(CRC) HANDBOOK SERIES

Handbook of Chemistry and Physics, 51st edition

Standard Mathematical Tables, 18th edition

Handbook of tables for Mathematics, 4th edition

Handbook of tables for Organic Compound Identification, 3rd edition

Handbook of tables for Probability and Statistics, 2nd edition

Handbook of Clinical Laboratory Data, 2nd edition

Manual for Clinical Laboratory Procedures, 2nd edition

Handbook of Laboratory Safety, 2nd edition

Handbook of Food Additives, 1st edition

Handbook of Biochemistry, selected data for Molecular Biology, 2nd edition

Handbook of Radioactive Nuclides, 1st edition

Handbook of Analytical Toxicology, 1st edition

Handbook of tables for Applied Engineering Science, 1st edition

* Handbook of Properties of Engineering Materials, 1st edition

* CRC-Fenaroli Handbook of Flavors, 1st edition

* Handbook of Environmental Pollution Control, 1st edition

* Handbook of Chromatography, 1st edition

* Handbook of Lasers, with Selected Data on Optical Technology, 1st edition

* Handbook of Laboratory Animal Science, 1st edition

* Handbook of Marine Sciences, 1st edition

* Handbook of Engineering in Medicine and Biology, 1st edition

* Handbook of Microbiology, 1st edition

* Currently in preparation.

HANDBOOK
of
LABORATORY
SAFETY

SECOND EDITION

Editor
NORMAN V. STEERE
Safety and Fire Protection Consultant
140 Melbourne Avenue, S.E.
Minneapolis, Minnesota 55414

Published by

THE CHEMICAL RUBBER CO.
18901 Cranwood Parkway, Cleveland, Ohio 44128

This book presents data on personal hazards and safety, obtained from authentic and highly regarded sources. Reprinted material is quoted with permission, and sources are indicated. A wide variety of references are listed. Every reasonable effort has been made to give reliable data and information but the editors and the publisher cannot cannot assume responsibility for the validity of all materials or for the consequences of their use. The objective is to furnish the best guides available relative to safety and hazards.

Preface to the Second Edition

The purpose of the *Handbook of Laboratory Safety* is to provide convenient information for hazard recognition and control. Fires, explosions and exposures to chemicals, biological agents and other hazards can usually be anticipated and measures taken to prevent such occurrences or minimize the consequences.

One of the new chapters in the Second Edition outlines several methods of organizing laboratory safety programs. The Handbook also emphasizes accident investigation and analysis, and the responsibility to report unexpected hazards encountered in experimental work.

Four new chapters have been added on Protective Equipment. They cover respiratory protective equipment, ear protection, eye hazards, and protective clothing. New chapters on Toxic Hazards include Principles and Procedures for Evaluating Toxicity of Chemicals, Mode of Action of Toxic Substances, Hazards of Isocyanates, and Chemical Cyanosis and Anemia Control. Chapters have been added on Protective Lockout and Tagging of Equipment, Grounding Electronic Equipment, Producing and Handling High Purity Water, and Laboratory Animal Housing. Two new chapters have been added to the section on Radiation Hazards.

Readers are invited to offer suggestions for changes in future editions and to report any errors which may be found. Comments and inquiries should be directed to the Publisher.

Comprehensive fire and health hazard data for 1094 chemicals are listed in tables at the end of the book. The tables include related physical properties of the chemicals and references to more detailed information available from other sources.

A great many people deserve credit for their contributions to this book, most notable the Advisory Board and the authors who prepared the text and the tables. Major impetus to prepare the book and valuable insights came from personnel and professional associates met during my experience at Michigan State University in the Office of Safety Services and at the University of Minnesota in the School of Public Health and the University Health Service. Understanding and support which were essential to completing the Second Edition were given by my wife Lyn.

Minneapolis, Minnesota   NORMAN V. STEERE
July 24, 1970

Advisory Board

Contributors

Ernest I. Becker, Ph.D.
Professor and Chairman
Department of Chemistry
University of Massachusetts
Boston, Massachusetts 02116

Mathew Braidech
Engineering Consultant
Wyckoff, N.J. 07481

Allen Brodsky, Sc.D., C.H.P.
Associate Professor of Health
    Physics
Graduate School of Public Health
University of Pittsburgh
Pittsburgh, Pennsylvania 15213

D. F. Bunch
Atomics International
Canoga Park, California 91304

Richard C. Charsha
"Freon" Products Division
Chambers Works
E. I. du Pont de Nemours & Co., Inc.
Deepwater, N.J. 08023

Daniel R. Conlon
Instruments for Research & Industry
Cheltenham, Pennsylvania 19012

William B. Cottrell
Director, Nuclear Safety Program
Oak Ridge National Laboratory
Oak Ridge, Tennessee 37830

Charles F. Dalziel
Professor Emeritus, Electrical
    Engineering
University of California
Berkeley, California 94720

Roger L. DeRoos
Instructor and Public Health
    Engineer
University of Minnesota
Minneapolis, Minnesota 55455

Caldwell N. Dugan
Division of Institutional Resources
National Science Foundation
Washington, D.C. 20550

Grace Mary Ederer
Associate Professor, Department
    of Laboratory Medicine
College of Medical Sciences
University of Minnesota
Minneapolis, Minnesota 55455

Theodore E. Ehrenkranz
Safety Engineer
Los Alamos Scientific Laboratory
Los Alamos, New Mexico 87544

John E. Evans, Jr., M.A., J.D.
Evans, Ivory and Evans
Pittsburgh, Pennsylvania 15219

K. Everett, B.Sc., A.R.I.C.
University Safety Officer
University of Leeds
Leeds LS2 9JT, England

Gari T. Gatwood
Environmental Health and Safety
    Engineer
Harvard University
Cambridge, Massachusetts 02138

Leon Goldman, M.D.
Professor and Chairman
Department of Dermatology
College of Medicine
Director, Laser Laboratory
Children's Hospital Research
    Foundation of the University of
    Cincinnati Medical Center
Cincinnati, Ohio 45229

F. A. Graf, Jr.
Research Scientist
Thiokol Chemical Corporation
Brigham City, Utah 84302

Roger Grimm
Blyth & Co., Inc.
New York, New York 10005

Everett Hanel, Jr.
U.S. Army Biological Laboratories
Fort Detrick
Fredrick, Maryland 21701

Scott A. Heider
Division of Institutional Resources
National Science Foundation
Washington, D.C. 20550

Peter Hornby
Operational Research Scientist
Institute for Operational Research
Coventry CV1 2FS, England

Harold Horowitz
Acting Director
Division of Institutional Resources
National Science Foundation
Washington, D.C. 20550

W. G. Hume, M.D.
Medical Division, Chambers Works
E. I. du Pont de Nemours & Co., Inc.
Deepwater, New Jersey 08023

Alvin B. Kaufman
Litton Systems Division
Litton Industries
Woodland Hills, California 91364

Edwin N. Kaufman
Senior Scientist
Douglas Aircraft Co.
Woodland Hills, California 91364

Howard L. Kusnetz, P. E.
Director, Div. of Occupational Injury
and Disease Control
Bureau of Occupational Safety and
Health
U.S. Public Health Service
Cincinnati, Ohio 45202

Adrian L. Linch
Medical Division, Chambers Works
E. I. du Pont de Nemours & Co., Inc.
Deepwater, New Jersey 08023

Herbert K. Livingston, Ph.D.
Professor
Wayne State University
Detroit, Michigan 48202

Gerald W. Marsischky
Director
Navy Ordnance Systems Command
Field Safety School
Crane Naval Ammunition Depot
Crane, Indiana 47522

Dennis G. Nelson
Product Development Supervisor
The 3M Company
St. Paul, Minnesota 55101

Richard J. Nocilla
Senior Health Physicist
Reynolds Electrical & Engineering
Company, Inc.
Las Vegas, Nevada 89102

G. Briggs Phillips, Ph.D.
Director, B-D Research Center
Becton, Dickinson and Company
Raleigh, North Carolina 27604

George G. Pinney
Manager of Quality Assurance
National Cylinder Gas Division
Chemetron Corporation
Oak Brook, Illinois 60521

C. F. Reinhardt, M.D.
Research Manager
Environmental Sciences Group
Haskell Laboratory
E. I. du Pont de Nemours & Co.,
Inc.
Wilmington, Delaware 19898

R. James Rockwell, Jr.
Directing Physicist
Laser Laboratory
University of Cincinnati Medical
School
Cincinnati, Ohio 45221

Robert S. Runkle
Safety Engineer
Becton, Dickinson Research Center
Raleigh, N.C. 27604

Gail D. Schmidt
Chief, Radioactive Materials Branch
Div. of Medical Radiation Exposure
Bureau of Radiological Health
U.S. Public Health Service
Rockville, Maryland 20852

David T. Smith
E. I. du Pont de Nemours & Co.,
   Inc.
Chambers Works
Deepwater, New Jersey 08023

Gail P. Smith, Ph.D.
Manager, International Research
Corning Glass Works
Corning, New York 14830

Ralph G. Smith, Ph.D.
Professor
School of Public Health
University of Michigan
Ann Arbor, Michigan 48104

Verity C. Smith
Vice President
Barnstead Still & Sterilizer Company
Roslindale, Massachusetts 02131

Eric W. Spencer
Safety Officer
Brown University
Providence, Rhode Island 02912

James G. Stearns
Safety Engineer
Los Alamos Scientific Laboratory
Los Alamos, New Mexico 87544

Herbert E. Stokinger, Ph.D.
Chief, Laboratory of Toxicology
   and Pathology
Bureau of Occupational Safety and
   Health
U.S. Public Health Service
Cincinnati, Ohio 45202

Paul W. Trott, Ph.D.
Division Safety Engineer
The 3M Company
St. Paul, Minnesota 55101

Barbara Tucker, M.T., (ASCP)
Chief Medical Technologist
Northwestern Hospital
Minneapolis, Minnesota 55407

Paul H. Woodruff
Manager, Chicago Office
Roy F. Weston, Environmental and
   Engineering Consultants
Wilmette, Illinois 60091

R. L. Wuertz, M.D.
Medical Division, Chestnut Run
E. I. du Pont de Nemours & Co.,
   Inc.
Wilmington, Delaware 19898

Michael G. Zabetakis, Ph.D.
Washington and Jefferson College
Washington, Pennsylvania 15301

John A. Zapp, Jr., Ph.D.
Haskell Laboratory
E. I. du Pont de Nemours & Co.,
   Inc.
Wilmington, Delaware 19898

ACKNOWLEDGEMENTS

The Editor and Publisher acknowledge the following journals, publishers and organizations for use of major portions of material appearing in the chapters listed below. Detailed acknowledgements appear in or at the end of the chapters and in the Introduction to the Tables of Chemical Hazard Information.

AMERICAN CONFERENCE OF GOVERNMENTAL INDUSTRIAL HYGIENISTS
12   Threshold Limit Values, in Tables of Chemical Hazard Information
12   Documentation of Threshold Limit Values

AMERICAN INDUSTRIAL HYGIENE ASSOCIATION
12   Hygienic Guides

AMERICAN INDUSTRIAL HYGIENE ASSOCIATION JOURNAL
1.7   Hydrofluoric Acid Burn Treatment
6.8   Chemical Cyanosis and Anemia Control
7.1   Laser Laboratory Design and Personnel Protection from High Energy Lasers
7.6   Determining Industrial Hygiene Requirements for Installations Using Radioactive Materials

AMERICAN INSTITUTE OF ARCHITECTS JOURNAL
3.3   Hoods for Science Laboratories
10.5   Laboratory Animal Housing

AMERICAN MEDICAL ASSOCIATION
1.6   First Aid Procedures
1.14   Occupational Health Programs

AMERICAN NATIONAL STANDARDS INSTITUTE
7.7   International Organizations Producing Nuclear Standards

AMERICAN SOCIETY OF HEATING, REFRIGERATING AND AIR-CONDITIONING ENGINEERS, INC.—ASHRAE Guide and Data Book
3.2   Exhaust Systems
3.4   Air Conditioning for Laboratories

APPLIED MICROBIOLOGY
10.3   Design of Facilities for Microbiological Safety

ARCHIVES OF ENVIRONMENTAL HEALTH, American Medical Association
6.3   Evaluating Toxic Exposures by Biospecimen Analysis
6.8   Chemical Cyanosis and Anemia Control
7.1   Laser Laboratory Design and Personnel Protection from High Energy Lasers
8.12   Controlling Hazards from Uses of the Plasma Torch

ARCHIVES OF INDUSTRIAL HEALTH, American Medical Association
6.6   Hazards of Isocyanates
6.8   Chemical Cyanosis and Anemia Control

CAMPUS SAFETY ASSOCIATION, National Safety Council
1.11   Safety Considerations in Research Proposals
8.9   Compressed Gas Cylinders and Cylinder Regulators
8.12   Controlling Hazards from Uses of the Plasma Torch
9.1   Prevention of Contamination of Drinking Water Supplies
11.1   Laboratory Considerations for Safety

Table of Contents

General

Responsibility For
Laboratory Safety

Norman V. Steere

Safety in the laboratory requires the same kind of continuing attention and effort that is given to research, teaching, and analytical techniques. Use of new or different techniques, chemicals, and equipment requires careful reading, instruction, and supervision, and may require consultation with other people with special knowledge or experience.

It should not be assumed that students, investigators, or technical staff have adequate information about laboratory safety. The information explosion makes it difficult to keep up-to-date on the possible consequences of exposures to laboratory chemicals, and the precautions needed to control the hazards of laboratory operations. Academic training or past job experience may not have dealt with hazards of the many new techniques and materials coming into use in every discipline and there is evidence that safety is not given adequate time or emphasis in the academic curriculum.

Many medical technologists and graduate chemistry students expressed the opinion in a survey[1] that not enough emphasis was placed on laboratory safety during their undergraduate college training, and the students also favored more emphasis on safety in graduate laboratories.

Representatives from the chemical industry have expressed privately and in open meetings the opinion that the college graduates they hire have generally not had adequate safety training, at any degree level.

Little if any time in chemistry courses is devoted to teaching the general principles of toxicology, according to a survey of a group of chemistry and chemical engineering professors.[2] Twelve of eighteen replied that they included no teaching on toxicology, and only two of the five who did, spent more than one hour. In contrast, most of the respondents believed that teaching about toxicity should begin in high school and continue through college and graduate school, but most students are expected to acquire toxicological knowledge solely by their own efforts. Early chemistry courses should integrate instruction on hazards and toxic potentials associated with the preparation and handling of compounds,[3] but it is even more important to concentrate efforts on providing chemical hazard information to the people who are working now in laboratories.

Efforts to prevent laboratory accidents by encouraging development of positive attitudes toward safety are valuable, but only if the emphasis is balanced by providing adequate and suitable information for understanding laboratory and chemical hazards and their consequences.

INFORMATION ON CHEMICAL HAZARDS AND TOXICITY

The central problems in any effort to provide chemical hazard and toxicity information to the people who need it in the laboratory, seem to be the absence of an organized, functioning, and available repository for the information, and a general disinclination to report hazards and toxicity information.

The "Handbook for Authors" published by the American Chemical Society for contributors to ACS journals states expressly, "Any unexpected hazards encountered with the experimental work must be noted and emphasized." If authors follow this direction it will be possible for the hazard information to be widely available and to be abstracted and stored for retrieval in the computer system of Chemical Abstracts Service, which has a tie with the computer system of the National Library of Medicine.

In addition to noting hazards in research reports, investigators should see that hazards, possible toxic effects, and necessary precautionary measures are recorded in theses and other documentation. As an example, the graduate student who suffered severe dermatitis during her work on formulation and characterization of organic tin compounds should have noted in her doctoral thesis that special handling and ventilation might be necessary to prevent dermatitis or other adverse reactions.

Hazard and toxicity information should be determined for chemicals described in doctoral theses, with a rigor similar to that used to determine precise chemical composition, melting and boiling points, structural formula, and other properties. Automatic equipment for determining flash point and differential thermal analysis is commercially available and should be on hand for regular use in university chemistry departments. Although evaluation of toxicity can be expensive if carried to the extent necessary for marketing a substance for household use, range-finding tests can be carried out at moderate prices and should become standard in research reports on new chemicals.[4]

Adverse reactions from laboratory exposures to chemicals and physical and biological agents should be reported. Reports on adverse reactions from occupational exposures are being solicited by the American Medical Association, Council on Occupational Health, Committee on Occupational Toxicology, for a registry on information from physicians in occupational medicine for dissemination to persons interested in occupational health.

Access to chemical hazard and toxicity information available in chemical company files and publications is usually not readily available for reasons which may include: lack of a convenient system for distributing such information, interest in maintaining competitive commercial position, and fear of misinterpretation of toxicity information. The information supplied by chemical manufacturers to the Food and Drug Administration is held as confidential, and the information accumulated by the National Advisory Center for Toxicology is available on a routine basis only to the USA Federal agencies that support the Center. A retrieval system for toxicity data is maintained by the Technical Information Section, Occupational Health Program, National Center for Urban and Industrial Health, U. S. Public Health Service, Cincinnati, Ohio. The Safety and Publications Committees of the American Chemical Society have requested the Chemical Abstracts Service to develop methods for computer storage and retrieval of hazard data, and this development will offer the best prospect of a useful central repository.

The basic premise for any rational efforts toward safety is that accidents are caused and that accidents and accidental injuries can be prevented. An operational definition that seems to be workable has been proposed by Lucille Huber[5]: "*An accident is the unplanned, occasional, but foreseeable consequence of one or more unsafe acts in combination with hazardous circumstances.*" Implicit in this definition is the important concept that accidents can not usually be attributed solely to "human factors" or to "mechanical factors," but that a chain or sequence of events and circumstances is necessary before most accidents can occur. The chain or sequence concept is important in selecting the accident or injury prevention measure that "will in the long run be most effective with the least sustained effort and the least chance of failure or breakdown," or the most effective measure for the lowest overall cost.

An example of the results of selecting injury prevention measures instead of accident prevention measures is the use of special protective clothing and face protection for handling small quantities of unstable explosives in the laboratory.

BASIC CAUSES OF ACCIDENTS

A proposed method of accident cause analysis described below is a modification of a system developed by Consumers Power Company of Michigan and used by Michigan State University and the University of Missouri. It is based on the philosophy that accidents are caused and can be prevented.

Some of the statements that accompany the list of basic causes include the following:

> List causes that apply. When more than one cause applies, show major cause first, then list contributing causes in order of importance.
>
> If correct basic causes are assigned, the corrective action will usually suggest itself.
>
> The best suggestions for preventing accidents may come from the injured person.

1.   Failure of Person in Charge to Give Adequate Instructions or Inspections

Failure to give necessary instructions.
Wrong instructions by person in charge.
Failure to provide a safety watcher.

Inspection not made before job started.
Inspection not made during job activity.
Other (explain).

2.   Failure of Person in Charge to Properly Plan or Conduct the Activity

Failure to plan safety into the activity.
Use of unsafe methods.
Use of inexperienced or unskilled people.
Failure to lead or maintain discipline.

Failure to enforce safety rules.
Failure to provide adequate manpower.
Other (explain).

3.   Improper Design, Construction or Layout

Inadequate engineering design or layout.
Faulty or inadequate design of equipment.
Unsafe housekeeping, poor ventilation, inadequate lighting and similar causes.

Inferior or faulty construction.
Other (explain).

4.   Protective Devices not Provided or Proper Equipment and Tools not Provided

Adequate protective devices not provided.
Failure to provide safe, suitable tools.

Failure to provide necessary materials.
Other (explain).

5.   Failure to Follow Instructions or Rules

Safe practice rules not followed.
Specific safe work procedure not followed.
Instructions not followed, warning unheeded.

Working without permission or authority.
Equipment or safety devices not inspected.
Failure to obey laws or regulations.
Other (explain).

6.   Failure to use Protective Devices Provided or Improper use of Equipment or Material

Protective devices not used.
Improper use of tools, or defects known.
Improper use of equipment.

Improper use of materials.
Failure of equipment by overloading.
Other (explain).

7. Physical Condition or Handicap

Chronic illness or disease.
Acute illness, nausea, fainting, and so forth.
Fatigue beyond capacity of individual.
Lack of skill or physical aptitude.
Physical handicap or lack of strength.
Conditions caused by allergy.
Under influence of intoxicants or drugs.
Other (explain).

8. Knowledge or Mental Attitude

Divided attention, failure to concentrate.
Lack of knowledge or understanding.
Lack of good judgment.
Unnecessary haste or hazardous short cuts.
Inability to work with others.
Temper, anger, or impulsive actions.
Excitement, fright, or involuntary reaction.
Other (explain).

9. Use of Devices with Unknown Defects

Faulty protective devices or equipment.
Defective tools (defects not known).
Unpredictable failure of tools or equipment.
Failure of material—defects not apparent.
Other (explain).

10. Agencies Outside the Organization (University, Company or Other Employer)

Accident caused by outside individuals (not reasonably preventable by the organization).
Equipment or conditions not under control of the organization.
Vandalism or willful damage.
Action of outside company.
Wild animals, insects, and so forth.
Wind, lightning, flood, and so forth.
Other (explain).

BASIC SAFETY HAZARDS

It may be helpful in analyzing the hazard potential of a work procedure, a laboratory environment, or a piece of equipment to consider some of the different kinds of hazards which can be present or which may occur. As a guide, some basic hazards and examples of the less common ones are listed, as adapted from the pamphlet "The Hazard Family," published by General Motors Corporation.

1. Pinch points
2. Catch points
3. Shear points
4. Squeeze points
5. "Run In" points
6. Flying objects
7. Falling objects
8. Sharp and pointed objects
9. Heavy objects
10. Hot and cold objects and environments
11. Slippery surfaces
12. Electricity
13. Radiation
14. Noise and vibration
15. Pressure extremes

PINCH POINTS are formed when two separated objects move and come in contact with on another. Examples:

> At the hinges of a door; between a rocker and the floor; pliers; riveters; spot welders; machine stops; clamping and holding fixtures; chucking and centering devices on machines.

CATCH POINTS are created by objects either stationary or in motion, which have sharp corners, splines, teeth or other rough shapes and surfaces capable of catching a person or his clothing. Examples:

> Rotating drills and reamers; spline shafts; milling machine cutters; broaches; fish hooks; nails sticking out; door knobs; keys and keyways.

SHEAR POINTS are created by two objects, one or both of which is in motion, as they pass one another. The two objects are close enough together when they pass to cause a shearing action on fingers, hands, feet and legs inadvertently placed in the shear point. Examples:

> Where a door closes into the door frame, or a drawer into a cabinet; shears; paper cutters; dies; reciprocating mechanisms.

SQUEEZE POINTS are created by two objects one or both of which is in motion as they move toward one another. The objects do not come into actual contact with one another but the distance between them reduces to the extent that a crushing injury will result if a person (or part of a person) is caught in the reduced area. Examples:

> Between two drawers of a cabinet or dresser; doorknobs too close to the edge of a door and the door frame; objects being moved by power conveyors creating squeeze points with fixed objects along the conveyor.

"RUN IN" POINTS are created by two objects in contact with and rotating toward one another. Examples:

> Belts and pulleys, as on vacuum pumps; chains and sprockets, as on a bicycle; gears in mesh; cables and drums; gears and racks.

INJURY CAUSATION AND PREVENTION

Haddon has developed an interesting and useful analysis of injury causation[6,7] which may be paraphrased as follows: Injury is produced when the human body receives energy in an amount or at a rate which either exceeds the injury threshold or interferes with the energy-exchange system of the whole body or parts of the body.

Haddon's first class of injuries includes those due to the delivery of energy in excess of local or whole-body tolerances or injury thresholds, and his illustrations are tabulated in Table 1.

Haddon's second classification of injuries includes those due to interference with normal body function—whole-body or local energy exchange; some examples are shown in Table 2.

RESPONSIBILITY AND REGULATIONS

Before discussing responsibility for laboratory safety within the teaching, research or service organization, brief consideration should be given to the interrelationships which make it difficult for any single laboratory or group of laboratories to be independent or self-sufficient in a comprehensive safety program. Every laboratory is largely dependent on the quality of the previous training and experience of its students or employees, on the testing and labeling of the chemicals which are available for purchase, on the design and maintenance of the physical facilities of the laboratories, on the reliability and safety of equipment provided, on the reasonableness of the laws and regulations which govern the design and operation of laboratories, and on the interest, judgment and support of the associations, companies, societies and other groups that can affect any of the dependent variables listed.

The various codes and standards which may affect or be applied to laboratory operations may not be appropriate, either because the laboratory hazards are greater than those for which the code or standard was written, or because the code or standard was written for commercial or industrial conditons and cannot properly be applied to laboratory conditions. In the latter situation, the answer seems to lie in the direction of having laboratory personnel represented on the code-making or standard-writing

TABLE 1

**Illustrations of Class I Injuries, Those Due to the Delivery of
Energy in Excess of Local or Whole Body Injury Thresholds**

Reprinted from "Textbook of Preventive Medicine" with permission of
Little, Brown and Company, Boston, Massachusetts.

Type of Energy Delivered	Primary Injury Produced	Examples and Comments
Mechanical	Displacement, tearing, breaking and crushing, predominantly at tissue and organ levels of body organization.	Injuries resulting from the impact of moving objects such as bullets, hypodermic needles, knives and falling objects; and from the impact of the moving body with relatively stationary structures as in falls, plane and auto crashes. The specific result depends on the location and manner in which the resultant forces are exerted. The majority of injuries are in this group.
Thermal	Inflammation, coagulation, charring, and incineration at all levels of body organization.	First, second and third degree burns. The specific result depends on the location and and manner in which the energy is dissipated.
Electrical	Interference with neuro-muscular function, and coagulation, charring and incineration, at all levels of body organization.	Electrocution, burns. Interference with neural function as in electroshock therapy. The specific result depends on the location and manner in which the energy is dissipated.
Ionizing Radiation	Disruption of cellular and sub-cellular components and function.	Reactor accidents, therapeutic and diagnostic irradiation, misuse of isotopes, effects of fallout. The specific result depends on the location and manner in which the energy is dissipated.
Chemical	Generally specific for each group or substance.	Includes injuries due to animal and plant toxins, chemical burns, as from KOH, Br_2, F_2, and H_2SO_4 and the less gross and highly varied injuries produced by most elements and compounds when given in sufficient dose.

TABLE 2

**Illustration of Class II Injuries, Those Due to Interference with
Normal Local or Whole Body Energy Exchange**

Reprinted from "Textbook of Preventive Medicine" with permission
of Little, Brown and Company, Boston, Massachusetts.

Type of Energy Exchange Interfered with	Types of Injury or Derangement Produced	Examples and Comments
Oxygen Utilization	Physiologic impairment, tissue or whole body death.	Whole body—suffocation by mechanical or chemical means. For example, by drowning, strangulation, CO and HCN poisoning. Local—"vascular accidents." (These, of course, also involve more than mere O_2 deprivation, since waste and nutrient exchange are also blocked.)
Thermal	Physiologic impairment, tissue or whole body death.	Injuries resulting from failure of body thermo-regulation, frostbite, death by freezing.

body, directly from organizations having laboratories or indirectly from scientific or technical associations.

The need for research personnel to participate in the development of codes and standards was expressed in an unpublished paper presented by Harry C. Hoy of Oak Ridge National Laboratory at a November 1966 meeting at the National Academy of Sciences—National Research Council:

"The problems of protection and safety resulting from such growth and complexity (in research) have gone far beyond those normally encountered by conventional safety experts. There is no previous knowledge concerning many of the operating parameters of these machines, such as voltages, current densities required, magnetic forces, vacuum handling ability, and temperature extremes, nor are there any standards by which to judge safety. In addition, it is usual that the rapidity of the change of requirements and conditions dictated by temporary experimental setups means the use of limited space and temporary equipment. The same separate, and yet interlocking, types of parameters are to be found in experimental situations in many of the research and development facilities. It is becoming increasingly evident from the size of experimental gear and the growing number of scientific personnel involved that unless adequate measures are taken to provide safe conditions, we will see a growing number of accidents.

"The basic problem of providing for safety, in either the production, operation, or research situation is the one of recognizing the hazard and eliminating, if possible, or isolating it. Many of these hazards are recognized and can be minimized by strict adherence to established codes. Unfortunately, many other conditions unique to research are not covered by these codes. The providing of the necessary standards, or more properly research safety guidelines, must in large measure be done by the most knowledgeable and experienced research people."

Even the research safety guidelines or laboratory safety rules to be used within a single college, hospital, or other organization should be developed with active participation of the groups that will be affected, not just to assure better acceptance but to get the benefit of the best possible ideas. If for any reason safety rules are developed unilaterally at the administrative level, it is strongly recommended that the rules be distributed as "proposed," with time and opportunity for laboratory personnel to ask questions and register objections; the delay will result in better rules with less need for exceptions.

RESPONSIBILITY WITHIN LABORATORIES

Responsibility for safety within the laboratories for an organization may be considered to exist at three different levels—individual, supervisory or instructional, and organizational or institutional.

The division of responsibility, in this or any other appropriate manner, needs to be clearly assigned and accepted, steps taken to see that the responsibilities are exercised, and the division reassessed if unexpected problems develop.

Each individual who works in a laboratory as a student or employee has a responsibility to learn the health and safety hazards of the chemicals he will be using or producing, and the hazards which may occur from the equipment and techniques he will employ, so that he may design his setup and procedures to limit the effects of any accident. The individual should investigate any accident which occurs, and record and report the apparent causes and the preventive measures which may be needed to prevent similar accidents.

The teacher or supervisor has the responsibility of giving all the necessary directions, including the safety measures to be used, and the responsibility of seeing that students or employees carry out their individual responsibilities. Whoever directs the activities of others has a concurrent responsibility to prevent accidental injuries from occurring as a result of the activities.

The organization or institution of which the laboratories are part, has a fundamental responsibility to provide the facilities, equipment, and maintenance to provide a safe working environment, or an organized program to make the improvements necessary for a safe working environment. Unless the institution or organization fulfills its responsibilities, it cannot expect its supervisors, employees or students to fulfill their responsibilities for laboratory safety.

In academic institutions with laboratories, there seems to have been a general lack of responsibility in some important areas which are outlined in the next paragraph. Whether important responsibilities have not been accepted or exercised is due to lack of adverse experience or to a general academic tendency toward freedom from supervision, the technological advances used in laboratories call for sophisticated attention to measures for controlling hazards and limiting injury frequency and severity. For example, the recent death of a university professor from an oxidizer explosion might have been prevented if he had been using quantity limits common in industrial research laboratories, and an oxidizer explosion injury to a graduate student could have been prevented if he had been using the kind of protective equipment commonly used in industrial research laboratories.

Some of the aspects of laboratory safety responsibility which need to be accepted and fulfilled include the following:

Preparation and practice of fire, emergency, and rescue procedures

Training in cardiopulmonary resuscitation

Labeling containers of chemicals

Providing appropriate means of collecting and disposing of waste chemicals and other hazardous wastes

Providing consultative assistance in laboratory safety and occupational health

Establishing workable policies on working alone

Requiring eye protection

Ventilating laboratory atmospheres

Monitoring laboratory atmospheres

Evaluating toxic exposures by biospecimen analysis by medical personnel

Providing appropriate immunization and protection against biohazards

Investigating and reporting laboratory accidents

Have on hand and maintain up-to-date toxicological and chemical hazard data for all chemicals to be used in the laboratory, if such information is available. This information should be readily accessible to all concerned; e.g., emergency personnel, safety officers and fire fighters.

As these responsibilities are more broadly and fully exercised, the frequency, severity and cost of laboratory accidents and injuries will decline.

REFERENCES

1. Allen, Joe E., "An Exploratory Study of the Attitudes of Laboratory Workers Toward Accident Prevention," Safety in the Chemical Laboratory, pp. 9–12, Division of Chemical Education Publishing Company, Easton, Pa., (1967).
2. Eckardt, R. E., Personal correspondence, Linden, N.J., (April 27, 1964).
3. Stokinger, Herbert E., Archives of Environmental Health, *Vol. 4*, p. 1, (1962).
4. Smyth, Henry F., The Place of the Range Finding Test in the Industrial Toxicology Laboratory, *J. Ind. Hyg. and Tox.*, Vol. 26, pp. 269–273, (October, 1944).
5. Huber, Lucile A., "The Investigation of Accidents," presented at the Eighth National Conference on Campus Safety, (June, 1961).
6. Haddon, William, Jr., Klein, David, and Suchman, Edward A., "Accident Research Methods and Approaches," pp. 537, Harper and Row, New York, N.Y., (1964).
7. Haddon, William, Jr., "The Prevention of Accidents," Textbook of Preventive Medicine; Clark and MacMahon, editors, Little, Brown and Company, Boston, Mass., (1967).

Organization for Safety in Laboratories

Ernest I. Becker and Gari T. Gatwood

For any institution, there must be management support if there is to be an effective safety program. Effort is needed to promulgate safety as a part of all activities to prevent accidental injury, suffering and financial losses.

The following program for safety is directed toward preventing interruption of an organization's primary activities by accidents, equipment failure, or poor communication about hazards. While safety accomplishment cannot easily be measured in dollars, it can be used as one means of measuring performance. The program is based on the basic premise that hazards and accidents are indications of mismanagement. Establishing safety as an integral part of an effective management system assists an organization in accomplishing its primary functions.

When programming safety for an organization, it is important to consider the type of motivation of the personnel in the organization. There can be important differences between the motivation of manufacturing employees and laboratory employees. The manufacturing employee is usually more closely supervised and will generally try to follow management practices, including safety. The self-supervising person who has only his final work observed will, in many cases, take great *risks* to arrive at his results. These motivational differences are reflected in the program below.

Three distinct variations in this program from conventional safety organization can be noticed. First, it has four levels of responsibility rather than the traditional three due to the fact that in universities and many research laboratories, supervisors do not have the same degree of communication with or responsibilities to top management as in industry. Secondly, there are safety review teams which must be given specific topics of study with a defined time limit, after which the teams are retired and not called upon again. Thirdly, the safety personnel (engineers or coordinators) are in an advisory position and work as consultants *only*. This point is important to stress because if the safety personnel are given any responsibility for approvals or disapprovals, the basic responsibility for safety (which belongs to management) shifts and individual motivation for safety is lost.

In cases where a full-time safety engineer has been employed at the laboratory level, care must be taken in appointing a safety committee. If the safety engineer is responsible to the Laboratory Director, as the organization chart suggests, and the committee chairman has a vested interest in the operation of the laboratory, there is likely to be friction between the committee chairman and the safety engineer. In such cases, it would be appropriate to appoint as chairman of the safety committee a person who already serves in an advisory capacity to the Director.

Except for the three differences noted, this proposed organization for laboratory safety is similar to conventional organization for industrial safety. In both cases an effective program depends on management recognition of the safety needs of the organization, and demonstrated support for integrating safety into the activities of the organization.

PROGRAM FOR SAFETY ORGANIZATION

These principles should be followed even when the program is adapted to medium or small organizations.

11

Company President or President of the University

1. Publishes for wide distribution a statement of policy concerning safety.
2. Assumes institutional responsibility for the general pattern of safety practices and their efficient administration.
3. Holds Deans, Directors of Laboratories and Department Heads accountable for the safety of all employees and personnel associated with the company or university.
4. Appoints the Corporate or University Safety Committee.

Corporate Safety Committee or University Safety Committee

1. Establishes, at the direction of the President, broad and consistent policies concerning safety throughout the company or university. Policies are directed toward meeting seven basic elements of a safety organization. (See Appendix A.)
2. Uses the Corporate or University Safety Department as the operational group of the committee.
3. Meets regularly from two to four times per year.

Corporate Level or University Safety Review Teams

1. Are appointed by the Corporate or University Safety Committee with the support of the President.
2. Are given assignments for in-depth studies of specific topics.
3. Make recommendations to the Corporate or University Safety Committee.
4. Are dissolved after the given time period and submission of recommendations (nominally three months).

Corporate or University Safety Department

1. Acts as the operational group of the Corporate or University Safety Committee.
2. Offers assistance in the form of safety engineering to any laboratory or university group requesting the same.
3. Performs periodic inspection of laboratories and makes recommendations for improvement of unsafe conditions.

Dean, Laboratory Director, or Departmental Head

1. Publishes statements of policy reflecting and amplifying the President's statement as it pertains to his area. Variation of the President's policy should be in implementation details only and not basic philosophy.
2. Has responsibility for the safety program pertinent to the personnel and facilities under his direction.
3. May appoint safety coordinator or engineer and/or safety committee as indicated and defines their duties.
4. Promulgates safety policies as formulated by the Corporate or University Safety Committee or others and holds senior faculty members, supervisors, or division heads responsible for implementation and enforcement of the safety policies.
5. Authorizes necessary expenditures for safety.

Safety Coordinator or Safety Engineer

1. Coordinates safety activities and maintains liaison with the Corporate or University Safety Consultant.

2. The extent to which other responsibilities will be used will vary with the need of the specific area. A list of these is presented in Appendix B.

Safety Committee

Responsibilities will be determined by the Dean, Laboratory Director, or Departmental Head who appointed them.

Local Safety Review Teams

1. Are appointed by the Local Safety Committee.
2. Are given short term assignments for in-depth studies of specific topics.
3. Make recommendations to the Local Safety Committee.

Senior Faculty Members, Supervisors, or Division Heads

1. Are responsible for enforcement of safety policies promulgated by Deans, Laboratory Directors, or Departmental Heads.
2. Should be aware of potential hazards in their research or other activities and should institute appropriate safety precautions.

Junior Faculty Member, Student, or Employee

1. Works in accordance with accepted safe practices.
2. Reports unsafe conditions and practices.
3. Observes safety rules and regulations.
4. Makes safety suggestions.
5. Does not undertake jobs he does not understand.
6. Is responsible for his personal safety.

APPENDIX A: SEVEN BASIC ELEMENTS OF A SAFETY ORGANIZATION

I. Management Leadership
  A. Assumption of responsibility
  B. Written declaration of policy
II. Assignment of Responsibility
  A. Operating Department Heads
  B. Safety Engineers or Coordinators
  C. Supervisors
  D. Committees
  E. Employees
III. Maintenance of Safe Working Conditions
  A. Inspections
  B. Engineering revisions
  C. Purchasing
  D. Supervisors
IV. Training
  A. Supervisors
  B. Employees
V. Accident Records
  A. Accident analysis
  B. Injury reports
  C. Measurement of results

VI. Medical and First Aid Systems
    A. Placement examinations
    B. First aid services
    C. Periodic health examinations
VII. Personal Responsibility of Employees
    A. Acceptance of responsibility
    B. Maintenance of interest

APPENDIX B: OTHER POSSIBLE ACTIVITIES OF SAFETY COORDINATOR AND/OR SAFETY ENGINEER

1. Keeps and analyzes accident records.
2. Conducts education activities for supervisors of all levels.
3. Conducts activities for stimulating and maintaining interest in safety of personnel in department.
4. Develops employee safety education programs.
5. Serves on departmental safety committee.
6. Supervises and appraises accident investigation.
7. Promotes, plans and directs a regular program of safety inspections.
8. Checks for compliance with applicable safety laws and codes.
9. Issues regular reports showing safety performance and accident trends.
10. Serves in staff capacity.
11. Reviews all drawings, projects, and work orders concerning safety.
12. Insures adequate preventive maintenance program.
13. Establishes and maintains emergency and disaster plans.
14. Establishes protective equipment requirements.
15. Establishes, promotes and administers safety procedures or standards.
16. Obtains assistance when required from Corporate or University Safety Consultant.

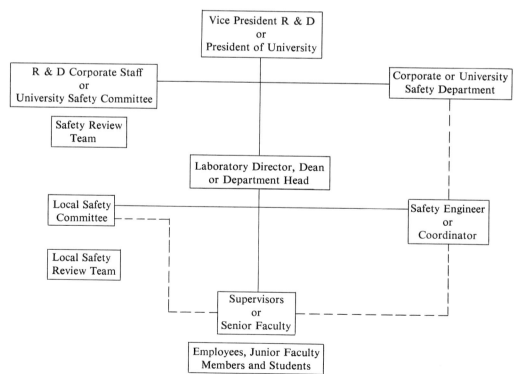

Fire, Emergency, and Rescue Procedures

Norman V. Steere

Fire in a laboratory building calls for prompt action to evacuate occupants safely and to extinguish the fire. Emergency procedures which are effective for fire evacuation can also be effective in case of toxic gas cylinder leaks and similar situations. Procedures which provide for fire rescue and fire fighting can be adapted to other emergencies requiring special equipment and training.

Many laboratories have inadequate provisions for meeting and controlling emergency situations which can arise, and many other laboratories have untested procedures that exist only on paper. The fact that an emergency has not arisen is no reason for taking this important planning for granted.

A realistic appraisal of the circumstances which can lead to emergencies in a laboratory will reveal many foreseeable and controllable problems. Some of the problems which can be expected to occur include:

> Fire involving one or more laboratories
> Chemical spills in corridors and laboratories
> Spills of radioactive materials
> Release of compressed toxic and corrosive gases
> Failure of power to laboratory hoods
> Escape of pathogens
> Explosions and injury of a person working alone or in an isolated area.

The general principles of emergency procedures will be outlined as a guide for preparation of individual specific plans of action for the full range of emergencies which can be anticipated.

The following section, "Primary Emergency Procedures," describes steps in the initial response which is the primary responsibility of people in the laboratory, requiring regular practice for proficiency but no special equipment. The section "Secondary Emergency Procedures" describes equipment and training needed to enable laboratory personnel to assist in rescue and damage control operations, which are the primary responsibility of a fire department.

Planning and preparation are necessary to assure that response to laboratory emergencies is prompt, correct, and effective. Damage and injury can be limited if emergency procedures are established and practiced regularly, and if adequate equipment and trained personnel are available for warning, rescue, and damage control.

PRIMARY EMERGENCY PROCEDURES

Laboratory personnel have sole responsibility for the primary emergency procedure, which consists of the following steps to prevent injury and limit the spread of the emergency:

> Alert personnel in the immediate vicinity of a fire or emergency
> Confine the fire or emergency
> Evacuate the building
> Summon aid

In all but minor situations, these steps take precedence over rescue, fire fighting, or damage control.

Even though small bench-top fires are commonly extinguished without evacuating the building or summoning the fire department, the first two steps of alerting and confining should always be taken. There should be an immediate readiness to evacuate the building and summon the fire department if such a small bench-top fire cannot be controlled.

There should be no criticism of a precautionary evacuation of the building. Fire departments are more often grateful than critical for being called to a fire which is out when they arrive. A fire which has gained headway before the fire department is called is much more difficult to extinguish.

Reporting a fire that has been extinguished will enable the fire department to be sure there are no smoldering embers and will provide a timely opportunity for fire department members to become familiar with the arrangement of the laboratory and the best means of access.

Equipment needed for the primary procedures includes an adequate system for signalling evacuation and a telephone or other means of summoning aid.

Alert Personnel in the Vicinity

Specific emergency procedures should include specific directions for alerting and directing occupants in the immediate vicinity of the emergency. Personnel in the vicinity should be informed of the nature and extent of the emergency, and the action expected of them. It may be very helpful to specify that, for example, one person should be directed to shut the door to the laboratory, one should actuate the evacuation alarm, and one should call the fire department and direct them to the scene of the emergency.

Persons who must work in isolated or unusually hazardous areas should be provided with some means, such as a compressed gas horn, for summoning help or giving warning.

Confine the Fire or Emergency

A fire or emergency situation should be confined in order to gain time for evacuation, and to limit the extent of damage. Methods of confinement could include shutting the sash on the hood and shutting the door to the laboratory.

Open transoms between laboratories and corridors provide the means for immediate spread of fire, gases, and air-borne contaminants, and thereby prevent effective confinement. Transoms should be sealed off with plaster or other fire-resistive construction.

Ordinary glass in doors or other openings between corridors and laboratories will fail rapidly during a fire, as will ordinary paneled or hollow-core wooden doors. Although the basic construction of walls between laboratories and corridors should be fire-resistant, there are fire-retardant coatings listed by Underwriters' Laboratories, Inc., which are intumescent and will provide a limited fire protection.

Open stairwells or stairwells with doors held open with fusible links will not provide an acceptable confinement of smoke and fire gases, since untenable conditions can develop so rapidly that people will be trapped before there is enough heat to melt fusible links.

An arrangement which will provide an acceptable measure of life safety in laboratory buildings would consist of stairwell doors held open with electromagnetic devices connected to a smoke detection system and to the evacuation alarm system, so that smoke or actuation of the alarm would close all the stairwell doors. Use of such a combined system seems to be a practical answer to providing convenient circulation without the threats to life safety of open stairwells, or stairwell doors wedged, propped, or wired open.

Evacuate the Building

Fire is only one of the emergency conditions which may require evacuation of a laboratory building. A broken bottle of formaldehyde in a main corridor or stairwell can present a formidable problem, as can large spills of many other chemicals. Leaking cylinders of toxic or corrosive gases are another type of emergency which may require quick evacuation to prevent many people from being injured or trapped.

Failure of power to laboratory hoods in which numerous containers of noxious chemicals are stored can create an immediate and urgent need to evacuate the entire building.

If building ventilation systems cannot be shut off by an accessible emergency switch, the entire building may have to be evacuated because of otherwise limited spills of radioactive materials or pathogens which can be picked up and spread by a recirculating ventilation system.

We recommend that laboratory buildings have at least three evacuation drills each year, with everyone leaving the building. Rapid evacuation of laboratory buildings with open stairwells is particularly important, since smoke and heat rise as quickly as if the open stairwell were a chimney. Untenable conditions which can trap people can develop in less than three minutes.

It is not reasonable to expect rapid and orderly evacuation of large buildings without the discipline and practice of regular exit drills.

Evacuation drills are practicable for laboratory buildings and have been carried out successfully in at least one large chemistry building. The Safety Committee of the department announced the day of the first drill, selected a time which minimized interruption of classes, and asked for total evacuation of the building, while recognizing that there might be unusual situations in which shutdown would cause irretrievable loss of data; the building was evacuated completely in less than $2\frac{1}{2}$ minutes.

There is no faultless method for being sure, quickly, that all occupants of the building have escaped. Search and rescue would be greatly simplified if missing persons could be accounted for as soon as the building was evacuated, but such accounting will be difficult even with good planning and almost impossible without.

As an approach to rapid accounting for occupants, evacuation procedures should be specific as to the point of rendezvous for each floor or laboratory area. By planning such rendezvous, it should be feasible to get a quick idea of who may be missing and thereby localize an initial search and increase chances of rescue. Evacuation drills will provide opportunities to practice such accounting methods, and to appraise operation of the evacuation and rendezvous procedures.

Evacuating a building but allowing people to re-enter before the emergency is under control is folly and a relinquishment of responsibility. Keeping occupants out of their buildings, from which they want to salvage research notes or equipment, is not easy. Keeping out reporters and other persons who have no right or reason to enter the building immediately is extremely difficult. Occupants must assume responsibility for controlling access as long as necessary or until relieved by police or other public safety personnel.

Summon Aid

The fire department is a valuable source of aid in many emergencies other than fire fighting. The fire department is, in most cases, equipped and ready to rescue, remove leaking gas cylinders, provide emergency lighting and ventilation, and perform a variety of other services. The fire department can give maximum assistance if their resources are known and if they have an opportunity to participate in pre-planning emergency assistance.

For example, the fire department should be directed to the scene of the emergency by meeting them at their expected point of response. Fire departments have received reports of fires where the person calling failed to mention the location of the building on fire, and they have responded to calls and been unable to find the problem or the person who called.

Calls for help from the fire department or any other group should not only tell where the problem is, but should describe what it is and what kind of help is needed. Giving an estimate of the size of the problem may help decide whether one or several fire companies will be sent, or whether one or several ambulances will be called.

SECONDARY EMERGENCY PROCEDURES

After laboratory personnel have carried out the primary procedures which are their basic responsibility, there should be action to bring the emergency under control. The control actions, or secondary emergency procedures, are generally the operational responsibility of the fire department and consist of the steps: (a) Rescue occupants, (b) Fight the fire, (c) Control the emergency.

Laboratory personnel definitely have a responsibility to inform and advise the fire department about such things as hazards and the problems which may arise from use of solid hose streams. We believe strongly that a group of laboratory personnel should be equipped and trained to function before and after the fire department arrives. Even if all fire departments were equipped with air masks, aluminized heat-reflective clothing, and special purpose extinguishers, laboratory personnel should be prepared to act immediately while the fire department is en route.

This section of the chapter describes briefly the kinds of equipment and training believed necessary to provide adequate capability for control of laboratory emergencies.

Rescue Occupants

Self-contained air masks with a 15-minute capacity should be provided (in pairs whenever possible) on all laboratory floors with chemical laboratories or radioactive materials, and especially near laboratories in which pathogenic or highly toxic materials are used. Large laboratory buildings should have two or three 30-minute air masks near the main entrance for administrative or supervisory personnel who will accompany the fire department on search and rescue operations. Self-contained masks are preferred because chemical concentrations may exceed the capacity of canister masks, and because of the possibility of oxygen deficiency in the atmosphere.

Aluminized heat-reflective suits are necessary to give amateur rescuers assurance and protection to effect a rescue from a laboratory in which a fire is out of control. Although a proximity suit or fire approach suit is not intended for entry into fires, such a suit can be used for entry during the first few minutes of a fire. Proximity suits of aluminized glass cloth were installed several years ago in Kedzie Chemical Laboratory at Michigan State University to provide staff and graduate students with protection for an immediate rescue from a burning laboratory. Self-contained air masks had been installed many years earlier. A rescue team of graduate students were able, with practice, to put on a mask and proximity suit in less than one minute.

Protection against gases which can be absorbed through the skin or which react with body moisture requires a gas-tight suit. Such a suit will not seem expensive if it provides protection in one serious emergency. Flashlights, ropes, and gloves round out the basic equipment needed for quick rescues in laboratories. More extensive equipment will be needed for rescue operations involved in disaster control.

All staff and graduate students should be made familiar with the function and operation of breathing masks, and with the purpose of aluminized heat-reflective clothing, and a group should be recruited for intensive training to include practice in search and rescue under simulated conditions of smoke, heat and limited visibility.

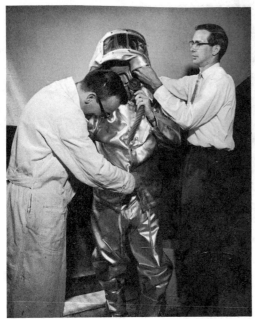

Courtesy Dow Diamond

FIG. 1. Emergency Equipment in a university laboratory. Capability for rescue and fire fighting is provided by an aluminized heat-reflective suit over a self-contained breathing neck.

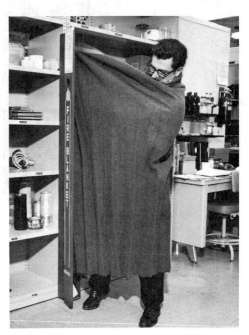

Courtesy of the 3M Company

FIG. 2. Illustration of one method of using a vertically-stored fire blanket—victim wraps himself in the blanket by turning.

Fight the Fire

In addition to the carbon dioxide fire extinguishers usually provided as the basic equipment in laboratories, there should be dry chemical extinguishers (sodium or potassium bicarbonate) for fighting flammable liquid fires which cannot be reached or controlled by CO_2. Water-type extinguishers are needed for waste basket fires and fires in wood, cloth, and other ordinary combustibles. Special purpose extinguishers should be available for controlling fires involving sodium, potassium, lithium, magnesium, or other metals and for electronic equipment and chemicals requiring special extinguishing agents. Consideration should be given to the purchase of wheeled dry chemical fire extinguishers which have great capacity for very large spill fires.

If fire hoses are part of the building equipment, the nozzles should be checked and replaced if they are the customary straight type. These produce a hard stream which is not as effective as a spray and which tends to break bottles and increase the fire fighting problems. Adjustable spray-type nozzles with shut-off should be provided for all hoses.

Training of a laboratory fire brigade should follow the guidelines which have been established by the National Fire Protection Association, and should include regular practice in the use of all types of fire extinguishers. Suppliers of fire extinguishers are an excellent source of training assistance.

Courtesy of the 3M Company

FIG. 3. A laboratory emergency station. The closet next to the emergency eyewash fountain has a self-contained air mask, rope, flashlight, three fire extinguishers—CO_2, dry chemical and water, and a stretcher (not showing).

Control the Emergency

This category of action may be so broad that training will have to be by dry runs rather than by actual practice. The emphasis here is on anticipating the problems which may arise, and acquiring the equipment and supplies needed for control measures before the need arises.

For example, large heavy plastic bags should be on hand for handling leaking gas cylinders, for holding waste generated in radioactive decontamination, and for a variety of other uses.

James Black at the U.S. Public Health Service suggested several years ago the technique of handling leaking gas cylinders by closing them in a heavy plastic bag with a tube, hose and clamp which can be released if the bag threatens to burst. The technique is simple, and has been used effectively many times to remove cylinders from buildings and transport the cylinders to disposal areas without damaging the vehicle or injuring the driver. The technique is one that should be known and prepared for in every laboratory with a cylinder of toxic or corrosive gas, because the cylinder may not fit into a hood or the leak rate may exceed the hood capacity for containment and dispersal.

More expensive than plastic bags are wet vacuum cleaners which can pick up liquids. We believe that every large laboratory building should have a wet vacuum for controlling liquid spills, and that the vacuum should be listed by Underwriters' Laboratories as explosion-proof.

Mention was made earlier that limited spills could involve the entire building if no switch were available to shut off a recirculating ventilation system. A similar problem could result if there were no accessible shutoff for a fresh air supply system which was picking up air-borne contaminants in heavy concentration. The answer to such problems seems to be an emergency shutoff switch located in a corridor where any responsible person could operate it without delay. The switch could be installed as a special sort of alarm box, and could be connected to the evacuation alarm system.

Although fire departments have smoke ejection fans which might be useful for emergency ventilation in case of spills, such fans may not be explosion-proof and do not provide adequate control for chemical spills. The kind of equipment which would do the best job would be a portable industrial fan with explosion-proof motor and flexible ducts for pick-up and exhaust.

EMERGENCY TEAM ORGANIZATION

It is recommended that the laboratory establish a group, which may be known as the Emergency Team, which will be trained to rescue occupants, use building fire equipment, and control emergencies before the arrival of and in cooperation with the Fire Department.

Organization

It is recommended that the Emergency Team consist of eight people from each floor of the building, and that the members from each floor include representatives of the groups which work there.

It is further suggested that each division head be given responsibility for initial selection, and replacement as necessary, of an appropriate number of representatives from his division from the floors on which they have their office or laboratory.

Responsibilities

It is recommended that members of the emergency team be given responsibilities for the following activities:

(1) Determine areas and operations in which it may be inadvisable to use water for fighting fires, in which some special type of extinguishing agent may be required, and in which special shut-down procedures or control measures may be necessary before, during or after fire-fighting operations. (The purpose of such determinations is to permit the laboratory to invite personnel of the Fire Department to the building so that such special situations may be explained, and pre-arranged plans drawn up for controlling emergencies.

(2) Assess laboratories for situations which seem likely to cause fire or other emergency.

(3) Develop plans for alerting personnel in the immediate vicinity of emergency situations.

(4) Determine the methods available for confining fires or emergencies to the laboratory, other work areas, and other rooms.

(5) Learn the operation of all types of fire extinguishing equipment available in the laboratory, and the types of fires for which such equipment is suited.

(6) Learn how to use the self-contained emergency breathing equipment available in the building.

(7) Learn the principles of search and rescue which may be needed in the laboratory.

(8) Become familiar with the operation of any emergency exhaust unit which may be available.

(9) Practice the use of fire extinguishers, breathing masks, search and rescue procedures, and any other emergency equipment which is available.

(10) Guard the stairwells and exterior doors of the building during exit drills and actual emergency situations, to prevent entry of unauthorized persons, who may interfere with emergency operations or be injured.

ACKNOWLEDGEMENT: Reprinted from the *Journal of Chemical Education, Vol. 41*, A369, May, 1964, with permission.

Laboratory First Aid

Norman V. Steere

Details of some important first aid procedures and cardiopulmonary resuscitation are described in the two chapters that follow, but here we want to emphasize the importance of first aid for chemical eye injuries, burns, and stoppage of respiration and circulation.

EMERGENCY TREATMENT OF CHEMICAL EYE INJURIES

Every person working with chemicals should know the correct method of emergency treatment of injuries in which chemicals enter the eye. The medically recommended method of emergency treatment of chemical eye injuries is to wash the injured eye thoroughly with plain water for 15 minutes—*without delay*! Contact lenses prevent thorough irrigation and must be removed for prevention of injury.

Dr. D. J. Kilian has made the following statement: "The most important part of the treatment of a chemical burn to the eye is what the individual does himself in the first few seconds of time. His part in taking care of the injury is far more important than the role that the doctor plays. If you allow that chemical to remain in the eye for even a few seconds too long, no matter what the doctor does, in many cases, you may end up with some permanent scarring. So, it should be perfectly obvious that the most important part of the whole treatment rests with the individual. That part of the treatment is very simple: *GET TO THE EYE BATH OR ANY SOURCE OF WATER* (whether it is a drinking fountain or a river by which you are working) *AND GET THE EYE THOROUGHLY WASHED OUT BEFORE THE CHEMICAL HAS ANY MORE CHANCE TO DO DAMAGE.* A burn to the cornea is like a burn to the skin. If you receive a light burn, such as a sunburn, when it heals you will have perfectly normal skin, but if you get a deep third degree burn, there will be permanent changes on the skin when the healing process is complete, regardless of anything that is done. The same thing is true of the cornea. If the burn is not deep, the cornea can return to normal. However, if you let the chemical stay in the eye too long, you may have damage that the most skilled eye-surgeon can not repair.

"Another point that is important is that alkaline materials, such as caustic and ammonia, are far more serious in the eye than are strong acid solutions, because acid will precipitate a protein barrier in the tissues and prevent the material from soaking in deeply. Alkaline materials do not precipitate this protein barrier, but they continue to soak down deep into the tissues as into a sponge. One has to wash and wash in order to get the strong alkaline materials out. So, be particularly cautious around strong bases or strong alkalies."

Contact lenses worn by persons working in laboratories can increase injury from chemical splashes because the wearer may not be able to remove the lenses to permit thorough irrigation, and a person giving first aid may not know that contact lenses are being worn or how to remove them. It is recommended that contact lenses not be worn in laboratories in which chemicals are handled, or that wearers be sure to use full eye protection at all times.

In the pamphlet "Use of Contact Lenses in Industry," published by the Council on Occupational Health of the American Medical Association, there are three paragraphs which are particularly applicable to wearing of contact lenses in laboratories:

Many physicians believe that the substitution of contact lenses for spectacles in industrial workers is contraindicated in workers whose eyes may be exposed to dusts, molten metals, or irritant chemicals. Small foreign bodies, which normally are washed away by tears, sometimes become lodged beneath contact lenses, where they may cause injury to the cornea. Similarly chemicals splashed into the eye may be trapped under a contact lens and cause extensive corneal damage before the lens can be removed and the eye adequately irrigated.

For effective protection for the eyes, the contact lens wearer should use in addition to his contact lenses the same approved face shields, conventional safety spectacles, or goggles for protection against job hazards as would any other worker on a similar job. Since removal of a contact lens for urgent irrigations after injury is made so difficult by spasm of the eyelids, the contact lens wearer is in even greater need of these protections than his counterpart who does not wear contact lenses, if the job carries high potential risk of eye injury.

Contact lenses are not in themselves protective devices and in fact may increase the degree of injury to the eyes. The same eye-protective devices used by other workers should be worn by contact lens wearers in similar employment.

All laboratories should have eyewash equipment installed and tested regularly to see that the water is not rusty and that pressure is adequate.

The color section shows the effects of prompt and delayed treatment of chemical eye injuries, and the effects of contact lenses in chemical splashes.

BURNS FROM FIRE AND CHEMICALS

Persons with burning clothing must be prevented from running, which fans the flames and can result in serious burns and death. Every person working in a laboratory, where a clothing fire may occur, should know how to use fire blankets, fire extinguishers and emergency showers. Chemical burns of all types should be immediately and thoroughly flushed with water to remove the chemical.

Cold water is considered to be an effective first aid measure for thermal burns, and if cold water is applied until pain subsides there may be more rapid healing. Applications of grease, tannic acids and other ointments, as a first aid procedure, on severe or extensive burns are to be avoided entirely.

RESPIRATION AND CIRCULATION STOPPAGES

We support the conclusion of the 1962 National Health Forum on Accident Prevention and Emergency Care that "adequate training programs for cardiopulmonary (heart-lung) resuscitation should be set up for physicians, dentists, nurses, medical students, rescue squads, and high-risk industries," and we believe that in most laboratories teams should be organized and trained to perform cardiopulmonary resuscitation.

Cardiopulmonary Resuscitation

INTRODUCTION

In May 1966, the work of an ad hoc Committee on Cardiopulmonary Resuscitation culminated in a Conference on Cardiopulmonary Resuscitation at the National Academy of Sciences-National Research Council (NAS-NRC). This study was undertaken in response to inquiries from the American National Red Cross and other national and federal agencies concerned with the need for standardized techniques of performance, training, and retraining requirements, and designation of the categories of persons to be taught mouth-to-mouth ventilation and external cardiac compression under present limitations on the supply of instructors. The ad hoc committee carefully reviewed and discussed these matters with representatives of over 30 national organizations attending the conference. The full proceedings of the conference will be published by the NAS-NRC. A summary of the recommendations of the ad hoc committee follows.

In November 1958, a Conference on Artificial Respiration was held at the National Academy of Sciences-National Research Council, and led to the publication in January 1959 of a "Statement on Emergency Artificial Respiration Without Adjunct Equipment." This statement unanimously endorsed the mouth-to-mouth and mouth-to-nose techniques of artificial respiration as the most practical methods of emergency ventilation without adjunctive equipment for an apneic person of any age. Since then, there has been worldwide acceptance and application of these techniques.

In July 1960, the clinical efficacy of an external manual technique for artificial circulation was reported. Since that time, well-documented experimental and clinical studies have established that the proper combination of artificial respiration and external cardiac compression can sustain a victim of sudden cardiac arrest for a reasonable period. Early experiences revealed both the benefits and the hazards of external cardiac compression and the need for its precise coordination with effective methods of artificial respiration by properly trained persons.

In an editorial in *Circulation* in September 1962,[1] closed-chest cardiopulmonary resuscitation was endorsed as a *medical* procedure. Subsequently, the method was reclassified as an *emergency* procedure in a second editorial in *Circulation* in May 1965.[2] This was endorsed by the American Heart Association, the American National Red Cross, the Industrial Medical Association, and the U.S. Public Health Service, which strongly recommended that the technique should be applied by "properly trained individuals of medical, dental, nursing and allied health professions and of rescue squads."

Since publication of the second editorial, the American Heart Association, the Public Health Service, and other organizations have inaugurated intensive training programs in cardiopulmonary resuscitation in response to the widespread interest and enthusiasm of highly motivated persons at all levels from first aid workers to professional medical personnel. Their experiences have indicated that clinical results vary widely and depend upon the exact technique taught, the effectiveness of training and periodic retraining, the personnel taught, the selection of cases, and numerous other factors. These considerations have guided the ad hoc Committee on Emergency Cardiopulmonary Resuscitation in formulating the following recommendations.

ABCD STEPS

Emergency cardiopulmonary resuscitation involves the following steps:

A—Airway opened

B—Breathing restored

C—Circulation restored

D—Definitive therapy

These should always be started as quickly as possible and always in the order shown. The recommended basic steps for performing the ABCs are shown in the Figure. Definitive therapy involves diagnosis, drugs, defibrillation (when indicated), and disposition. These definitive procedures are restricted to physicians or to members of allied health professions and paramedical personnel under medical direction. They are beyond the scope of this statement, which will be restricted to the ABCs of emergency cardiopulmonary resuscitation.

EXHALED-AIR VENTILATION (MOUTH-TO-MOUTH VENTILATION; MOUTH-TO-NOSE VENTILATION)

A and B are the basic steps of artificial ventilation and should always be applied first in emergency resuscitation. They constitute first aid measures which can be performed under almost any circumstances without adjunctive equipment or help from another person and regardless of the cause of the apnea.

Immediate Opening of the Airway

The most important single factor contributing to successful resuscitation is immediate opening of the airway. This is most easily and quickly accomplished by maximum backward tilt of the head. Sometimes an unconscious patient will be saved by this simple maneuver, which reestablishes an open airway and allows spontaneous breathing to resume. With the patient supine, the rescuer places one hand behind the patient's neck and the other on his forehead. He then lifts the neck and tilts the head backward. This stretches the neck and lifts the tongue away from the back of the throat, thereby relieving this anatomical obstruction of the airway. The head should be maintained in this position at all times. Obvious foreign material in the mouth or throat should be removed immediately with the fingers or by any other means possible.

Restoration of Breathing

If the patient does not resume spontaneous breathing after his head has been tilted backward, immediately begin artificial ventilation by either the mouth-to-mouth or the mouth-to-nose method. The first blowing effort will determine whether or not any obstruction exists. For mouth-to-mouth ventilation the patient's head is maintained in a position of maximum backward tilt with one hand behind the neck. In the unconscious patient this usually allows the mouth to drop open. The nostrils are pinched together with the thumb and index finger of the other hand. The rescuer then opens his mouth widely, takes a deep breath, makes a tight seal with his mouth around the patient's mouth, and blows in about twice the amount the patient normally breathes. He then removes his mouth and allows the patient to exhale passively. This cycle is repeated approximately 12 times per minute.* Adequate ventilation is ensured on every breath by (*a*) seeing the chest rise and fall, (*b*) feeling resistance of the lungs as they expand, and (*c*) hearing the air escape during exhalation.

* Caution: Unless the rescuer correctly gauges the rhythm at 12 to 15 times per minute, he is likely to hyperventilate himself, and the patient could be lost.

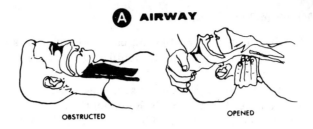

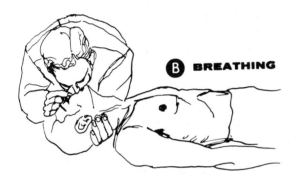

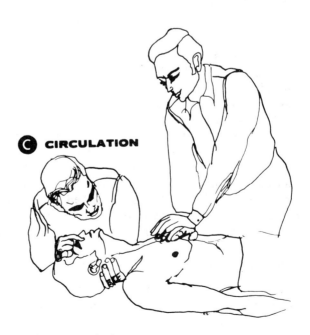

FIG. 1. Heart-Lung Resuscitation.

Mouth-to-Nose Ventilation

Mouth-to-nose ventilation can be used if it is impossible to open the patient's mouth, if it is impossible to ventilate through his mouth, if his mouth is seriously injured, if it is difficult to achieve a tight seal, or if the rescuer prefers the nasal route. For this technique the rescuer keeps the patient's head tilted back with one hand and uses the other hand to push the patient's lower jaw closed and seal his lips. He then takes a deep breath, seals his lips around the patient's nose, and blows in until he sees the patient's chest rise. He then removes his mouth, allows the patient to exhale passively, and repeats the cycle 12 times per minute. When mouth-to-nose ventilation is employed, it may be necessary to open the patient's mouth during exhalation, if this can be done, to allow the air to escape.

Incomplete Opening of Air Passages

Occasionally, there may be incomplete opening of the air passages even with properly performed "head-tilt." In such cases, further opening of the air passages can be achieved by displacing the patient's mandible forward so that his lower teeth are in front of his upper teeth, and by simultaneously holding his mouth open. This may be accomplished (a) by grasping the mandible between the thumb and index finger and lifting, or (b) by placing the fingers behind the angles of the lower jaw and pushing it forward.

Ventilation of Children

A and B are performed in essentially the same way for children, except that for infants and small children the rescuer covers the patient's mouth and nose with his mouth and blows gently, using less volume to inflate the lungs. Babies require only small puffs of air from the rescuer's cheeks. The rate of inflation should be 20 to 30 times per minute. The neck of an infant is so pliable that forceful backward tilting of the head may obstruct breathing passages; therefore, the tilted position should not be exaggerated.

Presence of Foreign Bodies

The presence of a foreign body should be strongly suspected if the rescuer is unable to inflate the lungs after proper head tilt and forward displacement of the mandible. The first blowing effort will determine whether or not any obstruction exists. An adult patient with this problem should be rolled onto his side quickly and firm blows delivered over his spine between the shoulder blades in an attempt to to dislodge the obstruction. The rescuer's fingers should then be swept through the patient's mouth to remove such material. Then exhaled-air ventilation should be resumed quickly. Sometimes, slow forceful breaths can be used to bypass a partial airway obstruction and inflate the lungs. A small child with an obstructive foreign body in the airway should be picked up quickly and inverted over the rescuer's forearm while firm blows are delivered over the spine between the shoulder blades. Then quickly resume ventilation.

Distension of the Stomach

Exhaled-air ventilation frequently causes distension of the stomach. This occurs often in children, and is not uncommon in adults. It is most likely to occur when excessive pressures are used for inflation or if the airway is not clear. Slight gastric distension may be disregarded. However, marked and easily discernible distension of the stomach may be deleterious because it promotes regurgitation, reduces lung volume by elevating the diaphragm, and may initiate vagal reflexes. Obvious gross distension should be relieved whenever possible. In the unconscious patient this can be accomplished by using one hand to exert moderate pressure over the patient's epigastrium between the umbilicus and the rib cage. If a second rescuer is available he can prevent recurrence of gastric distension by maintaining moderate pressure in this area. Special attention should be directed to lowering the patient's head or turning it to one side, or both, during this maneuver, to avoid aspiration of gastric contents.

Adjuncts to Exhaled-air Ventilation

No adjuncts are required for effective exhaled-air ventilation, and such devices usually do not increase the effectiveness of the method. Accordingly, there should be no delays caused by seeking such equipment or by putting it into use. If desired, a clean cloth or handkerchief can be used to cover the patient's mouth or nose, to overcome the esthetic and hygienic objections to direct oral contact. When available and properly employed, various devices (masks, tubes, airways) may be useful for professional medical and paramedical personnel.

Training People in Exhaled-air Methods

Every effort should be made to teach exhaled-air methods of artificial ventilation to as many members of the general populace as possible. The ad hoc committee urgently recommends that steps be taken to ensure training of the entire population from the fifth-grade level upward, by schools, clubs, local and national first aid groups, and medical, paramedical, and rescue organizations. Training should be according to the technique described above and in accordance with the training standards of the American National Red Cross. For optimum results it should include such media as lectures, demonstrations, posters, slides, and movies. Actual practice on life-like training manikins increases the efficiency of performance and is recommended, although it is not essential for satisfactory performance. When used, such manikins should provide obstruction of the airway when the head is not tilted back maximally, allow mouth-to-mouth and mouth-to-nose ventilation, and result in rise of the chest when the lungs are inflated.

ALTERNATIVE METHODS OF ARTIFICIAL VENTILATION

Regardless of the method used, the preservation of an open airway is essential. This can best be done by ensuring continued extension of the head and neck and forward displacement of the lower jaw. Some unconscious victims will be saved simply by establishing an open airway and permitting spontaneous breathing to resume.

Superiority of Exhaled-air Method

The ad hoc committee recommends exhaled-air ventilation as being unequivocally superior to all manual methods since it provides for control of the airway at all times, allows immediate assessment of airway obstruction, monitors ventilation on a breath-to-breath basis, and provides more actual ventilation than any of the manual methods. Furthermore, external cardiac compression cannot be used effectively in conjunction with manual methods. Manual methods should be used only in special circumstances that make it impossible or ill-advised to use exhaled-air methods. These include crushing facial injuries, conditions where the head is trapped, situations on the pole-top, etc.

Manual Methods

Those rescuers who cannot or will not use the exhaled-air techniques should use a manual method. No one manual method can be recommended as being unequivocally superior to the others. Some techniques are more effective than others, depending on the circumstances, and the rescuer should not be limited to the use of of a single manual method for all cases.

THE SUPINE CHEST-PRESSURE ARM-LIFT METHOD (Silvester) is generally preferred, especially when used with an improvised support under the shoulders. This tends to maintain an open airway because of backward tilt of the patient's head when he is in the supine position. Turning his head to one side is also useful in that it reduces the hazard of aspiration of vomitus. When a second rescuer is available, he should endeavor to maintain proper head and neck position or forward displacement of the mandible or both. When possible, the patient should be inclined to promote drainage of the lungs, with his head lower than his body.

THE PUSH-PULL PRONE MANUAL METHODS—Back-Pressure Arm-Lift (Holger Nielsen) and Back-Pressure Hip-Lift—can provide some ventilation, but the possibility of airway obstruction is always present, even with assistance of a second rescuer to keep the patient's head extended and to displace his mandible forward. The advantage claimed, that gravity alone will prove an open airway by causing forward displacement of the mandible with use of the prone position, cannot be substantiated in the unconscious patient. Furthermore, the prone position restricts movement of the thorax and abdomen, thereby restricting ventilatory volume.

CARDIOPULMONARY RESUSCITATION (CPR) (HEART-LUNG RESUSCITATION [HLR])

Restoration of Circulation

After three to five effective lung inflations, the carotid pulse should be checked. The operator maintains head-tilt with one hand; with index and middle fingers of the other hand he gently locates the larynx, and, after sliding laterally with the fingers flat, he palpates the carotid area. The pulse should be "felt," not "compressed." If apnea persists and there is unconsciousness, death-like appearance, and absence of carotid pulse, external cardiac compression should be started.

External Cardiac Compression

External cardiac compression consists of the application of rhythmic pressure over the lower half of the sternum. This compresses the heart and produces a pulsatile artificial circulation because the heart lies almost in the middle of the chest between the lower sternum and the spine. When properly performed, it can produce systolic blood-pressure peaks of over 100 mm Hg, with a mean blood pressure of 40 to 50 mm Hg in the carotid artery and carotid arterial blood flow of up to 35 percent of normal.

External cardiac compression must always be accompanied by artificial ventilation. Compression of the sternum produces some artificial ventilation but not enough for adequate oxygenation of the blood. For this reason artificial ventilation must be used whenever external cardiac compression is used.

Effective external cardiac compression requires sufficient pressure (80 to 120 pounds) to depress the patient's lower sternum $1\frac{1}{2}$ to 2 inches in an adult; the rate should be once a second. For external cardiac compression to be effective, the patient must be on a firm surface. If he is in bed, a board or improvised support should be placed under his back, but compression must not be delayed for that purpose. The rescuer stations himself at the side of the patient and places only the heel of one hand over the lower half of the sternum. Care must be exercised not to place the hand over the tip or xiphoid process of the sternum which extends down over the upper abdomen. He places his other hand on top of the first one and then rocks forward so that his shoulders are almost directly above the patient's chest. Keeping his arms straight, with elbows locked, he exerts adequate pressure almost vertically downward to move the lower sternum $1\frac{1}{2}$ to 2 inches in an adult. The preferred rate of 60 per minute is usually rapid enough to maintain blood flow and slow enough to allow cardiac refill; it is also practical in that it avoids fatigue and facilitates timing on the basis of one compression per second. The compressions should be regular, smooth, and uninterrupted, with compression and relaxation being of equal duration. Under no circumstance should compression be interrupted for more than five seconds.

When there are two rescuers, optimum ventilation and circulation are achieved by quickly "interposing" one inflation after each five chest compressions without any pause in compressions (5:1 ratio). One rescuer performs external cardiac compression while the other one remains at the patient's head, keeps it tilted back, and continues exhaled-air ventilation. Interposing the breaths without any pause in compressions is important, since every interruption in cardiac compression results in a drop of blood flow and blood pressure to zero. When there is only one rescuer, he must perform both artificial ventilation and artificial circulation using a 15:2 ratio, i.e., two quick lung inflations after each 15 chest compressions.

The Technique for Children

The technique is similar for children, except that the heel of only one hand is used for small children and only the tips of the index and middle fingers for babies. The ventricles of infants and small children lie higher in the chest and the external pressure should be exerted over the mid-sternum. The danger of lacerating the liver is greater in children because of the smallness and pliability of the chest and the higher position of the liver under the lower sternum and xiphoid. The force of compression should be sufficient to move the sternum about one fifth of the distance from the front to the back of the chest. In infants up to 1 year old this results in adequate blood-pressure peaks with 10 to 15 lb of pressure. Children up to 6 years old require 20 to 25 lb of pressure. The compression rate should be about 100 times per minute with breaths interposed after each five compressions. For infants and small children, backward tilt of the head arches the back off the horizontal. Firm support for external cardiac compression can be provided if the rescuer slips one hand beneath the patient's back while using the other hand to compress his chest. An alternate method for small infants is to encircle the chest with the hands and compress the mid-sternum with both thumbs.

Determining the Status of the Patient

The reaction of the pupils should be checked during cardiopulmonary resuscitation, since it provides one of the best clues to the overall status of the patient. A pupil which constricts when exposed to light indicates adequate oxygenation and blood flow to the brain. If the pupils remain widely dilated and do not react to light, serious brain damage is imminent or has occurred. Dilated but reactive pupils are less ominous. Normal pupillary reactions may be altered by the administration of drugs.

Periodic palpation of the carotid or femoral pulse should be employed to check the effectiveness of external cardiac compression or the return of a spontaneous effective heartbeat. The carotid pulse is more meaningful and more practical to use than the femoral or radial pulse.

Avoiding Complications

Complications can occur from the use of cardiopulmonary resuscitation. These include fracture of the ribs and sternum, laceration of the liver, and fat emboli. They can be minimized by careful attention to details of performance. Several important admonitions for proper performance of this technique are as follows: (a) Never compress over the xiphoid process at the tip of the sternum. This bony prominence extends down over the abdomen and pressure on it may cause laceration of the liver, which can be lethal. (b) Never let your fingers touch the patient's ribs when compressing. Keep just the heel of your hand in the middle of the patient's chest over the lower half of his sternum. (c) Never use sudden or jerking movements to compress the chest. The action should be smooth, regular, and uninterrupted, with 50 percent of the cycle compression and 50 percent relaxation. (d) Never compress the abdomen and the chest simultaneously. This traps the liver and may cause it to rupture.

Urgency

There must always be a maximum sense of urgency in starting external cardiopulmonary resuscitation. The outstanding advantage of this technique is that it permits the earliest possible treatment of cardiac arrest without any equipment and by other than professional medical personnel. The time between recognition of need and start of treatment should be measured in seconds.

A maximum sense of urgency must continue throughout the resuscitation procedures. Never interrupt heart-lung resuscitation for more than five seconds at a time for any purpose—moving the patient, changing rescuers, checking the pulse or electrocardiogram, injecting drugs, intubating the airway, applying mechanical devices, etc. With practice and attention to details this can usually be accomplished. The concept that the state of urgency no longer exists once artificial ventilation and artificial circulation have been established is completely erroneous. The longer it is necessary to continue cardiopulmonary resuscitation, the less likely it is to succeed. Even under optimum circumstances deterioration is progressive during heart-lung resuscitation, and the sense of urgency must be preserved until all efforts are abandoned.

When Internal Cardiac Compression Is Necessary

External cardiopulmonary resuscitation may be ineffective when certain conditions exist. These include crushing injuries of the chest, internal thoracic injuries, cardiac tamponade, tension pneumothorax, or severe emphysema with enlargement and fixation of the rib cage. If it can be determined that any of these conditions is present and if a physician is present with the necessary skill, equipment, and facilities in a hospital or other medical installation, he should open the chest, and perform internal cardiac compression in conjunction with artificial ventilation. A decision to use the internal method of cardiac compression may also be made under the proper circumstances by physicians who doubt the efficacy of the closed-chest method in particular cases or after prolonged application.

Deciding Not to Start CPR

Cardiopulmonary resuscitation should not be started when it is known or can be determined with a degree of certainty that cardiac arrest has persisted for more than five or six minutes (probably longer in drowning). If there is a question of the exact duration of the arrest, the patient should be given the benefit of the doubt and resuscitation started. If cardiopulmonary resuscitation is started in the absence of a physician, it should be continued until one is available to assume responsibility. Cardiopulmonary resuscitation is not indicated in a patient who is known to be in the terminal stage of an incurable condition.

When to Stop CPR

The decision to stop cardiopulmonary resuscitation is a medical one and depends upon an assessment of the cerebral and cardiovascular status. The best criteria of adequate cerebral circulation are the reaction of the pupils, the level of consciousness, movement, and spontaneous respiration. Deep unconsciousness, absence of spontaneous respiration, and fixed, dilated pupils for 15 to 30 minutes are indicative of cerebral death, and further resuscitative efforts are usually futile. Cardiac death may be assumed when there is no return of electrocardiographic activity after one hour of continuous cardiopulmonary support. In children, resuscitative efforts should be continued for longer periods than in adults since recovery has been seen even after prolonged unconsciousness.

TRAINING IN CARDIOPULMONARY RESUSCITATION

Heart-lung resuscitation is an emergency procedure which requires special ability in the recognition of cardiac arrest and special training in its performance. All training programs should adhere to the standards of the American Heart Association as detailed in its training manual. Experience has shown that, in addition to lectures, demonstrations, slides, and films, actual practice in both the ventilatory and the circulatory components of cardiopulmonary resuscitation is required on life-like manikins. An initial course in cardiopulmonary resuscitation should require a minimum of three hours of training for small groups of students and should include sufficient supervised, intensive manikin practice for every student to become proficient in detecting the presence or absence of the pulse and in performance of the individual steps and sequence of exhaled-air ventilation, external cardiac compression, and the combination of the two. Courses for instructors in cardiopulmonary resuscitation should, whenever possible, include practice in all of the ABCDs on experimental animals.

Retraining

Retraining or refresher courses which include manikin practice are required for all personnel. The exact frequency of such retraining may need to be regulated on the basis of the professional skill and experience of particular groups. However, the retraining requirements suggested at present for other than medical groups are twice the first year and annually thereafter.

Who Should Be Trained

All physicians, dentists, osteopaths, nurses, inhalation therapists, and rescue personnel should be carefully trained in heart-lung resuscitation. Training should be provided by either physicians or instructors who have had special instructor courses in cardiopulmonary resuscitation under such organized programs as those of the American Heart Association and the Public Health Service.

Cardiopulmonary resuscitation should not be taught to the general public at the present time. However, the committee recognizes that carefully controlled pilot projects should be carried out with highly motivated, select groups of lay personnel in order to determine the feasibility and effectiveness of such programs. Furthermore, instruction in heart-lung resuscitation for the general public cannot be inaugurated until sufficient numbers of trained instructors are available.

Certification

The committee also recognizes the desirability of certification in cardiopulmonary resuscitation set up under agencies with officially approved training-retraining programs. Such certification should indicate satisfactory completion of an approved cardiopulmonary resuscitation training course, or satisfactory training to serve as an instructor in cardiopulmonary resuscitation.

USE OF MECHANICAL EQUIPMENT IN CONJUNCTION WITH CARDIOPULMONARY RESUSCITATION

Attempts at reoxygenation of the lungs by exhaled-air methods or via mask should always precede attempts at tracheal intubation. Adequate lung inflations, interposed between external cardiac compressions, require high pharyngeal pressures which promote gastric distension. Therefore, the trachea should be intubated as soon as possible by trained personnel.

MANUALLY OPERATED SELF-INFLATING BAG-VALVE-MASK UNITS are recommended, since their use by trained personnel permits assessment and correction of ventilation volumes, airway obstruction, mask leak and gastric insufflation, and proper timing of inflations without interference with external cardiac compressions.

The highest possible inhaled oxygen concentration should be provided as soon as possible. By the use of a self-inflating bag-valve-mask unit, inhaled oxygen concentrations of over 50 percent can be obtained only by attaching to the intake valve a reservoir tube with a capacity equal to the tidal volume and an oxygen inflow rate of at least the minute volume. By attaching a demand valve to the bag intake, one can obtain 100 percent inhaled oxygen concentration.

Conventional pressure-cycled automatic ventilators or resuscitators are not recommended for use in conjunction with external cardiac compression because effective cardiac compression triggers termination of the inflation cycle prematurely, producing shallow and insufficient ventilation. The inflation flow rates of this type of equipment are usually inadequate.

OXYGEN-POWERED MANUALLY TRIGGERED VENTILATION DEVICES are acceptable if they can provide instantaneous flow rates of 1 liter/sec, or more, for adults. A safety-valve release pressure of about 50 cm of water should be provided. Ideally, they should permit the use of 100 percent oxygen and support of airway and mask with both hands. For use on infants and small children, specialized mechanical breathing devices producing lower flow rates are required.

EXTERNAL CARDIAC COMPRESSION MACHINES are adjuncts which may be used when prolonged resuscitation or transportation of the patient is required. They should be designed to approximate performance of the manual method. Their design should facilitate head-tilt and quick application of the machine, and minimize the danger of accidental malpositioning of the plunger during use. If automatic ventilation is also provided, it should be programmed to inflate the lungs after every fifth compression without a pause. When inflation is by face mask, gastric distension may occur and is an indication for gastric decompression and tracheal intubation.

External cardiac compression must always be started with the manual method first. When mechanical devices are used, one rescuer must always remain at the patient's head to monitor plunger action and ventilation, to check the pulse and pupils, and to support head-tilt and mask-fit manually unless the trachea is intubated.

ACKNOWLEDGMENT: This chapter is reprinted with permission from the *Journal of the American Medical Association*, Vol. 198, No. 4, pp. 373–379 (October 24, 1966).

Members of the ad hoc committee: Warren H. Cole, MD, *Chairman*; Larry H. Birch, MD, Butterworth Hospital, Grand Rapids, Mich.; James O. Elam, MD, Kansas City (Mo.) General Hospital; Archer S. Gordon, MD, Lovelace Clinic and Foundation, Albuquerque, N.M.; James R. Jude, MD, University of Miami (Fla.) School of Medicine; Peter Safar, MD, University of Pittsburgh School of Medicine; Leonard Scherlis, MD, University of Maryland School of Medicine, Baltimore; *Ex Officio Members*: Robert L. Flynn, MD, U.S. Public Health Service, Washington, D.C.; Robert M. Oswald, American National Red Cross, Washington, D.C.; J. Keith Thwaites, American Heart Association, New York; Leroy D. Vandam, MD, Peter Bent Brigham Hospital, Boston; Sam F. Seeley, MD, NAS-NRC staff, Washington, D.C.

This study was supported by the Division of Health Mobilization, Public Health Service, contract PH86-65-104, and by the American National Red Cross.

REFERENCES

1. The Closed Chest Method of Cardiopulmonary Resuscitation: Benefits and Hazards, editorial, *Circulation 26*:324 (Sept. 1962).
2. The Closed-Chest Method of Cardiopulmonary Resuscitation: Revised Statement, editorial, *Circulation 31*:641 (May 1965).

First Aid Procedures

American Medical Association

This chapter is an abridged and slightly modified version of the AMA First Aid Manual (1969 Revision) and is reprinted with the permission of the American Medical Association.

INTRODUCTION

First aid is the "immediate and temporary care given the victim of an accident or sudden illness until the services of a physician can be obtained."* The first objective is to save life:

Prevent heavy loss of blood Prevent shock
Maintain breathing   Send for a physician
Prevent further injury

The first aider must also:

Avoid panic
Inspire confidence
Do no more than necessary until professional help arrives

Common sense and a few simple rules are the keys to effective first aid.

This chapter presents general recommendations of the American Medical Association for first aid. It supplements but does not replace instruction in first aid techniques. Every person should receive basic instruction in first aid if possible.

HEAVY BLEEDING

Heavy bleeding comes from wounds to one or more large blood vessels. Such loss of blood can kill the victim in 3 to 5 minutes.

DON'T WASTE TIME... USE PRESSURE DIRECTLY OVER THE WOUND

DO: Place pad—clean handkerchief, clean cloth, etc.—over the wound and press firmly with your hand or both hands. If you haven't a pad or bandage, close the wound with your hand or fingers.

Apply pressure directly over the wound.

Hold the pad firmly in place with a strong bandage—neckties, cloth strips, etc.

Raise the bleeding part higher than the rest of the body unless bones are broken.

Keep victim lying down.

Call a physician.

DON'T: Never use a tourniquet to control bleeding except for an amputated, mangled, or crushed arm or leg, or profuse bleeding that cannot be stopped otherwise.

At this point you should look to the needs of other accident victims. Try to control bleeding and maintain breathing for as many of the victims as possible. Then, go back to the victim whose bleeding has been controlled and do the following:

DO: Keep the victim warm. Cover with blankets, coat, etc., and put something under him if he is on a cold or damp surface.

*American National Red Cross: First Aid Textbook (4th Ed.; Washington: Doubleday & Company, Inc., 1957).

If the victim is conscious and can swallow, give him plenty of liquids to
drink (water, tea, coffee).

DON'T: Do not give the victim alcoholic drinks.

If the victim is unconscious or if abdominal injury is suspected, do not give
him fluids.

MAINTAIN BREATHING

There is need for help in breathing when breathing movements stop or lips, tongue,
and fingernails become blue. When in doubt, begin artificial respiration. No harm can
result from its use and delay may cost the victim his life.

Artificial Respiration

DO: Start immediately. Seconds count.

Check mouth and throat for obstructions.

Place victim in position and begin artificial respiration.

Maintain steady rhythm of 15 breaths per minute.

Remain in position. After the victim revives, be ready to resume artificial
respiration if necessary.

Call a physician.

DON'T: Do not move the victim unless absolutely necessary to remove from danger.

Do not wait or look for help.

Do not stop to loosen clothing or warm the victim.

Do not give up.

Directions for mouth-to-mouth breathing and the manual method of artificial
respiration follow:

Mouth-to-Mouth Breathing for Adults

DO: Place victim half-way between a face-up and a side position.

Lift victim's neck with one hand and tilt his head back by holding the top of
his head with your other hand.

Pull the victim's chin up with the hand that was lifting the neck so that the
tongue does not fall back to block the air passage.

Take a deep breath and place your mouth over the victim's nose or mouth,
making a leak-proof seal.

Blow your breath into the victim's mouth or nose until you see the chest rise.
The air you blow into the victim's lungs has enough oxygen to save his
life.

Remove your mouth and let the victim exhale while you take another deep
breath.

As soon as you hear the victim breathe out, replace your mouth over his
mouth or nose and repeat procedure.

Repeat 15 times per minute.

Manual Method of Artificial Respiration

DO: Place victim in a face-up position.

Place something under victim's shoulders to raise them and allow the head to
drop backward.

Kneel above victim's head, facing the victim.

Grasp victim's arms at the wrists, crossing and pressing victim's wrists against the lower chest.

Immediately pull arms upward, outward, and backward as far as possible.

Repeat 15 times per minute.

If a second rescuer is present, he should hold the victim's head so that it tilts backward and the jaw juts forward.

Advantages to Mouth-to-Mouth Breathing

1. The victim does not need to be placed on the ground or in any special position. Mouth-to-mouth breathing can be given when the victim is in water or in cramped surroundings.

2. No special equipment is needed.

3. Rescuers can maintain mouth-to-mouth breathing for hours without fatigue, even when the victim is twice the size of the rescuer.

4. Hands are kept free to keep the head tilting backward and the jaw jutting forward. This prevents obstruction of the air passage, the most common cause of failure of any method of artificial respiration.

5. The rescuer can see, feel, and hear the effect of each lung inflation.

6. The rescuer can control the amount, rate, and pressure of air administered to the victim.

BREATHING STOPPED

A person who has stopped breathing will die if his breathing is not restored immediately. Even if breathing is restored, patients who have stopped breathing need hospitalization. Call an ambulance whenever feasible.

THESE THINGS CAN STOP BREATHING:

- Poisonous gases in the air or lack of oxygen
DO: Move victim to fresh air.

         Begin mouth-to-mouth breathing.

         Control source of poisonous gases if possible.

         Keep others away from the area.

DON'T: Do not enter an enclosed space to rescue an unconscious victim without first being equipped with a self-contained or air-supplied breathing apparatus.

- Electric Shock
DON'T: Do not touch victim until he is separated from current.

         Do not try to remove a person from an out-of-doors wire unless you have had special training for this type of rescue work.

DO: Call the power company and ask that the current be turned off. Or if you know how, pull the switch.

         Begin mouth-to-mouth breathing as soon as the victim is free of contact with current.

- Drowning

- Concussion resulting from explosions or blows to the head or abdomen

- Poisoning: Ingestion of sedative or chemical

- Constriction from cave-in

DO: Begin mouth-to-mouth breathing immediately

SHOCK

Shock usually accompanies severe injury or emotional upset. It may also follow infection, pain, disturbance of circulation from bleeding, stroke, heart attack, heat exhaustion, food or chemical poisoning, extensive burns, etc.

Signs of Shock

Cold and clammy skin with beads of perspiration on the forehead and palms of hands.

Pale face.

Complaint by the victim of a chilled feeling, or even shaking chills.

Frequently, nausea or vomiting.

Shallow breathing.

Save Life by Prevention of Shock

DO: Correct cause of shock if possible (e.g., control bleeding).

Keep victim lying down.

Keep his airway open. If he vomits, turn his head to the side so that his neck is arched.

Elevate victim's legs if there are no broken bones. Keep his head lower than trunk of the body if possible.

DO: Keep victim warm if weather is cold or damp.

Give fluids (water, tea, coffee, etc.) if the victim is able to swallow. The following formula can be used if available: 1 liter water, 5 grams salt, 2 grams sodium bicarbonate (1 quart water, 1 teaspoon salt, $\frac{1}{2}$ teaspoon baking soda (sodium bicarbonate).

Reassure victim.

DON'T: Never give alcoholic beverages.

Do not give fluids to unconscious or semi-conscious persons.

Do not give fluids if abdominal injury is suspected.

Prevention of shock should be considered with every injury and illness discussed.

POISONING

When to Suspect Poisoning

Odor of poison on the breath.

Discoloration of lips and mouth.

Pain or burning sensation in throat.

Unconsciousness, confusion, or sudden illness when access to poisons is possible.

Whenever bottles or packages of drugs or poisonous chemicals are found open in presence of children.

What to do Until You Contact a Physician

Speed is essential. Act before the body has time to absorb the poison. If possible, one person should begin treatment while another calls a physician or ambulance.

Save and give to the physician or hospital the poison container with its intact label and any remaining contents. If the poison is unknown, bring along the vomitus for examination.

The nature of the poison will determine the first aid measure to use.

SWALLOWED POISONS

DON'T: Do not induce vomiting if the victim

... is unconscious

... is in convulsions

... has symptoms of severe pain, burning sensation in mouth or throat, vomiting

... is known to have swallowed a petroleum product (kerosene, gasoline, lighter fluid), toilet bowl cleaner, rust remover, drain cleaner, lye, acids for personal or household use, iodine, styptic pencil, washing soda, ammonia water, household bleach.

DO: Call a physician immediately.

Begin mouth-to-mouth breathing if victim has difficulty breathing.

Give water or milk.

If safe (see above), induce vomiting. Induce vomiting by placing your finger at the back of the victim's throat or by use of 10 grams salt in 200 cc. of warm water (2 teaspoons in a glass of warm water) or use 30 cc. or one ounce of syrup of ipecac; one half ounce for a child.

When retching and vomiting begin, place the victim face down with head lower than hips. This prevents vomitus from entering the lungs and causing further damage.

How to Prevent Poisoning

Keep all drugs, poisonous substances, and household chemicals out of reach of children.

Do not leave discarded medicines where children or pets might get at them.

Do not store nonedible products on shelves used for storing food or food in laboratory areas.

Never give or take medicines in the dark.

Read labels before using chemical products.

Do not keep unneeded or unlabeled drugs and chemicals.

Never re-use containers of chemical substances.

Do not transfer poisonous substances to unlabeled containers.

INHALED POISONS

DO: Carry or drag victim (do not let him walk) to fresh air immediately.

Apply artificial respiration if breathing has stopped or is irregular.

Call physician.

Keep victim warm. ·

Keep victim as quiet as possible.

DON'T: Never give alcohol in any form.

Do not become a victim by exposure to the same poison.

SKIN CONTAMINATION (*Chemical Burns*)

DO: Drench skin with water (shower, hose, faucet).

Apply stream of water on skin while removing clothing.

Cleanse skin thoroughly with water. Speed in washing is most important in reducing extent of injury.

Burns from use of certain chemicals in technical areas require special first aid techniques. These techniques should be known by anyone working in such areas.

EYE CONTAMINATION (*Chemical or Other Foreign Material in Eye*)

DO: Immediately hold eyelids open and wash eyes with gentle stream of running water.

Delay of a few seconds greatly increases extent of injury.

Continue washing for at least 15 minutes; then take victim to a physician for treatment.

DON'T: Do not use boric acid ointments or any other chemicals. They may increase the extent of injury.

BURNS

Burns can result from heat (thermal burn) or from chemicals (chemical burn).

Every burn, even sunburn, can be complicated by shock, and the patient should be treated for shock.

Prevent shock ... Prevent contamination ... Control pain ... These are the objectives of first aid care for burns.

A person with "burn shock" may die unless he receives immediate first aid.

In "burn shock" the liquid part of the blood is sent by the body into the burned areas. There may not be enough blood volume left to keep the brain, heart, and other organs functioning normally.

Extensive Thermal Burn

DO: Place the cleanest available cloth material over all burned body areas to exclude air.

Have victim lie down.

Call physician. Keep victim lying down.

Place victim's head and chest a little lower than the rest of the body. Raise the legs if possible.

If the victim is conscious and can swallow, give him plenty of nonalcoholic liquids to drink (water, tea, coffee, etc.).

Move to hospital by ambulance immediately.

Small Thermal Burns

DO: Soak a sterile gauze pad or clean cloth in a baking soda solution: 2 tablespoons baking soda (sodium bicarbonate) to a quart of lukewarm water (24 to 30 grams sodium bicarbonate to 1 liter of lukewarm water).

Place pad over burn and bandage loosely.

Do not disturb or open blisters.

If skin is not broken, immerse burned part in clean, cold water or apply clean ice to relieve pain.

All burns, except where the skin is reddened in only a small area, should be seen by a physician or nurse.

Chemical Burns

DO: Immediately flush with water. Speed in washing is most important in reducing the extent of injury.

Apply stream of water while removing clothing.

Place the cleanest available material over the burned area.

If the burned area is extensive, have the victim lie down. Keep him down until a physician comes. Place his head and chest a little lower than the rest of the body (raise legs if possible). If he is conscious and can swallow, give him plenty of non-alcoholic liquids to drink.

All burns, except where the skin is reddened in only a small area, should be seen by a physician.

Burns from use of certain chemicals in technical areas require special first aid techniques. These techniques should be known by anyone working in such areas.

DON'T: Do not apply ointments, greases, baking soda or other substances to extensive burns.

CUTS AND ABRASIONS

In caring for minor wounds, it is most important to prevent infection.

DO: Immediately cleanse wound and surrounding skin with soap and warm water, wiping away from wound.

Hold a sterile pad firmly over the wound until the bleeding stops. Then change pad, and bandage loosely with a triangular or rolled bandage.

Replace sterile pad and bandage as necessary to keep them clean and dry.

DON'T: Never put mouth over a wound. The mouth harbors germs that could infect the wound. Do not breathe on wound.

Do not allow fingers, used handkerchiefs, or other soiled material to touch the wound.

Do not use an antiseptic on the wound.

HEAT AND COLD INJURIES

Heat Exhaustion

Symptoms:
Pale and clammy skin.
Pulse rapid and weak.
Victim complains of weakness, headache, or nausea.
Victim may have cramps in abdomen or limbs.

DO: Have victim lie down with his head level to or lower than his body.
Move victim to cool place, but protect him from chilling.
Give the victim salt water (5 grams salt to 1 liter of water) (1 teaspoon salt to 1 quart water) to drink if he is conscious.
Call a physician.

Heat Stroke

Symptoms:
Flushed and hot skin.
Pulse rapid and strong.
Victim often is unconscious.

DO: Call a physician.
Cool body by sponging it with cold water or by cold applications.
If the victim is fully conscious and can swallow, give him salt water; 5 grams salt to 1 liter of water (1 teaspoon salt to 1 quart water).

DON'T: Do not give alcohol in any form.

Frostbite

Symptoms:

Skin pink just before frostbite develops.

Skin changes to white or greyish-yellow as frostbite develops.

Initial pain, which quickly subsides.

Victim feels cold and numb; he usually is not aware of frostbite.

DO: Cover the frostbitten part with a warm hand or woolen material. If fingers or hand are frostbitten, have victim hold his hand in his armpit, next to his body.

Bring victim inside as soon as possible.

Place frostbitten part in warm water, about 42°C (108°F).

Gently wrap the part in blankets if warm water is not available or is impractical to use.

Let circulation reestablish itself naturally.

When the part is warmed, encourage the victim to exercise fingers and toes.

Give victim a warm, nonalcoholic drink.

DON'T: Do not rub with snow or ice. Rubbing frostbitten tissue increases the risk of gangrene.

Do not use hot water, hot water bottles, or heat lamps over the frostbitten area.

MOVING THE INJURED

Do not move an injured person before a physician or experienced ambulance crew arrives, unless there is real danger of his receiving further injury by being left at accident site. If possible, control bleeding, maintain breathing, and splint all suspected fracture sites before moving. If this is not possible, follow these general rules.

Pulling the Victim to Safety:

Pull victim head first or feet first, not sideways. Be sure head is protected.

Lifting the Victim to Safety:

If he must be lifted before a check for injuries can be made, every part of the body should be supported. The body should be kept in a straight line and should not be bent.

In carrying the injured person to an area where a stretcher can be manipulated, use either the two-, or three-man method, depending on the type and severity of the injury, the available help, and the physical surroundings (stairs, walls, narrow passages, etc.). The one- and two-man carry systems are ideal for transporting a person who is unconscious from asphyxiation or drowning, but are unsuited for carrying a person suspected of having fractures or other severe injuries. In these cases always use the three-man carry method.

An effective stretcher can be made by buttoning two shirts or a coat over two sturdy poles, or by wrapping a blanket in thirds about the poles. If the victim must be moved, a stretcher is the best means for transportation.

HISTORY OF INJURY OR ILLNESS

When other first aid measures have been taken and as time allows, the first aider should note the following information:

Identity of the victim.

Those persons the victim would like notified (including clergy).

Circumstances of the injury or illness.

Any special first aid measures taken (mouth-to-mouth breathing, administration of fluids, application of tourniquet, etc.).

Any disease or disability existing prior to the injury or illness (diabetes, heart trouble, allergies, etc.).

Hydrofluoric Acid Burn Treatment

INTRODUCTION

The treatment of hydrofluoric acid (HF) burns has been in a state of flux in the past, with numerous remedies being tried. The treatment recommended in this chapter is based upon that initiated in 1954 by E. E. Evans, M.D., who was the plant medical director at the Du Pont Chambers Works at that time. It is felt that this method of treatment is more satisfactory than that in prior use.

The approach given here centers around the use of certain high-molecular-weight quaternary ammonium compounds. The treatment consists basically in thorough and immediate flushing with water, followed by soaking in an iced solution of benzalkonium chloride (U.S.P., in a concentration of 0.1 to 0.133 %). It is imperative to treat HF burns immediately; any delay may greatly increase the severity of the burn.

The mechanism by which quaternary ammonium compounds alleviate the destructive action of HF has not been studied, but several routes of action have been postulated:

1. The quaternary ammonium nitrogen may exchange ionized chloride for fluoride ion to produce a nonionized fluoride complex in a manner similar to the sequestering action exerted by Versene for calcium and heavy metal ions.
2. The quaternary ammonium compound may directly alter the permeability of tissue cell membranes.
3. As a secondary effect, the compound may control invasive microorganism infection.
4. By reduction of surface tensions, better contact may be promoted between aqueous fluids and tissue components.

The use of cold makes a contribution by constricting lymph and blood vessels so as to delay or retard the passage of fluoride ion.

TREATMENT

Individuals who have had contact with HF should be showered immediately under a drenching spray of water. Contaminated clothing should be removed as rapidly as possible, even while the victim is in the shower. These things should be done at the site of the accident. It is essential that the exposed area be washed with a copious quantity of water for a sufficient period to remove all the HF from the skin or eyes. Speed in removing the patient from a contaminated atmosphere or removing HF from the affected area is of critical importance. After the initial shower, medical assistance should be obtained immediately.

On arrival at the medical facility, the patient is rapidly assessed for shock, and, if it is present, he is treated accordingly. If the patient's over-all condition does not contraindicate, he is given another shower. The affected areas are then soaked in iced aqueous or alcoholic benzalkonium chloride solution in a concentration of 0.1 to 0.133 %. If there is a significant delay in securing this solution, ice water or cold tap water may be used temporarily. Aqueous solutions are, of course, preferred in the vicinity of the eyes and mucous membranes; even then, care should be exercised, since benzalkonium chloride may be an eye irritant at the recommended concentration.

When the part to be treated can be positioned in an open vessel, the solution is brought to a depth sufficient to cover the part, and ice cubes are added. It should be

45

stated that immersion of a part of the body in an ice bath over a prolonged period of time may cause discomfort. Relief is readily obtained by removing the part from the solution every 10 minutes, waiting a few minutes, and then immersing it again.

If immersion is not practical, ice cubes are inserted between layers of gauze to form a compress which is then continually soaked with benzalkonium chloride solution. Towels are placed over the gauze to conserve cold and solution when possible. Experience indicates that these benzalkonium chloride soaks should be used for intervals varying from one to four hours, depending on the appearance and extent of the burn. A precaution for the use of these compresses should be mentioned. The ice should be cubed, not crushed, and should not be under any significant pressure to avoid a reaction similar to freezing or frostbite.

Should blisters form, complete debridement is necessary; all the white raised tissue should be cut away. The use of Elase ointment (fibrinolysin and deoxyribonuclease, combined [Bovine], Parke-Davis & Company) has proved quite effective in keeping the blistered areas free of debris.

After the soaks, HF ointment is applied to the burned area and a compression dressing is applied. The formula for the HF ointment is: 3 ounces of magnesium oxide powder, 4 ounces of heavy mineral oil, and 11 ounces of white petrolatum. Recently A&D ointment (a well-known vitamin ointment) or a topical steroid has frequently been used rather than the HF ointment because HF ointment hardens and is difficult to remove. Whichever ointment is used, it should be applied daily for several days, the exact time depending on the appearance of the burn.

Burns around the fingernails are extremely painful. They may require special treatment, to relieve the pain as well as to prevent infiltration of the HF into the deeper structures with the resulting destruction of tissues which may proceed to bone involvement. The nails may be split from the distal end of the nail bed to allow free drainage. Burns in this area should be soaked in iced benzalkonium chloride solution, as recommended above.

Eye Contact

If liquid HF has entered the eyes or if the eyes have been exposed to high concentrations of the vapor, they should be flushed with large quantities of clean water for 15 minutes. Repeated flushing may be required two or three times at intervals of 15 minutes.

The eyelids should be held apart during the irrigation to ensure contact of the water with all the tissues of the surface of the eyes and lids. Ice compresses should be applied intermittently for at least an hour when not irrigating. Pain can be relieved with two or three drops of 0.5% Pontocaine® solution or ointment. Further treatment can be instituted with one of the many eye solutions containing cortisone. The liquid is preferable to the ointment. If there is much blepharitis, a small amount of an ointment may be spread on the edge of the lid and into both angles. After local treatment, protection of the eye is secured with a compression patch until the inflammation has subsided. Specialist consultation should be obtained immediately if there is any doubt as to the degree of injury or of the professional skill available.

Fume Inhalation

High concentrations of fumes in the respiratory tract may cause burns more critical than those on exposed parts. Immediate removal to an uncontaminated atmosphere and prompt medical attention are required. To prevent the development of severe lung congestion (pulmonary edema), 100% oxygen inhalation should be started as soon as possible. Unpressurized inhalation with a respirator-type mask may be satisfactory. The use of positive pressure types of apparatus is predicated on the clinical findings.

Oxygen inhalation must be continued as necessary to maintain the normal color of the skin and mucous membranes. It may be advisable, even in borderline cases, to continue oxygen at half-hour intervals for three to four hours. If at the end of this period there are no signs of pulmonary edema, breathing is easy, and the color is good, oxygen may be discontinued.

These patients should be kept under observation for at least 24 to 48 hours. They should be kept warm and at complete rest throughout the treatment. Auxiliary treatment with bronchodilators and systemic steroids may be used as required.

Inflammatory reaction in the mouth, nose, and pharynx is difficult to treat; and, although ice applications may aid in reducing edema, specialized consultation may be necessary. Acute laryngeal edema is a complication which could be disastrous if not recognized. Adequate arrangements for tracheotomy should be incorporated in the treatment procedure, as laryngeal edema is an acute condition which cannot tolerate delay.

ACKNOWLEDGEMENT: This chapter is based on the article "Hydrofluoric Acid Burn Treatment," by C. F. Reinhardt, M.D., W. G. Hume, M.D., A. L. Linch, and J. M. Wetherhold, M.D., which was published in the *American Industrial Hygiene Association Journal*, *Vol. 27*, pp. 166–171, March–April, 1966. In that article Hyamine®, which is also a quaternary ammonium compound, was recommended for the treatment of HF burns. Because Hyamine® can no longer be obtained for the purpose of treating HF burns, it is now recommended that benzalkonium chloride be used.

This condensed and revised version of the original article is presented through the courtesy of C. F. Reinhardt, M.D., and A. L. Linch.

Containers and Labeling

Norman V. Steere

The volume, material and labeling of containers for chemicals have a significant bearing on the prevention and control of laboratory accidents. The duration and intensity of laboratory fires will be greatly increased if large volumes of flammable solvents become involved, which is likely to happen if there are many glass bottles or cans larger than one liter stored in the laboratory. Inadequate labeling of containers can be expected to result in problems if the contents cannot be identified quickly and positively, if the age of the contents which may peroxidize cannot be determined readily, or if the fire, health, and reactivity hazards of the contents must be found in a handbook or reference in the library.

CONTAINERS

VOLUME AND QUANTITY OF CONTAINERS. The volume of individual containers and the total volume of flammable solvents in the laboratory should be limited to the minimums practicable for research and operations. The argument against arbitrary restrictions of container size or total volume is interference with laboratory operations, but the argument against unrestricted volume and container size is the potential that intense or prolonged fire will interfere seriously with the operations of many laboratories.

While planning and anticipation of need can make it possible to reduce volumes stored in the laboratory to assure uninterrupted operation, there is fundamental need of an adequately supplied stockroom in the building, backed by a good institutional storehouse or purchasing system. Such stockrooms and supply systems are necessary for any successful effort to reduce quantities of chemicals stored in laboratories.

CONSTRUCTION OF CONTAINERS. There is probably no question that glass is the only acceptable material for containing a variety of chemicals, including many high purity solvents which are flammable. While ordinary metal or plastic containers may provide greater protection than glass from rupture by impact from mishandling, none of the three types of containers are certain to withstand an exposure to a laboratory fire.

Metal safety cans which will not rupture or release contents suddenly in a fire are available in terne plate, hot tin dip, monel or stainless steel, and in sizes from 473 ml. (one pint) to 18.9 liters (5 gallons U.S.). Safety cans approved by Underwriters' Laboratories, Inc. will leak at a limited rate if tipped over and have a spring-closing cap which will open to relieve internal pressure; safety cans approved by Factory Mutual Laboratories will pass the same tests and are equipped also with a flame-arresting metal screen in the pouring spout to prevent flashback if vapors are ignited.

Special safety cans are available for pouring, for waste storage and for convenient storage on shelves. Bottle carriers lined with polyurethane foam are available which will afford several minutes protection from a spill fire, and special pumps, self-closing faucets and flame-arresting vents for drums are available.

LABELING

The need for adequate labeling extends far beyond the immediate requirements of the individual user, since the individual user may not be present in case of fire or explosion when containers are broken or spilled, and he may not be around years later when the containers have deteriorated or otherwise lost their value. Therefore, wax pencil markings, abbreviations, formulas only, and code names or numbers should be avoided in favor of adequate labels.

Another argument in favor of the labeling systems described here is that good labels can serve to help educate students and technicians.

Basic Label Information

The recommendations of the Labels and Precautionary Information Committee of the Manufacturing Chemists' Association, Inc. are basic, and are included in the MCA Manual L-1. Recommendations for information that should be considered for inclusion on the label are:

1. Name of the chemical, preferably the chemical name, or the types of chemical.
2. A signal word to indicate severity of the hazard; e.g. Danger, Warning, or Caution.
3. Statement of hazards, with most serious first.
4. Precautionary measures to be taken, to avoid injury or damage, from hazards stated.
5. Instructions in case of contact or exposure if results are severe and immediate action may be necessary.

Such information is now commonly included on the labels of original packages of reagent chemicals (and consumer products) in the U.S.A.

Date and User Labeling

We recommend that labels for laboratory chemicals be dated when issued and marked with the name of the user, if it is issued to or formulated by an individual, particularly in case of solvents which tend to peroxidize or which are not common.

The reason for identifying chemical containers by name and date is to make it possible and easier to identify compounds which are out-dated, to relate quantity to rate of use, and to know to whom to turn for advice on handling or disposal.

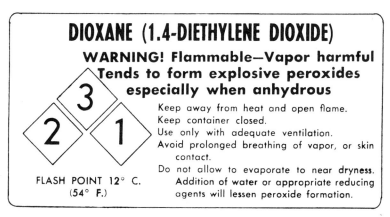

FIG. 1. Chemical Hazard Label.

Flash Point Labeling

Flammable and combustible solvents should be labeled to include their flash point, as a means of recognizing needs for ventilation, ignition source control or other precaution.

Hazard Labeling

Laboratories of Union Carbide Corporation, Western Electric, Inc., and the University of Minnesota are now using a very interesting and informative system of special labeling which deserves wider use in all types of laboratories. The system is described in National Fire Protection Association Manual No. 704M-1966, although the foreword states that the authors do not envision application of the guide to chemical laboratories.

The system identifies the order of severity of hazards from "4" to "0" (from severe hazard to little hazard) in the categories of health, flammability and reactivity. The information is given in a marking system that codes health hazard on the left in blue, flammability in the upper center in red, and reactivity on the right in yellow; the spatial arrangements are to be followed uniformly to inform persons who may be color blind, and one company adds the words that state the hazard. The bottom center space can be used for additional information such as radioactivity, unusual reactivity with water or air, and so forth.

ACKNOWLEDGEMENT: Reprinted from the *Journal of Chemical Education, Vol. 43*, A1057, December, 1966, with permission.

Chemical Waste Disposal

Norman V. Steere

Although the disposal of shock sensitive material such as ether peroxides may excite great concern, extreme precautions, and lengthy newspaper reports, the disposal of less hazardous laboratory wastes does not receive the routine attention needed.

In "The Care, Handling and Disposal of Chemicals," Gaston comments that the problem of waste disposal is "not being approached in a realistic manner by colleges and medical schools, no provision being made for sites where inflammables may be burned, or acids safely washed away."[1]

Gaston's book represents a big step toward filling the considerable gap that exists between chemical waste disposal needs and knowledge. Too frequently there is a dearth of any useful information on small-scale methods that can be used for disposing of chemicals without endangering personnel or polluting air or water supplies.

This chapter will suggest some approaches to methods which may be used for collection, handling, transport, and disposal of some chemicals that are flammable or detonable. The tables at the back of this book provide much of the information necessary to select a disposal method and the necessary precautions.

COLLECTION

If collection of chemical waste incurs a direct charge against laboratory budgets, almost all of the waste will be dumped down drains or otherwise disposed of without incurring measurable expense. Even when such collection is a "free" service, however, the collection system will likely not be used for all waste due to real or presumed inconvenience. The point is that any collection system must be convenient and without direct charge if the system is to be used for the disposal of most waste.

Collection of flammable wastes will be facilitated by a system for bringing labelled containers of waste to an area in each building that is convenient for truck pick-up and that is modified as necessary to provide fire-protected storage.

HANDLING

Waste solvents should be stored and handled in metal safety cans that are prominently labelled as to type of waste. Solvents which have contained sodium or similar material should be placed in separate waste containers from solvents which contain water, for obvious and important reasons. Grounding and other precautions to prevent static sparks or other sources of ignition are recommended.

Ethers which are suspected of containing peroxides, by visual observation of viscosity or crystal formation or by age, or which are found to contain peroxides in excess of limits which may be established as maximum, should be handled with the precautions accorded any material which friction or shock may detonate.

The precautions that should be taken for handling detonable material offer support for efforts to prevent formation of peroxide concentrations.

TRANSPORT

Transport of flammable solvent, leaking cylinders, and detonable material without incident requires planning and arrangement for suitable equipment. Beyond the provision of packing to prevent breakage of glass, rolling about of cylinders, and shock of peroxides, there is need for a vehicle or special trailer to provide separation between personnel and hazardous waste materials.

Transport of cylinders leaking toxic or corrosive gases and containers of explosive materials should proceed without delays or sudden stoppages, so that police escort or similar precautions should be provided.

FINAL DISPOSAL

Methods for final disposal of waste chemicals includes evaporation, neutralization, dilution, burial, storage, burning, and shock.

EVAPORATION. Solvents which are not combustible or which have flash points above ambient temperatures may be disposed of by evaporation if there is assurance that the vapors will not create problems. A method used for disposal of empty cans of anesthetic ether, by puncturing the end for ventilation, could be used for other empty solvent containers.[2] One university has developed a method of evaporating solvents from radioactive solutes by use of trays and large wicks.

NEUTRALIZATION AND DILUTION. Acid and alkaline materials can be carefully neutralized, and many soluble materials can be carefully diluted into a sewer or water course if the materials and concentrations will not damage the plumbing or injure biota, including man, which may be in the path of disposal. Gaston describes the construction of a useful pit for evaporation and dilution.[1]

BURIAL. Burial is a common method of disposal and one which may accomplish gradual dispersion of waste, or postponement of the problem. Burial methods should be planned to:

a. Prevent excessive quantities of injurious chemicals from reaching surface or ground waters,

b. Prevent uptake of toxic materials by crops grown over the burial site,

c. Provide enough earth cover to prevent scavengers from reaching the waste material,

d. Record the burial site so that the waste is not inadvertently dug up during subsequent grading or construction.

Some useful methods for disposing of empty pesticide containers and surplus pesticides have been described which could be used for other chemicals.[3,4]

STORAGE. The current procedures for disposal of radioactive waste in the U.S.A. is accumulation in storage, and periodic shipment to national storage centers. Carcinogenic chemicals which cannot safely be disposed of by other methods might be handled in a similar manner.

BURNING. Small quantities of flammable chemicals can be burned on the ground, or in shallow metal containers; larger quantities can be burned similarly in remote areas, preferably with a steady wind at one's back, or in incinerators.

Several chemical companies have incinerators which they use for burning waste chemicals, as do du Pont, 3M, the General Electric Research Laboratory in Schenectady, New York and the National Institutes of Health in Bethesda, Maryland.[5] Larger colleges and universities, and other institutions, which may not have access to an industrial, municipal or institutional incinerator but which have large quantities of solvent wastes should consider acquisition of special burners for disposal.

Obsolete or damaged cylinders for compressed gases may be disposed of by burial after the contents have been bled off through the valve or through holes in the sides of the cylinders made by bullets, well-aimed to prevent ricochet, or preferably by explosive charges placed correctly by qualified personnel.

SHOCK. Application of a physical shock can be the method for final disposal of shock-sensitive material such as the peroxides of ethers, and shock may be necessary for the opening of containers which will be disposed of by other methods.

The lack of a realistic or sophisticated approach to the problem of waste disposal by many institutions is typified by the common and frequently reported use of rocks to apply shock to glass containers of hazardous waste chemicals. Better methods, faster and safer, must be found and used.

A good example of an excellent disposal procedure, as well as the hazards of ether peroxides, is provided by this excerpt from a report on the disposal of a half-full 500-ml bottle of isopropyl ether by the Research Department of the Plastics Division of Imperial Chemical Industries Limited.[6]

"About five grams of a crystalline solid, presumably the peroxide, was just discernible at the bottom of the brown glass bottle, which was stoppered with a cork and had been stored in the dark. We did not try to open the bottle. When we broke it by remote control behind a safety wall in an outdoor disposal unit, the bottle exploded with a report that brought inquiries out of a nearby building."

RESPONSIBILITY

Each individual laboratory and project director has the responsibility for seeing that the laboratory's waste chemicals are safely collected, identified and stored for disposal, and that the institution is fully advised of the need for any special methods or facilities for proper disposal. Budget requests and research grant applications should include funds to pay the expenses of special neutralization, storage, shipping or other unusual costs of chemical waste disposal.

REFERENCES

1. Gaston, P. J., "The Care, Handling, and Disposal of Dangerous Chemicals," Institute of Science Technology, Northern Publishers, Aberdeen, (1964).
2. Cain, R. A., Disposing of Waste Ether Cans, *Anesthesiology, 24*, 255 (1963).
3. "Safe Disposal of Empty Pesticide Containers and Surplus Pesticides," Agricultural Research Service, U.S. Department of Agriculture, Washington, D. C., (1964).
4. Harein, P. K., "Waste Disposal," National Agricultural Chemicals Association, Washington, D. C., (1965).
5. Snow, D. I., and Fawcett, H. H., Occupational Health and Safety, Laboratory Planning, H. F. Lewis, Editor, Reinhold Publishing Corp., New York, (1962).
6. Pajaczkowski, A., personal communication, September 29, 1964.

Disposal of Hazardous Waste

Manufacturing Chemists' Association, Inc.

The Safety and Fire Protection Committee of the Manufacturing Chemists' Association, Inc. has prepared a safety guide on the recommended safe practices and procedures for disposal of hazardous waste, and the entire safety guide is reprinted here verbatim as a chapter with the permission of the Manufacturing Chemists' Association, Inc. The safety guide, known as SG-9, was adopted in 1961, and is available for 20 cents per copy from the Manufacturing Chemists' Association, Inc., 1825 Connecticut Avenue N.W., Washington, D.C., 20009.

PURPOSE

The purpose of this guide is to call attention to the hazards involved in the disposal of certain types of waste materials and to the legal difficulties and poor public relations which may result if proper disposal of these materials is not carefully planned and regularly carried out.

Any full discussion of the problem and methods of handling these materials is impractical within the limitations of this guide. It does, however, outline a basic method of approach which has been found to work reasonably well in a number of chemical organizations faced with a wide variety of waste disposal problems.

It should be noted that very serious problems of air and stream pollution, as well as serious hazards to plant personnel, may be created by the wastes resulting from a very small chemical operation. The basic methods outlined in this guide are applicable to the very smallest of chemical plant operations as well as those of considerable magnitude.

DEFINITION

While waste disposal includes the proper disposal of all unwanted materials, this guide is limited to suggestions for the handling of flammable liquids, toxic and corrosive materials and other substances which may create an air or stream pollution problem. Also included are suggestions in regard to the handling and disposal of unstable materials and leaking containers of dangerous gases or liquids.

RESPONSIBILITY

It is of the utmost importance to place the entire matter of waste disposal under the coordination control of a single individual or department specifically charged with that responsibility. The responsibility may be placed directly upon the plant safety organization or may be in some other operational unit. In any event, the safety organization shall act in a consultative capacity and shall have sufficient control power to insure the carrying out of established safe practices, to take care of emergency situations and to provide for the development of safe methods. To do this, the cooperative assistance of the medical division or industrial hygienists is also required. In small organizations, the various functions may all be combined in a single individual but his several functions and responsibilities should be recognized.

HAZARDS

In addition to the normal fire and explosion hazards of flammable liquids and their vapors and the expected hazards of toxic materials, certain other hazards must be anticipated in waste disposal handling. Some material may be severely corrosive to drainage piping. Other materials react with water or may cause violent reactions when mixed with other chemicals present in the plant. Certain materials, relatively non-hazardous in themselves, may be objectionable because of their effect on sewage disposal systems. The wide variety of these problems indicates the necessity of developing and maintaining a hazard index. Some assistance in developing this index may be obtained from the published materials in the attached bibliography. However, in many cases, hazards must be investigated by the plant itself.

Such a hazard index should be complete and include all materials handled at the plant or installation. It is usually desirable to develop a card file so that such information may be permanently and readily available. For this purpose, a suggested check-list of the type shown may appear on each card.

DISPOSAL INDEX

In addition to the card file or hazard index above described, the reverse side of this card should describe the proper method of disposal for this specific type of material. One form of such a card is a "Waste Disposal Index" on which the proper method for disposal may be described and will include a description of the standard method of disposal together with exceptions which may apply under special circumstances. Certain materials may be listed in such an index in general groups or types and it may be also necessary to list mixtures of materials. In any event, the book should provide a ready reference for the supervisor so that he will know immediately how to handle any of the types of materials or situations which he may encounter. The services of the waste disposal organization and its consultants are, of course, available for situations not covered directly in the index, but an attempt should be made to cover the great majority of situations by standard procedures.

METHODS

Some commonly used methods of disposal are briefly described in the following paragraphs.

BURNING IN THE OPEN. Many materials may be safely disposed of by burning in the open. Safety to personnel can be assured by making arrangements for igniting the material from a safe distance. Packing material which is commercially available in the form of paper tubes filled with excelsior may be used as lighting trains. Employees doing this work need flame-proofed clothing and gloves and it is also essential to provide safety showers.

In open field burning, however, combustion is usually incomplete so that the smoke produced may contain larger amounts of unburned carbon and appreciable amounts of irritant and toxic materials. Such smoke clouds not only create a severe smoke nuisance in the neighborhood, but may also create serious conditions by bleaching of plants, toxic effects on live stock and toxic and very annoying nuisances to residents. Such irritating and noxious smokes may be carried over a wide area and cause damage to paint on residential property or other structures.

BURNING IN INCINERATORS. While burning in incinerators under proper control can reduce most if not all of the smoke nuisance, the hazards created by flammable vapors or explosive solids in the incinerator structure may be quite serious. Chemical plant incinerators require expert design and the provision of automatic control to prevent the development of unsafe conditions. It is usually possible through the burning of relatively harmless wastes such as ordinary waste paper, to build up temperatures in the "combustion chamber" or the "after burner" which are high enough to ignite flammable liquid vapors immediately and to cause decomposition of irritant and toxic organic compounds. Sufficient temperature control and interlocking devices are required so that these materials cannot be introduced into the furnace unless a proper operating temperature, usually in the vicinity of 1700°F., is maintained. Great care is needed in the arrangement of flammable liquid tanks, piping and control valves. Safe clothing for workers, adequate showers and means of escape are positive requirements. Operations of this type must be under the responsible charge of competent operating personnel.

DISPOSAL THROUGH SEWERS.   While many water soluble materials may be readily flushed down sewers, great care is needed to avoid placing material in sewers which will create flammable vapor conditions, stream pollution problems, or which will upset the normal operation of sewage disposal plants. All disposal of wastes by this means must be in accordance with federal, state and local laws relating to stream pollution and to the type of material acceptable in the sanitary sewer system.

In instances where wastes are discharged directly into streams, special sewage treatment plants designed for the particular wastes encountered, may be necessary.

Great care must be taken to make sure that radioactive materials do not enter the normal waste disposal operation but are properly segregated and handled in accordance with recommendations of the Atomic Energy Commission.

REACTIVE WASTES.   Reactive chemicals which may cause a violent reaction when in contact with water or other common materials, require special treatment so that they may be made relatively harmless before disposal. Exact nature of this treatment will, of course, vary with the nature of the material.

EXPLOSIVE MATERIALS.   In general, standard procedures should be set up for the careful removal of shock-sensitive materials from locations within the plant to a disposal area where they may be safely destroyed, usually by burning. In some instances, it may be desirable to puncture containers by rifle shot or explosive charge. Great care, however, must be taken in such operations to prevent hazards created by ricochet of bullets and with proper permission from police authorities. Methods should be developed for picking up and safely transporting these materials to the disposal area.

DISPOSAL BY BURYING.   While many ordinary materials may be safely disposed of by burying or land-fill operations, it should be remembered that water soluble materials may eventually leach into streams and wells and create a hazardous situation. Buried materials may also create a severe hazard when dug up during later construction or other earth-moving operations. Land-fill operations are acceptable for many ordinary materials of organic origin but the use of this method should be carefully limited to those which will not cause hazards later on.

COMPLAINT INVESTIGATION

Responsibility should be assigned in regard to the handling of complaints which may arise from all types of plant operations, including waste disposal. All complaints in regard to odors, smoke and other nuisance conditions should be brought to the attention of one person or organization charged with investigating the origin of the nuisance. This work should be coordinated with that of industrial and public relations personnel and medical consultants.

REFERENCES

1. "Handbook of Chemistry and Physics," Weast, R. C. (editor), The Chemical Rubber Co., Cleveland, Ohio, (1966).
2. Hunter, D., "The Diseases of Occupations," Little, Brown, and Co., Boston, Massachusetts, (1962).
3. Lange, N. A., "Handbook of Chemistry," Handbook Publishers, Inc., Sandusky, Ohio, (1952).
4. Manufacturing Chemists' Association, "Air Pollution Abatement Manual," The Association, Washington, D. C., (1966).
5. Manufacturing Chemists' Association, Safety Data Sheets.
6. "Merck Index of Chemicals and Drugs," Stecher, P. G. (editor), Merck and Company, Rahway, New Jersey, (1960).
7. National Fire Protection Association, "Fire Hazaard Properties of Certain Flammable Liquids, Gases, Volatile Liquids," The Association, Boston, Massachusetts, (1965).
8. National Fire Protection Association, "Fire Protection Handbook," Tryon, G. H. (editor), The Association, Boston, Massachusetts, (1962).
9. National Safety Council, Safety Data Sheets.
10. Sax, N. I., "Dangerous Properties of Industrial Materials," Reinhold Publishing Corp., New York, (1963).

Safety Considerations
in Research Proposals

H. K. Livingston

Scientific research in the campus laboratories is one of the most exciting activities in the world of ideas, and one of the least orderly in the world of organization. Whether considered from the vantage point of the student, the professor, the dean, or the financial officer, university research defies generalization. Each professor has his independent ideas as to how his work should be organized, and each student is the end result of a long process designed to make him believe in the worth of the individual in science and the importance of preserving his individuality in his approach to his work. No two departments are organized alike, and funds come from a maze of sources varying from student breakage accounts to long-term research projects that represent large sums of money.

This chapter originated with the work of the Committee on Chemical Safety of the American Chemical Society, and most of the information available to this Committee dealt primarily with chemistry and chemical engineering. The American Chemical Society is properly concerned that the profession of chemistry be practiced safely, insofar as possible. For example, in determining if the bachelor's degrees granted by an American institution should be accredited by the ACS, representatives of the Society visit the campus and review a number of points, one of which is that "careful attention should be given to modern safety practices in the laboratory and in the storage and handling of chemicals."

A part of the over-all safety activities of the ACS is represented by the Committee on Chemical Safety which has been collecting information about safety hazards and problems that exist in chemical education beyond the bachelor's degree. Probably the exposure hours to chemical hazard on college campuses are at least as great for graduate students as they are for undergraduates. Most of the graduate student's laboratory time is spent in research. In the course of this work, we have established that there is one common point in all student-teacher relations involving academic research, and this we have called the research proposal. Once laboratory work has gone beyond the realm of the conventional undergraduate or graduate-level laboratory course, most laboratory work will be done on the basis of the research proposal, which we can define as follows:

A research proposal is a statement of an experimental program embodying a prospectus as to what may be discovered and an evaluation as to why the results may be desirable. Typically, a student outlines one or two specific experiments, points out how this may lead to a series of related experiments, and explains why these experiments may lead to surprising results that will lead to a well-regarded publication or otherwise "make a contribution" to the field in which he and his teacher are concentrating. Research proposals may be internal or external, formal or informal.

Traditionally the research proposal on the campus was always of the internal variety. A student outlined a program using available equipment or chemicals, or things he could buy or make himself, his professor described an area of research and suggested that the student select a program that was within his means. Only after 1940 did there begin to be any significant number of external proposals in which the professor, or the student working with his professor, proposed a line of experimentation to an outside agency, which would (if favorably impressed) make funds available via a research grant or research project that would finance the proposed research.

An informal research proposal may simply take the form of an oral colloquy between student and teacher, getting progressively closer to agreement, over a series of daily or weekly discussions, until finally it is realized that an agreement has been reached as to what is proposed. On the other hand a formal proposal will be in writing, frequently with several drafts preceding the final version.

SAFETY CONSIDERATIONS

The laboratory must be considered an unnatural environment for man. The temperatures and pressures encountered there cannot be dealt with "bare handed," things can happen at rates that are much too fast for human reaction times, and poisonous and noxious gases, liquids, and solids are commonplace. And yet in this environment the research student is expected to discover new science, which he can do only by running experiments that have never been run before.

Exploring the unknown in science means exploring the unknown in hazard. Certainly the objective is to make scientific discoveries, but to do so without physical discomfort to the research man and without damaging his research facilities or physically handicapping him in his future work.

The position of the ACS on this subject is made clear by its charter, granted by act of Congress in 1937, which lists among the objects of the Society "by its meetings, professional contacts, reports, papers, discussions, and publications to promote scientific interests and inquiry, thereby fostering public welfare and education ... and adding to the material prosperity and happiness of our people." Nothing is so sure to subtract from "material prosperity and happiness" as knowingly or unknowingly jeopardizing health or property values, which is what we mean when we speak of an experimenter as being unsafe.

This object of the ACS provides a good basis for defining safety considerations that enter into research proposals. Safety considerations are those mental processes that determine whether hazards to health or property values are likely to be involved in a proposed course of action, and evaluate steps that can be taken to minimize those hazards. The word hazard is used deliberately in this definition. Among the antonyms to safety, this appears to be the most expressive word, based on the definitions given in Fernald's "Standard Handbook of Synonyms, Antonyms, and Prepositions" (p. 231 of the 1947 edition), which are: "Hazard is the incurring of a possibility of loss or harm for the possibility of benefit. Danger may have no compensating alternative (of benefit). In risk the possibility of loss is the chief thought."

Thus, the possibility of benefit is the motivation for a research proposal; safety considerations assess the hazard involved.

REASONS FOR CONCERN FOR THE SAFETY OF RESEARCH STUDENTS

The above analysis implies that safety is desirable without elaborating on the reasons why safety considerations are important in research proposals. Humanitarian considerations demand that new knowledge not be gained at the expense of human well-being. The history of science includes examples of scientists who have harmed their health by deliberate self-experimentation, under circumstances that seemed to allow them no other course but to expose themselves to hazard if they were to obtain the information they needed. It does not condone deliberate experimentation that endangered the health of others, except where the hazards were fully understood and willingly accepted. Certainly students should not be involved in this type of hazard.

Probably at the root of this particular concern is the ethical challenge embodied in the phrase "Am I my brother's keeper?" However this question is answered, it is certainly unethical for a teacher to expose a student to a danger that is apparent only to the teacher. Safety considerations in universities and colleges properly start with this ethical consideration.

Legal requirements also enter into the question of safety in research, but they will not be dealt with in detail in this presentation. Certainly if humanitarian and ethical requirements are met, there are not likely to be any issues that will require legal action.

Sometimes safety considerations on the campus have taken a back seat to financial considerations. In the past, poor safety has sometimes been condoned because of "expediency," a euphemism for poverty. The financial support of the sciences today is such that anyone who feels an ethical or humanitarian responsibility to provide safety equipment can make a strong case for financial support for his needs, relying on the problem of legal liability if necessary, to clinch his argument. Hopefully we will never reach the point where competition for research dollars is such that a teacher will skimp on safety equipment just to "submit a low bid."

CURRENT SAFETY PRACTICE

Up to this point, this presentation has dealt separately with the terms research proposals and safety considerations, attempting to define the terms and show why they are important in colleges and universities. It is now proposed to combine the terms and review current practice regarding safety considerations in research proposals. The information needed for a review has been obtained by interviewing twenty scientists representing different points of view concerning research proposals. The interviews were based on the following question: "In your own experience, at what point do safety considerations enter into research proposals?" The scope of the experience of the scientists interviewed ranged from the direction of undergraduate research (the so-called B.S. thesis) to the direction of graduate research leading to the Ph.D. degree, as well as the direction of post-doctoral students. Also interviewed were a number of scientists engaged in the review of research proposals submitted for financial support by government agencies or by foundations engaged in supporting scientific research. The results of these interviews are summarized in Tables 1–3.

We were fortunate to obtain a specific statement regarding safety practices in one large chemistry department, which follows: "Senior thesis research is arranged by the student and faculty member concerned. It is felt that, at this level, the student cannot be expected to be familiar with all the hazards of his work. Accordingly the faculty members are expected to make the students aware of potential dangers. Obviously, as the work progresses, the student is not under the eye of the supervisor at all times, and accordingly the student must develop a sense of responsibility regarding his and others' safety. Frequent conferences with the instructor are used to evaluate the progress of the work and the proposed extensions of it. In this way the staff member supervising the work is directly involved in evaluating and approving projects."

"Graduate research is supervised in a similar manner. Here, of course, the direct supervision of the research lessens as the student matures. Part of this maturing process

TABLE 1
Research Proposals Classified by Type

Type of Proposal	Formality
Internal:	
B.S. Thesis Research	Almost always oral and informal.
M.S. Thesis Research	Usually oral and informal.
Ph.D. Thesis Research	Likely to be oral and informal unless external proposals are closely related.
Post-doctoral Research	Little information was obtained on this point, probably because much post-doctoral research originates from external research proposals (see below).
External:	Formal written research proposals are almost invariably required.

TABLE 2

Safety Considerations in Informal Proposals: Results of Interviews with Faculty Advisors

Item	Consensus from Interviews
Time at which safety is most likely to be discussed.	When proposal has definitely crystallized.
Frequency of discussion of safety with student prior to final formulation of proposal.	Rarely discussed in connection with his own research; more likely to be a subject of general discussion or in connection with his responsibility as an assistant in undergraduate laboratory courses.
Frequency with which safety is discussed at some time before experimental work is actually started.	Generally but not invariably.
Frequency with which safety is discussed after experimental work is started.	Not likely to be discussed unless student is observed violating good safety practice.
Degree of responsibility for safety felt by faculty advisor.	Perhaps not quite as great as for students in laboratories, since a student doing research has demonstrated greater ability than the average student.

TABLE 3

Safety Considerations in Formal Research Proposals:
Results of Interview with Representatives of Granting Agencies

Answers to Question:
  "In considering research proposals does your agency consider if research will be done safely?:"

Frequently	50%
Never	20%
First answer was "Never," but as the interview continued, the answer was changed to "Sometimes"	30%
"If frequently considered, by whom?"	
Agency staff	60%
Advisory panel	40%
"Major reliance for checking safety placed on:"	
Agency	40%
Institution at which research will be done	30%
Individual under whom research will be done	30%

is felt to be a critical evaluation of hazards involved in the work. In cases of doubt the student is expected to refer to one of the texts (in the Chemistry Library) on hazardous materials or laboratory safety. Final approval of potentially dangerous experiments rests with the supervisor."

It is difficult to determine the extent to which the increasing importance of formal research proposals on the campus is influencing practices with regard to informal proposals. It can be predicted that any safety considerations that become habitual with regard to formal proposals will gradually come to be an automatic part of informal proposals as well.

One must question the validity of the prevailing feeling that students who are qualified to do research are not as much of a responsibility to the faculty, as far as safety is concerned, as are the ordinary students taking laboratory courses. Evidence on this point is hard to obtain, but it is significant that a study of 148 accidents in chemistry laboratories in California high schools [R. D. Macomber, *Journal of Chemical Education*, 38, 367 (1961)], showed that, if minor accidents such as small cuts and burns were left out of consideration, most accidents occurred to students with the higher marks, A and B. It can be deduced from this study that the accident frequency per exposure hour was much higher when the student was carrying out experiments that were not in the laboratory manual. The more mature student may become less willing to take a chance with an experiment, but it is also possible that the tendency to experiment in areas of unknown hazard does not change as knowledge increases. It may be that the size of the danger area undergoes little change, and at any educational level the student needs to be told how to experiment safely in unknown areas.

It will be seen from Table 3 that there is no standard approach to safety in formal research proposals. In no case do the procedures for writing a proposal, as specified by the granting agency, require a statement on safety considerations. On the other hand, most agencies can cite one or more examples where safety considerations had a very important place in research proposals, in most cases at the instigation of the agency. These examples involved proposals to do research on explosives, with materials having an insidious toxicological character, or with high energy materials in metastable states.

An interesting feature of the interviews with representatives of granting agencies was the fact (see Table 3) that several, at first, stated that safety considerations did not come up in their agency's consideration of research proposals, but later in the interview the interviewee remembered cases where safety considerations had been a significant factor in the evaluation of the proposal.

The U. S. Atomic Energy Commission is in a special situation with regard to proposals submitted to it for research with man-made radioisotopes. Licenses from the U. S. Atomic Energy Commission or comparable state licensing agencies are required before any laboratory can work with such materials. Therefore a research proposal which states that radiological work will be carried on in a licensed laboratory automatically conveys the information that certain safety standards will be met. If it did not make such a statement, facilities would be inadequate to carry out the research. The arm of the U. S. AEC that deals with research grants therefore relies to a considerable extent on the licensing function of the Commission, whenever radiological safety is a concern.

CONCLUSIONS

In general, it is not current practice to require that a proposal to do laboratory research contain a statement on safety considerations. This fact may seem surprising. In the case of formal proposals the reason probably is that the legal liability of the granting agency is usually quite restricted, so such a statement is not felt to be required in what is, at least in part, a legal document. Without attempting to go into any legal matters, it can be stated that, when the expenditure of federal government funds is authorized, the extent of the governmental liability for damages that can be traced back to this authorization is in many cases limited to the size of the individual authorization itself, which is likely to be a rather small amount.

There is the definite feeling that since in sponsoring research the agency incurs little legal or financial liability, safety is someone else's problem—but because of humanitarian or ethical reasons, the agency gives some consideration to safety anyway.

Safety should not be approached through a search for liability—by trying to make somebody "it." If a serious accident occurs at a university, there is plenty of responsibility to share; neither advisor nor department head nor dean can (or does) tell himself "I wasn't responsible."

There is the now famous case of a barge carrying four huge cylinders of liquid chlorine, sunk by accident in the Mississippi River off Natchez, in which each of the parties concerned—the owner of the tugboat, the owner of the barge, and the owner of the chlorine—believed it had no legal liability. So the chlorine sat on the river bottom, to remain there until a cylinder failure would release it. Finally, the combined efforts of the Army and Navy salvaged the cylinders and relieved the concern of the residents of Natchez (downwind from where the barge sank).

On the campus, personal relationships are too close and the ethical responsibility of the teacher for the welfare and future of his students is too great to allow any question of legal liability to blur the importance of preventing accidents. The faculty and administration must be preserved from the weight of after-the-accident responsibility. This can be done by before-the-hazard precautions. How can this be done in the research laboratories? The following recommendations should be considered:

1. Discuss openly with faculty members directing laboratory research the nature of their ethical responsibility.

2. Encourage the practice of talking about safety considerations in informal research proposals, specifically at the time student and advisor agree on a proposal.

3. As a matter of university policy, have all formal research proposals contain a section entitled "Safety Considerations."

4. Make sure that the financial section of formal research proposals includes adequate funds to implement the suggestions outlined in the section on safety considerations.

ACKNOWLEDGEMENT: From Proceedings of the Eleventh National Conference on Campus Safety, July, 1964. Reprinted with permission.

Legal Liability
for Laboratory Accidents

John E. Evans, Jr.

The legal liability of teachers and schools for injuries resulting from laboratory accidents is no different than their liability for any type of accident that occurs to students or strangers in or about school property, albeit there is a greater risk of injury in the laboratory and workshop—so I feel justified in using the laboratory accident as a springboard for a discussion of the liability in general of teachers, schools, school districts, colleges, etc.

There are many and various fields of law but the one with which we are concerned in this chapter is the field of torts. A tort is an injury to one's person or property, from which injury the law recognizes that the injured individual has a corollary right of protection. Probably torts is the most social of all the areas of law, in that it affects more people in their day to day living than do contracts, criminal law, bankruptcy, or any of the other legal fields. In fact, over seventy-five per cent of all of the litigation in the United States arises from torts.

I would like to discuss first the individual tort liability of teachers, for this is fairly simple and there has not been great change in this law over the past years. Unlike the cases involving schools and colleges, the courts have not created an immunity for teachers which would insulate them from damages caused by their negligent acts. Instead, the courts have regularly applied, in the teacher cases, the basic common law rule that one who acts negligently or wrongfully to the injury of another person must bear the responsibility for damages caused by his wrongdoing. Actually, there have been relatively few cases where students or strangers have sued teachers, as such, for personal injuries. These cases which have arisen from laboratory accidents have generally involved injuries to children when explosions occurred during chemical experiments, or when students were 'fooling around' with chemicals, or when teachers had allegedly failed to properly warn or supervise hazardous experiments. In the cases which have been brought, the courts ruled that the teachers were legally liable if they were negligent; but, if the students were guilty of any carelessness on their own part (called contributory negligence) and if the child was of an age where he could be held responsible for his own acts, he was deprived of recovery for his injuries.

With relation to the legal liability of schools and colleges, anomalous but complete legal defenses have developed in our law. In the 19th Century the courts of this country created what I will call "twin immunities," wherein they ruled that both public schools and private charitable institutions of learning were free from all legal liability for the damages wrongfully inflicted by their employees. The courts, with general uniformity, held that in actions brought against schools and colleges they would not apply either the common law concept of *respondeat superior*—wherein an employer is responsible for damages wrongfully inflicted by his employees, while acting in the course of their employment—OR the basic legal concept that he who does a wrongful act to the injury of another must bear the responsibility in damages for the relief of the innocent victim.

Schools and colleges fall generally into two well defined groups, namely, those which are operated by the government; i.e., the public schools, and those which are private charitable institutions, which includes most of the balance. Each of these two groups was provided with its own separate type of legal immunity. I have called these twin immunities; but they were not identical twins, because a different rationale was generally used to justify the two legal creations.

In discussing these immunities, I am not talking about the various statutory laws which state legislatures have created and which courts must enforce, but rather I am discussing court-made law. The subject of these twin immunities has been a burning legal issue and a subject of controversy in this area of law during the past several years.

Referring first to public schools, why should not a school district be responsible for injuries to a student caused when its intoxicated school driver runs the bus into a utility pole? Of the several reasons assigned by the courts for granting immunity from liability in such a case as this, the one most frequently assigned has been that a school district is a branch of the soverign government, and it is above the law. When the Supreme Court of Illinois, in 1959, had before it a similar set of facts, it re-examined the basis for the old rule and it reversed its previous decisions which had created this governmental immunity. It said that we fought a revolution to overthrow the concept of 'Divine right of kings.' Yet, this legal immunity rests on the concept that the 'King can do no wrong.' Such thinking is not congenial to our democratic concept of justice. It has not even been the law in England for many years, where school districts are regularly held liable for wrongs caused by their employees. While today a majority of the states still give complete immunity to government schools, a strong trend has developed for eliminating this anomalous court-made protection. Illinois, Kentucky, Florida, California, Wisconsin, and at least five other states, have eliminated the governmental immunity and now hold that for such injuries wrongfully committed, the loss shall fall not on the innocent victim, but rather, on the employer of the wrongdoer.

The second of the twins, the so-called 'charitable immunity,' insulates charitable, non-governmental institutions of learning from responsibility for their wrongs. This defense was as firmly imbedded in our law 30 years ago as was the theory of governmental immunity, but the trend against it in more recent decisions has been strong. Today, the courts in well over half of the States have adopted new social and legal concepts and have reversed their decisions which gave private schools, colleges and other charities, legal immunity from liability for their torts. Pennsylvania and West Virginia eliminated this defense in 1965, and Ohio swept most of it away in the same year.

This change began in 1942, when Judge Rutledge (later a Justice of the United States Supreme Court) made a scholarly review of the bases for charitable immunity in the case of the *President and Directors of Georgetown College vs. Hughes.* He declared that it could not be rationally justified as a sound legal principle. The argument most frequently advanced on its behalf was that if the charities' money went to pay damages, its trust funds would be diverted from their intended use, to the detriment of the charitable undertaking.

In answer to this the majority of the courts have replied that an institution should be *just* before it is *charitable*; that most schools and colleges are protected with public liability insurance (it was really the insurance companies in many cases who were holding up the shield of charitable immunity); that there is no real danger of financial harm to private charities; and that there is no valid precedent, either in England or the common law, for disregarding our basic concept of justice that liability follows wrongful injury.

Conservative judges have opposed a change in status of school immunity and have argued in favor of *stare decisis*, i.e. against change in legal principles. An answer to the conservatives has been that if our law did not adapt itself to changing conditions, we would still be prosecuting for witchcraft, imprisoning for debt, and hanging for minor criminal offenses.

There are states wherein the courts were not willing to go the whole way of eliminating the defense of charitable immunity at one fell swoop. Instead, they whittled away at the doctine by holding that the school or charity would be liable only if it were negligent in the selection of its employees; or that the immunity would be retained for religious charities (as in Ohio); or that the school would be liable only if it were engaging in private or proprietary undertakings—such as when it opens its swimming pool to the public, or has paying spectators at its football games.

As a conclusion, I recommend that individual teachers be provided with the protection of liability insurance. Secondly, private schools and colleges are courting trouble if they fail to protect themselves in the same manner; and finally, government-sponsored institutions, even in those states where the doctrine of governmental immunity has not yet been discarded, would be wise to protect themselves with insurance —for the day may not be far off when they will have to bear the same social and legal responsibility as do individuals and corporations for the tortious wrongs of their employees.

AUTHORITIES CONSULTED

Pres. & Dirs. of Georgetown Col. vs. Hughes, 130 F. 2d 810 (D.C.-1942)
Guerrieri vs. Tyson, 147 Pa. Super. Ct. 239 (1942)
Flagiello v. Penna. Hosp., 417 Pa. 486 (1965)
Jones vs. Hawkes Hospital of Mt. Carmel, 175 Ohio St. 503 (1964)
Gibbons v. Y.M.C.A., 170 Ohio St. 280 (1960)
Adkins vs. St. Francis Hosp. of Charleston, 143 S. E. 2d 154 (W.Va.-1965)
Molitor vs. Kaneland Com. Unit. Dist. #302, 18 Ill. 2d 11; 163 N. E. 2d 89 (1965)
Darling vs. Charleston Community Mem. Hosp., 59 Ill. App. 2d 253; 200 N. E. 2d 149 (1965)
Bing vs. Thunig, 143 N. E. 2d 3 (N.Y.-1957)
31 *Amer. Trial Lawyers Journal* (1965)

1. 160 A.L.R. 7, 250
2. 25 A.L.R. 2d 1 and 203
3. 32 A.L.R. 2d 1163
4. 43 A.L.R. 2d 465
5. 86 A.L.R. 2d 489

Working Alone

Paul W. Trott

INTRODUCTION

"Working alone" has been defined, for safety considerations, as "the performance of any work by an individual who is out of audio or visual range of another individual for more than a few minutes at a time." Let us add one more sentence to this definition. "No other person is aware the individual is working alone, the nature of the work he is doing or the period of time the individual expects to work."

POLICY GUIDE

As a general policy the spirit of the statement "no one should ever work alone" should never be violated.

An individual should never work under conditions where emergency aid is not available. The availability of emergency aid, the degree of training and type of the emergency aid and the means of summoning the help depend on the nature of the hazard and the degree of exposure to the hazard.

POLICY ON WORKING ALONE

Always work under conditions where the availability of emergency aid is compatible with the nature of the hazard and the degree of exposure to the hazard. Examples of this philosophy are as follows:

Example 1. When a worker has a high degree of direct exposure to hazards such as high explosives or very toxic gases the emergency aid cannot be in the same location as the worker. This emergency aid must be at the nearest safe location and the worker monitored continuously by remote control. Closed circuit TV and/or intercom are satisfactory means of continuous monitoring.

Example 2. When a worker has a high degree of exposure to open handling of flammable liquids a second worker must be in the work area to provide immediate aid and summon additional emergency aid.

Example 3. When a worker is exposed to moving machinery such as a laboratory mill with adequate emergency stops, an alarm may be attached to the emergency stop to summon emergency aid.

Example 4. When the exposure is low and the nature of the hazard make an operation relatively safe—such as recording temperature on a fractionating column— a check procedure by telephone at definite intervals is adequate. Calls may be made' at one-half hour intervals or immediately before there is any change from the routine work.

Example 5. Even on routine jobs in non-hazardous locations—such as office and administrative employees, watchmen, and custodians—no worker should be alone longer than two hours without a check procedure. Phone checks, coffee breaks and lunch breaks may be used as times to accomplish a check. However, these check procedures at coffee breaks and lunch breaks should be formalized to prevent an injured or ill worker from being assumed absent from work.

Most work assignments can be routinely set up with emergency aid compatible with the degree of exposure and the nature of the hazard. Whenever there is a doubt concerning a work assignment and this philosophy, the supervisor and safety engineer should review the work assignment and jointly define the work assignment and the emergency aid compatible with the work assignment.

Occupational Health Programs

American Medical Association

This chapter is a slightly abridged version of a revised statement prepared by the Council on Occupational Health of the American Medical Association, approved by the Board of Trustees in April, 1960, adopted by the House of Delegates in June, 1960, and reprinted from the Journal of the American Medical Association, October 1, 1960, Vol. 174, pp. 533–536, Copyright, 1960, by American Medical Association. This abridged version is reprinted with the permission of the American Medical Association.

SCOPE, OBJECTIVES, AND FUNCTIONS

The term "occupational health program", as used here, means a program provided by management to deal constructively with the health of employees in relation to their work.

The term **occupational medicine** means that branch of medicine practiced by physicians in meeting medical problems and needs under occupational health programs.

Some employers, in the middle of the 19th century, established medical services for their employees in areas where satisfactory medical services were not readily available. Since 1911, workmen's compensation laws requiring employers to compensate employees or their heirs for occupational disability or death and to provide medical care for occupationally injured employees have been enacted in all states. In addition, most of these laws require employers to provide medical care for employees with occupational disease. These and other laws have given employers a greater incentive, as well as an obligation, to maintain safe and healthful working environments. The problems associated with the increasingly complex technology of industry, with ever new, potentially hazardous, physical and chemical agents, have served as an important stimulus to the development of occupational health programs. From these developments the earlier concept of curative occupational medicine has been broadened to include and emphasize prevention and health maintenance, and there has gradually emerged the type of occupational health program described here.

Organized medicine should exercise leadership in improving occupational health programs and in providing adequate occupational health services to all employees, including those in small establishments. The medical profession has a responsibility to bring to the attention of industrial management and labor the values of occupational health services. In assuming this responsibility, the profession should recognize that the considerations and recommendations contained in this statement should not be viewed as being limited to full-time, salaried physicians. They should serve to attract and guide physicians in private practice in rendering part-time or consultant services. It is specifically in the provision of occupational health services to small plants that physicians in private practice can make their greatest contribution. The county medical society should be prepared to help employers locate physicians willing and qualified to direct and serve occupational health programs.

General Considerations

Health maintenance is primarily the responsibility of the individual. However, the employer has an obligation to provide a safe work environment for his employees, and he has a valid interest in the prevention of loss of work time and of work efficiency resulting from his employees' ill health. Definitive diagnosis and therapy of non-occupational injury and illness are not responsibilities of the employer, but he may provide certain preventive health measures in a given situation where the employee, the employer and the community stand to benefit.

There are two types of health programs for those who work. The one that is dealt with here is the occupational health program that deals with the health of employees in relation to their work and is largely preventive. The other type is a medical care program for nonoccupational illnesses and injuries. These two types of programs differ in such respects as methods of financing and amounts and kinds of services. Failure on the part of employers, employees and physicians properly to distinguish between these two types of programs sometimes gives rise to misunderstandings and problems, particularly in those few situations in which the same professional personnel serves both programs.

Objectives

The objectives of an occupational health program are:
1. To protect employees against health hazards in their work environment.
2. To facilitate the placement and insure the suitability of individuals according to their physical capacities, mental abilities and emotional make-up in work which they can perform with an acceptable degree of efficiency and without endangering their own health and safety or that of their fellow employees.
3. To assure adequate medical care and rehabilitation of the occupationally ill and injured.
4. To encourage personal health maintenance.

The achievement of these objectives benefits both employees and employers by improving employee health, morale and productivity.

Activities

In order to attain these objectives the following activities are essential:

MAINTENANCE OF A HEALTHFUL WORK ENVIRONMENT. This requires that personnel skilled in industrial hygiene perform periodic inspections of the premises, including all facilities used by employees, and evaluate the work environment in order to detect and appraise health hazards, mental as well as physical. Such inspections and appraisals, together with the knowledge of processes and materials used, provide current information on health aspects of the work environment. This information will serve as the basis for appropriate recommendations to management for preventive and corrective measures.

HEALTH EXAMINATIONS

Preplacement examinations. These examinations serve to determine the health status of the individual in order to aid in suitable placement. Such examinations should include: (1) Personal and family medical history; (2) Occupational history, and (3) Physical examination, including appropriate laboratory procedures.

Periodic examinations. These health evaluations are similar to preplacement examinations and are carried out at appropriate intervals to determine whether the employee's health is compatible with his job assignment and to detect any evidence of ill health which might be attributable to his employment. Certain employees and groups may require examinations more frequently than others and additional procedures and tests, depending on their age, their physical condition, and the nature of their work.

All health examinations must be conducted by a physician with such assistance as he requires. The examination may be made in any properly equipped medical facility, at the place of work or elsewhere. However, to qualify as an appropriate part of the occupational health program, the examination must evaluate the health status of the employee in relation to his work.

The individual to be examined should be informed by appropriate means of the purpose and value of the examination. The physician should discuss the findings of the examination with the individual, explaining to him the importance of further medical attention for any significant health defects found.

Unrealistic and needlessly stringent standards of physical fitness for employment defeat the purpose of health examinations and of maximum utilization of the available work force.

DIAGNOSIS AND TREATMENT

Occupational injury and disease. Diagnosis and treatment in occupational injury and disease cases should be prompt and should be directed toward rehabilitation. Workmen's compensation laws and policies of medical societies usually govern the provision of medical services for such cases.

Nonoccupational injury and illness. Diagnosis and treatment in nonoccupational injury and illness cases are not responsibilities of an occupational health program with a few limited exceptions.

IMMUNIZATION PROGRAMS. An employer may properly make immunization procedures available to his employees under established principles.

MEDICAL RECORDS. The maintenance of accurate and complete medical records of each employee from the time of his first examination or treatment.

HEALTH EDUCATION AND COUNSELING. Occupational health personnel should educate employees in personal hygiene and health maintenance. The most favorable opportunity for reaching an employee with health education and counseling arises when he visits a health facility.

ORGANIZATION AND STAFFING

In order to provide a satisfactory occupational health program, it is essential that a qualified doctor of medicine be engaged to direct the program. The needs of a specific program will determine the amount of time required of this physician. Training and experience in occupational medicine are desirable. The medical director should have a major role in the development, interpretation and implementation of medical policy. He should administer the occupational health program and be directly responsible to a designated official at the policy making level of management.

The occupational health program should be tailored to each employee group according to its needs. These needs are determined by such factors as the number of employees in the group, the nature and extent of the hazards to which they are exposed, and the availability of community medical services. The personnel required in an occupational health program may include, in addition to physicians and nurses, persons skilled in various technological procedures of environmental appraisal and control, laboratory technicians, other specialized personnel and clerical help. The amount of time required of such personnel will depend upon the needs of the employee group to be served. The organizational relationships within the group, while varying with circumstances, should be such as to promote maximum collaboration and cooperation.

Nurses in occupational health programs should be graduates of accredited schools of nursing, registered and legally qualified to practice nursing where employed. Training and experience in occupational health are desirable. The nurse's professional duties should be clearly defined in writing by the physician.

In establishments which do not have a nurse, there should be one or more employees qualified in first aid available throughout the working hours. It is desirable, of course, in all employee groups to have a substantial proportion of the employees trained in first aid in the event of emergencies.

Cooperation among physicians, industrial hygienists, nurses, technicians and other occupational health personnel and management personnel responsible for the employment, safety and well-being of employees is essential. Occupational health personnel should cooperate also with voluntary and official community agencies providing health, safety, employment and welfare services.

Facilities

The extent of the facilities, including equipment, will depend upon the needs of the employees and the scope of the occupational health program.

SUMMARY

This statement provides only a small outline of the scope, objectives and functions of occupational health programs. More detailed information can be found in reference works on various aspects of occupational health and through consultation with persons whose professional training and experience have qualified them as specialists in this field. Guidance concerning the development and operation of occupational health programs also may be obtained from county and state medical societies, from the Council on Occupational Health of the American Medical Association, from the Industrial Medical Association, and from the American Industrial Hygiene Association.

Section 2

Protective Equipment

Respiratory Protective Equipment

It is important to recognize two distinct uses for respiratory protective equipment:

1. As emergency devices in atmospheres from which the wearer cannot escape without respiratory protection.

2. As non-emergency devices in atmospheres that are not immediately dangerous to the wearer, but are detrimental after prolonged or repeated exposure.

The table below lists types of respirators in each category, together with their limitations.

In addition to the information in the table, the following points should be noted about the various types of respiratory protection.

EMERGENCY RESPIRATORS

Self-contained breathing apparatus may be of the open- or closed-circuit types. The principal difference is in the handling of exhaled air. The open-circuit types vent directly to the outside atmosphere. The closed-circuit types pass it through a regenerator circuit that removes carbon dioxide and adds oxygen.

The most serious limitation of the open-circuit apparatus is the relatively short, and sometimes uncertain, supply of air or oxygen. Such units are nominally rated for 15 or 30 minutes, but the wearer's breathing rate, the working conditions, and the starting cylinder pressure will affect the actual service life.

Therefore, all open-circuit apparatus should be equipped with a low-pressure warning device, and the user should make his own service-life tests before wearing the apparatus in emergencies.

Usually, the only limitation to use of this apparatus is the possibility of poisoning by skin absorption.

It should be noted, however, that most open-circuit apparatus operate in the demand mode, creating a negative pressure in the facepiece during inhalation. With improperly fitted masks, inward contaminant leakage may occur. To overcome this, pressure-demand breathing has recently been introduced, wherein the facepiece exhalation valve and regulator are balanced to maintain a slight, positive pressure in the facepiece even during inhalation.

The closed-circuit apparatus is available with ratings up to two hours at present, with longer service lives soon to be available.

The closed-circuit apparatus requires more intensive training than the open-circuit types. Bureau of Mines personnel spend 30 hours training with such apparatus before they are qualified.

The **gas mask** operates with a negative pressure upon inhalation, and therefore the mask should be carefully fitted and the wearer should be alert for symptoms of leakage.

Both multi-purpose (formerly designated "universal") and special-purpose canisters are available.

The Bureau of Mines recommends that, where possible, canisters designed to remove specific gases or vapors be used.

Because gas masks cannot be used in oxygen-deficient atmospheres or in high concentrations of gases and vapors, they are being replaced in many cases by self-contained breathing apparatus and the third type of emergency respirator, the **hose mask with blower.**

The latter type incorporates a hand-operated or motor-driven blower, and the wearer may use up to 300 feet of large-diameter, wire-wrapped, oilproof hose. If the blower fails, the wearer can breathe through the large hose alone with no more resistance than is encountered in a gas mask.

It is important that a safety man be operating or standing by the blower to ensure that only *respirable* air is supplied.

A fourth type of emergency respirator was introduced rather recently, and combines an **air-line respirator with auxiliary self-contained cylinder.**

Air-line respirators alone are not regarded as safe for use in immediately dangerous atmospheres. Should the air supply fail for any reason, the wearer cannot get air through the valves and small-diameter hose. But with the auxiliary air cylinder, the ·wearer can breathe supplemental air while making his exit from the hazardous atmosphere.

NON-EMERGENCY RESPIRATORS

Respiratory protective devices intended only for use in non-emergency situations should ideally be used only as temporary measures while engineering controls for contaminants are being installed. They are, however, widely used in industry as a primary line of defense against inhalation hazards.

One of the **supplied-air respirators**, the air-line respirator with continuous flow, is particularly suitable, when equipped with helmets or hoods, for lead grinding or abrasive blasting.

Dispersoid respirators can be divided into three classes: those for use as protection against (1) pneumoconiosis-producing dusts and mists (such as silica and asbestos); (2) toxic dusts, metallic fumes, and mists not significantly more toxic than lead (zinc dust, lead fume, and chromic acid mist are examples); and (3) radionuclides and highly toxic dusts, fumes, and mists, significantly more toxic than lead (such as radionuclides and beryllium).

The third classification is subdivided into three groups, with a different type of dispersoid respirator for each: half-mask respirators are suggested for use in contaminant concentrations up to 10 times the threshold limit value, full facepieces up to 100 times the TLV, and positive-pressure dispersoid respirators up to 1000 times the TLV.

Nuisance dusts and mists are included with silica and asbestos in the first class. Bureau of Mines-approved respirators should be used even for nuisance materials such as coal, sawdust, and flour, since no one can foretell the possible effects of breathing such materials, and the legal complications that may arise from unapproved nuisance-dust respirators may be more costly and troublesome than the use of the approved respirator. (During 1966, there were at least six court cases arising from workers being provided unapproved respiratory protective devices.)

Chemical-cartridge respirators are designed for use in low concentrations of gases or vapors and are also frequently equipped with filters for protection against dusts, mists, and during paint spraying.

EMPLOYEE TRAINING

As is true with all protective equipment, respiratory protective devices will be effective only when acceptable to the worker.

The best way to achieve acceptance is to explain exactly why the respirator needs to be worn, and what might happen if the respirator were not worn.

Where non-emergency respirators are used, the wearer should be allowed to try on several styles from various manufacturers to choose the one best suited to him.

In addition to convincing a worker of the need for wearing a respirator, and providing him with a comfortable model, he must be trained to wear it correctly. Such training may vary from showing the user how to check the facepiece fit of a small dust respirator to training sessions of several hours for the more complicated self-contained apparatus.

During initial training, special emphasis should be placed upon facepiece fit. Several simple tests will disclose improper fit. The exhalation valve can be closed, and the wearer exhales (a slight positive pressure should build up in the facepiece without leakage of air at the seal). Or the filter or cartridge inlets or the breathing tubes are closed off, and the wearer inhales (the facepiece should collapse and remain collapsed for a few seconds).

These initial procedures can be carried a step further by allowing the wearer to check the fit he has obtained in an odorous atmosphere, such as that created by isoamyl acetate.

MAINTENANCE

Recently, the Bureau of Mines revised its recommended procedures for cleaning and disinfecting respirators to include new disinfectants replacing such materials as formalin and alcohol. They are:

1. Disassemble respirator.

2. Clean and disinfect facepiece and breathing tube in warm 49°C (120°F) cleaner-disinfectant solution. This may be prepared from quaternary-ammonium, detergent, and alkaline-salt compounds that are available from respirator manufacturers and other sources. Rinse thoroughly in warm water, since wearer may be allergic to quaternary-ammonium compounds.

or

Clean facepiece and breathing tube in a warm 49°C (120°F) liquid detergent solution, rinse in warm water, then immerse for 2 minutes in one of the following: (a) hypochlorite solution (50 ppm chlorine) prepared by diluting concentrated hypochlorite solution (5% chlorine) with water; (b) aqueous iodine solution (50 ppm iodine) prepared by dissolving a soluble iodine compound in water. Note: 50 ppm is equivalent to 0.005%. Chlorine and iodine solutions stronger than 50 ppm may damage rubber parts.

3. Air dry in a clean place.

4. Clean and inspect other respirator parts according to manufacturers' instructions.

5. Reassemble respirator and store in a clean, dry place.

Respirators for use in atmospheres NOT IMMEDIATELY DANGEROUS TO LIFE
(from which the wearer could escape without the aid of the respirator)*

Supplied-air respirators	Dispersoid respirators	Chemical-cartridge respirators
For atmosphere containing gases, vapors, and/or dusts, fumes, and mists Hose masks without blowers and air-line respirators (continuous, demand and pressure-demand flow) are in this classification Limitations: 1. Must be supplied with respirable air (do not use oxygen) 2. Air-line respirators with head and shoulder covering should be used for abrasive blasting 3. Length of air line	For atmospheres containing dusts, fumes, and mists Limitations: 1. Not for use in gases and vapors 2. Not for use in concentration of radioactive aerosols and organic phosphorus insecticides which are immediately dangerous to life 3. Service life limited by plugging of filter(s)	For atmospheres containing concentrations of gases and vapors up to the limits stated on the cartridge label Also for dusts, fumes, and mists, when equipped with filter(s) Limitations: 1. Up to 0.1% organic vapors 2. Eye irritation 3. Service life of cartridge(s)

*Use self-contained breathing apparatus in oxygen-deficient atmospheres or where a high concentration of gas or vapor may suddenly occur.

Respirators for use in atmospheres IMMEDIATELY DANGEROUS TO LIFE
(from which the wearer can *not* escape without the aid of the respirator)

Self-contained breathing apparatus*	Gas masks	Hose masks with blowers
For oxygen-deficient atmospheres or atmospheres containing gases, vapors and/or dusts, fumes, mists, and smokes Limitations: 1. Service life of apparatus 2. Sorption through skin 3. Weight and bulk of apparatus	For concentrations of gases and vapors up to the limits stated on the canister label Also for dusts, fumes, mists, and smokes, when equipped with filter Limitations: 1. Not for use in oxygen-deficient atmospheres 2. Sorption through skin 3. Service life of canister	For oxygen-deficient atmospheres or atmospheres containing gases, vapors, and dusts, fumes, mists, and smokes Limitations: 1. Attendant must be at blower 2. Blower must be located in respirable air 3. Sorption through skin 4. Length of hose (300 feet)

*Including devices that are combination self-contained breathing apparatus and supplied-air respirators.

ACKNOWLEDGEMENT: This chapter is a partial reprint of the article, "Lungs: Anatomy, Hazards, Monitoring, Protection," which appeared in *National Safety News*, vol. 95, 35–41 (March 1967). It is based on lectures by Dr. Carl A. Nau, Director, Institute of Environmental Health, University of Oklahoma Medical Center; Edward W. Poth, Jr., Chief, Engineering Section, Kelly Air Force Base, San Antonio, Tex.; Howard L. Kusnetz, Sanitary Engineer Director, State Services Branch, U.S. Public Health Service, Washington, D.C.; and Robert H. Schutz, Chief, Respiratory Section, Bureau of Mines, Pittsburgh. Reprinted by permission of the National Safety Council. This and other articles on industrial hygiene have been collected in "Fundamentals of Industrial Hygiene," National Safety Council, Chicago (1968).

Ear Protection

Airborne sound is rapid fluctuation of normal atmospheric pressure caused by a vibrating source. These air pressure fluctuations, or waves, can vary in intensity, harmonic content, frequency, and directionality.

The word "sound" is also used to indicate the sensation experienced when such pressure fluctuations strike the ear.

Since the brain's functions are performed by shunting tiny electrical impulses from one brain cell to another, all the sensory information it works with must be presented to it as a series of electrical impulses.

It is the job of the ear to translate the wave motions of sound (sense 1) to perceived sound (sense 2).

How does the ear effect this translation?

ANATOMY AND PHYSIOLOGY

The apparatus collectively called the ear consists of three distinct sections:

1. The outer ear (ear flap or pinna, and external auditory canal), which collects and funnels the sound waves to the eardrum stretched across the end of the canal.

2. The air-filled chamber of the middle ear, in which the vibrations of the eardrum are transmitted through a mechanical linkage of small bones to the third section of the ear. The three tiny bones are known collectively as the ossicles and separately as the malleus (hammer), incus (anvil), and stapes (stirrup). The hammer is vibrated by the eardrum, in turn pushes the anvil, which in turn moves the stirrup, which acts as a piston on the fluid in the inner ear.

3. The inner ear, containing the cochlea.* Fluid in this organ is set in motion by the mechanical vibration of the ossicles, and the nervous system in it converts the fluid vibrations into nerve impulses that are carried to the brain via the cochlear auditory nerve.

In summary: Wave motions in the air set up sympathetic vibration of the eardrum. This mechanical vibration is transmitted by the three bones in the middle ear to the fluid-filled chamber of the inner ear. In the process, the relatively large but feeble air-induced vibrations of the eardrum are converted to much smaller but more powerful mechanical vibrations by the three ossicles, and finally into yet stronger fluid vibrations. The wave motion in the fluid is sensed by the nerves in the cochlea, which transmit neural messages to the brain.

Since the principal industrial hazard to hearing is a noise-induced "wearing-out" of the nerve system in the inner ear, let's take a closer look at this subdivision of the hearing apparatus.

The Inner Ear

The portion of the inner ear we are concerned with here is the cochlea. This is a coiled tunnel (nature's way of fitting a long tunnel into a small space) resembling a snail shell.

Inside the tunnel are three fluid-filled canals. In the central canal nerve fibers from the auditory nerve end in hair cells, from which tiny fibers project into the cochlea fluid. There are some 20,000 hair cells in quadruple rows along the length of the cochlea.

*Part of the inner ear mechanism, the semicircular canals, is concerned with sense of balance, and is not germane to this discussion.

Typical Over-all and Octave Band Sound Pressure Levels in Decibels
(re 0.0002 dynes/cm^2)

Location	Remarks	Octave Bands in Hertz (Cycles per Second)								
		Over-all	20–75	75–150	150–300	300–600	600–1200	1200–2400	2400–4800	4800–9600
Circular cut-off wood saw	Cutting 1-inch birch	100	84	84	88	90	91	90	87	84
Planer—wood	Furniture parts	108	84	88	95	98	99	97	92	85
Jointer—wood	Furniture parts	98	85	88	90	90	89	86	81	75
Milling machine	Channel cut in steel	90	86	83	78	79	84	85	82	83
Turret lathe	3-inch hole, steel	91	81	85	78	80	84	84	82	81
Hand grinder	4-inch electric 1-inch steel	92	85	87	87	84	79	76	76	72
Internal combustion engine test	150 hp	102	86	90	98	91	94	92	90	90
Power house		116	109	115	91	79	73	68	68	64
Transonic wind tunnel	Mach 0.3	102	89	79	80	86	90	96	96	88
Supersonic wind tunnel	Mach 2.5	105	86	82	89	98	101	96	90	80
Fuel burner test room	1120 ft/sec	114	89	89	95	103	103	104	105	107
Jet test cell control room	7500 rpm	107	94	94	99	99	98	99	96	89
Aerosol generator	Spinning disk 50,000 rpm	101	75	60	62	72	100	80	89	86

Each cell is topped by some 20 of the hairlike sensory processes. These 400,000 "feelers" and their supporting cells are essential elements of sound perception.

The delicate projections, acted upon by sound waves in the fluid by the motion of the stapes, stimulate the fibers of the auditory nerve to give us what we call hearing.

The fluid in the inner ear may also be set in motion by vibrations transmitted through the skull bones, but only high intensity sounds can be transmitted in this manner.

Exactly how this sensory apparatus differentiates between the various qualities of sound, pitch, loudness, etc., is not perfectly understood. Enough is known, however, to give a general idea of the process.

Loudness, the strength of the sound, is undoubtedly a function of the intensity of the wave motion set up in the cochlear fluid.

How the sensory fibers distinguish pitch is still subject to different views. The sensory hairs at the base of the cochlea apparently react to the higher pitched sounds (up to 20,000 Hertz); those at the apex, or narrow end of the spiral, to the low pitch sound (down to about 20 Hz). In other words, there is a specific location along the length of the rows of hair cells where sensitivity to a given pitch will be greatest, and only the nerve fibers in that area are affected by a pure tone of that pitch.

Natural Defenses

The ear's natural defenses serve primarily to protect it from traumatic damage.

In the external auditory canal, hair and wax secretions serve to keep out small foreign objects.

The canal itself is bent, reducing the likelihood of an object penetrating through it and damaging the eardrum.

The eardrum and one of the three middle earbones are equipped with contractile muscles that react to intense sound, effectively damping the amount transmitted to the delicate inner ear.

Since both the outer auditory canal and the middle ear are air-filled, pressure differentials tend to deflect the eardrum that separates them. To accommodate pressure changes (such as from rapid changes in altitude) and thus protect the ear-drum from rupture, the middle ear is supplied with a safety valve, the eustachian tube. This tube connects the inner ear and the back of the throat. It is normally open sufficiently to equalize the pressure in the middle ear to minor atmospheric changes. In many individuals a conscious effort (yawning, chewing, etc.) is necessary to open the passage sufficiently to accommodate rapid pressure changes, such as experienced in skyscraper elevator riding and scuba diving.

HAZARDS

Although the human ear is subject to a number of types of disorders that can cause hearing loss, the major occupational hazard is sound itself—or rather excessive unwanted sound (noise).

The reader should be aware, however, of some of the non-job-related causes of hearing loss. Some 25 % of newly employed workers come to their jobs with a degree of hearing loss.

Non-noise-induced ear impairments can stem from:

1. Physical blockage of the auditory canals (with excessive wax, foreign bodies, etc.).

2. Traumatic damage, such as punctured eardrums or displacement of the ossicles.

3. Disease damage: childhood diseases (e.g., smallpox), infections of the inner ear, degenerative diseases, tumors, etc.

4. Hereditary or prenatal damages.

5. Drug-induced damages, such as from use of streptomycin, quinine.

6. Presbycusis, "natural" reduced hearing sensitivity due to aging.

Knowledge of these nonoccupational causes is especially important to the industrial medical staff, nurse, hygienist, or other person responsible for hearing tests, to avoid confusing them with occupational losses.

Noise

The young, healthy ear can detect sounds ranging from about 15 to 20,000 Hertz (cycles per second), and will detect sound pressures fairly near the arbitrary auditory threshold baseline throughout most of the spectrum of frequencies.

However, after prolonged exposure to extraordinary noise levels an irreversible impairment at various frequencies may occur.

The site of this impairment is the nervous system in the cochlea, and although the manner in which the damage is done is not known exactly, some aspects of it are clear:

1. The hearing loss occurs at a frequency about one half of an octave (an octave is the interval between frequencies having a ratio of two—as from 150 Hz to 300 Hz, or 2400 Hz to 4800 Hz) above the frequency of the noise to which the worker is exposed.

2. Temporary exposure to high noise levels will produce a temporary condition called auditory fatigue or temporary hearing loss, in which the hearing threshold will be shifted a number of decibels. This loss will be recovered after a period of time away from the exposure.

3. Long-term exposure to high noise levels (over a period of years) produces permanent hearing loss. According to one theory, the eventual irrecoverable loss will

be about the same as the temporary hearing loss experienced after a brief exposure to the same noise level.

4. It is believed that *any* exposure to noise with about 130 dB sound pressure level (such as a jet engine) is hazardous and should be avoided.

5. Apparently there is no hazard from exposures, no matter how prolonged, to levels up to about 80 dB (such as an office tabulating machine).

Hearing loss at the very high and very low frequencies is not a real handicap—but loss in the speech frequencies (roughly from 500 to 2000 Hz) is a significant impairment. Consequently, the usual criterion for determining whether a loss is compensable is that a loss of 15 dB or greater must have occurred in the speech frequencies. Actually, audiometric tests show that losses tend to occur earlier at higher frequencies, such as at 4000 Hz, and some authorities would like to see the criteria for compensation changed to encourage protection throughout the audible frequency range.

EVALUATING THE HAZARD

Since industrial concern with hearing loss is relatively recent—it is the most recent area of occupational disability to come under compensation laws—the risk criteria are much less clearly defined than they are, for instance, in the area of toxicity levels or radiation exposures.

Various criteria have been proposed since the problem was identified in the 1940's; there is still disagreement about which should be used.* The gray area is between 80 and 95 dB in the critical (speech frequency) octave bands. From the present information, it seems that below 80 dB even continuous exposure will not cause permanent impairment. Certainly continuous exposure to levels above 95 dB will cause hearing impairment. Risks at exposures of 85 dB are minimal. If the guideline is moved to 90 dB in the critical octave bands, you'll protect about 90% of the people exposed. If you go up to 95 dB, you'll protect only about 80 to 85% of the people exposed to the noise. This is due to differences in individual susceptibility. Some employers have attempted to control noise to provide personal protection to reduce all exposures to 80 dB. Others have pragmatically chosen damage risk levels as high as 90 dB.

Use of the various existing criteria, whichever one is chosen, requires careful monitoring and interpretation of the data. What frequencies constitute the noise? What are the sound pressure levels of the various frequencies? Is the noise continuous, intermittent, or varying? Is the sound energy concentrated in narrow frequency bands (more hazardous), or is it spread over wide bands? Do employees have rest periods from the exposure? What are the individual tolerances of the exposed workers? What prior hearing loss, if any, do the exposed workers have?

To make these determinations, two types of monitoring equipment are needed:

1. Noise level meters and noise analyzers, to identify the exposure.

2. Audiometric instruments to make tests before and periodically during exposure, to assess the effect of the exposures (and of protective measures) on workers.

Measuring Noise

There is a wide assortment of equipment available for noise measurements, including sound survey meters, sound level meters, octave band analyzers, narrow band analyzers, tape and graphic level recorders, impact sound level meters, and equipment for calibration of these instruments.

*See added section on Hearing Damage Risk Criteria at the end of this chapter.

For most noise problems encountered in the plant, the sound level meter and octave band analyzer will provide ample information.

Sound Level Meter

The sound level meter is the basic instrument for noise measurements. It comprises a microphone and an electronic circuit including an attenuator, an amplifier, three frequency response networks (weighting networks), and an indicating meter. The attenuator in the circuit controls the current within limits that can be handled by the indicating meter, which is calibrated in decibels. The value obtained is the root-mean-square sound pressure level expressed in decibels.

Three weighting networks (A, B, and C) are incorporated into the standard sound level meter. The purpose of these is to give a number that is an approximate evaluation of the total loudness level. Human response to sound varies with its frequency and intensity. The ear is less sensitive to the low and high frequencies at low sound intensities. At high sound levels there is little difference in human response at the various frequencies.

The three weighting networks provide a means for compensating for these variations in response. The A network is less sensitive to the low frequencies and is intended for use at sound pressure levels below 55 dB. The B network is an intermediate step for the range 55 to 85 dB. The C network has a flat response and is used for everything above 85 dB. The values obtained by the use of the three weighting networks will give a measurement of loudness, but for many problems additional information is needed, in which case the sound level meter must be used with a frequency analyzer.

Octave Band Analyzer

When the sound to be measured is complex, consisting of a number of tones spread over many octaves, the single value obtained from a sound level meter reading often is not sufficient for our purposes. It may be necessary to determine the sound pressure distribution according to frequency. The most practical and widely used analyzer is the octave band analyzer. As indicated by the name, the upper cutoff frequency for each band is twice the lower cutoff frequency.

Measurements can be made in one octave or multiples of octaves up to the complete spectrum with the second type of analyzer. The standard octaves are 37.5 to 75, 75–150, 150–300, 300–600, 600–1200, 1200–2400, 2400–4800, and 4800–9600 Hertz. At the present time there is a tendency to change the above octaves slightly, following the recommendations of the American National Standards Institute, and use octaves with center frequencies of 63–125, 250–500, 1000–2000, and 4000–8000 Hertz.

With both the sound level meter and octave band analyzer there is need of equipment for calibration. The manufacturers of these instruments have available acoustical calibrators. These usually consist of a signal generator of some type with a small speaker, which is placed over the microphone. These calibrators provide a means of calibration in both the laboratory and field.

Where to Monitor

With expanding mechanization, many potentially hazardous exposures to noise may be added to those which have been present for some time. As a rough guideline to potential noise problems, any noise situation that seriously hampers speech communication at close distances may produce hearing loss if an individual is exposed for long periods of time. Any such noise situation should be investigated with a sound level meter and octave band analyzer. Particular attention should be given to length of exposure. Many of the noisy operations may be of an intermittent nature.

Without noise measurements it is impossible to say that a particular operation or machine will produce a hazardous noise exposure (except for a small group of very intense noise producers, such as jet engines and rockets).

Audiometry

One of the biggest difficulties in setting up hearing protection programs is that there is no immediate tangible effect from control measures. Protective equipment doesn't prove its protectiveness by showing wear and tear on itself, as do other protective devices such as hard hats, glasses, and clothing.

So to evaluate hearing protection programs, a continuing check on the worker's hearing ability is necessary. This is called audiometric testing.

Hearing examinations have a number of benefits in addition to serving as the only gage of hearing conservation efforts. They can be a great force in employee relations, since they frequently identify correctible, nonoccupational hearing problems. And, of course, the audiometric record, when it includes a pre-employment test, is a defense against false compensation claims.

Audiometric testing equipment should be used under medical supervision. It consists of the audiometer (an instrument that produces pure tone sounds at different frequencies that can be gradually increased from below the auditory threshold until the test subject can hear them) and the test environment, which must be sufficiently quiet so as not to mask the test sounds.

Audiometers are either manual, in which case an operator does some dial twisting to increase the sound levels, or automatic. The acoustical environment can be specially constructed sound-protected areas, a prefabricated booth in which only the test subject is accommodated, or simply a very quiet room.

The result of testing will be an audiogram, a graph showing the test subject's hearing loss from the normal baseline as a function of frequency. The frequencies generally tested in industrial audiometry are 500, 1000, 2000, 4000, and 6000 Hz.

The main drawback of audiometry is that it is subjective. The person being tested is instructed to signal at the point where he can just hear a gradually increased sound. With manually operated audiometers, the operator's technique may affect the test results.

PROTECTION

Where noise level surveys show exposures above the accepted damage risk criteria, the exposure can be reduced by:
1. Environmental control
   a. Reducing the amount of noise produced at the source
   b. Reducing the amount of noise transmitted through air and building structures
   c. Revising operational procedures
2. Personal protection
   a. Plugs
   b. Muffs

The most satisfactory method of environmental control is to control the noise at the source. Unfortunately, this is not always possible. When the amount of noise produced at the source cannot be sufficiently reduced, a combination of control methods may be required to conserve hearing.

The engineering control of noise has not been sufficiently employed in industry. Few companies include noise level specifications when they purchase new equipment. Those who do, however, have found that suppliers can, in fact, often make machines much quieter. In many instances the supplier didn't realize his equipment was so noisy.

Noisy products are not only of concern to the purchaser—the doctrines of product liability are extending to noise, and compensation insurers are subrogating claims.

Engineering control of noise is sometimes fairly simple, particularly where machinery can be completely isolated. But there are many instances where engineering control to limit the noise to safe levels is impractical.

In such instances, hearing protection in the form of ear muffs or plugs can be used alone, or in combination with partial engineering methods.

Ear Protectors

There was a time when muff-type noise attenuators were recommended in preference to plugs, since muffs offer a few more decibels' protection than plugs. But experience in industrial hearing protection programs has indicated that over-protection can be the enemy of success, and that there are such strong psychological objections to wearing any hearing protection that sometimes a little less attenuation is a good trade-off for improved comfort and increased acceptability.

Here are the principal types of hearing protection devices, with comments on their relative merits and appropriate uses:

1. Muffs. Offer the highest attenuation. Should be mandatory for exposure to high intensity (above 95 dB) noise. Fitting not a problem. Easily adjusted. Cover whole pinna or ear flap, providing protection from noise that might be conducted through skull bones. *Disadvantages:* some workers complain of discomfort, heat, headaches; standard models cannot be worn with hard hats, welding shields, etc. Special models (back of the neck bands, and attached to hats) are available, however.

2. Plugs. Since, when properly fitted, plugs are more comfortable than muffs, they are to be preferred where there is 8-hour a day exposure to moderate (80–95 dB) noise levels.

Plugs are available in a variety of types—solid, sponge, or air-filled. Materials are rubber, neoprene, and plastics. All come in a variety of sizes. *Disadvantages:* not sufficient attenuation for really high-intensity exposures; require expert initial fit by medical department and proper daily insertion by the employee for maximum comfort and effectiveness.

3. Disposable malleable plug materials. Though frowned on in the past, recent studies have indicated that certain types of moldable materials have attenuation almost equal to the best plugs, and they have the advantage of being comfortable, providing a good fit, and having good employee reception. These include the waxed cotton type, and a new product known as "Swedish wool". The former was developed initially by an aircraft firm, and is now available under several trade marks. The latter was developed by a Swedish automotive company, and consists of extremely fine "glass down" fibers. Both types are especially good for problem employees who complain of discomfort with standard plugs. *Disadvantages:* slightly higher cost than plugs (since they are disposed of after a single use) and slightly less attenuation than some plugs.

MOTIVATION

There seems to be more resistance to wearing hearing protection than any other type of personal equipment. The two most common reasons: (1) employees may not be convinced that they need the protection—hearing loss is so gradual, even in intense exposures, that it does not make a dramatic impression, (2) many find hearing protectors "uncomfortable."

Evidence of these attitudes can be found in such practices as "springing" muffs so they don't seal against the head, clipping off the inner end of plugs and leaving only the outer end tab to fool the supervisor, and indifferent molding and insertion of malleable-type plug materials.

Some proven techniques for overcoming these attitudes have been found, and include:

1. Acquainting new employees with the hearing conservation program when they are being briefed on health insurance, first-aid facilities, safety glasses, hard hats, etc. In this way, they will tend to view wearing the hearing protective devices as a basic part of the job.

2. Using pre-employment audiograms to show the worker the present status of his hearing. A graphic presentation of the minor loss he may already have will help convince him that hearing loss is real, and encourage him to conserve the balance.

3. On-the-job fitting. Usually, fitting is done in a quiet dispensary or other place remote from the noisy exposure. Consequently, the employee gets no immediate idea of the protective value of the muff, plug, or malleable insert. The reason for dispensary fitting is obvious—a great many types of plugs and sizes are needed to find the combination that suits each individual, and these are not easy to transport about. But if on-the-job demonstrations can be arranged, they usually make a strong and lasting impression.

4. Group audiometric examinations that serve not to obtain audiograms, but to demonstrate to a group of workers the different degrees of hearing loss. Invariably such demonstrations will produce several older men in noisy jobs who can't hear the increasing signals until long after the other men have. There may be some who won't hear a 4000 Hz signal at all. This dramatizes hearing loss.

HEARING DAMAGE RISK CRITERIA

In Denver, at the 1969 annual meeting of the A.I.H.A. and A.C.G.I.H., the American Conference of Governmental Industrial Hygienists adopted threshold limit values for industrial noise. These values set the damage risk criteria for sound pressure levels.

Concurrent with this action, the Walsh-Healey noise safety regulation, establishing a similar set of threshold limit values, was put into effect with the signature of Secretary of Labor George P. Shultz. These damage risk criteria are shown below:

Duration per day, hours	Sound level, dBA	Duration per day, hours	Sound level, dBA
8	90	$1\frac{1}{2}$	102
6	92	1	105
4	95	$\frac{1}{2}$	110
3	97	$\frac{1}{4}$ or less	115
2	100		

Exposure to impulsive or impact noise should not exceed 140 dB peak sound pressure level.

ACKNOWLEDGEMENT: This chapter is reprinted with permission of the National Safety Council from *National Safety News*, vol. 95 (May 1967). It is based on lectures by Charles W. Reed, Director of Industrial Hygiene, Texas Employers Insurance Association; Herbert Jones, Assistant Chief, Engineering Section, U.S. Public Health Service; and Roger B. Maas, Hearing Consultant, Employers Mutual of Wausau. This and other articles on industrial hygiene have been collected in *Fundamentals of Industrial Hygiene*, National Safety Council, Chicago (1968).

Eye Hazards

Of all the major body organs prone to occupational injuries, the eye is perhaps the most vulnerable.

Although it does have some natural defenses, it has none to compare with the built-in self-cleansing mechanisms of the lungs, the recuperative power of the ear, or the healing properties of the skin.

Lacking a heavy armament of natural defenses, the eye requires artificial defenses, and immediate medical attention if injury occurs.

ANATOMY AND PHYSIOLOGY

Consideration of the structure of the human eye and how it can be affected by common occupational hazards will serve to underscore the wisdom of eye protection and efficient first aid.

The eyeball is housed in a cone of cushioning fatty tissue. It is thus insulated from the skull's bony eye socket. The latter has brow and cheek ridges projecting in front of the eyeball.

The eyeball itself is composed of some highly specialized tissue, tissue that does not react at all like other body tissue to injury. The cornea, lens, and humors, for instance, are clear. To maximize their transparency, they are nourished by very few blood vessels and, consequently, they do not heal rapidly. And the retinal tissue, the "photographic film" of the eye, is made up mostly of nerves, and nerves do not regenerate.

So the eye, being highly specialized tissue, is more likely to suffer permanent damage from injury than, for example, a finger.

This is not to say the eye does not have *some* natural defenses, however.

The bony ridges of the skull protect the eyeball from traumatic injury due to massive impact. A baseball, for instance, is too big to crush the eyeball; it will be stopped by the bony orbit.

The cushioning layers of conjunctiva and muscle around the eyeball tend to absorb impact. The fact that the eyeball can be displaced in its socket is a defense against injury. Related to this is the fact that the optic nerve is extra long to allow displacement of the ball without rupture of the nerve.

The eye is most subject to attack where it meets the world, at the corneal surface. Here the eye is equipped with an automatically actuated "windshield wiper and washer combination." The washers are the lacrimal glands, the wiper is the blink.

The function of the teary blink is to wash foreign bodies off the corneal or conjunctival surfaces before they can become imbedded. The triggering mechanism is irritation (the cornea has more nerves in it than practically any other part of the body).

The reflex blink can act like a closing door to shut out a foreign object heading for the eye, if the eye can see it coming and it is not coming too fast.

It will become apparent that protective equipment for the eye is simply apparatus to improve or extend these natural defenses: the bony ridge, the blink, and the tear glands.

These natural defenses might be adequate to protect against light, small foreign objects, and to wash away small quantities of mildly toxic liquids that might get into the eye, but they are no match for eye hazards such as the small, high-speed particle and the caustic burn.

PHYSICAL AND CHEMICAL HAZARDS

The eye is subject to several kinds of physical injury—from blunt objects, from sharp objects, and from foreign bodies.

Blows from blunt objects can produce direct pressure on the eyeball, or, if the object seals the rim of the bony orbit on contact, hydraulic pressure.

Such blows may cause contusion of the iris, lens, retina, or even optic nerve. Extreme blows might result in rupture of the entire globe. Contusions usually result in serious, irreversible injury. Hemorrhaging releases blood (which can be toxic to eye tissues), and physical dislocations of lens, retina, etc., are unlikely to repair themselves.

Lacerations of the cornea, lid, or conjunctiva may be caused by any sharp object, from the corner of a piece of typing paper to a knife.

Corneal lacerations frequently let the aqueous solution behind the cornea gush out—until the iris, which is like wet tissue paper, is pulled toward the laceration and plugs the wound. (The iris can be put back in place, the laceration sutured, and the eye may be nearly as good as new.)

Lacerations of the lid will heal, but the scar tissue may pull lids into unnatural positions. In addition to cosmetic deformity, this may result in exposure to future damage due to a holding-open of the lids and thus constant exposure to wind, dust, etc. Vertical lacerations are more serious in this respect.

Of physical injuries, *foreign bodies* are by far the most common.

Not all foreign bodies affect the eye in the same way. Low-mass, low-speed bodies are likely to lodge in or on the cornea or conjunctiva. High-speed objects are likely to lodge within the eye.

Foreign bodies affecting the conjunctiva usually are not too serious; they are located away from the business apparatus of seeing. They may result in redness and discomfort, but not vision damage. (Bodies on the conjunctiva, however, can be transferred to the cornea and become imbedded therein, if the man's reaction is to rub his eye; so even with minor irritations of the conjunctiva, a trip to the nurse is advisable.)

In industry, foreign bodies are likely to be corneal. In one study of 100 industry eye cases 77 involved the cornea, while only six were conjunctival. Fifteen affected the lids.

What can be expected from corneal foreign bodies?

1. Pain—Because the cornea is heavily endowed with nerves, an object sitting on the surface of the cornea will constantly stimulate the nerves.

2. Infection—Can be carried by the foreign particle in the form of bacteria or fungi, or from fingers used to rub the eye. (This used to be much more common; antibiotics have greatly reduced the problem.)

3. Scarring—Corneal tissue will heal, but the scars are optically imperfect, and may obscure vision.

What can be expected from intraocular foreign bodies?

1. Infection—Much less of a problem, though, than with low-speed, low-mass particles. Reason: the speed of small metallic particles often creates enough heat to effect sterilization. Wood particles, however, will not heat up, and if they penetrate the eye, can cause dangerous infection, usually leading to a marked vision reduction.

2. Damage via trajectory—Depending on its angle, point of entry, and speed, an intraocular particle may cause traumatic damage to the cornea, iris, lens, or retina, or all. Damage to the lens is especially serious, since, not being well supplied with blood, it is slow to heal. Also, *any* damage to the lens tends to act as a catalyst

for coagulation of its proteins, resulting in opacity and vision loss. Damage via trajectory will rupture numerous small blood vessels. The hemorrhaging from these is toxic to the eye because of the iron in the blood.

Several metals are of special concern as intraocular foreign bodies. Pure copper particles are almost certain to destroy the eye, because the toxic copper molecules become deposited in the lens, cornea, and iris (chalcosis). Copper alloys do not seem to have any toxic effects.

Iron, as already mentioned, has toxic effects (called siderosis), and even iron in the blood is sufficient to cause cell damage in eye tissues.

Pain cannot be counted on to alert a person that he has a foreign body *in* his eye. If it's on the cornea, blinking will probably cause constant irritation, but if the object has penetrated into the eyeball, there may be no acute pain. In one case, a machine shop employee complained of a mild distress in the eye, and vaguely remembered getting something in his eye several days before. Examination revealed a piece of metal had deeply penetrated the eye and the eye was eventually lost.

Burns

In many petrochemical and basic metals establishments, burns are a more serious threat to the eye than flying objects. Burns may result from chemicals (strong acids or alkalis), heat, or irradiation.

Chemical burns—Caustics are much more injurious to the eyes than acids. The medical prognosis of caustic burns always has to be guarded. The appearance of an eye on the first day may not be too bad, but it may deteriorate markedly on succeeding days. This is in contrast to acid burns, where the initial appearance is a good gage of the ultimate damage.

The reason for this is that strong acids tend to precipitate a protein barrier that prevents further penetration into the tissue. The alkalis do not do this, continuing to soak into the tissue as long as they are allowed to remain in the eye.

The end result of a chemical burn is usually a scar on the cornea. If this is not in front of the window in the iris, vision may not be greatly hampered. If the scar is superficial, a corneal transplant can alleviate burn damage. (Densely scarred corneal tissue cannot be repaired by transplants, contrary to popular misconception.)

If first-aid measures are not taken in time, and the chemical penetrates the anterior chamber of the eye, the condition is called iritis (bathing of the iris with the chemical agent.)

Thermal burns—Heat can destroy eye and eyelid tissue just as it does any other body tissue. But eye tissues may not recover so well as skin and muscle from such damage. The lids are more likely to be involved in burns than the eye itself, since involuntary closing of the eye is an almost certain response to excessive heat.

Irradiation burns—Two common radiation wave lengths can do harm to the eye:

1. Ultraviolet—Exposures to ultraviolet light are usually around welding operations. The welder is rarely involved, since he is too close to the arc to stand looking at it without the proper eyeshield. But welders' helpers and other bystanders frequently suffer from exposure. The toxic effects of UV radiation on the epithelium of the cornea is cumulative, and painful reaction may be delayed.

Unless at extremely high concentrations, UV does not penetrate the eye, being filtered out at the surface of the cornea.

2. Infrared—Unlike UV, infrared radiations pass easily through the cornea, and their energy is absorbed by the retina. With automation of basic metals operations, eye damage from infrared radiation is not nearly so common as it once was.

EVALUATING OCCUPATIONAL EYE HAZARDS

It does not take special training or engineering skills to identify most potential hazards to employees' eyes. Where people handle acids or caustics, where airborne particles of dust, wood, metal, or stone are possible, or where blows from blunt objects are likely, eye protection is needed.

Employees most directly involved with operations that produce such hazards are seldom overlooked by modern protective equipment programs, but all too often people on the perimeter of eye-hazardous operations are left unprotected, with costly results. Who has not heard of pieces of broken metal tools that found an eye after traveling great distances from the drill press, or of the welder's helper who failed to turn his back a few times and got painful corneal burns?

Even when a person's work may only occasionally bring him into proximity with eye hazards, the safest policy is to encourage eye protection all day long. The outdated technique of hanging a pair of community goggles near the grinding wheel is an example of the "eye-hazard job" approach to eye protection. The jobs that involved eye hazards were identified and eye protection was required for men doing these jobs, and only when they were doing the job. Another concept, which affords protection for the neighboring worker, the supervisor looking over a man's shoulder, or just a curious passerby, is the "eye-hazard area" approach. An "eye-hazard area" can be defined as "one wherein continuous or intermittent work on the part of anyone could cause an eye injury to anyone in the area." With proper enforcement and designation of areas, this approach is the most effective means of preventing eye injuries.

PROTECTIVE EQUIPMENT

All eye-protection equipment is designed to bolster one or more of the eye's natural defenses. Chippers' cup goggles extend the bony ridge protecting the eye socket and provide an auxiliary, more penetration resistant "cornea."

Chemical worker's goggles are better than a blinking eye lid for protection from chemical splashes.

The eyewash fountain takes over where the tear glands fall down.

There is a tremendous variety of eye protection available, from throw-away visitors' eye shields to trifocal prescription safety spectacles, from welders' helmets to clip-on anti-glare lenses. But the classic safety spectacles, with or without side-shields, will probably be adequate for 90% of general industrial work.

It is the most "sellable" eye protection on the market. Most workers find no real objections to wearing spectacles (when the fit is good and the prescription is right).

To go into specific recommendations for eyewear here would be to duplicate information easily available from suppliers or other sources. (See especially the 20-page section on eyewear in the October 1965 *National Safety News*.)

Worthy of mention, though, is the fact that unless the protective devices provided are *comfortable*, they will likely not be worn, even where eyewear is "compulsory." Many industrial ophthalmologists have seen injuries to workers who had goggles on their foreheads or spectacles in their hip pockets.

A sure way to guarantee undue discomfort is to require more protection than is really needed for the hazards involved, since level of protection and comfort are generally in inverse proportion.

First Aid

What should the employee do if, in spite of his protective eyewear (or through lack of it), he gets foreign material in the eye?

For propelled object injuries, he should get immediate medical attention. Even for foreign bodies on the corneal surface, self-help should be discouraged—removal of such particles is a job for a trained medical department person.

Chemical splash injuries require a different tack. Here the extent of permanent damage depends almost entirely on how the victim reacts. If the victim of a concentrated caustic splash gets quickly to an eyewash fountain and properly irrigates the eye continuously until medical help arrives, the chances are he will end up with a clear cornea or minimal damage.

Such irrigation should be with plain water from standard eyewash fountains, emergency showers, hoses, or any other available sources.

Water for eye irrigation should be clean, and within certain temperature limits for comfort. Tests have shown that 44.4°C (112°F) is about the upper threshold limit for comfort, but colder water, even ice water, apparently causes no harm and is not uncomfortable enough to discourage irrigation.

Use of substitutes for water (neutralizing solutions, boric acid solutions, mineral oil, etc.) is discouraged by nearly all industrial ophthalmologists, since in many instances such preparations can cause eye damage greater than no irrigation at all.

Contact Lenses

There has been much debate about the merits of wearing contact lenses in industry. Some chemical companies have adopted a rule that they cannot be worn on the job. On the other hand, some authorities think they are a blessing and can make work safer.

Probably the most important thing to keep in mind about contacts is that in the event of foreign bodies or chemicals in the eye, the contact *does not protect* the portion of the cornea it covers; liquids and even dusts will creep behind the lens. If contact lenses are worn they will require added eye protection, and wearers must be trained to remove their lenses before irrigation at eyewash fountains.

ACKNOWLEDGEMENT: This chapter is reprinted with permission of the National Safety Council from *National Safety News*, vol. 95, Feb. 1967. It is based on lectures by David Bishop, M.D., Chairman, Ophthalmology, Children's Memorial Hospital, Oklahoma City, Oklahoma; D. Jack Killian, M.D., Medical Director, Dow Chemical Co., Freeport, Tex.; and Clyde Harlin, Chief, Ground Safety, Tinker Air Force Base, Oklahoma City, Oklahoma. This and other articles on industrial hygiene have been collected in *Fundamentals of Industrial Hygiene*, National Safety Council, Chicago (1968).

Eye Protection

N. V. Steere

In all laboratories where wet or dry chemicals are used there is the hazard of splashes or dust particles entering the eye. There is also the hazard of flying particles or objects in laboratories where pressure or vacuum vessels are assembled if the vessel explodes or implodes. Work with electrical wiring or tools may result in eye injuries if an arc spits molten copper or if a tool throws a chip.

Explosions seem to be the cause of the most serious eye injuries. Spectacles with side shields should be worn as a basic protection against flying objects from explosions, and to afford a reasonable measure of protection against droplets which may be splashed. Although glass spectacles with side shields are more expensive than plastic goggles or eyeshields, the higher initial cost seems to be justifiable on the bases of comfort and wearability, optical quality, cleaning abrasion resistance, and durability.

If a hazard is present in a laboratory or during a course, so that eye protection is needed for some operations or some experiments, the best policy is to require protection to be worn at all times by all persons who work in or enter the laboratory. Hazardous operations and experiments may endanger others who come into the laboratory or who are unaware of the hazards. One example would be represented by the undergraduate who came into a different laboratory section to make up work he had missed in his regular section and, while others were doing their non-hazardous experiment, proceeded to try to prepare bromobenzene by a new method which resulted in explosion and injury.

If safety glasses are to be worn consistently by students or technicians, the instructor or supervisor should teach the practice by example. Even though visitors to the laboratory may be infrequent, they should be protected from the eye injury hazards which may exist. Requiring visitors to wear eye protection will impress both visitors and students.

Laser research or use presents some unique eye injury hazards that require special consideration. Glassblowing and welding present well-known hazards for which needed protection is readily available. Because eye injuries in "ordinary" laboratories are relatively rare, the hazards are not well-known and frequently not guarded against.

STANDARD FOR EYE AND FACE PROTECTION

The "U.S.A. Standard Practice for Occupational and Educational Eye and Face Protection, Z87.1-1968" details strength and quality standards for all forms of safety eyewear, excluding those relating to x-rays, gamma rays, high-energy particulate radiations, lasers, or masers. All safety eyewear for use in educational institutions, as well as in industry proper, should meet or exceed the requirements of the Z87 standard, or later revisions thereof. Manufacturers and distributors of eye and face protective devices should be required to furnish a statement on their business letterhead to the effect that all equipment furnished under a school or other purchase contract will in fact meet or exceed the specifications set out in the Z87 standard.

The Z87 document represents the views and recommendations of over 30 organizations concerned with the quality and use of eye and face protective devices. The new standard, as of September 18, 1968, replaced "The American Standard

Safety Code for Head, Eye, and Respiratory Protection, Z2.1-1959," both of which were developed under the guidance of the American National Standards Institute, Inc. (formerly known as the American Standards Association and the U.S.A. Standards Institute). The National Society for the Prevention of Blindness served as sponsor for the development of the Z87 standard.

The Z87 standard recommends specific types of eye and face protective devices for use in laboratories, as well as in other educational and occupational applications. The document also provides guidance on procedures for maintenance and disinfection of safety eyewear.

Glass or plastic safety lenses, which must be retained in industrial-quality frames, have permissible thickness ranging between 3.0 mm and 3.8 mm (0.11–0.15 in.); plastic lenses for flexible-fitting goggles must not be less than 1.3 mm (0.050 in.) in thickness. Face shields, which are not intended to provide basic eye protection, according to the Z87 standard, must be at least 1 mm (0.040 in.) in thickness.

All types of safety lenses must meet or exceed rigid impact and penetration tests. Safety spectacle lenses of glass or plastic, for example, must resist the impact of a 2.54-cm (1.00-inch) diameter steel ball, weighing approximately 68 g (2.4 ounces), dropped in a free fall from a height of 127 cm (50 inches) onto the horizontal upper surface of the lens.

Tests and experience have proved that pitted or scratched glass lenses have much less resistance to impact than lenses in good condition. The Z87 document recommends that all such lenses be removed from service.

TOUGHENING GLASS LENSES

Glass safety lenses for use by school and college personnel, as well as by industrial workers, are toughened to resist fracture by heating and surface cooling in a manner controlled to produce a high ratio of surface compression to internal tension. Reinforcing a glass lens through heat-tempering markedly increases its resistance to impact. Optical-quality plastic safety lenses are naturally strong, and require no hardening.

Figure 1 illustrates a method of cooling lens surfaces after furnace treatment, and shows schematically the areas in tension and in compression.

SPECIAL EYE AND FACE PROTECTION

If unusual hazards are inherent in certain laboratory operations, special or extra face and eye protection may be needed. There are chemical worker's goggles available that are vented for long periods of wear. Such protection would be desirable for acid pouring operations and other jobs with an appreciable hazard of splashes.

FACE SHIELDS

Face shields over basic eye protection are recommended for many exposures in chemical laboratories, as well as in woodworking; metal machining; spot welding, handling hot or corrosive metals; and buffing, wire brushing and grinding operations. Face shields over safety glasses are also recommended for protection when handling materials which are sensitive to friction, shock, and air.

Manufacturers and distributors of safety eyewear, usually listed in the classified section of the telephone directory, will make qualified representatives available to assist in the selection of proper eye and face protective devices.

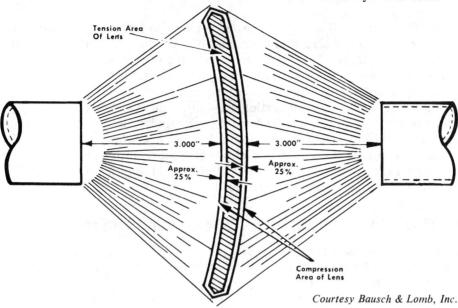

Tension Area
Of Lens

3.000"   3.000"

Approx.
25%   Approx.
25%

Compression
Area of Lens

Courtesy Bausch & Lomb, Inc.

FIG. 1. Sketch showing one method of surface cooling to toughen a safety lens, and areas of tension and compression. Tension area of lens, approximately 50%. Compression area of lens, approximately 50%.

LEGAL REQUIREMENTS FOR EYE PROTECTION

In 1963 Ohio became the first state to enact legislation requiring all students, teachers and visitors in educational laboratories and shops to wear industrial-quality eye protective devices. The Ohio school eye safety law was amended in 1965 to require eye-safety coverage in all schools, rather than just public institutions. As of

Courtesy Bausch & Lomb, Inc.

FIG. 2. Safety glasses with side shields provide protection against flying particles from the front and sides.

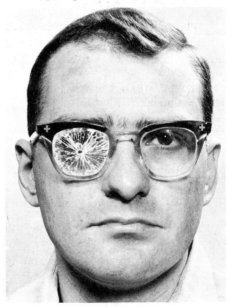

Courtesy Bausch & Lomb, Inc.

FIG. 3. Fractured safety lens is retained in the spectacle frame. Fracture pattern illustrates code requirement.

October 1969, a total of 30 states had enacted similar legislation, based upon a model school eye safety law developed and promulgated by the National Society for the Prevention of Blindness. The National Society reports its national campaign to prevent ocular damage and loss of sight among technical students, their teachers, and visitors is supported by the Child Welfare Commission of the American Legion, and safety, industrial, medical, and educational groups. The model school eye safety law specifies eye and face protective devices for every student and teacher in schools, colleges, universities, and other educational institutions, while participating in or observing instructional work in chemical-physical laboratories and industrial arts and vocational shops. Home economics laboratories, biology laboratories and geology field trips were not specifically cited in this law.

EMERGENCY TREATMENT OF CHEMICAL EYE INJURIES

It is important to note and be sure that every individual using laboratory facilities knows the correct method of emergency treatment of injuries in which chemicals enter the eye. The medically-recommended method of emergency treatment of chemical eye injuries is to *gently flush the injured eye thoroughly with regular tap water for* 15 *minutes—without delay*! Following this first-aid action, the injured individual should be taken for medical treatment, again without delay. Contact lenses present additional problems, should chemicals get into the eyes; they *must* be removed prior to irrigation of the eyes with water, otherwise irreparable damage to eyesight may occur. The importance of immediate flushing of eyes in the event that chemicals get into them is emphasized in the color illustrations following this chapter. Accident-prevention experts, as well as medical authorities, express serious concern over the use of contact lenses by personnel using shops and chemical laboratories. Many firms prohibit their in-plant use entirely.

LABORATORY EYE INJURIES

The importance of routine use of industrial-quality eye and face protection in laboratories has been demonstrated in many documented cases of severe eye injuries, loss of sight, and even total blindness.

ACKNOWLEDGEMENT: Reprinted from the *Journal of Chemical Education, Vol. 42*, A464, June, 1965, with permission. Revised in June, 1970.

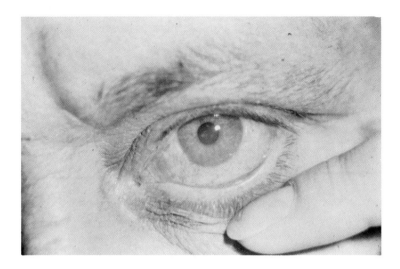

PLATE 1   DIAGNOSTIC AID TO DETECT INJURY FROM A CHEMICAL BURN: Following irrigation a drop of fluorescein dye is placed in the conjunctival sac, the eye closed, and the dye is allowed to remain for two minutes followed by further irrigation. In this case the lower one-third of the eye reveals injury from a chemical burn. In most instances the affected corneal tissue will stain green and the conjunctiva over the sclera will be yellow.

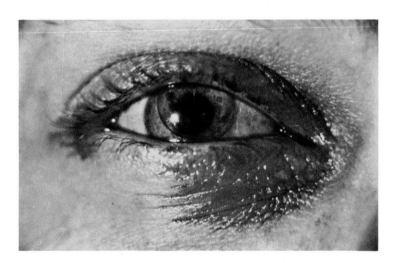

PLATE 2   CONTACT LENS RESTING ON THE SURFACE OF THE CORNEA: Just prior to taking this picture a drop of fluorescein dye was placed in the corner of the eye. It has crept under the contact lens around the border. This is what happens when a person gets a chemical in his eye. It is important to understand because several cases of chemical burns have been treated in which the individuals left their contact lenses in during irrigation, and the resulting burn was more serious. Contact lenses prevent the cornea from being adequately irrigated, therefore, they should be removed immediately after chemical exposure and before starting irrigation.

PLATE 3   FACTS CONCERNING CHEMICAL BURNS TO THE EYE: (1) Alkaline materials are more hazardous in the eye than strong acids. (2) Irrigation with neutralizing agents should not be used as a first aid procedure. This petrie dish contains egg white or egg albumin. A drop of strong sodium hydroxide was placed on the left, and, on the right, a drop of strong hydrochloric acid. Note on the right that the acid immediately reacted with the protein of the egg albumin to form an insoluble barrier. This barrier serves to prevent the acid from soaking deep into the tissue. The alkaline material on the left does not precipitate this barrier so it is free to soak deep.

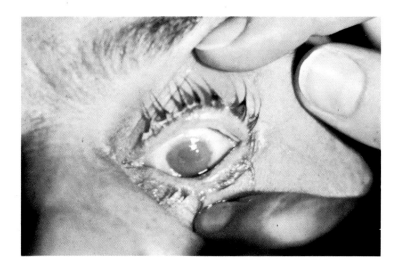

PLATE 4   THE DAMAGING EFFECT OF CHEMICALS IS ACCELERATED GREATLY BY HEAT: This man splashed hot chloroacetic acid in his eye, and the cornea was severely damaged. If the solution had been cold, the injury would not have been nearly as serious. Eye and face protective gear could prevent such injuries.

PLATE 5 THE EFFECTS OF ACID AND ALKALINE SOLUTIONS: The cornea in the front part of the eye, where the burn occurs, can be compared to the egg albumin in the photograph at the top of page 62. In the test tube on the right, the acid (red) produces an insoluble barrier when it contacts the cornea, while the alkaline material (blue) in the test tube on the left penetrates deep because it does not form a barrier. Alkaline materials are far more serious in the eye than strong acid solutions and require longer periods of time to leach out the damaging chemical by irrigation of the eye.

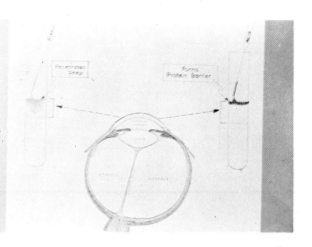

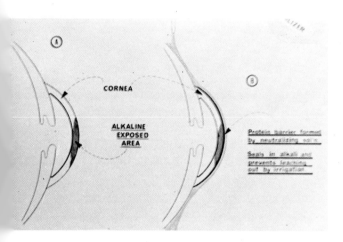

PLATE 6 HAZARD OF USING NEUTRALIZING SOLUTIONS: Assume that an alkaline material gets into the eye and soaks deep into the tissue as shown on the left side of the illustration. On the right, when the acidic neutralizer is used a protein barrier is formed over the cornea, sealing in the chemical and preventing the leaching out of the alkaline material by irrigation.

PLATE 7 IRRIGATE YOUR OWN EYES PROPERLY: Hold the eyelids apart with your fingers and roll your eyeball around as much as possible in order to wash all of the eye. If one of your co-workers happens to be the victim, assist in the irrigation procedure through holding—or forcing—his eyelids apart.

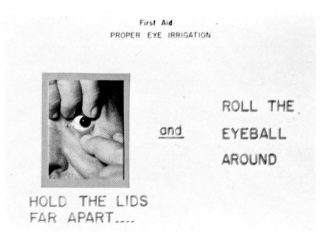

WATER HOSE

DRINKING FOUNTAIN

BEAKER of WATER

PLATE 8 EMERGENCY EYEWASH FOUNTAIN OR EYE BATH NOT AVAILABLE: Use a water hose, a drinking fountain, a beaker of water, or any other available source of water for the irrigation.

PLATE 9 EMERGENCY TREATMENT FOR EXTENSIVE CAUSTIC BURNS TO FACE AND EYES: The Y tube is being used from a gravity fed water source to provide a gentle flowing stream of water. The eyelids must be held apart to do an adequate job of irrigation. Preferably the victim should be lying down, and instructed to roll his eyeball around during irrigation so that the chemical may be flushed out of all of the interstices of the conjunctiva. The most important part of the treatment is the immediate and adequate irrigation with plain water. A delay of this procedure for 30 seconds can well mean the difference between no effect to the eye and permanent loss of vision. This man had no loss of vision in spite of the concentrated caustic, because the eye was immediately and adequately irrigated. The caustic burn to the face required extensive skin grafting over a period of two years.

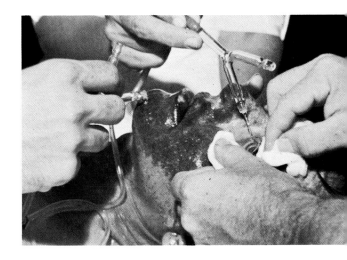

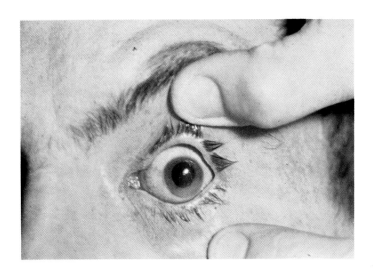

PLATE 10 IMPORTANCE OF PROMPT IRRIGATION IN PREVENTING PERMANENT EYE INJURY: While working alone, this man fell into a puddle of caustic. As a result of the fall, some splashed into his eye. By the time someone could get there and assist him to an eye shower, approximately 5 to 7 minutes had lapsed. By the time the eye was examined, within a few minutes, this opacity of the cornea was already present in the lower part of the eye. This man has a permanent wedge of scar tissue in the lower part of his eye because he could not get it irrigated in time. Safety eyewear would have eliminated eye damage, and the need for treatment.

PLATE 11 HYDROCHLORIC ACID SPLASHED IN THIS EYE DID NOT ALTER VISION BUT CAUSED FORMATION OF SCAR TISSUE: Treatment kept the cornea perfectly clear, but scar tissue began to form three weeks after the original injury and this photograph was taken several months later. The scar tissue continued to form and developed a large adhesion extending from the edge of the lower lid almost up to the iris (or colored part of the eye). The tissue became infected quite easily. Surgery was done to cut and remove this adhesion but it began to reform and grow several weeks later. Consequently this man had a life-long disability as the result of a chemical splash.

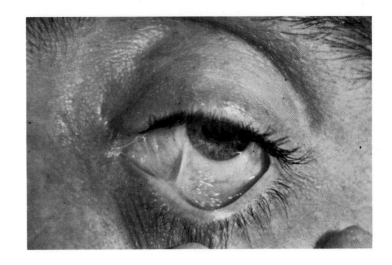

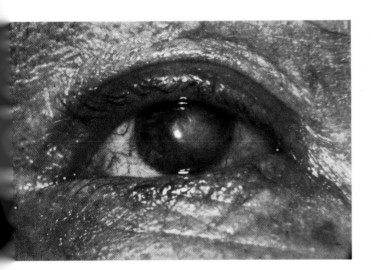

PLATE 12 PERMANENT DAMAGE WHICH CAN OCCUR FROM MANY CHEMICALS: Formation of scar tissue over the transparent cornea is like a window shade. The light rays can no longer go through the transparent cornea, and an individual's vision is impaired. This man splashed a drop of caustic in his eye but delayed irrigating and scar tissue formed over his cornea.

PLATE 13 SURGICAL REPAIR AND REMOVAL OF THE SCAR TISSUE IS POSSIBLE IN SOME CASES: This shows the same case after a corneal transplant was done. A circular window in the scar tissue has been removed and a piece of normal cornea has been grafted in its place. This type of operation is performed when someone "wills" or sacrifices their eye so that another may have restoration of their vision. It is necessary to have some reasonably good remaining cornea to receive the graft.

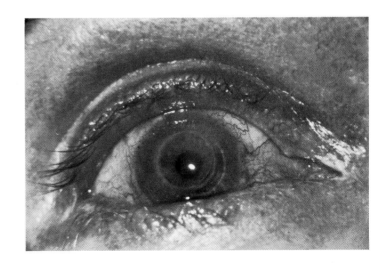

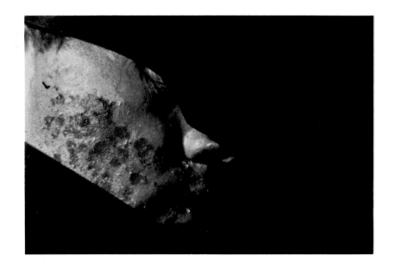

PLATE 14 INDIVIDUAL EXPOSED TO THE IN-
TENSE HEAT OF A SUDDEN ELECTRICAL ARC:
This reveals the excellent protection afforded
the eye region by the use of safety glasses. A
face shield over the primary protective device
would have minimized, if not eliminated,
facial damage.

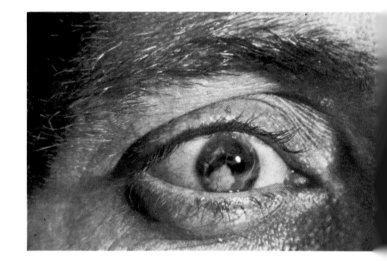

PLATE 15 A PERMANENT CORNEAL OPAC-
ITY RESULTING FROM A LABORATORY EX-
PLOSION: An injury from propelled glass-
ware.

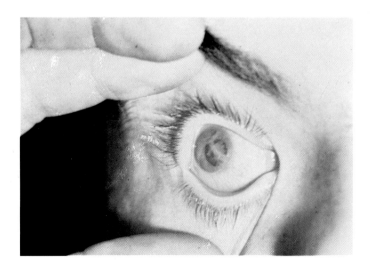

PLATE 16 COMPLETE LOSS OF VISION BE-
CAUSE OF FAILURE TO WEAR SAFETY
GLASSES: The scar tissue in the cornea and
the cataract in the lens were caused by a
traumatic injury from a propelled foreign
body; in this case a small piece of wood.

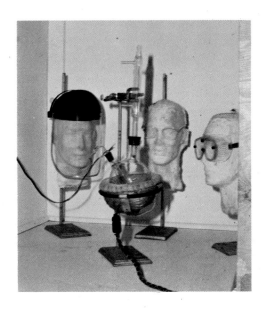

PLATE 17 EXPERIMENTAL SETUP FOR EVALUATING PROTECTION AFFORDED BY DIFFERENT TYPES OF SAFETY GLASSES AND FACE SHIELDS: Masks were constructed of papier mâché and sensitized with potassium thiocyanate solution, and flask containing ferric chloride was exploded by ignition of potassium chlorate and magnesium powder in the thermometer well. Tests were conducted by Professor Emeritus G. Norman Quam at Villanova University, under Public Health Service Research Grant AC-00129 from the Division of Accident Prevention. A 16 mm color/sound motion picture of Dr. Quam's research, "Eye and Face Protection in Chemical Laboratories," is being distributed by the National Society for the Prevention of Blindness, 79 Madison Avenue, New York, N.Y. 10016.

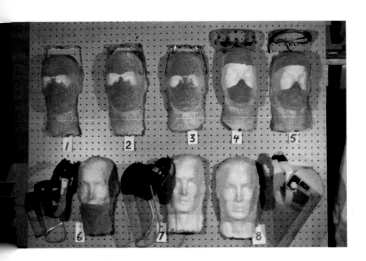

PLATE 18 EIGHT MASKS EXPOSED TO SIMULATED EXPLOSIONS: Eye and face protection afforded by different types of glasses, goggles and shields. Safety glasses 1, 2, and 3 are plain, perforated side shields, and solid side shields respectively. In each case, liquid spray bounced along the nose and cheeks into the eye socket. Goggles 4 and 5 did prevent damage by spray. All were effective in stopping solid missiles. Face shields 6, 7, and 8 varied in width and length of windows; each had a high crown. Number 6 had a 10-cm. window and did not protect mouth, chin and neck from missiles and splashes. The 20-cm. window of No. 7 was quite effective. The 24-cm. window and high crown of No. 8 gave complete protection of scalp, face, and neck.

PLATE 19 EVALUATION WAS MADE OF FACE PROTECTION AFFORDED FROM SIMULATED EXPLOSIONS AT THE SIDE: The plain glasses of No. 1 allowed spray to pass back of the lens. This was also true of the perforated side shields of No. 2 but the solid side shields of No. 3 did offer protection. The perforated and solid sides of goggles showed similar effects. The narrow 10-cm and 20-cm face shields (No. 6 and No. 7 respectively), exposed the ears in each case, while the very wide and very long shield of No. 8 gave good protection.

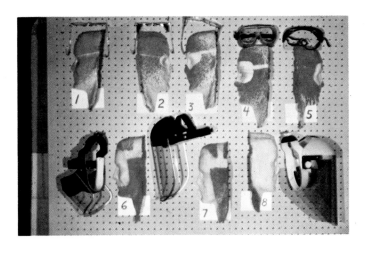

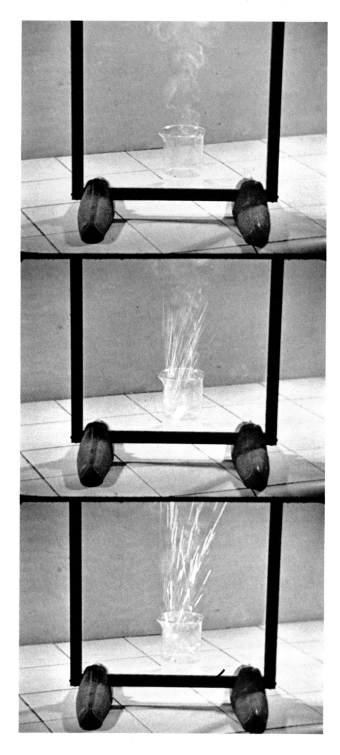

PLATE 20 METALLIC SODIUM BEHAVES ERRATICALLY IN CONTACT WITH WATER: A hydrogen explosion such as shown here can cause injury by violently spraying about burning sodium and sodium hydroxide. This is why it is so important to use the appropriate safety shielding, as shown here, when working with such chemicals. Photograph reprinted from the film, "Safety in the Chemical Laboratory", with permission from the Manufacturing Chemists' Association, Inc., and cooperation of Edward Feil Productions, Cleveland, Ohio.

PLATE 21 THIS IS A TYPICAL EXAMPLE OF A SECOND DEGREE BURN: One can readily see the fluid that has been formed between the two layers of skin. It is a general rule that if blisters like this form, it is probably a second degree burn. However, one cannot be absolutely sure, and in some instances it is virtually impossible to judge accurately between a second and third degree burn until many days have elapsed. Many burn experts feel that a better classification of burns is to call them partial thickness or full thickness burns, depending on whether the skin is completely destroyed or has enough of the small skin islands deep in the tissue to regenerate new skin.

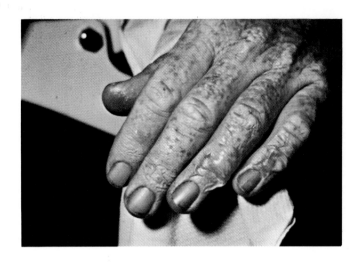

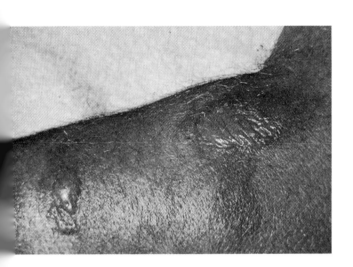

PLATE 22 THIS IS AN EXAMPLE OF A SECOND DEGREE BURN CAUSED BY EXCESS COLD: A burn is due to damage and coagulation of tissue and can occur from cold as well as heat. This man got up against a pipe that was full of liquid oxygen, which is several hundred degrees below zero; he got blisters just like those from a heat burn.

PLATE 23 AVULSION OF SKIN FROM THUMB AFTER A LABORATORY GLASSWARE ACCIDENT: Skin grafting was necessary for repair.

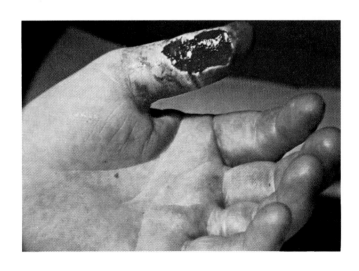

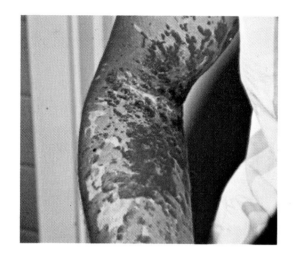

PLATE 24   CHEMICAL BURN TO SKIN FROM CONCENTRATED SULFURIC ACID: Picture taken one week after initial burn shows the pigmentation and dry eschar formation which are typical of many strong acid burns.

PLATE 25   SAME CASE FIVE WEEKS LATER, SHOWING SUPER-FICIAL SCARRING FROM THE ACID BURN: Strong mineral acid burns will not usually produce as serious an effect or as much scarring as strong alkaline burns. However, one notable exception is hydrofluoric acid, which has a unique destructive capability.

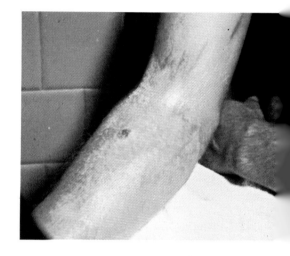

PLATE 26   SEVERE CONTACT DERMATITIS: This illustrates the potential problem in some people who develop skin sensitivity to certain chemicals.

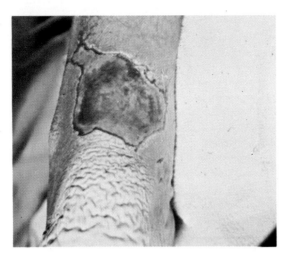

PLATE 27 SEVERE SODIUM HYDROXIDE BURN OVER THE ACHILLES TENDON OF A WORKER WHO GOT SOME FLAKE CAUSTIC IN HIS SHOE: That day he noticed only a very slight itching sensation and did not realize the significance of the exposure. Strong acids usually have good warning properties when they contact the skin, but many times strong alkaline solutions will give a minimum of discomfort when the skin is contacted.

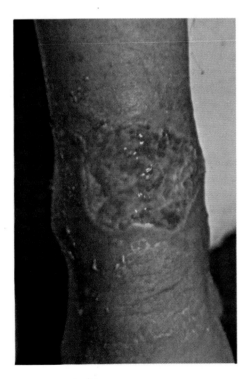

PLATE 28 THE SAME CASE ONE WEEK LATER: The burn has completely destroyed the full thickness of the skin and reveals the underlying subcutaneous tissue.

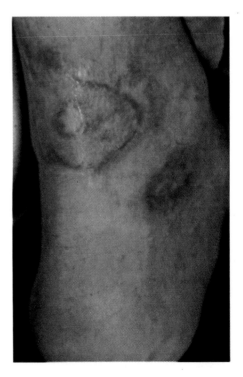

PLATE 29 SKIN GRAFTING WAS DONE: This resulted in a good take with no permanent disability. However, the man was unable to work for several months because of the extensive repair work involved.

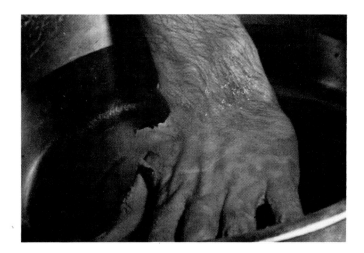

PLATE 30 THIS MAN ACCIDENTALLY HAD HIS HAND SPLASHED WITH SOME LIQUID BROMINE, A VERY CORROSIVE MATERIAL: You will notice how this case was treated. It was soaked in plain water to leach out all of the residual chemical possible. Neutralizers are of questionable value and in some instances do harm.

PLATE 31 THIS IS THE SAME CASE OF BROMINE SPLASH, SHOWING THE PALM OF THE HAND: It does not appear to be nearly as badly affected, because the skin of the palm is much thicker, and the amount of damage many times depends upon the thickness of the skin involved. One would expect the palm to have more of a protective action against chemicals.

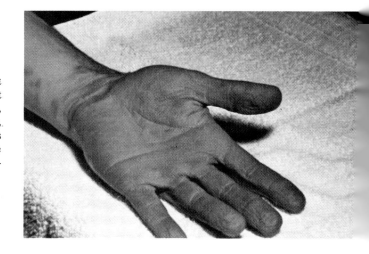

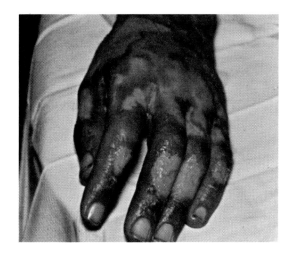

PLATE 32 THIS SHOWS THE BROMINE SPLASH CASE ON THE FOLLOWING DAY: Here one sees that the most severely burned area has been the upper part of the fingers and the back of the hand, where the skin is the thinnest.

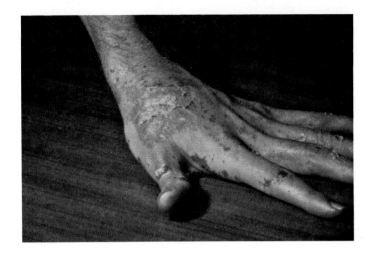

PLATE 33 THIS SHOWS THE SAME BROMINE
SPLASH CASE SEVERAL WEEKS LATER: This man
had an excellent result with no significant
permanent scarring, because of prompt treatment with plain water.

PLATE 34 A TYPICAL EXAMPLE OF THE PROB-
LEM PRODUCED WHEN CERTAIN ORGANIC CHEM-
ICALS ARE TRAPPED IN LEATHER: This man was
exposed to ethylene dibromide. He followed
the correct emergency procedure of immediately stripping off his clothing and showering.
He put on fresh clothing, but put his contaminated belt back on. The resulting burn appeared several hours later. Many chemicals
such as ethylene dibromide, methyl chloride,
methyl bromide, methylene chloride, ethylene
oxide and others have a similar effect when
trapped in leather, under wristwatches, or in
small bandages. Normally these chemicals, if
placed on the exposed skin, will evaporate
without causing any difficulty. However, when
their vapors are trapped next to the skin,
severe chemical burns can result.

PLATE 35 ANOTHER EXAMPLE OF THE PHE-
NOMENON WHEN ORGANIC CHEMICALS ARE
TRAPPED IN LEATHER: This man had only a
few drops of 1,3-dichloropropene drop on his
leather shoe. This gave no warning sensation.
Several hours later there was itching of the
foot, and the following day a large chemical
burn resulted. In such situations it is important
·to recognize these unusual properties and wear
proper footwear. In the event that leather
shoes are contacted by chemicals, they should
be thoroughly decontaminated before re-use.

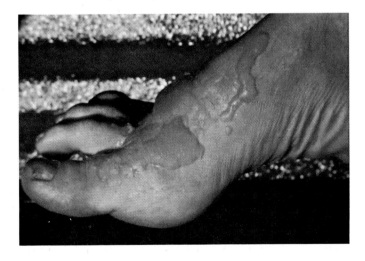

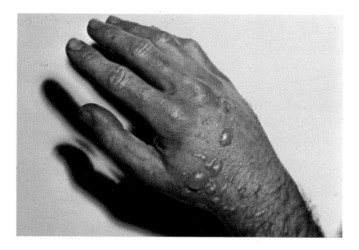

PLATE 36 A CHEMICAL BURN FROM ETHYL AZIRIDINYL FORMATE: This chemical is typical of many reactive organic chemicals that have poor warning properties. Skin contact did not produce any warning sensations and several hours later, slight itching and redness became apparent. By the end of the day blisters had begun to form, and decontamination at this time would do no good.

PLATE 37 SAME CASE 48 HOURS LATER, REVEALING THE EXTENSIVE SWELLING AND VESICULATION: Burns from these types of organic compounds are typically slow to heal, and frequently produce third degree burns.

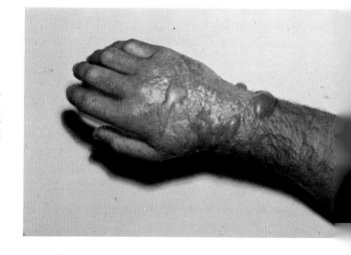

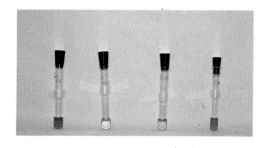

PLATE 38 THE FERROUS THIOCYANATE TEST FOR PEROXIDES IN ETHERS WAS CARRIED OUT ON TWO KNOWN CONCENTRATIONS AND ON SAMPLES OF TWO LABORATORY CONTAINERS OF p-DIOXANE: For comparison purposes, tube #1 contained 0.05% peroxide, and tube #2 contained 0.005% peroxide. The p-dioxane in tube #3 was 3 years old and was stabilized with p-benzylaminophenol, while the p-dioxane in tube #4 was new, but had not been stabilized.

110

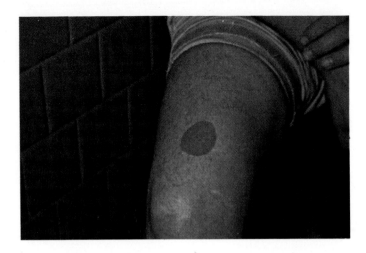

PLATE 39 THESE TWO PHOTOGRAPHS SHOW WHY ONLY SOAP AND WATER SHOULD BE USED FOR DECONTAMINATION OF PHENOL BURNS: The red area on the leg in this photograph simulates a phenol burn that is not large enough in area to cause a fatality from absorption, even if untreated.

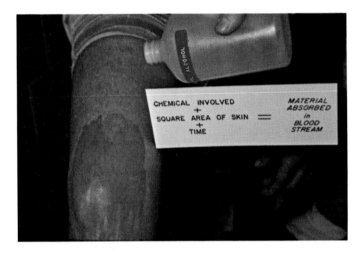

PLATE 40 THIS PHOTOGRAPH ILLUSTRATES WHAT MIGHT HAPPEN IF ALCOHOL WERE POURED ON THE PHENOL BURN: The alcohol would spread the phenol and increase tremendously the area of skin exposed to the phenol, because phenol is very soluble in alcohol. There have been several deaths from the injudicious use of alcohol in a manner such as this. The area of phenol was too small to produce a serious reaction in the body due to absorption; but because some well meaning person had read about using alcohol on phenol, he poured alcohol on the phenol, increased the surface area of absorption and produced a fatality.

Death due to improper first aid is possible, if proper decontamination procedures with soap and water are not set up ahead of time. Chemical involved plus square area of skin plus time equals the amount of material absorbed in the blood stream. Phenol is a unique chemical in this instance. Therefore it is strongly urged that you not use alcohol or advocate alcohol as a first aid measure for treatment of phenol burns; it may do a great deal more harm than good. Alcohol is used in medical departments in the treatment of phenol exposures, but in a very particular controlled way, with recognition of all the hazards that are involved. The recommended emergency first-aid treatment for phenol on the skin is to scrub with soap and water and get the phenol off just as fast as possible.

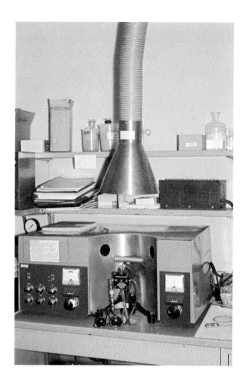

PLATE 41 DEMONSTRATION OF SPOT VENTILATION USED TO REMOVE GASES, VAPORS AND FUMES FROM AN ATOMIC ABSORPTION SPECTROPHOTOMETER. The cone shown is not the most effective shape of inlet, however.

PLATE 42 A SEPARATE VENTILATED ROOM TO REMOVE SOLVENT FUMES USED FOR A TISSUE PROCESSOR: Acetone is used in 1- or 2-liter quantities in the processor, and the room serves as fire protection for the laboratory area. Not shown in the photograph is a curb, which would prevent burning solvent from running out underneath the door.

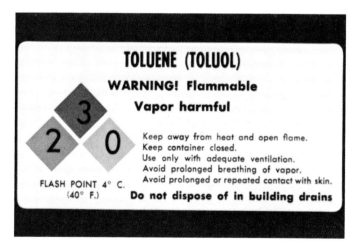

PLATE 43 CHEMICAL LABEL WITH HAZARD SIGNALS AND PRECAUTIONARY INFORMATION: The hazard signals are patterned after recommendations of the National Fire Protection Association. The precautionary information follows closely the recommendations of the Manufacturing Chemists' Association, Inc. Advice against disposal into the building drains is unique to labels used at the University of Minnesota.

Shields and Barricades For Chemical Laboratory Operations

David T. Smith

Whether personnel is performing experimental or routine laboratory work, it is important to provide protection from the hazards of explosion, rupture of apparatus and systems from overpressure, implosion due to vacuum, sprays or emissions of toxic or corrosive materials, or flash ignition of escaping vapors. In spite of planning and preparation, these types of mishaps can and do occur, and people have been injured by the results of the mishaps.

Analysis of the force and characteristics of the hazards involved can aid in selection of effective and economic solutions to the guarding problem. The decision is simplified by the fact that protection designed for the most severe exposure will suffice for lesser exposures, although overdesign may be uneconomic.

TYPES OF EXPOSURES

Although trinitrotoluene and other hazardous high energy materials are capable of detonation, most chemicals explode by chemical processes which are subsonic in speed and termed deflagrations. Because the time during which the destructive energy is liberated in deflagrations varies so widely, the total equivalent explosion force of various systems, as compared to a single material such as TNT, is of limited significance. TNT is capable of detonation and hence of developing its full potential instantaneously. Most chemicals, such as alcohol-air mixtures, are not. Even acetylene, dinitrotoluene, and nitrocellulose are caused to detonate only under severe conditions of confinement or pressure.

Accordingly, for most deflagrations the blast wave pressures are functions of the bursting pressure of their containers. The design of protection for chemical processes can be based on deflagration effects unless there is good reason to believe that a detonation can occur.

Missiles of great energy may be thrown by all types of explosions. They may consist of broken parts of apparatus, or even parts of damaged shields. For complete missile protection: (a) No unshielded line-of-sight path should be allowed between the exposing apparatus and any part of the observer's body; (b) No deflected missile path (ricochet) should be allowed where the angle of incidence with the deflecting surface exceeds 45 degrees (zero angle of incidence is a missile path perpendicular to the deflecting surface).

If adequate missile barriers are provided the only serious air-blast hazards in bench scale work besides failure of the shield or shield material are: (a) The possibility of ear damage; (b) Self-inflicted injury from involuntary or irrational reaction to the blast.

There is frequently the hazard of flash fire exposure, although the explosion hazard may be mild. While shielding of sufficient strength and completeness to protect from missile and blast hazards will normally be adequate for flame protection, complete enclosure in a ventilated hood is necessary to insure the containment of flames. Injuries from fires in laboratory work have most frequently been due to the direct proximity of the observer's body to the exposing apparatus. By limiting the mass of the sample, and avoiding the unshielded line-of-sight path between the exposing apparatus or location of emitted flammable vapors and any part of the observer's body, the hazard of flame burns can be minimized.

While combustible shield materials may contribute substantially to the fuel involved in a fire following an explosion or other mishap, slow burning materials may give good protection at the first impingement of flame and heat during the period when the observer retires to a safe location.

Elimination of unshielded line-of-sight paths between source of exposure and any part of the observer's body also provides essential protection from splash or spray of toxic or corrosive liquids, and ricochet protection minimizes splashes.

Less dramatic but equally important is the protection against drifting of toxic, corrosive, or flammable gases, vapors, and dusts from the source of exposure to the vicinity of the observer's body. A hood with controlled inward flow of air furnished by mechanical ventilation is required. While inward rates of airflow at the face of the hood in the order of 18.25 to 30.5 m per minute (60 to 100 fpm) are important to maintain as a minimum, it is more important to study the air flow at the hood opening with a smoke source under various actual operating conditions. Turbulence at the hood window opening may prove to be a problem. Flow of room air out of a window or door, or displacement of air by an air-conditioning system, may change the picture completely as to effective hood ventilation. Placement of materials and equipment in the hood also may have an effect—as a general rule, the further in from the hood face the better (with a 15.25-cm (6 in.) clearance, a good minimum distance).

As a second line of defense, personal protective clothing and equipment should be employed:

(a) Safety spectacles should be worn regardless of shielding and barricades. Side-shield spectacles offer far more protection than nonside shield ones and are preferred. Face shields or hoods may be specified for protection during periods when shielding is not effective.

(b) Permanently flame-retardant-treated cotton clothing and non-flammable synthetic fiber garments are available for use where flash fire or flame hazards are present.

(c) Shirt sleeves should be rolled down and buttoned at the wrists; shirt fronts should be completely buttoned. This gives protection not only from flame but also from flying particles, liquids, and dusts.

(d) Laboratory aprons, coats, special jackets, and other special garments, are available commercially for specific needs.

(e) Gloves should be used which provide complete hand protection, and consideration should be given to forearm protection as well.

MATERIALS OF CONSTRUCTION

While selection of materials of construction must be done jointly with consideration of design, the following general comments are applicable.

While observation of apparatus makes the use of transparent shields desirable, metal shields must be provided to protect against severe missile hazards. This applies when materials are in metal containers, or heavy missiles of ceramic or glass may develop. Mirrors, limited-area peepholes, and other devices may be used to facilitate observation in these cases. Transparent shields may be used for reactions in most laboratory glassware because the missiles developed will be very light and thus have relatively low energy.

ASTM test method D 256 measures the relative susceptibility of a sample to fracture by shock as the energy suspended by a pendulum breaks a sample notched in a standard manner in one blow. Another method is to drop balls of varying weights from varying heights on sheets rigidly supported, and measure the inch-pounds required for failure. Comparative values are indicated in Table 1.

Courtesy of the 3M Company

Fig. 1. Sliding shields for personnel protection when working at a glass vacuum system.

Considering its cost, transparency, high-tensile strength, resistance to bending loads, impact strength, and slow burning rate, methyl methacrylate appears to offer an excellent over-all combination of characteristics for laboratory shields up to the limit of its strength. Methyl methacrylate shields are not satisfactory for high temperature use (e.g., with Parr bombs). Metal shields serve best for high temperature exposures. Polycarbonate is much stronger than methyl methacrylate, self-extinguishing after ignition, but it is easily attacked by organic solvents. It currently is more expensive. Other plastics may be evaluated.

Ignition of 40-gram (0.14-oz) balls of nitrocellulose wrapped in paper masking tape at a distance of 15.25 cm (6 in.) from a 6.4-mm ($\frac{1}{4}$-in.) methyl methacrylate panel resulted in no damage.

Ignition of 3 grams (0.1 oz) of nitrocellulose in a 118-ml (4-oz) tightly stoppered glass jar at a distance of 15.25 cm (6 in.) from a 6.4-mm ($\frac{1}{4}$-in.) methyl methacrylate panel resulted in no damage.

Four grams (0.14 oz), similarly placed, resulted in the fracture of a similar panel. Five grams (0.18 oz), similarly placed, resulted in the fracture of a curved 6.4-mm ($\frac{1}{4}$-in.) methyl methacrylate panel.

TABLE 1
Shock Tests on Transparent Shields

Material	Thickness mm	Thickness (Inches)	Drop Ball kg/m	Drop Ball (inch-pounds)	ASTM D 256 (foot-pounds)
Double-strength glass	3.2	(1/8)	446	(25)	
Laminated glass	6.4	(1/4)	1964	(110)	
Plate glass	6.4	(1/4)	1964	(110)	
Wired glass	6.4	(1/4)	2000	(112)	
Tempered glass	6.4	(1/4)	10393	(582)	
Methyl methacrylate	6.4	(1/4)	19400	(1086)	(0.4 to 0.5)
Polycarbonate	6.4	(1/4)			(12.0 to 16.0)

Steel plate has proven to be about four times as effective as methyl methacrylate in shield work; i.e., use 3.2-mm ($^1/_8$-in.) steel plate for 12.8-mm ($^1/_2$-in.) methacrylate; 6.4-mm ($^1/_4$-in.) steel plate for 25.4-mm (1-in.) methyl methacrylate, and so forth.

Ordinary plate or rolled glass should not be considered for explosion shielding. Wire glass is undesirable for use in shields where there is severe blast effect because, if shattered, the wires may add to the missile damage. Tempered safety glass will shatter to small or grain-size particles which do not have sufficient weight or cutting edges to be a major missile hazard. However, if scratched ever so slightly, it may fail on the slightest bending or impact stress. For low-energy shielding, laminated safety glass offers excellent protection, is incombustible, and does not scratch easily.

Chemically-tempered glass has been developed by the Corning Glass Works in nominal, single thickness and laminated form, which offers greater impact resistance than equivalent thicknesses of conventional types of glass. The glass is known as Corning® CSG Glass.

Portable shields, including curved and weighted models, should be constructed of not less than 6.4-mm ($^1/_4$-in.) methyl methacrylate, polycarbonate or 6.4-mm ($^1/_4$-in.) laminated safety glass. Their use should be limited to work involving low missile hazards and low blast pressures. Their effectiveness against flame and splash hazards largely depends upon the angle of coverage.

An explosion of from four to six grams (0.14–0.21 oz) of nitrocellulose in a tightly stoppered 118-ml (4-oz) glass jar, at a distance of 15.25 cm (6 in.) from the shield, served to topple over each of two of the more popular type weighted and broad-case portable shields of methyl methacrylate. They will withstand detonation of 0.5 grams of TNT.

FIXED SHIELDS IN CONVENTIONAL HOODS

A good typical new installation for a ten-foot wide hood consists of six 18-inch wide panels with three on a forward track and three on an inner track, so that about half of the hood face is always closed and equipment can be reached around any 18-inch panel, which serves as a shield (36 to 41 cm or 14 to 16 in. is a more comfortable width.) Shielding panels should be made of material for expected uses.

Conventional hoods modified with 12.8-mm ($^1/_2$-inch) methyl methacrylate over-lapping horizontally sliding door shields and bottom guides, and 3.2-mm ($^1/_8$-in.) steel plate closure panels at exposed ends, may be used safely for nearly all work with de-flagration or pressure-failure type explosions involving bench scale laboratory quantities of materials.

They also will give adequate protection from detonations involving 2 grams (0.07 oz) TNT maximum per glass container, located 30.5 cm (12 inches) from the shield; six grams maximum per hood.

Rising-window hoods are not considered satisfactory unless additional shielding is provided. The space beneath curtain or rising-window type shielding presents the hazard of being left open for ventilation and manipulation. The most practical solution to the problem of existing rising-window hoods seems to lie in the installation of transparent horizontally-sliding door shields inside the rising windows.

SPECIAL BLAST HOODS

For detonatable materials with a TNT equivalent explosive energy over 2 grams (0.07 oz), and not more than 25 grams (0.9 oz), a special blast hood with 25.4-mm (1-inch) methacrylate sliding door shields, light plastic pressure-relief back, and 6.4-mm ($^1/_4$-inch) steel plate construction may be used.

REACTIONS IN SMALL AUTOCLAVES, ROCKER BOMBS, METAL CONTAINERS

Steel barricades of a minimum thickness of 6.4-mm ($\frac{1}{4}$-inch), and up to 19.2 mm ($\frac{3}{4}$ inch) or even higher, generally are required for adequate protection for explosions in metal containers. Equivalent protection should be provided for all walls and doors of the room which are exposed, and hinges and supports must be designed for equivalent strength. Indirect access by mazes is preferred to doors. Each installation should be individually designed. Complete protection must be afforded all inhabited surrounding areas.

Manipulation of work to avoid body exposure can be accomplished with special tongs and manipulative tools, and special ports or openings.

CAST ELASTOMER GLASSWARE ENCLOSURES

For transporting or storing high-energy materials that may detonate outside of the above barricades, so-called "tote barricades" which consist of a cast "elastomer" around glass laboratory-size containers have been developed. A 473-ml (one-pint) casting over a small container has contained a two-gram (0.07-oz) nitroglycerin blast; a 3.79-liter (one-gallon) casting over a similar container has contained a 15-gram (0.53-oz) nitroglycerin blast. The "tote barricades" tested were made of Adiprene, vacuum mixed to avoid bubbles.

These appear to be effective at temperatures from $-18°C$ (0°F) to normal room temperatures.

EAR PROTECTION

Designs previously discussed are based on blast and missile effects on the shield. Ear protection must be considered in addition.

A safe limit for air blast "overpressure" (pressure developed above atmospheric by the blast wave) to prevent damage to unprotected ears is 7.03 g/cm^2 (0.1 psi). This level is conservative because diffraction and reflection effects are unpredictable, and because individual ear response varies widely. The pressure from a single blast required to cause ear damage in healthy young persons probably is much higher, and there is considerable evidence that it is in excess of 70.3 g/cm^2 (1.0 psi).

Good earmuff protection will probably reduce air-blast over-pressure by 90% or more, so far as its effect on the ears. (Earmuffs have been observed to decrease considerably the tendency of observers to panic after a small blast.)

A substantial continuous enclosure around the blast with no vents into the observer's space will reduce the blast pressure to $\frac{1}{10}$ (20 dB), if the enclosure is not penetrated by the blast wave.

An intervening shield with side venting into the operator's area will reduce the blast pressure to $\frac{1}{5}$ or more of its original value, provided the operator is near enough to the shield that the nearest air line around the shield requires a diffraction of the sound wave around at least a 90-degree bend.

CONCLUSION

To conclude it should be stressed that:
1. There is a need for shielding chemical laboratory operations.
2. It is feasible to provide the shielding required.
3. It is much easier to design adequate shielding in a hood with horizontally-sliding windows than vertically-sliding windows.
4. Portable shields give very limited protection unless firmly anchored.

5. Metal shielding must be used where metal missiles may occur. Transparent materials can be used for protection against fragmented glass.

6. Line-of-sight and ricochet protection are both important.

7. The second line of defense, training of personnel to use the shielding, and to use available personal protective equipment, must be stressed.

ACKNOWLEDGEMENT: Reprinted from the *Safety Newsletter* of the National Safety Council, Chemical Section, with permission. Revised in 1969.

Personnel Protection from Ultraviolet Radiation

G. Briggs Phillips and Everett Hanel, Jr.

The general problem of protection of personnel from injury by ultraviolet radiation may be divided into two categories. Under the first category such visual aids as warning signs and indicator lights may be considered, while the second includes protective equipment to be worn by exposed personnel.

VISUAL AIDS

Some specific rules for the use of visual aids are listed below:

1. When ultraviolet lamps are controlled by manual switches, the switches should be located outside the room, preferably near the entrance door.
2. When manual switches are used, a small cobalt-blue indicator light should be mounted near the switch. The indicator light will serve as a constant reminder that the ultraviolet lamps are burning.
3. Warning signs must be used at every ultraviolet installation. The exact location of the signs and the message they convey will vary with different types of installations. In general it is desirable to post a sign outside a room housing an ultraviolet installation. The wording on the sign will coincide with the safety regulations recommended for the particular type of installation. The sentences listed below illustrate the type of message to be used on the signs:
    1. Caution—Ultraviolet lamps in use, protect your eyes.
    2. Caution—Ultraviolet lamps in use, do not enter.
    3. Caution—Turn off ultraviolet lamps before entering.
    4. Caution—Strong ultraviolet in use, protect your skin and eyes.
4. In some installations (such as ultraviolet door barriers) a danger pattern may be painted on the floor or walls to designate areas of high ultraviolet intensity.

PROTECTIVE EQUIPMENT

Ultraviolet radiation in the 2537A range has little penetrating effect. Ordinary glass completely absorbs the energy, as do most plastics, rubber, and similar materials. The penetration of ultraviolet through clothing will depend upon the closeness of weave of the fabric. Practical experience has shown that the skin is usually adequately protected by ordinary cotton laboratory clothing.

Eye Protection. While ordinary spectacles will in many instances offer adequate eye protection, it is recommended that safety glasses or goggles with solid side pieces be used. The side pieces prevent the entrance of the radiation when the source is to the left or right of the exposed individual. Cases of eye conjunctivitis have been known to occur when the individual wore ordinary spectacles.

Skin Protection. Installations requiring skin protection also require eye protection. The main portion of the body and the arms, and legs are protected by ordinary clothing. Rubber or cotton gloves can be used to protect the hands. A plastic personnel hood may be conveniently used to protect the eye, head and neck. In some cases face shields adequately protect the face and eyes. If a face shield is used, it is recommended that some type of cap be worn to protect the area of the upper part of the head. Personnel working in areas where respirators are required can be provided with a modified face shield.

In installations where personnel are exposed to high intensities for long periods of time, it has been necessary to wear safety goggles in addition to plastic personnel hoods; e.g., animal rooms with ultraviolet cage racks. This is because of penetration of the plastic by the longer ultraviolet wave lengths emitted in small quantities by the low pressure mercury vapor lamps.

Whenever plastic items, such as face shields, are used for ultraviolet radiation protection, tests should be made to assure that the formulation has zero transmission of 2537A. Lucite face shields, for example, have on occasion been found to transmit germicidal radiation.

Safety Showers

James G. Stearns

Two serious incidents, in which the water supply to safety showers was left turned off, led to the development of a standard for safety showers at the Los Alamos Scientific Laboratory, and an interest in development of a national standard for safety showers in laboratories and chemical-handling operations. This chapter combines a proposal presented by the author to a meeting at the 1965 National Safety Congress and the comments of members of the Campus Safety Association and Chemical Section of the National Safety Council, which sponsored the meeting.

The incidents which focused attention on safety shower standards resulted because shower heads were leaking and someone turned off the water supply—one incident involved an employee whose hair was on fire from a molten sodium spray and the second involved a chemistry student who had been sprayed on the chest and face with acid.

CRITERIA FOR SAFETY SHOWERS

The basic criteria proposed for requiring safety shower installations are hazards to personnel from acids, caustics, cryogenic fluids, clothing fires and other emergencies in which volumes of water are needed for protecting personnel by diluting, warming or cooling, flushing off chemicals or putting out clothing fires.

Criteria have been lacking for designing, installing and maintaining safety showers and this chapter sets forth criteria for consideration.

To assure reliable operation of safety showers the following specific requirements are recommended:

1. Locate shower in a conspicuous location, preferably in usual traffic patterns but no farther than 25 feet from laboratory entrances. (One shower in a corridor can serve several laboratories.)

2. Centerline of shower head to wall: 25 inches, minimum.

3. Shower head height: 7–8 feet (floor to base of head).

4. Actuating valves are to be operated by pulling on an easily located actuating device which may be one of three types: (a) ring and chain attaches to the lever or rocking arm of the valve; ring to be circular (8 inch minimum diameter) (b) a triangle and rod attached to the lever of the valve (c) a vertical and/or horizontal chain or cord configuration attached to the lever of the valve. The ring, triangle or overhead chain should be located a maximum 6′8″ above the floor, to be in reach but not infringing on headroom.

5. Actuating valves (3 choices given in order, all quick opening); (a) Slow-closing piston valve, 1 inch; (b) Self-closing globe valve, 1 inch; (c) Ball valve, ³/₄ to 1 inch.

6. Shut-off valve (for each head); (a) To be a 1¹/₂ inch O.S. and Y. or visible rising stem; (b) Not to be accessible to laboratory personnel; (c) To be clearly identified.

7. Shower head: deluge type.

8. Shower head material: brass (may be plated), plastic.

9. Locate showers a safe distance from electrical apparatus or power outlets.

10. Floor drains should be considered for conveniences in new building construction; however, traps may present problems if allowed to dry out. Floors may be left unsloped since adequate water flow will exceed drainage rate of any safe slope of the floor.

11. Shower locations should be marked on the floor with painted white three-foot

121

circles, colored tile or other marking that may be useful to assist injured persons in finding the shower. Signs designating shower locations are not necessary.

Courtesy of the 3M Company

FIG. 1. Safety showers in this laboratory are the type that will remain in operation until turned off, allowing a victim to use both hands to remove clothing splashed with chemicals.

12. If room permits, stall type and/or multiple head showers should be considered.
13. Shower head inlet: 1 inch I.P.S.
14. Supply: $1\frac{1}{2}$ inch line from main, potable water.
15. Flow rates: 30–60 gallons per minute from shower head.
16. Flow Pressure: 20–50 psi at shower head.
17. Piping to safety showers should be fed down from the ceiling, but if such is not practicable, shower supply plumbing may rise from the floor.
18. Exterior showers should be protected from freezing as should water lines supplying safety showers within buildings.

TESTING OF SAFETY SHOWERS

Under an engineering-maintenance program all showers should be tested every six months or oftener by operating the actuating valve to flush the lines and assure satisfactory operation. Any valve that leaks or does not operate properly should be replaced or repaired promptly. A record of tests should be kept. When maintenance on building water systems will require that any safety shower be inoperative, engineering-maintenance should definitely notify laboratory personnel in advance of outage and of restoration of service to showers.

The Los Alamos Scientific laboratory has more than 300 safety showers and each is tested every six months; two men spend approximately 400 hours per year performing these tests. Since our showers have been installed over the past 22 years, we have many unsatisfactory installations by our current standard. It is impracticable to make wholesale changes now; however, we have removed accessible valve handles, identified water supply valves, developed a list of shower locations, description and water pressures, and published the seriousness of inoperable safety showers.

We have four types of safety shower actuating valves in use. Because of ease of operation and maintenance we believe that the slow-closing piston valve more nearly fits our needs and as replacement or new installations are needed we use this type valve. We were forced to remove recently installed ratchet-type shower valves as they were not reliable.

Laboratory and safety personnel have been faced with questions about safety showers, and while safety shower manufacturers are most helpful and offer excellent detailed brochures, they naturally tend to favor their own product. The author believes that an independent standard developed by users, approved by national safety organizations and adopted as a national standard would be of value to all organizations that need safety showers.

Protective Clothing

Adrian L. Linch

INTRODUCTION

Most of the published information on protective clothing is in the form of technical data which delineate physical properties of a particular coating formulation or synthetic fiber, or promotional literature in which claims are mixed with facts. From such publications, one concludes that no standards have been established for the selection of chemically resistant materials, results and conclusions from meaningful laboratory tests have not been published, and the selection of a garment for a particular application can be mostly a hit-and-miss affair. This state of affairs initiated laboratory investigations at the Chambers Works of E. I. Du Pont de Nemours & Co., Inc., as early as 1941. With the opening of the Industrial Hygiene Laboratory at the Chambers Works in 1950, an aggressive program of protective clothing design, development and materials evaluation was initiated. Progress was summarized in 1955 in a chapter of "Modern Occupational Medicine" by Fleming, D'Alonzo and Zapp.[1] Further progress since 1955 is presented in this chapter.

LABORATORY EVALUATION OF COATED MATERIALS AND UNSUPPORTED PLASTICS

A study of the physical parameters which delineate such properties as loss of strength or change in visible appearance on chemical contact, chemical absorption expressed as weight gain or volume increase (swelling), degree of softening on chemical contact, tensile strength, breaking strength, elongation, tenacity, flex life, shrinkage, thermal aging, abrasion resistance and weatherability does not answer the one question of most vital concern: What is the resistance to penetration by the chemical for which protection must be supplied? What little information is available on chemical resistance is supplied in the form of qualifying expressions such as: not visibly affected, is resistant to, no effect, attacked, slight, none, good, excellent, fair, poor, etc. In no case coming to our attention is the source of information or reference to the test procedure given.

The penetration resistance factor must be divided into two parameters:
1. Time lapse for total resistance, or conversely stated: "Breakthrough Time."
2. Diffusion rate at "Breakthrough" expressed as g/100 m²/hour/mil (permeability) or mg/10 in²/hour.

In a limited number of cases, notably our Du Pont Film Department (for example "Tedlar"), permeability values for a small number of common solvents have been published. However, when toxic chemicals such as aniline and nitrobenzene are encountered, these penetration resistance factors must be determined in the laboratory.

Since no standards, test procedure or equipment for penetration resistance had been proposed, design of equipment became our first concern. The second phase was a development of a reliable analytical procedure. Then specification limits were established on the basis of these results. A modification of the test cup described by Schwartz and Goldman of the United States Public Health Service was adopted (Figure 1). This unit is composed of a flanged cup with matching flanged cover which

PERMEABILITY TEST CUP

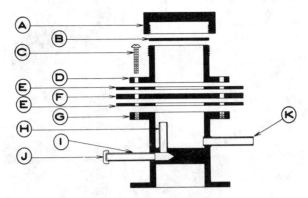

FIG. 1. Permeability test cup. (A) Threaded cover. (B) "Teflon" gasket. (C) Machine bolt (6) to secure flanges (#6-32-brass). (D) Upper flanged compartment. (E) Gasket (2). (F) Test specimen. (G) Lower flanged compartment. (H) Air inlet nozzle. (I) Air inlet tube. (J) Cap. (K) Air sampling tube (exhaust).

clamps the test piece against the cup flange to form a liquid-tight seal and reservoir for the test liquid. Inlet and outlet tubes were installed to permit passing a measured air stream through the bottom chamber to a midget impinger containing an appropriate liquid reagent to collect the vapor of the test material after diffusing through the test piece. The weight of material collected is determined by chemical analysis. The penetration rate after breakthrough is then calculated from the known time of sampling and the exposed area of the test piece. The equipment design and test procedure have been submitted to ASTM for approval.

Evaluation of corrosive chemical resistance was at first empirical and was based on the rate of destruction of a test piece exposed as the liner in a screw cap vial. Frequently color development in the test liquid, for example 96% H_2SO_4, was a good criterion of attack rate. The test was quantitized by securing the test piece over the open end of an inverted vial containing the test material and submerging in water to collect the fraction which diffuses through.

Retention of test liquid after an apparently satisfactory exposure test period is a property that should be determined also. In the case of acids, the test piece after thorough rinsing in distilled water is extracted for 24 hours in distilled water and then the liberated acid is titrated with standard alkali. This test has eliminated all vinyl plastic based materials from consideration as a component in safety apparel.

Monitoring of the air circulating within a garment that completely isolates the worker from his environment has been employed under conditions of extreme hazard (Figure 2). This technique has confirmed the safe operation of both the Neoprene and "Viton" "Air Suits" in the tetraethyllead (TEL) area sludge pits.

The question of shoe contamination can be decided often on the basis of an analysis of the socks worn inside the shoe. For example, aniline can be removed by a 30-minute dilute aqueous HCl extraction and analyzed by the standard diazotization and coupling technique adopted for air and urine specimens. If the test is negative, the cost of a new pair of shoes has been saved without jeopardizing the wearer's health. Otherwise, removal of core specimens destroys the shoe whether contaminated or not.

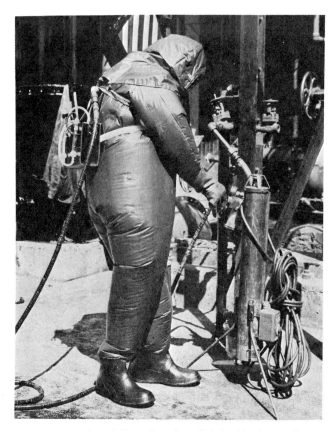

Fig. 2. Du Pont "Chem-Proof Air Suit" with air sampler installed to monitor conditioned interior air for toxic vapor leakage.

Other properties that must be evaluated in the laboratory include:

A. Flame resistance
   1. Ignition
   2. Sustained combustion
   3. Melting range
   4. Drip properties on melting
   5. Flame retardant coatings
B. Static properties
   1. Permanent static proofing
   2. Temporary static proofing applied after laundry
C. Substantivity
   1. Material not removed in laundry operation; e.g., benzidine
D. Durability
   1. Abrasion resistance
   2. Tenacity-coated fabrics, tendency to peel off
   3. Flex resistance
   4. Laundry properties
   5. Resistance to degradation by heat and light
   6. Tear and puncture resistance
E. Workability
   1. Elasticity (3 dimensional); Mylar is an example of 2 dimensional elasticity

      2. Seam strength
      3. Sealing properties
         (a) Cement
         (b) Heat
      4. Weight
         (a) Density of coating
         (b) Thickness vs. resistance
      5. Temperature effects

F. Comfort
      1. Thermal conductivity
      2. Surface roughness
      3. Drape and hand; soft vs. stiff
      4. Moisture absorption and evaporation

G. Safety
      1. Wet coefficient of friction, i.e., slippery when wet; gloves, shoe soles, etc.
      2. Flushing with water before garment removal
      3. Decontamination
      4. Heat transfer and resistance; reflective coatings; air layer
      5. Interference with body movements; bulk
      6. Proper fit; adequate fabrication design

MATERIALS AVAILABLE

Fabrics

Since this chapter is largely restricted to protective clothing, only a résumé of uncoated fabrics will be attempted. Briefly, the outstanding weaknesses of the commercially available fibers are:

Fiber	Weakness
Cotton and Rayon	Degraded by acids Substantive for some chemicals Fair durability
Wool	Degraded by alkalies Substantive for many chemicals Requires dry cleaning Fair durability

Polyethylene and -propylene
  Also available as nonwoven
  fabric—disposable
	Static build-up
	Requires flame retardant
	Durability not established

Nylon	Static build-up Melts when heated Requires flame retardant Nonabsorbent for water High heat transfer
"Orlon," polyacrylonitrile	As for nylon
"Dacron," polyester	As for nylon
"Dynel," modacrylic	As for nylon
"Nomex," polyamide	Nonabsorbent for water.

This "Nomex" fiber was developed to provide a high melting replacement for nylon in drag chutes to reduce the landing speed of jet planes. Field trials at Chambers Works have disclosed outstanding durability in addition to high melting point, permanent flame resistance, good static dissipation, excellent chemical resistance, low substantivity for chemicals in general, and good laundry response (suits have survived over 70 washings). The fabric is soft, somewhat elastic and highly resistant to tearing and ripping. The only complaints, and those have been weak, are high heat transfer, which produces a sensation of chill, clamminess, or over-heating and limited moisture absorption (between cotton and "Orlon" or "Dacron"), which sometimes produces discomfort in hot environments where perspiration is a problem. "Nomex" clothing is expected to outwear the best of conventional clothing by a factor of 4 or 5.

As might be expected, a transition zone between woven fabrics and impervious coatings has been developed. An inexpensive nonwoven, felt-like, lint-free polyethylene fabric suitable for the fabrication of disposable garments was developed by the Textile Fibers Department in 1964. This material would be attractive for clean rooms, acid areas and spray painting. A porous continuous polyethylene sheet—"Air Weave"—was developed by Industrial Products Company for use in raincoats which "breathe." The material passes gases, such as air, but repels liquids.

Coated Fabrics and Unsupported Sheet Plastics

The field of continuous coatings is of considerably greater importance in the field of protective clothing. It is obvious that in order to establish an effective barrier between the workman and his environment, one must either coat a fabric with an impervious film or fabricate the garment from a suitable continuous plastic sheet. In general, less elastomeric material is required to cover a woven fabric than is needed for unsupported film. This presents a cost saving, but more important, weight is reduced without sacrificing strength or chemical resistance.

Materials Available

Fabrics

Cotton
Rayons
Wool
Polyethylene
Polypropylene
Nylon
"Dynel," modacrylic
"Orlon," polyacrylonitrile
"Dacron," polyester
"Nomex" polyamide
Fluorocarbons

Coatings and Sheet

Cellophane
Polyethylene
Polypropylene
Vinyls:
    "Saran," polyvinylidene chloride
    "Orlon," polyacrylonitrile
"Mylar," polyester
"Teflon," polytetrafluoroethylene
"Tedlar," polyvinyl fluoride

Elastomers
Natural rubber
Butyl rubber
"Buna N"
Neoprene
"Viton"
"Hypalon".

Extensive laboratory investigation of commercially available elastomers and coating compositions indicated butyl rubber was the only material which would resist penetration by nitrobenzene, aniline and their derivatives for sufficient time to permit use in protective clothing. However, procurement proved to be a major obstacle to application. The very properties that render butyl rubber attractive for protective uses are those properties which are the most difficult to overcome in milling of coated fabrics, production of molded articles, or curing latex dipped ware. Chambers Works obtained some of the first butyl rubber-coated cotton produced at our Fairfield Works for fabrication of our "Chem-Proof Air Suit" in 1952. Not until 1959 were molded butyl storm rubbers produced and in 1956 molded butyl gloves became available as a result of the governmental procurement of protective garments for rocket fuel handlers. Up to these times, fabrication from butyl rubber-coated cotton had provided overshoes and gloves of very poor durability due to unsatisfactory cement adhesion in the seams. In 1963, a butyl-dipped cotton glove became available but some durability problems are currently being resolved. Butyl gloves produced by a latex dipping process in 1965 have not yielded consistent performance in laboratory tests.

A most significant improvement in the quality of butyl rubber-coated cotton appeared in 1962 with the development by the space agency of "rocket fuel handlers cloth." The original olive-drab butyl formulation exhibited rather poor acid resistance, which made necessary the stocking of neoprene garments for corrosive chemical service. However, the new black butyl coating demonstrated such outstanding acid resistance that all of Chambers Works' requirements for impervious protective clothing, except for severe TEL exposure, could be satisfied from this one source. In addition to outstanding chemical resistance, the material is much lighter, softer and drapes well.

Although the cup tests do not give either neoprene or butyl rubber even a fair rating for alkyl lead service, field use has demonstrated surprisingly effective protection from contaminated earth, water and high vapor concentrations. Care has been exercised to avoid gross contact with liquid TEL or TML. "Mylar" and "Viton" have exhibited unusual resistance to both TEL and Motor Mix. However, "Mylar's" lack of flexibility in the dimension normal to the surface has eliminated consideration for garment fabrication. Although one "Viton Air Suit" has been assembled, high cost renders routine application quite impractical. However, "Viton" coated neoprene gloves do show promise in this field. "Viton's" resistance to high temperature would recommend application for protection from hot liquid splashes such as molten lead and lead-sodium alloy in the TEL operations furnace areas.

Polyethylene has a limited application for its unusual resistance to hydrofluoric acid. However, this one property is not sufficiently important to require its use in favor of butyl rubber which will give adequate short term liquid HF protection.

A word about the "vinyls". Our experience both in laboratory testing and field service has been so uniformly poor that all vinyl formulations have been ruled out for use in protective clothing or safety devices at Chamber Works. As previously mentioned, vinyls retain acids after contact. Furthermore, resistance to aromatic nitro and amino compounds and chlorinated solvents is very poor, durability at best is only fair, and the change in resiliency with temperature is notoriously undesirable.

Fabrication of garments from "Teflon" may be possible, but the cost would limit its use to the most exotic applications. Our laboratory tests indicate some difficulty in obtaining thin sections that are free of pin holes, and seam sealing presents difficulties.

HEAD AND RESPIRATORY PROTECTION

As exposure potential increases beyond the level where protection of the hands and feet is no longer adequate, the environmental armor must be extended to include the head, neck and shoulders. If the design includes an inhalation air supply also, then respiratory protection can be provided without resorting to a separate face mask system. In fact, a properly designed cape or hood could replace air-supplied or canister gas masks completely and thus solve another troublesome problem; namely, how to provide vision-correcting lenses and provide adequate impact protection with respiratory protection.

Dissatisfaction with commercially available hoods initiated a design study in 1955. The first specification, two minutes under a safety shower without leaking, was not met by any submitted for evaluation. Since redesign offered the only solution, a decision to start from "scratch" was reached.

Any protective device intended to prevent head and shoulder contact with corrosive chemicals and inhalation of toxic fumes must provide dependable and expedient service with minimum weight, clear wide-angle vision, little or no restriction to head, shoulder and arm motion, and a supply of clean air sufficient not only for inhalation, but also for evaporation of perspiration and prevention of lens fogging. An air flow in the range of 85 l/min (3 cfm) will provide an adequate air sweep to prevent entrance of contaminated outside air under the cape skirt. Comfort to a great extent determines voluntary acceptance of any protective garment for optional precautionary use where exposure severity may not demand the wearing of chemical resistance hoods. Furthermore, all adjustments required for fitting before use must be simple, direct, and require a minimum of attention and effort. However, the final result must be secure to give the wearer the confidence needed to work freely under conditions of imminent exposure and either durable enough to survive severe physical abuse, or sufficiently inexpensive to be considered expendable after a few uses (Figure 3).

The following features were incorporated (Figure 4):

1. *"Hard hat" without modification used as the base.* Since head protection against falling missiles and bumping into fixed objects has become a universal requirement for most work areas, the most practical solution to the hood problem utilizes the "hard hats" already issued to the workmen. The problems associated with individual head band (8) adjustments and sanitation thus can be eliminated from the design, application and cost of the hood assembly.

2. *Vision and eye protection.* A semicircular, molded, clear, scratch resistant lens (6) that furnishes 180° vision without head motion and normal upward vision without head tilting is cemented into the face of the cape. This lens furnishes more than adequate mechanical impact protection. The use of safety glass behind this barrier is unnecessary unless the wearer requires prescription glasses to correct vision defects.

3. *Ventilation* is supplied through an all plastic air distributor ring (5) which circles the crown of the cape under the brim of the hard hat and air conditions the user's scalp as well as his neck and face. A series of closely spaced holes above the lens provides a downward flowing curtain of air, which effectively prevents fogging of the field of vision from condensation of water vapor. The air ring also locks the cape securely in place on the hard hat. Although an air supply is not necessarily required, use without ventilation is not encouraged.

4. *Air seal* is assured by a second inner skirt (2) attached to the inner surface of the cape skirt approximately midway between the air ring and the lower edge (1). This acts as a check valve to prevent fumes from entering when the wearer bends

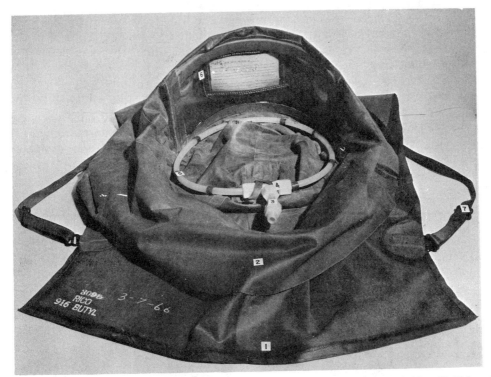

FIG. 3. Du Pont "Chemical Hazards Cape." (1) Outside skirt. (2) Inner wind-guard skirt. (3) Nylon air supply line connector. (4) Nylon "T" connector for perforated air ring. (5) Retaining loops for perforated air ring. (6) Plastic lens with pre-set curvature. (7) Tie-down strap.

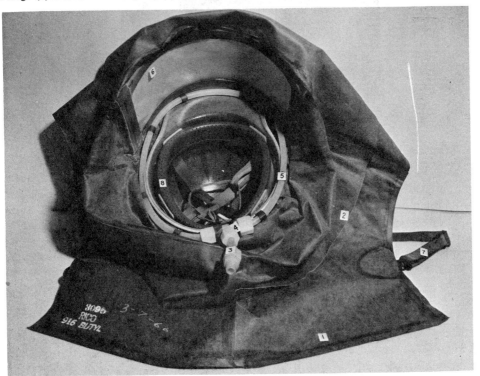

FIG. 4. Du Pont "Chemical Hazards Hood" assembled. (1) Outside skirt. (2) Inner wind-guard skirt. (3) Nylon air supply line connector. (4) Nylon "T" connector for perforated air ring. (5) Retaining loops for perforated air ring. (6) Plastic lens with pre-set curvature. (7) Tie-down strap. (8) Hard hat.

forward or tilts his head upwards and prevents the wind from raising the cape when employed in exposed locations. The air ring supplied 85 l/min (3 cfm) of air and 180 g/cm^2 (2.5 psi) back-pressure (3). The air venting under the skirt maintains a sweep that effectively blocks entrance of fumes.

5. *Weight.* The complete cape less hard hat weighs just under 2 lb.

6. *Material of construction.* The standard model is fabricated from butyl rubber coated cotton—rocket fuel handler's material—and is resistant to most chemicals. The pattern may be used for fabrication from polyethylene for a translucent cape for acid service, "Viton" A for TEL protection or an aluminized material (combined with a heat reflecting lens) for use where radiant heat is a problem (Figure 5).

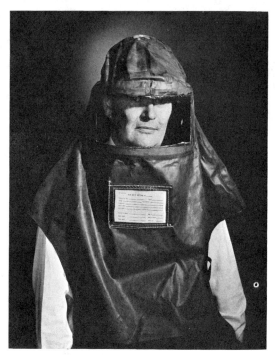

FIG. 5. Du Pont "Chemical Hazards Hood" in use.

DU PONT CHEMICAL HAZARDS SUIT

Next in order of development, based on extent of body coverage rather than chronological order, the old, familiar "acid suit" should be considered. The conventional outfit consisted of an "acid hood," with or without air supply, rubber jumper (waist length coat), rubber overalls, rubber gloves, storm rubbers or rubber safety shoes. Much dissatisfaction with the performance of these stiff, heavy and uncomfortable assemblies had been voiced from time to time. Furthermore, their use was restricted to corrosive alkalies and acids, and even here their utility was limited. Not until mid 1962, when the "rocket fuel handler's" coated fabric became available, was any serious attempt made to correct the deficiencies (Figure 6). With the substitution of the outstandingly better butyl rubber material for the obsolete rubber counterpart and correction of design deficiencies, the "Du Pont Chemical Hazards Suit" was developed. This suit provided not only increased protection for corrosive chemical service, but also permitted routine use in locations where severe hazard potentials were created by aromatic nitro and amino compounds, halogenated solvents and certain toxic organometallic compounds. The "Du Pont Chemical Hazards Cape" with its air supply system was adopted as an integral

FIG. 6. Du Pont "Chemical Hazards Suit" complete with ventilated "Chemical Hazards Hood."

component to provide respiratory protection where needed, as in hydrofluoric acid areas.

Although this "suit" covers the entire body, the armor is discontinuous where the jacket overlaps the coverall bib, the sleeves overlap the gloves, and the trousers overlap the foot protection. Furthermore, no air circulation is provided below the neckline. Therefore, these "suits" should be used on a light-duty precautionary basis and should not be worn for more than twenty minutes at a time. Enclosure of the human body in an unventilated impervious covering for periods beyond twenty minutes will impose stress that will at least reduce labor efficiency, if not be totally incapacitating. The garment has been well received by the man in the field and no accidents attributable to the design have occurred.

THE DU PONT "CHEM-PROOF" AIR SUIT

The ultimate in complete isolation from the work environment is provided by the "Chem-Proof Air Suit" (Figure 7).

Workmen engaged in the manufacture, transportation, or use of hazardous chemicals need complete body protection to prevent skin contact with toxic and corrosive materials as well as inhalation of injurious fumes during maintenance and repair work, unexpected operating incidents, equipment failure, and decontamination activities. Air conditioned full body coverage also permits efficient performance of routine job assignments in contaminated environments for extended periods under health-conserving conditions and in complete comfort. The "Air Suits", which offer little restriction to the wearer's movements, provide safe conduct into locations where exposure conditions would otherwise limit work to a few minutes or altogether prevent access with conventional protective clothing.

The impervious butyl rubber "Chem-Proof Air Suit" provides dependable health protection for work in areas contaminated with aniline, toluidines, xylidines,

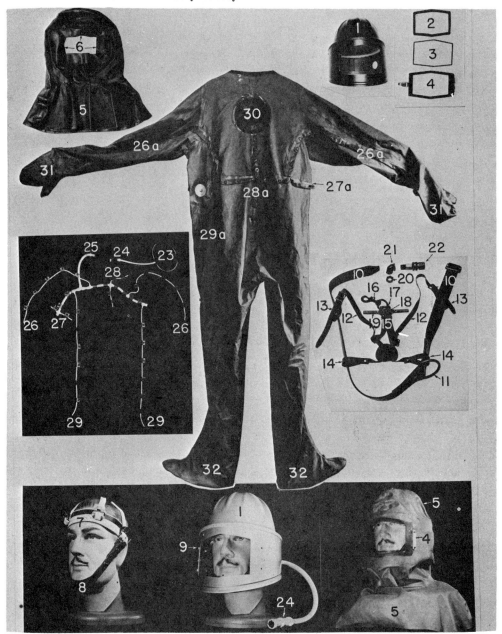

FIG. 7. The Du Pont "Chem-Proof Air Suit" with sandblaster's (or welder's) hood. (1) Sandblaster's helmet. (2) Rubber gasketed lens. (3) Heavy duty over-lens. (4) Metal retaining frame. (5) Cape. (6) Lens opening. (7) Hard hat head band. (8) Chin strap. (9) Lens assembly. (10) Safety belt. (11) "D" ring. (12) Shoulder straps. (13) Shoulder strap loops. (14) Stabilizer strap-extension of shoulder strap. (15) Support straps. (16) Swivel snap: secures "D" ring-11. (17) Ring on bulkhead fitting. (18) Bulkhead "T" connector—threaded side connection. (19) Air supply tubes. (20) Retaining nut. (21) Swivel elbow. (22) Quick-disconnect coupling. (23) Hood air supply ring. (24) Hood air supply connection. (25) Main body air supply line. (26) Arm branches—air distribution lines. (26a) Snap-tabs to secure air system. (27) Groin ventilation tubes. (27a) Groin ventilation tube retaining tabs. (28) Back ventilation tubes. (28a) Back ventilation tube retaining tabs. (29) Leg ventilation tubes. (29a) Leg ventilation tube retaining tabs. (30) Reinforcing patch for bulkhead fitting. (31) Butyl rubber gloves. (32) Molded butyl rubber soles on feet.

Note: A nylon safety belt which eliminates the support strap (15) and carries the support "D" ring (11) on the shoulder straps (12) and employs a figure 8 chain repair link to connect the bulkhead ring (17) to the "D" ring (11) directly is currently available from Industrial Products Company, Philadelphia, Penna.

chloroanilines, chlorotoluidines, and other aromatic amines or nitrobenzene, nitro-
toluenes, nitroxylenes, nitrochlorobenzenes, dinitrobenzene, or toluene and other
aromatic nitrocompounds, as well as corrosive acid and caustic alkalies (Figure 8).
The body of the 14-pound "Air Suit" is fabricated from a tough, lightweight, butyl
coated fabric with cemented seams which seal out both liquids and vapors from toe to
neck. A gusset over the chest provides freedom of action while putting on or removing the garment. Air circulation is provided by internal gum rubber, snap-in (a No's)
air lines (23, 29) extending to the wrists (26, 31), ankles (29, 32), the back (28), and
midriff (27). The air, after traveling from the extremities back to the neck opening,
exhausts under the hood cape skirt. Air pressure, equivalent to 7.6 to 25 g/cm^2
(3 to 10 inches of water), inflates the "Air Suit" balloon-like to provide freedom of
action impossible with close-fitting garments.

Fig. 8. Du Pont "Chem-Proof Air Suit" in use.

The hands are encased in sealed-on five finger butyl gloves (31) and the feet are
protected by the workman's steel-toe-cap safety shoes worn inside the suit (32). The
coated fabric over the shoes may be covered by low cut slipover slicker boots to
avoid excessive abrasion.

The "Du Pont Chemical Hazards Cape" (Figure 4) provides impact protection,
wide-angle vision, inhalation air supply and air conditioning from the perforated
distributor ring under the brim of the hard hat. The cape skirt is anchored in position
by straps which loop under the armpits. Fresh air for breathing and lens fog elimination flows downward from the air ring perforation directly above the lens.

Inhalation-quality tempered air is supplied from a one-half inch I.D. butyl hose,
which also serves as a lifeline for rescue if necessary. The air stream is divided through
a brass manifold (18), secured by a bulkhead seal in the upper back of the suit (30),
to the hood (23) and body air distribution lines (25). A nylon rescue harness (10),
which is worn inside the suit and is connected through a figure-8 linkage (16) between the belt's D-ring (11) and the manifold loop (17), carries most of the weight of
the air line and manifold assembly on the harness shoulder straps (13). The air hose
is secured to the manifold through a swivel L-connection (21) fitted with a quick-
disconnect snap-ring fitting (22).

Air flow is controlled by a piston-type regulator and passes through porous metallic and wool-felt filters before entering the delivery hose. Air, tempered to suit the worker's preference, should be supplied at 200 to 420 l/min (7 to 15 cfm), 21° to 32°C (70° to 90°F) and 40% to 60% relative humidity to attain maximum labor efficiency. For further detailed information, consult references 2–4.

Other variations of complete body protection have been designed and are currently in use. A transparent protective suit fabricated from a vinyl plastic film has been used extensively in atomic energy installations for protection from radioactive dust and gases. In locations where chemical penetration resistance is not a major factor, effective protection can be provided by inexpensive plastic garments at a cost that permits discarding the garment after exposure to hazardous materials rather than attempting to decontaminate. Capes and jackets, with or without an air supply, have been used where only partial body protection was needed. Unrestricted vision is a major advantage. Welding of radioactivity-contaminated or toxic metals such as cadmium and lead is an example of a health problem that can be solved with this type of protection.[5,6]

HEAT RESISTANT GARMENTS

No discussion of protective clothing would be complete without some mention of heat-resistant garments. Almost without exception an aluminized fabric is chosen to reflect radiant heat. A laminated aluminum foil is a more efficient reflector than a paint type coating. The base fabric usually is an asbestos fiber weave, although fiber glass and even cotton have been offered. Multicomponent laminations, in which a layer of neoprene is interposed between the aluminum and the fabric, provide additional protection from combustion products. As a result of the publicity given to asbestos as a source of pulmonary pathology, the use of asbestos in protective clothing undoubtedly will be re-evaluated in the near future. "Nomex" could well serve as a base in place of asbestos cloth for heat protection applications. Heat suits obviously should be air conditioned, but the problems associated with delivery of a sufficient volume of cool air has not been solved adequately. Self-contained suits that derive refrigeration from liquid air or a heat sink have reached field trial stage, but utility is limited by the amount of refrigeration a man can carry for more than a few minutes' operation. Air lines provided with sufficient insulation to deliver cooled air from an external source are not available in a form sufficiently flexible and devoid of the bulk usually associated with efficient thermal insulation to permit use by a man who expects to accomplish useful work.

The heat-resistant garment must be lined with an effective thermal insulator such as cotton or wool even when air circulation can be provided. The temperature of the outer fabric may reach levels that will produce thermal burns on contact with the exposed skin areas. In most applications the wearer must be provided with an inhalation air supply as a protection from air-borne thermal decomposition products present in the hot atmosphere. The lens of the hood must be provided with either a reflective metallic coating or fine mesh metal screen to protect the wearer from radiant heat and to prevent excessive heating of the lens, which may melt if plastic. Extra insulation must be provided for the soles of the feet, since the walking surfaces are almost certain to be excessively heated.

Portable Heat Exchangers

So much emphasis has been placed on air conditioning for the various safety garments discussed, some reference to portable heat exchangers should be furnished. Probably the most significant development in this field has been the vortex tube.

Although the principle was discovered many years ago, practical application was not discovered until after 1945.* The device derives its energy from the compressed air supply used to supply the user with a breathing air source. In principle, the device separates a high velocity compressed air jet into cold and hot streams by centrifugal force. The ratio of hot to cold rates will determine the temperature differential and can be as much as 111°C (200°F) with a 7 Kg/cm^2 (100 psi) air source. The device is compact, lightweight, employs no moving parts, and can be adjusted to supply air at any temperature within the range of the available pressure.

There are limitations involved, of course. The hot leg must be insulated to prevent thermal burns. The noise generated can reach very high levels, especially within a helmet or cape, and must be brought down to tolerable ranges with an efficient muffler. Although some effect can be obtained at pressures as low as 2.1 Kg/cm^2 (30 psi), best results require pressures above 3.5 Kg/cm^2 (50 psi). The vortex tube† is commercially available[7] and should be considered whenever there is need for a compact air conditioner that requires no external power.

A typical vortex tube operates on 570 to 700 l/min (20 to 25 cfm) of compressed air at 5.6 to 7 Kg/cm^2 (80 to 100 psi) and delivers 420 to 510 l/min (15 to 18 cfm) of air cooled 28° to 33°C (50° to 60°F) below the supply air temperature. The hot tube will operate under most conditions in the 66° to 93°C (150° to 200°F) range, but can reach 150°C (300°F) if a 75 % cold fraction is taken. With a 50 %–50 % split, the cold side may drop 56°C (100°F) below inlet temperature.

REFERENCES

1. Fleming, Allen J., D'Alonso, C. A., and Zapp, J. A., "Modern Occupational Medicine," Lea & Febiger, Philadelphia, Penna. (1955).
2. *Ibid.*, Chapter 8.
3. "Air Conditioned Safety Suit," *Chemical Engineering*, 60, 144–146 (Dec. 1953).
4. Brouha, L., Protecting the Worker in Hot Environments, *Industrial Hygiene Foundation*, *Transactions Bulletin No. 29, 20th Annual Meeting*, pp. 207–216 (Nov. 1955).
5. Snyder Manufacturing Co., New Philadelphia, Ohio.
6. Croley, J. J., Protective Clothing—Responsibilities of the Industrial Hygienist, *Amer. Ind. Hyg. Assoc. J.*, 27, 140 (March–April, 1966).
7. Encon Manufacturing Co., 4910 Augusta Street, Houston, Texas 77007.
8. Fleming, Allen J., D'Alonso, C. A., and Zapp, J. A., "Modern Occupational Medicine," 2nd ed., p. 98, Lea & Febiger, Philadelphia, Penna. (1960).

*G. J. Ranque, a French metallurgist was granted a U.S. patent in 1934 (U.S. Patent No. 1,952,281). Rudolf Hilsch, a German physicist, improved the device (Z. *Naturforsch 1* : 208–14, 1956). The result is often referred to as the Ranque-Hilsch, or simply the Hilsch tube.

†Most protective equipment dealers can supply the vortex tube.

Section **3**

Ventilation

Ventilation of Laboratory Operations

Norman V. Steere

Chemicals and micro-organisms which are toxic or pathogenic, or whose toxicologic properties are not known, should be controlled so that people in the laboratory do not absorb, ingest, or inhale the materials. In order to prevent inhalation of vapors, gases, and particulates (fumes, smokes, dusts, and aerosols) the contaminants must be contained or captured by hoods, enclosures, or spot ventilation.

Great flexibility at low initial investment and low operating costs can be provided by enclosures and spot ventilation. These two methods may be the only possible or economical means of improving existing laboratories, and they may be the best means of building ventilation flexibility into new laboratories.

LABORATORY HOODS

A laboratory hood is not performing its function unless it captures and retains the atmospheric contaminants generated within it. A hood is not intended to capture contaminants which become airborne elsewhere in the laboratory, nor is a hood generally designed to contain explosions.

Successful performance depends primarily on the velocity of air moving through the hood. Factors which affect the face velocity and air movement through the hood are cross currents, entrance shapes, thermal loading, mechanical action, exhaust slot design, and obstructions.

Successful performance of a hood may also depend upon its ability to confine a fire, to withstand corrosion,[3] to be readily cleanable if contaminated, and to collect certain contaminants such as radioisotopes and pathogens[4] before they enter the exhaust system.

We believe a laboratory hood intended for general use should be appropriate for use of flammable liquids and gases, and should be constructed of material which will withstand fire for several minutes so that the hood enclosure can maintain its integrity and confine the fire until it can be extinguished.[5]

Design of hoods which may become contaminated and have to be cleaned of radioactivity may be based on one of three current practices. Construction of stainless steel with welded joints and coved corners has been the generally accepted practice. A recent view is that equivalent results at lower cost can be obtained from ordinary hood materials covered with strippable coatings. We feel that stainless steel is better for college and university hoods in radioactive service, because we do not believe that strippable coatings can be maintained effectively or economically where laboratories are not numerous or concentrated or well-staffed. An evaluation of paint systems for use in radioisotope laboratories showed that strippable coating was ineffective for oily contaminants but that an epoxy paint, in routine use at one government laboratory, provided a readily cleanable surface.

Safe use of hot concentrated perchloric acid requires prevention of vapor contact with organic materials or vapor condensation where organic material may later come in contact. Safe use of perchloric acid may be accomplished by use in a separate hood with a duct washdown system and no exposed organic coating, sealing compound,[3] or lubricant; safe use may also be accomplished by use of a special scrubber unit, designed for handling exhaust from a large hood, or from a special small hood.[8] (For further detail see the chapter on handling perchloric acid.)

HOOD FACE VELOCITY

The velocity of air entering a hood at its face determines whether or not the hood will be safe, since an adequate face velocity is the basic requirement for capture and control of contaminants generated within a hood. A minimum face velocity of 100 lineal fpm (30m/min) for general laboratory hoods is a recommendation[1,7,9,10,11] with which we agree. Face velocities of 100 fpm (30m/min) will not be adequate to capture contaminants released with great force or more than very low velocity, as may be seen by reference to Table 1 on page 90. Since hoods with by-passes cannot be adjusted by laboratory personnel to provide higher than the minimum design velocities of the face, the only way higher capture velocities can be achieved in such hoods is to adjust the baffles or slots for concentrating the exhaust and to position the contaminant release as close as possible to the exhaust slot. Higher capture velocities can be achieved in ordinary hoods by closing the sash or positioning bench shields to reduce the open face area.

Horizontal sliding sash have a number of advantages that should be recognized.

1. Horizontal sliding sash allow adequate minimum face velocity across about $2/3$ the hood face area,[9] since the sash are stacked in about $1/3$ the face area, while the same face velocity should be provided over the entire face of hoods with vertical rising sash, since such sash are usually left all the way up.
2. Horizontal sliding sash provide better body shielding than vertical sliding sash, and are easier to work around, if not wider than about 38 cm (15 in.).
3. Horizontal sliding sash can be used as shielding to allow work at face velocities higher than the upper limits at which aerodynamic flow patterns would otherwise bring contaminants back to the hood user (See page 94 for discussion of upper limits).

Hoods for highly toxic materials require higher face velocities, ranging from 125 to 200 fpm (38 to 61m/min), as a means of minimizing outleakage which could be very hazardous.

Although face velocities much greater than 100 fpm (30m/min) may disturb gas flames, fine powder, or tissue slices in conventional hoods, the disturbances can be offset by use of electric heat, screening, or a hood design in which half of the air does not pass through the bottom slot.

Uniform flow of air into and through laboratory hoods will provide the best capture of contaminant,[7] just as uniform flow of air out of an air flow bench in a clean room provides the best exclusion of contaminants. Laminar air flow is expected to make a substantial contribution to the techniques of contamination control in clean rooms,[2] and laminar flow could contribute to contamination control in laboratories.

Cross-currents outside a hood can divert or nullify air flow into a hood and its capture ability, so it is important to locate hoods to minimize air currents from doors, windows, and air supply grilles.

One good and two poor locations for laboratory hoods are illustrated in Figure 1. Hood A will have a better chance of working correctly because it is in a location with minimum cross-currents from windows, doors, and pedestrian traffic.

Pedestrian traffic past a hood should be as little as possible since a walking rate of one mile an hour causes a cross-current velocity of 88 ft per minute; two miles an hour, 176 ft per minute. (1.6 km/hr is equivalent to 26m/min and 3.2 km/hr to 53m/min).

Figure 2a and b show how vapors eddy near the front of an ordinary hood and are easily drawn out by a passer-by. Figure 2c shows how an airfoil added to a hood will prevent eddies by providing a sweeping flow of air across the bottom of the hood.

Even when a hood has an adequate face velocity, an investigator can inhale toxic materials brought back by the eddies around his body. Clarke[3] cites an instance

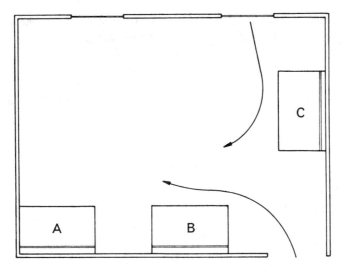

FIG. 1. Location of hoods. A is in the best location; B and C are poorly located.

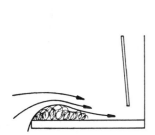

FIG. 2a. Vapors eddy near
the front of an ordinary
hood.

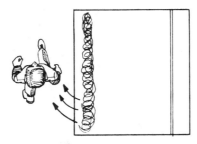

FIG. 2b. Eddying vapors are easily
drawn out by the cross-current from
a passer-by.

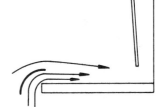

FIG. 2c. Adding an air-foil
provides a sweeping flow of
air that prevents eddies.

in which a person received a critical exposure from toxic chemicals in a hood with face velocity over 100 fpm (30m/min). Figure 3 shows how such an exposure can occur and how it can be prevented by a shield or by a section of horizontal-sliding sash.

TESTING PERFORMANCE OF LABORATORY VENTILATION SYSTEMS

Testing the face velocity of hoods or capture velocities of other laboratory exhaust systems can give false and misleading results if the tests are done on a single day and not correlated with the wind direction and other atmospheric conditions. As an example, tests of a hood on the day the wind was blowing away from the horizontal discharge vent on the roof could show adequate face velocities, while the face velocities could be very low or negative the next day when the wind was gusting against the discharge vent. Development of a set of standard test procedures for testing fume hoods in place has been needed and will now take place under a contract between Massachusetts Institute of Technology, U. S. Public Health Service and the American Society of Heating, Refrigerating and Air-Conditioning Engineers, Inc.

IMPROVEMENT OF EXISTING HOODS

Existing hoods not performing adequately can be improved at very modest expense by one or two simple measures.

The first step is to have the condition of the hood system checked to see that regular maintenance is provided. Points to check are obstruction of slots and the concealed space between slots, leaks or obstructions in ducts, and the condition of the fan. Since many hood and exhaust systems are not designed to provide convenient access, maintenance is often deferred and obstructions accumulate. System performance drops when room air leaks into ducts, and when fan impeller vanes are slowed down or reduced in area by corrosion.

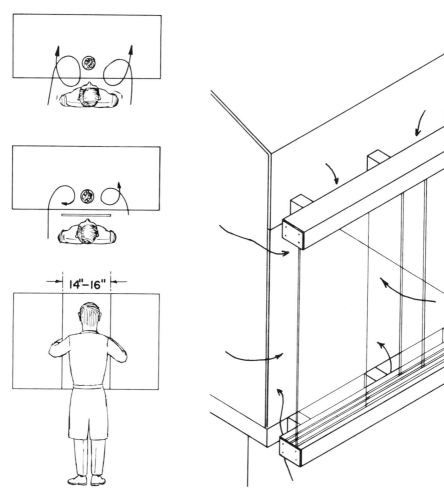

Fig. 3. Eddies which may allow escape and inhalation of contaminants can be prevented by a shield or a horizontal-sliding sash.

Fig. 4. Addition of horizontal-sliding sash is designed to provide improved performance with flexibility and continuous air flow.

Smoke candles have been used to see how effectively hoods capture and retain air-borne material, whether there are out-leaks in the duct system,[5] and also how well the roof outlet disperses the exhaust. Half-minute smoke candles for initial trials and three-minute candles for more extensive study are non-toxic and can be obtained from E. Vernon Hill & Co. or Superior Signal Co. Smoke candle tests should be supplemented with quantitative measurements of air velocities.

Increasing the air-moving capacity of a hood system is not the best second step in most cases, since a 20% increase in system capacity will require a 73% increase in horsepower.[6]

Hood performance can be improved readily by manipulating the air flow equation: $Q = Av$, where Q is the flow, A is the hood face area, and v is the face velocity. If Q remains essentially constant, a decrease in open face area will result in a proportional increase in face velocity. (This discussion applies mainly to sashless hoods, or to hoods with sash but no air by-pass. Improvement of hoods with air by-pass would require reduction of the by-pass capacity.)

Figures 4 and 5 show how the open face area of a sashless hood can be reduced by installing triple-grooved tracks for horizontal-sliding sash of laminated safety plate glass. Face area reduction can be varied from one-third to three-thirds. The track is spaced out from the hood to help keep the track clear and to allow air to enter the hood even with the sash covering the whole hood face. This inexpensive method of hood improvement with rabbetted oak tracks, increases face velocity and provides eddy, splash, and explosion protection as shown in Figure 3.

Immediate measures to decrease hood face area and increase face velocity can be taken by anyone using a hood, simply by placing a safety glass shield in the hood as near the face as possible.

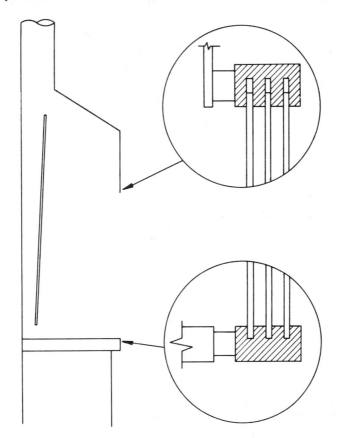

FIG. 5. Addition of horizontal-sliding sash will increase hood face velocity by decreasing face area open.

DESIGN OF NEW HOODS

Significant improvements have been made in the design of some new hoods,[9] and further developments may result from continued design and testing. Design and specification of new hoods should be based on critical and imaginative analysis of real and future ventilation needs, without the restrictions of unevaluated habit or custom. As an example, vertical-rising sash is conventional and sometimes insisted on without appraisal or comparison.

The need to provide effective hoods, to conserve conditioned air, and to balance air supply and exhaust suggest careful consideration before specifying new hoods. Vertical-rising sash are frequently left all the way up, so that the entire face should have the minimum velocity of 100 fpm (30m/min); some hoods with vertical-rising sash have by-passes which would apparently allow ready escape of contaminants in case of surge or sudden release.

Under no circumstances should corridors be designed to serve as plenums for air exhausted from laboratory areas, because air flowing into corridors can bring out and spread odors, technical contamination, air-borne pathogens, toxic gases and vapors, and fire.

Auxiliary Air Hoods

Auxiliary air hoods or supplementary air hoods are specially designed for use with a separate outdoor air supply which supplements the air exhausted from the laboratory. The purpose of the auxiliary supply of outdoor air is to reduce the conditioned air that might otherwise be exhausted from the room. Three designs for delivery of auxiliary air are illustrated in Figure 6. The first hood introduces the auxiliary air outside, overhead and in front of the sash. This auxiliary air is drawn into the sash opening as a part of the air exhausted from the room. The second hood introduces the air just inside the sash. The third hood introduces the air at the bottom of the back baffle.

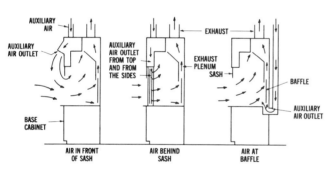

FIG. 6. Supplementary (auxiliary) air hoods.

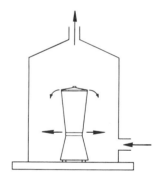

FIG. 7. Special enclosure for controlling aerosols and flammable vapors dispersed during blending.

Since the primary purpose of a hood is to capture and prevent leakage from the hood of contaminants released within the hood including the face opening, it is imperative that adequate velocity entering a hood be maintained at all times. If auxiliary air is introduced inside the hood, the total air volume drawn from the room is reduced and the face velocity drops below the safe value needed to prevent leakage of dangerous fumes through the face opening. For this reason, auxiliary air supplied within the hood is considered to be basically detrimental to the proper and safe operation of the hood.

Auxiliary air properly introduced outside the face opening will permit safe hood operation, but the outside air must be delivered at the proper location, velocity and temperature so as not to be objectionable to the hood user and so as not to disturb the normal hood face velocity. In cooler climates preheating may be required in winter; in warmer climates precooling (sensible only) may be needed in summer, especially if the operator is using the hood continuously. If the air is not comfortably tempered the operator may block the air outlets, seriously unbalancing the system and defeating the objective of reducing the air-conditioning load for the building.

Although auxiliary air supply systems are intended for reducing air-conditioning costs, under certain conditions the added cost for an auxiliary air supply system may be greater than the cost of the cooling capacity being saved, so that a careful economic study should be made before system design is selected.

OTHER ENCLOSURES

The amount of ventilation needed can be greatly reduced for laboratory operations that can be partially or completely enclosed. Glove boxes and dry boxes require little ventilation, while vacuum boxes and inert gas boxes require almost none.

A dry box or a glove box can be safely ventilated with an airflow of only 50 cfm for each square foot of open door area[10] ($15m^3/min/m^2$). A face velocity of 50 fpm or 15m/min is an effective capture velocity when face openings are small.

Figure 7 shows a special enclosure designed to prevent dispersion of toxic aerosols and to prevent an explosive concentration of flammable vapors which could be ignited by the blender motor.

Improved safety, great flexibility, and economy are advantages offered by glove boxes and other special enclosures. New laboratories should be designed to provide ventilation connections for such enclosures.

SPOT VENTILATION

Spot ventilation, exhausting contaminants near their point of origin, can prevent inhalation hazards from laboratory operations not suitable for enclosure because of bulk, access needs, or brief use.

Spot ventilation could be provided for every bench in laboratories where ventilation is imperative but adequate hood space cannot be provided. Such a system has been installed for welding fume control in the laboratory or shop in which welding is taught in Agricultural Engineering at the University of Minnesota. (Figure 8.) The instructor in charge of the laboratory will use the exhaust system to provide a valuable teaching experience; he has one flanged end removable so he can demonstrate how much better control is provided by a flange than other shapes.[10] Figure 9 attempts to show how a flange reduces entry losses and extends the reach of the exhaust pickup by reducing the volume exhausted.

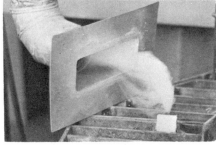

FIG. 8. Smoke candle illustrates effective capture by flanged spot ventilation of welding table; a similar system could be applied to laboratory spot ventilation.

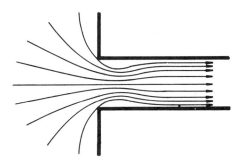

FIG. 9. Flanged end improves pick-up characteristics.

Figure 10 shows the most recent of three types of individual exhaust ventilators built into chemistry teaching laboratories at the University of Minnesota. Figure 11 shows a slotted exhaust recommended for spot ventilation[10] and used by Upjohn Company[4,11] and several university research laboratories for control of vapors from

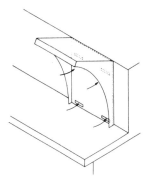

FIG. 10. Individual bench exhaust in a chemistry teaching laboratory.

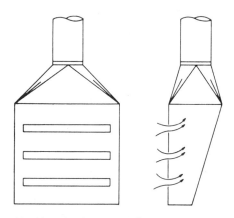

FIG. 11. Slotted exhaust used for spot ventilation of chromatography.

chromatography; the slotted exhaust is mounted on a track so that spot ventilation can be provided where needed.

A slotted exhaust can provide local or spot ventilation for exhausting vapors of moderate toxicity if the vapor-producing operations are done in reasonable proximity to the slot. Figure 12 shows a slotted exhaust system which was installed in a

FIG. 12.

neuropathology laboratory at the University of Minnesota when a laboratory hood was removed in remodeling; three work positions are ventilated instead of one and the technicians are now inhaling less vapors from their tissue work.

MARKING OF EXHAUST SYSTEMS

Controls, discharge outlets, fans, and ducts of hoods exhausting radioactive, pathogenic or highly toxic materials should be clearly marked to prevent shutdown or service without notification of hood users and personnel responsible for radiological or occupational health or safety.

ACKNOWLEDGEMENT: Reprinted from the *Journal of Chemical Education, Vol. 41*, A95, February, 1964, with permission.

REFERENCES

1. Barrett, James C., *Safety Maintenance, 119*, No. 1, 28 (1960).
2. Casey, Edward F., *J. Am. Assoc. for Contamination Control, 11*, No. 9, 5 (1963).
3. Clarke, John H., *National Safety News, 87*, No. 1, 28 (1963).

4. Jepson, G. L., in Sixth National Conference on Campus Safety—*Safety Monograph No. 9 for Colleges and Universities,* 127–130, National Safety Council, Chicago, 1959.

5. Ketcham, N. H., *Am. Ind. Hyg. Assoc. J., 19,* 324 (1958).

6. Morse, K. M., *National Safety Council Transactions, 14,* 42 (1957).

7. Schulte, H. F., Hyatt, E. C., Jordan, H. S., and Mitchell, R. N., *Am. Ind. Hyg. Assoc. Quart, 15,* 195 (1954).

8. Silverman, Leslie, and First, Melvin W., *Am. Ind. Hyg. Assoc. J., 23,* 463 (1962).

9. Walls, E. L., and Metzner, W. P., *Ind. Eng. Chem., 54,* No. 4, 42 (1962).

10. Committee on Industrial Ventilation, American Conference of Governmental Industrial Hygienists, "Industrial Ventilation," 7th ed., Edwards Brothers, Inc., Ann Arbor, Mich., (1962).

11. Michigan Department of Health, *Michigan's Occupational Health, 4,* No. 4, 1 (1959).

Exhaust Systems

American Society of Heating, Refrigerating and Air-Conditioning Engineers, Inc.*

ELEMENTS OF EXHAUST SYSTEMS

An exhaust system consists of (1) hoods or enclosures at sources of air contamination, (2) branch and main ducts through which an air stream transports the contaminant to air cleaning devices or to the atmosphere, (3) air moving equipment to produce the required air flow into hoods or enclosures, and (4) air cleaning equipment when required. See Chapter 11 in the 1969 GUIDE AND DATA BOOK for discussion of types, applications, and principles of operation of air cleaning equipment.

HOODS OR ENCLOSURES[1-10]

The most effective hood or enclosure is one that will require the minimum exhaust volume for effective contaminant control. The design must therefore be based upon a knowledge of the process or operation for which control must be obtained. The more complete the enclosure, the more economical and effective will be the installation.

Many designers give first consideration to a hood completely enclosing the operation and then provide necessary access and working openings. The familiar hoods, such as booths, sidedraft or downdraft hoods (with or without side shields) have been developed from this complete enclosure concept. Openings in hoods are kept to a minimum size and are placed away from the natural travel path of the contaminant, when possible. Doors should be provided for inspection and maintenance when needed.

Capture Velocities and Air Volume Exhausted. Only after the hood design has been determined can the exhaust volume requirements be calculated. With enclosures, volumes are calculated from the known open area of the hood and the selected capture or indraft velocity sufficient to prevent outward escape. Usual capture velocities for typical operations are listed in Table 1 and refer in the case of remote hoods to the air movement required at the zone of air contaminant generation. Required capture

TABLE 1
Minimum Air Velocities Required at Point of Origin to Capture Contaminant Effectively

Condition of Generation of Contaminant	Minimum Capture Velocity, fpm	Process
Released without noticeable movement	50–100	Evaporation of vapors, exhaust from pickling, washing, degreasing, plating, welding, etc.
Released with low velocity	100–200	Paint spraying in booth; inspection, sorting, weighing, packaging, low speed (less than 200 fpm) conveyor transfer points, blending, mixing, barrel filling.
Active generation	200–500	Foundry shakeout, high speed (over 200 fpm) conveyor transfer points, crushers, screens.
Released with great force	500–2000	Grinding, tumbling mills, abrasive cleaning.

*From ASHRAE Guide and Data Book, 1968, with permission.

velocities for any operation will vary with the magnitude of the air volume handled, with uncontrolled air movement in the area, and often with the location of the process or operation and size of the workroom. Large remote hoods exhausting large air volumes will provide effective control at lower maintained capture velocities than will small remote hoods handling lower exhaust volumes.

AIRFLOW-PRODUCING EQUIPMENT

The principal types of air-moving equipment are centrifugal fans, axial-flow fans, and venturi ejectors (see Chapter 4 of the 1969 ASHRAE GUIDE AND DATA BOOK). Gravity stacks have application for some systems handling higher air temperatures or steam where no air cleaning equipment is installed.

Centrifugal fans are most frequently used for the bulk of exhaust system applications due to the system pressures involved. Where air-cleaning equipment is included, fans are usually located on the clean-air side. The paddle wheel or modified paddle wheel designs are heavily constructed and have few blades to make them more suitable if wear, corrosion, or accumulations are a factor. The higher efficiency backward-curved blade designs find application where relatively clean, non-corrosive air is handled. Forward-curved blade designs have limited application due to the number, shape, and metal thickness of the short curved blades.

Belt-driven fans are recommended by many designers because a change in fan speed is often needed where future changes of the system may be involved, where added pressure losses from air cleaning equipment improvements are anticipated, or where the fan may be used at some later date for another application.

The axial-flow fan is used for systems having low-pressure losses. Propeller or disk designs develop pressures under 1 in. of water. The vane-axial designs develop higher pressures but seldom are used where pressures exceed 3 in. of water.

The venturi ejector[11] is an inefficient method of air movement, but has the advantage of causing airflow without having the exhaust air pass through the airflow-producing equipment. It minimizes the explosion or corrosion potentialities in certain types of systems.

Fans, motors, and drives should be located so that safe and easy access for periodic inspection, servicing, and maintenance is possible.

AIR-CLEANING EQUIPMENT

As discussed in detail in Chapters 10 and 11 of the 1969 ASHRAE GUIDE AND DATA BOOK, air-cleaning equipment should be considered in all systems where it can prevent property damage, neighborhood pollution, or reentry of polluted air to the working space, or where it can salvage usable material, reduce fire and explosion hazards, or make possible some recirculation of air to the working spaces. The present growing emphasis on air pollution control makes it desirable to remove all contaminants to the greatest practical degree based on reasonable cost and maintenance.

MAKE-UP AIR REQUIREMENTS

A correctly designed exhaust system may be ineffective during periods when windows and doors are closed. This condition should be anticipated and prevented by making provision for an adequate air supply. Poor performance of exhaust hoods is caused by: absence of air supply systems; cold drafts due to high indraft velocities through cracks and openings in building construction especially at windows and doors; reverse airflow through low-pressure systems of roof ventilators or general ventilation systems; downflow through heater vent stacks preventing exhaust of flue gases to the outdoors. Systems for conditioning make-up air do not necessarily increase heating requirements for the space, as cold air due to infiltration from the outdoors must be heated somehow to maintain comfortable temperatures in the area. The cost will depend on the ratio of exhaust air to that required for ventilation of the space occupied.

MAINTENANCE OF PERFORMANCE

Periodic inspection and checking of exhaust systems are necessary if control is to be maintained at the effective level of the original installation. Continued effectiveness depends on maintained design air volume flowing through the exhaust hoods. A checking procedure, therefore, must include some data to indicate at least relative airflow through the hoods. The static pressure or hood suction measurement will prove useful for such checking if data are available on air volumes and pressures at the time the system was installed. Testing and recording of such data for each new installation are of the utmost importance.

While hood suction readings have rightfully been discarded as a means of measuring airflow, they do offer a quick and accurate method of measuring relative airflow. If the hood suction is known while an exhaust system is functioning properly, its continued effectiveness can be assured so long as the hood suction is not reduced from its original value.

Difficulty from plugging may be encountered where heavy dust loads or moist air are encountered. Unless the hood design is altered or there are accumulations in the hood or branch pipe between hood and point of hood suction reading, the air volume exhausted from a hood cannot change without a change in hood suction reading.

The Pitot tube can be used for routine check purposes instead of the static pressure method but requires care in reading velocity pressures in an exact position with the tube paralleling the flow of air. The hood suction method of checking can, however, be more readily delegated to an assistant having no technical training. U-gages have been standard plant equipment long enough to eliminate any feeling of uncertainty in their use.

Since pressure readings vary as the square of the velocity or volume of flow, a slight change in flow is magnified by a comparison of gage readings. Normally, a reduction of volume or velocity of 10 to 15 percent will not be sufficient to reduce the effectiveness of the exhaust system. This range is equivalent to a reduction in static pressure readings of 19 to 30 percent.

A marked reduction in hood suction can often be traced to one or more of the following items:

1. Reduced performance by the exhaust fan caused by reduced speed due to belt slippage, wear on rotor or casing, or an accumulation of material in the rotor or casing obstructing the airflow.
2. Incorrect direction of exhauster rotation.
3. Reduced performance caused by defects in the exhaust piping, such as accumulations of material in branch or main ducts due to insufficient conveying velocities, condensation of oil or water vapors on duct walls, adhesive characteristics of material exhausted.
4. Leakage losses caused by loose cleanout doors, broken joints, holes worn in duct (most frequently in elbows), or poor connection to the exhauster inlet.
5. Losses in suction due to exhaust openings added to the system or due to a change of setting of blast gates in branch lines.
6. Increased pressure loss through the dust collector due to lack of maintenance, improper operation, wear, etc.

MATERIALS REQUIRED FOR CORROSION RESISTANCE

In many cases, exhaust systems including hoods, ducts, airflow-producing equipment, and air cleaning equipment will require protective coatings if carbon steel is used, or other materials of construction will be required.   Need for such materials will be determined by (a) the corrosion rates on interior or exterior duct surfaces, (b) protection against product contamination, and (c) explosion hazards.

Construction for corrosion protection is the most difficult due to the complex factors that influence rate of corrosion.   It is seldom possible to predict the concentration, composition, and dry-bulb and dew-point temperatures that will exist in actual operations.   A guide for selecting corrosive resistant materials based on a survey of actual experiences has been reproduced.[12]

REFERENCES

The material in this chapter is based largely on the recommendations of the American Conference of Governmental Industrial Hygienists as published in the *Industrial Ventilation Manual*, 1958 Edition.

1. Bloomfield, J. J. and DallaValle, J. M., The Determination and Control of Industrial Dust, *U. S. Public Health Service Bulletin 217*, (1935).
2. DallaValle, J. M., "Exhaust Hoods," Industrial Press, New York, (1952).
3. Hatch, T., Design of Exhaust Hoods for Dust Control Systems, *J. Ind. Hyg. and Toxicol.*, *18*: 595 (1936).
4. Hemeon, W. C. L., Air Dilution in Industrial Ventilation, *Heating and Ventilating*, (February, 1941).
5. Postman, B. F., Practical Application of Industrial Exhaust Ventilation for the Control of Occupation Exposures, *American Journal of Public Health*, *30*: 149 (1940).
6. Witheridge, W. N., "Principles of Industrial Process Ventilation," University of Michigan, Inservice Training Course, (October 1945).
7. Drinker, P., and Hatch, T., "Industrial Dusts," McGraw-Hill Book Co., New York, (1936).
8. Alden, J. L., "Design of Industrial Exhaust Systems," Industrial Press, New York, (1948).
9. Witheridge, W. N., Ventilation, *Ind. Hyg. and Toxicol.*, *1*, Chapter 10, F. A. Patty. ed.   Interscience Publishers, New York, (1948).
10. Stern, A. C., et al, Transport Velocities for Industrial Dusts. *Am. Ind. Hyg. Assoc. Quart.*, (December 1948).
11. McElroy, G. E., Design of Injector for Low Pressure Air Flow. *U. S. Bureau of Mines Technical Paper, 678*.
12. American Standard Safety Code for Ventilation and Operation of Open Surface Tanks, *American Standards Association*, Z 9.1-1951.

Hoods for Science Laboratories

Harold Horowitz, S. A. Heider and Caldwell N. Dugan

Architects have been giving vivid testimony to the significance of fume hoods and their associated exhaust equipment by the substantial number of recent science buildings in which they have moved service shafts housing exhaust ducts to the exterior wall as a means of expressing the functional characteristics of this type of building. Architects have been among the pioneers in the use of the chemical fume hood as a safety device in research laboratories and have been instrumental in bringing the technology of this means of environmental control to the present state of development. Nevertheless, after careful review of several hundred science buildings, the Architectural Services Staff of the National Science Foundation has concluded that much needs to be done to achieve wider dissemination and understanding of the basic principles of the fume hood's design and use. The objective of this chapter is to present a brief outline of principles for satisfactory use of fume hoods in science laboratory design, serving as an introduction to the subject for those architects who have not yet had the opportunity to study this important element of science building design.

A fume hood is an exhaust duct terminal, so conceived that it can enclose an experiment. The enclosure has one or more openable sides and is designed so it can transform the suction of the duct into a uniform movement of air across the face of the opening. Hazardous experiments involving toxic chemicals, and those with unpleasant odors, are conducted within the enclosure. The flow of air into the enclosure sweeps the toxic and odoriferous vapors and dusts into the duct to be exhausted out-of-doors, thus protecting the person working in front of the hood and also preventing the toxic and odoriferous materials from passing into the air of the laboratory.

THE IMPORTANCE OF FACE VELOCITY

Satisfactory performance of a fume hood requires that the airflow past the opening into the enclosure occur within minimum and maximum limits, since both are of great importance. They vary somewhat depending upon a number of factors relating to the design of the particular hood, its location in the laboratory and the degree of hazard of the experiments; therefore, the limits must be selected with judgment. The minimum face velocity must be great enough to insure that the direction of air movement at any point in the area of the open face of the hood will always be into the hood. It is desirable that the lowest possible amount of air be exhausted, consistent with safety requirements, because of the economic advantage of reducing losses of heated air in winter and cooled air in summer. Factors influencing the minimum safe face velocity for a particular hood are numerous.

The upper limit of air velocity for a fume hood is related to the aerodynamic flow pattern created by the air stream flowing past the scientist in front of the hood, past the experiment itself and out the exhaust opening. The scientist standing in front of a hood serves as a barrier to the air stream, and when the air velocity reaches a certain point, a low pressure or partial vacuum is created directly in front of him. The low-pressure zone extends into the fume hood increasingly as the face velocity is increased. A velocity can be reached where the low-pressure zone may extend into the area of the fume hood occupied by the experimental apparatus. When this condition is reached, the fumes generated by the experiment will fill the low-pressure zone and may contact the scientist's skin or be inhaled unless he is protected by a portable shield or a horizontally-sliding sash.

Within the two limiting extremes of face velocity the choice of air-flow rate is based on the desire to reduce to a minimum the total flow of air being exhausted and to maintain a safe environment. Safety is generally enhanced by increasing the rate of airflow to the point where the maximum airflow limitation is reached. It is usual to select fume hoods and to design the air-conditioning system to maintain the face velocity appropriate for the maximum hazards anticipated with the particular hood. Table 1 represents our general recommendations; but modifications in some situations may be warranted.

TABLE 1
Recommended Face Velocities

Degree of Hazard	Minimum Measured Velocity at Any Point Across Hood Face
Low toxicity levels	50 fpm
Average toxicity levels in research involving a wide range of materials	75 fpm
Low-level radioactive tracer materials with nominal toxicity hazards	100 fpm
Significant chemical toxicity levels and moderately radioactive materials	150 fpm
Higher levels of toxicity and highly radioactive materials	Consider the use of glove boxes and total enclosures if velocities in excess of 150 fpm are required

SELECTION OF FUME HOODS

Degree of Hazard. The degree of hazard of the experiments to be conducted within a fume hood has a great influence over the type that should be selected. It is important to establish the maximum degree of hazard anticipated before the choice of a hood is made. Conversely, it is important to know the limitations of each hood so that inconvenience or even tragedy, is not caused through improper matching of experimental work with hood capabilities.

Size of Experiment. For reasons of safety and economy it is desirable that hoods be as small as possible and exhaust the least possible amount of air from the laboratory. However, a hood must be sufficiently tall, wide and deep to house the particular type of experiments that will be undertaken. Experimental apparatuses vary greatly in size; therefore, fume hoods have been designed over a wide range of sizes and shapes. They may be as large as small rooms that are walked into, or they may be small, portable enclosures which are easily carried and placed in different positions on laboratory benches.

A common fault in the selection of fume hoods for science buildings is the tendency to select hoods of uniform size, usually a bench-type, as part of a conception of a modular laboratory arrangement. The program requirements for a science building should give consideration to the potential variation in size of experiments to be conducted so that an appropriate number of various size hoods, with a reasonable distribution within the building, can be obtained.

Fume Hood Construction Materials. A great many materials have been used for fume hood construction. Most of them have proven to be satisfactory when used correctly and properly selected with regard to the requirements of the experiments. It makes good sense to use the least expensive material that will do the job. Unfortunately, however, materials which are most versatile tend to be the most expensive. Table 2 relates fume hood materials with versatility for experimental work and economics.

TABLE 2

Hood Construction Materials

(Listed in decreasing order of cost)

Material	Limitations
a) Hood structure	
Chemical grade soapstone	Most versatile
Stainless steel.......................	May be attacked by some chemicals. Care should be used in selection
Monel metal........................	May be attacked by some chemicals. Care should be used in selection
Synthetic or cementitious "stones"....	Absorb water, tend to stain, may be attacked by some chemicals
Carbonized birch	Recommended only for light service
Aluminum	Recommended only for light service, readily attacked by alkaline materials
Reinforced plastics	Recommended only for light service. Resins vary in resistance to chemical attack and fire resistance. Care should be used in selection
Varnished wood	Limited to the lightest service where no possibility of fire exists and no use will be made of solvents and steam
b) Glazing materials	
Laminated glass.....................	Most versatile
Tempered glass	Not recommended where experiments may produce rapid thermal changes or explosion hazards exist. Although the glass shatters into fragments which are not sharp, failure results in loss of protection
Wire glass	Wires restrict vision
Plain glass	Not recommended
c) Coatings	
Epoxy	Too new for adequate experience but appears promising
Strippable paints....................	Not resistant to solvents and many chemicals. Use should be avoided except for special radioactive applications and when related to a decontamination program

DESIGN FACTORS

1) The depth of a hood is the most important dimension with respect to satisfactory operation. In general, the deeper the hood the more satisfactory it will be in providing uniform suction across the open face. The depth is also of great importance with respect to the possible size of apparatus and experimental setups that can be contained.

2) Generally, the lower the height of a fume hood (and the clear opening), the more satisfactory will be its airflow characteristics. A very tall hood with the exhaust duct connection near the top will present the greatest risk of escape for fumes because of the difficulty of securing uniform airflow at the base.

3) The design of the jamb is of utmost importance in preventing turbulence at the sides of the hood, which may result in fumes being swept out into the laboratory. In general, the jamb design should permit the smoothest possible airflow pattern.

4) The design of the sill should also provide for a smooth airflow to minimize turbulence. Sill designs which incorporate slots permitting air to enter and sweep across the working surface, irrespective of how close the operator may be standing, are valuable in insuring a safe hood. It is highly desirable that the sill be designed, by means of a slope or curvature near the front edge, to make it impossible for experimental equipment to be placed within 6 to 10 inches from the edge. Such design auto-

matically prevents fume-generating experiments from being positioned close to the front edge of the hood.

5) It is recommended that the sash tracks of a fume hood be installed some distance behind the front edge of the jamb and sill. This automatically insures that the experimental apparatus will be set back to this extent in the hood. In general, the further back the fume-producing apparatus is located in a hood, the greater the safety in the conduct of the experiments.

6) Uniformity of airflow across the face of the hood is of utmost importance. Averages can be deceptive, and a hood with an appropriate, average face velocity may actually provide below-minimum velocity at some points and above-maximum velocity at others. Such a hood can be unsafe. Each of the design factors influences uniformity of airflow across the face. Most important are the proportions of the hood—height, width, depth—and the relationship of the size and position of the exhaust-duct connection with the baffles. Unfortunately, no simple rule of thumb can be provided for guidance; the location of exhaust air ducts, the number of such ducts per hood, plus the type and design of baffles are of equal importance. For simple hoods with only a single exhaust air outlet near the top and with the usual type of slotted panel baffles, it is said that uniformity of airflow across the face will be increased as the hood depth increases, and will decrease with increasing height and width.

7) The function of baffles in a fume hood is to distribute the suction of the exhaust duct in such a way that uniform airflow through the face of the hood will result. The design and position of baffles and their openings are critical to satisfactory performance. Opinions on proper design of baffles vary widely among experts on the subject. At least two baffles, one at the top and one at the back of the enclosure, should be provided; they should be separated by air slots. Dividing the back baffle to form a center slot will improve velocity distribution at the hood face.

8) Many special design conditions occur with fume hoods and their associated components. These special situations should be explored when the program for a science building is being analyzed. Examples of special factors are: when radioactive materials are to be handled in a hood, the construction of the base cabinets and floor must be taken into consideration and the need for shielding blocks which may have a combined weight of 2000 pounds or more; when perchloric acid is to be used, the hood should be provided with washdown facilities and must be constructed so that its interior components and concealed air passages can be cleaned without undue jarring to avoid an explosion.

RELATION OF FUME HOODS TO AIR-CONDITIONING SYSTEMS

The advent of air-conditioning for laboratory buildings has greatly intensified the economic problems in the selection and utilization of fume hoods. The large volumes of expensive cooled air exhausted by fume hoods add substantially to the initial and operating costs of air-conditioning plants. Some mechanical engineers estimate that the initial extra cost of an air-conditioning plant is approximately equal to the installed cost of the fume hoods. In a science building with many fume hoods, the exhaust through this safety device can account for the greatest single cooling load requirement. (Of course, heated air is also lost through fume hoods during the winter, but the cost of heating is generally not as great as the cost of cooling and therefore is of less concern.) Many ideas, both simple and elaborate, have been explored to reduce the volume of conditioned air lost through fume hoods, thereby reducing the initial and operating costs of the air-conditioning system. Unfortunately, a number of these schemes have lost sight of the basic function of the fume hood—providing safety—and have tended to compromise the protection or comfort of the operator in the attempt to reduce air-conditioning costs.

The second important relationship of fume hoods to air-conditioning systems is the problem of maintaining satisfactory balance. In a building where an extensive number of fume hoods are being employed, the simplest approach to this has been to ignore the problem of balance and simply supply additional air throughout the system to make up for the losses through the hoods, on the assumption that about half the hoods may be operating at one time. The effectiveness of such a system seems to fluctuate depending on what actually occurs in use. Other approaches have been taken to control fume hoods and return air registers in a system to provide a constant volume of exhausted air, or to use constant-volume fume hoods as the sole exhaust outlet.

With respect to the air-conditioning system, fume hoods may be divided into three groups:

a) The standard hood is one in which the volume and velocity of air varies as the sash is raised or lowered, since no means is provided to compensate for the variable area of face opening. This type of hood causes the greatest difficulty with balance in an air-conditioning system. Laboratories containing such hoods must be supplemented with additional exhaust air openings to insure adequate laboratory ventilation.

b) The constant-volume hood incorporates an internal bypass feature permitting the same volume of air to be exhausted into the hood regardless of the position of the sash. Such hoods permit balancing of the air-conditioning supply and exhaust. Constant-volume hoods may not be manually controlled but operate continuously while they serve as the air exhaust outlet for the laboratory. It is often incorrectly assumed that this type of hood, because of its greater cost and more elaborate appearance, is safer than a standard hood. The special function of the constant-volume hood is related to the balance of the air-conditioning system rather than to safe operation and does not provide greater safety than can be achieved with a properly designed system using standard hoods.

c) The auxiliary air-supply hood attempts to reduce air-conditioning requirements by providing a separate supply of air that has not been cooled and dehumidified in the summer or fully heated in the winter. The supply of air for such a hood may be drawn from outdoors or from the service chases within the building, which are, in turn, supplied by air from attic or mechanical equipment rooms. Such hoods can substantially reduce the air-conditioning equipment capacity required to make up losses through fume hoods; operating costs can likewise be reduced. However, there are a number of disadvantages to the use of such hoods. One type of auxiliary air-supply hoods discharge untreated air just in front of the face of the hood, usually at the head, a scientist working at the hood must work in unconditioned air. The disadvantages are obvious, and the annoyance of scientists has been evidenced by their very human attempts to invent means of foiling the intended mode of operation. One such effort consists of securing cardboard over the outlets with adhesive tape, thus closing or reducing the auxiliary supply. Attempts to rectify this problem by partially cooling or heating the air supply, depending upon the season, substantially reduce any economic advantage of this type of hood. Another type of hood introduces the auxiliary air within the hood enclosure and is inherently unsafe because the face velocity is reduced below the rate necessary to capture fumes. This type of hood also poses several operational problems that can lead to significant hazards.

Another very important problem with auxiliary air-supply hoods is that the balance between the temperature of the auxiliary air and the air in the hood is critical. Unless balanced just right, much of the auxiliary air will not enter the hood but will enter the air-conditioned space of the laboratory, sweeping with it some of the contaminated air from the hood. Several manufacturers, universities, and public agencies are now attempting to improve upon the design of auxiliary air-supply hoods. For the present, however, it is our opinion that such hoods are usually an unwise selection.

Air supply outlets and returns in laboratories which contain fume hoods should be so located, in relation to the position of the hoods, that they do not cause strong drafts in the vicinity of the hood. If the fume hood is operating with a face velocity as low as 50 fpm, a relatively small draft of air can act to overcome and reverse the desired airflow into the hood and cause a hazardous condition.

CONTROLS

Piped services with outlets inside a fume hood should be controlled by handles in readily accessible locations outside the hood enclosure. If washdown facilities are provided for hoods to be used with perchloric acid, the control valve handle should also be located outside the hood enclosure.

Switches, rheostats and other control devices for electrical apparatuses and convenience outlets located within hoods should be located on the outside of the enclosure. The design of such controls should be approved for use in explosive environments.

Lighting controls, as with other electrical switches, should be located outside the hood enclosure and be of explosion-proof design.

When individual fan control is desired, the control device should be located outside the hood enclosure. Where several hoods are included in a single laboratory space, it is recommended that all hoods be controlled through a single switch, thus insuring that they will all operate simultaneously. It is undesirable to have some hoods operating while others are idle in a single laboratory space, as the idle hoods can serve as sources of make-up air for the hoods which are operating. When this occurs, the make-up air entering the laboratory through the idle hood may pick up sufficient contaminated material from the hood exhaust duct, the hood itself and any experimental apparatus within the hood to cause a hazardous condition in the laboratory.

Occasionally special controls are required for the adjustment of an experimental setup within the hood enclosure. It is desirable that the controls for such installation be located outside the hood in such a way that the sash may be completely lowered. Slots or other means should be provided so that the controls can be conveniently brought out of the enclosure. Should the fume hood be of the type with a sill incorporating slots for airflow, the slots may serve this purpose.

ADJUSTMENT FEATURES FOR AIRFLOW

A number of air-volume controls have been used with fume hoods. Among the more commonly encountered types are dampers in the exhaust duct and multispeed fan controls. We do not recommend the use of dampers as they require extensive maintenance and reduce efficiency of the exhaust system. Multispeed fan controls provide the fume hood operators with a degree of individual control; however, they may cause difficulty with the balance of the airconditioning system by frequent fluctuation of the supply air requirements. The recommended means of controlling air volume is to design the exhaust system carefully with respect to its airflow characteristics and provide the possibility of volume adjustment at the fan in such a way that adjustments can only be made by the operating personnel of the building's mechanical system.

The principal means of controlling the pattern of air distribution within a hood is through adjustable baffles. Such baffles are located between working areas of the hood and the point of connection with the exhaust duct. The baffles must be able to transform the suction of the duct into a uniform flow of air at the fume hood face. Baffles control air distribution by their position within the hood and by the size and position of slots in the baffles; usually the slots are adjustable. Occasionally the position of the baffle is also adjustable. The need for control through adjustments of the baffles increases with increasing height and width of hoods. Very large hoods require very sophisticated baffling systems to insure satisfactory airflow characteristics. The adjustment of the baffles is of extreme importance regardless of the characteristic of the baffle design. Therefore, a critical part of every fume hood installation should be careful adjustment of baffles and subsequent testing.

LOCATION OF THE FUME HOOD IN THE LABORATORY

Air Velocity Outside Fume Hood. The rate of airflow required to satisfy minimum face velocity requirements is quite small. Therefore, air movement outside the hood is of great importance in insuring safe operation. As already discussed, the location of room air outlets and returns may cause drafts sufficient to counteract and reverse the flow of air into a hood. Opening and closing of doors into a laboratory space will move a considerable volume of air and does have a marked influence on airflow patterns and pressure surges within a laboratory. This results in major air movements which can temporarily alter fume hood airflow characteristics. The effect of door openings is so great that considerable caution should be exercised in locating fume hoods in small laboratories where the relative volume of air movement generated by door action is substantial in relation to the total volume. In all laboratories it is desirable to locate fume hoods as far away as possible from door openings.

Windows which may be opened for ventilation can present even more serious interference with fume hood operation than doors. When fume hoods are located in buildings that are not airconditioned, or which require open-window ventilation during the spring and fall, hoods should not be located close to windows which may be open. However, it is preferable that hoods be located only in spaces provided with supply air through a mechanical ventilation system which can control the volume and location of incoming air.

Pedestrian traffic within a laboratory can also interfere with the operation of fume hoods. This can be readily recognized by recalling that a walking speed of one mile per hour is equal to 88 fpm, or 33 fpm greater than the minimum recommended face velocity for fume hoods. A person walking at 2 mph moves at a speed of 176 fpm, which is 26 fpm more than the maximum recommended face velocity for fume hoods. These are very moderate walking speeds; thus it should be assumed that any walking speed will probably exceed the face velocity, and persons walking by fume hoods will tend to produce some flow of contaminated air out of the hood. It is therefore undesirable to locate fume hoods along principal traffic lanes.

Relation to Work Areas. Fume hoods should be located conveniently in relation to benches and apparatuses that may be used by scientists in connection with the same experimental program. It is desirable that fume hoods be integrated with other laboratory furniture in such a way as to provide adequate bench space adjacent.

Proximity to Service Shafts. There are economic advantages in locating fume hoods close to service shafts which will contain the exhaust ducts and piped service lines. Such economic considerations have often been carried to the point where fume hoods have been located in undesirable positions in laboratories with respect to safety. It should be remembered that the function of the hood is to provide a safe working environment, and small savings in horizontal ducts and pipe runs are not worthwhile if the fume hoods cannot serve their intended function. When it becomes necessary to choose between safety and proximity to service shafts, the choice should always be for the location that will provide for proper function of the hood.

TESTING AND ADJUSTMENT

No standard testing procedure has yet been developed for the evaluation of fume hood performance. However, a number of procedures are being used which involve titanium tetrachloride smoke, smoke bombs, hot-wire anemometers and release of odoriferous materials (e.g. ammonia and mercaptans) within hoods. Testing services are available from independent consultants (general industrial hygienists), mechanical engineers and manufacturers' representatives.

At present there is an important need for the establishment of a standardized testing procedure for fume hoods and for fume exhaust system evaluation. Even though we are not satisfied with existing testing procedures, the necessity of adjusting hoods after they are installed and testing them prior to acceptance (as well as periodically during the life of the equipment) cannot be omitted. A satisfactory specification for a laboratory building should incorporate requirements for testing hoods with respect to airflow characteristics and face velocity. Hoods which do not provide satisfactory test results should not be accepted.

FUME HOOD EXHAUST SYSTEMS

Duct System

Ideally, every fume hood should be served by an independent duct. Such an arrangement insures maximum safety and flexibility. Economics, however, favor some degree of combination of ducts. Separate ducts should always be used with hoods for handling of radioactive materials, perchloric acid, strong oxidizing agents, or highly reactive chemicals of any sort. Aside from these cautions it is always permissible to combine ducts from several fume hoods in the same laboratory space, especially teaching laboratories where the work being done in the several hoods is under the control and supervision of an instructor. Hoods in several different laboratory rooms should never have their ducts combined, since it is not possible to predict the simultaneous use of hoods with materials that may be reactive when combined in the air stream. The combination of several hoods into a single duct also greatly complicates servicing problems, as any maintenance or repairs required for the duct or exhaust fan will shut down hood operation in all of the laboratories served by the combined duct.

Fume hood ducts differ from other air-handling ducts in that the materials that pass through them are often highly corrosive and very toxic. Consideration should be given to the fact that such ducts will have to be serviced or replaced during the life of the average laboratory building. Safety to personnel making repairs or replacement of ducts should not be overlooked.

Many satisfactory materials are available for fume hood ducts (see Table 3). Unfortunately, as in the case of fume hood construction materials, they vary in their

TABLE 3
Duct Construction Materials
(Listed in order of decreasing cost)

Material	Limitations of Use
Glazed ceramic pipe	Rarely used today because of installation problems
Epoxy coated stainless steel	Extensive experience not yet available but appears promising as most versatile material
Stainless steel	May be attacked by some chemicals. Care should be used in selection
Monel metal	May be attacked by some chemicals. Care should be used in selection
Synthetic or cementitious "stones"	Absorb moisture, may be attacked by some chemicals
Reinforced plastics	Various resins being used have different chemical and fire resistances. Care should be used in selection
Asphalt-asbestos coated steel	Limited solvent resistance. Care should be used in selection
Aluminum	Limited resistance to many chemicals. Care should be used in selection
Galvanized steel	Limited resistance to corrosion by wide variety of materials used in research
Black steel	Useful only with dry and uncorrosive dusts

cost and versatility of use. Expensive duct materials, such as stainless steel, are not completely versatile and are subject to attack by some chemicals. Therefore, considerable care must be exercised in evaluating the type of materials to be used in laboratories, the selection of duct materials and the features provided in the design of the building for service and replacement of ducts.

High-velocity air movement in the ducts is desirable to insure that dust and aerosol-size materials are not deposited in the joints, cracks or corners in the duct system. A minimum suggested design velocity is 2000 fpm. Higher conveying velocities are desirable. A minimum of turns, bends and other obstructions to airflow are desirable. Where perchloric acid is to be used, duct configuration should permit thorough washdown of duct surfaces.

Exhaust Fans and Outlets

Location of the Fan. Early in the history of the use of fume hoods, it was common to install exhaust fans directly above the hood. This has been found to be a very unsafe practice because it creates a high-pressure condition in the exhaust ducts and any leakage at joints or through pinholes caused by corrosion may result in distribution of contaminated air and toxic materials into the building. Unfortunately, some hoods are still being installed with fans located above the hood within the occupied space of the laboratory. This unsafe practice should be eliminated entirely, and we hope that building codes will eventually be revised to incorporate this requirement.

Locating the fan of a fume hood exhaust system in an attic or mechanical equipment room is less than wholly satisfactory, although a great improvement over a location directly at the hood. When the fan is located within an attic or a mechanical equipment room, there is a possibility of leakage into the building before the duct penetrates to the exterior where the fumes can be safely discharged. Attics and mechanical equipment rooms containing fume hood ducts in which the pressure is higher than in the surrounding space should be mechanically ventilated to the outdoors. Joints in ducts containing fumes under pressures should be sealed with compounds especially formulated for the purpose.

The preferred location for fans is outside the building, usually on the roof, and as close as possible to the terminal. Such a location insures that the ducts within the building operate at lower pressure than exists within the surrounding spaces in the building and that any leakage that does occur will be into the duct.

Location of the Discharge Terminal. When fume hoods are installed in existing buildings that have not been provided with adequate service shafts, the discharge will sometimes be made at the side of the building. Usually the duct will be carried through a panel substituted for a pane of glass in a window. This practice is unsafe and should be discontinued. The possibility of re-entrance of fumes through opened windows of the same or adjacent buildings is very great. Also, discharge at this position may result in noxious fumes collecting in the vicinity of the building rather than being swept rapidly away when the discharge coincides with being on the leeward side of the building.

Fume hood ducts are occasionally discharged into stacks which carry fumes to a point high above the building for discharge into the air. There are a number of architectural design problems associated with this type of discharge; however, the use of tall stacks may help solve the problem of safe disposal of especially toxic matter.

Occasionally fume hood exhaust ducts discharge their contents into areawells below grade. This may occur as a simple way of exhausting a fume hood located in a basement or subbasement space. This practice is not recommended, as toxic materials may accumulate in the areawells where they cannot be swept away by air movement. Fumes discharged in this way may also pass through on-grade areas where building personnel must walk and would be subjected to hazards. Occasionally rooftop wells are created as a result of screening, skylights or other architectural features on roofs. Discharge outlets of fume hood exhaust systems located within such depressed or screened areas are unsatisfactory, as this kind of arrangement tends to prevent the fumes from being swept away from the building.

The preferred location for fume hood exhaust duct discharge terminals is above the roof of a building. Ideally, the point of discharge should be above the transition zone between air moving freely past the building and the turbulent air restrained or trapped on the roof or lee side of the building.

Horizontal fan discharge outlets, fixed cap outlets, mushroom cap outlets and rotating cap outlets tend to prevent discharged noxious materials from being projected upward into the air stream which will move them away from the building. The preferred type of discharge terminal projects the fume hood exhaust air in a vertical direction at the highest possible velocity so that it can be captured by the free air stream above the turbulent zone influenced by the shape of the building. The only positive way of accomplishing this is to carry the discharge terminal to a sufficient height so that it is above the boundary between air flowing past the building rather than captured on the roof or in the leeward wake.

The function of the discharge terminal of a fume hood duct system is to project fumes away from the building in such a way that air intakes will not be contaminated and persons in the vicinity of the building or working on the roof will not be subjected to the hazard of breathing the discharged fumes. A common observation is that concern about rain entering the discharge end of the duct has led to a design decision negating this fundamental consideration. Discharge outlets of exhaust fans have been aimed downward to spread fumes along the surface of the roof or they may discharge horizontally which is nearly as bad. The fans may be positioned to discharge vertically but the opening of the duct terminal is covered by some type of weather cap, which reduces velocity and changes the direction of discharge so that the fumes are not projected away from the building. Such concern over rain penetration is needless as very simple means of protecting the fan can be found that do not compromise the effectiveness of the system. To begin with, it should be remembered that an appropriate terminal velocity of at least 2,000–3,000 fpm (609.6–914.4 m/min) is sufficient in itself to prevent rain penetration of the stack. The column of air leaving the terminal simply brushes the rain aside. If the fan is not operated continuously, there is the possibility of some rain penetration but this can easily be handled by a small drain hole at the bottom of the fan housing, or the use of one of a variety of details that have the effect of creating a drip within the duct to divert water and discharge it before entering the fan housing.

When radioactive materials are used, filters may be required to prevent discharge of these materials into the air. Scrubbers, burners and other types of air cleaners may also be used to treat fume hood discharges and reduce the potential hazard from toxic, biological and radioactive wastes. Filters and scrubbers can also partially reduce the need for concern over location of discharge terminals with respect to air intakes, experimental apparatus located on the roof and to personnel who may work there.

HAZARDS ASSOCIATED WITH FUME HOOD DISCHARGES

The problem of designing a satisfactory fume hood is not concluded with the hood, duct, fan and discharge terminal. Consideration must also be given to what happens when the fume hood exhaust leaves the system and enters the open air. As suggested in the previous section of this chapter there is considerable danger that the fume hood exhaust may circulate back into the building of its origin or into adjacent buildings. Special care should be taken to protect air intakes against contamination, and model tests in wind tunnels are often desirable for the study of complex situations. Corrosive fumes released a short distance above a laboratory roof can cause deterioration of scientific apparatus and rooftop mechanical equipment such as fans, and cooling towers. There may be considerable potential hazard to personnel, both scientific and building maintenance, who may have occasion to work on the roof of the building.

One of the most popular misconceptions is that an air intake at ground level will be adequately protected by a separation of several stories between it and fume hood discharge terminals on the roof. This conception is contrary to our understanding of airflow behavior around buildings and is not supported by observations of the performance of existing science buildings. One example is a university biochemistry research laboratory that is five stories high. The air intakes are located at ground level and fume hood outlets and an incinerator outlet are located on the rooftop. This building is a long slab with its major axis oriented in the northeast-southwest direction. The air intake is located near the center of the southeast elevation which turns out to be the prevailing leeward zone. The experience supports theoretical considerations completely as strong solvent and chemical odors are distributed throughout the building by the ventilating system with daily regularity. The five story separation here has not provided protection.

Studies of airflow around buildings both in the field and in wind tunnels show that a building forms an obstruction to airflow resulting in displacement of air. A cavity is formed on the leeward side of the obstacle in which relatively limited air movement takes place between the cavity and the surrounding regions. The shape of the cavity is influenced by the shape of the building and other nearby obstructions to airflow; the velocity of airflow and other meteorological conditions. Air intakes should not be located in positions that are likely to fall within the prevailing leeward side of a science building. However, this advice is really dependent upon the point at which the fumes are discharged from exhaust stacks. If the fumes are discharged at a height above the boundary of the cavity zone, then they will be carried away from the building and not contained within the cavity, however, if the height is not sufficient for the fumes to be discharged above the boundary they may be contained and held within the cavity. If the air intakes are also within the cavity region, they will draw in contaminated air.

There is no simple way to tell how high stacks should be to penetrate the cavity boundary. Photographs made of smoke moving past actual buildings, and simulation studies in wind tunnels show that the separation of the cavity from the air stream moving past may be sharply defined. However, the shape of the cavity and the height of the boundary for a given building shape will vary with wind speed and be influenced by such meteorological conditions as wind gust factors. There are a number of papers in the literature describing studies in which wind tunnel tests have been used for making a good approximation of the optimum height to insure dispersing fumes away from the vicinity of the point of discharge. Some authorities have formulated a rule of thumb which says that the height of the discharge point should be half again the height of the building. (By building height they are referring to the elevation of the leading edge or separation point at which the cavity begins to form.)

We are reluctant to support this rule of thumb because it is too conservative for many situations. It is increasingly conservative as buildings become taller and is also substantially influenced by the aerodynamics of the building shape and the relationship of the shape to the direction of the prevailing wind. Some smoke flow studies of low buildings, however, make the rule of thumb appear reasonably valid. It is interesting to think about this because one of the commonly made mistakes in the design of science buildings is the attempt to hide fans, stacks and other equipment on roofs of science buildings by means of parapets, wells, and other types of rooftop construction in the hope of giving the building a neater, more orderly appearance. Such obstructions to airflow simply aggravate the situation by making it necessary to raise the height of the stacks still further. If this is not done, fume dispersal from the rooftop area will be seriously compromised. It can lead to potentially hazardous conditions to all persons who must work on the roof while fume hoods are in operation since the parapets and wells tend to retain fumes to even a greater extent than if they had not been constructed. Solutions to this problem can be found readily once it is recognized that stacks are a necessary part of a science building ventilation system and the aesthetic design must incorporate some provision for accommodating them. A number of buildings have already been designed in which a series of stacks are visible, undisguised, but so treated and organized that their appearance is entirely acceptable as a component of the form. This approach makes much more sense than trying to cover them up with false roofs or parapets which defeat their own purpose by making it necessary to raise the stacks still higher.

One of the most important kinds of information when planning a science building with fume hoods is the pattern of wind direction distribution at the site. Most people have general notions of the prevailing wind direction, however, such general notions, usually expressed in such terms as — "Our prevailing winds are from the west, except in the winter when we have strong winds from the southeast."—are generally useless. What is needed is an indication of the prevailing percentage distribution of occurrences of wind from the various points of the compass. Even more important than knowing the direction the wind is coming from most often, is knowledge of the direction the wind comes from least often. It is the prevailing leeward zone of a science building that is the critical area with respect to the location of air intakes. By examination of a number of wind rose patterns for different sites, it can be seen that the prevailing leeward direction is rarely 180° opposed to the prevailing windward direction, which is the popular conception.

The quality of wind direction distribution information available at the present time is a great weakness in rational design of fume dispersal systems and location of building air intakes. Unless a proposed new science building is part of a college or university that is fortunate in having a Meteorology Department or some other academic unit or research group collecting wind distribution data for a period of years, it is necessary to fall back on the data available from the nearest Weather Bureau Station. This is far better than no information, but cannot be responsive to the variations introduced by local geographic circumstances. An alert university, industry or other user of fume hood systems, requiring construction and renovation of a substantial amount of facility space over the years, should make an effort to collect wind distribution data using recording equipment located in the areas where new construction is anticipated. Such information could become very valuable base data in science facility planning.

The problems associated with fume hood discharge, and airflow and gas diffusion around buildings require extensive research. However, considerable information has already been accumulated on these subjects and is available in the literature.

ADDITIONAL INFORMATION

This chapter has attempted to briefly outline the principles for satisfactory use of fume hoods for science laboratories. We have attempted to describe all of the elements of the system without becoming involved with details. Therefore, the information in this chapter should not be considered to be sufficient for design criteria or the preparation of specifications. A collection of literature on fume hoods and related subjects is available for reference in the offices of the Architectural Services Staff of the National Science Foundation in Washington. Architects who are planning science buildings are invited to visit and study this material. An extensive bibliography is available upon request.

ACKNOWLEDGEMENT: Reprinted from the *American Institute of Architects Journal, Vol. 26* 1965, with permission.

Air Conditioning for Laboratories

American Society of Heating,
Refrigerating and Air-Conditioning Engineers, Inc.

The information in this chapter is reprinted with permission from Chapter 10 of the ASHRAE GUIDE AND DATA BOOK 1968, the APPLICATIONS Volume, published for the profession by the American Society of Heating, Refrigerating and Air-Conditioning Engineers, Inc. The purpose of presenting this material is to further the understanding and cooperation necessary for building laboratory facilities that will meet the needs of teaching and research.

DESIGN CONDITIONS

Exhaust requirements for laboratory and testing facilities for the conveyance of contaminants to the atmosphere require the treatment of large quantities of outdoor air. Selection of outdoor design conditions, therefore, materially affects the size and cost of refrigeration and steam facilities. One degree Fahrenheit difference in outdoor design wet-bulb temperature involves approximately 0.375 tons of cooling capacity per 1000 cfm of outdoor air. Outdoor design data for each locality, if not available in standard design references, may be obtained from local weather bureau or airport weather records. For comfort air conditioning, apply standard comfort criteria to frequency of outdoor temperature occurrence. For spaces with specified maximum or maintained indoor states, maximum values of outdoor conditions should be applied.

Indoor conditions should be defined in terms of dry-bulb temperature, wet-bulb temperature, or relative humidity with specified tolerances for each value. Information should be obtained on whether a stated set of conditions represents limiting values or levels to be maintained. For variable temperature rooms, corresponding humidity requirements for the specified dry-bulb range should be obtained or an indication whether humidity levels are to be selectively available for any value within the specified dry-bulb range.

HOURS OF OPERATION

The hours and seasonal periods during which specific indoor states are required to be maintained should be specifically expressed since they determine whether the spaces should be served by individual ventilation, air-conditioning and refrigeration systems operable at any time of the day or year, or by central plants operating on common daily and seasonal schedules.

A system designed to deliver air at a dewpoint of 13°C (55°F) or less in order to continuously maintain and limit the dry-bulb temperature of an interior laboratory space requires mechanical refrigeration which operates whenever the outside dry bulb exceeds 13°C (55°F). It is erroneous to assume that, because a system serving such space operates on 100 percent outdoor air, it will always be capable of fulfilling conditions during the winter season without mechanical refrigeration. A dry-bulb temperature of 13°C (55°F) is frequently exceeded during the winter months.

THERMAL LOADS

Transmission, solar, lighting, and occupant room sensible and latent load components are supplemented in laboratories by equipment heat and vapor release. Such equipment consists of electric, steam or combustion heating devices, motors, electronic equipment and special apparatus. The evaluation of equipment heat gains to establish design loads requires equipment nameplate ratings, applicable load and use factors and the determination of overall diversity factors. Load factors relate actual equipment heat dissipation to nameplate rating. For motors, it is the ratio of actual power input to the motor rating. For heating devices such as ovens, furnaces and burners, it represents the ratio of energy required to maintain a mass at a temperature to the energy required to elevate it to that temperature. Thus a Bunsen burner may operate at partial capacity once a solution reaches its boiling point. Values of 50 to 75 percent of rated capacity are applicable to Bunsen burners. Values for ovens and furnaces may be obtained from manufacturers' data or computed from surface area and temperature data. Load factors for electronic equipment are generally 100 percent. Those for special apparatus should be obtained from manufacturers. Heat release from equipment located in fume hoods may be fully discounted, and equipment under canopy hoods or ventilated by special exhaust devices discounted 50 percent.

Equipment which is directly vented or water cooled should have appropriate reductions made in the heat released to the room.

Use factors relate to the portion of an hour during which equipment may be in use to the time it is actually operative. Many pieces of laboratory apparatus, such as vacuum pumps, are not in continuous use, and when operative, cycle frequently. The diversity factor is the percentage of all heat releasing equipment in a laboratory in simultaneous use. Use factors and diversity factors are best obtained from technical personnel engaged in the particular laboratory work.

The procedure for determining the simultaneous probable equipment heat release is to: (1) obtain nameplate ratings of all apparatus, (2) apply load and use factors to each, and (3) apply overall diversity factors.

Table 1 gives some representative overall net heat gains for bench-type laboratories.

TABLE 1

Check Figures for Heat Gain from Bench-Type Laboratory Equipment[a]

Equipment Loading	Typical Occupancy Examples	Net Internal Heat Gain from Laboratory Equipment, Total Btuh/Sq Ft[b]		Sensible Heat Ratio
		Base Maximum[c]	Peak Maximum[c]	
General Use	Chemistry, Biology, Physics	15	30	0.8–0.95
High	Physics, Physical Chemistry	30	60	0.8–0.95
Very High	Electronics	70	140	0.95–1.0

[a] Based on heat release for bench-type laboratory equipment in a 250 sq ft module. Large individual items should be evaluated separately.
[b] Includes net gain to space after deductions for diversity and hooding.
[c] Base maximum rates are design values for majority of spaces. Peak maximum rates are design values for few laboratory spaces with especially high concentrations of equipment and are usually handled by supplementary conditioning.

Because of equipment thermal gains, laboratory cooling loads are characteristically highly variable. Equipment loads often range up to 75 percent of total heat gains. For this reason, individual space control is required and if equipment is not evenly distributed throughout a space, further zone subdivision within a given space may be required to minimize excessive temperature gradients.

AIRFLOW RATES

Supply air quantities are established by room cooling requirements and load characteristics. Supply rates thus derived may be designated primary air rates. Supplemental supply required to make up deficiencies between room exhaust requirements and primary supply may be designated: (1) infiltrated supply if induced indirectly from other spaces, or (2) secondary supply if conducted to the room directly.

Exhaust air rates are established by requirements for the removal of heat, odor and airborne contaminants. Contaminants generated or present within research facilities are fumes, gases, pathogens, vapors, and radioactive particulates or ions. Many devices are employed to control the spread of airborne contaminants in laboratories, and all require controlled flow of large air quantities to entrain the pollutants and ultimately discharge them to the atmosphere. The most commonly employed general purpose enclosure is the fume hood. Air requirements for fume hoods range from 600 to 2000 cfm. Other special enclosures for the containment of highly toxic contaminants are also available. One variety is a small volume completely enclosed unit called a glove or dry box. Air requirements for these units are low and range from 5 to 10 cfm. Another is the massively shielded, specially constructed enclosure employed for high level radioactive work, called the hot cell or cave, wherein handling of substances is done with externally operated mechanical manipulators. Design rates for these units are approximately 1 air change per minute.

Generally, the entrapment of contaminants with air entails substantial exhaust volumes. These are consequently critical in establishing supply air rates, air flow patterns, and air systems design. Explicit and detailed qualitative and quantitative data pertaining to exhaust requirements is, therefore, essential for the design of air-conditioning and ventilating systems for laboratory and test facilities.

AIR BALANCE AND FLOW PATTERN

Control of the direction of airflow in many laboratory buildings is a necessary consideration for preventing the spread of airborne contaminants and protecting personnel from exposure to toxic and hazardous substances. In facilities involving work of such nature, the once-through principle of airflow is applicable. This method is based on the assumptions that: (1) 100 percent outdoor air is supplied and exhausted with no recirculation, (2) constant volume airflow is maintained with all exhaust facilities simultaneously operating at full capacity, and (3) any cross transfer of air between spaces is from areas of least contamination to those of highest contamination. In facilities where contamination isolation is not critical, such as electronics laboratories, air can be recirculated and control of interspace air patterns is not essential. Determinants for air pattern control are: (1) the type of contaminants handled or generated in each space, (2) the type, size and number of ventilated enclosures and auxiliary exhaust facilities in each space, and (3) permissibility of air transfer into or out of spaces.

Flow and air balance patterns applicable to various conditions of air transfer are:
1. No transfer to or from a space is permitted. The exhaust rate must be matched by the primary air rate or by a combination of primary and secondary air.
2. Only transfer into a space is permitted, with no outflow.
   a. Exhaust rate must equal the primary air rate.
   b. If the exhaust exceeds the primary air rate, then air may be infiltrated from adjoining spaces to make up the deficiency, providing the air is suitable in quality and psychrometric condition.
3. Only transfer out of a space is permitted with no inflow. The primary air rate may exceed the exhaust rate, and the excess supply air may be exfiltrated

providing its quality and condition permit. Otherwise, the room must be kept in balance.

For highly critical air balance conditions, an air lock provides a positive means of control. An air lock is an anteroom between a controlled and uncontrolled space with airtight doors electrically interlocked to prevent simultaneous operation. The air pattern in the air lock is designed to suit any of the foregoing laboratory space air balance requirements.

CONTAMINANT CONTROL

Contaminant control involves quantitative and qualitative control of air contaminants within prescribed limits. Contaminants are gases, vapors and particulates. Processes for contaminant removal are described in Chapter 11 of the 1967 HANDBOOK OF FUNDAMENTALS and Chapter 10 of the 1969 ASHRAE GUIDE AND DATA BOOK.

FUME HOODS

A laboratory fume hood is a ventilated enclosed work space consisting of side, back and top enclosure panels, a work surface or deck, a work opening called the face, and an exhaust plenum equipped with horizontal adjustable slots for the regulation of air flow distribution. The work opening may be unrestricted or may be equipped with operable glass doors for observation and shielding purposes. Doors may be: (1) vertically operable, (2) horizontally operable, or (3) vertically and horizontally operable. The combination door, which employs horizontally sliding sash within a vertical uplift frame, provides maximum access for setting up apparatus, reduced operating opening for the conservation of air, and horizontally moveable protective shielding. Hoods are equipped with a variety of accessories including filters, internal lights, service outlets, sinks, air bypass openings and airfoil entry devices. The location of the exhaust port distinguishes hoods either as updraft hoods, if the connection is at the top of the exhaust plenum, or as downdraft hoods, if located at the bottom. The up-draft design is most generally employed. The down-draft hood is applicable to work involving condensable contaminants, exhaust ductwork requiring wash-down facilities, or for radioactive work wherein the exhaust air must be discharged through high stacks remote from the building.

Supplementary air hoods are hoods designed for direct supply air connections to supplement and thereby reduce induced air volumes. They are designed to introduce supplementary air either directly into the hood or as a face air curtain.

Hood Air Requirements

Air requirements for hoods are a function of (1) operating face area and (2) design face velocity.

Operating Face Area. The operating face area of a hood is the opening area employed under operating conditions. It may be equal to or less than the maximum available face opening. For a hood equipped with a single uplift door or a pair of horizontally sliding doors, the operating face is equal to the maximum operable opening. For a combination horizontal-vertical door arrangement, the operating face can be considerably less than the maximum face opening, depending on the number of sliding panels and the track arrangement. A door with three sliding panels on separate tracks provides an operating area which is two-thirds the face area, whereas a three panel door with two tracks provides an operating area equal to one-third the face area. Definition of a hood's operating face area is requisite for the determination of its air requirements.

Design Face Velocity. A prime factor in determining the effectiveness of a fume hood in capturing and removing material emitted within it is the hood face velocity. The average face velocity is the total air passing across the face divided by the face area.

In operation, hood face velocities maintained at specific points on the hood face are often less than design because of velocity gradients over the hood face, depending on the hood's air distribution characteristics, and on maintenance and wear factors such as dirt accumulation in the ducts, fan belt slippage, and fan wheel corrosion and deterioration. Minimum face velocity is the minimum acceptable velocity at any point on the operating opening; for a well-designed, well-maintained hood this should not be less than 80 percent of the average design face velocity.

Table 2 gives design values for average and minimum face velocities as a function of the characteristics of the most dangerous material that the hood is expected to handle. Minimum values are prescribed in many states under labor codes. Maximum design face velocity is the maximum acceptable velocity at any point of the operating opening for any intermediate position of the hood door. Maximum velocities of 100 to 150 fpm will prevent disturbance to screened flames and most test materials.[1] Velocities up to 300 fpm are used for evaporation purposes.

TABLE 2
Laboratory Hood Design Face Velocities[2]

Nature of Materials Handled	Threshold Limits			Design Face Velocities, fpm	
	Gases Vapors	Dusts, Fumes, Mists	Mineral Dusts	Average[a]	Minimum[b]
Highly Toxic	Less than 0.1 ppm	Less than 0.1 mg/cu meter	—	150	125
Moderately Toxic General Laboratory Use	0.1 to 100 ppm	0.1 to 15 mg/cu meter	To 5 mppcf	100	80
Non-Toxic	Above 100 ppm	Above 15 mg/cu meter	Above 5 mppcf	60	50

[a]Total hood cfm divided by total face area.
[b]Lowest maintained velocity at any point across the hood face.

Hood Performance

Performance criteria for fume hoods are flow control, outfall and face velocity control.

Flow Control. Regulation of flow over the face opening of a hood is obtained by adjustment of the horizontal slots on the face of the hood plenum. One is provided at the bottom of the plenum to sweep the working surface. Another is located at the top to exhaust the canopy, and a third is frequently located midway on the plenum. These adjustable openings permit regulation of exhaust distribution for specific operations. Adjustment may be made for the collection of heavy fumes at the deck surface, for the capture of light fumes and hot gases at the canopy, or for uniform flow best suited to general purpose operations.

Outfall. Excluding external disturbances, outfall of fumes from hoods can be caused by: (1) eddy currents generated at hood opening edges, surface projections and depressions, and (2) thermal heads.

Eddy Currents. Corner and intermediate posts, deep deck lip depressions, sinks, and projecting serving fittings near the face produce air turbulence and potential outfall conditions. Plain entrance edges produce a vena contracta within 1 inch of the surface and up to a depth of 6 inches. Fumes generated in this area will be disturbed and possibly escape the hood enclosure. Airfoil shapes at the entry edges correct this condition and eliminate outfall possibilities. Sinks and service fittings should be located at least 6 inches beyond the hood face and deck lips should have minimal projections.

Air currents external to a hood can influence the air pattern of hoods and produce fume outfall.

Cross currents are generated by body movements, thermal convection, supply air movement and rapid operation of room doors and windows. Air motion at the hood face in excess of the hood face velocity disturbs face airflow and causes outfall. Terminal supply air velocity in the vicinity of hoods should be limited to 35 fpm. The location of hoods near heating elements or doors and windows which are frequently operated should be avoided.

Thermal Head. Thermal head is the pressure difference between the interior of the hood and the room due to the difference in density of gases at elevated temperatures in the hood and of air at room temperature outside the hood. If the thermal head exceeds the suction head at hood openings, outward leakage will occur. The suction pressure at a hood face operating at 100 fpm is 0.000624 in. of water. Thermal heads generated by two columns of air ranging from 6 to 12 in. in height and temperature differentials of $-1°C$ to $38°C$ (30 to 100°F) [room temperature 21°C (70°F) and hood temperature 38°C to 77°C (100 to 170°F)] range between 0.0004 to 0.0023 in. of water. The critical condition for development of thermal heads is when the hood door is closed coincident with high heat release. Hot gases accumulate near the top of the hood and leak through panel, lighting fixture and door frame joints. Effective control measures are: (1) bypass provisions arranged for continuous flow through the hood, (2) high face velocities (80 to 100 fpm) for optimum thermal dilution, (3) airtight construction of the canopy, gasketed lighting fixtures and weatherstrip door seals, (4) adjustment of plenum slots for maximum exhaust at the head of the hood under high thermal operations, and (5) temperature or pressure sensing alarms for safety indication.

Face Velocity Control. Variations in the resistance of a hood exhaust system reflect directly in face velocity variations. Two common causes are: (1) varying the face opening, and (2) filter resistance build-up.

Doors on hoods do not contribute to hood performance but are often required for shielding purposes. In order to control face velocities between prescribed design limits on hoods equipped with doors, a proportional air bypass device is required. Two types applicable to hoods with single uplift doors are: (1) a by-pass opening located in a front panel directly above the hood door, positioned so that the door functions as a damper as it is raised and lowered, and (2) a bypass damper opening to the hood canopy above the door which is mechanically linked to the door for proportional air bypass. Bypass devices applicable to both vertically or horizontally operable doors are: (1) a barometric damper responsive to the pressure variations produced in the hood as its face opening is varied, and (2) a by-pass damper operated by an air velocity sensitive controller.

Increases in fume hood exhaust system pressures due to filter loading can range from 50 to 100 percent of the clean filter condition when high efficiency filters are employed. Constant pressure regulation may be attained by an automatic pressure controlled damper in the duct system or by a similarly controlled bypass damper which admits air into the exhaust system when the filter is clean, and gradually closes as filter resistance builds up.

Performance Tests. Performance tests for hoods require: (1) measured exhaust air rate by a calibrated orifice, (2) a traverse of face velocity readings in the plane of the face opening for maximum, intermediate and minimum openings, and (3) heavy and light smoke tests under varying door positions.

EXHAUST SYSTEMS

Laboratory exhaust systems can be classified on the basis of hood characteristics and the method of system operation and control as: (1) constant volume, and (2) variable volume systems. These classifications can be further defined on the basis of the arrangement of the major system components such as the fans, plenums or duct mains, and branches as: (a) individual, (b) central, or (c) combination systems.

Constant Volume Systems. A constant volume system exhausts a fixed air quantity from each hood. Hoods served by such systems and equipped with doors must have individual bypasses for air volume and face velocity regulation. Since a system of this type will function with the same air volume being handled for any given set of conditions, the total number of exhaust hoods which can be installed throughout the facility must be limited by the capacity of the exhaust system and the ability of the supply system to provide make-up air. Constant volume systems are simple to balance and highly stable, and in most installations there is no need for continuous control of air balance during normal operation.

The constant volume system is flexible with respect to the number and location of hoods, but may incur high owning and operating costs because of the large air volumes handled. These costs may impose a practical limitation on the total number of hoods that can be installed in the building.

Variable Volume Systems. Since laboratory air is usually not recirculated, the cost of circulating, heating and cooling may be high if based upon the maximum air demand. In most laboratories, the installed hood capacity is seldom actively used at any one time, and a system which permits the application of a usage diversity factor to the installed hood capacity can reduce the size of the exhaust system required. Operating economies can be achieved by reducing the airflow during periods when the hoods are not in use or when they may be operated at less than full capacity. A reduction in exhaust air volume, coupled with constant hood face velocity when the hood face opening is partially closed, may be achieved by a control arrangement for the velocity-controlled hood. The sensing element responds to changes in hood face velocity and operates a motorized damper in the exhaust duct to maintain the face velocity within the desired range. When the hood is served by an individual exhaust fan, the operation of the damper will usually be sufficient to reduce the fan capacity, but it may be supplemented by a static pressure regulator in the duct operating inlet vanes, or a discharge damper at the fan. In large central systems, a modulating damper affords a satisfactory means for branch ducts, but it must be supplemented by static pressure regulators in the exhaust plenums, operating fan inlet vanes or discharge dampers, for system volume regulation.

Variable volume systems can provide initial and operating economies by limiting the normal operating air handled to less than the total required to exhaust all of the hoods and by permitting the shutting down of inactive hoods during off shifts. More freedom in the installation of hoods is possible, since the total number of hoods that may be provided is not directly dependent on the capacity of the exhaust system.

Variable volume systems, however, are difficult to balance, less stable in operation and more difficult to control than constant volume systems. Extensive, sensitive instrumentation and controls are required, and result in high initial and maintenance costs.

A particular operating problem is the regulation of the total simultaneous operating hood openings to match design hood usage factors. If the collective area of operating hood openings at any one time exceeds design opening diversity values, face velocity requirements will not be achieved. If, on the other hand, total hood openings are less than design values, bypass devices are required on hoods to maintain supply air rates, provide adequate thermal capacity, and assure air balance and flow patterns.

Individual Systems. Individual exhaust systems utilize a separate exhaust connection, exhaust fan, and discharge duct for each fume hood. This arrangement is extremely flexible, since the exhaust for the hood or space served by the fan or exhaust duct does not directly affect the operation of any other area of the building. The individual system permits selective operation of individual hoods merely by starting or stopping the fan motor. Shutdowns for repair or maintenance are localized. The unitary arrangement permits selective application of: (1) special exhaust air filtration, (2) special duct and fan construction for corrosive fumes, (3) emergency power connections to fan motors, and (4) off-hour operation. Individual exhaust fans are simple to balance and, when coupled with constant air volume, provide a stable and easily controlled system.

Although more fans are used than for central systems, the overall space requirements are usually less for individual systems because of the small, direct duct connections.

Where high discharge stacks are required because of extreme, generalized contamination of the effluent, individual systems may not be applicable.

Central Systems. Central exhaust systems consist of one or more fans, a common suction plenum, and branch connections to multiple exhaust terminals. Central systems are generally less costly than unitary systems in capital and maintenance cost, permit low cost stand-by exhaust fan provisions, and are applicable to remote high stack discharge requirements. They are more difficult to balance and present the difficulties of parallel fan operation when more than one fan is employed.

Filtration

Filtration facilities for exhaust systems are employed for the removal of hazardous and obnoxious air pollutants. A general application is the removal of radioactive particulates with disposable, very high efficiency, dry media absolute filters. The filter assembly for this purpose includes a pre-filter for coarse particle separation and a filter enclosure arranged for ready access and easy transfer of the contaminated filter to a disposal enclosure. For convenience of handling, replacement and disposal with minimum hazard to personnel the filter should be located outside the working area and on the suction side of the exhauster. A constant volume controller is required to maintain design exhaust rates as filter resistance increases with particulate accumulation.

Exhaust Fans

Fans handling contaminants should be located outside occupied areas of a building and close to the point of discharge to the atmosphere to avoid spread of contaminants through leakage by maintaining fume-collecting ducts outside of equipment areas under negative pressure. Collecting duct branches may be of conventional joint construction if appropriate allowance for leakage is added to the fan rating. Discharge ductwork should be of airtight construction.

SUPPLY AIR INTAKE LOCATIONS

In the arrangement of the supply air equipment, it is essential that the locations of the fresh air intakes be carefully considered, to avoid the possibility of recirculation of contaminated exhaust air.

Because concentration patterns are strongly influenced by wind direction, building shape and the location of the effluent source, exact air patterns are not predictable. Some recirculation will occur with flush exhaust vents and intakes within the cavity boundary, although depending upon the contaminants handled, sufficient dilution may take place to permit the use of such an arrangement. A rough but conservative approximation of the dilution effect that might be expected for exhaust vents flush with the roof, based on wind tunnel tests, can be made by the following equation.[4]

$$D = [4.66 + 0.147(L^2/A_e)^{1/2}]^2[V/V_e]$$

where

- D = dilution, defined as the ratio of gas concentration at the exhaust opening to the gas concentration at any point near the building, dimensionless.
- L = shortest air distance between exhaust opening and any point, feet.
- A_e = area of exhaust opening, square feet.
- V = wind speed, feet per minute.
- V_e = exhaust speed, feet per minute.

Less recirculation and more dilution can be expected as the exhaust stack is increased in height.[3] The results should be compared with the expected concentration of the exhaust contaminants at the point of discharge and the maximum concentration of these contaminants that can be tolerated in the supply air system.

Discharge of effluent through exhaust stacks which extend over the area of influence of the structure will assure additional dilution.

REFERENCES

1. Schulte, H. F. et al, Evaluation of Laboratory Fume Hoods, *Am. Ind. Hyg. Assoc. Quart.*, *15*:3, (September 1954).
2. Breif, R. S., Church, F. W., and Hendricks, N. V., Selection of Laboratory Hoods I, II, III, *Air Engineering* (September, October, November 1963).
3. Halitsky, James, Estimation of Stack Height Required To Limit Contamination of Building Air Intakes, *American Industrial Hygiene Conference,* (April 29, 1964).
4. Halitsky, James, Vent To Intake Short Circuit, *Air Conditioning, Heating and Ventilating,* p. 81 (July 1960).
5. Ruddy, J. M., Controlling Airborne Radioactivity During "Hot" Clean-up. *Air Conditioning, Heating, and Ventilating, 67* (February 1960).
6. Hale, R. J., Laboratory Ventilation. Paper presented at the Proceedings of the Sixth Hot Laboratory and Equipment Conference at the Nuclear Congress (March 19, 1958).
7. Ruddy, J. M., Designing Fume Hoods for Medium Level Radioactive Conditions, *Heating, Piping and Air Conditioning, 128* (March, 1958).
8. Harris, W. B., Laboratories for Handling Radioactive Materials, *Heating and Ventilating, 75* (November 1953).
9. York, J. E., Ventilation and Air-Conditioning for Laboratories, *Heating and Ventilating, 81,* (November 1953).
10. Ruddy, J. M., Air-Conditioning Rodent Quarters, *Heating and Ventilating, 95* (March 1952).
11. Turner, J. F., New Laboratory Fume Hoods Cut Air-Conditioning Load, *Heating, Piping and Air Conditioning, 113* (January 1951).
12. Building Research Advisory Board, Laboratory Design for Handling Radioactive Materials. *Conference Report No. 3* (November 27 and 28, 1951).
13. Clay, H. B., Controlling Fume Hood Exhaust in Atomic Energy Laboratories, *Heating, Piping and Air Conditioning, 77* (July 1960).
14. Peterson, J. E., and Peay, J. A., Laboratory Fume Hoods, *Air Conditioning, Heating and Ventilating* (May 1963).
15. Halitsky, James, Gas Diffusion Near Buildings, *ASHRAE Transactions, 69*:464 (1963).
16. Clarke, J. H., The Design of Exhaust Systems and Discharge Stacks, *Heating, Piping and Air Conditioning* (May 1963).
17. Economides, L., Design Criteria for Laboratory Air-Conditioning Systems, *ASHRAE Journal, 48* (September 1964).
18. Twietmeyer, H. E., Air-Conditioning of Research and Test Facilities, Load Characteristics, *ASHRAE Journal, 50* (September 1964).
19. Peterson, J., Air-Conditioning of Research and Test Facilities; Contamination Control, *ASHRAE Journal, 53* (September 1964).

Recirculation of Laboratory Atmospheres

Norman V. Steere

If conditioned air from offices, classrooms, and corridors of laboratory buildings, and other air generally uncontaminated by laboratory operations is recirculated, two special precautions should be observed to stop recirculation in case of emergency.

The first precaution consists of automatic equipment, fusible-link fire dampers on exhaust louvers and smoke detection equipment which can shut down the entire recirculation system for all kinds of buildings to prevent spread of smoke and fire.

The second precaution, recommended in the Laboratory Design Standards adopted by the Campus Safety Association, consists of a readily accessible control for laboratory fresh air supply or recirculation systems so that laboratory personnel can immediately stop the system in case of an emergency such as a spill or release of toxic, radioactive, or flammable materials. The Laboratory Safety Committee of the Association has urged installation of emergency shut-off controls on recirculating and fresh air supply systems because of incidents in which toxic or irritating materials were or could have been recirculated because ventilation controls were not accessible without special keys and special knowledge. It is suggested that the emergency control for the ventilation system be similar to a fire alarm pull station, properly labeled, and connected to the evacuation alarm system.

LABORATORY HOOD EXHAUST

Recirculation of the exhaust from laboratory hoods is not likely to be permissible or practical, except in special cases where the hoods have a restricted use and top-notch maintenance. Although the design of recirculating laboratory hoods is still in the experimental stage, there may be special situations where savings may be realized from recirculation of conditioned air *if*: 1. The products and by-products exhausted from the hood are known and *will always be known;* 2. the filters and other air-cleaning equipment are specific for the known contaminants; 3. the performance of air-cleaning equipment is automatically monitored; and 4. cleaners and monitors are systematically and skillfully maintained.

In the Industrial Ventilation Manual,[3] published by the American Conference of Governmental Industrial Hygienists, the question of acceptance of recirculating systems is related to the degree of health hazard of the contaminants being exhausted.

"It is the general policy of all official industrial health agencies not to recommend the recirculation of exhaust air if the contaminant is a material which may have a definite effect on the health of the worker. The reasons are as follows:

"1. Many types of air-cleaners do not collect toxic contaminants efficiently enough to remove the health hazard.

"2. Poor maintenance of the air-cleaner would result in the deliberate return of highly contaminated air to the breathing zone of the workers. Not being production equipment, air cleaners are too often poorly maintained.

"3. Improper operation of the air-cleaner, through mechanical failure or through ignorance or neglect on the part of the operators, would also result in the return of highly contaminated air."

MONITORING EXHAUST SYSTEMS

Monitoring operational performance would be absolutely essential if laboratory hood exhaust were going to be recirculated.

The first type of monitoring system should show that the exhaust fan motor has been turned on and that the fan is turning.

The second type is a manometer which shows the pressure drop across the filters or in the exhaust duct, and which can be used to judge the need for replacement of the filters or the fan.

The third type of monitoring system would be an automatic one which would signal when the level of contaminant passing through the cleaning system rises above a preselected value. Although such a monitoring system would be expensive initially, it would be the only sure and easy way to know when the contaminant concentration in the recirculated air has come up to the point where it would be necessary to service the cleaning system, stop recirculating, or stop operations.

Automatic monitoring would be more accurate than gas detection tubes and faster than the analysis of submarine atmospheres[6] which was done by mass spectrometric, infrared spectrometric, and gas chromatographic techniques. Accuracy and speed would be critical if exhaust were recirculated from hoods using materials as toxic as nitrogen dioxide, tolylene-2, 4-diisocyanate or beryllium.

Unfortunately, automatic monitoring is in its infancy and is available for relatively few gases; carbon monoxide, sulfur dioxide, methane, and some of the acid gases are among those which can be monitored automatically. Complexity of maintenance of automatic monitoring equipment is another significant factor to consider carefully before deciding on recirculation of laboratory hood exhaust as an economy measure.

CONCLUSION

While conditioned air from some parts of laboratory buildings may with precautions be recirculated, the exhaust from laboratory hoods, enclosures, and spot ventilators should not be recirculated. Even if the exhaust from some specific chemical operation in a laboratory could have the contaminants removed to absolutely safe levels by a perfectly maintained air-cleaning system, there is another important reason why such exhaust should not be recirculated.

The continually accelerating pace of research and development activities and technological progress make it impossible to predict what chemicals and reactions will be used in laboratories in the future, or what demands will be placed on laboratory hoods. Since knowledge of biological effects and detection instrumentation always lags behind chemical developments, the exhaust from laboratory hoods and enclosures should not be recirculated.

ACKNOWLEDGEMENT: Reprinted from *Research/Development Magazine,* June, 1965, by permission of the publisher, F. D. Thompson Publications, Inc.

REFERENCES

1. McConnaughey, W. E., Atmosphere Control of Confined Spaces, 390–399, "Man's Dependence on the Earthly Atmosphere," Proceedings of the First International Symposium on Submarine and Space Medicine, The MacMillan Company, New York, (1962).
2. Arnest, Richard T., Atmosphere Control in Close Space Environment (Submarine) *Report No. 367,* U. S. Naval Medical Research Laboratory, Submarine Base, New London, Connecticut (December 14, 1961).
3. Barnebey, H. L., Activated Charcoal for Air Purification, *ASHAE Journal* Section, *Heating, Piping and Air Conditioning,* 153–160 (March, 1958).
4. Sleik, Henry, and Turk, Amos, "Air Conservation Engineering," 2nd Edition, Connor Engineering Corporation, Danbury, Connecticut, (1953).
5. Schulte, John H., *Archives of Environmental Health,* 8:3, 438–452 (March, 1964).
6. Barrett, James C., *Safety Maintenance, 119*:1, 28 (1960).

Fire Hazards

Fire-Protected Storage
for Records and Chemicals

Norman V. Steere

Research records and flammable chemicals are not adequately protected against fire damage in most laboratories and research facilities.

Unprotected chemicals can provide fuel to increase the size and extent of damage of a small fire. Research notes, thesis material, unique chemicals, and financial and academic records may be irreplaceable or more difficult to reproduce than equipment or the building.

There are several ways of providing fire-protected storage, ranging from inexpensive methods of gaining additional minutes of protection to built-in systems and construction for providing hours of fire protection. We believe the potential for fire loss in laboratories calls for (1) realistic evaluation of possible losses, (2) immediate steps to minimize losses and (3) planning for systematic long-range improvements. While various innovations and expedients may be useful for temporary or partial protection, tested and approved equipment and construction will be more economical in the long run.

FIRE-PROTECTED STORAGE FOR RECORDS

Fire in a university laboratory building will usually result in an influx of graduate students trying to retrieve their research notes to avoid having to repeat a great deal of work. To avoid loss of valuable research material and interference with fire fighting operations, research safety considerations should include fire-protected storage.

Ordinary desks and file cabinets provide almost no protection for records, since wood will burn rapidly and metal will transmit enough heat in a few minutes to ignite or seriously char papers and other records.

Improvement of Existing Record Storage

When funds and space are limited, ordinary desks and file cabinets can be improved to provide a measure of fire-protected storage by use of special fire retardant paints.

Intumescent paints foam when heated and form an insulating barrier against heat and flames. Several such paints have been tested and are listed as fire retardant coatings by Underwriters' Laboratories, Inc., 207 East Ohio St., Chicago, Illinois 60611 in their "Building Materials List" and "Bi-Monthly Supplement."

Providing adequate fire protection for records will require a detailed analysis and a comprehensive plan based on determination of the value of various records, hazards to which the records are exposed and the present protection afforded.

"Standard for the Protection of Records," National Fire Protection Association (NFPA) standard No. 232-1963, $1, is recommended for every research department and organization. The publication details the need for management and protection of records, gives standards for fire-resistive vaults and file rooms and describes standards for fire-resistive safes and other record protection equipment.

"Standard for the Protection of Records" also has an excellent section on methods of salvaging records endangered or damaged by fire.

179

Although some large laboratories may need fire-resistive record vaults and file rooms, the discussion which follows will be limited to the standards for record protection equipment.

Record Protection Equipment

Equipment should be selected which will protect records from fires of the most destructive intensity and duration which may occur. If stairwells are open, if there is no automatic fire detection or sprinkler system, and if the building is of combustible construction, the probable maximum condition is complete destruction of all combustible contents and portions of the building.

Record protection equipment in wood-joisted or other nonfire-resistive buildings should be rated to withstand the impact following floor collapse, as well as the fire exposure.

The method described in NFPA No. 232 for selecting record protection equipment for a fire-resistive building is based on estimating the combustible material to which the equipment may be exposed, assuming uniform distribution, and dividing by floor area to obtain weight of combustibles per unit area. Flammable liquid weights are multiplied by two for the purpose of approximating their fuel contribution equivalent to ordinary combustibles.

Table 1, reproduced with NFPA permission, assumes complete burn-out of various combustible loadings in different arrangements and shows the class of equipment needed to assure record protection. Since Class E devices are less expensive than Class

TABLE 1
Record Protection Equipment for a Fire-Resistive Building

Total Combustible Contents per Floor (Including any Combustible Flooring, Partitions, and Trim)		Noncombustible Desks, Filing Cabinets, Lockers, and Other Closed Containers. Not Over 30% of Combustibles Exposed	Noncombustible Open-Front Shelving and Other Open Containers	Combustible Desks, Filing Cabinets, Shelving, Containers, Etc.
Lb Per Sq Ft of Floor Area	Kg/m^2			
Less than 5	24	Class E device	Class E device	Class D device
5 to 10	24–49	Class D device	Class D device	Class C safe or container
10 to 15	49–73	Class D device	Class C safe or container	Class B safe, container or 2-hr vault
15 to 20	73–98	Class C safe	Class B safe, container or 2-hr vault	Class B safe, container or 2-hr vault
20 to 30	98–146	Class C safe or container	Class B safe, container or 2-hr vault	Class A safe or 4-hr vault
30 to 35	146–171	Class B safe, container or 2-hr vault	Class A safe or 4-hr vault	Class A safe or 4-hr vault
35 to 45	171–220	Class B safe, container or 2-hr vault	Class A safe or 4-hr vault	6-hr vault
45 to 50	220–244	Class A safe or 4-hr vault	6-hr vault	6-hr vault
50 to 60	244–293	Class A safe or 4-hr vault	6-hr vault	6-hr vault with no combustible near door

D, D less than C and so on, it can be seen from the table that reducing the quantity of combustible contents in a laboratory and the amount exposed will reduce the cost of record protection. Replacing or fire-retardant coating of combustible desks, filing cabinets, shelving, and containers, will also reduce the cost of adequate record protection.

The only way to be sure that record protection equipment will meet the need for which it is purchased is to see that it bears the label of Underwriters' Laboratories, Inc., or other nationally recognized testing laboratory. In addition to initial tests on equipment which it labels and lists, Underwriters' Laboratories, Inc., also conducts periodic examinations and tests on samples selected at random from current production and/or stock.

Class D insulated filing devices listed by Underwriters' Laboratories, Inc., are effective in withstanding a standardized fire of controlled extent and severity for at least one hour, reaching 927°C (1700°F) before an interior temperature of 177°C (350°F) is reached at the center of the container or its insulated compartments during the period of fire exposure or the subsequent cooling period in the furnace. Class D devices also have to withstand a sudden heating without producing an explosion sufficient to cause an opening to their interior in a test in which the sample is inserted in a furnace preheated to 1093°C (2000°F) and kept at that temperature for 30 min. In both exposures papers loosely distributed and in contact with the filing faces must come through in usable condition, capable of ordinary handling without breaking and decipherable by ordinary means. Contents are not considered usable if they require special preparation to permit handling or are decipherable only by special photography or chemical processes.

Manufacturers of Class D-1 hour insulated filing devices listed by Underwriters' Laboratories, Inc., in January, 1965, include: John D. Brush & Co., Rochester, N. Y.; Diebold, Inc., Canton 2, Ohio; Herring-Hall-Marvin Safe Co., Div. of Diebold, Inc., Hamilton, Ohio; Meilink Steel Safe Co., Toledo 6, Ohio; The Mosler Safe Co., Hamilton, Ohio; Murphy Mfg. Co., Louisville 2, Ky.; Safe-Cabinet Div. of Remington Rand, Division of Sperry Rand Corp., Marietta, Ohio; Schwab Safe Co., Lafayette, Ind.; Shaw-Walker Co., Muskegon, Mich.

Prices for Class D files may range from approximately $200 for a 2-drawer file to $300 for a 4 drawer file, while prices for similar Class C files may range from $232 to $345.

Protectall Safe Corp., Hamilton, Ohio, and Safe-Cabinet Div., Marietta, Ohio, manufactures a Class E-one half hour fire test to 843°C (1550°F) and a sudden heating to 1093°C (2000°F) for 20 minutes.

Class C insulated record containers and safes, manufactured by many of the same companies have passed the 1 hr 927°C (1700°F) fire test before the interior temperature has reached 177°C (350°F) at 2.5 cm (1 in.) from walls or doors, have withstood an impact due to falling 9 m (30 ft) in the clear after being heated for 30 minutes and reheating in the inverted position for 30 minutes after impact without destroying the usability of papers stored inside. Other companies which manufacture Class C safes are: Acme Visible Records, Crozet, Va.; Gary Safe Co., Los Angeles 21, Calif.; Protectall Safe Corp., Hamilton, Ohio; Victor Safe & Equipment, Remington Rand Dealer Sales, Div. of Sperry Rand, North Tonawanda, N. Y.

Class B insulated record containers, and safes are tested to withstand a 2-hr fire test to 1010°C (1850°F), the sudden heating to 1093°C (2000°F) for 30 minutes and the 9 m (30 ft) impact test with heating for 45 minutes and reheating for 1 hour.

Class A safes, made by several of the companies mentioned and by LeFebure Corp., Cedar Rapids, Iowa, are tested to withstand a 4-hr fire test to 1093°C (2000°F) as well as the sudden heating test, and the impact test with 1 hour each, heating and reheating.

FIRE-PROTECTED STORAGE FOR CHEMICALS

Although fire will damage most chemicals or their containers, the greatest need of fire-protected storage is for chemicals which are easily ignited, difficult to extinguish or burn with great rapidity. Organic peroxides, azo compounds, pyrophoric metals, and flammable liquids are some of the types of chemicals which need separate and fire-protected storage.

It seems important to recommend that chemical storage practices be reviewed from time to time and the hazards evaluated. Sodium and potassium deserve better protection than results when they are stored according to the alphabet rather than the hazard. Placing all acids together may seem to be the simplest method of storage, except that 70–72% perchloric acid should not be next to glacial acetic, and picric acid should be in a third place.

The first step toward fire-protection storage of hazardous chemicals is to reduce quantities in the laboratory to the minimum practicable, which reduces the fire load and the amount of storage needed. We do not believe it is practicable to limit quantities in the laboratory to only one day's supply—but there should be a real effort to reduce quantities to workable minimums.

While it does not seem at all safe or workable for every laboratory to have to get all its flammable liquids each day, as some standards propose, neither is it reasonable to accumulate many liters or gallons of all the common solvents on open shelves. Only 7 liters (2 gal) of one flammable solvent recently fueled a fire that caused an estimated $60,000 of damage, and the damage would have been greater if 40 or 50 liters had been involved.

The NFPA Flammable Liquid Code calls for flammable or combustible liquid containers not larger than one quart (946 cc), which can generally be a workable size even in graduate organic research laboratories. However, if a laboratory process requires 20 liters (five gallons) of solvent for a single run, the solvent should not have to be in 20 single liter-size bottles. Some extractions or pilot plant operations may require such large quantities of solvent that one whole day's supply will be too much to store in the laboratory.

High shelves are unsuitable for storing bottles of combustibles because of the hazard of breakage and the more rapid build-up of heat in a fire. Open shelves and cabinets with glass doors are not suited for storage of chemicals which are hazardous if broken open by fire-fighting hose streams.

Many flammable liquids, particularly technical grade, can be used in metal containers which will not rupture under fire conditions. Safety cans approved by Factory Mutual Laboratories have an FM label and a flash-arresting screen in can openings. Teflon gaskets should be specified in preference to leather. Manufacturers of flammable liquid safety cans in the U.S. include Eagle Mfg. Co., 24th & Charles Sts., Wellsburg, West Virginia; Justrite Mfg. Co., 2061 N. Southport Ave., Chicago, Illinois 60614; The Protectoseal Co., 1920 S. Western Ave., Chicago, Illinois 60608; Advertising Metal Display Co., Chicago, Illinois 60650; and Safeway Safety Products Corp., Rockton, Illinois 61072.

The Protectoseal Company makes a very broad range of safety cans suitable for laboratory use, including stainless steel safety cans suitable for reagent grade solvents, sizes from pint to five gallon, special shapes for shelf storage, cans with flexible pouring spouts, and flammable waste disposal containers.

Protectoseal also makes a bottle carrier lined with polyurethane foam which has been used for holding bottles of flammable solvents and found to provide about six minutes protection from a spill fire.

Improvement of Existing Laboratory Storage for Flammable Liquids

The insulating value of a wooden door prevented involvement of two gallon glass bottles of benzene in the recent $60,000 laboratory fire mentioned earlier. Existing benches and cabinets used for laboratory storage of flammable liquids and other hazardous chemicals can be improved by intumescent fire retardant coatings or by other means which provide effective fire insulation.

Transite or asbestos board does not provide fire insulation. Although such material may withstand heat to which it is exposed, it transmits the heat rapidly by conduction.

Cabinets for Fire-Protection Storage

Although one national code requires flammable liquids in laboratories to be in double walled metal storage cabinets, if not in safety cans or containers less than a quart, other national standards and authorities attribute very little protection to such a cabinet.

Review of the fire performance standards for records protection equipment, listed earlier, supports the view that temperatures within uninsulated metal cabinets would be likely to exceed 177°C (350°F) in just a few minutes.

Consideration should be given to use of Underwriters' Laboratories listed insulated filing devices and record containers, for fire-protected storage of heat sensitive chemicals and flammable liquids. The well-developed standards and the testing program for record protection equipment seem to assure that such equipment would provide a known measure of fire resistance.

Special standards need to be developed for chemical storage cabinets and that special tests have to be devised and evaluated so that chemical storage cabinets can be built to meet the needs for fire-protected storage.

The Los Angeles Fire Department conducted some interesting tests early in 1959 to compare heat transmission of sections of various metal and wood cabinets, and adopted a standard (LAFD Standard No. 40, 1-1-60) for plywood cabinets for hazardous materials storage.

A typical cross section of five kinds of metal cabinet walls was made as follows: (1) A double wall, metal structure of 18 gauge CR steel, approximately 7 in. × 10 in. with 1.5 in. air space. (2) A similar cross section was made with a core of 5/8 in. sheetrock suspended midway in the 1.5 in. air space. (3) A similar cross section was used with a core of 1/2 in. Douglas Fir plywood untreated suspended in the air space. (4) Metal-walled structure insulated with 1 in. of 1 lb density fiberglass blanket in the 1.5 in. air space. (5) Metal-walled structure insulated with 1.5 in. of mineral rock wool density unknown.

The wood sections were set up in a similar manner and consisted of the following: (1) Two layers 1 in. Douglas Fir plywood; (2) One layer 1 in. Douglas Fir plywood; (3) Laminate formed by 1/2 in. plywood each side of 1/2 in. sheetrock.

Each of these structures was fastened to the opening of a furnace preheated to 704°C–788°C (1300° to 1450°F). A thermocouple was attached to the opposite face of the cross section mockup. The temperature and other observations were recorded.

Test results (Table 2) seemed to indicate that further testing of flammable liquid storage cabinets would be in order. Extensive tests were conducted later that year on wooden cabinets, and the results were so satisfactory that the Department recommended plywood cabinets in preference to non-insulated metal cabinets.

TABLE 2

Results of Tests on Cabinet Walls

Min.	5		10		15		20	
Metal	°F	°C	°F	°C	°F	°C	°F	°C
#1	430	221	500	260	510	266	650	343
#2	150	66	180	82	210	99	240	116
#3	130	54	140	60	160	71	170	77
#4	310	154	430	221	470	243	. . .	. . .
#5	270	132	433	223	466	241	. . .	. . .
Wood								
#1	100	38	100	38	100	38	100	38
#2	120	49	133	56	166	74	200	93
#3	90	32	100	38	110	43	130	54

It should be noted that while no metal cabinet manufacturers submitted samples for the tests, the Department did at a later date test and approve one manufacturer's metal cabinet for use in the City of Los Angeles.

The following are the Los Angeles Fire Department's specifications for the construction of Hazardous Materials Storage Cabinets as contained in the LAFD Standard No. 40, 1-1-60.

The Type A cabinet shall be used for the storage of dangerous chemicals. Either the Type A or the Type B cabinet may be used for the storage of hazardous materials other than dangerous chemicals (i.e., flammable and combustible liquids).

Cabinets shall be constructed of plywood.

The Type A cabinet shall not be less than 2 in. thick. The Type B cabinet shall not be less than 1 in. thick. No material shall be used having a thickness less than 1 in.

All joints shall be rabbetted and shall be fastened in two directions with flathead wood screws.

All doors shall be self-closing and equipped with substantial hinges. When more than one door is used there shall be a rabbetted overlap of not less than 1 in. Doors shall be equipped with a substantial locking or latching device.

The shelves in Type A cabinets shall be of wood not less than 1 in. thick. Shelves for Type B cabinets may be of either wood or metal.

The volumetric capacity of cabinets shall not be more than 50 cu ft.

The contents of all cabinets shall be indicated by appropriate warning signs, with letters 3 in. high on contrasting background.

The Los Angeles Fire Department standard is based on the absence of any record of fires originating within storage cabinets, and on a design philosophy of providing enough thermal insulation barrier to protect cabinet contents until the area is untenable and persons are not present to be injured when and if the cabinet fails. Coatings of fire retardant paint on the exterior of storage cabinets would provide additional fire protection, as would incorporation of a package system for automatic extinguishment by carbon dioxide or dry chemical.

Special Rooms

While outdoor locations or detached buildings are preferred for storage or handling of drums and bulk quantities of flammable liquids, most laboratory buildings need to have a well-supplied stockroom within the building.

Some industrial laboratories have outdoor storage for drums of flammable liquids that will not freeze, and several universities have separate storage buildings, more than 15 m (50 ft) from main buildings, for drums and case lots.

Every laboratory building that stocks over 185 liters (50 gal) of flammable or combustible liquids should have a special cutoff room to provide fire-protected storage for chemicals. Cutoff rooms for flammable and combustible liquid storage should not be below grade level in basements or sub-basements because of the difficulty of providing adequate drainage, ventilation, and explosion-venting.[1]

Cutoff storage rooms should be located at an exterior wall and be separated from laboratories and corridors by fire resistive construction, including walls, floors and automatic-closing fire doors, of at least one-half hour fire resistance. If the average weight of flammable and combustible liquids exceeds 73 kg/m^2 (15 psf) of floor area, the minimum fire resistance of construction should be increased by $^3/_4$ hr for each increase of 24 kg/m^2 (5 psf) of flammable and combustible liquids.[2]

The NFPA Flammable and Combustible Liquids Code,[3] which may be the basis of legal regulation in some jurisdictions, specifies that inside storage rooms (or cutoff rooms) shall have a fire-resistive rating of not less than 2 hr with noncombustible walls, floors and ceilings. The code also calls for 15 cm (6 in.) liquid-tight sills or ramps at openings to other rooms, proper ventilation, electrical wiring and equipment approved for Class I Division 2 Hazardous Locations if the room is used for flammable liquids with flash points below 38°C (100°F) and, where practical, large vents to provide fire and explosion relief.

Since explosions of common flammable liquid vapors under optimum mixture conditions may produce pressures in excess of 7 kg/cm^2 (100 psi),[1] the maximum practical amount of venting area should be installed on rooms used for storage and handling of flammable liquids. Vents may be doors, windows, explosion-venting sash, roof or wall panels or other devices arranged to open at internal pressures greater than 98 kg/m^2 (20 psf).

Positive mechanical exhaust ventilation should be provided for cutoff rooms in which flammable liquids with flash points below 38°C (100°F) are stored or transferred. Ventilation should be at the rate of 283 cubic meters (10,000 cu ft) of air for every 3.7 liters (gallon) of liquid vaporized, except carbon disulfide which would require 566 cubic meters (20,000 cu ft).

Automatic sprinklers are the basic protection for all flammable liquid occupancies,[1] to extinguish or limit fire, and should be provided to back up fixed automatic extinguishing systems utilizing foam, carbon dioxide or dry chemicals. Grating-covered troughs the full width of all doorways into the building may be provided instead of ramped curbs, and trapped drains should be provided from the troughs and the room to conduct to a safe outdoor discharge point the maximum sprinkler discharge in the cutoff room.

Portable extinguishing equipment should also be provided in convenient locations accessible under fire conditions. Building hoses for use by trained people should be equipped with adjustable spray nozzles, and wheeled dry chemical extinguishers should be considered, particularly for facilities on upper floors or remote from fully equipped fire departments.

Special Extinguishing Systems

For storage cabinets or small cutoff rooms there are available package systems of carbon dioxide and dry chemical extinguishing systems. Regular cutoff rooms require special extinguishing systems engineered to meet size and storage requirements.

ACKNOWLEDGEMENT: Reprinted with permission from the *Journal of Chemical Education, Volume 41,* A859, November, 1964.

REFERENCES

1. Factory Mutual Engineering Division, "Handbook of Industrial Loss Prevention." p. 41–1, McGraw-Hill Book Co., New York, (1959).
2. Factory Mutual Engineering Division, "Handbook of Industrial Loss Prevention," p. 2–4, McGraw-Hill Book Co., New York, (1959).
3. National Fire Protection Association, Flammable and Combustible Liquids Code. *NFPA No. 30-1963*, National Fire Protection Association, Boston.

Flame Extinction and Combustion Suppression

Mathew M. Braidech

EXTENDED DAMAGE AND LIFE LOSS CONSIDERATIONS

Reduction of property loss caused by the associated effects of fire other than by burning, such as spoilage by smoke, soot, odor and water damage, corrosion due to end-product gases, vapors and distillates, or thermal distortion of structures or equipment, are loss factors that should not be overlooked. Such damage may be quite extensive, since heat, smoke and gases can extend beyond the area actually involved in fire and travel to distant and remote areas in buildings. This problem is of particular interest where the damage susceptibility is high, as in the case of costly electronic precision equipment, such as data-processing computer installations, or certain readily perishable foodstuffs and pharmaceutical materials. Human safety and personal injury to persons involved in fire emergencies are always vital and preeminent considerations. They call for prompt detection and immediate alarm measures, as well as adequate means of egress and exit facilities, with proper pre-fire planning for safe fire-control action.

FIRE FACTORS CONCERNING EXTINGUISHING METHODS

A comprehensive fire protection program calls for a pre-fire hazard evaluation of the inherent hazard of the materials which might be involved and the degree of risk of initiation or fire expectancy, as well as the probable fire-severity potential. This involves the consideration of the following factors: fire-load (heat-release rate), the fuel-air (draft) arrangement, and any other exposure and fire-dimension factors favoring the rapid growth and uncontrolled spread of fire.

Some fires will be readily controlled by quenching or cooling below the ignition temperature of the burning material, some by diluting, smothering (blanketing) or inerting the surrounding atmosphere, and some by coating or encrusting with surface deposits or fusing sealants. Other fires will be controlled through disengagement of flame by blowing out (as in oil-well fires with explosives, to permit capping) or by inhibition of flame reactions with chemical agents, and in certain instances by actual physical removal of the fuel involved through auxiliary "dump tanks" or diversion piping to distant storage containers. Mechanisms for fire control are the basis for the established techniques for extinction of flame and suppression of combustion with water and steam, foam, vaporizing liquid, inerting gas and dry-powder chemical agents, either through manual application or fixed automatic installations.

CLASSIFICATION OF FIRES

For fire prevention and protection purposes all fires are commonly grouped into four basic classifications according to the nature of combustible material. The classifications were originally developed by Underwriters' Laboratories, Inc. and adopted by the National Fire Protection Association. Fires in each classification can be of different types, such as slow burning, fast burning, or ultra-fast burning (supersonic) with explosive violence. The established classifications have provided a serviceable guide in the selection of suitable methods of extinguishment and have also served to promote familiarization with their use limits.

Class A Fires

Class A fires occur in ordinary combustible solid materials, such as lumber, coal, rubber, textiles, paper, and fibrous materials except some synthetic fibers. They produce glowing embers or incandescence, and at the same time are accompanied by destructive distillation, with production of volatiles and flame, and the formation of carbonaceous material such as charcoal. Wetting, cooling and quenching with water or solutions containing large percentages of water, including chemically loaded streams, is the method of extinguishment to be used. Chemical additives such as wetting agents (technically termed surfactants and detergents) have demonstrated advantages with the reduction of the water surface-tension, thereby enhancing extinguishment by moistening water-repellent surfaces of the fuel or causing water penetration into the burning mass to check deep-seated, burrowing fires, as well as extending the gallonage-spread of water (in some instances even promoting aqueous emulsification of heavy combustible liquids to form temporary protective water-emulsion foam).

Class B Fires

Class B fires involve heavier combustible petroleum products (diesel fuels and lubricants) or the more flammable liquid hydrocarbons (gasoline, benzene, alcohol, etc.), energetic chemicals employed in rocket propellants, and other combustible liquids, greases, and solvents, where it is essential to exclude or dilute the atmospheric oxygen. Fire control or extinguishment with this class of burning materials is accomplished by blanketing (air-excluding and sealing combustible volatiles), inerting (dilution), or inhibition (interposing chemicals for flame disruption). These effects were respectively accomplished by (1) use of foam (chemical or mechanical), dry chemical (bicarbonate base) or multipurpose (chloride-phosphate base) agents; (2) inert gas (generally CO_2); or (3) by use of vaporizing liquids consisting of volatile halogenated hydrocarbons (chlorine-, bromine- or fluorine-treated), carbon tetrachloride, chloro-bromomethane, and their mixtures. Methyl bromide, bromotrifluoromethane and bromochlorodifluoromethane are examples of liquefied-gas extinguishing agents, becoming gaseous on their discharge into the atmosphere; this latter group has limited but special use in aircraft. The exact chemistry of extinguishment action of halogenated hydrocarbons is not known. However, it is believed that the interposing of these agents causes a "chain-breaking" of the ionized free radicals in the flame-combustion zone, which results in the inhibition and retardation of burning—in addition to the smothering effect.

NOTE: The extinguishing principle of dry chemical agents is also not fully understood, since with smothering action there is heat absorption in the combustion zone and heat reflection causing radiation blockage and volatile vapor feedback. These agents may not be lasting in their extinguishing effects as re-ignition can occur when hot metal surfaces are present, and some may also be incompatible with ordinary foams.

Foams (low expansion) of two types are available; chemical (wet) foam which is formed chemically on mixing alkaline salt and acid salt preparations and mechanical (dry) foam which is produced physically by turbulent mixing of air with water containing a small amount of foam-forming compounds, either of protein hydrolysate-type or nonprotein-type concentrates. Each, in turn, is further classified according to its suitability for various flammable liquids as a regular or special type, with a wide range of consistency and expansion ratios up to about 10:1 and as high as 50:1. The mechanical foams produce a tougher and more durable continuing blanket, resisting disruption due to solvents, wind or draft, or heat or flame. The newly developed high-expansion foams employ synthetic surface-action agents, termed "surfactants" (sometimes erroneously referred to as "detergent" foams). They feature a high expansion ratio up to 1000:1, with a tremendous generation rate, to 23,600 liters/sec or 50,000 cfm, making fast extinguishment possible through mass-volume fire attack by filling an entire building area or structure in a few minutes. Originally applied to Class A fires, this foam technique is now finding use for Class B fires, and is also gaining acceptance as an extinguishing procedure around electronic equipment, and in areas handling radioactive materials (avoiding spread of contamination by dispersion). It also tends to reduce water damage because there is minimal use of expanded water.

Another new surfactant technique is based on perfluorocarbon chemistry and is capable of spreading a very thin film over gasoline or oil surface to cut-off flammable vapor. The "Light Water" and a potassium dry chemical, known as "purple K," are applied separately from two nozzles handled by a single operator in a technique developed by the Naval Research Laboratory. The technique provides a compatible combination of "fire-knockdown" and "flash-back prevention" with spectacular success in spill fires, particularly where rescue of personnel is involved. These new foams should not be confused with the less stable wetting-agent foams. Synthetic surfactants hold much promise in the improvement of the three-dimensional "volume flooding" approach in fire control technology.

Class C Fires

Class C fires involve energized electrical equipment (transformers, switches, and motors) where the shock hazard due to the conductivity of the extinguishing media must be recognized and guarded against. However, with the equipment de-energized, extinguishants suitable for Class A fires may be used, unless flammable liquids are involved; otherwise, agents suitable for Class B are to be employed. When combinations of Class A and Class B fires are encountered, water spray or multi-purpose dry chemical extinguishing agents are applicable. Where delicate electronic equipment may be involved, due regard should be given for the possible corrosive and insulating effects of the various agents which might render such equipment inoperative.

Class D Fires

Class D fires involve combustible and reactive metals (either as massive solid parts, turnings, chips, or powder) which include magnesium, zirconium, titanium, lithium, thorium, uranium, or metallic sodium and potassium, their various alloy combinations, including metal hydrides, metal alkyds, and other organo-metallic compounds. Class D fires require control with special dry powder media, which will not react or combine adversely with the burning material. These special agents are not to be confused with the dry chemical bicarbonate-base extinguishing powders. These dry powder agents are generally proprietary formulations, prepared for specific materials requiring special application techniques, and may contain graphite, pulverized coke, hard pitch, shredded asbestos, vermiculite, cast-iron borings, talc, or even sand, mixed with low melting alkali-metal salts and various combinations of fluxing salts or resinous material. Extinguishment is due primarily to the smothering action of the flux-melt or a resinous insulation, resulting in a fused encrustation, which adhere as sticky coatings to the surfaces of the burning material to exclude atmospheric oxygen. These special agents obviate the violent and potentially dangerous reactions between some of the metals or their organic compounds and the conventional extinguishants. These preparations are generally approved for fire control of specific combustible metals.

FUNDAMENTALS OF MAJOR FIRE EXTINGUISHING SYSTEMS

Fire control and extinguishing methods may be classified into two major groups: (1) manual or portable hand-use and mobile devices, and (2) fixed automatic protective installations. The portable devices contain a limited supply of the extinguishant and are therefore designed for immediate use on fires in their early or incipient stages; they are auxiliary to all other methods, and not to be considered as an alternate to fixed systems. Fire extinguishing methods may also be classified based on the extinguishing media, into (a) water extinguishing methods, and (b) chemical extinguishing methods.

Despite developments with chemical agents in fire extinguishment, water and water-based solutions continue to be the most common and conventional extinguishing agents, particularly in providing more positive control of advanced and intensive massive fires. The ease of handling water and its ready availability in volume and pressure permit advantageous use of water in various ways, with hand hoselines or fixed nozzles, to be projected as solid streams to effect reach and to provide copious flow or deluge, or in the form of a water-curtain or sprinkler spray, fine mist or fog, or by generation into a foam. The general suitability of water for cooling, quenching, wetting, and penetration in deep-seated burning (enhanced by chemical additives or surfactants), as well as its diluting and emulsifying action and its capability of large-scale wetting down of exposed structures and surroundings to prevent ignition, reduce involvement, and retard spread of fire, make water virtually the universal extinguishant.

For small fires involving ordinary combustibles, water and water-based solutions used in manual application equipment or portable appliances, such as hand-pump tanks, soda-acid and stored pressure units, with chemically loaded streams and antifreeze formulations, appear to be by far the most common extinguishing systems for primary control of most fires. Water spray or fog are particularly effective on flammable liquid fires. Such application techniques are commonly used to protect flammable liquid and gas storage tanks, piping and equipment, and even electrical transformers.

Use of water with certain chemicals and reactive metals can present a dangerous reaction and explosion hazard; however, with an advance alert and suitable precautionary measures in this regard, use of water may be permissible with judicious application —to effect gradual reduction of temperature for slowing down exothermic runaway reactions, and for emergency protection of equipment and structures.

The most useful and widely used system of the automatic methods is the automatic sprinkler and its related methods, such as deluge, water curtains, and spray or fog. Automatic sprinkler systems are known for their simple "as-needed" prompt operation, with fast spray application over a given area. Some 30 different kinds of sprinkler heads with various adaptations are available for use in over 90 different occupancies classed either as light, ordinary, or extra-hazardous. Various type of heat-actuated release mechanisms, such as fusible (solder) links and pellets, non-solder frangible bulbs, or easily softened chemical compounds are employed as part of the sprinklers for automatic activation at different pre-determined temperature ranges— from 38°C to 182°C (100°F to 360°F) and higher. Sprinklers have discharge rates of 57 liters/min (15 gpm) at flowing pressures of 490–560 g/cm^2 (7–8 psi), and up to 133 liters/min (35 gpm) or more at pressures of 7 kg/cm^2 (100 psi) or more. A floor area of 9 m^2 (100 sq ft) can be covered by one sprinkler. In an average building with ordinary occupancy, one sprinkler is generally provided for an area of 9–18 m^2 (100–200 sq ft). The efficiency of spray sprinklers is shown by the relatively few heads that open in most fires: 5 or less in 76% of the fires, and 25 or less in 95% of the fires. Sprinkler systems not only minimize fire damage by quick extinguishment, but also minimize water damage by avoiding need for many large fire-hose streams.

In particular situations where water-damage susceptibility, dangerous water-reaction potentials, or electrical shock hazards may be prime considerations, chemical extinguishants may be used in specifically designed automatic installations, in the form of special foam agents, carbon dioxide, fluidized (free-flowing) special dry powders, and vaporizing liquids (with due regard for the possible toxic effects of halogenated hydrocarbons and their decomposition products). Foams and carbon dioxide are used either for localized application, as in processing equipment, or for the total area or room flooding, with appropriate warning signals.

More recently, increasing attention is being given to instant inerting systems for fire explosion prevention, involving flammable dusts, vapors, mists, or gases mixed with air or chemical supporters of rapid combustion. Such explosion-detonation suppression systems, timed with ultra-fast detectors (sensing either radiation or spectral effects) and low time-lag suppressors, are designed to arrest incipient fires and potentially explosive reactions in their "embryonic" stages, by automatic dispersal of an inerting or inhibiting or flame-dampening agent at extremely high speed (within milliseconds), thereby preventing the development of pressure waves that lead to destructive detonation.

Flammable Liquids

Flammable liquids present a serious fire and explosion hazard since they are easily ignited, difficult to extinguish, and burn with great rapidity. The more volatile liquids liberate heat up to 10 times faster than wood planking, and their vapors form explosive mixtures with air. Unvented containers may burst with explosive violence in exposure fires. Some liquids contain their own supply of oxygen and will burn in the absence of air, some are susceptible to spontaneous heating, and others may react violently with other materials, including water.

The danger from a flammable-liquid process or storage facility depends on such conditions as the quantity and flammability of the liquid, whether it is exposed to the air or is in a closed container or piping system, the probabilities of accidental leakage or overflow, the location in relation to important buildings, equipment, and outside ignition sources, the building construction, and the adequacy of fire protection.

The basic safeguards outlined in this chapter apply to unheated flammable liquids having flash points of 93°C (199°F) or less, and to other liquids if heated to within 28°C (82.4°F) of their flash points. The hazards of unheated liquids having flash points above 93°C (197°F) are usually adequately protected against by automatic sprinklers alone.

Automatic sprinklers are the basic protection against the hazards of all flammable liquids indoors and are the main factor in preventing serious damage. Although sprinklers cannot extinguish fires in low-flash-point liquids, they do prevent major damage to structures and equipment.

The seven basic safeguards for safe handling of flammable liquids are:

1. Isolate the hazards.
2. Confine the liquid.
3. Ventilate to prevent explosive mixtures.
4. Install explosion vents where needed.
5. Eliminate ignition sources.
6. Educate employees in hazards and safeguards.
7. Provide adequate fire protection.

The information in this chapter applies primarily to storage and handling of large quantities of flammable liquids, over 200 liters or 50 gallons, such as in laboratory stockrooms and storage rooms. However, the same basic principles may be applied to handling of any quantity of flammable liquids, including the liter size bottle.

ISOLATE THE HAZARD

The loss resulting from a fire in a container of flammable liquid outdoors is only the value of the liquid itself. If the container is in a small, detached building the loss will be increased only by the value of the building. If, however, the container of flammable liquid is located in a main building, the probability of serious fire damage is markedly increased, particularly if the building is of combustible construction or if valuable contents are exposed to damage. Fire in as little as 200 liters or 50 gallons of low-flash-point liquid can cause severe damage to a building and its contents.

Flammable-liquid storage facilities should be isolated either by distance or by fire cutoffs so that they do not expose important buildings and equipment, and in turn, are not likely to become involved in fires originating elsewhere.

Outdoor Locations Preferred

Locate outdoor flammable-liquid facilities at a safe distance from important structures. Where closer locations are necessary, they are acceptable only if exposed walls are of noncombustible construction, either blank or with openings protected.

Detached Buildings Acceptable

One-story buildings used exclusively for flammable-liquid facilities and located at a safe distance (usually 15 meters (49 feet) or more) from important structures may be of light noncombustible steel-frame construction. With less distance, the exposing wall of the flammable-liquid building should be a blank fire wall or the exposed main structure should be safeguarded as for outdoor locations.

Cutoff Rooms Acceptable

Cutoff rooms should be located at exterior walls of main buildings and isolated from occupied and valuable areas by fire walls or fire partitions. Walls of such cutoffs should have a fire resistance of $3^3/_4$ hours. Bond walls securely to floor and ceiling members to provide maximum explosion resistance. Wherever practical, avoid openings in the walls. Protect necessary openings with approved automatic-closing fire doors or other accepted means.

In multistory buildings, the preferred location for flammable-liquid rooms is in the first story, although upper-story locations are acceptable. Floors should be liquid-tight and should be provided with drainage. Floors and ceilings should have fire resistance equal to that of horizontal cutoffs. Existing combustible ceilings may be covered with fire-resistive sheathing. Because adequate drainage, ventilation, and explosion venting are difficult to provide in basement areas, flammable liquids should not be used or stored there.

CONFINE THE LIQUID

Flammable liquids in closed containers, equipment, and transfer systems offer only moderate hazard by themselves. In open containers or when they escape from equipment or piping they present an easily ignited fuel supply. The three primary objectives in the safe design and operation of flammable-liquid facilities are (1) to prevent the escape of liquid and vapors, (2) to provide rapid shutoff if liquid does escape, and (3) to confine the spread of escaping liquid to the smallest practicable area.

Storage or use of flammable liquids in large main areas of buildings should be avoided, and quantities in laboratories should be the minimum practicable. Manually operated pumps are recommended for small tanks and drums, rather than gravity transfer.

Provide ramped curbs or grating-covered troughs the full width of all doorways leading from flammable-liquid rooms to other indoor areas. Trapped drains of sufficient capacity piped to a safe outdoor discharge point should be provided to handle the maximum sprinkler discharge of the flammable-liquid area.

Above ground outside flammable-liquid facilities should be arranged so that escaping liquid will flow to a safe disposal location without approaching important buildings or equipment. Provide diversionary dikes or drains, taking advantage of natural grade, or surround the facilities with dikes of sufficient capacity to confine all the liquid.

VENTILATE TO PREVENT EXPLOSIVE MIXTURES

There is an explosion hazard wherever vapors are unconfined, as at open containers, at mixing, spraying, or dispensing locations, at any equipment where liquid surfaces are exposed or where there is normal minor leakage.

The preferred safety measure is to use closed equipment. Where this is not practical, ventilation should be provided to reduce the vapor concentration to a safe level.

Unheated flammable liquids having flash points above maximum ambient summer temperatures do not form explosive mixtures with air and are adequately safeguarded by natural-draft ventilation alone. Liquids having lower flash points form explosive mixtures and usually need positive exhaust ventilation to prevent development of hazardous vapor concentration. Adequacy of ventilation at all flammable-liquid processes under normal operating conditions should be checked with a flammable-vapor indicator. It is not ordinarily practical to remove the vapors from a major spill of low-flash-point liquids or heated liquids by ventilation.

Most flammable liquids are toxic in some degree and require ventilation to safeguard the health of operators. Ventilation that is safe from a health standpoint will usually be more than adequate to overcome any explosion hazard.

The amount of ventilation needed to prevent explosions is dependent on the rate at which vapors are produced and on the ability of specific vapors to form explosive mixtures with air. The volume of air made barely explosive by vapors from 3.8 liters (4 qts) of various common liquids is as given in Table 1.

TABLE 1
Flammable Liquids and Volume of Air made Explosive by 3.8 liters (1 gal)

Liquid	Cubic Meters of Air	Cubic Feet of Air
Acetone	49.7	(1,755)
Benzene (Benzol)	72.2	(2,550)
Butyl acetate	47	(1,660)
Butyl alcohol	56.6	(2,000)
Carbon disulfide	141.5	(5,000)
Ethyl acetate	40.5	(1,430)
Ethyl alcohol	44.2	(1,560)
Ethyl ether	49.5	(1,750)
Gasoline	65.1	(2,300)
Methanol	37.1	(1,310)
Pentyl acetate (Amyl acetate)	54.9	(1,940)
Pentyl alcohol (Amyl alcohol)	67.9	(2,400)
Toluene (Toluol)	71.6	(2,530)
Xylene (Xylol)	73.6	(2,600)

The vapors from 3.8 liters (4 qts) of gasoline mixed with about 65.1 m^3 of air produce a mixture that is barely explosive, and removal of this volume of air for each gallon evaporated will just maintain this explosive condition. In practice it is necessary to ventilate with about four times the volume of air made barely explosive in order to compensate for uneven distribution and to provide a proper factor of safety.

Flammable vapors are heavier than air and normally collect near floor level. Without proper ventilation they may travel along the floor and form explosive mixtures at locations remote from the point of origin. Ventilating systems must be designed to remove vapors from the lowest level at a rate that will prevent an unsafe vapor concentration. It is usually unnecessary to consider ceiling height or to arrange complete displacement of the entire room volume of air. The movement of ventilating air should be toward the vapor source from all directions. There should be no vapor concentration in air in excess of 25 percent of the lower explosive limit beyond 60 cm or 2 feet from the process or flammable-liquid container.

Natural-draft ventilation is usually adequate for above-grade locations of unheated flammable liquids having closed-up flash points above 45°C (113°F). It is adequate also for lower-flash-point liquids that are completely confined in closed vessels, unvalved piping, or equipment that permits no leakage of liquid or vapor, none of which are opened for charging, gaging, sampling, or other reason during regular operation.

Outlets for natural ventilation are best provided by permanent louvered openings to outdoors, located at floor level close to the source of vapors. Hooded roof ventilators or vent flues extending above the building are satisfactory. Inlets for fresh air should preferably be located near the ceiling level.

Even where mechanical ventilation is provided, it should be supplemented by natural-draft ventilation for idle or emergency periods.

Positive exhaust mechanical ventilation is needed to insure safety at storage or dispensing processes involving 143°C (289°F) and lower (closed-cup) flash-point liquids that are not entirely confined within the equipment or otherwise have exposed surfaces or unconfined vapors. This ventilation may also be needed for high-flash-point liquids which present large exposed surface areas or which are in the form of mist or spray.

Locate the suction intake close to the floor and to the source of vapor. Locate fresh-air inlets in exterior walls so as to obtain ventilation from outdoors, heating the air by unit steam heaters if necessary. If air is taken from other rooms, provide automatic-closing fire shutters at the inlet.

If the storage occupies a cutoff room, mechanical ventilation should be based on the floor area of the entire room. For processes in part of a larger area, it may be based on the area within draft curtains or floor curbs, or on the floor area within 3 m (10 feet) of exposed vapor sources.

Electrical equipment using flammable liquids should be interlocked to permit operation only when the ventilating system is running.

Do not manifold exhaust ducts from separate storage areas but run them separately to outdoors. Do not install dampers in exhaust ducts unless interlocked to close on actuation of special fire-extinguishing systems.

Approved flammable-vapor indicators are desirable for periodic checking of the ventilation. For the more hazardous processes it is advisable to provide a fixed flammable-vapor analyzer arranged to give a signal or to shut down operations if the vapor concentration exceeds a safe level.

INSTALL EXPLOSION VENTS WHERE NEEDED

Despite application of preventive measures, there are flammable-liquid processes at many locations where ventilation failure, mechanical damage, or other abnormal conditions may result in an explosion. Provision of adequate explosion vents will reduce the pressure developed by vapor explosions in buildings and equipment and thereby reduce structural damage.

Dispensing of flammable liquids having flash points below maximum ambient summer temperatures, and liquids which are heated within 14°C (57°F) of their flash points, which are in the form of mist, or which are otherwise capable of producing explosive concentrations in air, necessitate installation of explosion vents. Minor processes do not generally require explosion venting unless an explosive mixture will extend for several feet from the process if normal ventilation is interrupted. The amount of venting needed for a specific process is dependent mainly upon the maximum explosion pressure and the average rate of pressure rise of the exploding flammable vapor, and on the size and design strength of the building or equipment. The position of vents with relation to the point of origin of the explosion and the pressure required to burst or open the vent closure must also be considered.

Pressures produced by explosions of common vapors under optimum mixture conditions may exceed 7 kg/cm^2 (2.4 lb/in.^{2}). It is usually impractical to design structures or provide venting to prevent damage from the explosion of such mixtures if they occupy the entire volume of a room or item of equipment. Most explosions involve only a small part of the total volume of an enclosure and frequently occur near the limits of the explosive range. Such explosions are weak as compared to those of optimum mixtures, and proper explosion vents will prevent serious structural damage.

The maximum practical amount of venting area should be installed on equipment and buildings. Small enclosures require more venting in proportion to their size than large ones, because an explosive mixture is more likely to extend throughout their entire volume.

Vents for buildings may be unobstructed openings, louvers, roof vents, doors, windows, explosion-venting sash, roof or wall panels, or skylights, arranged to open under a pressure substantially below the strength of the structure. For equipment, vents may be diaphragms of waterproof paper or plastic-impregnated cloth, plastic sheets, metal foil, charging doors having spring or friction latches, or poppet relief valves. Saw-tooth or spear-point cutters may be used to quicken the opening of diaphragm-disk vents. Equipment vents should lead to outdoors through substantially constructed short ducts.

Locate indoor hazardous equipment close to exterior walls to facilitate venting to outdoors and provide separate explosion ducts and vents for each item of equipment. Single-story buildings are preferred as locations, although in multistory buildings corner locations or penthouses are acceptable.

Separate highly hazardous processing equipment by pressure-resisting walls. Keep explosion vents in operating condition and check them by a regular schedule of preventive maintenance.

ELIMINATE IGNITION SOURCES

The most common ignition sources of flammable-liquid fires and explosions are open flames, electrical equipment, overheating, hot surfaces, spontaneous heating, sparks and embers, static charges, and friction. Operations or equipment that create ignition sources should be eliminated from flammable-liquid areas, but if this is not possible, isolate the ignition source by a fire-resistive wall or partition. Do not introduce portable or temporary ignition sources except under extremely well-controlled conditions.

Open flames in common industrial use include those used for electric-arc and acetylene-flame welding, flame cutting, burning, brazing, torch soldering, heaters, furnaces, boilers, driers, ovens, and lead pots. They should be isolated from flammable-liquid locations.

Temporary open-flame processes should be strictly controlled by the welding-permit system under the direction of a responsible supervisor. This system forbids the use of any portable open flame equipment until a designated supervisor has inspected the job location to determine if all necessary safeguards are being carried out and has issued a signed permit to the operator. Work at, or in the vicinity of, hazardous flammable-liquid processes usually requires complete removal of the liquid and purging of the equipment with positive mechanical ventilation or with water or inert gas. When possible, equipment to be welded should be removed from the process area to a safe location.

Electrical equipment should be installed in accordance with recommendations for hazardous locations. Either electrical heating appliances should not be capable of heating to the autoignition temperature of the flammable liquid, or they should be equipped with temperature-limiting devices to prevent them from reaching that temperature.

Overheating of flammable liquids frequently starts fires at such processes as heating open containers, oil quenching of metals, and oven drying of specimens containing flammable vapors.

Steam is the preferred heating medium for flammable liquids. Provide heating equipment with automatic temperature controls and with emergency high-temperature-limit switches to shut off the heat supply if the liquid temperature exceeds a safe limit. Equip steam lines with pressure regulators and relief valves. Heating requires close supervision despite provision of automatic controls and safeguards.

Hot surfaces and radiant heat from equipment temperatures above the autoignition temperature of the flammable liquid are a potential ignition source. Ovens, furnaces, hot ductwork, and high-pressure steam lines should be designed and located so that dangerous concentrations of flammable vapors cannot come in contact with them.

Spontaneous ignition of residues and waste in the vicinity of flammable-liquid processes is a cause of fire. Many liquids, particularly vegetable and animal oils, combine readily with oxygen in air at ordinary temperatures and give off heat. Under conditions in which heat is produced faster than it is dissipated, the temperature of organic waste or rubbish increases, and ignition may occur. The reaction is accelerated by additional heat from hot surfaces.

Where materials subject to spontaneous heating are involved, equipment must be kept free of deposits both inside and outside. Maintain high housekeeping standards, store oily rags in covered metal containers, and remove all waste from the area regularly.

Sparks or embers from heaters, furnaces, incinerators, or vehicles; mechanical sparks from grinding or machining operations; and sparks from moving parts of equipment are common ignition sources. Such spark-producing equipment or operations should not be permitted in hazardous-vapor areas.

Smoking in flammable-liquid areas must be prohibited.

Static electricity should be eliminated; when possible, static-producing processes should be located outside of flammable-liquid areas.

Friction due to poor lubrication, poorly aligned or damaged moving parts, grinding processes, or machining is a common cause of fire. Design equipment with a safe amount of clearance between moving parts. Frequent inspections and good maintenance will prevent generation of high frictional heat.

Keep equipment in the best possible condition. Install spark-arresting screens on ember-producing devices. Use nonsparking metals for high-speed moving parts of machines, ventilating equipment, and hand tools. If industrial trucks must be used, they should be the special types recommended for such locations.

EDUCATE EMPLOYEES IN HAZARDS AND SAFEGUARDS

It is essential that operators and employees realize the danger of fire and explosion attendant on flammable-liquid processes and that they observe safe practices. A careless act can nullify the most complete fire-prevention safeguards.

Automatic protective devices must be in good working order. It is fundamental that employees should not tamper with equipment or make protective devices inoperative.

Employees should be trained to notice any defects in the arrangement or operation of process equipment or protective devices and should report them to management for immediate corrective measures. Supervisors also should be constantly on the alert to discover hazardous conditions and should never permit unsafe equipment to be operated even under the pressure of a production schedule.

Instructions outlining safe practice, proper operating procedure and procedure for emergency situations should be established and posted prominently.

Provide a continuous fire-prevention program for management and employees, including refresher programs for older employees. Organize fire squads and a fire brigade drilled in handling of flammable-liquid fires.

PROVIDE ADEQUATE FIRE PROTECTION

Burning flammable liquids can quickly cause destructive damage to combustible and exposed steel construction unless protective measures are taken. Automatic sprinklers are the basic protection for all flammable-liquid occupancies, although certain flammable-liquid hazards also require special fixed extinguishing systems. Automatic sprinklers should be provided to back up automatic-closing covers and fixed automatic extinguishing systems utilizing foam, carbon dioxide, or dry chemicals. Sprinklers should be installed at the ceiling, under hoods, under wide drainboards, and under other items of equipment where a fire may occur that would be shielded from ceiling sprinkler discharge.

Water discharged from automatic sprinklers and spray nozzles will usually extinguish fires in liquids having open-cup flash points of 66°C (151°F) and higher. This extinguishing action is mainly one of cooling.

Fires in heavier-than-water flammable liquids are readily extinguished by the action of water floating on top and excluding oxygen. Fires in water-miscible liquids may be extinguished by dilution.

Fires in most liquids with flash points below 66°C (151°F) cannot be extinguished by sprinklers. However, the cooling effect of sprinkler discharge will usually reduce temperatures below the autoignition point of ordinary combustibles and prevent serious damage to buildings and equipment.

Strong water supplies are needed for sprinklers and hose streams. In most situations, close sprinkler spacing and the extra-hazard schedule of pipe sizes are needed.

Equipment that necessarily exposes flammable-liquid surfaces should be fitted with automatic-closing covers when possible. Properly installed and maintained, such covers will extinguish fires in equipment by exclusion of air and will prevent flammable-liquid overflow caused by displacement by water. In the absence of automatic-closing covers, install standard fixed-pipe automatic extinguishing systems. Establish a testing and maintenance schedule that will assure proper operation of the system in event of fire.

Fixed automatic water-spray systems are suitable for equipment using flammable liquids, particularly those having flash points above 66°C (151°F) which have a specific gravity greater than water or which are water-miscible. For equipment with flammable liquids heated above 121°C (250°F) the spray system should extend well beyond the equipment on all sides in order to handle any spill fire that may result from foaming as steam is formed below the liquid surface.

Fixed automatic foam extinguishing systems are dependable when properly installed and maintained. The foam blankets the liquid for a considerable period of time and reduces the possibility of reflashing from hot metal surfaces. Some means must be provided to permit the build-up of a foam blanket, generally several inches in depth. Foam is not recommended for carbon disulfide, some ethers, and other very volatile liquids because their vapors may diffuse through the foam blanket and burn above it. A special foam system that will not cause boil-over is available for use with liquids heated above 121°C (250°F). A special mechanical (air) foam is available for some liquids such as alcohols, esters, and ketones that break down regular foam.

Fixed automatic carbon dioxide extinguishing systems are particularly adaptable to flammable-liquid equipment that is at least partially enclosed. Carbon dioxide has the advantage of eliminating damage and prolonged interruptions to production that would result from use of water or other extinguishing agents. The method of application, quantity of gas, and time of discharge are very critical because the extinguishing atmosphere may dissipate rapidly.

Fixed automatic dry-chemical extinguishing systems are also used for protection against flammable-liquid hazards. Ventilating equipment, conveyors, and other machinery that will spread a fire or dissipate the powder should be interlocked to shut down when the extinguishing system operates.

Portable extinguishing equipment—dry-chemical, carbon dioxide, or foam extinguishers—should be provided near all flammable-liquid facilities, and located to be accessible under fire conditions.

Small hose with spray nozzles is also recommended, particularly for fires in liquids that can be extinguished by water spray. Water extinguishers should be provided wherever ordinary combustibles are in the vicinity of the flammable-liquid process and where combustible residues may collect.

ACKNOWLEDGEMENT: Reprinted with permission from the "Handbook of Industrial Loss Prevention," Chapter 41, pp. 1–6, Factory Mutual Engineering Corporation, Factory Mutual System, New York, 1959.

Flammability Characteristics of Combustible Gases and Vapors

Michael G. Zabetakis

INTRODUCTION

Prevention of unwanted fires and gas explosion disasters requires a knowledge of flammability characteristics (limits of flammability, ignition requirements, and burning rates) of pertinent combustible gases and vapors likely to be encountered under various conditions of use (or misuse). Available data may not always be adequate for use in a particular application since they may have been obtained at a lower temperature and pressure than is encountered in practice. For example, the quantity of air that is required to decrease the combustible vapor concentration to a safe level in a particular process carried out at 200°C should be based on flammability data obtained at this temperature. When these are not available, suitable approximations can be made to permit a realistic evaluation of the hazards associated with the process being considered; such approximations can serve as the basis for designing suitable safety devices for the protection of personnel and equipment.

DEFINITIONS AND THEORY

Limits of Flammability

A combustible gas-air mixture can be burned over a wide range of concentrations —when either subjected to elevated temperatures or exposed to a catalytic surface at ordinary temperatures. However, homogeneous combustible gas-air mixtures are flammable, that is, they can propagate flame freely within a limited range of compositions. For example, trace amounts of methane in air can be readily oxidized on a heated surface, but a flame will propagate from an ignition source at ambient temperatures and pressures only if the surrounding mixture contains at least 5 but less than 15 volume-percent methane. The more dilute mixture is known as the lower limit, or combustible-lean limit, mixture; the more concentrated mixture is known as the upper limit, or combustible-rich limit, mixture. In practice, the limits of flammability of a particular system of gases are affected by the temperature, pressure, direction of flame propagation, gravitational field strength, and surroundings. The limits are obtained experimentally by determining the limiting mixture compositions between flammable and nonflammable mixtures.[88] That is,

$$L_{T,P} = \tfrac{1}{2}[C_{gn} + C_{1f}], \tag{1}$$

and

$$U_{T,P} = \tfrac{1}{2}[C_{gf} + C_{1n}], \tag{2}$$

where $L_{T,P}$ and $U_{T,P}$ are the lower and upper limits of flammability, respectively, at a specified temperature and pressure, C_{gn} and C_{1n} are the greatest and least concentrations of fuel in oxidant that are nonflammable, and C_{1f} and C_{gf} are the least and greatest concentrations of fuel in oxidant that are flammable. The rate at which a flame propagates through a flammable mixture depends on a number of factors including temperature, pressure, and mixture composition. It is a minimum at the limits of flammability and a maximum at near stoichiometric mixtures.[41]

200

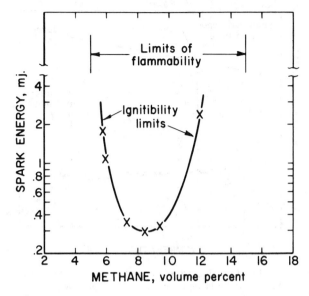

FIG. 1. Ignitibility curve and limits of flammability for methane-air mixtures at atmospheric pressure and 26°C.

The Bureau of Mines has adopted a standard apparatus for limit-of-flammability determinations.[17] Originally designed for use at atmospheric pressure and room temperature, it was later modified for use at reduced pressures by incorporating a spark-gap ignitor in the base of the 2-inch, glass, flame-propagation tube. This modification introduced a difficulty that was not immediately apparent, as the spark energy was not always adequate for use in limit-of-flammability determinations. Figure 1 illustrates the effect of mixture composition on the electrical spark energy requirements for ignition of methane-air mixtures.[28] For example, a 0.2-millijoule (mj) spark is inadequate to ignite even a stoichiometric mixture at atmospheric pressure and 26° C; a 1-mj spark can ignite mixtures containing between 6 and 11.5 volume-percent methane. Such limit-mixture compositions that depend on the ignition source strength may be defined as limits of ignitibility or more simply ignitibility limits; they are thus indicative of the igniting ability of the energy source. Limit mixtures that are essentially independent of the ignition source strength and that give a measure of the ability of a flame to propagate away from the ignition source may be defined as limits of flammability. Considerably greater spark energies are required to establish limits of flammability than are required for limits of ignitibility,[74] further, more energy is usually required to establish the upper limit than is required to establish the lower limit. In general, when the source strength is adequate, mixtures just outside the range of flammable compositions yield flame caps when ignited. These flame caps propagate only a short distance from the ignition source in a uniform mixture. The reason for this may be seen in figure 2 which shows the effect of temperature on limits of flammability at a constant initial pressure. As the temperature is increased, the lower limit decreases and the upper limit increases. Thus, since a localized energy source elevates the temperature of nearby gases, even a nonflammable mixture can propagate flame a short distance from the source. That is, a nonflammable mixture (for example, composition-temperature point *A*, fig. 2) may become flammable for a time, if its temperature is elevated sufficiently (composition-temperature point *B*).

Flammable mixtures considered in figure 2 fall in one of three regions. The first is to the left of the saturated vapor-air mixtures curve, in the region labeled "Mist." Such mixtures consist of droplets suspended in a vapor-air mixture; they are discussed in

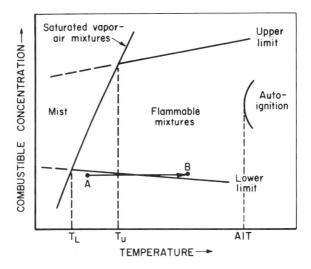

FIG. 2. Effect of temperature on limits of flammability of a combustible vapor in air at a constant initial pressure.

greater detail in the section on formation of flammable mixtures. The second lies along the curve for saturated vapor-air mixtures; the last and most common region lies to the right of this curve. Compositions in the second and third regions make up the saturated and unsaturated flammable mixtures of a combustible-oxidant system at a specified pressure.

In practice, complications may arise when flame propagation and flammability limit determinations are made in small tubes. Since heat is transferred to the tube walls from the flame front by radiation, conduction, and convection, a flame may be quenched by the surrounding walls. Accordingly, limit determinations must be made in apparatus of such a size that wall quenching is minimized. A 2-inch-ID vertical tube is suitable for use with the paraffin hydrocarbons (methane, ethane, etc.) at atmospheric pressure and room temperature. However, such a tube is neither satisfactory under these conditions for many halogenated and other compounds nor for paraffin hydrocarbons at very low temperatures and pressures.[67,88]

Because of the many difficulties associated with choosing suitable apparatus, it is not surprising to find that the very existence of the limits of flammability has been questioned. After a thorough study, Linnett and Simpson concluded that while fundamental limits may exist there is no experimental evidence to indicate that such limits have been measured.[42] In a more recent publication, Mullins reached the same conclusion.[46] Accordingly, the limits of flammability obtained in an apparatus of suitable size and with a satisfactory ignition source should not be termed fundamental or absolute limits until the existence of such limits has been established. However, as long as experimentally determined limits are obtained under conditions similar to those found in practice, they may be used to design installations that are safe and to assess potential gas-explosion hazards.

Industrially, heterogeneous single-phase (gas) and multi-phase (gas, liquid, and solid) flammable mixtures are probably even more important than homogeneous gas mixtures. Unfortunately, our knowledge of such mixtures is rather limited. It is important to recognize, however, that heterogeneous mixtures can ignite at concentrations that would normally be nonflammable if the mixture were homogeneous. For example, 1 liter of methane can form a flammable mixture with air near the top of a 100-liter container, although a nonflammable (1.0 volume-percent) mixture would result if complete mixing occurred at room temperature. This is an important concept, since layering can occur with any combustible gas or vapor in both stationary and flowing mixtures. Roberts, Pursall, and Sellers[57-61] have presented an excellent series of review articles on the layering and dispersion of methane in coal mines.

The subject of flammable sprays, mists, and foams is well-documented.[1,7,8,11,29,71,72,89] Again, where such heterogeneous mixtures exist, flame propagation can occur at so-called average concentrations well below the lower limit of flammability;[35] thus, the term "average" may be meaningless when used to define mixture composition in heterogeneous systems.

Ignition

Lewis and von Elbe,[41] Mullins,[45,46] and Belles and Swett[47] have prepared excellent reviews of the processes associated with spark-ignition and spontaneous-ignition of a flammable mixture. In general, many flammable mixtures can be ignited by sparks having a relatively small energy content (1 to 100 mj) but a large power density (greater than 1 megawatt/cm^3). However, when the source energy is diffuse, as in a sheet discharge, even the total energy requirements for ignition may be extremely large.[31,32,34,39,62,78] There is still much to be learned in this field, however, since electrical discharges are not normally as well defined in practice as they are in the laboratory.

When a flammable mixture is heated to an elevated temperature, a reaction is initiated that may proceed with sufficient rapidity to ignite the mixture. The time that elapses between the instant the mixture temperature is raised and that in which a flame appears is loosely called the time lag or time delay before ignition. In general, this time delay decreases as the temperature increases. According to Semenov,[64] these quantities are related by the expression

$$\log \tau = \frac{0.22E}{T} + B, \tag{3}$$

where τ is the time delay before ignition in seconds; E is an apparent activation energy for the rate controlling reaction in calories per mole; T is the absolute temperature, expressed in degrees, Kelvin; and B is a constant. Two types of ignition temperature data are found in the current literature. In the first, the effect of temperature on time delay is considered for delays of less than 1 second.[40,45] Such data are applicable to systems in which the contact time between the heated surface and a flowing flammable mixture is very short; they are not satisfactory when the contact time is indefinite. Further, equation (3) is of little help, because it gives only the time delay for a range of temperatures at which autoignition occurs; if the temperature is reduced sufficiently, ignition does not occur. From the standpoint of safety, it is the lowest temperature at which ignition can occur that is of interest. This is called the minimum spontaneous-ignition, or autoignition, temperature (AIT) and is determined in a uniformly heated apparatus that is sufficiently large to minimize wall quenching effects.[65,83] Figures 3 and 4 illustrate typical autoignition-temperature data. In figure 3 the minimum autoignition-temperature or AIT value for n-propyl nitrate is 170°C at an initial pressure of 1,000 psig.[87] Data in this figure may be used to construct a log τ versus $1/T$ plot such as that in figure 4. Such graphs illustrate the applicability of equation (3) to autoignition temperature data. The equation of the broken line in figure 4 is

$$\log \tau = \frac{12.3 \times 10^3}{T} - 25.1. \tag{4}$$

In this specific case, equation (4) is applicable only in the temperature range from 170° to 195° C; another equation must be used for data at higher temperatures. The solid lines in figure 4 define an 8 C° band that includes the experimental points in the temperature range from 170° to 195° C.

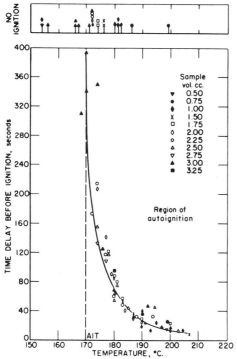

FIG. 3. Time delay before ignition of NPN in air at 1,000 psig in the temperature range from 150° to 210°C. (1-33 apparatus; type-347, stainless steel test chamber.)

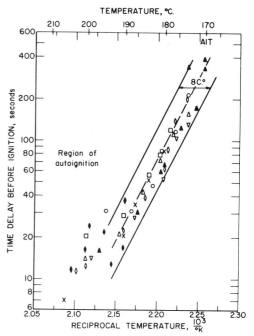

FIG. 4. Logarithm of time delay before ignition of NPN in air at 1,000 psig initial pressure. (Data from figure 3).

Formation of Flammable Mixtures

In practice, heterogeneous mixtures are always formed when two gases or vapors are first brought together. Before discussing the formation of such mixtures in detail, a simplified mixer such as that shown in figure 5 will be considered briefly. This mixer consists of chambers 1 and 2 containing gases A and B, respectively; chamber 2, which contains a stirrer, is separated from chamber 1 and piston 3 by a partition with a small hole, H. At time t_o, a force F applied to piston 3 drives gas A into chamber 2 at a constant rate. If gas A is distributed instantaneously throughout chamber 2 as soon as it passes through H, a composition diagram such as that given in figure 6 results; the (uniform) piston motion starts at t_o and stops at t_F. However, if a time interval Δt is required to distribute a small volume from chamber 1 throughout chamber 2, then at any instant between t_o and $t_F + \Delta t$, a variety of mixture compositions exists in chamber 2. This situation is represented schematically in figure 7. The interval of time during which heterogeneous gas mixtures would exist in the second case is determined in part by the rate at which gas A is added to chamber 2, by the size of the two chambers, and by the efficiency of the stirrer.

In practice, flammable mixtures may form either by accident or design. When they are formed by accident, it is usually desirable to reduce the combustible concentration quickly by adding enough air or inert gas to produce nonflammable mixtures. Under certain conditions, it may be possible to increase the combustible concentration so as to produce a nonflammable mixture. Such procedures are discussed in greater detail in the following section.

Flammable mixtures are encountered in production of many chemicals and in certain physical operations. These include gasfreeing a tank containing a combustible gas,[80] drying plastic-wire coating, and recovering solvent from a solvent-air mixture. When layering can occur, as in drying operations, it is not enough to add air at such a rate that the overall mixture composition is below the lower limit of flammability (assuming that uniform mixtures result). Special precautions must be taken to assure the rapid formation of nonflammable mixtures.[81] When a batch process is involved, an added precaution must be taken; a constituent at a partial pressure near its vapor pressure value may condense when it is momentarily compressed by addition of other gases or vapors. Accordingly, mixtures that are initially above the upper limit of flammability may become flammable. A similar effect must be considered when mixtures are sampled with equipment that is cooler than the original sample; if vapor condenses in the sampling line, the test sample will not yield accurate data. A flammable mixture sampled in this manner may appear to be nonflammable and thus create a hazardous situation.[82]

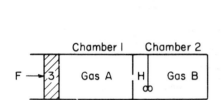

FIG. 5. Simplified mixer.

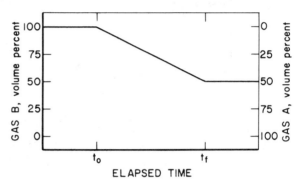

FIG. 6. Composition of gas in chamber 2, figure 5 (Instantaneous mixing).

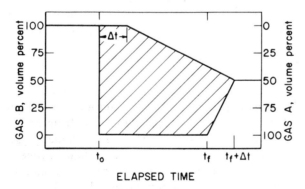

FIG. 7. Composition of gas in chamber 2, figure 5 (Delayed mixing).

A flammable mixture can also form at temperatures below the flash point of the liquid combustible either if the latter is sprayed into the air, or if a mist or foam forms. With fine mists and sprays (particle sizes below 10 microns), the combustible concentration at the lower limit is about the same as that in uniform vapor-air mixtures.[6,7,8,10,29,89] However, as the droplet diameter increases, the lower limit appears to decrease. In studying this problem, Burgoyne found that coarse droplets tend to fall towards the flame front in an upward propagating flame, and as a result the concentration at the flame front actually approaches the value found in lower limit mixtures of fine droplets and vapors.[10] With sprays, the motion of the droplets also affects the limit composition, so that the resultant behavior is rather complex. The effect of mist and spray droplet size on the apparent lower limit is illustrated in figure 8. Kerosine vapor and mist data were obtained by Zabetakis and Rosen,[89] tetralin mist data, by Burgoyne and Cohen,[10] kerosine spray data, by Anson;[1] and the methylene bistearamide data, by Browning, Tyler, and Krall.[7]

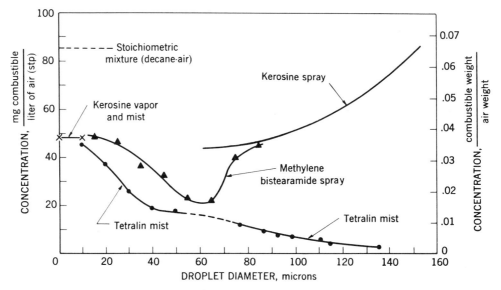

FIG. 8. Variation in the lower limits of flammability of various combustibles in air as a function of droplet diameter.

Flammable mist-vapor-air mixtures may occur as the foam on a flammable liquid collapses. Thus, when ignited, many foams can propagate flame. Bartkowiak, Lambiris, and Zabetakis found that the pressure rise ΔP produced in an enclosure by the complete combustion of a layer of foam of thickness h_f is proportional to h_f and inversely proportional to h_a, the height of the air space above the liquid before foaming.[2] That is

$$\Delta P \propto \frac{h_f}{h_a}. \tag{5}$$

Pressures in excess of 30 psi were produced by the ignition of foams in small containers.

Thomas found that an additional hazard could arise from production of foams by oxygen-enriched air at reduced pressures.[72] Air can become oxygen-enriched as the pressure is reduced, because oxygen is more soluble than nitrogen in most liquids.[33] Thus the presence of foams on combustible liquids are a potential explosion hazard.

A flammable foam can also form on nonflammable liquid if the foam is generated by a flammable gas mixture instead of air. Burgoyne and Steel, who studied this problem, found that the flammability of methane-air mixtures in water-base foams was affected by both the wetness of the foam and the bubble size.[12]

PRESENTATION OF DATA

Limit-of-flammability data that have been obtained at a specified temperature and pressure with a particular combustible-oxidant-inert system may be presented on either a triangular or a rectangular plot. For example, figure 9 shows a triangular flammability diagram for the system methane-oxygen-nitrogen. This method of presentation is frequently used because all mixture components are included in the diagram. However,

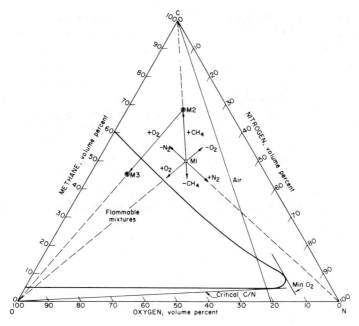

FIG. 9. Flammability diagram for the system methane-oxygen-nitrogen at atmospheric pressure and 26°C.

as the sum of all mixture compositions at any point on the triangular plot is constant (100%) the diagram can be simplified by use of a rectangular plot.[88] For example, the flammable area of figure 9 may be presented as illustrated in figure 10. As noted, the oxygen concentration at any point is obtained by subtracting the methane and nitrogen concentrations at the point of interest from 100 as follows:

$$\% \, O_2 = 100\% - \% \, CH_4 - \% \, N_2. \tag{6}$$

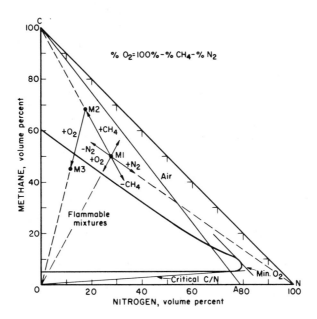

FIG. 10. Flammability diagram for the system methane-oxygen-nitrogen at atmospheric pressure and 26°C. (Data from figure 9).

With either type of presentation, addition of methane, oxygen, or nitrogen to a particular mixture results in formation of a series of mixtures that fall along the line between the composition point (for example, $M1$ in figures 9 and 10) and the vertices of the bounding triangle. For example, addition of methane ($+CH_4$) to mixture $M1$ yields initially all mixture compositions between $M1$ and C (100 pct CH_4). After a homogeneous mixture is produced, a new mixture composition point, such as $M2$, is obtained. Similarly, if oxygen is added ($+O_2$) to the mixture represented by point $M1$, all compositions between $M1$ and O (100 pct O_2) are obtained initially; if nitrogen is added, all compositions between $M1$ and N (100 pct N_2) are obtained initially. If more than one gas is added to $M1$, for example, methane and oxygen, the resultant composition point may be obtained by considering that the mixing process occurs in two steps. First, the methane is added to $M1$ and the gases are mixed thoroughly to give $M2$. Oxygen is then added to $M2$ with mixing to give a new (flammable) mixture, $M3$. If the methane and oxygen were added to a fixed volume at constant pressure, some of $M1$ and then of $M2$ would escape and mix with the surrounding atmosphere. In many instances this is an important consideration because the resulting mixtures may be flammable. For example, even if an inert gas is added to a constant-volume tank filled with methane, flammable mixtures can form outside the tank as the displaced methane escapes into the atmosphere. If the methane is not dissipated quickly, a dangerous situation can arise.

When a mixture component is removed by condensation or absorption, the corresponding composition point (for example, $M1$ in figures 9 and 10) shifts away from the vertices C, O, and N along the extensions to the lines $M1 - C$, $M1 - O$ and $M1 - N$, indicated in figures 9 and 10 by the minus signs. The final composition is determined by the percentage of each component removed from the initial mixture.

Mixtures with constant oxygen-to-nitrogen ratio (as in air), are obtained in figures 9 and 10 by joining the apex, C, with the appropriate mixture composition along the baseline, ON. Thus, the Air line, CA, (fig. 10) is formed by joining C with the mixture A (21 percent O_2 + 79 percent N_2). Using this latter point, A, one can readily determine the mixture compositions that are formed when mixture $M1$ is displaced from an enclosure and mixed with air. Initially, all mixture compositions between $M1$ and A would form. Since these would pass through the flammable mixture zone, a hazardous condition would be created. Similarly, if pure combustible CH_4 were dumped into the atmosphere (air), all mixtures between C and A would form. These would include the flammable mixtures along CA so that a hazardous condition would again be created, unless the combustible were dissipated quickly.

Mixtures with constant oxidant content are obtained by constructing straight lines parallel to zero oxidant line; such mixtures also have a constant combustible-plus-inert content. One particular constant oxidant line is of special importance—the minimum constant oxidant line that is tangent to the flammability diagram or, in some cases, the one that passes through the extreme upper-limit-of-flammability value. This line gives the minimum oxidant (air, oxygen, chlorine, etc.) concentration needed to support combustion of a particular combustible at a specified temperature and pressure. In figures 9 and 10, the tangent line gives the minimum oxygen value (Min O_2, 12 volume-percent) required for flame propagation through methane-oxygen-nitrogen mixtures at 26° C and 1 atmosphere.

Another important construction line is that which gives the maximum nonflammable combustible-to-inert ratio (critical C/N). Mixtures along and below this line form nonflammable mixtures upon addition of oxidant. The critical C/N ratio is the slope of the tangent line from the origin (figs. 9 and 10), 100 percent oxidant, to the lean side of the flammable mixtures curve. The reciprocal of this slope gives the minimum ratio of inert-to-combustible at which nonflammable mixtures form upon addition of oxidant. It is of interest in fire extinguishing.

An increase in temperature or pressure usually widens the flammable range of a particular combustible-oxidant system. The effect of temperature is shown in figure 11; two flammable areas, T_1 and T_2, are defined for a combustible-inert-oxidant system at constant pressure. The effect of temperature on the limits of flammability of a combustible in a specified oxidant was previously shown in figure 2. This type of graph is especially useful since it gives the vapor pressure of the combustible, the lower and upper temperature limits of flammability (T_L and T_U), the flammable region for a range of temperatures, and the autoignition temperature (AIT). Nearly 20 of these graphs were presented by Van Dolah and coworkers for a group of combustibles used in flight vehicles.[74]

The lower temperature limit, T_L, is essentially the flash point of a combustible, in which upward propagation of flame is used; in general, it is somewhat lower than the flash point, in which downward propagation of flame is used. Since T_L is the intersection of the lower-limit and vapor-pressure curves, a relationship can be developed between T_L, or the flash point, and the constants defining the vapor pressure of a combustible liquid. An excellent summary of such relationships has been presented by Mullins for simple fuels and fuel blends.[46]

At constant temperature, the flammable range of a combustible in a specified oxidant can be represented as in figure 12. Here the flammable range of JP–4 vapor-air mixtures is given as a function of pressure.[85] A more generalized flammability diagram of a particular combustible-oxidant system can be presented in a three dimensional plot of temperature, pressure, and combustible content—as illustrated in figure 13.[88] Here, composition is given as the ratio of partial pressure of the combustible vapor, p_{vapor}, to the total pressure, P. For any value of P, the limits of flammability are given as a function of the temperature. For example, at 1 atmosphere ($P = 1$), the flammable range is bounded by the lower limit curve $L_1 L_2 L_3 L_4$, and the upper limit curve $U_1 U_2$; all mixtures along the vapor pressure curve $L_4 U_3'' U_2$ are flammable. The flammable range is the same as that depicted in figure 2. At constant temperature (for example, T_1), the flammable range is bounded by the lower limit curve $L_1 P_{L1}$ and the upper limit curve $U_1 P_{U1}$; the broken curve $P_{L1} P_{U1}$ represents the low pressure (quenched) limit. The flammable range is the same as that depicted in figure 12. A similar range is defined at temperatures T_2, T_3, and T_4 which are less than T_1. However, at T_3 and T_4 the upper limit curves intersect the vapor pressure curves, so that no upper limits are found above U_3' and U_4'. In other words, all compositions along $U_3' U_3''$ and $U_4' L_4$ are flammable. The curve $L_4 P_L U_4' U_3' U_2$ defines the range of limit mixtures which are saturated with fuel vapor. Further since L_4 is the saturated lower limit mixture at one atmosphere, T_4 is the flash point.

Some of the points considered in this and the previous section are illustrated in figure 14.[80] This is the flammability diagram for the system gasoline vapor-water vapor-air at 21°C (70°F) and 100°C (212°F) and atmospheric pressure. The air saturation temperature, that is, the temperature at which saturated air contains the quantity of water given on the water vapor axis, is also included. For precise work, a much larger graph or an enlargement of the region from 0 to 8 percent gasoline vapor and from 0 to 30 percent water vapor would be used. However, figure 14 is adequate here. If water vapor is added to a particular mixture A, all mixture compositions between A and pure water vapor will form as noted (if the temperature is at least 212°F), and the composition point will shift towards the 100-percent-water-vapor point. If water vapor is removed by condensation or absorption, the composition point will move along the extension to the line drawn from A to the 100-percent-water-vapor point. The same applies to the other components, air and gasoline, as indicated earlier. Moreover, if more than one component is involved, the final composition point can be found by considering the effect of each component separately.

Figure 14 is of special interest since it can be used to evaluate the hazards associ-

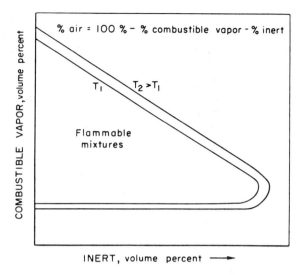

FIG. 11. Effect of initial temperature on limits of flammability of a combustible vapor-inert-air system at atmospheric pressure.

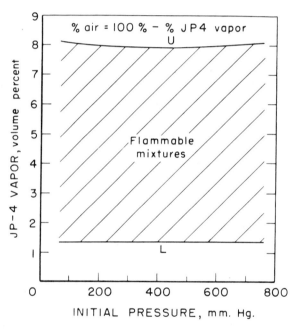

FIG. 12. Effect of initial pressure on limits of flammability of JP-4 (Jet Fuel Vapor) in air at 26°C.

ated with a gas-freeing operation. For example, mixture A represents a saturated gasoline vapor-air-water vapor mixture at 70° F. A more volatile gasoline than the one used here would give a saturated mixture with more gasoline vapor and less air in a closed tank; a less volatile gasoline would give less gasoline vapor and more air. In any event, if a continuous supply of air saturated with water vapor is added to a tank containing mixture A, all compositions between A and B (air plus water vapor) will be formed until all the gasoline vapor is flushed from the tank, and mixture B alone remains. If steam is used to flush mixture A from the tank, all compositions between A

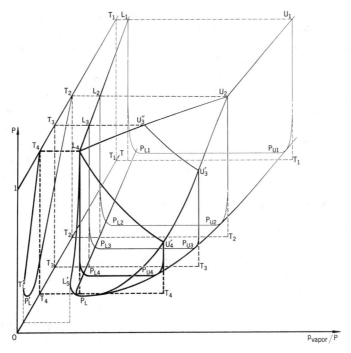

FIG. 13. Effect of temperature and pressure on limits of flammability of a combustible vapor in a specified oxidant.

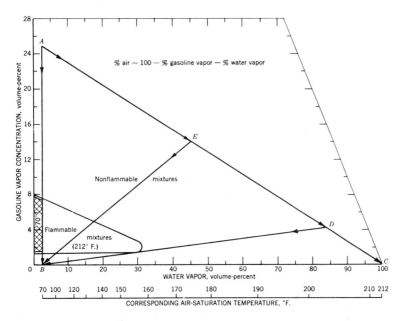

FIG. 14. Flammability diagram for the system gasoline vapor-water vapor-air at 70°F (21°C) and at 212°F (100°C) and atmospheric pressure.

and C will form until all the gasoline vapor has been flushed from the tank and only steam remains (at 212° F or higher). If the tank is permitted to cool, the steam will condense and air will be drawn into the tank giving mixtures along $C - B$. At 70° F, only air plus a small amount of water vapor will remain.

If hot water and water vapor at 175° F are used to flush mixture A from the tank, the mixture composition can only shift along AC to E. Mixtures between A and E that are flushed from the tank mix with air to give mixtures between points along AE and B. Again, as the water vapor in these mixtures condenses outside the tank, the composition of the resultant mixtures will shift away from the 100-percent-water-vapor point, C. The mixture in the tank will remain at E unless air is used to flush the tank, in which case mixture compositions between E and B will form. Again, if the water vapor within the tank condenses, the mixture composition will shift away from C. In any event, at this temperature (175° F), the addition of air to mixture E will lead to formation of flammable mixtures. Thus, mixture A cannot be flushed from a tank without forming flammable mixtures, unless steam or some other inert vapor or gas is used.

DEFLAGRATION AND DETONATION PROCESSES

Once a flammable mixture is ignited, the resulting flame, if not extinguished, will either attach itself to the ignition source or propagate from it. If it propagates from the source, the propagation rate will be either subsonic (deflagration) or supersonic (detonation) relative to the unburned gas. If it is subsonic, the pressure will equalize at the speed of sound throughout the enclosure in which combustion is taking place so that the pressure drop across the flame (reaction) front will be relatively small. If the rate is supersonic, the rate of pressure equalization will be less than the propagation rate and there will be an appreciable pressure drop across the flame front. Moreover, with most combustible-air mixtures, at ordinary temperatures, the ratio of the peak-to-initial pressure within the enclosure will seldom exceed about 8:1 in the former, but may be more than 40:1 in the latter case. The pressure buildup is especially great when detonation follows a large pressure rise due to deflagration. The distance required for a deflagration to transit to a detonation depends on the flammable mixture, temperature, pressure, the enclosure, and the ignition source. With a sufficiently powerful ignition source, detonation may occur immediately upon ignition, even in the open. However, the ignition energy required to initiate a detonation is usually many orders of magnitude greater than that required to initiate a deflagration.[14,90]

Deflagration

Where a deflagration occurs in a spherical enclosure of volume V with central ignition, the approximate pressure rise ΔP at any instant t after ignition is given by the expressions:

$$\Delta P = KP_1 \frac{S_u^3 t^3}{V} \leq P_m, \tag{7}$$

and

$$P_m = P_1 \frac{n_b T_b}{n_1 T_1} = P_1 \frac{\overline{M}_1 T_b}{\overline{M}_b T_1}, \tag{8}$$

where K is a constant, S_u is the burning velocity, P_1 is the initial pressure, P_m is the maximum pressure, T_1 is the initial temperature, n_1 is the number of moles of gas in the initial mixture, n_b is the number of moles of gas in the burned gases, $\overline{M}_1$ is the average molecular weight of the initial mixture, $\overline{M}_b$ is the average molecular weight of the burned gases, and T_b is the final (adiabatic) temperature of the products. With other enclosures, or with noncentral ignition, the flame front is disturbed by the walls before combustion is completed, so that calculated pressure cannot be expected to approximate actual pressure. Even with spherical enclosures, the flame front is not actually spherical, so that the walls tend to disturb the flame before combustion is complete.[38,41] A graph

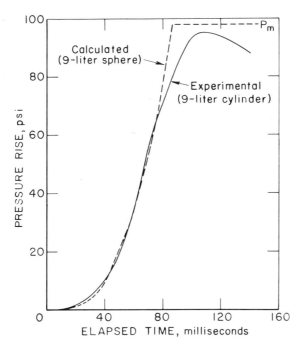

FIG. 15. Pressure produced by ignition of a 9.6 volume-percent methane-air mixture in a 9-liter cylinder (Experimental).

of the pressure developed by the combustion of a stoichiometric methane-air mixture (central ignition) in a 19.7 cm diameter, 9-liter cylinder is given in figure 15. The calculated pressure for a 9-liter sphere is included for comparison; K in equation (7) was evaluated from the experimental curve at 70 milliseconds. The calculated curve follows the experimental curve closely about 75 milliseconds, when the latter curve has a break. This suggests that the flame front was affected by the cylinder walls in such a way that the rate of pressure rise decreased, and the experimental curve fell below the calculated curve. Further, since the combustion gases were being cooled, the maximum pressure fell below the calculated value. The minimum elapsed time (in milliseconds) required to reach the maximum pressure appears to be about $75 \sqrt[3]{V}$ for the paraffin hydrocarbons and fuel blends such as gasoline; V is the volume in cubic feet in this case.

Detonation

Wolfson and Dunn[23,79] have expressed the pressure ratio P_2/P_1 across a detonation front as

$$\frac{P_2}{P_1} = \frac{1}{\gamma_2 + 1} (\gamma_1 M_1^2 + 1), \qquad (9)$$

where γ_2 is the specific heat ratio of the burned gases, γ_1 is the specific heat ratio of the initial mixture, and M_1 is the Mach number of the detonation wave with respect to the initial mixture. M_1 is given in terms of the temperatures T and molecular weights W of the initial and final mixtures by the expression:

$$\frac{(\gamma_1 M_1^2 + 1)^2}{\gamma_1 M_1^2} = \frac{(\gamma_2 + 1)^2 T_2 W_1}{\gamma_2 T_1 W_2}. \qquad (10)$$

Wolfson and Dunn have developed generalized charts that simplify the operations in-

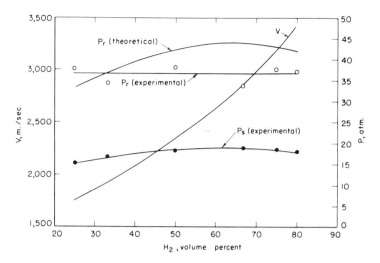

FIG. 16. Detonation velocity, V; static pressure, P_s; and reflected pressure, P_r. Developed by a detonation wave propagating through hydrogen-oxygen mixtures in a cylindrical tube at atmospheric pressure and 18°C.

volved in obtaining the pressure ratio as well as the density and temperature/molecular weight ratios across the detonation wave and the energy release in the detonation wave.

Many investigators have measured and calculated detonation and reflected pressures.[24,25,70] Figure 16 from the data of Stoner and Bleakney[70] gives the detonation velocity, the static or detonation pressure, and the reflected pressure developed by a detonation wave propagating through hydrogen-oxygen mixtures at atmospheric pressure and 18° C.

Blast Pressure

The pressures produced by a deflagration or a detonation are often sufficient to demolish an enclosure (reactor, building. etc.). As noted, a deflagration can produce pressure rises in excess of 8:1, and pressure rises of 40:1 (reflected pressure) can accompany a detonation. As ordinary structures can be demolished by pressure differentials of 2 or 3 psi, it is not surprising that even reinforced concrete structures have been completely demolished by explosions of near-limit flammable mixtures.

Jacobs and coworkers have studied the damage potential of detonation waves in great detail.[36,52] They have considered the principles involved in rupturing of pipes and vessels by detonations and the relevance of engineering and metallurgical data to explosions. More recently, Randall and Ginsburg[53] have investigated bursting of tubular specimens at ordinary and reduced temperatures. They found that the detonation pressure required to burst such specimens was, in general, slightly higher than the corresponding static-bursting pressure. Ductility of the test specimen appeared to have little effect on the bursting pressure, but ductility increased the strength of pipes containing notches or other stress raisers.

When a detonation causes an enclosure to fail, a shock wave may propagate outward at a rate determined by characteristics of the medium through which it is transmitted, and the available energy. If the shock velocity, V, is known, the resulting overpressure, $(P - P_o)$, is given by the expression[70]

$$P - P_o = P_o \left[\frac{2\gamma}{\gamma - 1} \right] \left[\frac{V}{a} - 1 \right], \tag{11}$$

where γ is the ratio of specific heats, and a is the velocity of sound in the medium through which the shock wave passes. The approximate damage potential can be assessed from the data in table 1.[73]

In conducting experiments in which blast pressures may be generated, special precautions must be taken to protect the personnel and equipment from blast and missiles. Browne, Hileman, and Weger[5] have reviewed the design criteria for suitable barricades. Other authors have considered the design of suitable laboratories and structures to prevent fragment damage to surrounding areas.[20,56,69,75]

TABLE 1

Conditions of Failure of Peak Overpressure-Sensitive Elements[73]

Structural Element	Failure	Approximate Incident Blast Overpressure	
		psi	g/cm²
Glass windows, large and small.	Usually shattering, occasional frame failure.	0.5–1.0	35.2–70.4
Corrugated asbestos siding.	Shattering.	1.0–2.0	70.4–140.8
Corrugated steel or aluminum paneling.	Connection failure, followed by buckling.	1.0–2.0	70.4–140.8
Wood siding panels, standard house construction.	Usually failure occurs at main connections, allowing a whole panel to be blown in.	1.0–2.0	70.4–140.8
Concrete or cinderblock wall panels, 8 or 12 inches thick (not reinforced).	Shattering of the wall.	2.0–3.0	140.8–211
Brick wall panel, 8 or 12 inches thick (not reinforced).	Shearing and flexure failures.	7.0–8.0	492–562

PREVENTIVE MEASURES

Inerting

In principle, a gas explosion hazard can be eliminated by removing either all flammable mixtures or all ignition sources.[9,84] However, this is not always practical, as many industrial operations require the presence of flammable mixtures, and actual or potential ignition sources. Accordingly special precautions must be taken to minimize the damage that would result if an accidental ignition were to occur. One such precaution involves the use of explosive actuators which attempt to add inert material at such a rate that an explosive reaction is quenched before structural damage occurs.[26,27] Figure 17 shows how the pressure varies with and without such protection. In the latter case, the pressure rise is approximately a cubic function of time, as noted earlier. In the former case, inert is added when the pressure or the rate of pressure rise exceeds a predetermined value. This occurs at the time t_i in figure 17 when the explosive actuators function to add the inert. As noted, the pressure increases momentarily above the value found in the unprotected case and then falls rapidly as the combustion reaction is quenched by the inert.

Flame Arrestors and Relief Diaphragms

Inert atmospheres must be used when not even a small explosive reaction can be tolerated. However, when the ignition of a flammable mixture would create little hazard if the burning mixture were vented, flame arrestors and relief diaphragms could be used effectively. The design of such systems is determined by the size and strength of the confining vessels, ducts, and so forth.

In recent studies of the efficiency of wire gauze and perforated block arrestors,[49,50] Palmer found the velocity of approach of the flame to be the major factor in determining whether flame passed through an arrestor. For these two types of arrestors, he found the critical approach velocity to be

$$V' = \frac{1.75\,k\,(T_h - T_o)}{m^{0.9}Q/x_o}, \tag{12}$$

and

$$V = \frac{9.6\,kA't\,(T_h - T_o)}{d^2Q/x_o}, \tag{13}$$

where k is the thermal conductivity of the gas; m is the mesh width; T_h is the mean bulk temperature of the flame gases through the arrestor; T_o is the initial temperature of the arrestor; Q is the heat lost by unit area of flame; x_o is the thickness of the flame propagating at the burning velocity, S; d is the diameter of an aperture; A' is the area of a hole in unit area of the arrestor face; and t is the arrestor thickness.

Equations (12) and (13) can be used to determine the mesh width or aperture diameter needed to stop a flame having a particular approach velocity. In practice, application of these equations assumes a knowledge of the flame speed in the system of interest. Some useful data have been made available by Palmer and Rasbash and Rogowski,[54,55] as well as by Jost[38] and Lewis and von Elbe.[41]

Tube bundles also may be used in place of wire screens. Scott found that these permit increased aperture diameters for a given approach velocity.[63]

In practice, it may be desirable to install pressure relief vents to limit damage to duct systems where flame may propagate. Rasbash and Rogowski[55] found that with propane- and pentane-air mixtures, the maximum pressure P_M (pounds per square inch) developed in an open-ended duct, having a cross section of 1 ft^2 is:

$$P_M = 0.07\,\frac{L}{D}, \text{ when } 6 \leq \frac{L}{D} \leq 48, \tag{14}$$

where L/D is the ratio of duct length to diameter. However, the presence of an obstacle (bend, constriction, etc.) in the path of escaping gases increased the pressure due to resistance to fluid flow by the obstacle. Location of a relief vent near the ignition source decreased the maximum pressure as well as the flame speed. For values of K (cross-section area of duct/area of vent) greater than 1, these authors found

$$0.8\,K \leq P_M \leq 1.8\,K, \tag{15}$$

where $2 \leq K \leq 32$, and $6 \leq \dfrac{L}{D} \leq 30$. To keep the pressure at a minimum either many small vents or a continuous slot was recommended rather than a few large vents. In addition, vents should be located at positions where ignition is likely to occur and should open before the flame has traveled more than 2 feet.

When possible, relief vents should be used with flame arrestors. The vents tend not only to reduce the pressure within a system following ignition but also to reduce the flame speed, thus making all arrestors more effective. Unfortunately, in certain large applications (for example, drying ovens), it is difficult to use flame arrestors effectively. In such cases, greater reliance must be placed on the proper functioning of relief vents. Simmonds and Cubbage[18,19,66] have investigated the design of effective vents for industrial ovens. They found two peaks in the pressure records obtained during the venting of cubical ovens (fig. 18). The first peak, P_1; the oven volume, V; the factor, K; and the weight per unit area (lb/ft^2) of relief, w, were related as fol-

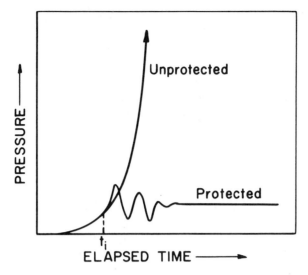

FIG. 17. Pressure variation following ignition of a flammable mixture in unprotected and protected enclosures.

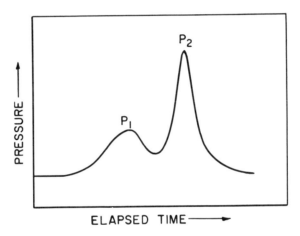

FIG. 18. Pressure produced by ignition of a flammable mixture in a vented oven.

lows for a 25 percent town gas [3]-air mixture:

$$P_1 V^{\frac{1}{3}} = 1.18kw + 1.57. \tag{16}$$

More generally,

$$P_1 V^{\frac{1}{3}} = S_o(0.3Kw + 0.4), \tag{17}$$

where S_o is the burning velocity of the mixture at the oven temperature.

The first pressure pulse was ascribed to the release and motion of the relief vent following ignition; the second pulse, to continued burning at an increased rate. The second pulse represents the pressure drop across the vent, and it is thus proportional to K. For small values of K it was found that

$$P_2 = K. \tag{18}$$

---

[3] Town gas contained approximately 52 pct hydrogen, 17 pct carbon monoxide, 15 pct methane; the balance was other hydrocarbons, 3 pct; nitrogen, 9 pct; carbon dioxide, 3 pct; and oxygen.

As with ducts, larger pressures were obtained when obstructions were placed in the oven.

In designing explosion reliefs for ovens, Simonds and Cubbage pointed out that (1) the reliefs should be constructed in such a way that they do not form dangerous missiles if an explosion occurs; (2) the weight of the relief must be small so that it opens before the pressure builds up to a dangerous level; (3) the areas and positions of relief openings must be such that the explosion pressure is not excessive; (4) sufficient free space must be utilized around the oven to permit satisfactory operation of the relief and minimize risk of burns to personnel; and (5) oven doors should be fastened securely so that they do not open in the event of an explosion.

Burgoyne and Wilson have presented the results of an experimental study of pentane vapor-air explosions in vessels of 60- and 200-cubic-foot volume.[13] They found the rates of pressure rise greater than could be predicted from laminar burning velocity data, so that the effect of a relief area in lowering the peak pressure was less than expected. All experiments were conducted at an initial pressure of 1 atmosphere. Vent data for use at higher initial pressures are summarized in an article by Block;[3] a code for designing pressure relief systems has been proposed in this article. Other authors have considered the effects of temperature and characteristics of the flammable mixture on vent requirements.[4,15,16,21,22,43,44,51,76]

FLAMMABILITY DATA

Flammability data are available in a number of publications including those of the National Fire Protection Association[48] and the Bureau of Mines.[91] Table A-1 of Bulletin 627, Bureau of Mines[91], gives a summary of limits and autoignition data for a number of individual gases and vapors in air at atmospheric pressure. Note especially that the limit of flammability data are for an ambient temperature of 25° C (77° F). At this temperature, the lower limit can be expressed as a simple function of the stoichiometric composition in air (C_{st}). In brief,

$$L_{25°(vol\ pct)} = K C_{st} \tag{19}$$

where K is 0.50 to 0.55 for a wide range of hydrocarbons and other flammable materials such as the alcohols, ethers, esters, aldehydes, and ketones. C_{st} may be obtained by the procedure given in Appendix B, Bureau of Mines Bulletin 627. On a weight basis,

$$L_{25°}(mg/l) \approx 48\ mg/l$$

where the limit is expressed in terms of the weight of combustible contained in a liter of air at STP (0°C and 760 mm Hg) (32°F and 29.9 mm Hg). Note that only the combustible content of the molecule is considered here. Thus, for the alcohols, ethers, esters, aldehydes, and ketones, we have

$$L_{25°}(mg/l) = 48\ \frac{M}{M - 16 \cdot n}\ mg/l$$

where M is the molecular weight and n is the number of oxygen atoms in the molecule.

At other temperatures,

$$L_t = L_{25°}\left(1 - \frac{t - 25°}{t_f - 25°}\right) \tag{20}$$

where L_t and $L_{25°}$ are the lower limits of flammability in air at $t°$ and 25° C (77° F) respectively (volume or weight basis), and t_f is the adiabatic flame temperature in °C. This equation assumes a constant flame temperature exists at the lower limit.[77] Another

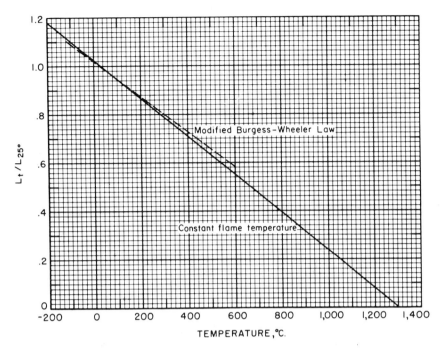

FIG. 19. Effect of molecular weight on lower limits of flammability of paraffin hydrocarbons at 25°C.

useful expression is that proposed by Zabetakis, Lambiris, and Scott;[86] this is the modified Burgess-Wheeler Law

$$L_t = L_{25°}\left[1 - \frac{0.75(t - 25°)}{L_{25°} \cdot \Delta H_c}\right] \qquad (21)$$

where ΔH_c is the heat of combustion in Kcal per mole. For example, Spakowski[68] has found $L_{25°} \cdot \Delta H_c$ to be 1,040 for the paraffin hydrocarbons so that

$$L_t = L_{25°}[1 - 7.21 \times 10^{-4}(t - 25°)] \qquad (22)$$

A plot of $L_t/L_{25°}$ (equations 20 and 21) for the paraffin hydrocarbons is given in figure 19. This figure may also be used to obtain the ratio $(\text{Min } O_2)_t/(\text{Min } O_2)_{25°}$. Also, when the heat release at the upper limit is equal to that at the lower limit, an expression similar to equation 21 can be written for U_t; since U_t increases with temperature, the correction term within the brackets would be added to 1 in this case. When cool flames are encountered, the upper limit is much higher than that predicted by assuming the heat release at the upper limit is equal to that at the lower limit. Further, many materials with a positive heat of formation, such as acetylene, propagate flame in the absence of air so that a true upper limit does not exist. However, an upper limit can be obtained by the addition of an inert gas or a flammable with a negative heat of formation.

Limits of flammability are not affected appreciably by small changes in pressure. However, a large pressure increase usually results in a widening of the flammable range. For example, an analysis of the data obtained by Jones and coworkers[30,37] shows the limits of flammability of a natural gas in air are given by the expressions:

$$L_{25°(\text{vol pct})} = 4.9 - 0.71 \log P(\text{atm}) \qquad (23)$$

and

$$U_{25°(\text{vol pct})} = 14.1 + 20.4 \log P(\text{atm}) \qquad (24)$$

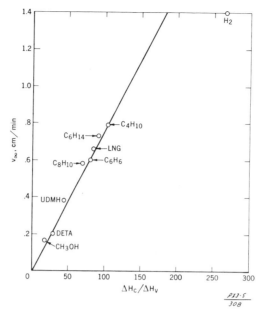

FIG. 20. Effect of temperature on lower limit of flammability of methane in air at atmospheric pressure.

with a standard error of estimate of 0.53 and 1.51 vol pct in L and U respectively. Similarly, for this same gas in a nitrogen-air atmosphere:

$$\text{Min } O_{2(\text{vol pct})} = 13.98 - 1.68 \log P(\text{atm}) \tag{25}$$

When the vapors above a pool of combustible liquid are ignited at atmospheric pressure and at temperature T_L (see figure 2), a flame propagates away from the ignition source. However, sustained burning usually does not occur until a slightly higher liquid temperature (fire point) is attained. Under wind-free conditions, the liquid regression rate, v, is

$$v = v_{\infty} (1 - e^{-kd}) \tag{26}$$

where v_{∞} and k are constants for a particular fuel burning in air, and d is the pool diameter; v_{∞} is given by the expression

$$v_{\infty} = 0.0076 \frac{\Delta H_c}{\Delta H_v} \text{ cm/min} \tag{27}$$

where ΔH_c is the net heat of combustion and ΔH_v is the sensible heat of vaporization. Figure 20 gives a summary of the v_{∞} values for a number of combustible including liquid hydrogen and liquefied natural gas (LNG). Equation 27 indicates that the linear burning rate (regression rate) is proportional to the fraction of the heat that must be fed back from the flame to the liquid pool to maintain vaporization. Thus, the burning rate increases with increase in heat of combustion and with decrease in heat of vaporization.

ACKNOWLEDGEMENT: The material in this chapter is taken from Bulletin 627, Bureau of Mines, Flammability Characteristics of Combustible Gases and Vapors, by Michael G. Zabetakis.

BIBLIOGRAPHY

1. Anson, D. Influence of the Quality of Atomization on the Stability of Combustion of Liquid Fuel Sprays. *Fuel, v. 32*, pp. 39–51, 1953.

2. Bartkowiak, Alphonse, Sotirios Lambiris, and Michael G. Zabetakis. Flame Propagation Through Kerosine Foams. *Combustion and Flame, v. 3*, pp. 347–353, 1959.

3. Block, Benjamin. Emergency Venting for Tanks and Reactors. *Chem. Eng.*, pp. 111–118, January 1962.

4. British Chemical Engineering. "Use of Bursting Discs at High Temperatures." p. 677, December 1957.

5. Browne, Howard C., Harold Hileman, and Lowell C. Weger. Barricades for High Pressure Research. *Ind. and Eng. Chem., v. 53*, pp. 52A–58A, 1961.

6. Browning, J. A., and W. G. Krall. "Effect of Fuel Droplets on Flame Stability, Flame Velocity, and Inflammability Limits." 5th Symp. (Internat.) on Combustion, Reinhold Pub. Corp., New York, pp. 159–163, 1955.

7. Browning, J. A., T. L. Tyler, and W. G. Krall. Effect of Particle Size on Combustion of Uniform Suspensions. *Ind. and Eng. Chem., v. 49*, pp. 142–148, 1957.

8. Burgoyne, J. H. Mist and Spray Explosions. *Chem. Eng. Prog., v. 53*, pp. 121-M—124-M, 1957.

9. ———. Principles of Explosion Prevention. *Chem. and Process Eng.*, pp. 1–4, April 1961.

10. Burgoyne, J. H., and L. Cohen. The Effect of Drop Size on Flame Propagation in Liquid Aerosols. *Roy. Soc. (London), Proc. v. 225*, pp. 375–392, 1954.

11. Burgoyne, J. H., and J. F. Richardson. The Inflammability of Oil Mists. *Fuel. v. 28*, pp. 2–6, 1949.

12. Burgoyne, J. H., and A. J. Steel. Flammability of Methane-Air Mixtures in Water-Base Foams. *Fire Res. Abs. and Rev., v. 4*, Nos. 1 and 2, pp. 67–75, 1962.

13. Burgoyne, J. H., and M. J. G. Wilson. "The Relief of Pentane Vapour-Air Explosions in Vessels." Proc. Symp. on Chem. Process Hazards, Inst. Chem. Eng. (London), pp. 25–29, 1960.

14. Cassutt, Leon H. Experimental Investigation of Detonation in Unconfined Gaseous Hydrogen-Oxygen-Nitrogen Mixtures. *ARS J.*, pp. 1122–1128, 1961.

15. Pane Relievers Vent Explosive Popoffs, Chem. Eng., pp. 70–72, Jan. 22, 1962.

16. Conison, Joseph. How to Design a Pressure Relief System. *Chem. Eng.*, pp. 109–114, 1960.

17. Coward, H. F., and G. W. Jones. Limits of Flammability of Gases and Vapors. *BuMines Bull. 503*, 155 pp., 1952.

18. Cubbage, P. A., and W. A. Simmonds. "An Investigation of Explosion Reliefs for Industrial Drying Ovens." Gas Council, Res. Communication GC23 (London), 46 pp., 1955.

19. ———. "An Investigation of Explosion Reliefs for Industrial Drying Ovens." Gas Council, Res. Communication GC43 (London), 41 pp., 1957.

20. Davenport, Donald E., George B. Huber, and Norman R. Zabel. Containment of Fragments From Runaway Nuclear Reactors. *Progress in Nuclear Energy, v. 4*, Pergamon Press, New York, pp. 484–503, 1961.

21. Davis, D. S. Emergency Vent Size for Tanks Containing Flammable Liquids. *British Chem. Eng. Nomogram No. 62, v. 6*, p. 644, 1961.

22. Diss, E., H. Karam, and C. Jones. Practical Way to Size Safety Disks. *Chem. Eng.*, pp. 187–190, 1961.

23. Dunn, Robert G., and Bernard T. Wolfson. Single Generalized Chart of Detonation Parameters for Gaseous Mixtures. *J. Chem. and Eng. Data, v. 4*, pp. 124–127, 1959.

24. Edwards, D. H., G. T. Williams, and J. D. Breeze, Pressure and Velocity Measurements on Detonation Waves in Hydrogen-Oxygen Mixtures. *J. Fluid Mech.*, pp. 497–517, November 1959.

25. Evans, Marjorie W., and C. M. Ablow. Theories of Detonation. *J. Chem. Rev., v. 129*, pp. 129–178, 1961.

26. Glendinning, W. G., and A. M. MacLennan. Suppression of Fuel-Air Explosions. *Nat. Fire Protect. Assoc. Quart., v. 45*, p. 61, July 1951.

27. Grabowski, G. J. Industrial Explosion Protection. *J. Am. Oil Chem. Soc., v. 36*, pp. 57–59, 1959.

28. Guest, P. G., V. W. Sikora, and Bernard Lewis. Static Electricity in Hospital Operating Suites: Direct and Related Hazards and Pertinent Remedies. *BuMines Rept. of Inv. 4833*, 64 pp., 1952.

29. Haber, F., and H. Wolff. Mist Explosions. *Zeit. angew. Chem., v. 36*, p. 373, 1923.

30. Hanna, N. E., M. G. Zabetakis, R. W. Van Dolah, and G. H. Damon. Potential Ignition Hazards Associated With Compressed-Air Blasting Using a Compressor Underground. *BuMines Rept. of Inv. 5223*, 33 pp., 1956.

31. Harris, D. N., Glenn Karel, and A. L. Ludwig. Electrostatic Discharges in Aircraft Fuel Systems. 41st Ann. API Meeting Preprint, 10 pp., 1961.

32. Herzog, R. E., E. C. Ballard, and H. A. Hartung. Evaluating Electrostatic Hazard During the Loading of Tank Trucks. 41st Ann. API Meeting Preprint, 7 pp., 1961.

33. Hildebrand, Joel H., and Robert L. Scott. "The Solubility of Nonelectrolytes." Reinhold Pub. Corp., New York, 3rd ed., p. 243, 1950.

34. Holdsworth, M. P., J. L. vander Minne, and S. J. Vellenga. Electrostatic Charging During the White Oil Loading of Tankers. 41st Ann. API Meeting Preprint, 30 pp., 1961.

35. Holtz, John C., L. B. Berger, M. A. Elliott, and H. H. Schrenk. Diesel Engines Underground. *BuMines Rept. of Inv. 3508*, 48 pp., 1940.

36. Jacobs, R. B., W. L. Bulkley, J. C. Rhodes, and T. L. Speer. Destruction of a Large Refining Unit by Gaseous Detonation. *Chem. Eng. Prog., v. 53*, pp. 565–573, December 1957.

37. Jones, G. W., R. E. Kennedy, and I. Spolan. Effect of High Pressures on the Flammability of Natural Gas-Air-Nitrogen Mixtures. *BuMines Rept. of Inv. 4557*, 16 pp., 1949.

38. Jost, Wilhelm. "Explosion and Combustion Processes in Gases." McGraw-Hill Book Co., New York, 621 pp., 1946.

39. Klinkenberg, A., and J. L. vander Minne. "Electrostatics in the Petroleum Industry." Elsevier Publishing Co., New York, 191 pp., 1958.

40. Kuchta, Joseph M., Sotirios Lambiris, and Michael G. Zabetakis. Flammability and Autoignition of Hydrocarbon Fuels Under Static and Dynamic Conditions. *BuMines Rept. of Inv. 5992*, 21 pp., 1962.

41. Lewis, Bernard, and Guenther von Elbe. "Combustion, Flames and Explosions of Gases." 2d ed. Academic Press, Inc., New York, 731 pp., 1961.

42. Linnett, J. W., and C. J. S. M. Simpson. "Limits of Inflammability." 6th Symp. (Internat.) on Combustion, Reinhold Pub. Corp., New York, pp. 20–27, 1957.

43. Lowenstein, J. G. Calculate Adequate Rupture Disk Size. *Chem. Eng., v. 65*, pp. 157–158, 1958.

44. Missen, R. W. Pressure Drop in Vapor-Relief Systems. *Chem. Eng.*, pp. 101–102, 1962.

45. Mullins, B. P. "Spontaneous Ignition of Liquid Fuels." Butterworths Sci. Pub. (London), 117 pp., 1955.

46. Mullins, B. P., and S. S. Penner. "Explosions, Detonations, Flammability and Ignition." Pergamon Press (London), 287 pp., 1959.

47. National Advisory Committee for Aeronautics. "Basic Considerations in the Combustion of Hydrocarbon Fuels With Air." Rep't 1300, ed. by Henry C. Barnett and Robert R. Hibbard, 259 pp., 1957.

48. National Fire Protection Association. "Fire-Hazard Properties of Flammable Liquids, Gases and Volatile Solids." NFPA 325, 60 Battery March St., Boston 10, Mass., 126 pp., 1960.

49. Palmer, K. N. The Quenching of Flame by Perforated Sheeting and Block Flame Arrestors. Proc. Symp. on Chemical Process Hazards, *Inst. Chem. Eng. (London)*, p. 51, 1960.

50. ———. "The Quenching of Flame by Wire Gauzes." 7th Symp. (Internat.) on Combustion, Butterworths Sci. Pub. (London), pp. 497–503, 1959.

51. Porter, R. L. Unusual Pressure Relief Devices. *Ind. and Eng. Chem., v. 54*, pp. 24–27, January 1962.

52. Randall, P. N., J. Bland, W. M. Dudley, and R. B. Jacobs. Effects of Gaseous Detonations Upon Vessels and Piping. *J. Chem. Eng. Prog., v. 53*, pp. 574–580, December 1957.

53. Randall, P. N., and I. Ginsburgh. Bursting of Tubular Specimens by Gaseous Detonation. *ASME, Trans., Paper No. 60-WA-12*, 9 pp., 1960.

54. Rasbash, D. J., and Z. W. Rogowski. Gaseous Explosions in Vented Ducts. *Combustion and Flame, v. 4*, pp. 301–312, December 1960.

55. ———. Relief of Explosions in Duct Systems. Proc. Symp. on Chem. Process Hazards, *Inst. Chem. Eng. (London)*, p. 58, 1960.

56. Rebenstorf, Melvin A. Designing a High Pressure Laboratory. *Ind. and Eng. Chem., v. 53*, pp. 40A–42A, 1961.

57. Roberts, A., B. R. Pursall, and J. B. Sellers. Methane Layering in Mine Airways. *Coll. Guard., v. 205*, pp. 535–541, October 1962.

58. ———. Methane Layering in Mine Airways. Part II, The Theory of Layering. *Coll. Guard., v. 205*, pp. 588–593, 1962.

59. ———. Methane Layering in Mine Airways. Part III. The Theory of Layering (Continued). *Coll. Guard., v. 205*, pp. 630–636, 1962.

60. ———. Methane Layering in Mine Airways. Part IV, Studying Layering Phenomena in a Laboratory Model. *Coll. Guard., v. 205*, pp. 723–732, 1962.

61. Roberts, A., B. R. Pursall and J. B. Sellers. Methane Layering in Mine Airways. Part V, Studying Layering Phenomena in a Laboratory Model. *Coll. Guard., v. 205*, pp. 756–763, 1962.

62. Rogers, D. T., and C. E. Schleckser. "Engineering and Theoretical Studies of Static Electricity in Fuels." Proc. 5th World Petrol. Cong., New York, sec. 8, paper 10, pp. 103–121, 1959.

63. Scott, George S., Henry E. Perlee, George H. Martindill, and Michael G. Zabetakis. "Review of Fire and Explosion Hazards of Flight Vehicle Combustibles." Aeronautical Systems Division Tech. Rept. 61–278, supp. 1, 43 pp., 1962.

64. Semenov, N. N. "Some Problems in Chemical Kinetics and Reactivity." Princeton Univ. Press, Princeton; N.J., v. 2, 331 pp., 1959.

65. Setchkin, Nicholas P. Self-Ignition Temperatures of Combustible Liquids. *J. Res. NBS, v. 53*, Res. Paper 2516, pp. 49–66, 1954.

66. Simmonds, W. A., and P. A. Cubbage. The Design of Explosion Reliefs for Industrial Drying Ovens. Proc. Symp. on Chem. Process Hazards. *Inst. Chem. Eng. (London)*, p. 69, 1960.

67. Simon, Dorothy M., Frank E. Belles, and Adolph E. Spakowski. "Investigation and Interpretation of the Flammability Region for Some Lean Hydrocarbon-Air Mixtures." 4th Symp. (Internat.) on Combustion, Williams & Wilkins Co., Baltimore, Md., pp. 126–138, 1953.

68. Spakowski, A. E. "Pressure Limit of Flame Propagation of Pure Hydrocarbon-Air Mixtures at Reduced Pressures." Nat. Advisory Committee for Aeronautics Res. Memo. E52H15, 33 pp., 1952.

69. Stenberg, John F., and Edward G. Coffey. Designing a High-Pressure Laboratory. *Chem. Eng.*, pp. 115–118, 1962.

70. Stoner, R. G., and W. Bleakney. The Attenuation of Spherical Shock Waves in Air. *J. Appl. Phys., v. 19*, pp. 670–678, 1948.

71. Sullivan, M. V., J. K. Wolfe, and W. A. Zisman. Flammability of the Higher-Boiling Liquids and Their Mists. *Ind. and Eng. Chem., v. 39*, p. 1607, 1947.

72. Thomas, A. "Flame Propagation Through Air-Fuel Foams." 6th Symp. (Internat.) on Combustion, Reinhold Pub. Corp., New York, pp. 701–707, 1957.

73. U.S. Atomic Energy Commission. "The Effects of Nuclear Weapons." Ed. by Samuel Glasstone, p. 163, April 1962.

74. Van Dolah, Robert W., Michael G. Zabetakis, David S. Burgess, and George S. Scott. Review of Fire and Explosion Hazards of Flight Vehicle Combustibles. *BuMines Inf. Circ. 8137*, 80 pp., 1962.

75. Weber, J. P., J. Savitt, and K. Krc. Detonation Tests Evaluate High Pressure Cells. *Ind. and Eng. Chem., v. 53*, pp. 128A–131A, 1961.

76. Weil, N. A. Rupture Characteristics of Safety Diaphragms. *J. Appl. Phys.*, pp. 621–624, 1959.

77. White, A. G. Limits for the Propagation of Flame in Inflammable Gas-Air Mixtures. III. The Effect of Temperature on the Limits. *J. Chem. Soc., v. 127*, pp. 672–684, 1925.

78. Winter, E. F. The Electrostatic Problem in Aircraft Fueling. *J. Royal Aeronautical Soc., v. 66*, pp. 429–446, 1962.

79. Wolfson, Bernard T., and Robert G. Dunn. Generalized Charts of Detonation Parameters for Gaseous Mixtures. *J. Chem. Eng. Data, v. 1*, pp. 77–82, 1956.

80. Zabetakis, M. G., Gasfreeing of Cargo Tanks. *BuMines Inf. Circ. 7994*, 10 pp., 1961.

81. Zabetakis, M. G., J. C. Cooper, and A. L. Furno. Flammability of Solvent Mixtures Containing Methyl Ethyl Ketone and Tetrahydrofuran in Air. *BuMines Rept. of Inv. 6048*, 14 pp., 1962.

82. Zabetakis, M. G., and H. H. Engel. Flammable Materials: A Lecture Demonstration. *BuMines Inf. Circ. 8005*, 17 pp., 1961.

83. Zabetakis, M. G., A. L. Furno, and G. W. Jones. Minimum Spontaneous Ignition Temperatures of Combustibles in Air. *Ind. and Eng. Chem., v. 46*, pp. 2173–2178, 1954.

84. Zabetakis, M. G., and Jones, G. W. The Prevention of Industrial Gas Explosion Disasters. *Chem. Eng. Prog., v. 51*, pp. 411–414, 1955.

85. Zabetakis, Michael G., George W. Jones, George S. Scott, and Aldo L. Furno. "Research on the Flammability Characteristics of Aircraft Fuels." Wright Air Development Center, Tech. Rept. 52–35, supp. 4, 85 pp., 1956.

86. Zabetakis, M. G., S. Lambiris, and G. S. Scott. "Flame Temperatures of Limit Mixtures." 7th Symp. (Internat.) on Combustion, Butterworths Sci. Pub. (London), pp. 484–487, 1959.

87. Zabetakis, M. G., C. M. Mason, and R. W. Van Dolah. The Safety Characteristics of Normal Propyl Nitrate. *BuMines Rept. of Inv. 6058*, 26 pp., 1962.

88. Zabetakis, M. G., and J. K. Richmond. "The Determination and Graphic Representation of the Limits of Flammability of Complex Hydrocarbon Fuels at Low Temperatures and Pressures." 4th Symp. (Internat.) on Combustion, Williams & Wilkins Co., Baltimore, Md., pp. 121–126, 1953.

89. Zabetakis, M. G., and B. H. Rosen. Considerations Involved in Handling Kerosine. *Proc. Am. Petrol. Inst., sec. 3, v. 37*, p. 296, 1957.

90. Zeldovich, Ia. B., S. M. Kogarko, and N. N. Simonov. An Experimental Investigation of Spherical Detonation of Gases. *Zhurnal Tekh. Fiziki (J. Tech. Phys. (USSR)), v. 26*, p. 1744, 1956.

91. Zabetakis, M. G., Bulletin 627, Bureau of Mines, "Flammability Characteristics of Combustible Gases and Vapors." GPO, Washington, D. C., 121 pp., 1965.

Section **5**

Chemical Reactions

The Safe Operation of Laboratory Distillations Overnight

Daniel R. Conlon

INTRODUCTION

Scientific laboratories differ greatly in the extent to which laboratory distillations are permitted to run unattended during the evening hours. Most scientists consider the familiar "distilled water" still to be free of trouble. Many laboratories have had such stills operating day and night for years with the very minimum of attention. However, other distilling units are often regarded as hazardous either because of their construction or because the material being distilled is toxic or flammable.

If these other stills are examined carefully, the sources of potential trouble can be eliminated, and these stills too can be safely operated into or through the evening hours.

FAILURE OF THE COOLING WATER

Most distillation units operate with water-cooled condensers; therefore, it is essential for safe operation that the water supply be dependable. However, water pressures change:

1. There are decreases in water pressure which may be caused by failure of a pump, by partial blocking of the mains, by increased consumption by other laboratories on the same water main, or by water-main breaks, lawn sprinkling or fire department water use.

2. There are increases in water pressure which may occur when auxiliary pumps are being switched on, usually at the beginning of the working day; when other laboratories turn off their water valves, usually at the end of the day; or at other times during the day or evening when water system pressures increase because of reduced consumption by many or major users.

Either type of pressure variation can cause trouble. Too low a pressure results in an inadequate water flow to the cooling condenser, thereby allowing distillate vapors to escape. Too high a pressure can cause the tubing connections to the condensers to expand and burst, which is doubly undesirable: the laboratory is flooded, and vapors may be released since there is no flow of cooling water to the condenser.

It is not at all difficult to protect against both hazards if the following four-point approach is used:

1. Install a simple pressure regulator[1] in the water main ahead of the valve that is used to adjust the flow to the condenser. The regulator can be adjusted to maintain an intermediate water pressure; it will then minimize the effect of both increases and decreases in the supply pressure.

2. Protect the pressure regulator with a suitable water filter, which will prevent particles of rust and other foreign materials from affecting its operation. Commercial filters with replaceable cartridges made of cotton or polypropylene twine are readily available.[2]

3. Monitor the flow of cooling water flowing *from* the cooling condenser to the drain. Monitoring this emergent water flow ensures that water has flowed *through* the condenser. Thus, it is inherently a more reliable approach than monitoring the main pressure.

4. Install a solenoid valve[3] in the water supply line and connect it electrically to

the water-flow monitor so that electrical power to this valve and to the still-pot heater will be turned off when the water-flow monitor responds to a water failure.

A number of water-flow monitoring devices have been described in the literature.[11-16] A modification of Houghton's device[13] is shown in Figure 1A, made of glass, and in 1B, made of metal. This design has the advantages that it imposes no back pressure on the apparatus being monitored; it possesses a short time delay so that adjustments in the flow can be readily made; and its flow-restricting capillary can be readily inspected.

Figure 2 shows a commercially available water-flow monitor[4] based on this same principle. The commercial unit has a built-in alarm plus a number of electrical receptacles, several of which are automatically turned off when the cooling water fails. The four-point approach recommended above is shown in Figure 3, items A, B, C, & D.

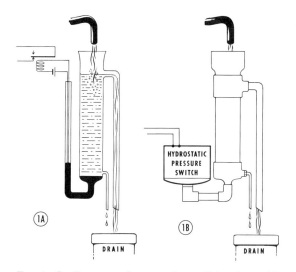

FIG. 1. Cooling water-flow monitors (1A glass, 1B metal). When the flow ceases, the column drains through the capillary, and the mercury contact or the pressure switch opens.

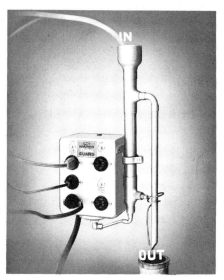

FIG. 2. Water-flow monitor equipped with alarm and convenience outlets.

FAILURE OF THE ELECTRICAL SUPPLY

It is important that the line voltage used for the distillation remain relatively constant since even moderate changes will affect the *rate* of distillation. However, electric voltages, like water pressures, change:

1. There are decreases in line voltage due to increased consumption by other laboratories, usually early in the day. This will be most noticeable if the lines are inadequate.

2. There are increases in line voltage when additional generating equipment is turned on, or when other laboratories turn off their electrically-operated equipment at the end of the day.

The following two-point approach will guard against difficulties from these variations:

1. As with water pressure, there is a convenient device for overcoming source fluctuations: the constant-voltage transformer. A 500-, 1000-, or 2000-watt unit would be suitable. A 2000-watt Sola constant-voltage transformer[5] costs approximately $250 and can accommodate several distillation columns. Such a unit is somewhat bulky (27 × 32 × 51 cm; or 10.6 × 12.6 × 20 in.) however, it may be mounted vertically under the other electrical controls. The insurance it provides against fluctuations in

line voltage greatly outweighs any inconvenience. These units are inherently reliable and trouble free.

2. A constant-voltage transformer cannot cope with a line voltage failure. Such failure in itself is not serious: the distillation merely stops. However, when the power is turned on again, "bumping" may well take place in the still-pot. If "bumping" is a serious problem, one should consider wiring a suitable relay into the still-pot heating circuit. Failure of the electrical supply should cause the relay to "drop out" and to stay out until it is manually reset.

An alternative way to achieve this same result is to plug the solenoid valve referred to previously into the water-flow monitor provided. A failure of the electric supply will cause the solenoid valve to close, shutting off the flow of cooling water and hence causing the water-flow monitor to de-energize its outlets. Since the solenoid valve is connected to one of these outlets, both it and the water-flow monitor are unable to reset themselves should the power failure be corrected. This approach has the advantage of having an inherent time delay—it will not be affected by momentary electrical cessations.

TERMINATION OF DISTILLATION

A question that often arises in deciding whether to leave an unattended distillation running during the night is how to stop the distillation if it is not to run until morning. There are several methods; the choice depends upon the nature of the distillation.

Timer Terminations. If the distillation is similar to the redistillation of a solvent, the still-pot and the still-head temperatures will not change greatly during the course of the distillation. Consequently, if, as has been recommended, a constant-voltage transformer is used to guard against variations in line voltage, the distillation rate will be reasonably constant. With experience, one can, therefore, estimate rather closely the amount of distillate that should accumulate during the night, and the distillation can be safely terminated with an electric timer. Such timers are familiar tools in most laboratories.

Temperature Terminations. If the boiling range of the material being distilled is wide rather than narrow, the rate of distillation will not remain constant. In this case, it will be better to terminate the distillation at a predetermined temperature rather than at a predetermined time. This is a common situation, not only with distillations left running for days on end but also with one-day runs that are not quite finished by "quitting time."

The scientist can readily terminate such a distillation through the use of a capacitance-actuated controller[6] which senses the position of the mercury in a thermometer located either in the still-pot or in the still-head. This is shown in Figures 3 and 4, items EE′ and FF′. When the still-pot heater is plugged into the controller, the heat to the still-pot will be cut off automatically when the temperature rises to the desired value. (A limit switch in the controller, which must be manually reset, keeps the power from turning on again when the temperature drops.)

Weight and Volume Terminations. If one prefers to terminate the distillation on the basis of the weight or the volume of the distillate, several other approaches may be used:

1. If the system need not be airtight and the product receiver can be supported on a pan balance, movement of the balance can be used to open an electrical circuit and terminate the distillation. When the balance moves, the heater is turned off (Fig. 5). The weights used on the balance should allow for the force used to actuate the switch.

This approach can be modified if the system must be airtight; flexible tubing can be run from the condenser to the product receiver mounted on the balance. It may be

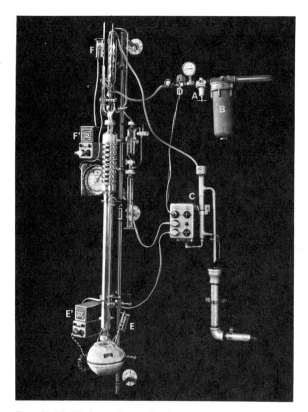

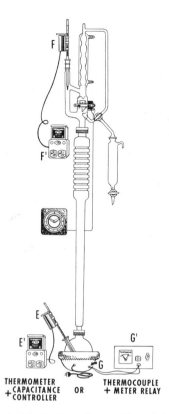

THERMOMETER
+CAPACITANCE   OR   THERMOCOUPLE
CONTROLLER   + METER RELAY

FIG. 3. Distillation column showing (A) water pressure regulator, (B) water filter, (C) emergent water-flow monitor, (D) solenoid valve in water line, and (EE′) (FF′) capacitance-actuated controllers for monitoring still-pot and still-head thermometers. (Glass equipment: Courtesy of Ace Glass Co., Vineland, New Jersey)

FIG. 4. Distillation column with still-head temperature being monitored for over-temperature or under-temperature by FF′, and still-pot being monitored for over-temperature by either EE′ or GG′. Note: If FF′ is not used for safety monitoring, it may be used to control the reflux ratio automatically by turning the reflux timer off whenever the still-head temperature exceeds the set point.

necessary to run two flexible lines—one for liquid flow, the other for vent. Care should be used in selecting the tubing used. It must not be attacked by the materials being distilled.

2. If a volume termination is preferred, there are several choices—these include the use of mercury contacts, capacitance-actuated controllers, and optical means of detecting liquid level. Figure 6 shows three examples schematically.

3. It is quite likely that automatic fraction collecting devices that have been developed for chromatographic purposes can be adapted for use with distillation columns. Termination of the distillation would be done on the basis of the number of fractions collected of a fixed volume.

POSSIBLE OVERHEATING OF THE STILL-POT

Once the foregoing factors have been taken into consideration, the more subtle hazards should receive attention. Laboratory distillations involving many liters of raw material are not likely to run dry overnight. However, it is conceivable that sooner or later this unlikely situation might develop because of some oversight. The distilling

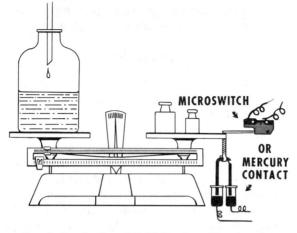

Fig. 5. Termination of distillation by weight of product.

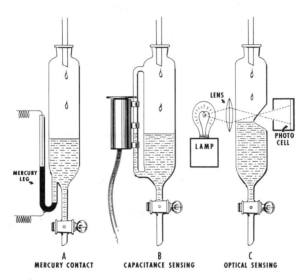

Fig. 6. Termination of distillation by volume of product.

apparatus can be monitored either by an over-temperature or an under-termperature approach:

1a. If a heating mantle is used with the still-pot, one can make use of the thermo-couple that is embedded in most heating mantles.[7] This can be attached to a suitable millivoltmeter equipped with an adjustable over-temperature sensing pointer (Fig. 4, CC'). These units, known as meter-relays, are commercially available.[8] They should be ordered complete with auxiliary relays capable of interrupting heater currents of several amperes.

1b. The still-pot itself can be monitored for over-temperature by the use of a thermoregulator plus an electronic relay or by a thermometer with a capacitance-actuated controller similar to that described above, but for monitoring rather than for termination (Figure 4, EE').

2. In some cases it is convenient to monitor the still-head temperature against a decrease in temperature. A decrease should occur if the liquid in the still-pot becomes quite small in volume; the *rate* of distillation will then decrease greatly and the still-head temperature will drop. This monitoring of the still-head temperature can also be

accomplished readily with either the thermocouple plus meter-relay[8] or with the still-head thermometer plus a capacitance-actuated controller[6] (Figure 4, FF').

BREAKAGE AND PHYSICAL SEPARATION

In any safety discussion it is important to consider the possible breakage of the glass equipment. Such breakage is highly unpredictable since it may result either from residual internal stresses in the glass, from stresses generated by improper external supports, or from an accidental blow.

Internal stresses in the glass apparatus may arise from inexperienced glass blowing or from careless commercial manufacture. *It is strongly recommended that all parts of the glass distillation equipment be carefully annealed and be checked for residual stress by means of polarized light.* Commercial polarized light units are available for this purpose,[9] or a stress-analysis unit can be improvised from two sheets of Polaroid and a light source. Some large research laboratories inspect with polarized light all glass apparatus which they receive.

The method used to support the distillation column is of the greatest importance in minimizing external stresses. Thought must be given to the design of the supporting framework, the method of clamping, and the design and location of the individual clamps.

A complete distillation column is usually rather tall and narrow. The column, head, and receivers can therefore be supported by a single support rod. The use of a rigid support rod and heavy-duty clamps and clamp holders is important in order to prevent one part of the apparatus from shifting with respect to the other parts. It is recommended that the rod be at least 1.90 cm (3/4 in.) in diameter and securely fastened to a wall. The following procedures are suggested:

1. The column itself is first firmly clamped either at its top, bottom, or at some enlarged portion. The clamp used must be strong enough to carry the full weight of the column and head.

2. A second clamp is then placed on the column. This is tightened only slightly; it is intended as a guide against lateral movement.

3. When the still-pot is attached to the bottom of the column, it can be held in place by a cradle supported by springs. The still-pot will then be free to align itself with the column.

4. The distillation head is placed on the top of the column and is held loosely in position by a clamp which allows it to move vertically but largely restricts lateral movement.

5. The receivers are then supported with care in order to prevent them from putting strain on the distillation head. Careful support of this part as well as of the rest of the apparatus is essential in order to prevent physical parting of the ground joints.

6. If the apparatus is extensive, or if the receivers when loaded will be heavy, then it may be desirable to introduce semi-flexible connecting links between the column and the receivers. One way of doing this is shown in Figure 7. The links illustrated are easily fabricated from ground-glass ball-and-socket joints.

In passing, mention should be made of an alternative method of supporting distillation equipment: namely, to mount the assembly on a rigid panel made of metal or other non-combustible material. In this case, a large number of supporting brackets may be used to clamp all parts of the assembly. A glassblower may seal the parts together after they are supported, thus eliminating any residual stresses due to the clamping.

Finally, breakage due to accidental blows should be prevented by (1) locating the apparatus in a corner of the laboratory, out of the main line of traffic, and by (2) the use of adequate safety shields. Such shields may be flat, rectangular, or semi-cylindrical. Those shown in Figure 8 are hinged and particularly convenient to use.[10]

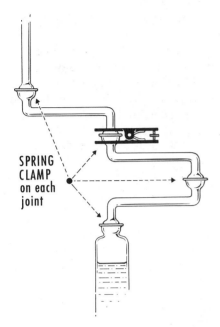

FIG. 7. Ball-and-socket links in transfer lines can flex and hence eliminate any stress present.

FIG. 8. Modern acrylic shields protect apparatus against external blows, yet are easily opened for manipulation.

OVERFLOW

Once a laboratory distillation is set up and allowed to run overnight, the operation may prove so convenient and satisfactory that the scientist forgets his initial caution. He might, at this time, forget to empty the receiver at the proper time, or might underestimate the distillation rate. A simple safeguard is to use a product receiver that is large enough to contain the entire charge to the distillation. If this is undesirable, one can use a smaller receiver, but connect it to a large overflow bottle. A five- or ten-liter bottle equipped with a two-hole rubber stopper and a drying tube in the vent line is a convenient safeguard.

Even after these precautions have been taken, it is worthwhile to mount the entire distillation apparatus in or over a metal tray large enough to contain all the liquid being distilled. A tray 8 cm ($3\frac{1}{4}$ in.) deep is usually adequate and is not too cumbersome.

MONITORING BY THE WATCHMAN

In large, well-staffed laboratories, the safety committee may insist that the night watchman, on his periodic trips through the laboratory, glance at any equipment that is running. Some scientists may feel that their "guard" has no feeling for scientific apparatus. Nevertheless, whenever apparatus is running at night, it is wise to have the apparatus so clearly labeled that an inexperienced, scientifically untrained guard could turn it off—preferably by turning no more than two or three conspicuously labeled switches and valves.

CONCLUSION

Many laboratories are quite cost conscious; other laboratories are hampered by shortages of personnel, while some are under great pressure to achieve results in a short period of time. Whatever the motivation, there are real advantages in being able to run laboratory operations including distillations, in an unattended condition, provided this can be accomplished in a safe manner.

In safety programs it is often difficult to agree on the exact sources of potential trouble and their relative importance. Hence, it is difficult at times to agree on which precautions are necessary and which are simply advisable. Nevertheless, there are few distillations that cannot operate unattended—not only during the night, but during the day as well—if careful plans are worked out along the lines suggested here.

ACKNOWLEDGEMENT: Reprinted from the *Journal of Chemical Education, Vol. 43*, A589 and A652, July and August, 1966, with permission.

REFERENCES

Components Cited

Note: These are typical components that may be used in the manner described in the text. Model numbers of electrical components correspond to 110 volts, 60 cycles. Manufacturers should be consulted for different voltage ratings. * indicates that there are a number of different models available depending upon the application intended.

1. Water Pressure Regulator, Catalog #1/4-N263. Watts Regulator Co., P. O. Box 810, Lawrence, Mass. 01842.
2. Water Filter, Model #L10U-3/4. Filterite Corp., Timonium, Md. 21093.
3. Solenoid Valve, Catalog #826225. Automatic Switch Co., Florham Pak, N. J. 07932.
4. WATER FLOW GUARD, Model WF-3. Instruments for Research & Industry, Cheltenham, Pa. 19012.
5. Constant Voltage Transformer, type CVS; for 2000 watts. Catalog #23-26-220. Sola Electric Co., Elk Grove Village, Ill. 60007.
6. THERM-O-WATCH Controller, Model L-6. Instruments for Research & Industry, Cheltenham, Pa. 19012
7. GLAS-COL Heating Mantles* (various sizes and styles). Glas-Col Apparatus Co., Terre Haute, Ind. 47803
8. COMPACK I,* Meter-Relay, Model 503K. Specify temperature range and thermocouple wire. API Instruments Co., Chesterland, Ohio 44026.
9. Bethlehem Apparatus Co., 825 Front Street, Hellertown, Pa. 18055
10. LAB-GUARD, Model H-16-30-1C. Instruments for Research & Industry, Cheltenham, Pa. 19012.

Literature Cited

11. Burford, H. C., Improved Flow-Sensitive Switch, *Journal of Scientific Instruments, 37,* 490, (1960).
12. Cox, B. C., Protection of Diffusion Pumps Against Inadequate Cooling, *Journal of Scientific Instruments, 33,* 148, (1960).
13. Houghton, G., A Simple Water Failure Guard for Diffusion Pumps and Condensers, *Journal of Scientific Instruments, 33,* 199, (1956).
14. Mott, W. E., and Peters, C. J., Inexpensive Flow Switch for Laboratory Use, *Review of Scientific Instruments, 32,* 1150, (1961).
15. Pike, E. R., and Price, D. A., Water Flow Controlled Electrical Switch, *Review of Scientific Instruments, 30,* 1057, (1959).
16. Preston, J., Bellows Operated Water Switch, *Journal of Scientific Instruments, 36,* 98, (1959).

Running Laboratory Reactions under Safe Control

Daniel R. Conlon

INTRODUCTION

Laboratory apparatus varies greatly in complexity—from the relatively simple thermostated ovens used for physical tests to quite complicated setups used for chemical operations and chemical reactions. Both scientists and safety engineers differ greatly in their attitudes toward apparatus and in their willingness to have apparatus running unattended at any time. Therefore, in some laboratories, practically every piece of equipment is shut down at the end of the working day. In others, numerous pieces of equipment and even chemical reactions run unattended. In general, the scientist finds it to his advantage to be able to have experimental equipment run unattended for either a few minutes, hours, or days.

The determining factors in deciding whether it is worthwhile to develop a set of conditions so that a laboratory apparatus can operate safely in the absence of the scientist are:

1. Is the operation repetitive and so demanding of close personal attention that it is quite tedious although not lengthy? (Exothermic reactions and vacuum-stripping operations are examples.)

2. Is it a lengthy experiment requiring many hours or days to run?

    a. If the experiment is lengthy, are there real disadvantages in interrupting the operation?

    b. If it were to be left running, could it be terminated automatically at some *desired* point?

    c. Could it be terminated at *any point* should some malfunction develop?

If the answer to these questions is yes, then one should examine the apparatus, determining which parts should be "watched" or monitored automatically. In such an examination, one usually finds there are very few units so complicated that they cannot safely be left running under automatic control; but also one finds there are few units so inherently safe that they do not need some safety monitoring.

This chapter considers both the simple apparatus and the complicated apparatus, both the endothermic reaction and the exothermic reaction, both the short-term operation and the long-term operation, both "control monitoring" and "safety monitoring."

SIMPLE ENDOTHERMIC EXPERIMENT

Figure 1 shows three apparatus types which from a safety point of view may be considered together: (A) the laboratory oven, (B) the constant temperature bath, and (C) the glass reaction flask. The assumptions made in grouping these together are: (1) only the variable of temperature need be "watched"; (2) should the temperature exceed a desired value, it can be detected by a monitor which automatically turns off the electric power; (3) since the operation is not exothermic, once the power is turned off, the apparatus will cool safely.

Apparatus such as ovens and constant temperature baths, (Figure 1, A & B) are not usually considered to be hazardous. Therein lies the danger; they are overlooked in safety inspections. This is particularly dangerous in the case of oil baths. If the oil bath thermostat fails, the oil will overheat and a fire may start. (Even if no fire results, valuable apparatus may be damaged and irreplaceable samples may be lost.)

Figure 1D represents schematically any experimental apparatus which is heated by an electric heater of either an external or an immersion type. Electrical power is supplied to the heater cyclically through a temperature controller connected to a sensing probe in the apparatus.

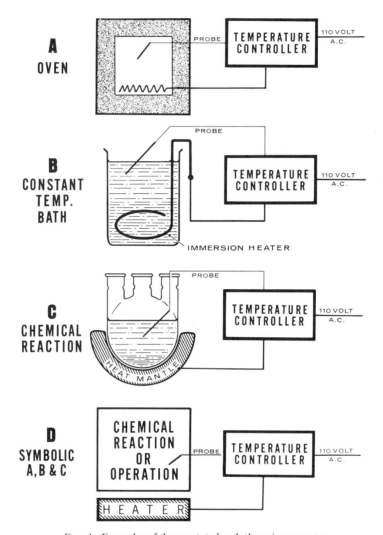

FIG. 1. Examples of thermostated endothermic apparatus.

As shown in Figure 2, to detect thermal malfunction of the apparatus in Figure 1, one may use a monitor to watch:
1. The temperature of the apparatus
2. The temperature controller
3. The temperature sensing probe
4. The power supplied cyclically to the heater
5. The temperature of the heater

One might reason that since the experiment is primarily concerned with the temperature of the apparatus (Item 1) this is what should be monitored. A good way of so monitoring is to use a secondary controller and probe that are independent of the primary temperature controller. If space is available, a relatively rugged bimetallic thermoregulator[1] is ideal for this purpose. It should be set to operate 5° to 10° above

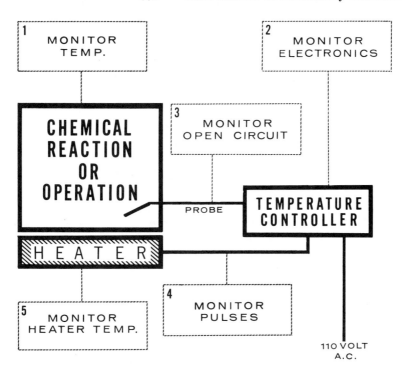

FIG. 2. Five possible ways to monitor an apparatus for malfunction of the temperature controller.

the normal temperature of the apparatus and can be electrically connected to take over control of the temperature if the primary controller fails. In that event, it would hold the apparatus temperature at the 5° to 10° higher level. It may, on the other hand, be wired to a relay circuit (Fig. 3) designed so that once the THERMOSWITCH opens, the relay will keep the power off until a push-button switch is reset. Figure 4 shows a commercially available "over-temperature guard" based on this design.[2]

With smaller apparatus it may not be convenient to use a bulky bimetallic thermo-regulator (although some are fairly small). If such is the case, one may use a thermo-couple connected either to a pyrometer controller[3-5] or to a meter relay.[6] A thermistor plus a bridge-type controller may also be used.[7]

If one prefers to insert no secondary probe into the apparatus, then one should either use an external sensor or should monitor items 2, 3, 4, or 5. In some cases (Item 2) the primary controller may be partially able to monitor itself. Pyrometer controllers are often designed so that if certain internal components fail, the controller will cut off the heater power. (This is known as *Fail Safe* design.) A somewhat different approach is to make use of a pyrometer that has two independent control points. The second control point can be set at a higher level to sound an alarm, open a circuit, or take remedial action. Admittedly, the pyrometric approaches are not 100% *Fail Safe*, but they do help to guard against trouble. This is also the case when one monitors Item 3— the sensing probe. This, again, is common practice with pyrometer controllers; they are usually designed so that the controller will turn off the heat if there is a thermo-couple break.

Item 4 calls for monitoring the ON and OFF power cycles to the heater. This can be accomplished with a so-called "reset," or "interval," timer or a "time delay" relay.[8] Such units are designed to open or to close a circuit after a period of seconds, minutes, or hours. When connected parallel with the heater, one of these can sound an alarm or turn off the heat if for any reason the electric power remains continuously ON. This approach has the advantage in that it monitors the apparatus without physical contact with the apparatus.

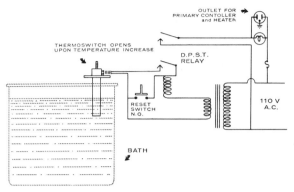

OVER-TEMPERATURE CUT-OFF FOR OIL BATHS

FIG. 3. Wiring diagram of an over-temperature relay circuit. Reset switch must be pushed to re-energize circuit.

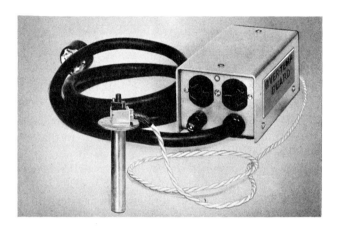

FIG. 4. Compact relay unit based on circuit shown in Figure 3.

Item 5 is sometimes the most convenient to monitor—particularly when one is using heating mantles into which the manufacturer has built a thermocouple. Again, one can attach the thermocouple either to a conventional pyrometer or to a meter relay.

Once the choice is made as to which approach (Items 1–5) one prefers, one then proceeds to the next decision: what action should the monitor take if an excessive temperature occurs?

1. Should the monitor permanently shut down the apparatus until an operator pushes a reset button?

2. Should the monitor allow the apparatus to continue to operate but at some slightly higher value?

3. Should the monitor give the disturbance a short time to disappear and take no action if the disturbance is only momentary?

4. Should the monitor make any record that a disturbance occurred?

The above choices apply not only to the monitors described above but also to many of the monitors described in the following sections. For some situations, the monitor should shut down the apparatus rather than exert corrective action because normal operating conditions *cannot* possibly be restored. In other cases, however, the monitor should take corrective action because normal operating conditions *can* readily be restored.

COMPLEX APPARATUS

While the scientist is almost always concerned with temperature, the apparatus is often more complicated and involves still more variables than the examples shown in Figures 1 and 2. Figure 5 illustrates schematically a more complicated setup. The ten circles symbolize the components and/or variables the scientist would naturally watch if he were personally present. They represent, in some cases, malfunction of the apparatus and, in other cases, the progress of the reaction. They are as follows:

1. The rate of the stirring
2. The viscosity of the reaction mixture
3. The rate at which reflux is rising into the reflux condenser
4. The flow of cooling water to any water-cooled part
5. The rate of distillation of product or by-product out of the reaction flask
6. The volume of product or by-product in the distillate receivers
7. The pressure in the apparatus: either hydrostatic pressure, gaseous pressure, or vacuum, depending upon the nature of the apparatus
8. The rate of flow of gas (or liquid) to or from the reaction
9. Other indicators of the progress of the reaction, depending upon the specific reaction
10. The time

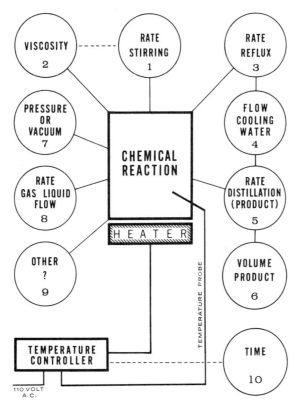

FIG. 5. Schematic illustration of a complex reaction apparatus.

It should be remembered that with each variable one usually has the choice of several different monitoring approaches. To illustrate:

1. The rotation of the stirrer may be monitored by (a) a meter relay which measures the electric current to the stirrer motor; or (b) a tachometer with either a switch output or a meter relay output, the tachometer being coupled either to the motor or to the stirrer shaft; or (c) an improvised tachometer such as is shown in Figure 6.

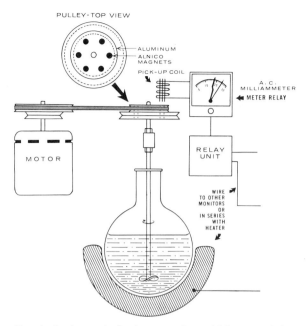

FIG. 6. An improvised stirrer monitor which uses Alnico magnets in an aluminum pulley to generate a rotating magnetic field. This changing magnetic field generates a current which is sensed by a meter relay connected to a pickup coil.

2. The viscosity of the mixture is often more difficult to monitor. Three possibilities can be suggested: (a) Sometimes a commercially available viscometer such as an Ultra-Viscoson[9] can be adapted to the apparatus either by immersion into the reaction area or into a side stream. (b) If the rate of stirring is sensitive to the viscosity (and it should be if one selects an A.C. motor having "poor regulation") then one can monitor viscosity by monitoring (1) as discussed above. (c) If the reactor is a continuous rather than batch-type reactor, the viscosity may be monitored by watching the back pressure as the reaction mixture is pumped.

3. The rate of reflux can be monitored quite readily by sensing the vapor temperature in the reflux condenser either in the region of the condensing zone or slightly above this zone. The temperature in these regions gives an indication of the reflux rate because an increase in the rate causes the condensing zone to move farther up into the condenser. This results in a temperature increase at any given point in the condensation zone. Such an increase is easy to monitor using either a thermoregulator, a thermocouple plus pyrometer, or a thermometer with a capacitance-actuated controller attached (Fig. 7). The monitor selected should be set 5° to 10° *below* the normal condensing temperature of the reflux vapors. The exact location of the sensing unit with respect to the condensing zone (and the manner in which the monitor is connected electrically) will depend on whether one wishes the monitor to turn off the heat cyclically whenever the rate of reflux gets too high, or to terminate the distillation should the rate drop too low or increase too much.

4. The flow of cooling water to the condensers or to any other part of the apparatus can be monitored with either a commercially available "WATER-FLOW GUARD"[10] or the glass apparatus also described in the previous chapter (Fig. 1A).

5. The rate of distillation is more difficult to monitor. One approach is discontinuous: to monitor the time required for successive samples of a given volume to accumulate. This could be done with either a relatively simple sample collector based on a syphon or one of the fixed-volume sample collecting and distributing devices

which has been developed for chromatographic purposes. To monitor the time required for successive samples to accumulate, one would again use one of the time-delay relays or timers. The timer, in this case, could monitor the time elapsed per sample as being too long or too short, depending on how the timer was connected. (To monitor both high and low rate would require two timers.)

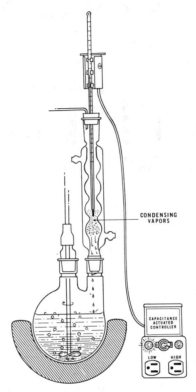

FIG. 7. Reflux rate monitoring with a thermometer located in condensing zone of reflux condenser.

6. The amount of product in a distillate receiver can be monitored either on a volume or weight basis as described in the previous chapter. It might also be monitored by counting the number of fixed volume fractions—if one were using a fraction collector as mentioned above.

7. The pressure within the apparatus can be monitored by use of either (a) commercially available pressure switches of the bourdon, bellows, or diaphragm type;[11] or (b) mercury manometers with contact probes; or (c) mercury manometers to which have been clipped capacitance-actuated controllers[12] which senses movement of the mercury levels. The two ways to use the mercury manometers are shown in Fig. 8A.

8. The rate of gas or liquid flow either *to* the reaction or *from* the reaction can be a very important variable. The monitoring of a clean flow can usually be accomplished with a flow meter which develops a pressure differential across an orifice or capillary. Monitoring can be based on adjustable or fixed position contacts in a mercury *U*-tube connected across the capillary. It also can be monitored with a capacitance-actuated controller attached to one leg of the *U*-tube (Fig. 8B).

9. This variable includes pH, color, presence of precipitate, etc. In general, it represents those variables that can only be monitored with rather sophisticated electronic devices. These may be either quite specialized units designed for a specific application or may be general-purpose laboratory tools that the scientist uses quite frequently.

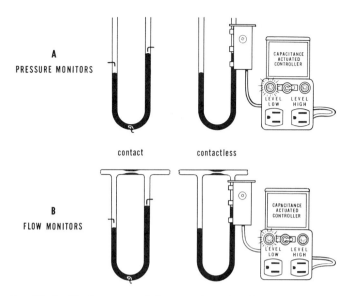

FIG. 8A & 8B. Pressure and flow monitors based on manometers
equipped either with contacts or capacitance-actuated controllers.

10. Time, the final variable on the list, can be monitored either (a) on the basis
of the time of day or (b) the elapsed time after a given stage of the reaction is
attained. Scientists are quite accustomed to using electric clocks and timers to shut
down apparatus at a predetermined time. They are less likely to be accustomed to the
many possible ways of using the time-delay timers. Such a device, if inter-connected
to one of the other monitors listed above, can terminate the reaction *n* hours after
the other monitor has indicated progress of the reaction. It is possible, for example,
to connect a time-delay timer to the primary temperature controller and have the timer
triggered by the first OFF cycle of the temperature controller. This will enable it to
control the time that the reaction was actually *at* temperature.

COUPLING OF MONITORS

With a complex apparatus, it may be desirable to interconnect a number of monitoring devices.

In general, the monitoring units referred to above all fall into several categories:
mercury contact systems, mechanical-electrical devices (such as pressure switches), and
electronic devices. These units generally have different electrical outputs:

1. The mercury contact systems are either "normally open" or "normally closed,"
and may be able to carry only a few milliamperes of current.

2. The electromechanical devices often have a microswitch with a single-pole
double-throw output that can be wired either "normally open" or "normally closed"
and is able to carry at least 5 amperes of current.

3. The electronic devices terminate in a variety of ways: (a) a low voltage (or low
current) control circuit, (b) an indicating meter that should be changed to a meter
relay, (c) a 110-volt power circuit either normally ON or normally OFF, (d) a single-
pole double-throw relay equivalent to the microswitch.

The output of any *single* monitor may be wired so that it interrupts the heat
to the reaction (or causes some other type of corrective action to take place). When
a number of monitors are being used, they can be wired together so that if any one
monitor is actuated the heater can be turned OFF. A series circuit can be used if the
output of every monitor is a "normally closed" relay or switch which opens upon

malfunction (Fig. 9). A parallel circuit would be used when every monitor has a "normally open" output which closes upon malfunction (Fig. 10).

If one has a hybrid system of monitors—some "normally open," some "normally closed," some supplying a low-voltage output, and some supplying 110-volt output—then in order to combine them, one may need to add relays to at least some of the monitors or groups of monitors so that they can be connected in a series or a parallel circuit.

Mention should be made of a more elaborate method of coupling monitors; the "sampling" technique in which the monitors are periodically rather than continuously checked. This may be accomplished with a stepping type switch[13] wired to the monitors. As it steps, it connects each monitor in turn to the alarm and control circuits.

EXOTHERMIC REACTIONS

Exothermic reactions, while usually not long-time reactions, at least so far as the exotherms are concerned, are particularly demanding of attention. When such reactions are running, the scientist may well feel he should not leave the laboratory for even a few minutes. This need not be, for many families of exothermic reactions are not dangerous and, once they are understood, can be entrusted to control by suitable monitors.

The monitors discussed under endothermic reactions may be used with exothermic reaction apparatus, both simple and complex, but with this difference: since the exothermic reaction evolves heat, the monitors may well have to cause *positive* cooling to be applied to the apparatus. Thus, instead of wiring the monitor (or monitors) in such a way that it merely interrupts power to the heater, *it should be connected to turn on a cooling device*.

The nature of the cooling device will depend upon the nature of the exothermic reaction. Mildly exothermic reactions can be controlled by blowing air at the reaction flask. This is accomplished quite readily by having the temperature monitor turn OFF the heater and turn ON an electric fan, or open a solenoid valve in an air tube connected to the laboratory air supply.

More active, but still "mild," exothermic reactions may require that the heater be removed from contact with the apparatus (so that any residual heat stored in the heater does not reach the reaction) and that simultaneously cooling air be turned ON. Still more active reactions require still more positive cooling. This can be accomplished if the cooling is applied in the form of an ice bath that can be raised automatically around the bottom of the reaction. (Heat to the reaction is applied from the top of the flask, by radiant heating.)

The steps described above can be applied automatically with a device known as Jack-O-Matic that is now commercially available.[14] When used in connection with a temperature-sensing system, Jack-O-Matic can either lower a heating bath from around a reaction flask and apply positive air-blast cooling, or turn off a radiant heater and raise an ice bath (Fig. 11). It controls temperature by repeating these operations cyclically as long as cooling is required. Thus, it is able to control automatically reactions that chemists have previously watched very closely. It is no longer necessary for the chemist to watch the thermometer, to lower the heating mantle, and to raise an ice bath around the reaction flask.

OTHER FACTORS

The foregoing notes discuss many of the points that often have to be considered in adding monitors and accessories to apparatus so that it can run unattended. The following are miscellaneous items that are less frequently encountered:

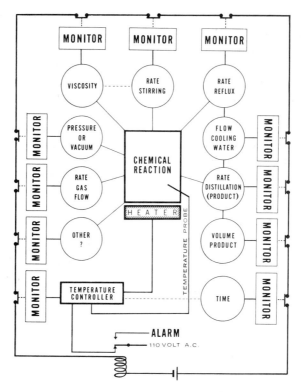

FIG. 9. A number of normally closed circuit monitors may be readily connected in series.

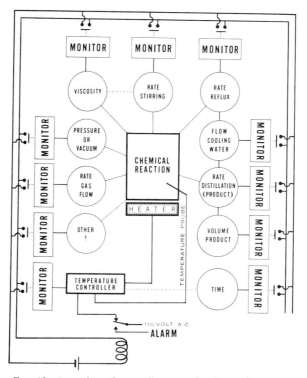

FIG. 10. A number of normally open circuit monitors may be connected in parallel.

Programmed Control.   Sometimes the reaction, instead of being carried out at constant temperature (or constant pressure, flow, etc.), should be carried out at increasing or decreasing levels.   A suitable controller for this purpose can usually be designed with the assistance of the laboratory instrumentation group and a research machinist. For example, a programmed temperature control that step-wise increases the temperature can be built around a group of constant-temperature regulators, each accurate to a few hundredths of a degree.   The regulators would be selected in sequence by a stepping switch controlled by a time-delay timer.   This arrangement would make it possible to maintain an apparatus at each of a series of fixed temperatures for a predetermined time.   (This type of programming is particularly useful in connection with physical test apparatus.)

Programmed control in which the variable is allowed to increase or decrease gradually is sometimes desired.   Programming controllers based on both linear and nonlinear cams have been used for many years for plant-type process control.[15]   Although these units are usually rather large physically, they can be and are used with laboratory ovens and in pilot-plant work.   They may also be used with laboratory reactions.

One can program-control other variables such as pressure or vacuum with plant-type cam controllers.   One can also program-control these variables in an improvised manner by coupling a slow-speed motor to one of the commercially available pressure or vacuum regulators.[16]   The motor will slowly drive the control point upscale.   If a limiting value is desired, it can be achieved by adding a suitable cam which activates a microswitch and turns off the motor.

Sequential Control.   Occasionally it may be desirable to program a sequence of operations.   Thus, as an example, a reaction can be programmed so that after a

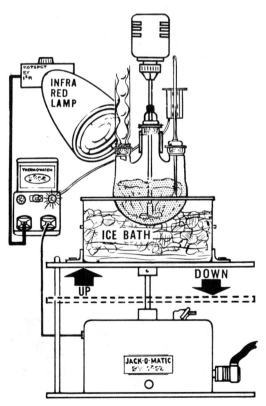

FIG. 11. Jack-O-Matic used to control exothermic reactions by raising and lowering ice baths.

predetermined number of hours a charge of reagent is added to the reaction flask; and after still another time period, still another charge is added. A different basis for sequential control is to monitor the reaction and use a given point in the development of the reaction, rather than time, to trigger a desired operation.

Foam-Level Control. In several rather different fields of laboratory work, excessive foaming in a laboratory operation can necessitate close personal attention and be a vexing problem. Vacuum stripping, although not a very long-time operation, is one example. As the lighter fractions are stripped from a mixture in a flask, foaming often occurs. If the stripping operation is not being watched quite carefully, part of the liquid in the still-pot may be swept over as foam into the condensing apparatus.

These boiling-type foams in glass flasks can usually be sensed either by probes coupled to capacitance-actuated controllers or by light beams using a light source and a light beam photocell. In either case, the foam-sensing device can be connected to a solenoid valve[17] which opens to admit a small amount of air or nitrogen to the apparatus (Fig. 12). The resulting pressure pulse will cause the foam to subside almost immediately. As a result, vacuum stripping can be carried out much more rapidly and efficiently.

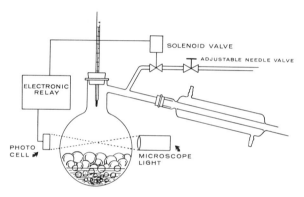

Fig. 12. Photoelectric monitoring and control of stripping foams.

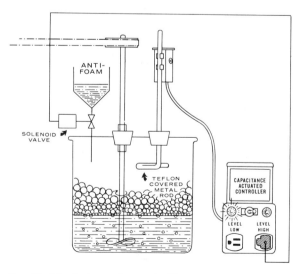

Fig. 13. Monitoring and control of fermentation foams with a probe and a capacitance-actuated controller.

A second field for foam monitoring and control is in microbiological preparations such as fermentations. Such processes are relatively slow compared with the foaming referred to above. Moreover, the fermentation foams are not so clean as the boiling foams. Finally, the fermentation apparatus is often made of stainless steel and not readily adaptable to optical sensing. Figure 13 shows foam-sensing, using Teflon-insulated probe and a capacitance-actuated control. The controller could either open a solenoid valve[18] and admit a small volume of "antifoam" or could turn on a small pump and inject "antifoam."

MONITORING THE MONITOR

One can consider two different categories of monitoring—"control monitoring" and "safety monitoring." In designing apparatus that is to run unattended, one is concerned with both. The following notes illustrate the differences:

1. "Safety monitoring" is always an addition to "control monitoring." In some cases, the "safety monitors" will be EXTRA monitors added to guard against failure of the "control monitors." In other cases the "safety monitors" will be extra monitors added to watch variables that were not being controlled.

2. "Safety monitoring" systems usually need not be as precise nor as fast in response as "control monitoring" systems. For example, in "control monitoring" the temperature of a bath, one may use a sensitive mercury thermoregulator and hold the temperature constant to a few hundredths of a degree. For "safety monitoring" the same bath it would be quite satisfactory to add the relatively insensitive, slow-responding, bimetallic thermoregulator.

3. Whenever possible, "safety monitors" should be unsophisticated devices whose mode of operation is visible, easily understood by the scientist, and readily checked. The mercury manometer is almost ideal from these points of view. It can be made into an excellent monitor for pressure, rate of pressure rise, rate of gas or liquid flow (or even flow of cooling water.) The electronic viscosity-measuring devices are examples of the opposite extreme of complexity. Admittedly, one is grateful for any method of measuring viscosity, but one wishes for devices that would be simple and readily checked.

4. Although the "safety monitor" will usually be a device that is inherently simple and insensitive, there are some situations, particularly involving exothermic reactions, where the "safety monitor" must be a highly sensitive device with a fast response time. Under such conditions, one may well be willing to use as the "safety monitor" a device of greater complexity. In fact, it can be a duplicate of the primary temperature controller.

5. One may be tempted, at times, to consider using the "safety monitor" as a "control monitor." This may even, at times, be quite reasonable, particularly when used to terminate a reaction. As an example, one can use a "safety monitor" attached to the stirrer motor to terminate a reaction when the reaction mixture has reached the desired viscosity.

However, one must guard against the opposite approach: considering that a "control monitor" eliminates the need for a "safety monitor." The five possible monitors shown earlier in Figure 2 were really "safety monitors" added to guard against failure of the primary controller.

6. Finally, this philosophy of guarding against all possible malfunction brings up the following question: If one always has to assume that the primary controllers can fail and therefore need monitoring, will the "safety monitor" sooner or later fail? Should one monitor the "safety monitor"? This is the age-old problem: Just how far should one go in monitoring the monitor, in guarding the guard, in policing the policeman?

Sometimes the solution to this problem is not difficult, since the variables are not independent. For example, if the temperature increases, other variables such as pressure, or reflux rate, will probably increase, and several different monitors will respond.

In considering monitors and their design, one should remember that the degree to which the monitors will be able to cope with both the "likely" and the "unlikely" malfunctions of the apparatus will depend upon the thought that has been given to their design and installation. In the final analysis, it is the good judgment of the scientist that is involved.

OTHER SAFETY RECOMMENDATIONS

The previous chapter made a number of specific recommendations relative to safe operation of glass distillation equipment. Some of these same recommendations also apply to the simpler physical testing apparatus and some to the more complex reaction apparatus. To review the recommendations briefly:

1. If electrical power fluctuations would affect the apparatus, these effects may be minimized by use of constant-voltage transformers.

2. Breakage of glass apparatus can be minimized by careful annealing, by inspecting the glass for strain, by proper support of the equipment, and by use of adequate safety shields.

3. Whenever overflow of a liquid product is a possibility, oversize receivers should be used. It sometimes is advisable to place large trays under the apparatus.

4. The apparatus should be conspicuously labeled so that an untrained person would have no trouble shutting it down in an emergency.

CONCLUSION

Although the suggestions made in this chapter are necessarily limited and will have to be modified to fit various local conditions, certainly there are great advantages in adapting laboratory apparatus so that it can be run safely unattended or with minimum attention. The advantages are: (a) Apparatus that is run continuously produces the desired results much more rapidly than would otherwise be the case. (b) The quality, such as color, of the experimental samples produced when running without interruption is often better. Likewise, the quality of physical tests that are run continuously rather than interruptedly is better. (c) Whenever the experiment can be run continuously, the scientist will have greater assurance that the run is valid and reproducible. In fact, when interrupted runs are made, the scientist may even have certain reservations about their validity. (d) Finally, whenever one is able to make use of the evening hours, one can greatly lower the research cost per experiment.

The benefits of safe unattended operation of laboratory equipment are so great that creative efforts to achieve these are most worthwhile. In fact, such efforts by individual scientists and their supporting instrumentation personnel have been extremely profitable and are one of the most promising fields for the improvement of research.

ACKNOWLEDGEMENT: Reprinted from the *Journal of Chemical Education, Vol. 43*, A589 and A652, July and August, 1966, with permission

REFERENCES

Components Cited

Note: The following list is but a partial guide to components and manufacturers. See buyers' guide published by various journals for still other manufactures. * indicates that there are a number of different models available depending upon the application intended.

1. THERMOSWITCH,* Fenwal, Inc., 113 Pleasant St., Ashland, Mass. 01721
2. OVER-TEMP GUARD. Instruments for Research & Industry, 108 Franklin Ave., Cheltenham, Pa. 19012.

3. VERSATRONIK Controller,* Models R 7161 B & C for single control point, Models R7161 D & E for two control points. Honeywell, Inc., 2701 Fourth Ave., S., Minneapolis, Minn. 55408.

4. GARDSMAN Controllers, Type J & JP for single control point, Type JPT-3 for two control points. West Instrument Corp., Schiller Park, Ill. 60176.

5. CAPACITROL Controllers, Series 270 or 470. Barber-Colman Co., Industrial Instruments Div., Rockford, Ill. 61111.

6. COMPACK I,* Meter Relay 503K. (Specify Temperature range and thermocouple wire.) API Instruments Co., Chesterland, Ohio 44026.

7. VERSA-TRAN Controller Model R 7079C. Honeywell, Inc., 2701 Fourth Ave. S., Minneapolis, Minn. 55408.

8. CYCL-FLEX Reset Timer,* Model HP5 Series. (Specify maximum time range desired—available from 5 sec. to 60 hours.) Eagle Signal Div., 726 Federal St., Davenport, Iowa 52803.

8A. Time Delay Relay Type 412 (Specify maximum time range) Cramer Div. Old Saybrook, Conn. 06475.

9. ULTRA-VISCOSON.* Bendix Cincinnati Div., 3130 Wasson Road, Cincinnati, Ohio 45241.

10. WATER-FLOW GUARD, Model WF-3. Instruments for Research and Industry, 108 Franklin Ave., Cheltenham, Pa. 19012.

11. Pressure and Vacuum Switches.* Bardsdale Valves, 5125 Alcoa Ave., Los Angeles, Calif. 90058.

12. THERM-O-WATCH Controller Model L-6. Instruments for Research & Industry, 108 Franklin Ave., Cheltenham, Pa. 19012.

13. Stepping Switch (Spring Driven or Direct Drive).* C.P. Clare & Co., 3101 Pratt Blvd., Chicago, Ill. 60645.

13. JACK-O-MATIC Model J-3M-3 or J-3M-L6. Instruments for Research & Industry, 108 Franklin Ave., Cheltenham, Pa. 19012.

14. ELECTRONIK 15 Circular Chart Program Controller.* Honeywell, Inc., Fort Washington, Pa. 19034. GARDSMAN Program Controller JGB. West Instrument Corp., Schiller Park, Ill. 60176.

15. Vacuum-Pressure Regulator, Series 44, and Pressure Regulator, Series 40. Moore Products Co., Springhouse, Pa. 19447.

16. Solenoid Valve Model #8262B1. Automatic Switch Co., Florham Park, N. J. 07932.

17. Solenoid Valve Model #8262B2. Automatic Switch Co., Florham Park, N. J. 07932.

Control of Peroxides in Ethers

Norman V. Steere

Ethyl ether, isopropyl ether, dioxane, tetrahydrofuran and many other ethers tend to absorb and react with oxygen from the air to form unstable peroxides which may detonate with extreme violence when they become concentrated by evaporation or distillation, when combined with other compounds that give a detonable mixture, or when disturbed by unusual heat, shock or friction.

Peroxides formed in organic compounds by autoxidation have caused many laboratory accidents, including unexpected explosions of the residue of solvents after distillation, and have caused a number of hazardous disposal operations. Some of the incidents of discovery and disposal of peroxides in ethers have been reported in the literature[1,2], some in personal communications and some in the newspapers. An "empty" 250-cc bottle which had held ethyl ether exploded when the ground glass stopper was replaced (without injury), another explosion cost a graduate student the total sight of one eye and most of the sight of the other, and a third explosion killed a research chemist when he attempted to unscrew the cap from an old bottle of isopropyl ether.[3]

Appropriate action to prevent injuries from peroxides in ethers depends on knowledge about formation, detection and removal of peroxides, adequate labeling and inventory procedures, personal protective equipment, suitable disposal methods, and knowledge about formation, detection and removal of peroxides.

FORMATION OF PEROXIDES

Peroxides may form in freshly-distilled and unstabilized ethers within less than two weeks, and it has been reported that peroxide formation began in tetrahydrofuran after three days and in ethyl ether after eight days.[4] Exposure to the air, as in opened and partially emptied containers, accelerates formation of peroxides in ethers[5,6], and while the effect of exposure to light does not seem to be fully understood, it is generally recommended that ethers which will form peroxides should be stored in full, air-tight, amber glass bottles, preferably in the dark.

Although ethyl ether is frequently stored under refrigeration, there is no evidence that refrigerated storage will prevent formation of peroxides, and leaks can result in explosive mixtures in refrigerators since the flash point of ethyl ether is $-45°$ C $(-49°$ F $)$.

The storage time required for peroxides as H_2O_2 to increase from 0.5 ppm to 5 ppm was less than two months for a tin-plate container, six months for an aluminum container, and over 17 months for a glass container.[7] The same report stated that peroxide content was not appreciably accelerated at temperatures about 11 degrees C $(20°$ F $)$ above room temperature. Davies has reported[8] the formation of peroxides in olefins, aromatic and saturated hydrocarbons and ethers particularly, with initial formation of an alkyl hydroperoxide which can condense on standing or in the presence of a drying agent to yield further peroxidic products. Davies refers to reports that the hydroperoxides initially formed (e.g. from isopropyl ether and tetrahydrofuran) may condense further, particularly in the presence of drying agents, to give polymeric peroxides and that cyclic peroxides have been isolated from isopropyl ether.

The literature contains an extensive report on autoxidation of ethyl ether[9].

Isopropyl ether seems unusually susceptible to peroxidation and there are reports that a half-filled 500-ml bottle of isopropyl ether peroxidized despite being kept over a wad of iron wool[10]. Although it may be possible to stabilize isopropyl ether in other ways, the absence or exhaustion of a stabilizer may not always be obvious from the appearance of a sample, so that even opening a container of isopropyl ether of uncertain vintage to test for peroxides can be hazardous[3]. Noller[5] comments that "neither hydrogen peroxide, hydroperoxide nor the hydroxyalkyl peroxide are as violently explosive as the peroxidic residues from oxidized ether."

DETECTION AND ESTIMATION OF PEROXIDES

Appreciable quantities of crystalline solids have been reported[1,2] as gross evidence of formation of peroxides, and a case is known in which peroxides were evidenced by a quantity of viscous liquid in the bottom of the glass bottle of ether. If similar viscous liquids or crystalline solids are observed in ethers, no further tests are recommended, since in four disposals of such material there were explosions when the bottles were broken.

Chemical and physical methods of detection and estimation of peroxides are described and cited by Davies[8,11], and he comments on several methods for detecting hydroperoxides and the problems of detecting dialkyl peroxides and polymeric alkylidene peroxides. Included in his citations[11] is a review by Criegee of the literature up to 1953 on the chemical methods for qualitative and quantitative analysis of organic peroxides.

Ferrous thiocyanate is reported to be a very sensitive method of detecting hydroperoxides. The ferrous thiocyanate test is as follows:

> A fresh solution of 5 cc of 1% ferrous ammonium sulfate, 0.5 cc of 1N sulfuric acid and 0.5 cc of 0.1 N ammonium thiocyanate are mixed (and if necessary, decolorized with a trace of zinc dust) and shaken with equal quantity of the solvent to be tested; if peroxides are present, a red color will develop. Examples of the red colors developed at two different peroxide concentrations are shown in the color section.

Acidified ferrous thiocyanate used as a spray reagent for paper chromotography will detect 15γ of a hydroperoxide or diacylperoxide, as cited by Davies[11].

The potassium iodide method of testing for peroxides in ethyl ether, as described by the Manufacturing Chemists Association[20], is as follows:

> Add 1 cc of a freshly-prepared 10% solution of potassium iodide to 10 cc of ethyl ether in a 25 cc glass-stoppered cylinder of colorless glass protected from light; when viewed transversely against a white background, no color is seen in either liquid.

If any yellow color appears when 9 cc of ethyl ether are shaken with 1 cc of a saturated solution of potassium iodide, according to one organization, there is more than 0.005% peroxide and the ether should be discarded[12].

A quantitative test for peroxide in tetrahydrofuran and dioxane is described as follows[12]:

> To 50 ml of the ether add 6 ml of glacial acetic acid, 4 ml of chloroform, and 1 g of potassium iodide; titrate with 0.1 N thiosulfate to find the per cent of peroxide, equal to

$$\frac{\text{ml of } Na_2S_2O_3 \times \text{Normality} \times 1.7}{\text{Weight of Sample}}$$

If the reaction is carried out in acetic acid, according to citations by Davies[11], air must be excluded by an inert gas or by the vapor of the boiling solvent to prevent autoxidation of the iodide; or acetic anhydride or hot isopropyl alcohol can be used as the reaction medium. (Avoid ignition of the flammable vapors of acetic acid, acetic anhydride and isopropyl alcohol).

A method for rapid detection of traces of peroxides in ethers has been developed from the use of N, N-dimethyl-*p*-phenylenediamine sulfate to detect gamma quantities of benzoyl and lauroyl peroxides[13].

Davies refers to oxidations by hydroperoxides which proceed quantitatively and can be used for their estimation, including the use of stannous and ferrous ions, and titanium sulfate, lead tetraacetate, phenolphthalein, 3-aminophthalic hydrazide, and cuprous dithiooxamide[11].

It is also reported[11] that dialkyl peroxides can be detected only after they have been hydrolyzed to hydroperoxides under strongly acid conditions and that no satisfactory method appears to have been developed for the estimation of alkylidene peroxides.[8]

Other methods cited by Davies[11] which may be used to detect organic peroxides include chromatography, ion exchange, polarography and absorption spectra.

INHIBITION OF PEROXIDES

No single method seems to be suitable for inhibiting peroxide formation in all types of ethers, although storage and handling under an inert atmosphere would be a generally useful precaution.

Some of the materials which have been used to stabilize ethers and inhibit formation of peroxide include the addition of 0.001% of hydroquinone or diphenylamine[9,11]; polyhydroxyphenols, aminophenols and arylamines. Addition of 0.0001 g pyrogallol in 100 cc ether reported preventing peroxide formation over a period of two years[8]. Water will not prevent formation of peroxides in ethers and iron, lead, and aluminum will not inhibit the peroxidation of isopropyl ether[14] although iron does act as an inhibitor in ethyl ether. Dowex-1® has been reported effective for inhibiting peroxide formation in ethyl ether[15], 100 ppm of 1-naphthol for isopropyl ether[16], hydroquinone for tetrahydrofuran[17] and stannous chloride or ferrous sulfate for dioxane[16]. Substituted stilbene-quinones have been patented as a stabilizer against oxidative deterioration of ethers and other compounds.

REMOVAL OF PEROXIDES

Reagents which have been used for removing hydroperoxide from solvents are reported[8] to include sodium sulfite, sodium bisulfite, stannous chloride, lithium tetrahydroalanate (caution: use of this material has caused fires)[19] zinc and acid, sodium and alcohol, copper-zinc couple, potassium permanganate, silver hydroxide, and lead dioxide.

Decomposition of ether peroxides with ferrous sulfate is a commonly used method; 40 g of 30% ferrous sulfate solution in water is added to each liter of solvent[20]. Caution is indicated since the reaction may be vigorous if the solvent contains a high concentration of peroxide.

Reduction of alkylidene or dialkyl peroxides is more difficult but reduction by zinc dissolving in acetic or hydrochloric acid, sodium dissolving in alcohol (note ease of ignition of hydrogen) or the copper-zinc couple might be used for purifying solvents containing these peroxides[8].

Addition of one part of 23% sodium hydroxide to 10 parts of ethyl ether or tetrahydrofuran will remove peroxides completely after agitation for 30 minutes; sodium hydroxide pellets reduced but did not remove the peroxide contents of tetrahydrofuran after two days[4]. Addition of 30% of chloroform to tetrahydrofuran inhibited peroxide formation until the eighth day with only slight change during 15 succeeding days of tests[4]; although sodium hydroxide could not be added because it reacts violently with chloroform, the peroxides were removed by agitation with 1% aqueous sodium borohydride for 15 minutes (with no attempt made to measure temperature rise or evolution of hydrogen).

A simple method for removing peroxides from high quality ether samples without need for distillation apparatus or appreciable loss of ether consists of percolating the solvent through a column of Dowex-1® ion exchange resin[15]. A column of alumina was used to remove peroxides and traces of water from ethyl ether, butyl ether, dioxane and petroleum fractions[21] and for removing peroxides from tetrahydrofuran, decahydronaphthalene (decalin), 1,2,3,4-tetrahydronaphthalene (tetralin), cumene and isopropyl ether[8].

Calcium hydride has been used for obtaining anhydrous and peroxide-free p-dioxane by refluxing followed by distillation[22] and the technique may also be applicable to other ethers. Sodium and potassium borohydrides have been used to reduce peroxide in tetrahydrofuran and bis(2-methoxyethyl)ether (diglyme) and to inhibit them for some time against further peroxidation[23].

Cerous hydroxide $Ce(OH)_3$ meets the need which has been expressed[19] for an insoluble solid which combines with peroxides and can be separated by filtration or decantation. Cerous hydroxide, prepared from a cerous salt solution by sodium hydroxide, changes from white to reddish brown within a minute or two after addition to an ether if peroxides are present, and removal of peroxides can be completed within 15 minutes[24]. Peroxyceric compound and unchanged cerous hydroxide can be removed by centrifugation and decantation (caution: flammable vapors may be ignited if the centrifuge is electrical and not explosion-proof.) Ramsey and Aldridge tested 19 high-grade commercially-available ethers and found peroxides in 12 of the ethers by the iodide test (2 ml of ether plus 5 ml of acidified potassium iodide solution plus 1 ml of starch solution). After treatment with cerous hydroxide each of 12 ethers which had contained peroxides gave negative potassium iodide tests.

The tests of 19 high-grade commercially-available ethers by Ramsey and Aldridge[24] is summarized as follows:

Ether	Quantities of Peroxides Found
Allyl ethyl ether	Moderate
Allyl phenyl ether	Moderate
Benzyl ether	Moderate
Benzyl n-butyl ether	Moderate
o-Bromophenetole	Very small
p-Bromophenetole	None
n-Butyl ether	Moderate
t-Butyl ether	Moderate
p-Chloroanisole	None
o-Chlorophenetole	None
Bis(2-ethoxyethyl)ether (Diethylene glycol diethyl ether)	Considerable
2-(2-Butoxyethoxy)ethanol (Diethylene glycol mono-n-butyl ether)	Moderate
p-Dioxane	Moderate
Ethyl ether[a]	None
Ethyl ether[b]	Considerable
Ethyl ether[c]	Moderate
Isopropyl ether	Considerable
o-Methylanisole	None
m-Methylphenetole	None
Phenetole	None
Tetrahydrofuran	Moderate

[a]Obtained from sealed tin can of anhydrous ether, analytical reagent, immediately after opening.

[b]Obtained from a partially filled tin can (well-stoppered) containing the same grade of anhydrous ether, originally, as that described in note a but of long standing.

[c]From a galvanized iron container used for dispensing ether from stockroom for research purposes.

After the removal of the peroxides the ethers were tested for the presence of cerium by the benzidine test, with negative results except for allyl ethyl ether and benzyl-*n*-butyl ether. The authors also noted that di-*t*-butyl peroxide does not liberate the iodine from acidified potassium iodide solution nor react with cerous hydroxide.

ACKNOWLEDGEMENT: Reprinted with permission from "The Chemistry of the Ether Linkage," 1967, Saul Patai, editor, Interscience Publisher, Inc. division of John Wiley & Sons, Inc.

REFERENCES

1. Douglass, I. B., *J. Chem. Educ., 40,* 469, (1963).
2. Steere, N. V., Control of Hazards from Peroxides in Ethers, *J. Chem. Educ., 41,* A575, (1964).
3. Accident Case History 603, *Manufacturing Chemists' Association,* reported in part in reference 2.
4. Fleck, E., Merck, Sharp & Dohme Company Memo, May 11, 1960.
5. Noller, C. R., "Chemistry of Organic Compounds," W. B. Saunders Co., Philadelphia, (1951).
6. Rosin, J., "Reagent Chemicals and Standards," 4th ed., D. Van Nostrand Co., Princeton, N. J., (1961).
7. Brubaker, A. R., personal communication, December 4, 1964.
8. Davies, A. G., Explosion Hazards of Autoxidized Solvents, *J. Roy. Inst. Chem.,* 386, (1956).
9. Lindgren, G., Autoxidation of Diethyl Ether and Its Inhibition By Diphenylamine, *Acta Chir. Scand., 94,* 110, (1946).
10. Pajaczkowski, A., personal communication, September 29, 1964.
11. Davies, A. G., "Organic Peroxides," Butterworths, London, (1961).
12. Brasted, H. S., personal communication, June 1, 1964.
13. Dugan, P. R., *Anal. Chem., 33,* 1630, (1961).
14. P. R. Dugan, *Ind. Eng. Chem., 56,* 37, (1964).
15. Feinstein, R. N., Simple Method for Removal of Peroxides From Diethyl Ether, *J. Org. Chem., 24,* 1172, (1959).
16. *Encyclopedia of Chemical Technology, Vol. 5* (Ed. R. E. Kirk and D. F. Othmer), Interscience Publishers, New York, 1950, pp. 871, 142.
17. *Encyclopedia of Chemical Technology, Vol. 6* (Ed. R. E. Kirk and D. F. Othmer), Interscience Publishers, 1950, p. 1006.
18. Jones, D. G., *Brit. Pat., 699,* 179; (1953); *Chem. Abstr., 49,* 3262f, (1955).
19. Moffett, R. B. and Aspergren, B. D., Tetrahydrofuran Can Cause Fire When Used As Solvent For LiAlH$_4$, *Chem. Eng. News, 32,* 4328, (1954).
20. *Chemical Safety Data Sheet—SD 29, Ethyl Ether,* Manufacturing Chemists' Association, Washington, D.C., (1956).
21. Dasler W. and Bauer, C. D., Removal of Peroxides From Ethers, *Ind. Eng. Chem. Anal. Ed., 18,* 52, (1946).
22. Birnbaum, E. R., personal communication, August 11, 1964.
23. *Manuals of Techniques,* Metal Hydrides, Inc., Beverly, Massachusetts, (1958).
24. Ramsey, J. B. and Aldridge, F. T., Removal of Peroxides From Ethers With Cerous Hydroxide, *J. Am. Chem. Soc., 77,* 2561, (1955).

Techniques for Handling High Energy Oxidizers

Dennis G. Nelson

Limiting the quantity of high energy oxidizers to be handled in glass apparatus in the laboratory is based on extensive tests which demonstrated that protective clothing would absorb the momentum imparted to glass fragments by explosion or detonation of quantities of 0.25 g or less, based on the molecular weight of the particular compounds being investigated. Protective clothing in this case consists of leather gloves, leather coat, face shield over safety glasses, and ear plugs.

Any research project can be carried out without injury to personnel, if adequate precautions are taken and if the design of protective shielding and equipment is based on tests rather than rough assumptions.

This chapter is based on a background of safety techniques accumulated by one company during seven years of research operations.

Due to the variety and initial uncertainty of the hazards involved in handling energetic solids, liquids and gases, the techniques must necessarily be versatile and comprehensive.

The majority of the safety techniques are applications of the principles:

Safety via miniaturization
Safety via dilution
Safety via remote protection
Safety via simplicity of operation
Safety via testing and analysis

Safety techniques have been developed for research, development and small-scale production, as well as complete chemical analysis and testing of high energy oxidizers. In addition, a continuous comprehensive safety program is maintained throughout the area which reviews current projects and sets up safety standards for new projects.

RESEARCH TECHNIQUES

The typical flow of a program is from research through development and small-scale production, and the project safety program is organized in the same manner. The bulk of safety data is accumulated at a very small scale to be later applied to larger scale operations. Figure 1 shows a schematic diagram of a nuclear-magnetic resonance tube, a basic tool for oxidizer research. These tubes are often used for screening reactions by combining reactants in the tubes and analyzing for reaction products. A quantity as small as 0.2 cm^3 of liquid in this tube with an oxidizer concentration of 10–20% is sufficient for detection.

Figure 2 shows one of the NMR detection units at The 3M Company. In particular, note the enclosed 1.2 cm thick Plexiglas glove box for the operator's protection during sample injection, analysis, and withdrawal. This unit is capable of detecting $\frac{1}{2}$ mmole (100–500 mg) of a fluorine-containing oxidizer. This minute quantity of explosive material, along with the relatively few manipulative analytical steps required, make this a safe and versatile tool for the basic researcher.

Slightly farther along in the research-development sequence is the 20-ml "sealless" reactor shown in Figure 3. These small units are useful in extending NMR data into

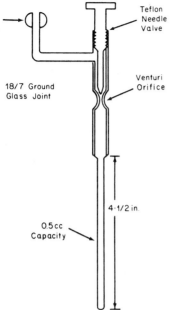

FIG. 1. Nuclear Magnetic Resonance Tube.

FIG. 2. Nuclear Magnetic Resonance Detector.

basic processing information: optimum reaction conditions, conversion rate, as well as identification and resolution of sensitive processing steps. The glass reactor body was made from a standard 2.54-cm Pyrex double strength flanged pipe cap. The 2.22-cm Teflon coated magnet and magnetic stirring unit are also standard commercial items. The Kel-F support rod prevents the magnet from being thrown out of position. This unit develops 200–300 rpm and is suitable for dispersing two-phase liquids, gas-liquid or dilute liquid-solid systems. Its environmental capabilities are from full vacuum to 1.4 atmospheres and from $-90°C$ to greater than $200°C$.

The relative size and simplicity of these units make them adaptable to oxidizer processing studies where relatively little safety data is available. For rapid processing studies these reactors may be safely placed in series such that they are individually barricaded but still accessible to an operator wearing protective clothing, as shown in Figure 4. In event of an explosion, normally only the glass components of the system are destroyed. In most instances operations may be quickly resumed after replacement of standard components.

The 300-ml stainless steel reactor shown in Figure 5 is the largest scale system still considered in the realm of research. The picture shows the reactor housed in its separate barricade which can be isolated from the rest of the system. A glass overhead system is still accessible to an operator clothed in protective gear.

ANALYTICAL TOOLS FOR NEW HIGH ENERGY OXIDIZERS

Essential to the safe operation of an oxidizer research and development program is a complete set of analytical tools. Both rapid in-process analyses and supplementary analytical techniques are required. Probably the most widely used tool for both in-process and supplementary analyses is the gas-liquid chromatograph. As with previous devices, the microliter sample size and the few manipulative steps required for gas-

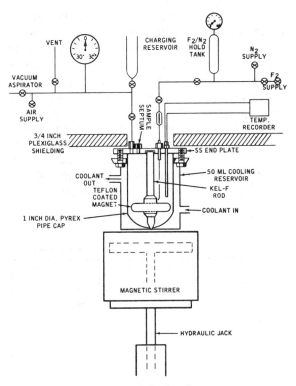

FIG. 3. 20 ml. Sealless Reactor.

FIG. 4. Operation of Multiple Sealless Reactor.

FIG. 5. 300 Milliliter Reactor.

liquid chromatograph analyses make this safe and versatile. A schematic diagram of a special gas-liquid chromatograph hookup is shown in Figure 6. This system has been designed so that both identification and isolation/purification of two condensible components is possible. The enclosed 1.27-cm Plexiglas box, appropriate shielding for the quantity of oxidizer involved (normally less than 50 mg), protects the operator while allowing full vision of the operation. The operator is able to monitor his product stream with the chromatograph and then switch the three-way valves from vent to trap when a desired product peak appears on the recorder chart. Following isolation in the liquid nitrogen cooled glass traps, a product may be expanded into a separate section of the box where it may be collected into bulbs or vented. An I.R. gas sample is also possible at this point. A Beckman Megachrom Preparative Gas Chromatograph, operating on this general principle, is also available for larger scale purification of samples, up to 50 g per day.

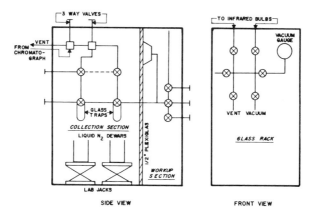

FIG. 6. GLC—Trapping System Schematic Diagram.

For development and small-scale production facilities, a series of quantitative tools are required that can be used remotely. Such a list of simple, versatile measurements that can be made without entering a bay where a dangerous chemical operation is in progress includes:

> Pressure-volume relationships for gases
> "In-line" graduates for liquids
> Titration systems for chemical reactions
> Remote micro-sampling systems for liquids and gases

Figure 7 shows the details of a liquid/gas oxidizer sampling system connected directly to the reactor inside the bay. This system was designed and constructed by 3M personnel for safe, reliable, remote samples of extremely hazardous reactor charges. The system was designed so that a maximum of 1 g of liquid oxidizer could be present in the sample box at any one time. The liquid sample line is cooled right up to the sample box to insure a sample representative of the reactor. Special washers around each of the orifices prevent any damaging shrapnel from escaping the box. Using this device, it is possible to take "in-process" samples at a safe level from a reactor containing up to 250 gm of hazardous oxidizers.

In addition to in-process analyses, a series of versatile and comprehensive supplemental analyses have been developed for analysis of gaseous liquid and solid oxidizers in the mg quantities shown in Table 1. All of these analyses can be run conveniently on a milligram scale such that conventional protective clothing; i.e., leather coat, face shield, ear plugs, and gloves, is sufficient for the operator's protection.

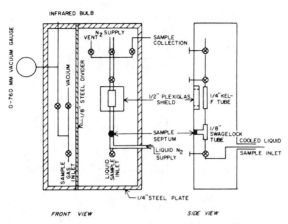

FIG. 7. Remote Sampling System—Schematic Diagram.

Table 1
Table of milligram-scale analysis

Analysis	Quantity
Oxidizing power	15 Milligrams
Infrared analysis	2
Equivalent weight	50
Elemental analysis	
Carbon	50
Fluorine	2
Nitrogen	20

DEVELOPMENT

In contrast to a huge rocket motor, $1\frac{1}{2}$ liters of propellant solution is a relatively small quantity, but in terms of potential destruction, is more than enough to create a meaningful human hazard. Beginning at the development stage, all subsequent oxidizer processing is done completely within remote location facilities. Almost without exception, a twin bay concept is used for all oxidizer development and small-scale production. The operator is separated from his material by 36-cm reinforced concrete walls. He views the operation through two thicknesses of 10-cm thick Plexiglas separated by a dead air space. Figure 8 shows a photograph of a typical reaction bay with the direct drive rods and flexible heavy duty cables which are connected to process valves. Figure 9 shows the other half of the twin bay where the operator conducts all the manipulations. In addition to the flexible cables and direct drive rods, the overhead switch panel provides power and control over all electrically driven equipment within the reaction bay. The services that are available within the reaction bay include: hot and cold water, steam, vacuum, pressure, inert gas, and fluorine.

Although process explosions are not frequent at the Development Laboratory, material is treated as though explosions occurred every day. Therefore, the facilities are planned for the worst to happen and to direct the explosion where it can do the least damage. Figure 10 shows a typical blast door which protects all personnel on the interior side of the reaction bay, and there is a flimsy blowout door on the other side of the bay which permits any sudden over-pressure to relieve itself where it can do no harm.

Fig. 8. Typical Reaction Bay.

Fig. 9. Typical Operator's Bay.

Fig. 10. Reaction Bay Blast Door.

Fig. 11. Cart and Cable Apparatus.

Since remote, twin bay facilities of the type described above are at a premium, they must be efficiently utilized. For this purpose the versatile cart and cable technique has been developed. Using this system, process equipment for specific applications is mounted on portable laboratory carts before transferring to reaction bays, thus expediting effective use of remote facilities. Figure 11 shows a photograph of such a system and cart before installation. Should an operation be terminated or process equipment damaged, the cart is removed from the bay and the next system is wheeled in. In most instances repairs and/or installations consume only a day or two of bay time.

Another concept that is applied wherever possible for safe yet efficient operation is the use of readily available, easily modified equipment. For example, conventional glassware is used wherever possible for processing, thus:

Minimizing damaging shrapnel in case of explosion.

Allowing rapid replacement of key equipment.

Similarly, conventional valves, fittings, tubing, and other processing equipment are used wherever feasible. In addition, process equipment and techniques are evolved from simplified systems as the characteristic hazards of particular oxidizers are determined. A fourth concept applied to oxidizer handling is the use of flow processes to replace batch processes, particularly where scale-up data is desired. With this method, large quantities of oxidizer solutions are never allowed to accumulate, but are continually transported and mixed through a series of pipes or tubes.

SAFETY BY DILUTION

A good rule developed over the years is to avoid whenever possible the handling of pure fluorochemical oxidizers or highly concentrated solutions of gaseous, liquid and solid oxidizers. The diluents and solvents commonly used at the Development Lab are inert gaseous diluents, fluorochemical solvents and conventional organic solvents. A complete history has been accumulated regarding the concentrations required for safe handling and the general processing of such solutions.

TESTING TECHNIQUES

An oxidizer testing program which runs concurrently with each significant development project is an essential part of the Development Laboratory's overall safety record. Before an oxidizer or intermediate may be handled by operator personnel, a thorough safety evaluation of the material is made. Determinations such as shock sensitivity, explosive limits of oxidizer solutions, vapor pressure determinations in solution, and solvent compatibility, as well as specialized tests where required, are common to each new oxidizer which comes into the laboratory. In addition, the Development Lab works closely with the United States Bureau of Explosives on test programs involving the shipment of oxidizer solution. Toxicity screening of oxidizers is a recent addition to the safety program.

SMALL-SCALE PRODUCTION

By most standards, oxidizer production equipment at the Development Laboratory seems quite small. Average oxidizer reactors range in size from about $1\frac{1}{2}$ liter (1.59 qts) capacity to approximately 40 liters (10.6 gal). The approach has been to expeditiously produce small quantities of fluorochemical oxidizers for propellant evaluation while retaining a high degree of versatility. As is found throughout the development facilities, the twin bay concept is used throughout production. With this facility a single operator may conduct two simultaneous operations on either side. Again the cart and cable technique is used as is the concept of using readily-available easily-modified equipment. Figure 12 shows the "Barricade within a Barricade" concept applied to small scale production. Moderate explosions have occurred within the 40 liter reactor without causing any other damage in the bay. With these kinds of facilities, up to 4.5 kg of high energy material may be prepared at one time.

Although most facilities at the Development Laboratory are designed for small quantities of materials, basic raw materials are handled in bulk quantity wherever possible. An example of this concept is shown in Figure 13: A double sealed, liquid

FIG. 12. Barricade Within a Barricade.

FIG. 13. 5000—Pound Liquid Fluorine Tank.

fluorine system capable of containing 2270 kg of liquid fluorine. The entire unit is below grade resting on two feet of coarse limestone which would act to neutralize any escaping material. An outer jacket of liquid nitrogen surrounds the fluorine, as well as a vacuum jacket beyond that. Primary and auxiliary shut-off valves are provided which may be operated remotely.

SAFETY PROGRAM

If, in spite of all the precautions described earlier, an accident does occur, the hallway outside the operator bays is complete with safety showers, fire stations, fire blanket, protective equipment, fresh air mask, and emergency exits. Figure 14 shows a close-up view of the emergency fire station complete with CO_2 cylinder, dry chemical extinguisher, water extinguisher, and an extinguisher for metallic fires. An emergency

FIG. 14. Emergency Fire Station.

horn is mounted on the wall for effective evacuation alarm. In addition to the material safety precautions, a vigilant safety inspection program is maintained. A rotating safety committee has been formed which meets once a month and more frequently as the situation demands. In fact, each major project is reviewed for its safety considerations and a monthly inspection of each building is accomplished. Close cooperation with the divisional safety engineer is maintained and specialized lectures and films are sought to maximize personnel's safety consciousness.

Although fluorochemical oxidizers are highly hazardous chemicals by almost any definition, vigilant safety techniques at the Development Laboratory make oxidizer processing a routine matter. With extreme cognizance of safety the oxidizer operation is actually less hazardous than many conventional chemical processes where safety is not emphasized to as great an extent.

ACKNOWLEDGMENT: This chapter was presented at the Seventh Annual Explosives Safety Seminar on High Energy Solid Propellants sponsored by the Armed Services Explosives Safety Board and hosted by Patrick Air Force Base at Cocoa Beach, Florida on August 26, 1965, and is reprinted with permission.

Handling Perchloric Acid and Perchlorates

K. Everett and F. A. Graf, Jr.

INTRODUCTION

Considerable interest has been taken in the explosion hazards to be encountered in the use of perchloric acid since a mixture of perchloric acid and acetic anhydride exploded in a Los Angeles factory in 1947, killing 15, injuring 400, and causing $2,000,000 damage. Literature surveys reveal that descriptions of explosions in laboratories using perchloric acid have been reported over a period of more than a century.

The most detailed available account of the chemistry of perchloric acid and a reference highly recommended to everyone who will be working with perchlorates is given by J. S. Schumacher in the American Chemical Society monograph "Perchlorates, Their Properties, Manufacture, and Uses."[1] Cummings and Pearson have reviewed the thermal decomposition and the thermochemistry and in a later report Pearson reviewed the physical properties and inorganic chemistry of perchloric acid.[2,3] Shorter, but very useful, reviews of the chemistry of perchloric acid also have been published[4,5] and several accidents which have occurred in France have been reviewed.[6] A review of the circumstances leading to the Los Angeles explosion and of 5 laboratory incidents involving perchloric acid has been published,[7] and a résumé of 5 serious accidents has been reported.[8] A summary of a number of accidents reported in the literature is appended to this chapter.

GENERAL DISCUSSION

Perhaps the most disturbing features of accidents involving perchloric acid are:
1. The severity of the accidents,
2. That the persons involved are, in the majority of cases, experienced workers.
Harris[7] concludes that the basic cause of accidents involving perchloric acid is due to contact with organic material, or to the accidental formation of the anhydrous acid. Smith[5] emphasizes the hazard of allowing strong reducing agents to come into contact with concentrated (72%) perchloric acid.

Properties of Perchloric Acid Solution

The Manufacturing Chemists' Association, Inc. has a Chemical Safety Data Sheet, SD-11, on perchloric acid solution which was adopted in 1947, and revised in 1965. Excerpts from MCA Safety Data Sheet SD-11 are reprinted in this chapter with permission. Properties of perchloric acid solutions are described in Table 1.

A.C.S. Reagent Grades of perchloric acid are 70–72% by weight $HClO_4$ and 60% by weight $HClO_4$.

The important physical and chemical properties of perchloric acid are that it is a water white liquid, it has no odor, the boiling point (of constant boiling mixture) is 203°C (397°F), and it is corrosive. Perchloric acid can be dangerously reactive. At ordinary temperatures 72% perchloric acid solution reacts as a strong nonoxidizing acid. At elevated temperatures (approximately at 160°C (320°F)), it is an exceedingly strong and active oxidizing agent, as well as a strong dehydrating reagent. Contact with combustible material at elevated temperatures may cause fire or explosion.

TABLE 1
Strengths of CP (A.C.S.) Perchloric Acid Solution

% HClO$_4$	Specific Designation	Sp Gr at 25°/4°C (77°/39°F)
60	—	1.5483
65	HClO$_4$ · 3H$_2$O	1.5967
66	—	1.6102
67	—	1.6237
68	—	1.6372
69	—	1.6507
70	—	1.6642
71	—	1.6777
72	—	1.6912
72.5	Constant Boiling (203°C)	1.6980
73	—	1.7047
73.6	HClO$_4$ · 2H$_2$O	1.7129
74	—	1.7182
75	—	1.7318

Relative Oxidizing Power of Perchloric Acid

Although no data are known to describe the change in the oxidizing power of perchloric acid with temperature and/or concentration, some observations describe the phenomenon sufficiently.

Cold perchloric acid, 70% or weaker, is not considered to have significant oxidizing power. The oxidizing power, however, increases rapidly as the concentration increases above 70%. Acid of 73+percent (which gives off fumes in even relatively dry air) is a fairly good oxidizer at room temperature. The monohydrate of perchloric acid (85% acid strength and a solid) is indeed a very good oxidizer at room temperature, as it will even react with gum rubber, whereas the 73% acid does not.

Temperature increases will also increase the oxidizing power of perchloric acid solutions; therefore, hot, strong perchloric acid solutions are very powerful oxidizing agents.

Hazards of Perchloric Acid

Hazards of perchloric acid solution are described in MCA SD-11 as follows.

1. Perchloric acid is strong, and contact with the skin, eyes, or respiratory tract will produce severe burns.

2. Perchloric acid is a colorless, fuming, oily liquid. When cold its properties are those of a strong acid but when hot, the concentrated acid acts as a strong oxidizing agent.

3. Aqueous perchloric acid can cause violent explosions if misused, or when in concentrations greater than the normal commercial strength (72%).

4. Anhydrous perchloric acid is unstable even at room temperatures and ultimately decomposes spontaneously with violent explosion. Contact with oxidizable material can cause immediate explosion.

The following are listed among the causes of fires and explosions involving perchloric acid.

1. The instability of aqueous or of pure anhydrous perchloric acid under various conditions.

2. The dehydration of aqueous perchloric acid by contact with dehydrating agents such as concentrated sulfuric acid, phosphorus pentoxide or acetic anhydride.

3. The reaction of perchloric acid with other substances to form unstable materials.

Combustible materials, such as sawdust, excelsior, wood, paper, burlap bags, cotton waste, rags, grease, oil and most organic compounds, contaminated with perchloric acid solution are highly flammable and dangerous. Such materials may explode on heating, in contact with flame, by impact or friction, or may ignite spontaneously.

Explosive Reactions

A chemical explosion is the result of a very rapid increase in volume, due to the evolution of gas or vapor, the reaction normally being exothermic. The force of the shock wave is governed by the rate at which the reaction takes place. The point is made by Burton and Praill[4] that, apart from the thermochemical aspects of explosions, it is the velocity of the decomposition which determines whether or not a reaction is explosive, and that the power of the explosive is governed largely by the pressure of the gases produced in the decomposition. Where the temperature of the explosion is several thousand degrees Centigrade, the power of the explosion is increased further by the thermal expansion of the gases.

When considering the hazards involved in the use of perchloric acid this point should be clearly recognized: many of the reported serious laboratory accidents involved only small quantities (< 1 gm) of reactant. (See Appendix to this chapter.)

The System: Perchloric Acid—Acetic Anhydride—Acetic Acid

Both Schumacher,[1] and Burton and Praill[4] examined the perchloric acid—acetic anhydride—acetic acid system in some detail, the former reproducing a triangular diagram for the system. Burton and Praill[4] quote the equation:

$$2.5 \, Ac_2O + HClO_4 \cdot 2.5H_2O = 5 \, AcOH + HClO_4 + 18.4 \, kcal$$

showing that there is a considerable evolution of heat when mixing the reagents, and that if excess acetic anhydride is present the solution may be considered as a solution of anhydrous perchloric acid in acetic acid. The most explosive mixture is given as the one in which complete combustion occurs.

$$CH_3COOH + HClO_4 = 2CO_2 + 2H_2O + HCl$$

These authors give the explosion temperature as 2,400°C and calculate that 1 gm of the mixture produces, instantaneously, about 7 liters of gas at the explosion temperature. They state finally that for this system all the investigated explosions have been due to the use of potentially dangerous mixtures, together with faulty equipment or technique.

In 1961, in a university laboratory, a young metallurgy student lost the sight of both eyes when an explosion took place while he was preparing an acetic anhydride—perchloric acid—water electro-polishing mixture. Turner and Bartlett[12] investigated the process and discovered that the reagents should be mixed in the following order.
    1. Add the perchloric acid to the acetic anhydride.
    2. Add water to the mixture, slowly and in small portions.

The Use of Magnesium Perchlorate as a Drying Agent

Several cases are on record in which magnesium perchlorate (Anhydrone) has exploded while being used as a desiccant (Appendix 1). Smith[5] regards the use of magnesium perchlorate for the drying of alcohol vapors as permissible, but Burton and Praill[4] warn that if magnesium perchlorate is to be used for drying organic liquids the purity of the drying agent should be determined, since the preparation may have left traces of free perchloric acid in the salt. Explosions involving magnesium perchlorate may have been caused by the formation of perchloric esters in the system. It should be noted that methyl and ethyl perchlorate are violently explosive compounds.

Anhydrous Perchloric Acid

The anhydrous acid is a much more dangerous material than the concentrated 72% acid. If stored for more than 10 days the acid is likely to develop a discoloration, and to be capable of spontaneously exploding. A principal hazard in the use of the anhydrous acid is the breakage of containers. Anhydrous perchloric acid will explode in contact with wood, paper, carbon and organic solvents. Anhydrous perchloric acid should only be made as required and should never be stored.

Miscellaneous Reactions

Burton and Praill[4] state that it is impossible to overemphasize that the simple alkyl esters of perchloric acid are extremely dangerous. They note that many explosions are on record resulting from the standard method of determining perchlorates or potassium as potassium perchlorate in the course of which an ethyl alcohol extraction is used. The same authors state that most organic perchlorate salts, with the exception of the diazonium salts, are safe unless they are over-heated or detonated, but mention that pyridine perchlorate can be detonated by percussion.

In the dry state, metal perchlorate—solvent complexes; e.g. the silver perchlorate –benzene complex, are explosive.

The spillage of concentrated 72% perchloric acid on suitably porous media; e.g. wood and other organic materials, produces mixtures which are extremely sensitive to impact and heat upon drying (Appendix 1).

The treatment of trivalent antimony or bismuth compounds with perchloric acid can be very hazardous (Appendix 1).[4]

Reducing Agents

In general, mixtures of strong reducing agents and concentrated 72% perchloric acid should be regarded as very hazardous.[5]

SAFE HANDLING OF PERCHLORATES

Schumacher[1] states that:

"Perchlorates appear to fall into two broad categories: those (1) more or (2) less sensitive to heat and shock. Included in the group of those qualitatively less sensitive are pure ammonium perchlorate, the alkali metal perchlorates, the alkaline earth perchlorates, and perchloryl fluoride. Among the more sensitive compounds are the pure inorganic nitrogenous perchlorates, the heavy metal perchlorates, fluorine perchlorate, the organic perchlorate salts, the perchlorate esters, and mixtures of any perchlorates with organic substances, finely divided metals, or sulfur. Any attempt to establish a more precise order of the degree of hazard to be expected from any given perchlorate seems unwarranted on the basis of data available. Each perchlorate system must be separately (and cautiously) evaluated.

TABLE 2
Materials Compatible with and Resistant to 72% Perchloric Acid[13]

Material	Compatibility	Material

Elastomers:
  Gum rubber
 	
  Viton B*
	each batch must be tested to determine compatibility
	slight swelling only
		Plastics:
		  Polyvinyl chloride
		  Teflon*
		  Polyethylene
		  Polypropylene
		  Kel-F†
		  Vinylidine fluoride

Metals and Alloys:
  Tantalum
  Titanium (chemically pure
    grade)
 	
  Zirconium
  Columbium (Niobium)
  Hastelloy C**
	excellent
	excellent
	
	excellent
	excellent
	slight corrosion rate
		Saran‡
		  Epoxies
		
		Others:
		  Glass
		  Glass-lined steel
		  Alumina
		  Carbon and graphite

*du Pont Trademark
†3M Company Trademark
‡Dow Chemical Corporation Trademark

"There do not appear to be any uniform recommendations for the safe handling of perchlorates which are generally applicable. A number of heavy metal and organic perchlorates, as well as hydrazine perchlorate (hydrazinium diperchlorate)* and fluorine perchlorate, are extremely sensitive and must be handled with great caution as initiating explosives. Mixtures of any perchlorates with oxidizable substances are also highly explosive and must be treated accordingly. For all of these, it is essential to avoid friction, heating, sparks, or shock from any source (as well as heavy metal contamination),* and to provide suitable isolation, barricades, and protective clothing for personnel."

However, "the more common ammonium, alkali metal, and alkaline earth perchlorates are considerably less hazardous."

Synthesis of new inorganic or organic perchlorates should only be undertaken by an experienced, cautious, investigator who is familiar with the literature.

A simple test to evaluate impact sensitivity can be conducted by placing a crystal or two of the perchlorate on a steel block and striking with a hammer. The degree of noise and relative impact to produce the explosion can be roughly correlated with the impact sensitivity.

A simple thermal stability test can be conducted by placing a crystal or two on a hot plate and observing the time to create a violent decomposition reaction. A gram of the material can be heated slowly in a loosely-closed vial for more exact determination of thermal stability.

RECOMMENDATIONS FOR THE SAFE HANDLING OF PERCHLORIC ACID

Several organizations have drawn up recommendations for the safe handling of perchloric acid, among them the Association of Official Agricultural Chemists, the Factory Mutual Engineering Division, and the Association of Casualty and Surety

*Author's note.

Companies. Graf[13] updated these recommendations in a recent paper. The recommendations from these and other sources are combined and summarized below.

Building Design

Perchloric acid should be handled in a masonry building with concrete or tile floors. Handling acid on wooden floors is dangerous, especially after the acid has dried. The wooden floor will then become sensitive to ignition by friction.

For conventional wooden wall construction, which is less desirable, it is highly recommended that a 6 in. concrete curb be provided for the walls to rest on. In this way, acid seepage under the wall is minimized.

FLOORS. Concrete of course, is not resistant to acids, and thus should be covered. Epoxy paints in general are resistant to room temperature perchloric acid spills; however, epoxy paint will peel off concrete if pools of water stand for several days. Therefore, the floor should have a gentle slope to a drain and contain no low spots.

No equipment of any kind should ever be bolted to a floor by using bolts that screw into the floor. Perchlorates can enter and form hazardous metallic perchlorates that can be initiated when the bolt is removed. Studs, firmly and permanently set into the floor, to which equipment can be bolted, are far safer.

Building Equipment

LABORATORY BENCHES. Laboratory benches should be constructed of metal (steel) and not wood to prevent acid absorption, especially at the bottom surface which rests on the floor and would be subject to the greatest exposure from acid spills. Bench tops of resistant and non-absorbent materials such as chemical stoneware, tile, epoxy composites, or asbestos composites are recommended.

SHELVES AND CABINETS. Shelves and cabinets of steel are highly recommended rather than wood.

HOODS. Hoods should be constructed of any of the materials listed above for bench tops, or of fluorocarbon covered steel* or other resistant material. Washdown capability is highly recommended to enable periodic removal of residues.

Although polyvinyl chloride is resistant to perchloric acid, the use of plastic duct work is the subject of a good deal of controversy in Britain. Many Fire Brigades do not like plastics because of their low melting point. Two separate instances of polypropylene ducts igniting and causing fire spread have been reported recently. The use of safety glass is normally advocated in hood sash, but where hydrofluoric acid conditions are likely to be found, transparent PVC may have to be accepted. To avoid the hazard of dust/perchloric acid buildup in the ducts, the fitting of corrosion resistant water wash nozzles in the lower part of the duct with a direct drain connection is becoming commonplace. A desirable feature of a fume cupboard for perchloric acid work is a flexible hose water spray which can serve as (a) a cupboard washdown facility, (b) a personnel shower/eye-wash spray, and (c) a fire hose spray.

EXHAUST BLOWERS. Exhaust blowers should be coated to prevent corrosion and should not be sealed into the hood with glycerine-litharge, since a fatal accident has occurred when blower removal was attempted. The exhaust blower should be lubricated with fluorocarbon lubricant. Ample blower capacity is a must for adequate ventilation of the heavy perchloric acid vapors. The blower should be accessible for periodic cleaning.

*Fluorocarbon covered steel hoods are now available from one of the scientific equipment manufacturers in the U.S.A.

EXHAUST DUCTS. Exhaust ducts should be as short as possible, not interconnected to other ducts, but routed directly outdoors. Polyethylene and polyvinyl chloride ducts are available for this service, but stainless steel eliminates fire hazards.

Laboratory Equipment

HEATING SOURCE. Hot plates (electric), electrically or steam heated sand bath, or a steam bath are recommended for heating perchloric acid. Direct flame heating or oil bath should not be used.

VACUUM SOURCE. Smith[5] describes a simple apparatus, using a water aspirator or pump, for drawing fumes from a reaction vessel. The use of this apparatus is to be commended in that the contamination of the fume cupboard duct with a dust/perchloric acid layer is avoided and the vapors are drawn into water and discharged safely to drain. A similar apparatus is marketed for carrying out Kjeldahl digestions.

Silverman and First[15] have described a self-contained unit which has been developed and field tested for collecting and disposing of chemical fumes, mists and gases. It is portable and compact and, when assembled, only requires connection to an electrical receptacle and water tap to be completely operational. Although originally designed for use in filter type radiochemical laboratory hoods for safe disposal of perchloric acid fumes arising from acid digestions, it may be used as a substitute for a permanent hood in a variety of locations. Collection and disposal of the acid at the source of emission is the guiding design principle.

The conventional vacuum pump has been used both in the laboratory and in a pilot plant handling large quantities of 72 percent perchloric acid. The pumps were protected by the use of well designed cold traps and desiccant columns as well as maintaining the practice of changing oil daily. The desiccant was routinely changed, also, as well as frequent thawing out and removing the cold trap contents.

GLASSWARE. The hazards that may ensue if an apparatus cracks or breaks due to thermal or mechanical shock are sufficient to make it desirable that quartz apparatus be considered, especially as it is necessary in many experiments to chill rapidly from the boiling point.

Glass to glass unions, lubricated with 72 percent perchloric acid, seal well and prevent joint freezing arising from the use of silicone lubricants. Rubber stoppers, tubes or stop-cocks should not be used with the acid, due to incompatibility.

STIRRERS. Pneumatically driven stirrers are recommended rather than the electric motor type. Repeated exposure of the motor interior to perchloric acid vapor could result in a fire, unless the motor is an explosion proof type, which is rare.

SUNDRY ITEMS. The choice of tongs for handling hot flasks and beakers containing perchloric acid mixtures should be given due thought. Since the use of radioactive materials has become commonplace, much thought has been put into the design of indirect handling equipment. The cheap, commonly used crucible tongs are most unsuitable for picking up laboratory glassware. If possible, tongs with a modified jaw design should be used to insure that a safe grip is obtained.

Perchloric Acid Storage

WITHIN THE LABORATORY. The maximum advisable amount of acid stored in the main laboratory should be no more than two 7 lb bottles. A 450 g (1 lb) bottle should be sufficient for individual use. Storage of perchloric acid should be in a fume hood set aside for perchloric acid use. The acid should be inspected monthly for discoloration; if any is noted, the acid should be discarded.

OUTSIDE OF THE LABORATORY. Perchloric acid should be stored on an epoxy painted metal shelf, preferably in a metal cabinet away from organic materials and flammable compounds. Discolored acid should be discarded. (See acid disposal recommendations.)

Acid Handling

73 % PERCHLORIC ACID OR LESS
1. Use goggles for eye protection whenever the acid is handled.
2. Always transfer acid over a sink in order to catch any spills and afford a ready means of disposal.
3. In wet combustions with perchloric acid, treat the sample first with nitric acid to destroy easily oxidizable matter.
4. Any procedure involving heating of the perchloric acid must be conducted in a ventilated hood.
5. No organic materials should be stored in the perchloric acid hood.
6. Do not allow perchloric acid to come in contact with strong dehydrating agents (concentrated sulfuric acid, anhydrous phosphorus pentoxide, etc.).
7. Perchloric acid should be used only in standard analytical procedures from well recognized analytical texts. (This does not apply to analytical research workers.)

73 to 85 % PERCHLORIC ACID
1. Same precautions as above.

ANHYDROUS PERCHLORIC ACID (GREATER THAN 85 %)
1. Only experienced research workers should handle anhydrous perchloric acid. These workers must be thoroughly familiar with the literature on the acid.
2. A safety shield must be used to protect against a possible explosion, and the acid must be used in an appropriate hood with a minimum of equipment present. No extraneous chemicals should be present in the hood.
3. A second person should be informed of the intended use of anhydrous acid and be in the same room with the research worker using this extremely strong oxidizer.
4. Safety goggles, face shield, thick gauntlets, and rubber apron must be worn.
5. Only acid freshly prepared should be used.
6. Dispose of the unused anhydrous acid at the end of each day.
7. Do not make any more anhydrous perchloric acid than is required for a single day's work.
8. Contact of the anhydrous acid with organic materials will usually result in an explosion.
9. Any discoloration of the anhydrous acid requires immediate disposal.

Acid Disposal

SPILLS. Perchloric acid spilled on the floor or bench top presents a hazard. It should not be mopped up, nor dry combustibles used to soak up the acid. The spilled acid should first be neutralized and then soaked up with rags or paper towel. The contaminated rags and paper towel must be kept wet to prevent combustion upon drying. They should be placed in a plastic bag and sealed and then placed in a flammable waste disposal can. If the spill can be rinsed down a chemical drain, neutralization of the wetted area is recommended followed by additional rinsing.

DISPOSAL. A small amount of perchloric acid can be dumped into a sink and flushed down with at least 10 times its volume of water.

DISMANTLING AN EXHAUST VENTILATION SYSTEM SUSPECTED OF CONTAMINATION WITH PERCHLORATES

Dismantling a laboratory exhaust system contaminated with shock-sensitive perchlorates is a hazardous operation, as evidenced by published and unpublished case histories. The procedures used by one university to reduce the hazards were described by Peter A. Breysse in the Occupational Health Newsletter (15(2 and 3) 1, February–March 1966) published by the Environmental Health Division, Department of Preventive Medicine, School of Medicine, University of Washington. The problem, procedures, and confirmation of perchlorate contamination were reported as follows.

A short time ago, the manager of Maintenance and Operations was requested to dismantle and relocate six laboratory exhaust systems. The possibility of perchloric acid contamination of these systems was considered. An investigation indicated that several laboratories serviced by the exhaust systems were utilizing or had in the past, used perchloric acid for wet ashing of tissues. Furthermore, the exhaust hoods were constructed with sharp corners and cracks, permitting the accumulation of contaminants not readily noticed or easily removed. The ducts were made of ceramic material and contained numerous joints as well as a number of elbows—areas conducive to perchlorate build-up. Organic compounds were also used—to pack the duct joints, as an adhesive for the flexible connectors, and as a sealing compound for the fan.

Recognizing the potential dangers of dismantling these systems, the following procedures were established and successfully carried out.

1. It was deemed desirable to dismantle the systems on the weekend when occupancy would be at a minimum.
2. The entire system was washed for 12 hours, just prior to dismantling, by introducing a fine water spray within the hoods, with the fans operating.
3. The fans were then hosed down.
4. Fan mounting bolts and connectors were carefully removed. Nonsparking tools were and should be used throughout.
5. The fans were immediately removed to the outdoors. As an added precaution during removal the fans were covered with a wet blanket.
6. After all the fans were taken outdoors one fan at a time was placed behind a steel shield for protection during dismantling. This fan was again washed down.
7. Plate bolts were evenly loosened to remove the plate without binding. If a fan puller is necessary, it should be nonsparking.
8. All disassembled parts were washed and cleaned. The gasket material contained on the flanges was scraped off with a wooden scraper.
9. Ordinarily, the ceramic ducts would be removed by breaking them apart with a sledge hammer. In this instance, the ducts were washed down again just prior to and during dismantling. A high speed saw was used to remove the duct work.

One of the flexible connectors and a piece of duct-joint sealing compound were collected and taken to the laboratory for examination. Qualitative analysis by X-ray fluorescence and chemical tests indicated the presence of perchlorates in both samples. While these procedures for dismantling and decontamination seem unduly severe, the uncertainty requires that they be followed.

Heating of perchloric acid should be accomplished within an exhaust system designed for this purpose. The exhaust hood should have smooth surfaces for ease of cleaning. Preferably the exhaust ducts should rise vertically in a straight line with the exhaust stack extended at least eight feet above the roof line to provide adequate dissipation. Nonabsorbent materials, such as stainless steel, should be utilized throughout the system and all organic sealers and compounds should be omitted. Since there is always the possibility that traces of perchloric acid will react with air-borne substances to form deposits within the system's confines, periodic washing of the entire structure is desirable. For greatest safety electric heating should be utilized with the controls located outside the hood. All spills should be cleaned up immediately and the hood should be posted—PERCHLORIC ACID HOOD, KEEP COMBUSTIBLES OUT.

SUMMARY AND CONCLUSIONS

The use of perchloric acid is becoming increasingly widespread and the properties of both the acid and its derivatives make it likely that the trend will continue. Perchloric acid may be used in safety, provided that its hazardous properties are clearly recognized, the purpose of the acid in a process is fully understood, and measures are taken to avoid known possibilities.

It is, however, clear that no one should attempt to use perchloric acid who is not fully conversant with the chemistry of the material and who has not made a careful appraisal of his operating conditions and techniques.

SAFE WORKING CONDITIONS

When assessing the minimum requirements to insure safe working conditions the following three questions are of importance.

1. Is the work involving the use of perchloric acid likely to be a continuing rather than an occasional and infrequent commitment?

2. Will the use of perchloric acid be accompanied by any form of heating? (Heat of reaction and frictional heat should not be overlooked in this context.)

3. Is it intended to use perchloric acid more concentrated than the 72% azeotrope?

Should the answer to all these questions be "no," then, in the absence of any other contra-indications, relaxation of the normal standards may be considered. If, however, a positive answer is received to any of these questions the working conditions should conform to the recommended standards. Any deviation from these standards should be made only with the agreement of the departmental safety supervisor.

APPENDIX 1—SOME ACCIDENTS INVOLVING PERCHLORIC ACID

1. Explosions may occur when 72% perchloric acid is used to determine chromium in steel, apparently due to the formation of mixtures of perchloric acid vapor and hydrogen. These vapor mixtures can be exploded by the catalytic action of steel particles.[1]

2. Two workers are reported to have dried 11,000 samples of alkali-washed hydrocarbon gas with magnesium perchlorate over a period of 7 years without accident. However, one sample containing butyl fluoride caused a purple discoloration of the magnesium perchlorate with the subsequent explosion of the latter.[1]

3. A worker using magnesium perchlorate to dry argon reported an explosion and warned that warming and contact with oxidizable substances should be avoided.[1]

4. An explosion was reported when anhydrous magnesium perchlorate used in drying unsaturated hydrocarbons was heated to 220°C.[1]

5. An explosive reaction takes place between perchloric acid and bismuth or certain of its alloys, especially during electrolytic polishing.[1,4]

6. Several explosions reported as having occurred during the determination of potassium as the perchlorate are probably attributable to heating in the presence of concentrated perchloric acid and traces of alcohol. An incident in a French laboratory is typical: an experienced worker in the course of a separation of sodium and potassium removed a platinum crucible containing a few decigrams of material and continued the heating on a small gas flame. An explosion pulverized the crucible, a piece of platinum entering the eye of the chemist.[6]

7. A violent explosion took place in an exhaust duct from a laboratory hood in which perchloric acid solution was being fumed over a gas plate. It blew out windows, bulged the exterior walls, lifted the roof, and extensively damaged equipment and supplies. Some time prior to the explosion, the hood had been used for the analysis of miscellaneous materials. The explosion apparently originated in deposits of perchloric acid and organic material in the hood and duct.[7]

8. A chemist was drying alcohol off a small anode over a bunsen burner in a hood reserved for tests involving perchloric acid. An explosion tore the exhaust duct from the hood, bent a portion of the ductwork near the fan, and blew out many panes of window glass.[7]

9. An employee dropped a 7-lb bottle of perchloric acid solution on a concrete floor. The liquid was taken up with sawdust and placed in a covered, metal waste can. Four hours later, a light explosion blew open the hinged cover of the can. A flash fire opened three sprinklers which promptly extinguished the fire.[7]

10. A 7-lb bottle of perchloric acid solution broke while an employee was unpacking a case containing three bottles. The spilled acid instantly set the wood floor on fire, but it was put out quickly with a soda-acid extinguisher.[7]

11. At a malleable iron foundry, perchloric acid had been used for about four years in the laboratory for the determination of the silicon contents of iron samples. A cast iron, wash-sink drain at the bench used for this purpose had corroded and the leaking acid had soaked into the wood flooring, which was later ignited while a lead joint was being poured. This fire was extinguished and part of the wood flooring was removed. Later in the day, at a point slightly removed from the location of the first fire, a similar fire occurred when hot lead was again spilled. This time the fire flashed with explosive violence into the exhaust hood and stack above the work bench. Laboratory equipment and records were wet down extensively, and damaged.

12. A stone table of a fume hood was patched with a glycerin cement and several years later, when the hood was being removed, the table exploded when a workman struck the stone with a chisel. The hood had been used for digestions with perchloric acid and, presumably, acid spills had not been properly cleaned up.[8]

13. A conventional chemical hood normally used for other chemical reactions, including distillation and ashing of organic materials, was also used during the same time for perchloric acid digestion. During a routine ashing procedure, the hot gases went up the 12 in. tubular transit exhaust duct and one of a series of explosions occurred that tore the duct apart at several angles and on the horizontal runs.[8]

14. During routine maintenance involving partial dismantling of the exhaust blower on a perchloric acid ventilating system, a detonation followed a light blow with a hammer on a chisel held against the fan at or near the seal between the rear cover plate and the fan casing. The intensity of the explosion was such that it was heard four miles away and of the three employees in the vicinity, one sustained face lacerations and slight eye injury; the second suffered loss of four fingers on one hand and possible loss of sight in one eye; the third was fatally injured with the 6 in. chisel entering below his left nostril and embedded in the brain.[8]

15. A 6-lb bottle of perchloric acid broke and ran over a fairly large area of wooden laboratory floor. It was cleaned up, but some ran down over wooden joists. Several years later a bottle of sulfuric acid was spilled in this same location and fire broke out immediately in the floor and the joists.[8]

16. A chemist reached for a bottle of perchloric acid stored on a window sill above a steam radiator. The bottle struck the radiator, broke, and the acid flowed over the hot coils. Within a few minutes the wooden floor beneath the radiator burst into flame.[8]

17. An explosion occurred when an attempt was made to destroy benzyl celluloses by boiling with perchloric acid.[11]

18. An explosion occurred as anhydrous perchloric acid was being prepared via sulfuric acid dehydration and extraction with methylene chloride when a stopper was removed from the separatory flask.[14]

ACKNOWLEDGEMENT: This chapter is based on and developed from a Safety Assessment prepared for use in the University of Leeds, England, and is reprinted with permission, as well as the work of Graf and associates at Thiokol Chemical Corporation, Brigham City, Utah. Revised in 1969.

REFERENCES

1. Schumacher, J. C., "Perchlorates: Their Properties, Manufacture and Uses." American Chemical Society, Monograph Series No. 146. Reinhold Publishing Corporation, New York, Chapman and Hall, Ltd., London, (1960).

2. Ministry of Aviation, *Perchloric Acid: A Review of its Thermal Decomposition and Thermochemistry*, Rocket Propulsion Establishment Technical Note No. 224, London: October 1963.

3. Ministry of Aviation, *Perchloric Acid: A Review of the Physical and Inorganic Chemistry*. Rocket Propulsion Establishment Technical Memo No. 352, London: March 1965.

4. Burton, H. and Praill, P. F. G., Perchloric Acid and Some Organic Perchlorates. *Analyst. 80.* 4, (1955).

5. Smith, G. F., The Dualistic and Versatile Reaction Properties of Perchloric Acid, *Analyst. 80.* 16, (1955).

6. Moureu, H. and Munsch, H., Sur Quelques Accidents Causés par la Manipulation de l'Acide Perchloriques et des Perchlorates. *Archives des Maladies Professionnelles de Médécine du Travail et de Sécurité Sociale. 12*, 157–159, (1951).

7. Harris, E. M., Perchloric Acid Fires. *Chem. Eng. 56:1*, 116–7, (1949).

8. Scheffler, G. L., University Health Service, University of Minnesota, private communication to Safcty Officer, Leeds University, London.

9. Jacob, K. D., Brabson, J. A., and Stein, C., *J. Assoc. Offic. Agric. Chemists, 43*, 171, (1960).

10. Factory Mutual Engineering Division, "Handbook of Industrial Loss Prevention," page 58–1. McGraw-Hill, New York. (1959).

11. Sutcliffe, G. R., *J. Textile Ind. 41*, 196T, (1950).

12. Bartlett, R. K. and Turner, H. S., An Unappreciated Hazard in the Preparation of Electropolishing Solutions: The Investigation of an Explosion. *Chemistry and Industry*, 1933–1934, (Nov. 20, 1965).

13. Graf, F. A., Safe Handling of Perchloric Acid. *Chem. Eng. Prog., 62:10*, 109–114, (1966).

14. Thiokol Chemical Corporation, Letter WLC-151/63, *Explosion Involving Anhydrous Perchloric Acid*, (11 April 1963).

15. Silverman, L. and First, M. W., "Portable Laboratory Scrubber for Perchloric Acid," *Industrial Hygiene Journal*, Nov.–Dec. 1962, pp. 463–472.

Section **6**

Toxic Hazards

Principles and Procedures for Evaluating Toxicity of Chemicals

The information in this chapter on toxicity was prepared by the Committee on Toxicology, Division of Chemistry and Chemical Technology, National Academy of Sciences—National Research Council and published as Publication 1138 by the National Academy of Sciences—National Research Council, Washington, D.C., in 1964 under the title "Principles and Procedures for Evaluating the Toxicity of Household Substances."

The complete publication is reprinted here by the courtesy of the National Academy of Sciences—National Research Council.

The Committee on Toxicology consisted of Arnold J. Lehman, Chairman, David W. Fassett, Horace W. Gerarde, Herbert E. Stokinger, and John W. Zapp.

PREFACE

Regulations issued August 12, 1961, pursuant to the Federal Hazardous Substances Labeling Act passed by Congress on July 12, 1960 (Public Law 86–613), require that the toxicity of chemicals contained in household substances shall be determined by the manufacturer. Maximal toxic doses for certain experimental animals, when administered by inhalation, oral ingestion, or percutaneous absorption, as well as the corrosive, irritant, and sensitizing actions, are clearly delineated in these regulations.

In order to facilitate such determinations and at the same time to promote a certain degree of standardization, the Committee on Toxicology of the National Academy of Sciences—National Research Council has outlined in some detail principles and procedures for evaluating the toxicity of household substances. It is hoped that this publication will be helpful to those who are concerned with providing data as required by the regulations noted above. For further details on procedure the reader is referred to the publications listed in the references.

The Committee on Toxicology acknowledges with gratitude the editorial preparation of this publication by the Advisory Center on Toxicology of the Academy-Research Council, under the direction of Harry W. Hays.

ORAL INGESTION

One of the most useful and frequently recorded parameters of biological activity is the acute single lethal dose by oral or parenteral routes. Medical literature for the past 80 years contains many thousands of values for acute lethal doses of chemical compounds, many of which were determined in connection with the development of new pharmacologic agents. Prior to the middle 1930's, the term "lethal dose" had no very specific meaning and usually was given as an approximate dose causing lethal effects in some or all of the animals treated. Principally as a result of the work of the pharmacologists, Trevan and Gaddum, and later of Bliss, it was shown that greater precision could be obtained through the application of statistical procedures.

It became apparent that the so-called LD_{50}, the dose killing approximately one-half of the population, could be determined with considerably greater accuracy than the dose killing only some other number of the population, and further that if the doses were spaced according to some geometric progression or some equal logarithmic interval, the values generally were symmetrically distributed. By plotting the log of the dose against some probability function, such as a probit, the data could be represented as a straight line having a certain position and certain slope. Elaborate statistical methods for fitting the best line to the values were developed by Bliss and others and became extremely useful in biological assays where it was desired to compare accurately one compound or sample with another.

Pharmacologists and toxicologists soon realized that the very large number of experimental animals used in such procedures was unnecessary when it was desired simply to obtain the approximate magnitude of the oral lethal dose. One of the early publications dealing with the simplification of this determination was that by Deichmann and Mergard, 1948. Although methods employing only a few animals per dose level give a fairly good idea of the position of the LD_{50}, they generally do not give a very accurate indication of the slope of the dose response curve. Nevertheless, a large portion of toxicologic data, particularly that from industrial laboratories, is now being expressed in terms of these so-called "range-finding" lethal doses.

The determination of an LD_{50} value, however, is only one of the purposes for performing oral toxicity studies. Of equal, or possibly more importance, is the description of the action of the compound. The range-finding LD_{50} tests actually facilitate this goal, since it is usually possible to make more careful observations on only two or three animals per dose level in terms of the time of onset and subsidence of symptoms, the nature of the toxic symptoms, and the length of time for recovery. It should be pointed out that the range-finding test, covering several doses in the zero as well as the 100 per cent response zones, provides the physician with considerably more useful and pertinent information in cases of poisoning than the highly precise single value of the more elaborately determined LD_{50}. Some of the factors that are of importance in determining an acute lethal dose and that should be included or considered in the description and evaluation of the results are discussed below.

Species and Strain of Animal

The animals most commonly used for LD_{50} determinations are rats and mice of varying strains. The principal advantage in using these animals is their relative uniformity and availability and the very large body of data that has been accumulated. The results obtained in these species may not, however, be uniformly reliable in predicting human responses, since both the rat and mouse may be resistant to certain types of chemical agents, e.g., aryl phosphates, phenols, amides, aromatic amines, etc., or may show responses quite different from those in other species. For example, morphine is a depressant in the rat, dog and man, but is stimulating in the mouse and cat. In the case of aryl phosphates, the rat is known to be resistant, while the chicken and cat are highly sensitive and thus would be the animals of choice in testing compounds of this type.

In some cases strain differences may be of importance. For example, in testing for carcinogenicity it is recognized that certain strains of mice are more susceptible than other strains to tumor formation. Strain differences are usually not so apparent in acute oral toxicity studies and therefore it is seldom necessary to determine an LD_{50} for two strains of laboratory animals. The species or strain differences may be due to differences in absorption, distribution, excretion and metabolism.

Sex

While various animal species may react differently to the same compound, there are also sex differences in response. A number of studies have shown female rats to be much more sensitive than males to certain compounds. The reason for these differences is still unknown, but it has been suggested that it may be due to differences in the activity of the metabolizing enzymes (Quinn, *et al.*, 1958). This sex difference in response to various compounds is seen often in rats and, although there appears to be no mandate for choosing either sex, it is important to indicate which sex has been used, and it would appear more conservative to use the more sensitive sex when there is a difference in response.

Weight and Age of Animals

There is considerable evidence to show that the weight and age of rats are important in the determination of the lethal dose of any compound. Studies on old rats with large deposits of fat may give values quite different from those of younger animals. To obtain more uniform results, it would seem desirable to select a group of animals with a fairly definite weight range. The usual range is 200–250 g for rats and 20–30 g for mice. Data on younger animals would seem to be of interest in connection with accidental poisoning.

Diet and Fasting of Animals

Free access to food and water prior to testing any compound may appreciably alter the LD_{50} value. This would be particularly true in case of materials rapidly absorbed and metabolized. While this is not generally a problem with single-dose studies, it is essential to state the nature of the diet used. Fasting animals overnight is now commonly practiced when doing acute toxicity studies. In view of the requirements in the regulation of the Hazardous Substances Labeling Act for administering large doses to rats, it is essential that fasting be of a sufficient period of time to make certain the stomach is empty in order to accommodate the large volumes of test compound required.

Preparation of Test Material

Differences in the preparation of test materials are probably responsible for many of the variations in the values given in the literature for any one compound. This is in part due to the vehicles used as solvents or for the dilution of oil-soluble materials. If the test material is a liquid, it should probably be given undiluted. When dilution is necessary and the test material is water-soluble, this would be the diluent of choice. If the material is oil-soluble, and vegetable, corn or cottonseed oils are used as the diluent, it should be recognized that the use of these solvents may alter the absorption as well as the response. In addition, solvents with known toxic properties should be avoided. With the use of microsyringes, it is possible to administer most liquids in sufficiently small volume to permit an accurate LD^{50} determination without dilution.

In the case of solids, it may be advisable to grind the material in a ball mill or mortar before attempting to put it in solution. If water-insoluble, it is better to make an aqueous suspension rather than to dissolve it in other solvents. Most suspensions are sufficiently stable to be administered orally, if this is done promptly. In cases where simple suspensions are not feasible, it may be necessary to use suspending agents, such as carboxymethylcellulose or guar gum.

Route of Administration

The oral route of administration is the one most commonly used for determining LD_{50}'s and is accomplished by the use of soft rubber or polyethylene tubing, or in some cases, a large blunt-tip needle. The maximum volume of aqueous solution and/or suspensions that should be given to a rat is probably in the neighborhood of 4 or 5 ml. Although as much as 10–12 ml can be given, the possibility of mechanical damage from large volumes should be considered. If the volume of 5 ml is taken as an upper limit, then 250-gram rats will require at least 25 % solutions or suspensions in order to reach a dose of 5 grams/kilogram. For materials that are not soluble in aqueous solution and must be administered in oily vehicles, 1.5–2.0 ml is generally the upper limit because of the laxative effect. The determination of the LD_{50} of insoluble solids at high dose levels poses difficult practical problems and the values may be difficult to interpret.

Number of Animals per Dose Level

It was indicated above that much valuable information can be obtained by so-called range-finding techniques, which should always precede any more elaborate determination of the LD_{50}. If it is desired to determine a slope, then a minimum of four or five animals per dose level should be used. Methods for calculating the LD_{50} are given by Weil (1952) and Litchfield and Wilcoxon (1949).

Dose Intervals

It is generally recognized that the dose intervals should increase by geometric or logarithmic progression. In a range-finding study it is customary to use one or two animals per dose and to use a factor of two between doses; e.g., 25, 50, 100, 200, 400, etc., mg/kg. The data will then consist of lower dose levels where no animals are killed, a middle zone in which there may or may not be lethal effects, and finally a zone in which lethal effects invariably occur. The mid-point of the distance between the uniformly zero and 100 per cent killed will generally be close to the LD_{50} determined with more accurate procedures. If more precise values are thought to be necessary, the number of animals should be increased and the dose interval chosen by some equal log interval. A factor of 1.26 is often used, equivalent to a 0.1 log interval, and an attempt made to give doses in the region of the LD_{50}.

Period of Observation

The time at which deaths occur or symptoms appear or subside may be important, particularly if there is any tendency for deaths to be delayed. It is characteristic that with certain compounds, such as alkylating agents, deaths may occur as late as the second week of observation, and in some cases even later. A 14-day observation period would seem, however, to be sufficient for most compounds.

Recording of Symptoms

Far too little attention has been paid to the accurate recording of symptoms during LD_{50} determinations. The preoccupation of toxicologists and pharmacologists with statistical precision has resulted in a tendency to lose sight of the fact that the physician treating a poisoning case is concerned with the symptoms produced. It is helpful to have check sheets for many of the standard types of symptoms, but it is essential that any stereotyped approach be avoided. The value of the LD_{50} determination will be directly proportional to the skill and experience of the toxicologist in the interpretation and recording of symptoms.

Weight Change of Animals

An indication as to the presence of a severe toxic effect may sometimes be gained by comparing the weights of the treated animals with those of controls. The weight of survivors at the end of a 14-day period should be recorded, and it is good practice to weigh the animals at least once during this period.

Necropsies

Necropsies of some of the surviving animals may be valuable in giving a clue as to the type of toxic effect produced by the test compound, and therefore should be a part of the general procedure. Gross inspection of the intestinal tract may reveal evidence of severe irritation. Pathologic changes of such organs as liver, kidney and spleen may be noted. If there is evidence of gross pathology, it may be of interest to determine the histopathology of the structures involved.

Evaluation of Acute Toxicity Data

While there have been a number of publications dealing with chronic exposure of animals in reference to prediction of chronic effects in humans, few have been concerned with the validity of such procedures for prediction of acute toxic effects in humans. Various reports on the tabulation of toxicity classes have been given by Hodge and Sterner (1949), Fassett (1959), and Hagan (1959).

It can probably be accepted as a general principle that the more highly toxic agents, excluding those whose action is primarily corrosive or irritant, are acting through some vital or specific enzyme systems which are of relatively uniform importance in man and lower animals. For example, there seems to be little doubt that compounds having an oral LD_{50} of about 50 mg/kg or less in the rat will often be highly toxic in a variety of other species, including man, and that the slope of the dose-response curve will generally be steep. As the lethal dose increases, however, the certainty of such predictions becomes less until at high LD_{50}'s, such as 5000 mg/kg, the exact value may have relatively little meaning in predicting human effects.

When such large doses of chemicals of low toxicity are ingested by man, they may cause symptoms or death by nonspecific means, such as changes in pH of body fluids, shock from irritation or perforation of viscera, aspiration of material from vomitus, etc.

If the material happens to be highly irritating, this may overshadow all the other toxic properties and result in a very low LD_{50} for concentrated solutions and a much higher LD_{50} for dilute solutions. Vomiting and diarrhea may also interfere with absorption of the material. The presence or absence of food in the stomach can alter the result either by influencing the rate of absorption or by protection against local irritant action. Finally, since toxic effects depend upon the influence of the material on biochemical processes, the advancement of knowledge in this area requires that the toxicologist continually attempt to interpret acute toxicity data in terms of probable biochemical mechanisms.

EVALUATION OF THE ASPIRATION HAZARDS OF LIQUIDS

Some liquids having a relatively low to moderate degree of oral toxicity are generally recognized as hazardous because they are readily aspirated into the lung, causing chemical pneumonitis. Kerosine, for example, has an oral LD_{50} of 30–40 ml/kg for the rat and yet a few drops aspirated into the lung will cause a rapidly fatal fulminating chemical pneumonitis. For kerosine, the ratio of the rat oral LD_{50} to the intratracheal LD_{50} is 140 to 1 (Gerarde, 1963).

The accidental aspiration of liquids from the mouth into the lungs is an acute incident that occurs in just a few seconds—the time required to take a breath. In this brief time the liquid flows from the back of the mouth through the glottis and into the respiratory tract. The volume of liquid aspirated is self-limiting in a conscious individual. As soon as liquid enters the lung, normal physiological reactions occur which oppose further entry of liquid. These responses are: (1) momentary reflex cessation of breathing and (2) the more active expulsive mechanism of coughing.

Although the aspiration hazard for certain liquids, hydrocarbons in particular, has been recognized for many years, there is very little information in the literature on methods for determining or measuring the toxicity and hazard by this portal of entry.

The direct instillation of liquid into the trachea with a syringe, for example, does not measure the aspiration tendency of a liquid, because it does not stimulate the conditions that prevail in clinical aspiration poisoning. It is well known in clinical medicine that it is dangerous to put liquids into the mouth of an unconscious patient because of the danger of aspiration in the absence of the swallowing reflex.

A method for measuring the aspiration hazard which is based on this well-known fact has been described recently. A liquid is placed into the mouth of an anesthetized animal (rat). In the absence of the swallowing reflex, a liquid of low viscosity and surface tension is readily aspirated.

Viscosity appears to be the most important physical property that determines the likelihood of entry into the lungs by aspiration. Surface tension, a measure of the spreading tendency of a liquid, is another property which, at least on theoretical considerations, would be expected to influence the aspiration tendency of a liquid.

The criteria for measuring aspiration tendency and toxicity are lung weights and mortality 24 hours after dosing. The method is as follows:

Male albino rats of Wistar strain and weighing from 200–300 g are anesthetized to the point of apnea in a covered wide-mouth jar (capacity one gallon) containing about one inch of wood shavings moistened with approximately one ounce of anhydrous diethyl ether. The animal is removed from the jar and placed on its back or side on the table top. The mouth is held open and the tongue pulled forward. With the animal's head elevated, 0.2 ml of the test material is delivered into the mouth with a Becton-Dickinson 1/2-ml syringe. This dose is a "mouthful" for the rat and is the maximum quantity that can be placed into the rat's mouth without danger of spilling. As breathing resumes and becomes regular, the nostrils are closed with the fingers at the end of the expiration phase in the breathing cycle. This is repeated until the liquid has been aspirated or the animal shows signs of regaining consciousness, which is usually preceded by a return of the swallowing reflex.

After dosing, the animals are observed for a minimum of four hours at intervals ranging from five minutes to a maximum of 30 minutes, depending on response. Lungs are removed and weighed as soon after death as possible. Twenty-four hours after dosing the survivors are sacrificed under ether anesthesia by exsanguination from the abdominal aorta. The lungs, dissected free from the heart, trachea and mediastinal structures, are blotted on a paper towel and weighed to the nearest centigram on a triple-beam or torsion laboratory balance. A rat with a lung weight of less than three grams, 24 hours after dosing, has minimal to moderate lung injury compatible with survival. For further details see Gerarde (1963).

PERCUTANEOUS ABSORPTION, EYE AND SKIN IRRITATION AND SENSITIZATION

The Federal Hazardous Substances Labeling Act is concerned primarily with the effects of substances on man. The alternative tests on animals described in the *Regulations* are meant to be used when human data and experience are lacking. The following is a discussion of some of the principles and procedures used in determining the toxicity of household substances.

Corrosive

Human tissue differs somewhat from the tissue of other mammals, but it is generally true that a chemical which is capable of destroying animal tissue by chemical action will be capable of destroying human tissue, and vice versa. The rabbit is most commonly used as the test animal for determining whether a substance is corrosive, but other mammalian species will serve as well. It is important that hair or fur be removed so that the test substance comes into intimate contact with the skin and that the effect on the skin can be accurately observed. The hair should be removed by clipping, shaving, or applying a depilatory. Clipping is generally preferred, since it will produce less damage to the skin.

It is important that enough of the test material be applied to the skin to produce injury. A dose of 0.5 ml of liquid or 0.5 g of solid is recommended. It seems likely, however, that in some instances smaller amounts would suffice to demonstrate a corrosive action.

The test material should remain in contact with the skin long enough to allow maximal reaction to occur. This can be accomplished by introducing the test substance under gauze, securing the patches with adhesive tape, and then overwrapping the area with an impervious material to retard evaporation. This would, for example, simulate contact with human skin under clothing. It is not intended that evaporation should be entirely prevented by this procedure, and the impervious overwrap should, therefore, not be tightly applied. Alternative procedures which will accomplish the same purpose should be acceptable.

If a substance is corrosive on normal intact skin, it is almost certain that it will also be corrosive on other tissues, e.g., mucous membranes or the eye, which are inherently more sensitive than intact skin. If a substance proves to be a strong primary irritant, but not corrosive, on normal intact skin, it should be tested for corrosive action on other tissues with which it might come in contact. Effects on the eye can be determined by applying a generous drop of liquid or about 0.1 g of powdered solid to the eye of an animal. The rabbit eye is perhaps most suitable for this purpose because of its size and the relative ease of observation of the results. As with the test on skin, the test substance must be left in the eye long enough to insure maximal reaction. Hence the eye should not be rinsed.

The mucous membranes of animals available for tests of corrosive substances are those of the mouth and the penile mucosa of the rabbit. The same principles apply as for tests on the eye.

Testing potentially corrosive substances on human beings is inherently dangerous and should not be deliberately undertaken for the routine evaluation of household substances.

Primary Irritation

By definition, primary irritation of the skin consists of a local inflammatory reaction which does not produce destruction of tissue or irreversible change ,at the site of contact. In man it would be recognized as erythema, with or without edema, of the inflamed area.

Irritation of animal skin is best detected by the use of albino animals, such as the guinea pig, rabbit, rat and mouse. Guinea pigs and rabbits are more commonly used for the test because it is possible, with these species, to extend the study to investigate the potential of the test substance for producing allergic skin sensitization.

A substance applied to the intact and abraded skin of six rabbits may be classified as a primary irritant if it produces a score of five or more (Draize, 1944). This score is based on the average degree of erythema and edema observed at 24 hours and 72 hours after application of the test material. It is the opinion of the Food and Drug Administration that any substance that would produce a score of five or more under the conditions of the prescribed test would indeed produce a substantial degree of primary irritation if applied to intact human skin. As described above under "Corrosive," the quantity of substance to be applied to shaved rabbit skin is 0.5 ml of liquid or 0.5 g of solid. It is to be applied under surgical gauze and overwrapped with an impervious material to aid in maintaining the patches in position and to retard, but not prevent, evaporation. After 24 hours, the bandages are removed and the resulting reactions evaluated according to a standard scoring system.

It seems likely that less elaborate procedures would also serve to characterize substances suspected of being primary skin irritants. Some laboratories prefer albino guinea pigs to rabbits because of their smaller size. Others have used rats, mice, or even a special strain of "hairless" mice which do not require shaving. As one investigator put it: "The rabbit is the standard animal for safety testing, not primarily because of any special property that would render it exclusively suitable for the purpose, but rather because of the admirable standardization of methods for the use of this animal by Draize, of the Food and Drug Administration."

It should be kept in mind that the primary purpose of all such tests is to predict whether a substance is capable of producing, on immediate, prolonged or repeated contact with normal living tissue, a local inflammatory reaction.

If the precautions discussed above under "Corrosive" are followed, i.e., if the substance is given ample opportunity to react maximally with the skin, it is usually not at all difficult to decide whether it does or does not produce a local inflammatory reaction at the site of contact.

There will, however, be borderline cases in which it is clear that a substance is not corrosive to animal skin, but doubtful whether it would produce an inflammatory reaction if applied to human skin. In such cases one might wish to carry out the "official" rabbit test prescribed in the *Regulations* (191.11), and determine whether the average score is five or more. Or, one might proceed more directly by applying a small amount, not more than one drop of liquid or 0.1 g of solid under a Band-Aid, to the skin of the upper arm or back of human volunteers. The tests on human skin would be decisive in answering the question of hazard posed by the Federal Hazardous Substances Labeling Act, but should only be carried out by persons skilled in the performance and evaluation of such tests. The subjects should be instructed to remove the patches immediately and to wash the site of application thoroughly if irritation is felt; otherwise the patches should remain in place for 24 hours.

Acute Dermal Toxicity

Some chemicals, when in contact with skin, will pass through the skin barrier and enter the circulation and produce general, rather than local, toxic effects. It is important, therefore, that household substances be tested for acute dermal toxicity as well as for skin irritation and sensitization, if contact with the skin is reasonably foreseeable.

The *Regulations* prescribe a procedure for evaluating acute dermal toxicity. It is essentially that described by Draize (1944) and was devised for the evaluation of substances, such as insect repellents, which might be applied fairly continuously to human skin. It requires that the test chemical be applied to intact and abraded skin of rabbits under an impermeable band, or sleeve, if liquid, and held in place for 24 hours, during which time the animal is immobilized in a suitable holder. Sufficient animals and graded doses are used to permit the determination of the lethal dose.

The prescribed test does not seem particularly appropriate for the evaluation of the dermal toxicity of household substances, for it is most unlikely that a chemical spilled on the skin or clothing will remain in continuous contact with the skin for 24 hours. Furthermore, the Regulation does not require the determination of an LD_{50}, but specifies only that a substance be considered highly toxic if it produces death in "half or more than half of a group of ten or more rabbits in a dosage of 200 mg or less per kilogram of body weight; or toxic if it produces death in half of a group of rabbits in a dosage of more than 200 mg/kg but not more than 2.0 g/kg of body weight."

From a practical point of view, there appears to be no necessity for confining a household substance under an impermeable bandage or cuff for 24 hours unless continuous human skin contact is reasonably foreseeable. In the case of liquids it should suffice to immobilize the animals only until the liquid has dried or been completely absorbed. Volatile liquids should be applied slowly enough to allow maximal opportunity for absorption. Solids should be kept in contact with the skin by a bandage which may or may not be impermeable, since the vapor pressure of solids is usually negligible. If the area of skin application is covered by a bandage that prevents the rabbit from ingesting the test material by licking, immobilization of the animal is unnecessary.

Eye Irritants

Albino rabbits are ideally suited for the study of compounds that may produce mild to severe eye irration. The relatively large size of the eye, the consequent ease of instillation of test material, and the ease of the evaluation of results commend the rabbit over other species for eye irritation tests.

Since rabbit eyes may frequently show evidence of injury sustained in the course of living, it is important that the eyes be examined before application of the test substance. Corneal injury may be detected by the instillation of a 5 per cent aqueous solution of fluorescein stain and flushed with distilled water twenty seconds after application. Any injured area should be grossly visible but the aid of a hand slit lamp will also be helpful. Animals showing corneal injury should be eliminated from the test.

For the purposes of the Hazardous Substances Labeling Act, a dose of 0.1 ml of liquid or 10.0 mg of fine dry powder is instilled into the conjunctival sac of one eye of each of six rabbits. The lids are held open for 20–30 seconds, and the animals are then returned to their cages. The other eye may be used as control or treated with the test material, but flushed with distilled water 20–30 seconds after application. While effects on washed eyes are not required by the *Regulations*, there would seem to be merit in knowing whether prompt washing would prevent injury in the case of accidental contamination with a household substance. The animals are observed for periods of 24, 48 and 72 hours and the changes noted in the cornea, iris and conjunctiva. In the initial eye tests if at any time during the first 72 hours, 4/6 rabbits show corneal opacity or iritis, the substance will be classified as a primary eye irritant. As for the conjunctiva if 4/6 have any of the following—2 redness, or 2 chemosis, or 2 discharge—the substance is also considered to be an eye irritant. On those occasions when only 1/6 show a reaction, the test is considered negative. However, if 2/6 or 3/6 rabbits show corneal opacity, iritis or conjunctivitis, a second test is performed using six more rabbits. In this instance if 3/6 animals show positive reactions to any of the 3 critical areas, the test is considered positive. When only 2/6 show reactions, a third set of rabbits will be required. In this instance it is necessary for only 2/6 rabbits to show reactions on the cornea, iris or conjunctiva as previously described for a positive test, 1/6 will be considered negative. Grading systems based on damage to the various structures and duration of injury are described by Friedenwald *et al.*, 1944, Draize *et al.*, 1944, and Carpenter and Smyth, 1946.

There would appear to be no justification great enough for attempting to evaluate potential eye irritants by deliberately applying them to the eyes of human volunteers.

Skin Sensitization

Skin sensitization is a phenomenon in which the skin, through prior contact with the sensitizing substance, acquires an increased sensitivity to that substance and responds to subsequent applications with erythema, with or without edema, and with or without blistering at the site of contact, whereas on first contact these local reactions did not occur or occurred only to a minimal degree. The reaction of the skin to poison ivy is typical of the human response to skin sensitizers. On first contact with poison ivy no irritation occurs. After an indefinite number of contacts, depending on the individual response, the skin acquires an increased sensitivity to the poison ivy, and thereafter responds with the typical reddening, itching, and pinpoint blisters. The allergic dermatitis produced by chemical substances does not differ in any important respect from that produced by poison ivy.

One difficulty in evaluating the potential of substances to produce skin sensitization is that the response of human skin to sensitizers differs markedly from that of many commonly used species of experimental animal. The rat and the mouse, for example, seldom exhibit skin-sensitization reactions, and hence would not be the animals of choice for the laboratory evaluation of potentially sensitizing agents. The albino guinea pig and rabbit do develop allergic reactions and can be used for the laboratory evaluation in such studies.

Among the many chemicals known to produce cutaneous sensitization are the aromatic nitroamino and halo compounds, e.g., *p*-phenylenediamine, picryl chloride and dinitrochlorobenzene; sulfur compounds (thioglycerol), diisocyanates (toluene diisocyanate), and certain metals such as nickel, mercury, and hexavalent chromium. It is seldom possible, however, to predict from the chemical structure of a substance whether it will produce an allergic reaction.

In testing a compound for sensitization, it is necessary that it penetrate the skin. In animals lacking sweat glands it is preferable to apply the material to abraded skin or to inject it intradermally.

A procedure which might be suitable for the evaluation of potential skin sensitizers using guinea pigs is described in some detail as follows:

Ten healthy, young adult, male, albino guinea pigs are selected for the test. Female guinea pigs, particularly if pregnant, do not respond as well as males. An area of skin between the shoulder blades is exposed first by clipping and then by shaving with an electric shaver. If the test substance is a liquid, approximately 0.1 ml is applied to the shaved area and gently massaged into the skin with a blunt glass rod. If a solid, the test substance may be dissolved in water or in a bland solvent such as alcohol, or it may be applied as a water paste. If marked primary irritation occurs, the sample is diluted with water or other bland solvent until a concentration is found that produces only slight irritation. This minimal reaction can then be compared with the final challenge reaction to the same concentration. An increased severity in response would indicate an acquired hypersensitivity, and further sensitization studies should be carried out.

For this purpose hair is removed, as described above, from a strip running from flank to trunk along each side of each animal. Repeat as necessary as the hair grows back. Starting at one end of one strip the first sensitizing application is made. If the substance is suitable for injection, 0.1 ml may be injected intradermally with a fine hypodermic needle. If not suitable for injection, it may be applied to lightly abraded skin and gently rubbed in with a glass rod. The light abrasions are made by scratching several times through the epidermis with a sharp needle or scalpel blade deeply enough to penetrate the epidermis but not deeply enough to cause bleeding.

The sensitizing injections, or applications to abraded skin, are made three times weekly on alternate days for three weeks, for a total of nine treatments. The succeeding injections or applications are made by moving along the shaved strips, choosing a new location for each treatment. The local reaction produced on each animal is recorded at 24 and 48 hours, since with strong sensitizers a build-up in severity of the reactions is observed even during the sensitizing treatments.

Following the ninth sensitizing treatment the animals are set aside for two weeks, after which they are challenged by a final application to intact skin between the shoulder blades and a final application to lightly abraded skin or intradermal injection. If the response to the final challenge applications or injection is greater in terms of intensity of local inflammatory response or number of animals responding, or both, there is a *prima facie* indication that the test substance is a skin sensitizer. However, since the sensitivity of guinea pig skin to primary skin irritants changes somewhat with the age of the animals, it is well to replicate the final applications on a group of control animals of the same age as the test group and to compare the two sets of responses. If they do not differ significantly, sensitization has not been demonstrated.

While the diet of the guinea pigs should be nutritionally adequate, they should not be given an excessive amount of Vitamin C (fresh greens), since this will tend to suppress the sensitization reaction. A standard commercial guinea pig ration should be sufficient if the manufacturer claims that it is a complete diet for guinea pigs.

It has been the experience of a number of laboratories that the above test, or suitable variants thereof, is quite useful in uncovering substances inherently capable of producing allergic skin sensitization in man. It does not, however, tell very much about the likelihood that sensitization will occur in a substantial number of people in the course of the use of the test substance for its intended technological purpose. About the only generalization that can be made is that strong sensitizers for guinea pigs are likely to cause a substantial number of sensitization reactions in man, whereas weak sensitizers for guinea pigs may or may not cause enough in man to be a problem. It is rare that a substance which *fails* to produce sensitization in guinea pigs when tested by the above technique will produce sensitization in man, if in this instance a sensitizing agent is defined as one that would produce an allergic dermatitis in more than one in 10,000 human users of the substance.

Skin sensitization tests can be carried out directly on human volunteers, but with some risk. If the volunteer becomes sensitized, he may for years thereafter react to the same substance, or a near chemical relative, as a sensitized individual. Most laboratories therefore use patch tests on at least 200 human volunteers to confirm the results of a *negative* test on guinea pigs. If negative results are obtained in these patch tests, it is reasonably safe to assume that the material is nonsensitizing. A discussion of the technique for conducting patch tests on humans has been described by Schwartz, 1960.

INHALATION

The *Regulations* of the Federal Hazardous Substances Labeling Act define, for the purpose of labeling substances that present an inhalation exposure hazard, two classes of substances, highly toxic and toxic (see Appendix 191.1 [e.2.], 191.1 [f.2]).

Toxic substances, falling within the category of air-borne hazard, are subject to the proviso that such concentrations are likely to be encountered by man when the substance is used in any reasonably foreseeable manner.

The following discussion is intended only as a guide to those who may be unfamiliar with the methods of acute inhalation toxicity testing and to provide assistance in determining procedures to be used in complying with the *Regulations*.

Principles and Procedures

Brief inhalation exposures of white rats offer in most instances the most appropriate method of forming an estimate of the toxicity of products which are included within the meaning of the Federal Hazardous Substances Labeling Act. When such exposures are performed in accordance with the conditions defined in the Act for respirable substances, the resulting data may be used as evidence of the toxicity class to which the product belongs and for which suitable labeling may be developed. Exceptions to the use of the white rat as the animal of choice are considered in the sections on acute oral and parenteral toxicity.

Physical State and Preparation of Exposure Atmosphere

The physical state of the substance requiring tests by inhalation procedure may be a gas, vapor, liquid or solid, or mixtures of these forms. The inhalation test procedures should use the product in the physical form that is employed under ordinary conditions of use as far as is technically possible. If the product is for use as an aerosol, this form should be used for the animal exposures, with proper regard for duplicating the particle size of the aerosol particles, whether it be a solid or a liquid aerosol. Methods for estimating the particle-size distribution of both liquid and solid aerosols are described by Fraser *et al.* (1959). If the particle size of a solid product is so coarse that an inhalation hazard is negligible as determined by particle-size analysis, no test need be made by inhalation. Sizes greater than 10 microns in diameter are considered to be beyond the respirable range and thus present a negligible inhalation hazard.

Although little difficulty should be experienced in attaining gas and vapor concentrations prescribed by the Act, the particulate concentrations (up to 200 mg/l by volume of mist or dust) imposed by the Act may prove impossible to attain in tests performed for brief intervals in small exposure chambers. In such cases, larger intermediate-size chambers should be used (Fraser *et al.*, 1959).

Gases may be metered into the exposure chamber from cylinders of the gas under pressure. Vapors may be metered similarly in the chamber from liquids with appreciable vapor pressure by means of a suitable syringe and pump mechanism (Machle *et al.*, 1939) or by metered dropping of the liquid on a heating element (Treon *et al.*, 1943), provided the liquid is stable to the heat applied. Liquids with low or negligible vapor pressure and mixtures of solids and liquids may be continuously dispersed at appropriate particle-size by a nebulizing device.

Solids may be dispersed by suitable dust-feed mechanisms, among which the Wright dust feed has proved useful. Often, however, dusts and powders require special dispersion devices, the characteristics of which must be adapted to the special properties of the dust. Various methods for aerosol distribution may be found in papers by Fraser *et al.* (1959) and Voegtlin and Hodge (1949).

Mixtures

Products or formulations that are mixtures of substances should be tested as mixtures. In the present state of knowledge, it is not possible to estimate the toxicity of a mixture from the toxicity and quantity of the individual components. In the case of mixtures, the analyzed air concentration shall represent the total ingredients and shall correspond to the specified limits of 200 ppm and 20,000 ppm v/v for gases and vapors, and 2 mg and 200 mg/1 for particulates. In determining the concentration of the total mixture where the components are chemically dissimilar, it is best to analyze for the toxicologically active ingredient existing in the largest proportion.

In the case of chemically similar mixtures, as petroleum distillates, halogenated hydrocarbons, ether, ester or glycol mixtures, a single analytical method usually exists which will determine the total concentration of the solvent in the product.

Animals

The information relative to species, strain, sex, weight and diet of animals as discussed in the sections on acute oral toxicity apply also to inhalation studies. The minimal number of animals for test at each dose level should be sufficient to give a statistically significant result. Four groups of ten animals each at doses which kill not less than 10 per cent and not more than 90 per cent of any group should suffice. Exposure tests should be repeated so that replicate animal mortality agrees within ± 10–15%.

Exposure Concentrations

It is difficult in most instances to attain and maintain exposure-chamber concentrations of test substances precisely at the specified limits of the Act. In this event one may use Haber's rule to correct for these conditions. In its simplest form this rule states that the animal response is constant when the product of the concentration, C, of a substance, multiplied by the time, t, during which the exposure is experienced is a constant; that is, $C \times t = K$. This rule holds with reasonable accuracy for a large number of toxic substances, provided variations of either variable (C or t) are not extreme. Certainly within the variations expected in the prescribed exposure conditions, this rule can be expected to hold.

For example, in applying the conditions of the rule to the tests under discussion, if it is found by analysis of the exposure-chamber air that the time-weighted average concentration of the test substance is 180 ppm, the exposure time should be extended to 1.11 hours to fulfill the requirements of the Act. Or if 21,500 ppm represents the exposure concentration, an exposure period of 0.93 hours is required to meet the conditions of the Act.

Sample and Analyses

The determination of the concentration of the substance in the test chamber should be made by quantitative analytical methods. Determination of the chamber-air concentration by loss of weight of the bulk sample per volume of air used throughout the exposure time does not provide sufficient accuracy for purposes of these tests. Because the accuracy of the determined concentration of the test substances in the exposure atmosphere cannot be more than that of the sampling procedure, efficient collection of the sample should be made. Both sampling and analytical procedures applied to the chamber air should provide an over-all error of estimation no greater than $\pm 10\%$. If a midget impinger is used for collection of the sample, proper attention should be given to the choice of solvent, the rate of air-flow, and the efficiency of the collection medium. Air-flow rates of choice for the midget impinger are 1–3 liters/minute when 10 ml solvent is used. If filtration is used for the collection of solids, membrane filters provide efficient collection of particles in the respiratory size range.

A discussion of air-sampling methods, including evaluation of the suitability of each, is given by Fraser *et al.* (1959).

The conventional method of expressing the concentration of air-borne particulates is in terms of milligrams per cubic meter of air. Tables are provided in most books on industrial hygiene and toxicology for converting milligrams per cubic meter to parts per million by volume of air, according to molecular weight of the substance.

Chamber Temperature and Humidity

The chamber temperature at which the tests are performed should be maintained preferably between 76° and 78°F. Excursions below and above these limits are not desirable since most substances exhibit a toxicity-temperature dependence of considerable magnitude for each 10° change in temperature (Fahrenheit).

The relative humidity range of 30 to 50% is, in most instances, satisfactory. Extremely high humidities are undesirable because of their effect on animal respiration, and in some instances, because of their effects on toxicity.

Air-Flow

The rate of flow of air through the exposure chamber should be adjusted in such a way as to fulfill two conditions: flow-rates that insure that the oxygen content of the exposure atmosphere is at least 16 per cent and a concentration of the test substance in the chamber at the inlet, which is substantially the same as that at the outlet. A sufficiently high flow-rate should be maintained to keep a uniform concentration throughout the chamber during exposure, e.g., 50 liter/min for a 100-liter chamber. Adsorption of substances in appreciable amounts on the hair of animals is not uncommon. When flow-rates through the chamber are minimal, hair adsorption, coupled with loss of test substances by inhalation, skin absorption and adsorption on the walls of the chamber, may materially reduce the concentration of the substance particularly at the lower test levels. High air-flow rates also help to maintain proper chamber temperatures by removing excess heat and moisture produced by the animals Certain substances also may be removed by combination with ammonia or other urinary components.

Exposure Chamber and Operation

Some degree of uniformity in exposure-chamber design and operation is desirable in order to assure the proper degree of uniformity in the test data. The description that follows is intended to illustrate the important features of an inhalation exposure unit and its operation. It is unimportant whether details of design and construction are identical to those described as long as the essential features are maintained.

Laboratory white rats are exposed in an all-glass cylindrical exposure chamber of a minimal size of 30-liter capacity ($12'' \times 16''$), as shown in Figure 1. The rats are placed on wire screens at three different levels in order to minimize crowding and maximize exposure to the substance under test. The jar is sealed with a glass or plastic top ground to fit the cylinder, and with ports suitable for the air inlet and sampling tube. Air is supplied from an essentially oil-free compressed air-line, bubbled through a potassium dichromate-sulfuric acid cleaning solution, thence through packed glass wool and through a desiccant, to remove any traces of aerosolized chromate mist. The air-flow is then directed into two channels equipped with calibrated flow meters. One channel delivers air through the source of vapor or gas, or liquid or solid particulate, at a sufficient rate to provide the desired chamber concentration of the substance when the make-up air permits the appropriate number of air changes per minute in the exposure chamber to maintain the conditions set forth under the above section on "Air-Flow." A control of the concentration is obtained by regulating the above relative rates of flow of the agent-laden air and the make-up air. Determination of the air-borne concentration of the test agent is made by drawing a known volume of chamber air through an appropriate sampler followed by analyses after the concentration has attained a steady state as discussed in "Sample and Analyses." At the recommended rates of flow approximately 5 minutes are required to reach a satisfactory percentage of the equilibrium concentration. Determination of exposure-chamber concentrations of the test agent should be made at intervals of 15 minutes throughout the duration of the exposure, or at intervals sufficiently frequent to insure a known and constant exposure concentration. Continuously recording, automatic sampling and analysis devices are becoming increasingly available (Fraser *et al.*, 1959). Corrections should be applied to the air-flows to maintain the concentration as determined by analysis, as close as possible to the stipulated limits of the test. Significant deviations from these values should be compensated for by altering the time factor to provide final Ct values of 200 ppm-hours and 20,000 ppm-hours, or the corresponding Ct values for the particulates. The ambient chamber-air temperature may be maintained within the range of 76° to 78°F by a dual air-conditioning and heating unit if required. The outlet of the exposure unit is vented into a well-ventilated hood. In exceptional circumstances it may be necessary to scrub the outlet air in a suitable absorbent before passing it into the ventilated hood and thence to the outside air.

Following attainment of the appropriate Ct values the animals are removed directly from the exposure unit and returned to their living quarters, provided with food and water, and observed for number of survivors for a period of 14 days. (See exceptions to the 14-day limit under section on "Chronic Response.")

Conditions Indicating Departures from Usual Tests

There are two conditions that might suggest departures from the ordinarily routine procedures described above: 1) instances in which it is not reasonably possible to attain the Ct value of the upper limits (20,000 ppm-hours, 200 mg/1-hours); 2) when borderline data are obtained. (Borderline data may be considered to be obtained when slightly less than LC_{50} occurs at an exposure at, or slightly below, 200 ppm, or slightly above 20,000 ppm or 200 mg/1.) To obtain supplementary data in the first instance, animal exposures may be made at substantially saturated vapor concentrations, provided this represents the exposure in use of the product. Exposure concentrations at saturated conditions should be analytically determined as usual. Tests made at saturated vapor conditions apply only to liquids or mixtures of liquids. However, if exposure to the product under normal conditions of use is by aerosolization, then the liquid should be nebulized. Many high-boiling liquids may be nebulized as a fog or mist to attain the higher level, viscosity permitting. Solids dissolved in liquids may be nebulized as fogs or fumes and thus the higher level can be attained.

It should be noted that the saturated vapor test may be the feasible means of test for products that are composed of a single volatile liquid or a liquid mixture containing only one volatile component and whose end use is not as a pressure-developed aerosol. If rats are not killed or exhibit no signs of injury in a one-hour exposure, this may be taken as an indication that the product has no vapor inhalation hazard.

In the second instance, the borderline case, a standard LC_{50} determination should be made. LC_{50} is defined as that concentration which is lethal to 50 per cent of a group of laboratory animals exposed by inhalation to a substance for a specified time. An LC_{50} value may be determined on the basis of four series of exposures. Beginning with a level in excess of 20,000 ppm, three additional levels are tested at geometrically decreasing concentrations which should differ by 0.75, 0.63, 0.56, or 0.5, depending on the "spread" in the effect of the limiting concentrations or the slope of the mortality curve with change in test concentration.

Calculation of the LC_{50} is made by the method of Weil (1952), providing the observed exposure concentrations are those geometrically calculated. If the actual concentrations are not identical to the geometrically determined concentrations, calculation of the LC_{50} value is made by the probit method of Miller and Tainter (1944).

Chronic Response from Acute Exposure

For proper labeling of a product within the intent of the Act, there should be concern for chronic as well as acute effects. A number of substances are known to cause chronic injury from overexposure to single and often small or infrequently repeated doses. The characteristics of substances that may give rise to chronic effects are delayed excretion and prolonged action. Substances which may be slowly excreted are those which have an affinity for certain tissues, such as bone, bone marrow, fat, liver, kidney and lungs. Among substances having a prolonged toxicologic effect are carcinogens, co-carcinogens, and mutagenic agents. Certain potent sensitizers, either chemical or photochemical, require but a single, and not particularly large, exposure to render an individual highly responsive to a subsequent exposure or even to one of similar, but not necessarily the same chemical structure.

Lung Clearance

It is apparent from much of the foregoing discussion that any material inhaled not only may be acutely injurious to the respiratory tract, but also may produce systemic poisoning. A considerable proportion of some inhaled substances, either gas or vapor, may be returned to the outside air and only a relatively small proportion remains in the lung or is absorbed into the blood and lymph. However, if the gas or vapor has an affinity for constituents of the respiratory tract, lung, or blood, the direction of lung clearance is reversed and little or no return to the outside air occurs.

Lung clearance of substances in other physical states, solid or liquid particulates, or colloidal matter, may be considered for all practical purposes not to involve loss from the lung by exhalation. Once deposited, such materials remain in the lung for varying periods of time. The degree to which they are retained will depend upon the relative affinity for circulating tissue fluids, the clearing mechanism of phagocytosis, direct mechanical passage through the lung, and the metabolic action of local pulmonary tissue enzymes. These substances or their metabolites ultimately find their way into the lymph or blood stream either wholly or in part, and thence to other tissues of the body. What ultimate organ or tissue is affected will depend upon the relative affinity of the substance, or its metabolites, for a particular tissue. For example, insoluble particles such as quartz or uranium oxides that escape the lung parenchyma are trapped in the pulmonary lymph nodes. Some, however, are solubilized and appear in the blood and urine.

A part, and often a large part, of an inhaled solid may be excreted via the gastrointestinal tract. This applies only to relatively insoluble, solid particulates and occurs by clearance of the respiratory tract and upper portions of the lung by ciliary action and by removal of a blood-absorbed substance or its metabolites via the bile. As much as 95 per cent of an inhaled solid particulate may be removed from the respiratory tract within 24 to 48 hours by the action of cilia.

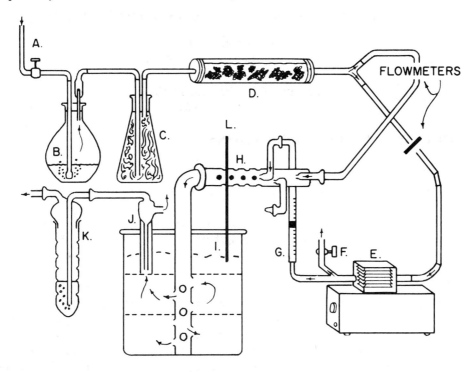

FIG. 1. GLASS INHALATION EXPOSURE UNIT

A. Compressed air line with reducing valve.
B. Potassium dichromate—sulfuric acid cleaning solution.
C. Glass wool filter.
D. Desiccant.
E. Source of vapor exposure.
F. Effluent release valve.
G. Rotameter.
H. Mixing chamber.
I. Exposure chamber.
J. Exhaust line.
K. Sampling tube connected to vacuum source.
L. Thermometer.

Appendix

Regulations under the Federal Hazardous Substances Labeling Act

Part 191, Chapter I, Title 21
Code of the Federal Regulations

Reprinted from Federal Register
of August 12, 1961

DEFINITIONS AND INTERPRETATIONS

§ 191.1 Definitions

(e) *Highly toxic substances.* "Highly toxic" is any substance falling within any of the following categories:

(1) Any substance that produces death within 14 days in half or more than half of a group of white rats each weighing between 200 grams and 300 grams at a single dose of 50 milligrams or less per kilogram of body weight, when orally administered.

(2) Any substance that produces death within 14 days in half or more than half of a group of white rats each weighing between 200 grams and 300 grams when inhaled continuously for a period of 1 hour or less in an atmospheric concentration of 200 parts per million by volume or less of gas or vapor or 2 milligrams per liter by volume or less of mist or dust, provided that such concentration is likely to be encountered by man when the substance is used in any reasonably foreseeable manner.

(3) Any substance that produces death within 14 days in half or more than half of a group of rabbits weighing between 2.3 kilograms and 3.0 kilograms each, tested in a dosage of 200 milligrams, or less, per kilogram of body weight when administered by continuous contact with the bare skin for 24 hours or less by the method described in § 191.10.

The number of animals tested shall be sufficient to give a statistically significant result and be in conformity with good pharmacological practices.

(4) Any substance determined by the Commissioner to be "highly toxic" on the basis of human experience.

(f) *Toxic substances.* "Toxic substances" is any substance falling within any of the following categories:

(1) Any substance that produces death within 14 days in one-half of a group of white rats each weighing between 200 grams and 300 grams, at a single dose of more than 50 milligrams per kilogram but not more than 5 grams per kilogram of body weight, when orally administered. Substances falling in the toxicity range between 500 milligrams and 5 grams per kilogram of body weight will be considered for exemption from some or all of the labeling requirements of the act under § 191.62, upon a showing that, because of the physical form of the substances (solid, a thick plastic, emulsion, etc.), the size of closure of the container, human experience with the article, or any other relevant factors, such labeling is not needed.

(2) Any substance that produces death within 14 days in one-half of a group of white rats each weighing between 200 grams and 300 grams when inhaled continuously for a period of 1 hour or less at an atmospheric concentration of more than 200 parts per million but not more than 20,000 parts per million by volume of gas or vapor or more than 2 milligrams but not more than 200 milligrams per liter by volume

of mist or dust, provided such concentration is likely to be encountered by man when the substance is used in any reasonably foreseeable manner.

(3) Any substance that produces death within 14 days in one-half of a group of rabbits weighing between 2.3 kilograms and 3.0 kilograms each, tested at a dosage of more than 200 milligrams per kilogram of body weight but not more than 2 grams per kilogram of body weight, when administered by continuous contact with the bare skin for 24 hours by the method described in § 191.10.

The number of animals tested shall be sufficient to give statistically significant results and be in conformity with good pharmacological practice.

(4) Any substance that is "toxic" (but not "highly toxic") on the basis of human experience.

(g) *Irritants.* The term "irritant" includes "primary irritant to the skin" as well as substances irritant to the eye or to mucous membranes.

(2) The term "primary irritant" means a substance that is not corrosive and that the available data of human experience indicate is a primary irritant; or which results in an empirical score of five or more when tested by the method described in § 191.11.

(3) *Eye Irritants.* A substance is an irritant to the eye mucosa if the available data on human experience indicate that it is an irritant for the eye mucosa, or when tested by the method described in § 191.12 shows that there is at any of the readings made at 24, 48, and 72 hours discernible opacity or ulceration of the cornea or inflammation of the iris, or that such substance produces in the conjunctivae (excluding the cornea and iris) a diffuse deep-crimson red with individual vessels not easily discernible, or an obvious swelling with partial eversion of the lids.

(h) *Corrosive.* A "corrosive substance" is one that causes visible destruction or irreversible alterations in the tissue at the site of contact. A test for a corrosive substance is whether, by human experience, such tissue destruction occurs at the site of application. A substance would be considered corrosive to the skin, if when tested on the intact skin or the albino rabbit by the technique described in § 191.11 the structure of the tissue at the site of contact is destroyed or changed irreversibly in 24 hours or less. Other appropriate tests should be applied when contact of the substance with other than skin tissue is being considered.

(i) *Strong sensitizer.* A "strong allergic sensitizer" is a substance that produces an allergenic sensitization in a substantial number of persons who come into contact with it. An allergic sensitization develops by means of an "antibody mechanism" in contradistinction to a primary irritant reaction which does not arise because of the participation of an "antibody mechanism." An allergic reaction ordinarily does not develop on first contact because of necessity of prior exposure to the substance in question. The sensitized tissue exhibits a greatly increased capacity to react to subsequent exposures of the offending agent. Thus, subsequent exposures may produce severe reactions with little correlation to the amounts of excitant involved. A "photodynamic sensitizer" is a substance that causes an alteration in the skin or mucous membranes, in general, or to the skin or mucous membrane at the site to which it has been applied, so that when these areas are subsequently exposed to ordinary sunlight or equivalent radiant energy an inflammatory reaction will develop.

TESTING PROCEDURES FOR HAZARDOUS SUBSTANCES

§ 191.10 Method of Testing Toxic Substances

The method of testing the toxic substances named in § 191.1 (e) (3) and (f) (3) is as follows:

(a) *Acute dermal toxicity (single exposure).* In the acute exposures the agent is held in contact with the skin by means of a sleeve for periods varying up to 24 hours. The sleeve, made of rubber dam or other impervious material, is so constructed that the ends are reinforced with additional strips and should fit snugly around the trunk of the animal. The ends of the sleeve are tucked, permitting the central portion to "balloon" and furnish a reservoir for the dose. The reservoir must have sufficient capacity to contain the dose without pressure. In the following table are given the dimensions of sleeves and the approximate body surface exposed to the test substance. The sleeves may vary in size to accommodate smaller or larger subjects. In the testing of unctuous materials that adhere readily to the skin, mesh wire screen may be employed instead of the sleeve. The screen is padded and raised approximately 2 centimeters from the exposed skin. In the case of dry powder preparations, the skin and substance are moistened with physiological saline prior to exposure. The sleeve or screen is then slipped over the gauze which holds the dose applied to the skin. In the case of finely divided powders, the measured dose is evenly distributed on cotton gauze, which is then secured to the area of exposure.

Dimensions of Sleeves for Acute Dermal Toxicity Test
(Test Animal Rabbits)

Measurements in Centimeters		Range of Weight of Animals (grams)	Average Area of Exposure (cm.²)	Average Percentage of Total Body Surface
Diameter at Ends	Over-all Length			
7.0	12.5	2,500–3,500	240	10.7

(b) *Preparation of test animals.* The animals are prepared by clipping the skin of the trunk free of hair. Approximately one-half of the animals are further prepared by making epidermal abrasions every 2 centimeters or 3 centimeters longitudinally over the area of exposure. The abrasions are sufficiently deep to penetrate the stratum corneum (horny layer of the epidermis), but not to disturb the derma—that is, not to obtain bleeding.

(c) *Procedures for testing.* The sleeve is slipped onto the animal, which is then placed in a comfortable but immobilized position in a multiple animal holder. Selected doses of liquids and solutions are introduced under the sleeve. If there is slight leakage from the sleeve, which may occur during the first few hours of exposure, it is collected and reapplied. Dosage levels are adjusted in subsequent exposures (if necessary) to enable a calculation of a dose that would be fatal to 50 percent of the animals. This can be determined from mortality ratios obtained at various doses employed. At the end of 24 hours the sleeves or screens are removed, the volume of unabsorbed material, if any, is measured, and the skin reactions are noted. The subjects are cleaned by thorough wiping, observed for gross symptoms of poisoning, and then observed for 2 weeks.

§ 191.11 Method of Testing Primary Irritant Substances

Primary irritation to the skin is measured by a patch-test technique on the abraded and intact skin of the albino rabbit, clipped free of hair. A minimum of six subjects are used in abraded and intact skin tests. Introduce under a square patch such as surgical gauze measuring 1 inch × 1 inch, two single layers thick,

0.5 milliliter (in case of liquids) or 0.5 gram (in the case of solids and semisolids) of the test substance. Dissolve solids in an appropriate solvent and apply the solution as for liquids. The animals are immobilized with patches secured in place by adhesive tape. The entire trunk of the animal is then wrapped with an impervious material such as rubberized cloth for the 24-hour period of exposure. This material aids in maintaining the test patches in position and retards the evaporation of volatile substances. After 24 hours of exposure, the patches are removed and the resulting reactions are evaluated on the basis of the designated values in the following table:

Evaluation of skin reactions	Value
Erythema and eschar formation:	
No erythema	0
Very slight erythema (barely perceptible)	1
Well-defined erythema	2
Moderate to severe erythema	3
Severe erythema (beet redness) to slight eschar formation (injuries in depth)	4
Edema formation:	
No edema	0
Very slight edema (barely perceptible)	1
Slight edema (edges of area well defined by definite raising)	2
Moderate edema (raised approximately 1 millimeter)	3
Severe edema (raised more than 1 millimeter and extending beyond the area of exposure)	4

The "value" recorded for each reading is the average value of the six or more animals subject to the test.

Readings are again made at the end of a total of 72 hours (48 hours after the first reading). An equal number of exposures are made on areas of skin that have been previously abraded. The abrasions are minor incisions through the stratum corneum, but not sufficiently deep to disturb the derma or to produce bleeding. Evaluate the reactions of the abraded skin at 24 hours and 72 hours, as described in this paragraph. Add the values for erythema and eschar formation at 24 hours and at 72 hours for intact skin to the values on abraded skin at 24 hours and at 72 hours (four values). Similarly, add the values for edema formation at 24 hours and at 72 hours for intact and abraded skin (four values). The total of the eight values is divided by four to give the primary irritation score.
Example:

	Exposure Time	Exposure Unit
Erythema and eschar formation:	Hours	Value
Intact skin	24	2
Intact skin	72	1
Abraded skin	24	3
Abraded skin	72	2
Subtotal		8
Edema formation:		
Intact skin	24	0
Intact skin	72	1
Abraded skin	24	1
Abraded skin	72	2
Subtotal		4
Total		12

Primary irritation score is $12 \div 4 = 3$

§ 191.12 Test for Eye Irritants

Six albino rabbits are used for each substance tested. One tenth of a milliter of the test substance is instilled in one eye of each rabbit; the other eye, remaining untreated, serves as a control. The treated eyes are not washed following instillation. Ocular reactions are read either with the unaided eye or with the aid of a hand slit lamp. Readings are made at 24 hours, 48 hours, and 72 hours after treatment.

FEDERAL REGISTER, vol. 28, No. 110, p. 5582, June 6, 1963

191.1 Definitions

(g) Irritants

(3) Eye irritants. A substance is an irritant to the eye if the available data on human experience indicate that it is an irritant to the eye, or if a positive test result is obtained when the substance is tested by the method described in 191.12.

2. It is proposed to amend 191.12 to read:

191.12 Test for eye irritants.

(a) Six albino test rabbits are used for each substance tested. The cages housing the animals shall be so designed as to exclude sawdust, wood chips, and other extraneous materials that might enter the eye. The eyes of the animals in the test group shall be examined before testing, and only those animals without observable eye defects shall be used. One tenth of a milliliter of the test substance, or in the case of solids or semisolids, 100 milligrams of the test substance, is allowed to fall on the everted lower lid of one eye of each rabbit; the upper and lower lids are then gently held together for 1 second before releasing, to prevent loss of material. The other eye, remaining untreated, serves as a control. The eyes are not washed following instillation. The eyes are examined at 24 hours, 48 hours, and 72 hours after instillation of the test material. An animal shall be considered as giving a positive reaction if there is, at any of the readings, discernible opacity of the cornea (other than a slight dulling of the normal luster), or ulceration of the cornea, or inflammation of the iris (other than a slight deepening of the folds (rugae) or a slight circumcorneal injection), or if such substance produces in the conjunctivae (excluding the cornea and iris) an obvious swelling with partial eversion of the lids, or a diffuse deep-crimson red with individual vessels not easily discernible.

(b) The test shall be considered positive if four or more of the animals in the test group exhibit a positive reaction. If one animal exhibits a positive reaction, the test shall be regarded as negative. If two or three animals exhibit a positive reaction, the test shall be rerun, using a different group of six animals. The second test shall be considered positive if three or more of the animals exhibit a positive reaction. If only one or two animals in the second test exhibit a positive reaction, the test shall be repeated with a different group of six animals. Should a third test be needed, the substance will be regarded as an irritant if two or more animals exhibit a positive response. Ocular reactions may be read with the unaided eye or aided by the use of a binocular loupe, the hand slit lamp, or any other expert means available. The diagnosis of corneal damage may be confirmed by instilling one drop of 2 percent fluorescein sodium ophthalmic solution into the treated eyes of two additional test rabbits. (The original group of six animals shall not be treated with fluorescein solution.) After flushing the excess fluorescein solution, the injured area of the cornea appears yellow, in contrast to the surrounding clear cornea. Fluorescein staining may be better visualized in a dark room under ultraviolet illumination.

Dated: May 31, 1963.

George P. Larrick,
Commissioner of Food and Drugs

[F.R. Doc. 63–5965; Filed, June 5, 1963; 8:47 a.m.]

REFERENCES

Carpenter, C. P. and Smyth, H. F., Jr., Chemical burns of the rabbit cornea, *Amer. J. Ophthal. 29*:136, 3–1372 (1946).

Deichmann, W. B. and Mergard, E. G., Comparative evaluation of methods employed to express the degree of toxicity of a compound, *J. Industr. Hyg. Toxicol., 30*, 373–378 (1948).

Draize J. H., Woodward, G. and Calvery, H. O., Methods for the study of irritation and toxicity of substances applied topically to skin and mucous membranes, *J. Pharmacol. Exptl. Therap., 82*, 377–390 (1944).

Fassett, D. W., Evaluation of toxicological data, *J. Occup. Med., 1*, 169–173 (1959).

Finney, D. J., "Probit analysis", Cambridge Univ. Press (1947).

Fraser, D. A., Bales, R. E., Lippmann, M. and Stokinger, H. E., Exposure chambers for research in animal inhalation, Public Health Monograph No. 57, U.S. Gov't Printing Office (1959).

Friedenwald, J. S., Hughes, W. F., Jr. and Herrmann, H., Acid-base tolerance of the cornea, *Arch. Ophthal., 31*, 279–283 (1944).

Gerarde, H. W., Toxicological studies on hydrocarbons. IX. Aspiration hazard and toxicity of hydrocarbons and hydrocarbon mixtures, *Arch. Environ. Health, 6*, 329–341 (1963).

Hagan, E. C., "Acute toxicity. In appraisal of the safety of chemicals in foods, drugs, and cosmetics," Association of Food and Drug Officials of the U.S., pp. 17–25 (1959).

Hodge, H. C. and Sterner, J. H., Tabulation of toxicity classes, *Amer. Industr. Hyg. Assoc. Quart., 10*, 93–96 (1949).

Litchfield, J. T. and Wilcoxon, F., A simplified method of evaluating dose-effect experiments, *J. Pharmacol. Exptl. Therap., 96*, 99–113 (1949).

Machle, W., Scott, E. W. and Treon, J., The physiological response to isopropyl ether and to a mixture of isopropyl ether and gasoline, *J. Industr. Hyg. Toxicol., 21*, 72–96 (1939).

Miller, L. C. and Tainter, M. L., Estimation of the ED_{50} and its error by means of logarithmic-probit graph paper, *Proc. Soc. Exptl. Biol. Med., 57*, 261–264 (1944).

Quinn, G. P., Brodie, B. B. and Shore, P. A., Drug-induced release of norepinephrin in cat brain, *J. Pharmacol. Exptl. Therap. 122*, 63A (1958).

Schwartz, L., Twenty-two years' experience in the performing of 200,000 prophetic patch tests, *Southern Med. J., 53*, 478–483 (1960).

Treon, J. F., Critchfield, W. E., Jr. and Kitzmiller, K. V., The physiological response of animals to cyclohexane, methycyclohexane, and certain derivatives of these compounds. II. Inhalation, *J. Industr. Hyg. Toxicol., 25*, 323–347 (1943).

Voegtlin, C. and Hodge, H. C., eds., "Pharmacology and toxicology of uranium compounds," McGraw-Hill Co., N.Y., (1949).

Weil, C. S., Tables for convenient calculation of median-effective dose (LD_{50} or ED_{50}) and instructions in their use, *Biometrics, 8*, 249–263 (1952).

Guide to Environmental
Exposure Limits

Richard J. Nocilla

As world-wide interest continues to move in the direction of developing better knowledge of the causes, effects, and methods of preventing environmental and occupational hazards, the complexities of these problems have also increased. The continuous contributions to the body of knowledge extant in the professions associated with occupational health stand as a tribute to man's efforts.[1,2] However, each refinement in the assessment of hazards has opened the door to a need for communicative and interpretative re-orientation, especially in the matter of quantitative evaluations.

The dynamic nature of environmental exposure evaluation certainly justifies the need for the present array of operating references.[3] Unfortunately, the use of terms like Tolerance Limits, Maximum Safe Concentrations, Toxic Limits, Maximum Safe Practice, Hygienic Standards, Recommended Good Practices, Threshold Limits, Hygienic Guides, Allowable Dose, and so forth, frequently fall into misuse and ultimately acquire a significance not intended originally. As a consequence, it becomes equally important for the person evaluating environmental exposures to maintain a familiarity with the accuracy of references[3] such as those promulgated by the American Conference of Governmental Industrial Hygienists, the U.S.A. Standards Institute (formerly the American Standards Association), the American Industrial Hygiene Association, National Academy of Science—National Research Council, the National Bureau of Standards, the International Committee on Radiation Protection, the Federal Radiation Council, the Atomic Energy Commission, and the Health Physics Society.[4,5,6,7,8,9,10]

The National Bureau of Standards handbooks provide guidance for concentrations of radioactive materials[10] and ionizing radiation exposure.[11,12] Judicious use of these guides depends upon a knowledge of the units, their measurement, supporting rationale[13] and governing regulations.[14]

Although there are numerous applicable regulations on ionizing radiation exposure,[15] in general for the U.S.A. the regulatory control for radioisotopes is as stipulated in Title 10 Code of the Federal Regulations Part 20, or Manual Chapter 0524 of the Atomic Energy Commission.

The Threshold Limit Values (TLV) which have been proposed as occupational exposure limits by the American Conference of Governmental Industrial Hygienists, (full text and values are included in Section 12, Tables of Chemical Hazard Information), represent an opinion of how much below the maximum exposure the time-weighted average exposure should be. It has been pointed out that TLV or MAC values (maximum allowable, acceptable or permissible values) are not fine lines dividing safe and unsafe exposures, and that exposure limits are significantly different if one is based on toxicity or hazard to health and another is based on discomfort with no health hazard.[16]

REFERENCES

1. Hatch, Theodore, Major Accomplishments in Occupational Health in the Past Fifty Years, *Amer. Ind. Hyg. Assoc. J. 25:*4, (March-April, 1964).
2. Brown, Harold V., The History of Industrial Hygiene: A Review with Special Reference to Silicosis, *Amer. Ind. Hyg. Assoc. J. 26:*3, (May-June, 1965).
3. Patty, Frank A., "Industrial Hygiene and Toxicology," *Vol. I,* 2nd Ed., pp. 164–171. Interscience Publishers, Inc., (1958).

4. "Documentation of Threshold Limit Values." American Conference of Governmental Industrial Hygienists, 1014 Broadway, Cincinnati, Ohio, (1962).

5. Stokinger, H. E., Standards for Safeguarding the Health of the Industrial Worker, *Pub. Health Reports, 70*:1, (1955).

6. Stokinger, H. E., Threshold Limits and Maximal Acceptable Concentrations. *AMA Arch. Envir. Health 4*:115 (1962).

7. Stokinger, H. E., International Threshold Limit Values, *Am. Ind. Hyg. J. 24*:469, (1963).

8. Am. Std. Assoc. Z37.23, (1962).

9. Stokinger, H. E., Threshold Limit Values and Maximal Acceptable Concentrations: Their Definition and Interpretation 1961 *Am. Ind. Hyg. Assoc. J. 23*:45, (1962).

10. U.S. Dept. of Commerce, "Maximum Permissible Body Burdens and Maximum Permissible Concentrations of Radionuclides in Air and in Water for Occupational Exposure," Handbook 69.

11. U.S. Dept. of Commerce, "Permissible Dose for External Sources of Ionizing Radiation," Handbook 59.

12. Federal Radiation Council, Washington D.C. Federal Register, "Radiation Protection Guidance for Federal Agencies," Reports Nos. 1 through 9, (May 18, 1960).

13. U.S. Dept. of Commerce, Nat'l. Bureau of Standards, Washington, D.C. (ICRU), Radiation Quantities and Units Handbook 84.

14. Title 10, Code of the Federal Regulations, Part 20, Standards for Protection Against Radiation, United States Atomic Energy Commission, (August 9, 1966).

15. Cottrell, William B., American Standards Assoc., "Compilation of National and International Nuclear Standards (excluding U.S. Activities)," Oak Ridge National Laboratory, Union Carbide Corporation, U.S. Atomic Energy Commission, Nuclear Safety Information Center, 2nd Edition, (1966).

16. Rowe, V. K., "The Significance and Application of Threshold Limit Data," National Safety Council Transactions (51st National Safety Congress) *12*:33–36 (1963).

Evaluating Toxic Exposures by Biospecimen Analysis

Ralph G. Smith

In recent years, a large measure of success has been achieved in making the chemist conscious of safety in the laboratory, and it is now commonplace to see safety devices, such as laboratory hoods, being used as they should in almost any laboratory. Nevertheless, there is one aspect of exposure to hazardous substances that must still be regarded as unsatisfactory, and it is doubtful if one laboratory in ten is fully conscious of the nature and magnitude of the problem.

I refer to the effects of repeated exposures, over a long period of time, and usually by inhalation, to low concentrations of substances which are capable of causing serious and often irreversible damage to the exposed individuals. Consider, for example, the infrared spectroscopist who utilizes carbon tetrachloride because it is transparent in his spectral region, and who daily dispenses small quantities of it to fill his cells. He knows that the liquid is non-flammable, and he probably has heard that excessive exposure to the vapors is harmful; in fact if he reads the label on most bottles he will be amply warned of the hazards of prolonged inhalation. It is probable, however, that he relies on the nose to warn him of excessive exposure, and doesn't know that the presently accepted threshold limit value for carbon tetrachloride is 10 ppm by volume, a concentration below the odor threshold for most persons. It is entirely possible for our spectroscopist to breathe enough carbon tetrachloride vapors over a period of weeks or months to cause adverse health effects, even though he hardly recalls a time when the odor was objectionable. A similar condition can result from many other solvent vapors, for mercury vapor, and for countless other substances which a laboratory worker may use repetitively.

The industrial hygienist has long been dealing with exactly this kind of problem in industry, and an extensive technology has developed in regard to evaluating exposures and controlling them. The most widely used approach has traditionally been the measurement of air concentrations of toxic substances, utilizing a number of sampling devices and direct reading instruments which have been developed for this purpose. The usual procedure is to go into the working environment at a certain time and take a number of rather short samples in what is hopefully termed the breathing zone of the workers. To anyone who has spent much time in industry or in chemical laboratories, it is quite apparent that the breathing zone is a rather difficult area to define, for in relatively few cases does the exposed worker stay in one place while breathing an unvarying concentration of some contaminant.

Certain measures can be adopted to improve the air sampling technique, and one of the more sophisticated is the installation of automatic air analysis equipment, which samples at a variety of points in the working environment and continuously records the data. In some instances, a further refinement is the installation of an alarm device which may be activated if concentrations become excessive. Even the simplest of such systems is initially rather expensive and requires a substantial annual budget for operation and maintenance, so that economic considerations alone tend to prevent widespread adoption of such systems. Furthermore, many situations in the laboratory simply do not lend themselves to this kind of fixed point air analysis, and could not be used even if cost were no object.

In recent times, a different sampling approach has become more popular, namely, the use of miniature devices which may be attached to the individual whose exposure is being studied, and which operate with lightweight pumps and battery pack. These devices are very useful and go a long way toward better defining the exposure of an individual in the performance of his normal duties, inasmuch as they follow him as he moves about.

No matter what method of air sampling is used, the purpose is to define the inhalation exposure, with the ultimate intention of comparing the analytical results to some standard values. In the United States, comparisons are normally made with reference to the Threshold Limit Value (TLV) list, put out annually by the Threshold Limit Value Committee of the American Conference of Governmental Industrial Hygienists.[1] This listing of chemicals, minerals, and assorted commercial substances implies that there exists a concentration of any substance included in the list which ordinarily will not produce adverse effects when the exposure period is defined as a normal work week. The values cited in the list are intended to be time-weighted average concentrations for a normal work day, and as such they are clearly an attempt to limit the total amount of a substance which an individual may breathe in any work period.

In actual practice, air sampling by any of the means previously cited does not normally yield time-weighted average data, and a considerable amount of auxiliary work may be required to obtain such information. Again, in many situations it is unfortunately true that it can be nearly impossible to get really good time-weighted averages, and air sampling fails to achieve its intended goal. In such situations, a completely different approach is called for, and in many instances this approach consists of sampling the exposed individual rather than his environment. Commonly this is done by examining samples of blood or urine, which are rather easily obtained, and less commonly by analyzing breath, sweat, tissue, feces, or other materials of biological origin.

A great many substances are absorbed after inhalation and exist essentially unchanged in the bloodstream, and may even be excreted unchanged in the urine. Other substances undergo changes after absorption and metabolites may be present in the blood or urine, while in still other instances a measurable change may result in the level of some naturally occurring biochemical substance. Frequently, these naturally occurring substances are enzymes, but many nonenzymic materials are also involved. It is obvious that under ideal circumstances biological sampling could be superior to air sampling, but unfortunately, the frequency with which ideal circumstances occur is not as great as might be wished. For example, certain materials when inhaled are found in the bloodstream at varying levels which bear some well established relationship to the total exposure of the individual, and it is possible to state that values below some level are normal, or at least not excessively elevated. Whenever a certain concentration is exceeded, by contrast, excessive exposure or incipient health damage may be indicated. In other instances the blood levels of many substances fail to show any correlation with exposure, or else the correlation is so uncertain as to be useless.

The chemical laboratory is obviously a work environment where exposure to almost any substance imagined or imaginable is possible. In the analytical laboratory, there will tend to be recurrent exposure to a number of reagents which are required by the analytical activity being performed. Thus, for example, many analytical laboratories will perform routine extractions with benzene or other solvents as a first step in the isolation of some organic substances, and there may be a constant low level of benzene in the general air of the laboratory, interspersed with higher concentrations for shorter periods of time as flasks are filled, or samples changed.

An organic laboratory will invariably take on a characteristic odor due to the class of chemicals with which a given research program is concerned, and in many cases the combination of high vapor pressure and toxicity may result in exposures which are very hazardous. Almost any laboratory can be expected to use mercury, and experience has demonstrated on numerous occasions that excessive exposure to this element is relatively commonplace. Many more examples of possible exposure of laboratory personnel to toxic substances could be cited, but it should be clear that the laboratory is a workplace where it is almost a certainty that such exposures will occur, and that they will very probably be nonrepetitive, unmeasured if not unmeasurable, and altogether frustrating to a responsible health official concerned with preventing health damage. The use of biological sample analyses in such circumstances is understandably attractive, and the remainder of this discussion is concerned with some applications of this approach. It must be kept in mind that in no way does the use of biological sample analysis preclude the evaluation of the hazards by air sampling, and for most situations air sampling accompanied by biological sample analyses proves to be very informative and helpful.

CHEMICAL EXPOSURES

It is not possible to attempt to list all of the chemical exposures which could be evaluated by biological sample analysis, and the following examples were selected for several reasons. First, it is believed that they are representative of several classes of substances which may be encountered; secondly, the individual substances chosen are themselves commonly encountered in many laboratories; and finally, a considerable body of knowledge exists concerning the interpretation of the results of these particular analyses.

With these thoughts in mind, the evaluation of exposure to mercury, benzene, carbon tetrachloride, organic nitro compounds, and organic phosphates will be briefly discussed. For many substances which are not mentioned, the chemist should be aware of sources of information to which he may turn when necessary, and several are deserving of note. A partial listing of informed agencies or organizations will include: the industrial hygiene section or the medical or toxicology departments of the employing organization itself, when the employer is a large corporation, university or institution; the state or local industrial hygiene agency; and the United States Public Health Service, Division of Occupational Health, in Cincinnati, Ohio. Often times it is satisfactory to consult an authoritative text or journal, and several which are particularly useful are cited in the reference list.[2-6] Many other texts and journals are available, and industrial hygiene organizations can readily identify them upon request.

Mercury

As stated earlier, mercury is to be found in virtually any laboratory, and more often than not, it is spilled and distributed in such a manner that hazardous situations may arise. The entire subject has been recently discussed[7] and remarks made therein need not be repeated. It is sufficient to make the observation that sampling and analysis of urine is uniquely useful when dealing with laboratory personnel exposed to mercury, and although it is not possible to equate urinary levels with illness, there is a clear relationship between elevated urinary levels and exposure by inhalation or skin absorption. Although there is no hard and fast urinary mercury level which can be considered as indicating hazardous conditions, any value above 0.30 mg/l is usually considered excessive, and lower values which are in excess of several hundredths of a mg/l indicate that the individual has been exposed to mercury in some manner or other.

There are a number of precautions to be observed in the collection of specimens and interpretation of urinary mercury values, and it is not recommended that laboratories unfamiliar with these matters attempt to do their own sampling and analyses. Instead, the appropriate industrial hygiene or medical organization should be consulted, and arrangements made for periodic sampling of this kind. In the event of elevated urinary levels, of course, a complete industrial hygiene survey is indicated, and consideration of the possibilities of health damage by a competent physician is wise. It should be noted in passing that at the present time there is no reason to analyze blood samples for their mercury content, inasmuch as most published studies have not shown any correlation between blood mercury levels and exposure. To summarize, the periodic sampling and analysis of urine specimens to determine mercury content is recommended as a means of preventing mercurialism in laboratory workers.

Benzene

The unique toxicity of benzene has been recognized by several decreases in its threshold limit value in recent years until at present the threshold limit value is 25 ppm, with the further recommendation by the committee setting the limit that at no time should the concentration be permitted to be in excess of 25 ppm. The fact that low concentrations of this order of magnitude are either odorless or not at all unpleasant to breathe means that concentrations in excess of the threshold limit value are readily tolerated by most individuals, but the effects of over-exposure to benzene are of such serious consequence that every possible precaution should be taken to minimize exposure to it. If possible, another solvent should be used in its place, but when benzene must be used, it should always be contained within a good laboratory hood or other well ventilated enclosure.

Absorbed benzene metabolizes rather quickly to phenol, which is excreted in the urine in sufficient quantity to substantially increase the normal phenol concentration. For many years, it has been common practice to measure the so-called ethereal sulfates by a simple procedure which distinguishes inorganic sulfates from organic sulfates in the urine. More recently, it has been shown that the determination of phenol as such is preferable to the sulfate determination,[8] but it is possible to interpret results no matter which of the two tests is employed. Normally, the inorganic sulfates will account for 85% or more of the total sulfates, and lesser quantities are indicative of exposure to benzene. By the same token, normal urine will usually contain less than 200 mg of phenol per liter of urine, and the increase in concentration is proportional to the exposure. If benzene is regularly used in any laboratory, these simple urine analyses should be made periodically as a method of detecting excessive exposure.

Carbon Tetrachloride

Carbon tetrachloride is considered every bit as insidious in its action as benzene, as evidenced by a threshold limit value currently set at 10 ppm, and its use should also be minimized and prevented if possible. At levels of 10 ppm carbon tetrachloride cannot be detected by odor, and hence the chemist may be quite unaware of its presence in the air.

Unlike benzene, carbon tetrachloride produces no simple metabolites which may be determined in urine or blood. Although some use may be made of direct measurements of carbon tetrachloride in the blood or breath, the means for performing such analyses are not readily available to most laboratories and interpretation of analyses is questionable. Carbon tetrachloride, therefore, is an example of a substance for which direct biological sample analysis is restricted to the measurement of variations in the levels of some normal constituent of the blood or tissue, with the purpose of detecting organic injury. Specifically in the case of carbon tetrachloride, it is the liver which is most likely to be damaged by repeated low level exposures, and if necessity demands that carbon tetrachloride be used with any frequency then the exposed individuals should be periodically examined by a physician who is aware of the hazard. Normally, the physician will supplement his examination by requesting that liver function tests be performed by the clinical laboratory, and the actual tests chosen will be a matter of medical judgement.

These remarks concerning carbon tetrachloride may be usefully applied to exposure to certain other chlorinated hydrocarbons, but care must be taken to avoid the misconception that all chlorinated hydrocarbons are toxicologically equivalent. For some, such as trichloroethylene, it is possible to measure metabolites in the urine to evaluate the exposure, whereas for many others reliance must be placed upon air sampling and medical examination, while in yet other cases breath analyses by gas phase infrared spectroscopy may be very useful. It is urged that the toxicological literature be consulted in the case of each substance whose use is anticipated so that adequate measures may be taken to minimize the hazards of use.

Organic Nitro Compounds

A large number of nitrogen-containing compounds are noted for their tendency to alter the normal composition of the blood by the formation of methemoglobin, a substance which, unlike hemoglobin, contains iron in the ferric state and is unsuitable for the transport of oxygen. Aniline and nitrobenzene are but two examples of commonly used chemicals with this property, and the hazards attending their use are all the more severe because of the ease with which they can pass through the unbroken skin. When using these chemicals and others like them, great care should be taken to minimize contact with skin and inhalation of the vapors, and it is advisable to establish a program of regular medical examination if such substances are used with frequency. Almost certainly the physician will wish to have periodic hematological tests performed, including methemoglobin determinations as well as other standard blood quality evaluations. Although it is true that not all nitrogen-containing compounds exert the same toxicological action, the literature should be checked before significant exposures to new compounds are permitted.

Organic Phosphate Compounds

Certain organic phosphate esters have become immensely important in recent years as a result of their insecticidal properties or their possible use as "nerve gases." It is highly improbable that laboratories doing research with such compounds will not be fully acquainted with the hazards attending their use, so these remarks will be kept very brief. Many of the organic phosphate esters are very toxic to man, and at very low levels specifically inactivate the enzyme cholinesterase. All persons who work with organic phosphates should be examined periodically. An important aspect of the examination is the determination of red blood cell and plasma levels of cholinesterase, for although these determinations do not specifically constitute a diagnosis of injury, they are most useful in giving evidence of exposure.

Conclusion

The examples cited above should make it clear that exposure to many toxic substances can be implied by the analysis of biological samples, and in some cases early diagnosis of intoxication is possible. Whenever such analyses can be performed, they have the virtue of integrating the individual's exposure to some degree, and of giving evidence of relatively brief high level exposures which could well go unnoticed with most air sampling programs. Furthermore, they reflect the extent to which absorption of the substance occurred, either after inhalation or by any other route of administration, and thus are potentially more informative than air samples. By no means, however, should complete reliance be placed on biological sampling to the exclusion of air sampling or other hygienic measures normally recommended for the handling of dangerous substances. When properly used in conjunction with other procedures, biological sample analysis can add much to a well conceived and executed program of health maintenance.

INTERPRETING BIOSPECIMEN ANALYSES

It is perhaps self-evident that a number of difficulties arise to confuse the interpretation of numerical values which at first glance seem so informative. Many questions can and should be asked about the validity and meaning of numbers obtained by analyzing biospecimens.

On the assumption that an accurate analytical value can be interpreted, what are the chances that any given result is accurate, and furthermore, what do we mean by "accuracy"? A definition of accuracy is easy enough and involves the nearness of any value to the truth which is presumably establishable if enough analyses are performed, or, in the case of "spiked" samples, is known prior to analysis. Biological samples are ordinarily much more difficult to analyze than air samples, and almost everyone recognizes the problem of the chemist seeking to isolate and quantitatively determine a few micrograms or nanograms of something in the presence of relatively huge amounts of organic matter and interfering ions. What is involved in an assessment of accuracy is a knowledge of the extent to which the result may be in error, coupled with an awareness of the significance of that much error in a particular concentration range. Thus, when the limit of detection of any method is approached, an error of 50% or even 100% may represent careful, meaningful analysis, whereas in the optimal range, equal care and skill may routinely yield results within 5% of the truth.

Unfortunately, reasonable accuracy is not often achieved even when allowance is made for analytical difficulties and awareness of the magnitude of a respectable error. A comparison of analytical values obtained on identical samples by different laboratories can be startling. Cross checking activities may result in reports which are in error by much more than the stated analytical uncertainty. In some cases, the results are so badly in error that the validity of the procedure designed to compare results had to be challenged, and when the procedure had been validated, one could only question the proficiency of the laboratory. An unintended modification of a good published method for mercury in urine may result in very low values only made known by comparing results with others. Serious errors can be committed by any laboratory but the probability of their occurrence can be minimized by good practice. It is good advice to place the greatest confidence in values obtained from laboratories that perform a given analysis relatively frequently with a reputation for integrity and proficiency. Keenan et al[9] cite replicate blood lead analyses by their own laboratory and ten collaborating laboratories which showed that in nearly every case the coefficients of variation were less than 10%, with no single value listed sufficiently in error to have been misleading in any way.

Even the best laboratory cannot give correct information on a bad sample, and unfortunately it is not at all difficult to collect unsatisfactory biological specimens. Much has been written about the precautions necessary to avoid contamination of a sample which at best may contain a few micrograms of something, but many who collect samples have not read such material, and submit worthless samples which may be grossly contaminated or otherwise incorrectly taken. There is rarely any need to question the value of a sample collected by, or under the supervision of, industrial hygiene laboratory personnel, but all samples submitted by others are certainly subject to proof that they were collected in a satisfactory manner. Sometimes the addition of an anticoagulant to blood may render it worthless for a given analysis, or a chemically unclean container may ruin a urine specimen. Many tissue samples have been leached of soluble substances by immersion in formalin.

When analytical data from other than industrial hygiene laboratories are examined, it is desirable to know by what means the analyses were performed before placing too much faith in the results. Many methods acceptable for establishing whether or not an acute poisoning episode occurred are not sufficiently sensitive for industrial hygiene work and a negative result from such methods means little. Equally misleading are very high results obtained by a method where the blanks were very high, or not even determined. Instances occur in which results are reported when blank values are several times higher than the increment of increase attributed to the specimen, a very unsatisfactory state of affairs.

As to the matter of attaching meaning to a numerical value on the assumption that it resulted from a carefully performed analysis in which the best available method was used, it is necessary to make clear that this comment is limited to a consideration of factors affecting the interpretation of analyses performed primarily for an industrial hygienic evaluation, or more specifically to assist in establishing whether or not workers are being overexposed to some toxic agent. The same considerations do not necessarily apply to the problems of the physician who must determine whether an individual patient is sick or well, and who requests certain analyses to assist him in making a judgment. It would be naive, however, to assume that biological analytical data will not be used for a variety of purposes, diagnostic, and even medicolegal, but the problems of interpretation differ, and are not treated here.

The most obvious question raised by an analytical result is whether it is normal or above-normal. Given a method of sufficient sensitivity, some of almost anything can be found in urine, blood or tissue, and its presence may even be necessary to support life. It is also true that some unsuspected factor in the environment of a person or persons thought to be normal may give rise to elevated values by comparison with other groups, so some definition of normalcy is required; this is not easy. Any search of the literature will turn up conflicting reports on almost any substance, and no doubt some of the differences are caused by varying degrees of normalcy. To paraphrase a well known quotation, "All men were created normal, but some were created more normal than others." The term "unexposed persons" is probably preferable to "normals," and could then be taken to exclude inhabitants of certain geographical areas where the water supply or soil is abnormally enriched in a component of interest. If the employees of a given plant reside in such an area, then the "normal" values for the region must be known. A range of values is always to be expected, even among the members of an unexposed group, and sometimes the range is surprisingly great and must be attributable to unknown intake, or metabolic variation. It can be very difficult to find reliable data on "normals," particularly if a less commonly performed analysis is involved. Texts such as those by Elkins,[4] Patty,[3] Fairhall,[10] Sunderman,[11] or Williams[5] are informative, as well as most pharmacology or toxicology texts; much information is given in the American Industrial Hygiene Association Hygienic Guide Series, but it is still difficult to establish the normal quantity of a less common metal or other substance to be expected in a biological specimen.

Beyond the normal or unexposed group levels are those values which are elevated but not excessive in relationship to some other level which is also rather ill-defined. Various terms for these higher levels are in use; Elkins[12] uses "maximum urinary concentrations," or MUCs; Vigliani[13] defines "maximum allowable biological concentration (of a toxic substance) which can be found in a biological medium without damage to health"; Zielhuis[14] speaks of maximum allowable limits, or MALs, and still other designations can be found. The American Industrial Hygiene Association Committee on Biochemical Assays has been seeking a satisfactory expression and thus far has preferred a term used by health physicists—"action levels," which presumably designates a level at which some kind of action should be taken. There are objections to any suggested definition because of the complexity of the circumstances that give rise to any value, as well as the uncertain and incomplete knowledge of the metabolic fate of many substances. A much higher degree of meaning can be attached to biological concentrations of a small group of substances in wide use, and which, historically, have been studied by many investigators, but this is not the case with the majority of substances which have received little study by comparison. This single factor, lack of sufficient data on exposed and control groups, is the greatest weakness of all but a few biological analytical values; only time and continuing study can improve this situation.

It is natural to try to assemble lists of maximum biological values comparable to the threshold limit value lists for air concentrations, but all such lists are much shorter than air lists, and the numbers given may signify substantially different things. It is presently quite improbable that a simple listing of values under a common heading for more than a few substances would be very useful, and it could prove troublesome by implying the same degree of internal consistency that characterizes a listing of air values. The American Industrial Hygiene Association Biochemical Assays Committee has thus decided, for example, that although all known biological values should be expressed, critical comment on their probable meaning and value is essential and should be referred to when seeking to interpret an analytical result. It is believed to be very important to have an idea of how values may fluctuate in relation to a host of variables, a partial listing of which includes the following:

1. Does a given value simply reflect exposure, or is it related to organic injury?
2. Does the value reflect the most recent exposure, or is it an index of long accumulated quantities?
3. To what extent does the unique metabolism of the individual affect the result, or how is the level related to general health, specific disease states, drinking habits, and other factors.
4. Is it possible that a higher value may reflect better excretory function, and actually be desirable?
5. If measurement of a metabolite or some naturally occurring substance is made in preference to the toxic agent in the environment, to what extent can variations occur which are not related to the exposure?

Practical difficulties are encountered when dealing with biological specimens which help to confuse efforts to interpret analytical results, even though they are of secondary importance and largely preventable. First there is the matter of expressing the result in some meaningful manner which involves the selection of concentration units. Unfortunately it is true that agreement on units has not been arrived at between countries, and often not even between different professional groups in the United States. Although conversions are usually possible, no good purpose is served by having the same analysis reported in the world literature as ppm, milligram per liter, microgram per 100 gm, milligram percent, or yet other units. In the particular case of tissue analyses, although less frequently performed, trouble arises over selecting a basis for calculating concentration, for there are at least three different possibilities: wet tissue weight, dry tissue weight, and ash weight. The worst situation frequently encountered, is the absence of any designation, presumably inferring a wet-weight basis for calculations. Tipton,[15] recognizing the problem, has published many of her results calculated, or calculable, in all three ways. Blood analyses can be performed on whole blood, serum, or plasma, and intercomparisons are difficult or impossible. Urinalyses can be expressed in a variety of units, and are sometimes further complicated by the practice of expressing 24-hour outputs while omitting actual concentration or volume figures. Even more troublesome is the problem of whether to correct a urinary value for dilution, or express it unchanged; either practice creates difficulties of interpretation. Other problems related to specific kinds of specimens could be recited but enough examples have been given to demonstrate clearly the need for careful consideration of the true meaning of numbers before interpretation is attempted.

Despite many serious and trivial problems which detract from the desirability of biospecimen analyses, their potential and actual usefulness far outweigh these short-comings, and more not less such analyses should be standard practice for the industrial hygienist. It is always possible to develop a satisfactory analytical method and have it mastered by a competent laboratory; results can be expressed in meaningful units on a properly obtained sample. It is not so easy to amass all desired knowledge on mode of toxic action and metabolic fate of a substance but, at the present pace of research in this area, new information becomes available at a very fast rate; yesterday's unsolved problems may be partially or completely answered in today's literature if it can be found.

Research effort should be accelerated to hasten the day when for every toxic agent in the environment a biospecimen analysis can be performed which will be as informa-tive to the industrial hygienist as blood or urinary lead values are at present. Most emphasis should be placed on blood and urine determinations for obvious reasons, but other types of analyses should also be developed to the greatest possible extent. Breath analyses have tremendous potential, as yet largely undeveloped, for estimating the ex-posure of an individual to volatile substances; the work of the Dow Chemical Company Group[16-18] is outstanding in this area. Analyses are performed on breath samples blown into plastic bags from which the material is released into infrared gas cells where meas-urements of great specificity and sensitivity are performed. The future of this approach together with gas chromatographic and yet other instrumental attacks insures that any number of analyses of volatile substances to be exhaled can be developed to help evaluate routine and accidental exposures. Some substances are present in the breath for days and even weeks after termination of exposure so the technique should be especially useful in arriving at estimated exposure levels for brief periods of time, as in accidents.

More attention should be given to tissue analyses also, primarily to tissue obtained at autopsy, but not excluding biopsy specimens. The former can be of considerable re-search value if related to the individual's occupational exposure to persistent materials. No doubt much more could be learned about the distribution of substances inhaled or absorbed over periods of many years than is presently known. Inevitably medicolegal problems can arise when such analyses are performed, but a large body of accumulated data is preferable to a situation where undue importance may be attributed to simply finding some of a substance in autopsy tissue.

SUMMARY

In summary, increased use of biospecimen analyses is advisable but only if coupled with increased efforts to interpret the results intelligently. Problems of interpretation are due to analytical uncertainties, biological variability, and lack of knowledge con-cerning metabolic behavior. Secondary problems are caused by the multiplicity of biological substances which can be analyzed and by lack of general agreement on uni-form concentration units in expressing results. When problems of interpretation are overcome, a wealth of useful information can be obtained from biological analyses, in-cluding the best possible assessment of total exposure to a toxic agent in some cases, and an indication of early damage to tissue in others. Increased research on the full utilization of breath analyses as well as blood and urine is recommended to develop profitable evaluation techniques in the future.

ACKNOWLEDGEMENT: Part of this chapter is reprinted with permission from the *Journal of Chemical Education, Vol. 32,* A42, January, 1966; and part is reprinted from the *Archives of Environmental Health, Vol. 10,* No. 4, pp. 604–618, April, 1965, copyright 1965 by the American Medical Association.

REFERENCES

1. "Threshold Limit Values for 1965." pub. by American Conference of Governmental Industrial Hygienists, 1014 Broadway, Cincinnati, Ohio 45202.

2. Patty, F. A., "Industrial Hygiene and Toxicology," *Vol. I,* second revised edition, Interscience Pub. Co., New York, (1958).

3. Patty, F. A., "Industrial Hygiene and Toxicology," *Vol. II,* second revised edition, Interscience Pub. Co., New York, (1962).

4. Elkins, H. B., "The Chemistry of Industrial Toxicology," second edition, John Wiley and Sons, Inc., New York, (1959).

5. Williams, R. T., "Detoxication Mechanisms," second edition, Chapman and Hall, Ltd., London, (1959).

6. "Hygienic Guide Series," pub. by American Industrial Hygiene Association, 14135 Prevost, Detroit, Michigan 48227.

7. Steere, N. V., Mercury Vapor Hazards and Control Measures, *J. Chem. Ed. 42:* A529, (July 1965).

8. Pagnotto, L. D., Elkins, H. B., Brugsch, H. G., and Walkley, J. E., Industrial Benzene Exposure from Petroleum Naphtha, *Amer. Ind. Hyg. Assoc. J. 22,* 417, (1961).

9. Keenan, R. G., et al, "USPHS" Method for Determining Lead in Air and in Biological Materials, *Amer. Ind. Hyg. Assoc. J. 24:* 581, (1963).

10. Fairhall, L. T., "Industrial Toxicology," second edition, Baltimore, Williams & Wilkins Co., (1957).

11. Sunderman, F. W., and Boerner, F., "Normal Values in Clinical Medicine," Philadelphia, W. B. Saunders Co., (1949).

12. Elkins, H. B., "Maximum Permissible Urinary Concentrations: Their Relationship to Atmospheric Maximum Allowable Concentrations, International Union of Pure and Applied Chemistry, Maximum Allowable Concentrations of Toxic Substances in Industry," 269, London, Butterworth & Co., Ltd., (1961).

13. Vigliani, E. C., "So-Called Maximum Allowable Biological Concentrations, International Union of Pure and Applied Chemistry, Maximum Allowable Concentrations of Toxic Substances in Industry," 285, London, Butterworth & Co., Ltd., (1961).

14. Zielhuis, R. L., "Maximum Allowable Limits in Biological Materials, Prevention of Inorganic Lead Poisoning, International Union of Pure and Applied Chemistry, Maximum Allowable Concentrations of Toxic Substances in Industry," 293, London, Butterworth & Co., Ltd., (1961).

15. Tipton, I. H. and Shafter, J. J., Statistical Analysis of Lung Trace Element Levels, *Arch. Environ. Health 8,* 58, (1964).

16. Stewart, R. D., et al, Human Exposure to 1,1,1,-Trichloroethane Vapor, Relationship of Expired Air and Blood Concentrations to Exposure and Toxicity, *Amer. Ind. Hyg. Assoc. J. 22,* 252, (1961).

17. Stewart, R. D., et al, Observations in Concentrations of Trichloroethylene in Blood and Expired Air Following Exposure of Humans, *Amer. Ind. Hyg. Assoc. J. 23,* 167, (1962).

18. Stewart, R. D., et al, Human Exposure to Carbon Tetrachloride Vapor, *J. Occup. Med. 3,* 12, 586, (1961).

Means of Contact and Entry of Toxic Agents

Herbert E. Stokinger

Of the various means of body exposure to toxic agents, skin contact is first in the number of affections occupationally related. Intake by inhalation ranks second, while oral intake is generally of minor importance except as it becomes a part of the intake by inhalation or when an exceptionally toxic agent is involved. For some materials, as might be inferred, there are multiple routes of entry.

SKIN CONTACT

Upon contact of an industrial agent with the skin, four actions are possible: (1) the skin and its associated film of lipid and sweat may act as an effective barrier which the agent cannot disturb, injure or penetrate; (2) the agent may react with the skin surfaces and cause primary irritation; (3) the agent may penetrate the skin, conjugate with tissue protein and effect skin sensitization; and (4) the agent may penetrate the skin through the folliculo-sebaceous route, enter the blood stream and act as a systemic poison.

The skin, however, is normally an effective barrier for protection of underlying body tissues, and relatively few substances are absorbed through this barrier in dangerous amounts. Yet serious and even fatal poisonings can occur from short exposures of skin to strong concentrations of extremely toxic substances such as parathion and related organic phosphates, tetraethyl lead, aniline and hydrocyanic acid. Moreover, the skin as a means of contact may also be important when an extremely toxic agent penetrates body surfaces from flying objects or through skin lacerations or open wounds.

INHALATION

The respiratory tract is by far the most important means by which injurious substances enter the body. The great majority of occupational poisonings that affect the internal structures of the body result from breathing air-borne substances. These substances lodging in the lungs or other parts of the respiratory tract may affect this system, or pass from the lungs to other organ systems by way of the blood, lymph, or phagocytic cells. The type and severity of the action of toxic substances depend on the nature of the substance, the amounts absorbed, the rate of absorption, individual susceptibility, and many other factors.

The relatively enormous lung-surface area (90 square meters total surface, 70 square meters alveolar surface), together with the capillary network surface (140 square meters) with its continuous blood flow, presents to toxic substances an extraordinary leaching action that makes for an extremely rapid rate of absorption of many substances from the lungs. Despite this action, there are several occupationally important substances that resist solubilization by the blood or phagocytic removal by combining firmly with the components of lung tissue. Such substances include beryllium, thorium, silica, and toluene-2,4-diisocyanate. In instances of resistance to solubilization or removal, irritation, inflammation, fibrosis, malignant change, and allergic sensitization may result.

Reference is made in the following material to various airborne substances and to some of their biologic aspects.

Particulate Matter: Dust, Fume, Mist, and Fog

Dust is composed of solid particulates generated by grinding, crushing, impact, detonation, decretitation, or other forms of energy resulting in attrition of organic or inorganic materials such as rock, metal, coal, wood, and grain. Dusts do not tend to flocculate except under electrostatic forces; if their particle diameter is greater than a few tenths micron, they do not diffuse in air but settle under the influence of gravity. Examples of dusts are silica dust and coal dust.

Fume is composed of solid particles generated by condensation from the gaseous state, as from volatilization from molten metals, and often accompanied by oxidation. A fume tends to aggregate and coalesce into chains or clumps. The diameter of individual particles is less than 1 micron. Examples of fumes are lead vapor on cooling in the atmosphere; and uranium hexafluoride (UF_6) which sublimes as a vapor, hydrolyzes, and oxidizes to produce a fume or uranium oxyfluoride (UO_2F_2).

Mist is composed of suspended liquid droplets generated by condensation from the gaseous to the liquid state as by atomizing, foaming or splashing. Examples of mists are oil mists, chromium trioxide mist, and sprayed paint.

Fog is composed of liquid particles of condensates whose particle size is larger than mists, usually greater than 10 microns. An example of fog is supersaturation of water vapor in air.

Gas and Vapor

A gas is a formless fluid which can be changed to the liquid or solid state by the combined effect of increased pressure and decreased temperature. Examples are carbon monoxide and hydrogen sulfide. An aerosol is a dispersion of a particulate in a gaseous medium while smoke is a gaseous product of combustion, rendered visible by the presence of particulate carbonaceous matter.

A vapor is the gaseous form of a substance which is normally in the liquid or solid state and which can be transformed to these states either by increasing the pressure or decreasing the temperature. Examples can include carbon disulfide, gasoline, naphthalene, and iodine.

Biologic Aspects of Particulate Matter

Size and surface area of particulate matter play an important role in occupational lung disease, especially the pneumoconioses. The particle diameter associated with the most injurious response is believed to be less than 1 micron; larger particles either do not remain suspended in the air sufficiently long to be inhaled or, if inhaled, cannot negotiate the tortuous passages of the upper respiratory tract. Smaller particles, moreover, tend to be more injurious than larger particles for other reasons. Upon inhalation, a larger percentage (possibly as much as 10-fold) of the exposure concentration is deposited in the lungs from small particles. In addition, smaller particles appear to be less readily removed from the lungs. This additional dosage and residence time act to increase the injurious effect of a particle.

The density of the particle also influences the amount of deposition and retention of particulate matter in the lungs upon inhalation. Particles of high density behave as larger particles of smaller density on passage down the respiratory tract by virtue of the fact that their greater mass and consequent inertia tend to impact them on the walls of the upper respiratory tract. Thus, a uranium oxide particle of a density of 11, and 1 micron in diameter will behave in the respiratory tract as a particle of several microns in diameter, and thus its pulmonary deposition will be less than that of a low density particle of the same measured size.

Other factors affecting the toxicity of inhaled particulates are the rate and depth of breathing and the amount of physical activity occurring during breathing. Slow, deep respirations will tend to result in larger amounts of particulates deposited in the lungs. High physical activity will act in the same direction not only because of greater number and depth of respirations, but also because of increased circulation rate, which transports the toxic amounts of certain hormones that act adversely on substances injurious to the lung. Environmental temperature also modifies the toxic response of inhaled materials. High temperatures in general tend to worsen the effect, as do temperatures below normal, but the magnitude of the effect is less for the latter.

Biologic Aspects of Gases and Vapors

The absorption and retention of inhaled gases and vapors by the body are governed by certain factors different from those that apply to particulates. Solubility of the gas in the aqueous environment of the respiratory tract governs the depth to which a gas will penetrate in the respiratory tract. Thus very little if any of inhaled, highly soluble ammonia or sulfur dioxide will reach the pulmonary alveoli, depending on concentration, whereas relatively little of insoluble ozone and carbon disulfide will be absorbed in the upper respiratory tract.

Following inhalation of a gas or vapor, the amount that is absorbed into the blood stream depends not only on the nature of the substance but more particularly on the concentration in the inhaled air, and the rate of elimination from the body. For a given gas, a limiting concentration in the blood is attained that is never exceeded no matter how long it is inhaled, providing the concentration of the inhaled gas in the air remains constant. For example, 100 parts per million of carbon monoxide inhaled from the air will reach an equilibrium concentration in the blood corresponding to about 13 percent of carboxyhemoglobin in 4 to 6 hours. No additional amount of breathing the same carbon monoxide concentration will increase the blood carbon monoxide level, but upon raising the concentration of carbon monoxide in the air a new equilibrium level will eventually be reached.

INGESTION

Poisoning of ingestion in the work place is far less common than by inhalation for the reason that the frequency and degree of contact with toxic agents from material on the hands, food, and cigarettes are far less than by inhalation. Because of this, only the most highly toxic substances are of concern by ingestion.

The ingestion route passively contributes to the intake of toxic substances by inhalation since that portion of the inhaled material that lodges in the upper respiratory tract is swept up the tract by ciliary action and is subsequently swallowed, thereby contributing to the body intake.

The absorption of a toxic substance from the gastrointestinal tract into the blood is commonly far from complete, despite the fact that substances in passing through the stomach are subjected to relatively high acidity and on passing through the intestine are subjected to alkaline media.

On the other hand, favoring low absorption are observations such as the following: (1) Food and liquid mixed with the toxic substance not only provide dilution but also reduce absorption because of the formation of insoluble material resulting from the combinatory action of substances commonly contained in such food and liquid; (2) these is a certain selectivity in absorption through the intestine that tends to prevent absorption of "unnatural" substances or to limit the amount absorbed; and (3) following absorption into the blood stream the toxic material goes directly to the liver, which metabolically alters, degrades, and detoxifies most substances.

ACKNOWLEDGEMENT: Reprinted from *Occupational Diseases,* Washington, D.C., U. S. Public Health Service Publication No. 1097, pp. 7–12, 1964.

Mode of Action of Toxic Substances

Herbert E. Stokinger

Toxic substances exert their effects by physical, or by chemical or physiologic (enzymatic) means, or by a combination of both.

The classification, as presented here, of the toxic mechanisms in the mammalian host has no precedent or any accepted basis other than that it appears to be inclusive, reasonable, and practicable. The classification has been developed to delineate two basic actions: the action of the toxic substance on the host, and the action of the host on the toxic substance. For it is the interplay of these two actions, together with the rate at which the body excretes the toxic substance, that determines what is called the toxicity of a substance.

The full toxic potential of most substances is not usually asserted because of destructive actions by the body and its mechanisms of elimination by urine, sweat, feces, and exhalations, or because of sequestration in inactive forms at certain tissue sites such as bone, skin, hair, and nails. If this were not so, synergistic or enhanced toxicities would never be manifest. Synergistic or enhanced toxicities arise from the development of unusual or enhanced concentrations of the toxic substance. This occurs when one or more of the usual means of elimination or reduction of the concentration of the toxic substance are blocked.

The following classification of toxic mechanisms must necessarily be based on prevailing knowledge, which varies greatly from discipline to discipline. In enzymology, for example, the state of knowledge is at the molecular and, in some instances, at the submolecular level. Such a situation obviously permits more exact definition of the governing mechanisms than is afforded by a discipline in which knowledge is at a cellular or organ level. Thus, a mechanism regarded at present as physical might be later labeled chemical or enzymatic to reflect the acquisition of new knowledge at a more intimate level. Indeed, when all mechanisms can be explained at the submolecular level, an entirely different classification will result. It is hoped that the present classification, believed appropriate within the limits of present knowledge, may not only provide greater insight into how chemicals act in the body, but also point to possible unsuspected relationships among the actions of diverse chemicals.

PHYSICAL MODES OF ACTION

Harmful substances that have a solvent or emulsifying action can produce after prolonged or repeated contact a dry, scaly, and fissured dermatitis. This effect is commonly attributed to the physical removal of surface lipid, but may also be caused by denaturation of the keratin or injury to the water barrier layer of the skin. Acidic or alkaline soluble gases, vapors, and liquids may dissolve in the aqueous protective film of the eye and mucous membranes of the nose and throat, and in sweat, causing irritation at these sites. Moreover, such insults may erode teeth and produce changes in hair structure.

On the inner surfaces of the body—the lungs and gastrointestinal tract—physical contact of unphysiologic amounts of substances may cause irritation. This may lead to inflammation or produce contraction, as in the reflex constriction of the respiratory passages upon inhalation of an irritant gas with resultant coughing, choking, or asphyxiation. In the upper gastrointestinal tract the effect may include vomiting, and further down in the tract the irritation may result in peristalsis and defecation.

Inert gases can exert serious and often fatal effects simply by physical displacement of oxygen, leading to asphyxia. Under pressure, inert gases such as nitrogen can produce compressed air illness by dissolving in unphysiologic amounts in the blood, lymph, and intercellular spaces, or may rupture delicate membranes such as the eardrum. Sudden decrease in pressure will result in decompression sickness. Less inert gases such as carbon dioxide and oxygen under greater than atmospheric pressure can lead to narcosis and other more serious effects, such as nerve and brain damage.

Physical adsorption of gases or vapors on solid or liquid particulates (aerosols) may, upon inhalation, lead to physiologic effects out of proportion to that anticipated from their inhaled concentration prior to adsorption. The action is known as synergism when the effect of gas and particulate exceeds the sum of the effects expected from either alone, or antagonism when the effect is less than expected. A physical theory has been developed to explain these abnormal actions. It is based on molecular properties of gases and accounts for the synergisms by postulating "adsorbed" layers of the gas on the particulate that, upon inhalation, carry to the sensitive lung tissue enormously increased concentrations of the gas that become localized point sources of contact. Synergism results when a rapid rate of desorption of the gas from particulate to the tissue occurs; antagonism, when the desorption rate is very slow or nonexistent.

An example of synergism is the inhalation of a mixture of sulfur dioxide and sodium chloride crystals (sodium chloride inhaled alone is inert), in which the effects on broncho-constriction are greater than that from the same concentration of inhaled gas.

An example of antagonism is the inhalation of welding fumes of nitrogen oxides and iron oxide particles; reduction of effect in this case is explained on the basis of a firmly combined layer of nitrogen oxides on the iron oxide particles.

Radioactive particles cause dislocation and breaking of chromosomal linkages, apparently from local energy release.

CHEMICAL OR PHYSIOLOGIC MODES OF ACTION

Substances that act chemically to produce injurious effects on the organs and tissues of the body do so by two basic means, either by depression or by stimulation of normally functioning pathways of metabolism. These two effects are brought about by a variety of mechanisms that are known in only a general way for most toxic substances, although the detailed mechanisms are known for a few important substances such as carbon monoxide, cyanide, arsenic, and uranium.

A single substance may have more than one pathway of action, or may act by stimulation of an enzyme system at a low concentration and by depression at a higher concentration. This is a characteristic response of many if not all toxic substances, better known examples of which are arsenic, cobalt, vanadium, chloroform, and benzene.

It is convenient to consider chemical mechanisms under the following categories: (1) primary mechanisms of injury which involve interactions of the toxic substance at the enzymatic level; (2) nonenzymatic interactions which involve more or less direct chemical combination or replacement of the toxic substances with a body constituent without enzyme intervention; and (3) secondary mechanisms of injury that may involve both enzymatic and nonenzymatic actions, resulting in injury only indirectly as a consequence of the presence of the toxic substance.

Primary Enzymatic Mechanisms

Most of the metabolic activity of the body is a result of the activity of enzymes, biologic catalysts formed by living cells throughout the body. Consequently, it is reasonable that the bulk of all toxic mechanisms should involve interference in some way with the normal enzyme activity.

Although the liver cells perform a major proportion of the metabolic activity of the body, enzymatic actions occur throughout the body without restriction to any particular organ site. Equally active but less diversified are all other tissues in the body, including lung, kidney, intestine, bone, brain, and nervous tissue. From this it may be inferred that enzymatic mechanisms may occur with the enzyme situated at nerve endings, within the nerve cell itself, or at cell surfaces.

Two groups of enzymes, phosphatases and dehydrogenases, are commonly involved in a large variety of toxic mechanisms because the two groups are included in a large number of important enzyme systems in the body.

It is important to observe that in "metabolizing" a toxic substance the enzyme is merely performing a function that it normally performs in metabolizing natural foodstuffs; no special enzymes exist to metabolize toxic substances.

Although substances are toxic for a variety of reasons, one is the frequent inability of enzymes to completely metabolize and destroy the toxic substance because of the rather high specificity or selectivity of the enzyme for the substance or substrate on which it acts.

Enzymes are proteins, highly complex interlocking chains of amino acids with specific spatial orientation of the chemical constituents. The orientation of the enzyme is such that it fits, much like the key to a lock, the substrate with which it combines prior to modifying it. Certain enzymes heretofore considered homogeneous in composition and in action may consist of several distinct components or isoenzymes, each still acting on the same substrate. Enzymes act with highest efficiency on substrates with chemical structures and configurations of natural foodstuffs, and only incompletely metabolize toxic agents which do not have the same precise structure.

Many enzymes have additional specificity in that they require a metal or a vitamin or both as cofactor(s) or activator(s). For example, the enzyme cocarboxylase, that splits carbon dioxide from certain organic acids, requires vitamin B_1 and magnesium ions as necessary constituents before it can function.

Because enzymes are proteins, they exhibit the physical and chemical properties of proteins. They undergo denaturation (1) by heat, as in burns, (2) by marked changes in acidity or alkalinity, as in contact with corrosive agents, or (3) by chemical denaturing agents, such as urea in high concentrations. These agents alike cause structural and configurational changes in the protein, and loss of the characteristic specificity and the catalytic activity of the enzyme.

Enzymes may become inactivated to varying degrees by less drastic means— among the enzymes requiring a specific metal as activator, any agent that will displace or inactivate the metal will render the enzyme inactive to the same degree. Metals with spatial requirements similar to the specific metal required by an enzyme may inactivate the enzyme—poisonous metals such as beryllium are believed to act in this way, and cyanide may combine with the iron of an iron-dependent enzyme to inactivate or inhibit the enzyme.

Another common way an enzyme may become inhibited is from competition with a substance whose structure is sufficiently similar to the natural substrate but does not quite fulfill the spatial requirements of the enzyme. This is probably the most common way in which toxic substances exert their effect on enzymes. An example is the drug sulfanilamide that competes with naturally occurring para-aminobenzoic acid.

A third way by which enzyme activity is inhibited is by accumulation of the product of the enzyme's activity. This is one of the natural ways by which body enzyme activity is regulated.

Like other catalysts, enzymes theoretically undergo no net change during the reactions they catalyze. Within a minute, one molecule of an enzyme can alter many thousand molecules of the substrate (turn-over rate). In no case does the enzyme contribute to the net energy requirements of the reaction and only those reactions that are energetically possible without an enzyme can occur in its presence. Enzymes merely accelerate a chemical reaction. They catalyze the backward as well as the forward direction of the reaction.

KNOWN ENZYMATIC MECHANISMS

Direct combination. The simplest way by which a toxic substance can alter enzyme action is by direct combination with active chemical groups on the enzyme. This is believed to occur with metals such as mercury and arsenic that combine so tightly with the active group of the enzyme that further action is blocked. If the enzyme or enzymes represent critical systems for which there is no shunt mechanism, then cells may die or function subnormally, resulting ultimately in injury to the cell, the organ, and the host. Similarly, nonmetallic substances such as cyanide can combine with and block the action of heavy metal-bearing enzymes because of the production of an inactive metal-cyanide enzyme. The blocking of this enzyme system to a significant degree results in the well-known fatal cyanide poisoning.

Another mechanism of poisoning by direct combination is illustrated by substances such as ozone and nitrogen dioxide, and possibly iodine and fluorine, that destroy enzymes by oxidation of their functioning groups. In these cases, specific chemical groups such as —SH and —SS— on the enzyme are believed to be converted by oxidation to nonfunctioning groups; or the oxidants may break chemical bonds in the enzyme, leading to denaturation and inactivation.

One of the more commonly encountered enzyme inhibition mechanisms in occupational exposures is that of the inhibition of the action of cholinesterase (acetylcholine esterase), an enzyme that regulates nerve-muscle action by destroying the muscle excitor acetylcholine. Acetylcholine is a powerful pharmacologic substance that can itself act as a poison if not destroyed after its release. The destruction is accomplished by the hydrolysis of the potential poison into its components, an acetyl group and choline. A large number of pesticides, chiefly organic phosphates, act in the body by blocking this same enzyme action, thus allowing excessive amounts of the muscle stimulator to accumulate. The excessive stimulation results in paralysis and prostration.

Competitive inhibition. One of the more usual toxic mechanisms involving enzymes is that of competition of the toxic substance with normal metabolites, or the cofactor(s) essential for enzyme action, for the site of action on the enzyme. This form of competition is highly effective and thus injurious only when the chemical structure of the competing toxic substance resembles that of the constituent normally used by the enzyme; the closer the structural similarity, the more effective the competition.

The successful competition of an unnatural or foreign toxic substance for the enzyme sites of action blocks normal action by preventing significant amounts of normal substances to be metabolized, or by preventing combination of a cofactor necessary for enzyme action. The cofactor can be a metal or a specific organic substance such as a vitamin.

Competitive inhibition has been shown to be the basis of many drug actions, as the action of sulfanilamide by reason of its close similarity to the B vitamin, para-aminobenzoic acid. Competitive inhibition is also the basis of the mechanism of action of a number of anticancer drugs, many of which are appreciably toxic, for example, the fluoropyrimidines.

Toxic mechanisms may operate also by metal-to-metal competition. For example, it is believed that the poisonous action of beryllium results from its capacity to compete effectively for the sites of combination of magnesium and manganese on critical body enzymes, by which action the enzyme is not longer able to function at its normal rate or may be inactivated completely.

An interesting example of a competitive mechanism is that recently found to explain the increased toxicity sustained following simultaneous exposure to two structurally similar economic poisons, malathion and EPN. Although EPN is highly toxic, malathion has a far lower order of toxicity, but when the two substances are present in the body together, malathion has a toxicity equalling that of EPN and the combined toxicity of both is far beyond expectation. EPN effectively competes for the same enzyme that hydrolyzes and would otherwise reduce the toxicity of malathion, and by inhibiting the enzyme action maintains the concentration of the toxic form of malathion at a high level in the body and consequently enhances the toxicity.

A number of other combinations of economic poisons are believed to produce enhanced toxicities by similar mechanisms, for example, the combinations malathion and Dipterex®, and Guthion® and Dipterex®. Other examples of competitive inhibition will undoubtedly be found.

Lethal Synthesis. Another means by which enzymes are involved in toxic mechanisms involves the synthesis of a new toxic product by enzyme action on the toxic substance originally taken into the body. The new product then exerts its toxic effect by interfering with normal metabolic processes.

A striking example of a substance acting in this manner is sodium fluoroacetate, the rat poison 1080®. Following its absorption into the body, an enzyme transfers the fluorine atom in fluoroacetate to citric acid, an important intermediate in the cycle of terminal metabolism. The converted fluorocitrate, unable to function to a significant degree in this important metabolic cycle, breaks the metabolic chain of activity, with the result that tissue respiration ceases, and death ensues.

Toxic Enzymes. A rather unusual type of toxic mechanism results when the toxic substance itself is an enzyme, such as snake and bee venoms and bacterial toxins. Although such substances exhibit a variety of toxic manifestations, for some of which the mechanisms are unknown, the venoms of bees and certain snakes possess enzymes (phosphatidases) that lyse red blood cells, destroying the oxygen-carrying power of the blood, as well as enzymes (proteolytic) that destroy cells and inhibit blood coagulation. In addition, bee venom contains a substance that inhibits dehydrogenases, enzymes important in the metabolism of many body functions.

Inducible Enzymes. All of the mechanisms discussed thus far have been depressant in action, but toxic substances may under certain conditions stimulate metabolic activity by inducing the physiologic synthesis of additional amounts of an enzyme.

Inducible enzymes are difficult to demonstrate in the mammalian host (even though a number have been demonstrated in bacteria and yeasts), but one instance of occupational health interest is presently known in detail and others will undoubtedly be found. High sucrose diets fortified with vitamins fed for 3 weeks to rats stimulate the enzymatic production of additional amounts of protein sulfhydryl groups in the kidney, which enables the rats to withstand otherwise lethal doses of mercury. The newly-formed sulfhydryl binds the mercury firmly and effectively reduces its toxic potential.

A mechanism exemplifying stimulation, probably mediated through inducible enzymes, is the increased production of serum alpha globulins when cobalt is absorbed into the body at relatively low levels of intake. At slightly higher levels of intake, cobalt stimulates the production of increased amounts of red blood cells—polycythemia—and associated with the polycythemia is increased production of hemoglobin. The exact mechanism of this stimulation is not known, but a new hormone (erythropoietin) whose production is stimulated by cobalt is believed involved. It appears that the action of erythropoietin is not entirely restricted to stimulating bone marrow to increased production of red cells, but may include stimulation of other centers as well.

Nonenzymatic Mechanisms

There are a number of occupationally important types of poisoning which proceed through mechanisms that do not involve the intervention of enzyme action but for which the energy is supplied by chemical action, so far as is known.

Direct Chemical Combination. Among the best known and understood mechanisms of poisoning is that of direct chemical combination of the toxic substance and a body constituent, as illustrated by carbon monoxide poisoning. In this instance, the gas combines rapidly and securely with hemoglobin, forming a new compound, carboxyhemoglobin, that cannot perform the usual function of hemoglobin to transport oxygen to the tissues. Hydrogen sulfide likewise unites with hemoglobin to convert it to sulfhemoglobin, a non-oxygen-carrying pigment, although this mechanism is not important in hydrogen sulfide poisoning.

Release of Body Constituents. A less well understood mechanism of injury, but on which there is nevertheless an enormous amount of indirect evidence, is the release by certain substances of natural body constituents in abnormal amounts that may lead to injury or death. Instances of this mechanism are numerous and involve the intake into the body of allergenic materials such as "hay fever" allergens or toluene-2,4-diisocyanate. Intake of these substances results in local release of histamine or histamine-like substances in large amounts, with the characteristic development of inflammation, edema, and other evidences of injury and allergic response.

A large number of amines are capable of histamine release; in these instances the mechanism involved is believed to be one of displacement, whereby the tissue-bound histamine is displaced and liberated by the unnatural amine. Similarly, any type of simple cellular damage results in the liberation of histamine-like substances.

There is accumulating evidence that release of hormones from nerves may be the common mechanism by which a number of chemical substances exert their toxic action. The example that follows (carbon tetrachloride) not only illustrates an action that releases body constituents, but also illustrates a highly indirect toxic action formerly believed to be a direct effect of a substance on an end organ.

Carbon tetrachloride has been shown to cause the massive discharge of epinephrine and related neurohumors from central sympathetic nerves. This discharge, possibly mediated by enzyme action, results in the stimulation of the nerve supply to the blood vessels of the liver to produce (1) restriction of the liver's blood flow, leading to reduced oxygen transport and ultimately the characteristic centri-lobular necrosis of the liver and (2) release of unesterified fatty acids from fat depots and their deposition in the liver to produce the well known "fatty" liver of carbon tetrachloride poisoning.

Chelation. A toxic mechanism that is increasingly being recognized to be one of the more common pathways of toxic action is chelation. Chelation is the term applied to the chemical combination of an organic structure and a metal whereby the metal is very firmly bound to the organic substance by both nonionic (organic) and ionic bonding. For example, the therapeutic agent ethylenedinitrilotetraacetic acid (EDTA) binds metals by chelation and many drugs and antibiotics are now believed to act by chelation. Chelating agents exert their effects in a number of ways:

1. By removal of biologically active metals that are normally bound in the cell or its components with resulting inactivation and cell damage. For example, treatment of lead poisoning with EDTA may in addition remove other metals, such as zinc, which is required for important functions in certain kidney enzymes (carbonic anhydrase).

2. By reacting with fixed intracellular metals.

3. By chelating firmly with a fixed tissue constituent. If the structure prior to chelation happens to be a critical one in a metabolic chain, ordinary function ceases and injury occurs as a result of the altered chelation structure. Chelating with a fixed tissue constituent is believed to be the mechanism by which boron, as borate, exerts its toxic action; borate is known to chelate with adjoining carbon atoms containing hydroxyl groups.

4. By increasing the absorption of a toxic agent. Instances are being recognized of toxicity resulting from abnormally increased amounts of absorption into the blood stream by a chelating compound. Iron, normally nontoxic when absorbed by the usual regulatory mechanism, may under unusual circumstances be absorbed in toxic amounts by the mechanism of chelation to form a soluble, easily absorbed substance.

STIMULATION OF IMMUNE MECHANISM. A mechanism whose toxic significance remains to be fully evaluated, but which nevertheless has been recognized for many years, is the stimulation of immune mechanisms as a result of the production of a new antigenic structure from the combination of a toxic substance with body constituents, usually protein. This mechanism is thought to be the basis of skin sensitivity resulting from contact with certain reactive organic substances, for example, the chloronitrobenzenes.

Another substance that illustrates this mechanism strikingly is toluene-2,4-diisocyanate and related aromatic isocyanates. These substances, upon inhalation, have unusual avidity for combining with body protein with resultant allergic sensitization of the respiratory tract.

Secondary Toxic Mechanisms

In the category of secondary toxic mechanisms are grouped those pathways of metabolism and mechanisms of injury not effected by the direct action of the toxic substance but that develop either (1) as a result of metabolic alteration of the toxic substance following its entrance into the body, or (2) as a consequence of an accumulation of toxic by-products from the initial direct action of the toxic substance. In the second instance, further injury occurs at a site in the body different from that of the original toxic action. Most if not all of the mechanisms considered here are performed by enzymes.

Detoxication (metabolic) Mechanisms. Mechanisms grouped here comprise all those metabolic activities that the body performs on a toxic substance in contradistinction to the actions that the toxic substance performs on the body. The latter actions were considered under Primary Enzymatic Mechanisms, and Nonenzymatic Mechanisms. Broadly, the so-called "detoxication" mechanisms are those performed by the body in the process of attempting to eliminate the toxic substance—oxidation, reduction, and synthesis. A few examples of each of these mechanisms will be given for well-known substances of a toxic nature.

The body does not always act to its own advantage when handling a foreign and generally a toxic substance, and peculiarly disadvantageous reactions result merely because the body is equipped with certain definitive pathways of metabolism derived from past utilization of food components. These pathways are the body's only resources when confronted with nonfood substances, and accordingly these mechanisms are used insofar as they can act on foreign substances bearing chemical structures similar to food substances. Whether this indiscriminate action by the body's enzymes results in an outcome favorable or unfavorable to the body depends only on the nature of the resultant modified foreign substance and not on any selective or guided action of enzymes.

Some examples of oxidation, reduction, and synthesis follow.

Oxidation is one of the most general metabolic activities of the body against foreign substances. It includes the oxidation of alcohols to aldehydes, aldehydes to acids, oxidation of hydrocarbon rings to phenols and quinones, alkyl groups to alcohols and acids, oxidative removal of ammonia from amines, oxidation of organic sulfur compounds, oxidative splitting of carbon ring compounds, removal of halogens from halogenated hydrocarbons, and a variety of other reactions, including the oxidation of certain metallic ions.

Secondary oxidative mechanisms are believed to play a dominant role in the toxicity of methanol. Oxidation of methanol to formaldehyde, which subsequently interferes with oxidative enzyme synthesis, is believed to be the pathway by which methanol exerts its injurious effect on the optic nerve leading to blindness. Ethyl alcohol, and presumably other alcohols, proceed through this metabolic pathway of oxidation to the corresponding aldehyde which is responsible, in part at least, for the toxic effects.

Perhaps one of the more important and interesting examples in which oxidative mechanisms play a decisive role in the ultimate toxic response is the oxidation of the carcinogenic hydrocarbon, 3,4-benzpyrene. Current theories of carcinogenesis consider some oxidized product, not the original hydrocarbon, to be a step in the process leading to tumor development. Several oxidized products of 3,4-benzpyrene have been identified following its entry into the body, including phenolic products and several quinones.

Similarly, the serious effect of the hydrocarbon benzene is believed to be the result of increasing oxidation of the benzene nucleus, first to phenol (monohydroxybenzene), then to dihydroxy- and trihydroxyphenol, which are considered responsible for the toxicity of benzene. Further oxidation to quinone may be involved, followed by further oxidative cleavage of the benzene ring to form the relatively nontoxic mucic acid.

It should be recognized that by no means do all metabolic alterations in the structure of toxic organic substances result in toxic by-products, since a great number of the metabolic products are detoxified in the process.

An important and striking example of the role of oxidative mechanisms in developing the toxicity of an organic substance is parathion. This substance containing sulfur in its molecule is relatively nontoxic until oxygen replaces the sulfur, forming paraoxon which is extremely toxic, inhibiting completely an important enzyme of nerve function, cholinesterase.

An example of oxidation among inorganic toxic substances is that of uranium. The tetravalent form is unstable to the body's oxidation-reduction potential, and is oxidized to the more toxic hexavalent form. The hexavalent form then combines with active sites (phosphate groups) on the surface of cells, blocking normal metabolic processes necessary for cell survival.

Much if not all of the toxicity of the long-recognized poisoning action of aniline arises not from aniline itself, but from its various oxidation products formed in the body. The more important of these are para-aminophenol and, by further oxidation, the quinoneimine, which is believed responsible for the methemoglobinemia that develops when aniline or other aromatic amines are absorbed into the body. The oxidized product of aniline oxidizes the ferrous iron of hemoglobin to the ferric form, resulting in methemoglobin, incapable of releasing oxygen.

Reduction is far less common a body function than oxidation. Nevertheless several types of foreign organic substances are metabolized by this pathway to produce one or more substances that are more injurious than the parent substance. Among certain of the inorganic metal ions, reduction is also the pathway of metabolism. Organic nitro-groups are reduced by stages to amines. Some aldehydes are reduced to alcohols. Unsaturated double bonds of carbon compounds may add hydrogen and thus become reduced, and many other examples could be listed.

In general, reduction tends to result in products that are less toxic than the original substance, e.g., reduction of aldehydes to alcohols, but reduction of nitrobenzene results in a para-aminophenol which is from 50–80 times more acutely toxic than nitrobenzene.

Among inorganic ions, pentavalent arsenic is relatively inactive in the body until reduced to the trivalent state. The physiologically active form of manganese is trivalent. If manganese is taken into the body in the form of pyrolusite, in which the manganese is tetravalent, reduction to the active form must occur, at least to that portion which is absorbed into the blood stream and later incorporated into active tissue components.

Synthesis, whereby the body contributes some tissue constituents in the conversion of the foreign substance to a new product, is one of the most common means the body has of disposing of the toxic agent. There are a dozen known synthetic mechanisms to accomplish this. Without listing them all, the addition of such substances as sulfate, sulfur, glucose, and protein derivatives to the toxic substance in general results in true detoxification and lessening of the injurious effects of the foreign substance.

The well-known synthesis of phenyl sulfate, which was one of the earliest synthetic mechanisms to be discovered (1876), converts highly toxic phenol to a substance that is practically nontoxic. Inorganic and organic forms of cyanide are synthesized to thiocyanate, a structure many times less toxic than cyanide. Certain toxic metal ions may react with sulfur of the body to be excreted as insoluble and nontoxic metal sulfides.

It should be pointed out that the synthetic detoxifying mechanisms are not entirely free of injury to the body. If synthesis is prolonged, the body may be injured by deprivation of vital amounts of some of its constituents.

Secondary Organ Involvement. A secondary mechanism of very general nature, and of considerable toxicologic importance, involves the indirect action of either the toxic agent or its metabolic by-products, or both. Once having injured a primary site, the substance(s) causes either the production or accumulation of deleterious products that in turn affect a secondary site.

A striking example of this secondary mechanism is the action of hexavalent uranium, which first injures the kidney in such a way as to prevent normal elimination of waste products such as urea, ammonia, and other substances. These products accumulate in the blood stream and injure the liver, resulting in fatty degeneration of this organ.

Similar indirect injury occurs to the heart when it is stressed unduly by restriction of blood flow from the lung, when the lung receives direct injury by some toxic substance.

There are numerous other examples, because the function of the body is so organized that there are few alterations of significant magnitude in an organ or tissue site that do not have repercussions in some other organ even at a remote site. The interlocking activities of the endocrine glands, with their respective hormones and their dependence on vitamins for normal function, is the basis for this entire group of secondary mechanisms.

An interesting example of the involvement of the highly sensitive interlocking endocrine systems is the simple inhalation of nonlethal concentrations of ozone, which produces alterations in the activities of the adrenal glands and disturbs the normal uptake of iodine by the thyroid gland, which in turn alters the activity of the thyroid-stimulating hormone of the pituitary body.

ACKNOWLEDGEMENT: Reprinted with permission from *Occupational Diseases, A Guide to Their Recognition*, U.S. Public Health Service Publication No. 1097, p. 13, Washington, D.C. (1964).

Hazards of Isocyanates

John A. Zapp, Jr.

Although organic isocyanates have been known, and many of their reactions recorded in the chemical literature, for at least a hundred years, it is only in the last two decades that their industrial potential has been appreciated and exploited. Perhaps the delay was due in part to the fact that much of the earlier work was concerned with monoisocyanates, whereas it is the polyisocyanates that have assumed importance as the building units of high molecular weight polymers having properties of adhesives, plastics, rigid or elastic foams, and synthetic rubber.

Like most synthetic polymers, the macromolecules based on polyisocyanates have proved to be physiologically inert. The isocyanate monomers which are used as starting materials in the polymerization are not physiologically inert. Although the acute toxicity of the commonly used polyisocyanate monomers is generally low, they are irritating to the skin, eyes, and respiratory tract; they may cause allergic skin sensitization; and they are known to cause in some people an allergic asthmatic reaction.

In view of the known properties of the isocyanates, the questions most frequently asked of the toxicologist are the following. 1. What hazards are involved in the production of the polymer? 2. Do any dermatitis hazards exist with respect to skin contact with the polymer? 3. Are there any special hazards when the polymer is exposed to high temperature?

This chapter is concerned chiefly with the hazards of isocyanates, but brief mention will be made of the dermatitis potential of polyurethane foam polymer and of the toxicity of pyrolysis products. The choice of polyurethane foam plastic as an example is rather arbitrary in the sense that the same hazards would be associated with the production of other forms of polymer based on polyisocyanates. The foam plastics are, however, the most widely used form at the present time. The hazards associated with all can be illustrated with this particular type.

FORMATION OF POLYURETHANE FOAM PLASTICS

Consider first the chemical reactions by which a polyurethane foam plastic might be formed. One starts with a polyol, which may be a trihydroxy resin like castor oil, or with a polyglycol. Equation 1 shows how such a polyol reacts with a diisocyanate (usually aromatic) in the absence of water to form a prepolymer in which the terminal hydroxy groups of the polyol are replaced with isocyanate groups.

$$2\text{-OCN-R-NCO} + \text{HO-R'-OH} \longrightarrow \qquad (1)$$
$$\text{OCN-R-NH·CO·O-R'-OCO·NH-R-NCO} \quad (I)$$

The original glycol now terminates with isocyanate groups (-NCO) and has acquired urethane groups (-R-NH·CO·O-R'-) in the process. Hence, the name polyurethanes is assigned to further polymerization products.

If water and catalyst are added to the prepolymer (I), a further reaction takes place, as shown in Equation 2.

$$2\text{-OCN}\textemdash\text{NCO (I)} + H_2O \longrightarrow \qquad (2)$$
$$\text{OCN}\textemdash\text{NHCONH}\textemdash\text{NCO (II)} + CO_2 \uparrow$$

The larger molecule still retains active isocyanate groups, which can react further with water and with each other or with monomeric diisocyanate to build up still larger molecules and additional carbon dioxide.

The evolution of carbon dioxide produces a foam, and as the molecular weight of the polymer increases, the products become more viscous and trap the evolved gas forming the stable foam. The reaction is also exothermic, and the heat evolved automatically cures the plastic and sets the foam. Terminal isocyanate groups on the polymer react in a variety of ways to form other end-groups.

In the actual formation of polyurethane foam plastics, an excess of diisocyanate monomer and water is usually present. The greater the excess, the more carbon dioxide produced, and the less rigid the foam. Recognizing that the reactions are exothermic, it can be seen that there is a possibility that some of the monomeric diisocyanate will be volatilized and escape from the mix into the surrounding atmosphere.

PHYSIOLOGICAL EFFECTS OF ISOCYANATES

Perhaps because of the relatively recent industrial importance of the isocyanates, there is very little recorded information on their physiological effects. One of the first important papers on the subject was that of Fuchs and Valade,[1] which appeared in 1951. This paper described a clinical study of toxic reactions among a group of workers exposed to toluene-2,4- and toluene-2,6-diisocyanate. After eight days to two months of exposure the workers complained of irritation of the eyes and throat, bronchial irritation, and then asthmatic attacks, which occurred mostly at night. The symptoms cleared up on rest away from the job, but reoccurred if the men returned to work. Fuchs and Valade commented, without giving specific references, that according to German investigators the diisocyanates attack mucous membrane and the epidermis and have been held accountable for asthmatic attacks. They then described experiments on dogs, rabbits, and guinea pigs, which showed that the mixed toluene diisocyanates were not highly toxic for animals by subcutaneous injection or skin absorption, but that they were respiratory irritants if inhaled.

An article by Reinl[2] in 1953 mentioned the strong physiological activity of isocyanates of high vapor pressure. This investigator noted that isocyanates in vapor form have a strongly irritant effect on all mucous membranes, particularly the respiratory organs and eyes. He mentioned the possible occurrence of asthmatic or bronchial difficulties related to what he termed isocyanate allergy.

During 1952 and 1953 we had occasion to carry out in the Haskell Laboratory some studies of the toxicity of toluene-2,4-diisocyanate (TDI). Since this material is widely used in polymerization reactions, and since we believe that the reactions of TDI are representative of those of other aromatic diisocyanates, a discussion of the results of these tests is pertinent to an evaluation of the hazards of isocyanates in polyurethane foam plastic production.

The acute oral toxicity of TDI was investigated by administering graded doses of the undiluted material by stomach tube to rats. The procedure followed was that of Deichmann and Le Blanc, in which a single rat is tested at each dose level and the smallest lethal dose in the series is characterized as the approximate lethal dose (ALD). For TDI, the ALD was 7500 mg/kg of body weight. Subsequent determination on a larger group of rats (10 at each of six doses) resulted in an LD_{50} of 5800 mg/kg, with 95% confidence limits of 7200 and 4600 mg/kg.

Pathological examination revealed a corrosive action on the stomach as well as a possible toxic effect on the liver. Subacute oral tests, in which each of a group of six rats received by stomach tube 1500 mg/kg/day (1/5 ALD), resulted in the death of three rats within a total of 10 treatments (2 ALD's). Pathological examination again revealed injury to the gastrointestinal tract and liver. These results suggested the possibility of cumulative effects.

The acute skin absorption toxicity of TDI was tested on rabbits and was again found to be low, since doses as large as 16,000 mg/kg of body weight failed to kill or produce anatomical injury to the internal organs. Severe local skin irritation, however, did result. Tests on guinea-pig skin confirmed the local irritancy of TDI, even when applied to intact skin as a 10 % solution in dimethyl phthalate. In addition, it was possible to produce allergic skin sensitization of guinea pigs with TDI.

The application of TDI to the rabbit eye resulted in marked irritation of the eyelids and mild damage to the corneal epithelium, unless the eyes were promptly and thoroughly flushed with water.

Acute inhalation studies revealed that a concentration (calculated) of 600 ppm for six hours was lethal to rats, whereas 60 ppm for six hours was not. The animals that died showed acute pulmonary congestion and edema.

Subacute inhalation studies with rats, guinea pigs, and rabbits showed that six exposures to analytical concentrations averaging 9 ppm were lethal to three of six rats, but were not lethal to guinea pigs and rabbits. Nevertheless, all survivors showed bronchitis and minimal bronchial pneumonia, whereas those rats that succumbed showed bronchitis and definite bronchial pneumonia.

Additional subacute inhalation tests with small animals established the following.
1. Ten six-hour exposures to analytical concentrations of 1 to 2 ppm of TDI produced no injury to rats, but after 30 six-hour exposures and killing of the animals microscopic evidence of tracheobronchitis was detected.
2. Two of the rats that had 30 six-hour exposures to analytical concentrations of 1 to 2 ppm of TDI were then exposed to 5 ppm. One died after the 8th and one after the 11th six-hour exposure. Both showed emphysema without pulmonary edema, and one showed definite bronchitis.
3. Five rats survived 79 six-hour exposures to analytical concentrations averaging 1.5 ppm of TDI. When killed, four showed bronchitis of varying degree.
4. Guinea pigs were killed after 23, 40, 57, 61, and 79 six-hour exposures to analytical concentrations of TDI averaging 1.5 ppm. All showed bronchitis with varying degrees of bronchial pneumonia.
5. One rabbit died after three and one after five exposures to analytical concentrations of TDI averaging 1.5 ppm. In addition to bronchitis, these animals showed gastroenteritis, which may have been coincidental. A third rabbit was killed after 19 exposures, a fourth after 52 exposures, and a fifth after 71 exposures. All showed bronchitis, the last showing, in addition, slight pulmonary edema.

Four male dogs were exposed 35 to 37 times over a period of four months to analytical concentrations of TDI averaging 1.5 ppm. The dogs reacted to exposure with lacrimation, coughing, restlessness, and spitting up of white frothy material. Because of these reactions exposures were limited from 30 minutes to 2 hours daily. No consistent objective signs of anatomical injury to the dogs could be detected during exposure. Blood pressure, heart rate, respiration rate, body weight, blood chemistry, hematology, and urine chemistry remained within normal limits. Coughing, lacrimation, and spitting up of frothy material continued throughout the entire course of exposure. When killed after the last exposure, all dogs showed mild congestion and inflammation of the trachea and large bronchi. A conspicuous feature was the presence of thick mucous plugs in some of the bronchial branches.

These animal tests substantiate the conclusions of Fuchs and Valade[1] and of Reinl[2] that the isocyanates are strong irritants of the eyes and respiratory tract. We did not, however, observe in our animals the asthmatic type of reaction which has been reported to occur in human beings and which suggests the possibility of inhalation sensitization in man.

Unpublished reports from the United States, as well as published reports from abroad, have confirmed the fact that asthmatic reactions do occur in men exposed to TDI and to other diisocyanates. Swensson, Holmquist, and Lundgren[3] reported three cases in 1955 of asthmatic reaction in men spray painting with polyisocyanate lacquer. In the same year Friebel and Lüchtrath[4] reported experiments on guinea pigs which were either injected intratracheally with TDI or were exposed repeatedly to an aerosol or vapor of TDI. They reported that the reactions in the guinea pigs resembled closely those observed in man, including asthmatoid respiration. They concluded, however, that the guinea pigs did not have a true allergic asthma, but rather a primary irritation which resulted in asthmatoid breathing. We have not observed this reaction in our series of guinea pigs. From the experimental and clinical observations which have just been described, it seems safe to conclude that toluene-2,4-diisocyanate is a material of low acute toxicity which, nevertheless, possesses a potential for producing considerable industrial morbidity. Exposure to TDI and other diisocyanates must be held to an absolute minimum if this morbidity is to be avoided.

DERMATITIS POTENTIAL OF POLYMERS

The polyurethane foam plastics and other isocyanate polymers could conceivably contain a small number of isocyanate groups if, for some reason, they escaped reaction with water or moisture in the air. It would seem unlikely that any such free isocyanate groups could exist on or near the surface of the polymer. Nevertheless, since TDI can be demonstrated to be a potential skin sensitizer for guinea pigs, we thought it advisable to conduct patch tests with the use of polyurethane foam plastic on human volunteers. In these tests, squares, 1 by 1 by 1/16 in., of foam plastic were applied by means of adhesive tape to the arms of 105 men and to the arms or legs of 104 women and were worn for six days. The skin under the patches was examined 24 hours after application and at the end of 6 days, when the patches were removed. Following a 10-day rest period, the patches were reapplied, removed after 24 hours, and the skin examined.

None of the 209 subjects showed skin irritation after the first 24 hours of contact. Three showed some erythema under the patch after six days. None of the 209 showed a reaction when the challenge patch was applied 10 days later. We conclude from this that the polyurethane foam plastic in these experiments was not a skin sensitizer.

PYROLYSIS PRODUCTS

In recent years we have come to recognize that plastics may yield toxic pyrolysis products when subjected to sufficiently high temperatures. We decided, therefore, to investigate polyurethane foam plastic in this respect and to compare its behavior with that of natural rubber latex, neoprene, and polyvinyl chloride foams. In our experiments the foam samples were heated in a glass tube by means of a multiple unit electric furnace, the temperature of which could be controlled to $\pm 2°C$. Air at a flow rate of 2 liters per minute was passed through the heating tube and into a glass chamber containing two or four white rats.

Table 1 shows the results when four rats were exposed to approximately 5 g of foam held at 200°C for six hours. Death in the animals exposed to polyvinyl chloride foam resulted from pulmonary congestion and edema.

TABLE 1

Pyrolysis of Elastomeric Foams at 200°C—Six Hours

Foam	Sample Wt., g	Weight Loss, per cent	Mortality
Polyurethane A..........	5.07	5.3	0/4
Polyurethane B..........	5.03	4.6	0/4
Polyurethane C..........	5.07	3.6	0/4
Neoprene...............	5.53	8.1	0/4
Rubber latex............	5.03	3.5	0/4
Polyvinyl chloride........	4.70	43.2	2/4

Table 2 shows the results of another series of experiments in which the foam temperature was held at 250°C for six hours. Polyvinyl chloride was not tested at this temperature in view of its behavior at 200°C. In the case of polyurethane A

TABLE 2

Pyrolysis of Elastomeric Foams at 250°C—Six Hours

Foam	Sample Wt., g	Weight Loss, per cent	Mortality
Polyurethane A..........	6.85	40.8	1/4
Polyurethane B..........	6.20	20.5	1/4
Polyurethane C..........	6.72	26.5	0/4
Neoprene...............	5.67	45.4	1/4
Rubber latex............	6.01	16.0	4/4

the one fatality occurred seven days after exposure, and autopsy showed extensive pneumonia with abscessation. This death may have been coincidental. Pulmonary congestion and edema were found in the rat dying after exposure to polyurethane B. The deaths occurring after exposure to neoprene foam and rubber latex foam were also associated with pulmonary congestion and edema.

Finally, 2 g samples of the foams were held at 560°C for 10 minutes, during which time decomposition was essentially complete for all except neoprene. The results are shown in Table 3. The rat dying after exposure to polyvinyl chloride was found

TABLE 3

Pyrolysis of Elastomeric Foams at 560°C—Ten Minutes

Foam	Sample Wt., g	Weight Loss, per cent	Mortality
Polyurethane A..........	2.00	100.0	0/2
Polyurethane B..........	2.00	97.3	0/2
Polyurethane C..........	2.00	100.0	0/2
Neoprene...............	2.00	75.4	0/2
Rubber latex............	2.00	93.5	0/2
Polyvinyl chloride........	2.00	95.7	1/2

to have pulmonary congestion and edema. All survivors were killed 9, 10, or 11 days after exposure, at which time no gross or microscopic pathological changes were detectable.

The pyrolysis experiments, described above, illustrate two things. First, all plastic foams tested, including natural rubber latex, yield toxic pyrolysis products if the temperature is sufficiently high. Second, the polyurethane foam plastics are not more hazardous with respect to pyrolysis products and critical pyrolysis temperature than other foam plastics in common use.

CONCLUSION

In conclusion, we believe that morbidity will result from exposure to toluene-2,4-diisocyanate or other similar diisocyanates. These materials are fairly strong irritants to skin, eyes, gastrointestinal tract, and respiratory tract. Asthmatic attacks will result in a significant proportion of people exposed to inhalation of the vapors, and skin sensitization may occur in a few exposed to vapor or liquid. This means that care must be taken in the handling of the diisocyanate monomers and during the production of the polyurethane foam plastics. We have suggested that the maximum concentration of TDI in the atmosphere should not exceed 0.1 ppm, basing this estimate on the positive response of animals to concentrations of 1 to 2 ppm. We know that 50% of a group of 24 men reported that the least detectable odor of TDI was apparent at 0.4 ppm. At 0.5 ppm irritation of the nose and throat occurred. These concentrations exceed the suggested maximum and point to the necessity for analytical monitoring of work areas.

The 1969 Threshold Limit Value (TLV) recommended by the American Conference of Governmental Industrial Hygienists (ACGIH) is 0.02 ppm for TDI and methylene-bis-(4-phenylisocyanate) (MDI).

Our results from patch testing 209 human volunteers suggest further that no dermatitis hazard should result from ordinary handling and skin contact with polyurethane foam plastic.

Finally, our pyrolysis experiments indicate that polyurethane foam plastics, like other foam plastics, can liberate toxic pyrolysis products, but that the polyurethane foam plastics are not more hazardous in this respect than other foam plastics in common use.

ADDENDUM ON ANALYTICAL METHODS

The analytical concentrations of TDI in the atmosphere referred to in this paper were determined by the method of K. E. Ranta of the Haskell Laboratory. In this method air is passed through a glass bubbler containing 15 ml of a reagent solution consisting of 200 ml of 1% aqueous sodium nitrite and 800 ml of ethylene glycol monoethyl ether (Cellosolve). In the presence of TDI a yellow-orange color develops on standing at room temperature, reaches a maximum by one hour, and is stable for at least another hour. The color densities are read in a photoelectric colorimeter, with use of a 420 mμ filter against a reagent blank, and are compared against standards containing known amounts of TDI. At a sampling rate of 0.5 liter of air per minute and with use of a single absorption tube, recovery of TDI from a dry atmosphere is 90% to 95%. In the presence of moisture, TDI hydrolyzes partially to give the TDI urea, 3,3'-diisocyanato-4,4'-dimethylcarbanilide. This gives essentially the same color with the nitrite-cellosolve reagent as TDI. Hence, in moist air our determination includes both TDI itself and TDI urea.

A rapid, sensitive colorimetric method for TDI was described by K. Marcali before the American Chemical Society at the Atlantic City meeting Sept. 20, 1956, and appeared in *Analytical Chemistry*.[5] Marcali's method is based on diazotization of the amine resulting from the hydrolysis of TDI. The TDI urea that might be formed in moist air does not produce the typical color and is not included in the diazometric determination.

This procedure was revised and adapted to a field kit, and extended to include methylene-bis-(4-phenylisocyanate) (MDI) in the region of the 0.02 ppm TLV adopted in 1961.

Attention is called to the fact that Marcali's method would be expected to give lower values for TDI in moist air than would our nitrite-cellosolve method. The toxicological importance of the TDI urea in inhalation exposures remains to be evaluated.

ACKNOWLEDGEMENT: Presented before the 44th National Safety Congress, Chicago, Oct. 24, 1956, and reprinted with permission from the *A. M. A. Archives of Industrial Health*, Vol. 15, pp. 324–330, April 1957, copyright 1957, by the American Medical Association.

REFERENCES

1. Fuchs, S., and Valade, P., Experimental and Clinical Study on Several Cases of Poisoning by Desmodur T (1, 2, 4- and 1, 2, 6-Toluylene Diisocyanates), *Arch. mal. profess.*, *12*, 191–196 (1951).
2. Reinl, W., Illnesses in the Manufacture of Polyurethane Plastics, *Zentralbl. Arbeitsmed. u. Arbeitsschutz*, *3*, 103–107 (July 1953).
3. Swensson, Å., Holmquist, C–E., and Lundgren, K–D., Injury to the Respiratory Tract by Isocyanates Used in Making Lacquers, *Brit. J. Indust. Med.*, *12*, 50–53 (Jan. 1955).
4. Friebel, H., and Lüchtrath, H., On the Effect of Toluylene Diisocyanate (Desmodur T) on the Respiratory Passages, *Arch. exper. Path. u. Pharmakol.*, *227*, 93–110 (Nov. 1955).
5. Marcali, K., Microdetermination of Toluene Diisocyanate in Atmosphere, *Anal. Chem.*, *29*, 552–558 (April 1957).
6. Grim, K. E., and Linch, A. L., Recent Isocyanate-in-Air Analysis Studies, *Amer. Ind. Health Ass. J.*, *25*, 285 (May–June 1964).

Mercury Vapor Hazards and Control Measures

Norman V. Steere

While there is a general recognition and common use of the high density and surface tension of mercury, there does not seem to be a corresponding appreciation of the vapor pressure and the toxicity, or of the difficulty of retrieving mercury spilled in a laboratory.

Mercury poisoning affected Pascal and Faraday,[1] and does occur currently with a variety of symptoms which depend on the intensity and duration of exposure. In perspective, the measures to prevent mercury poisoning do not need to be either heroic or hasty, since six months to two years of regular daily exposure would be required before symptoms would develop from ordinary spills of mercury, and symptoms will generally subside as soon as the exposure is discontinued. However, extreme concentrations from heated mercury could cause symptoms of acute mercury poisoning within several hours.

Mercury can corrode many metals and contaminate electrical instruments and other laboratory apparatus, as well as endanger health, so there are several arguments for adequate control measures. Consideration should be given to providing control measures because of the consequences of lost time or research effort if a laboratory has to be tediously decontaminated before it can again be routinely used.

The peculiar properties of mercury necessitate containment of mercury handling, special ventilation, special floor covering, routine monitoring, and special cleaning procedures, if mercury vapor is to be brought under reasonable control.

PROPERTIES OF MERCURY

Mercury has a great density and high surface tension, and such a low viscosity that pouring without splashing and spilling is almost impossible.[2] The droplets which are formed will roll and bounce when they hit the bench or floor, and may be broken into smaller droplets which cannot be seen by the unaided eye. Figure 1 shows droplets smaller than 0.02 mm in diameter which were found on a piece of flooring removed when it could not successfully be decontaminated, and Figure 2 shows a droplet smaller than 0.05 mm in diameter which was not removed when a piece of the same flooring was given special cleaning. Contrary to the concept that a sloped floor or bench will allow all mercury droplets to roll to a convenient sump for pickup, the author has frequently observed visible small droplets adhering to smooth vertical surfaces, on cupboard doors and on stainless steel surfaces of apparatus.

Mercury is insoluble in water, alkalies, common solvents, dilute hydrochloric acid and dilute sulfuric acid, so that cleaning up spilled mercury is not a simple problem. Mercury does dissolve in dilute nitric acid and hot, concentrated sulfuric acid and can be amalgamated with zinc dust, but these measures are too hazardous for routine use.

Biram reports that mercury exposed to the air slowly forms a skin, presumably of oxides, which may reduce the vapor pressure of the mercury below the tolerance level, but vibration will break the skin and allow atmospheric concentrations to increase.[2] The vapor pressure of mercury increases rapidly with increases in temperature, so that radiators, heating ducts, motors, ovens and other heating apparatus may greatly increase concentrations of mercury vapor if droplets get on or near such equipment.

The weight and viscosity of mercury may cause many standard laboratory containers and connections to break or leak, so tubing and connections should be designed to withstand the stress and not just be wired together.

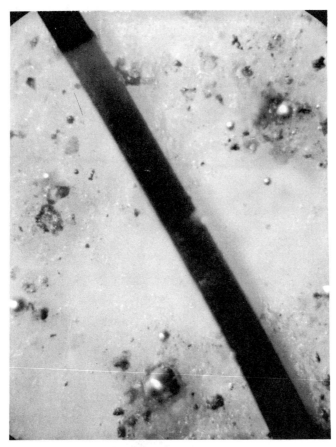

FIG. 1. Mercury Droplets on a Floor.
   An enlarged photograph of a section of floor covering removed from a laboratory, showing numerous small droplets of mercury. To show scale, a black hair $100\,\mu$ in width has been placed on the sample of floor covering.

Vapor Pressure of Mercury and Threshold Limits

Some of the vapor pressures for mercury shown by the "Handbook of Chemistry and Physics," 47th edition are as follows:

Temperature (°C)	Vapor Pressure (mm Hg)	Temperature (°C)	Vapor Pressure (mm Hg)
0	0.000185	28	0.002359
10	0.000490	30	0.002777
20	0.001201	40	0.006079
22	0.001426	50	0.01267
24	0.001691	100	0.273
26	0.002000		

At a room temperature of 25°C the equilibrium concentration of mercury vapor would be about 20 mg/m³ or 200 times the Threshold Limit Value of 0.1 mg/m³ recommended as the maximum atmospheric concentration for the normal work schedule by the American Conference of Governmental Hygienists. Compared to this "tolerance" level of 100 μg for the United States, Biram[2] in 1957 reported the British tolerance level to be 75 μg and the German tolerance level to be 1 μg/m^3, although the German level was based on work done by Alfred Stock who was known to be hypersensitive to mercury vapor.

FIG. 2. Mercury on Scratched Floor.
    A 100μ wide hair shows the scale of a mercury droplet that adhered to a piece of floor covering that had been removed and given special cleaning. The two lighter lines in the photograph are scratches in the surface of the floor covering.

SYMPTOMS AND SIGNS OF MERCURY POISONING

In Battigelli's 1960 review of the literature on mercury toxicity from industrial exposure, the characteristic signs listed for the diagnosis of mercurialism are gingivitis, emotional instability and tremor of the extremities, with muscular tremor as the eloquent sign.[3] A graphic example of the effects of muscular tremor is provided by the sketches of a direct current meter made by a repairman while suffering chronic mercury poisoning and 12 months later after complete recovery.[4] Figure 3 shows the two sketches.

The symptoms of metallic taste and sore mouth in women with chronic mercury poisoning are reported in a study[5] which includes analyses of blood and urine, atmospheric concentrations before and after control measures, and a summary of symptoms appearing most frequently in 29 persons with six months or more of mercury exposure.

The eyes of 51 direct current meter repairmen were examined by Locket and Nazroo[6] to find out if there were any eye changes following exposure to mercury vapor. The article reports the findings and shows four color photographs to demonstrate the brown reflex of the anterior lens capsule. The authors conclude that the brown reflex is related to length of time a worker has been exposed to mercury in any concentration, and that the reflex does not cause any visual symptoms or lead to more serious ocular disturbance. Severe pulmonary-tract irritation[7] or nephrotic syndrome[8] may be the predominant effects of acute mercury poisoning, if atmospheric concentrations of mercury vapor are high because the mercury has been heated or because of poor ventilation. Summaries of reports of mercury poisonings may be found in tables by Battigelli[3] and Bidstrup.[9]

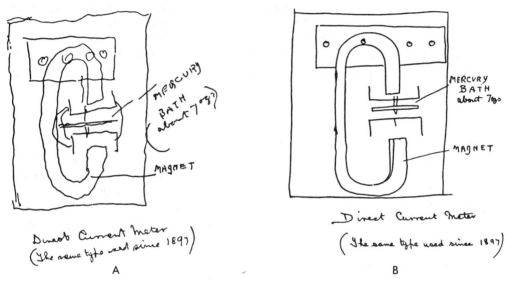

A   B

FIG. 3. Evidence of Tremor.
   Sketch "A" of a direct current meter by a repairman suffering chronic mercury poisoning from inhalation of mercury vapor shows the intensity of tremor. Sketch "B" made by the same man 12 months later, after removal from the exposure, shows the great contrast in muscular control.
   Sketches appear in "Toxicity of Mercury and Its Compounds," by P. Lesley Bidstrup, M.D., courtesy of Dr. Donald Hunter, and are reproduced here with the permission of the publisher, Elsevier Publishing Company, Amsterdam, Holland.

ABSORPTION, DISTRIBUTION AND EXCRETION OF MERCURY VAPOR

   The human organism is able to absorb and excrete substantial amounts of mercury, in some cases as high as 2 mg/day, without exhibiting any abnormal symptoms or physical signs.[10] Since humans not exposed to mercury vapor may have urinary excretion of up to 7.6 μg of mercury per day and since the mercury excretion of exposed workers may be high for the symptom-free and lower for those evidencing mercurialism,[10] urinary mercury levels have little or no diagnostic value. However, the measurement of mercury in 24-hour urine samples is a sensitive index of absorption of mercury.[9] Periodic tests of a group or sampling of workers who may be exposed to mercury vapor could be useful as a means of assessing whether an exposure exists and whether ventilation controls or improved cleaning methods are effective.

Absorption of Mercury Vapor

   The metabolism of inhaled mercury vapor in the rat was reported in 1962 by Hayes and Rothstein[11] in studies using a special exposure unit, Hg[203] as a tracer, and whole body counting techniques which allowed frequent estimation of body burden. They found that 86% of the mercury was cleared from inhaled air, deposited in the lung, oxidized to ionic form, and absorbed by the blood within a matter of hours. They report a study by Clarkson in 1961 which showed mercury in the blood to be bound 50% to the red blood cells and 50% to serum protein, with less than 0.1% filterable.

Distribution of Mercury in the Body

In their study of the metabolism of inhaled mercury vapor in the rat, Hayes and Rothstein found after exposures of 5 hours to a mercury level of 1.4 mg/m^3 that the metal was generally distributed in the body, but became highly localized in the kidney with an accumulation after 15 days of 70% or more of the body burden, or 150 times as high a concentration of the metal on a per gram basis as the other tissues. Assuming filterable mercury in blood at 0.1% they calculate the maximum amount of mercury that could be filtered by the kidneys in 24 hours after exposure at 0.86% of the body burden, and in comparison with the actual deposition of 16.7% of the body burden conclude that the extraordinary affinity of the kidney for mercury is not related to filtration phenomena.

Excretion of Mercury

In the rat inhalation studies, about 30% of the body burden of mercury cleared from the body rapidly with a half-time of 2 days, associated with a rapid fecal excretion, and the rest more slowly with a half-time of 20 days, with about equal rates of fecal and urinary excretion. The study showed that excreta during 15 days accounted for 42% of the initial body burden, and suggested that an additional reduction of 14% of initial body burden may in large part be accounted for by loss of mercury in the form of vapor from the animals.

Pharmacology of Mercury

The pharmacology of mercury and other heavy metals was reviewed in 1961 by Passow, Rothstein and Clarkson[12] in an article which mentions that mercury in the body has high affinity for sulfhydryl groups, chloride ions, and amines or simple amino acids, and that mercury inhibits urease, invertase and other enzymes carrying SH groups. Mercury can also block glucose uptake by erythrocytes and muscle, produce K$^+$ loss from the cells of all species, cause lesions of the central nervous system, and influence bioelectric phenomena by altering transmembrane potentials and by blocking nerve conduction.

MEASURES TO CONTROL MERCURY VAPOR HAZARDS

PERSONAL PROTECTIVE MEASURES. In laboratories where mercury is handled or manipulated regularly, as in gas measurements, clothing and shoes used in the laboratory should *not* be worn away from the laboratory, and there should be no smoking, drinking or eating in the laboratory. These precautions are necessary to limit the inhalation, ingestion and absorption of mercury during work and to prevent carrying mercury contamination to other areas. The precautions depend upon provision of the necessary laundry service, locker facilities, and facilities for washing, smoking and eating. A special soap which changes color from orange to deep purple in the presence of mercury on the skin is described by Bidstrup.[9]

CONTAINMENT OF OPERATIONS. Although most mercury handling operations in the laboratory seem to be out in the open, every consideration possible should be given to enclosing the operations in the smallest convenient space to control spills and simplify cleaning operations. Glove boxes, enclosures, and small ventilated rooms should be used, whenever they are feasible, to help control mercury vapor. Mercury handling operations should be located where foot traffic is a minimum.

MONITORING OF VAPOR CONCENTRATIONS. Evaluation of mercury vapor levels in the laboratory during operations and after cleaning procedures will be best in terms of speed and accuracy and long-term economy if direct-reading instruments are used which give nearly instantaneous response. There are portable battery-powered instruments which draw the air from the sampling point through a flexible tube to the instrument, where mercury vapor interference with 2,537 A ultraviolet light is translated into mg of mercury per m^3.

Other monitoring instruments are described by Biram[2] and in "Air Sampling Instruments for Evaluation of Atmospheric Contaminants," 2nd edition, 1962, from the Secretary of the American Conference of Governmental Industrial. Hygienists, 1014 Broadway, Cincinnati, Ohio.

Although a low-power microscope will not help evaluate mercury vapor, such a microscope is very effective in convincing laboratory personnel that the floor *does* have droplets of mercury on it that they didn't see before, that their shoes have enough mercury on the bottom so they shouldn't be worn home, and that a cigarette may have mercury droplets on the filter-tip. If a dissecting microscope is available, it is convenient to use.

FLOOR COVERING. We agree with Biram that much more work could be done on the flooring problem in laboratories. The general practice in the United States seems to be to cover laboratory floors, if at all, with 9 inch squares of asphalt or vinyl tile. While such a practice certainly makes it convenient to replace the floor covering economically in small worn or damaged spots, it does not make it convenient to remove mercury or other chemicals with which the floor may become contaminated, and the cracks between the tiles harbor many droplets of mercury.[13]

Based on recent experience with floor coverings in three laboratories, successful removal of microscopic droplets of mercury from laboratory floors depends first on a crack-free and *smooth* surface. Figure 2 shows a noticeable scratch made by very light foot traffic on inexpensive sheet flooring which was removed because it could not be decontaminated. Asphalt tile in another laboratory was finally removed after intensive cleaning efforts failed, and microscopic examination of the asphalt tile floor in the third laboratory shows roughness from wear that accounts for the difficulty in reducing mercury vapor concentrations at the surface.

Sheet linoleum with sealed joints and coved edges has been reported in a laboratory in Germany[2] and sheet polyvinyl chloride (PVC) is recommended for some laboratories in England. Another possible answer to the need for a smooth floor covering with good resistance to wear may be a suitable epoxy-based paint.

CLEANING AND DECONTAMINATION PROCEDURES. One of the most commonly used methods for picking up spilled mercury is the one described in 1938[14]: a glass tube of about 6 mm diameter drawn out to an opening of about 1 mm and connected by rubber tubing to a filter flask connected with a vacuum pump or aspirator, the flask acting as a trap. This method, sometimes substituting a hypodermic needle for the glass tubing, is a tedious one which is hardly adequate to cope with the construction and crowding which too often results when mercury work is done in laboratories not designed and built for such work and the resultant cleaning problems. We have seen laboratories where mercury droplets had bounced into places where they could not be seen and into places where they could not be reached by the pickup tube and needle. Especially when mercury droplets are so small they cannot be seen, this method fails to do the cleaning job necessary.

Because ordinary motor-driven vacuum cleaners are not designed to resist the corrosion of mercury they should not be used for such service, but the most important reason ordinary vacuum cleaners *must not* be used for picking up mercury is that the mercury droplets will be dispersed more finely throughout the laboratory. The only company known to manufacture a vacuum designed to pick up mercury while resisting corrosion and filtering the exhaust to avoid spreading contamination is the Acme Protection Equipment Company in South Haven, Michigan. This machine has a timer to show when the special Hopcalite filter should be changed.

Our experience with attempts to control mercury vapor with calcium polysulfide and with flowers of sulfur has indicated that such methods are not effective—when the floor was rubbed after the chemicals had been removed, the concentrations of mercury vapor approached the concentrations before the efforts. Since either rubbing or vibration, as occurs with foot traffic, will break the coating on mercury droplets and allow vaporization to continue, the critical process seems to be to remove as many of the droplets as possible.

VENTILATION FOR VAPOR CONTROL

None of the air exhausted from laboratories in which mercury is used should be recirculated to either the laboratory or to other parts of the building, in order to prevent the build-up of atmospheric concentrations of mercury vapor. Laboratories in which mercury is poured, handled or manipulated, or likely to be spilled or leaked, should have special exhaust ventilation. Three laboratories in the University of Minnesota Medical Center have exhaust inlets at floor level to help control vapor from mercury spilled during the frequent use of Van Slyke apparatus, and additional exhaust inlets have been found to be needed behind the equipment which has motor-driven agitators.

Local exhaust ventilation should be available as needed to control mercury vapor at its sources. One of the possible sources of mercury vapor which should have exhaust ventilation is the exhaust from the mechanical backing pump on a mercury diffusion pump, and any other vacuum pump which may throw mercury vapor.

CONCLUSION

Open surfaces of mercury can often be covered with water or oil to slow vaporization of the mercury, and vapor from mercury that cannot be closed or covered can be removed by suitably designed ventilation. Equipment and enclosures should be designed to localize splashes and spills, and steps can be taken to catch spills by pans, containers of water or polyurethane foam which seems to be useful in picking up large and middle-sized droplets.

Appropriate medical services should be provided as necessary to detect any mercury poisoning at an early stage, and monitoring equipment or services should be provided to prevent the development of any mercury poisoning in the laboratory.

Wherever mercury can be spilled in the laboratory there should be a smooth cleanable floor coated with non-skid wax listed by Underwriters' Laboratories, Inc., so that spilled mercury can be quickly picked up both thoroughly and with reasonable effort.

The literature contains some interesting and useful accounts of mercury hazards in medical laboratories[15], university laboratories[16], and other scientific laboratories[4,17].

ACKNOWLEDGEMENT: Reprinted from *Journal of Chemical Education, Vol. 42,* A529, July, 1965, with permission.

REFERENCES

1. Giese, A., *Science, 91*:476, (1940).
2. Biram, J. G. S., *Vacuum,* 5:77, (1957).
3. Battigelli, M. C., *J. Occupational Med.,* 2:337,394, (1960).

4. Shepherd, Martin, Schuhmann, Schuford, Flinn, Robert H., Hough, J. Walter, and Neal, Paul A., *J. Res. Natl. Bur. Std., 26*:357, (1941).
5. Benning, Dorothy, *Ind. Med. Surg., 27*:354, (1958).
6. Locket, S., and Nazroo, I. A., *Lancet, 2*:528, (1952).
7. McCord, Carey P., *Ind. Med. Surg., 31*:41, (1962).
8. Sher, David A., and Neff, W. S., *Minn. Med., 47*:1457, (1964).
9. Bidstrup, P. Lesley, "Toxicity of Mercury and Its Compounds," Elsevier Publishing Company, Amsterdam, Holland, (1964).
10. Goldwater, L. J., *Ann. N. Y. Acad. Sci., 65*:498, (1957).
11. Hayes, Alastair D., and Rothstein, Aser, *J. Pharmacol Exptl. Therap., 138*:1, (1962).
12. Passow, H., Rothstein, A., and Clarkson, T. W., *Pharmacol Rev., 13*:185, (1961).
13. Williams, Charles R., Eisenbud, Merril, and Pihl, Stanley E., *J. Ind. Hyg., 28*:378, (1947).
14. Goodman, Clark, *Rev. Sci. Inst., 9:*233, (1938).
15. Noe, Frances E., *New Eng. J. Med., 261*:1002, (1959).
16. Beauchamp, Isaac L., and Tebbens, Bernard D., *Am. Ind. Hyg. Assoc. Quart., 12*:171, (1951).
17. Kushakovskii, L. N., and Teplitskaia, R. T., *Gigiena i Sanit., 28*:76, (1963).

Chemical Cyanosis and Anemia Control

Adrian L. Linch, R. L. Wuertz and R. C. Charsha

INTRODUCTION

Cyanosis is an abnormal condition brought on by failure of the blood stream to deliver oxygen to the biochemical processes which sustain life. In order to understand the cyanotic state, some knowledge of normal conditions is required.

Normal State

Nearly all tissues of the human body require oxygen to convert biological fuels to energy. This oxygen, absorbed from the air in the lungs, must be transported to remote locations to sustain life. This critical function is carried on by the red pigment, hemoglobin, which, confined in erythrocytes, circulates freely as a suspension in the blood stream (Figure 1).

Hemoglobin (Hb) is composed of three major components: a porphyrin, a protein (globulin), and an atom of iron in the ferrous state. Four of these molecules associate to form hemoglobin, which has a molecular weight of about 65,000 and contains 0.34% iron. Normal blood contains approximately 15 g Hb per 100 ml. Hemoglobin combines with four molecules of oxygen to produce a loose complex (oxyhemoglobin—HbO_2) which readily dissociates on delivery to the vicinity of oxygen demand:

$$Hb + O_2 \rightleftharpoons HbO_2$$

$$\text{Dark red} \qquad\qquad \text{Bright red}$$

Cyanotic State

Cyanosis is a sign of tissue oxygen deficiency and occurs when the oxyhemoglobin level falls below the critical oxygen demand level.

A deficiency of oxygen in the blood stream can be produced by several conditions:

1. Insufficient respiratory oxygen pressure (normal partial pressure at sea level = 159 mm Hg). Anoxia, or lack of oxygen, is encountered in the rarefied atmosphere of high altitudes by pilots and mountain climbers or by workers in poorly ventilated environments where oxygen has been displaced by inert gas or consumed by combustion or fermentation. Interference with the mechanical processes of breathing such as encountered in asthma, pneumonia, and strangulation will deprive the blood of oxygen.

2. Impaired blood circulation. Conditions that prevent free circulation of the blood through the lungs or permit circulation to bypass in the heart will produce cyanosis through failure to oxygenate the available active hemoglobin.

3. Invasion of the blood stream by foreign chemicals that displace oxygen or block the reaction between hemoglobin and oxygen. These materials fall into three general classes: (a) the reactive gases and vapors absorbed through the lungs such as CO, H_2S, N_2O_3, HCN, nitrobenzene, aniline, etc.; (b) ingestion of compounds such as nitrates, nitrites, sulfides, and some medicinals such as sulfanilamide, acetanilide, etc.; and (c) the liquids and solids absorbed through the intact skin.

The association of chemical cyanosis with the manufacture of aromatic nitro and amino compounds has been a well-known occupational hazard almost from the

HEMOGLOBIN

first commercial production of nitrobenzene and aniline. Although the human bio-chemical reactions were not understood, reduced oxygen-carrying capacity of the blood producing general tissue anoxia after contact with aromatic compounds was recognized very early in the history of the dye industry.[1]

Cyanosis was a not uncommon industrial illness, and a number of deaths from contact with nitro compounds and amino compounds were reported.[1] Efforts to alleviate the condition were confined largely to the determination of methemoglobin (MHb) after the acute phase had developed and to supportive treatment for the victim. The subacute Hb depressing effect, anemia, produced by these compounds received even less attention. Many of the physical, physiological and biochemical effects became apparent only after extensive clinical and laboratory studies. The effects on work efficiency were vague, due principally to inadequate methods for controlling exposures through blood and urine analysis.

We have chosen to designate the over-all effects as the "cyanosis-anemia syn-drome", since cyanosis may appear alone, anemia may or may not follow, or anemia may occur without obvious cyanosis. Although methemoglobinemia is only one of the factors in the depression of hemoglobin oxygenation capacity under some con-ditions,[1] methemoglobin analyses provided the most expedient criteria for cyanosis severity classification.

The exposure severity peak was reached in Du Pont Chambers Works during 1940–1941, followed by a downward trend until 1951. Then the start of chloroaniline manufacture and greatly expanded production of the older well-established aniline derivatives sharply increased the frequency of the cyanosis-anemia syndrome. An effective control program introduced in 1956 produced a definite improvement[2-4] (Figure 2). Improvements and modifications in the microprocedures for the rapid determination of MHb and Hb have provided information that not only diagnosed cyanosis severity, but also provided the information necessary for effective preven-tion. Late in 1955, the National Research Council's recommended procedure and calibration standard for the cyanmethemoglobin procedure, which was later estab-lished as an official method for Hb, was introduced on a tentative basis for evaluation.

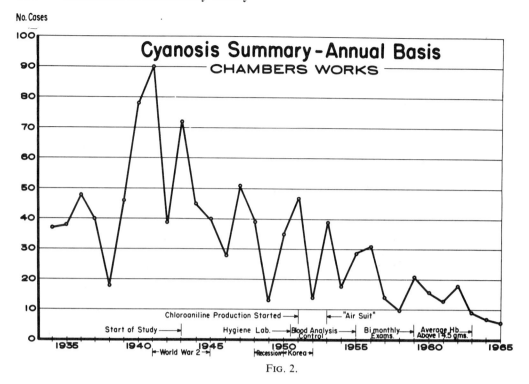

No. Cases

FIG. 2.

Then in 1957, with the adoption of the microhematocrit determinations for all routine blood specimens, another source of hemoglobin data became available for a cross comparison with the other cyanosis control methods. The oxygen evolution procedure for the determination of oxyhemoglobin was established at this time to provide a direct measurement of oxygen transport.

Absorption of fat-soluble materials such as aniline and nitrobenzene through intact human skin occurs with remarkable ease. Aniline absorption rates as high as 3.8 mg/cm^2/hour have been reported.[5] After the skin barrier has been passed, some of the contaminant is excreted unchanged through the kidneys, but a major portion is altered chemically to ionizable polar derivatives by the blood elements, the liver, and other tissues in an attempt to reduce the toxicity.

Most of the aromatic amino and nitro compounds probably are not cyanosis producers themselves, but oxidation-reduction enzyme systems promote conversion to known cyanogenic derivatives, which arise from either oxidation of the amine or reduction of the nitro group (Figure 3).

Aniline	Phenylhydroxylamine	Nitrosobenzene	Nitrobenzene
(1)	(2)	(3)	(4)

FIG. 3. Cyanosis Precursor Mechanism.

The reaction products of Hb with these metabolites of aromatic nitro (NC) and amino (AC) compounds have been loosely classified as "methemoglobin" from similarity of the absorption spectra to complexes in which the iron has been oxidized to the ferric valence state.[6] Investigation with radioactive tracer techniques disclosed at least six methemoglobin (MHb) precursor complexes containing ferrous iron, two forms containing ferric iron and two oxygenatable ferrous derivatives.[7] Total oxygenatable Hb is frequently less than would be expected from MHb analysis. Only the gasometric HbO_2 procedure will disclose the total inactive Hb.* (Abnormal Hb complexes associated with the onset and recovery of methemoglobinemia may be derived from the difference between the acid hematin and MHb (total oxidation) procedures for the determination of Hb.) Therefore, the determination of oxygenated hemoglobin (HbO_2) in the blood of chemical workers employed in cyanosis exposure areas should be included to provide the attending physician with a reliable indicator of oxygen transport loss.

The phenylhydroxylamines[6] are the most potent metabolites in the series. Restoration is undoubtedly due to the body's successful detoxification by conversion to the lower potential aminophenols and their esters, which can be found in the urine (Figure 4).

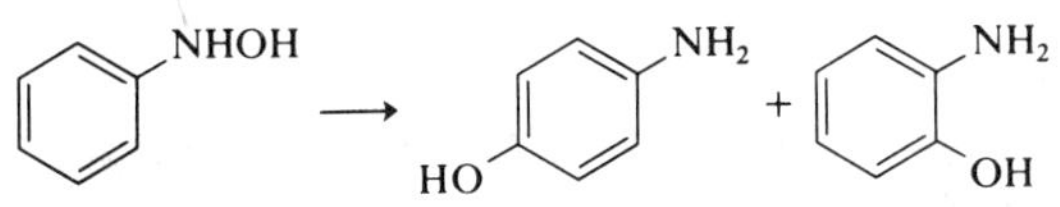

Phenylhydroxylamine   p-Aminophenol   o-Aminophenol

FIG. 4. Detoxification of phenylhydroxylamine in the body.

Some of the effects of industrial contact with cyanosis-producing chemicals have been described by Von Oettingen,[1] Bodansky,[6] Sroka,[9] Doctor,[10] Mangelsdorff,[11,12] and Halsted.[33] These references described a limited number of cases and the extent of the cyanosis-anemia syndrome was not disclosed. Various therapeutic regimens were recommended by these and other authors.[13,14] Based on the above reports and observation of employees with chemically induced cyanosis-anemia syndrome, a treatment program has been developed. Analysis of therapeutic results in all cyanosis-anemia syndrome cases reported from the manufacturing areas of the Du Pont Chambers Works during the 10-year period 1953–1962 forms the basis of our evaluation. Clinical data were gathered from the Chambers Works Medical Division records.

The discussion in this chapter will be limited to human experience only. Small animal response to cyanogenic agents was summarized excellently in 1962 by D. O. Hamblin.[14a] The systemic effects in man, such as metabolism alterations, jaundice and other evidence of liver damage, hematurea from kidney damage or glandular effects reported by other investigators,[14a] were not observed in this study of 215 case records.

ANALYTICAL METHODS

The methemoglobin procedure of Evelyn and Malloy[15] was modified and simplified for routine laboratory use. A quantitative procedure to evaluate some cyanosis precursors (HbC) was developed from the difference between acid hematin (Hb) and total MHb analyses.

Apparatus

A prism or diffraction grating spectrophotometer using 10–20 mm diameter cuvettes. The Bausch and Lomb "Spectronic 20" (diffraction grating) and 12 mm

---

* Two years experience with a "Micro-oximeter", which measures per cent HbO_2 directly by reflectance spectrophotometry, has confirmed this procedure as a reliable, accurate and rapid (less than 5 minutes) approach to the analysis of blood for oxygen capacity without chemical alteration.[8]

cuvettes have given excellent service. The necessary instruments for obtaining 0.1 ml of blood from an ear lobe puncture: 0.1 ml micro blood pipette, sterile cotton, gauze squares, and 70% alcohol for preparation of the skin surface and the lancet.

Reagents

M/60 PHOSPHATE BUFFER of pH 6.6.[15] Dissolve 6.7 g AR anhydrous disodium monohydrogen phosphate and 10.7 g AR anhydrous monopotassium dihydrogen phosphate in distilled water and dilute to one liter.

SAPONIN: 10% in distilled water.

POTASSIUM FERRICYANIDE: 10% in distilled water. Store in red, low actinic, glass-stoppered bottle. Make fresh monthly.

Procedure

One-tenth ml of freshly drawn blood is rinsed into a spectrophotometer cuvette containing 7.0 ml of M/60 phosphate buffer. The red blood cells are lysed with a drop of saponin solution and the components mixed by inverting the capped cuvette several times. Destruction of the cell stroma is necessary to avoid light scattering errors in the spectrophotometric procedure.[16]

The per cent transmittance is read at 630 mμ (%T_1) against a buffer blank containing a drop of saponin solution. The Hb is converted quantitatively to MHb with excess potassium ferricyanide solution to obtain the MHb value. One drop of 10% reagent is sufficient for full color development in three minutes. Per cent transmittance is read again at 630 mμ (%T_2) against the saponin blank. Ferricyanide reagent is not necessary in the blank, as it does not absorb at 630 mμ. In Figure 5, to

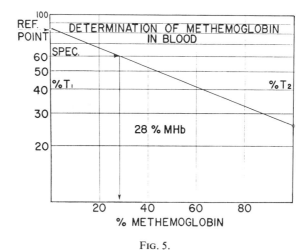

FIG. 5.

determine % MHb, connect the calibration chart's reference point (left hand vertical axis) by a straight line with %T_2 on the 100% MHb (HbM) right hand vertical axis. Extend %T_1 horizontally to the diagonal. A vertical to the MHb scale from this horizontal intersection determines the % MHb.

Calibration Chart

Blood specimens from 15–20 persons not under medication and not exposed to cyanosis-producing (cyanogenic) chemicals are processed by the foregoing procedure. The horizontal linear axis of semi-log graph paper is scaled 0 to 100% MHb and the log scale assigned %T values (see Figure 5). Assuming there is no MHb in the blood specimens, the average of $\%T_1$ results is plotted as the reference point on the left hand or 0% MHb axis. The $\%T_2$ scale on the right hand vertical (100% MHb) axis then provides an anchor point for each individual specimen. A line from the reference point to the $\%T_2$ anchor point is corrected thereby for variations in total hemoglobin concentration. In practice, a center ruled transparent straight edge is laid over the chart to connect the reference point and $\%T_2$ plot rather than draw individual curves.

Hamblin and Mangelsdorff[17] showed that the light transmission range at 630 mμ for normal hemoglobin was within a narrow range of $\pm 1\%$ T. Additional data may show that the reference point should be adjusted slightly downward or upward to avoid negative MHb readings, but this adjustment usually falls within the limits of the method's precision ($\pm 0.5\%$ T). The effective range of the spectral scale (20–90% T) can be utilized by varying the volume of buffer for any given spectrophotometer. The confidence limits lie in the range $\pm 0.5\%$ MHb, and less than 5% MHb is not considered significant.

Equivalence of Oxidized Hemoglobin (HbM) to Acid Hematin (Hb)

Since no change in molecular weight is involved, the amount of MHb in a specimen after complete oxidation by ferricyanide (HbM) is equivalent to the weight of Hb, and concentration values can be assigned to $\%T_2$. The acid hematin (Hb) procedure has been used in our Clinical Laboratory for more than seventeen years; therefore, $\%T_2$ was related to Hb %T for calibration purposes.[18] Under normal conditions, the two values are equivalent within $\pm 1\%$ T.[3] Blood specimens from personnel exposed to aromatic nitro and amino compounds frequently did not exhibit this equivalency. The percentage difference between these two Hb values, designated %HbC, is often a measure of MHb precursors after exposure to cyanosis-producing chemicals.[3]

$$\% \text{ HbC} = \frac{(g \text{ HbM} - g \text{ Hb}) \times 100}{g \text{ Hb}}$$

The algebraic sign is carried to determine whether negative HbC occurs with significant frequency. The origin of HbC remains a matter for conjecture.

Equivalence of HbM to Hb and total hemoglobin was obtained by titration of lysed whole blood in the buffer system with potassium ferricyanide. A specimen which contained 14.4 g/100 ml by both HbM and Hb procedures contained iron equivalent to 14.5 g of hemoglobin per 100 ml.[3] Linearity of the MHb curve down to 80% MHb was established also by this technique.

Acid Hematin (Hb)

A 0.02 ml blood specimen drawn from an ear lobe puncture served for both hemoglobin determination by light absorption at 525 millimicrons (mμ) after acid hydrolysis and red blood cell count (RBC) as described by West and others.[18,19] A precision of ± 0.2 g/100 ml was attained for this and the HbM methods. An erythrocyte count (RBC) is included in the West procedure.

Cyanmethemoglobin (CNMHb)

Fresh whole blood from an ear puncture was diluted in a ferricyanide-cyanide reagent and hemoglobin determined as CNMHb spectrophotometrically at 540 mμ as recommended by the National Research Council.[20] The precision appeared to be equivalent to the Hb and HbM procedures.

Microhematocrit[21]

Packed cell volume (PCV) was obtained by centrifuging blood specimens in 75 mm heparinized capillary tubes in an 8 inch diameter head at 10,000 rpm for five minutes. Hemoglobin was estimated by multiplying % PCV × 0.34.

Reduced Hemoglobin (Hb–R)[22]

Reduction of oxyhemoglobin with alkaline sodium hyposulfite in the presence of saponin,[23-25] was adapted from the procedure described by Van Slyke. Then total hemoglobin was determined as CNMHb by light absorption at 540 mμ.

Oxyhemoglobin (HbO$_2$)

Volume per cent of oxygen in a 0.04 ml blood specimen drawn from an ear puncture was obtained by the microgasometric procedure developed by Scholander and Roughton.[26] Multiplication of volume % by the factor 0.75 g/vol. % gave g HbO$_2$/100 ml of blood:[27]

$$\% \ HbO_2 = \frac{g \ HbO_2 \times 100}{g \ Hb}$$

Specimens containing $95 \pm 5\%$ HbO$_2$ were considered to be saturated.[28] The precision of the method has been shown to be within ± 0.1 g HbO$_2$.[26] Failure to obtain complete conversion of HbO$_2$ to reduced Hb has been encountered by other investigators in calibrating spectrophotometers.[29]

Frequent evidence of latent cyanosis has been observed in laboratory analytical results. Exposures sufficiently severe to produce high urinary excretion levels, 75–125 mg per liter nitrobenzene and its metabolites, have been detected (Figure 6). However, clinical observations and MHb analyses did not disclose abnormal hemoglobin conditions. Other cases have exhibited MHb above 10% at urinary excretion levels below 25 mg per liter.

Since the term methemoglobin (MHb) has been used rather loosely to describe what actually is a whole family of hemoglobin derivatives, the in-vitro work of Jackson and Thompson,[7] who used radioactive tracer techniques to identify six precursors (ferrous iron) of MHb, offered the most attractive solution to the cyanosis blood analysis dilemma (Figure 7). Para-radioiodine-tagged nitrosobenzene and phenylhydroxylamine combined firmly with Hb to produce a methemoglobinemia which disappeared in the course of an hour or two without loss of radioactivity from the red cells. The combining site was established at the iron atom of heme. None of the HbC types 1, 2, 5, 6, 7, and 8 absorbed light in the 630 mμ region. Type 3 could not be hydrolyzed to acid hematin. Types 6 and 8 could be oxygenated, but whether or not these derivatives gave up oxygen to the tissues in vivo is questionable. Jackson and Thompson also found two ferric iron forms with absorption maxima at 630 mμ (Types 3 and 4).

Examination of the absorption maxima (Figure 7) revealed that the routine Hb procedure for hemoglobin[19] would not disclose at least one (Type 3) and possibly two HbC forms at 525 mμ (Types 5 and 7) which are included in MHb and HbM results

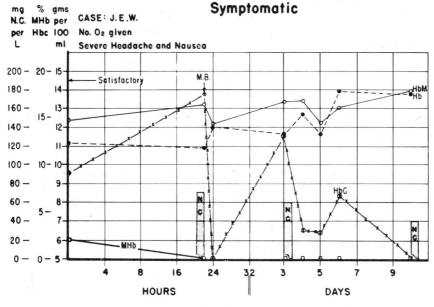

FIG. 6.

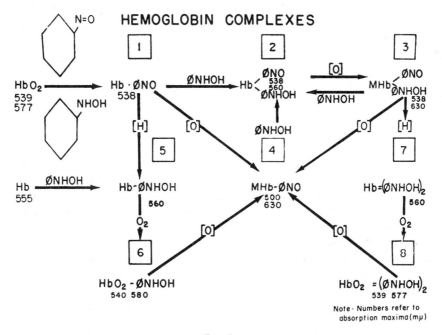

FIG. 7.

at 630 mμ. Loss of color by acid cleavage of the iron–aromatic nitrogen complex from the hemin molecule may account for low Hb values observed in many exposure cases. In normal blood taken from nonexposed personnel, differences between Hb and HbM results are within experimental deviation of ± 0.2 g, but HbM levels up to 20% higher than Hb have been found. This unstable condition often progresses to active cyanosis if exposure is not relieved, as HbC rarely rises above 20% before conversion to MHb occurs.

Figure 8 presents a chemical operator's hemoglobin anomalies during exposure to 2,5-dichloroaniline. For the first six days, HbM and Hb values agreed within experimental deviation (Area 1). MHb also was normal or less than 5%. Sometime during the next five days, an exposure occurred which produced significant concentrations of HbC (16%—Area 2). The absorption curve of the oxidized blood showed a

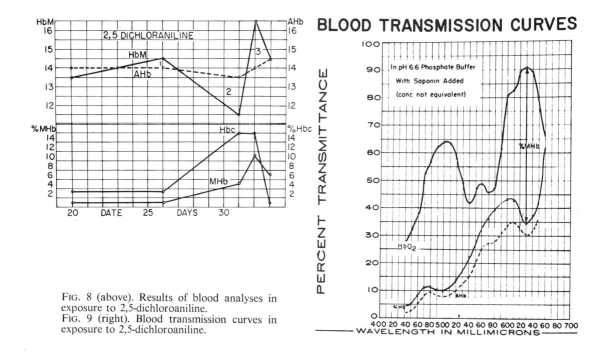

FIG. 8 (above). Results of blood analyses in exposure to 2,5-dichloroaniline.
FIG. 9 (right). Blood transmission curves in exposure to 2,5-dichloroaniline.

pronounced plateau in the 560 mμ region not found in normal MHb solutions (Figure 9). An HbC "reversal" occurred during the ensuing twenty-four hours (Area 3, Figure 8). The HbM values increased above Hb to a numerical value equal to the initial %HbC. Total MHb analysis rose to 11% as the precursor derivatives in the blood oxidized.

Figure 10. A further study of 2,5-dichloroaniline cyanosis with the additional aid from oxyhemoglobin values determined microgasometrically,[3] revealed considerably more HbC than indicated by 13% difference between Hb and HbM. HbM − HbO$_2$ difference was 24%, which in this case was a better indicator for HbC. Hb was not an accurate index for oxygenatable hemoglobin, nor did Hb differences alone reveal the exposure severity.

Figure 11. Superficially, HbO$_2$ would appear to be a reliable basis for cyanosis evaluation. However, the reverse condition has been met. The three procedures for hemoglobin (Hb, HbM and HbO$_2$) agreed, but 21% MHb by analysis was present from a meta-nitroaniline exposure. HbC Types 7 and 8 are strongly indicated as these derivatives oxidize readily on exposure to the atmosphere.[3] In less than eight hours, this condition developed into a typical cyanosis. At this time, methylene blue was administered when HbO$_2$ results revealed almost twice as much unoxygenated hemoglobin as HbC analysis indicated. Although methylene blue quickly restored oxygen capacity, no significant change in the difference between Hb and HbM hemoglobin levels occurred during the following nine hours. This persistence of high HbC concentration is typical of nitroaniline and dinitrobenzene exposures.

These results suggest that no one method of hemoglobin analysis will furnish sufficient information for accurate diagnosis of cyanosis. Rather, the interpretation

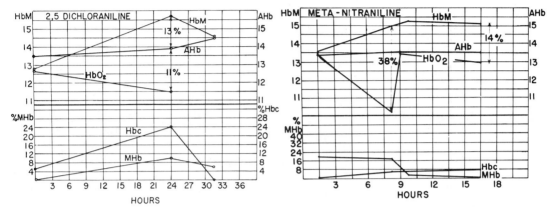

FIG. 10. Results of blood analyses in exposure to 2,5-dichloroaniline.

FIG. 11. Results of blood analyses in exposure to meta-nitroaniline.

of results from several different procedures including blood oxygen and an evaluation of rate of change is required for both exposure control and cyanosis treatment.

Oxyhemoglobin analysis disclosed what on the surface appeared to be discrepancies in some of the analytical results. Examination of the first blood specimen taken on admission disclosed the sum of HbO_2 and MHb exceeded 100 % in some patients with severe methemoglobinemia. For example in Figure 12, the sum was 113 %. Since the complexity of the reaction system of hemoglobin with the metabolites of aromatic nitrogen compounds has been delineated by Jackson and Thompson,[7] it is reasonable to assume that this reaction system is highly unstable and susceptible to minor environmental changes during the onset of cyanosis. Therefore, when the blood specimen is taken and hemolyzed in the presence of atmospheric oxygen, oxidation converts a portion of the HbO_2 to MHb. However, the reaction may be anaerobic, as sufficient oxygen is present in the blood itself to produce this oxidative effect. Further confirmation was found when specimens collected in the field were analyzed in the laboratory after a time lapse of one to two hours. Although these men worked in potential exposure areas, their urinary excretion did not account for the relatively high MHb results found in a number of instances.

In other cases, HbO_2 exceeded both hemoglobin results (Hb and HbM) by significant amounts (Figures 13, 14). Without exception, these conditions followed known absorption of aromatic nitrogen compounds and suggested the presence of hemoglobin derivatives that do not absorb light at 525 or 630 mμ.[30] In the work of Jackson and Thompson, background evidence for this explanation was found.[7]

In a third group, the sum of HbO_2 and MHb fell short of 90 %. In this condition, an inactive hemoglobin (HbI) with iron in the ferrous state analyzed as Hb and HbM. The cases illustrated by Figures 12 and 13 are representative of this condition which has been observed frequently when HbO_2 determinations are made during cyanosis episodes. At the point where the MHb in Figure 12 dropped to 5 %, the HbI was 17 %, and in Figure 13 the blood contained 33 % HbI at the HbO_2 low point. Again the system delineated by Jackson and Thompson could account for such conditions.[7,30]

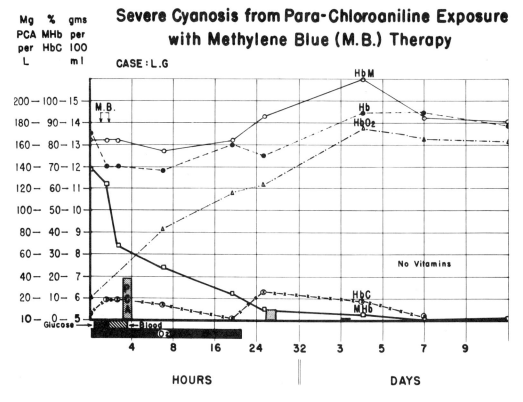

Severe Cyanosis from Para-Chloroaniline Exposure with Methylene Blue (M.B.) Therapy

CASE : L.G

FIG. 12.

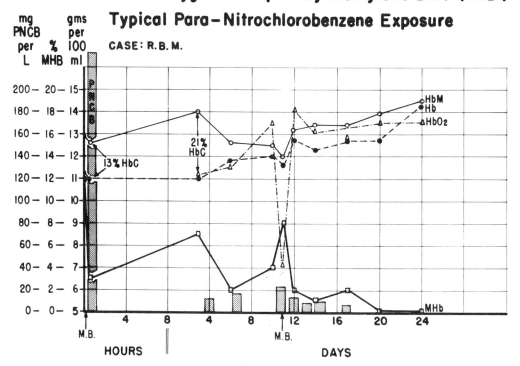

Restoration of Oxygen Transport by Methylene Blue (M.B.)
Typical Para-Nitrochlorobenzene Exposure

CASE: R.B. M.

FIG. 13.

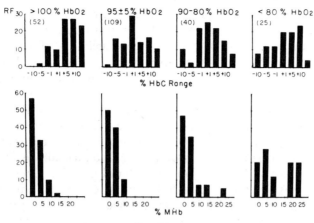

FIG. 14.

Evaluation of Hemoglobin Methods[30]

Hb-HbM-CNMHb-HbO$_2$ comparisons were made on 401 specimens drawn at random from the 13,675 sets of results obtained during the four-year period 1955 to 1959 (Figure 15).

The statistical methods used in compiling the results in this paper are summarized briefly:

Methods were compared only by differences in result when the methods to be compared were applied to the same blood sample at the same time. For any set of cases, these differences were treated as a statistical sample of a larger population. Methods were concluded to be different if a student's "t" test of the average difference versus the null hypothesis of no difference was significant, using, unless otherwise stated, a 95% confidence level.

Confidence limits were computed on the assumption of normal distributions. The above statistical methods were applied to original data in mg/100 ml units.

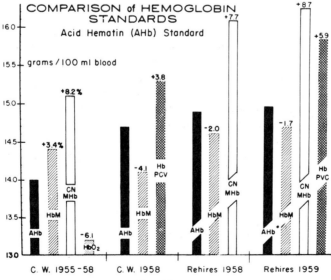

FIG. 15.

Conversions of both averages and confidence limits to percentages were made with reference to the average result by the method used as standard in each comparison within each set of specimens.

Results from the Hb, HbM and HbO_2 procedures were in close agreement when applied to nonexposure cases. On an over-all averaged basis, three-way agreements occurred with a frequency of 40–50 % and a reproducibility of ± 0.2 g/100 ml. A significant difference in Hb levels by all five methods was found between groups exposed and groups not exposed to aromatic nitrogen compounds. Seasonal differences also were observed.

Additional confirmation that the difference between Hb and HbM calculated as % HbC is a significant indication of chemical exposure and is often a precursor to methemoglobinemia was developed. Significant increases in Hb levels for recent cases as compared with Hb levels prior to 1958 confirmed the over-all benefit derived from the cyanosis control program introduced in 1955. Chemical exposures appeared to produce positive HbC; i.e., HbM results exceeded Hb values.

On an over-all basis, positive HbC occurred twice as often as negative values and 28 % of the specimens disclosed less than 1 % HbC. Although a near normal HbC distribution occurred in the absence of MHb, the proportion of positive HbC increased sharply as MHb concentrations exceeded 5 %. Deviations in the distribution of positive and negative HbC in the intermediate MHb zones stabilized beyond 20 % MHb where negative HbC all but disappeared.

The conversion factor, 0.34 g/100 ml/% PCV, appeared to be somewhat high when used to calculate Hb from microhematocrit determinations. However, nearly one-half of the specimens analyzed fell in the normal HbO_2 range (95 ± 5 %). On a frequency of occurrence basis, HbC and MHb levels increased as the % HbO_2 decreased, with a decided bias toward positive HbC values as HbO_2 dropped below 80 %. Oxygen evolution in 5 % of the cases indicated that both Hb and HbM methods had yielded low results. Significant quantities of MHb were found in the presence of excess HbO_2 in 3 % of the specimens. Inactive Hb not detectable as MHb was found in 73 % of the specimens which contained negligible amounts of HbC and less oxygen than required to saturate the non-MHb fraction. The frequency with which serious oxygen deficiency would remain undetected in the absence of HbO_2 determination would have been 1 in 35.

The CNMHb procedure for the determination of total Hb gave results that on the average exceeded the Hb, HbM, PCV-Hb and HbO_2 methods by amounts that precluded its use in an industrial cyanosis-anemia control program. Examination of individual cases also disclosed wide variances. The CNMHb levels were brought within Hb–HbM range by either hyposulfite reduction or saponin treatment before addition of the oxidizing reagent.

Many of the observed anomalies could be explained by the Hb complexes which are resistant to ferricyanide oxidation described by Jackson and Thompson and by a labilizing effect for acid cleavage after complexing with aromatic nitrogen metabolites. Similarities with other reported associated Hb systems from exposure of blood to highly active agents such as carbon monoxide and phenylhydrazine were observed.

DIAGNOSIS[2]

Cyanosis is a slate-gray or bluish discoloration best observed in daylight and is most likely to be seen in nail beds, ear lobes, lips, and mucous membranes. Cyanosis generally becomes apparent when the methemoglobin is 10 % or over. Thus methemoglobinemia of 10 % or more is considered as the criterion for the diagnosis of chemical cyanosis.[3,4,31] Under normal conditions, methemoglobin values of less than 5% are not considered significant.[3] HbC levels were not considered significant until a 10% level had been reached. Results above 10% indicated latent cyanosis conditions which also required close medical surveillance. Results above 20% required confirmation by microgasometric blood oxygen analysis. Methylene blue therapy was indicated in such cases. Hemoglobin concentrations below 13.0 mg/100 ml by either method were considered abnormal and too low to risk further exposure to cyanogenic or anemiogenic compounds. Two consecutive Hb results which, repeated weekly, fell below 12.0 g/100 ml required removal of the employee from further potential exposure until his average pre-exposure level was restored. A complete medical history and examination should be considered to detect non-occupational causes for this condition.

Potential cyanosis (P.C.) criteria were established to assist with the control program and included one or more of the following conditions:

1. MHb in the range 5%–10%. Close medical surveillance was considered necessary for these cases.
2. HbC over 14%.
3. Hb decline, over a period of 24 hours or less, greater than 13% calculated from maximum and minimum Hb and HbM results.
4. HbC of 10% or more, and either Hb result falls below 13.0 g/100 ml.

Urine specimens are collected from the cyanosis cases as soon as possible after admission and as frequently as practical during the recovery period. Aromatic nitro and amino compounds are determined by the procedure described by Koniecki and Linch.[32] If the causative agent is unknown or an unresolved mixture is present, the nitro compounds (NC) are calculated as nitrobenzene (NB) and the amino compounds (AC) are reported as aniline (mg NC/liter or mg AC/liter, Figures 6, 12, and 13). Urine specimens were classified abnormal if the NC or AC content exceeded 10 mg per liter. When the level exceeded 20 mg per liter, the employee was called in for a blood analysis and medical interview, if this had not been done at the time the urine specimen was collected. If the exposure was known to be exclusively nitrobenzene or dinitrotoluene, the threshold limit for recall was established at 50 mg per liter. At the 20 mg per liter limit, as much as 10% MHb was observed in the blood of the more susceptible employees when the exposure included the chloroanilines or dinitrobenzene. Results from these analyses help identify the causative agent in many cases, contribute information on its metabolism[3] and are used to supplement the cyanosis-anemia control program.[4]

A ten-year review of all cyanosis cases at the Du Pont Chambers Works disclosed 187 cases with methemoglobinemia of 10% or more (Tables 1 and 2). There were 28 additional cases thought to be cyanotic by clinical observation but whose MHb was less than 10%, so were not included in our statistics. These are mentioned only to point out the disparity which occasionally exists between the clinical impression of cyanosis and laboratory findings.

Most of the 187 cases fell in the 10–29% MHb range which was classified as mild methemoglobinemia. This group included 152 or 80% of the total. Moderate methemoglobinemia, defined as 30–50% MHb, was present in 32 or 18% of the cases. There were three cases of severe cyanosis with MHb over 50% (Table 1).

Signs of clinical cyanosis were noted in 139 or 74% of the 187 cases (Table 1, Figure 16). Forty-eight cases did not seem cyanotic, but had methemoglobinemia in the range 10–40%. Symptoms were recorded from 70 or 37% of the 187 men. Headache was the most frequent complaint. Other symptoms included weakness, fatigue, nausea, vertigo, chest pain, numbness in the extremities, abdominal pain.

malaise, aching joints, palpitation, aphonia, nervousness, air hunger, anuria, irrational behavior and syncope (Table 1). The 48 men who did not appear cyanosed clinically were all asymptomatic and would not have been diagnosed without supporting laboratory data (Figure 16).

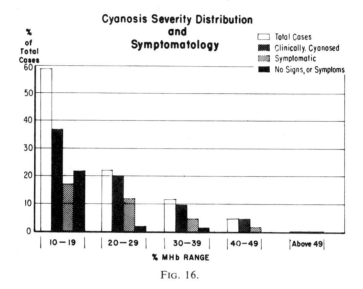

FIG. 16.

Examination of 157 cyanosis records, which included complete blood analyses, disclosed that 29 hemoglobin levels (18%) had dropped below 12.0 g/100 ml—the control limit below which further exposure is restricted.[1] (Table 3.) An additional 49 cases (31%) exhibited Hb concentrations in the 12.0 to 13.0 g/100 ml range, which is considered unsatisfactory for men assigned to cyanosis hazard areas.[4] This means that half of the employees involved in these cyanosis episodes incurred hemoglobin losses which depressed the concentration below 13.0 g/100 ml some time during the 7-day period following development of the methemoglobinemia maximum. Further examination for the magnitude of hemoglobin losses disclosed 117 cases (75%) in which a 10% or more decline was observed, and in 18 (12%), hemoglobin losses exceeded 20%.

TREATMENT[2]

Workers known to have had gross exposure to cyanosis-producing chemicals or who are suspected of being cyanosed or those found to have methemoglobinemia on routine examinations should have a thorough shower, including a shampoo and cleansing of the ear canals and nostrils. Cleaning under fingernails and toenails is especially important and may be facilitated with a scrub brush and orange stick. Contaminated clothing including shoes and socks should be discarded and replaced with fresh clothes and shoes.

Cyanosis victims should then be given bed rest, sweetened fruit juices are offered, and oxygen is administered by mask. Mild cases, with methemoglobinemia below 15%, usually improve without further treatment.

The methemoglobin level should be rechecked in one to two hours. Cases with a rising methemoglobin or those with an initial methemoglobin of 15% or more should

TABLE 1
Signs and Symptoms in Relation to Methemoglobinemia Severity (No. of Men = 143)

	None*	Mild		Moderate		Severe Above	Totals	
Methemoglobin range, %	0–9	10–19	20–29	30–39	40–49	50	No.	% of 187
Total No. of cases	28	111	41	22	10	3	187	100
No. of cases with cyanosis signs								
Degree — marked	0	0	0	0	0	2	2	1
Degree — moderate	0	2	4	6	6	0	18	10
Degree — mild	28	68	33	13	4	1	119	63
Total with signs	28	70	37	19	10	3	139	74
No. of cases with symptoms								
Headache	7	23	13	8	4	3	51	27
Weakness	2	9	8	1	0	1	19	10
Fatigue	1	4	7	1	0	0	12	6
Nausea	1	2	0	2	1	1	6	3
Vertigo	2	1	1	1	0	2	5	3
Chest pain	0	2	0	0	0	1	3	2
Numbness in extremities	1	2	0	0	0	1	3	2
Others†	0	2	3	3	0	2	10	5
Total with symptoms‡	10	32	22	9	4	3	70	37
No. of cases asymptomatic	18	79	19	13	6	0	117	63
No. of cases with neither signs nor symptoms	—	41	4	3	0	0	48	26

*Figures in this column excluded from line (horizontal) totals.
†Two each: abdominal pain and malaise; one each: aching joints, palpitation, aphonia, nervousness, air hunger, anuria, irrational behavior, and syncope.
‡Some cases gave more than one symptom; therefore, this total is not a summation of individual symptoms by columns.

TABLE 2
Cyanogenic Activity versus Structure

Structure	No. of Cases	% of Cases	Recovery Time, hours				Average MHb, %
			Without M.B.*		With M.B.*		
			<10%	<5%	<10%	<5%	
Nitrobenzene	12	6.4	77	32	6.2	—	13
Dinitrobenzene	19	10.2	172	181	9.3	20	19
With B-12			—	—	6.2	18	
Mononitrochlorobenzenes	10	5.3	17	27	4.9	11	16
Nitroanilines	6	3.2	21	23	8.8	30	24
Nitrotoluenes	2	1.1	—	—	—	—	14
Nitronaphthalene	2	1.1	—	—	—	—	10
Aniline	7	3.7	18	29	2.2	2.2	22
Toluidines	16	9.4	—	—	2.1	3.1	24
Monochloroanilines (mixed)	17	9.1	27	31	4.2	5.0	25
Para-chloroaniline	25	13.4	16	25	5.9	6.3	32
Dichloroaniline	2	1.1	—	—	—	—	16
Dinitrosobenzene	17	9.1	6.4	19	1.9	1.9	20
Unknown or mixed	52	27.0	18	18	1.9	6.0	23
Total—Average	187	100.0	40	56	4.2	8.9	—

*M.B. = methylene blue.

TABLE 3
Effects of Cyanosis on Hemoglobin Levels

Hb, g	No. of Cases	%		% Loss	No. of Cases	%
<12	29	18.4 ⎫ = 50%		<10	40	25.4
12–13	49	31.2 ⎭		10–20	99	63.1
13–14	52	33.2		>20	18	11.5
>14	27	17.2				
	157	100.0			157	100.0

receive methylene blue. This is available as a 1% solution* which is given by syringe very slowly intravenously at a rate not exceeding 2 ml per minute. The usual dose is 10 ml injected over a period of not less than five minutes. This amount is generally in keeping with the recommended dose of 1 to 2 mg (0.1–0.2 ml of 1% solution) per kilogram of body weight.[9,10,12,33-5] This usually results in a rapid fall of methemoglobin within an hour. Rarely is administration of more methylene blue necessary. When required, additional amounts are given in 5 to 10 ml doses at hourly intervals. Intravenous 5% glucose is reported to be beneficial. Failure of recovery to occur as rapidly as expected should be a signal to recheck the patient for residual chemical on skin or under nails or contaminated clothing. Repeat showering and cleansing may be indicated.

Since most cyanosis cases show some anemia, especially in the first few days after exposure, an oral hematinic, such as an iron-multivitamin tablet, should be given two or three times a day for several days or until the blood count has returned to normal. Vitamin B-12 (cyanocobalamine) may not only have a beneficial effect on the blood count, but also has been reported to augment recovery from the cyanosis phase.[2] Therefore, an intramuscular injection of 1 ml of cyanocobalamine (1000 mg/ml) should be administered early in the course of treatment.

During the recovery period of several days following a cyanosis-anemia episode, workers should be instructed to avoid heat, exertion, and ingestion of alcoholic beverages. Work assignments should be such as to preclude the possibility of any further exposure to cyanosis-producing chemicals until all blood values have returned to normal.

If recurrent episodes of cyanosis and/or anemia occur in the same employee in spite of standard safety precautions, individual susceptibility should be suspected. Such persons should be permanently removed from areas of potential exposure to cyanosis-producing chemicals.

RECOVERY[2]

The results of the preceding therapeutic program at the Chambers Works Medical Division are summarized in abbreviated Table 4.

Methylene blue lowered the average MHb reduction time (to less than 10% MHb) from 41 hours to 6.4 hours—an 84% shortening of the recovery period (Table 4). Note that methylene blue did not aggravate hemoglobin losses; in fact, the average duration of postcyanosis anemia was shortened from 95 to 42 hours.

The use of methylene blue in the treatment of chemical cyanosis was first reported by Williams and Challis[36] and Steele and Spink[37] in 1933. Its therapeutic efficiency

*William H. Rorer, Inc., Philadelphia, Pennsylvania, supplies 10 ml ampules of methylene blue.

was confirmed by others,[34,38-40] but its use was reserved generally for more severe cases of cyanosis.[6,10-12,33] Since 1955 we have used methylene blue to treat mild and moderate as well as severe cases, with marked reduction in recovery time. Blood analyses demonstrated not only accelerated reduction of MHb, but also restoration of oxygen transport. Oxygenation was restored also in several cases in which HbO_2 was depressed below 60% in the absence of significant concentrations of MHb (less than 10%). As a result of this more rapid recovery, most patients are able to leave the Medical Building in a few hours instead of being held for longer periods. Acute aniline poisoning has been treated by hemodialysis.[41]

TABLE 4
Cyanosis-Anemia Treatment and Recovery

Treatment, Methylene blue	No. of Cases	Methemoglobin		Hemoglobin	
		Average % MHb	Recovery, hours to <10%	% Loss	Regain, hours
Without	43	19	41	13.8	95
With	35	18	6.4	12.6	42
With + B-12	64	19	1.8	13.7	30

A total of 104 patients received methylene blue intravenously. None suffered any serious adverse side effects attributable to this therapy. Patients occasionally complained of a burning sensation along the course of the arm vein in which the injection was being given, but this subsided upon slowing the rate of injection. Rarely subcutaneous irritation and blue discoloration occurred around the site of injection due to extravasation of dye. One patient developed a superficial phlebitis associated with cellulitis of the forearm following inadvertent extravascular injection of dye; this subsided in a few days with the use of warm compresses. There were no instances of thrombophlebitis or cellulitis in patients where perivascular infiltration did not occur. No consistent effects of methylene blue on the pulse, blood pressure, or electrocardiogram were observed in the dosages used (5–10 ml of 1% methylene blue). Hemolytic effects were not noted at this dosage range. On the contrary, postcyanosis hemoglobin loss was arrested sooner when methylene blue was given, 8.3 hours versus 26 hours.

When vitamin B-12 was added to methylene blue therapy, the recovery time was reduced further from 6.4 to 1.8 hours—a 72% reduction in the time required for recovery by the methylene blue group or an over-all 95% improvement (41 hours to 1.8 hours for recovery) over the group treated with oxygen and iron-vitamin tablets alone (Table 4). In addition, vitamin B-12 further shortened the anemia phase to 30 hours.

The efficacy of methylene blue for accelerating the recovery from severe cyanosis can readily be appreciated by comparing Figures 17 and 18. Figure 17 illustrates a case of cyanosis-anemia syndrome produced by a para-toluidine splash. After the MHb rose from 48 to 71%, the patient convulsed and went into shock, requiring intravenous and intracardiac adrenalin for resuscitation. The methemoglobinemia subsided gradually over the next 24 hours and the patient recovered completely. This case was treated in our dispensary prior to the time when methylene blue was used.

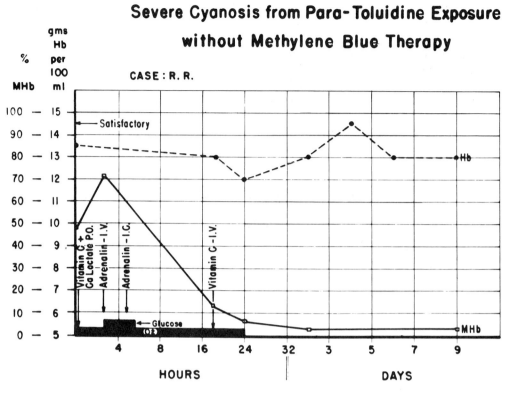

Fig. 17.

The next case (Figure 18) received 15 ml of 1% methylene blue (M.B.) in two divided doses. Note the steeper slope of the methemoglobin decline, falling from 78% to less than 10% in four hours with a compensating rapid rise in oxyhemoglobin (HbO$_2$). The exposure in this instance was to chloroaniline (CA). The patient was stuporous but not unconscious and responded so rapidly to methylene blue therapy that he was able to take a shower within a half hour after the first injection.

An example of differences in individual reaction to severe cyanosis due to chloroaniline (PCA) and treated with methylene blue (M.B.) is shown in Figure 12. This employee was brought into our dispensary by a truck driver when found weaving from side to side on the road while riding a plant bicycle. He appeared severely cyanosed and was irrational (MHb 69%). In addition to two 10 ml injections of methylene blue, this patient received two pints of whole blood. Clinical improvement coincided with the rapid decline in methemoglobinemia during the first three hours. During the remaining 24 hours the MHb decline was more gradual than in the previous case (Figure 18). Headache during this period was treated with oxygen inhalation and caffeine citrate. Note the reciprocal relationship of HbO$_2$ to MHb: as the MHb declined, HbO$_2$ increased.

The effect of methylene blue on the MHb declension is best illustrated in Figure 19, which shows two cyanosis episodes generated by chloroanilines in the same individual but differing in treatment. Methylene blue was not given in the first, but was used in the second instance. The severity of methemoglobinemia in both episodes was approximately the same (MHb 25 and 24% respectively), yet recovery was much more rapid when methylene blue was used. Note the MHb rise during the first four hours in the first episode. Experience accumulated from such cases formed the basis of our policy for more liberal use of methylene blue, even in the milder cyanosis cases,

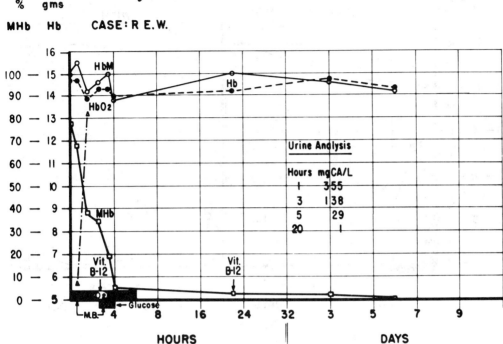

Severe Cyanosis from Chloroaniline (CA) Exposure
Methylene Blue (M.B.) + Vitamin B-12 Therapy

CASE: R E.W.

Urine Analysis

Hours	mgCA/L
1	355
3	138
5	29
20	1

HOURS   DAYS

Fig. 18.

to avoid the anticipated MHb rises that so frequently occurred, especially after chloroaniline and dinitrobenzene exposures.

Figure 20 illustrates methylene blue therapy in a patient with very mild initial cyanosis (MHb 10 %), whose response to the first 10 ml injection was poor. As the MHb was again slightly elevated (11 %) on the following day, another 10 ml dose was given. Progressive improvement followed. Relapses and prolonged low grade methemoglobinemia are typical of nitrobenzene (NB) absorption. A total of 15 relapsing cases in which the MHb returned to 10 % or above within a 24-hour period was found in this study. Of these, 11 (73 %) did not receive methylene blue. The average MHb was 15 % in the range of 10–35 %. Para-chloroaniline and dinitro-benzene each contributed four of these cases, nitrobenzene six, and meta-nitro-toluene one case.

Hemoglobin complexes (HbC) are believed to be precursors of MHb and present a condition of latent cyanosis.[3] Figure 6 illustrates the use of 10 ml methylene blue in an individual who complained of headache and nausea and presented laboratory evidence of potential cyanosis with an HbC of 17 %. No methemoglobinemia developed but urine analysis amply confirmed the decision to avert a cyanosis episode by such therapy. Urine analysis again revealed significant excretion of nitro com-pounds three days later. The accompanying relapse (HbC 13 %) was thought to be due to the continued wearing of contaminated shoes and jumper after other soiled clothing had been discarded. Note the gradual recovery of hemoglobin from 10.9 mg/100 ml to near satisfactory during the ten days after methylene blue administration.

The restoration of oxyhemoglobin (HbO_2) following methylene blue is demon-strated in Figures 12 and 18. Figure 13 further illustrates the rapid increase in HbO_2

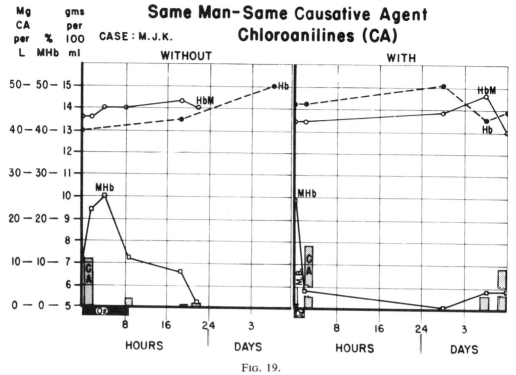

FIG. 19.

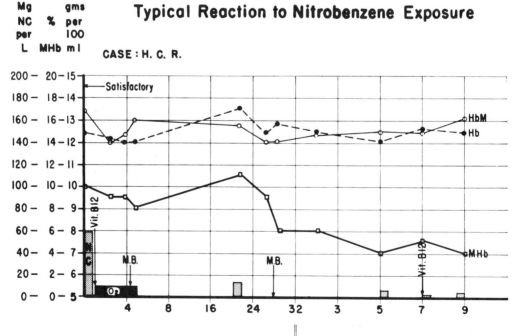

FIG. 20.

(from 7.1 to 14.1 g/100 ml) following 10 ml of 1 % methylene blue in treatment of a typical subcyanosis level case derived from para-nitrochlorobenzene (PNCB). This example illustrates the oxygen transport deficiency which can develop in the absence of true methemoglobinemia after exposure to some aromatic nitro compounds. Only by determination of HbO_2 can this type of alteration be revealed. Not only was the small but significant concentration of MHb reduced and the HbO_2 restored, but an improvement in total hemoglobin (Hb) level began immediately after methylene blue injection.

A random spot examination of employees reporting to the medical department from the major manufacturing areas disclosed several cases of elevated bilirubin levels. Examination of 14 cases disclosed 5 (36%) abnormals and 4 (29%) high-normal results.[27] The highest level reached, 4 times the upper limit of normal, appeared in the sequel of a severe PNCB exposure case (St–Table 5). In several instances (Ri, H, M, St), the abnormal bilirubin level was accompanied by low Hb concentration and incomplete oxygenation (So, Wr).

TABLE 5
Hemolytic Effects from Exposure to Aromatic Nitro and Amino Compounds

Case	MHb %	Hb g/100 ml	HbO$_2$ %	RBC ×10³	Bilirubin* mg/100 ml	HbM	HbC %
C	0	14.5	93	5.08	0.8	15.0	3.3
F	0	14.0	92	4.95	1.3	15.0	6.7
Q	0	16.0	91	5.70	0.8	15.4	3.8
WL	0	15.0	97	5.08	0.4	15.0	0
Sa	0	14.5	90	5.20	0.6	14.4	0
So	0	14.5	81	5.08	0.8	16.0	9.4
Ri	2	13.5	97	4.45	0.9	13.9	3.0
M	2	13.0	93	4.50	0.8	13.4	3.1
H	0	13.5	92	4.80	1.1	13.4	0
O	0	14.0	94	4.80	0.4	13.0	7.1
Re	0	12.5	88	4.45	0.5	13.0	3.9
Wr	0	15.0	86	4.90	2.7	14.4	4.0
Sm	2	13.5	95	4.90	0.2	13.6	0
St	19	13.5	—	4.75	3.1	14.4	6.3

*Normal serum bilirubin = 0.2–0.8 mg/100 ml (Ref. 27).

CYANOSIS POTENTIAL VERSUS CYANOGENIC AGENT STRUCTURE

Although the cyanogenic potential (cp) of nitrobenzene and aniline are both profoundly altered by the character, number and position of substituents in the benzene ring, physiological action has not been well documented. Contact with the substitution products has shown that dinitrobenzene, nitroanilines, chloronitrobenzenes and the chloroanilines possess considerably greater cyanosis potentials than would be expected from the activity of the parent compounds. The record has established the chloroanilines as particularly virulent cyanosis-producing agents.

Experience has shown that aniline derivatives produce cyanosis much more rapidly than do nitro compounds. However, the delay, or induction period, after exposure to some nitro compounds, such as meta-dinitrobenzene, and the prolonged recovery period present a more insidious hazard. Recovery from amine-type cyanosis is relatively rapid, usually a matter of hours to possibly a day or two, whereas the effects of nitro compounds persisted for a week or more in some cases.

A more detailed study disclosed a polarity factor as well as position effect exerted by the substituent group. As the polarity increases, cp decreases. Introduction of the hydroxyl (—OH) group significantly reduces cyanogenic potential, and is the mechanism by which the metabolic defense mechanism is able to alleviate the toxic stress (Figure 4). The presence of a carboxyl (—COOH) radical further reduces cp to a level that no longer presents a toxic threat. Sulfonation (—SO$_3$H) of the aromatic ring almost completely destroys physiological activity.

The solubility partition coefficient between lipoid components and water to a great degree determines the physiological effects from intact skin exposure. When the water solubility (hydrophilic effect) increases, the cp decreases. This behavior is in part due to a reduction in penetration rate through the skin, which preferentially absorbs fat soluble (hydrophobic) compounds. As an example, aniline penetrates rapidly, whereas the hydrophilic meta-phenylenediamine (MPD) is very poorly absorbed.[5] Transfer across the erythrocyte cell membrane also will be governed by lipoid solubility and polar strength. The persistence of low grade cyanosis and anemia after exposure to DNB and PNCB may be aggravated by reservoirs of nitro compound, which accumulate in adipose tissue and slowly diffuse back into the blood stream.

Transport across pulmonary membranes appears to be less selective, since the water soluble, polar compounds such as para-nitrophenol seem to be readily absorbed through the lungs. However, from an industrial health control standpoint, only a few derivatives present severe dusting problems (except nitroanilines) or sufficiently high vapor pressures (except aniline and nitrobenzene) to create inhalation problems.

Excretion rates through the kidneys are dependent on polarity and water solubility. The metabolic pathways not only introduce a hydroxyl group, but also esterify to further increase excretion potential. Most of the metabolites are excreted as sulfates, glucuronides, phosphates or mercapturic acids. Some unaltered aromatic amine or nitro compound is usually found in the urine from cyanosis cases. The aromatic amines are almost completely (over 90%) acetylated. Therefore, urine analysis must include a hydrolysis step. Aromatic nitro compounds are to varying degrees, depending upon the presence and position of other substituent groups, reduced to the corresponding aniline derivative. Partial reduction products from dinitro compounds can be detected by chromatographic separation on filter paper of the azo dyes obtained in acid hydrolyzed urine by coupling the diazotized amines to Chicago acid (8-amino-1-naphthol-5,7-disulfonic acid).

Hydroxylation occurs in the aromatic ring of both nitro and amino compounds at the 2 (ortho) and 4 (para) positions. However, under conditions employed for urine analysis, only the para-isomers are detected, since the ortho-aminophenols react with nitrous acid to produce a heterocyclic ring system (benzo-1, 2, 3-oxadiazole derivative), which does not couple to produce an azo dye.

Substituents in the aromatic ring not only determine methemoglobinemia severity, but also influence post-cyanosis recovery time and severity of hemoglobin loss. Data collected from twelve compounds which exhibited cyanogenic activity are summarized in Table 2, Cyanogenic Activity Versus Structure, and Figure 21, Causative Agent Potency.

Cyanogenic and anemiagenic potential series were developed from the sum of the rankings of each physiological factor taken in reverse order; that is, the worst offender was ranked 1. This sum was then divided by the number of factors for which information was available and these results again arranged in numerical ascending

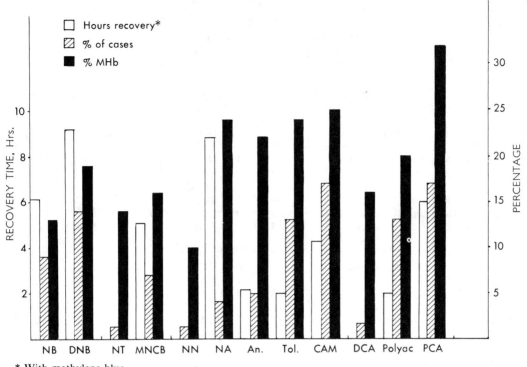

* With methylene blue

FIG. 21. Causative Agent Potency.

order which placed the most active agent in number one position. The factors evaluated were:

Factor	Calculation or Determinant	First Place Criteria
1. Frequency	% of total cases (187)	Max.
2. Severity	Average % MHb	Max.
3. Relapse rate	% of total cases (15)	Max.
4. Recovery rate	In hours—average	Longest
a. Without MB		
b. With MB		
5. Urinary excretion	mg/1 ether extractable at 10% MHb level	Min.
6. Frequency of severe cases	a. Over 50% MHb (3)	Max.
	b. 30–50% MHb (31)	Max.
7. Hb loss	Average %	Max.
8. Hb recovery	Time to regain loss—average	Max.
9. HbC	Average %	Max.
10. Time to restore Hb loss	To normal—average hours	Max.

Over-all Rank = Sum of factor rankings ÷ No. of factors in the sum
Cyanogenic Rank = Sum of rankings in Factors 1 through 6b ÷ No. of factors in the sum
Anemiagenic Rank = Sum of rankings in 7+8
Example: DNB = (2+6+2+1+2+4+2+7+4+7)÷10 = 3.7
Rank = 2 (Table 6)

The results from this study are summarized in Table 6.

As indicated in the Recovery section (Figure 13), para-nitrochlorobenzene (PNCB) is an especially insidious cyanogenic agent in that only a fraction of the nonoxygenatable Hb appears as MHb, and recovery is markedly delayed. Furthermore the anemiagenic potential is quite high (ranked No. 2). Examination of 10

TABLE 6

The Chemical Cyanosis-Anemia Syndrome

Relationship between Causative Agent Structure and Biochemical Potential (Ranked in descending order of relative hazard: No. 1 most, No. 13 least potent).

Rank	Cyanogenic Potential			Anemiagenic Potential			Over-all Potential		
	Initials	Product Name	Score	Rank	Initials	Score	Rank	Initials	Score
1	OCA	*ortho*-chloroaniline	—	1	NB	2.0	1	DNB	3.7
	MCA	*meta-* ,,	—	2	MNCB, PNCB	4.0	2	NB	3.9
	PCA	*para-* ,,	—	3	MNT	4.5	3	CAM	4.2
	CAM	mixed- ,,	2.3	4	PT	5.0	4	PT	4.6
2	DNB	dinitrobenzene	2.7	5	DNB	5.5	5	MNCB, PNCB	4.9
3	MNA	*meta*-nitroaniline	—	6	DCA	6.5	6	NA	5.5
	PNA	*para-* ,,	—	7	NA	6.5	7	MNT	5.7
	NA	nitroanilines	4.4	8	OT	7.5	8	MT	6.0
4	PT	*para*-toluidine	4.6	9	CAM	8.5	9	OT	6.3
5	NB	nitrobenzene	4.7	10	AN	9.5	10	AN	6.7
6	MT	*meta*-toluidine	4.7	11	Polyac	9.5	11	Polyac	6.8
7	ONCB	*ortho*-nitrochlorobenzene		12	MT	10.0	12	DCA	6.8
	PNCB	*para*-nitrochlorobenzene		13	NN	—	13	NN	7.0
	MNCB	mixed-nitrochlorobenzene	5.3						
8	AN	aniline	5.4						
9	—	"Polyac" = *para*-dinitrosobenzene	5.5						
10	OT	*ortho*-toluidine	5.8						
11	ONT	*ortho*-nitrotoluene	—						
	PNT	*para*-nitrotoluene	—						
	MNT	mixed- ,,	—						
	DNT	dinitrotoluenes	6.1						
12	NN	nitronaphthalene	6.7						
13	DCA	dichloroaniline (2,5 or 3,4)	7.4						

cases (Table 7) disclosed an average MHb of 16%, but total "inert" Hb calculated from Hb and HbO$_2$ differences averaged 42%. Severity of Hb losses is confirmed by an average Hb level of 12.5 g/100 ml and in one case (St) the serum bilirubin level reached 3.1 mg/100 ml (versus 0.8 mg/100 ml upper normal range) 8 days post-cyanosis (first peak).

Other observations noted:

1. In cases originating with *para*-toluidine exposure, HbC increases as MHb declines after methylene blue treatment.

2. Sharp Hb declines frequently set in as MHb declines after DNB exposure.

3. During episodes produced by PNCB or chloroanilines, Hb is usually depressed below the HbM level (Figures 12, 13).

Although insufficient data were available to assess the relative toxicities of the chloroaniline isomers directly from our human experience, work carried out by others with rats and guinea pigs with the acetylated derivatives indicated a decreasing order of toxicity: para—most, meta—intermediate, and ortho—least.[42]

TABLE 7
Cyanogenic and Anemiagenic Potential of PNCB

Case	Hb	HbO$_2$	% MHb	% Inert Hb	Urine mg/l	Days to Recover	No. of Relapses
K	13.2	7.1	11	46	24	14	1
McD	12.0	7.1	12	41	234	17	1
M	13.8	9.1	7	34	36	9	1
W	(15.9)	—	12	—	21	3	0
P	12.3	—	10	(13)*	30	0.1	1
	12.0	—	23	(14)*	21	4	0
McL	13.0	7.8	12	45	16	2+	0
Sc	12.3	—	38	(16)*	1170	0.1	1
C	12.0	—	15	(18)*	14	0.1	1
St	12.0	—	24	(21)*	74	14+	6

*Hb loss.

SUSCEPTIBILITY

Examination of the case histories for evidence of cyanosis susceptibility disclosed 143 employees contributed the 187 episodes classified by laboratory analysis. Thirty (21%) of this group were cyanosed 74 times (40% of the cases). Eight were classified as "chronic repeaters" who incurred three or more episodes each for a total of 30 cases (16%). Two of these men were cyanosed five times each (Table 8). These eight men have been removed permanently from areas of potential exposure to aromatic nitro and amino compounds.

In the major aromatic nitro and amino compound production area in which an average of 430 men were employed as operators and mechanics, 99 individuals or 23% of the crew were cyanosed once or more during this 10-year study. A breakdown of 133 episodes revealed that 54 or 41% of them occurred in 20 men or 5% of the crew.

Work published recently by others has indicated that the source of methemoglobinemia resides in the interference created by the AC and NC metabolites in the enzyme system responsible for maintaining the Hb iron in the ferrous state by the glucose oxidation pathway.[43-48] In those individuals whose erythrocytes have a diminished glucose phosphate dehydrogenase activity and glutathione levels, acute drug-induced hemolysis from absorption of aromatic amines and nitro compounds (nitrofurans, sulfanylamides, phenacetin, etc.) as well as methemoglobinemia has been observed. This mechanism may account for some relatively severe postcyanosis anemia episodes.[48] A simplified MHb reduction test has been proposed as a screening test to detect these individuals.[49]

Chemical cyanosis from aromatic nitrogen metabolites produced from the invasion of the blood stream by aromatic nitro and amino compounds appears to involve at least two pathways: (1) conjugation with the hemoglobin iron[7] and (2) interference with enzyme systems that maintain iron in the reduced (ferrous) state. In oxygenated blood there is a continuous tendency for the ferrous iron in the hemoglobin molecule to be oxidized to the ferric state, thus producing MHb. Fortunately an enzyme system exists in normal individuals that reduces MHb as rapidly as it is formed back to normal Hb. This is mediated by the enzyme, MHb reductase, which requires the coenzyme reduced diphosphopyridine nucleotide (DPNH), derived from anaerobic red cell glycolysis (Emden-Myerhof pathway)[43] (See Figure 22).

TABLE 8
Summary of Repeat Cases
(Individuals with Two or More Cyanosis Episodes)

	Men		Cases	
	Number	%	Number	%
Total	143	100	187	100
Repeaters—Total	30	21	74	40
Chronic Repeaters	4	—	3	—
	2	—	4	—
	2	—	5	—
Chronic—Total	8	6	30	16

In the presence of redox chemicals another coenzyme is involved, reduced triphosphopyridine nucleotide (TPNH), derived from another glycolytic pathway (hexosemonophosphate, HMP, or pentose pathway). DPNH and TPNH act as

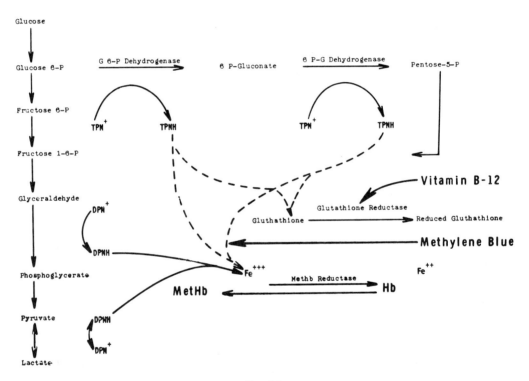

PATHWAYS OF CARBOHYDRATE METABOLISM
AND ENZYMATIC METHEMOGLOBIN REDUCTION

FIG. 22.

hydrogen ion donors for the reduction of MHb. When these systems are inactivated, the rate of MHb reduction slows and methemoglobinemia results. Reduced gluta-thione (GSH) normally present in the cell protects hemoglobin and sulfhydryl-containing enzymes against oxidative destruction, but this too becomes deficient when TPNH is depleted.

When metabolic processes in the red cell diminish to levels at which vital functions can no longer be carried out, degradation of cell protein may occur with the formation of Heinz bodies, and the lipoprotein of the cell membrane may become altered with resulting hemolysis. MHb formation is not necessarily a prelude to hemolysis, but the two conditions are frequently associated because of the concurrent effects of the etiological agent.

As to the mechanism of action of methylene blue, it is known that methylene blue produces marked acceleration of the pentose pathway by serving as an artificial electron donor working along with TPNH and MHb reductase to reduce ferric iron to the ferrous state, thus reducing MHb to normal hemoglobin.[46] The significance of the apparent augmentation of methemoglobin reduction by vitamin B-12 is unknown. Some clue to this action is found in animal experiments, which have shown that vitamin B-12 is involved in maintaining the activity of glutathione reductase, and as previously stated, reduced glutathione is necessary to prevent oxidation of sulfhydryl enzymes.[50]

CONTROL AND PREVENTION

An Industrial Hygiene program designed to control exposure to cyanogenic and anemiagenic aromatic nitro and amino compounds should include, in increasing order of importance:

1. Laboratory facilities to provide the blood and urine analyses required for early detection of exposure before physiological damage has occurred and for evaluation of the effectiveness of the program, i.e., quality control.
2. Recognition of potential cyanosis and blood level declines before methemoglobinemia or anemia develops.
3. Prompt removal from further exposure of those individuals who exhibit potential cyanosis, are definitely cyanosed, are afflicted with Hb levels depressed below 12.0 g/100 ml, and the "repeaters" who have demonstrated susceptibility to cyanosis or anemia.
4. Prevention of cyanosis and anemia by effective exposure control.

In the following discussion, a blood specimen was considered abnormal if:

1. Methemoglobin (MHb) or "hemoglobin complexes" (HbC) exceeded 9%.
2. Laboratory results indicate potential cyanosis.
3. The Hb value by either method dropped below 13.0 g/100 ml (caution limit).

Application of these two chemically different procedures for the determination of hemoglobin, acid hematin (Hb) and total oxidation methods (HbM), disclosed results significantly different (increased HbC) when applied to chemical exposure cases.

Experience gained from the application of these procedures to an exposure control program indicated these deviations provided a very sensitive indicator for the absorption of a variety of chemicals below levels that produced detectable physiological damage. Urine specimens were classified abnormal as defined under Diagnosis.

Operators and engineers assigned to areas where potential exposure to cyanogenic chemicals existed are examined routinely on a bimonthly schedule. These examinations include history taking, medical observation, blood pressure, pulse, cyanosis control, blood and urine analysis for nitro and amino compounds. All new transferees into such a control area are placed on a probationary "new man" program for sixty days to detect susceptible individuals. A new man receives a pre-employment or pretransfer examination, which is repeated at the end of three days, one week, three weeks, and the sixty-day control period.

Additional specimens are collected whenever unusually hazardous conditions or known exposure episodes have been encountered. All cyanosis and potential cyanosis cases, abnormal blood cases, urine results in excess of 20 mg N.C. or A.C. per liter, and frequency rates for abnormal bloods or urines over 20% (based on analysis of at least five specimens from the work area) are immediately reported to the area involved for corrective action.

Incidents and trends should be discussed with representatives from the Medical, Industrial Hygiene, Operating, and Engineering Divisions on a monthly basis to coordinate the exposure control program. Process changes, new products, production increases and maintenance problems are reviewed in advance to permit the institution of proper industrial hygiene controls and alert medical personnel to the detection of possible health problems early enough to avoid injury potential.

Since routine bimonthly tests from any given unit operation do not furnish on a daily basis sufficient data for an evaluation of cyanosis exposure control, Hb results are examined on a weekly and monthly basis. The building supervision is advised when the trend of abnormal specimens reaches 20%. When the average Hb level for a work crew falls in the range 14.0 to 15.0 g/100 ml (14.5 ± 0.5 g/100 ml), control over chronic exposure is considered satisfactory. When the Hb average for an entire crew (five or more individuals) exceeds 14.9 g, then the frequency rate for specimens containing 13.0 to 14.0 g becomes significant and may be used as an early warning of mildly chronic exposure conditions long before absorption severity can reach a critical level.

Application of the improved laboratory procedures for MHb and HbC disclosed information that could be employed for the detection of cyanosis conditions well enough in advance to permit alleviation of exposure before methemoglobinemia developed.[4] A brief review of the results collected from routine examination of operators and mechanics employed on Chambers Works in unit operations involved in the manufacture of nitrobenzene and aniline derivatives and supporting environmental data will serve to illustrate the development of successful Industrial Hygiene Control procedures.

Acute Phase

Although cyanosis has been a major industrial health problem in Chambers Works for many years, no fatalities attributable to this cause have occurred. However, excessive exposure without prompt and adequate medical attention can be fatal.

Within a few months following the establishment of the improved laboratory methods, both acute exposure conditions which required immediate correction and chronic subcyanosis absorption conditions were disclosed. Both the increased incidence and severity of cyanosis and the origin of a four-year hemoglobin decline appeared to coincide with the start of chloroaniline production, increased usage of *para*-nitroaniline, and accelerated production of dinitrobenzene (Figures 23, 24 and 25).

With recognition of the potency of chloroanilines new to the production areas, came realization that control measures satisfactory for the toluidines, phenetidines, *alpha*-naphthylamine, nitrotoluidines and anisidines were not effective enough for chloroaniline manufacture.

An incident which occurred in the iron reduction unit clearly illustrates this new problem. A graph of the hemoglobin history of a maintenance crew working on a chloroaniline stripper gave the first hint of trouble on May 23—24 when 100% of the blood specimens were abnormal. On May 28, two mild cases of cyanosis developed. On May 31, production was shifted to orthotoluidine. Immediately the percentage of

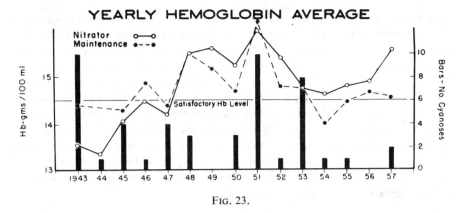

FIG. 23.

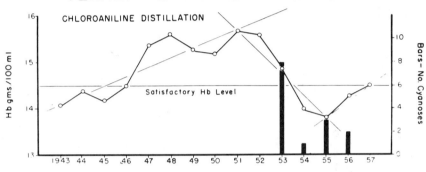

FIG. 24.

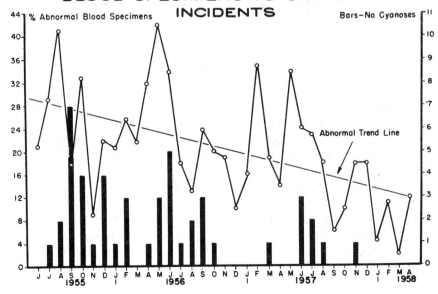

FIG. 25.

blood abnormals dropped below 15%. Since no precautions were taken to reduce exposure severity, this "break" indicated that *ortho*-toluidine is an aniline derivative possessing a relatively weak cyanosis potential. When chloroaniline production was started again June 12, blood abnormals rose to 60%. During the following day, a mild case of cyanosis developed; then on June 14, three severe cases of cyanosis were incurred. In each case, at least 24-hour warning was available from the high percentage of abnormal blood specimens collected from this crew. This experience furnished the basis for an exposure control system based upon a 20% upper limit for abnormal blood or urine specimens from a given area. Further experience confirmed this decision. A 20% control level would detect 70% of the cyanogenic conditions (Figure 26). If less than 12% of the specimens are abnormal, cyanosis would not be expected to develop in the group sampled.

ESTIMATED PROBABILITY
OF
CYANOSIS OCCURRENCE

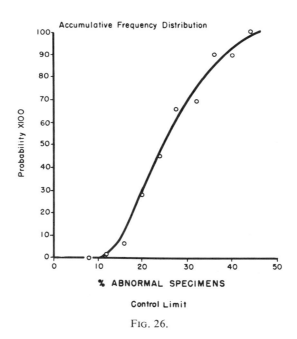

FIG. 26.

Study of the hemoglobin analyses compiled during 1955–1957 disclosed an approximate relationship between the percentage of abnormal specimens and the number of cyanosis cases for the entire production line on a monthly basis (Figure 25). This downward slope nearly parallels the decreasing cyanosis trend from the initiation of exposure control in mid-1955 through 1957.

Since heat has been recognized for some time as a major factor in the development of methemoglobinemia, the cyclic nature of the monthly cyanosis incidents within annual groups was examined for ambient temperature effects (Figure 25). A linear relationship, within surprisingly close tolerances between cyanosis frequency and ambient temperature, was derived from the data (Figure 27). At 30–35°, an apparent anomaly was found. However, consideration of the probable circumstances led to the conclusion that under these conditions, work would be carried out in heated buildings where there would be little thermal relationship to outside ambient

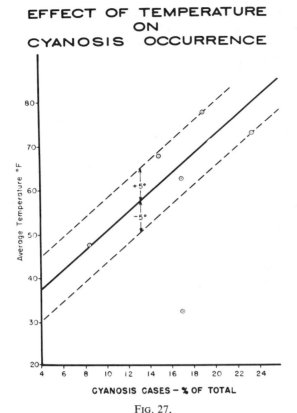

EFFECT OF TEMPERATURE
ON
CYANOSIS OCCURRENCE

FIG. 27.

temperature. Following a vertical from the anomalous point to the curve, an intersection in the 65–70°F interval, the expected normal winter indoor temperature, was obtained.

Temperature is undoubtedly one of the major influences responsible for the failure of severe exposures to aromatic nitrocompounds detected by urine analysis to produce HbC and MHb during winter months. Cyanosis from lesser exposures often develops rapidly under conditions of elevated ambient temperature in drier rooms, near steam stills, after hot showers at shift end, etc.[4] This temperature effect may be responsible for the "false positives" encountered in applying the control limits indicated by the probability curve.

Chronic Phase

Chemical cyanosis is usually a short term or acute condition which demands prompt action. With the help of relatively simple remedies, recovery is prompt on removal of exposure and seldom involves adverse aftereffects. However, repeated or chronic "subcyanotic" exposures over long periods of time lead to more serious damage to the blood-producing tissues and could produce prolonged hemoglobin deficiency or anemia. The period of progressive degeneration varies with individual susceptibility from a few months to years.

Plots of the hemoglobin (Hb) values, averaged on an annual basis for each of the unit operations and the maintenance crew, disclosed a steady rise from a low during the war years (1943–1946) to a peak level in 1951 (Figures 23 and 24). This rise was due in part to reduced tension following World War II, improved living habits, improved equipment maintenance, better process control, and the addition of new employees—literally introduction of fresh blood. In 1951, the production of chloro-anilines on a large scale was started. The depressing effect on Hb levels was apparent within two years, not only in the reduction operations, but also in all of the supporting units engaged in the manufacture and purification of nitrochlorobenzenes required for reduction.

The reaction of the chloroaniline distillation group, which had no direct contact with aromatic nitrogen compounds prior to 1951, is especially noteworthy. The superposed trend line not only illustrates the post-1951 Hb decline, but also shows a definite recovery trend after 1955 when exposure control measures became effective. The characteristic delay in Hb recovery after bringing cyanosis under control also is illustrated by the parallelism between the 1951–1955 declines for both cyanosis incidence and Hb concentration. However, in every case, the average Hb levels had returned to the satisfactory median, 14.5 g/100 ml, by 1957, largely as a result of the rigorous control measures developed and applied by the operating, engineering, and medical supervision.

The application of blood analysis to the solution of an exposure problem by evaluation of severity at subcyanosis levels is illustrated in Figure 28.

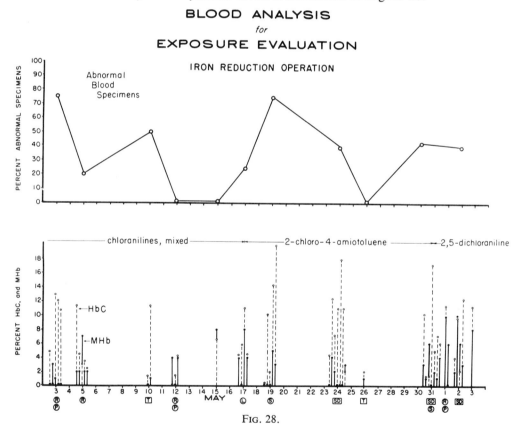

FIG. 28.

A plot of the MHb, HbC and % abnormals at frequent intervals for each operator on day shift furnished the following information:

1. Those pieces of equipment, or routine operations that repeatedly produce excessive exposure. This pinpointing confirmed the major sources of product loss for effective expenditure of maintenance, process revision, and design improvement funds to alleviate the unsatisfactory working environment, and conserve materials.

2. Evaluation of work habits of either individuals or shift crews.

3. Detection of individuals unusually susceptible to hemoglobin damage.

4. Estimation of the relative cyanosis-producing potentials of aniline derivatives. Chloroanilines, and 2-chloro-4-aminotoluene were approximately equivalent, whereas 2,5-dichloroaniline appeared to be somewhat less potent.

5. Air analysis (either nitro or amino compound), and blood analysis did not correlate within limits which justify use in exposure evaluation. These fat-soluble compounds are absorbed mostly by direct skin contact with contaminated surfaces, and steam-borne mists. Again, confirmation of the Industrial Hygiene principle "Man himself is the best sampler of his environment."

Prevention

Cyanosis can be prevented. Careful job analysis should be made to insure proper handling procedures, adequate equipment design for both operating and maintenance, and ventilation. Where appropriate, air analysis for high vapor pressure derivatives can be helpful, but, in general, results have been misleading more often than helpful because of low vapor pressure and absorption in bricks and other porous materials. However, the seriousness of mist from hot charges, leaking lines, steaming operations, etc., should not be underestimated as a source of gross skin exposure and contamination of work areas.

Field inspections of operations by the industrial physician and/or the industrial hygiene supervisor are of value in obtaining firsthand knowledge of the individual's work and his work environment. Specific problems, requiring an answer on the basis of preventive medicine or hygiene, can best be met by an on-site inspection.

Reduction of exposure can be attained by five practical procedures, arranged in order of increasing effectiveness:

1. Respiratory protection. Canister masks, air-supplied masks, and dust respirators alone have a very limited utility, since alleviation of skin absorption is the major problem. The hands and feet are responsible for about 50 % of skin absorption of cyanopathic materials.

2. Rotation of the members of the work crew. This example of minimal abnormal hemoglobin control limit required numerous daily blood analyses to determine which men were to be removed to areas of less exposure, and could be applied only to the construction department, where a variety of jobs was available.

3. Limited exposure duration. Limitation of working time in contaminated areas was practical only for maintenance crews and operators involved in unusual job incidents.

4. Use of butyl rubber protective equipment, such as gloves, overshoes, sleeves, and aprons consistently on a routine basis can reduce exposure within acceptable limits for all but the unusual incidents. No other fabricated commercially available elastomer approaches butyl rubber in resistance to penetration by aromatic nitro and amino compounds.

5. Complete body protection for severe exposure conditions. Application of the "Chem-Proof" Air Suit, an air-conditioned garment impervious to the neck line and provided with air-supplied helmet and cape which integrates head and body coverage as a single unit, has eliminated cyanosis at locations such as the chloroaniline distillation unit.

In case of skin contact through spills, splashes, leaks, equipment failure, and other accidents, prompt removal of clothing and drenching under a safety shower can avert cyanosis. Personal hygiene and a warm shower with plenty of soap vigorously applied at shift end will go a long way toward minimizing chronic exposure. All exposures—major, minor, or suspected—must receive immediate medical attention. Cyanosis control may be summed up: "If a man was exposed, product must have been lost."

The entire cyanosis-anemia syndrome has been summarized in a graphical flow chart form (Figure 29) to illustrate the complex interrelationships involved in the cyanosis cycle. Effective exposure control will prevent activation of this cycle at its source and protect those employees who for one reason or another may be unusually susceptible.

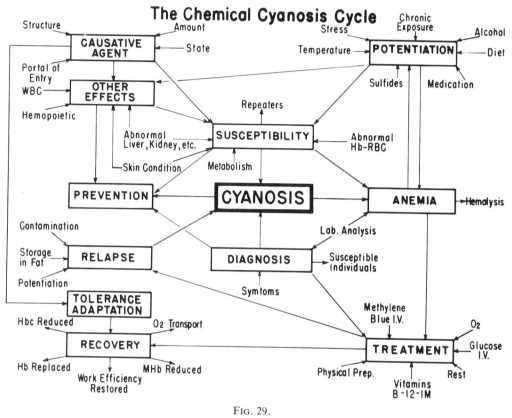

FIG. 29.

ACKNOWLEDGEMENT: Portions of this chapter are reprinted with permission from *Archives of Environmental Health*, vol. 1, p. 353 (1960); vol. 3, p. 165 (1961); and vol. 9, p. 478 (1964); *Archives of Industrial Health*, vol. 18, p. 422 (1958); and *American Industrial Hygiene Association Journal*, vol. 20, p. 396 (1959).

REFERENCES

1. Von Oettingen, W. F., The Aromatic Amino and Nitro Compounds, Their Toxicity and Potential Dangers: A Review of the Literature, *Public Health Bulletin No. 271*, U. S. Public Health Service (1941).
2. Wuertz, R. L., Frazee, W. H., Hume, W. G., Linch, A. L., and Wetherhold, J. M., Chemical Cyanosis-Anemia Syndrome, *Arch. Environ. Health*, 9, 478 (1964).
3. Evans, E. E., Charsha, R. C., and Linch, A. L., Evaluation of Chemical Cyanosis Through Improved Techniques for Hemoglobin Analysis, *A. M. A. Arch. Ind. Health*, 18, 422 (1958).
4. Wetherhold, J. M., Linch, A. L., and Charsha, R. C., Hemoglobin Analysis for Aromatic Nitro and Amino Compound Exposure Control, *A. I. H. A. J.*, 20, 396 (1959).
5. Piotrowski, J., Quantitative Estimation of Aniline Absorption Through the Skin in Man, *J. Hyg. Epid., Microbiol. Immun.*, 1, 23 (1957).
6. Bodansky, O., Methemoglobinemia and Methemoglobin Producing Compounds, *Pharmacol. Rev.*, 3, 144 (1951).

7. Jackson, H., and Thompson, R., The Reaction of Hemoglobin and Some of its Derivatives with *p*-Iodophenylhydroxylamine and *p*-Iodonitrosobenzene, *Biochem. J.*, *57*, 619 (1954).

8. Gambino, R. S., Goldberg, H. E., and Polanyi, M. L., Method for Determination of Oxygen Saturation by Means of Reflection Spectrophotometry, *Amer. J. Clin. Pathol.*, *42*, 364 (1964).

9. Sroka, K. H., Treatment of Industrial Diseases Due to Nitro and Amino Compounds of Benzene and its Homologs, *Seifen Oel und Fett Industrie*, *78*, 101 (1952).

10. Doctor, L., Classification and Treatment of Methemoglobinemia, *Quart. Bull. Northwest Univ. Med. School*, *27*, 134 (1953).

11. Mangelsdorff, A. F., Methemoglobinemia—Recognition, Treatment and Prevention, *Ind. Med.*, *21*, 395 (1952).

12. Mangelsdorff, A. F., Treatment of Methemoglobinemia, *A.M.A. Arch. Ind. Health*, *14*, 148 (1956).

13. Finch, C. A., Treatment of Intracellular Methemoglobinemia, *Bull. New England Med. Center*, *9*, 241 (1947).

14. Carnrick, M., Polis, B. D., and Klein, T., Methemoglobinemia: Treatment with Ascorbic Acid, *Arch. Intern. Med.* *78*, 296 (1946).

14a. Hamblin, D. O., Aromatic Nitro and Amino Compounds, Chapter 47, "Industrial Hygiene and Toxicology," 2nd ed., F. A. Patty, ed., Interscience Publishers, John Wiley & Sons, Inc., New York.

15. Evelyn, K. A., and Malloy, H. T., Microdetermination of Oxyhemoglobin, Methemoglobin and Sulfhemoglobin in a Single Sample of Blood, *J. Biol. Chem.*, *126*, 655 (1938).

16. Barer, R., Spectrophotometry of Clarified Cell Suspensions, *Science*, *121*, 709 (1955).

17. Hamblin, D. O., and Mangelsdorff, A. F., Methemoglobin and its Measurement, *Ind. Hyg. Toxicol.*, *20*, 523 (1938).

18. West, B. A., Determination of Hemoglobin and the Red Cell Count in the Same Sample of Blood with a Photoelectric Colorimeter, *Am. J. Med. Tech.*, *16*, No. 3 (1950).

19. Hunter, F. T., "The Quantitation of Mixtures and Hemoglobin Derivatives by Photoelectric Spectrophotometry," Charles C. Thomas, Springfield, Ill. (1951).

20. Cannon, R. K., Proposal for the Distribution of a Certified Standard for Use in Hemoglobinometry, National Research Council, Div. of Med. Sciences, *Science*, *122*, 59 (1955); *Amer. J. Clin. Pathol*, *30*, 211 (1958).

21. McInroy, R. A., Micro-Hematocrit for Determining Packed Cell Volume and Hemoglobin Concentration on Capillary Blood, *J. Clin. Pathol.*, *7*, 32 (1954).

22. Rostorfer, H. H., and Cormier, M. J., The Formation of Hydrogen Peroxide in the Reaction of Oxyhemoglobin with Methemoglobin-Forming Agents, *Arch. Biochem.*, *71*, 235 (1957).

23. Van Slyke, D. D., Miller, A., Weisiger, J. R., and Cruz, W. O., Determination of Carbon Monoxide in Blood and of Total and Active Hemoglobin by Carbon Monoxide Capacity, *J. Biol. Chem.*, *166*, 121 (1946).

24. Best, C. H., and Taylor, N. B., "Physiological Basis of Medical Practice," 4th ed., p. 49, Williams and Williams Co., Baltimore, Md. (1945).

25. Fricke, H., and Curtis, H. J., The Electrical Impedance of Hemolyzed Suspensions of Mammalian Erythrocytes, *J. Gen. Physiol.*, *18*, 821 (1935).

26. Roughton, F. J. W., and Scholander, P. F., Micro Gasometric Estimation of Blood Gases, *J. Biol. Chem.*, *148*, 541 (1943).

27. Wintrobe, M. M., "Clinical Hematology," 5th ed., Lea and Febiger, Philadelphia, Penna. (1961).

28. Kolmer, J. A., Spaulding, E. H., and Robinson, H. W., "Approved Laboratory Technic," Appleton-Century-Crofts, Inc., New York (1951).

29. Flood, F. T., Mandel, E. E., Owings, R. H., and Federspiel, C. F., Newer Standards in Hemoglobinometry, *J. Lab. Clin. Med.*, *43*, 897 (1954).

30. Wetherhold, J. M., Linch, A. L., Charsha, R. C., and Harris, H. C., Industrial Chemical Cyanosis Control—A Statistical Comparison of Results from Five Hemoglobin Procedures, *Arch. Environ. Health*, *3*, 165 (1961).

31. Wetherhold, J. M., Linch, A. L., and Charsha, R. C., Chemical Cyanosis: Causes, Effects and Prevention, *Arch. Environ. Health*, *1*, 353 (1960).

32. Koniecki, W. B., and Linch, A. L., Determination of Aromatic Nitro Compounds, *Anal. Chem.*, *30*, 1134 (1958).

33. Halsted, H. C., Industrial Methemoglobinemia, *J. Occup. Med.*, *2*, 591 (1960).

34. Finch, C. A., Methemoglobinemia and Sulfhemoglobinemia, *New England J. Med.*, *239*, 470 (1948).

35. Fleming, A. J., D'Alonzo, C. A., and Zapp, J. A., "Modern Occupational Medicine," 2nd ed., p. 406, Lea and Febiger, Philadelphia, Penna. (1960).

36. Williams, J. R., and Challis, F. E., Methylene Blue as Antidote for Aniline Dye Poisoning, *J. Lab. Clin. Med.*, *19*, 166 (1933).

37. Steele, C. W., and Spink, W. W., Methylene Blue in Treatment of Poisonings Associated with Methemoglobinemia, *New England J. Med.*, *208*, 1152 (1933).

38. Bodansky, O., and Gutman, H., Treatment of Methemoglobinemia, *J. Pharm. Exp. Therap.*, *90*, 46 (1947).

39. Walterskirchen, L., Oil of Mirbane Poisoning, *Wein. Klin. Wochenschr.*, *52*, 317 (1939).

40. Nadler, J. E., Green, H., and Rosenbaum, A., Intravenous Injection of Methylene Blue in Man with Reference to its Toxic Symptoms and Effect on Electrocardiogram, *Amer. J. Med. Sci.*, *188*, 15 (1934).

41. Lubash, G. D., Phillips, R. E., and Shields, J. D., Acute Aniline Poisoning Treated by Hemodialysis, *Arch. Ind. Med.*, *114*, 530 (1964).

42. Argus, M. F., Gryder, J. M., Seepe, T. L., Caldes, G., and Ray, F. E., Excretion and Distribution Studies with *o*-, *m*-, and *p*-Chloroacetanilide-Cl[36] in Rats and Guinea Pigs, *J. Amer. Pharm. Assoc.*, *47*, 516 (1958).

43. Huennekens, F. M., et al., Erythrocyte Metabolism: IV. Isolation and Properties of Methemoglobin Reductase, *J. Biol. Chem.*, *227*, 261 (1957).

44. Mahler, H. R., and Row, I., Methemoglobin Reduction through Cytochrome B-5, *Nature (London)*, *184*, 1651 (1959).

45. Stromme, J. H., and Eldjam, F., Role of Pentose Phosphate Pathway in Reduction of Methemoglobin in Human Erythrocytes, *Biochem. J.*, *84*, 406.

46. Boszormenyi-Nagy, I., et al., Effect of Methylene Blue on Metabolism of Adenine Nucleotides in Human Erythrocytes, *Arch. Biochem.*, *65*, 580 (1956).

47. Bockris, L., and Smith, R. S., Effect of Methylene Blue on Red Cell Glutathione, *Nature (London)*, *196*, 278 (1962).

48. Kellermeyer, R. W., Tarlov, A. R., and Brewer, G. J., Hemolytic Effect of Therapeutic Drugs, *J.A.M.A.*, *180*, 388 (1962).

49. Brewer, G. J., and Tarlov, A. R., Methemoglobin Reduction Test for Primaquine-Type Sensitivity of Erythrocytes, *J.A.M.A.*, *180*, 386 (1962). Dade Reagents, Inc., Miami, Florida; Cat. No. BS320 G-6-PD Reagent Kit.

50. Biswas, D. K., and Johnson, B. C., Glutathione Reductase and Dehydrogenase Activity in Vitamin B-12 Deficiency, *Arch. Biochem. and Biophysics*, *104*, 375 (1964).

Section 7

Radiation Hazards

Laser Laboratory Design and Personnel Protection from High Energy Lasers

Leon Goldman, R. James Rockwell, Jr., and Peter Hornby

INTRODUCTION

Because of the increasing importance of all phases of laser research, it is becoming necessary to develop laboratories which incorporate specific features for protection of personnel from laser radiation. In an attempt to develop a laser laboratory which not only accommodates existing laser systems but also has facilities for future laser developments, one must consider what safety features are necessary for lasers presently being used, as well as those now being developed.

The development of high-energy laser systems increases the problems of personnel protection. Although the long-term effects of even low-energy laser reactions in living tissue are not as yet known, many research laboratories now have high-energy laser systems. We have considered as high-energy lasers those with exit energies of 50 joules or more at wavelengths* of 4880 A to 5145 A, 10,600 A, and 106,000 A. Systems have been developed with outputs in the thousands of joules and with energy densities in the 100,000 joules/cm^2 range.

Increasing developments with CW lasers have produced increased outputs of high power. The argon CW laser (4880 A to 5145 A) now has outputs of 50 watts and the carbon dioxide laser (106,000 A) now goes over 60 kilowatts. Lasers operated in the Q-switched mode can now generate outputs as high as gigawatts. Ultra-short pulses in picoseconds are now available. With the generation of second and fourth harmonics, laser generation in the ultraviolet range is now possible. Also, pulsed and continuous wave (CW) ultraviolet lasers are available now. With such significant energy and power outputs, it is obvious that personnel working with lasers must be protected. In addition, new liquid or chemical lasers containing toxic materials may be a hazard in the laboratory. The development of divisions of laser chemistry also presents problems of safety for the laboratory. There is also now laser-triggered thermonuclear fusion, which presents hazards of radiation.

In brief, the protection program centers about the following laboratory design and operational features:

1. Personnel control
   a. Eye
   b. Exposed skin
   c. Inhalation
2. Area control
   a. Avoidance of specular reflectance
   b. Protection from laser plume and air contamination
   c. Proper ventilation
   d. Avoidance of electrical shock

It is not always possible to rigidly separate personnel control and area control, but the distinction is made to emphasize the need for the development of special laser laboratories, rather than the use of laser equipment in any available laboratory space. The construction of such special laboratories with their many safety precautions has been detailed before.[1,2]

*Wavelengths are given in Angstrom units, Å. One micron equals 10,000 Å.

381

Normally in a research laboratory, laser systems cannot be enclosed because of the continual need for changing optical arrangements and making adjustments during test programs. Thus, to protect the personnel from direct laser beam impacts, a warning system must be incorporated which warns of the charging of the laser power supply capacitors and allows the operating personnel sufficient time to leave the operating bench and turn away and cover their eyes before the actual firing. It has been the experience of the authors' research group that accidental premature discharge of the flash lamps can occur, particularly as the flash lamps become well used, so that a strict rule of operation should be "all work on the laser benches stops when warning lights appear." With CW lasers, doors to the laser areas should be locked.

EYE PROTECTION

The first concern in laser protection continues to be the eyes. Full protection of the eyes means not only from impact of the direct beam but also from the significant amount of reflection from surfaces. Eye protection is divided essentially into the following phases:
1. The design of laser systems to develop as much as possible the closed system technique.
2. The avoidance, as much as possible, of highly reflectant surfaces on and adjacent to the target areas.
3. Personnel protection in the form of protective devices for the eyes.
4. Constant reappraisal of all proposals for area and personnel protection.
5. Continued eye examinations and records for all operating personnel.

Continued eye examinations and good records are important especially to determine the long-term effects of exposure.

Since laser instrumentation production is still not assembly-line, investigative models will continue to be used, both in the research laboratory and in pilot experiments, even in the field of biomedical research. Many of these instruments will be open with the laser beam freely exposed on its course from the laser head to the target area. Area control means the development of dark, dull, nonreflecting surfaces in the laboratory and the use of extensive black, heavy felt drapes and the use of devices to attempt to focus on the target area and to observe the effects on the target area without direct observation of the target. These devices include reflectant focusing devices with diffusing screens and remote closed circuit television viewing. Light detectors placed about the laboratory can give some estimates of the hazards of reflected beams.

There is a large number of parameters involved in determining what the energy thresholds would be for causing retinal burns. Color of the retina, wavelength of incident light, degree of beam collimation, acceptance area of the eye (dark or light oriented), condition of the lens (i.e., how well it brings the laser beam to focus on the retina), energy density of the incident beam, and specific absorption characteristics of the eye other than the retina are but some of the variables involved. A generalized formula for determining the threshold energy for retinal damage which incorporates most of the parameters has been developed by Solon.[3] One can utilize this formula to develop threshold conditions for special cases. Rather than develop this formula, we will concentrate more on utilizing actual data collected at this laboratory on threshold energies and then determine under the worst conditions that could occur what quality of filter would be required to give adequate protection to the personnel.

Personnel protection for the eyes requires goggles which have sufficient protective material and which are so fitted that stray light cannot come in from any angle. This is especially important for high-energy laser work. This means the development essentially of a type of welder's goggles with an efficient filtering glass or the new protective plastics. We will recommend Schott BG-18 glass with an optical density of 8 for ruby, and 17.1 for neodymium, and neodymium YAG lasers.[4] It is difficult to make a general statement on what constitutes the maximum incident energy density permissible on the glass. However, we use the figure of 100 joules/cm^2 for our glasses, because at that value cracking and cratering become consistently evident. However, the glasses should be designed to give one-shot protection in the event of an accidental direct beam exposure. Swope and Koester[5] have recommended the use of a second filter with a lower absorption coefficient (Schott BG-38 glass) in front of the Schott BG-18 glass to present a high threshold for crazing and for breakage of the filter. Also, two BG-18 glass filters may be used together.

With the argon laser, eye protection is absolutely necessary. Protective glasses plus screens of amber plexiglass or ruby-tinted plastic are used for eye protection. In our work with argon gas lasers of 3 to 10 watts output, we have used amber-colored plexiglass, type-2442 sheets with an optical density of 2 to protect against the intense fluorescence developed at the target area in the tissue. These shields were recommended and are used by the argon gas laser research group at Bell Telephone Laboratories. Protection is also necessary against the ultraviolet radiation that is thrown off from the sides of the tube.

The carbon dioxide laser with its invisible beam is especially hazardous. The target area must be protected and hands, articles of clothing, and other objects should be kept out of the target area. Quartz glass, two 2-mm plates of fused quartz with an optical density over 35, will protect against the eye hazards of the invisible beam. For ultraviolet lasers, special ultraviolet protective glasses are required.

The helium-neon gas laser output at 6328 A is effectively filtered using the BG-18 glass. However, this and other gas lasers, together with junction diode lasers, generate many wavelengths in the infrared which because of their fairly low outputs do not cause any strong sensation when absorbed directly into the eye. Sufficiently large doses, however, can cause irreparable damage to the eye and as much caution should be exercised with these lasers as with the high energy lasers. The CO_2 laser with its output of 106,000 A, now above the kilowatt range, may cause burns of the surface of the eye or cornea since at this wavelength almost everything absorbs the laser beam. For pulsed lasers our laboratory has developed "failsafe" protective goggles in which a metal plate automatically covers both eyes during the charging period and opens automatically after firing. Recently, a "one pair" protective goggles with special filters has been proposed, but as yet, there is not enough data regarding the goggles to recommend them. Therefore, "one pair" goggles should not yet be used. It is imperative that personnel do not look down the barrel of any laser when the laser is operating or, in the case of a high energy laser, when there is a charge on the power supply capacitors. Helium-neon lasers are used commonly and carelessly today. It is emphasized that even low output helium-neon lasers can cause eye damage.

Laser protective glasses may be tested by biological techniques: a rabbit's eyes may be covered with the protective glass, subjected to laser irradiation and then examined for damage. Protection may be studied also by measuring experiments with various photosensitive devices. Schlickman and Kingston[6] have developed a dosimeter to measure the energy of the reflected laser pulse. This is used in an attempt to prevent eye damage by indicating radiation data which has reached the point where it can cause damage to the eye. Calibration is done by experiment on the eyes of rabbits. Finally, it may be done in the least desirable fashion by evaluation by the operating personnel after use of the protective glasses. Evaluation of the protective glasses by the operating personnel means that there should be detailed eye examinations by an ophthalmologist, preferably one who is familiar with laser technology. This eye examination should include a detailed examination and a fundus picture so that a record may be available. One individual who had worked considerably with lasers under poor industrial hygiene conditions was found to have a pigmented spot on the periphery of the fundus but no prior fundus examination was available.

Infrared color photography with special film from the Eastman Kodak Company (Kodak Ektachrome Infrared Aero film, Type 8443) has been used by us to study reflectance from target areas with laser impact. This type of photography done in a dark room using a wide-angle lens can show the detailed pattern of reflectance on personnel, including protective glasses and exposed skin.

There has been some concern about corneal opacities or cataracts caused by high-energy and high-peak-power Q-switching. Investigators who have developed cataracts are now under study to determine if these cataracts were caused by exposure to high output Q-switched ruby lasers. We have not yet observed cataracts in the experimental animal.

For biomedical purposes in direct impacts about the faces of patients, we have used extra black cloth coverings for the eyes, in addition to the protective goggles. A double felt curtain withstands up to 100 joules/cm² from a direct impact without appreciable transmission, at 6943 A (ruby wavelength). However, such material transmits on the order of 10% of the incident light at 10,600 A according to transmittance curves done recently by Buckley. This should be considered in protection programs with high-energy neodymium lasers. It is the experience of the patients who have had such impacts that some reddish light is still perceived as it is with the protective glasses alone (possibly by refraction of the light as it passes into the tissue). In addition, we are developing a protective shield of plastic covered with silver for the eyes similar to that used in X-ray radiation about the eyes. This protective shield is to be used in the eye itself.

It is recommended that as a rule personnel should not work in the dark since the difference in eye acceptance area between being in the dark and in the light is 16 : 1.

In summary, the eye protection program should be constant and a part of personnel protection in all phases. Even with low-energy lasers and with gas lasers, the question of eye protection must be considered and any programs must be continually re-evaluated in the light of the continuing studies on the immediate and delayed reactions of the eye to laser radiation.

Related to eye protection is the transmission of light to the eye or orbit by soft tissue, the transilluminating effect with lasers. This is important in accidents which may occur from impacts about the face, such as with premature firing of the flash tubes, and in treatment impacts of personnel about the head and neck of a patient, including laser dental surgery. We have attempted to study this in a patient by measures to avoid surface reflectance and, as mentioned above, with detailed photography, especially with the new Kodak colored infrared film. Black cloths and various other materials are used to attempt to protect the eyes and the face from the reflected beam, and the photographs are taken of the impact in darkness. The photographs of the protective areas are taken before, during and after impacts with either Kodachrome, Ektachrome or the new experimental Kodak infrared colored film. Exposure conditions for this film have been determined by Buckley.[7] In brief, in exposures about the forehead and the buccal cavity for laser dentistry, it appears that there is transillumination of the sinuses and also of the soft tissues. As yet, no permanent sequelae can be recognized. This is an important study and should be done before the biomedical applications of the laser become significantly more widespread and especially before the laser is used for actual transillumination procedures, such as the detection of cancer masses in tissues.

SKIN PROTECTION

Another part of the concern in personnel protection with the use of the high-energy laser is the exposure of the skin. The direct impact of high-energy lasers may cause considerable damage to the skin, especially where it is pigmented. Recently, studies have been done to measure this reflectance by a rapid scanning spectrophotometer. The skin has been checked in visible light and with examination under Wood's filter to detect any early pigmentation or dryness or scaling. There is, as yet, little data known of repeated exposures of exposed areas from reflectance. As yet, in our laboratory, with exposure periods varying up to more than seven years, no changes from chronic exposure to the skin have been found. In one research worker,[8] an individual with a hyper-reactive atopic skin, deliberate attempts were made to sensitize this individual. At present, even low-energy densities in the nature of 0.23 joules of exit energy and 15 to 20 joules/cm^2 energy density in a target area of 0.004 cm^2 produce significant reactions in the skin. Where patients are receiving impacts with the laser, personnel in the laboratory who will be called on to assist in positioning the patient and who will be exposed to numerous impacts, should attempt to keep their own hands and face protected. Black felt or leather gloves or felt coverings can be used on the hands. The face should be turned away from the target area. As indicated above, pictures of the impact in a dark room with infrared colored film reveal the extent of areas of exposure about the target zone. Only with long-term studies will it be possible to determine what the chronic radiation effects are on the exposed skin. (Laser radiation may ignite combustibles.)

In high-energy laser treatments of skin malignancies, the skin around the target area has been protected by single or multiple layers of cardboard. This protection should be applied tightly to the skin. To avoid reflectance, black cardboard should be used. Topical applications of dye mixtures, photochemic chemicals, and liquid crystals are also under study for protection of the skin.

The increased absorption of the laser beam in blood means also that direct impacts should be avoided on the superficial blood vessels in the skin of laboratory personnel. Detailed hematologic examinations in our laboratories have failed to reveal any anemia in laser personnel over a period of seven years.

AIR POLLUTION AND INHALATION HAZARDS

Air contamination is also a problem, especially as it relates to the use of liquid nitrogen coolants, nitrogen purges, spectroscopy and fragments from the laser plume. The concentration of nitrogen vapor could become significant in confined spaces with a consequent reduction of oxygen concentration. Preliminary measurements in our laboratory by the Occupational Health Field Headquarters of the Public Health Service have shown no low oxygen levels. The increasing use of water-cooled lasers in preference to the nitrogen-cooled types may reduce the frequency of the use of nitrogen. Liquid nitrogen produces burns when handled in a careless manner.

Ozone is produced at times about the flash lamps and concentrations of ozone could build up with high repetition rate lasers. Routine chest films of the staff of the Laser Laboratory have shown no findings.

A new field of chemistry is the study of the chemical lasers. When these contain toxic materials, the chemist must be aware of their hazards. An example of a chemical laser hazard is the new liquid laser containing selenium oxychloride, with an output of 10,600 A.

Raman laser spectroscopy and Brillouin scattering present potential and actual hazards with laser impacts of such materials as benzene, nitrobenzene, toluene and carbon disulfide. There may be air contamination from these toxic materials or new compounds formed after impact. With low-output instruments the air contamination may not be significant. Plume fragments from living tissues may disseminate viable cancer tissue in the air. High-power-output studies should be done under a hood or with adequate exhaust systems. Even with low-output systems, however, flammability of the solvent on impact may produce potential fire hazard. For over three years we have used traps over the laser head to contain plume fragments.

X-RAY GENERATION

In our experiments with peak power outputs of 100 to 150 megawatts, the search by our radiation physicist for X-ray generation from various targets, both metallic and non-metallic, failed to elicit any evidence of this, although such generation has been reported to us. The methods of our analysis were crude for this type of technique, i.e., essentially a Geiger counter and the use of X-ray sensitive film. Recently, in much more significant experiments, Schwartz[9] has reported no radiation traces from peak power outputs of 20 to 30 megawatts in a cloud chamber. However, this report did not analyze the interaction with metals.

ELECTRICAL SHOCK

In the research laboratory, the possibilities of electrical shock are great when space is at a minimum. Some high-energy systems may require upwards of 50 capacitors (4,000–10,000 volts). The energy stored at high voltage in these capacitors when dissipated through a human conductor will cause severe shock and massive thermal burns. This caution applies also to the cavities of some lasers which are maintained at the high voltage side of the system.

Shielding of the capacitors should be mandatory, not only to protect the personnel from accidental contact with the capacitors but also to protect them from the possibility of capacitor explosion which in its effect resembles a small hand grenade exploding. Often laboratory models of lasers have open electrical circuits which are hazards.[1] In the finished products, all electrical circuits should be housed in positively grounded cabinets to limit the possibility of shock from a floating ground potential.

LONG-TERM OBSERVATION PERIODS

With the late effects of the laser on living tissue as yet not known, and with the increasing development and use of the high-energy and high-power laser systems, it is necessary to set up long-term study programs. Impacts must be described in terms of the laser used, the target size, energy density, the pulse length, pulse spikes, lens systems and target area characteristics as regards vascularity, connective tissue, and so forth. Only in this manner can biomedical results observed at various laser centers be compared. In our series of over 600 patients treated with laser radiation in the last seven years, there has been no evidence of complications other than non-specific scarring. The increased reactivity of one research worker to low-energy laser radiation has been mentioned above.

As in any industrial process, personnel unfamiliar with the details of the process can get into the most difficulties. Those who use the high energy lasers should receive detailed instruction in the actual use and not rely on the details of an instruction manual. Instruction manuals for laser technology, in our experience, are often incomplete in details of protection and often in details of construction and assembly. For any laser installation, one individual trained in laser safety should be considered as the laser safety officer.

We are concerned also about the laser kits being produced for high school students and hobbyists. We are concerned that the kits are often used with little or no attention paid to the possible hazards to the eyes and other parts of the body and feel that clear instructions on protection should be included in these kits.

To summarize, current so-called safe energy levels may be listed. It must not be forgotten that the following factors may influence such energy levels:

1. The laser used
2. The target area
3. The character of the pulse (specifically the peak power and spiking characteristics)
4. The duration of the pulse

It should be emphasized, also, that the data with regard to the biological effect of lasers are incomplete at this time so that the following table of Tentative "Safe" Energy Levels for eye damage represents only the values presently used and should not be construed as being absolutely definitive. Brief mention has been made in this chapter of the high-power gas lasers such as the argon, YAG, including frequency doubled YAG, and carbon dioxide lasers (with outputs in the blue-green part of the visible spectrum and in the infrared). These lasers will undoubtedly take their place in the laser medical research field and, as they present a somewhat different protection problem, new techniques and protection equipment will have to be produced to counteract the potential hazards they represent.

CURRENT RECOMMENDED VALUES[10]

EYE			SKIN*			
1968 International Laser Safety Conference			American National Standards Institute Z-136 Standards Committee			
Pupil diameter, mm	Energy levels		Power level	Pulsed		Continuous wave	
	Q-Switch pulsed, j/cm^2	Non-Q-switched pulsed, j/cm^2	Continuous wave, w/cm^2	Exposure time, milli-second	Energy level j/cm^2 per pulse	Exposure time, second	Power level, w/cm^2
3	5.0×10^{-8}	5.0×10^{-7}	5.0×10^{-6}	>0.1	0.1	>1.0	0.1
7	1.0×10^{-8}	1.0×10^{-7}	1.0×10^{-6}	<0.1	0.01	$\leqslant 1.0$	1.0

*Values apply to wavelengths of 0.4 to 1000 micrometer (micron).

PROTECTION OF PATIENTS[11]

The medical laser laboratory must be equipped for exposure of patients who may require minor surgical procedures, either pre-impact or post-impact. In experiments where the laser head must touch the patient, the head must be kept sterile with nylon or glass coverings. Any manipulations about the exposed lesion demand sterile techniques with the accompanying masks, gloves, cameras, and sterile instruments. In the use of laboratory animals, any experiment involving surgical procedures such as exposure of the brain or circulatory system also demands careful surgical techniques. Therefore, all the occupational hazards of the operating room are now brought to the medical laser laboratory. In our laser laboratory, we have had exposure of patients under general anesthesia. The anesthetics used were those used in the presence of electrical spark procedures, such as halothane (Fluothane). When laser instruments are used in the operating room, the same precautions will hold as for any electrical instrumentation. In addition, there will be the need for eye and skin protection of operating room personnel. Also, in the exposure of infected animals and human tissues, the laser personnel must avoid bacterial or viral contamination of themselves or the laser applicator. Plume traps with an attached vacuum outlet are used on the laser head.

OUTLINE OF PROTECTIVE MEASURES IN THE LASER LABORATORY

Safety measures in a system for the protection of the operating personnel include:
1. A warning light system initiated on charging of any high-energy laser in the laser room, which causes red lights at the benches to flash intermittently in any adjoining rooms and in the laser room, and a large warning sign outside the laboratory to be illuminated.
2. A muted bell system which commences chiming on charging of the laser capacitor banks and continues until the laser is discharged.
3. Individual door locks triggered by the laser charging circuits which prevent opening of the doors except in an emergency.
4. A master override switch in the office or other adjoining rooms so that personnel can leave these rooms in an emergency whether the laser is firing or not. This switch is self-illuminated at all times.
5. Black curtains or screens which may be pulled around the laser benches to screen personnel from the laser flash.
6. Use of specifically designed antilaser flash goggles for personnel working close to the laser head, as well as for patients being exposed to the laser.
7. Avoidance of specular surfaces by rough finishing of the walls and painting with flat charcoal black paint.
8. Avoidance of any open electrical connections.
9. Proper air conditioning of the rooms.

There are perhaps some three or four years more of detailed experimentation work in the laboratory before the laser can be used routinely in clinical practice. At present, however, in the biology, physics and chemistry laboratories laser instrumentation is being used, often carelessly. It is emphasized again that as yet there is too little known of the effects in man of chronic exposure for the eyes, skin,[8] and other viscera. The safety program should be constantly surveyed and revised as new data and new lasers become available.

ACKNOWLEDGEMENT: Portions of this article dealing primarily with laboratory design are reprinted with permission from *Archives of Environmental Health*, Volume 10, pp. 493–497, March, 1965, copyright 1965 by the American Medical Association, and the major portion is reprinted with permission from the *American Industrial Hygiene Association Journal*, Volume 26, pp. 553–557, November–December, 1965. Revised in May 1970.

From the Laser Laboratory, Children's Hospital Research Foundation, of the Medical Center of the University of Cincinnati, Cincinnati, Ohio, supported by a grant from the John A. Hartford Foundation, Inc.

REFERENCES

1. Goldman, L., "Protection of Personnel Operating Lasers", *Amer. Jour. Med. Electronics*, 2, 4, 335–338, (Oct.–Dec., 1963).
2. Goldman, L. and Hornby, P., "The Design of a Medical Laser Laboratory", *Arch. Environ. Health*, 10, 493–497, (March, 1965).

3. Solon, L. R., "Occupational Safety with Laser (Optical Maser) Beams", *Arch. Environ. Health*, *6*, 414, (March, 1963).

4. Straub, H. W., "Protection of the Human Eye from Laser Radiation", Report from Harry Diamond Laboratories, Army Medical Command, (July 10, 1963).

5. Swope, C. H. and Koester, C. J., "Eye Protection Against Lasers", *Applied Optics*, *4*, 523–526. (1956).

6. Schlickman, J. J. and Kingston, R. H., "The Dark Side of the Laser", *Electronics*, (April 19, 1965).

7. Gibson, H. L., Buckley, W., and Whitmore, K. E., "New Vistas in Infrared Photography for Biological Surveys", *Jour. Bio. Photographic Assoc. 33*, 1–33, (1965).

8. Goldman, L. and Richfield, D. F., "The Effect of Repeated Exposures to Laser Beams: Case Report with Nine Months' Period of Observation", *Acta Derm-Venereol.*, *44*, 264–268, (1964).

9. Schwartz, J., Personal Communication.

10. Powell, Charles H., Bell, Herbert, Rose, Vernon, Goldman, Leon, and Wilkinson, Thomas K., "A Review of the Current Status of Laser Threshold Guides", *American Ind. Hyg. Assoc. J.*, in press.

11. Goldman, Leon, Rockwell, R. J., Fidler, James P., Altemeier, Wm. A., and Siler, V. E., "Investigative Laser Surgery: Safety Aspects", *Bio-Medical Engineering*, *4*, 415, (September, 1969).

12. Goldman, Leon and Rockwell, R. J., "Lasers in Medicine," Gordon and Breach Science Publishers Inc., New York, in press.

Basic Units of Radiation Measurement

Allen Brodsky

RELATIONSHIP BETWEEN PHYSICAL QUANTITIES AND BIOLOGICAL EFFECTS

As discussed in the sections on biological effects of radiation, a number of acute and long-term effects on animal and human species have been related to the physical energy absorbed from various types of ionizing radiation. However, the relative effectiveness of each type of radiation per unit energy absorbed in biological tissue has been found to vary not only with the type of radiation and its quantum energy, but also with the rate at which the energy is delivered, the kind of tissue, age and species of animal, the biological effect under consideration, and other experimental and epidemiologic variables. For mammals, beta radiation and the recoil electrons ejected from atoms by X or gamma radiation generally produce approximately the same order of magnitude of biological effects. On the other hand, heavier particles, such as the alpha particles emitted by certain radionuclides, lose their energy at higher rates of linear energy transfer (LET) along their paths and seem to produce somewhat higher damage per unit energy absorbed. In the case of alpha-emitting radionuclides, of course, since the characteristic alpha particles emitted do not have sufficient range to penetrate the dead layer of skin, biological damage is produced only when the radioactive material itself is distributed within an organ by inhalation or ingestion.

Thus, for purposes of radiation hazard evaluation, several units of radiation exposure and dose must be introduced to account for the several methods of measuring and assessing the effects of different types of radiation. Since most radiations of a given type, as emitted by most radionuclides, have average LET's within a narrow range, and consequently seem to have relative biological effectivenesses (RBE's) within a narrow range, characteristic simplifications can be introduced to limit the number of new units and the definitions required for most applications. In this chapter, only the most useful definitions and data are presented, and some basic references are given for further details.

BASIC UNITS OF RADIOACTIVITY, RADIATION MEASUREMENT, AND RADIATION DOSE

The following definitions of quantities and units will suffice in dealing with most problems in radiation protection (health physics) and dosimetry:

roentgen (r)—a unit for expressing exposure from X or gamma radiation in terms of the ionization produced in air, which can be measured by appropriate air ionization chambers and electrical instruments. It is defined as an exposure "such that the associated corpuscular emission per 0.001293 g of air produces, in air, ions carrying one electrostatic unit of quantity of electricity of either sign."[1] This simply means that an exposure (formerly called "exposure dose"[2]) of one roentgen generates recoil electrons per 0.001293 g of dry air within a small volume surrounding the point of measurement to the extent that these electrons produce 1 esu (electrostatic unit) of positive charge and 1 esu of negative charge (as ion pairs) in air. Since many electrons will leave the element of volume before losing all of their energy, a "standard air chamber" is necessary in order to most accurately measure the roentgen (see Figure 1). This chamber is specially designed to achieve an "electronic equilibrium" condition in which the ionization lost by electrons leaving that volume is compensated for by an equilibrium number of electrons entering the volume from a preceding volume of air.[1] Thus, the roentgen is a unit that expresses a point quantity, essentially the ionization density produced near the point of measurement in air. When applied to human exposure in a large beam or field of radiation, either the field must be uniform or the exposure in r must be measured at each point in the field. In summary, the roentgen may be remembered for practical purposes as the X or gamma exposure producing 1 esu (+ or −) per cc of air (at STP 0°C, 760 mm Hg).

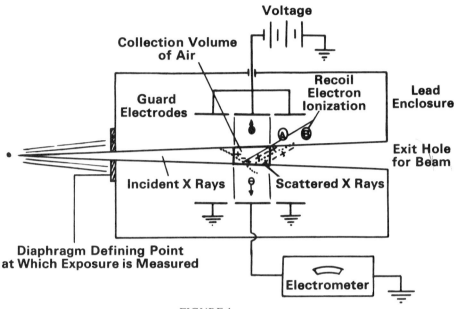

FIGURE 1.
Schematic Diagram of a "Standard Air Chamber".

rad—a unit of "absorbed dose" (D) in any medium for any kind of ionizing radiation.[1,2] It is simply defined as:

$$1 \text{ rad} = 100 \text{ ergs/gram.}$$

Integral absorbed dose—the integral $\int D dm$ over an organ or the whole body, where D is the variable absorbed dose within mass element dm. It is useful in obtaining an idea of the total energy absorbed by a region of tissue, which is more closely related to certain macroscopic biological changes than the dose only near one point. Common units are gram-rads, and 1 gram-rad = 100 ergs. The integral absorbed dose may be divided by the total mass of tissue in the region of interest to obtain the average tissue dose $\bar{D}$ in rads.

rep—a unit formerly used for similar purposes as the rad, but defined for any type of radiation in terms of the energy absorbed in tissue equivalent to that which would be absorbed from 1 r of X or gamma radiation. 1 rep has been defined variously as the absorption of energy ranging from 84 ergs/cm^3 to 93 ergs/gram of tissue.[3,4]

rem—a unit of "RBE-Dose" or "Dose Equivalent" (DE), used to express the estimated equivalent of any type of radiation that would produce the same biological end point as 1 rad delivered by X or gamma radiation. Thus,

$$DE \text{ (in rem)} = \text{Dose (in rad)} + QF \times DF,$$

where QF (formerly called RBE) accounts for the relative biological effectiveness of the radiation compared to X radiation for radiation protection purposes, and the DF (formerly designated as n) is the "relative damage factor"[5] or "distribution factor"[1,2] used to account for differences in the distribution of the rad dose to the organ of concern as a result of uneven uptake of the radionuclide, etc., as opposed to the QF or RBE factors reserved for more intrinsic biological characteristics of the emitted radiations.

gram-rem—the unit for integral absorbed dose when several types of radiation are involved and the equivalent doses of each are to be added after multiplication by appropriate relative biological effectiveness factors.

man-rem—a term used in estimating expected frequencies of disease in a population by determining the equivalent integral absorbed dose over the population. For somatic effects, the number of persons in the exposed population might be multiplied by the average dose equivalent to each person; for genetic effects, the average gonadal dose would be multiplied by the number of people exposed.

kerma (K)—the quotient E_K/m, where E_K is the sum of all kinetic energies of charged particles liberated by indirectly ionizing particles (e.g., neutrons) in a volume element, and m is the mass of matter in that volume element.[6]

Energy fluence (F)—as defined by the ICRU,[6] the quotient $\Delta E_F/\Delta a$, where ΔE_F is the sum of all the energies, exclusive of rest energies, entering a sphere of cross-sectional area Δa (e.g., in units of Mev/cm^2).

Energy flux density (intensity I)—the quotient $\Delta F/\Delta t$, where ΔF is the energy fluence in a small time interval Δt.

Particle fluence (Θ)—the quotient $\Delta N/\Delta a$, the number of particles flowing into a small sphere per unit cross-sectional area. This is the same quantity as nvt, used in the case of neutron-diffusion theory, where n = neutron density in neutrons/cm^3, v = average neutron velocity in cm/sec, and t = time in seconds over which the fluence is integrated. Another way of envisioning the meaning of nvt would be to imagine the total distance in cm traveled by all the neutrons present at a given moment per cm^3 of volume.

Particle flux density (ϕ)—the quotient $\Delta\Phi/\Delta t$, where $\Delta\Phi$ is the particle fluence in time Δt; units may be particles/cm^2-sec.

Mass attentuation coefficient (μ/ρ)—the fractional number of incident particles (or photons) interacting with a given material per unit mass thickness that they pass through; i.e., $\mu/\rho = dN/N\rho dl$, where dN/N is the probability of interaction per unit thickness dl of density ρ. Common units are cm^2/g (i.e., fractional number/g-cm^{-2}).

Mass energy-absorption coefficient (μ_{en}/ρ)—the fractional energy removed from incident indirectly ionizing particles (or photons) per unit mass thickness; i.e., $\mu_{en}/\rho = dE/E\rho dl$, where dE is the energy removed from the incident particles (not including rest energy or energy reirradiation as bremsstrahlung), E is the sum of the energies (excluding rest energies) of the incident particles, and ρdl is the mass thickness in g/cm^2, as above. Common units are again cm^2/g (i.e., fractional energy/g-cm^{-2}).

Curie (Ci)—the most common unit used to express the radioactivity (A) of a material; it is the amount of any radionuclide (or combination of radionuclides) in which there are 3.7×10^{10} nuclear transformations or disintegrations per second, or 2.22×10^{12} disintegrations per minute (dpm). Combined with appropriate prefixes, the symbol is commonly used for the following other scientific units:

$$1 \text{ megacurie (MCi)} = 10^6 \text{ Ci}$$
$$1 \text{ kilocurie (kCi)}\ = 10^3 \text{ Ci}$$
$$1 \text{ millicurie (mCi)} = 10^{-3} \text{ Ci}$$
$$1 \text{ microcurie } (\mu\text{Ci}) = 10^{-6} \text{ Ci}$$
$$1 \text{ nanocurie (nCi)}\ = 10^{-9} \text{ Ci}$$
$$1 \text{ picocurie (pCi)}\ \ = 10^{-12} \text{ Ci} = \mu\mu\text{Ci}.$$

The above designations are widely used, since we deal with a wide range of quantities of radioactivity in evaluating radiation exposure potentials.

In addition to the basic quantities and units defined above, other derived units will be introduced in subsequent sections of this chapter. Table 1 presents the other fundamental quantities and units listed by the ICRU.[2,6]

Table 1. QUANTITIES AND UNITS

No.	Name	Symbol	Dimensions	Units mksa	Units csg	Units Special
4	Energy imparted (integral absorbed dose)	—	E	J	erg	g-rad
5	Absorbed dose	D	EM^{-1}	J kg^{-1}	erg g^{-1}	rad
6	Absorbed-dose rate	—	$EM^{-1}T^{-1}$	J kg^{-1} s^{-1}	erg g^{-1} s^{-1}	rad s^{-1}, etc
7	Particle fluence or fluence	Φ	L^{-2}	m^{-2}	cm^{-2}	
8	Particle flux density	φ	$L^{-2}T^{-1}$	m^{-2} s^{-1}	cm^{-2} s^{-1}	
9	Energy fluence	F	EL^{-2}	J m^{-2}	erg cm^{-2}	
10	Energy flux density or intensity	I	$EL^{-2}T^{-1}$	J m^{-2} s^{-1}	erg cm^{-2} s^{-1}	
11	Kerma	K	EM^{-1}	J kg^{-1}	erg g^{-1}	
12	Kerma rate	—	$EM^{-1}T^{-1}$	J kg^{-1} s^{-1}	erg g^{-1} s^{-1}	
13	Exposure	X	—	—	—	r (roentgen)
14	Exposure rate	—	$QM^{-1}T^{-1}$	C kg^{-1} s^{-1}	esu g^{-1} s^{-1}	r s^{-1}, etc.
15	Mass attenuation coefficient	$\dfrac{\mu}{\rho}$	$L^2 M^{-1}$	m^2 kg^{-1}	cm^2 g^{-1}	
16	Mass energy-transfer coefficient	$\dfrac{\mu K}{\rho}$	$L^2 M^{-1}$	m^2 kg^{-1}	cm^2 g^{-1}	
17	Mass energy-absorption coefficient	$\dfrac{\mu_{en}}{\rho}$	$L^2 M^{-1}$	m^2 kg^{-1}	cm^2 g^{-1}	
18	Mass stopping power	$\dfrac{S}{\rho}$	$EL^2 M^{-1}$	J m^2 kg^{-1}	erg cm^2 g^{-1}	
19	Linear energy transfer	LET	EL^{-1}	J m^{-1}	erg m^{-1}	kev (um)$^{-1}$
20	Average energy per ion pair	W	E	J	erg	ev.
22	Activity	A	T^{-1}	s^{-1}	s^{-1}	Ci (curie)
23	Specific gamma-ray constant	Γ	$QL^2 M^{-1}$	C m^2 kg^{-1}	esu cm^2 g^{-1}	r m^2 h^{-1}, etc.
	Dose equivalent	DE	—	—	—	rem

From *NBS HANDBOOK 87*, p. 43.

PHYSICAL MEASUREMENTS OF DOSE AND THE BRAGG-GRAY PRINCIPLE

Since the standard air chamber described earlier is a relatively large, expensive, and sensitive instrument, and since measurements of the roentgen apply directly only to X or gamma rays, many other field and laboratory instruments have been devised for measuring the absorbed dose or dose equivalent from various types of radiation more directly, although in some cases less precisely. Instruments have also been designed to measure over wide ranges of intensity and energy. Some of the methodology has been reviewed in References 7 to 12.

Small-cavity chambers with air-equivalent or tissue-equivalent walls have often been used as secondary-standard instruments for measuring radiation exposure or absorbed dose. By use of the modified Bragg-Gray principle,[11] the ionization collected by an electrode within a small cavity in a suitably designed chamber can be related to the energy deposited per gram of wall material; also, the relationship may hold constant for a wide range of energies as long as the Bragg-Gray conditions are fulfilled. In its simplest form, the Bragg-Gray principle states that the ratio of the energy absorbed per gram of a medium to the energy absorbed per gram of gas in a small cavity in the medium is constant (almost independent of the initial energy of the recoil electrons produced in the medium). Since the energy to produce an ion pair in a gas (W_g) is also apparently independent of energy, we have the Bragg-Gray principle[13]

$$E_m = S_g^m \times W_g \times J_g,$$

where E_m is the energy absorbed per gram of wall material, S_g^m represents the relative mass stopping power ratio $(dE/\rho dl)_m/(dE/\rho dl)_a$ for electrons in the material and in air (i.e., the ratio of the rates of energy loss per unit path measured in g/cm^2), W_g expresses the average energy required to produce an ion pair in the gas (now usually taken as 34 ev/ion pair[11]), and J_g is the number of ion pairs produced per gram of gas in the cavity. Average mass stopping power ratios are given in Table 2 for electrons of various initial kinetic energies.[13,14]

The conditions to be met for the Bragg-Gray relation to hold[13] are briefly:

a) cavity dimensions must be small compared to the ranges of most secondary ionizing particles;

b) most ionizing particles should originate in the chamber walls, and very few primary interactions should occur in the gas;

c) the fluence of primary and secondary particles should be nearly uniform across the cavity;

d) the wall of the cavity should be thick enough so that all recoil charged particles traversing the cavity originate within the wall material, but the wall must not be so thick that it attenuates the primary nonionizing radiation appreciably.

Cavity chambers may be made with walls having an average atomic number $\overline{Z}$ simulating that of soft tissue, so that they can measure the rad dose more directly. More often, however, r chambers having air-equivalent walls are used, with appopriate wall thicknesses for the X- or gamma-ray energies to be measured (see Table 3 and Figure 2). Then the exposure or "air dose" measured in roentgens (r) can be converted to the appropriate absorbed dose in tissue, expressed in rads, by the equation[13]

$$D_{tissue}(rads) = 0.877 \times \frac{(\mu_{en}/\rho)_{tissue}}{(\mu_{en}/\rho)_{air}} \times R = f \times R.$$

Some f factors and values of $m\mu_{en} = \mu_{en}/\rho$ for various materials and gamma-ray energies are given in Table 4; for X-ray spectra, see Figure 3.

When exposure to X or gamma radiation is measured at or near the surface of the body with an air wall chamber of "equilibrium" thickness, the dose in rads at various depths in tissue must be corrected for attenuation by the use of depth dose curves or tables. However, the fractional dose at each depth depends not only on the quantum energy of the radiation, but on the area of the beam at the body surface, on the distance from source to skin, and on other factors.[2,13] A few representative depth dose data taken from Johns[13] are presented in Tables 5 to 13; the equivalent kilovoltage (effective kev) corresponding to each of the HVL specifications may be obtained from Figure 4. Table 14 provides data for converting photon fluences

Table 2.
MEAN MASS STOPPING POWER RATIOS RELATIVE TO AIR (S_m)

$$S_m = \frac{1}{T} \int_0^{T_0} S_m \, dT$$

Initial Electron Kinetic Energy (T_0), Mev	Including Density Effect		
	C	Water	Tissue
0.002	1.070	1.238	1.216
0.003	1.064	1.226	1.216
0.004	1.060	1.220	1.199
0.005	1.058	1.215	1.195
0.006	1.055	1.212	1.191
0.007	1.054	1.208	1.188
0.008	1.052	1.206	1.186
0.009	1.051	1.203	1.183
0.01	1.050	1.202	1.182
0.02	1.044	1.191	1.172
0.03	1.041	1.185	1.166
0.04	1.039	1.181	1.163
0.05	1.038	1.179	1.160
0.06	1.037	1.177	1.159
0.07	1.036	1.175	1.157
0.08	1.035	1.174	1.156
0.09	1.034	1.173	1.155
0.1	1.034	1.172	1.154
0.2	1.030	1.166	1.148
0.3	1.027	1.163	1.145
0.4	1.024	1.161	1.143
0.5	1.022	1.159	1.141
0.6	1.020	1.158	1.140
0.7	1.017	1.156	1.138
0.8	1.016	1.154	1.136
0.9	1.014	1.152	1.134
1	1.012	1.150	1.132
2	1.001	1.139	1.121
3	0.985	1.121	1.103
4	0.976	1.110	1.093
5	0.968	1.108	1.084
6	0.961	1.093	1.076
8	0.950	1.080	1.063
10	0.940	1.069	1.052

From: Johns, H. E., *The Physics of Radiology*, 2nd ed., p. 703. Charles C Thomas, Springfield, Illinois, 1964.

Table 3.
EQUILIBRIUM WALL THICKNESSES AND f FACTORS FOR SEVERAL PHOTON SPECTRA UP TO 3 Mev

Radiation	Peak Energy, Mev	Mean Energy, Mev	Equilibrium Wall, cm	f—rads/roentgens		
				Water	Bone	Muscle
^{137}Cs	0.66	0.66	0.2–0.3	0.975	0.933	0.966
^{60}Co	1.25	1.25	0.4–0.6	0.974	0.928	9.965
X rays	2.00	0.67	0.3–0.4	0.975	0.933	0.966
X rays	3.00	1.00	0.4–0.6	0.974	0.927	0.965

From: Johns, H. E., *The Physics of Radiology*, 2nd ed., p. 293. Charles C Thomas, Springfield, Illinois, 1964.

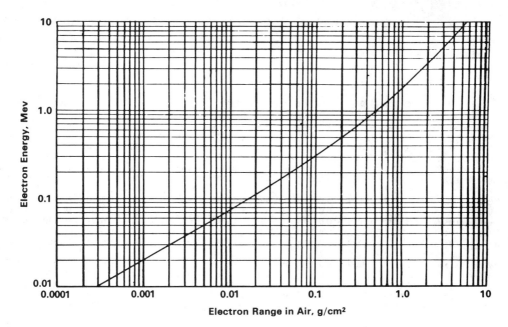

FIGURE 2.
Electron Range in Air (in g/cm²)
as a Function of the Electron Energy.
(From: Johns, H. E., *The Physics of Radiology*,
2nd ed., p. 291. Charles C Thomas,
Springfield, Illinois, 1964.)

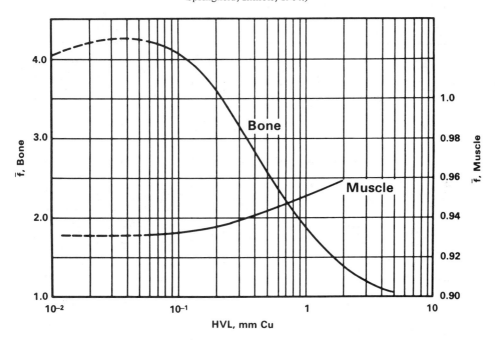

FIGURE 3.
Conversion Factors f̄ for X-Ray Spectra
of Various Half-Value Layers of Copper,
to Convert Roentgens to Rads for Bone and Muscle.
(From: Johns, H. E., *The Physics of Radiology*,
2nd ed., p. 282. Charles C Thomas,
Springfield, Illinois, 1964.)

Table 4. VALUES OF THE MASS ENERGY-ABSORPTION

Photon Energy, Mev	Mass Energy-Absorption										
	H	C	N	O	Na	Mg	Al	P	S	Ar	K
0.010	0.0099	1.9400	3.4200	5.5000	15.4000	20.9000	26.5000	40.1000	49.7000	62.0000	77.0000
0.015	0.0110	0.5170	0.9160	1.4900	4.4300	6.0900	7.6500	11.9000	15.2000	19.4000	24.6000
0.020	0.0133	0.2030	0.3600	0.5870	1.7700	2.4700	3.1600	5.0000	6.4100	8.3100	10.5000
0.030	0.0186	0.0592	0.1020	0.1630	0.4820	0.6840	0.8800	1.4500	1.8500	2.4600	3.1200
0.040	0.0280	0.0306	0.0465	0.0700	0.1940	0.2740	0.3510	0.5700	0.7310	0.9740	1.2500
0.050	0.0270	0.0226	0.0299	0.0410	0.0996	0.1400	0.1760	0.2820	0.3610	0.4840	0.6260
0.060	0.0305	0.0203	0.0244	0.0304	0.0637	0.0845	0.1040	0.1660	0.2140	0.2840	0.3670
0.080	0.0362	0.0201	0.0248	0.0239	0.0369	0.0456	0.0536	0.0780	0.0971	0.1240	0.1580
0.100	0.0406	0.0213	0.0222	0.0232	0.0288	0.0334	0.0372	0.0500	0.0599	0.0725	0.0909
0.150	0.0485	0.0246	0.0249	0.0252	0.0258	0.0275	0.0282	0.0315	0.0351	0.0368	0.0433
0.200	0.0530	0.0267	0.0267	0.0271	0.0265	0.0277	0.0275	0.0292	0.0310	0.0302	0.0339
0.300	0.0573	0.0288	0.0289	0.0289	0.0278	0.0290	0.0283	0.0290	0.0301	0.0278	0.0304
0.400	0.0587	0.0295	0.0296	0.0296	0.0283	0.0295	0.0287	0.0290	0.0301	0.0274	0.0299
0.500	0.0589	0.0297	0.0297	0.0297	0.0284	0.0293	0.0287	0.0288	0.0300	0.0271	0.0294
0.600	0.0588	0.0296	0.0296	0.0296	0.0283	0.0292	0.0286	0.0287	0.0297	0.0270	0.0291
0.800	0.0573	0.0288	0.0289	0.0289	0.0276	0.0285	0.0278	0.0280	0.0287	0.0261	0.0282
1.000	0.0555	0.0279	0.0280	0.0280	0.0267	0.0275	0.0269	0.0270	0.0280	0.0252	0.0272
1.500	0.0507	0.0255	0.0255	0.0255	0.0243	0.0250	0.0246	0.0245	0.0254	0.0228	0.0247
2.000	0.0464	0.0234	0.0234	0.0234	0.0225	0.0232	0.0227	0.0228	0.0235	0.0212	0.0228
3.000	0.0398	0.0204	0.0205	0.0206	0.0199	0.0206	0.0201	0.0204	0.0210	0.0193	0.0208
4.000	0.0351	0.0184	0.0186	0.0187	0.0184	0.0191	0.0188	0.0192	0.0199	0.0182	0.0199
5.000	0.0316	0.0170	0.0172	0.0174	0.0173	0.0181	0.0180	0.0184	0.0192	0.0176	0.0193
6.000	0.0288	0.0160	0.0162	0.0166	0.0166	0.0175	0.0174	0.0179	0.0187	0.0175	0.0190
8.000	0.0249	0.0145	0.0148	0.0154	0.0158	0.0167	0.0169	0.0175	0.0184	0.0172	0.0190
10.000	0.0222	0.0137	0.0142	0.0147	0.0154	0.0163	0.0167	0.0174	0.0183	0.0173	0.0191

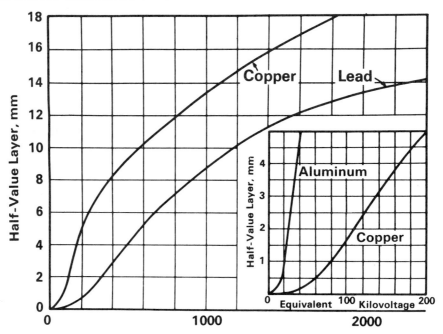

FIGURE 4.
Relationships Between HVL and Effective Kilovoltage.
(From: Johns, H. E., *The Physics of Radiology*,
2nd ed., p. 256. Charles C Thomas,
Springfield, Illinois, 1964.)

COEFFICIENTS AND THE FACTOR f

Coefficient $(m\mu_{en})$ cm^2/g									$f = 0.87_7 \left[\dfrac{(m\mu_{en}) \text{ medium}}{(m\mu_{en}) \text{ air}} \right]$		
Ca	Poly-styrene	Lucite	Poly-ethylene	Bakelite	Water	Air	Compact Bone	Muscle	$\dfrac{\text{Water}}{\text{Air}}$	$\dfrac{\text{Compact Bone}}{\text{Air}}$	$\dfrac{\text{Muscle}}{\text{Air}}$
89.8000	1.7900	2.9200	1.6600	2.4300	4.8900	4.6600	19.0000	4.9600	0.92_0	3.5_8	0.93_3
28.9000	0.4780	0.7880	0.4440	0.6510	1.3200	1.2900	5.8900	1.3600	0.89_7	4.0_0	0.92_5
12.5000	0.1880	0.3110	0.1760	0.2570	0.5230	0.5160	2.5100	0.5440	0.88_7	4.2_7	0.92_5
3.7500	0.0561	0.8920	0.0534	0.0743	0.1470	0.1470	0.0743	0.1540	0.87_7	4.4_3	0.91_9
1.5200	0.0300	0.0426	0.0295	0.0368	0.0647	0.0640	0.3050	0.0677	0.88_7	4.1_8	0.92_8
0.7640	0.0229	0.0288	0.0232	0.0259	0.0394	0.0384	0.1580	0.0409	0.90_6	3.6_1	0.93_4
0.4480	0.0211	0.0243	0.0218	0.0226	0.0304	0.0292	0.0979	0.0312	0.91_3	2.9_4	0.93_7
0.1910	0.0213	0.0226	0.0224	0.0217	0.0253	0.0236	0.0520	0.0255	0.94_0	1.9_3	0.94_5
0.1110	0.0228	0.0235	0.0241	0.0227	0.0252	0.0231	0.0386	0.0252	0.95_7	1.4_7	0.95_7
0.0488	0.0264	0.0267	0.0280	0.0261	0.0278	0.0251	0.0304	0.0276	0.97_1	1.0_3	0.96_4
0.0367	0.0287	0.0289	0.0305	0.0283	0.0300	0.0268	0.0302	0.0297	0.98_2	0.98_2	0.97_7
0.0319	0.0310	0.0311	0.0329	0.0305	0.0320	0.0288	0.0311	0.0317	0.97_7	0.94_7	0.96_3
0.0308	0.0317	0.0319	0.0337	0.0312	0.0329	0.0296	0.0316	0.0325	0.97_3	0.93_8	0.96_3
0.0304	0.0319	0.0320	0.0339	0.0314	0.0330	0.0297	0.0316	0.0327	0.97_4	0.93_3	0.96_8
0.0301	0.0318	0.0319	0.0338	0.0313	0.0329	0.0296	0.0315	0.0326	0.97_3	0.93_6	0.96_3
0.0290	0.0310	0.0311	0.0329	0.0305	0.0321	0.0289	0.0306	0.0318	0.97_4	0.92_9	0.96_5
0.0279	0.0300	0.0301	0.0319	0.0295	0.0311	0.0280	0.0297	0.0308	0.97_4	0.92_7	0.96_5
0.0253	0.0274	0.0275	0.0291	0.0270	0.0283	0.0255	0.0270	0.0281	0.97_3	0.92_9	0.96_6
0.0234	0.0252	0.0252	0.0267	0.0247	0.0260	0.0234	0.0248	0.0257	0.97_4	0.92_9	0.96_6
0.0213	0.0219	0.0220	0.0232	0.0216	0.0227	0.0205	0.0219	0.0225	0.97_1	0.93_7	0.96_3
0.0204	0.0197	0.0198	0.0208	0.0194	0.0205	0.0186	0.0199	0.0203			
0.0200	0.0181	0.0183	0.0191	0.0179	0.0190	0.0173	0.0186	0.0188			
0.0198	0.0170	0.0172	0.0178	0.0168	0.0180	0.0163	0.0178	0.0178			
0.0197	0.0153	0.0156	0.0160	0.0153	0.0165	0.0150	0.0165	0.0163			
0.0201	0.0144	0.0147	0.0149	0.0144	0.0155	0.0144	0.0159	0.0154			

to exposures expressed in roentgen units. Depth dose data and conversion factors for many other conditions may be found in Johns[13] and other references cited.

QUANTITATION OF DOSE–EFFECT RELATIONSHIPS

The analysis of possible mechanisms relating radiation dose delivered to tissue and the observed biological effects has been reviewed by Burch,[15] who emphasizes the need for quantitative methods of predicting long-term as well as early biological effects resulting from radiation exposure. Upton and Kimball[16] review data indicating the relationships between exposure and effects in animals, and Wald[17] reviews the various biomedical effects observed in humans exposed to radiation. Many other comprehensive reviews of radiobiological effects are available; they are cited in References 15 to 17 and, notably, in the United Nations Scientific Committee Reports.[18] The information now available shows certain definite biomedical effects of radiation on humans at high doses and dose rates, but continuing investigation is underway to provide improved understanding and predictability of the shape of the probabilistic dose–response relationships at low doses and dose rates. Therefore, recommended limits of long-term occupational and population exposure must be based on estimated upper limits of the probabilities of disease determined by conservative extrapolation of dose–response curves to low dose or dose rate regions. Since a comprehensive treatment of dose–response relations is beyond the scope of this handbook, only some examples of possible types of response relationships are presented below.

All survival after irradiation is often plotted as the logarithm of the surviving fraction versus dose (see Figure 5). These curves are often fitted to more or less smoothly varying data, showing that for large numbers of experimental units, such as bacterial or mammalian cells, an underlying probabilistic dose–response relationship can be explicitly and accurately defined.

Table 5.

DEPTH DOSE PERCENTAGES FOR CIRCULAR FIELDS FOR 70- AND 120-KVP
X RAYS, VARIOUS FILTRATIONS AND FOCAL-TO-SKIN DISTANCES (FSD)

A. HVL 1.0 mm Aluminum (Approximately 70 kvp with Inherent Filtration)

FSD = 15 cm

Area, cm²		0	3.1	7.0	12.5	28.3	50.0	100.0
Diameter, cm		0	2	3	4	6	8	11.3
Depth, cm	0.0	100	100	100	100	100	100	100
	0.5	61	74	79	81	84	86	87
	1.0	42	56	61	63	66	67	69
	2.0	23	32	36	39	41	42	44
	3.0	13	19	22	24	26	27	29
	4.0	8	12	13	15	17	19	20
	8.0	2	2	3	3	4	4	5

FSD = 20 cm

Area, cm²		0	3.1	7.0	12.5	28.3	50.0	100.0
Diameter, cm		0	2	3	4	6	8	11.3
Depth, cm	0.0	100	100	100	100	100	100	100
	0.5	62	75	80	82	84	86	88
	1.0	44	58	63	65	67	68	70
	2.0	24	34	38	41	43	44	45
	3.0	14	20	23	25	28	29	31
	4.0	9	13	15	16	18	20	21
	8.0	2	3	3	4	4	5	6

FSD = 30 cm

Area, cm²		0	3.1	7.0	12.5	28.3	50.0	100.0
Diameter, cm		0	2	3	4	6	8	11.3
Depth, cm	0.0	100	100	100	100	100	100	100
	0.5	63	76	81	83	85	88	89
	1.0	45	60	64	66	68	70	71
	2.0	25	36	40	42	44	46	48
	3.0	16	22	25	27	30	31	33
	4.0	10	14	16	18	20	22	23
	8.0	2	3	4	4	5	6	7

B. HVL 2.0 mm Aluminum (Approximately 120 kvp with Inherent Filtration)

FSD = 15 cm

Area, cm²		0	3.1	7.0	12.5	28.3	50.0	100.0
Diameter, cm		0	2	3	4	6	8	11.3
Depth, cm	0.0	100	100	100	100	100	100	100
	0.5	71	82	85	87	88	89	90
	1.0	52	65	69	72	74	76	77
	2.0	31	42	47	49	53	55	56
	3.0	20	28	32	34	38	40	42
	4.0	14	19	22	24	27	30	32
	8.0	3	5	6	7	9	10	11

Table 5. *(Continued)*

DEPTH DOSE PERCENTAGES FOR CIRCULAR FIELDS FOR 70- AND 120-KVP X RAYS, VARIOUS FILTRATIONS AND FOCAL-TO-SKIN DISTANCES (FSD)

FSD = 20 cm

Area, cm²		0	3.1	7.0	12.5	28.3	50.0	100.0
Diameter, cm		0	2	3	4	6	8	11.3
	0.0	100	100	100	100	100	100	100
	0.5	72	83	86	88	89	90	91
	1.0	54	66	71	73	76	77	78
Depth, cm	2.0	33	44	49	51	55	57	58
	3.0	22	30	34	36	40	42	44
	4.0	15	21	24	26	30	32	34
	8.0	4	6	7	8	10	11	13

FSD = 30 cm

Area, cm²		0	3.1	7.0	12.5	28.3	50.0	100.0
Diameter, cm		0	2	3	4	6	8	11.3
	0.0	100	100	100	100	100	100	100
	0.5	73	84	87	88	89	91	92
	1.0	55	68	73	74	77	79	80
Depth, cm	2.0	35	47	51	54	57	60	61
	3.0	24	33	37	39	43	45	47
	4.0	17	23	27	29	32	35	37
	8.0	5	7	8	9	11	13	15

C. HVL 3.0 mm Aluminum (Approximately 120 kvp, 1 mm Aluminum Filter)

FSD = 15 cm

Area, cm²		0	3.1	7.0	12.5	28.3	50.0	100.0
Diameter, cm		0	2	3	4	6	8	11.3
	0.0	100	100	100	100	100	100	100
	0.5	75	85	87	88	89	90	90
	1.0	58	70	74	76	77	78	80
Depth, cm	2.0	37	48	53	56	59	60	62
	3.0	24	33	37	41	45	46	48
	4.0	17	23	27	30	34	35	37
	8.0	4	6	8	9	11	13	14

FSD = 20 cm

Area, cm²		0	3.1	7.0	12.5	28.3	50.0	100.0
Diameter, cm		0	2	3	4	6	8	11.3
	0.0	100	100	100	100	100	100	100
	0.5	76	86	88	89	90	91	91
	1.0	60	72	75	77	79	80	81
Depth, cm	2.0	39	51	55	58	62	63	65
	3.0	27	35	40	43	47	49	51
	4.0	19	25	29	32	36	38	40
	8.0	5	7	9	10	12	14	16

Table 5. *(Continued)*

**DEPTH DOSE PERCENTAGES FOR CIRCULAR FIELDS FOR 70- AND 120-KVP
X RAYS, VARIOUS FILTRATIONS AND FOCAL-TO-SKIN DISTANCES (FSD)**

FSD = 30 cm

Area, cm²		0	3.1	7.0	12.5	28.3	50.0	100.0
Diameter, cm		0	2	3	4	6	8	11.3
	0.0	100	100	100	100	100	100	100
	0.5	77	86	88	90	91	92	92
	1.0	62	74	77	79	81	82	83
Depth, cm	2.0	41	54	58	61	65	66	67
	3.0	29	39	43	46	51	53	55
	4.0	21	28	32	35	40	42	44
	8.0	6	9	10	12	14	17	19

D. HVL 4.0mm (Approximately 140 kvp, 2.0 mm Aluminum Filter)

FSD = 15 cm

Area, cm²		0	3.1	7.0	12.5	28.3	50.0	100.0
Diameter, cm		0	2	3	4	6	8	11.3
	0.0	100	100	100	100	100	100	100
	0.5	78	87	89	90	91	92	93
	1.0	62	74	77	79	80	81	84
Depth, cm	2.0	40	52	56	59	62	63	67
	3.0	27	37	41	44	47	49	53
	4.0	19	26	30	32	36	38	42
	8.0	5	8	9	10	12	14	17

FSD = 20 cm

Area, cm²		0	3.1	7.0	12.5	28.3	50.0	100.0
Diameter, cm		0	2	3	4	6	8	11.3
	0.0	100	100	100	100	100	100	100
	0.5	79	88	89	90	92	93	94
	1.0	63	76	78	80	82	83	86
Depth, cm	2.0	43	55	59	62	64	66	70
	3.0	30	40	44	46	49	52	56
	4.0	21	29	32	35	38	41	45
	8.0	6	9	10	12	14	16	19

FSD = 30 cm

Area, cm²		0	3.1	7.0	12.5	28.3	50.0	100.0
Diameter, cm		0	2	3	4	6	8	11.3
	0.0	100	100	10 0	100	100	100	100
	0.5	80	90	91	92	93	94	95
	1.0	65	78	81	82	83	84	87
Depth, cm	2.0	45	58	62	65	68	69	73
	3.0	32	43	47	50	54	56	60
	4.0	24	32	36	38	42	45	49
	8.0	7	11	12	14	17	19	22

From: Johns, H. E., *The Physics of Radiology*, 2nd ed., pp. 705–706. Charles C Thomas, Springfield, Illinois, 1964.

Table 6.
DEPTH DOSE PERCENTAGES FOR CIRCULAR FIELDS, X RAYS OF HVL = 0.5 AND 1 mm COPPER

A. HVL 0.5 mm Copper, FSD 40 cm

Depth, cm	Area of Field, cm²								
	0	20	35	50	80	100	150	200	400
0	100.0	100.0	100.0	100.0	100.0	100.0	100.0	100.0	100.0
1	74.6	91.7	93.6	94.7	96.4	97.0	98.0	98.6	99.3
2	56.5	78.1	81.5	83.4	86.0	86.9	88.8	89.9	91.9
3	43.2	64.8	68.9	71.6	74.6	76.0	78.4	80.0	83.4
4	33.3	52.9	57.7	60.5	64.2	65.6	68.1	69.7	73.9
5	25.8	43.3	47.8	50.9	54.6	56.2	59.0	61.0	65.1
6	20.0	35.4	39.3	42.4	46.0	47.5	50.5	52.8	57.0
7	15.5	28.9	32.6	35.6	38.8	40.1	43.2	45.4	49.8
8	12.1	23.7	27.1	29.5	32.5	34.0	36.8	39.0	43.5
9	9.4	19.4	22.3	24.7	27.3	28.7	31.4	33.4	37.5
10	7.4	16.1	18.4	20.5	23.0	24.3	26.6	28.5	32.7
11	5.8	13.2	15.3	17.0	19.3	20.5	22.5	24.3	28.2
12	4.6	10.8	12.8	14.3	16.3	17.4	19.2	20.8	24.5
13	3.7	8.8	10.7	12.0	13.7	14.7	16.3	17.6	21.1
14	2.9	7.3	8.9	10.0	11.5	12.3	13.9	15.3	18.3
15	2.4	6.0	7.4	8.3	9.7	10.4	11.8	13.0	15.7
16	1.9	4.9	6.1	6.9	8.2	8.8	10.1	11.1	13.6
17	1.5	4.1	5.1	5.8	6.9	7.4	8.6	9.6	11.7
18	1.2	3.4	4.2	4.8	5.8	6.3	7.3	8.2	10.1
19	1.0	2.8	3.5	4.0	4.9	5.3	6.2	7.0	8.7
20	0.8	2.3	2.9	3.4	4.1	4.5	5.3	5.9	7.5

B. HVL 0.5 mm Copper, FSD 50 cm

Depth, cm	Area of Field, cm²								
	0	20	35	50	80	100	150	200	400
0	100.0	100.0	100.0	100.0	100.0	100.0	100.0	100.0	100.0
1	75.3	92.3	94.3	95.4	97.1	97.7	98.7	99.3	100.0
2	55.7	79.0	82.5	84.4	87.0	88.0	89.9	91.0	93.0
3	44.5	66.0	70.2	72.9	76.0	77.4	79.8	81.5	84.9
4	34.5	54.3	59.2	62.1	65.9	67.3	69.9	71.6	75.9
5	27.0	44.7	49.3	52.5	56.3	58.0	60.9	62.9	67.2
6	21.1	36.7	40.8	44.0	47.7	49.3	52.4	54.8	59.1
7	16.5	30.1	34.0	37.1	40.4	41.8	45.0	47.3	51.9
8	13.0	24.8	28.3	30.8	34.0	35.5	38.5	40.8	45.2
9	10.1	20.4	23.4	25.9	28.6	30.1	32.9	35.0	39.4
10	8.0	16.9	19.4	21.6	24.2	25.6	28.0	30.0	34.4
11	6.3	13.9	16.2	18.0	20.4	21.6	23.8	25.7	29.8
12	5.1	11.4	13.5	15.1	17.2	18.4	20.3	22.0	25.9
13	4.1	9.4	11.3	12.7	14.5	15.6	17.3	18.7	22.4
14	3.3	7.7	9.4	10.6	12.2	13.1	14.8	16.2	19.4
15	2.6	6.4	7.8	8.8	10.3	11.1	12.6	13.8	16.7
16	2.1	5.3	6.5	7.4	8.7	9.4	10.8	11.8	14.5
17	1.7	4.3	5.4	6.2	7.3	7.9	9.2	10.2	12.5
18	1.4	3.6	4.5	5.2	6.2	6.7	7.8	8.7	10.8
19	1.1	3.0	3.8	4.3	5.2	5.7	6.6	7.5	9.3
20	0.9	2.4	3.1	3.6	4.4	4.8	5.6	6.4	8.1

Table 6. *(Continued)*
DEPTH DOSE PERCENTAGES FOR CIRCULAR FIELDS,
X RAYS OF HVL = 0.5 AND 1 mm COPPER

C. HVL 1.0 mm Copper, FSD 40 cm

Depth, cm	Area of Field, cm²								
	0	20	35	50	80	100	150	200	400
0	100.0	100.0	100.0	100.0	100.0	100.0	100.0	100.0	100.0
1	78.3	93.5	96.2	97.5	99.2	100.1	101.3	101.9	102.3
2	61.7	82.1	87.2	89.0	92.0	93.0	94.7	95.6	97.1
3	49.0	71.1	75.9	79.0	83.1	84.7	87.1	88.9	91.4
4	39.0	60.5	65.5	68.8	73.2	75.2	78.2	80.3	84.2
5	31.1	50.9	55.8	59.3	63.9	65.6	69.1	71.3	75.5
6	25.0	42.8	47.4	50.7	55.1	57.1	60.3	62.6	67.4
7	20.0	35.8	40.1	43.2	47.4	49.3	52.7	55.1	59.9
8	16.1	29.8	33.7	36.5	40.5	42.6	45.7	48.1	53.1
9	13.0	24.9	28.5	31.0	34.7	36.7	39.9	41.9	46.9
10	10.4	20.8	24.9	26.4	29.6	31.4	34.4	36.4	41.5
11	8.4	17.4	20.3	22.4	25.3	27.0	29.6	31.6	36.4
12	6.7	14.6	17.1	19.0	21.5	23.1	25.6	27.5	31.8
13	5.4	12.2	14.4	16.0	18.4	19.7	22.0	23.9	27.8
14	4.4	10.2	12.2	13.6	15.7	16.9	19.0	20.7	24.3
15	3.5	8.5	10.2	11.5	13.5	14.5	16.3	17.8	21.3
16	2.8	7.1	8.6	9.7	11.5	12.4	14.0	15.4	18.6
17	2.3	6.0	7.2	8.3	9.8	10.6	12.1	13.3	16.3
18	1.9	5.0	6.1	7.0	8.3	9.0	10.4	11.5	14.3
19	1.5	4.2	5.2	5.9	7.1	7.8	8.9	9.9	12.5
20	1.2	3.5	4.4	5.0	6.1	6.7	7.7	8.5	10.9

D. HVL 1.0 mm Copper, FSD 50 cm

Depth, cm	Area of Field, cm²								
	0	20	35	50	80	100	150	200	400
0	100.0	100.0	100.0	100.0	100.0	100.0	100.0	100.0	100.0
1	79.0	94.2	96.9	98.2	99.9	100.8	102.0	102.6	103.0
2	63.0	83.2	88.3	90.2	93.2	94.2	95.9	96.9	98.4
3	50.5	72.5	77.4	80.5	84.7	86.3	88.8	90.6	93.5
4	40.5	62.0	67.2	70.6	75.1	77.1	80.2	82.4	86.4
5	32.5	52.5	57.5	61.1	65.9	67.6	71.2	73.5	77.8
6	26.3	44.4	49.1	52.5	57.1	59.2	62.5	64.9	69.8
7	21.3	37.3	41.8	45.0	49.4	51.4	54.8	57.3	62.3
8	17.3	31.2	35.2	38.2	42.4	44.6	47.8	50.3	55.5
9	14.0	26.1	29.9	32.5	36.4	38.5	41.8	43.9	49.3
10	11.3	21.9	25.2	27.8	31.2	33.1	36.2	38.3	43.6
11	9.1	18.3	21.4	23.7	26.7	28.5	31.3	33.4	38.5
12	7.4	15.4	18.2	20.1	22.8	24.4	27.1	29.1	33.8
13	5.9	12.9	15.3	17.0	19.5	20.9	23.4	25.3	29.5
14	4.8	10.8	13.0	14.4	16.7	17.9	20.2	21.9	25.8
15	3.9	9.1	10.8	12.2	14.3	15.4	17.4	18.9	22.7
16	3.2	7.6	9.1	10.3	12.2	13.2	14.9	16.4	19.8
17	2.6	6.4	7.7	8.8	10.4	11.3	12.9	14.2	17.3
18	2.1	5.3	6.5	7.4	8.9	9.6	11.1	12.3	15.2
19	1.7	4.5	5.5	6.3	7.6	8.3	9.5	10.6	13.3
20	1.4	3.7	4.7	5.4	6.5	7.1	8.2	9.1	11.6

From: Johns, H. E., *The Physics of Radiology*, 2nd ed., pp. 707–708. Charles C Thomas, Springfield, Illinois, 1964.

Table 7.
DEPTH DOSE PERCENTAGES FOR CIRCULAR FIELDS,
X RAYS OF HVL = 3 mm COPPER

A. HVL 3.0 mm Copper, FSD 50 cm

Depth, cm	Area of Field, cm²								
	0	20	35	50	80	100	150	200	400
0	100.0	100.0	100.0	100.0	100.0	100.0	100.0	100.0	100.0
1	82.3	94.7	96.5	97.4	98.6	99.0	100.0	100.5	101.4
2	68.0	85.8	88.2	89.8	91.7	92.7	94.3	95.4	97.6
3	56.2	75.0	78.8	81.0	84.1	85.4	87.5	89.2	92.4
4	46.4	64.8	69.1	71.8	75.4	77.0	79.8	81.8	85.9
5	38.6	56.0	60.0	63.0	66.8	68.6	71.6	73.9	78.4
6	32.0	47.7	52.0	54.9	58.8	60.9	64.0	66.4	71.0
7	26.5	40.8	44.8	47.8	51.8	54.0	56.9	59.4	61.4
8	22.0	34.9	38.7	41.5	45.5	47.6	50.4	53.0	58.2
9	18.4	29.7	33.3	36.0	39.8	41.7	44.6	47.2	52.2
10	15.4	25.3	28.6	31.1	34.7	36.6	39.5	41.8	46.8
11	12.8	21.7	24.6	26.9	30.3	32.0	34.8	37.2	41.9
12	10.7	18.5	21.1	23.2	26.4	27.9	30.6	32.7	37.3
13	9.0	15.7	18.2	20.0	22.9	24.4	26.9	28.8	33.3
14	7.5	13.4	15.7	17.3	19.9	21.2	23.6	25.4	29.5
15	6.3	11.5	13.4	15.0	17.3	18.5	20.7	22.4	26.3
16	5.3	9.8	11.5	12.9	15.0	16.4	18.2	19.7	23.4
17	4.5	8.4	9.9	11.2	13.4	14.0	15.9	17.4	20.8
18	3.7	7.2	8.5	9.6	11.1	12.2	14.0	15.4	18.5
19	3.1	6.1	7.3	8.3	9.9	10.7	12.3	13.6	16.5
20	2.6	5.2	6.3	7.2	8.6	9.3	10.8	11.9	14.6

B. HVL 3.0 mm Copper, FSD 60 cm

Depth, cm	Area of Field, cm²								
	0	20	35	50	80	100	150	200	300
0	100.0	100.0	100.0	100.0	100.0	100.0	100.0	100.0	100.0
1	82.9	95.3	97.1	98.0	99.2	99.5	100.6	101.1	102.0
2	68.8	86.7	89.2	90.8	92.7	93.7	95.3	96.4	98.7
3	57.3	76.2	80.1	82.3	85.4	86.8	88.9	90.6	93.9
4	47.5	66.1	70.5	73.2	76.8	78.5	81.3	83.3	87.4
5	39.8	57.5	61.4	64.5	68.3	70.2	73.2	75.5	80.1
6	33.2	49.1	53.4	56.4	60.3	62.5	65.6	68.1	72.8
7	27.6	42.2	46.2	49.3	53.3	55.6	58.6	61.1	66.2
8	23.1	36.2	40.0	42.9	47.0	49.1	52.0	54.7	60.0
9	19.4	30.9	34.6	37.3	41.2	43.2	46.2	48.9	54.0
10	16.3	26.4	29.8	32.3	36.1	38.0	41.0	43.3	48.5
11	13.6	22.7	25.7	28.0	31.5	33.3	36.2	38.6	43.4
12	11.4	19.4	22.1	24.2	27.5	29.1	31.9	34.1	38.8
13	9.6	16.5	19.1	20.9	24.0	25.5	28.1	30.1	34.8
14	8.1	14.2	16.5	18.1	20.9	22.2	24.7	26.6	30.9
15	6.8	12.2	14.2	15.7	18.2	19.5	21.8	23.6	27.6
16	5.8	10.4	12.2	13.6	15.8	17.0	19.2	20.8	24.6
17	4.9	8.9	10.5	11.8	13.9	14.8	16.8	18.4	21.9
18	4.1	7.6	9.1	10.2	12.1	12.9	14.8	16.3	19.6
19	3.5	6.5	7.9	8.9	10.5	11.4	13.1	14.4	17.5
20	2.9	5.6	6.8	7.7	9.2	9.9	11.5	12.7	15.6

Table 7. *(Continued)*

DEPTH DOSE PERCENTAGES FOR CIRCULAR FIELDS, X RAYS OF HVL = 3 mm COPPER

C. HVL 3.0 mm Copper, FSD 80 cm

Depth, cm	Area of Field, cm^2								
	0	20	35	50	80	100	150	200	400
0	100.0	100.0	100.0	100.0	100.0	100.0	100.0	100.0	100.0
1	83.8	95.9	97.8	98.6	99.7	100.1	101.1	101.6	102.5
2	70.0	88.0	90.3	92.0	93.9	94.8	96.5	97.6	99.7
3	58.6	77.9	81.7	83.8	86.9	88.2	90.4	92.1	95.4
4	49.0	68.1	72.3	75.2	78.7	80.4	83.3	85.3	89.5
5	41.3	59.5	63.4	66.5	70.4	72.3	75.3	77.7	82.3
6	34.7	51.1	55.5	58.6	62.6	64.8	68.0	70.5	75.1
7	29.2	44.1	48.3	51.4	55.5	57.8	60.9	63.6	68.9
8	24.5	38.1	42.0	44.9	49.1	51.4	54.4	57.1	62.6
9	20.7	32.7	36.5	39.2	43.3	45.3	48.4	51.2	56.4
10	17.5	28.1	31.5	34.1	38.0	40.0	43.1	45.7	50.9
11	14.7	24.2	27.3	29.7	33.4	35.2	38.2	40.7	45.8
12	12.5	20.8	23.6	25.8	29.2	30.9	33.8	36.1	41.0
13	10.5	17.8	20.4	22.2	25.5	27.1	29.9	31.9	36.8
14	8.9	15.3	17.7	19.4	22.3	23.7	26.3	28.2	32.8
15	7.6	13.2	15.2	16.9	19.4	20.8	23.2	25.0	29.4
16	6.4	11.3	13.2	14.7	17.0	18.2	20.5	22.1	26.3
17	5.4	9.7	11.4	12.8	14.9	15.9	18.1	19.7	23.5
18	4.6	8.3	9.9	11.1	13.0	13.9	15.9	17.5	21.0
19	3.9	7.2	8.5	9.6	11.4	12.3	14.1	15.5	18.8
20	3.3	6.2	7.4	8.4	9.9	10.7	12.4	13.7	16.7

D. HVL 3.0 mm Copper, FSD 100 cm

Depth, cm	Area of Field, cm^2								
	0	20	35	50	80	100	150	200	400
0	100.0	100.0	100.0	100.0	100.0	100.0	100.0	100.0	100.0
1	84.0	96.4	98.1	99.0	100.1	100.5	101.5	101.9	102.8
2	70.7	88.8	91.1	92.7	94.5	95.5	97.0	98.2	100.3
3	59.6	79.0	82.7	84.8	87.9	88.2	91.4	93.0	96.3
4	50.1	69.3	73.5	76.3	79.8	81.5	84.4	86.4	90.6
5	42.4	60.8	64.7	67.8	71.7	74.6	76.7	79.1	83.7
6	35.7	52.4	56.8	59.9	63.9	66.1	69.3	71.8	76.6
7	30.1	45.5	49.6	52.8	56.9	59.3	62.3	65.0	70.3
8	25.4	39.3	43.4	46.3	50.6	52.8	55.8	58.6	64.1
9	21.6	33.9	37.7	40.6	44.7	46.7	49.8	52.6	58.0
10	18.3	29.2	32.7	35.4	39.3	41.4	44.5	47.0	52.4
11	15.4	25.2	28.4	30.9	34.6	36.5	39.5	42.1	47.3
12	13.1	21.7	24.6	26.9	30.4	32.1	35.0	37.4	42.4
13	11.1	18.6	21.3	23.4	26.6	28.2	31.0	33.1	38.0
14	9.5	16.1	18.5	20.4	23.3	24.7	27.4	29.4	34.0
15	8.1	13.8	16.0	17.8	20.4	21.7	24.2	26.1	30.5
16	6.9	11.9	13.9	15.4	17.8	19.0	21.4	23.4	27.3
17	5.8	10.3	12.1	13.5	15.6	16.7	18.9	20.5	24.4
18	5.0	8.9	10.5	11.7	13.7	14.6	16.7	18.2	19.6
19	4.2	7.6	9.2	10.2	12.0	12.9	14.7	16.2	19.6
20	3.6	6.6	7.9	8.9	10.5	11.3	13.0	14.3	17.5

From: Johns, H. E., *The Physics of Radiology*, 2nd ed., pp. 714–715. Charles C Thomas, Springfield, Illinois, 1964.

Table 8.
DEPTH DOSE PERCENTAGES FOR CIRCULAR FIELDS,
X RAYS OF HVL = 4 mm COPPER

A. HVL 4.0 mm Copper, FSD 50 cm

Depth, cm	Area of Field, cm²								
	0	20	35	50	80	100	150	200	400
0	100.0	100.0	100.0	100.0	100.0	100.0	100.0	100.0	100.0
1	83.1	94.4	96.0	96.8	97.7	98.0	98.8	99.3	100.0
2	69.3	85.9	87.8	89.1	90.8	91.6	93.0	93.9	96.0
3	57.8	75.6	78.8	80.7	83.3	84.3	86.2	87.6	90.1
4	48.2	65.5	69.5	71.8	75.0	76.4	78.9	80.5	84.2
5	40.7	56.6	60.4	63.2	66.6	68.2	71.2	73.4	77.1
6	34.3	48.5	52.7	55.5	58.9	60.8	63.8	66.1	70.2
7	28.9	41.6	45.6	48.4	51.8	53.7	56.8	59.4	63.8
8	24.4	35.7	39.5	42.0	45.5	47.3	50.5	53.1	57.8
9	20.5	30.6	34.0	36.5	39.8	41.6	44.8	47.3	51.8
10	17.3	26.3	29.4	31.6	35.0	36.7	39.7	42.0	46.6
11	14.6	22.6	25.4	27.4	30.6	32.3	35.1	37.4	41.8
12	12.4	19.4	21.9	23.7	26.8	28.4	30.9	33.1	37.5
13	10.5	16.7	19.0	20.6	23.4	24.9	27.3	29.2	33.6
14	8.9	14.3	16.4	17.9	20.4	21.8	24.1	25.8	30.0
15	7.5	12.3	14.1	15.5	17.8	19.0	21.2	22.8	26.7
16	6.4	10.6	12.2	13.5	15.6	16.7	18.7	20.1	23.8
17	5.4	9.1	10.6	11.7	13.6	14.6	16.4	17.7	21.2
18	4.6	7.8	9.1	10.2	11.8	12.8	14.4	15.7	18.9
19	4.0	6.7	7.9	8.8	10.3	11.2	12.6	13.8	16.9
20	3.4	5.8	6.8	7.7	9.0	9.7	11.1	12.2	15.1

B. HVL 4.0 mm Copper, FSD 80 cm

Depth, cm	Area of Field, cm²								
	0	20	35	50	80	100	150	200	400
0	100.0	100.0	100.0	100.0	100.0	100.0	100.0	100.0	100.0
1	84.6	95.6	97.2	97.9	98.8	99.1	99.8	100.3	101.0
2	71.4	88.0	89.7	91.1	92.8	93.6	95.0	95.9	97.9
3	60.2	78.5	81.5	83.4	86.0	87.0	88.9	90.4	93.2
4	50.9	68.8	72.7	75.0	78.2	79.7	82.2	83.9	87.5
5	43.5	60.2	63.9	66.7	70.2	71.9	74.9	77.1	80.9
6	37.2	52.2	56.3	59.2	62.7	64.8	67.8	70.1	71.2
7	31.7	45.3	49.2	52.1	55.7	57.8	60.9	63.6	67.9
8	27.1	39.3	43.1	45.7	49.4	51.3	54.6	57.3	62.1
9	23.1	34.0	37.5	40.1	43.6	45.4	48.8	51.5	56.0
10	19.7	29.5	32.7	35.0	38.6	40.4	43.6	46.1	50.8
11	16.8	25.6	28.5	30.6	34.0	35.8	38.8	41.3	45.9
12	14.4	22.2	24.8	26.7	29.9	31.6	34.4	36.8	41.4
13	12.3	19.3	21.6	23.4	26.4	27.9	30.6	32.7	37.4
14	10.5	16.6	18.8	20.4	23.1	24.7	27.2	29.1	33.6
15	9.0	14.4	16.3	17.8	20.3	21.6	24.0	25.9	30.1
16	7.7	12.5	14.2	15.6	17.8	19.1	21.3	23.0	27.0
17	6.6	10.8	12.4	13.7	15.7	16.8	18.9	20.3	24.2
18	5.7	9.4	10.8	12.0	13.8	14.8	16.7	18.1	21.7
19	4.9	8.1	9.4	10.5	12.1	13.1	14.8	16.1	19.5
20	4.2	7.0	8.2	9.1	10.6	11.5	13.1	14.3	17.6

From: Johns, H. E., *The Physics of Radiology*, 2nd ed., p. 716. Charles C Thomas, Springfield, Illinois, 1964.

Table 9.

DEPTH DOSE PERCENTAGES FOR ^{137}Cs FOR SEVERAL FIELD SIZES,
SOURCE-TO-SKIN DISTANCES (SSD),
AND DIAPHRAGM-TO-SKIN DISTANCES (DSD)

A. Source 1.5 cm, SSD 15 cm, DSD = 0 (Courtesy of *Brit. J. Radiol.*)

Depth, cm	Circular Fields, Diameter, cm					Rectangular Fields, cm × cm		
	0	1	6	8	10	4 × 6	6 × 8	6 × 10
0.15	100.0	100.0	100.0	100.0	100.0	100.0	100.0	100.0
0.5	94.0	94.6	95.0	95.2	95.6	95.2	97.0	98.0
1	84.6	86.5	87.6	88.5	89.2	88.6	90.5	91.3
2	66.3	73.4	75.2	76.5	77.0	75.5	77.7	78.5
3	54.6	62.0	64.2	66.0	66.7	64.4	67.0	67.7
4	44.5	52.4	54.8	56.8	57.6	54.5	57.5	58.3
5	36.6	44.4	47.2	49.0	50.0	46.3	49.5	50.2
6	30.3	37.5	40.4	42.4	43.3	40.2	42.5	43.3
7	25.4	32.0	34.6	36.5	37.5	34.3	36.5..	37.2
8	21.2	27.5	29.8	31.5	32.4	29.5	31.5	32.0
9	17.7	23.6	25.7	27.5	28.6	25.3	27.3	28.0
10	15.0	20.4	22.5	24.2	25.3	21.9	23.8	24.4
11	12.8	17.4	19.5	21.3	22.4	19.0	20.9	21.5
12	11.0	15.2	17.0	18.8	19.7	16.7	18.4	18.9
13	9.4	13.2	14.9	16.6	17.6	14.6	16.3	16.7
14	8.0	11.4	13.1	14.7	15.6	12.8	14.4	14.7
15	6.8	9.9	11.5	13.0	13.8	11.3	12.6	12.9
16	5.9	8.6	10.1	11.5	12.3	9.8	11.1	11.3
17	5.1	7.6	8.9	10.2	10.9	8.6	9.8	10.0
18	4.4	6.6	7.8	9.0	9.7	7.5	8.6	8.8
19	3.8	5.8	6.9	7.9	8.6	6.5	7.6	7.7
20	3.3	5.1	6.0	7.0	7.6	5.7	6.7	6.8

B. Source 2.7 cm, SSD 35 cm, DSD 6 cm (Courtesy of *Brit. J. Radiol.*)

Depth, cm	Area of Field, cm²					
	0	25	50	100	200	400
0	0.0	35.0	50.0	65.0	75.0	80.0
0.15	100.0	100.0	100.0	100.0	100.0	100.0
1	88.1	93.4	94.2	94.5	94.9	95.0
2	76.4	85.1	86.4	87.1	87.6	88.2
3	66.4	76.8	78.8	80.0	80.7	81.6
4	57.7	69.1	71.6	73.2	74.2	75.3
5	50.1	62.1	64.8	66.7	68.0	69.3
6	43.6	55.6	58.4	60.6	62.2	63.6
7	37.9	49.7	52.5	54.7	56.7	58.3
8	33.0	44.4	47.2	49.3	51.5	53.3
9	28.8	39.6	42.4	44.5	46.7	48.7
10	25.3	35.3	38.1	40.2	42.3	44.5
11	22.3	31.5	34.2	36.3	38.5	40.6
12	19.7	28.1	30.7	32.8	35.0	37.2
13	17.3	25.1	27.5	29.6	31.9	34.0
14	15.2	22.4	24.6	26.7	29.0	31.0
15	13.4	20.0	22.0	24.1	26.3	28.3
16	11.8	17.8	19.7	21.7	23.9	25.8
17	10.4	15.9	17.7	19.6	21.7	23.6
18	9.2	14.2	15.9	17.7	19.7	21.6
19	8.1	12.7	14.3	16.0	17.9	19.8
20	7.2	11.3	12.8	14.4	16.2	18.0

Table 9. *(Continued)*
DEPTH DOSE PERCENTAGES FOR ^{137}Cs FOR SEVERAL FIELD SIZES, SOURCE-TO-SKIN DISTANCES (SSD), AND DIAPHRAGM-TO-SKIN DISTANCES (DSD)

C. Source, 3.2 cm, SSD 50 cm, DSD 10 cm (Courtesy of C. S. Simons et al. and *Am. J. Roentgenol.*)

Depth, cm	Area of Circular Field, cm^2					
	0	25	50	75	100	150
0.15	100.0	100.0	100.0	100.0	100.0	100.0
0.5	95.8	98.0	98.2	98.4	98.8	98.6
1	90.1	94.4	95.2	95.6	96.0	95.7
2	80.2	87.0	88.0	88.4	88.8	89.6
3	70.6	79.6	81.2	81.8	82.4	83.3
4	62.6	72.0	74.2	75.3	76.0	76.8
5	55.5	65.2	67.6	68.5	70.0	71.0
6	49.3	58.4	61.4	62.5	64.0	65.3
7	43.9	52.8	55.6	57.0	58.4	59.8
8	38.9	47.4	50.4	51.9	53.4	54.8
9	34.7	42.8	45.3	47.3	48.6	50.2
10	30.8	38.4	41.2	42.8	44.4	46.1
11	27.5	34.4	37.2	38.7	40.4	41.6
12	24.4	30.6	33.6	35.6	36.4	38.0
13	21.8	27.4	30.2	31.8	33.4	34.6
14	19.4	24.8	27.2	29.0	30.2	31.4
15	17.3	22.4	24.6	26.3	27.4	28.7
16	15.5	20.0	22.2	24.0	24.8	26.1
17	13.9	17.8	20.0	21.7	22.6	23.8
18	12.4	16.0	18.0	19.8	20.4	21.5
19	11.1	14.4	16.2	17.7	18.4	19.7
20	9.9	13.0	14.6	16.0	16.8	17.7

From: Johns, H. E., *The Physics of Radiology*, 2nd ed., pp. 719–720. Charles C Thomas, Springfield, Illinois, 1964.

Table 10.

DEPTH DOSE PERCENTAGES FOR ^{60}Co FOR SEVERAL FIELD SIZES AND SOURCE-TO-SKIN DISTANCES (SSD)

A. Average Photon Energy 1.25 Mev, HVL 11 mm Lead, SSD 50 cm

Depth, cm	Area of Field, cm^2					
	0	20	50	100	200	400
0.5	100.0	100.0	100.0	100.0	100.0	100.0
1	94.6	96.2	97.0	97.5	97.6	97.7
2	85.2	89.2	90.6	91.4	91.8	92.1
3	76.8	82.3	84.2	85.4	86.1	86.8
4	69.3	75.7	78.2	79.6	80.6	81.0
5	62.6	69.5	72.4	74.0	75.3	76.0
6	56.4	63.7	66.8	68.6	70.2	71.3
7	51.0	58.3	61.4	63.4	65.3	67.1
8	41.6	53.3	56.4	58.6	60.7	62.7
9	41.7	48.7	51.7	53.9	56.2	58.6
10	37.8	44.5	47.4	49.7	52.2	54.9
11	34.3	40.6	43.5	45.8	48.4	51.2
12	31.1	37.1	40.0	42.2	45.0	47.8
13	28.2	33.9	36.7	39.0	41.7	44.7
14	25.6	31.0	33.7	36.0	38.7	41.7
15	23.3	28.4	30.9	33.2	36.0	39.0
16	21.1	26.0	28.4	30.6	33.4	36.5
17	19.3	23.8	26.1	28.3	31.1	34.2
18	17.5	21.8	24.0	26.2	28.9	32.0
19	15.9	19.9	22.2	24.2	26.9	29.9
20	14.5	18.2	20.3	22.4	25.0	28.1

B. Average Photon Energy 1.25 Mev, HVL 11 mm Lead, SSD 60 cm

Depth, cm	Area of Field, cm^2					
	0	20	50	100	200	400
0.5	100.0	100.0	100.0	100.0	100.0	100.0
1	95.0	96.7	97.1	97.8	97.9	98.1
2	86.0	90.1	91.2	92.2	92.6	93.0
3	77.9	83.7	85.4	86.6	87.4	88.0
4	70.7	77.6	79.7	81.2	82.3	83.2
5	64.2	71.7	74.2	75.9	77.3	78.4
6	58.3	66.1	68.9	70.7	72.4	73.7
7	53.0	60.8	63.7	65.7	67.6	69.2
8	48.2	55.8	58.8	60.9	63.0	65.0
9	43.9	51.2	54.2	56.4	58.6	60.9
10	39.9	46.9	49.9	52.2	54.5	57.1
11	36.3	43.0	46.0	48.3	50.7	53.4
12	33.1	39.4	42.4	44.7	47.2	50.0
13	30.2	36.1	39.1	41.4	44.0	47.0
14	27.5	33.1	36.0	38.3	41.0	44.0
15	25.1	30.4	33.2	35.5	38.2	41.2
16	22.9	27.9	30.6	32.9	35.6	38.6
17	20.9	25.7	28.2	30.5	33.2	36.2
18	19.1	23.7	26.0	28.3	31.0	34.1
19	17.4	21.8	24.0	26.2	28.9	32.0
20	15.9	20.0	22.1	24.2	27.0	30.0

Table 10. *(Continued)*

DEPTH DOSE PERCENTAGES FOR ^{60}Co FOR SEVERAL FIELD SIZES AND SOURCE-TO-SKIN DISTANCES (SSD)

C. Average Photon Energy 1.25 Mev, HVL 11 mm Lead, SSD 80 cm

Depth, cm	Area of Field, cm^2					
	0	20	50	100	200	400
0.5	100.0	100.0	100.0	100.0	100.0	100.0
1	95.4	97.0	97.7	98.2	98.4	98.5
2	87.1	91.0	92.5	93.4	93.7	94.0
3	79.5	85.3	87.2	88.4	89.0	89.6
4	72.7	79.6	82.0	83.4	84.4	85.2
5	66.5	74.1	76.9	78.5	79.9	80.8
6	60.8	68.9	71.8	73.7	75.2	76.4
7	55.6	63.8	66.8	68.9	70.7	72.1
8	50.9	58.9	62.1	64.2	66.3	68.0
9	46.6	54.3	57.5	59.8	62.1	64.1
10	42.7	50.1	53.3	55.7	58.1	60.3
11	39.2	46.2	49.4	51.8	54.3	56.7
12	35.9	42.6	45.8	48.2	50.8	53.3
13	32.9	39.3	42.4	44.9	47.6	50.1
14	30.2	36.3	39.3	41.8	44.5	47.1
15	27.7	33.5	36.4	38.9	41.8	44.3
16	25.4	31.0	33.8	36.2	39.0	41.7
17	23.3	28.7	31.3	33.8	36.5	39.2
18	21.4	26.5	29.0	31.4	34.2	36.9
19	19.6	24.5	27.0	29.3	32.0	34.7
20	18.0	22.6	25.0	27.3	30.0	32.7

D. Average Photon Energy 1.25 Mev, HVL 11 mm Lead, SSD 100 cm

Depth, cm	Area of Field, cm^2					
	0	20	50	100	200	400
0.5	100.0	100.0	100.0	100.0	100.0	100.0
1	95.9	97.2	97.9	98.6	98.8	98.8
2	87.9	91.7	93.0	94.0	94.5	94.6
3	80.7	86.3	88.1	89.4	90.1	90.5
4	73.8	81.0	83.2	84.8	85.7	86.4
5	67.8	75.7	78.1	80.2	81.3	82.3
6	62.3	70.6	73.6	75.6	76.9	78.2
7	57.3	65.7	68.8	71.0	72.5	74.1
8	52.7	61.0	64.2	66.5	68.3	70.1
9	48.5	56.5	59.7	62.1	64.2	66.2
10	44.7	52.3	55.5	57.9	60.3	62.5
11	41.2	48.4	51.6	54.0	56.6	58.8
12	38.0	44.8	48.0	50.4	53.1	55.4
13	35.0	41.5	44.6	47.1	49.8	52.2
14	32.2	38.5	41.5	44.0	46.7	49.2
15	29.6	35.7	38.6	41.4	43.8	46.4
16	27.2	33.1	35.9	38.4	41.1	43.7
17	25.0	30.7	33.4	35.9	38.6	41.2
18	23.0	28.5	31.1	33.6	36.3	38.8
19	21.2	26.4	29.0	31.4	34.1	36.6
20	19.5	24.4	27.0	29.2	32.0	34.5

From: Johns, H. E., *The Physics of Radiology*, 2nd ed., pp. 721–722. Charles C Thomas, Springfield, Illinois, 1964.

Depth, cm	4 × 4	4 × 6	4 × 8	4 × 10	4 × 15	4 × 20	6 × 6	6 × 8	6 × 10	Rectangular 6 × 15
*	111.6	113.7	114.9	115.8	117.0	117.6	116.4	118.2	119.4	121.1
0	100.0	100.0	100.0	100.0	100.0	100.0	100.0	100.0	100.0	100.0
1	93.9	95.1	95.6	95.9	96.3	96.5	96.5	97.1	97.5	98.1
2	84.6	86.2	87.1	87.6	88.4	88.7	88.3	89.4	90.1	91.2
3	73.7	76.0	77.3	78.1	79.2	79.7	78.8	80.5	81.6	83.1
4	63.1	65.8	67.4	68.4	69.6	70.3	69.0	71.0	72.4	74.1
5	54.2	56.7	58.4	59.5	60.9	61.6	60.1	62.2	63.7	65.6
6	46.3	48.7	50.4	51.5	53.1	53.8	52.0	54.2	55.7	57.8
7	39.3	41.7	43.4	44.5	46.1	46.8	44.9	47.1	48.6	50.8
8	33.4	35.7	37.3	38.5	40.0	40.7	38.7	40.9	42.4	44.5
9	28.5	30.6	32.1	33.2	34.7	35.4	33.4	35.4	36.9	38.9
10	24.3	26.2	27.6	28.6	30.0	30.8	28.7	30.6	32.0	34.0
11	20.7	22.4	23.7	24.6	26.0	26.7	24.7	26.4	27.7	29.6
12	17.6	19.1	20.4	21.2	22.5	23.1	21.2	22.8	24.0	25.7
13	15.1	16.3	17.5	18.3	19.5	20.0	18.2	19.7	20.8	22.4
14	12.9	14.0	15.0	15.7	16.9	17.4	15.7	17.0	18.0	19.5
15	11.0	12.0	12.8	13.5	14.6	15.1	13.5	14.6	15.5	17.0
16	9.4	10.3	11.0	11.6	12.6	13.1	11.6	12.6	13.4	14.8
17	8.0	8.8	9.4	10.0	10.9	11.4	10.0	10.9	11.6	12.9
18	6.8	7.5	8.1	8.6	9.4	9.9	8.6	9.4	10.0	11.2
19	5.8	6.4	7.0	7.4	8.1	8.6	7.4	8.1	8.7	9.8
20	4.9	5.5	6.0	6.4	7.1	7.5	6.3	7.0	7.6	8.5

* The first line gives the dose at the maximum for 100 r of primary.

From: Johns, H. E., *The Physics of Radiology*, 2nd ed., p. 731. Charles C Thomas, Springfield, Illinois, 1964.

11.
RECTANGULAR FIELDS OF VARIOUS DIMENSIONS

Fields, cm × cm

6 × 20	8 × 8	8 × 10	8 × 15	8 × 20	10 × 10	10 × 15	10 × 20	15 × 15	15 × 20	20 × 20
122.1	120.4	121.9	124.1	125.3	123.7	126.2	127.7	129.6	131.5	133.7
100.0	100.0	100.0	100.0	100.0	100.0	100.0	100.0	100.0	100.0	100.0
98.3	97.9	98.3	99.1	99.3	98.9	99.7	100.0	100.6	101.0	101.4
91.5	90.7	91.6	92.8	93.3	92.6	93.9	94.5	95.6	96.2	96.8
83.6	82.4	83.7	85.4	86.3	85.1	87.0	88.1	89.5	90.8	92.3
75.0	73.4	74.9	77.1	78.1	76.7	79.1	80.3	82.1	83.8	85.7
66.6	64.8	66.5	68.8	70.0	68.4	71.1	72.5	74.5	76.2	78.3
58.8	56.8	58.6	61.1	62.3	60.6	63.5	65.0	67.0	68.9	71.3
51.8	49.7	51.5	54.1	55.3	53.6	56.5	58.1	60.1	62.0	64.2
45.6	43.4	45.2	47.7	49.0	47.2	50.1	51.7	53.7	55.6	57.9
40.0	37.8	39.6	42.0	43.3	41.5	44.3	45.9	47.9	49.8	52.0
35.0	32.8	34.5	36.9	38.1	36.3	39.1	40.6	42.6	44.5	46.7
30.6	28.5	30.0	32.3	33.5	31.7	34.4	35.8	37.7	39.6	41.7
26.7	24.8	26.1	28.3	29.4	27.7	30.3	31.5	33.3	35.1	37.2
23.3	21.5	22.7	24.7	25.8	24.2	26.6	27.8	29.4	31.1	33.1
20.3	18.6	19.7	21.6	22.6	21.1	23.3	24.5	25.9	27.5	29.4
17.7	16.1	17.1	18.9	19.8	18.4	20.4	21.6	22.9	24.4	26.2
15.5	13.9	14.9	16.5	17.4	16.0	17.9	19.0	20.2	21.6	23.2
13.5	12.0	13.0	14.4	15.3	13.9	15.7	16.7	17.8	19.1	20.7
11.8	10.4	11.3	12.6	13.4	12.2	13.8	14.7	15.7	16.9	18.5
10.3	9.0	9.8	11.1	11.8	10.7	12.1	13.0	13.9	15.0	16.4
9.0	7.8	8.5	9.7	10.3	9.3	10.6	11.4	12.3	13.3	14.5

Table
DEPTH DOSE PERCENTAGES FOR
FOR 80 cm AND 100 cm

A. FSD 80 cm

Depth, cm										Rectangular
	4 × 4	4 × 6	4 × 8	4 × 10	4 × 15	4 × 20	6 × 6	6 × 8	6 × 10	6 × 15
*	101.1	101.3	101.5	101.6	101.8	101.9	101.6	101.8	102.0	102.3
0								Surface dose 30 to 50%,		
0.5	100.0	100.0	100.0	100.0	100.0	100.0	100.0	100.0	100.0	100.0
1	96.8	97.0	97.2	97.3	97.4	97.4	97.4	97.6	97.7	97.8
2	90.6	91.2	91.5	91.6	91.8	91.8	91.9	92.2	92.5	92.7
3	84.7	85.5	85.9	86.1	86.4	86.4	86.5	86.9	87.3	87.6
4	79.0	79.9	80.4	80.6	81.0	81.1	81.1	81.7	82.1	82.5
5	73.5	74.5	75.1	75.3	75.7	75.9	75.9	76.6	77.0	77.5
6	68.1	69.2	69.9	70.1	70.5	70.7	70.7	71.5	71.9	72.5
7	62.9	64.1	64.8	65.1	65.5	65.7	65.7	66.5	67.0	67.6
8	58.0	59.2	59.9	60.3	60.8	61.0	60.8	61.7	62.2	62.9
9	53.5	54.7	55.3	55.8	56.3	56.6	56.2	57.1	57.7	58.5
10	49.3	50.5	51.1	51.6	52.2	52.5	52.0	52.9	53.5	54.4
11	45.5	46.6	47.3	47.8	48.4	48.6	48.4	49.0	49.6	50.5
12	41.9	43.0	43.7	44.2	44.8	45.4	44.5	45.4	46.0	46.9
13	38.6	39.7	40.4	40.9	41.4	41.8	41.1	42.0	42.7	43.6
14	35.6	36.6	37.3	37.8	38.4	38.7	38.0	38.9	39.6	40.5
15	32.9	33.8	34.5	35.0	35.6	35.9	35.2	36.1	36.7	37.6
16	30.4	31.3	32.0	32.4	33.1	33.4	32.6	33.5	34.1	35.0
17	28.1	29.0	29.6	30.0	30.7	31.0	30.2	31.1	31.6	32.6
18	26.0	26.9	27.4	27.9	28.5	28.8	28.0	28.8	29.4	30.0
19	24.0	24.9	25.4	25.9	26.5	26.8	26.0	26.7	27.4	28.2
20	22.1	22.9	23.5	23.9	24.5	24.8	24.0	24.8	25.4	26.2

B. FSD 100 cm

Depth, cm										Rectangular
	4 × 4	4 × 6	4 × 8	4 × 10	4 × 15	4 × 20	6 × 6	6 × 8	6 × 10	6 × 15
*	101.1	101.3	101.5	101.6	101.8	101.9	101.6	101.8	102.0	102.3
0								Surface dose 30 to 50%,		
0.5	100.0	100.0	100.0	100.0	100.0	100.0	100.0	100.0	100.0	100.0
1	97.1	97.3	97.5	97.6	97.7	97.7	97.7	97.9	98.0	98.2
2	91.4	91.9	92.2	92.4	92.5	92.6	92.6	92.9	93.1	93.4
3	85.8	86.5	86.9	87.2	87.3	87.5	87.5	87.9	88.2	88.6
4	80.2	81.2	81.7	82.0	82.2	82.4	82.4	83.0	83.4	83.8
5	74.8	76.0	76.6	76.9	77.2	77.4	77.3	78.1	78.6	79.0
6	69.7	70.9	71.6	71.9	72.3	72.5	72.4	73.2	73.8	74.3
7	64.8	66.0	66.7	67.1	67.5	67.7	67.6	68.4	69.0	60.6
8	60.1	61.3	62.0	62.4	62.9	63.1	62.9	63.8	64.4	65.1
9	55.7	56.9	57.6	58.0	58.5	58.8	58.4	59.4	60.0	60.7
10	51.5	52.7	53.4	53.8	54.4	54.7	54.2	55.2	55.8	56.6
11	47.7	48.8	49.5	49.9	50.5	50.8	50.3	51.3	51.9	52.7
12	44.1	45.2	45.9	46.3	46.9	47.2	46.7	47.7	48.2	49.1
13	40.8	41.9	42.6	43.0	43.6	43.9	43.3	44.3	44.9	45.8
14	37.8	38.9	39.5	40.0	40.6	40.9	40.2	41.2	41.8	42.7
15	35.0	36.1	36.7	37.2	37.8	38.1	37.4	38.3	38.9	39.9
16	32.5	33.5	34.1	34.5	35.2	35.5	34.8	35.6	36.3	37.2
17	30.1	31.1	31.7	32.1	32.8	33.1	32.3	33.1	33.8	34.7
18	27.9	28.8	29.4	29.8	30.5	30.8	30.0	30.8	31.5	32.4
19	25.8	26.7	27.3	27.7	28.4	28.7	27.9	28.7	29.3	30.2
20	23.8	24.7	25.3	25.7	26.4	26.7	25.9	26.7	27.3	28.2

* The first line gives the dose at the maximum for 100 r of primary.

Note that percentages are relative to 100% at 0.5 cm, the approximate wall thickness of a ^{60}Co r chamber. Doses at the surface 0.5-cm depth at the surface of the phantom.

From: Johns, H. E., *The Physics of Radiology*, 2nd ed., pp. 734–735. Charles C Thomas, Springfield, Illinois, 1964.

12.
^{60}Co FOR RECTANGULAR FIELDS, FOCAL-TO-SKIN DISTANCES

Fields, cm × cm

6 × 20	8 × 8	8 × 10	8 × 15	8 × 20	10 × 10	10 × 15	10 × 20	15 × 15	15 × 20	20 × 20
102.5	102.1	102.3	102.7	102.9	102.5	103.0	103.3	103.6	104.1	104.6
depending upon collimator										
100.0	100.0	100.0	100.0	100.0	100.0	100.0	100.0	100.0	100.0	100.0
97.8	97.8	98.0	98.1	98.1	98.2	98.3	98.3	98.4	98.4	98.4
92.8	92.7	93.0	93.2	93.3	93.3	93.6	93.6	93.9	93.9	94.0
87.7	87.6	87.9	88.3	88.5	88.3	88.8	88.9	89.3	89.4	89.6
82.7	82.5	82.9	83.4	83.6	83.4	84.0	84.2	84.7	84.9	85.2
77.7	77.4	77.9	78.5	78.8	78.5	79.2	79.5	80.1	80.4	80.8
72.7	72.4	73.0	73.7	74.0	73.6	74.4	74.7	75.4	75.8	76.4
67.9	67.5	68.1	68.9	69.2	68.8	69.8	70.1	70.8	71.4	72.1
63.3	62.7	63.4	64.3	64.7	64.1	65.2	65.7	66.5	67.2	68.0
58.9	58.2	58.9	59.9	60.4	59.7	60.9	61.4	62.3	63.1	64.0
54.8	54.0	54.8	55.8	56.3	55.6	56.9	57.4	58.4	59.2	60.2
51.0	50.1	50.9	52.0	52.5	51.7	53.1	53.7	54.7	55.6	56.6
47.4	46.5	47.3	48.4	49.0	48.1	49.5	50.2	51.2	52.1	53.2
44.1	43.2	44.0	45.1	45.7	44.8	46.2	46.9	47.9	48.8	50.0
41.0	40.1	40.9	42.0	42.6	41.8	43.1	43.9	44.9	45.8	47.0
38.1	37.2	38.0	39.2	39.8	38.9	40.3	41.0	42.0	43.0	44.2
35.5	34.5	35.3	36.5	37.1	36.2	37.6	38.8	39.3	40.3	41.5
33.1	32.1	32.8	34.0	34.6	33.7	35.1	35.8	36.8	37.8	39.0
30.8	29.8	30.5	31.7	32.3	31.4	32.8	33.5	34.5	35.5	36.7
28.7	27.7	28.4	29.6	30.2	29.2	30.7	31.4	32.3	33.4	34.6
26.8	25.7	26.4	27.6	28.2	27.2	28.6	29.4	30.3	31.4	32.6

Fields, cm × cm

6 × 20	8 × 8	8 × 10	8 × 15	8 × 20	10 × 10	10 × 15	10 × 20	15 × 15	15 × 20	20 × 20
102.5	102.1	102.3	102.7	103.0	102.5	103.0	103.4	103.7	104.1	104.6
depending upon collimator										
100.0	100.0	100.0	100.0	100.0	100.0	100.0	100.0	100.0	100.0	100.0
98.2	98.1	98.3	98.5	98.5	98.6	98.8	98.8	99.0	98.9	98.9
93.4	93.3	93.6	93.9	93.9	93.9	94.3	94.3	94.6	94.6	94.7
88.6	88.4	88.9	89.3	89.3	89.3	89.8	89.8	90.2	90.3	90.5
83.9	83.7	84.2	84.7	84.8	84.7	85.3	85.4	85.9	86.1	86.3
79.2	78.9	79.6	80.1	80.3	80.1	80.8	81.0	81.6	81.9	82.2
74.5	74.2	74.9	75.6	75.8	75.5	76.3	76.6	77.3	77.7	78.1
69.9	69.5	70.2	71.0	71.3	70.9	71.8	72.2	73.0	73.5	74.0
65.4	64.9	65.6	66.5	66.9	66.4	67.4	67.9	68.7	69.3	70.0
61.1	60.5	61.2	62.1	62.6	62.0	63.1	63.7	64.5	65.2	66.1
57.0	56.3	57.0	58.0	58.6	57.8	59.0	59.7	60.6	61.3	62.3
53.2	52.4	53.1	54.2	54.8	53.9	55.2	55.9	56.9	57.7	58.7
49.6	48.7	49.5	50.7	51.2	50.3	51.7	52.4	53.4	54.3	55.3
46.3	45.4	46.1	47.3	47.9	47.0	48.4	49.1	50.2	51.1	52.1
43.2	42.3	43.0	44.2	44.8	43.9	45.3	46.0	47.1	48.1	49.1
40.3	39.4	40.1	41.3	41.9	41.0	42.4	43.1	44.2	45.2	46.2
37.7	36.7	37.4	38.6	39.2	38.3	39.7	40.4	41.5	42.5	43.5
35.2	34.2	34.9	36.1	36.7	35.8	37.2	37.9	39.0	40.0	41.0
32.9	31.9	32.6	33.8	34.4	33.5	34.9	35.6	36.7	37.6	38.6
30.7	29.7	30.5	31.6	32.3	31.3	32.7	33.4	34.5	35.4	36.4
28.7	27.7	28.5	29.6	30.2	29.3	30.6	31.3	32.4	33.3	34.4

are 30 to 50% of the peak dose. The actual peak dose obtained is given in the first line relative to an r chamber reading at the nominal

Table 13.
TUMOR–AIR RATIOS FOR RECTANGULAR FIELDS FOR X RAYS OF HVL 3 mm COPPER AND ^{60}Co, RELATIVE TO THE AIR DOSE MEASURED IN A PARALLEL BEAM (INFINITE FSD)

A. HVL 3 mm Copper, FSD ∞ (Courtesy of *Am. J. Roentgenol.*)

Depth, cm	Field Size at Axis of Rotation, cm × cm												
	4 × 4	4 × 6	4 × 8	4 × 10	4 × 15	6 × 6	6 × 8	6 × 10	6 × 15	8 × 8	8 × 10	8 × 15	10 × 10	10 × 15
0	1.120	1.140	1.150	1.160	1.180	1.160	1.180	1.200	1.220	1.210	1.230	1.240	1.240	1.260
1	1.100	1.120	1.140	1.160	1.170	1.150	1.180	1.200	1.220	1.210	1.230	1.250	1.250	1.280
2	1.000	1.050	1.070	1.090	1.110	1.100	1.130	1.150	1.180	1.170	1.200	1.230	1.220	1.260
3	0.895	0.946	0.964	0.988	1.010	1.000	1.040	1.070	1.100	1.107	1.130	1.160	1.150	1.200
4	0.798	0.844	0.868	0.888	0.917	0.898	0.936	0.970	1.000	0.988	1.030	1.070	1.060	1.120
5	0.705	0.752	0.776	0.798	0.830	0.809	0.850	0.877	0.915	0.902	0.943	0.990	0.978	1.040
6	0.624	0.668	0.697	0.721	0.750	0.728	0.771	0.798	0.836	0.820	0.860	0.905	0.902	0.959
7	0.549	0.588	0.615	0.641	0.670	0.645	0.686	0.712	0.750	0.731	0.776	0.822	0.815	0.876
8	0.476	0.546	0.543	0.565	0.595	0.566	0.607	0.634	0.674	0.655	0.700	0.745	0.736	0.796
9	0.419	0.450	0.479	0.500	0.529	0.500	0.539	0.565	0.600	0.580	0.622	0.669	0.660	0.722
10	0.363	0.394	0.417	0.438	0.468	0.439	0.473	0.497	0.535	0.513	0.549	0.596	0.587	0.645
11	0.317	0.345	0.366	0.384	0.412	0.382	0.417	0.436	0.473	0.450	0.489	0.531	0.520	0.574
12	0.273	0.299	0.319	0.337	0.364	0.335	0.365	0.381	0.417	0.398	0.434	0.474	0.461	0.512
13	0.239	0.261	0.279	0.296	0.320	0.294	0.320	0.337	0.370	0.350	0.381	0.419	0.407	0.455
14	0.206	0.228	0.244	0.259	0.281	0.257	0.281	0.298	0.326	0.308	0.335	0.371	0.360	0.405
15	0.182	0.199	0.214	0.228	0.248	0.226	0.247	0.260	0.288	0.270	0.295	0.329	0.318	0.361
16	0.158	0.174	0.187	0.198	0.218	0.198	0.217	0.230	0.255	0.238	0.259	0.290	0.281	0.320
17	0.138	0.152	0.165	0.175	0.192	0.174	0.190	0.202	0.225	0.210	0.228	0.258	0.249	0.287
18	0.120	0.133	0.143	0.153	0.169	0.151	0.166	0.177	0.200	0.185	0.200	0.228	0.220	0.256
19	0.105	0.116	0.126	0.135	0.149	0.133	0.146	0.156	0.176	0.163	0.176	0.202	0.194	0.226
20	0.092	0.102	0.110	0.118	0.132	0.117	0.129	0.137	0.156	0.143	0.157	0.180	0.172	0.201

Table 13. (*Continued*)

TUMOR–AIR RATIOS FOR RECTANGULAR FIELDS FOR X RAYS OF HVL 3 mm COPPER AND ⁶⁰Co, RELATIVE TO THE AIR DOSE MEASURED IN A PARALLEL BEAM (INFINITE FSD)

B. ⁶⁰Co, FSD ∞ (Courtesy of *Am. J. Roentgenol.*)

Depth, cm	\	\	\	\	\	Field Size at Axis of Rotation, cm × cm								
	4 × 4	4 × 6	4 × 8	4 × 10	4 × 15	6 × 6	6 × 8	6 × 10	6 × 15	8 × 8	8 × 10	8 × 15	10 × 10	10 × 15
0.5	1.011	1.013	1.014	1.016	1.019	1.016	1.019	1.020	1.023	1.021	1.023	1.026	1.026	1.029
1	0.993	0.998	1.000	1.010	1.020	0.990	0.996	1.000	1.005	1.005	1.010	1.115	1.020	1.025
2	0.950	0.958	0.961	0.965	0.968	0.966	0.973	0.978	0.984	0.981	0.986	0.992	0.992	0.998
3	0.910	0.916	0.925	0.930	0.936	0.928	0.941	0.946	0.952	0.948	0.954	0.962	0.959	0.968
4	0.866	0.878	0.884	0.889	0.893	0.893	0.903	0.908	0.916	0.913	0.920	0.926	0.927	0.936
5	0.823	0.835	0.845	0.850	0.854	0.852	0.869	0.871	0.878	0.872	0.882	0.890	0.888	0.900
6	0.780	0.794	0.802	0.808	0.815	0.811	0.822	0.829	0.839	0.836	0.844	0.854	0.853	0.867
7	0.739	0.752	0.761	0.768	0.774	0.769	0.783	0.788	0.800	0.794	0.804	0.815	0.811	0.829
8	0.693	0.708	0.716	0.723	0.732	0.726	0.740	0.748	0.759	0.754	0.764	0.778	0.775	0.792
9	0.653	0.668	0.676	0.683	0.694	0.685	0.700	0.708	0.723	0.716	0.726	0.740	0.736	0.756
10	0.613	0.628	0.637	0.644	0.653	0.646	0.661	0.670	0.683	0.676	0.687	0.703	0.699	0.717
11	0.577	0.591	0.600	0.608	0.616	0.608	0.622	0.632	0.649	0.639	0.650	0.669	0.662	0.682
12	0.541	0.555	0.563	0.570	0.580	0.573	0.587	0.597	0.611	0.602	0.614	0.632	0.628	0.648
13	0.509	0.521	0.531	0.538	0.549	0.538	0.552	0.565	0.580	0.568	0.580	0.599	0.595	0.615
14	0.476	0.489	0.498	0.505	0.516	0.507	0.522	0.532	0.547	0.537	0.549	0.568	0.563	0.584
15	0.448	0.461	0.469	0.476	0.488	0.476	0.491	0.503	0.518	0.505	0.518	0.536	0.533	0.553
16	0.420	0.432	0.440	0.448	0.459	0.449	0.463	0.473	0.487	0.478	0.489	0.508	0.503	0.525
17	0.398	0.409	0.416	0.421	0.433	0.421	0.436	0.448	0.460	0.450	0.460	0.480	0.475	0.499
18	0.371	0.383	0.391	0.397	0.408	0.398	0.411	0.421	0.434	0.425	0.436	0.454	0.449	0.472
19	0.351	0.361	0.370	0.375	0.386	0.376	0.387	0.398	0.410	0.399	0.411	0.430	0.421	0.448
20	0.330	0.340	0.348	0.354	0.364	0.354	0.366	0.374	0.386	0.378	0.387	0.404	0.399	0.425

Note the higher backscatter percentages for the lower-energy X rays.

From: Johns, H.E., *The Physics of Radiology*, 2nd ed. Charles C Thomas, Springfield, Illinois, 1964.

Table 14.
ENERGY FLUX PER ROENTGEN, PHOTON FLUX PER ROENTGEN,
AND EXPOSURE DOSE RATE PER MILLICURIE
AS A FUNCTION OF PHOTON ENERGY

hn Photon Energy, Mev	Energy-Absorption Coefficient $(\mu_{en})_{air}$, cm^2/g	Energy Flux per Roentgen, ergs/cm^2/r	Photon Flux per Roentgen, N/cm^2/r	Roentgens per Hour per Millicurie at 1 cm
0.010	4.6600	18.8	11.7×10^8	9.020
0.015	1.2900	68.0	28.30	3.740
0.020	0.5160	170.0	53.10	2.000
0.030	0.1470	597.0	124.00	0.853
0.040	0.0640	1,370.0	214.00	0.495
0.050	0.0384	2,284.0	286.00	0.371
0.060	0.0292	3,003.0	313.00	0.339
0.080	0.0236	3,716.0	290.00	0.365
0.100	0.0231	3,796.0	237.00	0.447
0.150	0.0251	3,494.0	146.00	0.728
0.200	0.0268	3,272.0	102.00	1.090
0.300	0.0288	3,045.0	63.40	1.670
0.400	0.0296	2,963.0	46.30	2.290
0.500	0.0297	2,953.0	36.90	2.870
0.600	0.0296	2,963.0	30.80	3.440
0.800	0.0289	3,035.0	23.70	4.470
1.000	0.0280	3,132.0	19.60	5.420
1.500	0.0255	3,439.0	14.30	7.400
2.000	0.0234	3,748.0	11.70	9.050
3.000	0.0205	4,278.0	8.91	11.900
4.000	0.0186	4,715.0	7.36	14.400
5.000	0.0173	5,069.0	6.33	16.700
6.000	0.0163	5,380.0	5.60	18.900
8.000	0.0150	5,847.0	4.57	23.200
10.000	0.0144	6,090.0	3.80	27.900

From: Johns, H. E., *The Physics of Radiology*, 2nd ed., p. 694. Charles C Thomas, Springfield, Illinois, 1964.

These types of curves often are fitted by the n-target model

$$S_D = 1 - (1 - e^{-\lambda D})^n,$$

where n is termed the extrapolation (or target) number, and λ the probability that unit radiation dose D will "hit" any of the n targets.

Although useful for describing much of the accumulated data, this model is of limited use in describing the influence of many factors on radiosensitivity, as pointed out by Burch,[15] who has developed more elaborate multiple-hit models. Dose–response curves determined in terms of chromosome breaks in the plants *Tradescantia* and *Vicia* show that two-break aberrations increase nearly as the square of the dose for radiation of low LET at high intensity, but increase at a lower rate and almost linearly with the dose at low intensity.[16] For high LET radiation (see the neutron curve of Figure 6), the two-break aberrations increase almost linearly with dose and independently of intensity.

Figure 7 illustrates the type of dose–response relationships that may be observed in studying genetic mutations at the animal level. One may note that, although large confidence intervals must be attached to each data point as a result of statistical limitations in observing the relatively rare mutations with an animal population of limited size, a definite increase in mutation rate with dose is, nevertheless, observed. Also, high dose rates again appear more effective than low dose rates.

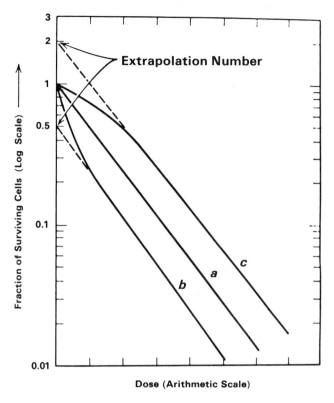

FIGURE 5.
Types of Mammalian Cell Survival Curve.
[Logarithm of fraction of cells S surviving the (acute) irradiation against dose D]
(From: Morgan, K. Z. and Turner, J. E., eds., *Principles of Radiation
Protection—A Textbook of Health Physics*, p. 377. John Wiley and Sons,
New York, New York, 1967.)
Curve *a*: $S = 2D$; observed when homogeneous cells are irradiated by
particles (such as low-energy natural α particles) of high LET. Curve
b gives an extrapolation number of less than unity (0.5 in example);
observed when heterogeneous mixtures of cells (of low and high radiosensitivity) are irradiated. Curve *c* gives an extrapolation number of
more than unity (2 in example); observed when most types of mammalian
cells are irradiated by, for example, ^{60}Co γ rays—that is, by radiation
of low average LET.

When a response such as mortality (1 − survival) of animals is plotted as a function of radiation dose in rad, curves such as those in Figure 8 are obtained. The S-shaped curve typical of many chemically toxic agents[19] is observed here for radiation.

When plotting this type of mortality data for mice exposed to high-energy protons (440 and 730 Mev) and to X rays, Bradley et al.[20] found that a probit mortality (or probability) versus log (dose) scale linearized the dose–response curves and made them approximately parallel for given strains exposed to different LET radiations (see Figure 9). The point where each line crosses the 50% mortality level is called the LD_{50} for that type of radiation, that age, strain and species of animal, and that time interval of observation. Thus, whatever the basic mechanisms at the cellular and subcellular levels, they may often combine to produce an effect represented at the animal or human level by the empirical log-normal dose–response function.[19,21] This is the functional shape that the probit-type plot is designed to linearize for convenience in bioassay of toxic agents or pharmaceuticals, and statistical methods of curve fitting and analysis are available for these functions.[19]

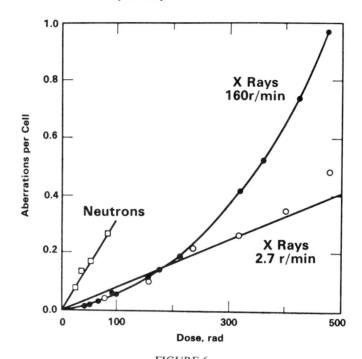

FIGURE 6.
Dose Curves for Two-Break Chromosomal Aberrations.
(From: Morgan, K. Z. and Turner, J. E., eds.,
Principles of Radiation Protection—A Textbook of Health Physics, p. 404.
John Wiley and Sons, New York, New York, 1967.)
The X-ray data are replotted from Sax (1941), the neutron data from Giles
(1943). The neutron doses were given in *n* units. and have been con-
verted to rad by multiplication with the factor 2.5 (this conversion is very
approximate and should not be relied upon).

A summary of the log-normal function, its properties, and simplified procedures for its application are presented in Reference 21. Briefly, suppose that x is the dose of a population of animals and that $P(\ln X \leq \ln x)$ is the probability of animals having "sensitivities" or lethal effects at doses $X \leq x$; then P may be considered as the (expected) fraction of animals dying in a group administered a dose x, and a log-normally distributed P would be given by the function

$$P(\ln X \leq \ln x) = \frac{1}{\sqrt{2\pi}\sigma_g} \int_{\ln X = \ln 0 = -\infty}^{\ln X = \ln x} \exp\left[-(\ln X - \ln \mu_g)^2/2\,\sigma^2_g\right] d(\ln X).$$

Here, σ_g is the standard deviation ($\sqrt{\mathrm{Var}(\ln X)}$) in the frequency distribution of $\ln X$, $\overline{\ln}$ is the mean value of $\ln X$, and μ_g is the "geometric mean of X". The above expression is the form for the cumulative integral over a normal distribution of $\ln X$, and has been shown to be a special case of a family of logarithmic distributions.[22] The LD_{50} value where a probit-type mortality plot (Figure 9) crosses the 50% mortality line would be an estimate of μ_g of the log-normal dose–response function (if the data indicate that a straight line on such a plot is applicable). The parameter σ_g is the standard deviation in $\ln X$ (the natural logarithm of the randomly distributed dose required to kill an animal), and this parameter may also be easily estimated from the probit plot. The estimated value of σ_g can be obtained from the equation

$$\sigma_g = \ln x_{0.8413} - \ln x_{0.50} = \ln(x_{0.8413}/x_{0.50}),$$

or

$$\sigma_g = \ln x_{0.50} - \ln x_{0.1587} = \ln(x_{0.50}/x_{0.1587}),$$

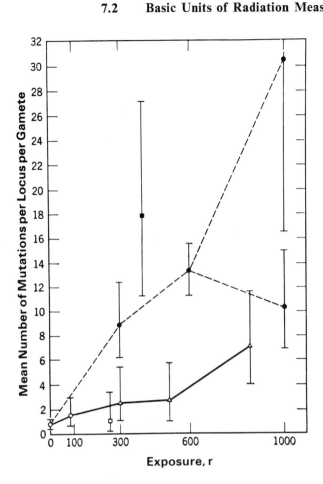

FIGURE 7.

Exposure Curves for Specific-Locus Mutation in the Mouse.
(From: Morgan, K. Z. and Turner, J. E., eds.,
Principles of Radiation Protection—
A Textbook of Health Physics, p. 416.
John Wiley and Sons, New York, New York, 1967.)
90% confidence intervals shown. Solid points represent
results with acute X rays (80 to 90 r/min). Open points
are chronic γ-ray results (triangles and square, 90/r week;
circles, 10 r/week). Squares are mutation rates
in females; all other points are mutation rates in males.
The point for zero dose is the sum of all male controls.
The top 1000-r point represents results of a single exposure;
the lower represents results of successive exposures
to 600 and 400 r respectively, separated by an
interval of 15 weeks. (Russel et al., 1960.)

where $x_{0.8413}$ is the value of x where the " % mortality versus log X line" crosses the 84.13% mortality value, and $x_{0.50} = \mu_g$ is the value of x where the line crosses the 50% mortality value. Now the "standard geometric deviation"[21] or geometric mean standard deviation[22] is defined as

$$s_g = x_{0.8413}/x_{0.50} = x_{0.50}/x_{0.1587},$$

and the average dose $\bar{x}$ required to kill an animal is then given by the relationship[21]

$$\log_{10} \bar{x} = \log \mu_g + 1.1513(\log_{10} s_g)^2.$$

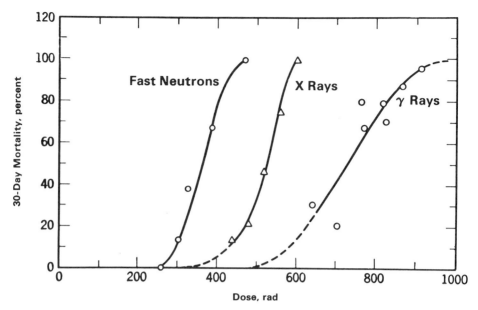

FIGURE 8.
Thirty-Day Mortality of Mice Exposed to X-Rays, Fast Neutrons, and Gamma Rays.
(The mice were 9 to 12 weeks old at the time of whole-body irradiation.)
Reproduced by permission of Academic Press, Inc. and A. C. Upton et al.
(From: Morgan, K. Z. and Turner, J. E., eds.,
Principles of Radiation Protection—
A Textbook of Health Physics, p. 429.
John Wiley and Sons, New York, New York, 1967.)

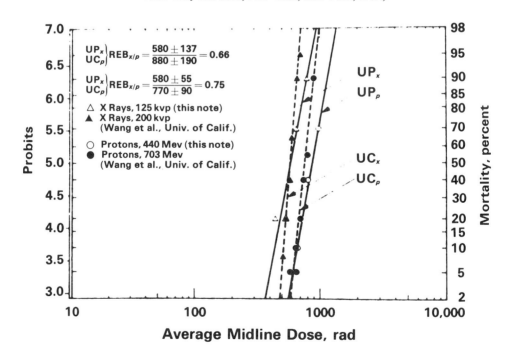

FIGURE 9.
Relationship Between Probit of Mortality and Average Midline Dose
of X and Proton Radiation to Mice.
(From: Bradley, F. J., Watson, J. A., Doolittle, D. P.,
Brodsky, A., and Sutton, R. B. *Health Phys.*, 10: 72, 1964.)

The particular significance of μ_g and s_g for a log-normal dose–response relationship is that $\mu \overset{x}{\div} s_g$, and not $\bar{x} \pm \sigma_x$, gives the interval in which 68 percent of the population "sensitivities" will lie, if X is interpreted as the dose to which a fraction of the population is sensitive enough to die (or manifest some other biological end point). The concept of s_g may also be useful, for example, in predicting that less than 2.5 percent of the population are likely to require doses greater than $s_g^2 \mu_g$ to die; in other words, if the log-normal function holds at higher doses, less than about 2.5 percent of the population would survive more than s_g^2 times the LD_{50}. On

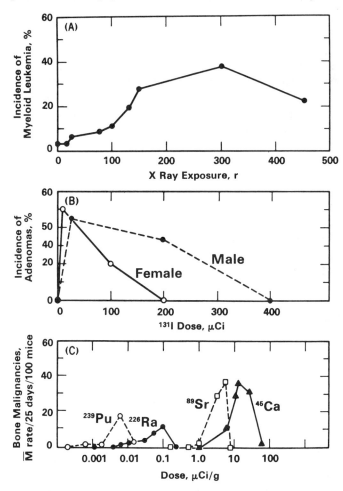

FIGURE 10.
Incidence of Various Neoplasms in Relation to Radiation Dose.
(From: Morgan, K. Z. and Turner, J. E., eds.,
Principles of Radiation Protection—
A Textbook of Health Physics, p. 434.
John Wiley and Sons, New York, New York, 1967.)

the other hand, 2.5 percent of the population would be affected by doses less than the LD_{50} divided by s_g^2. Thus, the S-shaped or log-normal dose–response relationship would predict that at low doses some finite, although small, effect is still possible.

The S-shaped curve has also been suggested by some of the animal data on carcinogenesis. However, when dose levels are so high that cell lethality is produced, the rising portion of the S-shaped curve may not be evident (see Figure 10). However, a stochastic two-stage model of carcinogenesis,[23] involving a second event conditioned on a specific prior event followed by the cell growth stage of Arley,[24] predicts a dose–response curve shape so similar to the log-normal that it could not be distinguished from it by any reasonably sized experiment. Thus, the variation in the dose X at which the biological end point (e.g., an observed tumor by time T) may occur could be interpreted as resulting largely from ordinary chance variations in the action of a radiation dose on an individual animal rather than from real individual variations in sensitivity between animals. Some purely stochastic variation of this kind, as well as real variations in individual susceptibility, may also be expected to contribute to the range of doses over which lethality may occur. In any case, however, the type of dose–response relationship represented by any of these S-shaped curves would yield a predicted incidence of cancer at low doses or dose rates that would be lower[23] than the incidence obtained by linear extrapolations through the origin from cancer incidences observed at higher doses or dose rates. Leukemogenesis, on the other hand, may indeed be linearly related to dose.[18]

Thus, from such considerations, conservative estimates of biomedical effects at the population level may be made for purposes of setting standards for radiation protection. Some order-of-magnitude estimates of the biomedical effects of radiation per unit dose are given below for use in assessing the potential hazards of "maximum credible accidents" or inadvertent exposures deemed possible in individual applications.

ESTIMATION OF HUMAN RISKS OF LOW-LEVEL RADIATION EXPOSURE

Lethal Dose (Whole Body). $LD_{50}^{30} = 300$ to 500 rem when received within a short period of time (say, less than 1 day) and without therapeutic treatment.[17]

Shortening of Life Span. Estimated at 1 to 4 days per roentgen of exposure early in life, but the exact magnitude is not well established.[17,18]

Bone Cancer. About 4 to 8 cases per million population per 70-year period per rem dose received over a 70-year life span, assuming 10 percent of the natural incidence is a result of background radiation;[15,18] the frequency at low doses may actually be much less than this upper-limit estimate.

Leukemia. An average of about 2 cases per million adults per rem average exposure to the entire population considered per year-at-risk following the exposure, although the incidence-versus-time curve peaks after a latent period that probably varies according to dose level;[23,25] however, for children irradiated *in utero* or in preconception gametal stages, recent evidence of Saxon Graham et al.[26] tends to confirm the earlier data of McMahon,[27,28] indicating a higher sensitivity by perhaps a factor of 50 to 70.

Cell Cancer. Probably no more than 250 cases per million population per 70-year period per rem dose to the population.[24]

Cataracts. More than 1,000 rads of X or gamma radiation to the lens of the eye are required, or more than 100 rads from neutrons, to cause an appreciable increase in cataract incidence.[17,18]

Genetic Effects. A single-generation "doubling dose" to double the natural mutation rate has been estimated to be 40 rems to the entire population ($40 \times 200,000,000$ man-rems to the United States population, for example).[29] Considering the natural mutation rate, a doubling dose of 40 rems to every member of a population might cause about 1 out of every 200 births in the next generation to result in a death or failure to reproduce.[24] The total numbers of induced extinctions in all future generations might be a maximum of about 40 million for a constant population size of 200 million. If the 40-rem dose were delivered to the population each generation for much longer than the average life of a mutant gene, i.e., much longer than 40 generations or 1,200 years,[30] then at equilibrium there would be an additional 40 million abortive conceptions, stillbirths, neonatal deaths, or failures to reproduce per generation per 200,000,000 population. Of course, the proportion of each of these is uncertain. In addition,

Other Radiation Effects. Some additional nonspecific health impairment or loss of vitality might result from cell loss following somatic mutations, although, if the destruction of cells is at a low rate, regeneration may prevent organ failure or ill health.[31] These additional effects, particularly by degenerative cardiovascular and renal diseases,[15] may have a smaller relative increase per rem above natural incidence levels; but since they are more prevalent, they may somewhat exceed in absolute numbers the excess deaths from radiation-induced cancer. Nevertheless, the total excess mortality induced by radiation exposure would not be expected to be more than the order of magnitude produced by genetic and carcinogenic effects.[24]

ACKNOWLEDGEMENT: This chapter is reprinted with permission from "CRC Handbook of Radioactive Nuclides", Yen Wang, Editor, The Chemical Rubber Co., Cleveland, Ohio (1969).

REFERENCES

1. International Commission on Radiological Units and Measurements, Report of the ICRU. *NBS Handbook 78*. U.S. Government Printing Office, Washington, D.C., 1959.
2. International Commission on Radiological Units and Measurements, Clinical Dosimetry—Recommendations of the ICRU. *NBS Handbook 87*, p. 38. U.S. Government Printing Office, Washington, D.C., 1963.
3. Parker, H. M., Health Physics, Instrumentation, and Radiation Protection. *Advance. Biol. Med. Phys.*, 1: 243, 1948.
4. Roesch, W. C. and Attix, H. F., eds., *Radiation Dosimetry*, 2nd ed., Vol. 1, Ch. 1. Academic Press, New York, 1968.
5. International Commission on Radiological Protection, Report of Committee 2. *Health Phys.*, 3: 1–380, 1960.
6. International Commission on Radiological Units and Measurements, Report 10a, Radiation Quantities and Units. *NBS Handbook 84*. U.S. Government Printing Office, Washington, D.C., 1962.
7. Attix, F. H. and Roesch, W. C., eds., Instrumentation. *Radiation Dosimetry*, Vol. 2. Academic Press, New York, 1966.
8. Hine, G. J. and Brownell, G. L., eds., *Radiation Dosimetry*. Academic Press, New York, 1956.
9. Price, W. J., *Nuclear Radiation Detection*, 2nd ed. McGraw-Hill, New York, 1966.
10. Morgan, K. Z. and Turner, J. E., eds., *Principles of Radiation Protection—A Textbook of Health Physics*. John Wiley and Sons, New York, 1967.
11. National Committee on Radiation Protection and Measurements, NCRP Report No. 27, Stopping Powers for Use with Cavity Chambers. *NBS Handbook 79*. U.S. Government Printing Office, Washington, D.C., 1961.
12. National Committee on Radiation Protection and Measurements, Measurement of Neutron Flux and Spectra for Physical and Biological Applications, Recommendations of the NCRP. *NBS Handbook 72*. U.S. Government Printing Office, Washington, D.C., 1960.
13. Johns, H. E., *The Physics of Radiology*, 2nd ed. Charles C Thomas, Springfield, Illinois, 1964.
14. International Commission on Radiological Units and Measurements, Report of the ICRU. *NBS Handbook 62*, U.S. Government Printing Office, Washington, D.C., 1956.
15. Burch, P. R. J., Radiation Physics. *Principles of Radiation Protection—A Textbook of Health Physics*, pp. 366–397, Morgan, K. Z. and Turner, J. E., eds. John Wiley and Sons, New York, 1967.
16. Upton, A. C. and Kimball, R. F., Radiation Biology. *Principles of Radiation Protection—A Textbook of Health Physics*, pp. 398–447, Morgan, K. Z. and Turner, J. E., eds. John Wiley and Sons, New York, 1967.
17. Wald, N., Evaluation of Human Exposure Data. *Principles of Radiation Protection—A Textbook of Health Physics*, pp. 448–496, Morgan, K. Z., and Turner, J. E., eds. John Wiley and Sons, New York, 1967.
18. United National Scientific Committee on the Effects of Atomic Radiation, Various Reports. United Nations, New York, 1958, 1962, 1964, 1966. (Continuing reviews are underway.)
19. Finney, D. J., *Probit Analysis*. Cambridge University Press, New York, 1962.
20. Bradley, F. J., Watson, J. A., Doolittle, D. P., Brodsky, A., and Sutton, R. B. *Health Phys.*, 10: 71–74, 1964.
21. Schubert, J., Brodsky, A., and Tyler, S., The Log-Normal Function as a Stochastic Model of the Distribution of Strontium-90 and Other Fission Products in Humans. *Health Phys.*, 13: 1187–1204, 1967.
22. Espenscheid, W. F., Kerker, M., and Marijevic, E. *J. Phys. Chem.*, 68: 3093, 1964.
23. Brodsky, A., A Stochastic Model of Carcinogenesis and Its Implications in the Dose–Response Plane. Presented at the Health Physics Society Meeting in Los Angeles, 1966. Abstract in *Health Phys.*, 12: 1176, 1966. Detailed treatment of the model in Brodsky, A., *A Stochastic Model of Carcinogenesis as Applied to Skin Tumors in Mice* (Dissertation). University of Pittsburgh, 1966.
24. Brodsky, A. *Am. J. Public Health*, 55(12): 1971–1992, 1965.
25. Cobb, S., Miller, M., and Wald, N., On the Estimation of the Incubation Period in Malignant Disease, I. *J. Chron. Dis.* 9(4); 385–393, 1959.

26. Graham, L. S., Levine, M. L., Lilienfeld, A. M., Schuman, L., Gibson, R., David, J. E., and Hempleman, L. H., Preconception Intrauterine and Postnatal Irradiation as Related to Leukemia. Presented at the Meeting of the American Public Health Association, New York, October 8, 1964.
27. McMahon, B. *J. Nat. Cancer Inst.*, 28: 1173, 1962.
28. United Nations Scientific Committee on the Effects of Atomic Radiation, Report. United Nations, New York, 1962.
29. National Academy of Sciences, The Biological Effects of Atomic Radiation. *Summary Reports*. National Research Council, Washington, D.C., 1956.
30. Muller, H. J., Radiation and Human Mutations. *Sci. Amer.*, 193: 58–68, 1955.
31. Henshaw, P. S. *Health Phys.*, 1: 141–151, 1958.

Radiation Protection and Regulations

GUIDES FOR LIMITING ORGAN AND WHOLE-BODY DOSE

The history and rationale behind the limits of exposure recommended by the ICRP, NCRP and FRC have been reviewed recently by Morgan.[1] We shall list—for reference only—some of the more important guides for limiting exposure from sources external and internal to the body. These guides were generally arrived at by the consensus of individual scientists, considering the types of dose–response data illustrated in previous sections. As stated by Morgan,[2] "The goal is to avoid all unnecessary radiation exposure, to receive exposure even within the limits discussed above when—and only when—the expected benefits exceed the likely harmful consequences of radiation exposure." Of course, the "harmful consequences" may be estimated by the reader from the data in the preceding sections; in most cases he will probably find the probabilities of harm small compared to those of many other occupations and the normal risks of daily life,[3] if radiation exposures are kept well below the recommended guide levels.

Table 15 presents the basic recommended limits of occupational exposure to various parts of the body for a period of 13 weeks and for one year, and cumulative limits according to age for certain organs. Although the quarterly limits may differ somewhat from those recommended by the various committees and agencies, the more important annual limits are practically identical. The most important rule to remember is that the whole-body dose should remain within the cumulative limit 5(N-18) rems, where N is the age after the 18th birthday. If this limit is met for workers exposed in a general field of external radiation, and if no appreciable additional extremity exposure or intake of radioactive material is received (which is often the case), then the other annual limits will automatically be met. If the exposure is fairly uniformly distributed throughout the year by administrative controls, then the quarterly limits are also easily met. In any case, regulatory limits must also be met as set forth in the Code of Federal Regulations, Title 10, Part 20, and in other applicable federal, state, and local regulations, which will be discussed further on.

A summary and comparison of recommended limits for workers and for the general public is presented in Table 16, as adopted from Morgan.[1] The groups B and C are no longer separately recommended by the ICRP.[13] Also, the safety factor of 1/3 for the average dose to the "critical" population group is not suggested, but left up to individual countries to decide under specific circumstances. (The factor 1/3 may be reasonable, in the absence of other data, for use in evaluating exposure distributions from radionuclides distributed to the environment; for weapons-test fallout, for example, less than 0.4 percent of the population in the 0-to-25 age group receive exposures from ^{90}Sr in excess of 3 times the average exposure.[16]) In general, no member of the public should receive more than 10 percent of the annual whole-body dose permitted for radiation workers.

Furthermore, when large populations may be exposed to radioactivity dispersed from the environment, consumer items, etc., the limits on genetic dose presented in Table 17 must be considered. As indicated in the previous sections, the total number of mutants produced throughout all future generations is related to the man-rem product, regardless of how the

Table 15.
PRESENTLY (1967) RECOMMENDED DOSE EQUIVALENTS* TO BODY ORGANS OF OCCUPATIONAL WORKERS EXPOSED TO IONIZING RADIATION

Body Organ	Maximum Dose Equivalent In Any 13 Weeks, rem/13 weeks	Annual Permissible Dose Equivalent, rem/year	Accumulated Dose Equivalent To Age N, rem‡
Red bone marrow†	3[4-6]	5[4,6,7]	$5(N-18)$[4-7]
Total body	3[4,6-9]	5[5,6,8,9]	$5(N-18)$[4,6-9]
Head and trunk	3[4,6]	5[4,5]	$5(N-18)$[4,6,7]
Gonads	3[4-6,8,9]	5[4-9]	$5(N-18)$[4-9]
Lenses of eyes	3[4-6,8-11] 4[12] 8[13]	5[4-11] 15[13]	$5(N-18)$[4-11]
Skin	8[5,9,11] 10[4,9,15] 15[13]	30[4-6,8,9,14,15]	
Thyroid	8[5,9,11] 10[4] 15[13]	30[4-6,8,9]	
Bone	15[13]	30[8,9]	
Feet, ankles, hands, forearms	20[5] 25[4,6,9] 38[13]	75[4-6,9,14]	
Other single organs	4[5,8] 5[4] 8[13]	15[4-6,8,9,14]	

* The recommended permissible dose equivalent for the occupational worker is in addition to doses from medical and background exposure.

† Referred to in earlier reports as active "blood-forming organs."

‡ The 5(N − 18) accumulated dose equation may put a ceiling on the 13-week dose, but doses should be kept as far below even this cumulative limit as is feasible.

From: Morgan, K. Z., Maximum Permissible Exposure Levels—External and Internal. *Principles of Radiation Protection*, Ch. 14, Morgan, K. Z. and Turner, E. J., eds., 1967. By permission of John Wiley and Sons, New York.

Table 16.
DOSE-EQUIVALENT LEVELS RECOMMENDED IN EARLY REPORTS OF THE ICRP, AND FACTORS TO BE APPLIED TO OCCUPATIONAL VALUES

Exposed Group	Dose Equivalent to Gonads or Total Body		Dose Equivalent to Most Other Organs	
	Dose Rate	Factor	Dose Rate	Factor
A. Occupational worker	3.00 rem/qtr 12.00 rem/yr or 5.00 rem/yr (av)	1	8.0 rem/qtr 15.0 rem/yr (av)	1
B. Work in vicinity of controlled area	1.50 rem/yr	3/10	1.5 rem/yr	1/10
C. Visit area occupationally	1.50 rem/yr	3/10	1.5 rem/yr	1/10
D. Whole population	0.50 rem/yr 5.00 rem (av) to age 30 or 0.17 rem/yr (av) 0.05 rem/yr (av)	1/10 1/30 1/100	1.5 rem/yr 15.0 rem (av) to age 30 or 0.5 rem/yr (av)	1/10 1/30

From: Morgan, K. Z., Maximum Permissible Exposure Levels—External and Internal. *Principles of Radiation Protection*, Ch. 14, Morgan, K. Z. and Turner, E. J., eds., 1967. By permission of John Wiley and Sons, New York.

mutants may be distributed among immediately succeeding generations. Since the average occupational genetic dose is actually less than 0.3 rem per year distributed over more than 2×10^5 employees,[17] it represents fewer man-rem than the natural background genetic dose of about 0.1 rem per year received by 2×10^8 persons in the United States.[18] Thus, occupational exposure adds less than about 0.3 percent to the naturally received population genetic dose. However, as the nuclear-energy industry expands, and as radioisotopes are increasingly used in consumer products, it will become necessary to account for and limit all of the man-made sources of population exposure. A more detailed discussion of this problem and an illustrative method for such population exposure accounting have been presented elsewhere.[3] It has been suggested that all consumer items ever distributed together should not expose the public to an average genetic dose of more than about 0.01 rem per year (0.3 rem per generation of 30 years).[3]

Table 17.
PERMISSIBLE GENETIC DOSE TO THE POPULATION AT LARGE
SUGGESTED IN THE EARLY REPORTS OF THE ICRP TO SERVE AS A GUIDE
(Dose Equivalent in rem to Age 30)

Medical			4.5 rem/generation
Background			3.0 rem/generation
Other Sources			
Population at large			
Internal	1.5		
External	0.5		
Subtotal		2.0	
Other groups			
Occupational	1.0		
Special groups	0.5		
Reserve	1.5		
Subtotal		3.0	
Total from Other Sources		5.0	
Total from All Sources			12.5 rem/generation

From: Morgan, K. Z., Maximum Permissible Exposure Levels—External and Internal. *Principles of Radiation Protection*, Ch. 14, Morgan, K. Z. and Turner, E. J., eds., 1967. By permission of John Wiley and Sons, New York.

REGULATORY LIMITS OF EXPOSURE

In regulatory agencies there has been a need to establish some legal boundaries to allowable limits of exposure, which may be inspected conveniently and which may serve as a basis for enforcement of remedial measures or legal action, if necessary. Thus, although regulations established by different nations and states will generally be consistent with ICRP recommendations (Table 15) in the degree of protection afforded, there may be slight differences in detailed requirements, which should be noted by the user of radioactive material. For example, for ease in examining the majority of licenses, the regulations of the United States Atomic Energy Commission, 10 CRF 20,[30] reduce the allowable quarterly limit of radiation exposure below the values in Table 15 when retrospective verification of an employee's previously accumulated exposure is inconvenient or impossible. This may be in some cases more lenient rather than more restrictive, when codified, than the recommendations of the ICRP, since the maximum recommended annual exposure given in Table 15 may in some cases already have been exceeded. Yet it could be unreasonable to insist that an employer determine previous exposure histories that may not readily be available. Thus, according to 10 CFR, Section 20.101 (a):

" (a) Except as provided in paragraph (b) of this section, no licensee shall possess, use, or transfer licensed material in such a manner as to cause any individual in a restricted area to receive in any period of one calendar quarter from radioactive material and other sources of radiation in the licensee's possession a dose in excess of the limits specified in the following table:

Rems per Calendar Quarter

1. Whole body; head and trunk;
   active blood-forming organs;
   lenses of eyes; or gonads $1\frac{1}{4}$
2. Hands and forearms; feet and ankles $18\frac{3}{4}$
3. Skin of whole body $7\frac{1}{2}$ "

On the other hand, Section 20.101 (b) spells out additional record-keeping requirements for those employers who occasionally must avail themselves of the 3-rem-per-calendar-quarter limit, in order to ensure that the $5(N - 18)$ cumulative lifetime limit (Table 15) is not grossly exceeded in the long run.

Additional regulatory considerations resulted in the publication in 10 CFR 20 of limiting permissible concentrations of radionuclides in air or water, which are generally taken from ICRP recommendations,[23] but which may differ slightly for some nuclides.

Other requirements of 10 CFR 20 deal with the following: exposure limits for minors; permissible levels of radiation in unrestricted areas (two maxims—neither more than 2 mrem in one hour, nor more than 100 mrem in any seven consecutive days); orders requiring furnishing of bioassay services; surveys; personnel monitoring; caution signs, labels, and signals; instruction of personnel and posting of notices of employee rights; storage of licensed materials; disposal of waste; records of surveys, radiation monitoring, and disposal; reports and notification required for accidents of varying degrees of seriousness. Before ordering radioactive materials, copies of this regulation should be examined together with other pertinent state or local regulations. Additional specific requirements may be added by the AEC or other regulatory body, where necessary, to a license to use radioactive materials. Information on obtaining a license to use by-product materials, source materials containing fissionable elements, or special nuclear materials may be obtained in the United States from:

Director of Regulation
U.S. Atomic Energy Commission
Washington, D.C. 20545

In about 19 states of the U.S.A., agreements have been made between the respective states and the USAEC, so that each "agreement state" now regulates the use of materials originating from the federal atomic-energy program as well as other radiation sources not under AEC jurisdiction, such as radium, X-ray machines, and high-energy particle accelerators. However, state regulations are generally fairly consistent with those of the AEC under the conditions of the agreements between these bodies. Moreover, licensing control of the larger production and utilization facilities (including nuclear reactors) is retained by the AEC.

In addition to AEC and state regulations, the user of radioactive materials should also consult the regulations of the U.S. Department of Transportation (Code of Federal Regulations, Title 49) and recent regulations from the Department of Labor, which are related to the Walsh-Healey Act. There are, however, certain small and sufficiently safe quantities of various radioactive materials that have been exempted from regulatory control by the USAEC for use in teaching, demonstration, and calibration of sensitive radiation detectors. Information on exempt and/or "generally licensed" radioactive materials may be obtained from many nuclear chemical and pharmaceutical companies.

RECOMMENDATIONS FOR
LIMITING EXPOSURES IN EMERGENCIES

Recommendations for limiting the radiation exposure of employees as well as members of the public when unplanned and unexpected situations arise that release radioactive materials to uncontrolled areas have been promulgated by various national and international organizations, such as the ICRP, NCRP, FRC, and the British Medical Research Council. Some of the recommendations have been summarized in various texts.[21,27,33-36] The calculations used to estimate the acceptable emergency exposures to inhalation or land contamination by gross fission-product mixtures have been recalculated with more recent data[37] and more conservative biological assumptions. Many considerations are required in an emergency in order to balance the risk of exposure of the individual against the need for some emergency action to save either life or property, and thus comprehensive discussion of this subject within the scope of this handbook is difficult. Generally, additional references containing summarized data needed for making emergency decisions should be on hand.[21,23,31,33,35-38]

Morgan[1] summarized some limits for "planned special exposures and emergency exposures", giving the basic dose limits of the various organs of the body in situations where there are good reasons for allowing persons to exceed the limits recommended in previous sections of this chapter. The summarized recommendations, taken from ICRP,[13] include:

a) a limit of 2 R_{50} committed in any single event to any body organ for a planned special exposure, where R_{50} represents the annual permissible dose equivalent for the respective organ under consideration;

b) a limit of 5 R_{50} committed in a lifetime to any body organ in planned radiation exposures;

c) a maximum permissible intake of any radionuclide for planned special exposures corresponding to the intake that would result from breathing at the MPC for 2 years;

d) a maximum permissible intake from all planned special exposures equivalent to breathing a nuclide at MPC for 5 years;

e) in addition, planned special exposures are not permitted if a single exposure exceeding $R_{50}/2$ has been received in the previous 12 months or if the worker previously—at any time—received an abnormal exposure exceeding 5 R_{50}.

Also, planned special exposures "are not permitted to organs of reproductive capacity". They are not permitted to gonads, total body, or red bone marrow if—as a consequence—the individual's cumulative lifetime limit of $5(N - 18)$ rem would be exceeded.

Thus, the external planned special exposure limits in a single event for the respective body organs, rounded off to R_{50}, would be as follows.

1. Gonads, total body, and red bone marrow: 2 R_{50}, if taken as 12 rem, would correspond to the routinely permitted exposure if the individual has not exceeded $5(N - 18)$, and 5 R_{50} (equivalent to 25 rem) for planned special exposures over a lifetime.

2. Thyroid, skin, and bone: $R_{50} = 30$ rem, 2 $R_{50} = 60$ rem, and 5 $R_{50} = 150$ rem.

3. Hands, forearm, feet, and ankles: $R_{50} = 75$ rem, 2 $R_{50} = 150$ rem, and 5 $R_{50} = 375$ rem.

4. For all other body organs (including lens of the eye): $R_{50} = 15$ rem, 2 $R_{50} = 30$ rem, and 5 $R_{50} = 75$ rem.

After inhalation of fission-product mixtures, the revised acceptable emergency doses (AED's) chosen by Brodsky[37] are closer to those given above than to the previous values used by Cowan and Kuper[39] in the calculations quoted in NCRP Report No. 29.[31] The later calculations[37] should result in estimates of acceptable emergency exposures in curie-seconds per cubic meter, or in curies inhaled, that would "not be likely to produce clinically observable injury either acutely or after many years following an incident."[37] Since a considerable degree of synergism is known to occur between radiation exposures to different body organs, the individual-organ AED's were further chosen so that the fractional AED's contributed to separate parts of the body, as developed by the various nuclides, could be assumed to be roughly additive in determining the total fraction or multiple number of AED's contributed by a particular fission-product mixture.

ACKNOWLEDGEMENT: This chapter is reprinted with permission from "CRC Handbook of Radioactive Nuclides", Yen Wang, Editor, The Chemical Rubber Co., Cleveland, Ohio (1969).

REFERENCES

1. Morgan, K. Z., Maximum Permissible Exposure Levels—External and Internal. *Principles of Radiation Protection—A Textbook of Health Physics*, Ch. 14, Morgan, K. Z. and Turner, J. E., eds. John Wiley and Sons, New York, 1967.
2. Morgan, K. Z., in *Principles of Radiation Protection—A Textbook of Health Physics*, Ch. 14, p. 532, Morgan, K. Z. and Turner, J. E., eds. John Wiley and Sons, New York, 1967.
3. Brodsky, A. *Amer. J. Public Health*, 55(12): 1971–1992, 1965.
4. Federal Radiation Council, Report No. 2, *Radiation Protection for Federal Agencies*. Federal Radiation Council, Washington, D.C., 1961.
5. International Commission on Radiological Protection, Recommendations of the International Commission on Radiological Protection. *ICRP Publ. 2*. Pergamon Press, London, England, 1959.
6. National Committee on Radiation Protection and Measurements, Maximum Permissible Radiation Exposures to Man. *Suppl. NBS Handbook 59*. U.S. Government Printing Office, Washington, D.C., 1958.
7. National Committee on Radiation Protection and Measurements, Maximum Permissible Radiation Exposures to Man. *NBS Handbook 59*. U.S. Government Printing Office, Washington, D.C., 1957.
8. International Commission on Radiological Protection, Report of Committee 2 on Permissible Dose for Internal Radiation. *ICRP Publ. 2*. Pergamon Press, London, England, 1959, and *Health Phys.*, 3: 1, 1960.,
9. National Committee on Radiation Protection and Measurements, Maximum Permissible Body Burden and Maximum Permissible Concentrations of Radionuclides in Air and Water for Occupational Exposure. *NBS Handbook 69*. U.S. Government Printing Office, Washington, D.C., 1959.
10. International Commission on Radiological Protection, Report on Decisions at the 1959 Meeting of the International Commission on Radiological Protection (ICRP), Addendum to ICRP Publ. 15. *Radiology*, 74: 116, 1960.
11. International Commission on Radiological Protection, Report of Committee 3 on Protection Against X Rays up to Energies of 3 Mev and Beta and Gamma Rays from Sealed Sources. *ICRP Publ. 3*. Pergamon Press, London, England, 1960.
12. International Commission on Radiological Protection, Recommendations of the International Commission on Radiological Protection. *ICRP Publ. 6*. Pergamon Press, London, England, 1966.
13. International Commission on Radiological Protection, Recommendations of the International Commission on Radiological Protection. *ICRP Publ. 9*. Pergamon Press, London, England, 1966.
14. National Committee on Radiation Protection and Measurements, Maximum Permissible Amounts of Radioisotopes in the Human Body and Maximum Permissible Concentrations in Air and Water. *NBS Handbook 52*. U.S. Government Printing Office, Washington, D.C., 1953.
15. National Committee on Radiation Protection and Measurements. *Radiology*, 75: 122, 1960.
16. Schubert, J., Brodsky, A., and Tyler, S., The Log-Normal Function as a Stochastic Model of the Distribution of Strontium-90 and Other Fission Products in Humans. *Health Phys.*, 13: 1187–1204, 1967.
17. Brodsky, A., *Unpublished Data*.
18. United Nations Scientific Committee on the Effects of Atomic Radiation, *Various Reports*. United Nations, New York, 1958, 1962, 1964, 1966. (Continuing reviews are underway.)
19. International Commission on Radiological Protection, Report of Committee 2. *Health Phys.*, 3: 1–380, 1960.
20. International Commission on Radiological Protection, *Recommendations of the International Commission on Radiological Protection*. Pergamon Press, Macmillan Co., New York, 1959.
21. Brodsky, A. and Beard, G. V., compilers and eds., *A Compendium of Information for Use in Controlling Radiation Emergencies, TID-8206* (Rev.). U.S. Atomic Energy Commission, Washington, D.C., 1960.
22. Society for Nuclear Medicine, MIRD Report. *J. Nucl. Med.*, Suppl. 1: 5–39, 1968.
23. International Commission on Radiological Protection, Report of Committee 2. *Health Phys.*, 3: 1–380, 1960.

24. International Commission on Radiological Protection, Report of the ICRP Task Group on Lung Dynamics. *Health Phys.*, 12: 173–207, 1966.
25. Fitzgerald, J. J., Brownell, G. L., and Mahoney, F. J., *Mathematical Theory of Radiation Dosimetry*. Gordon and Breach, New York, 1967.
26. Cember, H., *Introduction to Health Physics*. Pergamon Press, London, England, 1968.
27. Morgan, K. Z. and Turner, J. E., eds., *Principles of Radiation Protection—A Textbook of Health Physics*. John Wiley and Sons, New York, 1967.
28. Hine, G. J. and Brownell, G. L., eds., *Radiation Dosimetry*, 1st ed. Academic Press, New York, 1956.
29. Roesch, W. C. and Attix, F. H. *Radiation Dosimetry*, 2nd ed., Vol. 1, Ch. 1, Attix, H. F. and Roesch, W. C., eds. Academic Press, New York, 1968.
30. U.S. Atomic Energy Commission, *USAEC Rules and Regulations, Title 10, Code of Federal Regulations*, Part 20, Standards of Protection Against Radiation, Revised August 9, 1966. U.S. Government Printing Office, Washington, D.C.
31. National Committee on Radiation Protection and Measurements, Exposure to Radiation in an Emergency. *NCRP Report No. 29*. NCRP Publications Office, Washington, D.C., 1968.
32. National Committee on Radiation Protection and Measurements, Medical X-Ray and Gamma-Ray Protection for Energies up to 10 Mev. Equipment Design and Use, *NCRP Report No. 33*. NCRP Publications Office, Washington, D.C., 1968.
33. National Research Council, *Radiobiological Factors in Manned Space Flight*, Langham, W. H., ed. National Academy of Sciences, Washington, D.C., 1967.
34. Lanzl, L. H., Pingel, J. H., and Rush, J. H., *Radiation Accidents and Emergencies in Medicine, Research, and Industry*. Charles C Thomas, Springfield, Illinois, 1965.
35. U.S. Atomic Energy Commission, *The Effects of Nuclear Weapons*. U.S. Government Printing Office, Washington, D.C., 1962.
36. Martin, T. L. and Latham, D. C., *Strategy for Survival*. The University of Arizona Press, Tucson, Arizona, 1963.
37. Brodsky, A., Criteria for Acute Exposure to Mixed Fission-Product Aerosols. *Health Phys.*, 11: 1017–1032, 1965.
38. Bureau of Ships, Principles of Radiation and Contamination Control. *Procedures and Guidelines Relating to Nuclear-Weapons Effects*, Vol. 2. Department of the Navy, Washington, D.C.
39. Cowan, F. P. and Kuper, J. B. H. *Health Phys.*, 1: 76, 1958.

Safe Handling of Radioisotopes

International Atomic Energy Agency

INTRODUCTION

Scope

This chapter is provided as a guide to the safe handling of radioisotopes. It is hoped that it should be helpful particularly to small scale users who may not have direct access to other sources of information.

Large scale users and those with specialized experience may prefer to adopt other procedures which are known to provide equivalent or even superior protection. Other published guides can be recommended when appropriate to specialized fields of application. It is presumed that those using radioisotopes in the practice of their profession (such as radiologists), will supplement the recommendations of the chapter by application of their normal professional training.

The chapter does, of course, not prevent the application of more stringent and more extensive instructions that may possibly be in force in some countries.

This chapter contains a series of recommendations which should be interpreted with scientific judgement in their application to a particular problem. The choice of wording is intentionally precise and the user must understand its implication before departing from any recommendation.

As most natural objects contain some radioactive material, it is clear that the provisions of the chapter are not intended to apply below a certain limiting degree of radioactivity. This lower limit can be taken as a concentration of 0.002 microcuries per gram of material, or a total activity in the working area less than 0.1 microcuries. These limits are based on the most dangerous radioisotopes so that the use of somewhat higher limiting levels of activity is permissible provided the isotopes present are not the most dangerous. A guide to quantities of the less toxic isotopes which may be handled without special precautions is provided in column one of Table 2 and the provisions of the two paragraphs containing Tables 1 and 2. In general, the relaxation of controls must be based on an assessment of the possibility of hazard, taking into account the nature of the material, operations and working facilities.

Treatment of all radioisotopes as potentially dangerous, however, is recommended for its training value and the protection it offers against misidentification.

Definitions

In general, technical terms are used with their accepted scientific meanings.

A few definitions of significant terms follow:

IONIZING RADIATION: electromagnetic or corpuscular radiation capable of producing ions directly or indirectly in its passage through matter (for instance: alpha rays, beta rays, gamma rays, X-rays, neutrons).

SEALED SOURCE: a source of ionizing radiations that is firmly bonded within material or sealed in a cover of sufficient mechanical strength which excludes the possibility of contact with the radioisotope and the dispersion of the radioactive material into the environment under foreseeable conditions of use and wear.

UNSEALED SOURCE: any other radioactive source.

EXTERNAL RADIATION: radiation received by the body from radioactive sources external to it.

INTERNAL RADIATION: radiation received by the body from radioactive sources within it.

DOSE: a measure of the quantity of radiation delivered to a specified absorber.

RADIOACTIVE CONTAMINATION: the undesired presence of radioactive substances in or on any material.

ADEQUATE PROTECTION: protection against external radiations and against intake of radioactive material such that the radiation dose received by any person from sources external and/or internal to the body does not exceed the maximum permissible levels set for exposure by the competent authority.

INSTALLATION: any accommodation or facility where radioactive substances are produced, processed, used or stored.

ENCLOSED INSTALLATION: an installation in which the radiation source and all objects exposed thereto are within a permanent enclosure:

    (a) to which no person has access, or within which no person (except those undergoing treatment) is permitted to remain during irradiation; and

    (b) which affords under all practical operating conditions adequate protection for all persons outside the enclosure.

OPEN INSTALLATION: an installation which, due to operational requirements; e.g. the use of mobile equipment, does not meet the conditions specified for "enclosed installation."

COMPETENT AUTHORITY: a national or international authority whose jurisdiction in the field of problems concerned applies to the activities of the installation considered.

CONTROLLED AREA: area in which exposures may exceed the permissible levels for non-occupationally exposed persons and therefore requires the supervision of a radiological officer.

The terms "Workers" or "Personnel" are used in the sense of including all persons potentially exposed to radiation or radioactive substances as a result of their occupation.

Maximum Permissible Levels for Exposure to External Radiation and to Radioactive Contamination

Pending the issuing by the International Atomic Energy Agency of regulations on maximum permissible levels for exposure to external radiation and to radioactive contamination, it will be generally acceptable to this Agency if all work performed in installations using radioactive isotopes obtained through the International Atomic Energy Agency is in conformity with maximum permissible levels fixed by the competent authority.

As in most countries the setting of maximum permissible levels has been done on the basis of recommendations of the International Commission for Radiological Protection, for countries where such maximum permissible levels have not been fixed the recommendations of the International Commission for Radiological Protection of 1954, as subsequently amended in 1956 and 1958, are recommended as a common basis until such time as the regulations of the International Atomic Energy Agency may be issued.

A generally accepted maximum permissible level is often not available with respect to certain specific problems, particularly for surface contamination or waste disposal. The problem involved and useful working guides are given in the applicable paragraphs of the chapter.

Organization

PRINCIPLES. Good radiation safety practice depends on an effective health and safety organization. Experience shows that even the most competent worker cannot be relied upon to keep in mind all health and safety requirements while preoccupied with the successful prosecution of his work. Responsibilities and duties must be set out clearly to assure safety.

Responsibility of the Authority in Charge of the Installation

The authority in charge of the installation is customarily held responsible for the radiological safety of both the workers and the general public. To meet those responsibilities the authority should ensure that the following actions are taken:

Health and safety rules (in conformity with this chapter) should be prepared for the areas in which radioactive material is to be handled.

All necessary operating instructions should be provided.

Suitable installation and equipment should be provided.

Provisions should be made for necessary medical supervision of the workers and for suitable medical casualty service.

Only persons medically suitable and adequately trained or experienced should be allowed to work with radioactive material.

All workers liable to exposure to ionizing radiation in the course of their work should be instructed about the health hazards involved in their duties.

Suitable training with reference to health and safety should be provided for all staff.

A person technically qualified to advise on all points of radiation safety should be employed or otherwise provided. In this chapter he will be referred to as the "radiological health and safety officer" although various titles are customary in different countries.

The authority in charge of the installation should consult the radiological health and safety officer on all points of radiation safety.

Appropriate means should be taken to ensure that all persons who may be exposed to radiation hazards know the name of the radiological health and safety officer and how to get in touch with him. Any necessary alternates should be provided.

Duties of the "Radiological Health and Safety Officer"

The radiological health and safety officer's duties will vary somewhat according to the organizational structure of the group with which he is working and the degree of the hazard of the class of work undertaken. In general he will assist the authority in charge to carry out the latter's responsibilities for radiation protection. In the accomplishment of his duties, the "radiological health and safety officer" should call for advice or help upon professionally competent persons whenever necessary.

His work will usually include the following duties:

Any necessary administrative, technical and medical instructions concerning the radiation hazards and safe working practices relevant to the nature of the installation and work should be provided to all employees whose duties involve the handling of radioactive material and to all other employees who are not regularly employed in such work but who may occasionally be exposed to radiation and radioactive material. These instructions should be written, understandable, practicable and, whenever possible, posted.

All persons working with radioactive materials should be instructed in the use of all necessary safeguards and procedures and all visitors should be informed of pertinent precautions to be taken. All persons, workers, personnel and visitors should be supplied with such auxiliary devices as may be necessary for protection. The "radiological health and safety officer" should ensure that every visitor has a proper authorization and should recommend that no unnecessary visit is made.

Radioactive material (including that in patients, animals and equipment) should be prevented from leaving the jurisdiction of the authority in charge under circumstances that may subject other persons to radiation in excess of the limits prescribed by the competent authority. The "radiological health and safety officer" should ensure that the proper arrangements for safe waste disposal are made.

Any area, inside or outside the installation should be ensured against subjection to radiation levels or concentrations of radioactive material exceeding the maximum permissible levels indicated by the competent authority for such a type of area.

The appropriate authorities (for instance the Fire Department) should be notified of the existence of any conditions or situations that, while not normally considered a radiation hazard, may become a hazard under special or unusual circumstances.

Measures should be taken to ensure that no modification of equipment or installations which might lead to unforeseen radiation hazards is made without provision of appropriate safeguards.

Measures should be taken to ensure that no radioactive material is dealt with by unauthorized people in the installation.

Suitable alternates or other means should be provided to ensure that necessary advice is available at all times in case of an emergency and the particular safety measures to be taken in such cases are provided for.

It should be established that suitable records are kept.

It should be established that the necessary tasks of monitoring, medical supervision and protection measures are carried out and properly co-ordinated.

Duties of the Worker

The operating instructions provided should be known.

Health and safety rules for the worker's area should be known and followed.

The safety equipment provided should be used properly.

The worker should protect both himself and others by acting carefully and working safely.

Any accident or unusual incident or any personal injury, however slight, should be reported.

Workers exposed to radiation hazards should immediately report any significant ailment and any suspected overexposure to external radiation, or any suspected introduction of radioactive material into their systems.

Medical Supervision of Workers

GENERAL CONSIDERATIONS.   Medical supervision of persons employed in radiation work should be based on the experience that in any properly run radiation laboratory, radiation accidents will be secondary to normal industrial accidents. Preemployment and routine medical examinations should be primarily those desirable in good industrial medical practice but certain medical requirements are specific to radiation work and should supplement regular industrial medical practice. Opportunities for observation of genuine symptoms of radiation injuries will be extremely rare unless very bad working conditions prevail. Undesirable working conditions may be present to a considerable degree before any clinical symptoms of radiation damage appear.

Young persons should not be occupationally exposed to radiation. In many countries the minimum age is taken as eighteen years.

Special attention should be directed toward protecting women of reproductive age.

If X-ray examinations are carried out, care should be given to keep to a minimum the exposure involved.

MEDICAL EXAMINATIONS BEFORE EMPLOYMENT.   No person should be employed in work involving a possible radiation hazard unless within the period of 2 months preceding his first employment in that work he has undergone a medical examination.

It is recommended that this medical examination on recruitment include the following:

1. A complete medical examination, as given normally in pre-employment examinations, including a personal history covering family, medical and occupational background, as well as the usual clinical tests;

2. Special investigations of those organs and functions which are considered as particularly vulnerable to radiation hazards according to the class of work undertaken; e.g., by hematological examinations, dermatological examinations, ophthalmological examinations, pulmonary examinations, gynecological examinations, and neurological examinations.

MEDICAL EXAMINATIONS DURING EMPLOYMENT. All persons employed in work involving radiation hazard should undergo medical examinations.

The routine examinations should be carried out every twelve months, or such other periods as the competent authority may require. They should include the general examinations practiced in industrial medicine and also special examinations desirable because of the hazards of external radiation and contamination in each particular case. The special examinations given above should be carried out at appropriate intervals. In the case of suspected over-exposure or internal contamination, the physician should specify any required program of examinations. In case of internal radioactive contamination, the radio-toxicological examinations yield information on the nature and extent of such contamination, by means of measurements and analyses carried out directly on the organism and indirectly on the excreta (urine, feces, exhaled air). In addition, in cases of inhalation of aerosols or radioactive dust or gases, the examination of the lungs should include the investigation of combined mechanical, chemical or radioactive effects.

In the case of workers handling unsealed radioactive isotopes, tests are useful from time to time to determine the total body burden; in many cases monitoring of the excreta (more particularly of the urine, or in case of radium of the radon in the breath) will permit an assessment of the body burden.

In certain circumstances, a more elaborate test can be adopted to determine the body burden, by measuring the gamma-radiation (or Bremsstrahlung) emitted by the body. If it is possible to measure such a body burden the dose of radiation received should be estimated and noted on the personnel record and taken into consideration by the physician.

MEDICAL CASUALTY SERVICE. The form of medical casualty service provided will depend on the availability of medical staff within the establishment.

First aid advice and equipment should be immediately available throughout the working area. The scope of first aid treatment attempted should be based on medical advice.

Arrangements for referring casualties and personnel contamination problems to medical services at an appropriate stage should be clearly defined and known.

Determination of Radiation Exposure of Personnel

GENERAL CONSIDERATIONS. The essential aim of radiological protection is to prevent injury from ionizing radiations. Its basis is respect for the recommended maximum permissible doses, but it also calls for systematic observation, to detect any irradiation or irradiation effect. This observation must include both physical and medical control.

Symptoms following irradiation are at present detectable only for relatively high doses. This lack of sensitivity in the clinical examination is aggravated by the lack of specificity of the injuries observed, and the often considerable latent time between irradiation and the manifestation of its effects. This in no way reduces the necessity for systematic medical examinations to detect any radiation-induced effects but makes it essential to complement them by rigorous control of the doses received.

Present physical or radiochemical techniques allow the measurement of very low radiation doses and quantities of radioelements. This sensitivity is very helpful, as it permits the detection of irradiations considerably lower than those considered permissible. The methodical application of these techniques should therefore be regarded as essential. These techniques may be classified as follows:

Personnel Monitoring. (a) External Radiation Monitoring in which radiation measuring devices are worn by the worker; (b) Internal Contamination Monitoring in which suitable instruments may be used or the body wastes may be sampled and analyzed, to determine the presence and quantity of radioactive material within the body.

Area Monitoring. The determination of radiation levels and air contamination in the working area. (a) Measurement by the use of radiation measuring instruments and devices; (b) Calculation based on the amount of radioactive material present, its form and the nature of the processes in which the workers will be exposed.

Determination by Personnel Monitoring

MONITORING FOR EXTERNAL RADIATION EXPOSURE WITH PERSONNEL DOSIMETERS. This simple and convenient method should be used for the measurement of external radiation exposure of all workers in the controlled area.

The preferred device is the film dosimeter which permits measurement of the accumulated radiation dose over a period. This film also provides a permanent means of checking the accumulated external radiation exposure record which should be kept for each individual. Similar film dosimeters should be used on the hands, wrists or other extremities when these are exposed to higher radiation fields than is the trunk of the body.

Pocket ionization chambers, luminescent individual radiation detectors and thimble chambers supplement the above film dosimeters and are particularly useful where an immediate and sensitive measurement is needed in connection with a specific task.

In the use of both film dosimeters and ionization chambers for personnel monitoring, serious errors may occur unless standard procedures are adopted.

MONITORING FOR INTERNAL CONTAMINATION. The difficulties of this monitoring are very real due to the complicated and specialized nature of the techniques involved. In the case of monitoring of body wastes by radiochemical analysis, there are further difficulties in interpreting results.

Monitoring by Instruments. Whole body or gamma spectrometry radiation detectors may be used to determine the presence and quantity of radioactive material in the body. However these instruments are expensive and their operation and the interpretation of results is very specialized. It is unlikely that the small users would have such instruments, though in special cases their use could be arranged through other institutions.

Monitoring by Analysis of Body Wastes. A routine program of urine analysis should be drawn up for workers exposed to the possibility of significant internal contamination. The frequency of urine sampling should be evaluated on the basis of an appraisal of the nature and quantity of the isotopes involved and the operations necessary in the particular process.

In the event of suspected internal contamination, if appropriate, a special series of samples should be collected and analyzed. The biological half-life and the period of body retention should be borne in mind in scheduling these samples. Where appropriate, urine analysis should be supplemented by fecal analysis, nose swabbing, examination of stomach washings and radon breath tests.

Determination by Area Monitoring

MONITORING BY INSTRUMENTS. The use of ionization chambers, pocket ionization chambers and film dosimeters exposed under conditions similar to those in which the workers will be exposed, enable the dose to an individual over any particular time to be inferred.

Measurements of contamination present in the air or drinking water can be used to estimate possible body uptake. However, considerable errors will occur, especially if the measurements are not representative due to the presence of particles of high specific activity.

MONITORING BY CALCULATION. Knowledge of the total radioactive material present, its nature, the processes and the working conditions in a laboratory enable the estimation of possible exposure of personnel. However, considerable experience and technical skill are demanded for such estimates.

MONITORING OF THE AREA. In addition to personnel monitoring to determine the exposure history of individuals, general area monitoring is carried out to determine the need for protective action. Monitoring should be done periodically or continuously with due regard to the external and internal radiation hazards for the purpose of determining the possibility of exposure of persons, workplaces and articles.

MONITORING OF RADIATION FROM EXTERNAL SOURCES. All places around radioactive sources emitting penetrating radiation where persons can be exposed to radiation, not neglecting adjoining rooms or places outside the building, should be monitored for radiation. This should be done before starting a project, after any significant modification of the set-up and also periodically during work.

Portable ionization chambers, pocket ionization chambers, GM-counters, scintillation counters (in some cases also film dosimeters) may be used. All instruments used for monitoring should be calibrated and checked regularly, for which a radiation standard should be available. Duplication of instruments is desirable in some cases.

MONITORING OF CONTAMINATION ON SURFACES OF ROOMS AND EQUIPMENT. Everything used for work with radioactive materials may be subject to wide-spread contamination. This includes surfaces of working places, walls of fume hoods or glove boxes, floor or walls of working rooms, clothing, equipment, and all other surfaces.

Contamination by radioactive substances of working surfaces, clothing and equipment can be a hazard to health and also may interfere with the work being carried out.

It is not yet possible to recommend definite permissible levels for surface contamination and contamination of clothing and equipment. However, the inexperienced user may adopt any one of a number of proposed levels accepted in certain countries.

If it is known that contamination is permanently fixed, monitoring can be based on the consideration of permissible external radiation levels.

It is necessary to carry out a systematic monitoring of contamination of all places and equipment that have been in contact with radioactive materials. Such monitoring must be performed at least when work has been completed but, if necessary, also several times during work.

Monitoring should be performed both with the help of dosimetric instruments and by smear tests. Thin windowed GM-counters are suited for examining the smear samples taken; the presence of alpha emitters may make an alpha scintillation monitor or equivalent device desirable.

When alpha or soft beta emitters are used, the walls of beakers, bottles, pipettes, and other containers may absorb most of the radiation so that monitoring from outside of these containers might be insufficient.

Experimental animals, their excreta and the premises, such as cages, where they are kept should be monitored.

MONITORING OF CONTAMINATION OF THE AIR. In cases where radioactive aerosols, gases or powders (dust) are handled or produced the air must be monitored for contamination.

A reliable system of monitoring of the air after filtering, before releasing it into the open, should be carried out in cases when the activity released could exceed levels set for such outside places by the competent authority.

For monitoring aerosols the airborne substances are either deposited by electrostatic precipitation, impactors or by filtration.

Some radioactive gases can only be monitored after collection by chemical or other means.

It will often be desirable to identify the radioactive contamination by radiochemical analysis or physical means.

MONITORING OF CONTAMINATION OF WATER. A simple monitoring method (e.g. dipping a GM-counter or scintillation counter into the water) in many cases will prove to be unsatisfactory and a more elaborate procedure for monitoring must be effected.

A reliable assessment of the contamination of waters to be released to public drains or sewers, in accordance with the section on "Disposal to drains and sewers," is necessary. Sampling may prove necessary, in which case the radioactive substances dissolved may require concentrating (for instance by ion exchange or evaporation) before activity measurements can be carried out.

MONITORING OF SKIN AND CLOTHING. Monitoring of hands, clothing and particularly shoes should always be carried out when working with unsealed sources. No person should leave the working place (room) without checking for contamination.

Monitoring for contamination of the skin and clothing should be performed by appropriate means. A thin window GM-counter may be sufficient in many cases. When alpha contamination may occur independently of beta and gamma radiation an alpha-selective monitor should also be provided.

SEALED SOURCES

Choice and Design of Sealed Sources

A source used to produce radiation field should be sealed in a suitable container or prepared in a form providing equivalent protection from mechanical disruption. The following characteristics are desirable and consistent with the work being carried out:

The activity of the source used should be a minimum.

The energy or penetrating power of the emitted radiation should not be greater than that necessary to accomplish the task with a minimum total exposure.

If possible, the radioactive material in the source should be of low toxicity and in such a chemical and physical form as to minimize dispersion and ingestion in case the container should be broken.

Sealed sources should be permanently marked to permit individual identification and facilitate determination of nature and quantity of radioactivity without undue exposure of the worker.

Sealed sources or appropriate containers should be regularly examined for contamination or leakage (smear tests, and/or electrostatic collection may be used). The interval between examinations should be determined by the nature of the source in question.

Mechanically damaged or corroded sources should not be used and should immediately be placed in sealed containers. They should be repaired only by a technically skilled person, using suitable facilities.

Methods of Use of Sources

Sources should always be handled in such a way that proper location is possible at all times. Inventories should be kept.

If any person has reasons for believing that a source has been lost or mislaid, he should notify the "radiological health and safety officer" immediately.

If the loss is confirmed, the competent authority should be notified without delay.

Sources should be handled in such a way that the radiation dose to personnel is reduced to a minimum by such methods as shielding, distance and limited working time.

Sources should be handled in such a way as to avoid hazards to all personnel including those not involved in the operations. Attention should be paid to people in adjacent areas including rooms above and below. Areas subject to high radiation levels should be clearly marked and, if necessary, roped off.

Beams of radiation arising from a partially shielded source should be clearly indicated. Care should be taken to insure that such a beam is stopped at the minimum practical distance by suitable absorbing material. Monitoring procedures should be planned to take into account the sharp collimation of radiation fields which may occur.

When practical, sealed sources should be used in enclosed installations from which all persons are excluded during irradiation.

Sources should not be touched by hands. Appropriate tools should be used, for instance, long handled, lightweight forceps with a firm grip. If needed, even more elaborate means of protection have to be considered, such as master slave manipulation, or other methods.

Work with radioactive materials should be planned to permit as short an exposure as possible. The extent of protection provided by limiting working time can easily be lost if unexpected difficulties occur in the work, so that dummy runs should preferably be performed whenever it is possible.

Although work should be planned to limit exposure time to a safe figure, if sufficient shielding cannot be provided and time of exposure must be controlled, this should be carried out in a systematic way, preferably with time keeping and warning services outside the responsibility of the actual worker.

Shielding

Adequate shielding should be provided.

For beta rays the protection of eyes, face and body can be accomplished, by transparent plates of moderate thickness.

For gamma rays the protection of head and body may be effected by screens of adequate shielding effect for the source in question.

In addition to shielding for direct radiation, shielding may be necessary to give adequate protection against the back-scattering from the floor and ceiling (proper care should be taken against direct radiation through the supporting structure).

Bricks used for shielding sources should overlap to prevent penetration of the radiation at the joints:

Shielding should, as far as possible, be near the source.

As there are many possibilities for error in shielding calculation, the adequacy of shielding should always be tested by direct measurements.

UNSEALED SOURCES

General Operations with Unsealed Sources

The provisions of the sections on "Methods of Use of Sources" and "Shielding" with respect to protection from external radiation also apply to unsealed sources. In respect to marking, the corresponding indications should be put on the associated container of the unsealed source.

All operations should be planned to limit spread or dispersal of radioactive material. To this end all unnecessary movement of persons or materials should be avoided.

Areas in which radioactive work is carried out should be designated, marked and monitored. At the boundaries of such areas monitoring and control measures should be set up if so required by the levels present. In larger establishments such check points should be established not only between areas subject to radioactive contamination and those not, but also between active areas subject to different usage. In this way an accidental escape of radioactive material is limited to a restricted local area and one may avoid difficult and expensive decontamination.

Equipment, glassware, tools and cleaning equipment for use in any particular active area should not be used for work in inactive areas and should be suitably marked. Special consideration should be given to avoiding contamination of major items of equipment which might need to be transferred for economic reasons.

Equipment should not needlessly be brought from inactive areas to active ones. Contaminated equipment, and apparatus should not be released from the controlled area for repair, until the level of activity has been reduced to the safe limits set up by the "radiological health and safety officer."

The use of new techniques should first be approved by the responsible person and be tried out with inactive materials or with material of low activity, before being put into operation.

Planning should allow adequate time for the operations required.

Precautions to be taken for handling unsealed sources depend on the degree of radiotoxicity and on the quantity of the substance being used. Contamination hazards such as risks from external contamination, skin penetration, ingestion or inhalation should be considered in addition to the usual radiation hazards of sealed sources. It is recommended that the "radiological health and safety officer" issue working instructions taking into account working conditions and acceptable risks.

A radioisotope can be classified in one of the four following categories of radiotoxicity per unit activity:

1. very high radiotoxicity
2. high radiotoxicity
3. moderate radiotoxicity
4. slight radiotoxicity

Hazards arising out of the handling of unsealed sources depend on factors such as the types of compounds in which these isotopes appear, the specific activity, the volatility, the complexity of the procedures involved, and of the relative doses of radiation to the critical organs and tissues, if an accident should occur giving rise to skin penetration, inhalation or ingestion. Taking these factors into account, the broad classification is given in Tables 1 and 2. Table 1 gives the list of the main isotopes in each of the above mentioned categories. The various types of laboratories or working places required are indicated in Table 2.

Modifying factors should be applied to the quantities indicated in the last 3 columns of Table 2, according to the complexity of the procedures to be followed. The following factors are suggested, but due regard should be paid to the circumstances affecting individual cases:

Procedure	*Modifying factor*
Storage (stock solutions)	× 100
Very simple wet operations	× 10
Normal chemical operations	× 1
Complex wet operations with risk of spills; simple dry operations	× 0.1
Dry and dusty operations	× 0.01

TABLE 1

Classification of Isotopes According to Relative Radiotoxicity per Unit Activity

(The isotopes in each class are listed in order of increasing atomic number)

Class I	Class II	Class III			Class IV
Very high toxicity	High toxicity	Moderate toxicity			Slight toxicity
Sr-90 + Y-90	Ca-45	Na-22*	Ga-72*	I-132*	H-3
Pb-210* + Bi-210 (Ra D + E)	Fe-59*	Na-24*	As-74*	Cs-137 plus	Be-7*
Po-210	Sr-89	P-32	As-76*	Ba-137*	C-14
At-211	Y-91	S-35	Br-82*	La-140*	F-18
Ra-226 + 55 per	Ru-106 plus	Cl-36	Rb-86*	Pr-143	Cr-51*
cent daughter	Rh-106*				
products*			Zr-95* plus	Pm-147	
Ac-227	I-131*		Nb-95*	Ho-166*	Ge-71
U-233*	Ba-140*	K-42*	Nb-95*	Lu-177*	Tl-201*
Pu-239	La-140	Sc-46*	Mo-99	Ta-182*	
Am-241*	Ce-144 plus	Sc-47	Tc-98	W-181*	
Cm-242	Pr-144*	Sc-48*	Rh-105*	Re-183*	
	Sm-151	V-48*	Pd-103 plus	Ir-190*	
	Eu-154*	Mn-52*	Rh-103	Ir-192*	
	Tm-170*	Mn-54*	Ag-105*	Pt-191	
	Th-234* plus	Mn-56*	Ag-111	Pt-193*	
	Pa-234*	Fe-55	Cd-109 plus	Au-196*	
	natural uranium*	Co-58*	Ag-109*	Au-198*	
		Co-60*	Sn-113*	Au-199*	
		Ni-59	Te-127*	Tl-200	
		Cu-64*	Te-129*	Tl-202	
		Zn-65*		Tl-204	
				Pb-203*	

*Gamma-emitters.

TABLE 2

Toxicity Levels and Workplaces

Radio toxicity of isotopes	Minimum significant quantity	Type of laboratory or working place required		
		Type C Good Chemical Laboratory	Type B Radioisotope Laboratory	Type A High Level Laboratory
Very high	0.1 μCi	10 μCi or less	10 μCi— 10 mCi	10 mCi or more
High	1.0 μCi	100 μCi or less	100 μCi—100 mCi	100 mCi or more
Moderate	10 μCi	1 mCi or less	1 mCi— 1 Ci	1 Ci or more
Slight	100 μCi	10 mCi or less	10 mCi— 10 Ci	10 Ci or more

As in the case of sealed sources, unsealed sources should be handled with equipment providing protection against external radiation.

Manipulations should be carried out over a suitable drip tray, or with some form of double container which will minimize the importance of breakages or spills. It is also useful to cover the working surfaces with absorbent material to soak up minor spills. The absorbent material should be changed when unsuitable for further work and considered as radioactive waste.

Shielding should be provided as near the container of radioactive substance as possible.

Handling-tools and equipment used should be placed in nonporous trays and pans with absorbent disposable paper, which should be changed frequently. Pipettes, stirring rods and similar equipment should never be placed directly on the bench or table.

After use, all vessels and tools should be set apart for special attention when cleaning.

Choice of Radioactive Material and Suitable Processes

When a choice between several isotopes of varying toxicities is possible one of relatively low toxicity should be used.

Materials of low specific activity should be used if possible.

The working methods should be studied and procedures adopted to avoid as much as possible the dispersal of radioactive material, in particular through the formation of aerosols, gases, vapors or dusts.

Wet operations should be used in preference to dry ones.

Frequent transfers should be avoided.

The quantity of radioactive substances necessary for a specific purpose should always be chosen as small as possible.

Choice and Design of Work Places

GENERAL CONSIDERATIONS. Special consideration should be given to the choice of fire-proof construction for the buildings. As a rule the choice of location of premises should be such that there is small risk of landslide or flood.

The reservation of special working places for the handling of radioisotopes is recommended. This rule may be considered as optional for work with quantities corresponding to column C of Table 2; but it should be compulsory for work with quantities corresponding to column B and A of Table 2.

As far as possible, the active areas should be planned and utilized in such a manner as to separate widely the different levels of activity.

The radioisotopes working areas should be marked.

FLOORS, WALLS, WORKING SURFACES. The floors, walls and working surfaces should be such as to be easily kept clean.

For Type C working places a linoleum covered floor and working surfaces covered with non-absorbent material and with disposable covers is an example of what would be considered satisfactory. The working surfaces must be able to support the weight of the necessary shieldings against the gamma radiations.

For Type B working places the walls and the ceilings should be covered with a washable, hard, non-porous paint; the floor with such materials as linoleum, rubber tiles or vinyl. The junction of floors and walls should be rounded off in order to facilitate the cleaning. Corners, cracks and rough surfaces should be avoided. When working with gamma emitters the floor and the working surfaces should be able to support the weight of the shielding.

Type A working places should be specially designed by an expert. In general, Type A laboratories will use glove boxes or other completely enclosed systems.

Walls and floors should be free from unnecessary obstacles and all unecessary objects should be removed from the working surface.

SINKS. Sinks should be provided in the working area of Type B or C laboratories. In general the usual type of sink, with a smooth white glaze finish, without blemishes, will suffice. It is desirable to have the sinks connected directly to the main pipe; connections to open channels should be avoided, and also any unnecessary devices which might accumulate slime. Taps should be designed for operation by foot, knee, or elbow, rather than by hand. A suitable waste disposal system as discussed later in this chapter should be provided.

FURNITURE. The furniture should be reduced to a minimum and easily washable. Dust collecting items such as drawers, shelves and hanging lamps should be as few as possible.

LIGHTING. The working premises should be adequately lighted.

VENTILATION. Provision for adequate ventilation should be included in the original design of the premises.

Routes of entry and exit for the ventilating air should be clearly defined under all conditions of use, including open and closed positions of doors and windows and various operating arrangements of the fume hoods. In small laboratories it may be possible to provide the needed flow of air simply by the exhaust systems of the fume hoods, but in such a case special attention must be given to inflow of fresh air into the laboratory under all conditions, by such means as providing adequate louvres in the doors of rooms.

Consideration should be given to any need to treat or filter incoming air. In cold climates the problem of heating the intake air fora large group of fume hoods should not be overlooked as this may be a major problem.

Siting of inlet and exhaust vents should be such as to prevent any recirculation of exhausted air. The need to filter air exhausted from work places and fume hoods will depend on the nature of the work, the position of the exhaust vent relative to surroundings and potential nuisance value of particulates settling in the surrounding neighborhood.

Fume hoods should produce a regular airflow without any eddies. The speed of the airflow should be such that there can be no escape of air into the working place from the fume hood under typical operating conditions including opening of windows and doors, suction of other fume hoods. This can be checked by smoke tests. It is recommended that the fan be placed on the exhaust side of any filter in the system. The gas, water and electrical appliances should be operated from the outside of the fume hood. The inside of the hood and the exhaust ducts should be as easy to clean as possible.

Protective Clothing

Protective clothing appropriate to the radioactive contamination risks should be worn by every person in the controlled area, even if only very small quantities of radioactive materials are manipulated.

In Type C working place the personnel should wear simple protective clothing such as ordinary laboratory coats or surgical coats. In Type A or B working places protective clothing or devices should be provided according to the nature of the work. When working with experimental animals, clothing proof against teeth or claws may be desirable and protection of the face against blood or body fluid splashings should be provided.

In Type A and B working places the protective clothing should be clearly identified, for example by a different color. It should not, in any case, be worn outside the controlled area.

The working clothes and town clothes should be kept in separate cubicles or changing rooms. When changing from one to the other, one should be careful to avoid cross contamination risks.

Rubber gloves should be worn when working with unsealed radioactive substances.

Rubber gloves are provided to protect against contamination of the skin and are of no value for protection from penetrating radiation.

Care should be taken not to needlessly contaminate objects by handling them with protective gloves, in particular light switches, taps, door knobs, and other surfaces. The gloves should be either taken off or a piece of non-contaminated material (paper), which should be disposed of afterwards with the contaminated residue, should be interposed.

Contaminated gloves should be washed before taking them off.

A method of putting on and removing rubber gloves without contaminating the inside of the gloves should be used. This procedure is such that the inside of the glove is not touched by the outside, nor is any part of the outside allowed to come in contact with the bare skin. It is desirable to use gloves for which the inside and outside are distinguishable.

Personal Protective Measures

No unsealed radioactive sources should be manipulated with the unprotected hand.

No solution should be pipetted by mouth in any isotope laboratory.

It is recommended that special precautions be taken to avoid punctures or cuts, especially when manipulating the more dangerous radioisotopes.

Anyone who has an open skin wound below the wrist (protected by a bandage or not) should not work with radioactive isotopes without medical approval.

The use of containers, glassware, and other equipment with cutting edges should be avoided.

Glass blowing by mouth should be avoided in places where unsealed radioactive substances are utilized. Glass blowing, welding, brazing, soldering, and other operations should never be permitted on equipment contaminated with radioactive materials unless it is done in specially ventilated facilities, and unless special techniques are used to prevent the inhalation of radioactive dust and fumes.

Only self-adhesive labels should be used in controlled areas. Labels requiring to be wetted should be avoided.

The following should not be introduced or used in working places containing unsealed sources:

1. Food or beverages (where necessary, drinking fountains should be provided in the vicinity).
2. Smoking items or snuff tobacco.
3. Handbags, lipsticks and other cosmetics, or items used to apply them.
4. Handkerchiefs, other than those mentioned below.
5. Utensils for eating or drinking.

Disposable paper towels and paper handkerchiefs or the equivalent should be provided for the workers. Special containers should be placed in the working places, in which these towels and handkerchiefs should be discarded after use. These should be treated as radioactive residue.

Hands should be washed thoroughly before leaving the controlled area (special attention should be given to the nails, in between fingers and outer edges of the hands).

Showers should be taken when recommended by the "radiological health and safety officer." Monitoring of hands, shoes and street clothing, if worn at work, may also be necessary before leaving the controlled areas.

Control of Air Contamination

Radioactive contamination of the air of the working places should be reduced as much as possible. All operations likely to produce radioactive contamination of the air through the production of aerosols (in particular the heating of radioactive solutions), smoke or vapors, should be done in an air-tight enclosure kept below atmospheric pressure (glove box) or in a fume hood.

The aim should be to improve collective protection, and so avoid resort to individual protection such as respirators, compressed air masks and frog suits. Conditions which require extended use of respirators should be discouraged.

When the contamination level cannot be maintained below the levels set out by the "radiological health and safety officer," individual protection against contamination must be furnished to the people using the working places.

Respirators should be of a form approved by a recognized testing laboratory for the class of service required, and the practical safe limits of use should be known and respected.

Respirators should be capable of standing up to the conditions of use and should be checked and tested periodically.

Respirators should be individually fitted and tested for tightness of fit by attempting to breathe with the inlet closed off.

Users of respirators must accustom themselves to the discipline necessary in their use, otherwise more harm than good is likely to be done by introducing contamination under the facepiece.

In difficult cases when the use of respirators would not give adequate safety (for instance with radioactive gases), air-line hoods may provide the only reliable form of protection. In such cases attention should be given to the purity of the air and its proper supply.

Special Uses of Unsealed Sources

RADIOACTIVE ISOTOPES IN ANIMAL EXPERIMENTS. The general provisions for work with unsealed sources should be followed. However, the unwarranted spread of contamination by animals or from animal excreta requires special consideration in the design of cages and rooms.

Excreta, body constituents from biopsies and autopsies and animal cadavers should be considered as radioactive wastes. Possible hazards of spread of contamination through decomposition process should be prevented; e.g., by deepfreezing, use of disinfectants or sealed plastic containers. Special provisions for collection of excreta and decontamination of cages should be made. The radioactive animals or their cages should be marked with labels indicating the nature and amount of radioisotopes used and the time of administration.

No uncontrolled exchange of animals, instruments, cages, or other items between active and inactive laboratories should be allowed.

Precautions should be taken to prevent the possibility of contaminated wounds produced by handling the animals and the contamination from radioactive aerosols or splashings produced by animals' movements, coughing or other activities.

For work with radioactive animals, the modifying factor to be applied in Table 2 should normally be taken as 0.1.

Presence of vermin as potential vectors of contamination should be considered.

STORAGE OF SOURCES

Place of Storage

When not in use, radioactive sources should be kept in a place of storage assigned for this purpose only.

The place of storage should be adequately shielded.

Only authorized personnel should be allowed to introduce or remove sources from the place of storage which should be secure against tampering.

The place of storage should be in a room provided with a suitable means of exit that can be operated from the inside.

The place of storage should be chosen so as to minimize risk from fire.

The places where sources are stored should be inspected regularly and checked for possible contamination.

Conditions of Storage

All radioactive sources should be clearly labelled, giving information on the activity and nature. It may be found desirable to include the name of the person who is responsible for the source. In the event that a number of sources are normally in use in a fume-hood or other working area, as in analytical work, the marking might be of a general nature to apply to the whole working area. Any source involving hazards greater than those listed in the general warning should be specially marked.

The containers for beta-emitting isotopes should have adequate thickness to reduce the primary radiation to a safe level. Considerable *Bremsstrahlung* may arise from high intensity sources and additional shielding should be provided if necessary.

Gamma-emitting sources should be stored in such a way as to limit the radiation exposure from other sources when any one source is being handled.

When either sealed or unsealed sources are liable to release a radioactive gas, their place of storage should be efficiently vented to the open air by mechanical means before it is opened.

Special equipment should be provided for storing unsealed sources of radioactive substances to prevent not only external irradiation hazards, but also radioactive contamination hazards.

In Type C working places the sources may be stored in special cupboards providing adequate protection.

In Type B working places it is better to use a special secure receptacle which provides adequate protection and could be ventilated if necessary.

Storage Operations

Records should be kept of all stored radioactive sources.

The records should give clear information on type of source, activity and time of removal and return as well as the name of the person responsible for the source during its absence from the store.

Periodic inventories should be performed.

The removal of sources from the store and the time for which they are removed should be checked to provide adequate control.

Thermally unstable solutions containing radioactive materials in nitric acid or other oxidizing solutions containing even traces of organic material and stable solutions with alpha-activity in excess of 5 mc or beta-activity in excess of 50 mc should always be stored in vented vessels.

Bottles and containers should be chosen to open easily.

Solutions having a high alpha-activity in excess of 1 mc/ml should not be stored in thin walled glass bottles, since irradiation might weaken the glass. All glass vessels must be expected to fail without apparent cause.

Bottles containing radioactive liquids should be placed in vessels large enough to hold the entire contents of the bottles in case of breakage.

Special precautions are required when opening vessels containing radioactive liquids liable to catch fire, explode or froth.

TRANSPORTATION OF RADIOACTIVE MATERIAL

Transportation Within An Establishment

The amount of radioactive material moved should be limited to that required.

Transportation should be done in adequately shielded and closed containers. The containers should be constructed to prevent accidental release of the source material in case of upset.

If radioactive material in liquid or gaseous form or in powder, or other dispersible solid form is in a shatterable container it should be transported in an outer non-shatterable container. With liquid sources the container should be provided with absorbing material able to retain all the liquid in case of breakage.

Suitable means should be provided for the transfer of the source to and from the transport container.

The transport container should be clearly marked with warning signs.

Containers in transit should bear a transportation tag showing necessary information for safety such as: (a) nature of contents; (b) physical condition; (c) activity in curies; (d) dose rate of radiation at contact of the outer surface of the container; (e) dose rate of radiation at a specific distance; (f) kind of packing (when applicable).

In case of unsealed sources, the transportation tag should, in addition, certify that the outsides of the container and carrier are free from contamination.

It is recommended that the transportation tag should be disposed of only when the source is in the charge and under the complete physical control of a person who is aware of the nature of the radioactive material and of the radiation hazards involved.

Emergency procedures should be planned to cover accidents to radioactive material in transit.

Any loss of radioactive materials during transport should at once be reported to the "radiological health and safety officer."

Suitably trained workers should be in charge of all transportation of hazardous quantities of radioactive material inside an establishment.

ACCIDENTS

Identification of Accidents

Any unplanned happening which could affect radiation safety is considered an accident from the point of view of this chapter. The most essential and often the most difficult problem in coping with accidents is the recognition that an accident has occurred.

Precautionary Measures

All work should be carried out according to some prearranged plan. Any departure from the plan should be followed by a reassessment of the radiation hazards involved. Appropriate accident instructions should be prepared and posted and the staff should thoroughly understand them.

Accident instructions should avoid any rigid restrictions on conditions of application. A serious situation can develop from a wide variety of causes ranging from a simple spread of radioactive contamination to such natural causes of disaster as fire, flood or earthquakes.

All planning of measures to be taken in case of accidents should give priority to human safety according to need and urgency. Responsibility for protecting the public must take precedence.

The staff should be thoroughly familiar with the position and method of use of the protection and first-aid equipment for emergencies. Practice drills are essential. Equipment should be checked regularly to ensure that it is in good working order.

The public health authorities, fire service, and other services or authorities (guards, police . . .) should be kept informed of special radiation hazards and of the radiation safety measures to be taken.

No person should undertake dangerous work without someone standing by who can assist in case of trouble.

Emergency facilities and staff should be available when work is being carried out with unsealed sources requiring the use of Type A or B working places.

First aid measures in conformity with medical advice should be available.

Actions Common to All Accidents

The control of measures for dealing with any accident should be the responsibility of one individual. The individual concerned or his alternates should be clearly indicated in the accident instructions and should be available to all concerned.

All accidents should be fully reported. This report may have an important bearing on staff health and legal responsibilities and may assist the "radiological health and safety officer" in making detailed study with a view to avoiding similar accidents in the future.

All accidents should be investigated and appropriate measures should be taken to prevent repetition of the accident.

Accidents Involving Radioactive Contamination

Radioactive materials may be accidentally released by a spill, by a failure of equipment or by rupture of a sealed source. The actions which may be appropriate to prevent wide-spread contamination and exposure of personnel in such a case, and the order in which such actions should be taken, will depend upon the circumstances involved. For example, in the case of a small spill of liquid it may be desirable to contain and clean up the contamination immediately without severely affecting the routine activities in the room in which the spill occurs. However, the release of a relatively large quantity of a radioactive powder or aerosol in a room will require immediate action to contain the contamination in the room and evacuation of the room by all personnel, followed by elaborate monitoring and decontamination procedures. Actions frequently desirable in cases of accidental releases of radioactivity are listed below. Although some effort is made to list these actions in the approximate order in which they are likely to be appropriate, some of these actions will be appropriate only in cases of large releases or under special conditions.

Persons in the vicinity of the spill or release who are liable to either external or internal contamination as a result of the accident should be given appropriate information immediately.

The protection of personnel and the containment of the radioactive material in the room in which the accident occurs should be given primary consideration. If the spill or release is of such a nature that, in the judgement of the person immediately responsible for the work, it is advantageous to take immediate action to contain the material or limit its release, such action may be appropriate.

Persons directly contaminated by a wet spill should immediately remove clothing affected and thoroughly wash the hands and other contaminated areas of the body.

If an inhalation hazard exists, all persons not involved in carrying out planned safety procedures should vacate the contaminated area immediately.

Evacuation or other action should be accomplished with the minimum required movement about the room. Movement should not start until the individual is aware of the situation and has determined the purpose of his movement. Areas of known or suspected contamination should be avoided.

The "radiological health and safety officer" or his representative in the area should be given all available information on the nature and extent of the release.

If evacuation of the room is required, it will generally be desirable to shut off all mechanical ventilation and to close all outside openings. However, there may be local conditions which require consideration. For example, if the release occurs in or near a fume hood, it may be disadvantageous to take any action which would discontinue ventilation by the hood.

If considerable contamination of the air is suspected, inhalation of radioactive material should be minimized by holding the breath or by use of such respiratory protection as may be available.

After all persons are out of the room, it may be desirable to prevent further escape of radioactive material from the room by sealing doors and other closures with adhesive tape.

Persons suspected of inhaling or ingesting considerable quantities or radioactive material should seek or be given immediate attention as discussed in "Decontamination of Personnel."

Except in case of injury or other urgent need, persons who have vacated the contaminated area should not leave the immediate vicinity until they have been monitored and necessary precautions, such as the removal of shoes or outer clothing, taken to limit further spread of radioactivity.

The extent of the area of contamination should be determined and the area roped off, with appropriate warning or guards.

The person in charge of measures for dealing with accidents, or a designated alternate, should arrange for immediate decontamination of personnel as required.

Safe and efficient decontamination of the working area and equipment will generally require careful planning based on an evaluation of all factors involved. In general, the following sequence of procedures will be involved:

1. Locate and contain the contamination.
2. Assess the contamination and plan clean-up operations.
3. Reduce the contamination by appropriate methods.
4. Assess the residual contamination and repeat the procedure as necessary.

Personnel carrying out decontamination procedures should be provided with appropriate instructions and equipment for their own protection and for the protection of other personnel.

DECONTAMINATION

Decontamination of Personnel

The "radiological health and safety officer" should set up instructions and facilities (materials and equipment) for normal decontamination and first aid procedures in conformity with the paragraph on "medical casualty service." The staff should be fully acquainted with these procedures.

MEASURES TO BE TAKEN IN CASE OF INTERNAL CONTAMINATION. Radioactive contamination of personnel can be internal through ingestion, inhalation, wounds or skin penetration. If anyone suspects internal contamination in case of an accident during work, it should be immediately reported to the "radiological health and safety officer."

Internal contamination is essentially a medical problem, parallel in some ways to the absorption of chemical toxins. Special corrective procedures should therefore combine with normal medical practice under medical advice and supervision.

Aims of the corrective procedures are: (a) try to eliminate as much of the internally introduced contaminant still remaining in the mouth, gastro-intestinal or respiratory tract, as quickly as possible and try to prevent or reduce its uptake into the bloodstream and tissues; (b) try to prevent fixation of the contaminant in the body or try to increase its excretion from the body.

For the first of these aims it is sometimes necessary that the contaminated person or another non-medical person takes immediate action (in the first seconds or minutes) for instance, to promote the mechanical elimination of the contaminant by vomiting or expectoration.

In case of contaminated small open wounds, cuts, punctures, or other injuries, the wound should be immediately washed and bleeding encouraged if necessary, and referred to the medical officer.

For the second of the aims indicated above any further procedure of internal decontamination; e.g., more complicated chemical or physico chemical methods, is a matter of medical treatment. It should be undertaken as soon as possible but only under medical supervision.

MEASURES TO BE TAKEN IN CASE OF EXTERNAL CONTAMINATION OF PERSONNEL. External contamination on the person can be a hazard in three ways:

1. It may cause injury from local exposure of the skin.
2. It may penetrate the intact skin (especially in the presence of certain organic solvents).
3. It may eventually be transferred into the body by ingestion or inhalation.

The danger of loose activity being eventually carried into the body is by far the most critical hazard, so that decontamination procedures are primarily concerned with loose contamination.

As a rule, except for decontamination of hands, or except in cases of emergency as agreed upon by the "radiological health and safety officer," all mild decontaminating procedures described in the two paragraphs below should be carried out under supervision of the "radiological health and safety officer." Attempts to remove contamination which resists mild procedures should only be made under medical supervision.

The immediate washing of contaminated areas with water and soap is the method of choice for removing loose contamination, subject to certain elementary precautions: (a) tepid water, not too hot, should be used; (b) soap should not be abrasive or highly alkaline; (c) washing can be helped by scrubbing with a soft brush only and in such a way as not to abrade the skin; (d) the skin should be washed for a few minutes at a time, then dried and monitored.

Washing could be repeated if necessary (as indicated by monitoring) providing there is no indication of the skin getting damaged.

If this procedure fails, only mild detergent approved by the "radiological health and safety officer" might be used, although repeated applications of detergents to the same area of the skin, hands for instance, might injure the skin and make it penetrable.

Use of organic solvents or of acid or alkaline solutions should be avoided.

Special attention should be paid to proper decontamination of creases, folds, hair and of such parts of the hands as finger nails, inter-finger space and the outer edges of the hands.

Care should be taken to avoid as much as possible the spreading of the contamination to uncontaminated parts of the body and to avoid internal contamination. If there is a risk of such a spread, an attempt should first be made to remove the contamination locally with absorbent material, and, if necessary, with a proper masking of the adjacent non-contaminated areas of the skin. A non-contaminated open wound should be protected.

After each decontamination operation, the treated place should be dried with a fresh non-contaminated towel or swab, and monitored. All towels and swabs, used in the decontamination process should be treated as contaminated material.

While decontaminating the face, special care should be taken not to contaminate the eyes or lips.

Decontamination of the eyes should be undertaken immediately. Not only the radioactive isotope is to be considered, but also the chemical nature of the contaminant and eventual complications due to foreign bodies and mechanical or chemical irritants. Additional irritation of the eyes by decontamination procedures should be avoided. Immediate irrigation of the eyes with a copious amount of water or with appropriate medically approved solutions is recommended. These solutions and a suitable vessel for eye washing should be provided for first-aid. After this first procedure every case of contamination of the eyes should be submitted to medical control and further treatment.

Attempts to remove contamination which resists washing should only be made under medical supervision.

Decontamination of Equipment

DECONTAMINATION OF GLASSWARE AND TOOLS. The decision to decontaminate material must take into account the continuing value of the material compared to the cost of decontamination.

Where the half-life of the contaminating element is short, it may be desirable to store tools and glassware for decay of activity rather than to attempt decontamination.

Decontamination of equipment should generally be done as soon as possible after its use. In many cases this will prevent the contamination from getting fixed and from being ultimately more difficult to deal with. It will often be found that surfaces that have been kept moist are easier to clean.

The cleaning of contaminated glassware and tools should be done with great care by informed persons in a well ventilated hood set aside in the laboratory for that purpose, or in special decontamination areas.

If it is necessary to dismantle any equipment prior to decontamination procedures, careful monitoring should be carried out during the operation.

Glassware can be cleaned by any of the normal chemical agents, of which chromic-acid solution is probably the most useful. Other cleaning agents are concentrated nitric acid, ammonium citrate, penta sodium triphosphate and ammonium bifluoride.

Metal tools and similar equipment should be washed with a detergent combined with brisk brushing to dislodge trapped contamination. Contamination resisting this treatment may be washed in stronger agents including dilute nitric acid or a 10% solution of sodium citrate or ammonium bifluoride. Other cleaning agents can be chosen based on the material of construction of the equipment and the likely chemical nature of the contaminant. Stainless steel could be treated with sulphuric or, as a last resort, hydrochloric acid.

If the decontamination causes any corrosion of the metal, future decontamination will be more difficult to remove and a coat of glossy paint on the decontaminated surface is desirable. Contamination prevention by the use of strippable coatings or plastic covers is useful. A coat of paint may provide adequate protection from soft emitters which prove resistant to decontamination.

The uptake of radioactive substances by glassware may be reduced by a preliminary treatment with the corresponding inactive chemical.

In some cases immersion in solutions of the non-radioactive isotope of the contaminant may be tried, although this is a slow procedure.

The solutions used for cleaning should not be returned to the stock bottles between uses.

Laboratory equipment should be surveyed for residual contamination following decontamination procedures. If the residual contamination indicates that the level of activity remains greater than that specified as permissible, equipment should not be re-used and should be regarded as radioactive waste.

DECONTAMINATION OF WORKING AREAS, BENCHES, AND OTHER SURFACES. As soon as possible after contamination of working areas, benches, and other surfaces has occurred or has been detected, decontamination should be carried out by suitably equipped and informed persons.

All surfaces should be cleaned by wet methods if possible, as the use of dry methods may create a dust hazard. For porous materials of construction which prove unsuitable for cleaning by wet methods, vacuum cleaning with proper filtration of the rejected air might be attempted; in any case special precautions in using dry techniques are necessary.

Cleaning tools should be assigned to the area in which the operations are being performed and not removed or used elsewhere without careful decontamination.

Paintwork can be cleaned with soap (or detergent) and water or, in extreme cases, removed with a paint remover. Polished linoleum can be cleaned with soap and water, followed, if necessary, by the removal of the wax polish by means of a solvent.

If the contamination is by alpha or soft beta emitters, the radiation may possibly be controlled by painting over. The use of two coats with the undercoat in a contrasting color is useful to indicate any wearing away of the protective coat. This method of contamination control should be used with caution with respect to future possible uses of the installation.

If after attempted decontamination adequate protection cannot be assured, the contaminated rooms or premises should be abandoned and contaminated removable objects disposed of in accordance with the requirements of the competent authority. Access to these abandoned areas should be forbidden to unauthorized persons and such areas should be identified by an appropriate and recognizable warning sign.

DECONTAMINATION OF CLOTHING, HOSPITAL LINEN OR SIMILAR ITEMS. In any handling of contaminated clothing appropriate precautions should be taken to prevent or control contamination of the worker and of the surrounding areas by the formation of aerosols. The sorting of contaminated garments will often need to be carried out in a fume hood. Care must be taken to prevent air-borne contamination from clothing placed in storage.

Contaminated clothing and linen should not be released to public laundries without the approval of the "radiological health and safety officer."

With short-life radioactive contamination, storage is recommended until the activity has fallen to safe levels.

It will usually be desirable to wash the contaminated clothing in specially provided laundering facilities; the area where decontamination goes on should be monitored. Personnel in charge of these facilities should be provided with protective coats and suitable gloves.

Contaminated garments should be segregated into batches of differing degrees of activity to avoid cross-contamination.

Routine washing of moderately contaminated clothing may be carried out according to schedules recommended for commercial laundry practice. However, it may be advantageous to substitute a standard detergent (chosen on the basis of economy) for the soap because of the tendency of latter to form deposits which may fix the activity in the fabric.

Clothing with resistant contamination or high levels of activity is dealt with by longer periods of washing and especially by repeated rinsings.

Rubber gloves and other rubber goods and plastics usually decontaminate readily. Such items should first be washed with an ordinary laundry formula. If this does not prove effective, rubber items can be washed in dilute nitric acid or agents chosen in the light of the nature of the contamination. This should be followed by a wash using scouring powder and a thorough rinse in running tap water.

If the clothing, linen, or similar items cannot be decontaminated to a safe level, it should be regarded as radioactive waste.

RADIOACTIVE WASTE CONTROL AND DISPOSAL

Waste Collection

In all working places where radioactive wastes may originate, suitable receptacles should be available.

Solid waste should be deposited in refuse bins with foot-operated lids. The bins should be lined with removable paper bags to facilitate removal of the waste without contamination.

Liquid waste should, if no other facilities for liquid waste disposal exist, be collected in bottles kept in pails or trays designed to retain all their contents in the event of a breakage.

All receptacles for radioactive wastes should be clearly identified. In general, it will be desirable to classify radioactive wastes according to methods of disposal or of storage, and to provide separate containers for the various classifications used. Depending upon the needs of the installation, one or more of the following bases for classification of wastes may be desirable:

1. Gamma radiation levels (high, low)
2. Total activity (high, intermediate, low)
3. Half-life (long, short)
4. Combustible, non-combustible

For convenient and positive identification, it may be desirable to use both color-coding and wording.

Shielded containers should be used when necessary.

It is generally desirable to maintain an approximate record of quantities of radioactive wastes released to drainage systems, to sewers, or for burial. This may be particularly important in the case of long-lived radioisotopes. For this purpose it is desirable or necessary to maintain a record of estimated quantities of radioactivity deposited in various receptacles, particularly those receiving high levels of activity or long-lived isotopes. Depending upon the system of control used by the installation, it may be desirable to provide for the receptacle to be marked or tagged with a statement of its contents.

Radioactive wastes should be removed from working places by designated personnel under the supervision of the "radiological health and safety officer."

Waste Storage

All wastes which cannot be immediately disposed of in conformity with requirements of the competent authority have to be placed in suitable storage.

Storage may be temporary or indefinite. Temporary storage is used to allow for decrease of activity, to permit regulation of the rate of release, to permit monitoring of materials of unknown degree of hazard or to await the availability of suitable transport. Indefinite storage in special places has to be provided for the more hazardous wastes for which no ultimate disposal method is available to the particular user.

Storage condition should meet the safety requirements for storage of sources.

The storage site should not be accessible to unauthorized personnel. (Control of animals should not be overlooked.)

The method of storage should prevent accidental release to the surroundings.

Appropriate records should be kept of the storage.

Disposal of Wastes to the Environment

GENERAL CONSIDERATIONS.   Disposal of radioactive wastes to the environment should be made in accordance with the conditions established by the "radiological health and safety officer" and by the competent authority.

The ways in which radioactive materials may affect the environment should be carefully examined for any proposed waste disposal method.

The capacity of any route of disposal to safely accept wastes depends on evaluation of a number of factors, many of which depend on the particular local situation. By assuming unfavorable conditions with respect to all factors it is possible to set a permissible level for waste disposal which will be safe under all circumstances. This usually allows a very considerable safety factor. The real capacity of a particular route of waste disposal can only be found by a lengthy study by experts.

The small user should first try to work within restrictive limits which are accepted as being safe and which will usually provide a workable solution to the problem of waste disposal. Such a restrictive safe limit is provided by keeping the level of activity at the point of release into the environment below the permissible levels for nonoccupationally exposed persons recommended by the International Commission on Radiological Protection for activity in drinking water or in air. This rule should be superseded if the competent authority provides any alternative requirements or if local studies by experts provide reasonable justification for other levels.

DISPOSAL TO DRAINS AND SEWERS.   The release of wastes into drains does not usually need to be considered as a direct release into the environment. Hence, a restrictive safe limit will usually be provided if the concentrations of radioactive waste material based on the total available flow of water in the system, averaged over a moderate period (daily or monthly), would not exceed the maximum permissible levels for drinking water recommended by the International Commission on Radiological Protection for individuals occupationally exposed. This would provide a large safety factor since water from drains and sewers is not generally to be considered as drinking water. However, in situations where the contamination affects the public water supply, the final concentrations in the water supply should be to the levels set for non-occupationally exposed persons. Some present studies suggest that if the contamination affects water used for irrigation the final concentrations in the irrigating water should be a factor of at least ten below the levels set for occupational exposure and the possible build up of activity in the irrigated lands and crops should be carefully surveyed.

Finally, prior to release of wastes to public drains, sewers and rivers, the competent authorities should be informed and consulted to ascertain that no other radioactive release is carried out in such a way that the cumulated release may result in a hazardous situation.

Radioactive wastes disposed to drains should be readily soluble or dispersible in water. Account should be taken of the possible changes of pH due to dilution, or other physico-chemical factors which may lead to precipitation or vaporization of diluted materials.

In general, the excreta of persons being treated by radioisotopes do not call for any special consideration. (This, however, does not apply to the unused residues of medical isotope shipments.)

Wastes should be flushed down by a copious stream of water.

The dilution of carrier-free material by the inactive element in the same chemical form is sometimes helpful.

Maintenance work on active drains within an establishment should only be carried out with the knowledge and under the supervision of the "radiological health and safety officer." Special care should be given to the possibility that small sources have been dropped into sinks and retained in traps or catchment basins.

The release of waste to sewers should be done in such a manner as not to require protective measures during maintenance work of the sewers outside the establishment, unless other agreement has been reached with the authority in charge of these sewers. The authority in charge of the sewer system outside the establishment should be informed of the release of radioactive wastes in this system; mutual discussion of the technical aspects of the waste disposal problem is desirable to provide protection without unnecessary anxiety.

DISPOSAL TO THE ATMOSPHERE.   Release of radioactive waste in the form of aerosols or gases into the atmosphere should conform with the requirements of the competent authority.

Subject to the preceding paragraph concentrations of radioactive gases or aerosols at the point of release into the environment should not exceed the accepted maximum permissible levels for non-occupationally exposed persons as set forth in the appropriate current national or international standards which have been established for maximum permissible levels for exposure to external radiations and for radioactive contamination of air and water. If higher levels are required and protection is based on an elevated release point from a stack, such levels can only be set after examination of local conditions by an expert.

Even if activity below permissible levels is achieved at the release point for an aerosol, a hazard of nuisance may still arise from fall-out of coarse particles. Therefore, the need for filtration should be assessed.

Used filters should be handled as solid wastes.

BURIAL OF WASTES.   Burial of wastes in soil sometimes provides a measure of protection not found if the wastes are released directly into the environment. The possibilities of safe burial of waste should always be appraised by an expert.

Burial under a suitable depth of soil (about one meter) provides economical protection from the external radiation of the accumulated deposit.

A burial site should be under the control of the user with adequate means of excluding the public.

A record should be kept of disposals into the ground.

INCINERATION OF WASTES.   If solid wastes are incinerated to reduce the bulk to manageable proportions, certain precautions should be taken.

The incineration of active wastes should only be carried out in equipment embodying those features of filtration and scrubbing as may be necessary for the levels of activity to be disposed of.

Residual ashes should be prevented from becoming a dust hazard, for example by damping them with water, and should be properly dealt with as ordinary active waste.

ACKNOWLEDGEMENT: This chapter is a reprint of a major part of "Safe Handling of Radioisotopes," published by the International Atomic Energy Agency, Vienna 1, Kaerntnerring, Austria, 1958, and is reprinted with permission.

Dose to Various Body Organs From Inhalation or Ingestion of Soluble Radionuclides

D. F. Bunch

At the National Reactor Testing Station (NRTS), a great number of calculations are performed each year by the various contractors to estimate the radiological consequences from operational releases, minor and major incidents, and potential accidental releases. Much of this is repetitious in that each person making the calculation must perform the necessary mathematics to solve the various equations used in calculating dose. In almost all cases, the mathematical and biological parameters are those recommended for the "standard man" by the International Commission on Radiological Protection (ICRP).[1,2] To eliminate the need for this repetition and to establish more uniform practices in calculations, a computer program was written and estimates of dose were prepared for the isotopes and major organs listed in the ICRP reports. These estimates take the form of dose conversion factors, such that a rapid and reasonable estimate of dose may be made. This chapter gives information for dose from the ingestion or inhalation of soluble radionuclides.

Calculation of Dose

Since, for the most part, close accuracy is not desired or even warranted, the parameters used are those recommended by ICRP for continuous exposure. The general expression used is:

$$\text{Dose} = \frac{AfET_E}{m}\left(1 - \exp\frac{-1.26 \times 10^4}{T_E}\right) \times \frac{1.6 \times 10^{-6} \times 3.2 \times 10^{15}}{0.693 \times 10^2} \quad (1)$$

where

f = fractional uptake by ingestion or inhalation to the organ of interest
E = effective energy = $\Sigma EF(\text{RBE})n$
T_E = effective half-time of material in organ of interest
m = mass of organ
1.6×10^{-6} = erg/MeV
3.2×10^{15} = dis/day/curie
0.693×10^2 = erg/g/rad.

The exponential term assumes a 50-year post-exposure period to correct for certain isotopes that do not reach equilibrium in this time. The term "A" is defined as:

$A = 1$ to calculate rem per curie inhaled or ingested (2)

$A = 1 B$ to calculate rem per curie-sec/m^3 where B = breathing rate in m^3/sec (curie-sec/m^3 is the time integrated concentration of airborne radioactivity) (3)

$A = 1/f$ to calculate rem per curie in the organ. (4)

452

The derivation, assumptions, and limitation of Equation (1) have been discussed in detail in References 1, 3, and 4, and these should be referred to for more detailed information (see especially Reference 3). It should be emphasized that these calculations should not be applied to the general population and, further, that derived doses are only approximations.

Application

Sample Calculation 1. If it is known that one microcurie of I-131 has been inhaled, the estimation of dose may be made as follows:

(1) If the thyroid is the organ of interest, the conversion factor from curies inhaled to dose in rem (from Table 2) is 1.48 E + 06 or 1.48×10^6 rem/curie inhaled

(2) Dose = $1.48 \times 10^6 \dfrac{\text{rem}}{\text{Ci inhaled}} \times 10^{-6}$ curie = 1.48 rem $\approx$ 1.5 rem.

Sample Calculation 2. If the air concentration is 10^{-6} μCi/cc of I-135 and the individual will be exposed for eight hours, the estimation of dose may be made as follows:

(1) If the exposure is occupational, assume the high breathing rate factor (Ci-sec/m^3)

(2) If the thyroid is the organ of interest, the conversion factor is 4.28E + 01 or 4.28×10^1 rem/Ci-sec/m^3

(3) 10^{-6} μCi/cc = 10^{-6} Ci/m^3

(4) Dose = 10^{-6} Ci/m$^3 \times 8$ hours $\times 3600 \dfrac{\text{sec}}{\text{hour}} \times \dfrac{4.28 \times 10^1 \text{ rem}}{\text{Ci-sec/m}^3} \approx 1.2$ rem.

Sample Calculation 3. If one curie of Sr-89 were released from a stack and the dose at 16 km (10 miles) were desired, it could be calculated as follows:

(1) Relative axial concentration at 10 miles (Pasquill, Class F—strong inversion from Figure 1) is 2×10^{-5}m^{-2}. This must be divided by the wind speed (4 m/sec) to obtain relative dilution of 5×10^{-6} sec/m^3

(2) Bone dose conversion factor for Sr-89 = 9.42E + 01 or 9.42×10^1 rem/Ci-sec/m^3 (lower breathing rate)

(3) Dose = 1 Ci $\times 5 \times 10^{-6} \dfrac{\text{sec}}{\text{M}^3} \times 9.42 \times 10^1 \dfrac{\text{rem}}{\text{Ci-sec/m}^3} = 4.7 \times 10^{-4}$ rem.

Notes on Format—Table 2

Column 1: The identification format is Z.A. as in tritium 01.003. (Z = atomic number; A = atomic weight; M = metastable)

Column 2: This is $\Sigma EF(\text{RBE})n$ in MeV.

Column 3: This is T_E in days.

Column 4: Weight of organ of interest, assuming standard-man parameters; e.g., Bone = 700 g.

Column 5: rem/curie inhaled.

Column 6: rem/curie-sec/m^3 for breathing rate typical of active portion of day, 10 m^3/8 hours.

Column 7: rem/curie-sec/m^3 for average breathing rate, 20 m^3/24 hours.

Column 8: rem/curie in organ.

Column 9: rem/curie ingested.

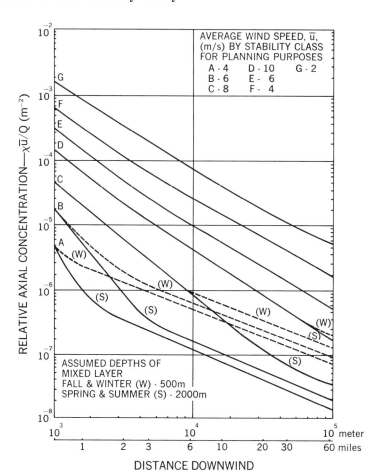

FIG. 1. Average relative axial concentration by stability [5].

Alternate Data

Since it may be desirable to use parameters other than those used in these conversion factors, a nomogram has been prepared and included as Figure 2 so that any or all of the parameters may be varied to obtain different conversion factors. This nomogram also is based on Equation (1) with the assumption that $e^{-1.26 \times 10^{4} T_E}$ is near zero. The nomogram may be used as follows:

    (1) *rem/Ci-sec/m³*

        (a) Draw a line from the time integrated concentration through the assumed breathing rate to derive curies inhaled.

        (b) Draw a line from curies inhaled through f_a to derive curies in organ.

        (c) Draw a line from curies in organ through the effective energy, mark the point, and draw a line through the effective half-life to the next column.

        (d) Draw a line from this column through the organ weight to derive dose in rem.

    (2) *rem/curie inhaled*

        Done in the same fashion as (1) except that step (a) is eliminated; begin at column 3. To determine rem/curie ingested, substitute f_w for f_a.

    (3) *rem/curie in organ*

        Same as (1); eliminate steps (a) and (b) and begin at column 5.

For the short-lived isotopes, the use of a single exponential model may grossly overestimate the actual dose. The conversion factors for these isotopes should be corrected by $T_R/(T_R + T_U)$ where T_R is the radiological half-life and T_U is

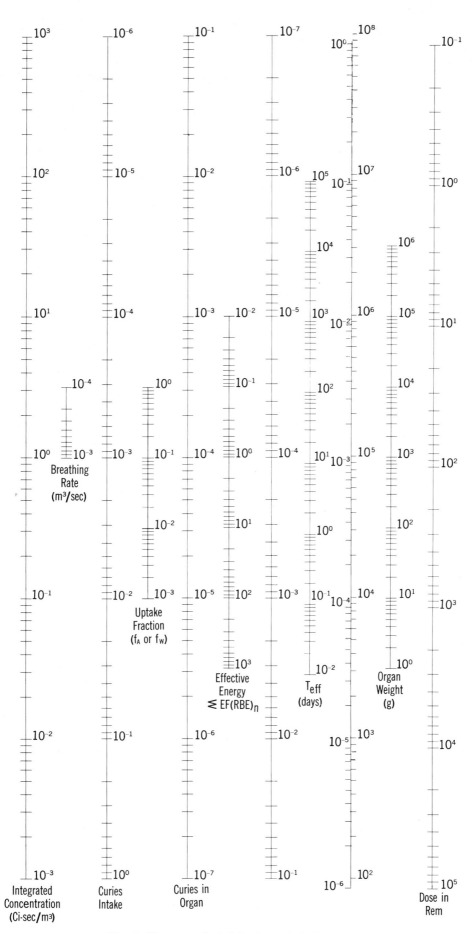

FIG. 2. Nomogram for infinity dose calculations.

the half-time for uptake (0.25 day for iodine). The corrected iodine factors are shown in Table 1. It can be seen that many of these isotopes would not constitute a significant hazard if mixed fission products were released. In addition, for most of the organs a few isotopes constitute 80–98$^+$ percent of the exposure. Therefore, for estimating dose from mixed fission products it is not necessary to make a calculation for the contribution of every dose.

TABLE 1
Thyroid Conversion Factors

	ICRP II				Corrected for Uptake Decay		
	rem/Ci inhaled	rem/Ci-sec/m^3 (2.32 × 10^{-4}m^3/sec)	rem/Ci-sec/m^3 (3.47 × 10^{-4}m^3/sec)	rem/Ci organ	rem/Ci inhaled	rem/Ci-sec/m^3 (2.32 × 10^{-4}m^3/sec)	rem/Ci-sec/m^3 (3.47 × 10^{-4}m^3/sec)
I-131	1.48 × 10^6	343	514	6.3	1.44 × 10^6	330	500
I-132	5.35 × 10^4	12.4	18.5	0.23	1.5 × 10^4	3.5	5.2
I-133	4 × 10^5	92.8	139	1.8	3.1 × 10^5	72	110
I-134	2.5 × 10^4	5.8	8.7	0.11	3.3 × 10^3	0.75	1.1
I-135	1.24 × 10^5	28.8	43.0	0.54	6.6 × 10^4	15	23

$T_R(T_U + T_R)$ where $T_U = 0.25$ day

TABLE 2
Dose Conversion Factors

Identification (Z. A.)	Energy (MeV)	T(1/2) Days	Weight in Grams	Inh. Dose (rem/Ci)	Dose/Conc rem/Ci-s/m^3 Hi-Rate	Dose/Conc rem/Ci-s/m^3 Lo-Rate	Organ Dose (rem/Ci)	Ing. Dose (rem/Ci)
THYROID	**20 GRAMS**							
24.051	8.40E−03	2.66E+01	20.0	1.90E+02	6.57E−02	4.29E−02	8.26E+05	3.72E−00
49.114M	9.20E−01	7.20E−00	20.0	2.45E+03	8.47E−01	5.53E−01	2.45E+07	1.96E+01
49.115M	1.60E−01	1.90E−01	20.0	1.12E+01	3.89E−03	2.53E−03	1.12E+05	8.99E−02
49.115	1.70E−01	8.40E−00	20.0	5.28E+02	1.82E−01	1.19E−01	5.28E+06	4.22E−00
50.113	1.60E−01	4.30E+01	20.0	7.12E+02	2.46E−01	1.60E−01	2.54E+07	1.27E+02
50.125	9.30E−01	8.40E−00	20.0	8.09E+02	2.79E−01	1.82E−01	2.89E+07	1.44E+02
51.122	5.90E−01	1.60E−00	20.0	2.79E+01	9.66E−03	6.30E−03	3.49E+06	3.14E−00
51.124	5.70E−01	3.80E−00	20.0	6.41E+01	2.21E−02	1.44E−02	8.01E+06	7.21E−00
51.125		4.00E−00	20.0	1.66E+01	5.76E−03	3.75E−03	2.08E+06	1.86E−00
52.125	1.10E−01	7.80E−00	20.0	1.20E+03	4.17E−01	2.72E−01	3.17E+06	7.93E+02
52.127M	3.00E−01	8.30E−00	20.0	3.50E+03	1.21E−00	7.90E−01	9.21E+06	2.30E+03
52.127	2.40E−01	3.70E−01	20.0	1.24E+02	4.31E−02	2.81E−02	3.28E+05	8.21E+01
52.129M	6.80E−01	7.10E−00	20.0	6.78E+03	2.34E−00	1.53E−00	1.78E+07	4.46E+03
52.129	6.00E−01	5.10E−02	20.0	4.30E+01	1.48E−02	9.70E−03	1.13E+05	2.83E+01
52.131M	6.90E−01	1.10E−00	20.0	1.06E+03	3.69E−01	2.40E−01	2.80E+06	7.02E+02
52.132	7.40E−01	2.40E−00	20.0	2.49E+03	8.63E−01	5.63E−01	6.57E+06	1.64E+03
53.126	1.60E−01	1.21E+01	20.0	1.64E+06	5.69E+02	3.71E+02	7.16E+06	2.14E+06
53.129	6.80E−02	1.38E+02	20.0	7.98E+06	2.76E+03	1.80E+03	3.47E+07	1.04E+07
53.131	2.30E−01	7.60E−00	20.0	1.48E+06	5.14E+02	3.35E+02	6.46E+06	1.94E+06
53.132	6.50E−01	9.70E−02	20.0	5.36E+04	1.85E+01	1.21E+01	2.33E+05	6.99E+04
53.133	5.40E−01	8.70E−01	20.0	3.99E+05	1.38E+02	9.02E+01	1.73E+06	5.21E+05
53.134	8.20E−01	3.60E−02	20.0	2.51E+04	8.69E−00	5.66E−00	1.09E+05	3.27E+04
53.135	5.20E−01	2.80E−01	20.0	1.23E+05	4.28E+01	2.79E+01	5.38E+05	1.61E+05
75.183	3.40E−02	2.90E−00	20.0	1.27E+03	4.41E−01	2.88E−01	3.64E+05	1.27E+03
75.186	3.60E−01	1.70E−00	20.0	7.92E+03	2.74E−00	1.78E−00	2.26E+06	7.92E+03
75.187	1.20E−02	3.00E−00	20.0	4.66E+02	1.61E−01	1.05E−01	1.33E+05	4.66E+02
75.188	8.00E−01	5.70E−01	20.0	5.90E+03	2.04E−00	1.33E−00	1.68E+06	5.90E+03
85.211	6.10E+01	3.00E−01	20.0	1.80E+06	6.22E+02	4.05E+02	7.82E+07	2.34E+06

TABLE 2 (Continued)

Identification (Z. A.)	Energy (MeV)	T(1/2) Days	Weight in Grams	Inh. Dose (rem/Ci)	Dose/Conc rem/Ci-s/m³ Hi-Rate	Dose/Conc rem/Ci-s/m³ Lo-Rate	Organ Dose (rem/Ci)	Ing. Dose (rem/Ci)
KIDNEY	**300 GRAMS**							
4.007	1.20E−02	3.70E+01	300.0	8.21E+02	2.84E−01	1.85E−01	1.09E+05	6.57E−00
21.046	5.00E−01	4.00E+01	300.0	2.46E+04	8.53E−00	5.56E−00	4.93E+06	9.86E−00
21.047	2.00E−01	3.30E−00	300.0	8.14E+02	2.81E−01	1.83E−01	1.62E+05	3.25E−01
21.048	1.10E−00	1.80E−00	300.0	2.44E+03	8.44E−01	5.51E−01	4.88E+05	9.76E−01
23.048	7.00E−01	1.32E+01	300.0	2.27E+04	7.88E−00	5.14E−00	2.27E+06	1.82E+03
24.051	1.20E−02	2.66E+01	300.0	5.35E+01	1.85E−02	1.20E−02	7.87E+04	1.02E−00
27.057	4.50E−02	9.20E−00	300.0	8.16E+01	2.82E−02	1.84E−02	1.02E+05	6.12E+01
27.058M	4.80E−02	3.70E−01	300.0	3.50E−00	1.21E−03	7.90E−04	4.38E+03	2.62E−00
27.058	2.20E−01	8.40E−00	300.0	3.64E+02	1.26E−01	8.22E−02	4.55E+05	2.73E+02
27.060	5.60E−01	9.50E−00	300.0	1.04E+03	3.63E−01	2.36E−01	1.31E+06	7.87E+02
29.064	1.70E−01	5.10E−01	300.0	4.27E+02	1.47E−01	9.65E−02	2.13E+04	2.13E+02
30.065	1.10E−01	9.30E+01	300.0	3.02E+04	1.04E+01	6.83E−00	2.52E+06	1.00E+04
30.069M	4.70E−01	5.80E−01	300.0	8.06E+02	2.79E−01	1.82E−01	6.72E+04	2.68E+02
30.069	3.70E−01	3.60E−02	300.0	3.94E+01	1.36E−02	8.89E−03	3.28E+03	1.31E+01
31.072	8.90E−01	5.50E−01	300.0	6.03E+02	2.08E−01	1.36E−01	1.20E+05	2.41E−00
32.071	1.00E−02	6.00E−00	300.0	1.18E+02	4.09E−02	2.67E−02	1.48E+04	4.44E−00
33.073	3.60E−02	6.70E+01	300.0	1.60E+03	5.55E−01	3.62E−01	5.94E+05	1.78E+02
33.074	3.40E−01	1.70E+01	300.0	3.84E+03	1.33E−00	8.68E−01	1.42E+06	4.27E+02
33.076	1.10E−00	1.10E−00	300.0	8.05E+02	2.78E−01	1.81E−01	2.98E+05	8.95E+01
33.077	2.40E−01	1.60E−00	300.0	2.55E+02	8.84E−02	5.77E−02	9.47E+04	2.84E+01
34.075	7.20E−02	1.01E+01	300.0	5.38E+03	1.86E−00	1.21E−00	1.79E+05	7.17E+03
40.093	1.90E−02	9.00E+02	300.0	2.37E+04	8.24E−00	5.37E−00	4.76E+06	9.53E−00
40.095	4.60E−01	5.90E+01	300.0	3.34E+04	1.15E+01	7.55E−00	6.69E+06	1.33E+01
40.097	1.50E−00	7.10E−01	300.0	1.31E+03	4.54E−01	2.96E−01	2.62E+05	5.25E−01
41.093M	3.80E−02	6.40E+02	300.0	2.99E+04	1.03E+01	6.76E−00	5.99E+06	1.19E+01
41.095	2.00E−02	3.35E+01	300.0	8.26E+02	2.85E−01	1.86E−01	1.65E+05	3.30E−01
41.097	6.00E−01	5.10E−02	300.0	3.77E+01	1.30E−02	8.51E−03	7.54E+03	1.50E−02
42.099	4.80E−01	1.50E−00	300.0	9.31E+03	3.21E−00	2.09E−00	1.70E+05	1.02E+04
43.096M	4.20E−01	3.60E−02	300.0	1.86E+01	6.45E−03	4.20E−03	3.72E+03	1.86E+01
43.096	4.70E−01	3.50E−00	300.0	2.02E+03	7.01E−01	4.57E−01	4.05E+05	2.02E+03
43.097M	9.00E−02	1.60E+01	300.0	1.77E+03	6.14E−01	4.00E−01	3.55E+05	1.77E+03
43.097	2.00E−02	2.00E+01	300.0	4.93E+02	1.70E−01	1.11E−01	9.86E+04	4.93E+02
43.099M	2.60E−02	2.50E−01	300.0	8.01E−00	2.77E−03	1.80E−03	1.60E+03	8.01E−00
43.099	9.40E−02	2.00E+01	300.0	2.31E+03	8.02E−01	5.23E−01	4.63E+05	2.31E+03
44.097	7.70E−02	1.30E−00	300.0	1.23E+03	4.27E−01	2.78E−01	2.46E+04	1.48E+02
44.103	2.20E−01	2.40E−00	300.0	6.51E+03	2.25E−00	1.46E−00	1.30E+05	7.81E+02
44.105	8.40E−01	1.80E−01	300.0	1.86E+03	6.45E−01	4.20E−01	3.72E+04	2.23E+02
44.106	1.30E−00	2.48E−00	300.0	3.97E+04	1.37E+01	8.97E−00	7.95E+05	4.77E+03
45.103M	5.40E−02	3.80E−02	300.0	5.06E−00	1.75E−03	1.14E−03	5.06E+02	3.03E−00
45.105	1.90E−01	1.44E−00	300.0	6.74E+02	2.33E−01	1.52E−01	6.74E+04	4.04E+02
46.103	6.10E−02	1.10E+01	300.0	4.96E+03	1.71E−00	1.12E−00	1.65E+05	3.31E+03
46.109	4.20E−01	5.60E−01	300.0	1.74E+03	6.02E−01	3.92E−01	5.80E+04	1.16E+03
47.105	2.20E−01	8.00E−00	300.0	2.17E+03	7.50E−01	4.89E−01	4.34E+05	8.68E+01
47.110M	6.50E−01	1.00E+01	300.0	8.01E+03	2.77E−00	1.80E−00	1.60E+06	3.20E+02
47.111	3.70E−01	4.00E−00	300.0	1.82E+03	6.31E−01	4.11E−01	3.65E+05	7.30E+01
48.109	9.80E−02	1.84E+02	300.0	1.11E+05	3.84E+01	2.50E+01	4.44E+06	1.11E+03
48.115M	6.10E−01	3.80E+01	300.0	1.42E+05	4.94E+01	3.22E+01	5.71E+06	1.42E+03
48.115	5.60E−01	2.20E−00	300.0	7.59E+03	2.62E−00	1.71E−00	3.03E+05	7.59E+01
49.113M	1.90E−01	7.30E−00	300.0	3.42E+03	1.18E−00	7.72E−01	3.42E+05	2.73E+01
49.114M	9.30E−01	2.70E+01	300.0	6.19E+04	2.14E+01	1.39E+01	6.19E+06	4.95E+02

TABLE 2 (Continued)

Identification (Z. A.)	Energy (MeV)	T(1/2) Days	Weight in Grams	Inh. Dose (rem/Ci)	Dose/Conc rem/Ci-s/m³ Hi-Rate	Dose/Conc rem/Ci-s/m³ Lo-Rate	Organ Dose (rem/Ci)	Ing. Dose (rem/Ci)
KIDNEY (Continued)								
49.115M	1.90E−01	1.90E−01	300.0	8.90E+01	3.08E−02	2.00E−02	8.90E+03	7.12E−01
49.115	1.70E−01	6.00E+01	300.0	2.51E+04	8.70E−00	5.67E−00	2.51E+06	2.01E+02
52.125M	1.40E−01	2.00E+01	300.0	2.07E+04	7.16E−00	4.67E−00	6.90E+05	1.38E+04
52.127M	3.20E−01	2.30E+01	300.0	5.44E+04	1.88E+01	1.22E+01	1.81E+06	3.63E+04
52.127	2.40E−01	3.90E−01	300.0	6.92E+02	2.39E−01	1.56E−01	2.30E+04	4.61E+02
52.129M	7.80E 01	1.60E+01	300.0	9.23E+04	3.19E+01	2.08E+01	3.07E+06	6.15E+04
52.129	6.80E−01	5.10E−02	300.0	2.56E+02	8.87E−02	5.79E−02	8.55E+03	1.71E+02
52.131M	8.10E−01	1.20E−00	300.0	7.19E+03	2.48E−00	1.62E−00	2.39E+05	4.79E+03
52.132	9.60E−01	2.90E−00	300.0	2.06E+04	7.12E−00	4.64E−00	6.86E+05	1.37E+04
55.131	2.10E−01	8.00E−00	300.0	3.10E+03	1.07E−00	7.01E−01	4.14E+05	4.14E+03
55.134M	1.10E−01	1.30E−01	300.0	2.64E+01	9.15E−03	5.97E−03	3.52E+03	3.52E+01
55.134	4.60E−01	4.00E+01	300.0	3.40E+04	1.17E+01	7.68E−00	4.53E+06	4.53E+04
55.135	6.60E−02	4.20E+01	300.0	5.12E+03	1.77E−00	1.15E−00	6.83E+05	6.83E+03
55.136	2.90E−01	1.00E+01	300.0	5.36E+03	1.85E−00	1.21E−00	7.15E+05	7.15E+03
55.137	3.70E−01	4.20E+01	300.0	2.87E+04	9.94E−00	6.48E−00	3.83E+06	3.83E+04
56.131	1.40E−01	4.90E−00	300.0	4.73E−00	1.63E−03	1.06E−03	1.69E+05	8.46E−01
56.140	1.20E−00	5.10E−00	300.0	4.22E+01	1.46E−02	9.53E−03	1.50E+06	7.54E−00
58.141	1.80E−01	3.00E+01	300.0	6.66E+03	2.30E−00	1.50E−00	1.33E+06	2.66E−00
58.143	8.20E−01	1.33E−00	300.0	1.34E+03	4.65E−01	3.03E−01	2.69E+05	5.38E−01
58.144	1.30E−00	1.91E+02	300.0	3.06E+05	1.05E+02	6.91E+01	6.12E+07	1.22E+02
59.142	8.10E−01	8.00E−01	300.0	7.99E+02	2.76E−01	1.80E−01	1.59E+05	3.19E−01
59.143	3.20E−01	1.35E+01	300.0	5.32E+03	1.84E−00	1.20E−00	1.06E+06	2.13E−00
60.144	2.00E+01	6.56E+02	300.0	3.23E+07	1.11E+04	7.30E+03	3.23E+09	1.61E+04
60.147	3.10E−01	1.11E+01	300.0	8.48E+03	2.93E−00	1.91E−00	8.48E+05	4.24E−00
60.149	9.70E−01	8.30E−02	300.0	1.98E+02	6.87E−02	4.48E−02	1.98E+04	9.92E−02
61.147	6.90E−02	3.83E+02	300.0	3.25E+04	1.12E+01	7.35E−00	6.51E+06	1.30E+01
61.149	4.20E−01	2.20E−00	300.0	1.13E+03	3.94E−01	2.57E−01	2.27E+05	4.55E−01
62.147	2.30E+01	6.56E+02	300.0	1.86E+07	6.43E+03	4.19E+03	3.72E+09	7.44E+03
62.151	4.20E−02	6.45E+02	300.0	3.34E+04	1.15E+01	7.54E−00	6.68E+06	1.33E+01
62.153	2.50E−01	1.95E−00	300.0	6.01E+02	2.08E−01	1.35E−01	1.20E+05	2.40E−01
63.152	7.10E−01	3.80E−01	300.0	4.99E+02	1.72E−01	1.12E−01	6.65E+04	1.99E−01
63.152	2.50E−01	1.13E+03	300.0	5.22E+05	1.80E+02	1.17E+02	6.96E+07	2.09E+02
63.154	7.60E−01	1.18E+03	300.0	1.65E+06	5.73E+02	3.74E+02	2.21E+08	6.63E+02
63.155	8.30E−02	4.38E+02	300.0	6.72E+04	2.32E+01	1.51E+01	8.96E+06	2.69E+01
65.160	4.00E−01	6.60E+01	300.0	4.88E+04	1.68E+01	1.10E+01	6.51E+06	1.95E+01
67.166	6.90E−01	1.10E−00	300.0	9.36E+02	3.23E−01	2.11E−01	1.87E+05	3.74E−01
68.169	1.90E−01	9.30E−00	300.0	2.17E+03	7.53E−01	4.91E−01	4.35E+05	8.71E−01
68.171	4.60E−01	3.10E−01	300.0	1.75E+02	6.08E−02	3.96E−02	3.51E+04	7.03E−02
69.170	3.40E−01	9.20E+01	300.0	3.85E+04	1.33E+01	8.70E−00	7.71E+06	1.54E+01
69.171	3.00E−02	2.26E+02	300.0	8.36E+03	2.89E−00	1.88E−00	1.67E+06	3.34E−00
70.175	1.50E−01	4.10E−00	300.0	1.97E+03	6.82E−01	4.45E−01	1.51E+05	7.58E−01
71.177	1.60E−01	6.70E−00	300.0	6.61E+02	2.28E−01	1.49E−01	2.64E+05	2.64E−01
72.181	2.50E−01	4.30E+01	300.0	1.32E+04	4.58E−00	2.99E−00	2.65E+06	5.30E−00
73.182	4.50E−01	8.80E+01	300.0	7.32E+04	·2.53E+01	1.65E+01	9.76E+06	2.93E+01
76.185	2.50E−01	4.80E−00	300.0	5.92E+03	2.04E−00	1.33E−00	2.96E+05	1.48E+03
76.191M	3.90E−02	5.20E−01	300.0	1.00E+02	3.46E−02	2.25E−02	5.00E+03	2.50E+01
76.191	1.10E−01	3.80E−00	300.0	2.06E+03	7.13E−01	4.65E−01	1.03E+05	5.15E+02
76.193	3.80E−01	1.00E−00	300.0	1.87E+03	6.48E−01	4.23E−01	9.37E+04	4.68E+02
77.190	1.20E−01	9.70E−00	300.0	4.01E+03	1.39E−00	9.07E−01	2.87E+05	1.29E+03
77.192	5.00E−01	3.00E+01	300.0	5.18E+04	1.79E+01	1.16E+01	3.70E+06	1.66E+04
77.194	8.10E−01	7.80E−01	300.0	2.18E+03	7.54E−01	4.92E−01	1.55E+05	7.01E+02

TABLE 2 (Continued)

Identification (Z. A.)	Energy (MeV)	T(1/2) Days	Weight in Grams	Inh. Dose (rem/Ci)	Dose/Conc rem/Ci-s/m³ Hi-Rate	Dose/Conc rem/Ci-s/m³ Lo-Rate	Organ Dose (rem/Ci)	Ing. Dose (rem/Ci)
KIDNEY (Continued)								
78.191	2.20E−01	2.90E−00	300.0	4.72E+03	1.63E−00	1.06E−00	1.57E+05	1.57E+03
78.193M	2.30E−02	3.20E−00	300.0	5.44E+02	1.88E−01	1.22E−01	1.81E+04	1.81E+02
78.193	1.40E−02	6.00E+01	300.0	6.21E+03	2.15E−00	1.40E−00	2.07E+05	2.07E+03
78.197	5.00E−01	5.60E−02	300.0	2.07E+02	7.16E−02	4.67E−02	6.90E+03	6.90E+01
78.197	2.30E−01	7.40E−01	300.0	1.25E+03	4.35E−01	2.84E−01	4.19E+04	4.19E+02
79.196	1.50E−01	5.50E−00	300.0	1.83E+03	6.33E−01	4.13E−01	2.03E+05	6.10E+02
79.198	4.10E−01	2.70E−00	300.0	2.45E+03	8.50E−01	5.54E−01	2.73E+05	8.19E+02
79.199	1.20E−01	3.10E−00	300.0	8.25E+02	2.85E−01	1.86E−01	9.17E+04	2.75E+02
80.197M	1.80E−01	9.40E−01	300.0	9.18E+03	3.17E−00	2.07E−00	4.17E+04	1.08E+04
80.197	4.30E−02	2.30E−00	300.0	5.36E+03	1.85E−00	1.21E−00	2.43E+04	6.34E+03
80.203	1.50E−01	1.10E+01	300.0	8.95E+04	3.09E+01	2.02E+01	4.07E+05	1.05E+05
81.200	1.30E−01	9.70E−01	300.0	7.46E+02	2.58E−01	1.68E−01	3.11E+04	7.15E+02
81.201	1.10E−01	2.10E−00	300.0	1.36E+03	4.73E−01	3.08E−01	5.69E+04	1.31E+03
81.202	2.40E−01	4.40E−00	300.0	6.25E+03	2.16E−00	1.41E−00	2.60E+05	5.99E+03
81.204	2.50E−01	7.00E−00	300.0	1.03E+04	3.58E−00	2.33E−00	4.31E+05	9.92E+03
82.203	6.90E−02	2.16E−00	300.0	1.47E+03	5.08E−01	3.31E−01	3.67E+04	3.67E+02
82.210	1.00E+01	4.94E+02	300.0	4.87E+07	1.68E+04	1.09E+04	1.21E+09	1.21E+07
82.212	8.10E+01	4.40E−01	300.0	3.51E+05	1.21E+02	7.93E+01	8.79E+06	8.79E+04
83.206	5.80E−01	3.10E−00	300.0	3.54E+04	1.22E+01	8.00E−00	4.43E+05	1.33E+03
83.207	3.30E−01	6.00E−00	300.0	3.90E+04	1.35E+01	8.81E−00	4.88E+05	1.46E+03
83.210	1.90E+01	3.00E−00	300.0	1.12E+06	3.89E+02	2.53E+02	1.40E+07	4.21E+04
83.212	8.20E+01	4.20E−02	300.0	6.79E+04	2.35E+01	1.53E+01	8.49E+05	2.54E+03
84.210	5.50E+01	4.60E+01	300.0	1.24E+07	4.31E+03	2.81E+03	6.24E+08	2.49E+06
89.227	6.20E+01	6.00E+03	300.0	2.41E+08	8.35E+04	5.45E+04	8.05E+10	8.05E+04
89.228	5.50E+01	2.60E−01	300.0	1.05E+04	3.66E−00	2.38E−00	3.52E+06	3.52E−00
90.227	6.10E+01	1.84E+01	300.0	2.76E+06	9.57E+02	6.24E+02	2.76E+08	1.38E+03
90.228	5.60E+01	6.78E+02	300.0	9.36E+07	3.23E+04	2.11E+04	9.36E+09	4.68E+04
90.230	4.80E+01	2.20E+04	300.0	1.13E+09	3.92E+05	2.56E+05	1.13E+11	5.67E+05
90.231	1.40E−01	1.07E−00	300.0	3.69E+02	1.27E−01	8.33E−02	3.69E+04	1.84E−01
90.232	4.10E+01	2.20E+04	300.0	9.70E+08	3.35E+05	2.18E+05	9.70E+10	4.85E+05
90.234	9.00E−01	2.41E+01	300.0	5.35E+04	1.85E+01	1.20E+01	5.35E+06	2.67E+01
91.230	1.50E+02	1.77E+01	300.0	1.31E+06	4.52E+02	2.94E+02	1.31E+08	5.22E+02
91.231	7.90E+01	5.10E+04	300.0	2.17E+09	7.52E+05	4.90E+05	2.17E+11	8.70E+05
91.233	1.50E−01	2.74E+01	300.0	1.01E+04	3.50E−00	2.28E−00	1.01E+06	4.05E−00
92.230	3.50E+02	8.70E−00	300.0	2.10E+07	7.27E+03	4.74E+03	7.51E+08	8.26E+03
92.232	1.10E+02	1.50E+01	300.0	1.13E+07	3.94E+03	2.57E+03	4.07E+08	4.47E+03
92.233	5.00E+01	1.50E+01	300.0	5.18E+06	1.79E+03	1.16E+03	1.85E+08	2.03E+03
92.234	4.90E+01	1.50E+01	300.0	5.07E+06	1.75E+03	1.14E+03	1.81E+08	1.99E+03
92.235	4.60E+01	1.50E+01	300.0	4.76E+06	1.64E+03	1.07E+03	1.70E+08	1.87E+03
92.236	4.70E+01	1.50E+01	300.0	4.86E+06	1.68E+03	1.09E+03	1.73E+08	1.91E+03
92.238	4.30E+01	1.50E+01	300.0	4.45E+06	1.54E+03	1.00E+03	1.59E+08	1.75E+03
92.240	1.10E−00	5.66E−01	300.0	4.30E+03	1.48E−00	9.70E−01	1.53E+05	1.68E+02
93.237	4.90E+01	6.40E+04	300.0	1.03E+09	3.58E+05	2.33E+05	1.38E+11	4.14E+05
93.239	2.10E−01	2.33E−00	300.0	9.05E+02	3.13E−01	2.04E−01	1.20E+05	3.62E−01
94.238	5.70E+01	1.60E+04	300.0	1.22E+10	4.24E+06	2.76E+06	1.22E+11	4.90E+06
94.239	5.30E+01	3.20E+04	300.0	1.36E+10	4.71E+06	3.07E+06	1.36E+11	5.44E+06
94.240	5.30E+01	3.20E+04	300.0	1.36E+10	4.71E+06	3.07E+06	1.36E+11	5.44E+06
94.241	2.50E−00	4.20E+03	300.0	2.46E+08	8.51E+04	5.55E+04	2.46E+09	9.84E+04
94.242	5.10E+01	3.20E+04	300.0	1.31E+10	4.53E+06	2.95E+06	1.31E+11	5.24E+06
94.243	3.70E−01	2.08E−01	300.0	9.49E+01	3.28E−02	2.14E−02	1.89E+04	1.13E−02
94.244	4.80E+01	3.20E+04	300.0	7.52E+08	2.60E+05	1.70E+05	1.50E+11	9.02E+04

TABLE 2 (Continued)

Identification (Z. A.)	Energy (MeV)	T(1/2) Days	Weight in Grams	Inh. Dose (rem/Ci)	Dose/Conc rem/Ci-s/m³ Hi-Rate	Dose/Conc rem/Ci-s/m³ Lo-Rate	Organ Dose (rem/Ci)	Ing. Dose (rem/Ci)
KIDNEY (Continued)								
95.241	5.70E+01	2.30E+04	300.0	1.02E+09	3.53E+05	2.30E+05	1.36E+11	4.09E+05
95.242	7.20E+01	1.80E+04	300.0	1.01E+09	3.50E+05	2.28E+05	1.34E+11	4.05E+05
95.242	7.80E+01	6.67E−01	300.0	8.08E+04	2.79E+01	1.82E+01	1.08E+07	3.23E+01
95.243	5.40E+01	2.70E+04	300.0	1.00E+09	3.47E+05	2.26E+05	1.34E+11	4.02E+05
95.244	4.40E+01	1.80E−02	300.0	1.46E+03	5.06E−01	3.30E−01	1.95E+05	5.86E−01
96.242M	7.80E+01	1.61E+02	300.0	1.54E+07	5.35E+03	3.49E+03	3.09E+09	6.19E+03
96.243	6.00E+01	8.40E+03	300.0	4.82E+08	1.67E+05	1.08E+05	9.65E+10	1.93E+05
96.244	6.00E+01	5.20E+03	300.0	3.50E+08	1.21E+05	7.91E+04	7.01E+10	1.40E+05
96.245	5.60E+01	2.40E+04	300.0	6.77E+08	2.34E+05	1.52E+05	1.35E+11	2.70E+05
96.246	5.60E+01	2.40E+04	300.0	6.77E+08	2.34E+05	1.52E+05	1.35E+11	2.70E+05
96.247	5.50E+01	2.40E+04	300.0	6.64E+08	2.30E+05	1.50E+05	1.32E+11	2.65E+05
96.248	5.20E+01	2.40E+04	300.0	5.46E+09	1.89E+06	1.23E+06	1.09E+12	2.18E+06
SKELETON 7000 GRAMS								
4.007	8.50E−03	4.80E+01	7000.0	3.45E+02	1.19E−01	7.78E−02	4.31E+03	2.76E−00
6.014	2.70E−01	4.00E+01	7000.0	2.28E+03	7.89E−01	5.15E−01	1.14E+05	2.85E+03
9.018	1.40E−00	7.80E−02	7000.0	4.61E+02	1.59E−01	1.04E−01	1.15E+03	6.11E+02
15.032	3.50E−00	1.41E+01	7000.0	1.66E+05	5.77E+01	3.76E+01	5.21E+05	1.98E+05
16.035	2.80E−01	7.61E+01	7000.0	4.50E+03	1.55E−00	1.01E−00	2.25E+05	6.75E+03
20.045	4.30E−01	1.62E+02	7000.0	3.68E+05	1.27E+02	8.30E+01	7.36E+05	3.97E+05
20.047	2.60E−00	4.90E−00	7000.0	6.73E+04	2.32E+01	1.51E+01	1.34E+05	7.27E+04
21.046	9.00E−01	2.40E+01	7000.0	1.14E+04	3.94E−00	2.57E−00	2.28E+05	4.56E−00
21.047	8.90E−01	3.10E−00	7000.0	1.45E+03	5.04E−01	3.29E−01	2.91E+04	5.83E−01
21.048	1.60E−00	1.70E−00	7000.0	1.43E+03	4.97E−01	3.24E−01	2.87E+04	5.75E−01
23.048	1.20E−00	1.44E+01	7000.0	7.30E+03	2.52E−00	1.64E−00	1.82E+05	5.11E+02
26.055	6.50E−03	6.65E+02	7000.0	1.37E+03	4.74E−01	3.09E−01	4.56E+04	4.56E+02
28.059	7.70E−03	8.00E+02	7000.0	1.30E+04	4.50E−00	2.93E−00	6.51E+04	9.76E+03
28.063	1.10E−01	7.79E+02	7000.0	1.81E+05	6.26E+01	4.08E+01	9.05E+05	1.35E+05
28.065	5.30E−00	1.10E−01	7000.0	1.23E+03	4.26E−01	2.78E−01	6.16E+03	9.24E+02
30.065	9.40E−02	2.06E+02	7000.0	9.21E+03	3.18E−00	2.07E−00	2.04E+05	3.07E+03
30.069M	2.10E−00	5.80E−01	7000.0	5.79E+02	2.00E−01	1.30E−01	1.28E+04	1.93E+02
30.069	1.90E−00	3.60E−02	7000.0	3.25E+01	1.12E−02	7.34E−03	7.23E+02	1.08E+01
31.072	2.60E−00	5.60E−01	7000.0	1.15E+03	3.99E−01	2.60E−01	1.53E+04	4.61E−00
38.085M	3.40E−02	4.90E−02	7000.0	1.69E−01	5.85E−03	3.82E−03	6.04E+01	1.26E−01
38.085	9.10E−02	6.48E+01	7000.0	1.74E+04	6.03E−00	3.93E−00	6.23E+04	1.30E+04
38.089	2.80E−00	5.04E+01	7000.0	4.17E+05	1.44E+02	9.42E+01	1.49E+06	3.13E+05
38.090	5.50E−00	6.40E+03	7000.0	3.84E+07	1.32E+04	8.67E+03	3.20E+08	2.88E+07
38.091	7.40E−00	4.00E−01	7000.0	7.53E+03	2.61E−00	1.69E−00	2.68E+04	5.65E+03
38.092	8.00E−00	1.10E−01	7000.0	2.60E+03	9.01E−01	5.87E−01	9.30E+03	1.95E+03
39.090	4.40E−00	2.68E−00	7000.0	2.36E+04	8.19E−00	5.34E−00	1.24E+05	9.34E−00
39.091M	3.00E−00	3.50E−02	7000.0	2.10E+02	7.29E−02	4.75E−02	1.11E+03	8.32E−02
39.091	2.90E−00	5.80E+01	7000.0	3.37E+05	1.16E+02	7.62E+01	1.77E+06	1.33E+02
39.092	6.90E−00	1.50E−01	7000.0	2.07E+03	7.19E−01	4.69E−01	1.09E+04	8.20E−01
39.093	6.50E−00	4.20E−01	7000.0	5.48E+03	1.89E−00	1.23E−00	2.88E+04	2.16E−00
40.093	9.50E−02	1.00E+03	7000.0	9.75E+04	3.37E+01	2.19E+01	1.08E+06	3.90E+01
40.095	1.10E−00	5.95E+01	7000.0	6.22E+04	2.15E+01	1.40E+01	6.91E+05	2.49E+01
40.097	6.20E−00	7.10E−01	7000.0	4.18E+03	1.44E−00	9.45E−01	4.65E+04	1.67E−00
41.093M	1.20E−01	7.87E+02	7000.0	9.98E+04	3.45E+01	2.25E+01	9.98E+05	3.79E+01
41.095	3.70E−01	3.38E+01	7000.0	1.32E+04	4.57E−00	2.98E−00	1.32E+05	5.02E−00

TABLE 2 (Continued)

Identification (Z. A.)	Energy (MeV)	T(1/2) Days	Weight in Grams	Inh. Dose (rem/Ci)	Dose/Conc rem/Ci-s/m³ Hi-Rate	Dose/Conc rem/Ci-s/m³ Lo-Rate	Organ Dose (rem/Ci)	Ing. Dose (rem/Ci)
SKELETON (Continued)								
41.097	2.40E−00	5.10E−02	7000.0	1.29E+02	4.47E−02	2.92E−02	1.29E+03	4.91E−02
43.096M	3.90E−01	3.60E−02	7000.0	1.48E−01	5.13E−05	3.34E−05	1.48E+02	1.48E−01
43.096	3.50E−01	3.70E−00	7000.0	1.36E+01	4.73E−03	3.08E−03	1.36E+04	1.36E+01
43.097M	3.70E−01	2.00E+01	7000.0	7.82E+01	2.70E−02	1.76E−02	7.82E+04	7.82E+01
43.097	1.90E−02	2.50E+01	7000.0	5.02E−00	1.73E−03	1.13E−03	5.02E+03	5.02E−00
43.099M	2.00E−02	2.50E−01	7000.0	5.28E−02	1.82E−05	1.19E−05	5.28E+01	5.28E−02
43.099	4.70E−01	2.50E+01	7000.0	1.24E+02	4.29E−02	2.80E−02	1.24E+05	1.24E+02
44.097	1.30E−01	2.40E−00	7000.0	6.95E+01	2.28E−02	1.48E−02	3.29E+03	7.91E−00
44.103	6.20E−01	1.20E+01	7000.0	1.57E+03	5.44E−01	3.54E−01	7.86E+04	1.88E+02
44.105	3.50E−00	1.90E−01	7000.0	1.40E+02	4.86E−02	3.17E−02	7.03E+03	1.68E+01
44.106	6.50E−00	1.50E+01	7000.0	2.06E+04	7.13E−00	4.65E−00	1.03E+06	2.47E+03
45.103M	1.90E−01	3.80E−02	7000.0	1.52E−00	5.28E−04	3.44E−04	7.63E+01	7.63E−01
45.105	9.50E−01	1.39E−00	7000.0	2.79E+02	9.65E−02	6.30E−02	1.39E+04	1.39E+02
47.105	1.60E−01	1.70E+01	7000.0	3.73E+02	1.29E−01	8.43E−02	2.87E+04	1.43E+01
47.110M	1.10E−00	2.60E+01	7000.0	3.93E+03	1.35E−00	8.87E−01	3.02E+05	1.51E+02
47.111	1.80E−00	6.00E−00	7000.0	1.48E+03	5.13E−01	3.34E−01	1.14E+05	5.70E+01
49.113M	6.80E−01	7.30E−02	7000.0	2.09E+01	7.26E−03	4.73E−03	5.24E+02	1.78E−01
49.114M	4.50E−00	2.60E+01	7000.0	4.94E+04	1.71E+01	1.11E+01	1.23E+06	4.20E+02
49.115M	7.40E−01	1.90E−01	7000.0	5.94E+01	2.05E−02	1.34E−02	1.48E+03	5.05E−01
49.115	8.50E−01	5.70E+01	7000.0	2.04E+04	7.08E−00	4.62E−00	5.12E+05	1.74E+02
50.113	7.00E−01	5.30E+01	7000.0	3.13E+04	1.08E+01	7.08E−00	3.92E+05	7.84E+03
50.125	4.80E−00	8.70E−00	7000.0	3.53E+04	1.22E+01	7.97E−00	4.41E+05	8.82E+03
51.122	2.30E−00	2.70E−00	7000.0	1.96E+03	6.81E−01	4.44E−01	6.56E+04	1.96E+02
51.124	6.90E−01	3.80E+01	7000.0	8.31E+03	2.87E−00	1.87E−00	2.77E+05	8.31E+02
51.125	2.80E−01	9.00E+01	7000.0	1.77E+03	6.12E−00	4.00E−00	5.89E+05	1.77E+03
52.125M	5.10E−01	2.00E+01	7000.0	3.66E+03	1.26E−00	8.27E−01	1.07E+05	2.48E+03
52.127M	1.50E−00	2.30E+01	7000.0	1.24E+04	4.28E−00	2.79E−00	3.64E+05	8.38E+03
52.127	1.20E−00	3.90E−01	7000.0	1.68E+02	5.81E−02	3.79E−02	4.94E+03	1.13E+02
52.129M	3.20E−00	1.60E+01	7000.0	1.84E+04	6.36E−00	4.15E−00	5.41E+05	1.24E+04
52.129	2.80E−00	5.10E−02	7000.0	5.13E+01	1.77E−02	1.15E−02	1.50E+03	3.47E+01
52.131M	2.60E−00	1.20E−00	7000.0	1.12E+03	3.87E−01	2.53E−01	3.29E+04	7.58E+02
52.132	3.10E−00	2.90E−00	7000.0	3.23E+03	1.11E−00	7.29E−01	9.50E+04	2.18E+03
55.134M	1.70E−02	9.30E−00	7000.0	5.01E+01	1.73E−02	1.13E−02	1.67E+03	6.68E+01
55.131	4.90E−01	1.30E−01	7000.0	2.02E+01	6.98E−03	4.55E−03	6.73E+02	2.69E+01
55.134	9.90E−01	1.20E+02	7000.0	3.76E+04	1.30E+01	8.50E−00	1.25E+06	5.02E+04
55.135	3.30E−01	1.40E+02	7000.0	1.46E+04	5.06E−00	3.30E−00	4.88E+05	1.95E+04
55.136	7.20E−01	1.19E+01	7000.0	2.71E+03	9.40E−01	6.13E−01	9.05E+04	3.62E+03
55.137	1.40E−00	1.38E+02	7000.0	6.12E+04	2.11E+01	1.38E+01	2.04E+06	8.16E+04
56.131	1.10E−01	9.80E−00	7000.0	2.16E+03	7.49E−01	4.88E−01	1.13E+04	3.98E+02
56.140	4.20E−00	1.07E+01	7000.0	9.02E+04	3.12E+01	2.03E+01	4.75E+05	1.66E+04
57.140	2.70E−00	1.68E−00	7000.0	4.79E+03	1.65E−00	1.08E−00	4.79E+04	1.91E+04
58.141	8.10E−01	3.10E+01	7000.0	1.99E+04	6.88E−00	4.49E−00	2.65E+05	7.96E−00
58.143	3.80E−00	1.33E−00	7000.0	4.00E+03	1.38E−00	9.04E−01	5.34E+04	1.60E−00
58.144	6.30E−00	2.43E+02	7000.0	1.21E+06	4.19E+02	2.73E+02	1.61E+07	4.85E+02
59.142	3.90E−00	8.00E−01	7000.0	3.29E+03	1.14E−00	7.44E−01	3.29E+04	1.31E−00
59.143	1.60E−00	1.36E+01	7000.0	2.30E+04	7.95E−00	5.19E−00	2.30E+05	9.20E−00
60.144	1.00E+02	1.50E+03	7000.0	1.42E+08	4.93E+04	3.21E+04	1.58E+09	5.54E+04
60.147	1.40E−00	1.12E+01	7000.0	1.49E+04	5.16E−00	3.36E−00	1.65E+05	5.80E−00
60.149	4.70E−00	8.30E−02	7000.0	3.71E+02	1.28E−01	8.37E−02	4.12E+03	1.44E−01
61.147	3.50E−01	5.70E+02	7000.0	1.89E+05	6.56E+01	4.28E+01	2.10E+06	7.38E+01
61.149	1.90E−00	2.20E−00	7000.0	3.97E+03	1.37E−00	8.97E−01	4.41E+04	1.54E−00

TABLE 2 (Continued)

Identification (Z. A.)	Energy (MeV)	T(1/2) Days	Weight in Grams	Inh. Dose (rem/Ci)	Dose/Conc rem/Ci-s/m³ Hi-Rate	Dose/Conc rem/Ci-s/m³ Lo-Rate	Organ Dose (rem/Ci)	Ing. Dose (rem/Ci)
SKELETON (Continued)								
62.147	1.15E+02	1.50E+03	7000.0	1.64E+08	5.67E+04	3.70E+04	1.82E+09	6.38E+04
62.151	1.30E−01	1.44E+03	7000.0	1.78E+05	6.16E+01	4.01E+01	1.97E+06	6.92E+01
62.153	1.10E−00	1.96E−00	7000.0	2.05E+03	7.09E−01	4.62E−01	2.27E+04	7.97E−01
63.152	2.90E−00	3.80E−01	7000.0	1.04E+03	3.62E−01	2.36E−01	1.16E+04	4.19E−01
63.152	4.50E−01	1.14E+03	7000.0	4.88E+05	1.68E+02	1.10E+02	5.42F+06	1.95E+02
63.154	2.70E−00	1.19E+03	7000.0	3.05E+06	1.05E+03	6.89E+02	3.39E+07	1.22E+03
63.155	2.80E−01	4.39E+02	7000.0	1.16E+05	4.04E+01	2.63E+01	1.29E+06	4.67E+01
64.153	2.30E−01	1.91E+02	7000.0	5.10E+04	1.76E+01	1.15E+01	4.64E+05	2.08E+01
64.159	7.50E−01	7.50E−01	7000.0	6.54E+02	2.26E−01	1.47E−01	5.94E+03	2.67E−01
65.160	1.10E−00	6.80E+01	7000.0	1.18E+05	4.10E+01	2.67E+01	7.90E+05	4.74E+01
66.165	1.50E−00	9.70E−02	7000.0	2.30E+02	7.98E−02	5.20E−02	1.53E+03	9.22E−02
66.166	3.90E−00	3.40E−00	7000.0	2.10E+04	7.27E−00	4.74E−00	1.40E+05	8.41E−00
67.166	3.40E−00	1.10E−00	7000.0	6.32E+03	2.18E−00	1.42E−00	3.95E+04	2.53E−00
68.169	5.80E−01	9.30E−00	7000.0	8.55E+03	2.95E−00	1.93E−00	5.70E+04	3.42E−00
68.171	2.00E−00	3.10E−01	7000.0	9.83E+02	3.40E−01	2.21E−01	6.55E+03	3.93E−01
69.170	1.70E−00	1.13E+02	7000.0	3.24E+05	1.12E+02	7.33E+01	2.03E+06	1.32E+02
69.171	1.50E−00	4.10E+02	7000.0	1.04E+06	3.59E+02	2.34E+02	6.50E+06	4.22E+02
70.175	7.10E−01	4.10E−00	7000.0	4.61E+03	1.59E−00	1.04E−00	3.07E+04	1.78E−00
71.177	7.60E−01	6.75E−00	7000.0	9.21E+03	3.18E−00	2.08E−00	5.42E+04	3.68E−00
72.181	7.40E+01	4.30E+01	7000.0	1.34E+06	4.65E+02	3.03E+02	3.36E+07	5.04E+02
73.182	1.00E−00	8.20E+01	7000.0	4.33E+04	1.49E+01	9.78E−00	8.66E+05	1.73E+01
74.181	4.70E−02	8.50E−00	7000.0	8.44E+01	2.92E−02	1.90E−02	4.22E+03	2.95E+01
74.185	6.80E−01	8.00E−00	7000.0	1.15E+03	3.97E−01	2.59E−01	5.75E+04	4.02E+02
74.187	1.40E−00	9.00E−01	7000.0	2.66E+02	9.21E−02	6.01E−02	1.33E+04	9.32E+01
75.183	5.50E−02	3.30E−00	7000.0	9.59E−00	3.31E−03	2.16E−03	1.91E+03	9.59E−00
75.186	1.80E−00	1.82E−00	7000.0	1.73E+02	5.99E−02	3.90E−02	3.46E+04	1.73E+02
75.187	6.20E−02	3.50E−00	7000.0	1.14E+01	3.96E−03	2.58E−03	2.29E+03	1.14E+01
75.188	3.90E−00	5.90E−01	7000.0	1.21E+02	4.20E−02	2.74E−02	2.43E+04	1.21E+02
81.200	9.50E−02	9.70E−01	7000.0	2.53E+01	8.76E−03	5.71E−03	9.74E+02	2.43E+01
81.201	4.40E−01	2.10E−00	7000.0	2.53E+02	8.78E−02	5.73E−02	9.76E+03	2.44E+02
81.202	9.40E−01	4.40E−00	7000.0	1.13E+03	3.93E−01	2.56E−01	4.37E+04	1.09E+03
81.204	1.30E−00	7.00E−00	7000.0	2.50E+03	8.65E−01	5.64E−01	9.62E+04	2.40E+03
82.203	5.10E−02	2.17E−00	7000.0	9.35E+01	3.23E−02	2.11E−02	1.16E+03	2.33E+01
82.210	2.90E+01	2.40E+03	7000.0	5.85E+07	2.02E+04	1.32E+04	7.31E+08	1.46E+07
82.212	4.10E+02	4.40E−01	7000.0	1.52E+05	5.27E+01	3.44E+01	1.90E+06	3.81E+04
83.206	4.30E−01	4.30E−00	7000.0	1.50E+02	5.20E−02	3.39E−02	1.95E+04	5.86E−00
83.207	2.40E−01	1.32E+01	7000.0	2.57E+02	8.92E−02	5.81E−02	3.34E+04	1.00E+01
83.210	4.00E+01	3.60E−00	7000.0	1.17E+04	4.05E−00	2.64E−00	1.52E+06	4.56E+02
83.212	4.11E+02	4.20E−02	7000.0	1.40E+03	4.86E−01	3.17E−01	1.82E+05	5.47E+01
84.210	2.80E+02	2.00E+01	7000.0	1.77E+06	6.14E+02	4.00E+02	5.92E+07	3.55E+05
88.223	2.80E−00	1.17E+01	7000.0	6.92E+04	2.39E+01	1.56E+01	3.46E+05	5.19E+04
88.224	2.80E−00	3.64E−00	7000.0	2.15E+04	7.45E−00	4.86E−00	1.07E+05	1.61E+04
88.226	1.10E−00	1.60E+03	7000.0	7.43E+05	2.57E+02	1.67E+02	1.85E+07	5.57E+05
88.228	1.90E−00	2.10E+03	7000.0	1.68E+06	5.82E+02	3.79E+02	4.20E+07	1.26E+06
89.227	1.00E+03	7.20E+03	7000.0	5.03E+09	1.74E+06	1.13E+06	6.28E+10	1.88E+06
89.228	9.70E+02	2.60E−01	7000.0	2.13E+05	7.37E+01	4.81E+01	2.66E+06	7.99E+01
90.227	9.90E+02	1.84E+01	7000.0	3.46E+07	1.19E+04	7.82E+03	1.92E+08	1.34E+04
90.228	9.70E+02	6.93E+02	7000.0	1.27E+09	4.42E+05	2.88E+05	7.10E+09	4.97E+05
90.230	2.40E+02	7.30E+04	7000.0	5.28E+09	1.82E+06	1.19E+06	2.93E+10	2.05E+06
90.231	5.60E−01	1.07E−00	7000.0	1.14E+03	3.94E−01	2.57E−01	6.33E+03	4.43E−01
90.232	2.70E+02	7.30E+04	7000.0	5.94E+09	2.05E+06	1.34E+06	3.30E+10	2.31E+06

TABLE 2 (Continued)

Identification (Z. A.)	Energy (MeV)	T(1/2) Days	Weight in Grams	Inh. Dose (rem/Ci)	Dose/Conc rem/Ci-s/m³ Hi-Rate	Dose/Conc rem/Ci-s/m³ Lo-Rate	Organ Dose (rem/Ci)	Ing. Dose (rem/Ci)
SKELETON (Continued)								
90.234	4.50E−00	2.41E+01	7000.0	2.06E+05	7.13E+01	4.65E+01	1.14E+06	8.02E+01
91.230	1.60E+03	1.77E+01	7000.0	6.58E+06	2.26E+03	1.49E+03	5.98E+07	2.68E+05
91.231	7.50E+02	7.30E+04	7000.0	1.00E+10	3.49E+06	2.27E+06	9.17E+10	4.12E+07
91.233	4.10E−01	2.74E+01	7000.0	1.30E+04	4.51E−00	2.94E−00	1.18E+05	5.34E+01
92.230	1.80E+03	1.95E+01	7000.0	1.03E+07	3.59E+03	2.34E+03	3.71E+08	4.08E+03
92.232	1.20E+03	3.00E+02	7000.0	1.06E+08	3.68E+04	2.40E+04	3.80E+09	4.18E+04
92.233	2.50E+02	3.00E+02	7000.0	2.22E+07	7.68E+03	5.01E+03	7.92E+08	8.72E+03
92.234	2.40E+02	3.00E+02	7000.0	2.13E+07	7.37E+03	4.80E+03	7.61E+08	8.37E+03
92.235	2.30E+02	3.00E+02	7000.0	2.04E+07	7.06E+03	4.60E+03	7.29E+08	8.02E+03
92.236	2.30E+02	3.00E+02	7000.0	2.04E+07	7.06E+03	4.60E+03	7.29E+08	8.02E+03
92.238	2.20E+02	3.00E+02	7000.0	1.95E+07	6.75E+03	4.40E+03	6.97E+08	7.67E+03
92.240	5.30E−00	5.87E−01	7000.0	9.20E+02	3.18E−01	2.07E−01	3.28E+04	3.61E+01
93.237	2.50E+02	7.30E+04	7000.0	3.36E+09	1.16E+06	7.59E+05	3.05E+10	1.37E+06
93.239	9.80E−01	2.33E−00	7000.0	2.65E+03	9.18E−01	5.99E−01	2.41E+04	1.08E−00
94.238	2.80E+02	2.30E+04	7000.0	2.15E+09	7.45E+05	4.86E+05	2.87E+10	8.61E+05
94.239	2.70E+02	7.20E+04	7000.0	2.47E+09	8.56E+05	5.58E+05	3.29E+10	9.89E+05
94.240	2.70E+02	7.10E+04	7000.0	2.47E+09	8.55E+05	5.57E+05	3.29E+10	9.88E+05
94.241	1.40E+01	4.50E+03	7000.0	4.69E+07	1.62E+04	1.05E+04	6.25E+08	1.87E+04
94.242	2.50E+02	7.30E+04	7000.0	2.29E+09	7.93E+05	5.17E+05	3.05E+10	9.17E+05
94.243	2.00E−00	2.08E−01	7000.0	8.79E+02	3.04E−01	1.98E−01	4.39E+03	1.05E−01
94.244	2.40E+02	7.30E+04	7000.0	7.16E+09	2.48E+06	1.61E+06	3.57E+10	8.59E+05
95.241	2.80E+02	5.10E+04	7000.0	2.08E+09	7.20E+05	4.69E+05	3.30E+10	8.26E+05
95.242M	3.60E+02	3.20E+04	7000.0	2.09E+09	7.25E+05	4.73E+05	3.33E+10	8.32E+05
95.242	4.00E+02	6.67E−01	7000.0	1.49E+05	5.16E+01	3.37E+01	2.37E+06	5.92E+01
95.243	2.70E+02	7.10E+04	7000.0	2.07E+09	7.18E+05	4.68E+05	3.29E+10	8.23E+05
95.244	2.50E+02	1.80E−02	7000.0	2.99E+03	1.03E−00	6.76E−01	4.75E+04	1.18E−00
96.242	4.00E+02	1.62E+02	7000.0	5.13E+07	1.77E+04	1.15E+04	6.85E+08	2.05E+04
96.243	3.00E+02	1.10E+04	7000.0	1.78E+09	6.17E+05	4.02E+05	2.37E+10	7.13E+05
96.244	3.00E+02	6.10E+03	7000.0	1.26E+09	4.38E+05	2.85E+05	1.68E+10	5.06E+05
96.245	2.80E+02	7.20E+04	7000.0	2.56E+09	8.87E+05	5.79E+05	3.42E+10	1.02E+06
96.246	2.80E+02	7.10E+04	7000.0	2.56E+09	8.86E+05	5.78E+05	3.41E+10	1.02E+06
96.247	2.70E+02	7.30E+04	7000.0	2.47E+09	8.57E+05	5.59E+05	3.30E+10	9.90E+05
96.248	2.60E+02	7.30E+04	7000.0	2.07E+10	7.18E+06	4.68E+06	2.77E+11	8.30E+06
96.249	2.70E+01	4.40E−02	7000.0	9.41E+02	3.25E−01	2.12E−01	1.25E+04	3.76E−01
97.249	2.00E+01	2.89E+02	7000.0	1.22E+07	4.22E+03	2.75E+03	6.11E+07	1.46E+03
97.250	2.90E+01	1.34E−01	7000.0	8.21E+03	2.84E−00	1.85E−00	4.10E+04	9.85E−01
98.249	3.00E+02	5.10E+04	7000.0	7.08E+09	2.44E+06	1.59E+06	3.54E+10	8.49E+05
98.250	3.10E+02	3.50E+03	7000.0	2.23E+09	7.71E+05	5.03E+05	1.11E+10	2.67E+05
98.251	2.90E+02	5.80E+04	7000.0	6.94E+09	2.40E+06	1.56E+06	3.47E+10	8.33E+05
98.252	1.10E+03	7.94E+02	7000.0	1.84E+09	6.38E+05	4.16E+05	9.23E+09	2.21E+05
98.253	3.70E+02	1.80E+01	7000.0	1.40E+07	4.87E+03	3.17E+03	7.04E+07	1.68E+03
98.254	1.89E+04	5.60E+01	7000.0	2.23E+09	7.74E+05	5.05E+05	1.11E+10	2.68E+05
99.253	3.70E+02	2.00E+01	7000.0	1.56E+07	5.41E+03	3.53E+03	7.82E+07	1.87E+03
99.254M	6.60E+02	1.60E−00	7000.0	2.23E+06	7.72E+02	5.03E+02	1.11E+07	2.67E+02
99.254	6.20E+02	4.77E+02	7000.0	6.25E+08	2.16E+05	1.41E+05	3.12E+09	7.50E+04
99.255	3.80E+02	3.00E+01	7000.0	2.41E+07	8.33E+03	5.43E+03	1.20E+08	2.89E+03
100.254	6.60E+02	1.35E−01	7000.0	1.88E+05	6.51E+01	4.25E+01	9.41E+05	2.26E+01
100.255	3.80E+02	8.96E−01	7000.0	7.19E+05	2.49E+02	1.62E+02	3.59E+06	8.63E+01
100.256	1.80E+04	1.11E−01	7000.0	4.22E+06	1.46E+03	9.53E+02	2.11E+07	5.06E+02

TABLE 2 (continued)

Identification (Z.A.)	Energy (MeV)	T(1/2) Days	Weight in Grams	Inh. Dose (rem/Ci)	Dose/Conc rem/Ci-s/m³ Hi-Rate	Dose/Conc rem/Ci-s/m³ Lo-Rate	Organ Dose (rem/Ci)	Ing. Dose (rem/Ci)
LIVER	**1700 GRAMS**							
4.007	1.60E−02	4.50E+01	1700.0	7.83E+02	2.71E−01	1.76E−01	3.13E+04	6.26E−00
15.032	6.90E−01	8.00E−00	1700.0	9.61E+03	3.32E−00	2.16E−00	2.40E+05	1.20E+04
21.046	6.40E−01	2.50E+01	1700.0	2.78E+04	9.63E−00	6.28E−00	6.96E+05	1.04E+01
21.047	2.10E−01	3.10E−00	1700.0	1.13E+03	3.92E−01	2.55E−01	2.83E+04	4.25E−01
21.048	1.10E−00	1.70E−00	1700.0	3.25E+03	1.12E−00	7.34E−01	8.14E+04	1.22E−00
23.048	9.00E−01	1.31E+01	1700.0	1.02E+04	3.55E−00	2.31E−00	5.13E+05	6.15E+02
25.052	9.60E−01	4.50E−00	1700.0	1.31E+04	4.55E−00	2.97E−00	1.88E+05	3.76E+03
25.054	2.30E−01	2.30E+01	1700.0	1.61E+04	5.57E−00	3.63E−00	2.30E+05	4.60E+03
25.056	1.30E−00	1.10E−01	1700.0	4.35E+02	1.50E−01	9.83E−02	6.22E+03	1.24E+02
26.055	6.50E−03	3.68E+02	1700.0	4.16E+03	1.44E−00	9.39E−01	1.04E+05	1.35E+03
26.059	4.20E−00	4.17E+01	1700.0	3.04E+05	1.05E+02	6.88E+01	7.62E+06	9.91E+04
27.057	5.30E−02	9.20E−00	1700.0	4.24E+02	1.46E−01	9.57E−02	2.12E+04	1.48E+02
27.058M	5.90E−02	3.70E−01	1700.0	1.90E+01	6.57E−03	4.28E−03	9.50E+02	6.65E−00
27.058	2.90E−01	8.40E−00	1700.0	2.12E+03	7.33E−01	4.78E−01	1.06E+05	7.42E+02
27.060	7.20E−01	9.50E−00	1700.0	5.95E+03	2.06E−00	1.34E−00	2.97E+05	2.08E+03
28.059	7.70E−03	5.00E+02	1700.0	5.02E+03	1.73E−00	1.13E−00	1.67E+05	3.35E+03
28.063	2.10E−02	4.92E+02	1700.0	1.34E+04	4.66E−00	3.04E−00	4.49E+05	8.99E+03
28.065	1.20E−00	1.10E−01	1700.0	1.72E+02	5.96E−02	3.89E−02	5.74E+03	1.14E+02
29.064	1.90E−01	5.30E−01	1700.0	1.31E+02	4.54E−02	2.96E−02	4.38E+03	8.76E+01
30.065	1.50E−01	6.60E+01	1700.0	4.74E+04	1.63E+01	1.06E+01	4.30E+05	1.50E+04
30.069M	5.00E−01	5.80E−01	1700.0	1.38E+03	4.80E−01	3.13E−01	1.26E+04	4.41E+02
30.069	3.70E−01	3.60E−02	1700.0	6.37E+01	2.20E−02	1.43E−02	5.79E+02	2.02E+01
31.072	1.10E−00	5.30E−01	1700.0	1.59E+03	5.53E−01	3.60E−01	2.53E+04	6.34E−00
32.071	1.00E−02	4.60E−00	1700.0	1.00E+01	3.46E−03	2.25E−03	2.00E+03	4.00E−01
33.073	4.10E−02	6.70E+01	1700.0	9.56E+02	3.30E−01	2.15E−01	1.19E+05	1.07E+02
33.074	3.80E−01	1.70E+01	1700.0	2.24E+03	7.78E−01	5.07E−01	2.81E+05	2.53E+02
33.076	1.10E−00	1.10E−00	1700.0	4.21E+02	1.45E−01	9.50E−02	5.26E+04	4.74E+01
33.077	2.40E−01	1.60E−00	1700.0	1.33E+02	4.62E−02	3.01E−02	1.67E+04	1.50E+01
34.075	9.40E−02	2.00E+01	1700.0	4.09E+03	1.41E−00	9.23E−01	8.18E+04	4.91E+03
37.086	6.60E−01	1.44E+01	1700.0	1.65E+04	5.72E−00	3.73E−00	4.13E+05	2.06E+04
37.087	9.00E−02	6.30E+01	1700.0	9.87E+03	3.41E−00	2.22E−00	2.46E+05	1.23E+04
40.093	1.90E−02	3.20E+02	1700.0	6.14E+03	2.12E−00	1.38E−00	3.06E+05	2.15E−00
40.095	5.70E−01	5.30E+01	1700.0	2.63E+04	9.09E−00	5.93E−00	1.31E+06	9.20E−00
40.097	1.60E−00	7.10E−01	1700.0	9.88E+02	3.42E−01	2.23E−01	4.94E+04	3.46E−01
41.093M	3.80E−02	6.88E+02	1700.0	2.27E+04	7.87E−00	5.13E−00	1.13E+06	1.02E+01
41.095	2.60E−01	3.36E+01	1700.0	7.60E+03	2.63E−00	1.71E−00	3.80E+05	3.42E−00
41.097	6.40E−01	5.10E−02	1700.0	2.84E+01	9.83E−03	6.41E−03	1.42E+03	1.27E−02
42.099	4.80E−01	2.66E−00	1700.0	3.53E+03	1.22E−00	7.97E−01	5.43E+04	4.34E+03
43.096M	6.00E−01	3.60E−02	1700.0	1.41E−00	4.87E−04	3.18E−04	9.40E+02	1.41E−00
43.096	6.40E−02	3.80E−00	1700.0	1.58E+01	5.49E−03	3.58E−03	1.05E+04	1.58E+01
43.097M	9.00E−02	2.30E+01	1700.0	1.35E+02	4.67E−02	3.05E−02	9.01E+04	1.35E+02
43.097	2.00E−02	3.00E+01	1700.0	3.91E+01	1.35E−02	8.84E−03	2.61E+04	3.91E+01
43.099M	9.40E−02	3.00E+01	1700.0	1.84E+02	6.36E−02	4.15E−02	1.22E+05	1.84E+02
43.099	3.50E−02	2.50E−01	1700.0	5.71E−01	1.97E−04	1.28E−04	3.80E+02	5.71E−01
45.103M	5.50E−02	3.80E−02	1700.0	1.27E−00	4.40E−04	2.87E−04	9.09E+01	7.27E−01
45.105	1.90E−01	1.40E−00	1700.0	1.62E+02	5.60E−02	3.65E−02	1.15E+04	9.26E+01
46.103	6.30E−02	9.00E−00	1700.0	7.40E+02	2.56E−01	1.67E−01	2.46E+04	4.93E+02
46.109	4.20E−01	5.50E−01	1700.0	3.01E+02	1.04E−01	6.80E−02	1.00E+04	2.01E+02
47.105	2.90E−01	1.10E+01	1700.0	1.06E+03	3.69E−01	2.41E−01	1.38E+05	4.16E+01
47.110M	8.40E−01	1.42E+01	1700.0	3.99E+03	1.38E−00	9.02E−01	5.19E+05	1.55E+02
47.111	3.80E−01	5.00E−00	1700.0	6.36E+02	2.20E−01	1.43E−01	8.27E+04	2.48E+01

TABLE 2 (continued)

Identification (Z.A.)	Energy (MeV)	T(1/2) Days	Weight in Grams	Inh. Dose (rem/Ci)	Dose/Conc rem/Ci-s/m³ Hi-Rate	Dose/Conc rem/Ci-s/m³ Lo-Rate	Organ Dose (rem/Ci)	Ing. Dose (rem/Ci)
LIVER (continued)								
48.109	1.00E−01	1.40E+02	1700.0	1.15E+05	4.00E+01	2.61E+01	6.09E+05	1.15E+03
48.115	6.10E−01	3.50E+01	1700.0	1.76E+05	6.10E+01	3.98E+01	9.29E+05	1.76E+03
48.115	5.80E−01	2.20E−00	1700.0	1.05E+04	3.65E−00	2.38E−00	5.55E+04	1.05E+02
49.113M	2.10E−01	7.30E−02	1700.0	2.66E+01	9.23E−03	6.02E−03	6.67E+02	1.86E−01
49.114M	9.40E−01	2.60E+01	1700.0	4.25E+04	1.47E+01	9.60E−00	1.06E+06	2.97E+02
49.115M	2.00E−01	1.90E−01	1700.0	6.61E+01	2.28E−02	1.49E−02	1.65E+03	4.63E−01
49.115	1.70E−01	5.80E+01	1700.0	1.71E+04	5.93E−00	3.87E−00	4.29E+05	1.20E+02
50.113	2.30E−01	4.30E+01	1700.0	1.20E+03	4.17E−01	2.72E−01	4.30E+05	2.15E+02
50.125	9.40E−01	8.40E−00	1700.0	9.62E+02	3.32E−01	2.17E−01	3.43E+05	1.71E+02
51.122	6.70E−01	2.60E−00	1700.0	3.79E+01	1.31E−02	8.55E−03	7.58E+04	4.54E−00
51.124	9.20E−01	2.30E+01	1700.0	4.60E+02	1.59E−01	1.03E−01	9.21E+05	5.52E+01
51.125	2.10E−01	3.60E+01	1700.0	1.77E+02	6.14E−02	4.00E−02	3.55E+05	2.12E+01
52.125M	1.40E−01	2.00E+01	1700.0	2.43E+03	8.43E−01	5.50E−01	1.21E+05	1.21E+03
52.127M	3.20E−01	2.30E+01	1700.0	6.40E+03	2.21E−00	1.44E−00	3.20E+05	3.20E+03
52.127	2.40E−01	3.90E−01	1700.0	8.14E+01	2.81E−02	1.83E−02	4.07E+03	4.07E+01
52.129M	8.30E−01	1.60E+01	1700.0	1.15E+04	3.99E−00	2.60E−00	5.78E+05	5.78E+03
52.129	7.30E−01	5.10E−02	1700.0	3.24E+01	1.12E−02	7.31E−03	1.62E+03	1.62E+01
52.131M	9.70E−01	1.20E−00	1700.0	1.01E+03	3.50E−01	2.28E−01	5.06E+04	5.06E+02
52.132	1.10E−00	2.90E−00	1700.0	2.77E+03	9.60E−01	6.26E−01	1.38E+05	1.38E+03
55.131	2.40E−02	9.00E−00	1700.0	4.70E+02	1.62E−01	1.06E−01	9.40E+03	6.58E+02
55.134M	1.50E−01	1.30E−01	1700.0	4.24E+01	1.46E−02	9.57E−03	8.48E+02	5.94E+01
55.134	5.70E−01	8.10E+01	1700.0	1.00E+05	3.47E+01	2.26E+01	2.00E+06	1.40E+05
55.135	6.60E−02	9.00E+01	1700.0	1.29E+04	4.47E−00	2.91E−00	2.58E+05	1.80E+04
55.136	3.50E−01	1.14E+01	1700.0	8.68E+03	3.00E−00	1.95E−00	1.73E+05	1.21E+04
55.137	4.10E−01	8.90E+01	1700.0	7.94E+04	2.74E+01	1.79E+01	1.58E+06	1.11E+05
56.131	1.90E−01	1.15E+01	1700.0	1.61E+01	5.59E−03	3.64E−03	9.51E+04	2.85E−00
56.140	1.40E−00	1.26E+01	1700.0	1.30E+02	4.51E−02	2.94E−02	7.67E+05	2.30E+01
57.140	1.10E−00	1.68E−00	1700.0	3.21E+03	1.11E−00	7.26E−01	8.04E+04	1.20E+04
58.141	1.80E−01	2.90E+01	1700.0	1.36E+04	4.71E−00	3.07E−00	2.27E+05	5.68E−00
58.143	8.50E−01	1.32E−00	1700.0	2.93E+03	1.01E−00	6.61E−01	4.88E+04	1.22E−00
58.144	1.30E−00	1.46E+02	1700.0	4.95E+05	1.71E+02	1.11E+02	8.26E+06	2.06E+02
59.142	8.10E−01	8.00E−01	1700.0	1.41E+03	4.87E−01	3.18E−01	2.82E+04	5.64E−01
59.143	3.20E−01	1.32E+01	1700.0	9.19E+03	3.18E−00	2.07E−00	1.83E+05	3.67E−00
60.144	2.00E+01	1.31E+02	1700.0	1.48E+07	5.12E+03	3.34E+03	1.14E+08	5.70E+03
60.147	3.20E−01	1.00E+01	1700.0	1.81E+04	6.26E−00	4.08E−00	1.39E+05	6.96E−00
60.149	9.90E−01	8.30E−02	1700.0	4.64E+02	1.60E−01	1.04E−01	3.57E+03	1.78E−01
61.147	6.90E−02	3.83E+02	1700.0	2.30E+04	7.95E−00	5.19E−00	1.15E+06	6.90E−00
61.149	4.40E−01	2.20E−00	1700.0	8.42E+02	2.91E−01	1.90E−01	4.21E+04	2.52E−01
62.147	2.30E+01	1.87E+02	1700.0	1.68E+07	5.82E+03	3.80E+03	1.87E+08	6.55E+03
62.151	4.20E−02	1.86E+02	1700.0	3.06E+04	1.05E+01	6.90E−00	3.40E+05	1.19E+01
62.153	2.60E−01	1.94E−00	1700.0	1.97E+03	6.83E−01	4.45E−01	2.19E+04	7.68E−01
63.152	7.10E−01	3.80E−01	1700.0	7.04E+02	2.43E−01	1.59E−01	1.17E+04	2.93E−01
63.152	3.30E−01	1.24E+02	1700.0	1.06E+05	3.69E+01	2.41E+01	1.78E+06	4.45E+01
63.154	8.60E−01	1.24E+02	1700.0	2.78E+05	9.63E+01	6.28E+01	4.64E+06	1.16E+02
63.155	9.50E−02	1.05E+02	1700.0	2.60E+04	9.01E−00	5.87E−00	4.34E+05	1.08E+01
64.153	9.90E−02	1.56E+02	1700.0	2.01E+04	6.97E−00	4.55E−00	6.72E+05	8.06E−00
64.159	3.30E−01	7.50E−01	1700.0	3.23E+02	1.11E−01	7.29E−02	1.07E+04	1.29E−01
66.165	3.90E−01	9.70E−02	1700.0	3.29E+01	1.13E−02	7.43E−03	1.64E+03	9.88E−03
66.166	7.80E−01	3.40E−00	1700.0	2.30E+03	7.98E−01	5.21E−01	1.15E+05	6.92E−01
67.166	6.90E−01	1.10E−00	1700.0	6.60E+02	2.28E−01	1.49E−01	3.30E+04	1.98E−01
68.169	2.20E−01	9.20E−00	1700.0	6.60E+02	2.28E−01	1.49E−01	8.81E+04	2.64E−01

TABLE 2 (continued)

Identification (Z.A.)	Energy (MeV)	T(1/2) Days	Weight in Grams	Inh. Dose (rem/Ci)	Dose/Conc rem/Ci-s/m³ Hi-Rate	Dose/Conc rem/Ci-s/m³ Lo-Rate	Organ Dose (rem/Ci)	Ing. Dose (rem/Ci)
LIVER (continued)								
72.181	2.90E−01	4.30E+01	1700.0	5.97E+04	2.06E+01	1.34E+01	5.42E+05	2.44E+01
73.182	5.60E−01	8.80E+01	1700.0	1.71E+05	5.93E+01	3.87E+01	2.14E+06	6.43E+01
74.181	8.70E−02	3.90E−00	1700.0	2.95E+02	1.02E−01	6.66E−02	1.47E+04	8.86E+01
74.185	1.40E−01	3.80E−00	1700.0	4.63E+02	1.60E−01	1.04E−01	2.31E+04	1.38E+02
74.187	4.40E−01	8.00E−01	1700.0	3.06E+02	1.06E−01	6.91E−02	1.53E+04	9.19E+01
75.183	1.00E−01	1.17E+01	1700.0	2.54E+02	8.80E−02	5.74E−02	5.09E+04	2.54E+02
75.186	3.70E−01	2.98E−00	1700.0	2.39E+02	8.30E−02	5.41E−02	4.79E+04	2.39E+02
75.187	1.20E−02	1.40E+01	1700.0	3.65E+01	1.26E−02	8.25E−03	7.31E+03	3.65E+01
75.188	8.50E−01	6.80E−01	1700.0	1.25E+02	4.35E−02	2.83E−02	2.51E+04	1.25E+02
76.185	2.90E−01	5.20E−00	1700.0	6.56E+02	2.27E−01	1.48E−01	6.56E+04	2.62E+02
76.191	4.90E−02	5.20E−01	1700.0	1.10E+01	3.83E−03	2.50E−03	1.10E+03	4.43E−00
76.191	1.20E−01	4.10E−00	1700.0	2.14E+02	7.40E−02	4.83E−02	2.14E+04	8.56E+01
76.193	3.80E−01	1.00E−00	1700.0	1.65E+02	5.72E−02	3.73E−02	1.65E+04	6.61E+01
77.190	1.60E−01	8.30E−00	1700.0	4.04E+03	1.39E−00	9.13E−01	5.78E+04	1.32E+03
77.192	6.00E−01	2.00E+01	1700.0	3.65E+04	1.26E+01	8.25E−00	5.22E+05	1.20E+04
77.194	8.10E−01	7.70E−01	1700.0	1.90E+03	6.57E−01	4.28E−01	2.71E+04	6.24E+02
78.191	3.10E−01	2.60E−00	1700.0	8.42E+01	2.91E−02	1.90E−02	3.50E+04	2.80E+01
78.193M	3.20E−02	3.20E−00	1700.0	1.06E+01	3.70E−03	2.41E−03	4.45E+03	3.56E−00
78.193	1.90E−02	2.00E+01	1700.0	3.96E+01	1.37E−02	8.95E−03	1.65E+04	1.32E+01
78.197M	5.10E−01	5.60E−02	1700.0	2.98E−00	1.03E−03	6.73E−04	1.24E+03	9.94E−01
78.197	2.40E−01	7.20E−01	1700.0	1.80E+01	6.24E−03	4.07E−03	7.52E+03	6.01E−00
79.196	2.10E−01	5.50E−00	1700.0	5.02E+02	1.73E−01	1.13E−01	5.02E+04	2.01E+02
79.198	4.40E−01	2.70E−00	1700.0	5.17E+02	1.78E−01	1.16E−01	5.17E+04	2.06E+02
79.199	1.30E−01	3.10E−00	1700.0	1.75E+02	6.06E−02	3.95E−02	1.75E+04	7.01E+01
80.197M	1.90E−01	9.30E−01	1700.0	6.92E+02	2.39E−01	1.56E−01	7.69E+03	8.46E+02
80.197	5.20E−02	2.30E−00	1700.0	4.68E+02	1.62E−01	1.05E−01	5.20E+03	5.72E+02
80.203	1.70E−01	1.04E+01	1700.0	6.92E+03	2.39E−00	1.56E−00	7.69E+04	8.46E+03
81.200	1.80E−01	9.20E−01	1700.0	1.44E+02	4.98E−02	3.25E−02	7.20E+03	1.44E+02
81.201	1.20E−01	1.90E−00	1700.0	1.98E+02	6.86E−02	4.47E−02	9.92E+03	1.98E+02
81.202	2.70E−01	3.50E−00	1700.0	8.22E+02	2.84E−01	1.85E−01	4.11E+04	8.22E+02
81.204	2.50E−01	5.00E−00	1700.0	1.08E+03	3.76E−01	2.45E−01	5.44E+04	1.08E+03
82.203	9.40E−02	2.17E−00	1700.0	2.04E+02	7.06E−02	4.60E−02	8.87E+03	5.68E+01
82.210	1.00E+01	1.50E+03	1700.0	1.50E+07	5.19E+03	3.38E+03	6.52E+08	4.17E+06
82.212	8.30E+01	4.40E−01	1700.0	3.65E+04	1.26E+01	8.25E−00	1.58E+06	1.01E+04
83.206	8.00E−01	4.50E−00	1700.0	6.26E+03	2.16E−00	1.41E−00	1.56E+05	2.35E+02
83.207	4.50E−01	1.49E+01	1700.0	1.16E+04	4.03E−00	2.63E−00	2.91E+05	4.37E+02
83.210	1.30E+01	3.80E−00	1700.0	8.60E+04	2.97E+01	1.94E+01	2.15E+06	3.22E+03
83.212	8.30E+01	4.20E−01	1700.0	6.06E+04	2.09E+01	1.36E+01	1.51E+06	2.27E+03
84.210	5.50E+01	3.20E+01	1700.0	3.83E+06	1.32E+03	8.64E+02	7.66E+07	7.66E+05
89.227	6.20E+01	1.90E+03	1700.0	6.65E+08	2.30E+05	1.50E+05	5.12E+09	2.56E+05
89.228	5.60E+01	2.60E−01	1700.0	8.23E+04	2.85E+01	1.85E+01	6.33E+05	3.16E+01
90.227	6.10E+01	1.84E+01	1700.0	4.88E+05	1.69E+02	1.10E+02	4.88E+07	2.44E+02
90.228	5.60E+01	6.91E+02	1700.0	1.68E+07	5.82E+03	3.80E+03	1.68E+09	8.42E+03
90.230	4.80E+01	5.70E+04	1700.0	2.36E+08	8.17E+04	5.33E+04	2.36E+10	1.18E+05
90.231	1.60E−01	1.07E−00	1700.0	7.45E+01	2.57E−02	1.68E−02	7.45E+03	3.72E−02
90.232	4.10E+01	5.70E+04	1700.0	2.01E+08	6.97E+04	4.55E+04	2.01E+10	1.00E+05
90.234	9.00E−01	2.41E+01	1700.0	9.44E+03	3.26E−00	2.13E−00	9.44E+05	4.72E−00
91.231	6.30E+01	5.80E+04	1700.0	4.03E+08	1.39E+05	9.11E+04	3.10E+10	1.55E+05
91.233	1.80E−01	2.74E+01	1700.0	2.79E+03	9.65E−01	6.29E−01	2.14E+05	1.07E−00
93.237	4.90E+01	5.40E+04	1700.0	3.11E+08	1.07E+05	7.03E+04	2.39E+10	1.19E+05
93.239	2.20E−01	2.33E−00	1700.0	2.90E+02	1.00E−01	6.54E−02	2.23E+04	1.11E−01

TABLE 2 (continued)

Identification (Z.A.)	Energy (MeV)	T(1/2) Days	Weight in Grams	Inh. Dose (rem/Ci)	Dose/Conc rem/Ci-s/m³ Hi-Rate	Dose/Conc rem/Ci-s/m³ Lo-Rate	Organ Dose (rem/Ci)	Ing. Dose (rem/Ci)
LIVER (continued)								
94.238	5.70E+01	1.60E+04	1700.0	1.08E+08	3.74E+04	2.44E+04	2.16E+10	4.32E+04
94.239	5.30E+01	3.00E+04	1700.0	1.18E+08	4.10E+04	2.67E+04	2.37E+10	4.74E+04
94.240	5.30E+01	3.00E+04	1700.0	1.18E+08	4.10E+04	2.67E+04	2.37E+10	4.74E+04
94.241	1.00E−00	4.10E+03	1700.0	8.51E+05	2.94E+02	1.92E+02	1.70E+08	3.40E+02
94.242	5.10E+01	3.00E+04	1700.0	1.14E+08	3.95E+04	2.57E+04	2.28E+10	4.56E+04
94.243	2.50E−01	2.08E−01	1700.0	8.60E+01	2.97E−02	1.94E−02	2.26E+03	1.01E−02
95.241	5.70E+01	3.40E+03	1700.0	7.24E+08	2.50E+05	1.63E+05	8.22E+09	2.88E+05
95.242M	6.80E+01	3.30E+03	1700.0	7.06E+08	2.44E+05	1.59E+05	8.02E+09	2.81E+05
95.242	7.50E+01	6.67E−01	1700.0	1.60E+05	5.56E+01	3.63E+01	1.82E+06	6.40E+01
95.243	5.40E+01	3.50E+03	1700.0	7.04E+08	2.43E+05	1.58E+05	8.00E+09	2.80E+05
95.244	2.00E+01	1.80E−02	1700.0	1.37E+03	4.77E−01	3.11E−01	1.56E+04	5.48E−01
96.242	7.80E+01	1.54E+02	1700.0	5.22E+07	1.80E+04	1.18E+04	5.22E+08	2.90E+04
96.243	6.00E+01	2.50E+03	1700.0	6.48E+08	2.24E+05	1.46E+05	6.48E+09	2.59E+05
96.244	6.00E+01	2.10E+03	1700.0	5.47E+08	1.89E+05	1.23E+05	5.47E+09	2.18E+05
96.245	5.60E+01	3.00E+03	1700.0	7.20E+08	2.49E+05	1.62E+05	7.20E+09	2.88E+05
96.246	5.60E+01	3.00E+03	1700.0	7.20E+08	2.49E+05	1.62E+05	7.20E+09	2.88E+05
96.247	5.50E+01	3.00E+03	1700.0	7.07E+08	2.44E+05	1.59E+05	7.07E+09	2.82E+05
96.248	5.20E+01	3.00E+03	1700.0	5.81E+08	2.01E+05	1.31E+05	5.81E+09	2.32E+05
BRAIN 1500 GRAMS								
15.032	6.90E−01	1.35E+01	1500.0	2.02E+03	6.99E−01	4.56E−01	4.59E+05	2.43E+03
29.064	2.10E−01	5.30E−01	1500.0	2.19E+01	7.59E−03	4.95E−03	5.49E+03	1.64E+01
LUNGS 1000 GRAMS (SOL.)								
14.031	5.90E−01	1.10E−01	1000.0	3.36E+02	1.16E−01	7.58E−02	4.80E+03	4.32E+02
24.051	1.40E−02	2.66E+01	1000.0	2.75E+02	9.53E−02	6.21E−02	2.75E+04	5.51E−00
26.055	6.50E−03	8.19E+02	1000.0	2.36E+03	8.17E−01	5.33E−01	3.93E+05	7.87E+02
26.059	4.20E−01	4.45E+01	1000.0	8.29E+03	2.87E−00	1.87E−00	1.38E+06	2.76E+03
43.096M	3.80E−01	3.60E−02	1000.0	4.55E−01	1.57E−04	1.02E−04	1.01E+03	4.55E+01
43.096	6.40E−01	2.30E−00	1000.0	4.90E+01	1.69E−02	1.10E−02	1.08E+05	4.90E+01
43.097M	9.00E−02	4.70E−00	1000.0	1.40E+01	4.87E−03	3.17E−03	3.13E+04	1.40E+01
43.097	2.00E−02	5.00E−00	1000.0	3.33E−00	1.15E−03	7.51E−04	7.40E+03	3.33E−00
43.099M	3.50E−02	2.40E−01	1000.0	2.79E−01	9.67E−05	6.31E−05	6.21E+02	2.79E−01
43.099	9.40E−02	5.00E−00	1000.0	1.56E+01	5.41E−03	3.53E−03	3.47E+04	1.56E+01
51.122	6.70E−01	2.70E−00	1000.0	1.07E+03	3.70E−01	2.41E−01	1.33E+05	1.20E+02
51.124	9.20E−01	3.80E+01	1000.0	2.06E+04	7.15E−00	4.67E−00	2.58E+06	2.32E+03
51.125	2.10E−01	9.00E+01	1000.0	1.21E+04	4.17E−00	2.72E−00	1.51E+06	1.35E+03
55.131	2.40E−02	9.30E−00	1000.0	3.77E+01	1.31E−02	8.50E−03	1.64E+04	4.92E+01
55.134M	1.70E−01	1.30E−01	1000.0	3.76E−00	1.30E−03	8.48E−04	1.63E+03	4.90E−00
55.134	5.70E−01	1.20E+02	1000.0	1.16E+04	4.02E−00	2.62E−00	5.06E+06	1.51E+04
55.135	6.60E−02	1.40E+02	1000.0	1.57E+03	5.44E−01	3.54E−01	6.83E+05	2.05E+03
55.136	3.50E−01	1.91E+01	1000.0	7.08E+02	2.45E−01	1.59E−01	3.08E+05	9.24E+02
55.137	4.10E−01	1.38E+02	1000.0	9.62E+03	3.33E−00	2.17E−00	4.18E+06	1 25E+04
56.131	1.90E−01	1.16E+01	1000.0	8.97E−00	3.10E−03	2.02E−03	1.63E+05	1.63E−00
56.140	1.40E−00	1.28E+01	1000.0	7.29E+01	2.52E−02	1.64E−02	1.32E+06	1.32E+01
81.200	1.80E−01	9.50E−01	1000.0	3.66E+01	1.26E−02	8.28E−03	1.26E+04	3.41E+01
81.201	1.20E−01	2.00E−00	1000.0	5.15E+01	1.78E−02	1.16E−02	1.77E+04	4.79E+01
81.202	2.70E−01	4.00E−00	1000.0	2.31E+02	8.01E−02	5.23E−02	7.99E+04	2.15E+02
81.204	2.50E−01	6.00E−00	1000.0	3.21E+02	1.11E−01	7.26E−02	1.11E+05	2.99E+02
HEART 300 GRAMS								
29.064	1.70E−01	5.30E−01	300.0	8.88E+01	3.07E−02	2.00E−02	2.22E+04	6.66E+01

TABLE 2 (continued)

Identification (Z.A.)	Energy (MeV)	T(1/2) Days	Weight in Grams	Inh. Dose (rem/Ci)	Dose/Conc rem/Ci-s/m³ Hi-Rate	Dose/Conc rem/Ci-s/m³ Lo-Rate	Organ Dose (rem/Ci)	Ing. Dose (rem/Ci)
SPLEEN 150 GRAMS								
4.007	1.20E−02	4.90E+01	150.0	1.45E+02	5.01E−02	3.27E−02	2.90E+05	1.16E−00
23.048	7.00E−01	1.37E+01	150.0	1.23E+04	4.25E−00	2.77E−00	4.73E+06	9.46E+02
26.055	6.50E−03	3.88E+02	150.0	7.46E+03	2.58E−00	1.68E−00	1.24E+06	2.48E+03
26.059	3.40E−01	4.19E+01	150.0	4.21E+04	1.45E+01	9.51E−00	7.02E+06	1.40E+04
27.057	4.50E−02	9.20E−00	150.0	1.14E+02	3.95E−02	2.58E−02	2.04E+05	8.57E+01
27.058	4.80E−02	3.70E−01	150.0	4.90E−00	1.69E−03	1.10E−03	8.76E+03	3.67E−00
27.058	2.20E−01	8.40E−00	150.0	5.10E+02	1.76E−01	1.15E−01	9.11E+05	3.82E+02
27.060	5.60E−01	9.50E−00	150.0	1.46E+03	5.08E−01	3.31E−01	2.62E+06	1.10E+03
29.064	1.70E−01	4.20E−01	150.0	1.05E+03	3.65E−01	2.38E−01	3.52E+04	7.04E+02
31.072	8.90E−01	5.40E−01	150.0	5.92E+02	2.05E−01	1.33E−01	2.37E+05	2.37E−00
34.075	7.20E−02	1.60E+01	150.0	1.98E+03	6.88E−01	4.48E−01	5.68E+05	2.55E+03
37.086	6.60E−01	1.32E+01	150.0	1.28E+04	4.46E−00	2.90E−00	4.29E+06	1.71E+04
37.087	9.00E−02	4.50E+01	150.0	5.99E+03	2.07E−00	1.35E−00	1.99E+06	7.99E+03
40.093	1.90E−02	9.00E+02	150.0	1.46E+04	5.07E−00	3.31E−00	9.78E+06	5.87E−00
40.095	4.60E−01	5.90E+01	150.0	2.00E+04	6.94E−00	4.53E−00	1.33E+07	8.03E−00
40.097	1.50E−00	7.10E−01	150.0	7.88E+02	2.72E−01	1.77E−01	5.25E+05	3.15E−01
41.093M	3.80E−02	7.56E+02	150.0	2.83E+04	9.80E−00	6.39E−00	1.41E+07	1.13E+01
41.095	2.00E−01	3.38E+01	150.0	6.66E+03	2.30E−00	1.50E−00	3.33E+06	2.66E−00
41.097	6.00E−01	5.10E−02	150.0	3.01E+01	1.04E−02	6.81E−03	1.50E+04	1.20E−02
45.103M	5.40E−02	3.80E−02	150.0	3.54E−00	1.22E−03	7.99E−04	1.01E+03	2.02E−00
45.105	1.90E−01	1.42E−00	150.0	4.65E+02	1.61E−01	1.05E−01	1.33E+05	2.66E+02
46.103	6.10E−02	8.00E−00	150.0	8.42E+02	2.91E−01	1.90E−01	2.40E+05	4.81E+02
46.109	4.20E−01	5.50E−01	150.0	3.98E+02	1.37E−01	9.00E−02	1.13E+05	2.27E+02
49.113M	1.90E−01	7.30E−02	150.0	3.42E+01	1.18E−02	7.72E−03	6.84E+03	2.73E−01
49.114	9.30E−01	2.40E+01	150.0	5.50E+04	1.90E+01	1.24E+01	1.10E+07	4.40E+02
49.115M	1.90E−01	1.90E−01	150.0	8.90E+01	3.08E−02	2.00E−02	1.78E+04	7.12E−01
49.115	1.70E−01	4.80E+01	150.0	2.01E+04	6.96E−00	4.54E−00	4.02E+06	1.61E+02
52.125M	1.40E−01	2.00E+01	150.0	5.24E+03	1.81E−00	1.18E−00	1.38E+06	3.45E+03
52.127M	3.20E−01	2.30E+01	150.0	1.37E+04	4.77E−00	3.11E−00	3.63E+06	9.07E+03
52.127	2.40E−01	3.90E−01	150.0	1.75E+02	6.07E−02	3.95E−02	4.61E+04	1.15E+02
52.129M	7.80E−01	1.60E+01	150.0	2.33E+04	8.09E−00	5.27E−00	6.15E+06	1.53E+04
52.129	6.80E−01	5.10E−02	150.0	6.50E+01	2.24E−02	1.46E−02	1.71E+04	4.27E+01
52.131M	8.00E−01	1.20E−00	150.0	1.79E+03	6.22E−01	4.06E−01	4.73E+05	1.18E+03
52.132	9.60E−01	2.90E−00	150.0	5.21E+03	1.80E−00	1.17E−00	1.37E+06	3.43E+03
55.131	2.10E−02	9.10E−00	150.0	3.58E+02	1.23E−01	8.08E−02	9.42E+04	4.71E+02
55.134	1.30E−01	1.30E−01	150.0	3.16E+01	1.09E−02	7.14E−03	8.33E+03	4.16E+01
55.134	4.60E−01	8.80E+01	150.0	7.58E+04	2.62E+01	1.71E+01	1.99E+07	9.98E+04
55.135	6.60E−02	9.80E+01	150.0	1.21E+04	4.19E−00	2.73E−00	3.19E+06	1.59E+04
55.136	2.90E−01	1.15E+01	150.0	6.25E+03	2.16E−00	1.41E−00	1.64E+06	8.22E+03
55.137	3.70E−01	9.70E+01	150.0	6.72E+04	2.32E+01	1.51E+01	1.77E+07	8.85E+04
56.131	1.40E−01	6.10E−00	150.0	5.89E−00	2.04E−03	1.33E−03	4.21E+05	1.05E−00
56.140	1.20E−00	6.40E−00	150.0	5.30E+01	1.83E−02	1.19E−02	3.78E+06	9.47E−00
72.181	2.50E−01	4.10E+01	150.0	1.51E+05	5.24E+01	3.42E+01	5.05E+06	6.57E+01
73.182	4.50E−01	7.60E+01	150.0	4.21E+04	1.45E+01	9.51E−00	1.68E+07	1.68E+01
77.190	1.20E−01	9.70E−00	150.0	3.44E+03	1.19E−00	7.77E−01	5.74E+05	1.14E+03
77.192	5.00E−01	3.00E+01	150.0	4.44E+04	1.53E+01	1.00E+01	7.40E+06	1.48E+04
77.194	8.10E−01	7.80E−01	150.0	1.87E+03	6.46E−01	4.22E−01	3.11E+05	6.23E+02
78.191	2.20E−01	2.90E−00	150.0	7.55E+02	2.61E−01	1.70E−01	3.14E+05	2.51E+02
78.193M	2.30E−02	3.20E−00	150.0	8.71E+01	3.01E−02	1.96E−02	3.63E+04	2.90E+01
78.193	1.40E−02	6.00E+01	150.0	9.94E+02	3.44E−01	2.24E−01	4.14E+05	3.31E+02

TABLE 2 (continued)

Identification (Z.A.)	Energy (MeV)	T(1/2) Days	Weight in Grams	Inh. Dose (rem/Ci)	Dose/Conc rem/Ci-s/m³ Hi-Rate	Dose/Conc rem/Ci-s/m³ Lo-Rate	Organ Dose (rem/Ci)	Ing. Dose (rem/Ci)
SPLEEN (continued)								
78.197	5.00E−02	5.60E−02	150.0	3.31E−00	1.14E−03	7.48E−04	1.38E+03	1.10E−00
78.197	2.30E−01	7.40E−01	150.0	2.01E+02	6.97E−02	4.54E−02	8.39E+04	6.71E+01
79.196	1.50E−01	5.50E−00	150.0	6.10E+02	2.11E−01	1.37E−01	4.07E+05	2.03E+02
79.198	4.10E−01	2.70E−00	150.0	8.19E+02	2.83E−01	1.84E−01	5.46E+05	2.73E+02
79.199	1.20E−01	3.10E−00	150.0	2.75E+02	9.52E−02	6.21E−02	1.83E+05	9.17E+01
80.197M	1.70E−01	9.00E−01	150.0	7.54E+02	2.61E−01	1.70E−01	7.54E+04	1.50E+03
80.197	4.30E−02	2.10E−00	150.0	4.45E+02	1.54E−01	1.00E−01	4.45E+04	8.90E+02
80.203	1.50E−01	8.20E−00	150.0	6.06E+03	2.09E−00	1.36E−00	6.06E+05	1.21E+04
83.206	5.80E−01	3.90E−00	150.0	2.90E+03	1.00E−00	6.54E−01	1.11E+06	1.11E+02
83.207	3.30E−01	1.00E+01	150.0	4.23E+03	1.46E−00	9.55E−01	1.62E+06	1.62E+02
83.210	1.70E+01	3.00E−00	150.0	6.54E+04	2.26E+01	1.47E+01	2.51E+07	2.51E+03
83.212	8.20E+01	4.20E−02	150.0	4.41E+03	1.52E−00	9.96E−01	1.69E+06	1.69E+02
84.210	5.50E+01	4.20E+01	150.0	1.13E+07	3.94E+03	2.57E+03	1.13E+09	2.27E+06
85.211	6.10E+01	3.00E−01	150.0	2.08E+05	7.15E+01	4.70E+01	1.04E+07	3.12E+05
PANCREAS 70 GRAMS								
25.052	5.60E−01	2.80E−00	70.0	1.49E+04	5.16E−00	3.36E−00	1.65E+06	4.97E+03
25.054	1.30E−01	5.60E−00	70.0	6.92E+03	2.39E−00	1.56E−00	7.69E+05	2.30E+03
25.056	1.10E−00	1.10E−01	70.0	1.15E+03	3.98E−01	2.59E−01	1.27E+05	3.83E+02
27.057	4.00E−02	9.20E−00	70.0	3.11E+02	1.07E−01	7.02E−02	3.89E+05	2.33E+02
27.058M	3.90E−02	3.70E−01	70.0	1.22E+01	4.22E−03	2.75E−03	1.52E+04	9.15E−00
27.058	1.70E−01	8.40E−00	70.0	1.20E+03	4.17E−01	2.72E−01	1.50E+06	9.05E+02
27.060	4.40E−01	9.50E−00	70.0	3.53E+03	1.22E−00	7.97E−01	4.41E+06	2.65E+03
30.065	8.40E−02	2.30E+01	70.0	1.83E+04	6.35E−00	4.14E−00	2.04E+06	6.12E+03
30.069M	4.50E−01	5.70E−01	70.0	2.44E+03	8.44E−01	5.50E−01	2.71E+05	8.13E+02
30.069	3.70E−01	3.60E−02	70.0	1.26E+02	4.38E−02	2.85E−02	1.40E+04	4.22E+01
37.086	6.50E−01	1.43E+01	70.0	2.26E+04	7.81E−00	5.10E−00	9.82E+06	2.94E+04
37.087	9.00E−02	6.00E+01	70.0	1.31E+04	4.54E−00	2.96E−00	5.70E+06	1.71E+04
TESTES 40 GRAMS								
14.031	5.90E−01	1.10E−01	40.0	4.08E+01	1.41E−02	9.21E−03	1.20E+05	5.16E+01
16.035	5.60E−02	7.64E+01	40.0	7.75E+03	2.68E−00	1.75E−00	7.91E+06	1.02E+04
30.065	5.60E−02	1.28E+02	40.0	3.58E+03	1.23E−00	8.08E−01	1.32E+07	1.19E+03
30.069M	4.30E−01	5.80E−01	40.0	1.24E+02	4.30E−02	2.81E−02	4.61E+05	4.15E+01
52.125M	1.10E−01	2.00E+01	40.0	4.47E+03	1.54E−00	1.01E−00	4.07E+06	3.05E+03
52.127M	3.10E−01	2.30E+01	40.0	1.45E+04	5.01E−00	3.27E−00	1.31E+07	9.89E+03
52.127	2.40E−01	3.90E−01	40.0	1.90E+02	6.58E−02	4.29E−02	1.73E+05	1.29E+02
52.129M	6.90E−01	1.60E+01	40.0	2.24E+04	7.77E−00	5.07E−00	2.04E+07	1.53E+04
52.129	6.00E−01	5.10E−02	40.0	6.22E+01	2.15E−02	1.40E−02	5.66E+04	4.24E+01
52.132	7.30E−01	2.90E−00	40.0	4.30E+03	1.49E−00	9.72E−01	3.91E+06	2.93E+03
30.069	3.70E−01	3.60E−02	40.0	5.06E+04	1.76E+01	1.15E+01	3.38E+07	6.76E+04
PROSTATE 20 GRAMS								
24.051	8.40E−03	2.66E+01	20.0	1.90E+02	6.57E−02	4.29E−02	8.26E+05	3.72E−00
30.065	5.60E−02	1.30E+01	20.0	5.38E+04	1.86E+01	1.21E+01	2.69E+06	1.61E+04
30.069M	4.30E−01	5.60E−01	20.0	1.78E+04	6.16E−00	4.02E−00	8.90E+05	5.34E+03
30.069	3.70E−01	3.60E−02	20.0	9.85E+02	3.40E−01	2.22E−01	4.92E+04	2.95E+02
ADRENALS 20 GRAMS								
14.031	5.90E−01	1.10E−01	20.0	1.68E+02	5.81E−02	3.79E−02	2.40E+05	2.04E+02

TABLE 2 (continued)

Identification (Z.A.)	Energy (MeV)	T(1/2) Days	Weight in Grams	Inh. Dose (rem/Ci)	Dose/Conc rem/Ci-s/m³ Hi-Rate	Dose/Conc rem/Ci-s/m³ Lo-Rate	Organ Dose (rem/Ci)	Ing. Dose (rem/Ci)
OVARIES 8 GRAMS								
14.031	5.90E−01	1.10E−01	8.0	2.40E+01	8.30E−03	5.41E−03	6.00E+05	3.00E+01
30.065	5.60E−02	7.40E+01	8.0	4.59E+03	1.59E−00	1.03E−00	3.83E+07	1.53E+03
30.069M	4.30E−01	5.80E−01	8.0	2.76E+02	9.57E−02	6.24E−02	2.30E+06	9.22E+01
30.069	3.70E−01	3.60E−02	8.0	1.47E+01	5.11E−03	3.33E−03	1.23E+05	4.92E−00
85.211	6.10E+01	3.00E−01	8.0	2.93E+05	1.01E−02	6.62E+01	1.95E+08	3.91E+05
SKIN 2000 GRAMS								
14.031	5.90E−01	1.10E−01	2000.0	7.20E+01	2.49E−02	1.62E−02	2.40E+03	7.20E+01
16.035	5.60E−02	8.24E+01	2000.0	1.28E+03	4.42E−01	2.88E−01	1.70E+05	1.70E+03
43.096M	2.10E−02	3.60E−02	2000.0	1.39E−01	4.83E−05	3.15E−05	2.79E+01	1.39E−01
43.096	8.30E−03	3.00E−00	2000.0	4.60E−00	1.59E−03	1.03E−03	9.21E+02	4.60E−00
43.097M	7.10E−02	9.00E−00	2000.0	1.18E+02	4.08E−02	2.66E−02	2.36E+04	1.18E+02
43.097	1.10E−03	1.00E+01	2000.0	2.03E−00	7.04E−04	4.59E−04	4.07E+02	2.03E−00
43.099M	2.20E−03	2.40E−01	2000.0	9.76E−02	3.37E−05	2.20E−05	1.95E+01	9.76E−02
43.099	9.40E−02	1.00E+01	2000.0	1.73E+02	6.01E−02	3.92E−02	3.47E+04	1.73E+02
49.113M	1.30E−01	7.30E−02	2000.0	1.75E+01	6.07E−03	3.96E−03	3.51E+02	1.26E−01
49.114M	9.00E−01	2.60E+01	2000.0	4.32E+04	1.49E+01	9.76E−00	8.65E+05	3.11E+02
49.115M	1.40E−01	1.90E−01	2000.0	4.92E+01	1.70E−02	1.11E−02	9.84E+02	3.54E−01
49.115	1.70E−01	6.70E+01	2000.0	2.10E+04	7.28E−00	4.75E−00	4.21E+05	1.51E+02
75.183	1.20E−03	1.90E+01	2000.0	1.09E+02	3.79E−02	2.47E−02	8.43E+02	1.09E+02
75.186	3.60E−01	3.30E−00	2000.0	5.71E+03	1.97E−00	1.28E−00	4.39E+04	5.71E+03
75.187	1.20E−02	2.50E+01	2000.0	1.44E+03	4.99E−01	3.25E−01	1.11E+04	1.44E+03
75.188	7.80E−01	6.90E−01	2000.0	2.58E+03	8.95E−01	5.84E−01	1.99E+04	2.58E+03
FAT 10000 GRAMS								
6.014	5.40E−02	1.20E+01	10000.0	1.82E+03	6.30E−01	4.11E−01	4.79E+03	2.39E+03
55.131	2.40E−02	9.30E−00	10000.0	3.79E−00	1.31E−03	8.57E−04	1.65E+03	4.95E−00
MUSCLE 30000 GRAMS								
30.065	3.20E−01	2.18E+02	30000.0	1.54E+04	5.35E−00	3.49E−00	1.72E+05	5.16E+03
30.069	6.40E−01	5.80E−01	30000.0	8.24E+01	2.85E−02	1.85E−02	9.15E+02	2.74E+01
30.069	3.70E−01	3.60E−02	30000.0	2.95E−00	1.02E−03	6.67E−04	3.28E+01	9.85E−01
37.086	7.00E−01	1.51E+01	30000.0	8.86E+03	3.06E−00	2.00E−00	2.60E+04	1.17E+04
37.087	9.00E−02	8.00E+01	30000.0	6.03E+03	2.08E−00	1.36E−00	1.77E+04	7.99E+03
55.134M	2.60E−01	1.30E−01	30000.0	2.50E+01	8.65E−03	5.64E−03	8.33E+01	.00E−99
55.134	1.10E−00	1.20E+02	30000.0	9.76E+04	3.37E+01	2.20E+01	3.25E+05	.00E−99
55.135	6.60E−02	1.40E+02	30000.0	6.83E+03	2.36E−00	1.54E−00	2.27E+04	.00E−99
55.136	6.50E−01	1.19E+01	30000.0	5.72E+03	1.98E−00	1.29E−00	1.90E+04	.00E−99
55.137	5.90E−01	1.38E+02	30000.0	6.02E+04	2.08E+01	1.35E+01	2.00E+05	.00E−99
56.131	3.80E−01	1.15E+01	30000.0	8.94E−00	3.09E−03	2.01E−03	1.07E+04	1.61E−00
56.140	2.30E−00	1.27E+01	30000.0	5.98E+01	2.06E−02	1.34E−02	7.20E+04	1.08E+01
81.200	4.00E−01	9.30E−01	30000.0	2.38E+02	8.25E−02	5.38E−02	9.17E+02	2.20E+02
81.201	1.70E−01	1.90E−00	30000.0	2.07E+02	7.16E−02	4.67E−02	7.96E+02	1.91E+02
81.202	3.80E−01	3.80E−00	30000.0	9.26E+02	3.20E−01	2.08E−01	3.56E+03	8.54E+02
81.204	2.50E−01	5.50E−00	30000.0	8.81E+02	3.05E−01	1.99E−01	3.39E+03	8.13E+02
TISSUE 43000 GRAMS								
1.003	1.00E−02	1.20E+01	43000.0	2.06E+02	7.14E−02	4.66E−02	2.06E+02	2.06E+02

TABLE 2 (continued)

Identification (Z.A.)	Energy (MeV)	T(1/2) Days	Weight in Grams	Inh. Dose (rem/Ci)	Dose/Conc rem/Ci-s/m³ Hi-Rate	Dose/Conc rem/Ci-s/m³ Lo-Rate	Organ Dose (rem/Ci)	Ing. Dose (rem/Ci)
TOTAL BODY 70000 GRAMS								
1.003	1.00E−02	1.20E+01	70000.0	1.26E+02	4.38E−02	2.86E−02	1.26E+02	1.26E+02
4.007	3.50E−02	4.10E+01	70000.0	3.79E+02	1.31E−01	8.55E−02	1.51E+03	3.03E−00
6.014	5.40E−02	1.00E+01	70000.0	4.28E+02	1.48E−01	9.66E−02	5.70E+02	5.70E+02
9.018	8.90E−01	7.80E−02	70000.0	5.50E+01	1.90E−02	1.24E−02	7.33E+01	7.33E+01
11.022	1.60E−00	1.10E+01	70000.0	1.39E+04	4.82E−00	3.14E−00	1.86E+04	1.86E+04
11.024	2.70E−00	6.00E−01	70000.0	1.28E+03	4.44E−01	2.89E−01	1.71E+03	1.71E+03
14.031	5.90E−01	1.10E−01	70000.0	4.66E+01	1.61E−02	1.05E−02	6.86E+01	5.83E+01
15.032	6.90E−01	1.35E+01	70000.0	6.20E+03	2.14E−00	1.40E−00	9.84E+03	7.38E+03
16.035	5.60E−02	4.43E+01	70000.0	1.96E+03	6.80E−01	4.43E−01	2.62E+03	2.62E+03
17.036	2.60E−01	2.90E+01	70000.0	5.97E+03	2.06E−00	1.34E−00	7.97E+03	7.97E+03
17.038	2.30E−00	2.60E−02	70000.0	4.74E+01	1.64E−02	1.06E−02	6.32E+01	6.32E+01
20.045	8.60E−02	1.62E+02	70000.0	8.10E+03	2.80E−00	1.82E−00	1.47E+04	7.36E+03
20.047	1.40E−00	4.90E−00	70000.0	3.98E+03	1.37E−00	9.00E−01	7.25E+03	3.62E+03
21.046	1.30E−00	2.20E+01	70000.0	7.55E+03	2.61E−00	1.70E−00	3.02E+04	3.02E−00
21.047	2.60E−01	3.10E−00	70000.0	2.13E+02	7.36E−02	4.80E−02	8.52E+02	8.52E−02
21.048	2.20E−00	1.70E−00	70000.0	9.88E+02	3.41E−01	2.23E−01	3.95E+03	3.95E−01
23.048	1.90E−00	1.16E+01	70000.0	6.05E+03	2.09E−00	1.36E−00	2.32E+04	4.65E+02
24.051	2.50E−02	2.66E+01	70000.0	1.75E+02	6.08E−02	3.96E−02	7.03E+02	3.51E−00
25.052	2.10E−00	4.20E−00	70000.0	2.79E+03	9.67E−01	6.31E−01	9.32E+03	9.32E+02
25.054	5.10E−01	1.62E+01	70000.0	2.62E+03	9.06E−01	5.91E−01	8.73E+03	8.73E+02
25.056	1.90E−00	1.10E−01	70000.0	6.62E+01	2.29E−02	1.49E−02	2.20E+02	2.20E+01
26.055	6.50E−03	4.63E+02	70000.0	9.54E+02	3.30E−01	2.15E−01	3.18E+03	3.18E+02
26.059	8.10E−01	4.27E+01	70000.0	1.09E+04	3.79E−00	2.47E−00	3.65E+04	3.65E+03
27.057	9.00E−02	9.20E−00	70000.0	3.50E+02	1.21E−01	7.90E−02	8.75E+02	2.62E+02
27.058M	9.90E−02	3.70E−01	70000.0	1.54E+01	5.35E−03	3.49E−03	3.87E+01	1.16E+01
27.058	6.10E−01	8.40E−00	70000.0	2.16E+03	7.49E−01	4.88E−01	5.41E+03	1.62E+03
27.060	1.50E−00	9.50E−00	70000.0	6.02E+03	2.08E−00	1.35E−00	1.50E+04	4.51E+03
28.059	7.70E−03	6.67E+02	70000.0	2.17E+03	7.51E−01	4.90E−01	5.42E+03	1.62E+03
28.063	2.10E−02	6.52E+02	70000.0	5.78E+03	2.00E−00	1.30E−00	1.44E+04	4.34E+03
28.065	1.40E−00	1.10E−01	70000.0	6.51E+01	2.25E−02	1.46E−02	1.62E+02	4.88E+01
29.064	2.50E−01	5.30E−01	70000.0	5.46E+01	1.88E−02	1.23E−02	1.40E+02	3.92E+01
30.065	3.20E−01	1.94E+02	70000.0	1.96E+04	6.81E−00	4.44E−00	6.56E+04	6.56E+03
30.069M	6.40E−01	5.80E−01	70000.0	1.17E+02	4.07E−02	2.65E−02	3.92E+02	3.92E+01
30.069	3.70E−01	3.60E−02	70000.0	4.22E−00	1.46E−03	9.53E−04	1.40E+01	1.40E−00
31.072	1.80E−00	5.40E−01	70000.0	2.56E+02	8.88E−02	5.79E−02	1.02E+03	1.02E−00
32.071	1.00E−02	9.20E−01	70000.0	2.52E−00	8.74E−04	5.70E−04	9.72E−00	9.72E−02
33.073	6.10E−02	6.00E+01	70000.0	1.04E+03	3.61E−01	2.35E−01	3.86E+03	1.16E+02
33.074	5.60E−01	1.65E+01	70000.0	2.63E+03	9.12E−01	5.95E−01	9.76E+03	2.93E+02
33.076	1.30E−00	1.10E−00	70000.0	4.08E+02	1.41E−01	9.21E−02	1.51E+03	4.53E+01
33.077	2.40E−01	1.60E−00	70000.0	1.09E+02	3.79E−02	2.47E−02	4.05E+02	1.21E+01
34.075	2.00E−01	1.01E+01	70000.0	1.49E+03	5.17E−01	3.37E−01	2.13E+03	1.92E+03
35.082	1.80E−00	1.30E−00	70000.0	1.85E+03	6.41E−01	4.18E−01	2.47E+03	2.47E+03
37.086	7.00E−01	1.32E+01	70000.0	7.32E+03	2.53E−00	1.65E−00	9.76E+03	9.76E+03
37.087	9.00E−02	4.50E+01	70000.0	3.21E+03	1.11E−00	7.24E−01	4.28E+03	4.28E+03
38.085M	9.80E−02	4.90E−02	70000.0	8.65E−00	3.00E−03	1.94E−03	2.16E+01	6.49E−00
38.085	3.30E−01	6.47E+01	70000.0	9.02E+03	3.12E−00	2.03E−00	2.25E+04	6.77E+03
38.089	5.50E−01	5.03E+01	70000.0	1.16E+04	4.04E−00	2.64E−00	2.92E+04	8.77E+03
38.090	1.10E−00	5.70E+03	70000.0	2.36E+06	8.16E+02	5.32E+02	5.90E+06	1.77E+06
38.091	1.90E−00	4.00E−01	70000.0	3.55E+02	1.23E−01	8.04E−02	8.88E+02	2.67E+02
38.092	2.60E−00	1.10E−01	70000.0	1.20E+02	4.18E−02	2.72E−02	3.02E+02	9.07E+01

TABLE 2 (continued)

Identification (Z.A.)	Energy (MeV)	T(1/2) Days	Weight in Grams	Inh. Dose (rem/Ci)	Dose/Conc rem/Ci-s/m³ Hi-Rate	Dose/Conc rem/Ci-s/m³ Lo-Rate	Organ Dose (rem/Ci)	Ing. Dose (rem/Ci)
TOTAL BODY (continued)								
39.090	8.90E−01	2.68E−00	70000.0	6.30E+02	2.18E−01	1.42E−01	2.52E+03	2.52E−01
39.091	9.30E−01	3.50E−02	70000.0	8.60E−00	2.97E−03	1.94E−03	3.44E+01	3.44E−03
39.091	5.90E−01	5.80E+01	70000.0	9.04E+03	3.12E−00	2.04E−00	3.61E+04	3.61E−00
39.092	1.60E−00	1.50E−01	70000.0	6.34E+01	2.19E−02	1.43E−02	2.53E+02	2.53E−02
39.093	1.70E−00	4.20E−01	70000.0	1.88E+02	6.52E−02	4.25E−02	7.54E+02	7.54E−02
40.093	1.90E−02	4.50E+02	70000.0	2.54E+03	8.82E−01	5.75E−01	1.02E+04	1.02E−00
40.095	1.10E−00	5.55E+01	70000.0	1.61E+04	5.58E−00	3.64E−00	6.45E+04	6.45E−00
40.097	2.10E−00	7.10E−01	70000.0	3.94E+02	1.36E−01	8.89E−02	1.57E+03	1.57E−01
41.093M	3.80E−02	6.30E+02	70000.0	6.32E+03	2.18E−00	1.42E−00	2.53E+04	2.53E−00
41.095	5.10E−01	3.35E+01	70000.0	4.51E+03	1.56E−00	1.01E−00	1.80E+04	1.80E−00
41.097	8.70E−01	5.10E−02	70000.0	1.17E+01	4.05E−03	2.64E−03	4.69E+01	4.69E−03
42.099	5.10E−01	1.80E−00	70000.0	6.62E+02	2.29E−01	1.50E−01	1.02E+03	8.15E+02
43.096M	3.00E−01	3.60E−02	70000.0	5.70E−00	1.97E−03	1.28E−03	1.14E+01	5.70E−00
43.096	1.40E−00	8.00E−01	70000.0	5.92E+02	2.04E−01	1.33E−01	1.18E+03	5.92E+02
43.097M	9.00E−02	9.90E−01	70000.0	4.70E+01	1.62E−02	1.06E−02	9.41E+01	4.70E+01
43.097	2.00E−02	1.00E−00	70000.0	1.05E+01	3.65E−03	2.38E−03	2.11E+01	1.05E+01
43.099M	8.00E−02	2.00E−01	70000.0	8.45E−00	2.92E−03	1.90E−03	1.69E+01	8.45E−00
43.099	9.40E−02	1.00E−00	70000.0	4.96E+01	1.71E−02	1.12E−02	9.93E+01	4.96E+01
44.097	1.50E−01	2.00E−00	70000.0	8.56E+01	2.96E−02	1.93E−02	3.17E+02	9.51E−00
44.103	4.40E−01	6.20E−00	70000.0	7.78E+02	2.69E−01	1.75E−01	2.88E+03	8.65E+01
44.105	1.20E−00	1.90E−01	70000.0	6.50E+01	2.25E−02	1.46E−02	2.41E+02	7.23E−00
44.106	1.40E−00	7.20E−00	70000.0	2.87E+03	9.95E−01	6.49E−01	1.06E+04	3.19E+02
45.103M	5.50E−02	3.80E−02	70000.0	7.73E−01	2.67E−04	1.74E−04	2.20E−00	4.41E−01
45.105	2.00E−01	1.33E−00	70000.0	9.84E+01	3.40E−02	2.22E−02	2.81E+02	5.62E+01
46.103	6.40E−02	3.90E−00	70000.0	9.23E+01	3.19E−02	2.08E−02	2.63E+02	5.27E+01
46.109	4.20E−01	5.10E−01	70000.0	7.92E+01	2.74E−02	1.78E−02	2.26E+02	4.52E+01
47.105	6.30E−01	4.40E−00	70000.0	7.61E+02	2.63E−01	1.71E−01	2.93E+03	2.93E+01
47.110	1.70E−00	4.90E−00	70000.0	2.28E+03	7.92E−01	5.16E−01	8.80E+03	8.80E+01
47.111	4.00E−01	3.00E−00	70000.0	3.29E+02	1.14E−01	7.44E−02	1.26E+03	1.26E+01
48.109	1.10E−01	1.40E+02	70000.0	4.07E+03	1.40E−00	9.18E−01	1.62E+04	4.07E+01
48.115M	6.10E−01	3.50E+01	70000.0	5.64E+03	1.95E−00	1.27E−00	2.25E+04	5.64E+01
48.115	7.10E−01	2.20E−00	70000.0	4.12E+02	1.42E−01	9.31E−02	1.65E+03	4.12E−00
49.113M	2.90E−01	7.30E−02	70000.0	5.59E−00	1.93E−03	1.26E−03	2.23E+01	4.47E−02
49.114M	9.70E−01	2.40E+01	70000.0	6.15E+03	2.12E−00	1.38E−00	2.46E+04	4.92E+01
49.115M	2.60E−01	1.90E−01	70000.0	1.30E+01	4.51E−03	2.94E−03	5.22E+01	1.04E−01
49.115	1.70E−01	4.80E+01	70000.0	2.15E+03	7.46E−01	4.86E−01	8.62E+03	1.72E+01
50.113	3.20E−01	2.70E+01	70000.0	2.55E+03	8.84E−01	5.77E−01	9.13E+03	4.56E+02
50.125	9.40E−01	7.50E−00	70000.0	2.08E+03	7.21E−01	4.70E−01	7.45E+03	3.72E+02
51.122	8.20E−01	2.60E−00	70000.0	6.08E+02	2.10E−01	1.37E−01	2.25E+03	6.76E+01
51.124	1.60E−00	2.30E+01	70000.0	1.05E+04	3.63E−00	2.37E−00	3.89E+04	1.16E+03
51.125	4.30E−01	3.60E+01	70000.0	3.65E+03	1.26E−00	8.23E−01	1.35E+04	4.05E+02
52.125M	1.50E−01	1.20E+01	70000.0	7.23E+02	2.50E−01	1.63E−01	1.90E+03	4.75E+02
52.127M	3.20E−01	1.30E+01	70000.0	1.67E+03	5.78E−01	3.77E−01	4.39E+03	1.09E+03
52.127	2.40E−01	3.80E−01	70000.0	3.66E+01	1.26E−02	8.26E−03	9.64E+01	2.41E+01
52.129M	1.10E−00	1.00E+01	70000.0	4.41E+03	1.52E−00	9.97E−01	1.16E+04	2.90E+03
52.129	9.80E−01	5.10E−02	70000.0	2.00E+01	6.94E−03	4.53E−03	5.28E+01	1.32E+01
52.131M	1.60E−00	1.15E−00	70000.0	7.39E+02	2.55E−01	1.66E−01	1.94E+03	4.86E+02
52.132	1.90E−00	2.60E−00	70000.0	1.98E+03	6.86E−01	4.47E−01	5.22E+03	1.30E+03
53.126	2.30E−01	1.21E+01	70000.0	2.20E+03	7.63E−01	4.97E−01	2.94E+03	2.94E+03
53.129	8.90E−02	1.38E+02	70000.0	9.73E+03	3.36E−00	2.19E−00	1.29E+04	1.29E+04
53.131	4.40E−01	7.60E−00	70000.0	2.65E+03	9.17E−01	5.98E−01	3.53E+03	3.53E+03

TABLE 2 (continued)

Identification (Z.A.)	Energy (MeV)	T(1/2) Days	Weight in Grams	Inh. Dose (rem/Ci)	Dose/Conc rem/Ci-s/m³ Hi-Rate	Dose/Conc rem/Ci-s/m³ Lo-Rate	Organ Dose (rem/Ci)	Ing. Dose (rem/Ci)
TOTAL BODY (continued)								
53.132	1.70E−00	9.70E−02	70000.0	1.30E+02	4.52E−02	2.95E−02	1.74E+02	1.74E+02
53.133	8.40E−00	8.70E−01	70000.0	5.79E+03	2.00E−00	1.30E−00	7.72E+03	7.72E+03
53.134	1.50E−00	3.60E−02	70000.0	4.28E+01	1.48E−02	9.66E−03	5.70E+01	5.70E+01
53.135	1.30E+02	8.00E−01	70000.0	8.24E+04	2.85E+01	1.86E+01	1.09E+05	1.09E+05
55.131	2.90E−02	8.75E−00	70000.0	2.01E+02	6.96E−02	4.54E−02	2.68E+02	2.68E+02
55.134M	1.90E−01	1.30E−01	70000.0	1.95E+02	6.77E−02	4.41E−02	2.61E+01	2.61E+01
55.134	1.10E−00	6.50E+01	70000.0	5.66E+05	1.96E+02	1.27E+02	7.55E+04	7.55E+04
55.135	6.60E−02	7.00E+01	70000.0	3.66E+04	1.26E+01	8.26E−00	4.88E+03	4.88E+03
55.136	6.50E−01	1.10E−00	70000.0	5.66E+03	1.96E−00	1.27E−00	7.55E+02	7.55E+02
55.137	5.90E−01	7.00E+01	70000.0	3.27E+05	1.13E+02	7.38E+01	4.36E+04	4.36E+04
56.131	3.80E−01	9.80E−00	70000.0	1.10E+03	3.81E−01	2.48E−01	3.93E+03	1.96E+02
56.140	2.30E−00	1.07E+01	70000.0	7.28E+03	2.52E−00	1.64E−00	2.60E+04	1.30E+03
57.140	1.90E−00	1.68E−00	70000.0	8.43E+02	2.91E−01	1.90E−01	3.37E+03	3.37E−01
58.141	2.10E−01	3.00E+01	70000.0	1.66E+03	5.76E−01	3.75E−01	6.66E+03	6.66E−01
58.143	9.70E−01	1.33E−00	70000.0	3.40E+02	1.17E−01	7.69E−02	1.36E+03	1.36E−01
58.144	1.30E−00	1.91E+02	70000.0	6.56E+04	2.27E+01	1.48E+01	2.62E+05	2.62E+01
59.142	8.50E−01	8.00E−01	70000.0	1.79E+02	6.21E−02	4.05E−02	7.18E+02	7.18E−02
59.143	3.20E−01	1.35E+01	70000.0	1.14E+03	3.94E−01	2.57E−01	4.56E+03	4.56E−01
60.144	2.00E+01	6.56E+02	70000.0	3.46E+06	1.19E+03	7.82E+02	1.38E+07	1.38E+03
60.147	4.00E−01	1.11E+01	70000.0	1.17E+03	4.05E−01	2.64E−01	4.69E+03	4.69E−01
60.149	1.10E−00	8.30E−02	70000.0	2.41E+01	8.34E−03	5.44E−03	9.65E+01	9.65E−03
61.147	6.90E−02	3.83E+02	70000.0	6.98E+03	2.41E−00	1.57E−00	2.79E+04	2.79E−00
61.149	5.40E−01	2.20E−00	70000.0	3.13E+02	1.08E−01	7.08E−02	1.25E+03	1.25E−01
62.147	2.30E+02	6.56E+02	70000.0	3.98E+07	1.37E+04	8.99E+03	1.59E+08	1.59E+04
62.151	4.20E−02	6.45E+02	70000.0	7.15E+03	2.47E−00	1.61E−00	2.86E+04	2.86E−00
62.153	3.00E−01	1.95E−00	70000.0	1.54E+02	5.34E−02	3.48E−02	6.18E+02	6.18E−02
63.152	8.80E−01	3.80E−01	70000.0	8.83E+01	3.05E−02	1.99E−02	3.53E+02	3.53E−02
63.152	6.60E−01	5.59E+02	70000.0	9.75E+04	3.37E+01	2.20E+01	3.90E+05	3.90E+01
63.154	1.30E−00	5.72E+02	70000.0	1.96E+05	6.79E+01	4.43E+01	7.86E+05	7.86E+01
63.155	1.60E−00	3.14E+02	70000.0	1.32E+05	4.59E+01	2.99E+01	5.31E+05	5.31E+01
64.153	1.70E−01	1.65E+02	70000.0	7.41E+03	2.56E−00	1.67E−00	2.96E+04	2.96E−00
64.159	3.60E−01	7.50E−01	70000.0	7.13E+01	2.46E−02	1.61E−02	2.85E+02	2.85E−02
65.160	8.50E−01	6.60E+01	70000.0	1.48E+04	5.12E−00	3.34E−00	5.93E+04	5.93E−00
66.165	5.10E−01	9.70E−02	70000.0	1.30E+01	4.52E−03	2.95E−03	5.22E+01	5.22E−03
66.166	7.90E−01	3.40E−00	70000.0	7.09E+02	2.45E−01	1.60E−01	2.83E+03	2.83E−01
67.166	7.00E−01	1.10E−00	70000.0	2.03E+02	7.04E−02	4.59E−02	8.14E+02	8.14E−02
68.169	3.70E−01	9.30E−00	70000.0	9.09E+02	3.14E−01	2.05E−01	3.63E+03	3.63E−01
68.171	6.50E−01	3.10E−01	70000.0	5.32E+01	1.84E−02	1.20E−02	2.13E+02	2.13E−02
69.170	3.40E−01	1.07E+02	70000.0	9.61E+03	3.32E−00	2.16E−00	3.84E+04	3.84E−00
69.171	3.00E−02	3.42E+02	70000.0	2.71E+03	9.38E−01	6.11E−01	1.08E+04	1.08E−00
70.175	1.60E−01	4.10E−00	70000.0	1.73E+02	5.99E−02	3.91E−02	6.93E+02	6.93E−02
71.177	1.70E−01	6.70E−00	70000.0	3.01E+02	1.04E−01	6.79E−02	1.20E+03	1.20E−01
72.181	5.00E−01	4.30E+01	70000.0	5.68E+03	1.96E−00	1.28E−00	2.27E+04	2.27E−00
73.182	1.10E−00	7.60E+01	70000.0	2.20E+04	7.64E−00	4.98E−00	8.83E+04	8.83E−00
74.181	2.00E−01	1.00E−00	70000.0	6.34E+01	2.19E−02	1.43E−02	2.11E+02	2.11E+01
74.185	1.40E−01	1.00E−00	70000.0	4.44E+01	1.53E−02	1.00E−02	1.48E+02	1.48E+01
74.187	6.80E−01	5.00E−01	70000.0	1.07E+02	3.73E−02	2.43E−02	3.59E+02	3.59E+01
75.183	2.40E−01	6.40E−00	70000.0	8.11E+02	2.80E−01	1.83E−01	1.62E+03	8.11E+02
75.186	3.80E−01	2.50E−00	70000.0	5.02E+02	1.73E−01	1.13E−01	1.00E+03	5.02E+02
75.187	1.20E−02	7.00E−00	70000.0	4.44E+01	1.53E−02	1.00E−02	8.88E+01	4.44E+01
75.188	9.40E−01	6.40E−01	70000.0	3.17E+02	1.10E−01	7.17E−02	6.35E+02	3.17E+02

TABLE 2 (continued)

Identification (Z.A.)	Energy (MeV)	T(1/2) Days	Weight in Grams	Inh. Dose (rem/Ci)	Dose/Conc rem/Ci-s/m³ Hi-Rate	Dose/Conc rem/Ci-s/m³ Lo-Rate	Organ Dose (rem/Ci)	Ing. Dose (rem/Ci)
TOTAL BODY (continued)								
76.185	5.10E−01	2.00E−00	70000.0	3.23E+02	1.11E−01	7.30E−02	1.07E+03	1.07E+02
76.191M	6.00E−02	4.50E−01	70000.0	8.56E−00	2.96E−03	1.93E−03	2.85E+01	2.85E−00
76.191	1.60E−01	1.80E−00	70000.0	9.13E+01	3.15E−02	2.06E−02	3.04E+02	3.04E+01
76.193	3.80E−01	8.00E−01	70000.0	9.64E+01	3.33E−02	2.17E−02	3.21E+02	3.21E+01
77.190	3.70E−01	7.50E−00	70000.0	8.80E+02	3.04E−01	1.98E−01	2.93E+03	2.93E+02
77.192	1.10E−00	1.58E+01	70000.0	5.51E+03	1.90E−00	1.24E−00	1.83E+04	1.83E+03
77.194	8.10E−01	7.60E−01	70000.0	1.95E+02	6.75E−02	4.40E−02	6.50E+02	6.50E+01
78.191	7.00E−01	2.70E−00	70000.0	5.99E+02	2.07E−01	1.35E−01	1.99E+03	1.99E+02
78.193M	7.50E−02	3.00E−00	70000.0	7.13E+02	2.46E−02	1.61E−02	2.37E+02	2.37E+01
78.193	4.30E−02	2.40E+01	70000.0	3.27E+02	1.13E−01	7.38E−02	1.09E+03	1.09E+02
78.197M	5.50E−01	5.60E−02	70000.0	9.76E−00	3.37E−03	2.20E−03	3.25E+01	3.25E−00
78.197	2.60E−01	7.30E−01	70000.0	6.01E+01	2.08E−02	1.35E−02	2.00E+02	2.00E+01
79.196	4.60E−01	5.40E−00	70000.0	7.87E+02	2.72E−01	1.77E−01	2.62E+03	2.62E+02
79.198	5.80E−01	2.60E−00	70000.0	4.78E+02	1.65E−01	1.07E−01	1.59E+03	1.59E+02
79.199	1.80E−01	3.10E−00	70000.0	1.76E+02	6.12E−02	3.99E−02	5.89E+02	5.89E+01
80.197M	3.00E−01	9.10E−01	70000.0	1.81E+02	6.28E−02	4.10E−02	2.88E+02	2.16E+02
80.197	9.70E−02	2.10E−00	70000.0	1.35E+02	4.69E−02	3.06E−02	2.15E+02	1.61E+02
80.203	2.50E−01	8.20E−00	70000.0	1.36E+03	4.72E−01	3.08E−01	2.16E+03	1.62E+03
81.200	4.00E−01	9.20E−01	70000.0	1.86E+02	6.45E−02	4.21E−02	3.89E+02	1.75E+02
81.201	1.70E−01	1.90E−00	70000.0	1.63E+02	5.67E−02	3.69E−02	3.41E+02	1.53E+02
81.202	3.80E−01	3.50E−00	70000.0	6.74E+02	2.33E−01	1.52E−01	1.40E+03	6.32E+02
81.204	2.50E−01	5.00E−00	70000.0	6.34E+02	2.19E−01	1.43E−01	1.32E+03	5.94E+02
82.203	2.20E−01	2.17E−00	70000.0	1.46E+02	5.06E−02	3.30E−02	5.04E+02	4.03E+01
82.210	5.20E−00	1.20E+03	70000.0	1.91E+06	6.61E+02	4.31E+02	6.59E+06	5.27E+05
82.212	8.20E+01	4.40E−01	70000.0	1.10E+04	3.82E−00	2.49E−00	3.81E+04	3.05E+03
83.206	1.80E−00	2.80E−00	70000.0	1.38E+03	4.79E−01	3.12E−01	5.32E+03	5.32E+01
83.207	1.00E−00	5.00E−00	70000.0	1.37E+03	4.75E−01	3.10E−01	5.28E+03	5.28E+01
83.210	1.00E+01	2.50E−00	70000.0	6.87E+03	2.37E−00	1.55E−00	2.64E+04	2.64E+02
83.212	8.30E+01	4.20E−02	70000.0	9.58E+02	3.31E−01	2.16E−01	3.68E+03	3.68E+01
84.210	5.50E+01	2.50E+01	70000.0	4.07E+05	1.40E+02	9.18E+01	1.45E+06	8.72E+04
85.211	6.10E+01	3.00E−01	70000.0	1.67E+04	5.78E−00	3.78E−00	2.23E+04	2.23E+04
88.223	2.80E+02	5.90E−00	70000.0	6.98E+05	2.41E+02	1.57E+02	1.74E+06	5.23E+05
88.224	2.80E+02	2.30E−00	70000.0	2.72E+05	9.42E+01	6.14E+01	6.80E+05	2.04E+05
88.226	1.10E+02	9.00E+02	70000.0	4.18E+07	1.44E+04	9.44E+03	1.04E+08	3.13E+07
88.228	2.30E+02	2.30E+02	70000.0	2.23E+07	7.73E+03	5.04E+03	5.59E+07	1.67E+07
89.227	2.00E+02	6.00E+03	70000.0	2.78E+08	9.62E+04	6.28E+04	1.11E+09	1.11E+05
89.228	2.30E+02	2.60E−01	70000.0	1.58E+04	5.46E−00	3.56E−00	6.32E+04	6.32E−00
90.227	2.00E+02	1.84E+01	70000.0	9.72E+05	3.36E+02	2.19E+02	3.89E+06	3.89E+02
90.228	2.30E+02	6.91E+02	70000.0	4.20E+07	1.45E+04	9.47E+03	1.68E+08	1.68E+04
90.230	4.80E+01	5.70E+04	70000.0	1.43E+08	4.96E+04	3.23E+04	5.73E+08	5.73E+04
90.231	1.80E−01	1.07E−00	70000.0	5.09E+01	1.76E−02	1.14E−02	2.03E+02	2.03E−02
90.232	6.20E+01	5.70E+04	70000.0	1.85E+08	6.40E+04	4.18E+04	7.40E+08	7.40E+04
90.234	9.10E−01	2.41E+01	70000.0	5.79E+03	2.00E−00	1.30E−00	2.31E+04	2.31E−00
91.230	2.90E+02	1.77E+01	70000.0	2.70E+05	9.38E+01	6.12E+01	1.08E+06	1.08E+02
91.231	1.40E+02	4.10E+04	70000.0	4.01E+08	1.38E+05	9.05E+04	1.60E+09	1.60E+05
91.233	3.20E−01	2.74E+01	70000.0	2.31E+03	8.01E−01	5.22E−01	9.26E+03	9.26E−01
92.230	3.50E+02	1.47E−00	70000.0	1.35E+05	4.70E+01	3.06E+01	5.43E+05	5.43E+01
92.232	2.80E+02	1.00E+01	70000.0	7.40E+05	2.56E+02	1.67E+02	2.96E+06	2.96E+02
92.233	5.00E+01	1.00E+02	70000.0	1.32E+06	4.57E+02	2.98E+02	5.28E+06	5.28E+02
92.234	4.90E+01	1.00E+02	70000.0	1.29E+06	4.48E+02	2.92E+02	5.18E+06	5.18E+02
92.235	4.60E+01	1.00E+02	70000.0	1.21E+06	4.20E+02	2.74E+02	4.86E+06	4.86E+02

TABLE 2 (continued)

Identification (Z.A.)	Energy (MeV)	T(1/2) Days	Weight in Grams	Inh. Dose (rem/Ci)	Dose/Conc rem/Ci-s/m³ Hi-Rate	Dose/Conc rem/Ci-s/m³ Lo-Rate	Organ Dose (rem/Ci)	Ing. Dose (rem/Ci)
TOTAL BODY (continued)								
92.236	4.70E+01	1.00E+02	70000.0	1.24E+06	4.29E+02	2.80E+02	4.96E+06	4.96E+02
92.238	4.30E+01	1.00E+02	70000.0	1.13E+06	3.93E+02	2.56E+02	4.54E+06	4.54E+02
92.240	1.30E−00	5.85E−01	70000.0	2.09E+02	7.23E−02	4.71E−02	8.03E+02	8.03E−00
93.237	4.90E+01	3.90E+04	70000.0	1.39E+08	4.82E+04	3.14E+04	5.57E+08	5.57E+04
93.239	2.90E−01	2.33E−00	70000.0	1.78E+02	6.17E−02	4.03E−02	7.14E+02	7.14E−02
94.238	5.70E+01	2.20E+04	70000.0	1.44E+08	4.99E+04	3.26E+04	5.78E+08	5.78E+04
94.239	5.30E+01	6.40E+04	70000.0	1.60E+08	5.54E+04	3.61E+04	6.40E+08	6.40E+04
94.240	5.30E+01	6.30E+04	70000.0	1.59E+08	5.53E+04	3.60E+04	6.39E+08	6.39E+04
94.241	2.30E−00	4.50E+03	70000.0	2.56E+06	8.88E+02	5.79E+02	1.02E+07	1.02E+03
94.242	5.10E+01	6.50E+04	70000.0	1.54E+08	5.34E+04	3.48E+04	6.17E+08	6.17E+04
94.243	3.70E−01	2.08E−01	70000.0	2.03E+01	7.03E−03	4.59E−03	8.13E+01	2.44E−03
94.244	4.80E+01	6.50E+04	70000.0	1.80E+08	6.12E+04	3.99E+04	7.09E+08	2.12E+04
95.241	5.70E+01	1.80E+04	70000.0	1.36E+08	4.72E+04	3.08E+04	5.46E+08	5.46E+04
95.242M	7.30E+01	1.50E+04	70000.0	1.38E+08	4.77E+04	3.12E+04	5.52E+08	5.52E+04
95.242	8.00E+01	6.67E−01	70000.0	1.18E+04	4.09E−00	2.67E−00	4.74E+04	4.74E−00
95.243	5.40E+01	2.00E+04	70000.0	1.33E+08	4.61E+04	3.01E+04	5.33E+08	5.33E+04
95.244	4.40E+01	1.81E−02	70000.0	2.10E+02	7.28E−02	4.74E−02	8.41E+02	8.41E−02
96.242	8.00E+01	1.61E+02	70000.0	3.40E+06	1.17E+03	7.68E+02	1.36E+07	1.36E+03
96.243	6.00E+01	8.40E+03	70000.0	1.03E+08	3.57E+04	2.33E+04	4.13E+08	4.13E+04
96.244	6.00E+01	5.20E+03	70000.0	7.51E+07	2.59E+04	1.69E+04	3.00E+08	3.00E+04
96.245	5.60E+01	2.40E+04	70000.0	1.45E+08	5.01E+04	3.27E+04	5.80E+08	5.80E+04
96.246	5.60E+01	2.40E+04	70000.0	1.45E+08	5.01E+04	3.27E+04	5.80E+08	5.80E+04
96.247	5.50E+01	2.40E+04	70000.0	1.42E+08	4.92E+04	3.21E+04	5.69E+08	5.69E+04
96.248	5.20E+01	2.40E+04	70000.0	1.17E+09	4.06E+05	2.64E+05	4.68E+09	4.68E+05
96.249	5.20E−00	4.40E−02	70000.0	6.04E+01	2.09E−02	1.36E−02	2.41E+02	2.41E−02
97.249	3.80E−00	2.89E+02	70000.0	2.90E+05	1.00E+02	6.54E+01	1.16E+06	3.48E+01
97.250	5.70E+01	1.34E−01	70000.0	2.01E+03	6.98E−01	4.55E−01	8.07E+03	2.42E−01
98.249	6.00E+01	4.70E+04	70000.0	1.75E+08	6.06E+04	3.95E+04	7.01E+08	2.10E+04
98.250	6.20E+01	3.50E+03	70000.0	5.57E+07	1.92E+04	1.25E+04	2.23E+08	6.69E+03
98.251	5.90E+01	5.30E+04	70000.0	1.74E+08	6.04E+04	3.94E+04	6.99E+08	2.09E+04
98.252	2.10E+02	7.93E+02	70000.0	4.40E+07	1.52E+04	9.93E+03	1.76E+08	5.28E+03
98.253	7.30E+01	1.80E+01	70000.0	3.47E+05	1.20E+02	7.83E+01	1.38E+06	4.16E+01
98.254	3.80E+03	5.60E+01	70000.0	5.62E+07	1.94E+04	1.26E+04	2.24E+08	6.74E+03
99.253	7.30E+01	2.00E+01	70000.0	3.85E+05	1.33E+02	8.70E+01	1.54E+06	4.63E+01
99.254	1.30E+02	1.60E−00	70000.0	5.49E+04	1.90E+01	1.24E+01	2.19E+05	6.59E−00
99.254	1.20E+02	4.76E+02	70000.0	1.50E+07	5.22E+03	3.40E+03	6.03E+07	1.81E+03
99.255	7.50E+01	3.00E+01	70000.0	5.94E+05	2.05E+02	1.34E+02	2.37E+06	7.13E+01
100.254	1.30E+02	1.35E−01	70000.0	4.63E+03	1.60E−00	1.04E−00	1.85E+04	5.56E−01
100.255	7.50E+01	8.96E−01	70000.0	1.77E+04	6.14E−00	4.00E−00	7.10E+04	2.13E−00
100.256	3.60E+03	1.11E−01	70000.0	1.05E+05	3.65E+01	2.38E+01	4.22E+05	1.26E+01

LEGAL NOTICE

This report was prepared as an account of Government sponsored work. Neither the United States, nor the Commission, nor any person acting on behalf of the Commission:

A. Makes any warranty or representation, express or implied, with respect to the accuracy, completeness, or usefulness of the information contained in this report, or that the use of any information, apparatus, method, or process disclosed in this report may not infringe privately owned rights; or

B. Assumes any liabilities with respect to the use of, or for damages resulting from the use of any information, apparatus, method, or process disclosed in this report.

As used in the above, "person acting on behalf of the Commission" includes any employee or contractor of the Commission, or employee of such contractor, to the extent that such employee provides access to, any information pursuant to his employment or contract with the Commission, or his employment with such contractor.

ACKNOWLEDGEMENT: This chapter is reprinted from "Dose to Various Body Organs from Inhalation or Ingestion of Soluble Radionuclides" which was prepared in the Health and Safety Division of the U.S. Atomic Energy Commission, Idaho Operations Office, AEC Research and Development Report, Health and Safety, TID-4500, Issued: August 1966.

REFERENCES

1. International Commission on Radiological Protection, *Radiation Protection.* Recommendations of the International Commission on Radiological Protection. ICRP-2. Report of Committee II on Permissible Dose for Internal Radiation. New York: Pergamon Press, (1959).
2. International Commission on Radiological Protection, *Radiation Protection.* Recommendations of the International Commission on Radiological Protection. ICRP-6. Amended 1959 and Revised 1962. New York: Pergamon Press, (1962).
3. Vennart, J. and Minski, M., Radiation Doses From Administered Radionuclides, *J. Brit. Radiol. 35,* pp. 372–387 (June 1963).
4. Loevinger, R., Holt, J. G., and Hine, G. J., "Internally Administered Radioisotopes" Chapter 17. Radiation Dosimetry. G. J. Hine and G. L. Brownell (ed.), New York: Academic Press, Inc., (1956).
5. Markee, Jr., E. H., A Simplified Method of Estimating Environmental Hazards from Accidental Airborne Release of Radioactive Material, *NRTS Meteorological Information Bulletin Number 2,* (April 1966).

Limits for Radioactive Surface Contamination

G. D. Schmidt

The control of surface contamination has been a common procedure of radiation control programs since the early days of the atomic industry. These procedures include both the direct measurement of total surface contamination using portable survey meters, and smear or wipe tests for sampling the "removable" surface contamination. Smears have proven effective in detecting and measuring the extent of radioisotope spills and contamination tracked into a clean area. The smear samples are usually counted in sensitive laboratory instruments and provide measurement of "removable" contamination, which may be related to the extent of possible internal exposure (due to ingestion, inhalation, wound contamination, and so forth).

Standards for the permissible amounts of radioactive surface contamination have been established by governmental authority in the United States only for radioactive material shipments and by the Department of Defense. However, surface contamination limits have been adopted by most nuclear organizations on an individual basis. These limits have not necessarily been derived on the basis of the associated personnel hazards, but rather may be minimum levels obtainable without economic hardship. The general philosophy adopted in the United States has been to limit surface contamination to "the lowest practicable level." In some cases, the values adopted appear to be extremely conservative; however, the facility has been able "to live with" these values. Surface contamination limits have also been set at a low level because of a desire to limit (prevent) the contamination of sensitive areas and materials (such as in low level counting rooms and cross contamination of experiments).

In order to summarize the available information and guidance on contamination levels, a review was undertaken in 1965 of adopted guides and the data relating to personnel hazards from surface contamination. A compilation has been made of the surface contamination limits used in the United States and in other countries, based on literature and a questionnaire sent to various organizations. The limits for use in occupational radiation areas and release to non-controlled areas (unrestricted or general public use) are summarized in Table 1.

The contamination categories used in Table 1 involve an arbitrary assignment of the actual data, which included limits for more than 30 specific categories. Generally similar values are used for many of these categories, and it is felt that only the following categories merit specific guides:

1. *Occupational: Basic Guide*—The "basic guide" is applicable for use in a radiation area where control over the entry and activity of individuals is exercised for the purposes of radiation protection. The basic guide is applicable for floors, benches, tools, and equipment. A similar guide is also generally used for protective clothing, as seen in Table 1.

2. *Occupational: Clean Area*—The "clean area" guide is for use in areas where radioactive materials are generally not utilized (or only low levels), but the area is subject to contamination from radiation areas. These may include non-radiation laboratories, service areas and areas sensitive to radioactive contamination.

3. *Non-controlled—Skin and Personal Clothing*—This guide applies to personal clothing, shoes, hands, and skin of the whole body.

TABLE 1
Summary of Contamination Guides, Recommendations, and 1965 Facility Survey*

Application	Alpha (dpm/100 cm²)		Beta/Gamma	
	Total	Removable	Total (mR/hr)	Removable (dpm/100 cm²)
Occupational				
Basic guides	200–22,000	N.D.–22,000	0.1–1.0	N.D.–220,000
U.S. practice	300–2,000	N.D.–540	0.1–1.0	N.D.–2,200
Clean areas	N.D.–2,200	N.D.–2,200	N.D.–0.5	N.D.–660,000
U.S. practice	N.D.–1,000	N.D.–54	N.D.–0.5	N.D.–220
Protective clothing	N.D.–22,000	N.D.–22,000	0.1–2.5	80–220,000
U.S. practice	N.D.–5,000	N.D.–300	0.1–2.5	80–1,000
Non-controlled				
Skin and clothing	N.D.–2,200	N.D.–2,200	0.05–2.0	N.D.–22,000
U.S. practice	N.D.–1,500	N.D.–5	0.05–2.0	N.D.–80
Material				
U.S. practice	N.D.–5,000	N.D.–500	N.D.–0.3	N.D.–200

N.D.—Non-detectable
* The ranges of contamination limits presented have been compiled from the responses of various organizations in the U.S. and other countries as well as published guides and recommendations.

4. *Non-controlled—Release of Materials*—This guide applies to equipment, tools, vehicles, and facilities for use by the general public without any radiation protection precautions.

It is observed that the contamination limits for the same category used by different organizations vary by factors of up to 1000. There are basically two different limits used: (1) that recommended by Dunster[1,2] and Barnes[3] and used by most foreign organizations, and (2) the "lowest practicable limit," generally used in the United States. The basic guide recommended by Dunster for readily removable contamination is 22,000 dpm/100 cm² for alpha activity.

The literature contains only a few cases where work in contaminated areas has resulted in a detectable body burden. The article by Eisenbud, Blatz, and Barry[4] reported that significant radium burdens were observed only when the contamination was greater than about 100,000 dpm alpha activity per 100 cm² of surface. Schultz and Becher[5] showed a correlation between the amount of surface contamination, the airborne concentrations, and the urinary excretion of uranium. A contamination level of 40,000 dpm/100 cm² was correlated with the maximum permissible air concentration for uranium. The body burden data on three families who had lived in a radium contaminated residence was reported by Evans.[6] Radium was processed in the basement until 1941. Surveys of the residence in 1964 showed extensive alpha contamination levels in the range of 10,000 to 100,000 dpm/100 cm² with peak levels greater than 1,000,000 dpm/100 cm². The results of radium body burden measurements were negative with the exception of those persons who participated in the radium processing. The negative findings included individuals who lived in the house as young children.

The results obtained from a survey in the United Kingdom of twenty-three luminising establishments, which provided seventy-five workers for body radioactivity measurements, are summarized and discussed by Duggan and Godfrey.[7] They found that widespread contamination of about 6,600 dpm ^{226}Ra/100 cm² (loose) and 66,000 dpm ^{226}Ra/100 cm² (total) in allegedly controlled areas is associated with undesirably high radium body burdens in the workers (i.e., of the order of the maximum permissible body burden (MPBB) of 0.1 μCi ^{226}Ra). Duggan and Godfrey recommend levels lower than those of the U.K. Ministry of Health, based upon operational experience. However, the authors do report that radiation hygiene was generally not of the highest standard.

A study of contamination in aircraft instrument repair facilities, which in some cases were involved in the stripping of radium luminous dial instruments, is reported by the Bureau of Radiological Health and Eberline Instrument Corp.[8] Radium body burdens of 40 % of the MPBB were observed in two individuals from one facility which had alpha surface contamination levels of 1,650,000 dpm/100 cm² maximum total; 50,000 dpm/100 cm² average total and 2,700 dpm/100 cm² maximum removable.

Since 1965* there have appeared in the literature additional recommended guides and adopted levels for surface contamination, which are summarized in Table 2.

TABLE 2
Referenced Contamination Guides

Facility or Reference	Alpha (dpm/100 cm²)		Beta/Gamma (mR/hr & dpm/100 cm²)		Application
	Total	Removable	Total	Removable	
DOT, 49 CFR 173[18]*		220		2,200	Package
		220	<0.5	2,200	Vehicles
PHS No. 999-RH-36[8]	25,000 (max)	500			Radium dial stripping
	5,000 (av)				
DSAM 4145.8[17]	1,000	200	2.0	200	Radiation laboratory
	1,000	1,000	2.0	2.0	Contamination clothing
	600	30	0.25	100	Non-radiation laboratory
	200	200	0.2	0.2	Personal clothing
	200	0	0.06	0	Skin-body
U.K. Ministry of Health[15]†		22,000		220,000	Radiation area
		2,200		2,200	Non-radiation area (body, personal clothing)
Duggan and Godfrey[7]‡	6,600	660			Radium dial painting

*For natural uranium, natural thorium, and depleted uranium use values higher by a factor of 10.
†The guide is the maximum permissible levels of contamination not known to be fixed to the surface.
‡Value is for dpm ^{226}Ra per 100 cm²; note that one might expect to observe 2–3 alpha daughters associated with radium contamination.

Included in Table 2 are the values adopted in 1964 by the U.K. Ministry of Health[15] which are based on the recommendation of Dunster.[2] Similar values were recently adopted by the U.K. Department of Employment and Productivity.[16] The values adopted by the U.K. Ministry of Health allow significantly higher levels of contamination than any of the other referenced guides. The values adopted by the Department of Defense in DSAM 4145.8[17] are quite typical of U.S. practice. The U.S. Department of Transportation,[18] however, has set limits for contamination of packages offered for transport intermediate between U.S. practice and the U.K. Ministry of Health levels.

*Recommendations concerning radioactive surface contamination, prior to 1965, have been made by Saenger,[9] General Dynamics,[10] Los Alamos Scientific Laboratory,[11] American Standards Association,[12] National Committee on Radiation Protection,[13] and the International Atomic Energy Agency.[14]

Based upon the adopted radioactive surface contamination guides and the operational experience relating internal exposure to actual surface contamination, it appears that the limits used in the United States are conservative. Therefore, the suggested radioactive surface contamination guides in Table 3 appear reasonable and should not represent a health hazard.

TABLE 3
Suggested Radioactive Surface Contamination Guides

Application	Alpha (dpm/100 cm²)		Beta/Gamma	
	Total	Removable	Total (mR/hr)	Removable (dpm/100 cm²)
Occupational				
Basic guide	25,000 (max) 5,000 (av)	500	1.0	5,000
Clean area	1,000	100	0.5	1,000
Non-controlled				
Skin, personal clothing	500	N.D.	0.1	N.D.
Release of material	2,500 (max) 500 (av)	100	0.2	1,000

The recommendations, however, are subject to the following conditions and interpretations:

1. These limits are to be used as guides, and in practice professional judgment should be used by the radiological physicist to determine the acceptability of the actual contamination.

2. The values are for the most hazardous radionuclides, and a factor of 10 increase in the levels may be allowed for other radionuclides. The most hazardous radionuclides (high radiotoxicity) are generally taken as those in Class 1 of the IAEA classification[14] which includes, for instance, ^{90}Sr, ^{226}Ra, ^{239}Pu, and ^{241}Am.

3. Although it is felt that the recommended values should not result in a health hazard, good radiation protection practice dictates that a reasonable effort be made to keep actual contamination levels below these values.

4. Compliance with contamination guides should not be used as evidence that exposure of persons to internal sources of radiation is within the prescribed standards. Biological sampling or whole body counting should be used to ascertain internal exposures.

5. *For release of material to the general public:*
   (a) A reasonable effort should be made to minimize the contamination (i.e., the application of additional decontamination procedures have little effect on the contamination levels).
   (b) Surfaces of premises or equipment likely to be contaminated that are inaccessible for measurement shall be presumed to be contaminated in excess of the above limits and not released.

6. Total contamination, as used here, refers to that contamination measured with an appropriate survey instrument probe in direct contact with the surface. Removable contamination refers to activity removed from the surface by wiping the surface with a filter paper or other similar material moistened with a detergent solution (or appropriate solvent); the wipe is dried, counted and the activity removed per 100 cm² of surface calculated.

REFERENCES

1. Dunster, H. J., Contamination of Surfaces by Radioactive Materials: The Derivation of Maximum Permissible Levels, *Atomics*, (August 1955).

2. Dunster, H. J., Surface Contamination Measurements as an Index of Control of Radioactive Materials, *Health Physics*, 8, 353–356, (1962).

3. Barnes, D. E., "Basic Criteria in the Control of Air and Surface Contamination," Health Physics in Nuclear Installations; Report of Symposium Organized at the Danish Atomic Energy Center of Riso, (May 1959).

4. Eisenbud, M., Blatz, H. and Berry, E. V., How Important is Surface Contamination? *Nucleonics*, 12, No. 8, (1954).

5. Schultz, N. B. and Becher, A. F., "Correlation of Uranium Alpha Surface Contamination, Air-Borne Concentrations, and Urinary Excretion Rates," *Health Physics*, 9, 901–909, (1963).

6. Evans, Robley D., Long-term Effects of Residence in a Highly Contaminated House, M.I.T. 952–1, *Radium and Mesothorium Poisoning and Dosimetry and Instrumentation Techniques in Applied Radioactivity*, pp. 15–24, (May 1964).

7. Duggan, M. and Godfrey, B., "Some Factors Contributing to the Internal Radiation Hazard in the Radium Luminising Industry," *Health Physics*, 13, 613–623, (June 1967).

8. U. S. Department of Health, Education, and Welfare, Public Health Service, Bureau of Radiological Health, Rockville, Md. and Eberline Instrument Corp., Santa Fe, New Mexico, "Evaluations of Radium Contamination in Aircraft Instrument Repair Facilities," Public Health Service Publication No. 999-RH-36.

9. Saenger, E. L., Medical Aspects of Radiation Accidents, *Handbook for Physicians, Health Physicists and Industrial Hygienists*, (February 1963).

10. General Dynamics Corp., "Health Physics Handbook," OSP-279 Health Physics Office Dept. 3, General Dynamics Corp., Fort Worth, Texas, (April 1963).

11. Los Alamos Scientific Laboratory, Los Alamos, New Mexico, "General Handbook for Radiation Monitoring", compiled and edited by Jerome E. Dummer, Jr., LA-1835, U. S. Atomic Energy Commission, Superintendent of Documents, U. S. Government Printing Office, Washington, D.C. 20402, (Nov. 1958).

12. American Standards Association, "Radiation Protection in Nuclear Fuel Fabrication Plants," N 7.2, American Standards Association Inc., New York, N.Y. 10016, (July 8, 1963).

13. National Committee on Radiation Protection, "Safe Handling of Radioactive Materials," National Bureau of Standards Handbook 92, U. S. Department of Commerce, Superintendent of Documents, U. S. Government Printing Office, Washington, D.C. 20402, (March 1964).

14. International Atomic Energy Agency, "Safe Handling of Radioisotopes," Safety Series No. 1, Vienna, (March 1962).

15. United Kingdom Ministry of Health, "Code of Practice for the Protection of Persons against Ionizing Radiations arising from Medical and Dental Use," (1964).

16. United Kingdom Department of Employment and Productivity, "Ionizing Radiations: Precautions for Industrial Users," Safety, Health, and Welfare, New Series No. 13, HMSO, (1969).

17. Department of Defense, Defense Supply Agency, "Radioactive Commodities in the DoD Supply Systems," DSAM 4145.8. Superintendent of Documents, U. S. Government Printing Office, Washington, D.C. 20402, (Dec. 1967).

18. Code of Federal Regulations, Title 49, Chapter 1—Department of Transportation, Part 173 Shippers, subsection 173.393 (h) General packaging requirements. Federal Register, Vol. 33, No. 194, Part II, Superintendent of Documents, U. S. Government Printing Office, Washington, D.C. 20402, (Oct. 4, 1968).

Determining Industrial Hygiene Requirements for Installations Using Radioactive Materials

Allen Brodsky

INTRODUCTION

The purpose of this chapter is to present some guidelines that have proved useful in evaluating radiation safety requirements for various operations over a wide range of types, quantities, and forms of radioactive material. These guidelines are illustrated by the development of a table that first groups the commonly used radionuclides in eight groups corresponding to the relative magnitudes of their maximum radiotoxicities. Then the table presents curie quantities above which the need for certain safeguards should be examined. For reasons given below, the eight groups are chosen to correspond to the eight orders of magnitude over which the estimated maximum doses per curie range when the radioactive material is delivered in a single intake by inhalation. However, safeguards against external exposure are based on the external dose rate for each radionuclide.

Although the use of single intakes by inhalation may seem to apply only to accidental situations, it also turns out to yield about the same ordering of radionuclides that one would select on the basis of permissible concentrations for continuous exposure under routine operations. Also, since the calculations of dose by inhalation take into account an appreciable fraction of material assumed to pass through the gastrointestinal tract within the first few days after exposure, the relative grouping of radionuclides is also consistent with the relative ingestion toxicity, within one order of magnitude.

Some words of caution should be inserted at this point. Since the generalized calculations in this chapter cannot take into account additional safety considerations, or operating conditions that may be designed into a particular facility, the guides given herein should not ordinarily be used in themselves as the sole criteria for selecting safeguards. The numbers given in the table are chosen to represent (in general) the magnitudes at which each safeguard should be considered if materials are to be handled in the most dispersible form, or if processes are varied or unpredictable in regard to the unwanted release of material from process locations. The values in the table may be scaled upward, for example, for situations in which the radioactive material is normally diluted with other materials that would prevent an intake of enough radionuclide to produce serious exposures, or for processes normally enclosed in a manner known to prevent the escape of significant quantities. Some suggested scaling factors are presented with the table of safeguards to take into account the degree of hazard contributed by the nature of the operation. On the other hand, in unusual circumstances where, for example, good ventilation control is absent and the process is known to disperse the radioactive material toward the operator, operation in chemical hoods might be required at even lower curie levels than those indicated in the table. Principles of ventilation design to deal with these lower levels of radioactivity would be the same as those used in general industrial hygiene practice.

The methods of hazard estimation and the numbers given in this chapter are intended primarily for use as guides by health physicists or industrial hygienists experienced in radiation control, and require addition of the other considerations that apply to specific operations, particular materials, or the possible chemical reactions, in order to avoid the application of larger and more expensive factors of safety than are necessary for the protection of health. Further, although the methods described in this chapter have been based on the general limits of radiation exposure recommended by the National Committee for Radiation Protection and Measurement (NCRP) and the Federal Radiation Council (FRC), the use of these methods does not automatically guarantee that all other applicable safety or regulatory requirements will be met. The industrial hygienist or health physicist must balance the combination of safeguards in order to (1) protect health, (2) meet additional regulatory requirements, and (3) ensure that the first two objectives are met with minimum cost to the employer.

CLASSIFICATION OF RADIONUCLIDES INTO GROUPS

A number of different classifications have been suggested for purposes of estimating hazards of given radionuclides.[1] Many of these classifications recognize the differences in relative toxicity per curie of the various radionuclides, but divide them arbitrarily into only three or four groups for purposes of selecting safeguards. Since the maximum dose per curie inhaled varies over about nine orders of magnitude or more for the radioactive materials of interest, the author has found that the use of only three groups may result in applying similar safeguards to two radionuclides that differ by a factor of approximately 1000 in relative radiotoxicity. This means that either the occasional introduction of unnecessary factors of safety of about 1000 would be encouraged by such guidelines, or else dangerous reductions in the factor of safety may occur. Although the uncertainties[2] in present methods of dose estimation are recognized, the author believes that relative maximum radiotoxicities per curie are generally reliable to within a factor of 10 for most radionuclides of concern.[3] Also, the author has generally found that dose estimates obtained by using present ICRP-NCRP methods[2] tend to be on the safe side when compared with information given in the literature. A considerable part of the wide variation among radionuclides in dose per curie inhaled can be seen to result from the wide range in types and energies of radiation emitted, the wide range in physical half-lives, and the wide ranges in linear energy transfer per unit path through tissue. The uncertainties in these physical parameters are generally small compared to uncertainties in biological factors such as fractional uptake in the critical organ, the biological rate of elimination, or the relative biological effectiveness of the energy deposited per gram of tissue. The uncertainty of these biological factors may range up to factors of 10 to 100.[3] Thus the radionuclides will be arranged in order of dose per curie (values in Table 1 will be given in the inverse units of curies per 15 Rem for convenience in showing the minimum activity that must be inhaled to give a significant dose). On the assumption that the relative ordering of the maximum dose per curie inhaled (under the worst chemical and physical conditions) is reliable on the average to within a factor of 10, the radionuclides are then arranged in groups that cover only a factor of 10 in relative radiotoxicity. Exceptions to this grouping are made in the case of natural uranium and natural thorium, which have such low specific activities that even in the pure state many milligrams of material would have to be inhaled to produce an appreciable dose (see Table 1).

The total dose to the critical organ in 50 years per microcurie intake, assuming instantaneous uptake in the critical organ from a single exposure of duration short compared to the effective half-life of the radionuclide in the body, was calculated

TABLE I. Radiotoxicity Versus Levels Above Which Various Safeguards May Be Required (See Notes on page 500)

Group	Radionuclide	Half-Life (Physical) (days)	Specific Activity (curies per gram)	External Gamma Dose Rate (R/hour at one meter per curie)	A — Critical organ (receiving highest proportion of permissible dose from solubilized activity [after inhalation; other than GI tract])	B — Body burden (for permissible continuous dose rate in critical organ) (microcuries)	C — Effective half-life in critical organ (days)	D — SOLUBLE Permissible concentration in water (40 hrs/wk) (µCi per ml)	E — SOLUBLE Permissible concentration in air (40 hrs/wk) (µCi per cc)	F — INSOLUBLE Permissible concentration in water (40 hrs/wk) (µCi per ml)	G — INSOLUBLE Permissible concentration in air (40 hrs/wk) (µCi per cc)	Single inhalation to give 15 REM to critical organ (curies per 15 REM)	Single inhalation to give 15 REM to lung ('insol' materials) (curies per 15 REM)
I	H-3	4.5×10^3	9.78×10^3	<0.0002	Total water	10^3	12	1×10^{-1}	5×10^{-6}	1×10^{-1}	5×10^{-6}	6.15×10^{-2}	—
	C-14	2.0×10^6	4.61	<0.01	Fat (as CO_2)	300	12	2×10^{-2}	4×10^{-6}	—	5×10^{-5} (as CO_2)	2.88×10^{-2}	—
II	Br-82	1.5	1.06×10^6	—	Total body	10	1.3	8×10^{-3}	1×10^{-6}	1×10^{-2}	2×10^{-7}	7.47×10^{-3}	5.3×10^{-3}
	Cr-51	27.8	9.2×10^4	—	Total body	800	26.6	5×10^{-2}	1×10^{-5}	5×10^{-2}	2×10^{-6}	8.84×10^{-2}	2.3×10^{-3}
	Fe-55	1.1×10^3	2.51×10^3	—	Spleen	10^3	388	2×10^{-2}	9×10^{-7}	7×10^{-2}	1×10^{-6}	2.17×10^{-3}	2.3×10^{-3}
III	S-35	87.1	4.28×10^4	<0.01	Testes	90	76.5	2×10^{-3}	3×10^{-7}	8×10^{-3}	3×10^{-7}	7.23×10^{-4}	6.9×10^{-4}
	Au-198	2.7	2.44×10^5	0.25	GI(LLI)	—	26 (Total body)	2×10^{-3}	3×10^{-7}	1×10^{-3}	2×10^{-7}	7.25×10^{-4}	5.3×10^{-4}
	Ca-47	4.9	5.9×10^5	—	Bone	5	4.9	1×10^{-3}	2×10^{-7}	1×10^{-3}	2×10^{-7}	2.59×10^{-4}	4.6×10^{-4}
	I-132	0.097	1.05×10^7	—	Thyroid	0.3	0.097	2×10^{-3}	2×10^{-7}	5×10^{-3}	9×10^{-7}	4.5×10^{-4}	—
	Ce-141	32	2.8×10^4	—	Bone	30	31	3×10^{-3}	4×10^{-7}	3×10^{-3}	2×10^{-7}	7.06×10^{-4}	4.2×10^{-4}
	Mixed fission†† products	††	$<4 \times 10^{11}$	—	Bone, Lung	—	—	—	—	—	—	1.7×10^{-4}	—
	Sr-85	65	2.37×10^4	—	Total body	60	64.7	3×10^{-3}	2×10^{-7}	5×10^{-3}	1×10^{-7}	2.0×10^{-3}	2.7×10^{-4}
	La-140	1.68	5.61×10^5	0.95	GI(LLI)	9	1.68	7×10^{-4}	2×10^{-7}	7×10^{-4}	1×10^{-7}	4.2×10^{-3}	2.6×10^{-4}
	Nb-95	35	3.93×10^4	—	Total body	40	33.5	3×10^{-3}	5×10^{-7}	3×10^{-3}	1×10^{-7}	3.6×10^{-3}	2.3×10^{-4}
	Zn-65	245	8.2×10^3	—	Prostate	60 (Total body)	13 (Prostate) / 194 (Total body)	3×10^{-3}	1×10^{-7}	5×10^{-3}	6×10^{-8}	2.6×10^{-4}	1.5×10^{-4}
	Co-58	72	3.13×10^4	0.65	Total body	30	8.4	4×10^{-3}	8×10^{-8}	3×10^{-3}	5×10^{-8}	8.4×10^{-3}	1.3×10^{-4}
	Fe-59	45.1	4.92×10^4	—	Spleen	20	41.9	2×10^{-3}	1×10^{-7}	2×10^{-3}	5×10^{-8}	3.0×10^{-3}	1.3×10^{-4}
IV	Hf-181	46	1.62×10^4	<0.01	Spleen	4	41	2×10^{-3}	4×10^{-8}	2×10^{-3}	7×10^{-8}	9.94×10^{-5}	1.92×10^{-4}
	Pm-147	920	9.25×10^2	<0.01	Bone	60	570	6×10^{-3}	6×10^{-8}	6×10^{-3}	1×10^{-8}	8.9×10^{-5}	2.3×10^{-4}
	P-32	14.3	2.85×10^5	—	Bone	6	14.1	5×10^{-4}	7×10^{-8}	7×10^{-4}	8×10^{-8}	8.7×10^{-5}	2.1×10^{-4}
	Ba-140	12.8	7.3×10^4	1.54	Bone	4	10.7	8×10^{-4}	1×10^{-7}	7×10^{-4}	4×10^{-8}	1.4×10^{-4}	8.6×10^{-5}
	Th-234	24.1	2.32×10^4	0.0019	Bone	4	24.1	5×10^{-4}	6×10^{-8}	5×10^{-4}	4×10^{-8}	8.5×10^{-5}	7.3×10^{-5}
	Kr-85	3.9×10^3	3.96×10	0.51	Total body	—	—	—	1×10^{-5}(sub)	—	$3 \times$	6.9×10^{-2}	5.8×10^{-5}
	Ir-192	74.5	9.16×10^3	—	Kidney	6	30	1×10^{-3}	1×10^{-7}	1×10^{-3}	3×10^{-8}	3.2×10^{-4}	6.9×10^{-5}
	Cl-36	1.2×10^8	3.21×10^{-2}	—	Total body	80	29	2×10^{-3}	4×10^{-7}	2×10^{-3}	2×10^{-8}	2.7×10^{-3}	5.3×10^{-5}
	Y-91	58	2.50×10^4	—	Bone	5	58	8×10^{-4}	4×10^{-8}	8×10^{-4}	3×10^{-8}	5.0×10^{-5}	7.3×10^{-5}
	Ta-182	112	6.2×10^3	—	Liver	7	88	1×10^{-3}	4×10^{-8}	1×10^{-3}	2×10^{-8}	1.1×10^{-4}	5.0×10^{-5}
	Ca-45	164	1.77×10^4	<0.01	Bone	30	162	3×10^{-4}	3×10^{-8}	1×10^{-3}	3×10^{-8}	4.3×10^{-4}	2.6×10^{-5}

RADIONUCLIDE AND GROUP		PHYSICAL PROPERTIES			RADIOBIOLOGIC PROPERTIES							RELATIVE RADIOTOXICITY	
					SOLUBLE					INSOLUBLE			
Radionuclide Group		Half-Life (Physical) (days)	Specific Activity (curies per gram)	External Gamma Dose Rate (R/hour at one meter per curie)	Critical organ (receiving highest proportion of permissible dose from solubilized activity [after inhalation; other than GI tract]) A	Body burden (for permissible continuous dose rate in critical organ) (microcuries) B	Effective half-life in critical organ (days) C	Permissible concentration in water (for continuous use by employees 40 hrs per week) (microcuries per milliliter) D	Permissible concentration in air (for continuous occupation exposure 40 hrs per week) (micro-curies per cc) E	Permissible concentration in water (for continuous use by employees 40 hrs per week) (micro-curies per ml.) F	Permissible concentration in air (for continuous occupational exposure 40 hrs per week) (micro-curies per cc) G	Single inhalation in curies to give 15 REM to critical organ (curies per 15 REM)	Single inhalation in curies to give 15 REM to lung (" insol materials) (curies per 15 REM)
IV	Ce-144	290	3.18×10^{3}	0.20	Bone	5	243	3×10^{-4}	1×10^{-8}	3×10^{-4}	6×10^{-9}	1.4×10^{-5}	1.5×10^{-5}
(*cont.*)	I-126	13.3	7.8×10^{4}	—	Thyroid	1	12.1	5×10^{-5}	8×10^{-9}	3×10^{-3}	3×10^{-7}	1.4×10^{-5}	7.3×10^{-4}
	Eu-154	5.8×10^{3}	1.45×10^{2}	—	Kidney	5	1.18×10^{3}	6×10^{-4}	4×10^{-9}	6×10^{-4}	7×10^{-9}	1.3×10^{-5}	1.6×10^{-5}
	I-131	8	1.24×10^{5}	0.25	Thyroid	0.7	7.6	6×10^{-5}	9×10^{-9}	2×10^{-3}	3×10^{-7}	1.2×10^{-5}	7.3×10^{-4}
	Tm-170	127	6.08×10^{3}	0.004	Bone	9	113	1×10^{-3}	4×10^{-8}	1×10^{-3}	3×10^{-8}	3.8×10^{-5}	7.5×10^{-5}
V	I-129	6.3×10^{9}	1.62×10^{-4}	—	Thyroid	3	138	1×10^{-5}	2×10^{-9}	6×10^{-3}	7×10^{-8}	2.3×10^{-6}	1.6×10^{-4}
	Tc-99	7.7×10^{7}	1.71×10^{-2}	—	GI(LLI)	—	—	1×10^{-2}	2×10^{-6}	5×10^{-3}	6×10^{-8}	9.1×10^{-6}	—
					Kidney	10							
VI	Ra-223	11.7	5.0×10^{4}	—	Bone	0.05	11.7	2×10^{-5}	2×10^{-9}	1×10^{-4}	2×10^{-10}	3.9×10^{-6}	5.3×10^{-7}
	Po-210	138.4	4.5×10^{3}	<0.00005	Spleen	0.03	42	2×10^{-5}	5×10^{-10}	8×10^{-4}	2×10^{-10}	1.3×10^{-6}	5.0×10^{-7}
	Th-227	18.4	3.17×10^{4}	—	Bone	0.02	18.4	—	—	—	—	5.5×10^{-7}	4.6×10^{-7}
	Sr-90	1×10^{4}	1.44×10^{2}	<0.01	Bone	2	6.4×10^{3}	*1×10^{-5}	*1×10^{-9}	1×10^{-3}	5×10^{-9}	3.9×10^{-7}	1.3×10^{-5}
	Pb-210	7.1×10^{3}	8.8×10	—	Kidney	0.04	494	4×10^{-6}	1×10^{-10}	5×10^{-3}	2×10^{-10}	3.2×10^{-7}	5.3×10^{-7}
	Cm-242	162.5	3.34×10^{3}	<0.01	Liver	0.05	154.3	7×10^{-4}	1×10^{-10}	7×10^{-4}	2×10^{-10}	3.0×10^{-7}	4.6×10^{-7}
	U-233	5.9×10^{7}	0.01 (with 80 ppm U^{232})	0.0002 (with 20 ppm U^{232})	Bone	0.05	300	9×10^{-4}	5×10^{-10}	9×10^{-4}	1×10^{-10}	7.0×10^{-7}	2.7×10^{-7}
	U-235 (+1% U-234)	2.6×10^{11}	2.15×10^{-6}	<0.002	Kidney	0.03	15	8×10^{-4}	5×10^{-10}	8×10^{-4}	1×10^{-10}		
					Bone	0.06	300	—	—	—	—	1.1×10^{-6}	2.6×10^{-7}
	U-238 and U–nat'l	1.6×10^{12}	3.34×10^{-7}	<0.002	Kidney	0.005	15	1×10^{-3}	7×10^{-11}	1×10^{-3}	1×10^{-10}	1.9×10^{-7}	3.0×10^{-7}
	Th-232 and Th–nat'l	5.1×10^{12}	1.11×10^{-7}	<0.0002	Bone	0.01	7.3×10^{4}	*5×10^{-5}	*3×10^{-11}	1×10^{-3}	3×10^{-11}	2.25×10^{-9}	2.6×10^{-8}
VII	Sm-147	4.8×10^{13}	1.95×10^{-8}	—	Bone	0.1	1.5×10^{3}	2×10^{-3}	7×10^{-11}	2×10^{-3}	3×10^{-10}	7.7×10^{-8}	6.9×10^{-7}
	Nd-144	7.3×10^{17}	4.97×10^{-15}	—	Bone	0.1	1.5×10^{3}	2×10^{-3}	8×10^{-11}	2×10^{-3}	3×10^{-10}	7.7×10^{-8}	7.3×10^{-7}
	Ra-226	5.9×10^{5}	1.00	—	Bone	0.1	1.6×10^{4}	4×10^{-7}	3×10^{-11}	9×10^{-4}	*5×10^{-11}	4.9×10^{-8}	1.5×10^{-8}
	Cm-244	6.7×10^{3}	8.2×10	—	Bone	0.1	2.1×10^{3}	2×10^{-4}	9×10^{-12}	8×10^{-4}	1×10^{-10}	1.1×10^{-8}	2.3×10^{-7}
VIII	Am-243	2.9×10^{6}	1.85×10^{-1}	—	Bone	0.05	7.1×10^{4}	1×10^{-4}	6×10^{-12}	8×10^{-4}	1×10^{-10}	7.6×10^{-9}	2.7×10^{-7}
	Am-241	1.7×10^{5}	3.21	0.039	Bone	0.05	5.1×10^{4}	1×10^{-4}	6×10^{-12}	8×10^{-4}	1×10^{-10}	6.6×10^{-9}	2.7×10^{-7}
	Np-237	8×10^{8}	6.9×10^{-4}	—	Bone	0.06	7.3×10^{4}	9×10^{-5}	4×10^{-12}	9×10^{-4}	1×10^{-10}	5.2×10^{-9}	2.7×10^{-7}
	Ac-227	8×10^{3}	7.2×10	—	Bone	0.03	7.2×10^{3}	6×10^{-5}	2×10^{-12}	9×10^{-3}	3×10^{-11}	3.0×10^{-9}	6.9×10^{-8}
	Th-230	2.9×10^{7}	1.97×10^{-2}	0.009	Bone	0.05	7.3×10^{4}	5×10^{-5}	2×10^{-12}	9×10^{-4}	1×10^{-11}	2.8×10^{-9}	2.3×10^{-8}
	Pu-242	1.4×10^{8}	3.9×10^{-3}	—	Bone	0.05	7.3×10^{4}	1×10^{-4}	2×10^{-12}	9×10^{-4}	4×10^{-11}	2.5×10^{-9}	8.5×10^{-8}
	Pu-238	3.3×10^{4}	2.68×10	<0.02	Bone	0.04	2.3×10^{4}	1×10^{-4}	2×10^{-12}	8×10^{-4}	3×10^{-11}	2.2×10^{-3}	7.3×10^{-8}
	Pu-240	2.4×10^{6}	2.27×10^{-1}	<0.001	Bone	0.04	7.1×10^{4}	1×10^{-4}	2×10^{-12}	8×10^{-4}	4×10^{-11}	2.0×10^{-9}	8.5×10^{-8}
	Pu-239	1.9×10^{6}	6.17×10^{-2}	<0.00001	Bone	0.04	7.2×10^{4}	1×10^{-4}	2×10^{-12}	8×10^{-4}	4×10^{-11}	2.0×10^{-9}	8.5×10^{-8}

TABLE I. Radiotoxicity Versus Levels Above Which Various Safeguards May Be Required (continued) (See Notes on page 500)

RADIONUCLIDE AND GROUP	FACILITIES AND EQUIPMENT								SITE	PROCEDURES							
Radionuclide Group	Chemical hood required (curies)	Glovebox required (curies)	Glovebox inside hot cell or cave (Based on gamma dose rates) (curies)	1 Absolute filter (in the exhaust from active atmosphere) (curies)	2 Absolute filters in series (in the active exhaust) (curies)	Continuous general air sampler with alarm (in work areas) (curies)	Continuous exhaust stack monitor and alarm (to protect public) (curies)	Building containment or controlled leak rate (to protect public) (curies)	Radius of low population zone X (in m.) dist. beyond which cloud dose is less than 15 Rem for Q curies released, where $Q = fC$, for C curies in process	Personnel monitoring and/or appropriate shielding vs. external gamma radiation (curies)	Occasional excretion radio-assay spot checks of operating personnel (curies)	Routine excretion assay of all operating personnel (curies)	Emergency dosimeters worn to measure high external doses (curies)	Routine environmental monitoring of site and community (curies)	Pre-planned written emergency procedures and drills (curies)	Written routine operating procedures (curies)	Written pre-operational analysis of maximum credible accidents (including doses to people) (curies)
I — H-3, C-14	1	10	—	10	10^4	10^4	10^5	10^6	$X = 0.47Q^{2/3}$	—	10	10^2	—	1,000	10^4	10^6	10
II — Br-82, Cr-51, Fe-55	0.1	1	5	1	10^3	10^3	10^4	10^5	$2.2Q^{2/3}$	0.5	1	10	50	100	10^3	10^5	1
III — S-35	10^{-2}	0.1	100	0.1	10^2	10^2	10^3	10^4	$10Q^{2/3}$	10	0.1	1	1,000	10	10^2	10^4	0.1
Au-198			4							0.4			40				
Ca-47, I-132, Ce-141			1							0.1			10				
Mixed fission products†, Sr-85, La-140, Nb-95, Zn-65			1							—			—				
Co-58, Fe-59	10^{-2}	0.1	2	0.1	10^2	10^2	10^3	10^4	$10Q^{2/3}$	0.2	0.1	1	20	10	10^2	10^4	0.1
IV — Hf-181	10^{-3}	10^{-2}	100	10^{-2}	10	10	10^2	10^3	$47Q^{2/3}$	10	10^{-2}	0.1	1,000	1	10	10^3	10^{-2}
Pm-147			100							10			1,000				
P-32			0.5							0.05			5				
Ba-140			500							50			5,000				
Th-234			2							0.02			20				
Kr-85																	
Ir-192			100							10			1,000				
Cl-36																	
Y-91			—							—							
Ta-182			100							10			1,000				
Ca-45			100														
Sr-89			3							0.3			30				

TABLE 7.6 ... in Radioactivity Versus Levels Above Which Various Safeguards May Be Required (*continued*) (See Notes on page 500)

Group	Radionuclide	Chemical hood required (curies)	Glovebox required (curies)	Glovebox inside hot cell or cave (Based on gamma dose rates) (curies)	1 Absolute filter (in the exhaust from active atmosphere) (curies)	2 Absolute filters in series (in the active exhaust) (curies)	Continuous general air sampler with alarm (in work areas) (curies)	Continuous exhaust stack monitor and alarm (to protect public) (curies)	Building containment or controlled leak rate (to protect public) (curies)	SITE — Radius of low population zone X (in. dist.) beyond which cloud dose is less than 15 REM for Q curies released, where Q = fC, for C curies in process	Personnel monitoring and/or appropriate shielding vs. external gamma radiation (curies)	Occasional excretion radio-assay spot checks of operating personnel (curies)	Routine excretion assay of all operating personnel (curies)	Emergency dosimeters worn to measure high external doses (curies)	Routine environmental monitoring of site and community (curies)	Pre-planned written emergency procedures and drills (curies)	Written routine operating procedures (curies)	Written pre-operational analysis of maximum credible accidents (including doses to people) (curies)
IV (cont.)	Ce-144, I-126, Fu-154 / I-131 / Tm-170	$10^{-3} \leftrightarrow 10^{-3}$	$10^{-2} \leftrightarrow 10^{-2}$	5 / 4 / 250	$10^{-2} \leftrightarrow 10^{-2}$	$10 \leftrightarrow 10$	$10 \leftrightarrow 10$	$10^{2} \leftrightarrow 10^{2}$	$10^{3} \leftrightarrow 10^{3}$	$47Q^{2/3} \leftrightarrow 47Q^{2/3}$	0.05 / 0.4 / 25	$10^{-2} \leftrightarrow 10^{-2}$	$0.1 \leftrightarrow 0.1$	5 / 40 / 2,500	$1 \leftrightarrow 1$	$10 \leftrightarrow 10$	$10^{3} \leftrightarrow 10^{3}$	$10^{-2} \leftrightarrow 10^{-2}$
V	I-129, Tc-99	10^{-4}	10^{-3}	—	10^{-3}	1	1	10	10^{2}	$220Q^{2/3}$	100	10^{-3}	10^{-2}	10^{4}	0.1	1	10^{2}	10^{-3}
VI	Ra-223, Po-210, Th-227, Sr-90, Pb-210, Cm-242, U-233	$10^{-5} \leftrightarrow 10^{-5}$	$10^{-4} \leftrightarrow 10^{-4}$	20,000 / 100 / 100 / 5,000 (500 kg)	$10^{-4} \leftrightarrow 10^{-4}$	$0.1 \leftrightarrow 0.1$	$0.1 \leftrightarrow 0.1$	$1 \leftrightarrow 1$	$10 \leftrightarrow 10$	$1,000Q^{2/3} \leftrightarrow 1,000Q^{2/3}$	10 / 10 / 1 / 50	$10^{-4} \leftrightarrow 10^{-4}$	$10^{-3} \leftrightarrow 10^{-3}$	1,000 / 1,000 / 100 / 5,000	$10^{-2} \leftrightarrow 10^{-2}$	$0.1 \leftrightarrow 0.1$	$10 \leftrightarrow 10$	$10^{-4} \leftrightarrow 10^{-4}$
VI	U-235 (+% U-234), U-238 and U-nat'l, Th-232 and Th-nat'l	10^{-5}	10^{-4}	—	10^{-4}	0.1	0.1	1	10	$1,000Q^{2/3}$	—	10^{-4}	10^{-3}	—	10^{-2}	0.1	10	10^{-4}
VI	Sm-147, Nd-144, Ra-226, Cm-244	10^{-6}	10^{-5}	—	10^{-5}	10^{-2}	10^{-2}	0.1	1	$4,700Q^{2/3}$	—	10^{-5}	10^{-4}	—	10^{-3}	10^{-2}	1	10^{-5}
VIII	Am-243, Am-241, Np-237, Ac-227, Th-230, Pu-242, Pu-238, Pu-240, Pu-239	$10^{-7} \leftrightarrow 10^{-7}$	$10^{-6} \leftrightarrow 10^{-6}$	25 / 100 / 50 / 1,000 (Based on criticality)	$10^{-6} \leftrightarrow 10^{-6}$	$10^{-3} \leftrightarrow 10^{-3}$	$10^{-3} \leftrightarrow 10^{-3}$	$10^{-2} \leftrightarrow 10^{-2}$	$0.1 \leftrightarrow 0.1$	$22,000Q^{2/3} \leftrightarrow 22,000Q^{2/3}$	0.25 / 1 / 0.5 / 10 (Based on criticality)	$10^{-6} \leftrightarrow 10^{-6}$	$10^{-5} \leftrightarrow 10^{-5}$	25 / 100 / 50 / 1,000 (Based on criticality)	$10^{-4} \leftrightarrow 10^{-4}$	$10^{-3} \leftrightarrow 10^{-3}$	$0.1 \leftrightarrow 0.1$	$10^{-6} \leftrightarrow 10^{-6}$

from the equation

$$\text{Dose (Rem/}\mu\text{Ci)} = \int_{t=0}^{t=50\times365} I_0 e^{-0.693t/T_{1/2}} \, dt \tag{1}$$

$$= 1.44 I_0 T[1 - \exp(1.265 \times 10^4/T)]$$

where I_0 = initial dose rate to the critical organ per microcurie intake, or $f_a R/q f_2$ (with the standard symbols from reference 2); f_2 = the fraction of radionuclide in the organ of reference divided by that in the total body (Table 12 of reference 2); q = body burden listed beside the corresponding critical organ in Table 1 of reference 2; T = effective half-life in the body in days, from Table 12 of reference 2; and R = permissible dose rate for continuous exposure in Rem per day for the body organ concerned, obtained from reference 2 as 0.1/7 Rem/day for irradiation of the whole body, 0.08 Rem/day for bone, 0.1/7 Rem/day for the gonads, 0.6/7 Rem/day for the thyroid and skin, and 0.3/7 Rem/Day for other parts of the body. The expression in brackets in equation (1) is essentially 1, except for the few bone-seeking radionuclides that have effective half-lives that are not short compared to 50 years. For strontium-90, with an effective half-life of 6.4×10^3 days, the factor in brackets becomes 0.861. For purposes of this chapter, the same relative dose from daughter products is assumed for single intake or for continuous exposure from radionuclides that have radioactive daughters building up in the body. The contribution from daughters is thus taken into account by using the total body burdens, q, to give dose rates, R. These body burdens were calculated by the ICRP Committee, taking into account daughter products that build up in the body.

Since for insoluble materials an average of $12\frac{1}{2}\%$ of the material inhaled may remain in the lung, with a half-life of 120 days, the lung must also be taken into consideration as a possible critical organ. Single intake doses based on the lung dose were calculated from the equation:

Dose to the lung per microcurie (μCi) inhaled (insoluble)

$$= \frac{\text{Permissible dose rate per week}}{(\text{MPC}_{\text{air}} \text{ based on continuous exposure to lung}) \times 1.4 \times 10^8 \text{ cc/wk}} \tag{2}$$

$$= 2.14 \times 10^{-9}/\text{MPC}_{\text{air}} \text{ based on lung}$$

since the equilibrium dose rate from continuous exposure is the same as the average dose rate from a series of single intakes of the same total quantity of radioactivity per week, for effective half-lives short compared to 50 years.

The smallest value in curies per 15 Rem was selected as a basis for ordering the radionuclides in order to derive initial criteria that would be safe even for processes in which the radionuclides might occur in chemical forms that could give the highest potential internal exposures. This method provides a safe starting point from which other correction factors may be applied where they are determinable in specific circumstances. The final relative ordering of radionuclides for purposes of selecting safeguards against internal exposure is given in Table 1.

METEOROLOGICAL CALCULATIONS

In Figure 1, the minimum curie quantities that would need to be released at zero point in order to produce an estimated dose of 15 Rem to an adult standing X meters downwind are plotted for some of the radionuclides frequently in use. The ordinates at a distance of 10 meters show that the radionuclides do conveniently group themselves into eight groups corresponding to eight orders of magnitude of

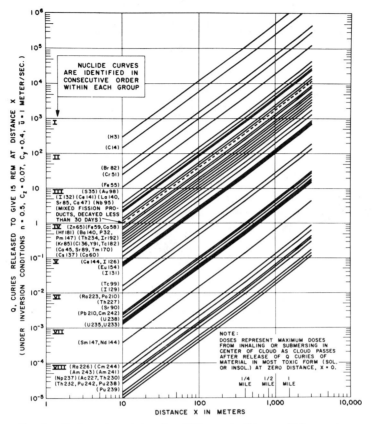

FIG. 1.   Curies released to give 15 Rem at distance X for various nuclides.

relative maximum hazard of various radionuclides under different proximities of the potential point of release and likely points of exposure. Although the meteorological calculations as well as the single intake doses used to obtain Figure 1 would yield considerable factors of safety under most actual situations, the use of even raw guides such as Figure 1 when additional information about materials and procedures cannot be specified in advance may often result in less safety overdesign than the use of "professional judgment" alone without any consideration of relative radiotoxicity. Figures 2 and 3 have also been used in estimating exposures and exposure durations from released materials. A discussion is presented in the Appendix of the calculations used in obtaining Figures 1 through 3, and the bases for estimating the fractions of released material that may be inhaled or deposited under various circumstances.

DERIVATION OF SAFEGUARD GUIDE LEVELS

In the introduction, arguments were presented for beginning the estimation of hazards and safeguard requirements by a consideration of the relative internal or external doses per curie of the radionuclides involved under the various possible modes of exposure. This section will outline the derivation of the base-line levels in Table 1 above which the need for various safeguards should be examined. These levels were derived with the aid of the basic information on relative radiotoxicity and the methods of estimating general hazard potential discussed in earlier sections. The industrial hygienist will ordinarily need to consider additional factors, such as specific activity, chemical dilution and form, volatility, types of unit operation, probability of accidents indicated by process history, or other aspects of the particular industrial

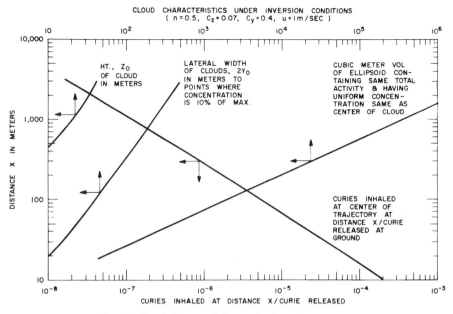

FIG. 2. Cloud characteristics under inversion conditions.

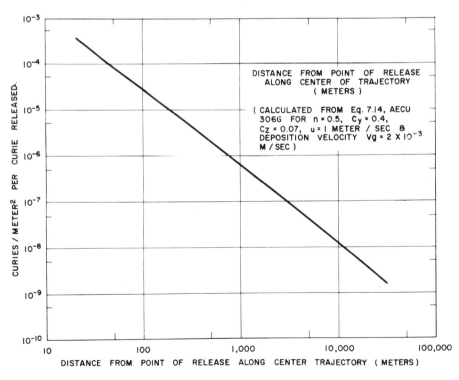

FIG. 3. Typical deposition of cloud activity plotted versus distance, with activity deposit in curies per square meter per curie released given as the probable upper limit along the center of the trajectory under inversion conditions.

or laboratory operations, in order to determine which combination of safeguards is actually required at what curie level.

Radionuclide Properties and Radiotoxicity Groups

Some of the properties of the radionuclides encountered in industry and their relative radiotoxicities are listed on the left-hand side of Table 1. These data are presented here not only for convenient reference, but also because the repeated use of such a table with the magnitudes of important variables revealed helps to generate additional perspective regarding the relative importance of various factors in hazard evaluation. In Table 1 other radiobiologic properties such as MPC's are also listed for convenient reference and to illustrate that the radionuclides will generally order themselves into approximately the same groupings if the permissible concentrations for continuous inhalation are used as the criterion for relative radiotoxicity. Also, examination of the relationships between permissible intakes for ingestion and inhalation will show that the inhalation hazard is generally the limiting consideration in the following selection of safeguard levels. The maximum radiation doses from inhaled substances were obtained from ICRP data[2] as described above. Specific activities and gamma dose rates were either calculated from half-lives and decay schemes[2,5] or taken from compilations.[6,7] Values in the gamma dose-rate column preceded by < are generally calculated upper limits to the bremsstrahlung dose rate, with self-absorption of the X-radiation and absorption by any container neglected.

In a previous paper,[4] Figure 1 was plotted from the single intake doses calculated by Fairbairn and Dunning,[7] whose values were obtained in a similar fashion to these given here. Fairbairn and Dunning gave values generally within a factor of 2 of those given in Table 1, except for ^{3}H, ^{14}C, ^{32}P, and ^{230}Th. In this chapter, the effective lung dose from inhaled insoluble particles containing ^{3}H or ^{14}C is assumed to be much less than the dose calculated from a uniform distribution of absorbed energy throughout lung tissue. The maximum ^{3}H beta-ray range is only about 6 microns in tissue,[8] and the average range would be about 1 micron.[9] Absorption within the insoluble particle or phagocytic cell and the mucous linings of the tracheobronchial and alveolar surfaces would be expected to reduce sharply the dose to sensitive tissue.[10] However, since the mucoid film on the alveolar surfaces is only about 0.2 micron thick, these assumptions may not be applicable in the case of fine particulates that remain dispersed throughout the lung for long effective half-lives.

Estimation of Safeguard Requirements

CHEMICAL HOOD. A value of 0.1 μCi (10^{-7} Ci) was selected for the most toxic group as the lowest quantity requiring that operations potentially capable of dispersing the radioactive material be conducted within a hood. This is the value given in reference 11 as the minimum significant quantity for dusty operations, and it is the value in reference 11, Table 2, requiring chemical hoods for radioisotope laboratory operations (10 μCi), multiplied by the factor of 0.01 for dry and dusty operations. Also, it is about one-thirtieth of the approximate level at which some laboratories have found it generally desirable to require drybox operations.

For the most hazardous nuclides in Table 1, 0.1 μCi is only about two and one-half times the maximum permissible body burden.[2] In Group 8 of Table 1, all the important nuclides have long effective half-lives in the body. Thus, it seems reasonable to attempt to minimize the probability of inhaling a significant fraction of the permissible body burden.

Since both the permissible continuous intake values[2,12] and the single intake values in Table 1 range over about eight orders of magnitude, the values requiring hood operation have been scaled up a factor of 10 each for Groups 7 to 1.

For some of the uranium and thorium nuclides and their natural mixtures, the extremely low specific activity even in the carrier-free state almost guarantees a reduction in the probability of intake[1] by inhalation. Therefore, these nuclides have been placed in higher categories than would be indicated purely on the basis of dose per curie. The low specific activity also limits the external dose rates from these elements in bulk form even in multicurie quantities, and this further limits the external radiation safeguards needed for large quantities. Such an *a priori* consideration of specific activity[1] was not considered valid for any of the other nuclides in Table 1.

GLOVEBOX. The level where gloveboxes may begin to be desirable was set at 1 μCi for Group 8, based primarily on experience[13] with ^{239}Pu. This level of 1 μCi comes to 10 times the quantity where operations within a chemical hood may be required, and is about 25 times the permissible body burden for the most hazardous nuclides. Scaling upward was done on the same basis as for chemical hoods.

GLOVEBOX INSIDE HOT CELL OR CAVE (See also Personnel Monitoring). The point at which operations should be conducted within a hot cell or cave, as well as a glovebox (for loose material), was based primarily on the external gamma, X-ray, or bremsstrahlung dose rate from the quantity of material as a point source. Since this external dose rate varies widely even within the same internal toxicity category, a separate number has been listed where applicable in Table 1, appropriate to the individual nuclide. This level is based on the quantity that could give approximately 1 R/hr at 1 meter under some conditions of geometry. Of course, for materials such as natural uranium and natural thorium, the specific activity is so low in any case that such high external radiation levels are not possible (as a result of self-absorption of the radiation within the material).

Although a hot cell or cave, if provided, would give a certain additional amount of containment against dispersal, this was not factored into consideration in other parts of this table, since the degree of containment is too uncertain and dependent on other facility design parameters. This is one of the many other factors that must remain subject to the judgment of the reviewing industrial hygienist or health physicist in deciding what combination of safeguards and what overall degree of containment must finally be imposed on the particular operations to be conducted.

Of course, for the well-designed glovebox[13] we may assume for our purposes that less than about 10^{-8} of the material within the box will escape from the box on the average as a result of ordinary operations, including small glove leaks. The reviewing health physicist or industrial hygienist must ensure that proper glovebox design and ventilation are provided and that proper procedures for changing gloves, removing materials or equipment, and other activities, will be carried out to meet the expected containment requirements.

ONE ABSOLUTE FILTER (in Exhaust from Active Atmosphere). The micrograms per square meter times factors of 10^{-6} to 10^{-4} gives generally the air concentrations in micrograms per cubic meter above a contaminated ground area where ordinary human activities are conducted.[6,14] Recent work by Jones and Pond[15] and by Stewart[16] has also produced resuspension factors in this range, and Stewart recommends the factor of 10^{-6} as an appropriate average value for use in hazard evaluation both in the laboratory and in the field. Assuming that in a glovebox with finite floor area and controlled air flow a factor of 10^{-6} may be applied, then we might expect that the

$$\text{Air concentration } (\mu\text{Ci/ml}) = 10^{-12} \times \mu\text{Ci/m}^2 \qquad (3)$$

Thus, the number of microcuries at which no absolute filter should be required was taken as the quantity that could, if spread over 1 square meter of the glovebox surface, produce an air concentration averaging about the permissible concentration for occupational exposure.[2] This would require a filtered air exhaust whenever the exhaust air leaving the box may be expected to exceed occupational MPC, which is thus nominally taken as a level of operations at which $10^{12} \times MPC_{air}$ microcuries or more are handled. Additional dilution as the exhaust air flows to the stack, and atmospheric dilution, would thereby ensure that concentrations in controlled areas would be well below maximum permissible levels. Of course, appropriate monitoring would be required, as shown below, to ensure that build-up of significant contamination in the surrounding environment does not occur, and that 10 CFR 20 requirements are met.[17]

The above resuspension factor of 10^{-6} has been compared with the value of 35 $\mu g/m^2$ for permissible ^{239}Pu contamination in controlled areas.[6] It is found that 35 $\mu g/m^2$ divided by 10^{12} gives the occupational MPC of ^{239}Pu in air, so these figures seem to be in good agreement.

For Groups 7 to 1, the values have been scaled upward by factors of 10 each, to take into account the increasing MPC's of these groups.

TWO ABSOLUTE FILTERS IN SERIES (*in Exhaust from Active Atmosphere*). This level was set as one thousand times the quantity requiring one absolute filter in the exhaust stream. Opinions gathered by the author indicate that a reduction in air concentrations by a factor of 1000 may be assumed for each filter over long operating times when the filter is properly tested and installed.[18]

Additional filters beyond two may be required for facilities handling very large quantities of material in one batch.[18] The attenuation factors for each filter beyond the second would have to be based on further information on the specific absolute filters to be used and the particle size characteristics of the material handled.

CONTINUOUS ROOM AIR SAMPLER WITH ALARM. To arrive at a reasonable level at which we should consider requiring a continuous air sampler with alarm, in addition to the routine air sampling and breathing zone sampling program that may be required, the suggestion by the NCRP in Handbook 69[12] that an individual could be exposed if necessary to 1200 times the MPC of a nuclide for one hour, if previous records and monitoring results showed that this would not cause the individual to exceed the quarterly dose limit, was considered. It was also considered that it would be easy to construct a device to detect 1000 MPC (40-hour occupational) of any radionuclide within one hour.

Then, the resuspension factor in the section above was used to calculate the number of microcuries spread per square meter that would produce an air concentration of 1000 MPC. This gave, for ^{239}Pu, a quantity of

$$10^{12} \times 10^3 \times 2 \times 10^{-12} = 2000 \,\mu Ci/m^2, \qquad (4)$$

which was rounded off to 1 mCi, and it was assumed that this quantity of material spread over any area of less than 1 square meter would not be likely to produce the same concentration as this quantity per square meter spread over an infinite plane surface.[16]

It should be noted that this requirement would not take the place of routine monitoring and other safeguards, but would be required primarily for the purpose of detecting unexpected incidents in time to prevent individuals from receiving biologically serious overexposures.

Continuous Exhaust Stack Monitor and Alarm. This item would also be required when appropriate to prevent serious results from unexpected incidents and would not take the place of routine effluent sampling and monitoring, although the same instrumentation could be arranged to serve both purposes in some respects.

The level at which consideration should be given to requiring a stack monitor and alarm, in addition to a routine sampling and monitoring program, was established on the basis that less than 1/10,000 of the material released if a filter should fail would be inhaled by an individual standing at 15 meters from the stack opening (see Figure 2). Multiplying the level of 1 μCi, at which an absolute filter in the exhaust might be required, by 10,000 gives a level of 10 mCi as the minimum amount of ^{239}Pu spread on a 1-square-meter drybox floor that could produce a concentration of the order of MPC at 15 meters from the stack if the filter should fail. Also, if the entire 10 mCi of ^{239}Pu should be released up the stack in a single incident, the inhaled dose of ^{239}Pu to an individual 15 meters from the stack could be of the order of 1 μCi, which would deliver 7100 Rem to the bone over 50 years; however, the individual would be unlikely to inhale this much material unless directly downwind from the stack and at the same height as the center of the cloud of activity. The maximum dose would be 710 Rem at 66 meters, or 71 Rem at 280 meters. These numbers also apply only to extremely unfavorable meteorological conditions, as discussed previously.

The other groups of nuclides were scaled up according to the maximum dose per curie, as before.

BUILDING CONTAINMENT (Airtight). If it is assumed that there is a 1 % release from a drybox and that 1 % of that released from the drybox escapes an ordinary building, then $10^4 \times 10^{-5} = 0.1$ Ci of ^{239}Pu could give a total bone dose of 15 Rem to a person standing 10 meters away (Figure 1). Thus, above this level of operation, the advisability of requiring sealed entrances, and other protective features, should be considered, depending on the type of operations, the site, and other factors. As before, the quantities in the less toxic groups were scaled up according to the relative magnitude of the possible dose per curie (adjusted for specific activity in the case of natural uranium and natural thorium).

SITE RADIUS OF LOW POPULATION ZONE. The proposed radius for the low population zone for each group is given by the equation:

$$X = A_i Q^{2/3} \tag{5}$$

where X is the distance from the point of release in meters; Q is the quantity of the material released $= fC$; C is the number of curies in process within one building at one time; f is the assumed fraction of material in process that is released (for example, 1 for noble gases, 0.5 for halogens and other gases, and 0.01 for nuclides incorporated in materials in solid or liquid form under operating conditions.)[19] For Groups 1 through 8, A_i becomes 0.47, 2.2, 10, 47, 220, 1000, 4700, and 22,000 respectively. This corresponds to the equation giving the distance at which Q curies of the most hazardous radionuclide in the group will give a dose of less than 15 Rem to a person standing directly downwind from the point of release under inversion conditions: $n = 0.5$, $C_z = 0.07$, $C_y = 0.4$; and wind speed $u = 1$ m/sec. The equations are consistent with the diffusion equation specified in 10 CFR 100 and include the assumption that all the radioactive material remains in the cloud as it travels downwind.[19] The value of 15 Rem was selected as the one-year occupational dose limit used by the NCRP-ICRP Committee in its calculations, based on individual body organs other than the skin and thyroid, or whole body.[2] It was used for all the nuclides in the table, since this permissible level is applicable to most of them, and differences in permissible levels for other nuclides are smaller than the uncertainties in dose estimation.

The dose of 15 Rem may be compared to the whole-body dose of 25 Rem and the thyroid dose of 300 Rem used in 10 CFR 100 for defining the low population zone. It is noted that for Groups 2 through 8, the internal dose to the critical organ is the limiting consideration, except for ^{85}Kr, which is limited by the external beta dose. A skin dose of 30 Rem rather than 15 Rem would not change the group to which ^{85}Kr belongs. In the previous version of this chapter,[4] the 15-Rem dose for mixed gross fission products of 180-day irradiation time was obtained by assuming, as in reference 20, that a dose of 25 Rem to the whole body is equivalent to an exposure to 10 Ci/m^3 for one second. Further estimates[23] for various reactor operation and decay times indicate that 1 Ci-sec/m^3 produces a combined internal irradiation from various fission product nuclides that would be approximately equivalent to a 25 R external gamma exposure, within a factor of 2, for 0 to 3 years of operation and decay times of less than 30 days. An exposure of 1 Ci-sec/m^3 would actually deliver an estimated combination of 0.28 R to the whole body, 36 Rem to the lungs over several months, about 40 to 50 Rem to the bones within 50 years, about 7 Rem to the gastrointestinal tract within two days, about 10 Rem to the thyroid within several weeks, and an undetermined amount of radiation to the blood and other organs.[23] Although the relative effectiveness of such a combination of doses is not well known, it would not be expected to differ from the gross effect of a 25 R whole-body exposure by more than an order of magnitude. Since this equivalent dose includes an external dose of 0.28 R, the use of this equivalent automatically limits the external gamma radiation levels, when used to establish safeguards against the release of radioactive material. Since pressure-resistant containment structures are not usually required in nuclear material licensing, the "worst case" was used in developing these criteria by assuming that all the material to be released is released within a short span of time. This makes the dose from cloud passage the important consideration, since there is no fixed gamma radiation source[20] for any appreciable period of time.

PERSONNEL MONITORING AND/OR SHIELDING. The level at which we might wish to require personnel monitoring and/or shielding, rather than rely on the judgment of supervisors alone to control exposure times and other factors, was based on the quantity of material that could produce a dose rate of 10 mR/hr at 1 meter. Lower or higher levels than this might be more appropriate, depending on the experience of operating personnel, the type of operation, and other variables. The level selected here is intended to avoid a dogmatic requirement for personnel monitoring at dose rates low enough that well-trained personnel could maintain appropriate exposure limits by administrative procedures. However, the author actually prefers the philosophy that either shielding or personnel monitoring be provided whenever recordable occupational exposures may be received, unless economic considerations are prohibitive and compliance with permissible exposure limits can be otherwise assured.

OCCASIONAL BIOASSAY SPOT CHECKS FOR OPERATING PERSONNEL. These levels were the same as those requiring glovebox operation, for similar reasons.

CONTINUED ROUTINE BIOASSAY. The level selected at which consideration should be given to requiring routine bioassay (unless containment is considered absolutely foolproof) is set at 10 times the level for occasional bioassay checks. It is also several hundred thousand times the quantity that could give an internal dose of 15 Rem. For ^{239}Pu, the quantity becomes 10 μCi, which is the quantity that, if released in a cloud, could give a dose of about 15 Rem to the bone to a person standing 10 meters away, under the inversion parameters given in Figure 1.

BIOASSAY SAMPLING FREQUENCY. The bioassay sampling frequency depends on the degree of hazard, since in no case would it be desirable to wait so long between samples that serious internal exposures could go undetected. However, an upper limit on the length of time between samples should be set also on the basis that sufficient material will be excreted at the time(s) of sampling so that any preceding intermittent or continuous exposures averaging above MPC can be detected. Thus, the maximum length of time between bioassay samples should be set on the basis that a single intake of the material at the beginning of the sampling period that would lead to an average exposure exceeding MPC should be detectable at the end of the sampling period. To establish this time, both the number of curies to deliver the permissible average dose and the effective biological half-life of the radionuclide in the body must be considered. Suggested sampling intervals[4] have been removed from Table 1, since further information on elimination of inhaled substances is needed.[2,3,10]

A minimum length of time between samples may be based on the upper limit of the dose that could be received in the interval, with consideration given to the other safeguards and emergency procedures that would be established in cases where high potential exposures are possible. Also, some consideration should be given to making a more accurate measurement of repeated acute doses from fractions of the radionuclide that are eliminated with shorter biological half-lives, than of those governing calculation of maximum permissible continuous intake,[2] in situations where these fractions could possibly contribute the major part of the cumulative dose over a long period of time.

Thus, in general, the maximum time interval is limited by the ability to detect the quantity excreted at the end of the interval as the result of an intake at the beginning of the interval that would deliver an average dose during the interval exceeding permissible levels. The minimum time interval would be set, in the most hazardous operations, so that an individual would not be likely (in the event of safeguard failures) to inhale a serious dose of material before the condition is detected.

EMERGENCY DOSIMETER PROGRAM. Where the radionuclide emits gamma or X-rays of sufficient penetrating power to be measured by the ordinary personnel dosimeter, the quantity of nuclide that can produce up to 10 R/hr at 1 meter has been chosen at the level at which the dosimetry should be capable of measuring up to lethal doses of radiation. Special consideration will have to be given to neutron-emitting materials, and to situations where criticality is possible.

ENVIRONMENTAL SAMPLING AND MONITORING. A lost batch of 100 μCi is enough to contaminate about 3 million square feet (or 0.1 square mile) to the minimum level of 60 dpm per square foot detected at Los Alamos after nine years of operation[24] releasing about 1 Ci of ^{239}Pu. The other groups were scaled up by factors of 10 as before, 100 μCi taken as the level at which environmental monitoring be considered for ^{239}Pu. This approach is generally sufficiently conservative from ecological as well as immediate inhalation or ingestion considerations as a result of the usually small fraction of material that would be dispersed either routinely or accidentally.

Of course, the frequency and extent of the environmental monitoring program would depend on many factors, such as the maximum possible quantity released, the stack height, the half-life of the radionuclide and the sensitivity of analytical techniques.

PREPLANNED EMERGENCY PROCEDURES. This is set at the same level as that where a continuous air monitor and alarm are required. The two safeguards are obviously related—when the alarm sounds, personnel should be prepared to act appropriately.

WRITTEN ROUTINE OPERATING INSTRUCTIONS. Written operating instructions for personnel should be required where the potential magnitude of the hazard is sufficient to require sealed building entrances and exitways, unless facilities within the building are obviously so foolproof that no individual inside or outside the building could receive serious exposure even in the event of the maximum credible accident. Thus, the values are the same as those listed under building containment.

WRITTEN PREOPERATIONAL ANALYSIS OF MAXIMUM CREDIBLE ACCIDENTS. The level at which a written analysis of the maximum credible accident is required is taken to be the same as that at which a glovebox is required, since this is the approximate level at which serious overexposures in the event of an accident begin to have a significant probability if appropriate safeguards fail or are not provided. In all applications, of course, the applicant should analyze the potential hazards and indicate safeguards provided to minimize them. The extent and detail with which the additional evaluation of "maximum credible accidents" should be given would of course depend on the number and the degree of hazard of the possible types of accidents.

DISCUSSION

This chapter has attempted to illustrate methods of estimating potential hazards and safeguard requirements from the types and quantities of radioactive material to be handled. No recommendations on specific equipment or industrial hygiene procedures have been included. There are already many good references[7,11,13,14,18,25-29] on appropriate procedures or proper design considerations that can be applied once the need for certain types of procedures or facilities is established.

The guidelines given in Table 1 refer only to those safeguards that are subject to quantitative assessment within an order of magnitude or two. Thus, the table is only an aid to, and cannot be a substitute for, the professional judgment of an industrial hygienist experienced in health physics and in balancing the relative proportions of various safeguards designed into a specific industrial or laboratory facility. Furthermore, an examination of the items included in Table 1 will show that the safeguards listed are neither mutually exclusive nor independent of one another. In many instances, combinations of several of the protective measures in various relative proportions can best be used to achieve a desired level of protection.

The unit for quantity of radioactive material in Table 1 has been the curie rather than the gram, since measurements of radioactivity and quantities shipped to users are usually reported in curies or in other units proportional to the curie. Except for natural uranium (mainly ^{238}U) or natural thorium (^{232}Th), no *a priori* allowance has been made for the specific activity of the material in listing the relative radiotoxicity, as others have done in estimating the relative inherent hazard of the radionuclides.[1,7,11,28,30] For the purposes of this chapter, we have considered that the probability of intake is not likely to be related to specific activity for any radionuclides in which inhaled quantities of only 10 mg or less can produce internal doses of 15 Rem or more. In most industrial or laboratory operations, the specific activity of the radioactive material in process and the probabilities or potentials for intake will depend not on the specific activity of the pure nuclide, but on variables specific to the individual operation—such as the quantities and kinds of chemical reagents with which the radionuclide is mixed, the potential mechanisms of dispersal, and local ventilation characteristics. For all the radionuclides listed in Table 1 except natural uranium and natural thorium, the amount of inhaled nuclide needed to give a serious internal dose is far less than 10 mg and is usually in the submicrogram range (see columns "Specific Activity" and "Relative Radiotoxicity" in Table 1). These mass quantities are so small that regardless of their exact magnitude they could become attached to a single dust particle.[10,26] Thus, the specific radioactivity of a nuclide cannot be included in setting generalized guidelines such as those in Table 1, but it must be considered by the industrial hygienist along with all the other specific operational characteristics of an industrial process in selecting the optimum and most economic combination of industrial hygiene facilities and procedures.

APPENDIX

The values in Figure 1 were obtained by calculating the integrated time-concentration exposures in units of curie-seconds per cubic meter at various distances, X, along the center-line trajectory by using a modified form of Sutton's diffusion equation:[31]

$$(\text{Ci-sec/m}^3) = \frac{2Q_0}{u\pi C_y C_z X^{(2-n)}} \exp\left(\frac{-4VX^{n/2}}{nu\pi^{1/2}C_z}\right) \qquad (A.1)$$

where Q_0 is the number of curies released as a point source at $X = 0$. Conservative parameters representing inversion weather conditions were selected consistent with those given in proposed 10 CFR 100[19]: $n = 0.5$, $C_z = 0.07$, $C_y = 0.40$, and $u = 1$ m/sec wind velocity. The particle settling velocity, V, was taken to be 2×10^{-3} m/sec, corresponding to a particle about 4 to 5 microns in diameter,[31] of density 2.5 g/cm³. Figure 1 shows that the exponential settling term is not an important factor in determining cloud concentrations at distances less than two miles for particles of size smaller than about 4 microns. Dense particles of plutonium would have a higher settling velocity for a given particle size, but the probable smaller size of PuO fumes escaping in an accidental release would compensate for this. Thus, the settling velocity of 2×10^{-3} m/sec is believed to be a reasonably conservative choice for maximizing cloud concentrations.[16,31] (See also reference 16, in which a similar choice has been made.)

After the chosen parameters are inserted, equation A.1 reduces to

$$(\text{Ci-sec/m}^3) = (22.7\ Q_0/X^{1.5}) \exp\left(-0.129\ X^{0.25}\right) \qquad (A.2)$$

By the use of equation A.2, the probable limit to the number of curies inhaled at distance X downwind per curie released at ground level zero point was plotted in Figure 2. This curve was used to obtain the curves for each nuclide in Figure 1 and is also useful in general hazard evaluation work for other nuclides or other modes of exposure than those assumed for Figure 1.

For convenience in estimating the total area or number of people that may be affected by an accidental release of radioactivity, the cloud width out to the points where the concentration is 10 % of that in the center of a Gaussian cloud, the cloud height to 10 % concentration, and the equivalent cloud volume are also plotted in Figure 2, for the case of a cloud at ambient temperature with the center line of the trajectory remaining at ground level. The cloud volume is calculated as the volume of an ellipsoid that would give the same inhaled dose, if it traveled along the same trajectory as the diffusing cloud at a wind speed of 1 m/sec, to a person breathing at the center of the trajectory, under the conditions that the ellipsoid contained the same total activity as the cloud and had a uniform concentration of activity equal to the concentration at the center of the Gaussian cloud. Thus, the curve of cloud volume versus distance gives an easy and conservative estimate of volumes over which the activity may be assumed to be dispersed at various distances and, together with wind speed, may be used to estimate the maximum exposure to individuals at various distances as well as the average exposure and the numbers of persons exposed. The cloud width and cloud height were calculated from the equations for $2y$ and z_0 in reference 31. The cloud volume was calculated from the equation

$$V = (\pi^{1/2} X^{(2-n)/2})^3 C_y^2 C_z \qquad (A.3)$$

Ground contamination levels outside of nuclear facilities, for either accidental or routine release of radioactive material, were estimated from the following equation:[32]

$$S = \frac{2Q_0 V}{u\pi C_y C_z X^{(2-n)}} \exp\left(\frac{-4 V X^{n/2}}{nu\pi^{1/2} C_z}\right) \qquad (A.4)$$

where S is the curies per square meter deposited along the center of the trajectory at distance X meters from the release point of Q_0 curies at ground level, V is the average deposition velocity[16] of the particles in meters per second, u is the wind speed in meters per second, and the other symbols are diffusion parameters that characterize weather conditions, as in equation $A.1$. In Figure 3, the number of curies deposited per square meter per curie released at zero point is plotted as a function of the distance X in meters along the center of the cloud trajectory, for inversion conditions[19] in which $n = 0.5$, $C_y = 0.4$, and $C_z = 0.07$, and the wind speed, u, is only 1 m/sec. Figure 3 tends to overestimate contamination levels in curies per square meter, but the cloud dimensions from similar meteorological conditions in Figure 2 would tend to be correspondingly diminished for any given curie level released. Thus, the area of most immediate concern and the likely upper limits of dose rate or ingestion rate tend to be emphasized by Figures 2 and 3.

The above methods of estimating dispersion and deposition of radioactive materials released outside nuclear facilities have been used to determine base-line protection factors provided by atmospheric dilution in the selection of appropriate safeguard levels in Table 1. Although the values in Figures 1 through 3 tend to overestimate population exposures, they often suffice in evaluating most facility requirements to determine whether various safeguards are unnecessary or required, since the relative hazard potentials often fall either well below or well above the orders of magnitude over which the meteorological factors alone may vary. However, in situations where the balance between safety and economic considerations becomes marginal, the industrial hygienist may be obligated to consider the effects of more representative meteorological conditions, particularly when dealing with methods for controlling long-term, relatively low-level population exposures from routine plant operations. Various methods[31,33-40] have been developed for estimating meteorological dilution factors under different weather conditions, for different source configurations other than a point release, and for clouds of material released above ambient temperature or from elevated stacks. Consideration of these other conditions will generally lead to estimated population exposures lower than those that would be obtained from Figures 1 through 3. Even the single change to neutral or average meteorological conditions[33,36] would give maximum cloud concentrations one-tenth or less of those implied in Figures 1 through 3. When a detailed picture of likely average or single-incident diffusion patterns are desired for emergency planning or hazard evaluation, an experienced meteorologist should be consulted.

For estimating inhalation exposures from accidents within buildings, the Fickian equations for molecular diffusion would in most cases[37] not be applicable, since there would be interference from normal room air currents, exhaust ventilation, "splash" of particles from operations such as cutting or grinding, or explosions and fires that are often associated with accidental releases.[25,26]

However, estimates of the upper limits of possible inhalation exposure are often useful. For example, when immediate alarm and evacuation procedures are in effect, an individual more than a few feet from the point of release would generally not remain for more than a minute in the vicinity of the accident. For an unsuspected instantaneous point release, the maximum quantity of activity inhaled can be estimated by assuming that the released material is immediately dispersed in a hemisphere of radius r feet equal to the distance between the point of release and the breathing zone. If the worker is assumed to remain at a fixed position, then the hemisphere may be assumed to be ventilated by at least a rate $A = 15\pi r^2$ cu ft/min, where a minimum convection velocity of 15 ft/min is adopted as recommended by Hemeon.[25] The concentration of a contaminant introduced at time $t = 0$ into an air space of volume V cubic feet is

$$X = X_0 e^{-At/V} \qquad (A.5)$$

where X is the concentration at time t minutes, X_0 is the initial concentration of the material dispersed instantaneously throughout volume V cubic feet, and A is the rate of ventilation of volume V in cubic feet per minute. Thus, by assuming a maximum breathing rate of 3.5 cu ft/min (100 liters/min) the number of curies, C, inhaled by a worker remaining r feet from a point release would be less than

$$C = 3.5 \int_{t=0}^{\infty} (C_0/V)e^{-22.5t/r}\,dt \qquad (A.6)$$
$$= 0.074\ C_0/r^2$$

where C_0 is the number of curies released instantaneously (less than 1 second), $V = (2/3)\pi r^3$ is the volume of the hemisphere up to the breathing zone, and r is the distance in feet from the point of release to the breathing zone. It is noted that the mean duration of the activity in such a hemisphere, on the assumption of further dilution by natural convection only, would be only $r/22.5$—which is a small fraction of a minute for distances of a few feet. Also, equation $A.6$ shows that someone more than three feet from the point of release would probably inhale less than 1 % of the escaping material. Since the usual air velocities outdoors would be greater than 150 ft/min, an individual more than three feet from an outdoor release would probably inhale less than 1/1000 of the total quantity released (see also Figure 2).

For the continuous release of C_0 curies per minute, the build-up equations given by Hemeon[25] can be integrated in similar fashion to give the inhaled dose. Similar calculations would apply for a worker moving in a large ventilated room in which the radioactive material is introduced, except that the volume, V, would be that of the room and the ventilation rate, Q, in the derivation of equation $A.6$ would be the general ventilation rate of the room.

NOTES TO TABLE 1 (page 484):

*Permissible concentrations are taken from the Code of Federal Regulations, Title 10, Part 20, August, 1966, since licensees of the U. S. Atomic Energy Commission must comply with pertinent requirements of these regulations. Other values under "Radiobiologic Properties" are taken from the report of Subcommittee 11, International Commission on Radiological Protection (ICRP), *Health Physics*, Vol. 3, 1960. In general, the permissible concentrations in the Federal Regulations are based on the MPC's of the ICRP. However, for reasons presented in the regulations there are minor differences (less than about an order of magnitude) for the isotopes ^{90}Sr and ^{232}Th).

†Taken for gross fission product mixtures, separated from reactor fuel less than 30 days after reactor shutdown, as calculated in A. Brodsky, "Criteria for Acute Exposure to Mixed Fission Product Aerosols," *Health Physics*, 11, 1017–1032, 1965.

NOTE: The Curie levels for protection against inhalation and contamination are intended for dry, dusty materials that may be easily dispersed in concentrated form. For liquids, or where active material will be diluted with other materials, the above safeguard levels may be raised by factors of 10 or more. For simple storage of stock solutions, or where operations are conducted only in such a way that no materials of specific activity greater than 0.1 millicurie per gram can be dispersed as an aerosol, the above levels concerned with protecting against intake of loose radioactive materials may be multiplied by 100 or more, depending on the nature of the material and the particular combination of safeguards selected (see text).

The reader should be sure to check values in the table against the most recent recommendations of the NCRP and any local or federal regulations that may pertain to his operation.

ACKNOWLEDGEMENT: This chapter is reprinted with permission from *American Industrial Hygiene Association Journal*, 26, 294 (1965).

REFERENCES

1. Morgan, Karl Z., Snyder, W. S., and Ford, M. R., Relative Hazard of the Various Radioactive Materials. *Health Phys. 10*, 171–182 (March 1964).
2. *Report of Committee II on Permissible Dose for Internal Radiation.* Pergamon Press, London, England (1959), *Health Phys. 3*, 1 (1960).
3. Snyder, W. S., Range of Uncertainty of MPC Values. *In Hearings before the Joint Committee on Atomic Energy*, May 1960, 338–381, U. S. Government Printing Office, Washington, D.C.
4. Brodsky, A., *Determining Industrial Hygiene Requirements for Installations Using Radioactive Materials*, Presented at the 1964 Conference of the American Industrial Hygiene Association in Philadelphia (April 29, 1964).
5. Strominger, D., Hollander, J. M., and Seaborg, G. T., *Table of Isotopes, UCRL–1928* (2nd Rev.); also *Rev. Mod. Phys. 30*, No. 2, Part II, (1958).
6. Brodsky, A., and Beard, G. V., *A Compendium of Information for Use in Controlling Radiation Emergencies*, U. S. At. Energy Comm. Publ. *TID-8206 (Rev.)* 52–100, (September 1960).
7. Fairbairn, A., and Dunning, N. J., The Classification of Radioisotopes for Packaging. In *Regulations for the Safe Transport of Radioactive Materials—Notes on Certain Aspects of the Regulations, Safety Series No. 7*, 25–78, International Atomic Energy Agency, Vienna (1961).
8. Slack, L., and Way, K., *Radiations from Radioactive Atoms in Frequent Use*, U. S. At. Energy Comm. Rept. M-6965, National Research Council (February 1959).
9. Cronkite, E. P., Fliedner, T. M., Killman, S. A., and Rubini J. R., *Tritium in the Physical and Biological Sciences*, 2, 191, International Atomic Energy Agency, Vienna (1962).
10. Hatch, T. F., and Gross, P., *Pulmonary Deposition and Retention of Inhaled Aerosols*, AIHA-AEC Monograph, Academic Press, New York, 9–18, 74–78, 132–135, (1964).
11. *Safe Handling of Radioisotopes, Safety Series No. 1*, 99, International Atomic Energy Agency, Vienna (1958).
12. Maximum Permissible Body Burdens and Maximum Permissible Concentrations of Radionuclides in Air and in Water for Occupational Exposure, Recommendations of the National Committee on Radiation Protection. *Nat. Bur. Std. (U. S.) Handbook 69*, (June 5, 1959).
13. Garden, N. B., *Report on Glove Boxes and Containment Enclosures*, U. S. At. Energy Publ., *TID-16020* (June 20, 1962) (available from Office of Technical Services, Department of Commerce, Washington, D.C.).
14. Bradley, F. J., Nuclear Plant Engineering and Maintenance. *Production Handbook*, 24–55 to 24–80, Gordon B. Carson, Ed., 2nd Ed., The Ronald Press, New York (1958).
15. Jones, I. S., and Pond, S. F., *Some Experiments to Determine the Resuspension Factor of Plutonium from Various Surfaces, AERE-R4635*, Health Physics and Medical Division, UKAEA Research Group, Atomic Energy Research Establishment, Harwell, England (May 1964); also see A. Fairbairn, The Derivation of Maximum Permissible Levels of Radioactive Surface Contamination of Transport Containers and Vehicles. In *Regulations for the Safe Transport of Radioactive Materials—Notes on Certain Aspects of the Regulations, Safety Series No. 7*, 79–82, International Atomic Energy Agency, Vienna (1961).
16. Stewart, K., *The Resuspension of Particulate Material from Surfaces*, p. 15. Presented at the International Symposium on Surface Contamination, Gatlinburg, Tennessee, June 1964. UKAEA, AWRE, Aldermaston, Berkshire, England.
17. *Standards for Protection against Radiation*, Title 10, Code of Federal Regulations, Part 20, effective January 1961, U. S. Atomic Energy Commission, Washington, D.C.
18. *High Efficiency Particulate Air Filter Units*, U. S. At. Energy Comm. Publ. (August 1961) (available from Office of Technical Services, Department of Commerce, Washington, D.C.).
19. *Reactor Site Criteria*, U. S. Atomic Energy Commission Regulation 10 CFR Part 100, 27 Federal Regulation 309 (April 12, 1962).
20. Cowan, F. P., and Kuper, J. B. H., Exposure Criteria for Evaluating the Public Consequences of Catastrophic Accidents in Large Nuclear Plants. *Health Phys. 1*, 76–84 (June 1958); also C. K. Beck et al., *Theoretical Possibilities and Consequences of Major Accidents in Large Nuclear Power Plants*, U. S. At. Energy Comm. Rept. WASH-740 (March 1957).
21. Brodsky, A., *Acceptable Emergency Doses from Fission Product Clouds*. Presented at the Fall Symposium of the Baltimore-Washington Chapter of the Health Physics Society, Bethesda, Maryland (October 12, 1963) (available from Graduate School of Public Health, University of Pittsburgh, Pittsburgh, Pennyslvania).

22. Suzuki, M., A Biological Consideration on Variability of Doses Received by Exposure to Radio-active Aerosol Released in a Hypothetical Big Reactor Accident and Evaluation of Exposure Hazard. *Reactor Safety and Hazard Evaluation, II*, Symposium in Vienna, 14–18 May 1962, International Atomic Energy Agency, Vienna (August 1962).

23. Brodsky, A., Criteria for Acute Exposure to Mixed Fission Product Aerosols. *Health Phys.* (Sept.–Oct., 1965).

24. Jordan, H. S., and Black, R. E., Evaluation of the Air Pollution Problems Resulting from Discharge of a Radioactive Effluent. *Amer. Ind. Hyg. Assoc. J. 19*, 20–25 (1958).

25. Hemeon, W. C. L., *Plant and Process Ventilation*, 2nd Ed., 217–244, The Industrial Press, New York.

26. Drinker, P., and Hatch, T., *Industrial Dust*, 32, 243–257, McGraw-Hill Book Company, New York (1954).

27. Blatz, H., *Radiation Hygiene Handbook*, McGraw-Hill Book Company, New York (1959).

28. *Safe Handling of Radioisotopes—Health Physics Addendum, Safety Series No. 2*, International Atomic Energy Agency, Vienna (1960); see also *Medical Addendum, Safety Series No. 3*, International Atomic Energy Agency, Vienna (1960).

29. Safety Standard for Non-Medical X-Ray and Sealed Gamma-Ray Sources, Part I. General, *Nat. Bur. of Std. (U. S.) Handbook 93* (Jan. 3, 1964); see also Control and Removal of Radioactive Contamination in Laboratories, *Natl. Bur. Std. (U. S.) Handbook 48*, (1951) and other handbooks of this series.

30. Duhamel, F., and Lavie, J. M., La Limitation des Quantités de Substances Radioactives Manipulés en Transportes. *Health Phys. 10*, 453–468 (July 1964).

31. Wexler, H., et al., *Meteorology and Atomic Energy, AECU 3066*, U. S. Weather Bureau-USAEC Document (July 1955) (available from U. S. Government Printing Office, Washington, D.C.).

32. Wexler, H. et al., ibid, p. 93.

33. Pack, D. H., Meteorological Aspects of Nuclear Emergencies. *Reactor Safety and Hazards Evaluation*, 9–17, a Training Course Manual published by the Division of Radiological Health, Public Health Service, R. A. Taft Sanitary Engineering Center, Cincinnati 26, Ohio (February 1963).

34. Gifford, F. G., Use of Routine Meteorological Observations for Estimating Atmospheric Dispersion. *Nucl. Safety 2, No. 4*, 47–51 (1961).

35. Cramer, H. E., Engineering Estimates of Atmospheric Dispersal Capacity. *Amer. Ind. Hyg. Assoc. J. 20*, 183–189 (1959).

36. Pasquill, F., "Atmospheric Diffusion", D. Van Nostrand Co., Ltd., London, England, 192–198, 204–213 263–270 (1962).

37. Sutton, O. G., "Micrometeorology," McGraw-Hill Book Company, New York, 133–140, 273–288 (1953).

38. Smith, M. E., and Singer, I. A., Diffusion and Deposition in Relation to Reactor Safety Problems. *Amer. Ind. Hyg. Assoc. Quart., 18*, 4, 319–330 (December 1957).

39. Healy, J. W., *Calculations of Environmental Consequences of Reactor Accidents*, Report HW-54128, General Electric Co., Hanford, Washington (Dec. 11, 1957).

40. Fitzgerald, J. J., Hurwitz, Jr., H., and Tonks, L., *Method for Evaluating Radiation Hazards from a Nuclear Incident*, Report KAPL-1045, Knolls Atomic Power Laboratory, Schenectady, N.Y. (Mar. 26, 1954).

International Organizations
Producing Nuclear Standards

William B. Cottrell

Early in 1963 Section Committee N6, Reactor Safety Standards of the American Standards Association (now the American National Standards Institute) established Subcommittee N6.9 to maintain an indexed catalog of standards of interest to the nuclear industry for the U.S.A. and other countries. In order to fulfill its designated functions, Subcommittee N6.9 first compiled and published the compilation of U.S.A. nuclear standards, and next undertook a survey of nuclear standards of other countries. This chapter reprints part of the fifth edition of the Compilation of National and International Nuclear Standards (ORNL-NSIC-63); the fourth edition of the Compilation of U.S.A. Nuclear Standards was published in 1967 as ORNL-NSIC-43.

Report ORNL-NSIC-63 contains a description of the organizations involved. Readers interested in additional descriptive information are referred to the bibliography of this report and to the "Selected Bibliography of Radiation Protection Organizations" published in 1963 by the American Conference of Governmental Industrial Hygienists. An additional reference is the "Atomic Handbook," edited by John W. Shurtall, published by Morgan Brothers, Ltd., London.

While committee members have undertaken to contact all organizations and committees actively engaged in the generation of nuclear standards, the nature of the work is such that information is soon outdated. Therefore, the subcommittee cannot guarantee the complete accuracy of the report.

International Organization	Standard Organization	Informant, Title, Address
1. Bureau Veritas	Maritime Technical Committee; Subcommittee for Nuclear Energy	P. Weiss, Secretary Bureau Veritas 31 rue Henri Rochefort Paris 17, France
2. CERN (European Organization for Nuclear Research)		Prof. Victor F. Weisskopf M.I.T. Cambridge, Massachusetts 02139 United States of America
		Prof. B. Gregory Director General CERN 1211 Meyrin Geneva 23, Switzerland
3. EURATOM (European Atomic Energy Community)		M. Pierre Chatenet, President European Atomic Energy Community 51 rue Belliard Brussels, Belgium
		M. Jean Rey, President Commission of the European Communities 24, av. de la Joyeuse Entree Brussels 4, Belgium

International Organization	Standard Organization	Informant, Title, Address
4. IMCO (Inter-governmental Maritime Consultative Organization)		T. Busha, Head External Relations and Legal Matters Section Intergovernmental Maritime Consultative Organization 22 Berners Street London, W.1., England
5. IAEA (International Atomic Energy Agency)		Hon. Sigvard Eklund, Director General International Atomic Energy Agency Kärntnerring 11 Vienna 1, Austria
6. ICRP (International Commission on Radiological Protection)		E. Eric Pochin, Chairman International Commission on Radiological Protection University College Hosp. Med. School University Street London, W.C.1, England
		F. D. Sowby, Scientific Secretary International Commission on Radiological Protection Clifton Avenue Sutton, Surrey, England
	A. Committee 1. Radiation Effects	H. G. Newcombe, Head Biology Branch Atomic Energy of Canada Ltd. Chalk River, Ontario, Canada
	B. Committee 2. Internal Exposure	K. Z. Morgan, Director Health Physics Division Oak Ridge National Laboratory P.O. Box X Oak Ridge, Tennessee 37830
	C. Committee 3. External Exposure	Bo Lindell, Director Statens Strålskyddsinstitut Karolinska ajukhuset Stockholm 60, Sweden
	D. Committee 4. Application of Recommendations	H. Jammet, Chef du Department de la Protection Sanitaire au Commissariat a l'Energie Atomique B.P. No. 6 Fontenay-aux-Roses (Seine), France
	E. Committee 5. Radioactive Waste Handling and Disposal	
	F. Committee 6. Committee on RBE	
	G. Joint ICRP and ICRU Study	
7. ICRU (International Commission on Radiation Units and Measurements)		Dr. L. S. Taylor, Chairman International Commission on Radiation Units and Measurements Suite 402 4201 Connecticut Avenue, N.W. Washington, D.C. 20008

International Organization	Standard Organization	Informant, Title, Address
7. ICRU (International Commission on Radiation Units and Measurements), *continued*	IA. Fundamental Physical Parameters and Measurement Techniques	Dr. F. W. Spiers, Chairman
	IB. Medical and Biological Applications	Dr. H. Vetter, Chairman Austria
	IIB. X-rays, Gamma rays and Electrons	Dr. W. Pohlit, Chairman Germany
	IIC. Heavy Particles	Dr. W. K. Sinclair, Chairman
	IID. Medical and Biological Applications (Therapy)	Dr. W. J. Meredith, Chairman
	IIE. Medical and Biological Applications (Diagnosis)	Dr. Olle Olsson, Chairman
	IIF. Neutron Fluence and Kerma	Dr. A. H. W. Aten, Jr., Chairman
	IIIA. Protection Instrumentation and Its Application	Dr. L. Larsson, Chairman
8. IEC (International Electrotechnical Commission)		L. Ruppert, General Secretary International Electrotechnical Commission 1 rue de Verembe Geneva, Switzerland
	A. Technical Committee No. 45, Measuring Instruments Used in Connection with Ionizing Radiation	A. Rys, Chairman IEC Technical Committee No. 45 Inspection Generale, CEA 29 rue de la Federation Paris 15, France
	Working Group No. 1, Classification and Terminology	A. Rys (As above)
	Working Group No. 2, Safety of Radiation Instruments	S. J. Dagg, Chairman Central Electricity Generating Board 20 Newgate Street, London, E.C.1, England
	Working Group No. 3, Interchangeability	W. Böhme, Chairman Siemens A.G. Rheinbrückenstrasse 50 75 Karlsruhe, Germany
	Working Group No. 4, Reactor Instrumentation	
	Working Group No. 5, Measuring Instruments Used in Prospecting and Mining	P. Fabre, D.P.R.M. Direction of Productions, CEA B.P. No. 4 92-Chatillon sous Bagneux France
	Working Group No. 6, Electrical Measuring Instruments Using Sealed Radiation Sources	M. A. Goldmann, Chairman Comitato Nazionale Energie Nucleare C.N.E.N. Via Belisario, 15 Roma, Italy

International Organization	Standard Organization	Informant, Title, Address
8. IEC (International Electrotechnical Commission), *continued*	Working Group No. 7, Testing Methods	M. L. Strackee Hoofd. Afd. Radioactiviteit Rijksinsituut voor de Volksgezondheit Sterrenbos 1 Utrecht, Netherlands
	Working Group No. 8, Health Physics Instruments	
	Working Group No. 9, Radiation Detectors	L. Costrell, Chairman National Bureau of Standards Washington, D.C. 20234
	1. Subcommittee No. 45A Reactor Instrumentation	S. H. Hanauer, Chairman University of Tennessee Knoxville, Tennessee 37916
	Working Group A1, Reactor Instrumentation Principles	J. L. Petrie, Secretariat Walden House 24 Cathedral Place London, E.C.4, England
	Working Group A2, Reactor Instruments	J. Furet, D.E.G. CEA, Saclay–BP. No. 2 91 Gif-sur-Yvette, France
	2. Subcommittee No. 45B, Health Physics Instrumentation	Prof. Brunello Rispoli, Chairman Via Belisario 15 Roma, Italy
	B. Technical Committee No. 62, Medical X-ray Equipment	Secretariat: Germany
9. ISO (International Standards Organization)		Olle Sturen, Secretary-General International Standards Organization 1 rue de Verembe 1211 Geneva 20, Switzerland
	A. Technical Committee No. 85, Nuclear Energy	Secretariat U.S. of America Standards Institute 10 East 40th Street New York, N.Y. 10016
	Subcommittee No. 1, Terminology, Definitions, Units, and Symbols	Secretariat U.S. of America Standards Institute 10 East 40th Street New York, N.Y. 10016
	Subcommittee No. 2, Radiation Protection	M. R. Frontard, Director General Association Francaise de Normalisation 23 rue Notre-Dame des Victoires 75, Paris 2e, France
	Subcommittee No. 3, Reactor Safety	H. A. R. Binney, C.B., Director British Standards Institution 2 Park Street, London, W.1, England
	Working Group No. 1, Siting	
	Working Group No. 2, Meteorological Aspects	

International Organization	Standard Organization	Informant, Title, Address
9. ISO (International Standards Organization), *continued*	Working Group No. 3, Containment Structures	L. P. Zick Chicago Bridge and Iron Co. 901 West 22nd Street Oak Brook, Ill. 60521
	Working Group No. 4, Effects of Nuclear Radiation in Steels	
	Working Group No. 5, Criticality Safety	Secretariat: United Kingdom
	Working Group No. 6, Nuclear Reactor Pressure Vessels	G. Wiesenach, Chairman Germany
	Subcommittee No. 4, Radioisotopes	l'Ing. J. Wodzicki, President Polski Komitet Normalizacyjny U1, Swietokrzyska 14 Warsaw 51, Poland
	B. Technical Committee No. 12, Quantities, Units, Symbols, Conversion Factors and Conversion Proposals	Secretariat: Denmark Dansk Standardiseringsraad Aurehøjvej 12 Kobenhavn Hellerup, Denmark
10. Lloyds		The Secretary Lloyds Register of Shipping 71 Fenchurch Street London, E.C.3, England
11. OECD (Organization for Economic Co-operation and Development); European Nuclear Energy Agency (ENEA)	Committee on Reactor Safety Technology	Mr. F. R. Farmer, Chairman Safety Division United Kingdom Atomic Energy Authority Risley, Warrington, Lancs. England
		Henri B. Smets, Secretariat Committee on Reactor Safety Technology 38 Boulevard Suchet Paris 16, France
12. United Nations		U Thant, Secretary-General United Nations United Nations Building New York, New York 10017
	A. Food and Agriculture Organization (FAO)	G. Wortley, Head Pesticide Residues and Food Protection Section Joint FAO-IAEA Division of Atomic Energy in Agriculture IAEA, Kärtner Ring 11 Vienna, Austria

International Organization	Standard Organization	Informant, Title, Address
12. United Nations, *continued*	B. International Labor Organization (ILO)	Dr. Luigi Parmeggiani, Chief Occupational Safety and Health Branch Conditions of Work and Life Department International Labor Organization CH 1211 Geneva 22, Switzerland
	C. United Nations Scientific Committee on the Effects of Atomic Radiation (UNSCEAR)	Francesco Sella, Secretary United Nations United Nations Building New York, New York 10017
	D. World Health Organization (WHO)	Dr. W. H. P. Seelentag Chief Medical Officer Radiation Health World Health Organization Avenue Appia, 1211 Geneva 10, Switzerland
	E. World Meteorological Organization (WMO)	Dr. K. Langlo, Chief Technical Division World Meteorological Organization Palais des Nations Geneva, Switzerland

ACKNOWLEDGEMENT: This chapter is reprinted from "Compilation of National and International Nuclear Standards" by William B. Cottrell, Chairman of USA Standards Institute Subcommittee, N6.9, which includes W. B. Allred, J. P. Blakely, D. Davis, J. B. Godel, J. E. McEwen, Jr., M. Novick, C. Roderick, and J. C. Russ.

LEGAL NOTICE

This report was prepared as an account of Government sponsored work. Neither the United States, nor the Commission, nor any person acting on behalf of the Commission:

A. Makes any warranty or representation, expressed or implied, with respect to the accuracy, completeness, or usefulness of the information contained in this report, or that the use of any information, apparatus, method, or process disclosed in this report may not infringe privately owned rights; or

B. Assumes any liabilities with respect to the use of, or for damages resulting from the use of any information, apparatus, method, or process disclosed in this report.

As used in the above, "person acting on behalf of the Commission" includes any employee or contractor of the Commission, or employee of such contractor, to the extent that such employee or contractor of the Commission, or employee of such contractor prepares, disseminates, or provides access to, any information pursuant to his employment or contract with the Commission, or his employment with such contractor.

Electrical and Mechanical Hazards

Protective Lockout and Tagging of Equipment

Gari T. Gatwood

The development of a technique that can protect an individual working on exposed electrical circuits, open process lines, machinery or other potentially hazardous tasks has been a difficult assignment for many safety engineers. Such a technique is particularly difficult to develop for use in laboratories where management, supervision and acceptance of procedures are very limited.

The objectives of protective lockout and tagging are two-fold and the differences are very important. First, there should be a procedure which provides an assurance for a person who intends to work on a piece of equipment that no one can energize or activate that piece of equipment while he is working on it. This is accomplished by the use of a personal padlock attached to the control center, main valve or power source. Secondly, there needs to be a procedure for indicating that a piece of equipment is out of service and could be dangerous if activated. This indication should be done with a tag attached at the same location as the padlock. In the first case (lockout), if the equipment is activated a person *will be* injured, while in the second case (tagging), a person *may be* injured.

In industry this problem is most frequently handled by: (1) providing the capability of padlocking switches and valves in the "off" position; (2) establishing a lockout procedure; (3) educating the personnel to proper use of the procedure; and (4) enforcing the procedure.

Much of the equipment used in laboratories does not lend itself to the use of padlocks, and to modify it adequately would either be very costly or virtually impossible. Laboratory personnel, particularly in research and university laboratories, are not inclined to use restrictive procedures or close supervision. As a result, lockout and tagging procedures become difficult to establish.

The following procedure was developed with all of these things in mind and is successfully being used in a large university laboratory. In this case, special tapes were designed to serve as locks where padlocks could not be used; strong emphasis was placed on education; and the laboratory management was convinced of the importance of the procedure.

This procedure can be adapted for use in most laboratories where management truly recognizes the need. Most of the equipment—padlocks, lockout devices, and tags—is commercially available. However, the tapes will have to be made up on a special order. The author prefers a cloth tape for strength.

PROTECTIVE LOCKOUT AND TAGGING

Purpose

1. This procedure provides a uniform method for tagging and locking out machinery or equipment that is being worked on or is otherwise unsafe. It will prevent the possibility of setting moving parts in motion, energizing electrical circuits or opening valves while work is being performed. All personnel are required to comply with these procedures.

2. The requirements of this procedure shall apply to circuit breakers, switches or other power source controls; air or hydraulic valves controlling the operations of equipment; and valves controlling the flow of liquids or gases.

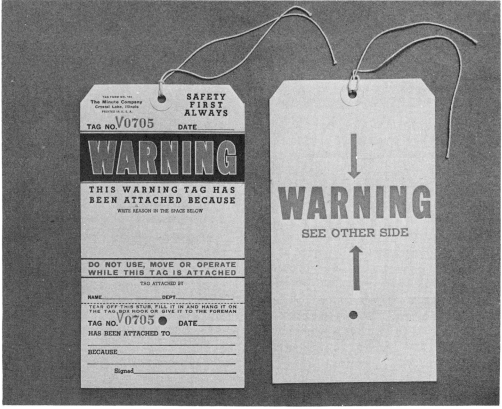

Fig. 1. The two-piece warning tag.

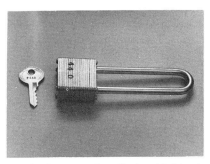

Fig. 2. A personal padlock with identifying number.

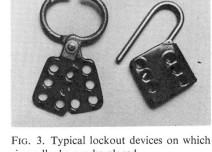

Fig. 3. Typical lockout devices on which six padlocks can be placed.

Fig. 4. A piece of adhesive cloth lockout tape.

Fig. 5. A roll of 2500 cloth lockout tapes.

Definitions

TAGS.   For the purpose of this procedure, tags are two-piece cards with strings to be attached to a piece of equipment being held out of service. A tag is used to identify a piece of equipment that, if activated, could cause injury or damage.

LOCK.   A lock is an issued personal padlock identified by a number and used to physically prevent the operation of a control, switch, valve or other equipment while being worked on. Each personal lock has only one key, which is kept on the person of the user.

LOCKOUT DEVICES.   Lockout devices are mechanical devices that can be held closed by attaching a lock through any of six pairs of lockout holes. A lockout device enables from one to six men to individually lock out a switch, valve, or other properly fitted piece of equipment.

LOCKOUT TAPE.   Lockout tape is a specially-printed adhesive cloth tape used to hold a personal lock in place. This tape is to be used only in those cases where a personal lock *cannot* be directly attached.

Responsibility

DIVISION HEADS.   Each Division Head is responsible for the following:

1. Maintaining a supply of tags, lockout devices, lockout tape, and personal padlocks at his shop or office.
2. Instructing each man under his supervision in the procedural use of tags, personal padlocks, lockout tape, and lockout devices.
3. Supervising the procedural use of tags, personal padlocks, lockout tape, and lockout devices.
4. Maintaining a register of tag stubs in current use and a record of issued personal locks.

FIG. 6. Proper application of lockout device, padlock and tag to secure main electrical power circuit.

FIG. 7. Lockout tape as used to lock out a low pressure gas system.

SAFETY OFFICE. The Safety Office is responsible for the following:

1. Maintaining a supply of tags, personal locks, lockout devices, and lockout tape.
2. Instructing supervisors and experimenters in the procedural use of tags, personal locks, lockout devices, and lockout tape.

ALL USERS OF TAGS AND LOCKOUT EQUIPMENT.

1. Each person normally engaged in maintenance, installation, or operation of equipment requiring tagging and lockouts must have in his possession or have easily accessible to him the following:
   (a) tags
   (b) personal padlock with key
   (c) lockout device
   (d) lockout tapes
2. Each person is responsible for the proper use of his lockout equipment.
3. Each man attaching his personal lock to a device must keep the key to the lock on his person.

Procedure

1. When a piece of equipment is to be held out of service, a tag must be applied. Before work is performed on the equipment, each person involved shall attach his own tag and personal lock. If a personal padlock cannot be used directly, it must be attached with the lockout tape to the switch or control.

2. The tag must be signed by the person attaching it together with the reason for its use and the names of additional people authorized to remove it. Only persons whose names appear on the tag are authorized to remove the tag.

FIG. 8. The lockout tape and padlock hold the circuit breaker in the "off" condition.

FIG. 9. Chain and padlock insures "off" position of water line.

FIG. 10. A process line which has been chained and padlocked.

3. If a tag must be removed while the named persons are not available, only the Division Head of the division applying the tag jointly with the person responsible for the laboratory may remove the tag. They may do this only after a complete examination of the equipment and an attempt to consult with all of the individuals named on the tag. Such removal is to be reported to the Safety Office immediately in writing by the persons removing the tag.

4. In addition to tags, personal padlocks or lockout tapes are to be used for the protection of the individual working on the equipment. Under all circumstances these padlocks or tapes have to be removed by him at the end of the project or working day, whichever comes first. Only he is authorized to remove his own personal padlock or tape.

5. If a personal padlock or tape of an absent person must be removed, this can only be done after receiving specific authorization from the Laboratory Director.

6. Whenever a tag is used, its bottom half will be filled in and returned to the issuing supervisors. Tags and lockout equipment are available through supervisors to all laboratory personnel.

Violations

1. The prime purpose of this procedure is to protect people when working on equipment; secondly, to prevent operation of dangerous equipment. In order for a man to have confidence that a personal padlock or lockout tape will protect him, these procedures must not be violated.

2. The following violations are considered a serious offense that must unequivocally be reported to the Laboratory Director:

(a) Operation of equipment in violation of a tag or lock.

(b) Unauthorized removal of a tag.

(c) Unauthorized removal of a personal lock, lockout device, or lockout tape.

ACKNOWLEDGEMENT: This chapter is copyrighted by Gari T. Gatwood and is used with special permission.

FIG. 11. A chemical process line locked out with the use of lockout tapes.

FIG. 12. Use of lockout tapes on switch panel.

Grounding Electronic Equipment

Electronic Industries Association
2001-I Street, N.W. Washington, D.C. 20006

The information in this chapter is reprinted with permission from a safety data sheet published in 1966 by the Electronic Industries Association, Safety and Health Committee, Industrial Relations Department.

The introduction explains that EIA safety data sheets are designed to serve the interest of its member companies by making users more fully aware of the potential accident and health hazards associated with various materials, equipment and processes and offering suggested precautions to follow when working with them.

Further qualifications given in the safety data sheet are that the information and recommendations contained in the safety data sheet do not constitute an official code or standard. They have been compiled from sources believed to be reliable and to represent the best current opinions on the subject, and there is no warranty expressed or implied with reference to the accuracy or completeness of the safety data sheet, or the results to be obtained from the use of the data contained therein. Also, it should not be assumed that all acceptable safety measures are contained in the safety data sheet, and the Electronic Industries Association assumes no responsibility in connection therein.

SCOPE

This chapter outlines the recommended minimum safety and fire protection requirements for grounding of electronic equipment, but does not include grounding requirements for lightning protection or for suppression of static electricity accumulation.

DEFINITIONS

GROUNDING. Grounding is the act of connecting circuits or equipment to the earth or a common ground plane (aerospace vehicle frame) through a continuous low impedance electrical path.

ELECTRONIC EQUIPMENT. Packaged electrical devices, fixed or portable, for the generation, amplification and control of electron flow, as distinguished from electrical devices primarily engaged in the generation, transformation, transmission or utilization of electrical power.

EQUIPMENT GROUND. The electrical connection of non-current-carrying metal parts of equipment to ground.

GROUND. A conducting connection, whether intentional or accidental, between an electrical circuit or equipment and earth, or to some conducting body which serves in place of earth.

GROUNDED. Grounded means connected to earth or to some conducting body which serves in place of earth.

GROUNDED CONDUCTOR. A conductor which is intentionally grounded, either solidly or through a current limiting device.

GROUNDING CONDUCTOR. A conductor used to connect an equipment, device or wiring system with a grounding electrode or electrodes.

GROUNDING SYSTEM. The complete electrical connection including leads, attachments and hardware necessary for a functional ground.

PORTABLE EQUIPMENT. Equipment which is actually moved or can easily be moved from one place to another in normal use.

HAZARDS

Lack of or improper use of electrical grounding systems not provided with double insulation can result in:
A. Fatal or non-fatal electrical shock, burns, and secondary injuries to personnel.
B. Fire in electronic equipment and facilities.
C. Possible faulty operation of electronic equipment by generation of spurious signals.
D. Disruption of operations.

SAFETY CRITERIA

EQUIPMENT GROUND—electrically connect and ground metallic non-current-carrying parts enclosing electronic equipment in the same area to prevent differences of potential and, in turn, eliminate hazards.

EARTH CONNECTIONS—(connections of electronic grounding circuits to the earth)
1. Consider soil moisture and temperature when designing a connection. The temperature has a direct effect on the moisture. Soil temperatures below freezing or hot enough to dry the soil cause the connection resistance to increase. Soils with low moisture content (less than 15–20 percent by weight) are very poor electrical conductors.
2. Extend connection electrodes into the soil far enough to eliminate the above variables. One or a few electrodes extending far into the soil are better than many short electrodes near the surface.
3. Make the ground resistance connection low enough to blow the fuse or trip the circuit breaker protecting the circuit; in no case should the connection exceed 25 ohms.
4. Use conductors that are corrosion resistant and chemically compatible with the soil in which they are buried.
5. Make contact surface area between the connection parts and earth sufficiently large to prevent heating from high current densities during high current flow faults.
6. Use parts that are strong enough to resist mechanical damage during installation and earth movements. Digging into a part with a hand shovel should not damage the connection.
7. Check utility piping systems used for earth connections to assure that the system is a continuous metallic connection, without non-metallic couplings and joints which would prevent current flow.

Electronic equipment designed and manufactured in accordance with standards of Underwriters' Laboratories and similar agencies is generally provided with adequate grounding means or fully insulated for safety.

Design and Construct Electronic Equipment as Follows

1. Bond non-current-carrying metal parts of electronic equipment together and provide means for grounding.
   a. Where signal circuits permit, bond chassis together and connect to ground.
   b. Bond component parts to chassis.

      c. Bond exposed non-current-carrying metal racks and cabinets together and ground.

2. Isolate or insulate internal electrical circuits from the equipment enclosure when the circuits operate at:

      a. Potentials of 30 volts or more, and at

      b. Frequencies of 50 kilocycles or less.

3. External circuits need not be isolated from the equipment enclosure provided the circuit is capable of operating only outside the above limits, i.e., more than 30 volts and less than 50 kilocycles.

4. The recommended minimum safe isolation impedance between an isolated circuit and equipment enclosure is 400 ohms per maximum circuit rating volt. It is suggested that multiple or general purpose test equipment load (not energy source) circuit terminals be labeled with the designed maximum permitted potentials.

5. A placard of non-conductive material is recommended for multiple purpose test equipment where external connections are individually made through terminals or jacks. The colors of the placard should be in accordance with Standards of American National Standards Institute ("Specifications for Industrial Accident Prevention Signs" USASI Z35.1).

6. Isolated circuits should not be capable of delivering more than 2.5 (two point five) milliamperes of current from the circuit to the equipment enclosure when circuit grounds are disconnected.

7. The above requirements also apply to equipment subassemblies or removable drawers. Circuits internal to these assemblies need not be isolated.

8. Avoid the use of grounding circuits or equipment enclosures in place of internal circuit conductors wherever possible. Do not use grounding circuits or equipment enclosures as a substitute for external circuit conductors. Intentional current flow through the grounding system may disrupt the purpose of the grounding system; that is, to eliminate the differences of potential.

9. Ground isolated circuits by means accessible and disconnectable from outside the equipment enclosure to permit disconnecting for testing and special applications.

10. The National Electrical Code should be used to color all external power carrying conductors supplying electronic equipment. Neutral conductor should be white. The mating internal connections of electronic equipment to an outside power source should be color coded to match.

Equipment Enclosure Grounding Methods

1. Ground equipment with an electrical conductor or equivalent metallic wiring enclosure, as outlined in the National Electrical Code. Size the grounding conductor so that it will carry the fault current without overheating. The equipment ground conductor should be green. The mating internal connection should be color coded to match.

2. Place the grounding conductor for portable equipment in the same cable as, but electrically separate from, the equipment incoming power circuit conductors.

3. Connect equipment to the same ground as that used for the equipment power source.

4. Use receptacles, plugs, and connectors having an identified grounding circuit terminal for portable equipment grounding. Use the grounding terminal only for attaching the grounding conductor.

5. Plugs which disconnect external power cords containing the equipment ground conductor should be compatible with power outlet receptacles and should be listed by Underwriters' Laboratories, or other recognized testing organizations.

CONTROL OF HAZARDS BY PROCEDURE

Facility Grounding Systems

1. Inspect and test facility grounding systems intended for grounding electronic equipment after completion of construction, before use, and periodically thereafter for compliance with requirements. Test intervals should be short enough to check for variations caused by weather conditions.
2. Visually inspect wire joints for loose connections, conductor damage, and corrosion.
3. Test grounding systems, using a ground tester. Verify the integrity of earth ground at the utilization locations.

Electronic Equipment

1. Inspect equipment ground connections before using. Where plug terminals are visible on an assembled plug, make sure the identified grounding terminal has the grounding conductor attached.
2. Test equipment for grounding provisions. The test should indicate that the metallic non-current-carrying parts of the equipment are grounded.
3. Test equipment external circuits for compliance with safety criteria. NOTE: Disconnect circuit grounding means before performing tests. Connect meter directly between the circuit under test and the equipment enclosure.
4. Identify equipment failing tests with a "DANGER—DO NOT USE OR OPERATE" tag. The color of the tag should be in accordance with ANSI Standards ("Specifications for Industrial Accident Prevention Signs," USASI Z35.1).

NOTE: Figures in the data sheet not included here are: 1. Electronic Equipment Ground; 2. Electronic Equipment Grounding; 3. Earth Connections; 4. Equipment Circuits; 5. Equipment Maximum Potential Warning Placard; 6. Preferred Grounding of Internal and External Circuits; 7. Testing for Equipment Ground; 8. Testing for Hazardous Circuit Connections to Equipment Cabinet.

Appendices not included here are: I, Safety Criteria, Ground Resistance Test Methods: Triangulation Method; II, Safety Criteria, Ground Resistance Test Methods: Fall-of-Potential Method; III, Safety Criteria, Ground Resistance Test Methods: Ratio Method.

REFERENCES

1. National Electrical Code, *NFPA No. 70 (ASA C1)*, National Fire Protection Association International.
2. Electronic Computer Systems, *NFPA No. 75*, National Fire Protection Association International.
3. Dimensions of Caps, Plugs and Receptacles, *USASI C73*, American National Standards Institute.
4. Specifications of Industrial Accident Prevention Signs, *USASI Z35*.1, American National Standards Institute.
5. Grounding of Portable Electric Equipment, *Data Sheet D-299*, National Safety Council.
6. Accident Prevention Handbook—Ground Safety, *Para. 0412.3 Electrical Machines, AFM 127-101*, Department of the Air Force.
7. Data Processing Equipment, *Electronic, Electrical Appliance and Utilization Equipment List*, Underwriters' Laboratories, Inc.
8. Office Appliances and Business Equipment, *UL 114, Standards for Safety*, Underwriters' Laboratories, Inc.
9. Grounding and Bonding Equipment, *UL 467 (ASA C33.8), Standards for Safety*, Underwriters' Laboratories, Inc.
10. Attachment Plugs and Receptacles, *UL 498, Standards for Safety*, Underwriters' Laboratories, Inc.

11. General Purpose Wiring Devices, *Publication No. WD 1*, National Electrical Manufacturers Association.

12. Electrical Grounding, *Industrial Hazards Control Bulletin No. S-13*, The Boeing Company.

13. Basic Electrical Standards for Industrial Equipment (Formerly JIC Electrical Standards), *Manufacturing Standards*, General Motors Corporation.

14. Standards for Electrical-Electronic Test Equipment, *Industrial Hygiene and Safety Pamphlet No. 15*, North American Aviation.

15. General Specification-Support Equipment, *DSC-SP-1000*, Defense and Space Center, Westinghouse Electric Corporation.

16. Standard Test Equipment Grounding—Operational Procedure, *Specification No. A-257*, Westinghouse Electric Corporation.

17. Grounding of Cord-connected Electric Appliances, *No. Z-87*, *Special Hazards Bulletin*, Association of Casualty and Surety Companies.

Deleterious Effects of Electric Shock

Charles F. Dalziel

A survey of the effects of electric current on man as the shock current is increased in magnitude may be helpful in focusing attention on the several hazards inherent in the use of electricity. Such knowledge is of assistance in establishing reasonable limits for the allowable leakage from insulation in the design of appliances and hand tools. Quantitative knowledge of maximum tolerable currents is essential for the proper design of grounding mats for electric power stations and for fixing the maximum output from devices having exposed electrodes. Such information is helpful in the formation and explanation of safety codes, and in analyzing accidents. The information may also be of value to the physician in explaining what may have happened to his patient. Such knowledge is important in stressing the necessity of safe rescue and quick resuscitation, and is the basis for the development of electric defibrillators for victims suffering from ventricular fibrillation. Appreciation of the deleterious effects of electric shock is necessary for maintaining vigilance in matters pertaining to electrical safety, and for educating people in the safe use of electrical appliances and equipment.

Man is very sensitive to electric current because of his highly developed nervous system. Although minute electric shocks are generally considered annoying and objectionable rather than harmful, such shocks constitute an ominous warning of the presence of potentially hazardous conditions. The device or appliance in question should be disconnected immediately and the cause ascertained by a person competent in such matters.

Upon increasing the alternating current, using the condition of copper wires held in the hands, the sensations of tingling give way to contractions of the muscles. The muscular contractions and accompanying sensations of heat increase as the current is increased. Sensations of pain develop and voluntary control of the muscles that lie in the current pathway becomes increasingly difficult. Finally a value of current is reached for which the subject cannot release his grasp of the conductor. The maximum current a person can tolerate when holding an electrode in the hand and still let go of the energised conductor by using the muscles directly stimulated by that current is called his "let-go current." Let-go currents are important as experience has shown that an individual can withstand, with no ill after-effects, repeated exposure to his let-go current for at least the time required for him to release the conductor. It is noted that the current flowing between the hands is sufficient to affect many muscles of the body. Currents only slightly in excess of one's let-go current are said to "freeze" the victim to the circuit. Such currents are very painful, frightening and hard to endure for even a short time. Failure to interrupt the current promptly is accompanied by a rapid decrease in muscular strength due to the pain and fatigue associated with the accompanying severe involuntary muscular contractions, and it would be expected that the let-go ability would decrease rapidly with duration of contact. Prolonged exposure to currents only slightly in excess of a person's let-go limit may produce exhaustion, asphyxia, collapse and unconsciousness followed by death.

In tests, 134 men and 28 women subjects held and then released a test electrode consisting of a number 6 or 8 copper wire. The circuit was completed by holding the other hand on a flat brass plate or by clamping a conducting band wrapped with saline soaked cloth on the upper arm. After one or two preliminary trials to accustom the subject to the sensations and muscular contractions produced by commercial alternating current (60 Hertz), the current was increased to a certain value and the subject was commanded to let go of the wire. If he succeeded, the test was repeated at a current of slightly higher value. If he failed, a lower current was used and the values were again progressively increased until the subject could no longer release the test electrode. The end point was checked by several trials, and the highest value was taken as the individual's let-go value in order to eliminate the effects of fatigue. The experimental points plotted in Figure 1 were obtained with hands wet with salt water solution to secure uniform conditions, and to reduce the sensation of burning caused by high current densities at tender spots, and at the instant of releasing the test electrode. Other tests were made with dry hands, hands moist from perspiration and hands dripping wet from weak acid solutions. The effect of the size of the electrodes was also investigated. It was found that the location of the indifferent electrode, the moisture conditions at the points of contact and the size of the electrodes had no appreciable effect on the individual's let-go current. It is believed that results obtained from tests in which hands wet with saline solution grasp and then release a small copper wire may be used to predict let-go currents of a specified degree of safety within an accuracy sufficient for most practical purposes. Sixty-cycle let-go currents were obtained from 134 men and 28 women, and the average value of let-go threshold was established at 16 and 10.5 milliamperes for men and women, respectively. The ratio of the let-go threshold of women to men is thus approximately 2/3, and this ratio is frequently used in estimating let-go currents for women for other frequencies and wave forms.

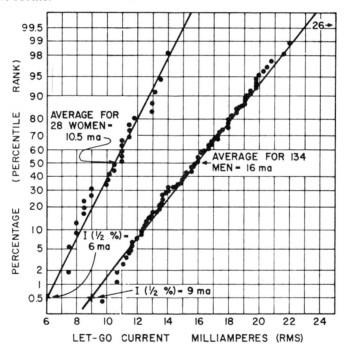

FIG. 1. Let-go current distribution curve for men and women, 60-Hz commercial alternating current.

Tests using gradually increasing direct current produce sensations of internal heating rather than severe muscular contractions. Sudden changes in the current magnitude produce powerful muscular contractions, and interruption of the current always produces a very severe shock. The muscular reactions when the test electrode was released at the higher values were objectionable and sooner or later all subjects declined to attempt higher currents. Tests were conducted on 28 men, and in each case little difficulty was experienced in releasing the electrode. The maximum a subject could tolerate and release was termed his "release current" since this represents a psychological limit rather than the physiological limit of the let-go tests.

Although the deleterious effects of electric shock are due to the current actually flowing through the human body, in accidents the voltage of the circuit is usually the only electrical quantity known with certainty. While current and voltage are related by Ohm's law, the great variances in skin and contact resistances are so unpredictable that let-go voltages are relatively meaningless. On very high voltage circuits, the skin and contact resistances break down instantly and thus they play only a minor role in limiting the current received by a victim. However, on the lower voltages the resistances at the contact locations become of increasing importance, and these resistances are of paramount importance on very low voltage circuits. Obviously, wet contacts create a most dangerous condition for receiving an electric shock, and let-go voltages under these conditions may be of limited interest.

The muscular contractions resulting when 20 or more milliamperes flow across the chest are sufficient to stop breathing during the period the current flows, and the reactions at the instant of current interruption during the d-c release tests occasionally threw the subject a considerable distance. However, normal breathing returns automatically upon interruption of the current and no adverse after-effects were produced by such experiences. The muscular reactions during accidents frequently cause fractures, and the contractions resulting when a victim grasps bare overhead wires may be sufficient to freeze him suspended to the circuit in spite of his struggles to drop free. In many accidents a victim frees himself by breaking the conductor, or his body weight may assist him in interrupting the circuit; however, fortuitous circumstances must not be relied upon to assure safety to human life.

The resistance of the human body has a negative characteristic, the resistance decreasing with increasing current, voltage or time. A value of 500 ohms is frequently used for the body circuit resistance between major extremities for analyzing severe accidents where deep burns are present or for sparking contact with energized high-voltage conductors. A somewhat higher value, say 1,000 to 1,500 ohms, is probably more realistic for low or medium voltage accidents involving moist or firm contacts, but with skin intact.

The minimum current likely to produce an effect on the heart known as ventricular fibrillation has for many years been recognized as the most dangerous electric shock hazard, because once fibrillation is established it is not likely to cease naturally before death. Unfortunately, such cases do not respond to resuscitation and the skill and equipment needed to apply the only known remedy, a controlled counter-electric shock, in the small time during which it might be effective, is not generally available for use in the field.

Currents considerably in excess of those just necessary to produce ventricular fibrillation may cause cardiac arrest, respiratory inhibition, irreversible damage to the nervous system, and serious burns. Hence, if accidental shock currents can be kept below the fibrillating threshold, death from this cause, and death or serious injury from accidents involving still higher currents will also be avoided.

Ventricular fibrillation is caused by relatively small currents flowing in the heart. For short shocks the probability of fibrillation increases with increasing current up to a certain value and then decreases, with the probability of fibrillation becoming small at high current values. Fibrillation is due to over-stimulation rather than damage to the heart; however, when fibrillation occurs the normal pumping action of the heart ceases, and death usually follows in a few minutes. The importance of establishing the minimum current just necessary to produce ventricular fibrillation, even if only approximate, is thus of great importance.

It is obvious that shocks involving currents likely to produce ventricular fibrillation cannot be performed on man and the only recourse is to extrapolate results obtained from animal experimentation to man. In addition to the questionable validity of relating results of experiments made on animals to man, there exists relatively little quantitative data available for such purpose. The first comprehensive work on this subject was entitled "Effect of Electric Shock on the Heart" by Ferris, King, Spence, and Williams, published in 1936.[1]

Recent work by Dr. W. B. Kouwenhoven of the Johns-Hopkins University[2,3] involving capacitor discharges on dogs indicates that, although it appears more difficult to produce ventricular fibrillation in dogs with a current pathway between one front foot and the opposite hind legs using capacitor discharges than with 60-cycle alternating currents, fibrillation was produced in 9 out of 35 tests. He states that a discharge of

TABLE 1
Summary of Human Accidents on Impulse Currents.
Calculated Electric Shock Quantities Received by Victims.

No.	Location/year		Time Constant Microseconds	Voltage kv	Current A	Quantity mc	Energy ws	Remarks
						Surge Discharges		
1	England	1936	0.99	750	1,250	1.2	385	Lichtenberg figures behind ears and on chest. Suffered shock.
2	U.S.A.	1950	2.5	60	120	0.3	9	No trace of discharge on body. Headache for three days.
3	Japan	1942-3	6.0	50	100	0.6	15	Semiconscious and dizzy for short time.
4	Japan	1950	7.8	960	1,600	12.5	5,000	Lost sight one eye. Suffered pain and shock. No trace of discharge on body.
5	France	1948	8.3	228	456	3.8	429	Lichtenberg figures. Intense muscular reactions, and temporary paralysis of hand.
6	Japan	1942	62.5	80	160	10	400	Partially paralysed for three hours.
7	U.S.A.	1937	100	500	1,000	100	25,000	Lichtenberg figures. Intense muscular reactions and pain. Deep burns. Paralysed for 16 hours. Current pathway from abdomen to feet.
8	Sweden	1948	1,200	25	42	50	520	Unconscious and paralysed for short period.
9	Japan	1942-3	1,200	5	8	10	21	Burn on sole of foot.
10	Switzerland	1944	3,200	17.5	30	96	720	Unconscious. Wounds on arm and hand.
11	U.S.A.	1958	3,750	20	40	15	1,500	State of shock for three hours. Current pathway hand to hand, no burns.
12	Switzerland	1950	4,070	17.5	16	66	264	Two men in series, both had wounds.
13	Sweden	1949	11,180	6	12	134	402	Burns on heels.
14	Sweden	1950	35,000	2	4	140	140	Concussion due to fall.
15	England	1959	57,000	4	8	456	912	Suffered deep burns requiring one month to heal. Current pathway left upper arm to lower front of thorax.
16	U.S.A.	1954	106,000	0.5	1	106	26.5	Fell to floor, was shaken and pale but uninjured. Current pathway between the hands, no burns.
						Oscillatory Discharges		
17	Sweden	1952		$\frac{1}{2}$ to 1			24	Electrocuted. Small burn on finger.

The above are based on an assumed body and contact resistance equal to 500 ohms. For detailed analysis see Reference (10).

50 watt-seconds or less had no effect on the heart; however, discharges from 80 to 150 watt-seconds produced ventricular fibrillation, and in some cases, cardiac standstill. In view of the above one might be tempted to suggest that the danger threshold for impulse shocks for man be raised to some value in excess of 50 watt-seconds. However, considering the very limited data available it is believed that a value of 50 watt-seconds is more appropriate at this stage of our knowledge on the subject.

A summary of 17 serious accidents involving impulse discharges is given in Table 1. A study of the various cases reveals that the most severe injuries were experienced on discharges having the greatest energy. In accidents involving arcing or sparking contact, discharges having a short-time constant often produce Lichtenberg figures, but these disappear in a short time. In contrast, discharges involving long-time constants and arcing contact often result in wounds or deep burns. Accidents involving wet or firm contact usually leave no trace on the body.

Impulse shocks of less than 50-watt-sec., although startling and disagreeable, may not be harmful. Both field and laboratory experience with capacitor and inductive types of electric fence controllers indicates that impulse shocks having an energy content of about 0.25 watt-second, while harmless, are definitely very objectionable. It is suggested therefore that an objectionable impulse shock threshold be established at 0.25 watt-second. Such a threshold may have very practicable applications in industry and the laboratory. For example, while it is likely that a technician might tolerate shocks of this magnitude say once a week without comment, it is possible that a daily dose of a half-dozen or more such shocks might produce both violent complaint and possible permanent deleterious nervous effects.

Unfortunately, quantitative information regarding shock intensities necessary to cause other serious effects remain largely unknown. For example, the minimum current required to produce unconsciousness lies somewhere between the let-go and fibrillating thresholds. Higher currents passing through the chest or vital nerve centres may produce paralysis of the breathing mechanism, an effect called respiratory inhibition. Much higher currents, such as those used in electrocution of criminals, may raise the body temperature sufficiently to cause immediate death. Currents sufficient to blow fuses and trip circuit breakers often create awesome destruction of tissue, and may produce very severe shock and irreversible damage to the nervous system.

TABLE 2.

Quantitative effects of electric current on man.

Effect	Milliamperes					
	Direct Current		Alternating Current			
			60-Hertz		10,000 Hertz	
	Men	Women	Men	Women	Men	Women
Slight sensation on hand	1	0.6	0.4	0.3	7	5
Perception threshold, median	5.2	3.5	1.1	0.7	12	8
Shock—not painful and muscular control not lost	9	6	1.8	1.2	17	11
Painful shock—muscular control lost by $\frac{1}{2}\%$	62	41	9	6	55	37
Painful shock—let-go threshold, median	76	51	16	10.5	75	50
Painful and severe shock—breathing difficult, muscular control lost by $99\frac{1}{2}\%$	90	60	23	15	94	63
Possible ventricular fibrillation						
Three-second shocks	500	500	675	675		
Short shocks (T in seconds)			$116/\sqrt{T}$	$116/\sqrt{T}$		
Capacitor discharges	50*	50*				

*Energy in watt-seconds

Burns suffered in electrical accidents are of great concern. These burns may be of two types, electric burns and thermal burns. Electric burns are the result of the electric current flowing in the tissues. Typically, electric burns are slow to heal, but they seldom become infected. Thermal burns are the result of high temperatures in close proximity to the body, such as produced by an electric arc, vaporized metals or hot gases released by the arc, by overheated conductors caused by short circuits, or by explosions. These burns are similar to burns and blisters produced by any high-temperature source. Currents of the let-go level, if they flow for an appreciable time, are more than sufficient to produce deep burns, and both types of burns may be produced simultaneously. Any serious burn should receive prompt medical attention.

No discussion of electric shock would be complete without mention of rescue and resuscitation for victims of serious electric shock accidents. Rescue the victim from the circuit promptly and safely. In many cases the victim may remain in contact with the circuit because of his inability to let go of the energized conductor, or due to unconsciousness. Apply *immediately* an approved method of artificial respiration if the victim is not breathing, or if he appears not to be breathing, and apply an approved method of external heart massage if circulation appears to have stopped. Dispatch assistants for medical assistance and a mechanical respirator. Continue resuscitation without interruption until the victim revives, until rigor mortis sets in, or until he is pronounced dead by a physician. Many victims of serious electric shock accidents recover, perhaps after extensive burns have healed, with no serious permanent aftereffects. (See chapter on Cardiopulmonary Resuscitation)

SUMMARY OF THE LETHAL EFFECTS OF ELECTRIC CURRENT ON MAN

1. If long continued, currents in excess of one's let-go current may produce collapse, unconsciousness, and death.
2. Currents flowing through the chest, the head or nerve centers controlling respiration may produce respiratory inhibition. Respiratory inhibition is dangerous because paralysis of the respiratory organs may last for a considerable period even after interruption of the current, and the approved method of artificial resuscitation *must* be applied promptly to prevent suffocation.
3. Ventricular fibrillation is caused by moderately small currents which produce over-stimulation of the heart rather than physical damage to that vital organ. When fibrillation occurs the rhythmic pumping action of the heart ceases and death usually follows in a few minutes.
4. Heart standstill may be caused by relatively high currents.
5. Relatively high currents may produce fatal damage to the central nervous system.
6. Relatively high currents may produce deep burns, and currents sufficient to materially raise body temperature produce immediate death.
7. Victims who have been revived sometimes die suddenly without apparent cause. This is thought to be due to (a) aggravation of pre-existing conditions, (b) the result of hemorrhages affecting vital centers, or (c) the effects of shock to the nervous system. Delayed death may be also due to burns or other complications.

ACKNOWLEDGEMENT: *Presented at a Meeting of Experts on Electrical Accidents and Related Matters, Sponsored by the International Labour Office, World Health Office and International Electrotechnical Commission, Geneva, Switzerland, October 23–31, 1961. Reproduced with permission of the author.

LITERATURE CITED

1. Ferris, L. P., King, B. G., Spence, P. W. and Williams, H. B., Effect of Electric Shock on the Heart, *A.I.E.E. Transactions (Electrical Engineering), 55:*5, pp. 498–515, (May 1963).
2. Kouwenhoven, W. B., Effect of Capacitor Discharges on the Heart, *A.I.E.E. Transactions (Power Apparatus and Systems), 75:*23, Part III, pp. 12–15, (April 1956).
3. Milnor, W. R., Knickerbocker, G. G. and Kouwenhoven, W. B., Cardiac Responses to Transthoracic Capacitor Discharges in the Dog, *Circulation Research, 6:*1, pp. 60–65 (January 1958).

FOR FURTHER REFERENCE

1. Thompson, Gordon, Shock Threshold Fixes Appliance Insulation Resistance, *Electrical World, 101:24,* pp. 793–95, New York, (June, 1933).
2. Dalziel, C. F., Ogden, E., Abbott, C. E., Effect of Frequency on Let-Go Currents, *A.I.E.E. Transactions (Electrical Engineering), 62:*12, pp. 745–50, (December, 1943).
3. Dalziel, C. F., Effect of Waveform on Let-Go Currents, *A.I.E.E. Transactions (Electrical Engineering), 62:*12, pp. 739–44, (December, 1943).
4. Dalziel, C. F., Lagen, J. B., and Thurston, J. L., Electric Shock, *A.I.E.E. Transactions (Electrical Engineering), Vol. 60,* pp. 1073–79, (December, 1941).
5. Dalziel, C. F. and Massoglia, F. P., Let-Go Currents and Voltages, *A.I.E.E. Transactions (Applications and Industry), 75:*24, Part II, pp. 49–55, (May, 1956).
6. Dalziel, C. F., Dangerous Electric Currents, *A.I.E.E. Transactions (Electrical Engineering), 65:*8–9, pp. 579–85, (August–September, 1946).
7. Kouwenhoven, W. B., Chesnut, R. W., Knickerbocker, G. G., Milnor, W. R. and Sass, D. J., A-C Shocks of Varying Parameters Affecting the Heart, *A.I.E.E. Transactions (Communications and Electronics), 78:*42, pp. 163–69, (May, 1959).
8. Dalziel, C. F., Threshold 60-Cycle Fibrillating Currents, *A.I.E.E. Transactions (Power Apparatus and Systems),* No. 50, pp. 667–673, (October, 1960).
9. Dalziel, C. F., and Lee, W. R., Lethal Electric Currents, *IEEE Spectrum, 6:*2, 44–50 (Feb. 1969).

Electrical Equipment, Wiring and Safety Procedures

T. E. Ehrenkranz and G. W. Marsischky

ELECTRICAL SAFETY PROCEDURES

There are many precautionary procedures for working around and on electrical equipment, and the procedures vary with the kind of equipment and the extent of use. The following procedures have been compiled as a guide for preparation of specific directions for individuals, or groups of employees, with the selection, emphasis and additional amplification based on your needs.

Equipment producing a "tingle" should be reported promptly for repair. "Shorts" become progressively worse and can become extremely hazardous, especially where contact may readily be made against the metal frame-work of an exhaust hood or the damp floor and bench surfaces of a "walk-in" box. Do not rely on grounding to mask a defective circuit nor attempt to correct a fault by insertion of another fuse, particularly one of larger capacity.

Keep the use of extension cords to a minimum and cords as short as possible. Be sure insulation and wire size of extension cords are adequate for the voltage and current to be carried and the environmental conditions in which the cord is to be used. (See article 400, National Electrical Code.)

Work on electrical devices should be done after the power has been disconnected or shut off and suitable precautions taken to keep the power off during the work. On portable equipment the power cord should be unplugged and secured so the electricity cannot be accidentally turned on by someone else. On fixed equipment the main power switch should be shut off and locked with a padlock; shutting off and blocking starter switches is not a positive control nor safe enough since an internal short can bypass the switch and start equipment.

If it is necessary to work on live electric equipment, the person doing so should be fully knowledgeable and have a second person present who is trained in rescue, first aid and cardiopulmonary resuscitation. Never work on live equipment alone.

Use only tools and equipment with nonconducting handles when working on electrical devices.

Treat all electrical devices as if they are live.

Drain capacitors before working near them or removing the device from service, and keep the short on the terminals during the work since some of the charge may return due to a dielectric effect.

Never touch another person's equipment or electrical control devices unless instructed to do so. He may be working on the equipment out of your line of sight.

After servicing electrical equipment do not turn the power back on until you are sure all persons have moved to a safe location and are aware the switch is to be activated.

Enclose all electric contacts and conductors so that no one can accidently come into contact with them.

Mark all high voltage equipment with signs stating the approximate voltage with letters at least three inches high if possible.

Wear safety glasses or a face shield where sparks or arcing may occur. When working with large vacuum tubes or cathode ray tubes which may implode, wear safety glasses and a face shield and use a table shield if possible.

Never use metallic pencils or rulers, or wear rings or metal watchbands when working on electrical equipment.

Cleaning solvents for electrical equipment should be carefully chosen. The use of carbon tetrachloride or benzene should be avoided because of their toxicity. Highly toxic and/or highly flammable cleaning solvents require special ventilation, storage and control procedures.

Laboratory wiring should be done by electricians. Electronic equipment wiring should be done by trained technicians or electronic engineers, or by students under professional supervision.

Never handle electrical equipment when hands, feet, or body are wet or perspiring, or when standing on a wet floor.

With high voltages regard all floors as conductive and grounded unless covered with well maintained and dry rubber matting of suitable type for electrical work.

Whenever possible, use only one hand when working on circuits or control devices.

When it is necessary to touch electrical equipment (for example, when checking for overheated motors), use the back of the hand. Thus, if accidental shock were to cause muscular contraction, you would not "freeze" to the conductor.

Avoid using or storing highly flammable liquids near electrical equipment.

Keep in mind that on some equipment the interlocks disconnect the high voltage source when a cabinet door is opened but power for control circuits remains on.

ELECTRICAL APPLIANCES AND EQUIPMENT

All electrical appliances and equipment should show listing and approval by Underwriters' Laboratories, Inc., or other nationally recognized testing laboratory unless the device is one for which test standards have not been established.

All electric appliances and equipment should be properly grounded in accordance with the National Electrical Code.

A safety manual for electrical equipment design should be devised so that the persons in charge of research groups will be aware of the minimum requirements for construction and purchasing of electric equipment for laboratories, including the requirements recommended in the section on New Electrical Equipment.

All current-carrying parts of any electrical devices should be enclosed.

All "home-made" electrical apparatus should be inspected and approved by a competent electrician before being placed in service.

All electric appliances, equipment, and wiring should be approved for use at the maximum temperature it can reasonably be expected to be subjected to in use.

Equipment for Use in Hazardous Atmospheres

Electric appliances, equipment and devices that are intended for use in areas which contain or may contain flammable vapor-air mixtures, should be constructed so that they meet requirements of the National Electrical Code, Article 500.

Due to the expense of explosion-proof equipment, if at all possible the operating conditions should be altered so that electrical equipment is not subjected to hazardous atmospheres. This can usually be accomplished with a little planning so that switches, lights, motors, and electric contacts are not located in the hazardous area, but are placed in remote or purged locations.

The use of intrinsically safe systems for recording and controlling may eliminate the need for at least part of the explosion-proof equipment. Intrinsically safe systems are those approved for use in hazardous atmospheres without enclosures due to their low operating power, and are constructed as recommended by the Instrument Society of America.

Automatic purging of motor housings with nitrogen during hazardous processes can effectively prevent ignition of flammable mixtures, but purging is complicated and should be done according to the standards of Underwriters' Laboratories, Inc., the Instrument Society of America, or other nationally recognized authority.

Refrigerators

Refrigerators approved for use in hazardous locations should be acquired for the storage of flammable liquids. Some responsible safety men approve the revamping of ordinary domestic refrigerators for this purpose, but such revamping is seldom complete. To adequately protect a refrigerator or freezer, all electrical contacts, switches, thermostat switches lights, etc., must be removed from inside the compartment and the motor placed on top of the refrigerator or the entire refrigerator compartment made vapor tight. Since the revamping of a domestic refrigerator is expensive and the presence of revamped and ordinary domestic refrigerators in the same area may prove misleading to the laboratory personnel, the provision of special explosion-proof refrigerators is preferred. Any refrigerator or freezer not meant for the storage of flammable liquids should be so marked in red letters near the handle.

Ovens

All ovens should be so designed and installed that even if the thermostatic controls fail, nearby combustible materials will not be ignited. Ovens with open flame or exposed electric heating elements, switches or thermostatic controls should not be used for drying materials wet with flammable liquids. The drying of materials wet with flammable liquids should be conducted only in steam, hot water, remotely heated or nitrogen purged ovens. Exceptions may be made for the use of remotely heated electric ovens if the oven temperature can be adequately maintained at a level well below the flash point of the flammable liquid involved.

Electrical Power Tools

1. Portable electric tools should be grounded if:
   a. The tool is not completely encased in insulating material and approved as "double insulated."
   b. Working in a damp area or on any wet or grounded surface.
   c. Conducting particles or moisture might bridge the insulation.
   d. They might overheat in the course of normal operation, resulting in deterioration of insulation.
2. Stationary power tools should be adequately grounded.
3. When pig-tail, clip, or similar ground connector is used, connect it first and disconnect it last. Make the connection firmly. If a clip is to be used, one with a strong spring and insulated handle should be selected.
4. When extension lights are required in damp places or inside metal vessels, the use of a flashlight on a 6V to 12V lamp with a stepdown isolation transformer is desirable. 110V leads should be at a distance from the place of work.
5. The following is a suggested periodic check for portable electric equipment:
   a. Examine equipment, cord, and plug.
   b. Standing on a dry, insulated floor and away from grounded objects, turn the switch on and off and see if it works properly.
   c. Watch for sparks while bringing the tool loosely against a grounded object with switch on as well as off.
   d. If the device has a two-prong plug, reverse the plug and repeat b and c.

e. If shocks or sparks or other defects are observed, have the tool repaired.
   Note: Equipment may be checked for grounding continuity with portable tool
      testers.

6. Frayed cords must be replaced.

Electronic Equipment

1. Equipment, whether of commercial manufacture or not, should be furnished with a circuit diagram, operating instructions, and an explanation of the associated hazards and the operation of safety devices. Door interlocks should be installed (preferably in sight) to de-energize high-voltage circuits when doors are opened.

2. In general, chassis and exposed metal parts should be grounded and bonded together. Conductors used for such purpose should be adequate for maximum anticipated fault current.

3. High-voltage connectors should be of the type on which the body of the connector contacts its mating part before the high-voltage conductor makes contact.

4. If it is necessary to work on energized components, the decision to use grounds can be at the discretion of the persons doing the work. However, if grounds are not used, the entire area should be "ground-free" and denoted as such. If grounding is desirable for safety but "noise" is objectionable, then consideration should be given to the use of a "quiet" laboratory ground, which is a separate shielded grounding conductor (independent from power ground) to provide a zero potential, free from pick-up. If such a conductor is used, it should be prominently marked.

5. Bench tops in electronic laboratories should be nonconductive, and only a minimum of connected equipment should be on the bench tops.

6a. Conductors used to dissipate the charge on capacitors or capacitive circuits should contain sufficient copper to dissipate the calculated maximum energy without excessive temperature rise, which endangers soldered joints, etc. The length of the insulated wand should be ample for the reach and the voltages involved. Connections made with a spring clip are not recommended.

6b. Shorting conductor wands can be of rigid plastic, or dry hardwood painted with clear shellac. Conductor must be a bare conductor, but transparent plastic tubing may be slipped over it. The end of conductor should terminate in a copper hook so that it can be left hanging on a terminal of a discharged capacitor during repair work as extra protection.

7. Many electronic tubes now contain radioactive materials (Table 1). Some of the special application tubes contain quantities which are hazardous when the tubes are broken and the material enters the body by inhalation, ingestion, or through cuts in the skin. Some of these tubes contain small amounts of active material which in a single tube is not a health problem under normal conditions.

TABLE 1
Examples of Radioactive Materials in Electronic Tubes

Type	Manufacturer	Isotope	Microcuries
OA2	Raytheon	Cobalt-60	0.0067
OA2-WA	CBS-Hytron	Nickel-63	0.01–0.05
OA2-WA	Raytheon	Cobalt-60	0.0067
OB2	Raytheon	Cobalt-60	0.0067
OB2-WA	CBS-Hytron	Nickel-63	0.01–0.05
OB2-WA	Raytheon	Cobalt-60	0.0067

8. Selenium (in rectifiers) should be handled with caution because of its toxicity.

9. When a selenium rectifier burns out or arcs over, ventilate the area to remove fumes, particularly in the case of large rectifiers.

10. Measurement of filament voltage on power rectifier tubes should preferably be through permanently installed voltmeters rather than portable instruments. The meters should be rated for the maximum peak inverse voltage, and should be protected with a glass or plastic shield.

11. Metal-cased meters should only be installed on grounded panels. Note: Although it is assumed that voltmeters and ammeters will be connected to the ground side of the series resistor and the grounded side, respectively, the above two precautions are nevertheless recommended.

12. Filaments on some high voltage vacuum rectifier tubes, for example, 1B3-GT or IV2, are capable of producing a fatal shock if the filament circuits are ungrounded.

13. Rectifier tubes operating in circuits in which peak inverse voltages are around 16,000 or over produce X-rays, for which shielding should be provided.

14. The possibility of high-frequency burns is not confined to rf equipment, but can exist in large audio-frequency equipment due to high frequency parasitic oscillations.

15. Before handling cathode-ray tubes, short high-voltage terminal to outer coating and ground. Store and carry them with care to prevent breakage, and wear eye protection.

16a. Only insulated or grounded shafts should protrude through chassis panels.

16b. As an added precaution use short, well recessed setscrews, preferably of nylon.

17. Whenever technical requirements permit, install current-limiting resistors in series with the output of power supplies.

18. Bleeder resistors should be installed across filter capacitors in power supplies, but even with this precaution capacitors should be manually discharged before working on them.

19. If power must be on while adjusting equipment, these precautions should be observed to minimize hazards:
   a. Use insulated test prods
   b. Have another person who is cognizant of hazards and familiar with artificial respiration near you
   c. Stand on an insulating mat
20. Principal high voltage danger points:
   a. Transformer terminals
   b. Rectifier-tube plate caps
   c. Filter capacitor terminals
   d. Filter choke
   e. Rf tuning capacitors and coils
   f. Fuse panels
   g. Zero adjusting screws of meters (depending on mode of installation)
   h. Cathode-ray tube terminals
21. Do not rely completely on relays because contacts may weld together. Two relays in series are an added safety feature
22. If it is necessary to disable an interlock, tag it for the time it is inoperative.
23. Fail-safe considerations:
   a. Safety circuits should be designed with normally open relays
   b. When possible, install fuses in series with filter capacitors
   c. Run control circuit wiring so that a short circuit causes a fail-safe situation

IMPULSE CURRENTS

1. Large capacitors should be installed in barricaded locations so that all personnel are protected from bursting capacitors.

2. Keep capacitors short-circuited when not in use.

3. Floor areas around high voltage or impulse current generating equipment where operators or observers are likely to stand should be covered with a suitable kind of rubber matting. Matting should be properly maintained, kept dry, and tested per ASTM D-178-24 if visibly deteriorated.

4. Never put your hand on or near a condenser bank or anything attached to a condenser unless a hard "ground" wire has first been attached. In all but less than 1000-J condensers this has to be preceded by discharging through a resistive "ground". Examples of the number of joules dissipated per foot of copper conductor, (approximately 400°C (720°F) temperature rise) are listed in Table 2.

TABLE 2
Energy Dissipated Per Foot of Conductor

Wire Sized	Joules
0000	44,200
000	35,100
00	27,800
0	22,200
2	13,900
4	8,700
6	5,475
8	3,450
10	2,160
12	1,370

MICROWAVE AND RADIO FREQUENCY EQUIPMENT

1. Electronic equipment generating rf will induce voltages in resonant circuits which may produce burns, sparks, and hazard to ammunition, volatile liquids, and gases.

2. High-frequency circuits may cause burns when contacted or closely approached.

3. Stay clear of high frequency fields as they produce heat at a high rate. Rings and watches should not be worn nearby.

4. Areas in which microwave power density of more than 0.01 W/cm^2 is detected or suspected should be considered hazardous and posted with warning signs. Personnel should not be permitted in such areas while power density as measured or calculated exceeds 0.01 W/cm^2. Personnel exposure to lower levels should be held to a minimum. The 0.01 W/cm^2 figure has been considered a good figure but is being revised downward and a time factor incorporated by Committee A-95 of the U.S.A. Standards Institute, subject to revision. According to one source, $0.001.$ W/cm^2 is the upper limit of exposure for the eyes, particularly around 3000 Mc.

5. If the output of generators producing average power densities greater than 0.01 W/cm^2 has to be discharged, dummy loads, water loads, or other suitable materials should be used to absorb the energy output.

6. Where test procedures require free-space radiation, the radiating device should be so located that the energy beam is not directed toward personnel. In positioning such devices the probable directions of reflected beams should also be considered.

7. Microwaves over 3000 Mc are usually reflected or absorbed by skin and can be felt by heating of surface tissue. Between about 1000 to 3000 Mc they can penetrate the skin and the fat layer subject to individual variations. Frequencies below 1000 Mc penetrate the deep tissues without subjective awareness of heating. Different parts of the body vary in susceptibility to these effects. Eyes and those organs which cannot readily dissipate heat are most vulnerable.

8. Where microwave generating equipment such as klystrons and magnetrons operate, spurious X-rays are generally present. In general, power supplies, oscilloscopes, electron microscopes, and other equipment, operating over 10 kV should be checked for X-rays with survey instruments which are sensitive in the energy range in which the equipment is being operated.

9. Where power levels are high, screening or absorptive enclosures should be used around components.

9a. A nomograph for estimating the shielding effectiveness of a grid of wires is shown in Figure 1.

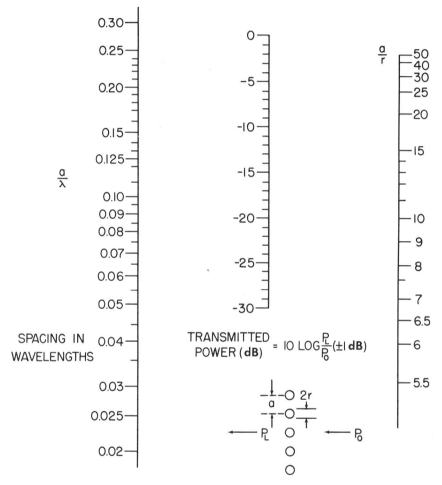

FIG. 1. Nomograph for estimating the shielding effectiveness of a grid of wires. Reprinted from Proceedings of the IRE, 1961, "Some Technical Aspects of Microwave Radiation Hazards," W. W. Mumford, with the permission of The Institute of Electrical and Electronics Engineers, Inc.

10. Do not punch holes for power cables into rf equipment enclosures. Use shielded cables for power, with cable terminating in male-female connection and shielding connected to cabinet.

11. Air vents and other openings should be covered with copper screen electrically connected and tightly fastened to the enclosure.

12. In addition to rf hazards, be alert to high dc potentials on components.

13. Guard against sharp edges or points which can emit corona discharges and cause burns.

14. High-frequency heating equipment should incorporate the following safety features:

a. High-frequency leads should be guarded.

b. Minimum suggested clearances for non-metallic or metallic guarding:

2,000V peak or 150 A rf:	5.1 cm (2 in.)
9,000V peak or 200 A rf:	7.62 cm (3 in.)
18,000V peak or 320 A rf:	2.70 cm (5 in.)

c. Minimum suggested lengths of non-conducting hose ($^3/_8$ in.) for cooling water:

2,000V rf peak:	50.8 cm (20 in.)
9,000V rf peak:	31.44 cm (36 in.)
18,000V rf peak:	127 cm (50 in.)

d. Keep in mind that high-frequency burn is not the only hazard; there may be high voltage dc on the work coil.

e. For large, heavy-duty, high-frequency furnaces individual operating instructions should be prepared and posted, covering a complete operating cycle, emergency procedures, and principal maintenance procedures.

Electrical Motors

All motors and motor installations should conform to the standards set forth in Article 430 of the National Electrical Code.

Care should be taken to insure that 3-phase motors are properly wired for the actual voltage levels existing in each facility. Line voltages nominally considered as 220V by the layman may vary from 208V to 240V in fact. Three-phase motors are equipped with an identification diagramming the proper wiring method for the various voltages. Failure to use the right wiring procedures can lead to overheating and early motor failure.

A competent electrician should inspect and service all motors on at least an annual basis. This will not only prolong the service life of the motors, but will allow for detection of possible electrical faults before failure actually occurs.

All motors, especially if they are located in a hazardous area or perform a fire protection or safety related service such as the exhausting of contaminants, should have a quick disconnect switch accessible from the corridor or a location out of the operating area.

ELECTRICAL POWER DISCONNECTION

1. Every motor or major component should have a disconnect installed ahead of the starter and within sight from the component it protects. This disconnect should be capable of being padlocked in disconnected position while the motor or component is being worked on. Closely grouped equipment may be served with one disconnect.

2. Do not open slow-opening disconnects under load, especially if the load is inductive. If, in an emergency, it should become necessary to open a disconnect under load, a long pole or rope could be used. The face must be averted to avoid flash-burn of eyes.

3. A disconnect should be padlocked in the open position by the person who plans to work on a motor. He should retain the key until finished and secure a tag to the lock showing his name and the time and date.

4. When additional people need to keep open a circuit already disconnected, they should place their own individual lock and tag.

5. Circuits shall not be energized until all locks and tags have been removed by all the people who placed them.

6. The use of explanatory signs or tags on work in progress or interrupted work is to be encouraged.

7. Open ends of wires disconnected from equipment shall be taped.

8. When an air-break switch is open, check to make certain that safe clearance is obtained on all phases. The position of the handle is not to be depended upon as evidence that the switch is open.

9. When operating circuit breakers:

   a. Use only one hand; get in the habit of keeping your other hand in your pocket.
   b. Keep clear of everything except operating handles.
   c. Turn your face away before operating the breaker.
   d. Don't operate isolating switches under load.
   e. Before closing the breaker determine that the equipment is in condition to be energized, that all tags have been removed by the persons who placed them, that protective devices are operative, and that all persons concerned have been notified that the circuit is to be energized.

ELECTRICAL SYSTEM WIRING

Adequacy of Wiring System

The general wiring in a new building should be designed to provide for future expansion. Expansion should be possible without major remodeling or installation of new conduits in feeder circuits. Many universities now plan for 100 percent expansion of their electrical demands during the life time of the building. In research areas, the inability to expand circuit capacities and the number of circuits can have a serious effect on the safe and economical use of the building. While miniaturization may reduce future power loads, the increasing demand for sophisticated electronic equipment to measure, test, supervise, and initiate research functions in the laboratory will probably outweigh the gains made by miniaturization.

Care should be taken to insure that the purchase of new electrical equipment or minor remodeling will not place too high a demand upon the existing electrical supply. Never increase fuse size to compensate for overloading circuits. Fuses, or over-current protection devices, are installed on the basis of their ability to protect a given wiring system and should not be changed except on the advice of a competent and qualified electrician after a thorough evaluation of the entire wiring system affected by the change.

Grounding

All laboratory electrical services should be grounded. Before final acceptance or use of any new laboratory facility, the entire electrical system should be tested to insure that an adequate and continuous grounding system has been installed. Grounding should be done in accordance with the standards set forth in the latest edition of the National Electrical Code.

Many older buildings do not have grounded electrical systems. Since the lack of a safe route to ground for electrical current produced by failures within systems creates the most severe personnel hazard, the provision of an adequate grounding system for the entire electrical system should be given high priority. Where it is impractical to provide an adequate and continuous ground for the entire electrical wiring system, operational rules should be established that will ensure that all potentially hazardous equipment is effectively grounded; e.g., all electrical systems operating at 50 volts or more and having exposed metal or current conducting parts that are normally energized or may become energized due to failure within the system should be grounded.

Equipment grounding may be accomplished in a number of ways. Probably the most common method is by attaching a ground connection to a continuous metallic system such as water pipes. Inspections of water pipe grounding systems often reveal them to be inadequate due to poor electrical bonding to the system or due to the lack of electrical continuity in the water pipe system. The use of water pipe grounding systems must be closely supervised by competent electricians to insure that adequate grounding is achieved. Grounding should not be left to the discretion of the individual laboratory worker or supervisor, but should be made mandatory for all hazardous equipment.

A second means of grounding is by installation of a separate grounding system for hazardous equipment which requires more initial expense, but provides a more reliable and more efficient grounding system. In this system, existing pipe or conduit chases are utilized to carry a grounding bus-bar or cable to each room or laboratory which may foreseeably need such protection. The grounding system is then carried into the areas to be protected simply by installing a No. 4 or larger bare conductor in an easily accessible location. Connection to the grounding system is made by attaching clamps at the desired locations. This system has the advantage of flexibility in service and insures an adequate and effective grounding system. Where it is impractical to provide either of the above grounding systems, a competent electrical engineer should be employed to design a suitable grounding system. For installation of ground rods and further information on grounding, see Article 250 of the National Electrical Code.

Electrical Outlets on the Floor

If the laboratory facilities are equipped with floor outlet boxes, a system of routine periodic inspection and repair should be instituted to prevent the accumulation of water, solvents, acids or other foreign materials in the under floor conduit. Preventive maintenance may require the installation of new gaskets and special guards to prevent damage to the floor boxes. The floors around floor outlet boxes should never be flooded when mopping; damp mopping will aid in preventing unnecessary damage.

Wiring Design and Maintenance

All wiring used in laboratory buildings should be suitable for use in dry and wet locations and should be suitable for use where temperatures may reach 75°C (167°F).

All existing wiring systems in laboratory buildings or buildings converted to laboratory use should be carefully inspected and tested at least annually. Where excessive deterioration is apparent, the wiring in affected areas should be replaced with wiring meeting present standards.

Wiring in Laboratory Hoods

Where possible, outlets, switches and lighting fixtures should not be installed within laboratory fume hoods. The location of lighting fixtures behind enclosed and gasketed (vapor tight) panels of safety glass and the location of outlets and fixtures outside of the fume hood proper have been successfully used. NOTE: Only explosion-proof electrical equipment should be used within a hood enclosure in which flammable concentrations of vapor and air may be present. Wiring in hoods should be at a minimum because of corrosion.

Existing hoods which have exposed light fixtures, switches or outlets should be restricted to use with nonflammable materials or be revamped so that all fixtures, switches and outlets are effectively protected from exposure to corrosive agents and explosive concentrations of flammable mixtures. This can be effectively accomplished by placing approved "enclosed and gasketed" enclosures around light fixtures and opening the unexposed side of the fixture to atmosphere and relocating switches and outlets to the front edges of the hood enclosure.

Number and Location of Electrical Outlets

A sufficient number of electrical outlets should be provided and located so that the installation of extension cord wiring will not be necessary. Research laboratories should have at least four double outlets for each individual. The outlets should be so spaced that there is no more than 1.83 m (6 ft) from any point on any wall, bench or workspace to the nearest outlet.

Outlet Configurations

Outlets serving nominal 110V, 220V, 440V or other ac voltages should be, as far as possible, according to NEMA configurations for general purpose non-locking plugs and receptacles. Now USASI Std. (ASA) Locking Type plugs can be designated so that no one can mistakenly connect 110V equipment to 220V or 440V circuits. All dc circuits should be clearly labeled and segregated from ac circuit connectors.

Emergency Power Supply

Emergency power should be provided for operating exhaust fans, exit lighting, exit signs, emergency evacuation alarms, at least one elevator and other emergency or fire and safety supervisory equipment. The emergency power source should be capable of automatic starting and manual starting from a remote location. Where desired, especially marked and identified outlets in research laboratories may also be connected to the emergency power system to prevent loss of experimental products or data.

Where applicable, unit battery-supplied emergency lights and a separate battery supply for the emergency evacuation alarm system may suffice.

ELECTRICAL PANELS AND EQUIPMENT

Electric Control Panels

The exact location of the corridor panel board for the electrical circuits to your work area should be known, to allow remote shut-off of apparatus in an emergency situation. It would be helpful to have panel and breaker switches which control room circuits inside or outside of the main door of each laboratory. They should be labeled. Main breakers should be *prominently* labeled.

New Electrical Equipment

For nonfixed electrical equipment supplied by a power cord, the purchase requisition should contain the statement that "Exposed noncurrent carrying metal parts shall be grounded through a grounding type cord and plug" whenever any of the following apply:

1. Hand-held motor-operated equipment.
2. Equipment used with or around moisture such as water baths, physiotherapy equipment, stirrers, and water pickup machines.
3. Readily movable equipment, such as centrifuges, ovens, and hot plates, which may be used with or around moisture, other grounded equipment or fixed grounds such as water pipes, sinks, and metal hoods.
4. Equipment which is explosion-proof.
5. Autotransformers.

6. Any other apparatus which may be procured with grounding-type cord and plug as a standard item.

Exceptions to the above paragraph are "double-insulated" tools. These are rated as "double-insulated" by the Underwriters' Laboratories and recognized by the National Electrical Code. In addition to the "functional" insulation needed for electrical operation they are encased in an insulating material of ample electrical and mechanical strength. Such tools are particularly suitable for use in wet or "massive-ground" locations. Other typical hand-held electrical tools that generally need not be grounded are electric draftsmen's erasers, and some marking tools. There has been little adverse experience with these because they are conservatively designed and are generally not used in hazardous locations.

Because of the possibility of errors in the required rewiring of ordinarily ungrounded appliances, such rewired appliances should be checked to see that the grounding conductor is not wired so that it normally carries current, and to see that the switch (if any) is in the "hot" conductor.

ACKNOWLEDGEMENT: Parts of this chapter reprinted from the *LASL Health and Safety Manual,* University of California, Los Alamos Scientific Laboratory, Los Alamos, New Mexico.

Explosion-Proof Electrical Equipment

T. E. Ehrenkranz

Article 500 of the National Electrical Code (NEC) divides "hazardous locations" into Classes, Groups and Divisions and specifies required electrical equipment in each category. The principal agency which tests and rates electrical equipment for such locations is the Underwriters' Laboratories, Inc., (UL). Electrical equipment can also be specified by the National Electrical Manufacturers Association (NEMA) Type numbers.

Definitions

There are Class I, Class II and Class III hazardous locations according to the NEC. In each Class, the NEC also recognizes a Division 1 location where hazardous atmospheric concentrations can exist under normal operating conditions, and Division 2 where, in general, such hazardous conditions can exist only infrequently through an accident or an abnormal situation. Division 2 may include areas adjacent to Division 1 locations.

The National Electrical Code defines Class I, Divisions 1 and 2 as follows:

Class I, Division 1. Locations (1) in which hazardous concentrations of flammable gases or vapors exist continuously, intermittently, or periodically under normal operating conditions, (2) in which hazardous concentrations of such gases or vapors may exist frequently because of repair or maintenance operations or because of leakage, or (3) in which breakdown or faulty operation of equipment or processes which might release hazardous concentrations of flammable gases or vapors, might also cause simultaneous failure of electrical equipment.

This classification usually includes locations where volatile flammable liquids or liquefied flammable gases are transferred from one container to another, interiors of spray booths and areas in the vicinity of spraying and painting operations where volatile flammable solvents are used; locations containing open tanks or vats of volatile flammable liquids; drying rooms or compartments for the evaporation of flammable solvents; locations containing fat and oil extraction apparatus using volatile flammable solvents; portions of cleaning and dyeing plants where hazardous liquids are used; gas generator rooms and other portions of gas manufacturing plants where flammable gas may escape; inadequately ventilated pump rooms for flammable gas or for volatile flammable liquids; the interiors of refrigerators and freezers in which volatile, flammable materials are stored in open, lightly stoppered, or easily ruptured containers, and all other locations where hazardous concentrations of flammable vapors or gases are likely to occur in the course of normal operations.

Class I, Division 2. Locations (1) in which volatile flammable liquids or flammable gases are handled, processed or used, but in which the hazardous liquids, vapors or gases will normally be confined within closed containers or closed systems from which they can escape only in case of accidental rupture or breakdown of such containers or systems, or in case of abnormal operation of equipment, (2) in which hazardous concentrations of gases or vapors are normally prevented by positive mechanical ventilation, but which might become hazardous through failure or abnormal operation of the ventilating equipment, or (3) which are adjacent to Class I, Division 1 locations, and to which hazardous concentrations of gases or vapors might occasionally be communicated unless such communication is prevented by adequate positive-pressure ventilation from a source of clean air, and effective safeguards against ventilation failure are provided.

This classification usually includes locations where volatile flammable liquids or flammable gases or vapors are used, but which, in the judgment of the Code enforcing authority, would become hazardous only in case of an accident or of some unusual operating condition. The quantity of hazardous material that might escape in case of accident, the adequacy of ventilating equipment, the total area involved, and the record of the industry or business with respect to explosions or fires are all factors that should receive consideration in determining the classification and extent of each hazardous area.

Class I Locations

Class I includes Groups A, B, C and D. Group A is for atmospheres containing acetylene, Group B for atmospheres containing hydrogen, or gases or vapors of equivalent hazard such as manufactured gas, Group C for atmospheres containing ethyl ether vapors, ethylene, or cyclopropane, and Group D is for atmospheres containing gasoline, hexane, naphtha, benzine, butane, propane, alcohol, acetone, benzol, lacquer solvent vapors, or natural gas. The approximate flammable limits in air for Group A (acetylene) is 2.5%–81% by volume; for Group B (hydrogen) it is 4%–75%; for the Group C gases the spread is from 2%–30%, and for Group D it is only 1%–17%.

In general, electrical equipment for Class I, Division 1 locations does not have gas-tight enclosures. Enclosures are built so that they can resist an internal explosion if gas seeps inside and is ignited by a spark, and that the resulting flame is quenched so that a gas mixture outside the enclosure will not be ignited. For example, a Class I, Group C and D cast iron junction box is hydrostatically tested to four times the internal peak pressure that is attainable with a stoichiometric mixture. Its bolted metal-to-metal joints, if, for example, 1.9 cm (^{3}/$_4$ inch) wide, have a maximum clearance of 0.05 mm (.002 inches) or, if 2.54 cm (1 inch) wide, a maximum of 0.09 mm (.0035 inches) so that only cooled combustion products can issue. If an item is not to be hydrostatically tested by UL, then a calculated factor of safety of 5 must be demonstrated, unless the material is rolled steel for which UL accepts lower factors of safety. It is important to note that for Groups A and B ground joints as described above are not acceptable; only threaded joints are allowed. With Class I fixtures surface-temperature limitations are usually easily met, but in the case of lighting fixtures surface temperature is a controlling factor.

Class II Locations

Class II locations contain electrically conducting or combustible dusts. This Class includes: Group E for atmospheres containing metal dust, including aluminum, magnesium and their commercial alloys, and other metals of similarly hazardous characteristics, Group F for atmospheres containing carbon black, coal or coke dust, and Group G for atmospheres containing flour, starch, or grain dusts. Fixtures approved for this Class must be so constructed that they are "dust-ignition-proof." This means that enclosures exclude ignitible amounts of dust; or amounts which would affect equipment performance and permit an interior ignition to be communicated to the exterior of the enclosure; and that surface temperatures are below the ignition temperatures of the combustible dusts, about 166°C (330°F) to 99°C (390°F).

Class III Locations

Class III designates areas which contain ignitible fibers. It has no groups but is classified into two divisions. Equipment approved for such locations may have a maximum surface temperature of 166°C (329°F), or 120°C (248°F) if subject to overloading. Covers must minimize entrance of fibers and escape of sparks.

Wiring

Wiring for Classes I, II and III in general is in rigid conduit or MI (mineral insulated) cable with approved end fittings. There are also approved types of flexible conduit. The NEC prescribes the use of special conduit seals to prevent the passage of combustible gases from one fixture to another or from a hazardous area into a non-hazardous area. These seals and associated breather and drain fittings must be installed according to the manufacturer's recommendations.

Some typical NEMA Types are the following:

Type I   General purpose.
Type 4   Watertight (excludes water from a hose-stream).
Type 7, with letter (Corresponds to Class I with same Group letter; air break equipment).
Type 8, with letter (Corresponds to Class I with same Group letter; oil-immersed equipment).

The Type 4 is especially useful in areas where the main service requirement is for equipment to withstand periodic washing-down.

Intrinsic Safety

The NEC also recognizes the principle of "intrinsic safety". Equipment approved as "intrinsically safe" for use in specific hazardous atmospheres is defined as incapable of releasing sufficient electrical energy to cause ignition under normal or abnormal conditions. Such equipment does not need the expensive, specialized enclosures which characterize "explosion-proof" equipment. This approach to the explosion hazard is not yet widely utilized in the USA, but in Great Britain, for example, there is a formidable list of lamps, instruments, communications equipment and test apparatus certified "intrinsically safe" for atmospheres which correspond to our Class I Groups C and D.

Other ways to avoid the high cost of "explosion-proof" installations are to locate as much of the electrical equipment as possible outside the hazardous area, or to locate conventional equipment in an enclosure which is under a constant purge of clean air, or preferably, nitrogen.

Equipment for Use with Hydrogen

Around hydrogen facilities one is faced with one of two problems: the unavailability, or the high cost of Class I Group B equipment. When the above schemes are not practical, there are two schools of thought. One is to use Group C or D equipment; the other is to use "enclosed and gasketed" equipment. (The latter used to be called "vapor-proof" and more recently "vapor-tight" but these terms are now obsolete and not recognized by the NEC or UL.) In spite of the fact that "explosion-proof" equipment is not gas tight, and "enclosed and gasketed" equipment "breathes" upon heating and cooling, the above alternatives are often acceptable and the following reasoning is offered:

The ignition energy and the quenching distance for hydrogen, as for gases in general, are lowest at stoichiometric proportions, then increase rapidly. Likewise, explosion pressures are highest in these regions. Since equipment for each Group is designed for safety at stoichiometric mixtures, a Group C or D fixture may be entirely safe up to 10–15% hydrogen. As for "enclosed and gasketed" fixtures, even in a high concentration of hydrogen they are not likely to "breathe" enough of the hazardous atmosphere for the internal concentration to rise to 4% hydrogen, unless the fixture exposure is over a protracted period.

Maintenance of Equipment

Suggestions for maintenance of electrical equipment for hazardous locations:

a) Turn off and lock-out power before opening housings. This includes lighting fixture servicing.

b) When ground joints need cleaning use stiff bristle brush (not wire brush) and a high flash point solvent such as kerosene. Wipe dry and apply light oil or special lubricants for corrosive locations as recommended by manufacturer. Be sure that there are no nicks or grit in the ground-joint surfaces.

c) Use original bolts when replacing covers, and tighten them all evenly. There must be no missing bolts.

d) Perform (b) every time cover is removed.

e) Do not paint over nameplates.

f) Avoid painting e.p. fixtures containing heat-sensitive devices as dark paint will alter their functioning.

g) Repairs to e.p. motors technically negate the UL label, and should preferably be undertaken by the manufacturer, especially extensive work such as rewinding.

h) Some drains in sealed conduit work automatically, but those which don't must be periodically drained manually.

i) Some people specify a watertight joint as one made with a sealant, and made up to be "wrench-tight."

ACKNOWLEDGEMENT: Reprinted from *LASL* Health and Safety Newsletter No. 64-3, July 30, 1964, Los Alamos Scientific Laboratory, Los Alamos, New Mexico.

REFERENCES

1. National Electrical Code, (1966).
2. Standards Nos. 886 and 844, Underwriters' Laboratories.
3. "Electrical Accidents and Their Causes," H.M.S. Stationery Office, (1960).
4. K. A. Woodard, "The Application of Commercial Electrical Equipment . . . ," Stearns-Roger Manufacturing Company, (1954).
5. J. R. Petree, NEC, UL and Practical Factors Applying to Class I Groups A and B.
6. Crouse-Hinds Company, Personal Communication, (1961).
7. von Elbe and Scott, ASD-TDR-62-1027.
8. Lewis and von Elbe, "Combustion, Flames and Explosions of Gases," Academic Press, (1951).
9. W. F. Hickes, Intrinsically Safe Electronic Instrumentation, *ISA Journal,* 8, 7, (1961).
10. C. F. Kisselstein, Keeping the 'Proof' in Explosion-Proof, *ISA Journal,* June, (1961).
11. L. G. Matthews, Union Carbide Corporation, personal communication.

Glass

Gail P. Smith

DEFINITION OF GLASS

Glass is indispensable in the laboratory. To use it to greatest effectiveness it is necessary to understand its nature and properties. We will start with some definitions of the material:

The Dictionary Definition

"An amorphous inorganic usually transparent or translucent substance, consisting typically of a mixture of silicates or sometimes borates or phosphates formed by fusion of sand or some other form of silica or by fusion of oxides of boron or phosphorus with a flux (as soda, potash) and a stabilizer (as lime, alumina) and sometimes metallic oxides or other coloring agents so that a mass is produced that cools to a rigid condition without crystallization and that may be blown, cast, pressed, rolled, drawn, or cut into various forms." (Websters Third New International Dictionary)

The Scientist's Definition

"A glass is an inorganic substance in a condition which is continuous with, and analogous to, the liquid state of that substance, but which, as the result of having been cooled from a fused condition, has attained so high a degree of viscosity as to be for all practical purposes rigid." (G. W. Morey)

The Engineer's Definition

"An inorganic product of fusion which has cooled to a rigid condition without crystallizing." (A.S.T.M.)

GLASS COMPOSITION AND STRUCTURE

"Glass" is a generic term for a large family of materials. About 700 different compositions are in commercial use. The atoms in a glass form an extended three-dimensional network which has no symmetry, no long-range order. Nearly all the glasses used in the laboratory are based on silica as the structure-determining oxide. The silica tetrahedron has four atoms of oxygen and one atom of silicon; the oxygen atoms are shared with, and bonded to, neighboring silicon atoms. We can classify silica glasses into several large groups, according to their composition: silica glass, soda-lime glass, lead-alkali glass, borosilicate glass, and aluminosilicate glass. In addition, there are glasses which contain no silica.

Silica Glass

Silica glass is the most important of the single-oxide glasses. In many respects pure vitreous silica can be considered to be an ideal glass. It has a high use-temperature, a very low coefficient of thermal expansion, and very low ultrasonic absorption. It is an excellent dielectric. Its chemical durability and stability to weathering are very good. It is resistant to radiation. Silica glass is used in ultrasonic delay lines, as supersonic wind tunnel windows, in various optical systems, and as crucibles for growing germanium or silicon crystals. Its high use-temperature results from its extremely high viscosity at high temperatures which makes it a difficult glass to produce or to fabricate.

Some of the difficulties of manufacture of silica glass—fused silica—have been overcome in two ways: (1) a method of producing SiO_2 glass, other than that of melting selected crystals of quartz, was devised—a vapor deposition of the reaction product of $SiCl_4$ and water; and (2) a glass was developed containing at least 96% SiO_2 whose properties are comparable in many respects to those of SiO_2 itself. Products made from this material are sold under the trademark VYCOR*, and are made by leaching a glass which is primarily SiO_2 and B_2O_3, after fabrication, to remove all but the last traces of B_2O_3. The article is then fired to re-close its structure, during which all dimensions are considerably shortened. The new glass is nearly as refractory and nearly as low in expansion as SiO_2.

Soda-lime Glass

For nearly all commercial glasses, fluxes are added to reduce the high viscosity inherent in silica glass and to bring glass manufacture into the range of industrially accessible temperatures and refractories. The most usual flux is soda, Na_2O. The addition of alkali "softens" the glass, i.e., reduces its viscosity at high temperatures, by breaking Si—O bonds. The addition of alkali also increases solubility, and sodium silicate glasses form the basis of the soluble silicate industry. Stabilizing oxides are added to decrease greatly the solubility of the sodium silicates. Calcium oxide is a cheap and effective stabilizer. In the soda-lime-silica system, the optimum glass with respect to cost, durability and ease of manufacture, is the composition silica-72%, soda-15%, lime and magnesia-10%, alumina-2%, and miscellaneous oxides-1%. The miscellaneous oxides result from, and enable the use of, cheap raw materials. With only slight variations, this composition is used throughout the world to produce flat glass, containers, and incandescent lamp envelopes. It accounts for about 90% of all glass made.

Lead-alkali Glass

Lead-alkali glasses contain a substantial content of PbO which for the most part replaces the CaO in soda-lime glasses. They have a high density and increased index of refraction which makes the glass more brilliant and consequently desirable for art ware sometimes referred to as "crystal." Addition of lead also permits reduction of alkali content to bring about lower electrical conductivity—for this reason lead glasses are in common use in the electronics field. The use of lead also results in a relatively high dielectric constant and low power factor, desirable for electrical capacitors. Lead glasses are also used for radiation shielding windows, optical glasses, and for very low-melting "solder" glasses. Lead glasses have a long working range and low annealing temperatures, making them easily manipulated in the fire. Accordingly, the glassblower speaks of lead glasses as "sweet."

Borosilicate Glass

Addition of fluxes, as discussed above, to the silica network always increases its expansion coefficient. This effect is reduced by addition of B_2O_3, another glass-former. Thus, stable and workable glasses can be produced which retain enough of the low-expansion characteristic of SiO_2 glass to be classed as heat-resisting. This fact, together with their generally good chemical durability, makes borosilicates desirable for such uses as cooking or baking ware, laboratory glassware, reagent bottles, telescope mirrors. In the making of complicated laboratory apparatus, the low expansion coefficient reduces strain development and risk of breakage during fabrication. Code 7740 glass is an example of a borosilicate glass: products made from it are sold under many trademarks, including HYSIL, PHOENIX, DURAN 50, SIMAX, KIMAX, K-33, as well as the familiar PYREX®.

Aluminosilicate Glass

When alumina is added to an alkali silicate glass, an increase in viscosity results. Alumina also improves structural stability and chemical durability. Alkaline-earth aluminosilicates form the basis for glasses requiring high use-temperature and/or good dielectric properties. Other aluminosilicate glasses are used for top-of-stove ware and fiber glass.

Non-silicate Glasses

In addition to the major silica types, there are many less common but interesting glasses—phosphates, borates, germanates, tellurates and nonoxide glasses such as fluorides and sulfides. Although the amount of these glasses is never very great, many of them are very important in optics and other technical fields.

PHYSICAL PROPERTIES OF GLASS

Table 1 (pg. 334) lists some properties of glasses representative of each of the classes based on silica. But to properly use this table, and to understand the properties and hence the use of glasses, some discussion of the physical properties of glasses is necessary.

Viscosity

At ordinary temperatures the viscosity of glass is so high that it can be considered to be infinite. As the temperature is raised the viscosity decreases and the glass gradually assumes the character of a viscous liquid. Four points on the viscosity-temperature curve have been developed to represent the viscosity of the glass at important points for the glassworker, the strain, annealing, softening, and working points. These points, or reference temperatures, are listed in Table 1, Column 11, and are defined as follows:

STRAIN POINT. A temperature at which the internal stress is substantially relieved in a matter of hours. The strain point corresponds to a viscosity of approximately $10^{14.5}$ poises.

In general the strain point represents the extreme upper limit of serviceability for annealed glass.

ANNEALING POINT. A temperature at which the internal stress is substantially relieved in a matter of minutes. The annealing point corresponds to a viscosity of approximately $10^{13.0}$ poises.

In an annealing operation the glass is heated somewhat above the annealing point and slowly cooled to somewhat below the strain point. The temperature range, strain point to annealing point, is called the annealing range. Distortion of the glass becomes a problem above the annealing point.

SOFTENING POINT. A temperature at which a uniform fiber, of a specified diameter and length, elongates under its own weight at a rate of 1 mm per minute when heated in a specified manner. For glass of density near 2.5 gm/cm^3 this temperature corresponds to a viscosity of $10^{7.8}$ poises.

At the softening point the glass deforms very rapidly and starts to adhere to other bodies.

WORKING POINT. The temperature where the glass is soft enough for hot working by most of the common methods. Viscosity at the working point is approximately 10^4 poises.

Strength

For all practical purposes and at normal temperatures, glass can be considered to be a perfectly elastic and, therefore, a perfectly brittle material, although some internal friction and a slight delayed elastic effect may be observed. Thus, it does not plastically deform before failure and it fractures only from tensile stresses, never from shear or compression. The stress-strain curve for glasses is a straight line up to the breaking point.

The intrinsic strength of all glasses is extremely high, possibly as much as 220,000 to 280,000 kg/cm^2 or three or four million pounds per square inch. Glass fibers have supported tensile stresses of over 70,000 kg/cm^2 or a million psi; specially prepared bulk glass will withstand 18,000 kg/cm^2 or a quarter of a million psi, but commercially produced glassware may fail under a stress less than 140 kg/cm^2 or 2000 psi. Table 2 shows some average values of the breaking strength of glass in different forms.

TABLE 2
Breaking Stress of Annealed Glass

Condition of Glass	Average Breaking Stress	
	kg/cm^2	psi
Fine fibers		
Freshly drawn	2000–30000	30000–400000
Annealed	700–3000	10000–40000
Rods, acid etched and lacquered,		
6.35 mm ($1/4$ in. diam)	20000	250000
Blown Ware		
Inner surfaces	1000–3000	15000–40000
Outer surfaces	300– 700	4000–10000
Drawn window glass	600–1400	8000–20000
Polished plate glass	600–1100	8000–16000
Drawn tubing	400–1000	6000–15000
Pressed ware	200– 600	3000– 8000
Ground or sand-blasted surface	100– 300	1500– 4000

This large difference is caused principally, if not entirely, by changes in the condition of the surface. Commercially produced glass articles contain in their surfaces a number of flaws, most of them tiny, which result from handling during manufacture. These flaws give rise to high local stresses when a load is applied. The effect of surface condition is so powerful that it masks any effect of change in composition except in extreme cases. In a commercially acceptable glass item, failure will always originate in the surface, for only a severe internal flaw can cause a failure to originate in the interior of the glass.

This has strong implications for the handling of glass in the laboratory. Sharp and localized impacts; scratches on the surface, especially the interior surface of a container; localized and sharp heating are to be avoided as much as possible. These usual and normal precautions result from the fact that glass is perfectly elastic, and therefore cannot accommodate local high stresses by local plastic deformation.

Strengthening of Glasses

Many attempts have been made to increase the usable strength of glass, but only one has been generally successful—prestressing; that is, strengthening the glass article by putting all its surfaces under compression. Any applied load must first neutralize this built-in compression before it can exert tension on the surface. Thus, the strength of the glass article will equal approximately the sum of the compression and the ultimate strength of the glass surface when it is free from stress.

THERMAL STRENGTHENING. Heat-treating, or tempering, has been the method traditionally employed to prestress glass. A glass article is heated until it is almost soft enough to deform under its own weight; i.e., its viscosity is so low that it can support no internal stress. Then it is suddenly quenched, usually with a blast of cold air but occasionally in a bath of oil or molten salt. As they contact the quenching medium, the surfaces cool, shrink, and harden. At this point the interior of the article is still hot enough to flow and accommodate the shrinkage. As the temperature of the interior decreases, the whole item shrinks and forces the surface into compression. Figure 1 represents a fair picture of the residual stress in the glass when thermal equilibrium is established. Figure 1 also shows that a much larger applied tensile stress is required to rupture the glass than if the surface compression were absent; i.e., if the glass were annealed.

Under certain conditions tempering can produce compression of the order of 2100 kg/cm^2 or 30,000 psi in a surface. The amount of stress is influenced by the coefficient of thermal expansion and thickness of the glass as well as the tempering schedule employed. Compositions with high coefficients of expansion develop the highest stresses, but in some of the lowest-expansion glasses (fused silica, for instance) only small stresses can be generated. Appreciable stress cannot be readily induced in glass less than 2.5mm or 0.1-inch thick, and complex shapes generally cannot be efficiently tempered.

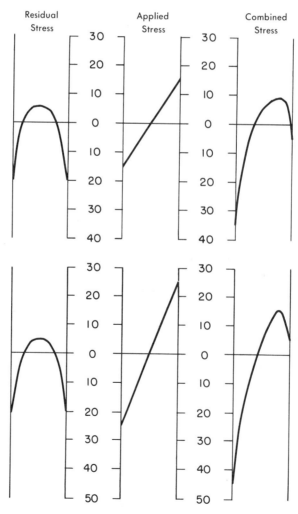

FIG. 1

Although tempering can double, and even quadruple, the strength of a glass article, it does exert two penalties.

(1) The maximum service temperature of a tempered article is lower than that of an annealed—stress-free—article. The energy stored in glass by tempering causes the glass structure to deform at a lower temperature, to release the tempering stress. Usually the reduction in maximum service temperature is around 200°C (380°F), as may be seen in Table 1. In spite of this restriction, tempering is used extensively for car windows, boiler gauge glasses, pipe lines, doors for buildings, and cooking ware.

(2) The other penalty is that tempered glass usually has stored within it sufficient strain energy so that when the compressive layer is penetrated, the article will fracture. This means that usual cold-forming operations on glasses (grinding, sawing, drilling), cannot normally be performed on tempered ware.

CHEMICAL STRENGTHENING.   The limitations on thermal tempering or thermal strengthening, imposed by considerations of size, shape, maximum temperature gradient possible, and other factors, can often be avoided by the newer methods of chemical strengthening. In these methods, the composition or state of the surface is made different from that of the interior: by an over-lay at high temperature of one glass on another of higher expansivity; by crystallization of the surface, resulting in a desired lower expansivity; and/or by exchange of ions at the surface of the glass. This latter method will produce a surface layer which is either (a) of lower expansion coefficient, or (b) in high surface compression if the diameter of the substituting ion is larger than that of the one originally present (as when sodium replaces lithium in a lithium-containing glass).

Generally, the surface compressed layer is thinner, surface compression is much higher, interior tension is lower, and the stored energy—the energy released on rupture —is lower in chemically strengthened glass than it is in thermally strengthened glass.

But the shape of the stress pattern indicates that chemically strengthened glass will be more sensitive to deep accidental scratches or bruises than will thermally strengthened glass, and that the glass should be handled so as to avoid such injury.

WORKING STRESSES.   With an adequate safety factor provided, the prolonged working stress for annealed glass, almost independent of composition, is taken as 70 kg/cm^2 or 1,000 psi and for tempered, or thermally strengthened glass as 140 to 280 kg/cm^2 or 2,000 to 4,000 psi; depending on the piece in question. The corresponding strength for commercially available chemically strengthened glass, with short-time breaking load near 3500 kg/cm^2 or 50,000 psi, is in the region of 1100 kg/cm^2 or 15,000 psi.

Elasticity

For all ordinary pusposes it can be assumed that glass is perfectly elastic up to the point of fracture. The Young's Modulus of elasticity varies from 440,000 to 910,000 kg/cm^2 or from 6,000,000 to 13,000,000 psi, but most commercial glasses have values between 560,000 to 700,000 kg/cm^2 or 8,000,000 and 10,000,000 psi. Values are listed in Table 1.

Poisson's ratio covers a fairly narrow range, from about 0.16 for fused silica to 0.28 for high-lead glasses.

Hardness

The term "hardness" has two meanings to the glass worker. Where used with reference to viscosity, it refers to a glass with high viscosity at high temperatures—the high silica glasses are "hard" and the lead-alkali glasses "soft." Where used with reference to the mechanical properties of glass, it has the conventional mechanical meaning.

TABLE I PROPERTIES OF GLASSES

1	2	3	4	5	6 Corrosion Resistance			7 Thermal Expansion 10⁻⁷in./in./°C		8 Upper Working Temperatures (Mechanical Considerations Only)				9 Thermal Shock Plates 6″ ×	
Glass Code	Type	Color	Principal Use	Forms Usually Available	Weathering	Water	Acid	$0-300°C$ $32-572°F$	Room Temp.-Setting Point	Annealed Normal Service °C	Annealed Extreme Limit °C	Tempered Normal Service °C	Tempered Extreme Limit °C	Annealed $\frac{1}{8}″$ Thk. °C	Annealed $\frac{1}{4}″$ Thk. °C
0010	Potash Soda Lead	Clear	Lamp Tubing	T	2	2	2	93	100	110	380	—	—	65	50
0080	Soda Lime	Clear	Lamp Bulbs	B M T	3	2	2	92	103	110	460	220	250	65	50
0120	Potash Soda Lead	Clear	Lamp Tubing	T M	2	2	2	89	98	110	380	—	—	65	50
1720	Aluminosilicate	Clear	Ignition Tube	B T	1	1	3	42	52	200	650	400	450	135	115
1723	Aluminosilicate	Clear	Electron Tube	B T	1	1	3	46	54	200	650	400	450	125	100
1990	Potash Soda Lead	Clear	Iron Sealing	—	3	3	4	124	136	100	310	—	—	45	35
2405	Borosilicate	Red	General	B P U	—	—	—	43	51	200	480	—	—	135	115
2475	Soda Zinc	Red	Neon Signs	T	3	2	2	93	—	110	440	—	—	65	50
3320	Borosilicate	Canary	Tungsten Sealing	—	³1	³1	³2	40	43	200	480	—	—	145	110
6720	Soda Zinc	Opal	General	P	²⁻	1	2	80	92	110	480	220	275	70	60
6750	Soda Barium	Opal	Lighting Ware	B P R	²⁻	2	2	88	—	110	420	220	220	65	50
6810	Soda Zinc	Opal	Lighting Ware	B P R	²⁻	1	2	69	—	120	470	240	270	85	70
7040	Borosilicate	Clear	Kovar Sealing	B T	³3	³3	³4	48	54	200	430	—	—	—	—
7050	Borosilicate	Clear	Series Sealing	T	³3	³3	³4	46	51	200	440	235	235	125	100
7052	Borosilicate	Clear	Kovar Sealing	B M P T	³2	³2	³4	46	53	200	420	210	210	125	100
7056	Borosilicate	Clear	Kovar Sealing	B T P	2	2	4	51	57	200	460	—	—	—	—
7070	Borosilicate	Clear	Low Loss Electrical	B M P T	³2	³2	³2	32	39	230	430	230	230	180	150
7250	Borosilicate	Clear	Seal Beam Lamps	P	³1	³2	³2	36	38	230	460	260	260	160	130
7570	High Lead	Clear	Solder Sealing	—	1	1	4	84	92	100	300	—	—	—	—
7720	Borosilicate	Clear	Tungsten Sealing	B P T	³2	³2	³2	36	43	230	460	260	260	160	130
7740	Borosilicate	Clear	General	B P S T U	³1	³1	³1	33	35	230	490	260	290	180	150
7760	Borosilicate	Clear	General	B P	2	2	2	34	37	230	450	250	250	160	130
7900[1]	96% Silica	Clear	High Temp.	B P T U M	1	1	1	8	7	800	1100	—	—	1250	1000
7913	96% Silica	Clear	High Temp.	B P R S T	1	1	1	8	7	900	1200	—	—	—	—
7940	Fused Silica	Clear	Ultrasonic	U	1	1	1	5.5	7	900	1100	—	—	1250	1000
8160	Potash Soda Lead	Clear	Electron Tubes	P T	2	2	3	91	100	110	380	—	—	65	50
8161	Potash Lead	Clear	Electron Tubes	P T	2	1	4	90	97	110	390	—	—	—	—
8363	High Lead	Clear	Radiation Shielding	L C	3	1	4	104	112	100	200	—	—	—	—
8871	Potash Lead	Clear	Capacitors	—	2	1	4	102	113	125	300	—	—	55	45

10	11				12	13	14			15			16			17	18
Thermal Stress Resistance °C.	Viscosity Data				Impact Abrasion Resistance	Density grams per C.C.	Young's Modulus		Poisson's Ratio	Log₁₀ of Volume Resistivity			Dielectric Properties at 1 Mc and 20°C			Refractive Index Sod. D Line (.5893 Microns)	Glass Code
	Strain Point °C.	Anneal-ing Point °C.	Softening Point °C.	Working Point °C.			(10⁶lb./sq. in)	(10⁶kg/cm²)		25°C. 77°F	250°C. 482°F	350°C. 662°F	Power Factor	Dielectric Const.	Loss Factor		
19	395	435	625	985	0.8	2.86	8.9	0.63	.21	17.+	8.9	7.0	.16%	6.7	1.%	1.539	0010
17	470	510	695	1005	1.2	2.47	10.0	0.70	.24	12.4	6.4	5.1	.9	7.2	6.5	1.512	0080
20	395	435	630	980	0.8	3.05	8.6	0.60	.22	17.+	10.1	8.0	.12	6.7	.8	1.560	0120
28	670	715	915	1190	2.0	2.52	12.7	0.89	0.25	—	11.4	9.5	.38	7.2	2.7	1.530	1720
25	670	710	910	1175	2.0	2.64	12.5	0.88	0.25	—	13.5	11.3	.16%	6.3	1.0%	1.547	1723
14	330	360	500	755	—	3.47	8.4	0.59	.25	—	10.1	7.7	.04	8.3	.33	—	1990
⁴37	500	530	770	1085	—	2.50	9.9	0.70	0.21	—	—	—	—	—	—	1.507	2405
⁴17	440	480	690	1040	—	2.59	10.0	0.70	—	—	7.8	6.2	—	—	—	1.511	2475
⁴40	500	540	780	1155	—	2.27	9.4	0.66	0.19	—	8.6	7.1	.30	4.9	1.5	1.481	3320
19	510	550	775	1010	—	2.58	10.2	0.72	.21	—	—	—	—	—	—	1.507	6720
⁴18	445	485	670	1040	—	2.59	—	—	—	—	—	—	—	—	—	1.513	6750
⁴23	490	530	770	1010	—	2.65	—	—	—	—	—	—	—	—	—	1.508	6810
37	450	490	700	1080	—	2.24	8.6	0.60	.23	—	9.6	7.8	.20	4.8	1.0	1.480	7040
39	460	500	705	1025	—	2.24	8.7	0.61	.22	16.	8.8	7.2	.33	4.9	1.6	1.479	7050
41	435	480	710	1115	—	2.28	8.2	0.58	.22	17.	9.2	7.4	.26	4.9	1.3	1.484	7052
34	470	510	720	1045	—	2.29	9.2	0.65	.21	—	10.2	8.3	.27	5.7	1.5	1.487	7056
66	455	495	—	1070	4.1	2.13	7.4	0.52	.22	17.+	11.2	9.1	.06	4.1	.25	1.469	7070
48	490	540	780	1190	3.2	2.24	9.2	0.65	.20	15.	8.2	6.7	.27	4.7	1.3	1.475	7250
21	340	365	440	560	—	5.42	8.0	0.56	.28	—	10.6	8.7	.22	15.	3.3	—	7570
49	485	525	755	1140	3.2	2.35	9.1	0.64	.20	16.	8.8	7.2	.27	4.7	1.3	1.487	7720
53	515	565	820	1245	3.1	2.23	9.1	0.64	.20	15.	8.1	6.6	.50	4.6	2.6	1.474	7740
52	480	525	780	1210	—	2.23	9.1	0.64	—	17.	9.4	7.7	.18	4.5	.79	1.473	7760
202	820	910	1500	—	3.5	2.18	10.0	0.70	.19	17.	9.7	8.1	.05	3.8	.19	1.458	7900¹
211	820	910	1500	—	3.5	2.18	9.6	0.67	.19	—	9.7	8.1	.04	3.8	0.15	1.458	7913
290	990	1050	1580	—	3.6	2.20	10.5	0.74	.16	—	11.8	10.2	.001	3.8	.0038	1.459	7940
⁴18	395	435	630	975	—	2.98	—	—	—	—	10.6	8.4	.09	7.0	.63	1.553	8160
22	400	435	600	860	—	4.00	7.8	0.55	.24	—	12.0	9.9	.06	8.3	0.50	1.659	8161
19	300	315	380	460	—	6.22	7.4	0.52	.27	—	9.2	7.5	.19	17.0	3.2	1.97	8363
17	350	385	525	785	—	3.84	8.4	0.59	.26	—	11.1	8.8	.05	8.4	.42	—	8871

TABLE I (Continued)

1	2	3	4	5	6			7		8				9		
					Corrosion Resistance			Thermal Expansion 10⁻⁷in./in./°C		Upper Working Temperautres (Mechanical Considerations Only)				Thermal Shoc Plates 6″ x		
Glass Code	Type	Color	Principal Use	Forms Usually Available				0–300°C 32–572°F	Room Temp.- Setting Point	Annealed		Tempered		Anneale		
					Weath- ering	Water	Acid			Normal Service °C.	Extreme Limit °C.	Normal Service °C.	Extreme Limit °C.	⅛″ Thk. °C.	¼″ Thk. °C.	T
9010	Potash Soda Barium	Grey	TV Bulbs	P	2	2	2	89	102	110	380	—	—	—	—	
9700	Borosilicate	Clear	u v Trans- mission	T U	³1	³1	³2	39	39	220	500	—	—	150	120	
9741	Borosilicate	Clear	u v Trans- mission	B U T	³3	³3	³4	39	49	200	390	—	—	150	120	

COLUMN 1

[1]Glasses 7910,7911 & 7905, for special U.V. and Infrared Applications.

COLUMN 5

B—Blown Ware   P—Pressed Ware   S—Plate Glass
M—Multiform   R—Rolled Sheet   T—Tubing and Rod
U—Panels   LC—Large Castings

COLUMN 6

[2]Since weathering is determined primarily by clouding which changes transmission, a rating for the opal glasses is omitted.
[3]These borosilicate glasses may rate differently if subjected to excessive heat treatment.
See page 338 for further discussion of corrosion resistance.

COLUMN 7

See Text Pages 336 and 337.

COLUMN 8

Normal Service: No breakage from excessive thermal shock is a sumed.
Extreme Limits: Glass will be very vulnerable to thermal sh
Recommendations in this range are based
mechanical stability considerations only. T
should be made before adopting final des
These data approximate only.

COLUMN 9

These data approximate only.
Based on plunging sample into cold water after oven heatir
Resistance of 100°C means no breakage if heated to 110°C a
plunged into water at 10°C. Tempered samples have over twi
the resistance of annealed glass.

Mechanical hardness numbers are empirical; the value depends on the test method. Many hardness tests have been devised for glasses, the most common of which are: scratch hardness tests, grinding or abrasion hardness tests, penetration hardness tests. The method to be used depends on the application.

The hardness of glass cannot be measured by the Brinell or Rockwell methods developed for metals because the high localized pressures fracture the specimens. Knoop, Vickers, scratch, and abrasion tests are usually used to evaluate the hardness of glass. There exists good correlation between the indentation hardness and the elastic (Young's) modulus: glasses with highest indentation hardness also have highest moduli.

On the Mohs scale of scratch hardness glasses lie between apatite (5) and quartz (7). Some common materials that are hard enough to scratch glass include agate, sand, silicon carbide, hard steel and emery. Glasses are harder than mica, mild steel, copper, aluminum and marble.

Impact abrasion resistance of glasses is evaluated by measuring their resistance to sandblasting under standard conditions. Values recorded are relative only, showing resistance as compared to soda lime glass. Data for several glasses are listed in Table 1.

Thermal Expansion

When the temperature is raised, all commercial silica-based glasses expand. In general, the change is smaller than with most ordinary substances; but, because glass is perfectly elastic, and therefore does not locally yield, the expansion is often quite important both in connection with heat shock resistance and in connection with rigid seals to other materials such as metals, ceramics, and other glasses.

10	11				12	13	14			15			16			17	18
Thermal Stress Resist-ance °C.	Viscosity Data				Impact Abrasion Resistance	Density grams per C.C.	Young's Modulus		Poisson's Ratio	Log_{10} of Volume Resistivity			Dielectric Properties at 1 Mc and 20°C			Refractive Index Sod. D Line (.5893 Microns)	Glass Code
	Strain Point °C.	Anneal-ing Point °C.	Softening Point °C.	Working Point °C.			$(10^6\text{lb.}/ \text{sq. in})$	$(10^6\text{kg/} \text{cm}^2)$		25°C. 77°F	250°C. 482°F	350°C. 662°F	Power Factor	Dielectric Const.	Loss Factor		
18	405	445	650	1010	—	2.64	9.8	0.69	.21	—	8.9	7.0	.17	6.3	1.1	1.507	9010
45	520	565	805	1200	—	2.26	9.6	0.67	.20	15.	8.0	6.5	—	—	—	1.478	9700
55	410	450	705	—	—	2.16	7.2	0.51	.23	17.+	9.4	7.6	—	—	—	1.468	9741

COLUMN 10

Resistance in °C is the temperature differential between the two surfaces of a tube or a constrained plate that will cause a tensile stress of 1000 p.s.i. on the cooler surface.

COLUMN 11

See page 330. These data subject to normal manufacturing variations.

COLUMN 12

Data show relative resistance to sandblasting.

COLUMN 15

Data at 25° extrapolated from high temperature readings and are approximate only.

ALL DATA SUBJECT TO NORMAL MANUFACTURING VARIATIONS

For most glasses, the curves are initially nearly linear but the curvature is positive, indicating a higher rate of expansion as the annealing zone is approached. The quantity usually referred to by the term "expansion" but more properly called "expansion co-efficient" or "expansivity" is listed in Table 1 as "Thermal Expansion Coefficient," and is the average slope of the initial portion of the expansion curve. To be more precise, it is the average change of length per unit length per °C between 0°C and 300°C. For sealing applications, values are given in Table 1 for the average expansion coefficient from room temperature to the setting point—arbitrarily defined as 5°C above the strain point. A comparison of the expansion coefficients to the setting point between a material of known expansion and a particular glass will give a good estimate of sealing compatibility. However, for precise predictions of stresses due to expansion differences, complete expansion curves for the sealing materials should be compared.

Thermal Stresses

The coefficient of thermal expansion is the determining factor in the thermal shock resistance of glasses. Two different conditions are to be discussed: steady-state thermal stresses and transient thermal stresses.

STEADY-STATE THERMAL STRESSES. The importance of stresses due to steady-state thermal gradients depends greatly on the degree of constraint imposed by some parts of the item upon others or by the external mounting. Thus under no constraint and with uniform temperature gradient through the thickness, very large temperature difference can be tolerated. Under complete constraint, the stress depends on the temperature difference and on the glass properties (elastic and thermal) and can be calculated. The formula is:

$$S = \frac{\alpha E \Delta T}{2(1 - \mu)}$$

where

S = maximum stress (tension on cooler surface, compression on hotter surface)
α = coefficient of linear thermal expansion
E = Modulus of elasticity
μ = Poisson's ratio, and
ΔT = temperature differential between the two surfaces

It is important to know the temperature difference that approaches the potential danger point of $S = 70$ kg/cm^2 or 1,000 psi for annealed glasses. When complete constraint is imposed:

$$\Delta T_{1000} = \frac{2000 (1 - \mu)}{E \alpha}$$

The face-to-face temperature differentials that will cause a tensile stress of 70 kg/cm^2 or 1,000 psi on the cooler face are listed in Table 1, Column 10, for tubes and constrained plates.

For Code 7740 glass the listed figure is 48°C. Therefore a furnace sight glass in a fully constraining frame with an inner surface temperature of 148°C and an outside face temperature of 100°C will be under a tensile stress of 70 kg/cm^2 or 1,000 psi at the outside surface. It must be remembered that temperature differential means temperature difference between the two glass surfaces, exclusive of gradients at surface itself. In air, particularly, an appreciable difference exists between surface temperature of the glass and of the air moving past it.

TRANSIENT THERMAL STRESSES. When a glass article is suddenly cooled, such as by removal from a hot oven, tensile stresses are introduced in the surfaces and compensating compressional stresses in the interior. Conversely, sudden heating leads to surface compression and internal tension. In either case the stresses are temporary (transient) and disappear on attainment of temperature uniformity.

Since glass will fail from tension at the surface, the temporary stresses from sudden cooling are much more damaging than those resulting from sudden heating, assuming of course, that all surfaces are heated or cooled at the same time.

The transient thermal stresses increase with expansion coefficient and with glass thickness. They also depend upon the shape of the article and on the method of chilling or heating. Thus, a complicated shape would be more severely stressed than a simple one. Sudden chilling by immersion in cold water is more rigorous than by blowing with cold air.

Column 9 of Table 1 illustrates the most extreme case: direct plunging into cold water. Cooling into less severe media, such as air, permits much higher temperature differences than those listed.

CORROSION OF GLASSES

The corrosion resistance of commercial glasses varies with the glass composition. Borosilicate glass, Code 7740, and high silica glass, Code 7900, are extremely corrosion resistant, particularly in acid or neutral solutions.

Factors affecting the rate of corrosion for a given glass are type and concentration of corroding liquid, temperature, contamination of the liquid by the glass and the degree of agitation of the liquid. Agitation removes the products of decomposition from the vicinity of the glass surface and permits more corrosive liquid to contact the glass.

Corrosion may be evidenced in many ways. Most common are clouding of the surface and contamination of solutions in contact. A few other less common effects include increase of electrical surface leakage, surface hydration, loss of strength and surface discoloration. Loss in weight may also occur and is frequently the basis for testing. However, it cannot always be correlated to the above effects. For example, some glasses remain clear even after severe corrosion.

The corrosion of a silicate glass by the following aqueous solutions generally decreases in the order listed: HF, NaOH, Na_2CO_3, NH_4OH, sea salt, H_3PO_4, certain organic acids and chelates, HCl and H_2SO_4, HNO_3, H_2O and neutral salts. Organic solutions are generally less corrosive than water.

Resistance to Water

The resistance of most commercial glasses to corrosion by water is very high. The corrosion that does occur is seldom a matter of simple solution of the glass; it involves ion exchange, hydration, and selective attack. Column 6, Table 1 classifies the various glasses in their resistance to water. Glasses rated 1 will rarely give trouble at temperatures below 100°C. For instance, Code 7740 glass loses only about 125×10^{-7} mm or 5×10^{-7} inches per day when exposed to boiling water. Those rated 2 may show corrosion under adverse conditions. Those rated 3 may require special consideration in their use.

Corrosion increases with temperature and this factor can be significant when dealing with super-heated water.

Resistance to Weathering

Weathering of glass is defined as its corrosion by atmospheric gases such as water, carbon dioxide and other gases. The effect is usually evidenced by clouding or streaking of the surface or electrical surface leakage. Weathering resistance closely follows the ratings in Column 6, Table 1. Glasses rated 1 will virtually never show weathering effects; those rated 2 may occasionally be troublesome, particularly if weathering products cannot be removed; those rated 3 require more careful consideration.

Resistance to Acids

Most silicate glasses are highly resistant to all acids, except hydrofluoric. The rate of corrosion decreases with time, because acids, except HF, attack primarily the alkalis in the glass, leaving a porous, silica rich surface layer which reduces further corrosion. Column 6, Table 1 classifies various glasses as to their resistance to acids. Glasses rated 1 will lose less than 25×10^{-6} mm or 1×10^{-6} inches for one day when exposed to 5% hydrochloric acid at 95°C. those rated 2 between 25×10^{-6} and 25×10^{-5} mm or between 1×10^{-6} and 1×10^{-5} inches, those rated 3 between 25×10^{-5} and 25×10^{-4} mm (1×10^{-5} and 1×10^{-4} inches), and those rated 4 greater than 25×10^{-4} mm or 1×10^{-4} inches. In general the rate at which other acids, except hydrofluoric and hot phosphoric, attack these glasses will be comparable.

The injunction follows: never put hydrofluoric acid or hot phosphoric acid in glass containers.

Resistance to Alkali

Compared to that by acids, the corrosion of glass by strong alkalies is relatively high. The mechanism differs from attack by water or acid in that the entire structure of a silicate glass is dissolved, and the rate of attack is more nearly linear with time.

Corrosion by a given alkali solution will not vary over wide ranges with respect to different glasses as in the case of acid attack. Most glasses will lose between 75 and 330×10^{-4} mm or between 3 and 13×10^{-4} inches per day in 5% NaOH at 95°C. Code 7280 glass is the most alkali resistant glass and will lose only about 125×10^{-5} mm or 5×10^{-5} inches per day in 5% NaOH at 95°C.

REFERENCES

Glass structure and properties

1. Condon, E. U. and Odishaw, H., eds., "Glass," Chapter 8 by Lillie, H. R., Handbook of Physics. McGraw-Hill Book Co., New York, (1958).
2. Hutchins, J. R., and Harrington, R. V., "Glass," Encyclopedia of Chemical Technology, 2nd edition, Interscience Publishers, New York, (1967).
3. Kingery, W. D., "Introduction to Ceramics," John Wiley & Sons, New York, (1960).
4. McKenzie, J. D., ed., "Modern Aspects of the Vitreous State," Butterworth, Washington, (1964).
5. Morey, G. W., "The Properties of Glass," *2nd edition,* Reinhold, New York, (1954).
6. Phillips, C. J., "Glass, The Miracle Maker," Pitman Publishing Co., New York, (1948).
7. Shand, E. B., "Glass Engineering Handbook," McGraw-Hill Book Co., New York, (1958).
8. Stanworth, J. E., "Physical Properties of Glass," Oxford University Press, London, (1953).

Laboratory procedures

1. Parr, L. M., "Laboratory Glass-Blowing," Chemical Publishing Co., New York, (1957).
2. Robertson, A. J. B., et al, "Laboratory Glass-Working for Scientists," Academic Press, New York, (1957).
3. "Guide for Safety in the Chemical Laboratory," D. Van Nostrand, New York, (1954).
4. Lewis, E. J., Proper Care Will Prolong the Life of Chemical Glassware, *Chemical and Engineering News 21:* 552, (April 25, 1943).
5. Laboratory Glassware, *Data Sheet No. 23 (revised),* National Safety Council, Chicago, (1964).
6. Laboratory Glass Blowing with Corning's Glasses, *Bulletin B-12,* Corning Glass Works, Corning, New York, (1961).
7. How to Use and Care for Your Corning Organic Chemistry Kit, *Bulletin CK-2,* Corning Glass Works, Corning, New York, (1964).

Acid Cleaning of Glassware

Barbara Tucker

The clinical or research laboratory has many complex safety problems, including the common one of safety factors involved with procedures used in the acid cleaning of glassware. The major component in most acid cleaning solutions is sulfuric acid and many laboratory personnel have experienced at least one minor acid burn. One survey[1] of three years experience in a university hospital clinical laboratory showed that 25% of the laboratory injuries were caused by acid or alkali, and a Manufacturing Chemists' Association publication[2] lists fourteen case histories of injuries caused by sulfuric acid.

It has been suggested[3] that a job hazard analysis check list be used to determine the safe performance of any job by an individual, and this technique has been applied to acid cleaning of laboratory glassware. This job hazard analysis consists of a detailed listing of key job steps, the tools or equipment used, the potential health and injury hazards which may be involved, and the safety practices or equipment which may be needed to control the hazards.

Job Hazard Analysis for Sulfuric Acid Cleaning of Laboratory Glassware

1. Job description: Provide chemically clean glassware.
2. Job location: A sink within the laboratory area.
3. Key job steps:
   a. Prepare acid cleaning solution—sodium dichromate and concentrated sulfuric acid.
   b. Prepare glassware for acid cleaning by rinsing, soaking, or other means.
   c. Immerse glassware in acid or provide contact of the acid by rotation on the interior of the glassware.
   d. Rinse glassware with tap water.
   e. Rinse glassware with deionized or distilled water.
   f. Oven dry the glassware.
   g. Dispose of contaminated acid.
4. Tools or equipment used:
   a. Carrier for acid bottle.
   b. Safe container for acid solution.
   c. Carrier for pipettes.
   d. Basket or rack for test tubes.
   e. Porcelain or stainless steel dipper for acid solution.
   f. Washer for rinsing pipettes.
5. Potential health and injury hazards:
   a. Sulfuric acid can splash or spill in routine handling or if containers break and cause severe burns on contact with body tissues.
   b. Sulfuric acid mist can present an inhalation hazard if inhaled in concentrations greater than 1 milligram/cubic meter for an eight-hour exposure.
   c. Sulfuric acid can cause a fire hazard if hydrogen is liberated by the action of the acid on metals.
   d. Sulfuric acid can be highly corrosive to many metals and alloys, so that sink traps and waste lines can create a safety hazard to worker and plumber.
   e. Sudden dumping of several liters of concentrated sulfuric acid into drain

lines can cause the release of toxic hydrogen sulfide gas into laboratories through dry traps in waste lines or floor drains.

6. Safety practices, apparel and equipment:
   a. Safety goggles
   b. Face shield
   c. Rubber gloves
   d. Rubber apron

The job hazard analysis outlined above provides a basis for a thorough study of the entire procedure. Several physical modifications could be introduced which would facilitate safer performance by the worker.

The first of these physical modifications would be to provide a lead-lined or stainless steel double sink. One half of this double sink should be considerably deeper to house the acid container and automatic pipette washer, as shown in Figure 1. A perforated copper tubing connected to a cold-water supply is fastened several inches below the top of the sink to provide a constant stream of water to flush away any excess acid solution. The trap and waste line should be of Pyrex Brand drainline construction.

The second physical modification to make acid cleaning of glassware safer would be the use of the "Movable Safety Shield," as shown in Figure 2.

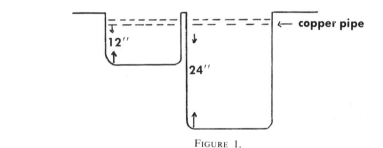

FIGURE 1.

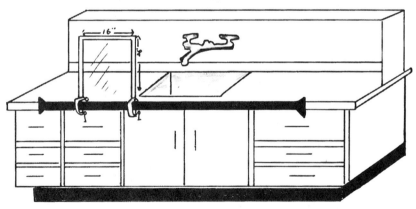

FIGURE 2.

The shield is mounted on a track and is designed to be of any appropriate linear measurement. The entire track can be permanently mounted to a bench or semi-permanently secured by the use of clamps. The worker should roll the safety shield between himself and the acid cleaning solution at all times.

Additional physical requirements should be as follows:

1. The suggested level of ventilation for all laboratory areas is a minimum level of ten changes of air per hour.
2. The lower level of light available for all laboratory areas is 200 foot candles.

3. A safety shower and floor drain should be located immediately behind the operator.

4. A spray or hose should be available with easy access at the sink area.

5. An eye wash fountain should be readily available.

REFERENCES

1. Ederer, G. M. and Tucker, B., Accident Surveys and Safety Programs in Two Hospital Clinical Laboratories, *Amer. J. Med. Techn., 26:*219–229, (1960).

2. "Case Histories of Accidents in the Chemical Industry," *Volume One* Manufacturing Chemists' Association, Inc., Washington, D. C. (1962).

3. DeReamer, R., "Modern Safety Practices." John Wiley & Sons, Inc., New York, p. 47–52, (1958).

4. "Chemical Safety Data Sheet SD-20," Manufacturing Chemists' Association, Inc., Washington, D.C., (1963).

5. Rappaport, Arthur E., "Manual for Laboratory Planning and Design." The College of American Pathologists, Chicago, p. 75, (1960).

6. Laboratory Design Considerations, Part II: Safety in the Chemical Laboratory, Norman V. Steere, Editor: *J. Chem. Ed., 42:*A666, (1965).

Instrument and Equipment Hazards

Edwin A. Wynne*, Barbara Tucker, Grace Mary Ederer and Norman V. Steere

The purchase or fabrication of laboratory equipment with adequate controls and safeguards, the installation of the equipment in safe locations with adequate ventilation, and the use of the equipment within the limitations of the original design or with appropriate modifications are important steps towards insuring safety in the laboratory.

One example of an apparatus that has proven to be a serious hazard is the fraction collector. Fraction collectors have been the cause of fires when they have been used in operations different from those for which they were originally designed. Most fraction collectors were designed for use with aqueous solvents, but serious fires have resulted when flammable solvents were used in the same apparatus and solvent vapors were ignited by the sparking of open electrical contacts. The Forest Products Laboratory in Madison, Wisconsin, solved the problem by adding a small exhaust fan to remove the solvent vapors. With this safety feature added, a fire was prevented when the transport mechanism jammed; the solvent vapor was dispersed by the fan.

Commercially available instruments and laboratory equipment are generally safer and more reliable than laboratory-fabricated instruments. However, safe design which would permit safe operation is not always given adequate attention and consideration by manufacturers. For example, vacuum pumps and condensers are often produced without the minimum safeguard of a guard to cover the belt. Without the protection of a belt guard, a worker can easily be injured by getting his clothing or fingers caught between the belt and the wheel. Belt guards should be required on new vacuum pumps and compressors. This requirement should be stated in the purchase order. Belt guards may be purchased or fabricated and installed on existing models.

Laboratory equipment that requires high voltages, high pressure, or operates at high speeds or high temperatures, requires special safeguards to permit persons working with them, or in the vicinity of them, to perform their work without harm. A recent death from a high voltage device is a case in point.[1]

Electrical and many other safety standards have been established for several types of laboratory equipment. It is possible to specify purchase of equipment that has been tested and labeled by such nationally recognized testing laboratories as Underwriters' Laboratories, Inc., Factory Mutual Laboratories, and Underwriters' Laboratories of Canada, Inc. Some of the types of laboratory equipment that are listed and labeled as meeting safety tests of the Underwriters' Laboratories, Inc. are:

1. Bench lights
2. Colony counters
3. Emergency lighting equipment
4. Explosion-proof appliances for hazardous locations
5. Furnaces
6. Heaters
7. Hot plates
8. Incubators
9. Ovens
10. Shakers
11. Stirrers
12. Water baths

A performance evaluation program for new instruments for clinical laboratories has been proposed by Bradley E. Copeland, M.D., of the Commission on Continuing Education of the American Society of Clinical Pathologists, at New England Deacon-

*Deceased.

ess Hospital in Boston, Massachusetts.[2] Instrument evaluations have been invited for an Instrument Evaluation Registry by Matthew Patton, M.D., Chairman of the Subcommittee on Instruments of the Standards Committee of the College of American Pathologists. This subcommittee has prepared the pamphlet, "A Suggested Guide for Manufacturers for Preparation of Manuals of Operation for Laboratory Instruments." Dr. Patton is at Sacred Heart General Hospital in Eugene, Oregon.

Technical literature should be obtained from manufacturers prior to purchasing instruments or equipment, and most companies will provide schematics of their instruments. All manufacturers will supply operating instructions. It would be wise to study the schematic diagram of the instruments to determine repairability and safety aspects. After instruments or equipment are purchased, there should be a check-out procedure using the operating instructions. A good axiom in using equipment is to observe its limitations and not force or employ it for work for which it was not designed.

A list of hazards and safeguards which should be considered in purchasing, setting up, and using laboratory equipment is presented. This list is not complete, and additions should be made when new knowledge is acquired.

INSTRUMENTS USING GAS FROM CYLINDERS OR PIPING SYSTEMS

Use of gases, pressure bottled or piped, requires normal precautions for determining positive closure of all fittings to prevent leaks. The list below will help in determining safety check points on equipment such as flame photometers, automatic analyzers, spectrophotometers, and similar equipment:

1. Adequate safeguards to prevent inadvertent ignition.
2. Suitable check valves to prevent gases from surging back into gas line.
3. Flame arrestors, where necessary, to prevent flash back of flame into gas lines.
4. Purchase or use of fuel in approved gas cylinders, within the maximum size and quantity limits established by local fire authorities, particularly in hospital laboratories.
5. Equipment for holding gas cylinders stable should be required and used with equipment requiring gas cylinders.
6. Gas-tight fittings and hoses rated for the application, if permanent piping is not practicable.
7. Pressure limiting devices to prevent overpressure to the system.
8. Durable and appropriate piping or tubing; e.g., no oil in oxygen lines, no copper in acetylene lines, gas hoses provided to meet service specifications.
9. Atomic absorption spectrophotometers, in particular, need to have suitable exhaust connections to prevent build-up in work areas of hazardous concentrations of toxic gases, vapors or fumes emerging from the flame unit.
10. Flames should be guarded by built-in shields which will attenuate any explosion, eruption or injurious radiation, and protect the person using the equipment.

CENTRIFUGES

In selecting a centrifuge, carefully consider location, type and use. If a table-top type is acquired, make certain that it is securely anchored. Instruct all users on the importance of balancing each time the centrifuge is used. Other important check points are:

1. Adequate shielding against accidental "fly-aways"—double walls provide greater protection.
2. Prevention of "walking," with suction cups or wheel brakes.
3. Accessibility of parts, particularly for rotor removal.

4. Top equipped with disconnect switch which shuts off rotor if top is inadvertently opened.

5. Safeguards for handling flammables and pathogens, including positive exhaust ventilation, to a safe location from which recirculation is very unlikely.

6. Positive locking of head.

7. Electrical grounding.

8. Location where vibration will not cause bottles or equipment to fall off shelves.

OVENS, FURNACES, HOT PLATES, MANTLES AND OTHER HEATING EQUIPMENT

Caution personnel to use asbestos gloves and tongs when placing or removing samples into or from ovens, furnaces and hot plates. Mantles should be used only with suitable variable powerstats so that rated wattage is not exceeded; if a higher wattage heater is necessary, a higher rating should be provided rather than using the full input of the powerstat. Additional checkpoints are:

1. Blow-out panels or magnetic latches. Latches are more desirable if they open at pressures just above one atmosphere.

2. Forced draft, well designed convection, or inert-gas purging to prevent explosive concentrations.

3. Controls marked in definite units.

4. Reliable and well-maintained thermostatic controls.

5. Double controls for use with hazardous materials.

6. Explosion-proof contactors, controls and switches for equipment to contain or be used in potentially explosive atmospheres.

7. Controls that fail on the safe side.

8. Higher wattage ratings provided by design for 220V, rather than 110V.

9. Electrical grounding of all exposed metal or electrically conductive cabinets and parts.

10. Sealed thermostats to prevent possible ignition of flammable solvents.

11. On-off switch with pilot light is a must for hot plates.

12. Hot plate wiring should not be exposed.

13. Hot plates should be designed for laboratory use (not the hardware store type) to prevent hot spots.

14. Electrical controls on mantles should have thermostats to prevent excessive heating.

15. Facilities for an inert-gas purging of mantles in case of spillages.

16. Bright indicator light on heating equipment with time switches, to show when equipment has been set to turn on later (intentionally or inadvertently).

ELECTRONIC EQUIPMENT

Select equipment adequate for the uses for which it has been designed. Always follow manufacturers' directions for installation or request the manufacturer's technical representative to install equipment. Plan to acquire and use proper protective glasses where ultraviolet fluorescent equipment is to be used. If X-ray, neutron activation, or similar processes are employed, insure that personnel are instructed in the hazards of stray radiation. Additional checkpoints are:

1. Protection against the possibility that a zero voltage setting may give full line potential, as can occur with variable auto-transformers.

2. Clear distinction between instrument circuits and building electrical circuits.

3. A desirable safety feature would be the introduction of a double pole-switch and/or a polarized or keyed system to prevent insertion of an improper plug.

4. Electric motors and controls should be chosen for specified jobs: explosion-proof for use with flammables; nonexplosion-proof types for general use.

5. Equipment should be fused, and have under-voltage protection in event of a partial power failure.

6. Equipment should have a combination of magnetic and thermostatic branch circuit protection.

7. Lamp holders and switch terminal insulation should be ceramic, particularly where used in locations of high relative humidity.

8. Powerstats should be properly shielded and supplied with built-in fusing.

9. Temperature limit switches should be provided to prevent overheating.

10. Electrical immersion heaters should be protected against overheating through failure of thermostatic controls, by use of an automatic cutoff and by a low water level cutoff.

11. Gas chromatograph electronics should be insulated—particularly the "hot" side.

12. Gas chromatographs should be ventilated adequately.

13. All cords for heating units should have insulation approved by Underwriters' Laboratories, Inc.

14. X-ray equipment should be shielded to prevent stray radiation.

15. Neutron-guns should have built-in shielding and adequate directions for use.

16. Lasers should be well protected to prevent shock, accidental discharge and stray radiation.

MISCELLANEOUS

1. Pressure release controls should be provided for opening autoclaves.

2. High-pressure types of autoclaves should have adequate explosion protection, and all controls should be of the remote type, so the operator can be behind a protective wall.

3. Calorimeter bombs should be adequately shielded as protection against explosions.

4. Safety shields required for equipment should be sized and designed to provide adequate coverage, stability and resistance to penetration and convenience of use.

5. Microtomes—Positive lock to prevent unexpected operation; guard to keep operator's body parts out of the path of long knives which may project beyond the sectioning area.

6. Fraction collectors—Ventilated to exhaust leaks or spills of flammable liquids which may be used; explosion-proof construction, if possible, or design to keep ignition sources out of area where flammable vapors may accumulate; consider location in an area or enclosure where the fire damage potential will be minimized; signal to indicate cycle completion.

7. Tissue processors; e.g., Auto-Technicon® and Tissuematon®—Consider location in a separate room or closet with special exhaust ventilation, and with curbing to confine any spill of the flammable liquids used in the equipment.

8. Paraffin dispensers and vacuum infiltrators for paraffin—Automatic over-temperature shutoff in series with thermostatic control so that thermostat failure will not result in overheating and fire.

9. Electrophoresis apparatus—Proper grounding and electrical interlocks are necessary to prevent personnel from possible contact with any voltages which could cause shock or death.

10. Chromatography equipment—Carefully controlled ventilation to remove any excess or air-borne vapors that are toxic, narcotic, irritating or a nuisance. Excessive exhaust rates on chromatography jars will cause temperature differentials that introduce gross errors by alteration of migration rates.

11. Equipment for distillation, zone melting and similar equipment to be used on a continuous run should have "fail-safe" attachments to prevent serious consequences in case of water pressure changes or electrical failure.

12. Refrigerators—Many laboratory explosions have resulted when ordinary domestic refrigerators have been used for storage of flammable liquids and leaking vapors have reached one of the many ignition sources within such refrigerators. One means of preventing such explosions has been twofold: to mark all ordinary refrigerators as unsuitable for flammable liquid storage, and where such storage is necessary, to modify ordinary refrigerators by removing all lights, switches, heating units, open thermostats and other ignition sources. Several universities have standard purchase specifications requiring such modifications by suppliers of refrigerators for laboratories. Another means of preventing refrigerator explosions has been the development of explosion-proof refrigerators by three manufacturers. Such refrigerators are tested and listed as explosion-proof by Underwriters' Laboratories, Inc., and the entire refrigerator would be safe in most explosive atmospheres. A question remains whether the expense of such explosion-proof refrigerators is justified in laboratories where there are many sources of ignition and very few pieces of laboratory equipment can be purchased that are explosion-proof.

13. Environmental chambers, incubators and walk-in coolers—Walk-in coolers, incubators and environmental chambers should be equipped so that persons cannot be trapped inside, so that persons inside can signal an emergency and summon help, and so that toxic or flammable concentrations can be prevented by regular or special means. Even in cold chambers it is practicable to provide routine exhaust ventilation to prevent development of hazardous atmospheres from normal activities in the chamber. Standard equipment for cold rooms at one university includes inside alarm signals and gasket heaters to prevent doors from freezing shut.

SUMMARY

Safety check lists have been presented which should be considered when purchasing laboratory equipment. Many of the precautions listed may become unnecessary as the manufacturers of scientific apparatus improve design and build in safeguards in response to demand and the specific requirements of users. These precautions and safeguards recommended for purchasing, installing and using instruments and equipment are not inclusive, and the authors invite reports of experience with new equipment and with new uses of existing equipment.

REFERENCES

1. Spencer, E. W., Ingram, V. M., and Levinthal, C., Electrophoresis: An Accident and Some Precautions, *Science, 152,* 1722–23 (1966).
2. Copeland, Bradley E., *J. Ass. Advan. Med. Instrum., 1,* 28–31 (1966).

Compressed Gas Cylinders and Cylinder Regulators

George Pinney

The use of compressed gases in educational, medical, and industrial laboratories presents many hazards not encountered in non-laboratory use. The problems include the wide variety of flammable, toxic, and radioactive materials and their mixtures with properties that are frequently unfamiliar to the researcher, and the propensity of laboratory personnel to modify, adapt, and repair cylinder valves and regulators themselves, rather than to leave such work to their suppliers or specially trained personnel. Incorporating a cylinder into an experimental apparatus so that foreign materials can enter the cylinder or so that the cylinder may be subjected to extreme pressures is an extremely hazardous practice that is unfortunately fairly common.

COMPRESSED GAS

The regulations of the U. S. Department of Transportation define a compressed gas as one having a pressure in the container of 2.8 kg/cm^2 (40 psia) or greater at 21°C (70°F) or, regardless of the pressure at 21°C (70°F), having an absolute pressure exceeding 7.31 kg/cm^2 (104 psia) at 54°C (130°F). In addition, any liquid flammable material having a Reid vapor pressure exceeding 40 psia at 38°C (100°F) is classified as a compressed gas.

The regulations define the minimum pressure in a cylinder, but the maximum pressure can range from a low of a few atmospheres for poisonous or radioactive materials to 500 atmospheres or above for the non-condensible gases.

The maximum working pressure for the common 3A and 3AA seamless steel cylinders is stamped permanently into the metal of the cylinder near the valve. This pressure may be exceeded by 10% if certain additional requirements are met.

Regardless of the pressure rating of the cylinder, the pressure of the gas in the cylinder will depend to a great extent on its physical state. So-called permanent gases have a pressure more or less in direct proportion to the remaining contents. Gases which are liquefied in the cylinder, such as carbon dioxide, propane, ammonia, and others, will exert their own vapor pressure as long as any liquid remains or their critical temperatures are not exceeded.

The cylinder sizes encountered are equally varied, ranging from the common lecture bottle 5 cm × 38.1 cm (2 in. × 15 in.) and weighing about 1.4 kg (3 lb), through the large industrial type, 23 cm × 140 cm (9 in. × 55 in.) and weighing up to 64 kg (140 lb).

General Rules

Compressed gas cylinders can be used in laboratories with complete safety if the following general rules are complied with completely during cylinder receiving operations, storage, transportation to the laboratory or other use point, usage, and empty cylinder disposal.

Know Contents of Cylinder. No cylinder should be allowed to remain on your premises unless its contents can be quickly and completely determined by the wording on the cylinder or a tag securely attached to the cylinder. If the tag becomes detached or the label is defaced, the cylinder should not be used, but should be marked "contents unknown" and placed in the empty stock for return to the supplier. Do not remove this identification from empty cylinders as this might present a hazard to your supplier. Do not rely on a color code as this will vary from supplier to supplier. An example of this is that one company paints hydrogen cylinders red while another supplier paints his oxygen cylinders red. Also, colors appear different under artificial light; and, finally, many persons are color blind.

Know Properties of Contents.   Knowledge of the properties of cylinder contents is primary to a laboratory operation due to the unusual uses to which gases may be put, as well as the uncommon gases or gas mixtures used. Not only should the flammability, corrosiveness, or oxidation potential be known but the physiological properties must be kept in mind—such as toxic, anesthetic, or irritating qualities. Two examples are that toxicity and flammability of carbon monoxide must both be kept in mind, while for hydrogen sulfide its toxicity and ability to desensitize the sense of smell should be recognized.

Handle Cylinders Carefully.   Cylinders are primarily shipping containers and as such are constructed to be as light as possible consistent with safety and durability. Rough handling or abuse, such as using a cylinder for a roller to move heavy equipment or using one as a hammer, could seriously weaken the cylinder and render it unfit for further use.

Store Cylinders Appropriately.   Store and use in ventilated areas away from heat or ignition sources. Store flammables away from other gases. Limit the quantity stored in one location. Cylinders containing gases under high pressure could very quickly render an area unsafe if the large volume of gas should be released. Most cylinders, except those in toxic gas service, are equipped with safety devices of the rupture disk type. The pressure-relief devices may function prematurely if cylinders are heated to a temperature in excess of 52°C (125°F) and release the entire content of the cylinders. Also, cylinders containing low vapor pressure liquids could become liquid-full at elevated pressures and burst. If a cylinder must be heated, this should be done in a very well-thermostated water bath heated to not more than 52°C (125°F). In one industrial laboratory a researcher, in an effort to increase the flow rate of propane from a cylinder, turned an infrared heat lamp on it from a distance of about 10 cm (4 in.). This lamp was aimed at a fusible metal plug which shortly released and jetted a stream of liquid propane on the heat lamp. For some reason the gas did not ignite even though the researcher had left the room. If the gas had ignited, the resultant explosion and fire could have leveled the entire laboratory.

Fasten Cylinders Securely in Use or Storage.   If a cylinder should fall or roll off a bench it might break off the regulator, releasing a large quantity of gas and causing the cylinder to pin-wheel, which can break a person's legs; another danger is that the valve could shear off and the cylinder might "rocket" like a projectile due to release of pressure.

Transport Cylinders Safely.   Transport large cylinders only on a wheeled cart. Do not slide or roll them, since it is easy to lose control of a cylinder while rolling or dragging it no matter how much practice a man might have. If one falls it could land on the foot, severely crushing the toe bones. When a cylinder is dragged, the valve protective cap may pull off and strike the man in the face. Mishandling of cylinders in transit is the cause of many pulled muscles and back injuries.

Do Not Tamper with Cylinders.   Never tamper with any part of a valve such as the safety nut or stem packing nut. A leak at either place is potentially hazardous; such cylinders should be marked as leakers and removed to an open area until picked up by the supplier. Do not put unmarked leaking cylinders in the empty pile, as it is contrary to Department of Transportation Regulations to ship leaking containers by common carrier. There have been fatalities in laboratories caused by unfamiliarity with

Courtesy of the 3M Company

FIG. 1. Illustration of a method of securing compressed gas cylinders in storage.

Courtesy of the 3M Company

FIG. 2. Illustration of a specially built cart for transporting six large cylinders of compressed gas. (Cart in background is loaded with safety cans used for handling flammable liquids.)

valves. In one instance, the safety nut was confused with an outlet cap, which is frequently installed on the outlet, and the safety nut was completely removed. Unfortunately, the gas was carbon monoxide. You will note that the safety nut connects directly to the valve inlet and once removed, the flow of gas cannot be stopped.

Do Not Strike Arcs on Cylinders.   Do not strike an electric arc on cylinders. This rule is applied primarily to industrial use, where inert gases are used for shielded arc welding. It is very tempting to test the arc on the large metal surface. Arc burns, however, not only are stress raisers, but, due to metallurgical changes, could cause the heat-affected portion of the cylinder to become brittle.

Use Compressed Gases with Appropriate Equipment.   Use cylinders only with equipment suitable for the contents and do not force the connection or use homemade adaptors. The importance of this rule cannot be overemphasized. Accidents have occurred because of attaching flammable gas regulators to oxygen cylinders, improperly identifying the contents of a cylinder, and so forth. USASI Standard B57.1 lists the various standard connections for compressed gases. There are 38 connections listed, classified into four thread divisions: left and right hand thread and internal and external thread, plus some pipe threads and yoke type connections. The various gases are assigned to connections so that hazardous interconnections cannot be made. Upon application to the Compressed Gas Association, recommendations are made for the connection to be used for newly developed or newly marketed materials. Generally speaking, left hand threads are reserved for flammable gases and right hand threads for non-flammables. There are a few exceptions made necessary by previous practice. Almost always, however, hazardous connections cannot be made except by homemade adaptors or by forcing the connection.

Do Not Use Cylinders Without a Regulator. Cylinders contain pressures greater than the pressures which most laboratory equipment can withstand. The inadvertent closing of a vent valve or stop cock or the plugging of a line or mercury trap could cause a violent failure of the apparatus. There are fine needle valves available which can reduce the flow of gas from the high pressure cylinder to a few bubbles a minute. Such valves are not regulators and the design of any equipment used with them must keep this fact in mind.

Close Cylinder Valves When Not in Use. Do not stop the gas flow from cylinders overnight by backing off on the regulators. Even the best of regulators can develop seat leaks and allow excessive pressures to develop in using equipment. Closing the valve will eliminate this hazard. If this rule is followed meticulously, any question as to the position of a cylinder valve in an emergency is removed. Finally, no foreign materials can enter the cylinder if through leakage or other malfunction the cylinder pressure should become lower than the pressure in some other part of the apparatus.

Close Valves on Empty Cylinders and Mark the Cylinder "Empty." If cylinders are returned to the supplier with the valve open, the interior will become contaminated with atmospheric air and moisture. Such cylinders cannot be used for high purity gases without an extensive reconditioning. If the cylinder had contained such materials as anhydrous hydrogen chloride, or chlorine, this resultant humid atmosphere would corrode the cylinder very rapidly. Empty cylinders should be so marked and stored separately to avoid returning full cylinders to the supplier or sending empties to laboratory or other use point.

Never Attempt to Refill a Cylinder. It is very tempting to refill your own small cylinders from large ones by interconnecting them with high pressure tubing. There are a number of reasons why this practice is hazardous. The cylinder being filled may, unknown to you, have a lower working pressure than the large cylinder. Too rapid a filling can result in extremely high cylinder temperatures which could damage the valve. The cylinder being filled may contain a residue of some reactive material. For cylinders containing liquids, the DOT prescribes filling weights which result in a vapor space in the cylinder at temperatures and pressures up to that at which the safety device functions. If these weights are exceeded, the cylinders may become liquid-full at room temperatures and fail.

At each refilling your supplier gives all cylinders a careful visual inspection for defects, some of which are not obvious, and gives those cylinders which require it a hydrostatic pressure test at legally required intervals. All other cylinders must be given extremely thorough internal and external visual inspection at stated intervals. This inspection and testing can be best performed by one having a thorough knowledge of cylinders.

Finally, some laboratory and speciality gases are available in lightweight minimum-design cylinders which are classified as non-refillable by the DOT. In other words, for safety reasons such a cylinder must be discarded after use the same as the common aerosol spray cans.

Further Information

Much valuable information on the subject of cylinders and safety is contained in the various pamphlets of the Compressed Gas Association, the Chlorine Institute, the National Safety Council, and other national organizations. Pamphlets available from the Compressed Gas Association include those on Acetylene, Ammonia, Sulfur Dioxide, Oxygen, Hydrogen, and Carbon Dioxide. The pamphlets are separated into groups on cylinders, valves, safety devices, and gases. A complete list of all publications is available from the CGA at 500 Fifth Avenue, New York City, New York. Local suppliers of compressed gases can be valuable sources of safety information.

Incidentally, the most satisfactory way of avoiding some hazards which have been described is to keep compressed gas cylinders out of the laboratory completely. This does not rule out use of compressed gases as the cylinders themselves can be placed in a properly selected area and tubing run from regulators to the point of use. One analytical laboratory requires about 25 cylinders of various gases and not one of the cylinders is in the laboratory itself.

Summary of General Rules

1. Know cylinder contents. Never remove or deface contents identification label.
2. Know properties of contents.
3. Handle cylinders carefully.
4. Store cylinders in ventilated area away from heat or ignition sources.
5. Fasten cylinders securely in use, transit, or storage.
6. Transport large cylinders only on a wheeled cart.
7. Never tamper with any part of a valve such as the safety nuts or packing nuts.
8. Do not strike an electric arc on cylinders.
9. Use cylinders only with equipment suitable for the contents. Do not force connections or use homemade adaptors.
10. Do not use cylinders without a regulator.
11. Close cylinder valves when not in use.
12. Close valves on empty cylinders and mark the cylinder "empty."
13. Never attempt to refill a cylinder.

ACKNOWLEDGEMENT: From Proceedings of the Twelfth National Conference on Campus Safety, June, 1965. Revised in 1969.

REFERENCES

Pamphlets issued by the Compressed Gas Association:
"Safe Handling of Compressed Gases", *P-1.*
"Characteristics & Safe Handling of Medical Gases," *P-2.*
"Safe Handling of Cylinders by Emergency Rescue Squads," *P-4.*
"Suggestions for the Care of High Pressure Air Cylinders for Underwater Breathing," *P-5.*
"Compressed Gas Cylinder Valve Inlet and Outlet Connections," (ASA Standard B57.1), *V-1.*
"Pin-Index Safety System for Flush-Type Cylinder Valves," *V-2.*
"Recommendations for the Disposition of Unserviceable Containers," *C-2.*
"Method of Marking Portable Compressed Gas Container to Identify the Material Contained," *C-4.*
"Standards for Visual Inspection of Compressed Gas Cylinders," *C-6.*
"Safety Relief Device Standards," *S-1.1, S-1.2 and S-1.3.*
"Acetylene," *G-1.*
"Anhydrous Ammonia," *G-2.*
"Storage and Handling of Anhydrous Ammonia," (ASAK 61.1), *G-2.1.*
"Sulfur Dioxide," *G-3.*
"Oxygen," *G-4.*
"Hydrogen," *G-5.*
"Carbon Dioxide," *G-6.*
Handbook of Compressed Gases, Compressed Gas Association, Reinhold Publishing Corp., New York, 1966,

Cold Traps

Alvin B. Kaufman and Edwin N. Kaufman

Cold traps are used in instrumentation and elsewhere to prevent the introduction of vapors or liquids into a measuring instrument from a system, or from a measuring instrument (such as a McLeod gauge) into the system. A cold trap provides a very-low-temperature surface on which such molecules can condense, and improves pump-down by one or two magnitudes.

However, cold traps improperly employed can impair accuracy, destroy instruments or systems, and be a physical hazard. For example, many of the slush mixtures used in cold traps are toxic or explosive hazards, and this is not indicated in the literature.

The authors became aware of the deficiencies in tunnel instrumentation, where it was necessary to measure pressures in the micron to 760-torr* region. The instrumentation system used Statham gauges for ambient pressure down to 100 to 150 torr or about 2–3 psia and NRC Alphatron gauges for pressures to 5×10^{-2} torr. To prevent calibration shifts and contamination of the NRC transducers by oil fumes from the vacuum pump and possible wind-tunnel contaminants, a cold trap was placed in the line.

The cold trap was filled with liquid nitrogen, and the valve to the tunnel line shut off. When the valve was opened, cold gas shot out, shown by condensation; the overpressure developed in the system destroyed the Statham strain-gage bridge, although it was not sufficient to rupture the transducer diaphragm. As no satisfactory explanation was forthcoming, a glass cold trap was procured and set up in a dummy system. The cause of the phenomenon soon became apparent: air in the trap and system lines was becoming liquefied in the trap. When the valve was opened, this liquid air was being blown into the warmer lines by atmospheric pressure; the resultant volatilization of liquid into gas was practically an explosion.

Nevertheless, cold traps are often the only satisfactory means of removing contaminants, although in ordinary experimental work the charcoal trap is occasionally acceptable. A charcoal trap will remove oil and condensable vapors so that pressures to 10^{-8} torr or better may be secured, but it presents a serious restriction on pumping speed and requires bakeout when it has become charged with oil and vapors. Molecular sieve traps place similar restrictions on pumping speed.

The errors introduced by water vapor, when measuring low pressures, depend on the vacuum gauge used. The presence of water vapor also affects the magnitude of vacuum that can be achieved. The equilibrium point of a dry-ice-acetone slush is $-78°C$ ($-108.4°F$), which, although sufficient to trap mercury vapor effectively, does not remove water vapor; a temperature of at least $-100°C$ ($-148°F$) is required to eliminate water vapor, or, alternatively, exposure to anhydrous phosphorus pentoxide (P_2O_5). This material is usually rejected for field use because of possible biological, fire and explosive hazards: in absorbing water it produces heat, and reacts vigorously with reducing materials. Slush mixtures using liquid air and liquid oxygen were considered and dropped, either because of the explosive hazard or toxicity of the vapors,[2] or because they were not cold enough. Table 1 lists many common thermal transfer and coolant fluids with their hazards and limitations.

Carbon dioxide is adequately caught by traps cooled by liquid air or nitrogen, its vapor pressure at the liquid-air temperature being 10^{-6} to 10^{-7} torr. Methane, ethyl-

*The torr is equal to a pressure of 1 mm of mercury at standard condition.

ene and carbon monoxide have considerably higher vapor pressures and are not effectively trapped by even a liquid-nitrogen trap.

Vapor pressure of most standard roughing pump oils is 10^{-3} to 10^{-4} torr at 25°C (77°F), $\frac{1}{5}$ of this value at 0°C, and negligibly small below the temperature of dry ice. Fractionating oils currently used in vacuum pumps have very low vapor pressures, ranging from 20×10^{-6} to 10^{-7} torr, and pose no problem for most work. Nevertheless, gases produced by thermal decomposition of the oil may contaminate the vacuum unless trapped.

TABLE 1

Thermal Transfer Fluids[1] Used with Instrumentation Cold Traps

Element[2]	Temperature[3]		Hazard[4]			Remarks
	°C	°F	Inhalation Toxicity	Skin Toxicity	Explosive or Fire	
Glycerine 70% by weight water 30%	−38.9	−38.0	None	Slight	Slight	Vapor pressure 2.5×10^{-3} torr @ 50°C
Ethyl alcohol, dry ice	−78	−108	Moderate	Slight	Dangerous	
Ethylene glycol 52.5% by volume, water 47.5%	−40	−40	None	Slight	Slight	
Chloroform, dry ice. (Trichloromethane)	−63.5	−82	Extreme	Slight	Slight	Vapor pressure 100 torr @ 22°C
Liquid SO_2	−75.5	−103	Extreme	Extreme	None	Very dangerous
Methanol (Methyl alcohol), dry ice	−78	−108	Slight	Slight	Dangerous	Vapor pressure 100 torr @ 22°C Ingestion very dangerous
Acetone, dry ice	−78	−108	Moderate	Slight	Dangerous	Vapor pressure 400 torr @ 39°C
Methyl bromide, dry ice	−78	−108	Extreme	Extreme	Moderate	Very dangerous
Fluorotrichloromethane (Freon 11) dry ice	−78	−108	Slight	Slight	None	
Methylene chloride, dry ice	−78	−108	Moderate	Moderate	Slight	Very dangerous to the eyes. Vapor pressure 380 torr @ 22°C
Calcium chloride	−42	−44	Slight	Slight	None	
Ethyl methyl ketone	−78	−108	Moderate	Moderate	Dangerous	

[1] Transfer fluids will freeze solid and become colder if subject to temperatures lower than their freezing point. A slush mixture is secured by lowering temperature, such as by the introduction of limited quantities of dry ice until the mixture is quasi-frozen.

[2] These materials are often sold under tradenames (listed in the Handbooks of Chemistry and Physics). In general, any combination of elements shown was selected for the coldest slush mixture obtainable.

[3] If the refrigerant is dry ice, the transfer fluid will not go below −78°C (−108.4°F), the temperature of solid CO_2.

[4] The consensus is that many of these liquids while hazardous at room temperature, are not hazardous when cooled, since their evaporation at low temperatures is fairly low. For utmost safety those noted dangerous should not be employed unless venting or other special precautions are taken. For greater detail see Reference 2.

VIRTUAL LEAKS

If the cold trap is chilled too soon after the evacuation of the system begins, gases trapped will later re-evaporate, when the pressure reaches a sufficiently low value. The evaporation of the refrigerated and trapped gases is not rapid enough to be evacuated by the system, but is enough to degrade the vacuum, producing symptoms very similar to those of a leak.

To avoid these virtual leaks, keep the traps warm until a vacuum of about 10^{-2} torr is obtained. The tip of the trap is then cooled until ultimate vacuum is reached, at which time the trap may be immersed in the coolant to full depth.

SAFETY PRECAUTIONS

If liquid nitrogen is the coolant, liquid air can condense in the trap, inviting explosion. Liquid air, comprising a combination primarily of oxygen and nitrogen, is warmer than liquid nitrogen. Depending on the nitrogen content, air liquefies anywhere from $-190°C$ ($-310°F$) ($5°C$ warmer than liquid nitrogen) to $-183°C$ ($-297.4°F$) (liquid oxygen). If liquid nitrogen is used, the trap should be charged only after the system is pumped down lest a considerable amount of liquid oxygen condenses, creating a major hazard.

Handle any liquid gas carefully; at its extremely low temperature, it can produce an effect on the skin similar to a burn. Moreover, liquefied gases spilled on a surface tend to cover it completely and intimately, and therefore cool a large area.

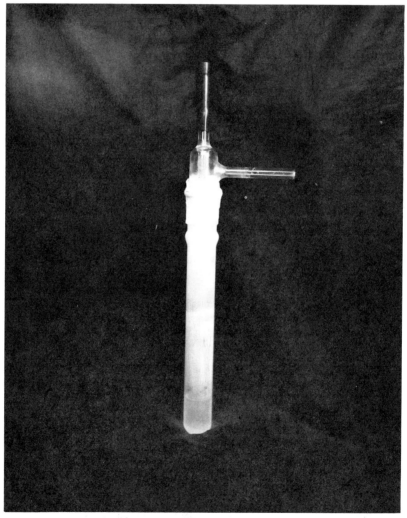

FIG. 1. This photograph illustrates the liquefaction of air in an instrumentation trap with its exposure to liquid nitrogen.

The evaporation products of these liquids are also extremely cold and can produce burns. Delicate tissues, such as those of the eyes, can be damaged by an exposure to these cold gases which is too brief to affect the skin of the hands or face.

Eyes should be protected with a face shield or safety goggles (safety spectacles without side shields do not give adequate protection). Gloves should be worn when handling anything that is or may have been in contact with the liquid; asbestos gloves are recommended, but leather gloves may be used. The gloves must fit loosely so that they can be thrown off quickly if liquid should spill or splash into them. When handling liquids in open containers, high-top shoes should be worn with trousers (cuffless if possible) worn outside them.

Stand clear of boiling and splashing liquid and its issuing gas. Boiling and splashing always occur when charging a warm container or when inserting objects into the liquid. Always perform these operations SLOWLY to minimize boiling and splashing.

Should any liquefied gas used in a cold trap contact the skin or eyes, immediately flood that area of the body with large quantities of unheated water and then apply cold compresses. Whenever handling liquefied gases, be sure there is a hose or a large open container of water nearby, reserved for this purpose. If the skin is blistered, or if there is any chance that the eyes have been affected, take the patient immediately to a physician for treatment.

Oxygen is removed from the air by liquid nitrogen exposed to the atmosphere in an open Dewar. Store and use liquid nitrogen only in a well ventilated place; owing to evaporation of nitrogen gas and condensation of oxygen gas, the percentage of oxygen in a confined space can become dangerously low. When the oxygen concentration in the air becomes sufficiently low, a man loses consciousness without warning symptoms and will die if not rescued. The oxygen content of the air must never be allowed to fall below 16%.

The appearance of a blue tint in liquid nitrogen is a direct indication of its contamination by oxygen, and it should be disposed of, using all the precautions generally used with liquid oxygen. Liquid nitrogen heavily contaminated with oxygen has severe explosive capabilities. In addition, an uninsulated line used to charge Dewars will condense liquid air; liquid air dripping off the line and revaporizing causes an explosive hazard during the charging operation.

If the cold trap mixture is allowed to freeze, and the cold trap becomes rigid, slight movement in other parts of the apparatus could result in breakage of the trap or other glassware.

If a gas trap has to be lifted out of the Dewar cold bath for inspection, it will be difficult to reinsert into the slush. Therefore, it is preferable to use a liquid that will not freeze at −78.5°C.

ACKNOWLEDGEMENT: Reprinted with permission of Rimbach Publications, Pittsburgh, Pennsylvania, "Instrument and Control Systems," Vol. 36, pp. 109–111, July, 1963.

REFERENCES

1. Strong, J., Neher, H. V., Whitford, A. E., Cartwright, C. H., and Hayward, R., "Procedures in Experimental Physics," Prentice-Hall, Inc., New York, (1938).
2. Sax, N. Irving, "Dangerous Properties of Industrial Materials," Reinhold Pub. Co., New York, (1961).
3. Dushman, Lafferty, "Scientific Foundation of Vacuum Techniques," John Wiley & Sons, New York, (1962).

Cryogenic Safety

Eric W. Spencer

Cryogenics may be defined as low temperature technology, or the science of ultra-low temperatures. To distinguish between cryogenics and refrigeration, a commonly used measure is to consider any temperature lower than $-73.3°C$ ($-100°F$) as cryogenic. Although there is some controversy about this distinction, and some who insist that only those areas within a few degrees of absolute zero may be considered as cryogenic, the broader definition will be used here.

Low temperatures in the cryogenic area are primarily achieved by the liquefaction of gases, and there are more than twenty-five which are currently in use in the cryogenic area; i.e., gases which have a boiling point below $-73.3°C$ ($-100°F$). However, the seven gases which account for the greatest volume of use and applications in research and industry are helium, hydrogen, nitrogen, fluorine, argon, oxygen, and methane (natural gas).

Cryogenics is being applied to a wide variety of research areas, a few of which are: food processing and refrigeration, rocket propulsion fuels, spacecraft life support systems, space simulation, microbiology, medicine, surgery, electronics, data processing, and metalworking.

TABLE 1
Properties of Cryogenic Fluids

Gas	Normal B.P. °C	°K	Vol. expansion to Gas	Flammable	Toxic	Odor
Helium-3	−269.9	3.2	757 to 1	No	No[c]	No
Helium-4	−268.9	4.2	757 to 1	No	No[c]	No
Hydrogen	−252.7	20.4	851 to 1	Yes	No[c]	No
Deuterium	−249.5	23.6	. . .	Yes	Radioactive	No
Tritium	−248.0	25.1	. . .	Yes	Radioactive	No
Neon	−245.9	27.2	1438 to 1	No	No[c]	No
Nitrogen	−195.8	77.3	696 to 1	No	No[c]	No
Carbon monoxide	−192.0	81.1	. . .	Yes	Yes	No
Fluorine	−187.0	86.0	888 to 1	No	Yes	Sharp
Argon	−185.7	87.4	847 to 1	No	No[c]	No
Oxygen	−183.0	90.1	860 to 1	No	No[c]	No
Methane	−161.4	111.7	578 to 1	Yes	No[c]	No
Krypton	−151.8	121.3	700 to 1	No	No[c]	No
Tetrafluoromethane	−128	145	. . .	No	Yes	No
Ozone	−111.9	161.3	. . .	Yes	Yes	Yes
Xenon	−109.1	164.0	573 to 1	No	No[c]	No
Ethylene	−103.8	169.3	. . .	Yes	No[c]	Sweet
Boron trifluoride	−100.3	172.7	. . .	No	Yes	Pungent
Nitrous oxide	−89.5	183.6	666 to 1	No	No[c]	Sweet
Ethane	−88.3	184.8	. . .	Yes	No[c]	No
Hydrogen chloride	−85.0	188.0	. . .	No	Yes	Pungent
Acetylene	−84.0	189.1	. . .	Yes	Yes	Garlic
Fluoroform	−84.0	189.1	. . .	No	No[c]	No
1,1-Difluoroethylene	−83.0	190.0	. . .	Yes	No[c]	Faint Ether
Chlorotrifluoromethane	−81.4	191.6	. . .	No	Yes	Mild
Carbon dioxide	−78.5[b]	194.6	553 to 1	No	Yes	Slight Pungent

[a] $0°K = -273.16°C; -459.69°F$.

[b] Sublimes.

[c] Nontoxic, but can act as an asphyxiant by displacing air needed to support life. As with most chemicals, even harmless materials can be toxic or poisonous if taken in sufficient quantities under the right circumstances.

Cryogenic fluids (liquefied gases) are characterized by extreme low temperatures, ranging from a boiling point of $-78.5°C$ ($-109°F$) for carbon dioxide to $-269.9°C$ ($-454°F$) for helium. Another common property is the large ratio of expansion in volume from liquid to gas, from approximately 553 to 1 for carbon dioxide, to 1438 to 1 for neon. Table 1 contains a more complete summary of the properties of cryogenic fluids.

HAZARDS

There are four principal areas of hazard related to the use of cryogenic fluids or in cryogenic systems. These are: flammability, high pressure gas, materials, and personnel. All categories of hazard are usually present in a system concurrently, and must be considered when introducing a cryogenic system or process.

The flammability hazard is obvious when gases such as hydrogen, methane, and acetylene are considered. However, the fire hazard may be greatly increased when gases normally thought to be non-flammable are used. The presence of oxygen will greatly increase the flammability of ordinary combustibles, and may even cause some non-combustible materials like carbon steel to burn readily under the right conditions. Liquefied inert gases such as liquid nitrogen or liquid helium are capable, under the right conditions, of condensing oxygen from the atmosphere, and causing oxygen enrichment or entrapment in unsuspected areas. Extremely cold metal surfaces are also capable of condensing oxygen from the atmosphere.

The high pressure gas hazard is always present when cryogenic fluids are used or stored. Since the liquefied gases are usually stored at or near their boiling point, there is always some gas present in the container. The large expansion ratio from liquid to gas provides a source for the build-up of high pressures due to the evaporation of the liquid. The rate of evaporation will vary, depending on the characteristics of the fluid, container design, insulating materials, and environmental conditions of the atmosphere. Container capacity must include an allowance for that portion which will be in the gaseous state. These same factors must also be considered in the design of transfer lines and piping systems.

Materials must be carefully selected for cryogenic service because of the drastic changes in the properties of materials when they are exposed to extreme low temperatures. Materials which are normally ductile at atmospheric temperatures may become extremely brittle when subjected to temperatures in the cryogenic range, while other materials may improve their properties of ductility. The American Society of Mechanical Engineers' Boiler and Pressure Vessel Code, Section VIII—Unfired Pressure Vessels may be used as a specific guide to the selection of materials to be used in cryogenic service. Some metals which are suitable for cryogenic temperatures are stainless steel (300 series and other austenitic series), copper, brass, bronze, monel, and aluminum. Non-metal materials which perform satisfactorily in low temperature service are Dacron, Teflon, Kel-F, asbestos impregnated with Teflon, Mylar, and Nylon. Once the materials are selected, the method of joining them must receive careful consideration to insure that the desired performance is preserved by using the proper soldering, brazing, or welding techniques and materials. Finally, chemical reactivity between the fluid or gas and the storage containers and equipment must be studied. Wood or asphalt saturated with oxygen has been known to literally explode when subjected to mechanical shock. When properties of materials which are being considered for cryogenic uses are unknown, or not to be found in the known guides, experimental evaluation should be performed before the materials are used in the system.

Personnel hazards exist in several areas where cryogenic systems are in use. Exposure of personnel to the hazards of fire, high pressure gas, and material failures previously discussed must be avoided. Of prime concern is bodily contact with the extreme low temperatures involved. A very brief contact with fluids or materials at cryogenic temperatures is capable of causing burns similar to thermal burns from high temperature contacts. Prolonged contact with these temperatures will cause embrittlement of the exposed members because of the high water content of the human body. The eyes are especially vulnerable to this type of exposure, so that eye protection is necessary.

While a number of the gases in the cryogenic range are not toxic, they are all capable of causing asphyxiation by displacing the air necessary for the support of life. Even oxygen may have harmful physiological effects if prolonged breathing of pure oxygen takes place.

There is no fine line of distinction between the four categories of hazards, and they must be considered collectively and individually in the design and operation of cryogenic systems.

GENERAL PRECAUTIONS

Personnel should be thoroughly instructed and trained in the nature of the hazards and the proper steps to avoid them. This should include emergency procedures, operation of equipment, safety devices, knowledge of the properties of the materials used, and personal protective equipment required.

Equipment and systems should be kept scrupulously clean and contaminating materials avoided which may create a hazardous condition upon contact with the cryogenic fluids or gases used in the system. This is particularly important when working with liquid or gaseous oxygen.

Mixtures of gases or fluids should be strictly controlled to prevent the formation of flammable or explosive mixtures. As the primary defense against fire or explosion, extreme care should be taken to avoid contamination of a fuel with an oxidant, or the contamination of an oxidant by a fuel.

As further prevention, when flammable gases are being used, potential ignition sources must be carefully controlled. Work areas, rooms, chambers, or laboratories should be suitably monitored to automatically warn personnel when a dangerous condition is developing. When practical, it would be advisable to provide for the cryogenic system or equipment to be shut down automatically as well as to sound a warning alarm.

When there is a possibility of personal contact with a cryogenic fluid, full face protection, an impervious apron or coat, cuffless trousers, and high-topped shoes should be worn. Watches, rings, bracelets, or other jewelry should not be permitted when personnel are working with cryogenic fluids. Basically, personnel should avoid wearing anything capable of trapping or holding a cryogenic fluid in close proximity to the flesh. Gloves may or may not be worn, but if they are necessary in order to handle containers or cold metal parts of the system, they should be impervious, and sufficiently large to be easily tossed off the hand in case of a spill. A more desirable arrangement would be hand protection of the potholder type.

When toxic gases are being used, suitable respiratory protective equipment should be readily available to all personnel. They should thoroughly know the location and use of this equipment.

STORAGE

Storage of cryogenic fluids is usually in a well insulated container designed to minimize loss of product due to boil-off.

The most common container for cryogenic fluids is a doublewalled, evacuated container known as a Dewar flask, of either metal or glass. The glass container is similar in construction and appearance to the ordinary "Thermos" bottle. Generally, the lower portion will have a metal base which serves as a stand. Exposed glass portions of the container should be taped to minimize the flying glass hazard if the container should break or implode.

Metal containers are generally used for larger quantities of cryogenic fluids, and usually have a capacity of 10 to 100 liters (2.6 to 26 gallons). These containers are also of double-walled evacuated construction, and usually contain some adsorbent material in the evacuated space. The inner container is usually spherical in shape because this has been found to be the most efficient in use. Both the metal and glass Dewars should be kept covered with a loose-fitting cap to prevent air or moisture from entering the container, and to allow built-up pressure to escape.

Larger capacity storage vessels are basically the same double-walled containers, but the evacuated space is generally filled with powdered or layered insulating material. For economic reasons, the containers are usually cylindrical with dished ends, which approximates the shape of the sphere but is less expensive to build. Containers must be constructed to withstand the weights and pressures that will be encountered, and adequately vented to permit the escape of evaporated gas. Containers should also be equipped with rupture discs on both inner and outer vessels to release pressure if the safety relief valves should fail.

Cryogenic fluids with boiling point below that of liquid nitrogen (particularly liquid helium and hydrogen) require specially constructed and insulated containers to prevent rapid loss of product from evaporation. These are special Dewar containers which are actually two containers, one inside the other. The liquid helium or hydrogen is contained in the inner vessel, and the outer vessel contains liquid nitrogen which acts as a heat shield to prevent heat from radiating into the inner vessel. The inner neck as shown in the illustration, should be kept closed with a loose fitting, non-threaded brass plug which prevents air or moisture from entering the container, yet is loose enough to vent any pressure which may have developed (Fig. 1). The liquid nitrogen fill

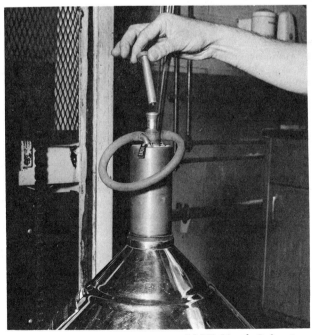

M.I.T. Lincoln Laboratory Photo

FIG. 1.   Dewar container with a loose-fitting, non-threaded brass plug.

and vent lines should be connected by a length of gum rubber tubing with a slit approximately 2.54 cm (1 in.) long near the center of the tubing. This prevents the entry of air and moisture, while the slit will permit release of gas pressure. Piping or transfer lines should be double-walled evacuated pipes to prevent the loss of product during transfer.

Most suppliers are now using a special fitting to be used in the shipment of Dewar vessels. Also, there is an automatic pressure relief valve, and a manual valve to relieve pressure before removing the device. Dewar vessels of this type must be strictly and regularly maintained to prevent the loss of product, and to prevent an ice plug from forming in the neck.

The liquid nitrogen outer jacket should be kept filled to maintain its effectiveness as a radiant heat shield. The cap must be kept on at all times to prevent entry of moisture and air, which will form an ice plug. The liquid helium fill (inner neck) should be reamed out before and after transfer, and at least twice daily. Reaming should be performed with a hollow copper rod, with a marker or stop to prevent damaging the bottom of the inner container. Some newer style Dewar vessels are equipped with a pressure relief valve, and pressure gauge for the inner vessel.

Transfer of liquids from the metal Dewar vessels should be accomplished with special transfer tubes or pumps designed for the particular application. Since the inner vessel is mainly supported by the neck, tilting to pour the liquid may damage the container, shortening its life, or creating a hazard due to container failure at a later date. Piping or transfer lines should be so constructed that it is not possible for fluids to become trapped between valves or closed sections of the line. Evaporation of the liquid in a section of line may result in pressure build-up and eventual explosion. If it is not possible to empty all lines, they must be equipped with safety relief valves and rupture discs. When venting storage containers and lines, proper consideration must be given to the properties of the gas being vented. Venting should be to the outdoors to prevent an accumulation of flammable, toxic or inert gas in the work area.

ACKNOWLEDGEMENT: Reprinted with permission from the *Journal of the American Society of Safety Engineers.* Vol. 11 (8). 15–19, August, 1963, Chicago, Illinois.

REFERENCES

"Cryogenics," Marsh & McLennan, Inc., Chicago, Ill., (1962).

"Industrial Gas Data," Air Reduction Sales Co., Acton, Massachusetts.

"Matheson Gas Data Book." 47th edition, The Matheson Co., Inc., East Rutherford, New Jersey, (1961).

"Precautions and Safe Practices for Handling Liquid Hydrogen," Linde Company, New York, (1960).

"Precautions and Safe Practices for Handling Liquefied Atmospheric Gases," Linde Company, New York, (1960).

Braidech, Mathew M. "Hazards/Safety Considerations in Cryogenic (Super Cold) Operations," Conference of Special Risk Underwriters', New York, (1961).

Hoare, Jackson, and Kurti, "Experimental Cryophysics," Butterworths, London, (1961).

MacDonald, D. K. C., "Near Zero, An Introduction to Low Temperature Physics," Anchor Books, Doubleday & Co., Inc., New York, (1961).

Neary, R. M., "Handling Cryogenic Fluids," Linde Company, New York, (1960).

Scott, Russell B., "Cryogenic Engineering," D. Van Nostrand Company, Inc., Princeton, New Jersey, (1959).

Timmerhaus, K. D., editor "Advances in Cryogenic Engineering," 7, Plenum Press, New York, (1961).

Vance, R. W., and Duke, W. M., Editors, Applied Cryogenic Engineering," John Wiley & Sons, Inc., New York, (1962).

Zenner, G. H., "Safety Engineering as Applied to the Handling of Liquefied Atmospheric Gases," in Advances in Cryogenic Engineering, 6, Plenum Press, New York, (1960).

"Cryogenic Safety," A Summary Report of the Cryogenic Safety Conference, Air Products, Incorporated, Allentown, Pennsylvania, (1959).

Controlling Hazards from Uses of
the Plasma Torch

Howard L. Kusnetz and Roger Grimm

The plasma torch is a device which has the ability to develop extremely high temperatures and as such is expected to be of increased use in research and in industry.

Recent literature references, requests to the Division of Occupational Health for information on potential health hazards, and information from equipment manufacturers lead us to believe that our estimates of the number of users of plasma torches have been somewhat conservative, and that usage of plasma devices, together with their potential health hazards, will become a major item of concern to persons evaluating occupational exposures. Prevention of possible harmful effects may be a difficult task because few environmental health studies have been made and knowledge of the full extent of the potential health hazards is limited.

Plasma is defined as a state of matter in which individual atoms are separated into free electrons and positive ions and plasmas occur in nature as well as in man-made devices. Lightning and the aurora borealis are familiar examples of natural plasmas, and the glowing area of an electric arc is a man-made plasma. True plasmas, those in which complete ionization occurs, exist only at extreme temperatures. Since they will vaporize any known material, the plasmas cannot be contained within any device for unlimited periods of time. The processes described in this chapter will be concerned with partial ionization of gas on a continuous basis.

Plasma was described by Sir William Crookes in 1879 as a "fourth state of matter." Essentially, it is a highly ionized stream of gas molecules. A plasma torch is a device capable of heating gas to extremely high temperatures. Unlike a conventional flame, no combustion is involved, hence plasma temperatures are not limited by internal heats of reaction. Maximum plasma temperature becomes, rather, a function of the energy input to the torch. The gas, which acquires energy from an electric arc, is heated, dissociated and ionized, and plasma is formed. In the process of recombination, the energy which had been acquired is released in the form of large quantities of heat. Where the arc is totally contained within the torch, striking from the cathode to the cooled nozzle, the mode of operation is said to be "non-transferred." However, where the work piece is used as the anode, the electron and ion bombardment adds to the heat transferred by the plasma. In this "transferred arc" mode of operation, even higher temperatures are realized.

The type of arc stabilization may also affect the rate of heat transfer, and does affect the noise levels resulting from use of the torch. Arc stabilization refers to the passage of the gas through the orifice of the torch. Higher enthalpies are available with the laminar flow or low pressure torch than with the vortex or high pressure type of device. Commercial plasma jets rated up to 5,000 kw are available and a 10,000 kw plasma generator has been reported in the literature.

Techniques of plasma generation and containment are not recent developments. Reid, in a patent dated 1908, proposed the use of plasma in iron ore treatment.[1] Use of plasmas as a heat source for high temperature furnaces was discussed by Mathers[2] in a patent dated 1911. Irving Langmuir designated the name "plasma" for this phenomenon in his development of plasma theory in the 1920's[3] in which he related the kinetic energy of charged particles to the temperatures at which they exist.

It was not until after 1945 that reliable commercial equipment was developed in the plasma torch field. An increase in interest in extreme temperatures and greater availa-

579

ability of highly refractory materials for critical component parts contributed to development of plasma torches. Today, the demand for information concerning high-temperature conditions, effects, and operations is skyrocketing. Information is desired for application in such areas as generation of electric energy, the heating effects of controlled nuclear and termonuclear reactions, rocket propulsion, and space vehicle reentry testing.

PHYSICAL DESCRIPTION AND PRINCIPLE OF OPERATION

The processes in the arc, where the electric energy is transferred, are extremely complex.[4,5,6,7] The torch consists mainly of a chamber housing the anode and cathode, and an orifice through which the plasma is emitted. A source of dc power is needed for starting and maintaining the arc. Equipment operating at 250 volts dc with power rating up to 1,000 kw is available. For more routine applications, power levels of 40 to 80 kw are used.

Two types of plasma torches are in industrial use at present. These are the non-transferred arc and the transferred arc torches. Uses of the non-transferred arc torch include metallizing, surface hardening, welding, crucible and furnace melting, crystal growing, thermo-chemical studies, particle spheroidization and vaporization, materials testing, piercing, rocket propulsion, and space vehicle reentry simulation. Temperatures up to 16,650°C (30,000°F) may be reached with the non-transferred mode of operation. The transferred arc torch is used primarily for cutting purposes. Here, the material being cut becomes the torch anode. Temperatures almost twice that of the non-transferred arc are possible. The increased enthalpy is due primarily to heating of the work-piece caused by electron impingement.

The plasma process begins when the dc power supply is turned on. The voltage drop across the gap between the electrodes extracts the first few electrons from the surface of the cathode. These electrons accelerate and collide with molecules of air and atoms of the gas in the gap. Energy is transferred to these molecules and atoms with some being ionized. The gas temperature begins to increase with a steady current of electrons toward the anode and a counter-current of positive ions toward the cathode. As the conductivity of the system rises, the anode heats up sufficiently to emit anode material into the electric field. It is at this point that a plasma arc is produced.

The temperature of the plasma is increased by magnetohydrodynamic and thermal effects which tend to increase the density of the plasma and the frequency of collisions. A "thermal pinch effect" is caused by the tangential injection of water or gas into the torch chamber, and gases which have been used include argon, helium, nitrogen, hydrogen, and compressed air. The injected fluid cools the chamber walls and the gases in the outer regions of the plasma. Since this cooling lowers the ionization and the conductivity of the gases near the walls, the current in the discharge tends to concentrate in the hotter central region of the plasma. When the current density in the center of the discharge reaches a high enough level, a second effect, the "magnetic pinch," occurs. This effect is analogous to the attraction between two conducting wires with their currents flowing in the same direction. In the plasma the electrons and the positive ions are respectively attracted by their own self-induced magnetic fields. A further constriction of the discharge results. These magnetohydrodynamic forces and the pressure which is built up in the chamber cause the plasma to be ejected through the orifice.

INCREASE IN USAGE OF PLASMA TORCHES

The extent of the use of plasma torches may be determined from a perusal of the literature and from manufacturer's data. In 1960, the Applied Science and Technology Index listed 25 articles under "Plasma Torch." In 1961, there were 34 references; in

1962, 32; and in 1963 there were more than 50 such references in addition to listings under "Plasma Physics." It is interesting to note the number of titles containing the word, "New." Although many of the references pertain to research projects, the number of torches in routine commercial use is growing, and many of the research results presage problems of controlling occupational exposures.

Manufacturers' estimates in 1963 were that there were about 1,000 non-transferred arc torches and about 300 units of transferred arc metal-cutting equipment in use. These numbers were expected to double by 1965 and to triple by 1970. For the non-transferred arcs about 60–70 percent are estimated to be in routine use, primarily for metallizing or spraying operations. About 15–20 percent are used for general purpose research studies, such as arc phenomena; this would include home-built units. The balance are used for materials testing and for high temperature work, including high temperature syntheses. Reference to the home-built units should be emphasized. Plasma torches are not difficult or complicated to build or operate. A recent *Scientific American* article described the project of a high school boy who built and operated his own torch.

POTENTIAL HEALTH HAZARDS

Increased use of the plasma torch will mean that a large number of workers will be exposed to the potential health hazards associated with the device, so that the hazards should be evaluated thoroughly before mass utilization of the torch is achieved.[8]

The high pressure or vortex mode which may utilize gas velocities up to Mach 3 in certain operations presents a severe noise hazard. Sound levels in the octave bands above 1,200 cycles per second were in excess of 110 dB in the operations we observed, and from personal communications we have become aware of sound levels in excess of 130 dB. Personal protection is indicated for those in the immediate vicinity and isolation of the operation is indicated to protect other employees, but these remedies may not be simple to implement. At operations we observed, certain employees refused to wear the protective devices since the use of ear muffs, together with the necessary welding helmet, is an uncomfortable combination.

Torch operation in an enclosed room should protect other employees—providing that the nature of the operation can be limited to a given area, which is not always possible. Further, total enclosure may increase the heat load and fume and gas concentration, thus making the task of ventilation more difficult.

A second major potential hazard from the torch is exposure to a broad portion of the electromagnetic spectrum. Radiations above and below the visible light spectrum are present. The infrared, or heat, contribution is immediately apparent to anyone in the vicinity of the torch operation, but the shorter wavelength components are even more invidious. If we consider a 5 Mw torch operating at a potential of 250 volts, we can calculate that the wavelength at the lowest end of the electromagnetic spectrum will be about 50 A. Indications in the literature are that operations at voltages as high as 4,000 volts are possible. Theoretically, for such voltage the lower end of the resultant energy spectrum would be about 3 A. For reference, recall that the UV range is from about 30A–4,000 A; X-rays are from 10^{-4} A to 10^3 A. The possibility of X-radiation is not remote. This is especially true with the transferred arc mode where electron and ion bombardment of the work-piece may very well result in secondary emission.

The U. S. Public Health Service, in cooperation with the New Hampshire State Department of Health, conducted an engineering and medical survey of plasma torch operations. The investigations were designed to provide base line data on hazard potentials and some of the findings are summarized here. Noise exposures encountered in spraying and cutting operations with the standard high velocity, high pressure torch were generally high. Overall noise levels between 101 and 121 dB. (re: 0.0002 μ bars) were encountered. Most of the energy was concentrated in the upper octave bands, 1,200-2,400 cps, 2,400-4,800 cps, and greater than 4,800 cps. The high sound pressure level is due to the design of the nozzle which causes the plasma to be emitted with high velocity. It is mandatory that devices for protection against hearing loss be worn in noise fields of this intensity.[9,10] Ear plugs, ear muffs, or both should be used.

An alternate mode of operation utilizing low velocity, low pressure flow (laminar flame) is also in use. Measurements made upon this torch showed an over-all sound pressure level of 78 dB. The highest reading in any octave band was 73 dB. Use of the low velocity nozzle will help to reduce substantially the noise problem associated with the high velocity plasma torch.

The exact range of wavelengths in the electromagnetic spectrum which is produced by the torch is still unknown. It is theorized that the torch could produce radiation in the range of 50 A to 100,000 A, which includes a portion of the X-ray band, ultraviolet, visible light, and infrared. Ultraviolet radiation was measured with an Archer-Reed ultraviolet meter which was calibrated for 2,537 A. At times, readings in excess of 250-μw per square centimeter, the limit of our instrument, were reached. Ultraviolet output varied with torch operation, energy input, metal involved and gas used. Severe erythema of exposed skin or "flash" injury to the conjunctivae may be produced by exposure to ultraviolet light. Further, persons being treated with certain drugs may react unduly to some wavelengths of ultraviolet radiation. Actinic or senile keratoses as well as actinic skin degeneration can be augmented by exposure to ultraviolet.

Measurements of ionizing radiation were made by the use of a 1.4 mg per square centimeter, thin-window G.M. tube. The measurements indicated that some radiation was being produced during transferred arc operation. The highest value recorded was 0.012 milliroentgen per hour, which is within acceptable limits. It must be understood, however, that this device is not adequate for measuring ultrasoft (grenz) rays. Thus, it is not possible to exclude the possibility of these rays in the torch emissions until further tests are made. Grenz rays are capable of producing effects upon the skin which result in reddening and hyperpigmentation.

The high temperatures available with plasma equipment—particularly when applied to thermochemical syntheses, or metallizing operations, open the door to a flood of new hazards. Materials available in powder form for metallic spraying include carbides, oxides, and borides. Materials such as HCN or cyanogen have been made with the torch at yields that indicate commercial feasibility. The plasma torch will vaporize anything in the periodic table. Formation of new compounds by vaporphase technology will give us heretofore unthought-of materials for which toxicity data are lacking. New intermetallic compounds, new ceramics, and compounds ordinarily unstable or metastable at normal temperatures can be manufactured with the new plasma technology. Titanium and magnesium nitrides are examples.

CONTROL OF HAZARDS

As a result of the hazard potential of the electromagnetic radiation by the torch, several recommendations are proposed. All persons working in proximity to torch operations should wear protective clothing which may be made of closely woven cotton fabric. In addition, users should wear full face shields containing filters for eye protection against the ultraviolet, intense visible, and infrared emissions from torch operations. Welders' gloves should be a part of the standard protective equipment. Based on presently available data, these protective devices should be adequate. If it is found that the radiations are capable of penetrating these fabrics, heavier protective equipment will be needed, and better shielding of the torches will have to be designed.

An example of the need for proper use of personal protective equipment is demonstrated by a recent occurrence, not associated with our survey, involving the transferred arc torch and low momentum flame. Argon was being used, and a very intense flame, 20 cm to 30 cm long, was being produced. Several observers, who were given hand-held welders' shields, faced away from the flame and received a 10-second exposure to the side of the face and to the eye. By evening, those exposed required strong analgesic to control the severe pain which had developed. By contrast, workers wearing No. 12 welders' lenses and directly observing this same flame for periods of up to 30 minutes experienced no difficulty or after-effects.

With operations such as metal-cutting and metal-spraying, particulate matter will escape into the working environment unless adequate control methods are used. In a process involving barium, for example, air-borne concentrations of 0.23–0.72 mg per cubic meter were found when operations were carried on outside an exhaust hood. When the operation was moved into the hood, the concentration dropped to 0.002 mg per cubic meter. There have been several reports from the Navy of cases of metal fume fever among men spraying with the plasma torch.[11] For most metallizing operations an exhaust-ventilated metal spray booth should be used. The high velocities of the plasma stream necessitate either complete enclosure or high capture velocities.

For cutting operations, systems for stationary as well as movable work must be provided. A portable exhaust hood can provide sufficient local exhaust ventilation when work is performed away from a specific cutting area. In a fixed area where transferred arc operations are performed, a stationary hood should be provided. A suggested design would be a hood comprising two components: (1) a booth with a face velocity of 61 m/min. or 200 fpm and (2) a quench trough below the hood with a slot above the trough for collecting particulate after quenching. A slot velocity of 610 m/min. (2000 fpm) and a face velocity of 61 m/min. (200 fpm) should be maintained. The booth should be designed so that as much as possible of the cutting operation will be located within the hood.

The use of standard exhaust ventilation techniques may not be possible with the laminar flame. Air velocities as low as 30 m/min. (100 fpm) were able to deflect the flame. No definitive recommendations regarding controls for this mode of operation can be given without further study.

As in the inert-gas-shielded arc welding technique, the high temperature and ultraviolet energy in the flame leads to the formation of oxides of nitrogen and ozone from the normal components of the air. For example, with the transferred arc torch being operated at a 55 kw input, using a 70% A-30% H_2 mixture, and with no exhaust ventilation, concentrations of NO_2 and O_3 were 6 ppm and 0.9 ppm, respectively. When the same gas mixture was used, and the ventilation system was turned on, the O_3 concentration was found to be 0.09 ppm. No NO_2 was found. This occurred even with a power boost to 80 kw.

For these gases, properly designed local exhaust ventilation, similar to that now being used with other welding procedures, should control the NO_2 and O_3 output.

CONCLUSIONS

Investigations of the potential hazards associated with the plasma torch operation have shown that exposure to ozone, oxides of nitrogen, particulates, noise, and radiant energy exist. The atmospheric contaminants generated by plasma torch operation do not differ greatly from those generated by more common metal-cutting or metallizing operations, but new types of contaminants with unknown toxicities may be generated.

The design of exhaust ventilation equipment for plasma torch operation and the design of personal protective devices to shield the worker against a combination of noise, heat, ultraviolet and visible light, X-radiation, and respirable contaminants is a formidable challenge, and control of hazards from uses of the plasma torch lies in the areas of development of appropriate measurement and protection techniques based on detailed studies of potential health hazards at installations utilizing plasma equipment.

We thank the New Hampshire State Department of Health for its assistance, particularly Mr. R. S. Dumm of that department for his participation in the study. Acknowledgment is also made of the contributions of Drs. D. J. Birmingham and T. H. Milby on the medical aspects.

ACKNOWLEDGEMENT: Reprinted from "Plasma Torch" by Roger Grimm and Howard Kusnetz, Archives of Environmental Health, March, 1962, Vol. 4, pp. 295–300. Copyright 1962, by American Medical Association; and "Controlling the Hazards of the Plasma Torch" by Howard Kusnetz, Proceedings of the Campus Safety Association, 1964.

REFERENCES

1. Reid, J. H., U.S. Patent No. 896248
2. Mathers, E. A., U.S. Patent No. 1,002,721
3. Langmuir, I., "Phenomena, Atoms, and Molecules," New York, Philosophical Library, Inc. (1950).
4. Hellund, E. J., "The Plasma State," New York, Reinhold Publishing Corp., (1961).
5. Linhart, J. G., "Plasma Physics," New York, Interscience Publishers, Inc., (1960).
6. Giannini, G. M., The Plasma Jet, *Sci. Amer. 197*:80 (Aug., 1957).
7. Browning, J. A., "Techniques for Producing Plasma Jets," Publication 905-59, New York, American Rocket Society
8. Speicher, H. W., Plasma Jet, *Arch. Environ. Health 2*:278 (March, 1961).
9. Rosenblith, W. A., Stevens, K. N., and Staff of Bolt, Bernaek, and Newman, Inc.; "Handbook of Acoustic Noise Control," *Vol. II*, Noise and Man, U.S.A.F.W.A.D.C. Techn. Rep. No. 52-204, (1953).
10. U.S. Air Force, "Hazardous Noise Exposure," U.S.A.F.O.S.G. Reg. No. 160-3 (Oct., 1956).
11. Bergtholdt, C.P.I.; Recent Welding Practices at Naval Facilities, *Arch Environ. Health 2*:257 (March, 1961).
12. "Industrial Ventilation," A.C.G.I.H. Manual; Print No. VS-23, Lansing, Mich., Committee on Industrial Ventilation, P.O. Box 453 (1960).

Water Supply

Prevention of Contamination of Drinking Water Supplies

Paul H. Woodruff

The safety of drinking water supplies is usually taken for granted, but there are conditions under which contamination can enter potable water systems with a resulting health hazard.

In 1942, in a Pittsburgh factory employing some 500 people, a new water connection was being made from the building's plumbing to the city water supply and the building's water was shut off for a short period to make the connection. Employees meanwhile continued to use water in the building on the lower floors. In this way, a partial vacuum was drawn in the building's plumbing, and toilets were siphoned back into the drinking water. Several hundred employees came down with intestinal disturbances within a few hours.

Another instance occurred in 1947 at a school in Milford, Nebraska. There had been a small fire in the school and the school's fire system used a sewage polluted river as a water supply for fire fighting. Inadvertently, someone had left a valve open which connected the fire water supply with the drinking water system in the school. Contaminated river water was pumped into the school's plumbing and 50 students came down with intestinal disorders.

These situations can, should, and must be avoided.

WATER SUPPLY SYSTEM

A review of the elements in a typical water supply system is helpful in appreciating the opportunities for contamination to occur.

A water supply system includes a source of surface or ground water supply, conduits which bring the raw water to a pumping station, treatment processes, and high pressure pumps, to move the water into the distribution system and perhaps into treated water storage.

Contamination of the water supply can occur at any point in the system and the effect is usually more serious if it occurs after the water is treated.

The apparent causes of water supply contamination are as many as the sources, but basically water contamination results from two conditions:

1. A cross-connection between a potable system and a non-potable system, and
2. Pressure conditions such that the non-potable material is forced into the potable water supply.

CROSS CONNECTIONS

Figure 1 illustrates schematically the simplest and most obvious type of cross-connection, a potable supply separated from a non-potable supply only by a valve. This is the type of cross-connection that is easiest to locate.

Table 1 lists some typical types of plumbing fixtures where contamination by cross-connections of the direct type is possible. Familiar equipment such as air conditioning apparatus, and sterilizers, could contribute contamination by cross-connection to the potable supply.

The second type of cross-connection exists in a much more insidious form. Table 2 lists typical types of plumbing fixtures that indirectly connect potable water

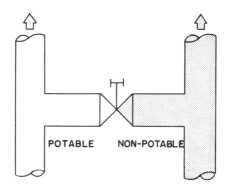

Fig. 1.   Valved connections between potable water and non-potable fluid.

with a non-potable system. These fixtures have in common one essential factor and that is the ability of the inlet fixture to be flooded with non-potable liquid that in many cases will eventually be drained away to waste.

Pressure which allows the non-potable liquid to reach a potable water supply can be caused by many conditions. However, these pressure conditions can be divided into two general classes, backsiphonage and backflow. In backsiphonage, the driving force is atmospheric pressure, meaning the potable water will be at some pressure less than atmospheric (under a partial vacuum). Backflow occurs when a non-potable system is under greater pressure than the potable system.

Backsiphonage. Following are several examples of contamination occuring by backsiphonage. Figure 2 shows a sterilizer which is connected to the city water system. The sterilizer is allowed to cool after use without opening the air vent; as a result a vacuum is created in the sterilizer and suction is created on the inlet to the sterilizer. In the next room a hospital attendant is washing out bed pans. A hose is connected to the sink with the discharge end left under water in the basin. Suction from the sterilizer pulls contents of the sink over into the sterilizer. Meanwhile, people on the lower floors flush toilets creating a demand on the plumbing system greater than the undersized main will supply, and thus contaminated contents of the sterilizer are drawn back into the plumbing system in the building. A correction for this situation would be vacuum breakers on both the sink and the sterilizer.

Figure 3 illustrates another set of circumstances where the fault is a rubber hose in a sink. In this case, the cause of reversed flow may be different. Two adjacent multi-storied buildings are served by the same water main which often lacks adequate pressure. A booster pump in the adjoining building supplies additional pressure when needed. However, the pump at times will pull more water than the main can supply, thereby creating a partial vacuum on the suction side of pump and the circumstances necessary for backsiphonage. Correction for this condition would be a vacuum breaker on the sink and a vacuum cut-off switch on the booster pump.

Figure 4 illustrates another common condition which can cause a contaminated water supply. In this case, a dishwasher is installed with solid connections to both the waste line and the hot and cold water supply. The first two floors of this building are supplied by city pressure but because of relatively low pressure in the city main, the upper floors are supplied by a booster pump drawing water from the city main. During periods of exceptionally low pressure in the city main, the booster pump creates a partial vacuum on the suction side of the pump. This, in turn, will siphon contents of the dishwater and waste line down to the pump suction. To correct this situation, the dishwasher piping should be modified to insure both supply and discharge through an air gap. In addition, a low pressure switch should be installed on the booster pump.

TABLE 1
Partial List of Plumbing Hazards and Fixtures
With Direct Connection

Air Conditioning, Air Washer	Cooling System
Air Conditioning, Chilled Water	Dishwasher
Air Conditioning, Condenser Water	Fire Standpipe or Sprinkler System
Air Line	Hydraulic Equipment
Aspirator, Laboratory	Laboratory Equipment
Aspirator, Medical	Lubrication, Pump Bearings
Autoclave and Sterilizer	Photostat Equipment
Auxiliary System, Industrial	Pump, Pneumatic Ejector
Auxiliary System, Surface Water	Pump, Prime
Auxiliary System, Unapproved Well Supply	Pump, Water Operated Ejector
Boiler System	Sewer, Sanitary
Chemical Feeder, Pot-Type	Sewer, Storm
Chlorinator	Swimming Pool
Coffee Urn	Weedicide and Fertilizer Sprayer

TABLE 2
Partial List of Plumbing Hazards and Fixtures
With Submerged Inlets

Bathtub	Lavatory
Bedpan Washer, Flushing Rim	Lawn Sprinkler System
Brine Tank	Photo Laboratory Sink
Cooling Tower	Sewer Flushing Manhole
Drinking Fountain	Slop Sink, Flushing Rim
Floor Drain, Flushing Rim	Slop Sink, Threaded Supply
Garbage Can Washer	Steam Table
Ice Maker	Urinal, Siphon Jet Blowout
Laboratory Sink, Serrated Nozzle	Water Closet, Flush Tank, Ball Cock
Laundry Machine	Water Closet, Flush Valve, Siphon Jet

A fairly common situation is shown in Figure 5, where a submerged inlet tank is used for processing. Fire elsewhere in the building activates the fire spinkler system, the fire pump pulls directly from the city main which is not sufficiently large to supply water for fire fighting and water to feed the process tank. A vacuum is created on the suction side of the fire pump and the contents of the processing tank will be siphoned back into the building's plumbing. An air gap is needed in the feed line to the process tank.

Backflow. Circumstances in which contamination occurs because of backflow are common in occurrence. Figure 6 shows a boiler that is cross-connected to the city supply for purposes of filling the system and for boiler water make-up where the boiler water is chemically treated for corrosion and scale control. Due to some unforeseen circumstances, the city pressure is reduced below the boiler water pressure as could happen if there were a major fire in a neighborhood. The valve separating the boiler water from the city water leaks, or has perhaps been left partially open and the chemically treated water will be pumped out into the city mains. In this case, an air gap or

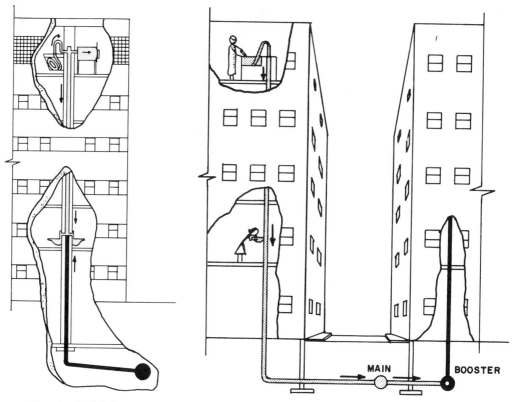

FIG. 2. Backsiphonage.

FIG. 3.   Backsiphonage.

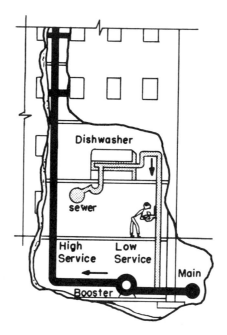

FIG. 4.   Backsiphonage.

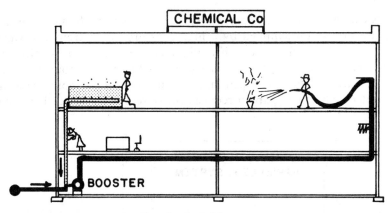

Fig. 5.   Backsiphonage.

reduced-pressure-principle backflow preventer should be used to prevent contamination of the city water.

A private well is shown in Figure 7 being used to irrigate a lawn; a septic tank located adjacent to the well is polluting the well. A cross-connection has been made from the city main to the well, probably for purposes of priming the pump. If and when the valve separating the two systems is left open, or leaks, polluted water from the well will be pumped into the building's water supply system. The connection between the well and the city main should be broken to prevent contamination.

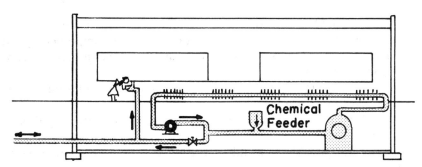

Fig. 6.   Backflow.

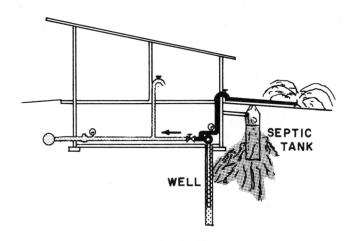

Fig. 7.   Backflow.

Fire systems are often improperly installed. Figure 8 shows a sprinkler water system for fire protection supplied by a high pressure pump from a sewage polluted river. For "safety," the fire system has also been connected to the city supply with only a single valve separating the two systems. If this valve should be left open, or leak, the river water would be pumped into the potable water supply. This installation should be protected by an air gap, or reduced-pressure-principle backflow preventer.

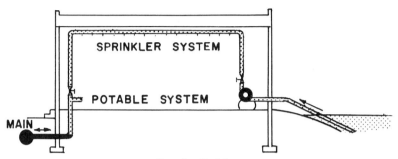

FIG. 8.   Backflow.

PROTECTIVE MEASURES

Safety practices which are up-to-date with regard to protection of potable water supplies should include:

1. An adequate plumbing code. (Check with your Local or State Health Department.)

2. Insistence on all plumbing work being done by licensed plumbers.

3. Periodic reminders to all personnel of the hazards of improper operation of equipment which is attached to the potable water supply. Vacuum breakers should not be removed from laboratory faucets.

4. Periodic inspections of plumbing and plumbing fixtures to insure all equipment is properly installed and maintained.

However, at the heart of preventing contamination of water supplies are two fundamental rules:

1. Never, if it can possibly be avoided, directly connect a non-potable system to a potable water system.

2. If a direct cross-connection is absolutely necessary (rare, if ever) then the use of properly installed backflow preventers and vacuum breakers are necessary.

Several common types of devices in use to prevent backsiphonage and backflow are illustrated in Figures 9, 10 and 11.

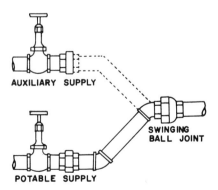

FIG. 9.   Swing connection.

The vacuum breakers shown in Figures 10 and 11 are only useful for preventing contamination caused by backsiphonage and, of course, must be maintained and checked periodically if they are to serve their purpose.

A better method of eliminating the possibility of contamination by cross-connection is to install an air gap. Figure 12 illustrates the technique for supplying water to a non-potable purpose through a surge and a booster pump. Figure 13 illus-

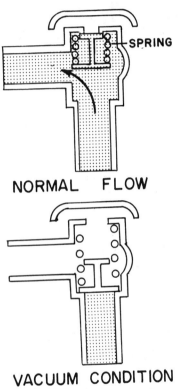

NORMAL   FLOW

VACUUM CONDITION

FIG. 10. Operation of a vacuum breaker.

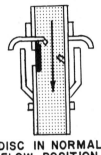

DISC IN NORMAL FLOW POSITION

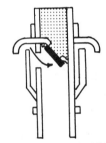

DISC IN VACUUM BREAKING POSITION

FIG. 11. Pressure-type vacuum breaker. (Not applicable where back-pressure may occur).

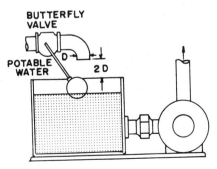

FIG. 12. Surge tank and booster pump.

trates a dual water supply system for protection, again utilizing an air gap. With an air gap, of course, neither backflow or backsiphonage is possible. Figure 14 is a schematic illustration of the proper air gap on a lavatory installation.

Figures 15 and 16 show a recommended method of discharging into an indirect waste system through an air gap, again insuring that backflow is impossible even if the sewer should be under pressure.

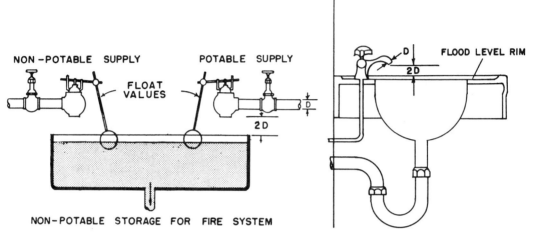

FIG. 13.   Fire system make-up tank for a dual water system.

FIG. 14.   Air gap on lavatory.

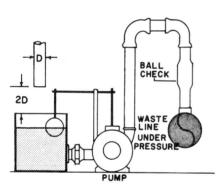

FIG. 15. Air gap to sewer subject to back-pressure.

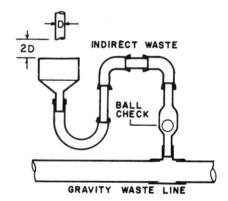

FIG. 16. Air gap to sewer subject to back-pressure.

Editor's Note.   If reduced-pressure-principle backflow preventors are used a strict maintenance schedule must be set up. These units must be tested each year by a certified plumbing inspection and dismantled for repair each five years.

ACKNOWLEDGEMENT: Illustrations were adapted from U. S. Public Health Service Publication No. 957, "Water Supply and Plumbing Cross-Connections," and the text was presented at the 11th National Conference on Campus Safety.

Producing and Handling
High Purity Water

Verity C. Smith

Because water is both an oxidizing agent and a reducing agent, it will dissolve almost anything to some extent. Purity of water is a relative term and is usually defined in relationship to some property of the water such as the specific resistance or the amount of a certain contaminant such as copper that might be present. The physical and chemical properties of water, such as the freezing point, index of refraction, molecular composition, specific resistance, color and turbidity can be described and measured; however, the most widely used measurement, specific resistance, is the most practical method of determining the quality of a given water sample. This measurement, of course, is complicated by the fact that the resistance of theoretically 'pure' water at 25°C (77°F) is 18.3 megohms, whereas the resistance of the same water is 87.4 megohms at 0°C (32°F) and 1.28 megohms at 100°C (212°F).

Since this resistance is based on the activity of the molecules dissolved in the water, it will vary with each type of contaminant. This means the absolute determination of the quantity of impurity could be made only if one knew the type of contaminant present and knew what the activity of that contaminant was. In any case it is generally assumed that water having a resistance of 1,000,000 ohms at 25°C (77°F) is of excellent purity and water having a resistance above 10,000,000 ohms is in the extreme purity range. However, the resistance measurement only indicates the ionizable impurities present and should always be considered in the relationship to ionized salts only. 18,000,000-ohm water, which can be obtained quite readily from a mixed-bed demineralizer, for example, would in almost all cases be pyrogenic and would be unfit for use in making up intravenous solutions.

Once one decides on the type of water needed as an end product, the method of production of this water can be considered in relation to the quantities involved, the contaminants that must be removed from the water, and the money available to do the job.

There are two practical methods of producing pure water, namely by distillation or by ion exchange. The processes of distillation remove the water from the impurities by converting the liquid to the gas phase, and then recondensing it as distilled water. Ion exchange or demineralization is based on the removal of the impurities from the water by means of synthetic resins which have an affinity for dissolved, ionized salts.

When considering the quantity required, it is very important that the figure be based on a daily or weekly consumption, rather than a gallon per minute or gallon per hour usage. Generally speaking it is much cheaper to design the equipment on the basis of a 24-hour operation with the idea that the purified water can be stored for use over an eight hour period. In the case of a relatively large installation increased storage capacity can produce a saving because the equipment can run over the weekend and produce enough water to help the requirements for the five days that follow.

WATER CONTAMINANTS

The next consideration, the contaminants to be removed, becomes more complex and it is generally necessary to consult with someone familiar with water purification in order to insure a proper solution to the problem. The production of high purity water means trying to remove essentially all of certain types of impurities that might be a detriment to the particular operation in question. For example, if water is being prepared for pharmaceutical use for intravenous injection, it is essential that the water be freed of pyrogens, which are large organic molecules that cause patients to go into shock; however, it is not necessary that the same water be free of many salts in the ppm range since sodium chloride and other chemicals are added to the water so that the water will be compatible with the patients' blood.

TABLE 1

Classification of Pure-water Contaminants and Possible Methods of Removal

Classification	Examples	Possible Methods of Removal
Liquid Impurities		
Miscible organic	Alcohol, detergents, etc.	Fractional distillation and absorption
Miscible inorganic	Hydrogen peroxide	Distillation, chemical decomposition
Immiscible organic	Oils	Filters, absorption, solvent extraction, distillation
Immiscible inorganic	Mercury, bromine	Generally not a problem
Solid Impurities		
Molecular organic	Sugars, plastics, pyrogens, viruses	Distillation
Molecular inorganic	Salts	Ion exchange or distillation
Colloidal organic	Graphite, fiber particles, skin, ion exchange resins	Distillation, perhaps coagulation
Colloidal inorganic	Silica, clay, metal oxides	Distillation, perhaps coagulation
Suspended organic	Biologicals of all kinds	Distillation, filtration, coagulation
Suspended inorganic	Metal oxides, mud	Distillation, filtration, coagulation
Gaseous impurites		
Volatile organic	Methane, CO_2	Distillation, degasification, ion exchange
Volatile inorganic	Ammonia, oxygen, HCl	Distillation, degasification, ion exchange

The cost of production, of course, depends upon the complexity of the train of equipment and the energy required to operate it.

To give an idea of the problems involved in the removal of impurities, the types of contaminants that may be present in water are shown in Table 1. On the left are listed the classification of impurities; in the center, examples of this impurity; and on the right, methods or combinations of methods that may be used to remove the impurities.

Liquid Impurities

In the first case we have the miscible organics such as alcohols and detergents which may be removed at least partially by fractional distillation and absorption. Some of these impurities, incidentally, are almost impossible to remove completely, especially if they boil at about the same temperature as water. The most obvious miscible inorganic is hydrogen peroxide, which is generally not a problem and can be removed by distillation after chemical treatment. The most obvious immiscible organic, of course, is oil in its various forms; and although this seemingly is not a problem, it can become serious if distribution systems, pumps and containers are not properly cleaned or maintained in any given system. Various types of oils can be removed by filtration, absorption or solvent extraction as well as by distillation. The immiscible inorganics such as bromine and mercury are generally not a problem.

Solid Impurities

Solid impurities are the most plentiful and, of course, are the ones usually considered by the user. Molecular organic impurities such as sugars, plastics, pyrogens, and biologicals, can be removed by distillation. Unfortunately they cannot be dependably removed by ion exchange methods. The molecular inorganic impurities, which include all of the dissolved salts, can be removed very easily by ion exchange or distillation. The colloidal organics, including impurities such as particles of graphite, fiber particles, and ion exchange resins, can be removed by distillation and can be reduced by various coagulation steps under certain conditions. Likewise the colloidal inorganics such as silica, clay, and metal oxides, can be removed by distillation and reduced by coagulation. Suspended organic matter, which covers all of the biological impurities, can be removed by distillation, filtration, or in some cases, coagulation. They may also be removed chemically by strong oxidizing agents. Suspended inorganic impurities such as metal oxides, sand, mud, and similar materials, can be removed readily by filtration, coagulation or distillation.

Gaseous Impurities

Gaseous impurities, especially in trace amounts, can become quite difficult to remove, especially in instances when the gas is quite soluble in the water. Volatile organics such as methane are very insoluble and are not usually found in water. On the other hand, carbon dioxide is present in the atmosphere and is always present in water in contact with the atmosphere. It may be removed by distillation, degasification or ion exchange, since it is an ionized molecule when dissolved in water. Volatile inorganics such as ammonia and hydrogen chloride can be removed readily by ion exchange; however, the inert gases such as oxygen, nitrogen, and hydrogen, must be removed by degasification or distillation.

ION EXCHANGE

The references mentioned to distillation and demineralization must be considered further, but it should be remembered that they can be used in various combinations and that pretreatment steps can be used with either process in all sorts of combinations which are too numerous to outline in detail. First, we will consider the demineralization step, which employs the principle of ion exchange. This is a process first discovered many years ago when it was noticed that water passing through certain types of sand was suitable to drink, even though it may have come from the ocean. The original ion exchange materials were naturally occurring green sands and were used almost exclusively in the softening industry. As research on the subject of ion exchange was intensified, along with the development of synthetic resins, it was found that organics could be synthesized that would have ion exchange properties far exceeding those of the naturally occurring materials. These resins are broken down into two classifications, namely cation resins that remove the positive or metallic ions, and those that remove the negative ions, known as anion resins.

The synthetic resins which are in use are, with some exceptions, made of styrene cross-linked with divinyl benzene before being treated with activating chemicals such as sulfuric acid to make them cation acceptors or amines to make them anion acceptors.

Figure 1 shows the simple chemical equation of the ion exchange process, where *Rz* stands for the resin molecule. The hydrogen ion attached to the resin is exchanged for the sodium ion of the sodium chloride, forming a sodium resin salt which is insoluble. In the case of regeneration the concentration of the acid is increased so that the sodium ions are pushed off into solution and the resin is again returned to the acid or hydrogen form. Figure 2 is a diagram of the operation of cation resin in the hydrogen cycle on the left-hand side, and in the sodium cycle on the right-hand side.

Service:

$$RzSO_3H + NaCl \leftrightharpoons RzSO_3Na + HCl$$

Equilibrium shifts———$\rightarrow$

Regeneration:

$$RzSO_3H + NaCl \leftrightharpoons RzSO_3Na + HCl$$

$\leftarrow$———Equilibrium shifts

FIG. 1. Chemical equation of the ion exchange process.

In other words, if the resins are regenerated with an acid, other cations in the water come through and displace the hydrogen ions, which means the effluent contains the acid of the salts entering. On the right-hand side the cation resin was originally regenerated with sodium salt so that any divalent ions such as calcium and magnesium entering at the top, having a greater affinity for the resin, attach themselves to the active sites displacing the sodium which comes out the bottom, resulting in nothing but sodium salts at the outlet. This process is referred to as softening.

Figure 3 shows the operation of an anion resin on the left side and an operation of a column containing a mixture of cation and anion on the right side. This, of course, is the so-called 'mixed-bed' which removes both the cations and anions very effectively.

The average ion exchange purification system consists of a two-bed demineralizer, a multiple-bed demineralizer or a mixed-bed unit, or any combination thereof. The two-bed is the simplest demineralizer to operate, since the regeneration merely involves passing acid through the first column, with a subsequent rinse, followed by caustic, through the second column, and a subsequent rinse. Because the anion resins are more effective in removing acids than anions as such, with few exceptions the water is always passed through the cation column first. The water from the two-bed generally

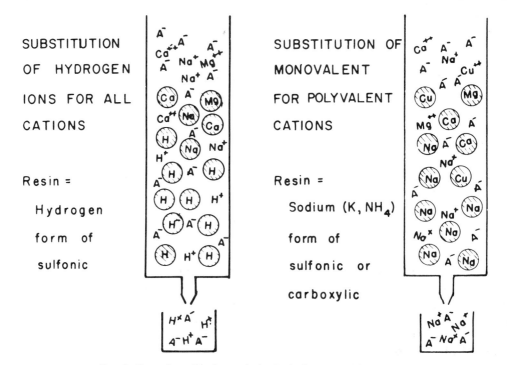

FIG. 2. Operation of cation resin in the hydrogen and sodium cycle.

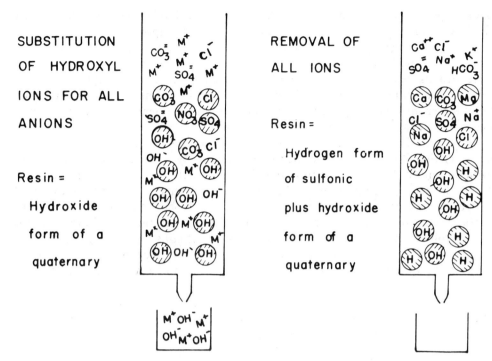

FIG. 3. Operation of an anion resin and a column containing a mixture of cation and anion resins.

will have a resistance averaging above 150,000 ohms; however, the cutoff point is usually set at 50,000 ohms, which is equivalent to 10 ppm impurities expressed as sodium chloride. This means that units of this sort are quite adequate for many washing operations and for feed to stills or devices such as weather simulating instruments.

If water of higher quality is desired, the slightly more complicated mixed-bed should be used. Disadvantages of the mixed-bed are the fact that the resins must be separated prior to regeneration and that the regeneration is not as efficient as in the case of resins contained in individual columns. The mixed-bed does have the advantage, of course, of being less expensive for a given flow rate, since there is only one column and less resin for a given diameter. The mixed-bed may be complete with fully automatic controls which are, of necessity, quite complicated and expensive. These units produce water having a resistance between one and ten million ohms, though under certain conditions, especially when new resins are employed, the resistance may be as high as 18 million ohms. It should be remembered that these resistances are based on 25°C (77°F). It used to be common to see specifications calling for 26 million ohm water from a mixed-bed, however, this figure is based on 18°C (64.4°F) rather than 25°C (77°F) and is equivalent, considering the temperature difference.

DISTILLATION OF WATER

Distillation is an older, more conventional and perhaps more easily understood method of producing pure water. This process should be quite simple, since we all boil water in our homes every day and should be able to condense it quite easily on a cool surface. On the contrary, it turns out to be very difficult if it is to be done properly and economically.

The stills must be designed so that the water in the gas phase separates from the liquid phase without carrying over particulate contamination as entrainment. As N.

E. Dorsey points out in "Properties of Ordinary Water Substances," it is exceedingly difficult to obtain water free of suspended particles. The evaporator of the still should be of the right diameter and designed to give the optimum conditions at the interface of the heated liquid, and the baffling system must be designed so that the velocity is sufficient to insure the removal of any of the entrained particles. From this point on, the purified steam must be protected from all the classifications of contaminants, and the condenser must be designed to insure an optimum removal of volatile impurities which might normally recondense in the distillate.

Because of the requirement of nature that 540 calories be added to every gram of water to change it into steam, the source of heat for any still is always an important consideration. Stills with capacities larger than 38 liters (10 gallons) per hour are generally heated by steam because of the cost of electricity and the complication of using gas. In the case of large stills it may pay to conserve the latent heat of the steam and use it over one or more times to boil the water in the several evaporations that may follow. These stills are called multiple-effect stills and are operated under pressure and/or vacuum so that there is a slight temperature drop in going from one evaporator to another. This is the principle of operation of many of the large saline water stills.

In certain situations, laboratories are more interested in dependable purity than in the cost of operation, in which case the first still feeds the distillate to a second still, the second still to the third still, etc. Generally speaking, the use of this type of equipment is limited to the hospital field, since they do not want to take any chance of contamination of a given batch of water due to a temporary condition such as a boil-over. The multiple stills also have the advantage of producing a better grade of water since, to a limited extent, the water produced from a still varies in purity with the purity of the feed-water; hence double distillation is almost always employed wherever extreme purities are desired. Needless to say, the cost of operation of the multiple stills is in direct proportion to the number of stills employed. Conversely, in the case of the multiple-effect stills noted previously, the operating cost diminishes more or less in direct proportion to the number of effects.

There are other distillation methods for producing various grades of distilled water, some of them having a high degree of complexity but having the advantage of conserving on energy. These include such stills as vapor compression stills, centrifugal rotating heat exchanger stills, falling film evaporator stills, and combinations of these with the multiple-effect principle. Since such units are basically designed for the production of potable water, they are usually made of materials that are as inexpensive as possible, which means that the water produced contains impurities picked up from the metallic components of the equipment. An exception to this rule is the stainless steel vapor compression still which has been used in the pharmaceutical industry quite successfully. Because of high cost of energy in areas outside of the United States, small vapor compression stills are popular and have been employed in the laboratory field for some time. On the other hand, stainless steel does impart impurities to the water and the water may or may not come under the *broad* "high purity" classification.

Another deterrent to the production of a quality distillate is the removal of impurities that are carried over in a vapor which has been contaminated with foam. Generally speaking, this condition exists whenever the water being fed to the still is high in total solids, or when the still has not been drained, thus allowing the contents of the evaporator to become quite concentrated. Properly designed baffle systems minimize the effect of carry-over and should be recommended whenever the problem is suspected. Not only do conditions of this sort present a problem as far as the purity is concerned, but they also mean that the maintenance of the equipment is increased, since the still must be cleaned frequently in proportion to the salt concentration in the evaporator and the hours during which the still is operated.

STORAGE AND DISTRIBUTION

Once water of required purity has been produced, there is the difficulty of storing the water and transmitting it to the point of use without reducing purity. There are two basic considerations: first, the material of contact of the tanks and distribution system, including pumps, valve packing and gaskets, and piping, and second, the atmosphere maintained over the water being stored. Gravity distribution, even at considerable extra expense, avoids the problems of pumping. The piping should be sized so that there will be an adequate supply of water at all of the distilled water drawoff faucets, but because of the relatively infrequent and spasmodic use of distilled water from any given faucet, it can be assumed that the flow of water in any particular pipe will never be great at any time. Therefore, the principle of sizing pipes similar to that used in the design of other piping systems does not necessarily apply to the design of piping used in distilled water distribution. If the use of water is substantial, it is better to have the water come from the faucet at a relatively low pressure to cut down on the use, than to have a high pressure available at all times.

The cheapest and most widely used materials for water storage and piping can be classified generally as the plastics, since polyethylene bottles and carboys are used to handle distilled water in practically every laboratory. There are new plastics available at the rate of about one per year and, of course, some are undoubtedly better than others as far as resistance to solution by ultrapure water is concerned. On the other hand, the plastics all seem to have the same general disadvantages of a relatively porous surface, which can harbor biological impurites, and the tendency to contaminate the water with organics and inorganics. In some cases plasticizers or the plastic molecules themselves become degraded by the water and enter the effluent stream. In other cases inorganic impurities picked up by the plastic during its manufacture or processing are dissolved out of the plastic surface by the water.

The lowering of the resistance of water from approximately 18,000,000 ohms to approximately 8,000,000 to 10,000,000 ohms can occur by passage of water through a length of 100 feet of polyvinyl chloride pipe. We know that polyethylene is relatively porous and that carbon dioxide and other gases from the atmosphere can permeate the plastic and lower the resistance of water stored within a carboy made of the material. On the other hand, experiments have shown that Teflon® has good resistance to attack and will preserve the quality of the pure water to an amazing degree. Unfortunately, Teflon® is expensive and is not available as pipe or as fittings, except in small sizes, where it can be purchased as tubing. Teflon® also has the disadvantage of the physical property of cold flow, which means you cannot fabricate an elbow from Teflon® and expect that it will stand up over a period of time under pressure.

If we consider tanks and distribution systems designed primarily for ultra-pure water, we find that metals have been used almost exclusively, since engineers and architects have been reluctant to put plastic piping within the walls of buildings. When considering metal systems, one is usually limited to aluminum, stainless steel or tin as a material of contact with the water. The only other material used is the glass-lined steel tank and, in some cases, glass pipe. Manufacturers of glass-lined vessels will not recommend their use for high purity water, since the glass used for the lining of vessels is one that is particularly soluble in high purity water, especially if that water happens to be hot. The use of glass pipe is a little more prevalent, especially in the pharmaceutical industry where silica as a contaminant is not serious, and where the advantage of inspection is very desirable. We do know, however, that the glass is soluble to the extent that even the manufacturer of the pipe has implied that the impurities imparted to the water, especially during the first six months of operation, would probably be detrimental to use of the water for manufacture of certain types of transistors.

Aluminum piping which has been used in many installations, should be recommended only when the user appreciates the fact that he is going to have some aluminum in the water coming from the system. The quantity of aluminum and the amount of impurities such as copper that might be given off by the aluminum pipe and fittings will vary with the type of aluminum used, and might or might not be a problem in any particular instance. Aluminum will dissolve to the extent of from 10 to 20 ppb when in contact with pure water. Stainless steel, generally speaking, is less desirable than aluminum, since more people are bothered by the impurities of iron, nickel and chromium given off by the stainless, than they are by the impurities from aluminum. Stainless steel is also more expensive and somewhat difficult to handle during the installation of the system.

Both aluminum and stainless steel have been and are being used for storage tanks, since they are readily available at a relatively low cost. If metal tanks are to be specified, the fabricator should grind the interior welds to the equivalent of a 2-B finish. It might be of interest to note that the contamination imparted to the water in any storage tank is quite often greater than the impurities from the distribution system, since the water remains in the storage tank for many hours or days. If the distribution system is known to contaminate the water, the operator may minimize the effect by merely running the first slug to waste before taking his quantity for use.

The critical user of ultra-pure water usually demands a tank and distribution system fabricated in such a way that the only material in contact with the water is pure block tin, or perhaps a silver tin alloy. Because pure tin is relatively soft by itself, it must be either coated on a stronger metal or bonded to that metal. Tin-lined pipe and tin-lined tanks are fabricated by bonding a 1.6 mm (1/16″) sheet of tin to steel, or perhaps by lining a piece of red brass pipe with 1.6 mm (1/16″) of tin, or we have the so-called tin-coatings which are generally applied by heat and fluxing agents to copper or copper-base alloys. Since the cost of these two processes varies tremendously, it is important that the difference be noted. Strangely enough, in some cases, we believe now that the hot tinning done properly will last longer than a tin-lined or cladded surface, particularly if the item is subjected to wide temperature variations, as would be the situation in the operation of a distilled water cooler.

INSTALLATION

No matter what type of material is used in the fabrication of the tank and distribution system, it is very important that the supervisor of any installation be made aware of the care required in the handling of the components to make sure that the proper type of threads are used, the proper pipe compounds are employed, and of course, make sure that an optimum condition of cleanliness is maintained. In one instance a hospital research wing was completely piped in polyvinyl chloride pipe on the basis of the fact that the PVC had been used in a previous installation and was adequate. Unfortunately, the pipefitters installing the new system used liberal quantities of cutting oil when they threaded the pipe and found that they could not clean the system properly since the oil penetrated into the relatively porous surface of the pipe and was detected by the laboratory instruments as long as two years after the installation.

CONCLUSION

If one is concerned with a pure water system for his laboratory, it is important that he first establish his present and probable future needs, then consider the method of production and distribution. Attention must be given to the raw water available, since a successful installation in one location may not be suitable at another. In any case it should be stressed that there is a problem in producing high purity water and the water which comes from the faucet marked "DW" in the laboratory, and no matter what systems are used, the purity of water is only relative.

Biological Hazards

Animal Care and Handling

Roger De Roos

Bioscience research conducted with the use of laboratory animals has contributed a great deal to mankind, including a significant reduction in deaths from communicable diseases, a growing understanding of cancer, heart disease, and mental illness, and development of surgical methods for better treatment of congenital, acquired, and malignant diseases.[1]

The importance of healthy animals for experimental research has been well recognized by the scientists, since the experimental animals must be free of disease for the research results to be accurately interpreted. For the maintenance of their health, adequate hygiene and care are of utmost importance. The diets of the animals must be nutritionally adequate and free of pathogenic organisms. The water supply must be of satisfactory chemical and bacteriological quality. The environmental conditions must be such that the animals are not subjected to disease-causing dosages of pathogenic organisms. To maintain the concentration of pathogenic organisms at a low level, such factors as vermin control, removal of animal wastes, maintaining clean facilities and cages, and maintenance of satisfactory air quality must all be taken into consideration.

The control of man as he works with and takes care of animals is also of great importance in keeping animals free of disease since under some conditions diseases can be transmitted from man to animals. Salmonellosis is one example where human carriers have been incriminated as the source of infection for animals, and other examples are influenza, staphylococcal infections, tuberculosis, and infectious hepatitis. Therefore, personnel who work with the animals should take the necessary precautions to prevent the transmission of human infections to animals, avoiding contact with animals when personnel are ill and practicing good personal hygiene whenever working with animals.[2]

Since man may contract diseases from animals, personnel who take care of or work with animals should try to minimize the chance of pathogenic organisms from the animals being transmitted to them. Animals which are obviously ill should be isolated from the normal animal quarters and added precautions should be followed in their care and in maintenance of their quarters. However, an infected animal will not necessarily be conspicuously ill and animals which appear to be perfectly healthy are usually the ones from which man contracts an infection. In many instances the animal host maintains the disease parasite in peaceful co-existence, not being adversely affected by its presence, so that even apparently healthy animals may be a reservoir of infection.[3]

Animal caretakers and research personnel must be continuously alert to and make use of techniques and environmental control measures which will limit the spread of infection in animal quarters from animal to animal, man to animal, or from animal to man. A discussion of the infectious process is given in the next few paragraphs to create a better understanding of how an environmental control program will aid in preventing spread of disease, in an animal-quarters facility or in a microbiological laboratory. In the infectious disease process there are six primary links, as follows:[4]

1. A causative or etiological agent
2. A reservoir or source of the causative agent
3. A mode of escape from the reservoir
4. A mode of transmission from the reservoir to the potential new host
5. A mode of entry into the new host
6. A susceptible host

The total environmental control program should hinge on the breaking of one or more of the links of the infectious disease process, for disease will not result if one of the links is broken even though the remaining five factors necessary to cause disease are present.

DISEASE ORGANISMS.   When considering the etiological agents which may be present, attention should be given to the possible types of organisms, and such characteristics of the organisms as infectivity, virulence, pathogenicity, and the organism's resistance to environmental factors outside of the host. For example, if one is investigating a pseudomonas infection in a mouse colony, it would be very helpful to know that pseudomonas organisms do not survive readily under dry conditions. It might also be helpful to know that the pseudomonas organisms are resistant to quaternary ammonium disinfectants, but are destroyed more easily when chlorine is used as a disinfectant.

RESERVOIR OF DISEASE ORGANISMS.   The etiological agents which cause infection require pre-existing life for their growth and perpetuation. The place where these organisms perpetuate is called the reservoir of infection. With few exceptions, the reservoir for diseases which attack man is made up of human and animal sources. As was previously mentioned, a large portion of the reservoir consists of sub-clinical infections and carriers who are not obviously ill. People or animals who are obviously ill with a disease, the frank cases, constitute the remainder of the reservoir, with the exception of pathogenic plant forms such as fungi and molds. If the size of the reservoir can be decreased, the opportunity for transmission will also be decreased. The use of antibiotics is one method for decreasing the size of the reservoir, by decreasing the numbers of animals harboring the disease organisms, and another method would be the use of "disease-free" stock for the conduct of laboratory animal experimentation.

ESCAPE OF ORGANISMS.   Organisms may escape from animals in many ways; they may be exhaled from the respiratory tract; excreted through the intestinal or urinary tracts; exit through an open sore which is discharging from the surface of the body; shed from the skin or hair; or may be liberated from the body by outside forces. An example of mechanical escape, an outside force liberating the organism from the body, would be a biting or sucking insect which draws infected blood and in turn transmits the infection to man or other animals.

The mode of escape may have a definite bearing on the approach to the control program. For example, with leptospirosis special care should be exercised in handling of urine from diseased animals, and it is important to emphasize the importance of proper handling of wastes removed from the cages of animals with leptospirosis.

TRANSMISSION OF ORGANISMS.   Transmission of organisms to another host may occur either directly or indirectly. Direct transmission occurs through very close association between the reservoir and the new host. An example of this would be the transmission of respiratory staphylococcal infection from an animal to man— the animal sneezes and the organisms are breathed in by the attendant who is caring for the animal.

Pathogens may also be transmitted indirectly on inanimate vehicles, and in this instance there is not as close an association between the reservoir and the new host. For indirect transmission to take place, the organisms must be able to survive for some period of time outside the host, allowing them to be transferred by a vehicle from one location to another. The vehicle which is involved in the transfer may be of either an animate or inanimate nature. Cockroaches may act as animate vehicles in the transfer of salmonella organisms from the feces of one animal to the food of another animal, in which case the most obvious control measure would be to break the chain of infection by eradicating cockroaches in the animal-quarters facility.

Animal feed may act as an inanimate vehicle of infection, because the raw materials have been contaminated with pathogenic organisms, or the feed becomes contaminated by handling procedures during preparation. If the feed is properly heat treated to destroy pathogenic organisms subsequent to formulation, and is not re-contaminated before being fed to the animals, this link in the chain of infection will be broken and the animal feed will no longer act as an inanimate vehicle of transmission.

ENTRY OF DISEASE ORGANISMS. The mode of entry into the body for organisms which cause disease may be through the respiratory tract, gastrointestinal tract, or through a break in the skin. A break in the skin may provide entry into the body for a staphylococcal infection when a person with a scratch handles an animal with a staphylococcal lesion and the organisms are transferred by direct contact. Entry of organisms can often be prevented by use of protective equipment, such as rubber gloves, ventilated cabinets, or other devices.

SUSCEPTIBILITY OF HOST TO DISEASE ORGANISMS. After an etiological agent has gained entry into a new host, whether or not disease will develop depends upon the susceptibility of the host. Under normal conditions the body of the new host will have certain defensive mechanisms which will tend to ward off invasion of pathogenic organisms. These defense mechanisms are called "resistance." When there is enough resistance to protect against the average infecting dose of a pathogenic organism, the person is "immune" to the disease. This is not to say that if the same person were exposed to an overwhelmingly large dose of the pathogenic organism that the disease would not develop.

Rabies is an example of disease where the host can be made less susceptible by immunization.[4]

CONCLUSION

The preceding paragraphs contain the principles which may be utilized in approaching the problem of controlling communicable disease in animal quarters. It will be helpful if these principles are given careful consideration in the establishment of administrative and housekeeping procedures, in the evaluation of new equipment, and in the design of housing facilities. In each of these steps, continuous attention should be given to ways and means of breaking one or more links in the infectious disease chain.

REFERENCES

1. Dragstedt, Lester R., Ethical Considerations in the Use and Care of Laboratory Animals, *J. Med. Ed., 35*:1, (January, 1960).
2. Meyer, K. S., "Evolution of the Problems of the Occupational Disease Acquired from Animals," Conference on Occupational Diseases Acquired from Animals, 1964. The University of Michigan, School of Public Health, Continuing Education Series—No. 124, pp. 4–55.
3. Clarkson, M. R., "Development of Protective Measures," Conference on Occupational Diseases Acquired from Animals, 1964. The University of Michigan, School of Public Health, Continuing Education Series—No. 124, pp. 370–75.
4. Anderson, Gaylord W., Arnstein, Margaret G., Lester, Mary R., "Communicable Disease Control," 4th Edition, p. 14–46, MacMillan Co., New York, (January, 1962).

Prevention of Laboratory-Acquired Infections

G. Briggs Phillips

Almost since microbiology began as a science, accidental infections resulting from laboratory manipulation of pathogenic microorganisms were recognized and recorded. Louis Pasteur finally disproved the theory of spontaneous generation in 1861, and in the 1870's began his studies with disease-producing organisms. Robert Koch solved the problem of growing pure bacterial cultures in the laboratory in 1881, and in the following two years discovered the etiologic agents of tuberculosis and cholera. The organisms producing typhoid fever were identified in 1880, and five years later, in 1885, two cases of occupationally-acquired typhoid fever were recorded in the German Imperial Health Service. In 1893 another case of laboratory-acquired typhoid fever was recorded in Germany and a case of tetanus was reported in France. In 1903 the first recorded case of blastomycosis following an accidental self-inoculation occurred.

Today, more than 80 years later, the problem of accidentally-acquired laboratory disease still exists. It is not at all unusual to see reports of laboratory infections in the current medical literature. Moreover, through the years the frequency of reports of laboratory infections appears to have increased as the science of microbiology has expanded.

Through the years, several hundred publications have mentioned approximately 6000 laboratory infections. The largest single collection of cases was that published in 1951 by Sulkin and Pike.[1] This survey listed 1342 laboratory infections occurring in the U.S. during a 20-year period. Included were infections caused by 69 different disease agents, resulting in 39 deaths. Through a committee of the American Public Health Association, Sulkin has continued to tabulate reported cases; the total number of cases now stands at 2348 with 107 deaths.[2]

While these reports and surveys illustrate that there can be a microbiological safety problem in infectious disease laboratories, they give little concrete information in terms of frequency or severity rates. In fact, probably only a fraction of the laboratory-acquired infections are ever reported in the literature, and it sometimes is difficult to prove whether or not a disease actually was acquired during laboratory work. An even more elusive factor is the occurrence of accidental infection in laboratory workers who do not show recognizable clinical symptoms.

Microbiological safety in its simplest form relates to the precise control of the microbial elements in any particular environment. Its application in laboratories where pathogenic cultures or infected animals are being used will help to prevent infections in laboratory workers.

A second reason for microbiological environmental control is to protect the validity of the experiment. In the absence of suitable controls, laboratory results can be confounded by accidental or unintentional transfer of infections microorganisms from animal to animal or from test tube to test tube.

A third reason for microbiological environmental control is to protect man from infection by laboratory animals not known to be infected. Human infections may result from any laboratory use of animals. For example, laboratory animals used by a psychology department for behavioral studies can present a human infectious hazard if they carry an unrecognized disease and if microbiological environmental control is inadequate. Tuberculosis in monkeys could be an example of an unrecognized disease. For example, since 1934 there have been 18 human cases of B virus infection identified with the handling of monkeys or their tissues.[3] Most of the cases were fatal; the causative virus apparently occurs naturally in monkeys without producing ill effects. Another example is the observation of an abnormal incidence of hepatitis among persons handling apparently healthy subhuman primates, chiefly chimpanzees—from 1953 to 1962 some 69 human cases have been documented in which primate to human transfer was suspected.[4] A third potential hazard occurs to human handlers who have close contact with monkeys that have acquired tuberculosis.[5]

This chapter is directed primarily toward controlling microbiological hazards in the infectious disease laboratory to prevent occupational infections. We should realize, however, that other applicable areas exist.

First we will review the known and probable causes of laboratory infections. Then we will consider the available methods for prevention, dealing with five approaches to laboratory safety and giving some specific recommendations for each. Finally, we will discuss the educational needs and the future prospects for microbiological laboratory safety on the campus.

CAUSES OF LABORATORY INFECTIONS

The true causes of accidents lie in a combination of circumstances rather than the simple, direct effect of one or two external agencies. This is why cause analyses can conveniently follow an epidemiological approach, wherein the total interrelationships and interactions of the host, the accident agencies, and the environment are considered.

In the infectious disease laboratory there are a number of ways in which the major elements present interact with each other. The host or person affects and is affected by the environment, physical agents, infectious agents, animals, and even insects in various ways. Of course it is not the interactions themselves that are responsible for accidents; the interactions are normally required for carrying out laboratory functions. But accidents do happen when the sequence is wrong, when the timing is wrong, when the amount is too much or too little, when the wrong choice is made, or where there is a combination of any of these or other factors.

We have already noted that naturally infected monkeys and chimpanzees can transmit diseases to humans in the laboratory. There are, in fact, well over 100 diseases of animals that could conceivably be passed to man. About a dozen of these zoonoses are known to have been transferred from naturally infected animals to man in the laboratory—lymphocytic choriomeningitis, infectious hepatitis, cat scratch fever, Newcastle disease, psittacosis, monkey B virus, leptospirosis, tuberculosis, malaria, amebidsis, shigellosis, and streptococcal and staphylococcal infections.

Because naturally infected animals can cause laboratory infections we may also expect animals challenged in the laboratory with infectious disease agents to be capable of transmitting infection to humans. Although there are no definitive data showing the frequency with which this happens, we do know that the hazard exists. Moreover, a sizable amount of supportive evidence has been provided by animal cross-infection studies.[6] In these studies control animals are usually caged together with infected animals or in adjoining cages. Then periodic tests are made to determine if the controls themselves become infected. The frequency with which cross-infection of this type occurs provides good presumptive evidence of the hazards that may exist for man in the same environment.

Infections from animals may be caused by bites or scratches, from contact with contaminated cage debris, or from breathing of air-borne organisms from sick or coughing animals. If animals have been used in aerosol experiments, infectious organisms on their fur may be released to the air to infect animal attendants.

The specific primary causes of laboratory infections fall into two groups. One group, which includes about 20 percent of the total, consists of recognized accidents that result in the infection of laboratory personnel. The second group, which includes approximately 80 percent consists of accidents whose causes are often classified as "unknown" because there were no previously recognized or recorded accidents or incidents that could be shown to have been responsible for the infections. Although this is a somewhat unique situation, that it is true is shown by the fact that the percentage of "unknown" causes has remained reasonably consistent in the various infection surveys.

TABLE 1
Known and Unknown Causes of Laboratory Infections

Data Source	Percentage of Accidents	
	Known Cause	Unknown Cause
Paneth, 1915[7]	61	39
Sulkin & Pike, 1951[1]	16–20	84–80
Schafer, 1950[8]	16	84
Survey of 18 countries, 1959	14	86
Fort Detrick, 1955–57	35	65
Fort Detrick mechanical and chemical lost-time injuries	100	0

The five most frequently recognized causes of laboratory infections are: (a) accidental oral aspiration of infectious material through a pipette, (b) accidental inoculation with syringes and needles, (c) animal bites, (d) sprays from syringes, and (e) centrifuge accidents. Together these caused about 12 percent of some 3700 laboratory infections. A general estimate of the percent of accidents due to each cause is shown in Table 2. Some other commonly recognized causes of laboratory infections are: (a) cuts or scratches from contaminated glassware, (b) cuts from instruments used during animal autopsy, and (c) the spilling or spattering of pathogenic cultures on floors, table tops, and other surfaces.

TABLE 2
Common Causes of Laboratory Infection

Accident Cause	Percent of 3700 Infections
Oral aspiration through pipettes	4.7
Accidental syringe inoculation	4.0
Animal bites	1.4
Spray from syringes	1.2
Centrifuge accidents	0.8

Because unsafe acts or unsafe conditions have not been identified in approximately 80 percent of the recorded laboratory infections, some laboratory procedures and equipment have been suspected of creating hazards. Indeed this suspicion has been confirmed by a number of studies in which the amount of microbial aerosol produced by various laboratory techniques has been measured. Most of the usual techniques have been tested in studies done in this country[9,10] and in England.[11,12] They show that most common laboratory techniques carried out in the ordinary manner will produce infectious air-borne particulates. At least one study has shown that these particulates are of a size which will readily penetrate to the human lung if they are breathed.[11] Of course, these results only suggest possible means of laboratory infection. The type of microorganism, its probable infectious dose, its environmental resistance, the resistance of the host, and many other factors would have to be evaluated to accurately define a hazard.

Laboratory infections caused by accidents that can be identified with unsafe acts and conditions or with procedures and techniques that unsuspectingly release infectious aerosols to the laboratory environment illustrate that laboratory safety is a problem of environmental control. The microbe must remain in its environment (test tube, flasks, etc.) and the microbiologist must be externalized from the organism's environment. Although this solution appears simple and straight forward, its application is complex. Microbes capable of causing human infection are not readily detectable in the usual sense; the infecting dose may be odorless, tasteless and invisible to the eye.

Statistical studies of accidents and infections at several large laboratory institutions provide some additional epidemiological data of probable significance in infection prevention. These are summarized as follows:

In general, more infections are associated with manipulating cultures than with handling animals.

Laboratory technicians, students, trained professional personnel, and animal handlers, those most closely associated with the infectious operations, are in the greatest danger of becoming infected.

Among those who work directly with pathogens, it is the younger persons with less formal training who are more apt to become infected.

Inhalation of infectious aerosols is by far the most frequent mode of laboratory infection.

The physical form of an infectious agent is related to its hazard level. Dried or lyophilized cultures and infected eggs are more dangerous to handle than liquid cultures or infected blood specimens.

PREVENTION OF LABORATORY INFECTIONS

As with any type of preventive effort, the prevention of laboratory-acquired infections should begin with an assessment of the extent of the problem or an estimate of the probable extent of future problems. The potential hazards may result from research being carried on with infectious cultures, from the use of cultures or infected tissues in classroom demonstrations, from clinical diagnostic procedures, or from the use of animals in laboratory situations. In any case, there must be an understanding and agreement on the dangers by administrators and laboratory directors. Even when the microbiological hazards are understood, the philosophy of scientific freedom characterized by academic life can often work to oppose the inspections, investigations, and regulatory requirements of a good safety program. Moreover, one is often faced even today with the martyr-to-science complex in which laboratory scientists feel that being infected is part of the job. Particularly in a school situation, legal and moral considerations make this view unacceptable.

Once there is an adequate assessment of the potential microbiological hazards and management is committed to a preventive program, there should be evolved a precise personnel policy regarding occupational health. Management should make a series of policy decisions relating to the goals of the safety program and how it is to operate. An adequate list of policy questions has been formulated for those concerned with the construction of laboratory facilities for infectious disease work.[13] Many of these questions apply to the safety policy as a whole. For example:

What level of occupational infection is acceptable to management? Is it desired to attempt to prevent all work-incurred infections, including subclinical infections that can be detected only serologically? Or is management's aim to prevent only those infections that are likely to result in incapacitating illnesses, or only those for which there is no treatment?

To what extent is the control of microbiological hazards to be extended to protect persons in areas peripheral to the laboratory? Public relations, economic and legal considerations are involved here.

What type of supporting medical program is to be provided for persons at risk in the laboratory?

A program for controlling microbiological hazards should begin with a clear concept of the goals, an understanding of the nature of the hazards, and an expression of the policies to be followed in achieving control. By this action, administration establishes responsibility for safety control, includes planning for accident control in all phases of laboratory work, and makes it clear that no job will be considered so important that it cannot be done safely.

In the main, the cardinal tenets of microbiological control will be education, engineering, and enforcement. The detailed implementation of safety control can be discussed by considering five important elements. Each element's use is determined by the extent of the microbiological hazards and by management's policy concerning them. Management aspects, vaccination, and safe techniques and procedures are discussed below; equipment and building design are referred to in the next chapter.

Management Aspects

Some of the programming and policy responsibilities have already been discussed. Management at various levels must also concern itself with the proper selection of laboratory employees. Proper selection refers not only to technical competence but also to the fact that it may be undesirable to employ persons with certain physical conditions for work with some types of infectious agents, because of medical decisions.

Management should likewise be concerned with providing safety training for laboratory personnel, formulating safety regulations, and establishing methods for adequately reporting and investigating accidents.

The management approach should also attempt to control human factors in accident causation and strive to provide an atmosphere wherein personnel may develop attitudes conducive to safe performance. Practical experience has shown that such an approach is essential for an accident and infection prevention program to be successful. Good laboratory management includes good safety management since safety is an essential part of any productive enterprise.

Vaccination

Vaccination of laboratory personnel is recommended when a satisfactory immunogenic preparation is available. Good immunity is conferred after vaccination against smallpox, tetanus, yellow fever, botulism, and diphtheria. The new living vaccine for tularemia gives excellent protection. Other vaccines such as those for psittacosis, Q fever, Rift Valley fever, and anthrax have been or are being tried experimentally with varying degrees of success.

Vaccines have not yet been developed for a number of human diseases which have been known to occur in laboratory workers—dysentery, blastomycosis, brucellosis, coccidioidomycosis, glanders, histoplasmosis, infectious hepatitis, leptospirosis, and toxoplasmosis. We generally evaluate the efficiency of vaccines for laboratory workers on the basis of their effectiveness in preventing disease in the general population but the laboratory worker may be exposed to infectious microorganisms at a higher dose level than would be expected from normal public exposure, and the exposure may be by a route different from that normally expected; e.g., respiratory infection with the tularemia or anthrax organism.

Safe Techniques and Procedures

Sound fundamental laboratory techniques, well supervised and conscientiously carried out, can do much to achieve environmental control and reduce the hazards of infection. Many procedural rules are obvious because their aim is to prevent direct contact with harmful microbes. Others may be less well understood because their purpose is to prevent airborne contamination of the workers' environment at a level where such contamination is not easily or readily detected. Infectious aerosols are like dangerous radiations, except that the former are more difficult to monitor. A list of procedural rules that are widely applicable in infectious disease laboratories follows.[14]

1. Never do direct mouth pipetting of infectious or toxic fluids, use a pipettor.

2. Plug pipets with cotton.

3. Do not blow infectious material out of pipettes.

4. Do not prepare mixtures of infectious material by bubbling expiratory air through the liquid with a pipette.

5. Use an alcohol-moistened pledget around the stopper and needle when removing a syringe and needle from a rubber-stoppered vaccine bottle.

6. Use only needle-locking hypodermic syringes. Avoid using syringes whenever possible.

7. Expel excess fluid and bubbles from a syringe vertically into a cotton pledget moistened with disinfectant, or into a small bottle of cotton.

8. Before and after injecting an animal, swab the site of injection with a disinfectant.

9. Sterilize discarded pipettes and syringes in the pan where they were first placed after use.

10. Before centrifuging, inspect tubes for cracks. Inspect the inside of the trunnion cup for rough walls caused by erosion or adhering matter. Carefully remove all bits of glass from the rubber cushion. A germicidal solution added between the tube and the trunnion cup not only disinfects the surfaces of both of these, but also provides an excellent cushion against shocks that otherwise might break the tube.

11. Use centrifuge trunnion cups with screw caps or equivalent.

12. Avoid decanting centrifuge tubes; if you must do so, afterwards wipe off the outer rim with a disinfectant. Avoid filling the tube to the point that the rim ever becomes wet with culture.

13. Wrap a lyophilized culture vial with disinfectant-wetted cotton before breaking. Wear gloves.

14. Never leave a discarded tray of infected material unattended.

15. Sterilize all contaminated discarded material.

16. Periodically, clean out deep-freeze and dry-ice chests in which cultures are stored to remove broken ampoules or tubes. Use rubber gloves and respiratory protection during this cleaning.

17. Handle diagnostic serum specimens carrying a risk of infectious hepatitis with rubber gloves.

18. Develop the habit of keeping your hands away from your mouth, nose, eyes, and face. This may prevent self-inoculation.

19. Avoid smoking, eating, and drinking in the laboratory.

20. Make special precautionary arrangements for respiratory, oral, intranasal, and intratracheal inoculation of infectious material.

21. Give preference to operating room gowns that fasten at the back.

22. Evaluate the extent to which the hands may become contaminated. With some agents and operations, forceps or rubber gloves are advisable.

23. Wear only clean laboratory clothing in the dining room, library, and other non-laboratory areas.

24. Shake broth cultures in a manner that avoids wetting the plug or cap.

EDUCATION IN MICROBIOLOGICAL SAFETY

In spite of the considerations, approaches, and equipment discussed thus far, human factors in accident prevention occupy a dominant position and the educational process is essential in safety.

As we are speaking primarily about safety in the campus situation, education takes on a double significance. The responsibility for the safety of the individual working in the infectious disease laboratory rests with the teaching institution that provided his initial training in laboratory procedures. Endowing the student with heuristic desires and technical knowledge is not enough. He must be taught how to use the instruments and apparatus of the laboratory. He must, in the learning process, be made to understand the importance of the manipulations and impressed with the notion that a good scientist is also a safe scientist. Too often school authorities have avoided the need for microbiological safety education and solved infectious hazards problems simply by forbidding the use of pathogenic agents in the school's laboratories.

It is true that not all microbiologists handle or need to handle infectious organisms in their work. But it is academic "buck-passing" to tell the student taking a course in infectious diseases that if he is required to handle pathogens later in his career his employer or someone else will give him the proper instructions. If microbiologists and others who work with infectious microorganisms are to be given every opportunity to protect themselves from acquiring occupational diseases, should not safety education in the hazards associated with handling highly virulent microbes be included in the college curriculum? If safe behavior is a concept of life and important contributions to attitudinal outlook are formed early, is it realistic to wait until after the completion of professional training to institute education in safety?

CONCLUSIONS AND FUTURE PROSPECTS

As has been noted the use of animals and infectious agents often leads to occupational disease among laboratory people. From the point of view of each safety administrator, it is important that an evaluation be made of actual, potential, or future microbiological hazards before deciding if and how much of a prevention effort is required.

It has been principally during the last two decades that attention has been given to the problem of correcting or reducing laboratory-acquired illnesses. Former traditions of personal sacrifice are gradually becoming outdated by economic, moral, and legal pressures. Also, in the last few years, it has become eminently clear that laboratory determinations will be accurate only if controlled to the extent that concurrent culture cross-contamination or animal cross-infection can be prevented. This has prompted research helpful in developing techniques and methods which reduce human infectious risks in the laboratory. The most important single conclusion from this research is that preventing the release of accidental microbial aerosols at the laboratory working surface through careful techniques and through the use of containment devices is the best way of achieving microbiological environmental control.

The specific tools for controlling microbiological hazards are:

The required management supports and administrative techniques of reporting, analyzing, selecting, regulating, and training.

The use of correct techniques.

The use of safety equipment.

Properly designed laboratories.

Vaccination of personnel.

On a campus, it is essential that the school discharge its responsibility for the safety education of students who are exposed to infectious disease hazards. Increased national expenditures for education and increased emphasis on microbiological research portends an increasingly greater demand for microbiological safety programs in laboratories in order to protect potentially exposed students, researchers, and scientists.

Education in laboratory safety methodology requires, as a background, an adequate body of facts about laboratory hazards, their prevention, and most particularly, their causes. In a time when there is a clearly recognized shortage of educators and teachers it is appropriate that scientific methodology be applied in efforts to control and reduce laboratory infections among teachers, researchers, and students.   Moreover, future demands on the educational system signal a need for research information on this subject.

Of significance is the trend toward team research in which persons trained in fields other than microbiology use infectious cultures as tools in the solving of life-science problems. Should the effort to isolate and identify virus strains as the etiological agents of certain cancers be successful to any degree, the need to protect research workers handling such strains would come under consideration. Perhaps it should be considered now. Likewise, in the space satellite research program it has been recognized that uncontrolled transfer of microbes between planets is undesirable. In the medical field, a more immediate hospital problem is that of the spread of staphylococci and other infections among hospital personnel and patients.   The principles of environmental control applicable in laboratory microbiological safety are helpful in solving these problems.

Before microbiological safety control can be successfully integrated into needed areas in colleges and universities, it will be necessary, through safety education, to impart knowledge about these hazards to the university staff and through them to the students.

ACKNOWLEDGEMENT: Reprinted from the *Journal of Chemical Education*, *Vol. 42*, January and February, 1965, with permission.

REFERENCES

1. Sulkin, S. E., and Pike, R. M., *Am. J. Public Health, 41,* 769, (1951).
2. Sulkin, S. E., *Bacteriol. Rev., 25,* 203, (1961).
3. Love, F. M., and Jungherr, E., *J. Am. Med. Assoc., 179,* 804, (1962).
4. Held, J. R., "Sub-human Primates in the Transmission of Human Hepatitis," Presented at the Sixth CDC Biennial Veterinary Conference, Atlanta, Georgia, (August 6–10, 1962).
5. Ruch, T. C., "Diseases of Laboratory Primates," W. B. Saunders Company, Philadelphia, Pennsylvania, (1959).
6. Kirchheimer, W. F., Jemski, J. V., and Phillips, G. B., Cross Infections Among Animals of Diseases Transmissible To Man, *Proc. Animal Care Panel, 11,* 83, (1961).
7. Paneth, L., *Med. Klin. 11,* 1938–1939, (1915).
8. Schafer, W., *Arch. Hyg. Bakteriol., 132,* 15, (1950).
9. Barbeito, M. S., Alg, R. L., and Wedum, A. G., *Am. J. Med. Technol., 27,* 318, (1961).
10. Reitman, M., and Wedum, A. G., Microbiological Safety, *Public Health Rep., 71,* 659, (1956).
11. Tomlinson, A. J. H., *Brit. Med. J. 2,* 15, (1957).
12. Whitwell, F., Taylor, P. J., and Oliver, A. J., *J. Clin. Pathol., 10,* 88, (1957).
13. Wedum, A. G., and Phillips, G. B., *J. Am. Soc. Heating, Refrig. and Air-Conditioning Engineers,* (Feb., 1964).
14. Wedum, A. G., Laboratory Safety in Research With Infectious Aerosols, *Public Health Rep., 76,* 619, (1964).
15. Phillips, G. B., "Microbiological Safety in U. S. and Foreign Laboratories," Safety Division, U. S. Army Biological Laboratories, Frederick, Md., (Sept. 1961).

Design of Facilities for Microbiological Safety

G. Briggs Phillips and Robert S. Runkle

Incorporation of microbiological safety measures in the design of biomedical laboratory facilities is needed for one or more of the following reasons:

1. To prevent the uncontrolled escape of infectious materials from the building to safeguard the health of the surrounding community.

2. To assist in the prevention of accidentally-acquired infections among building personnel.

3. To prevent the unintentional spread of diseases among animals by animal-to-animal or man-to-animal transfer.

4. To prevent false laboratory results due to cross-contamination of microbiological cultures.

The initial step that should be taken in the design process for a microbiological research laboratory is an analysis of the research activities to be undertaken, the hazards associated with the research and with each operation, and a functional analysis of the relationships that will exist between each activity. This analysis should enable the laboratory director and the architect/engineer to estimate the extent of the hazardous operations and to concentrate and minimize the amount of containment equipment required, thereby realizing economic savings.

Before determining what safety measures to incorporate in the design of infectious disease laboratories, much research and study was necessary to understand how laboratory workers become infected, how micro-organisms might escape and spread within a building or escape to the surrounding community, how animal cross-contamination occurs and other similar problems.[1-3] From the results of a number of laboratory hazards studies,[4-8] two concepts emerged that have proved successful in designing biologically safe laboratories. The first was the concept of primary and secondary barriers for the containment of infectious materials within the laboratory and the second concept provided the laboratory designer a logical division of major functional zones within a typical laboratory building.

THE PRIMARY-SECONDARY BARRIER CONCEPT

Enclosures, barriers, or any containment devices that immediately surround the infectious or potentially infectious material may be designated as primary barriers. They are the first line of defense (other than the test tubes, flasks, etc.) for preventing escape and possible spread of infectious microorganisms. Examples of primary barriers are ventilated microbiological cabinets, Figure 1; closed ventilated animal cages, Figure 2; closed centrifuge cups, Figure 3; and safety blendor bowls, Figure 4.

The secondary barriers in a laboratory are the features of the building that surround the primary barriers. These provide a separation between infectious areas in the building and the outside community, and between individual infectious areas within the same building. Examples of secondary barriers are (1) floors, walls and ceilings, (2) ultraviolet air locks and door barriers, (3) personnel change rooms and showers, (4) differential pressures between areas within the building, (5) provisions for filtering or decontaminating potentially contaminated exhaust air, and (6) provisions for treatment of potentially contaminated liquid wastes. These and other secondary barriers provide supplementary microbiological containment, serving mainly to prevent the

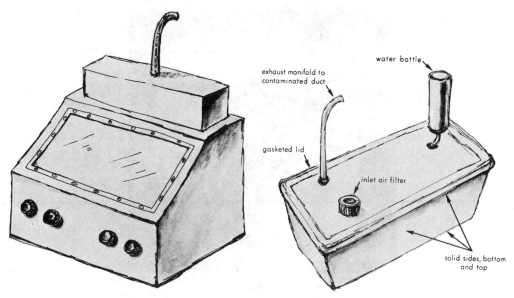

FIG. 1. Safety cabinets.   FIG. 2. Ventilated animal cages.

FIG. 3. Safety centrifuge cup.

escape of infectious agents if and when a failure occurs in the primary barriers. Figure 5 is a graphic representation of the functions of primary and secondary barriers.

Actually, the more effective the primary barriers are, the less need there is for emphasis on secondary barriers. Therefore, during the design phase of any infectious disease laboratory, it is both important and economically necessary to first determine and select the primary containment devices to be used, thereby reducing the complexity and cost of the secondary barriers.

FIG. 4. Safety blendor bowl.

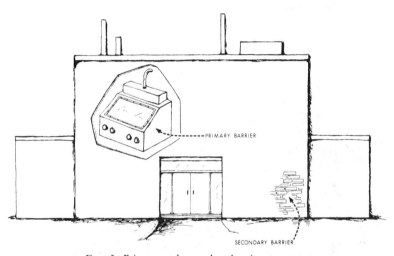

FIG. 5. Primary and secondary barrier concept.

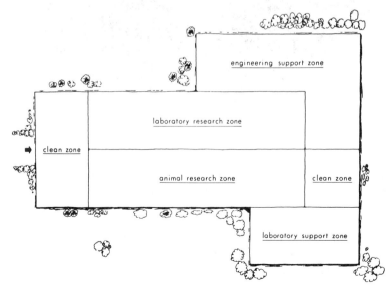

FIG. 6. Five functional zones of hypothetical laboratory.

FUNCTIONAL ZONES OF A LABORATORY BUILDING

Figure 6 illustrates five functional zones that might exist in a hypothetical laboratory building for research with infectious disease agents. Obviously there can be numerous physical arrangements of these zones, but the typical arrangement shown will illustrate their relationship to each other and provide a basis for understanding the design requirements for biological safety.

The Clean Zone

The clean zone of a laboratory building (Figure 7) contains the entrance area, the office area, conference room and library, and those functional rooms where administrative operations, conferences, reading, writing and other tasks not involving infectious materials are carried out. Also, within this zone are the transitional rooms through which personnel and materials enter and leave the potentially infectious parts of the building. These transitional rooms preserve the integrity of the secondary barrier when people and materials enter and leave the infectious areas. In addition, the necessary shipping, receiving and clean storage areas are in the clean zone. The non-contaminated mechanical equipment space, although a clean area, will be discussed later, under the engineering support zone.

Personnel should enter and leave the infectious areas through clean and contaminated change rooms, illustrated in the lower right-hand portion of Figure 7. The clean change room should provide lockers to store street clothing, storage shelves for laboratory clothing and appropriate toilet and wash facilities. An air lock with an ultraviolet (UV) door barrier and ceiling-mounted UV lamps separates the clean from the contaminated change room. Adjacent to this or between the two change rooms should be located a shower room for use when leaving the infectious areas. The contaminated change room should contain a storage rack for laboratory shoes, a bag for discarding laboratory clothing upon exit and suitable toilet facilities. The use of UV lamps in the shoe storage rack and clothing discard bag can be an effective secondary barrier.[9,10]

Transitional arrangements for materials and supplies generally are needed at two locations. At the front of the building provisions are usually required for transferring books, data sheets and similar items between clean offices and offices in the infectious area. Typically this may include a small through-the-wall ethylene oxide gas chamber for the cold sterilization of heat sensitive materials[11] and an UV exposure apparatus for decontaminating sheets of paper passed out of the infectious area,[12] as illustrated in Figure 8. Transitional rooms at the rear of the laboratory are needed for receiving laboratory and animal room equipment and supplies and for removing equipment, trash and other items from the infectious areas. As shown in Figure 9, a typical arrangement at the rear of the building will consist of an UV air lock for the inward passage of supplies, and clean and contaminated receiving rooms separated by large through-the-wall autoclaves that are also operable with mixtures of ethylene oxide gas. The use of small viewing windows and speaking diaphragms facilitates communication and operation in the front and rear transitional rooms.

The Laboratory Research Zone

The laboratory research zone, shown in Figure 10, contains the laboratories where infectious microbiological operations, exclusive of animal work, are performed. This zone is separated at least by a corridor from the zone where infected animals are used. In addition to laboratory rooms, this zone may contain potentially contaminated offices adjacent to offices in the clean zone, necessary toilets and change

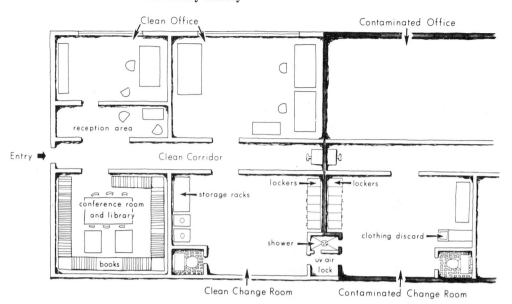

FIG. 7. Clean zone, clean and contaminated change rooms.

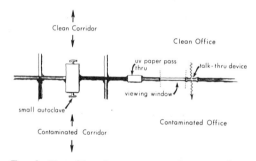

FIG. 8. Transitional arrangements between clean
and contaminated offices.

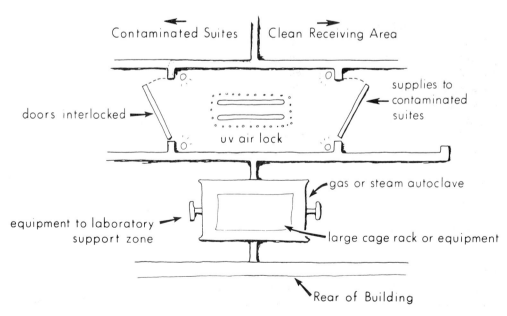

FIG. 9. Transitional arrangements between clean receiving area and contaminated research area.

rooms, constant temperature rooms for incubation and refrigeration, and instrument rooms for centrifuges and other research apparatus.

One of the most important tasks in planning the laboratory research zone is the selection of the primary barrier ventilated cabinets. Although cabinets are available

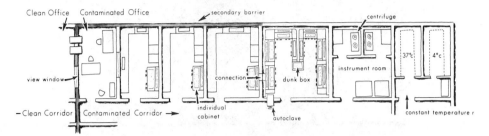

FIG. 10 Laboratory research zone.

in many shapes and sizes and constructed of a variety of materials, there are only two basic types: partial barrier cabinets and absolute barrier cabinets.[13] Figure 11 shows a partial barrier cabinet that provides an inward sweep of air through the open panel and away from the operator. This unit may also be used with a panel and gloves attached as shown in Figure 12. Absolute barrier cabinets are similar in appearance, as shown in Figure 13, but are constructed gas-tight and operated at a constant negative air pressure. The selection of partial or absolute barrier cabinets will have a significant effect upon sizing of the air-handling equipment, because the partial barrier cabinets require 850 l/min (300 cfm) supply and exhaust, while the absolute barrier cabinets require only 85 to 280 l/min (3–10 cfm) per 1.8-meter (6-foot) unit. Obviously the selection of the types of cabinets for use with the infectious cultures will vary according to an assessment of the risk of the types of laboratory operations together

FIG. 11. Partial barrier safety cabinet.

FIG. 12. Partial barrier safety cabinet—gloves attached.

with the particular microorganisms to be used. In reference to the hypothetical laboratory research zone, Figure 10, it is assumed that 75 per cent of the research area will be for low to medium risk work requiring partial barrier cabinets and 25 per cent will need to be equipped with absolute cabinet systems for performing high-hazard operations with minimum risk. The absolute cabinet system terminates with a double-doored autoclave and a germicidal liquid bath is also available for passing materials in and out of the system. At least some of the other laboratory rooms should have free-standing, single-door autoclaves. This reduces the need to transport potentially infectious materials in the corridors. In determining spatial arrangements for safety cabinets and the most effective and efficient room sizes, the bench concept developed by Norman[14] is appropriate, except that the barrier cabinets replace the open bench in many of the activities.

FIG. 13. Absolute barrier safety cabinet system.

Within the laboratory research zone there are a number of design considerations that will directly influence the containment by the secondary barriers. The most important of these are:

1. Microbial filtration of non-recirculated air exhausted from the laboratory rooms. High-efficiency microbial filters[15] with 99 per cent retention of 1.0 micron diameter particles are appropriate.

2. Microbial filtration of exhaust air from absolute barrier safety cabinets and other apparatus where aerosols of infectious microorganisms are intentionally generated. Ultrahigh efficiency filters[15] with 99.95 per cent retention of 0.3 micron diameter particles are acceptable in these applications. In some instances, it may also be desirable to utilize an electric or gas fired air incinerator in addition to the ultra-high efficiency filter for ultimate protection.

3. Air pressure balance within the zone should maintain the laboratory rooms negative to adjoining halls and the partial barrier cabinets negative to the laboratories. Differential air pressures are established by controlling the direction of air movement. Thus a typical laboratory can be maintained at negative pressure to surrounding areas by exhausting larger quantities of air than are supplied. Obviously the balance of such a system can be significantly affected by the "pumping action" of doors, traffic, and other factors. The entire laboratory research zone should be maintained at a negative pressure to the non-contaminated zones.

4. Paints and coatings used on walls, wall curbings, ceilings, floors and other surfaces must be resistant to flowing steam and disinfectants used in the decontamination process and frequent washings. Walls, wall curbings and ceilings should be free of cracks and the coatings flexible enough to span minor shifts in the structural system. A monolithic covering is often used on the floors.

5. Casework and other installed equipment, when possible, should be sealed to floors and walls to limit possible spread of contamination. Equipment not sealed to walls or floors should be movable or mounted on wheels to facilitate decontamination and cleaning.

6. Lighting fixtures, pipes, conduit and other services that penetrate the secondary barrier wall must be designed and installed to preserve the biological separation between the contaminated and clean zones. Electrical conduits should be internally sealed. Fixtures should be sealed to the wall or ceiling or mounted away from the surface for ease of cleaning.

7. To limit personnel traffic, liberal use should be made of viewing windows and speaking diaphragms in doors and/or walls to laboratories, particularly doors to walk-in refrigerator and incubator rooms.

8. In some rooms it may be desirable to install ceiling-mounted UV fixtures operated from the corridor when the room is unoccupied. These fixtures aid in reducing non-specific microbial contamination and are useful in case of an accidental spill of infectious material.

The Animal Research Zone

When the animal research zone is used for small to moderate-sized laboratory animals, the rooms are usually located in the same building as the laboratory research zone. As shown in Figure 14, this zone would typically include rooms for animal inoculation and autopsy, and infected animal holding rooms equipped with primary-barrier isolation equipment, such as ventilated cage and UV cage racks. In some instances it may be desirable to locate aerosol exposure equipment, such as the Henderson apparatus, in a room adjoining the animal room.

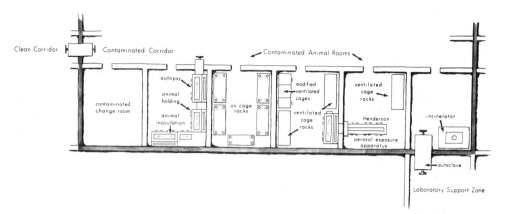

Fig. 14. Small animal research zone.

If not available elsewhere, the animal area should have an incinerator for disposing of animal carcasses, and a large autoclave for sterilizing cages and passing them into a cage washing room.

In designing the animal zone, the type and degree of animal isolation should receive early consideration. It should always provide for safe working conditions for personnel and prevent undesired animal cross-infection. The caging system selected will affect the cost of the facility and ventilated cages will have an impact upon the size of the air-handling equipment.

Dust filters should be installed at the air exhaust ducts of the animal rooms to prevent excessive loading of the downstream microbial filters with animal hair and dander. It is desirable to have these filters decontaminated *in situ* and changed by laboratory personnel. The recommended minimum ventilation rate[16] for the rooms is 15 changes per hour of filtered, non-recirculated, draft-free air of the relative humidity and temperature appropriate to the animal species. Since the walls and floors of animal areas are decontaminated and washed frequently and exposed to urine and other wastes, careful selection should be made of wall paints and finishes.

Other design considerations of importance in the animal zone are: waterproof lighting fixtures in wash areas, floor drains, adequate storage areas, cage and rack washing equipment, and automated animal care activities. Although the nature of some disease agents in small animals is such that no special provisions are needed to prevent cross-infection or to protect animal handlers, many infectious agents require animal isolation equipment for protecting personnel. In situations where a high degree of isolation is required and where animal-to-animal separation is needed, infected animals are held in small, individually-ventilated cages as shown in Figure 15. In some instances, instead of closed ventilated cages, animals may be housed in cages under an UV barrier[17] illustrated in Figure 16. Non-portable ventilated animal compartments have been found effective in some infectious disease laboratories. Each of the Horsfall units[18] shown in Figure 17 holds one or two animal cages and is equipped with a viewing window, and inlet and outlet air filters.

If the animal research zone uses infected large animals such as horses and cattle, it may be in a remote wing or suite, or may even be located in a completely separated facility. In any case, since it is very difficult to provide primary barriers around large animals, greater emphasis is ordinarily placed on the secondary barriers. A typical suite for housing infected large animals, as shown in Figure 18, would contain holding rooms with movable stanchions and partitions, and rooms or areas for animal inoculation, surgery, autopsy and incineration of carcasses. In case the epizootic diseases being studied are transmissible to man, protection for operating

Fig. 15. Ventilated animal cage rack (with UV screen).

Fig. 16. Non-ventilated animal cage rack (with UV screen).

FIG. 17. Modified Horsfall isolation cages.

personnel must be provided by protective garments and respirators or ventilated suits, if needed. The large animal zone should have change rooms and transitional rooms for movement of food and equipment as previously described.

The Laboratory Support Zone

In many infectious disease laboratories the zone for laboratory support is best located outside the contaminated research zones. From an economic and functional standpoint, this location reduces the amount of secondary barrier area required and provides an opportunity of grouping similar support functions in one area. An admitted disadvantage of locating the laboratory support zone outside the contaminated research zone is that wash room and animal room personnel cannot move between the contaminated and clean zones without a change of clothes and showering, and therefore additional personnel may be required.

As shown in Figure 19, a typical laboratory support zone may include rooms for washing and sterilizing glassware and animal cages; preparing culture media; storing equipment, glassware and animal cages; and repairing various laboratory items. In some instances the support zone may contain animal rooms for the quarantine and

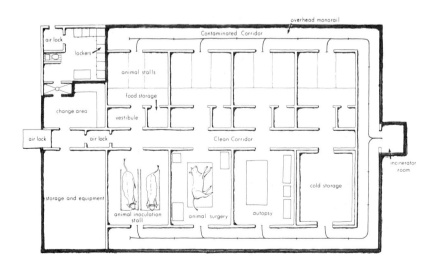

FIG. 18. Large animal research zone.

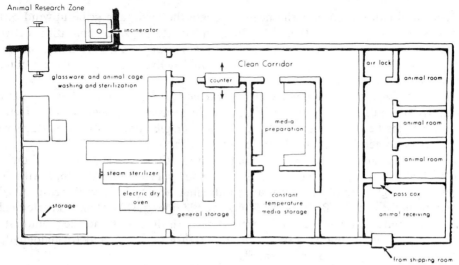

FIG. 19. Laboratory support zone.

acclimatization of animals before they are passed into the infectious area for use. Careful attention must be given to the design of the ventilation system for this zone because of the many heat-generating and odor-producing procedures which are carried out in the washing area. Also, because of the great amount of water involved in washing operations, floors, walls and other surfaces should be resistant to moisture. The room for culture media preparation should have a controlled movement of filtered supply air. To limit the ingress of microorganisms, air in the room and cabinetry always should be maintained at positive pressure to the other laboratory support rooms.

The Engineering Support Zone

The fifth functional zone in the hypothetical laboratory provides engineering support for the entire facility. Included here are the necessary pipes, ducts, pumps, blowers and filters, the liquid waste treatment system, and most of the air-handling systems. Special engineering arrangements should maintain the integrity of the secondary barrier when penetration of the wall is made by pipes, wires, and ducts. As much of the engineering support equipment as possible should be located outside contaminated zones to reduce the necessity of entrance of maintenance personnel. The zone can be comprised of space located on the grounds adjacent to the building, or in the attic, basement or other building space. Because of the large amount of engineering equipment needed, as much as half the total building area is sometimes required for this zone.

The equipment required to heat, cool, filter and distribute the supply and exhaust air for the various building zones will occupy a large part of the engineering support zone. This equipment may represent an investment amounting to nearly one-half the cost of the facility.

Wherever a device or item of equipment is essential to maintenance of a microbiological barrier and is subject to failure, "fail-safe" design features are required. This aspect will be accomplished through the use of redundant fans and motors, interlocked supply and exhaust fans to prevent pressurization of infectious areas, emergency electrical generators, filters in series with air incinerators, and automatic cycling of autoclaves after the door on the infectious laboratory side is opened. The inclusion of the above items in a laboratory should be determined after an assessment of the risks associated with the research activity has been made.

One vital consideration is the need to locate the air-intake grille upwind of the exhaust stack. In addition, there should be adequate physical separation between both to prevent cross-contamination. In the engineering support zone, air ducts going to each room in the infectious zone should deliver non-recirculating, filtered and conditioned air.

The supply ducts need not be of airtight construction, but exhaust air ducts coming from the infectious zone must be airtight to assure no leakage of infectious microorganisms before air reaches the exhaust filter plenum. Galvanized ducts with taped, epoxy-coated joints have been found satisfactory.

The exhaust air plenums in a building are often sized to serve several laboratory rooms of about the same hazard level. They may be equipped either with high efficiency spun-glass mats, or with ultrahigh-efficiency units for filtering air discharged from the rooms. Because the filters in a plenum must be changed periodically, provisions should be made for decontaminating both the filters and the plenum itself before entry by maintenance personnel. A mixture of steam and formaldehyde is frequently used as the decontaminant.[19]

In virus laboratories, inlet air is often passed through ultrahigh-efficiency filters. This high quality air is required to prevent accidental contamination of experimental cultures by endogenous organisms or by other organisms utilized in the research program. In tissue culture cubicles or other areas requiring this high quality air supply, the use of a high volume recirculated air system using ultra-high efficiency filtration should be considered, if no hazards to operating personnel are created. This system will result in significant economies by reusing conditioned air and will be acceptable if proper separation by zones of similar usage is employed.

An important part of the engineering support zone is the central control board illustrated in Figure 20. Here, read-outs of all systems in the area can be readily monitored by engineering personnel. Such boards should have visual and audible alarms that will automatically signal the failure of any part of the system when preset limits of environmental factors such as temperature are exceeded. Another important item is a stand-by electrical generator that is used in the event commercial power supply is interrupted. While it may not be possible to provide a generator large enough to supply full electrical requirements, the stand-by current should at

Fig. 20. Central engineering control board.

least be sufficient for ventilated animal cages, ventilated cabinets, freezers, re-frigerators, incubators and emergency lighting. The stand-by units may be mounted in trailers to provide easy portability or incorporated in the engineering support zone.

In some laboratories, facilities must be provided for treating contaminated liquid wastes. Two basic systems are usually employed: the batch and the continuous flow system. However, if the hazards associated with the program are minimal and adequate space is available, a sanitary drain field can be utilized. Regardless of the system, in the laboratory no infectious culture fluids should be knowingly poured into drains without prior sterilization.

The batch system shown in Figure 21 is used to collect potentially infectious

Fig. 21. Batch-type liquid waste treatment system.

effluents from infected animal areas by gravity flow. When a tank is about one-half full of liquid, the drain lines are closed and the liquid is sterilized by adding steam and holding for a period of time. Usually, a second tank is automatically put into service at this time. All piping to the tank should have welded joints to assure no possibility of leakage. A concrete curb should be provided to contain the liquid in the event of a rupture in the system. The necessity for microbiological monitoring of the effluent from the system should be considered in the design process to insure that adequate valves, sampling ports and other means are incorporated in the design for removing samples.

For larger volumes of effluent a continuous flow arrangement, shown in Figure 22, that utilizes injected steam to raise the liquid to a proper temperature may be used. In this case the effluent–steam mixture flows through a series of retention tubes and is cooled through a heat-exchanger before being discarded. All liquid waste treatment systems can be monitored and controlled from a central panel where flow arrangements, effluent volumes, treatment times, and temperatures are visually indicated.

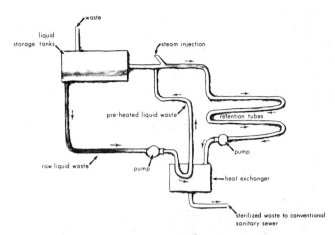

FIG. 22. Continuous flow waste treatment system.

ACKNOWLEDGEMENT: Reprinted from *Applied Microbiology*, March, 1967, with permission. This chapter is an abridged corollary to the film "Laboratory Design for Microbiological Safety," which may be borrowed free of charge, from the Public Health Service, Communicable Disease Center, 1600 Clifton Road, Atlanta, Georgia.

REFERENCES

1. Kirchheimer, W. F., Jemski, J. V., Phillips, G. B., Cross-Infection Among Experimental Animals by Organisms Infectious for Man. *Proc. Animal Care Panel, 11*, 83–92 (1961).

2. Phillips, G. B., Broadwater, G. C., Reitman, M., Alg, R. L., Cross Infections Among Brucella-Infected Guinea Pigs. *J. Infect. Dis., 99*, 56–59 (1956).

3. Phillips, G. B., Jemski, J. V., Brant, H. G., Cross-Infection Among Animals Challenged with *Bacillus Anthracis. J. Infect. Dis., 99*, 222–226 (1956).

4. Wedum, A. G., Assessment of Risk of the Human Being in the Microbiological Laboratory. (to be published)

5. Chatigny, M. A., Protection Against Infection in the Microbiological Laboratory: Devices and Procedures. *Advances in applied microbiology. 3:* 131–92 (1961).

6. Wedum, A. G., Bacteriologigal Safety. *Am. J. Public Health, 43*, 1428–1437 (1953).

7. Wedum, A. G., Laboratory Safety in Research with Infectious Aerosols, *Public Health Rep., 79*, 619–633 (1964).

8. Phillips, G. B., Microbiological Hazards in the Laboratory, *J. Chem. Ed., 42:*43–8; 117–30 (1965).

9. Phillips, G. B., Hanel, E., Jr. Use of Ultraviolet Radiation in Microbiological Laboratories. *U. S. Government Res. Rep. 34:* p. 122 (1960).

10. Wedum, A. G., Hanel, E., Jr., Phillips, G. B., Ultraviolet Sterilization in Microbiological Laboratories. *Public Health Rep., 71,* pp. 331–336 (1956).

11. Phillips, C. R., Gaseous Sterilization, p. 746–765. *In* G. F. Reddish (ed.) Antiseptics, Disinfectants, Fungicides and Chemical and Physical Sterilization. Philadelphia. Lea and Febiger (1957).

12. Phillips, G. B., Novak, F. E., Applications of Germicidal Ultraviolet in Infectious Disease Laboratories: II. An Ultraviolet Pass-Through Chamber for Disinfecting Single Sheets of Paper. *Appl. Microbiol. 4,* 95–96 (1956).

13. Phillips, G. B., et al., Microbiological Barrier Equipment and Techniques, a state of the art report. Boston. Am. Assoc. for Contamination Control (1966).

14. Norman, J. D. The Bench Concept in Laboratory Planning. *Health Laboratory Science. 1:* 179–84. (1964).

15. Decker, H. M., Buchanan, L. M., Hall, L. B., Gardner, G. D., Jr., Air Filtration of Microbial Particles. (Public Health Service Publications 953). Washington, D. C. (1962).

16. Guide for Laboratory Animal Facilities and Care. Revised, (P.H.S. publication 1024). Washington, (1965).

17. Phillips, G. B., Reitman, M., Mullican, C. L., Gardner, G. D., Jr., Applications of Germicidal Ultraviolet in Infectious Disease Laboratories: III. The Use of Ultraviolet Barriers on Animal Cage Racks. *Proc. Animal Care Panel, 7*, 235–244 (1957).

18. Horsfall, F. L., Bauer, J. H., Individual Isolation of Infected Animals in a Single Room. *J. Bacteriol. 40*, 569–580 (1940).

19. Glick, C. A., Gremillion, G. G., Bodmer, G. A., Practical Methods and Problems of Steam and Chemical Sterilization. *Proc. Animal Care Panel, 11*, 37–44 (1961).

Infectious Hazards of Common Microbiological Techniques

Studies of the potential sources of infection have centered on the hazards associated with common laboratory techniques. The experimental method used has been described previously.[1,2] Essentially, this method involves sampling air with the sieve-type air sampler during standard bacteriological operations such as pipetting, centrifuging, inoculating and lyophilizing cultures, and autopsy of animals. The operational area is surrounded by samplers, each of which draws air at the rate of 28 liters per minute (1 cfm) through 340 small openings, thereby impinging organisms on the surface of a petri dish agar plate 1–2 mm below the openings. After a suitable incubation period [36–48 hours at 30°C. (86°F)] for bacteria and 4–16 hours at 30°C. (86°F) for bacteriophage) colonies or plaques are counted in a Quebec colony counter.

Contamination of the environment is determined also by swabbing surfaces with cotton moistened with nutrient broth. The swabs are streaked on agar plates, which are then incubated as are the air-sampler plates.

Three easily identified organisms were used in these studies: (a) *Serratia indica*, a red pigmented vegetative rod; (b) *Bacillus subtilis* var. *niger,* designated *B. globigii* in Fort Detrick laboratories; and (c) coliphage T_3. *S. indica* and *B. subtilis* spores were sampled on corn-steep, molasses agar, and coliphage T_3 on tryptose phosphate glucose agar.[2]

TABLE 1
Aerosols Produced by Common Microbiological Techniques

Technique	Number of operations	Number of colonies appearing on sampler plates		
		Average	Minimum	Maximum
Agglutination, slide drop technique (one slide)	60	0.3	0	0.66
Animal injections (*Serratia indica*):				
1. 10 shaved guinea pigs injected intraperitoneally with 0.5 ml culture, no disinfectant	3	15	15	16
2. Same as (1) but injection site disinfected before and after injection with 1 percent tincture of iodine	3	0	0	0
Autopsy, guinea pig:				
1. Immediately after 1 ml *S. indica* culture injected intraperitoneally...............	2	4.5	3	6
2. Immediately after 10 ml culture injected intracardially.........................	6	3	1	6
3. Grinding tissue 2 minutes in mortar and pestle with 2 ml sterile broth:				
Guinea pig liver as in (1), 10 ml inoculum.........................	10	1.8	0	8
Guinea pig heart as in (2)	6	19.5	0	103
Centrifuging:				
1. Pipetting 10 ml *S. indica* culture into 50 ml tube..............................	100	0.6	0.1	1.2
2. Pipetting 30 ml culture into 50 ml tube...	100	1.2	0	5.5
3. Removal of one cotton plug after centrifuging................................	100	2.3	0.8	5.0

Technique	Number of colonies appearing on sampler plates			
	Number of operations	Average	Minimum	Maximum
Centrifuging (continued)				
4. Removal of one rubber cap after centrifuging...............................	80	0.025	0	0.25
5. Decanting supernatant into flask	10	17.6	0	115
6. Siphoning supernatant from 10 tubes, each containing 30 ml centrifuged culture...................................	100	3	0	24
7. Adding 30 ml saline to one tube of packed centrifuged cells and resuspending by mixing by alternate sucking and blowing with a pipette	100	4.5	0.7	12.8
8. One 50 ml tube breaking in centrifuge but all 30 ml culture staying in trunnion cup	10	4	0	20
9. As in (8) but culture splashing on side of centrifuge	10	1,183	80	1,800
10. Swabbing outside of centrifuge tubes after filling, centrifuging, taking off supernatant, and resuspending	10	(1)	....	....
One drop of S. indica **culture falling** 3 inches onto:				
1. Steel surface	200	1.3	0.02	4.7
2. Painted wood...........................	200	0.3	0.01	0.6
3. Kem-rock	100	0.04	0.00	0.05
4. Dry hand towel.........................	100	0.16	0.00	0.35
5. Dry paper towel	200	0.11	0.00	0.45
6. Dry wrapping towel.....................	100	0.02	0.00	0.05
7. Towel wet with 5 percent phenol.........	100	0.02	0.00	0.05
8. Pan of 5 percent phenol	100	0.00	0.00	0.00
Inoculating loop:				
1. Streaking one agar plate with one loopful of S. indica broth culture..............	10	0.6 4.6^2	0	20
2. Streaking one agar plate with one loopful of agar culture........................	15	0.26	0	0.7
3. Loopful of broth culture striking edge of tube	15	0.60	0	2.3
4. Inserting one hot loop into 100 ml culture in a 250 ml Erlenmeyer flask...........	550	8.7	0.68	25
5. Inserting one cold inoculating loop into 100 ml culture in a 250 ml Erlenmeyer flask	250	0.08	0	0.22
Hypodermic syringe and needle [withdrawing 1 ml phage suspension from rubber-capped vaccine bottle and making tenfold dilutions in rubber-capped vaccine bottles (10^{-1} to 10^{-9}), pledget does not always protect fingers against contamination]:				
1. Cotton pledget around needle	90	2.3	0	10
2. Ethanol soaked cotton pledget...........	90	0	0	0
Lyophilization:				
1. Breaking one ampule containing 2 ml of lyophilized S. indica culture in milk plus broth menstruum by dropping on the floor, first 10 minutes	10	2,029	1,939	2,040
2. Same as (1), 50-60 minutes after breakage.	10	741	162	1,447
3. Opening one lyophile tube by filing and breaking tip	20	86	4	256

TABLE 1 (Continued)

Aerosols Produced by Common Microbiological Techniques

Technique	Number of colonies appearing on sampler plates			
	Number of operations	Average	Minimum	Maximum
Lyophilization (continued)				
4. Same as (3), but wrapped in 70 percent ethanol soaked cotton pledget..........	50	0.08	0	0.8
5. Transferring one dry inoculum from one lyophile tube by wire loop.............	50	1.0	0	5
6. Same as (5), but shaking powder into broth tubes...........................	20	5.4	0	30
7. Same as (5), but wet inoculum transferred by syringe and needle after reconstituting with one ml broth	10	4.4	0	17
Petri dish plates:				
1. Preparation of pour plate, pipetting one ml inoculum of S. indica into plate without blowing, and adding melted agar and mixing	15	2.6	0.2	5
2. Streaking one smooth agar plate with 0.1 ml; spread with glass rod	50	0.06	0	0.4
3. Streaking one rough agar plate with one loopful of broth culture	10	25.1	7	73
4. Same as (3), but using 0.1 ml and glass rod	50	8.7	2	25
Pipettes (also see centrifuging.):				
1. Inoculating 50 ml broth in 125 ml Erlenmeyer flask with 1 ml culture (S. indica)	5	1.2	0	2
2. Mixing 7 ml broth culture by alternate suction and blowing, without forming bubbles	5	0.2	0	1
Plug, stopper, or cap removed from culture container of 1–10 dilution of 24-hour broth culture of S. indica:				
1. Escher rubber stopper removed from 5-oz square dilution bottle immediately after shaking up and down	15	5.0	0	20
2. Same as (1), stopper removed after 30 seconds wait..........................	15	2.5	0	12
3. Plastic screw cap removed from 8-oz prescription bottle immediately after shaking..................................	15	4.0	0	13
4. Cotton plug removed from 250 ml Erlenmeyer flask immediately after rotary shaking (dry plug)	15	5.0	0	16
5. Same as (4), but wet plug................	5	10.2	0	35
High speed blendor, S. indica culture mixed 2 minutes:				
1. Screw-capped, no rubber gasket (1 minute)	10	8.7	0	31
2. Screw-capped, rubber gasket, worn bearing..................................	10	61.0	12	126
3. Loose fitting plastic cover	15	518	77	> 1,246
4. Removing tight cover immediately after mixing	15	(1)	(1)	(1)
5. Removing tight cover 1 hour after mixing .	15	8.2	5	33

[1] Colonies too numerous to count.
[2] Two technicians.

Wide variations from average determinations of contamination hazards associated with laboratory procedures are possible (see Table 1). These variations often seem to depend on minor changes in technique peculiar to the individual testing a particular procedure. A reported count of two, for example, means that two colonies grew on the agar sampling plates. It has been reported that most bacteria in the air occur in clumps.[3] Also, the efficiency of the sieve sampler in recovering aerosolized particles of heterogeneous size and composition under varying humidities is not easily nor precisely determinable. In the presence of bacterial aerosols of known concentrations, efficiencies have varied from 43 to 73 percent. Therefore, the reported number of colonies is significantly smaller than the actual number of bacteria.

It is evident that certain procedures create larger amounts of aerosols than others. Grinding tissue with mortar and pestle, decanting the supernatant after centrifugation, resuspending packed cells, inserting a hot loop in a culture, withdrawing a culture sample from a vaccine bottle, opening a lyophile tube, streaking an inoculum on a rough agar surface, and shaking and blending cultures in high-speed mixers appear to be potentially dangerous to the technician if the micro-organisms are infectious. Accidents during centrifugation or handling of dried cultures caused extensive contamination of the laboratory. Practically every manipulation in the microbiological laboratory creates aerosols, and these aerosols are probably the source of many laboratory infections.

ACKNOWLEDGEMENT: This chapter appeared originally as part of an article on Microbiological Safety by Morton Reitman, Ph.D., and A. G. Wedum, M.D., Ph.D., in *Public Health Reports 71* (7), 659–665, (1956).

REFERENCES

1. Reitman, M., Frank, M. A., Sr., Alg, R. L., and Wedum, A. G., Infectious Hazards of the High Speed Blendor and Their Elimination By A New Design. *Appl. Microbiol. 1:* 14–17 (1953).
2. Reitman, M., Moss, M. L., Harstad, J. B., Alg, R. L., and Gross, N. H., Potential Hazards of Laboratory Techniques. I. Lyophilization. *J. Bact. 68:* 541–544 (1954).
3. Du Buy, H. G., Hollaender, A., and Lackey, M. D., A Comparative Study of Sampling Devices For Air-Borne Microorganisms. *Pub. Health Rep. Suppl. No. 184.* 40 pp. Washington, D. C., U. S. Government Printing Office, (1945).

Laboratory Animal Housing

Robert S. Runkle

In order to discuss the subject of animal housing facilities, three types of animals must be recognized. Those raised in an open colony with normal sanitation measures are classified as conventional animals. This type of animal is of sufficient quality for most present applications. Specific pathogen-free animals are of a higher quality than conventional animals. They should be tested for certain pathogens or organisms. Germ-free animals are those which are maintained free of detectable microorganisms. In addition to the distinctive qualities of these three types of animals, the physical barriers that are erected between the outside environment and the laboratory animal become increasingly complex as one changes from conventional to germ-free animals. This chapter is limited to a discussion of facilities for housing conventional animals.

In designing animal facilities for research laboratories, it must be remembered that emphasis on different species of animals will change from year to year. Therefore, the facilities should whenever possible have a built-in flexibility to allow a maximum of different uses for any one room or area.

In any discussion of animal housing facilities, three specific areas must be recognized: production and breeding; quarantine facilities; and experimental and holding areas.

The sections that follow deal with facilities for production and breeding of conventional laboratory rodents, quarantine of dogs, cats and primates, and for holding rodents and/or larger animals under experimentation.

PRODUCTION AND BREEDING FACILITIES

Animals bred in standard production are mice, guinea pigs, rats, hamsters, and rabbits. The design of housing for these species is very similar and, in many cases, an organization will alternate the use of its animal rooms to accommodate all of these animals. When laboratory facilities include production buildings, they should be kept in the vicinity of the research complex. This greatly reduces the difficulties of transportation for animals and personnel. The area may be separated from other buildings and confined to one portion of the property so that unwanted personnel may be excluded from the buildings, and noises and odors from the animals can be isolated.

There are several methods of predicting the area needed in a production colony, depending on the type of facility to be built. Growth curves for the research institution and existing animal production, if obtainable, should be analyzed and extrapolated. The research program will, of course, affect the numbers of animals required and all available information should be obtained from the scientific directors.

Although there are many different arrangements for breeding and production colonies, two comparatively new arrangements have proven effective in increasing production rates and reducing the possibility of contamination. However, these may not be feasible in small-scale operations.

The "clean and refuse" corridor system, as illustrated in Figure 1, utilizes dead-end service corridors, which greatly reduce the transfer of contaminatives. The "clean corridor" provides circulation for operations such as food and bedding delivery to the breeding rooms, and removal of animals for research. The "refuse corridor" provides a means of controlling contamination through such operations as the removal of dirty cages, dead animals or used bedding.

The second system, "separate building concept," as illustrated in Figure 2, utilizes many small buildings, each containing one or more breeding rooms, mechanical equipment, and storage space. The small buildings are located around a central service building, which acts as a receiving building for supplies, an office area, and holds cage and rack washing equipment.

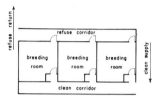

FIG. 1. "Clean and refuse" corridor layout of breeding rooms.

There are several distinct areas to be included in a production facility. Of these, the breeding room itself is the most important. The size of this room may vary between 28 and 65 sq. meters (300 and 700 square feet). Each room should contain a sink, workbench, storage space, and electric outlets. If the rooms are too small, the above items will be unnecessarily duplicated. If the rooms are too large, epizootics may wipe out a large portion of the animals, or even the entire colony. It would be difficult to place a monetary value on the time and effort of the researcher and technician, if the colony were to be lost during an experiment. A small vestibule for hand-washing before the caretaker enters the room is helpful in preventing cross-contamination. Figure 3 illustrates a possible arrangement of a breeding room. Provision of an area for caretakers to change clothes and shower before entering the actual breeding rooms is an important means of preventing entrance of contamination.

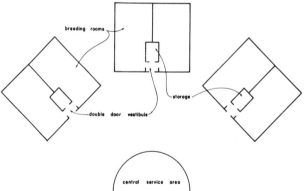

FIG. 2. "Separate building concept" of breeding facilities.

The type of caging system will affect room size and layout.* Standardization of cages and cage racks will facilitate planning of the building and washing equipment. The possibility of using fixed racks with removable cages should be investigated. If cages with grid bottoms and waste pans that can be flushed are used, labor and handling costs can be reduced substantially. This system has been used successfully for rabbits, rats, and monkeys, as illustrated in Figure 4.

*University of California report of Statewide Animal Care Panel; available from Institute of Laboratory Animal Resources, National Research Council.

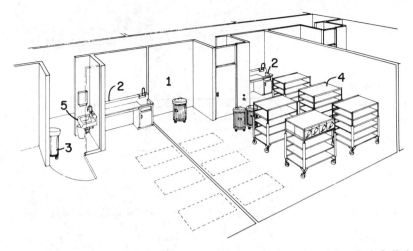

FIG. 3. Breeding rooms. 1. Storage space alcove 2. Sink and work bench 3. Soiled linen receptacle in vestibule 4. Animal cage racks on casters 5. Lavatory with waste paper receptacle and paper towel dispenser.

The need for storage space should not be underestimated. Feed, bedding, cleaning equipment, racks and cages can occupy considerable space. The cage washing area should be conveniently located with respect to the breeding rooms. Space should be provided for soiled as well as clean racks and cages. A large loading dock leading directly into an enclosed short-term storage room is needed for receiving

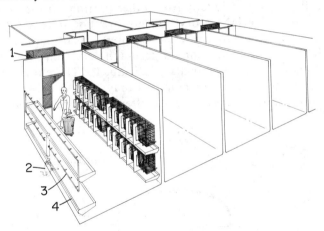

FIG. 4. Typical module showing automated features. 1. Vestibule (screened vestibule only for primates) 2. Garbage disposal in floor under each rack 3. Automatic watering device 4. Flushing pans.

shipments of feed, cages, and other supplies.* While this dock can be used for shipment of animals, it is more desirable to provide a separate dock for this purpose. Addition of a pathology laboratory and autopsy facility will permit early detection of disease in the colony. Provision should be made for employee shower, locker, and lunch rooms. See Figure 5 for a diagram of area relationship.

QUARANTINE FACILITIES

With a few minor modifications, indicated in Figure 5, the same arrangement can be used for a quarantine facility.

*See Table 3.

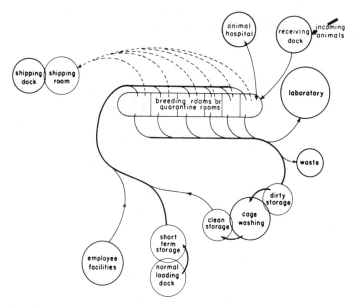

Fɪɢ. 5(a). Relationship of areas in a production or quarantine facility.

If no breeding stock is maintained, an area should be reserved for inspecting and conditioning the animals received at an institution. This area should preferably be separated physically and operationally from other portions of the animal colony.

Quarantine rooms for laboratory rodents are not different from rooms in experimental areas, which are to be described later. Small animals, such as mice, rats and other laboratory rodents, should be purchased from dealers who are well known and the quality of their animals predictable.* Therefore, this section will describe facilities for dogs, cats and monkeys only. These animals require a long period of quarantine; and, because they are often transported many miles and may

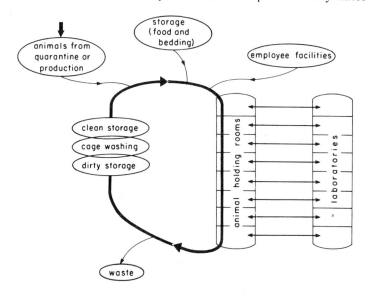

Fɪɢ. 5(b). Relationship of areas in a holding facility adjacent to research laboratories.

*University of California report of Statewide Animal Care Panel; available from Institute of Laboratory Animal Resources, National Research Council.

be in poor condition of health when received, medical and surgical treatment may be needed.

In quarantine facilities the movement of animals is from the receiving dock to conditioning rooms to holding areas. Materials are moved from the receiving dock to storage. Separate covered platforms for shipping and receiving animals should be provided. The receiving platform should be located adjacent to the quarantine area. There should also be adequate storage space adjacent to the receiving dock. A separate preparation room for animals to be shipped is also desirable.

Dogs and cats are usually obtained from city pounds or from dealers who collect the animals from such organizations. Provision must be made for examining and dipping or spraying the animals as they enter the quarantine building. This room should be completely washable with a well-sloped floor and floor drain. The room should contain cages for animals to shake and dry in after being dipped or sprayed.

Dogs that are on long-term holding should be provided with some type of exercise area. If space is available, an indoor-outdoor run arrangement is a very convenient method of housing the dogs. It is best to allow at least 0.92 sq. meters (ten square feet) for a 13.6 Kg to 15.9 Kg (thirty to thirty-five pound) dog in an indoor run. It is undesirable to house more than four dogs in a runway. However, when dogs are placed in a run together, they should be of similar size and of the same sex. Large gang cages in which many dogs are housed together are more difficult to keep clean and fighting becomes more of a problem. Figure 6 shows the relationship of the interior run to exterior runs.

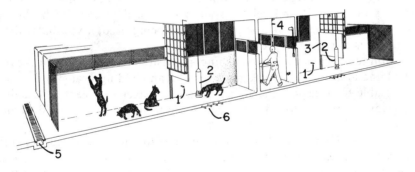

FIG. 6. Dog runs. 1. Automatic drinking nipple 2. Automatic feeder 3. Outside door (full width of dog run) 4. Corridor control (for dog run door) 5. Drain 6. Piping for radiant heating.

In a quarantine building housing all types of dogs, a solid partition should be provided on the lower 1.21 meters to 1.52 meters (four to five feet) of the runs so that waste cannot be transferred from one pen to the next. This also prevents much of the barking which is caused by visual contact between animals. A grate-covered waste gutter should be provided at the outside end of the outdoor runs. The runs may be modified for holding cats by putting a wire covering over the top of the outside run. Hot water radiant heat coils in the indoor run will dry the floors quickly after washing and help keep the animals healthy. The climate in which the facility is located will determine whether facilities for snow removal from the outside runways should be provided.

A quarantine period, which will vary among institutions, is required for primates. During this time outdoor runs are not required; in fact, individual cages are preferable. These may be either movable cages with solid sides, top and grid bottom, or the stainless steel or galvanized wire type which are placed over a stainless steel waste trough. This latter method of caging has become very popular in the short time it has been in use. Automatic watering devices and automatic flushing devices for the waste pan cut down caretaker cleaning time considerably.

An important aspect of the quarantine area is a multipurpose operating room. This should be available for minor surgery and for required treatment of animals brought from pounds or other sources. Provision should be made for employee locker, shower, and lunch rooms.

EXPERIMENT AND HOLDING AREAS

Rooms of this type may be required to house almost any type of animal under experimentation in close proximity to research laboratories. The individual details and arrangement of the room are dependent upon the type of research and species of the animal housed. Three basic arrangements are used: room adjacent to the laboratory; animal rooms in one area on each floor; and/or a separate animal building adjacent to the laboratory building connected with a walkway. The first system is not economical, because the rooms are dispersed over a large area and facilities such as storage and mechanical equipment are unnecessarily duplicated. However, the system may need to be used when the scientist must visit his animals several times a day.

Thought should be given to the use of separate elevators and corridors for animal traffic, if animals are housed in a laboratory building. In designing this type of facility, a number of precautions must be observed. Care should be taken in waterproofing the floor to prevent water from leaking into laboratories on the lower floors. Cracks in the wall and ceiling should be sealed to exclude harborage for insects or disease-carrying agents. If primates are being used, a vestibule with outer and inner doors is required to prevent escape.

In addition to the cage and rack washing, provision must be made for the removal of soiled bedding and the supply and storage of clean bedding, water and feed. Very often it is feasible to transport the soiled racks and cages to a central washing area where the bedding is removed and racks, cages and water bottles are washed and then reassembled.

The second arrangement, while eliminating the duplication of facilities inherent in the first system, does present the same construction problems. A possible way of eliminating some problems is to locate the animal quarters, cage and rack washing, and storage rooms on the ground floor. This method would eliminate vertical traffic and minimize possibility of cross-contamination.

The most economical and practical arrangement is the third, provided that it is satisfactory to the investigators involved. When all animal facilities are grouped in one building, duplication is at a minimum, cross-contamination problems are reduced (since the animal buildings and the laboratory are separated by a natural barrier of air), and service and storage areas can be centrally located.

The inclusion of an operating suite and supporting areas, required for corrective and experimental surgery, in the separate animal building will facilitate the work of the scientists. It should include a preparation room, operating room, recovery area, examination and treatment room, X-ray and autopsy room, pharmacy, and pathology laboratory. Provisions should also be made for employee locker, shower, and lunch rooms. Figures 7 through 11 illustrate the different areas of an operating suite.

PHYSICAL DESIGN FACTORS

After the information has been assembled concerning the function of the animal facilities and the design of the major areas, the following factors should be evaluated for each area: architectural; lighting and electrical; mechanical; and automation.

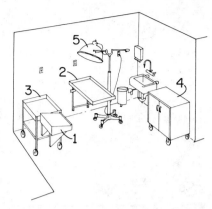

FIG. 7. Examination alcove. 1. Wall-hung writing shelf 2. Wall-hung examination table (removable) 3. Soiled instrument cart 4. Clean instrument storage 5. Movable examination light.

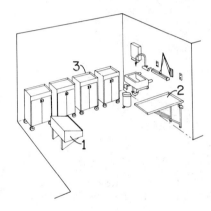

FIG. 8. Treatment alcove. 1. Wall-hung writing shelf 2. Wall-hung examination table (removable) 3. Treatment carts.

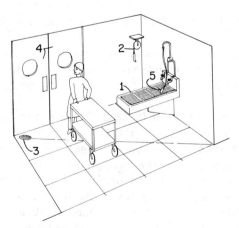

FIG. 9. Preparation alcove. 1. Preparation table 2. Hair cutter 3. Floor drain 4. Doors to operating room 5. Spray hose with mixing valve.

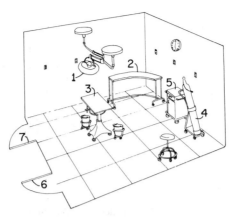

FIG. 10. Small operating room. 1. Operating lights 2. Instrument table 3. Operating table 4. Oxygen tank 5. Storage cabinet with casters 6. Door to sub-sterile room 7. Door from scrub-up.

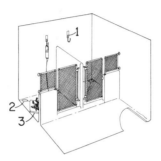

FIG. 11.   Recovery   room. 1. Hook for blood plasma 2. Floor drain 3. Pallet.

The interior of the animal room must be easily cleanable and have many of the same properties as the laboratory. The floor and wall surfaces must be more durable than the laboratory finishes, because stronger detergents are used to remove the dirt which collects in animal rooms. Cage and equipment traffic is also more severe than personnel traffic.

The floor is one of the most important aspects of animal room sanitation, both because of construction and maintenance considerations. Bacteria, which attach themselves to dust particles and settle to the floor, will incubate if a suitable medium is present.

Perhaps one of the most widely used flooring materials in animal rooms is concrete. It is a very satisfactory material if it is troweled smooth and treated with a surface hardener and sealer. Construction specifications should be very exacting about surface designation and workmanship. Terrazzo is somewhat more durable. However, it is also more costly to install and repair. In corridors where traffic is very heavy, consideration should be given to the installation of terrazzo. Quarry tile, used in many installations, is resistant to organic acids and salts in animal wastes. Two disadvantages are the difficulty in cleaning because of the joints and noise as wheeled vehicles are moved through the corridors. Some of the plastic-type surfaces which are troweled over concrete in a thin layer resist wear and cracking somewhat better than plain concrete. The resiliency of the material also makes it quieter than the harder surface finishes. However, the greatest problem is obtaining good supervision to insure proper application. Membrane waterproofing should be used whenever animal facilities are housed on a supported slab.

The wall surface should be smooth, hard, impact-resistant, free from joints, and resistant to urine and cleaning compounds. Many materials provide these qualities, but compromises usually have to be made. Glazed tile has been used for many years in areas where a high degree of cleanability is required. It is durable, but will fracture if subjected to impact from vehicles. The use of an epoxy grout will render the joints impervious.

Film materials that are applied with adhesives and overlapped to form a tight joint give an impervious surface. These materials are flexible so that the danger of cracking or joint separation is reduced. Many of the new synthetic coatings, designed to be applied over concrete or cinder blocks, produce a durable surface at a moderate cost. Care must be exercised in the selection of a product, since there is considerable variation in their properties. The concrete block and mortar should be free of any unstable compounds that might cause a separation of the synthetic coating. Cement plaster on cinder or concrete block wall will result in a fairly durable smooth wall. Any type of hung or false ceiling presents a location for harborage of objectionable organisms. This can be overcome by using a flat concrete ceiling with an application of a sealer or paint. A finishing operation may be required to produce a smooth surface. With this type of ceiling, air ducts may be surface-mounted and caulked at the ceiling, to reduce dirt-collecting surfaces. However, it is best if ducts can be eliminated from the room and placed in the corridor ceiling for easy access without disturbing the animal population.

Doors should be at least 1.08 meters (3 feet 6 inches) wide by 2.13 meters (7 feet) high to accommodate cage racks and service carts. Push plates and recessed door pulls should be used rather than door knobs, because of the problem of their becoming damaged or broken. In planning corridor widths, consideration should be given to cage and rack dimensions. An investigation should be made as to the feasibility of using sliding doors. The use of sliding doors, suspended from a track, mounted either on the corridor side or the room side of the wall, would allow elimination of the area required for swing of a normal door and would reduce damage to door hardware from moving carts.

Doors in corridors and supply aisles may be equipped with photoelectric mechanisms for automatic operation. Gaskets should be installed at the bottom of doors to prevent entry of wild rodents or pests. Since gaskets will render the door airtight, exhaust louvers in corridor doors or walls will facilitate accurate balancing of air pressure, which is quite important in isolating animal room odors and possible contaminations. Doors located in areas that are hosed down frequently should have some means of protecting the doors against rust. A stainless steel kick plate on both sides and the bottom is one solution.

A curb suitable to protect the wall from impact of wheeled vehicles could be installed in animal rooms and corridors of buildings specifically designed for animal housing. A curb 15.2 cm (6 inches) high and projecting 9.1 cm (4 inches) from the wall with a sloped top to eliminate dust-catching surfaces should be sufficient. If desired, a large radius cove may be used instead of curbing. A curbing or cage bumper should always be used with plaster walls.

An important consideration is control of noise. This is especially important when animal facilities are located in the same building as laboratories or office space.

Noises in an animal facility generally originate from two sources: the animal itself and transportation of the cages and racks from one area to another. The main problem is encountered with the barking of dogs or chattering of primates. Where animal noise would create a problem, an attempt should be made to isolate the animals in rooms. Some laboratories practice debarking as a temporary measure; however, this is not a permanent solution to noise control. Keeping doors closed reduces noise transmission. The use of rubber tires and a smooth resilient surface in heavy traffic areas will reduce a good percentage of the noise. The architect and engineer should realize, however, that probably the most effective means of handling noise transmission is to separate the animal quarters physically from any area that would be disturbed by noises.

Lighting and Electrical Factors

Adequate electrical outlets should be provided, preferably of the grounded, waterproof type. All rooms should be provided with normal 110 V service, and 220 V service should be available in areas that use heavy-duty equipment.

Light fixtures of the fluorescent type are preferred and should be watertight to permit hosing. They should be either surface-mounted and caulked or recessed to eliminate a horizontal, dust-catching surface. The light intensity design range should be 650 to 1080 lumens/sq. meter (60 to 100 foot candles). The use of multiple switching in each room will allow the change in intensity for different functions. A cycling device for the lighting system in the interior breeding room is very useful. Switches should be located outside of the animal room for convenient access and protection from moisture. Sunlight is not required for small animal breeding rooms and is undesirable from the standpoint of the added heat load. Direct sunlight on small animals confined in cages may cause temperatures to rise above a safe level.

A generator should be provided as an auxiliary supply if a power failure occurs. This should be connected to automatically energize the circuits handling the mechanical equipment for all the production and breeding rooms where a controlled atmosphere is present, as well as operating rooms and other key areas.

Mechanical Factors

The following plumbing requirements should be thoroughly investigated when planning the animal facility. Hot and cold water will be required for all facilities; however, many institutions will have additional needs for demineralized water, steam for sterilization, compressed air, gas and vacuum piping, and special sewering provisions.

Use of floor drains in small-animal rooms will depend on whether it is desired to hose down the rooms regularly or to sweep and damp-mop them with a disinfectant. However, it is cheaper to install floor drains during initial construction than to add them at a later date if the room's function is changed. Rooms with floor drains are more flexible in that they can be used for small and large animals or for procedures where larger volumes of water may be required in cleaning.

If drains are installed but not needed immediately, a removable gasketed cover should be placed over the drain. If drains are likely to become clogged with waste feed and bedding, a flushing floor drain or a garbage disposal unit in connection with the drain is advisable. Even when floor drains are omitted, materials that are easily washed should be provided, since damp-mopping of floors and walls with hot water and detergent or wet vacuuming is necessary for adequate germicidal protection. In rooms where fixed cage racks are used in conjunction with flush-type waste pans, a drain line must be used to dispose of waste water. Use of a closed drain system, coupled with membrane waterproofing, will greatly reduce the possibility of leakage between floors when animals are housed over other research functions. Adequate provision should be made for clean-outs in a closed drain system.

Floor drains are usually essential in rooms in which cages are hosed down. This type of cage sanitation is very efficient and dependable, and the animal house must be designed to accommodate such operations. All solid wastes may be flushed from the cage into an open gutter and then into a drain. It is suggested that the drains in monkey and dog rooms should be 15.2 cm (6 inches) in diameter of the flushing type with special hair traps to avoid clogging.* Adequate and easily accessible cleanouts should be provided.

Although climatic conditions generally determine whether to provide cooling systems for human occupants, facilities for rodents should be provided with complete air-conditioning, regardless of the geographic location of the facility.

The size and type of air-conditioning system will depend on the locality, the available utilities, and the heating and cooling requirements of the facility. Individual controls should be installed for each animal room or zone to accommodate various species. Unless the institution provides night watchmen in the animal quarters, a dual thermostat arrangement or alarm system should be installed to safeguard against heating or cooling emergencies. Small animals that have been raised under ideal conditions with controlled temperature and relative humidity are very susceptible to change. If a power failure occurs and is not corrected in a short period of time, the breeding cycle of the animals will be upset or the whole colony may be lost.

Table 1 summarizes the recommendations of a few reliable sources for the optimum ranges of relative humidity and temperature. It is recognized that the stated ranges are more critical for the small animals and the monkey than for the cat and dog. The relative humidity in the larger-animal areas is usually higher, because of the frequent hosing rather than through the mechanical air-handling system. However, in an ideal design, the relative humidity should be mechanically controlled in the larger-animal areas.

The architect and engineer should assure themselves that the scientific personnel who will use the facility have evaluated their criteria and selected optimum temperature and humidity range for each area. Since the figures in Table 1 are averages, specific applications may need different ranges. However, there is sufficient evidence to suspect that, when temperature and/or humidity vary outside the stated ranges, the animals will be more susceptible to disease and will have a lower breeding rate.

In choosing the equipment to air-condition an animal facility, the architect/engineer should consider not only the heat generated by the animals, but also the air

TABLE 1
Summary of Ranges of Temperature and
Relative Humidity

Species	°C	°F	%RH
Rat	18.3–22.8	65–73	45–55
Mouse	20–23.9	68–75	50–60
Guinea pig	18.3–23.9	65–75	45–55
Rabbit	15.6–23.9	60–75	40–45
Hamster	20–23.9	68–75	40–55
Dog	18.3–23.9	65–75	45–55
Cat	21.1–23.9	70–75	40–45
Monkey	16.7–29.4	62–85	40–75

*U. S. Public Health Service, Medical School Facilities, Planning Considerations and Architectural Guide. Washington, D.C., Government Printing Office, 1961, p. 75.

volume required to keep the odor below an objectionable level. Table 2 summarizes useful design criteria.

It must be realized, however, that the figures given in Table 2 are based on frequent cage-washing procedures and normal animal-holding densities. These procedures contribute to the reduction of objectionable odors, the control of enzootic conditions and a reduction in required rate of air change. If the cages are to be washed more or less than the one-and-a-half to two times per week that Table 2 is based upon, an adjustment should be made accordingly in the required air volume/animal. The graph, mentioned in footnote *c*, Table 2, will give heat emission for any animal whose weight is known. Therefore, if animals to be used vary greatly from weights listed, the heat/hour/animal should be recalculated.

TABLE 2
Criteria for Design of Mechanical Systems

Animal	Weight[a]		Air volume/animal		Total heat/hr/animal[c]	
	kg	lb	(1/min)	(CFM[b])	(kg-cal)	(BTU)
Mouse	0.022	0.0484	4.16	0.147	0.15	0.6
Hamster	0.11	0.2640	11.5	0.406	0.63	2.5
Rat	0.25	0.5500	23.1	0.815	1.1	4.3
Guinea pig	0.35	0.7700	32.7	1.155	1.4	5.6
Chicken	up to 3.2	7.	297	10.5	7.6	30.0
Rabbit	up to 3.6	8.	340	12.0	8.6	34.0
Cat	up to 3.6	8.	340	12.0	8.6	34.0
Monkey	up to 5.4	12.	510	18.0	11	43.0
Dog	up to 27.	60.	2550	90.0	38	150.0

[a]Recommended average weights as reported by California University, statewide Animal Care Committee.
[b]Computed by multiplying average weights times CFM/lb (from *Heating, Piping and Airconditioning 10*:289–291, Apr. 1938, and US National Institute of Health, Environmental Services Branch, Survey of Animal Rooms, Clinical Center, unpublished report, March 1956).
[c]See graph, page 1-163, paragraph—Heat Emission of Animals, in "Handbook of Airconditioning, Heating and Ventilating," New York, Industrial Press, 1959.

Supply air for animal rooms should ideally be 100 per cent fresh air, not recirculated. If air is recirculated, more efficient filtration for removal of odors and contaminants will be required. The fresh air does not need sterilization, but will generally need filtering; however, provision should be made, if possible, to add high efficiency filters at a later date, when the required experimental conditions are revised. If the animal quarters are contiguous with the other research activities, the air supply system should be separate from other parts of the building. The air should, if possible, be introduced into the rooms near the ceiling and exhausted near the floor to remove the heavier-than-air ammonia fumes. The supply of air to the animal rooms should be introduced into the rooms in as low a velocity as possible, since higher-velocity air tends to increase the chance of the animals' catching disease.

In some installations, the exhaust ducts from animal rooms in which there is loose hair present have no grills over the exhaust duct. This allows hair to be drawn up the duct and blown out. If conditions of the surrounding area do not allow hair to be exhausted, then a filter or incinerator should be used at the room exhaust register inside the exhaust duct. If this is not originally installed, space should be provided for installation at a later date. The exhaust from the animal portion of the building should definitely be separate from the main exhaust header to prevent contamination of laboratory space in time of power failure. The exhaust outlets should be located so they will not interfere or cause objectionable odors in critical areas. The fresh air intake should be located above ground level on the prevailing wind side of the building and as far away as possible from any exhaust to prevent reentry of contaminated air.

MECHANIZATION

Much of the cost in caring for animals is the labor and delay in cleaning or repairing facilities. Industrial engineering methods can be used to study present or proposed operating procedures and determine the areas in which it would be most advantageous to automate equipment and material movement. The architect and engineer, when designing an animal facility, should attempt to incorporate as many labor-saving devices as are economically and practicably feasible.

One of the newer innovations for the above-mentioned operations is the vacuum extractor used for removing waste material from cages. This system may be used in several types of facilities and at different points in the equipment processing procedure. In conventional colonies, dirty cages may be removed from the animal room and brought to the cage-washing area. The vacuum system can then be used as a means of removing waste from the cages and transporting it from the cage washing area to a storage bin or incinerator, where it is destroyed. The vacuum system may also be used to clean cages within the animal room, as well as the room itself, thus eliminating the need to transport soiled bedding and animal waste through the building corridors. A large-diameter vacuum hose has even been used to remove mouse cages from a colony room directly to an incinerator.

In cage-washing areas, the operations of soiled bedding removal, cage washing, water bottle-washing and refilling, bedding replacement and fresh food supply can be integrated so that all functions are carried out progressively with a minimum of manual labor. Automatic cage and rack-washers can operate at much higher water temperatures than human hands can stand and, therefore, can kill a much larger percentage of the microorganisms at these high temperatures. The cages and racks are also washed faster; and, since the machines are automatic, there is no question of variation from the set standard of cleanliness. By using the vacuum extractor at the beginning of the assembly line to remove soiled bedding and an automatic bedding dispenser at the clean end, the cage would be handled a minimum number of times. Machines have been developed to perform the complete process of water-bottle cleaning and filling. The automatic washers clean and sterilize the bottles more thoroughly because they are designed specifically for this operation. An important consideration in the design of washing facilities is provision of adequate space for spare cages, both clean and dirty.

The addition of an efficient incinerator to the animal facility would eliminate the cost of removing waste material. There would be less chance of spreading contamination outside the reservation if the diseased material is destroyed in the incinerator. Handling of infected material can be eliminated if a vacuum extractor is used to transfer the material directly into the incinerator.

In any operation the equipment-handling and service area should be centralized in the animal colony. If expansion is planned or is possible, a location should be selected for the central services so that the area will not be isolated or inconvenient if expansion or extension does take place.

Other areas in which automation has either been used or is thought to be feasible are: automatic watering devices for animal cages, automatic flush pan to catch droppings, overhead storage of bedding for automatic dispensing, specially treated corrugated paper as a cage liner and automatic weight sorter for separating animals in different weight ranges. As more experience is gained in the field of animal care, many different labor-saving devices will be developed. Although the architect/ engineer cannot be expected to develop these devices, he should check with groups who have had experience in these areas. (A list of such groups is available upon request from the Research Facilities Planning Branch, NIH.)

Average food and bedding requirements for various species of laboratory animals may be useful as a planning aid. Although scientifically and statistically valid information is difficult to obtain, Tables 3 and 4 may serve as guidelines.

TABLE 3[a]

Approximate Bedding Consumption

Species	Animals/month (average)	Bedding/month (average) kg	Bedding/month (average) lb	Bedding/animal/month (average) g	Bedding/animal/month (average) lb
Mouse	109,000	9,430	20,800	86.2	0.19
Rat	38,300	11,700	25,900	304	0.67
Guinea pig	10,350	21,000	46,200	2,020	4.46
Rabbit	3,400	7,170	15,800	2,110	4.65
Hamster	4,850	1,800	4,000	374	0.824
Dog	700	none	none	none	none
Cat	160	450	1,000	2,840	6.25
Monkey	1,100	1,800	4,000	1,650	3.63

[a]From production records, Animal Production and Animal Hospital Sections, Laboratory Aids Branch, DRS, NIH.

TABLE 4[a]

Approximate Feed Consumption

Species	Daily g	Daily avdp oz	Monthly g	Monthly lb
Mouse	5	0.18	180	0.4
Rat	15	0.53	450	1
Guinea Pig	30	1.0	900	2
Rabbit	150	5.3	4,500	10
Hamster	10	0.35	300	0.7
Dog	500	18	15,000	33
Cat	150	5.3	4,500	10
Monkey	300	10	9,500	21

[a]Ralston Purina Co., St. Louis, "Purina Laboratory Animal," 1962 (-), p. 32.

ACKNOWLEDGEMENT: This chapter is reprinted with permission from *Journal of the American Institute of Architects*, vol. 44, pp. 55–58, 77–80, Mar.–Apr., 1964, copyright 1964 by the American Institute of Architects.

ANIMAL FACILITIES BIBLIOGRAPHY

Selected References: 1938–1962

Compiled by Gertrude Fox, Librarian

1. Adams, S. S., Some Problems of Animal House Design and Equipment, *J. Anim. Tech. Assn.*, *11*, 39–43 (Sept. 1960).
2. Barker, E. V., Design and Construction of Animal Quarters for Medical Education and Research, *J. Med. Educ.*, *35*, 15–23 (Jan. 1960).
3. "Basic Care of Experimental Animals," Rev. ed., Animal Welfare Institute, New York, p. 68 (1958).
4. Brewer, N. R., Housing for Research Dogs, *Fed. Proc.*, *20*, 917–918 (Dec. 1961).
5. Brewer, N. R., Comp. References on Design of Animal Quarters, *Bull. Med. Res.*, *11*, 6–8 (Jan./Feb. 1957). Selected references 1942–57. (Reprints available from Institute of Laboratory Animal Resources, National Academy of Sciences, Washington, D.C.)
6. Brewer, N. R., and Penfold, T. W., Thoughts Concerning the Design of Animal Quarters, *Proc. ACP*, *11*, 281–290 (Dec. 1961). Presented at the Ninth Annual Meeting of the Animal Care Panel, Dec. 3–5, 1958, Chicago, Illinois.
7. Buxton, A. H., Preclinical Animal House, School of Medicine, U. of Western Australia, *J. Anim. Tech. Assn.*, *11*, 88–9 (March 1961).

8. California, University of, "Report of the Statewide Animal Care Committee." (Available from Institute of Laboratory Animal Resources, National Academy of Sciences, Washington, D.C.)

9. California, University of, School of Veterinary Medicine, Davis, California, Third Annual Progress Report, AEC Project No. 6: "The Effects of Continual Sr^{90} Ingestion during the Growth Period of the Beagle and its Relation to Ra^{226} Toxicity," A. C. Anderson et al., 84 p., (Sept. 1960) pp. 8–16.

10. Care and Diseases of the Research Monkey, *Ann. N.Y. Acad. Sci.*, *85*, 735–992 (May 12, 1960).

11. Cass, J. S., Campbell, I. R., and Lange, L., A Guide to the Production, Care and Use of Laboratory Animals, *Fed. Proc.*, *19*, *Pt. 3*, *Suppl. 6*, 1–196 (Dec. 1960). Annotated list of references.

12. Charles, R. T., The Design and Management of Animal Colonies in Radiobiological Research, *J. Anim. Tech. Assn.*, *5*, 31–34 (Sept. 1954).

13. Clarke, E. M. W., A New Unit for Infected Small Animals, *J. Anim. Tech. Assn.*, *11*, 27–30 (June 1960).

14. Cochin, R., Experimental Animal Unit, Perth Chest Hospital, W. Australia, *J. Anim. Tech. Assn.*, *11*, 76–7 (March 1961).

15. Cohen, B. J., Caring for Research Animals in Hospital Laboratories, *Hosp.*, *35*, 43–48 (Dec. 16, 1961).

16. Cohen, B. J., Organization and Functions of a Medical School Animal Facility, *J. Med. Educ.*, *35*, 24–33 (Jan. 1960).

17. "Comfortable Quarters for Laboratory Animals," Animal Welfare Institute, New York, 35 p. (1958).

18. Coppa-Zuccari, G., Le Centre de Virologie de l'Université de Milan, *Produits Pharmaceutiques*, *16*, 216–20 (May 1961). Includes animal facilities.

19. Crossman, R. F., and Elsea, R., Air Conditioning Laboratory Animal Quarters, *Air Cond. Heat. & Vent.*, *58*, 71–76 (July 1961).

20. Cumming, C. N. W., The History, Growth and Development of Commercial Laboratory Animal Production, *Biomedical Purview*, *1*, 21–28 (Fall 1961).

21. De Ome, K. B., The Economies of Animal Colony Operation, *Proc. ACP*, *8*, 113–127 (Sept. 1958).

22. Dolowy, W. C., Medical Research Laboratory of the University of Illinois, *Proc. ACP*, *11*, 267–280 (Dec. 1961).

23. Earl, A. E., Care and Use of Dogs for Research, *J. Anim. Tech. Assn.*, *5*, 52–60 (Dec. 1954).

24. Farris, E. J., "The Care and Breeding of Laboratory Animals," Wiley, New York, 515 p. (1950).

25. "Isolated Ideas Concerning the Design of Animal Quarters," Discussion at Animal Care Panel Meeting, 6 p. (Nov. 1957). (Available from Institute of Laboratory Animal Resources, National Academy of Sciences, Washington, D.C.)

26. Kallman, R. F., The Maintenance of an Experimental Mouse Colony in a University Medical School Department, *Proc. ACP*, *11*, 73–82 (April 1961).

27. Kennard, M. A., Ruch, J. R., and Milton, J. F., The Housing, Care and Surgical Handling of Laboratory Primates, *Yale J. Biol. Med.*, *18*, 443–471 (May 1946).

28. Lack, C. R. et al., Saving Labor in the Animal House, *J. Anim. Tech. Assn.*, *8*, 81–2 (March 1958).

29. Lau, K. K., and Harrison, J. L., An Animal House in the Tropics, *J. Anim. Tech. Assn.*, *10*, 150–9 (March 1960).

30. MacPherson, A. D. H. et al., The Architects Approach to Design, *J. Anim. Tech. Assn.*, *4*, 67–70 (March 1954).

31. Marois, P., Boulay, G., and di Franco, E., Colony Husbandry of the Monkey, *Canad. J. Comp. Med. Vet. Sci.*, *25*, 291–5 (Nov. 1961).

32. Marquette University School of Medicine, Milwaukee, "Animal Research Facility," The University, 8 p. (1961).

33. Medical Research Council (Great Britain), Laboratory Animals Bureau, "The Design of Animal Houses," Collected Papers, *2*, 1–75 (1954).

34. Munkelt, F. H., Air Purification and Deodorization by Use of Activated Carbon, *Refrig. Eng.*, *56*, Recommended Ventilation Requirements for Animal Laboratories, 8 p. insert after p. 222 (Sept. 1948).

35. Munkelt, F. H., Odor Control in Animal Laboratories, *Heating, Piping, & Air Cond.*, *10*, 289–291 (April 1938).

36. National Research Council, Institute of Laboratory Animal Resources, "Standards for the Breeding, Care, and Management of Syrian Hamsters," Washington, D.C. (1960).

37. Newhall, L. E. J., Some Aspects of the Management of an Experimental Animal Wing, *J. Anim. Tech. Assn.*, *12*, 47–53 (Sept. 1961).

38. Pereira, W. P., Housing and Management of Experimental Animals, *J. Anim. Tech. Assn.*, *3*, 49–52 (Dec. 1952).

39. Pipher, W. V., "Programming and Design of a Centralized Medical Research Laboratory," Skidmore, Owings and Merrill, New York, 6 p.

40. Report of a Symposium on the Organization and Administration of an Animal Division. Carshalton, England, Laboratory Animals. Centre. Collected papers 7, 1–107 (1958).

41. Reyniers, J. A., The Production and Use of Germ-free Animals in Experimental Biology and Medicine, *Amer. J. Vet. Res.*, *18*, 678–687 (July 1957).

42. Rickell, M., The Care of Baboons, *J. Anim. Tech. Assn.*, *12*, 60–62 (Dec. 1961).

43. Short, D. J., Automatic Rabbit Battery, *J. Anim. Tech. Assn.*, *11*, 13–21 (June 1960).

44. Smith, R. T., The New Animal Unit at Parke-Davis, Hounslow, *J. Anim. Tech. Assn.*, *12*, 43–6 (Sept. 1961).

45. Softly, A., The Animal Rooms, Royal Perth Hospital, Western Australia, *J. Anim. Tech. Assn.*, *11*, 63–5 (Dec. 1960).

46. Staley (A. E.) Mfg. Co., Decatur, Ill., "A Practical Guide on the Care of Laboratory Animals," 3rd ed., 45 p. (1958).

47. Stone, W. C., Management Practices in a Normal Monkey Colony, *Proc. ACP*, *12*, 99–106 (June 1962).

48. Strock, C., ed., "Handbook of Air Conditioning, Heating, and Ventilating," The Industrial Press, New York, p. 1–163 and 1–293 (1959).

49. Taylor, W. R., Nelson, C. E., and McMaster, W. W., Diagnostic X-ray Suites for the General Hospital, *Arch. Record*, *126*, 217–224 (Oct. 1959). Although the article deals with X-ray suites for hospitals, the information may be adapted to animal facilities.

50. Thorp, W. T. S., The Design of Animal Quarters, *J. Med. Educ.*, *35*, 4–14 (Jan. 1960).

51. U. S. Army and Air Force, "Care and Management of Laboratory Animals," Washington, GPO, 112 p (1958).

52. U.S.N.I.H., Division of Research Grants, "The Dog in Medical Research," Revised. Washington, GPO, 15 p. (1961).

53. U.S.P.H.S., Division of Hospital and Medical Facilities, Planning the Laboratory for the General Hospital, *Arch. Record*, *129*, 161–8 (Feb. 1961). This article deals with hospital planning, but contains material useful in planning animal facilities.

54. Weiss, C., Care of Guinea Pigs Used in Clinical and Research Laboratories, *Amer. J. Clin. Path.*, *29*, 49–53 (Jan. 1958).

55. West Foundation, "Housing for Experimental Animals," Brant Lake, N.M., 19 p.

56. Worden, A. N., and Lane-Petter, W., eds., "UFAW Handbook on the Care and Management of Laboratory Animals," Universities Federation for Animal Welfare, London, England, 951 p. (1957). (Available from Animal Welfare Institute, New York, N.Y.)

Laboratory Design
and Equipment

Laboratory Design Considerations for Safety

Campus Safety Association

These considerations proposed by the Laboratory Safety Committee of the Campus Safety Association received approval at the 13th National Campus Safety Association Conference on June 24, 1966.

Introduction

This set of considerations has been prepared from considerable hard-won experience to provide persons working with the designing of new or remodeled laboratory facilities with a suitable reference guide to design safety.

The following items should be used as guides only. These considerations do not spell out the details nor do they cover all of the facets of the many design problems. In most instances, various alternatives are offered and the selection of the method and degree of protection required for your circumstances is left to your judgement. It is suggested that where specialized protective equipment is to be used an engineer or engineering firm be employed with experience with the equipment to be installed. The various state and insurance rating agencies and manufacturers or distributors of equipment will often do preliminary work on specialized equipment without obligation.

The title of these considerations deliberately deletes any reference to the type of laboratory since recent experience has indicated a need for most of the items covered in all laboratory buildings. Reference to existing standards of design or installation are made in the various sections of these guides so that you will be able to acquire the design details. It has been impossible to provide design detail in this set of standards for reasons of desired brevity.

Automatic Systems for Fire and Explosion Protection

Automatic fire and explosion detection and/or protection equipment should be installed in those areas of all laboratory buildings which present special hazards. Due to the nature of the work performed in most laboratory facilities and the nature of the chemicals employed, the hazards of fire and explosion are much greater than in other campus buildings. Even a small fire in laboratory buildings can cause large losses. Laboratory fires have destroyed valuable research equipment, products and data.

As research work increases on campuses, the concentration of values also increases and the inherent hazard increases. To maintain control of this fast rising spiral of hazard and values, it is suggested that the use of the following types of protective equipment be considered in all new or remodeled laboratory buildings. The decision as to which type of equipment and how complete an installation is desired or necessary is left to the individual designers. Table 1 lists several, but not all, potential hazard areas, and Table 2 provides a list of technical design standards for various protective devices.

In many hazardous areas, it will be desirable to provide more than one type of protection. For example, flammable liquid rooms should have automatic detection, automatic extinguishing and explosion venting equipment.

655

TABLE 1

Hazardous Areas Needing Automatic Detection and/or Protection Devices

Flammable liquid storage, repackaging or dispensing rooms.
Chemical stockrooms.
Compressed gas cylinder storage or manifold rooms.
General storage, equipment storage and janitor closets.
Radioactive isotope storage or production areas.
Mechanical equipment rooms, boiler rooms, electrical rooms, transformer vaults, air-handling equipment and battery rooms.
Vertical chutes, ducts, pipe chases which pierce more than one floor.
Horizontal concealed spaces that pass through fire walls.
Laboratories used for handling infectious, carcinogenic or pathogenic agents.
Laboratories below grade level or without windows.
Laboratories used for special hazard experimentation, such as hydrogenation, use of explosives, rocket propellants, highly toxic gases, etc.

TABLE 2

Technical Design Standards for Automatic Systems

Installation of Sprinkler Systems, NFPA No. 13.
Carbon Dioxide Extinguishing Systems, NFPA No. 12.
Dry Chemical Extinguishing Systems, NFPA No. 17.
Foam Extinguishing Systems (high expansion foam), NFPA No. 11.
Water Spray Systems for Fire Protection, NFPA No. 15.
Explosion Venting Guide, NFPA No. 68.
Inerting for Fire and Explosion Prevention, NFPA No. 69.
Installation Standpipe and Hose Systems, NFPA No. 14.
Explosion Suppression Systems, NFPA Handbook.
Factory Mutual Laboratories Engineering Manual, NFPA Fire Protection Handbook.
Fire, Smoke and Combustible Gas Detection Equipment should be installed in accordance with its individual listing in the Fire Protection and Hazardous Location Equipment Listings of Underwriters Laboratories, Inc.

Emergency Alarm Systems

Due to the high fire hazards, explosion hazards and possible escape of toxic, radioactive or bacteriological agents, every laboratory building should be equipped with an emergency evacuation alarm system which sounds within the building.

Since quick response by professional fire fighters or other emergency personnel is highly desired to minimize damage and danger to the surrounding area, remote signaling devices connecting the building alarm system to the local fire or police department or campus security department should be given consideration.

The following emergency signaling or reporting equipment should be provided:

A. Remote signaling systems connecting the emergency evacuation alarm, any automatic detection and/or protection devices to a centrally supervised panel at the fire department, police department or campus security office in accordance with the standards of NFPA No. 72, Proprietary, Auxiliary and Local Protective Signaling Systems.

Where remote signaling systems are impractical, emergency telephones should be provided inside the building where they will be accessible when the buildings are closed. The emergency telephones should connect directly to the emergency offices so that there will be no need to dial or locate a telephone number.

B. Consideration should be given to providing outside emergency telephones for use by night watchmen, passers-by, or persons who have evacuated the building.

C. Annunciator panels should be provided for all automatic detection, extinguishing and manual alarm systems in a location easily accessible to the fire department to indicate the location of any detection, extinguishing or manual alarm devices which have operated.   Such annunciator panels are important for prompt response to sites of emergencies.

D. Building air conditioning fans should be automatically shut down or switched to total exhaust by the operation of any automatic or manual fire or gas detection, protection or alarm device. Laboratory exhaust system and fume hoods should continue to operate unless manually shut down by the personnel in charge of the laboratory.

E. Stairway or other firedoors that are normally held open by electromagnetic devices should be released by the action of any automatic detection, extinguishing or manual alarm device.

F. All fume hoods and spot ventilation equipment should be provided with manual shutoff devices so that they can be shut off in case of fire, if such action does not cause additional hazards. NOTE: Automatic shutdown is not recommended for this equipment.

Special Facilities for Chemical Storage, Handling and Disposal

Special provisions should be made for the handling, storage and disposal of flammable liquids, compressed gas cylinders and hazardous chemicals. The more rapid our scientific advances, the more hazardous laboratory operations become. Our new and remodeled laboratory buildings should be designed with the expected increase in hazard in mind.

A. Flammable liquid storage, repackaging or dispensing should be done in separate buildings, separate but adjacent rooms or in special cut-off rooms designed in accordance with NFPA No. 30, Flammable and Combustible Liquids Code or the Factory Mutual Laboratories Engineering Manual. Bulk storage should be in a remote area.

B. Provision should be made for easy distribution of materials to the laboratories so that excessive storage in laboratories can be prevented. Ideally, such facilities should provide for effective security and distribution control.

C. Provision should be made for safe storage within laboratories in ventilated, fire resistive storage cabinets and safety containers.

D. Provision should be made for the collection, disposal and destruction of flammable liquids, hazardous chemicals and biological wastes within each laboratory facility. The nature of this problem is such that individual analysis of the expected problem is required before a solution can be found. In many instances, a system of collection and remote destruction is the most feasible. NOTE: In some areas, outside firms can be employed to pick up hazardous wastes.

E. Provisions should be made for storing compressed gas cylinders of a flammable and/or toxic nature outside of the building in an area protected from the weather and easily accessible to the freight elevator. See NFPA No. 58, Storage and Handling of Liquefied Petroleum Gases, for a guide in designing storage facilities. Segregated storage facilities will be needed for certain gases that should not be stored in the same area or enclosure. NOTE: The increasing use of tank farms and gas manifold systems for the distribution of the primary laboratory gases via a piped system introduces hazards many architects, engineers and university personnel may overlook. Such systems must be carefully engineered to prevent accidental release of explosive or toxic gases in populated areas or buildings. The general guidelines set forth in NFPA Pamphlets 51, 54, 56, 565, 566, 567, and 58 should be used in determining the safety of such installations. Gas masks, self-contained breathing apparatus, proximity suits and special fire extinguishing equipment should be provided in locations where they will be accessible under emergency conditions.

F. Special hazard laboratories should be provided for work which must be unattended at times or which is so inherently hazardous that it cannot be conducted in the normal laboratory. Special hazard labs should be equipped with automatic fire and explosion protection and detection devices. The rooms should be located on the outer perimeter of the building and have a direct means of egress to the outside. Hydrogenation and high pressure autoclave laboratories represent the type of laboratory which needs special consideration. Such laboratories must be carefully designed with attention given to explosion relief, structural integrity, control of entrance and egress and location within the main structure.

G. A special, well ventilated and protected area should be established for the storage of potentially explosive chemicals such as organic peroxide which should not be stored in flammable liquids storage rooms. Such storage should provide for the segregation of potentially explosive chemicals into reasonably small quantities.

Safety Equipment

A. Water-type fire extinguishers (Class A) should be provided in accordance with the standards of NFPA No. 10, Installation, Maintenance and Use of Portable Fire Extinguishers. In addition to the Class A fire extinguishers, Class B:C fire extinguishers should be provided for fighting flammable liquid and electrical fires. Multi-purpose, Class AB:C fire extinguishers can be used instead of providing two different types of extinguishing units. There should be at least one 30 lb Class B:C or AB:C fire extinguisher on each floor, and one 5 lb Class B:C or AB:C fire extinguisher in each laboratory. Special Class D fire extinguishers should be provided where the use of metals or metal hydrides indicates a need for them.

B. Standpipe and fire hose equipped with adjustable fog nozzles should be installed in accordance with NFPA No. 14, Installation of Standpipe and Hose Systems.

C. Consideration should be given to installing two 30-minute, self-contained breathing masks for alternate floor levels for use in emergency rescue operations. The air masks should be set in permanent cabinets. This requirement is in addition to any specific gas masks that may be provided for use within specific hazardous environments. Gas masks with canisters or cartridges are not acceptable for use in oxygen-deficient areas or where concentrations of contaminants exceed the listing on the particular mask.

D. Consideration should be given to providing a proximity suit or suits for fire fighting and rescue operations especially where the local fire department is located some distance away or does not have this equipment.

E. At least one safety shower and eyewash fountain or hose should be installed in each laboratory, or near the entrance to each laboratory. There should be no more than a 50-foot travel distance to such devices from any point in the laboratory. Care should be taken to locate these devices and the actuating mechanisms where they will not be blocked.

F. Each laboratory should be equipped with at least one fire blanket.

G. Each laboratory should have access to a first aid kit. The contents of the first aid kit should be determined by the campus physician. Where hospital facilities and emergency transportation is available, it is recommended that only sterile bandages and compresses be provided.

H. Respirators, ventilated suits, or similar devices should be provided for protection of persons entering areas where infectious disease agents or radioactive isotopes may be or are air-borne under normal conditions or due to accidents.

I. Safety bulletin boards should be placed in the building for communication of safety information and regulations.

Facilities for Infectious Agents and Animals

Animal handling and infectious agent laboratories, and handling or housing areas should be designed with the prevention of human infection in mind. This is especially important in research areas where the level of possible infection is rapidly increasing. All areas of possible contamination should be well segregated from public areas and provided with a separate positive ventilation system which can be equipped with bacteriological filters. Any make-up air necessary for the ventilation of contaminated areas should be introduced from the outside. Table 3 lists some of the hazards normally encountered and suggested preventive design features. Close cooperation between laboratory personnel, safety personnel, environmental health personnel and designers is imperative if adequate protection is to be provided.

TABLE 3
Infectious Hazards and Prevention

Hazard	Preventive Design
Air-borne, aerosol, infectious agents as in laboratories or animal housing rooms.	Ultraviolet lights and/or spot ventilation. e.g. Ventilated animal cages or cabinets, glove boxes.
Surface borne infectious agents from spills or normal contamination.	Ultraviolet lights, where surface is directly exposed, and easily cleanable walls, floors, ceilings and equipment, e.g. glazed tile, ceramic tile and stainless steel. Floor drains and wall hydrants should be provided
Cross contamination.	Segregation of high hazard areas by intervening safe areas or decontamination rooms equipped with showers and lockers. Well defined segregation of ordinary hazard areas with change rooms between them and public areas.
Disposal of infectious waste or carcasses.	Autoclaves for sterilization and an incinerator for disposal.
Contaminated equipment.	Autoclaves large enough for cages and enough autoclaves to handle all of the equipment and material without removing it from the contamination area and through public areas. After the material has been sterilized, it should be placed in special containers, sealed and incinerated.
General equipment and supplies.	Each area housing animals used for the handling, introduction or testing of infectious agents should be equipped with autoclaves, a stock of disinfectant, respirators, and ventilated hoods and/or suits. All floor drains in such areas should be segregated from the main drainage and be equipped with sterilization equipment. A sufficient number of refrigerators should be provided for infectious agents so that they will not be left on tables or benches.

Ventilation

In order to meet the needs of both safety and economy, laboratory ventilation systems must effectively remove air-borne toxic and flammable materials and at the same time exhaust a minimum volume of air. Make up air should be supplied to laboratories to replace the air removed by exhaust systems so that such systems work properly, and air exhausted from laboratories should not be recirculated.

LABORATORY HOODS. Laboratory hoods are the most commonly used means of removing gases, dusts, mists, vapors and fumes from laboratory operations, preventing toxic exposures and flammable concentrations, but they are often misused and frequently specified when not really needed. Spot ventilation and special enclosures are other means of capturing or containing gases, vapors and particles effectively while exhausting minimum volumes of laboratory air.

To be effective, a laboratory hood and its associated components must confine contaminants within the hood, remove them through ductwork, and disperse them so they do not return to the building through the fresh air supply system.

Storage of chemicals and gas cylinders in laboratory hoods is a common practice which may be both wasteful and dangerous. If chemicals and cylinders are stored in hoods to guard against leaks, then the hoods should operate 24 hours per day to assure constant exhaust. In addition to the expense of such long operation, storage of chemical containers in hoods will restrict performance by reducing work space and by obstructing air flow so that velocities may be too low to retain toxic materials, within the unit.

SPECIAL STORAGE CABINETS. Hood performance should be assisted by providing ventilated chemical storage cabinets which will require less air flow and which can be exhausted in a separate system 24 hours every day.

HOOD LIMITATIONS. Laboratory hoods are not generally designed to contain explosions of a high order, even though laboratory personnel may think so. For general use, laboratory hoods should be constructed of materials to withstand fire which may occur from flammables, solvents and gases; an enclosure should be able to maintain its integrity and confine a fire until the fire can be extinguished.

Successful performance of laboratory hoods depends primarily on the velocity of air moving through the hood which is affected by cross-currents, entrance shapes, thermal loading, mechanical action of particles, exhaust slot design, and obstructions in the hood. Successful performance also depends on corrosion resistance, cleanability if contaminated, and in some cases the collection of contaminants, such as radioisotopes and pathogens.

Design criteria have been established for hoods in radioactive service, including scrubbers and filters with gauges to show pressure drop across the filter as it loads up.

PERCHLORIC ACID HOODS. Hoods in which hot concentrated perchloric acid is to be used should be used only for that service, should have no exposed organic coating, lubricant, or sealing compound, and should be equipped with a special wash-down system in the ducts and back. Safe use may also be accomplished by use of a special scrubber unit which effectively removes acid vapors and mists. Because of the hazards of reactions between hot acid vapors and organic materials, gas and oil should not be used for heating perchloric acid. Perchloric acid hoods and ducts should be clearly labeled to prevent confusion.

HOOD FACE VELOCITIES. An adequate velocity of air entering a hood at its face is the basic requirement for capture and control of contaminants generated within a hood, and an entering face velocity of 100 lineal feet per minute for normal openings of general units is recommended. Hoods for highly toxic materials may require face velocities ranging from 125 to 200 feet per minute.

Since cross-currents outside a hood can interfere with the operation by counteracting the capturing velocity, it is important to minimize air currents from doors, windows, pedestrian traffic and air supply grilles.

DESIGN OF NEW HOODS. Design and specification of new hoods should be based on critical analysis of present ventilation needs and an imaginative analysis of future needs, without restricting design requirements to past custom. Some of the questions that can be asked before specifying new hoods include the following: (a) Should hoods incorporate automatic fire extinguishing systems? (b) Should hood benches be 36 inches above the floor, or lower to accomodate tall equipment and to be more convenient and safe for short people? (c) Are there enough hoods to meet present and anticipated needs? (d) Are hoods specified when other enclosures would do a better job at less cost?

Each laboratory hood should generally have its own exhaust fan, switch and pilot light; multiple hoods in a laboratory should be interconnected in some manner so that one may not pull air down through a hood not operating. A good practice is to include all portions of hood design, construction and installation in one lump contract so that the effectiveness of the entire system can be judged. Letting individual contracts for hoods, duct work and electrical work weakens your bargaining position.

ENCLOSURES. The amount of exhaust air needed for laboratory operation is reduced if the operation can be moved from a hood to a partial or complete enclosure such as a glove box, requiring little ventilation, or a vacuum or inert gas box, which requires almost none.

A glove box can be safely ventilated with an air flow of only 50 cubic feet per minute for each square foot of open door area.

Improved safety, great flexibility, and economy are advantages offered by glove boxes and other special enclosures. New laboratories should be designed to provide ventilation connections for such enclosures.

SPOT VENTILATION. Spot ventilation—exhausting contaminants near their point of origin—can prevent inhalation hazards from laboratory operations not suitable for enclosure because of bulk, access needs or brief use. Ventilation could be provided for every bench in laboratories where ventilation is imperative but adequate hood space cannot be provided.

New laboratories should be designed to provide special local ventilation and exhaust connections for enclosures.

ACCESS FOR SERVICE. Fan and duct designs should provide convenient access ports for inspection and cleaning and should be designed with no constricting sharp bends that require excessive fan horsepower for adequate hood performance.

Provision for control of inhalation hazards is one of the most important functions of a laboratory building and design of the building should allow streamlined ductwork and shaftways large enough for future additions of ductwork.

DUCT MATERIAL. Material for hood exhaust ducts should resist corrosion by chemicals and moisture to which the ducts will be exposed and fire should not be able to spread from one duct to another.

DUCT VELOCITY. A minimum duct velocity of 2000 feet per minute is recommended for vapors and gases; a velocity of 2500–3000 feet per minute is suggested to scavenge condensed moisture. Transport velocities for particulate materials will range from 3500–4500 feet per minute. Good practice would be to determine the appropriate velocity from the Industrial Ventilation Manual.

FAN LOCATION AND EXHAUST POINT. The best location for hood exhaust fans is on the roof of the building to place the duct system under negative pressure, assuring that any leaks which may develop will not allow contaminants to escape into the building.

Consideration should be given to contamination or air pollution problems which could result from direct discharge of hood exhaust to the atmosphere. There is an increasing need to provide filters, collectors, condensers, scrubbers or other air cleaning equipment.

A laboratory ventilation system with the best hoods and best transport system is a failure if the exhaust returns to the laboratory through windows or the fresh air system. Discharge outlets and discharge velocities should be designed so that exhaust is effectively dispersed and nuisance prevented.

PRECAUTIONS FOR SHUTDOWN OF AIR SUPPLY.   If conditioned air from offices, classrooms and corridors of laboratory buildings, and other air generally uncontaminated by laboratory operations is recirculated, two special precautions should be observed to stop recirculation in case of emergency.

The first precaution consists of automatic equipment, fusible link fire dampers on exhaust louvers and smoke detection equipment which can shut down the entire recirculation system to prevent spread of smoke and fire.

The second precaution recommended consists of a readily accessible control for laboratory fresh air supply or recirculation systems so that laboratory personnel can immediately stop the system in case of an emergency such as a spill or release of toxic radioactive, or flammable materials. The emergency control for the ventilation system should be similar to a fire alarm pull station, properly labelled and connected to the evacuation alarm system.

MONITORING HOOD OPERATION.   Monitoring the operational performance of laboratory hoods is desirable. The first thing to know is whether the hood is operating. A pilot light can be provided to show that the exhaust fan motor has been turned on, or a device can be built to assure that the motor is turning, or a gauge can be installed to indicate that the fan has vanes and is drawing air.

If filters are part of the exhaust system or fan vanes are liable to corrode seriously, a manometer which shows the pressure drop across the filters or in the exhaust duct can be used to judge the need for replacement of the filters or the fan.

NO RECIRCULATION.   The exhaust from laboratory hoods, enclosures, and spot ventilators should not be recirculated. The continually accelerating pace of research and development activities and technological progress make it impossible to predict what chemical and reactions will be used in laboratories in the future or what demands will be placed on laboratory hoods.

Illumination

Adequate lighting should be provided to conform with current recommendations of the Illuminating Engineering Society:

Office and Lecture Rooms	100 footcandles
Laboratories	200 footcandles
Stairways, Washrooms and Other Service Areas	20 footcandles
Shops	100 footcandles
Precision Manual Arc Welding and Extra Fine Bench and Machine Work	1000 footcandles

An emergency power source should be provided so that exit paths and exit signs will be available even if the commercial power source fails. The emergency power source can also supply fire detection, protection or alarm devices, other emergency equipment, exhaust ventilation and special outlets where electrical power loss may lead to disastrous failure.

Radioisotopes

A radioisotope laboratory should be designed like any other work area, primarily for convenience, efficiency and safety in normal operations. But in addition to maintaining control of the hazards inherent in planned materials and processes, it is also important that the laboratory be designed to reduce the likelihood of accidents and to ameliorate the consequences and promote return to normalcy if an accident does occur.

With the exception of provisions for radiation shielding, the requirements of a radioisotope laboratory are essentially not different from the requirements of laboratories handling other dangerous materials such as highly toxic chemicals or contagious pathogens, In many ways, radioactive materials differ only in DEGREE from other toxic agents in that: (1) incorporation of radioactivity in normally harmless materials can render them highly toxic, (2) radioactivity never gives warning by virtue of disagreeable taste or odor, (3) radioactivity cannot be destroyed or neutralized, (4) radioactive material, on a weight or volume basis can be many times more toxic than ordinary toxic materials.   However, the toxicity hazards are controlled by essentially the same methods of controlling inhalation, ingestion and contact:

A. By promotion of cleanliness and good housekeeping.
   1. Eliminate dust-catchers, e.g. overhead pipes, ledges, grooves, corners.
   2. Provide smooth, non-porous easily cleaned surfaces on equipment, furniture, floors and walls.
   3. Provide a separate change area (for high hazard labs) for exchange of street clothes for laboratory clothing.
B. By control of ventilation and air circulation.
   1. Establish air flow patterns from cold areas to hot areas, i.e. higher level areas, if possible should operate at a slightly negative pressure with respect to lower level areas.
   2. Use hoods or dry-boxes for operations likely to create air-borne contamination.
   3. Place blowers on hood ducts at exit end of duct so as to maintain negative pressure in duct.
   4. Place filters, if used, preferably at entrance end of duct.
   5. Utilize a common duct and blower when two or more hoods are in one room. If separate ducts and blowers are used, the blowers should be wired to a common switch to insure simultaneous operation (continuous operation is preferred) otherwise downdrafts and cross contamination are probable.
   6. Design room ventilation supply free of strong drafts—especially near hood faces.
   7. Separate exhaust outlets a sufficient distance from ventilation intakes.
C. By control of waste disposal.
   1. Select corrosion resistant and break resistant components for sinks and plumbing.
   2. Install hold-up tank where necessary.
   3. Select waste containers to minimize handling of waste. Choose disposable containers or those with disposable liners.
   4. Provide special features or equipment when necessary for handling of animal waste and carcasses in radiobiological laboratories.
D. Shielding *may* be required for:
   1. Work areas.
   2. Storage.
   3. Hood ducts and filter boxes.
   4. Plumbing.
   5. Doors and walls common to other areas.
E. Shielding sometimes requires additional structural support incorporated into walls, floors and furniture.

Design of hot labs is a specialized subject on which the available references should be consulted.

Egress

Each laboratory or other potentially hazardous area should have two exits as remote from each other as possible. Laboratory benches and other equipment should be arranged to facilitate egress, and visibility from the corridor through small wired-glass panels. Doors from labs to corridors should be at least 36 inches wide, have school type hardware and be so designed that they can swing safely in the direction of egress without swinging more than 12 inches into the corridor. Maximum distances to exits should be 75 feet.

Fire Resistance

Walls, doors and windows between laboratories and corridors should have a minimum fire resistance of $3/4$ hour, as determined by fire test ratings.

Water Supply

Laboratory water supplies should be suitably designed so that the drinking water supply will not become contaminated. (If there are separate supplies, the potable and non-potable outlets need to be clearly identified.) All plumbing fixtures equipped for hose connections should have anti-siphonage devices installed.

Consideration should be given to providing an emergency water supply wherever pressure failure would create hazards; e.g. condensers, cooling water, eyewashers.

Miscellaneous Design Features

Office and individual laboratories should be separated with offices nearer the corridor.

Each laboratory should have one sink supplied with hand washing materials so that persons using the lab can conveniently remove toxic materials before putting on street clothes or going out of the laboratory for drinking coffee, eating or smoking.

The number of researchers working in a single research laboratory space should be limited with a preferable minimum of two and preferable maximum of four.

Laboratory hoods should be equipped with safety glass sash—horizontal sliding panels no wider than 16 inches are preferred. Laboratory benches should be designed so that safety glass shielding is facilitated, since all reactions under pressure or vacuum should be shielded. Double laboratory benches, where persons may work opposite each other, should be provided with safety shielding; safety glass would minimize interference with lighting and would permit observation of reactions which may be temporarily unattended.

Safety centers where electrical, gas, steam, air and other utilities can be shut off should be located outside the laboratory rooms.

Provisions for hanging coats so that they do not obstruct safety equipment or egress should be made. Hanging coats in halls is not recommended.

Steam outlets on benches should be aimed downward. Valves for gas, steam and air should be of the type that indicates at a glance if the valve is open. One gas outlet in each lab should be fitted with a check valve to prevent oxygen being introduced into gas lines in glass blowing operations. All valves should be consistently color-coded and where possible shape-coded to reduce confusion in use.

ACKNOWLEDGEMENT: This chapter is reprinted with special permission of the Executive Committee of the Campus Safety Association.

Membership of the Campus Safety Association Laboratory Safety Committee
that participated in development of the
Laboratory Design Considerations for Safety

G. W. Marsischky, Chairman

William A. Burgess
Harvard University

Robert Dennis
University of Michigan

Charles W. Easley
California Institute of Technology

Chandler Eaton
Harvard University

John M. Fresina
Massachusetts Institute of Technology

John C. Martin
University of Illinois

R. J. Nocilla
Reynolds Electrical & Engineering Co., Inc.

G. N. Quam
Villanova University

Norman V. Steere
University of Minnesota

Correspondence about these considerations should be addressed to the current Chairman, William A. Burgess, Harvard University, or to another member of the Laboratory Safety Committee.

Section **12**

Tables of Chemical Hazard Information

Introduction to Tables of Chemical Hazard Information

(For Condensed Key to Tables, see Inside Back Cover.)

The Tables of Chemical Hazard Information are a unique compilation of fire and health hazard data on over one thousand chemicals. Chemicals listed include those used in the laboratory with reported hazards, ones for which carcinogenicity reports were found, and those with exposure limits recommended or proposed by the American Conference of Governmental Industrial Hygienists. The hazard data were compiled from a wide variety of publications, including a privately circulated manual, and several other sources of information, as described in the sections which follow.

The Tables are organized to group health hazard data and references on left pages, and fire hazard data and related properties on right pages. For a flammable solvent, this arrangement would provide information in left to right sequence for:

hazards of low-level long-term exposures (Columns 3 and 4);

hazards of concentrated short-term exposures (Column 5);

hazards of higher concentrations where fire is a hazard (Column 8–11), and

properties which may be of use in assessing hazards or in evaluating methods for disposing of waste.

In some cases, hazard information is listed across one or more columns, in a general hazard category, when the information is from sources other than those listed for the column, or when the hazard needs to be described in terms other than that of the column heading.

Attention is called to the two different numbering systems used in Columns 5 and 7, and the possibility of adapting data from one column if similar data needs to be estimated for the other.

COLUMN 1. CHEMICAL NAME AND FORMULA

The chemical names chosen for listing hazard data in the tables conform as closely as possible with the system of nomenclature adopted by the International Union of Pure and Applied Chemistry and used by the Chemical Abstracts Service, American Chemical Society. Reference was made to the SOCMA Handbook of Commercial Organic Chemical Names published by the Chemical Abstracts Service of the American Chemical Society, and valuable advice on nomenclature was received from Dr. Mary A. Magill, Nomenclature Division of Chemical Abstracts Service, Dr. Ernest I. Becker, University of Massachusetts, and Dr. W. E. Noland, University of Minnesota.

Names are shown in the uninverted form, and some of the more common names are cross-referenced. The single intentional exception to Chemical Abstracts nomenclature is the use of common names for the pesticides and other economic poisons which are listed (chemicals not commonly used in the laboratory, but listed because they are included in the ACGIH Threshold Limit Values).

Chemical formulas are from the Handbook of Chemistry and Physics, the 1965 SOCMA Handbook of Commercial Organic Names or other listed sources.

COLUMN 2. CHEMICAL ABSTRACTS REGISTRY NUMBER

The Chemical Abstracts Registry Number shown was taken from the 1965 SOCMA Handbook of Commercial Organic Chemical Names or provided by the Chemical Abstracts Service. The unique Registry Number identifies organic compounds for which the structure is defined and serves to identify the compound throughout the CAS Registry System, a computer-file of structural information that is now being built. Even if nomenclature changes, the Registry number will remain the same.

COLUMN 3. A.C.G.I.H. THRESHOLD LIMIT VALUES 1966

Values for gases and vapors are given in parts of vapor or gas per million parts of air by volume at 25°C and 760 mm Hg, followed in parentheses by approximate milligrams per cubic meter of air. Values for particulates and aerosols are shown only in parentheses, in milligrams per cubic meter. As explained in detail below in the full text of Threshold Limit Values for 1966, reprinted by permission of the American Conference of Governmental Industrial Hygienists, the values are time-weighted average limits unless preceded by C, which is a recommended ceiling limit. The letter S, following values, stands for Skin and refers to the paragraph "Skin" Notation, and T stands for Tentative Values.

The Threshold Limit Values refer to air-borne concentrations of substances and represent conditions under which it is believed that nearly all workers may be exposed, day after day, without adverse effect. Because of wide variation in individual suscepti-bility, exposure of an occasional individual at or even below the threshold limit may not prevent discomfort, aggravation of a pre-existing condition, or occupational illness.

Threshold limits should be used as guides in the control of health hazards and should not be regarded as fine lines between safe and dangerous concentrations. Ex-ceptions are the substances given in Appendix A and certain of the substances given a "C" listing. The values not given a "C" listing refer to time-weighted average con-centrations for a normal workday. The amount by which these concentrations may be exceeded for short periods without injury to health depends upon a number of factors such as the nature of the contaminant, whether very high concentrations even for short periods produce acute poisoning, whether the effects are cumulative, the frequency with which high concentrations occur, and the duration of such periods. All must be taken into consideration in arriving at a decision as to whether a hazardous situation exists. Enlightened industrial hygiene practice inclines toward controlling exposures below the limit rather than maintenance at the limit.

Threshold limits are based on the best available information from industrial experience, from experimental human and animal studies, and when possible, from a combination of the three. The basis on which the values are established may differ from substance to substance; protection against impairment of health may be the guiding factor for some, whereas reasonable freedom from irritation, narcosis, nui-sance or other forms of stress may dominate the basis for others. The Committee holds to the opinion that limits based on physical irritation should be considered no less binding than those based on physical impairment; growing bodies of evidence indi-cate that physical irritation may promote and accelerate physical impairment.

DOCUMENTATION OF THRESHOLD LIMIT VALUES. A separate com-panion piece to the TLVs is issued by ACGIH under this title. This publication gives the pertinent scientific information and data with reference to literature sources, that were used to base each limit. Each documentation also contains a statement defining the type of response against which the limit is safeguarding the worker. For a better understanding of the TLVs it is essential that the Documentation be consulted when the TLVs are being used.

CEILING VS TIME-WEIGHTED AVERAGE LIMITS. Although the time-weighted average concentration provides the most satisfactory, practical way of moni-toring air-borne agents for compliance with the limits, there are certain substances for which it is inappropriate. In the latter group are substances which are predominantly fast acting and whose threshold limit is more appropriately based on this particular response. Substances with this type of response are best controlled by a ceiling "C" limit that should not be exceeded. It is implicit in these definitions that the manner of sampling to determine compliance with the limits for each group must differ; a single brief sample, that is applicable to a "C" limit, is not appropriate to the time-weighted limit; here, a sufficient number of samples are needed to permit a time-weighted average concentration throughout a complete cycle of operations or throughout the work shift.

Whereas the ceiling limit places a definite boundary which concentrations should not be permitted to exceed, the time-weighted average limit requires an explicit limit to the excursions that are permissible above the listed values. The magnitude of these excursions may be pegged to the magnitude of the threshold limit by an appropriate factor shown in Appendix C. It should be noted that the same factors are used by the Committee in making a judgement whether to include or exclude a substance for a "C" listing.

"SKIN" NOTATION. Listed substances followed by the designation "S" (Skin) refer to the potential contribution to the over-all exposure by the cutaneous route including mucous membranes and eye, either by air-borne, or more particularly, by direct contact with the substance. Vehicles can alter skin absorption. This attention-calling designation is intended to suggest appropriate measures for the prevention of cutaneous absorption so that the threshold limit is not invalidated.

MIXTURES. Special consideration should be given also to the application of these values in assessing the health hazards which may be associated with exposure to mixtures of two or more substances. A brief discussion of basic considerations involved in developing threshold limit values for mixtures, and methods for their development, amplified by specific examples are given in Appendix B.

"INERT" OR NUISANCE PARTICULATES. A number of dusts or particulates that occur in the working environment ordinarily produce no specific effects upon prolonged inhalation. Some insoluble substances are classed as inert; e.g., iron and steel dusts, cement, silicon carbide, titanium dioxide, cellulose; others may be soluble (starch, soluble oils, calcium carbonate) but are of such a low order of activity that in concentrations ordinarily encountered, do not cause physiologic impairment; still others may be rapidly eliminated or destroyed by the body (vegetable oils, glycerine, sucrose). In the case of the insoluble substances, there may be some accumulation in the respiratory passages. In the case of the soluble substances, this accumulation will ordinarily be temporary but may interfere to some extent with respiratory processes. Hence, it is desirable to control the concentrations of such particulates in the air breathed by any individual, in keeping with good industrial hygiene practice.

A threshold limit of $15mg/m^3$, or 50 mppcf, whichever is less, is recommended for substances in these categories and for which no specific threshold limits have been assigned. This limit, for a normal work day, does not apply to brief exposures at higher concentrations. Neither does it apply to those substances which may cause physiologic impairment at lower concentrations but for which a threshold limit has not yet been adopted. Some "inert" particulates are given in Appendix D.

SIMPLE ASPHYXIANTS—"INERT" GASES OR VAPORS. A number of gases and vapors, when present in high concentrations in air, act primarily as simple asphyxiants without other significant physiologic effects. A TLV may not be recommended for each simple asphyxiant because the limiting factor is the available oxygen. The minimal oxygen content should be 18 percent by volume under normal atmospheric pressure (equivalent to a partial pressure, pO_2 of 135 mm Hg). Atmospheres deficient in O_2 do not provide adequate warning and most simple asphyxiants are odorless. Several simple asphyxiants are listed in Appendix E. Some asphyxiants present an explosion hazard. Account should be taken of this factor in limiting the concentration of the asphyxiant.

These limits are intended for use in the field of industrial hygiene and should be interpreted and applied only by persons trained in this field. They are not intended for use, or for modification for use, (1) as a relative index of toxicity, by making a ratio of two limits, (2) in the evaluation or control of community air pollution or air pollution nuisances, (3) in estimating the toxic potential of continuous uninterrupted exposures, (4) as proof or disproof of an existing disease of physical condition, or (5) for adoption by countries whose working conditions differ from those in the U.S.A.

PHYSICAL FACTORS. It is recognized that such physical factors as heat, ultraviolet and ionizing radiation, humidity, abnormal pressure and the like may place added stress on the body so that the effects from exposure at a threshold limit may be altered. Most of these stresses act adversely to increase the toxic response of a substance. Although most threshold limits have built-in safety factors to guard against adverse effects of moderate deviations from normal environments, the safety factors of most substances are not of such a magnitude as to take care of gross deviations. For example, continuous work at temperatures above 90°F, or over-time, extending the work-week more than 50%, might be considered gross deviations. In such instances judgement must be exercised in the proper adjustments of the threshold limit values.

ANNUAL REVIEW. These values are reviewed annually by the Committee on Threshold Limits for revision or additions, as further information becomes available.

"NOTICE OF INTENT". At the beginning of each year, proposed actions of the Committee for the forthcoming year are issued in the form of a "Notice of Intent." This Notice provides not only an opportunity for comment, but solicits suggestions of substances to be added to the list. The suggestions should be accompanied by substantiating evidence for a tentative limit.

AS LEGISLATIVE CODE. The Conference does not consider the Threshold Limit Values appropriate matter for adoption in legislative codes and regulations, and recommends against such use. If, however, the list is so used, the intent of the concepts contained in the Preface should be maintained and provisions should be made to keep the list current.

REPRINT PERMISSION. The pamphlet "Threshold Limit Values for 1966" may be reprinted provided that written permission is obtained from the Secretary-Treasurer of the Conference and that it be published in its entirety.

TENTATIVE VALUES. Substances for which a "T" or tentative value is shown comprise those for which a limit has been assigned for the first time or for which a change in the "Recommended" listing has been made. In both cases, the assigned limits should be considered trial values that will remain in the tentative listing for a period of at least two years, during which time definitive evidence and experience is sought. If acceptable at the end of two years, these substances and values will be

Respirable Dusts Evaluated by Count

Substance: mppcf*

SILICA

Crystalline

Quartz, Threshold Limit calculated from the formula	$\dfrac{250**}{\%SiO_2 + 5}$
Cristobalite, Threshold Limit calculated from the formula	
Amorphous, including natural diatomaceous earth	20

SILICATES (less than 1% crystalline silica)

Asbestos	5
Mica	20
Soapstone	20
Talc	20
Portland Cement	50
GRAPHITE (natural)	15
"Inert" or Nuisance Particulates	50 (or 15 mg/m^3 which-
see Appendix D	ever is the smaller)

Conversion factors
   mppcf × 35.3 = Million particles per cubic meter
                = particles per c. c.

 * Millions of particles per cubic foot of air, based on impinger samples counted by light-field technics.
 ** The percentage of crystalline silica in the formula is the amount determined from air-borne samples, except in those instances in which other methods have been shown to be applicable.

moved to the RECOMMENDED list. Documentation for tentative values are available for each of these substances.

RADIOACTIVITY. For permissible concentrations of radioisotopes in air, see U. S. Department of Commerce, National Bureau of Standards, Handbook 69, "Maximum Permissible Body Burdens and Maximum Permissible Concentrations of Radionuclides in Air and in Water for Occupational Exposure," June 5, 1959. Also, see U. S. Department of Commerce National Bureau of Standards, Handbook 59, "Permissible Dose from External Sources of Ionizing Radiation," September 24, 1954, and addendum of April 15, 1958.

Appendix A

Specific information listed in Appendix A is given directly in the tables.

Appendix B; Threshold Limit Values For Mixtures

When two or more hazardous substances are present, their combined effect, rather than that of either individually, should be given primary consideration. In the absence of information to the contrary, the effects of the different hazards should be considered as additive. That, is if the sum of the following fractions,

$$\frac{C_1}{T_1} + \frac{C_2}{T_2} + \cdots \frac{C_n}{T_n}$$

exceeds unity, then the threshold limit of the mixture should be considered as being exceeded. C_1 indicates the observed atmospheric concentration, and T_1 the corresponding threshold limit, (See Example 1A.a.)

Exceptions to the above rule may be made when there is good reason to believe that the chief effects of the different harmful substances are not in fact additive, but independent as when purely local effects on different organs of the body are produced by the various components of the mixture. In such cases the threshold limit ordinarily is exceeded only when at least one member of the series $\left(\frac{C_1}{T_1} \text{ or } \frac{C_2}{T_2} \text{etc.}\right)$ itself has a value exceeding unity, (See Example 1A.b.)

Antagonistic action or potentiation may occur with some combinations of atmospheric contaminants. Such cases at present must be determined individually. Potentiating or antagonistic agents are not necessarily harmful by themselves. Potentiating effects of exposure to such agents by routes other than that of inhalation is also possible, e.g. imbibed alcohol and inhaled narcotic (trichloroethylene). Potentiation is characteristically exhibited at high concentrations, less probably at low.

When a given operation or process characteristically emits a number of harmful dusts, fumes, vapors or gases, it will frequently be only feasible to attempt to evaluate the hazard by measurement of a single substance. In such cases, the threshold limit used for this substance should be reduced by a suitable factor, the magnitude of which will depend on the number, toxicity and relative quantity of the other contaminants ordinarily present.

Examples of processes which are typically associated with two or more harmful atmospheric contaminants are welding, automobile repair, blasting, painting, lacquering, certain foundry operations, diesel exhausts, etc. (Example 2.)

Examples

1A.   General case, where air is analyzed for each component.
    a. ADDITIVE EFFECTS

$$\frac{C_1}{T_1} + \frac{C_2}{T_2} + \frac{C_3}{T_3} + \cdots \frac{C_N}{T_N} = 1$$

Air contains 5 ppm of carbon tetrachloride (TLV, 10), 20 ppm of ethylene dichloride (TLV, 50) and 10 ppm of ethylene dibromide, (TLV, 25).

$$\frac{5}{10} + \frac{20}{50} + \frac{10}{25} = \frac{65}{50} = 1.3$$

Threshold limit is exceeded.

b. INDEPENDENT EFFECTS

Air contains 0.15 mg/m^3 of lead (TLV, 0.2) and 0.7 mg/m^3 of sulfuric acid (TLV, 1).

$$\frac{0.15}{0.20} = 0.75; \qquad \frac{0.7}{1} = 0.7$$

Threshold limit is not exceeded.

1B.   Special case when source of containment is a mixture and atmospheric composition is assumed similar to that of original material, i.e. vapor pressure of each component is the same at the observed temperature.

a. ADDITIVE EFFECTS, approximate solution.

  1. A mixture of equal parts (1) trichloroethylene (TLV, 100), and (2) methyl chloroform (TLV, 350).

$$\frac{C_1}{100} + \frac{C_2}{350} = \frac{Cm}{Tm}$$

Solution applicable to "spot" solvent mixture usage, where all or nearly all, solvent evaporates.

$$C_1 = C_2 = \tfrac{1}{2}\,Cm$$

$$\frac{C_1}{100} + \frac{C_1}{350} = \frac{2C_1}{Tm}$$

$$\frac{7C_1}{700} + \frac{2C_1}{700} = \frac{2C_1}{Tm}$$

$$Tm = 700 \times \tfrac{2}{9} = 155 \text{ ppm}$$

1B.b. General Exact Solution for Mixtures of N Components With Additive Effects and Different Vapor Pressures.

(1) $\dfrac{C_1}{T_1} + \dfrac{C_2}{T_2} + \cdots + \dfrac{C_N}{T_N} = 1;$

(2) $C_1 + C_2 + \cdots + C_N = T;$

(2.1) $\dfrac{C_1}{T} + \dfrac{C_2}{T} + \cdots + \dfrac{C_N}{T} = 1.$

By the Law of Partial Pressures,

(3) $C_1 = ap_1,$

and by Raoult's Law,

(4) $p_1 = F_1 p_1{}^\circ.$

Combine (3) and (4) to obtain

(5) $C_1 = aF_1 p_1{}^\circ.$

Combining (1), (2.1) and (5), we obtain

(6) $\dfrac{F_1 p_1{}^\circ}{T} + \dfrac{F_2 p_2{}^\circ}{T} + \cdots + \dfrac{F_N p_N{}^\circ}{T} =$

$\dfrac{F_1 p_1{}^\circ}{T_1} + \dfrac{F_2 p_2{}^\circ}{T_2} + \cdots + \dfrac{F_N p_N{}^\circ}{T_N}$

and solving for T,

(6.1) $T = \dfrac{F_1 p_1{}^\circ + F_2 p_2{}^\circ + \cdots + F_N p_N{}^\circ}{\dfrac{F_1 p_1{}^\circ}{T_1} + \dfrac{F_2 p_2{}^\circ}{T_2} + \cdots + \dfrac{F_N p_N{}^\circ}{T_N}}$

or

(6.2) $T = \dfrac{\sum\limits_{i=1}^{i=n} F_1 p_1{}^\circ}{\sum\limits_{i=1}^{i=n} \dfrac{F_1 p_1{}^\circ}{T_1}}$

T = Threshold Limit Value in ppm.

C = Vapor concentration in ppm.

p = Vapor pressure of component in solution.

p° = Vapor pressure of pure component.

F = Mol fraction of component in solution.

a = A constant of proportionality. Subscripts 1, 2, ... n relate the above quantities to components 1,2,...*n*, respectively. Subscript *i* refers to an arbitrary component from 1 to *n*. Absence of subscript relates the quantity to the mixture.

Solution to be applied when there is a reservoir of the solvent mixture whose composition does not change appreciably by evaporation.

Exact Arithmetic Solution of Specific Mixture

	Mol. wt.	Density	T	$p°$ at 25°C	Mol Fraction in half-and half solution by volume
Trichloroethylene (1)	131.4	1.46 g/ml	100	73 mm Hg	0.527
Methyl chloroform (2)	133.42	1.33 g/ml	350	125mm Hg	0.473

$$F_1 p_1° = (0.527)(73) = 38.2$$
$$F_2 p_2° = (0.473)(125) = 59.2$$

$$T = \frac{38.2 + 59.2}{\frac{38.2}{100} + \frac{59.2}{350}} = \frac{(97.4)(350)}{133.8 + 59.2} = \frac{(97.4)(350)}{193.0} = 177$$

$T = 177$ ppm   (Note difference in TLV when account is taken of vapor pressure and mole fraction in comparison with above example where such account is not taken).

2. A mixture of one part of (1) parathion (TLV, 0.1) and two parts of (2) EPN (TLV, 0.5).

$$\frac{C_1}{0.1} + \frac{C_2}{0.5} = \frac{Cm}{Tm} \; ; C_2 = 2C_1 \qquad\qquad \frac{7C_1}{0.5} = \frac{3C_1}{Tm}$$

$$Cm = 3C_1$$

$$\frac{C_1}{0.1} + \frac{2C_1}{0.5} = \frac{3C_1}{Tm} \qquad\qquad Tm = \frac{1.5}{7} = 0.21 \text{ mg/m}^3$$

1C. TLV for Mixtures of Mineral Dusts.

For mixtures of biologically active mineral dusts the general formula for mixtures may be used. With the exception of asbestos, pure minerals are assigned TLV of 2.5, 20 or 50.

For a mixture containing 80% talc and 20% quartz, the TLV for 100% of the mixture "C" is given by:

$$TLV = \frac{1}{\frac{0.8}{20} + \frac{0.2}{2.5}} = 8.4 \text{ mppcf}$$

Essentially the same result will be obtained if the limit of the more (most) toxic component is used provided the effects are additive. In the above example the limit for 20% quartz is 10 mppcf.

For another mixture of 25% quartz 25% amorphous silica and 50% talc:

$$TLV = \frac{1}{\frac{0.25}{2.5} + \frac{0.25}{20} + \frac{0.5}{20}} = 7.3 \text{ mppcf}$$

The limit for 25% quartz approximates 8 mppcf.

Appendix C, Bases For Assigning Limiting "C" Values

By definition a listed value bearing a "C" designation refers to 'ceiling' value that should not be exceeded; all values should fluctuate below the listed value. In general the bases for assigning or not assigning a "C" value rest on whether excursions of concentration above a proposed limit for periods up to 15 minutes may result in a) intolerable irritation, b) chronic, or irreversible tissue change, or c) narcosis of sufficient degree to increase accident proneness, impair self rescue or materially reduce work efficiency.

In order for the Committee to decide whether a substance is a candidate for a "C" listing, some guidelines must be formulated on the permissive fluctuation above the limit in terms of the seriousness of the response in the categories a, b, c, given above. For this the factors given in the table below have been used by the Committee. For both technical and practical reasons, the factors have been pegged to the concentration in an inverse manner. It will be noted that as the magnitude of the TLV increases a correspondingly decreased range of fluctuation is permitted; not to decrease the factor for TLV's of increasing magnitude would permit exposures to large absolute quantities, an undesirable condition, a condition that is minimized at low TLV's. Moreover, larger factors at the lower TLV's are consistent with the difficulties in analyzing and controlling trace quantities.

TLV RANGE ppm* or mg/m^3	Test TLV Factor	Examples
0 to 1	3	Toluene diisocyanate-TLV, 0.02 ppm, if permitted to rise above 0.06 ppm may result in sensitization in a single subsequent exposure. "C" listing recommended on category b.
1 + to 10	2	Manganese-TLV, 5mg/m^3, contains little or no safety factor. All values should fluctuate below 5mg/m^3. "C" listing recommended on category b.
10 + to 100	1.5	Methyl styrene-TLV 100 if encountered at levels of 150 ppm will prove intensely irritating, "C" listing recommended on category a.
100 + to 1000	1.25	Methyl chloroform-TLV 350ppm, at 438 ppm for periods not exceeding 15 minutes is not expected to result in untoward effects relating to category c. No "C" listing recommended.

*Whichever unit is applicable

Permissible Excursions for Time-Weighted Average (TWA) Limits

As stated in the preface, the same factors may be used as guides for reasonable excursions *above* the limit for substances to which the time-weighted average applies. The time-weighted average implies that each excursion *above* the limit is compensated by a comparable excursion below the limit. Thus, a value of 6 ppm of HF is permissible for periods not exceeding 15 minutes, provided an equivalent decrease below the limit of 3 ppm obtains.

Appendix D

Some "Inert" or Nuisance Particulates*

Alundum (A1$_2$0$_3$)	Graphite (synthetic)	Marble
Calcium carbonate	Gypsum	Plaster of Paris
Cellulose	Vegetable oil mists (except castor,	Rouge
Portland Cement	cashew nut, or similar irritant	Silicon Carbide
Corundum (A1$_2$0$_3$)	oils)	Starch
Emery	Limestone	Sucrose
Glycerine Mist	Magnesite	Tin Oxide
		Titanium Dioxide

*When toxic impurities are not present.

Appendix E

Some Simple Asphyxiants—"Inert" Gases and Vapors.

Acetylene	Ethane	Helium	Methane	Nitrogen
Argon	Ethylene	Hydrogen	Neon	Nitrous Oxide

Threshold Limit Committee of the American Conference of Governmental Industrial Hygienists

COLUMN 4. PRINCIPAL EFFECTS OF INHALATION EXPOSURES ABOVE THRESHOLD LIMIT VALUES

The information shown in this column is the personal judgment of Dr. Ralph G. Smith, Wayne State University, a member of the Threshold Limit Value Committee of the American Conference of Governmental Industrial Hygienists.

The purpose of this column is to help users realize that Threshold Limit Values cannot be compared on a numerical basis if the principal of first major effects of overexposure are of different types. A second purpose of this column is to help users appreciate the effects of exposures above the Threshold Limit Values, and the value of referring to the full details in the Documentation of Threshold Limit Values.

In almost every case these conclusions were arrived at by referring to the Documentation of Threshold Limit Values, Revised Edition, Copyright 1966 by American Conference of Governmental Industrial Hygienists, and usually were taken from the last sentence of the justification, which ordinarily gives a summary of the basis for the value selected. It is obvious that any selection is somewhat arbitrary and many substances may be toxic, irritating, and produce narcosis, etc., but it is believed that the designation selected is most appropriate.

TOXIC. The principal action of the substance is damage to some essential function, usually chronic effects produced by repeated exposures in excess of TLV concentrations.

IRRITANT. Principal action is irritation of tissue with which the substance comes into contact. This includes the entire respiratory system, or skin or eyes.

NUISANCE. At concentration several times the TLV no known systematic effects or irritation is produced. Most substances are considered physiologically inert.

CARCINOGENIC. Usual definition, substance is definitely known to produce cancer.

FUME FEVER. The principal action of the substance is the production of a fever following initial or interrupted exposure which ordinarily clears up within 24 hours.

NARCOSIS. Depression of the central nervous system tending to produce sleep or unconsciousness.

COLUMN 5. RELATIVE HAZARD TO HEALTH FROM CONCENTRATED SHORT-TERM EXPOSURES

The system used for describing Relative Hazard to Health is based on the privately circulated "Manual of Hazards to Health from Chemicals" compiled by Henry F. Smyth, Jr. in cooperation with the Medical Department, Union Carbide Corporation, and the Chemical Hygiene Fellowship, Mellon Institute, Pittsburgh, Pennsylvania. The manual is copyrighted by Union Carbide Corporation and information is reprinted here with permission.

(Statements from other sources are listed in this column, without hazard grades or in parentheses.)

The information in the tables is intended for the use of people who have responsibilities for teaching, research operations, engineering, and manufacturing. It should be used as a guide to assist in planning equipment, handling methods, and protective devices and procedures.

Hazard is defined as the probability that injury will result from a particular way of using a material. It is distinct from the toxicity of the material, defined as the capacity of the material to produce injury. Statements about hazard are meaningful only when the contemplated manner of use of a material is described. The judgements expressed herein refer to health hazards in chemical procedures which are conducted under informed supervision and with reasonable care. They consider only accidental single exposures as opposed to repeated or substantially continuous exposures.

When crude mixtures are being handled the hazards may be greater than for the partially purified or substantially pure materials considered herein.

The estimates in the tables are based upon experiments with small animals modified by the experience of industrial physicians. The judgments expressed are necessarily tentative and subject to revision as experience accumulates.

The listing of relative health hazards is *not* a good source for toxicological data. Many entries are predictions based on structural analogies, not measurements. The five ranges of relative hazard which are listed are too broad to indicate differences which may be significant for some judgments, and at times have been modified for unusual physical properties, or to incorporate human experience. In particular, the entries do not predict the safety of technical or consumer products, where contacts may be quite different from those of laboratory operations.

Specific suggestions for safe handling methods, first aid, or therapy are the responsibility of industrial hygiene or medical personnel.

Symbols Used to Describe Relative Hazard to Health

1. No residual injury is to be expected from accidental exposure even if no treatment is applied.
2. Minor residual injury may result from some accidental exposures if no treatment is applied.
3. Minor residual injury may result in spite of prompt treatment.
4. Major residual injury may result in spite of prompt treatment.
5. Major residual injury is likely in spite of prompt treatment.
x. Entry based upon analogy with a closely similar structure, or on other estimate believed sound.
y. Entry based on a non-standard test (a test different from the usual test described for each type of contact).
z. Entry based upon human experience, superseding animal data.
* See supplemental note.

The exposures for which relative hazards are predicted are those of accidental brief contact which may occur, but are not inevitable, during operations in the laboratory use of chemicals. They are essentially "once in a lifetime" contacts; that is to say, it is expected that sufficient time will elapse after one contact so that the exposed person will have returned completely to his initial state before the next contact. When the potentiality of a material for chronic toxicity or sensitization is known to make such a return unusually slow, a supplemental note is appended to the relative hazard symbol.

It must be borne in mind that the relative hazard shown is that of a material meeting specifications for purity. The hazard of a less pure grade may be different.

Blanks in the tabulation indicate either that there is insufficient information to justify an entry, or that physical properties of material make the particular contact very unlikely.

Information Supplemental to the Relative Hazard Grades

The tabulated numbers do not communicate all of the pertinent judgments about the relative health hazards of handling the listed materials, because (1) some of the entries are not based on the standardized laboratory tests, described below, hence their validity differs from those based on standard tests, (2) some of the materials may cause conspicuous effects which may be alarming but are unlikely to be injurious, (3) some of the materials are so toxic that even the handling precautions appropriate to hazard grade 5 materials are inadequate for protection, and (4) some of the materials may cause injurious effects, particularly after prolonged or repeated contact, which dictate more careful handling than would be justified by the hazard grade, which is based upon injuries likely after a single contact.

The coding used to describe relative hazard allows one to form some idea of the validity of the judgments, and is intended for the guidance of medical and industrial hygiene personnel. The symbol 'x' is shown where the judgment is based only upon analogy with a material of closely similar structure, 'y' is shown where physical characteristics of the material, or other reasons, forced reliance on a non-standard experimental test, and 'z' is shown when experience with humans has yielded a judgment of hazard differing from that based on animal experiment. All entries not including 'x', 'y', or 'z' are based on the standardized tests on laboratory animals described below.

All of the statements of supplemental information are reproduced below, in alphabetical order, with such elaboration as appears desirable. The statements usually include the word "may" because effects of materials are related to quantity and time of contact; below some quantity, no effect results. When the word "contact" appears in a statement, it may refer to swallowing, inhalation, skin, or eye contact depending upon which of the five routes of contact bears the asterisk.

The particular effect indicated may never have been known to have been caused by the material, but is shown because of structural analogies with a known causative material. It is impossible to be certain that information is shown for every material which may produce a significant effect.

Supplemental Effects

ACNE-LIKE ERUPTION. Acne-like eruption may follow prolonged contact. This effect is unlikely to be an injury, but the eruption may leave permanent scar tissue when it heals.

CARCINOGENIC IN ANIMALS. Cancer has resulted in animals; prolonged contact may affect humans. It is not certain that the experimental production of cancer in animals indicates that human cancer can result, but it is prudent to reduce human contact to the minimum possible.

CHRONIC TOXICITY. Repeated contact is more hazardous. The tabulated hazard grades are based upon acute toxicity, the result of a single contact separated from the next contact by a sufficient time so that the two are not additive. The entry marks those materials from which an unperceived effect results from one contact, sufficiently persistent so that contact a day or more later adds, and eventually chronic poisoning may result; and structural analogs of such materials.

DEATH/HOURS INHALATION. Prolonged breathing of high concentration has killed a human. This entry might be applied to almost any highly volatile material, but is reserved for one of only moderate volatility whose unusual metabolic pathway results in widely different kinds of effects, depending on quantity contacted. Here, prolonged means a few hours.

BREATH/ODOR RELIANCE. Odor is unreliable warning. This entry would be justified for every odorous material, but it is reserved for materials where it is known that death has resulted from relying upon odor to warn of danger.

EXTREME TOXICITY. So toxic that a microdrop in the eye has killed a rabbit. The quantity required has been as little as 0.005 ml, penetration has been so rapid that death has resulted within a few minutes. Materials so marked call for greater handling precautions than do the typical hazard grade 5 material.

EYE BURN WITHOUT PAIN. Eye burn possible with no pain at time of contact. So-called delayed eye burns are particularly likely to result in permanent disability, because absence of pain at the time of contact reduces the likelihood that eye-washing will be undertaken early enough to limit eye injury. Contact with either vapor or fluid has injured.

EYE PIGMENT. Long repeated contact may result in permanent eye pigmentation. The effect referred to has been noted only after several years of daily contact with airborne material. It alters a person's appearance, but has not resulted in disability.

GAS OR VAPOR PENETRATES SKIN. Gas or vapor penetrates the skin freely. This entry marks air borne materials for which impervious protective clothing may be as necessary to an exposed person as is a respirator.

IRRITATION EYE, NOSE, THROAT. Irritation of eye, nose, and throat may result. Ordinarily the irritation referred to constitutes nothing more than discomfort, with injury not probable unless the contact is sufficient so that the discomfort is almost intolerable, or unless the contact is prolonged or often repeated. Irritation which is usually trivial, such as may be expected from air borne dust no matter how bland, is not indicated.

LUNG INJURY FROM PROLONG. Prolonged contact may injure lung. Here, prolonged means several years.

METAL FUME FEVER. Metal fume fever may follow breathing very fine fume. This effect results from a physical action in the lung, not a toxic action. It may be temporarily disabling, but has not produced permanent disability.

NOSE AND LUNG INJURY. Prolonged contact may injure nose and lung. Here, prolonged means a few weeks for the nose, many years for the lung.

PERMANENT INJURY. Permanent injury from non-fatal contact increases hazard. The materials bearing this entry have permanently disabled persons who appeared to recover, with the help of medical care, from the severe effect of a single contact.

PHOTOSENSITIZATION. Photosensitization may follow exposure. Materials so marked have left a victim temporarily so sensitive to sunlight that he has received a dangerous sunburn when an unexposed companion was unaffected.

POLYMER FUME FEVER. Polymer fume fever may follow breathing very fine fume. See "metal fume fever" above.

REACTION WITH HCl—HAZARD 5. Reaction with hydrochloric acid yields toxic irritant of hazard 5 for all contacts. Because of the hazard of the reaction product, any mixture with any amount of free hydrochloric acid at any temperature, should be considered to be in hazard grade 5.

SENSITIZATION OF SKIN, RESP. TRACT. Sensitization of skin or respiratory tract may result. Materials so marked may elicit in the skin of many people an allergic sensitization so that later contact with a very small quantity results in annoying or disabling dermatitis; or in the respiratory tract so that later contact results in disabling or fatal asthma-like interference with breathing. It is said that any material whatsoever can sensitize someone. Only those known to have sensitized a considerable proportion of frequent contactors, and close structural analogs of these, are shown.

SKIN STAIN. Harmless stain may result (color is often specified). This entry is not used for dyes, whose staining ability should be obvious. It is used for substances which may react with skin to yield a colored product, after contacts which are not sufficient to cause injury. Ordinarily the color is removed only by normal replacement of old skin cells by new ones.

TEMPORARY VISUAL DISTURBANCE. Temporary visual disturbance may result. The disturbance referred to had not been permanent, but it may be alarming and shows unwisely excessive contact.

WOUNDS NEED MEDICAL CARE. Wounds require prompt medical attention to facilitate orderly healing. Unless every fragment is removed from even a trivial wound, healing may not take place for months.

Frequent Phenomena not Covered by Supplemental Information

All halogenated materials pyrolyze in varying degree in intense ultraviolet, electric discharges, flames, or on hot surfaces, to yield lung injuring gases of hazard grade 5.

In this manual skin irritation hazards refer to contact of materials with uncovered skin, such as hands or cheeks. All materials are more irritating when they are held against the skin by a covering, such as clothing or finger rings. Covering will increase the hazard grade by one or two units. The increase is most noticeable with volatile materials, it is most likely to produce disabling results when feet within shoes are contacted.

All materials are more irritating to the skin and more injurious to the eyes when they are above room temperature.

All compressed gases and fluids boiling below room temperature are capable of injuring the skin by freezing. In extreme instances this frost bite may result in very severe permanently disabling injury.

Hazard grades for breathing refer to vapors from substances at room temperature. Breathing vapor from hot material (or material aerosolized by high speed machinery, or by chilling hot vapors) may involve higher concentrations with or without degradation products, hence will be more hazardous.

A large number of materials, possibly essentially all, are more hazardous to swallow soon after alcoholic beverages have been consumed.

Most of the pairs of materials tested upon animals have been found to be additive. It is not known that contacts with mixtures which are likely in the groups using this manual are more hazardous than one would expect from the sum of the materials. This does not mean that when two materials in hazard grade 1 are mixed, the hazard grade of the mixture is 2. For a variety of reasons it is likely to be grade 1.

Experimental Results Used to Predict Relative Hazard

When human experience with a material has not been sufficient to allow estimates of relative hazard of general industrial handling, observations upon experimental laboratory animals have been relied upon. The definitions of relative hazard in terms of experimental observations are shown below.

EYE CONTACT. The most severe injury to the cornea of the eyes of five male albino rabbits following instillation of various volumes of undiluted fluid material, or of an excess of solutions of a material in glycol, deobase or water. (Excess, as used below, means the greatest amount which can reach the eye by accident. In the rabbit test an excess is 0.5 ml.)

Relative hazard 1.	0.5 ml (an excess) does not cause severe injury.
Relative hazard 2.	0.005 ml does not cause severe injury.
Relative hazard 3.	0.005 ml or excess 40% solution causes severe injury.
Relative hazard 4.	Excess 5% solution causes severe injury.
Relative hazard 5.	Excess 1% solution causes severe injury.

When solubilities have limited the concentrations applied to the rabbit eye, or when the material was applied only as a solid, the letter 'y' indicates that correction has been applied to make the prediction consistent with the above.

BREATHING VAPORS. Mortality among 6 male albino rats weighing 90 to 120 grams, inhaling vapors substantially saturated at room temperature and observed for 14 days thereafter.

Relative hazard 1.	8 hours inhalation kills 0, 1, 2 or 3 of 6 rats
Relative hazard 2.	2 or 4 hours inhalation kills 2, 3 or 4 of 6 rats
Relative hazard 3.	1/4 to 1 hour inhalation kills 2, 3 or 4 of 6 rats
Relative hazard 4.	2 or 5 minutes inhalation kills 2, 3 or 4 of 6 rats
Relative hazard 5.	2 minutes inhalation kills 5 of 6 rats

The test described is strictly a test of hazard, the likelihood that a material free to evaporate at room temperature, will produce a vapor concentration which will be injurious to breathe for a short period. The relative hazard of breathing dusts and some volatile materials has been estimated from other tests, believed to be consistent with the above. When the hazard is based on mortality among 6 rats weighing 90 to 120 grams inhaling known vapor concentrations for 4 hours and observed for 14 days thereafter, the experimental results are:

Relative hazard 1y.	4 hours inhalation of 128,000 ppm kills 0 to 4 of 6 rats
Relative hazard 2y.	4 hours inhalation of 16,000 ppm kills 0 to 4 of 6 rats.
Relative hazard 3y.	4 hours inhalation of 2,000 ppm kills 0 to 4 of 6 rats.
Relative hazard 4y.	4 hours inhalation of 250 ppm kills 0 to 4 or 6 rats.
Relative hazard 5y.	4 hours inhalation of 250 ppm kills 5 to 6 of 6 rats.

SKIN PENETRATION. LD_{50} is the dosage killing half of a group of male albino rabbits weighing about 3 kilograms, within 14 days following administration by 24 hours' contact with about 40% of the body surface.

Relative hazard 1.	LD_{50} more than 20 ml/kg. body weight.
Relative hazard 2.	LD_{50} 2 to 20 ml/kg.
Relative hazard 3.	LD_{50} 0.2 to 1.99 ml/kg.
Relative hazard 4.	LD_{50} 0.02 to 0.19 ml/kg.
Relative hazard 5.	LD_{50} less than 0.02 ml/kg.

IRRITATION OF UNCOVERED SKIN. The most severe reaction observed following contact of 0.01 ml. of undiluted chemical, or of a 40% solution of a chemical, with the clipped belly of 5 male albino rabbits

Relative hazard 1.	Undiluted causes only capillary injection
Relative hazard 2.	Undiluted causes only slight erythema
Relative hazard 3.	Undiluted causes erythema and slight edema
Relative hazard 4.	Undiluted causes necrosis
Relative hazard 5.	10% solution causes necrosis

When the material has not been applied to the rabbit belly in undiluted form or when a somewhat different test of irritation was relied upon, the letter 'y' is appended to the numerical hazard grade. Correction in such non-standard tests has been made for the estimated reduction in irritation due to dilution, so that the entries are extrapolated to irritation expected from undiluted material.

It is important to stress the fact that irritation and all other effects on the skin are more severe when a material is covered, as when it contacts the skin on contaminated clothing. This increase is most dramatic when the material is between the skin and impervious gloves, or when it is inside shoes.

SWALLOWING. LD_{50} is the dosage killing half of a group of male albino rats weighing 90 to 120 grams, within 14 days following administration through a stomach tube. In general a smaller dosage would be required to kill a man.

Relative hazard 1.	LD_{50} more than 10 gm/kg. body weight
Relative hazard 2.	LD_{50} 1 to 10 gm/kg
Relative hazard 3.	LD_{50} 0.1 to 0.99 gm/kg
Relative hazard 4.	LD_{50} 0.01 to 0.099 gm/kg
Relative hazard 5.	LD_{50} less than 0.01 gm/kg

COLUMN 6. REFERENCES TO SUPPLEMENTARY INFORMATION ON TOXICITY, FLAMMABILITY AND OTHER HAZARDS

The column on supplementary information lists a variety of sources which provide more detailed hazard or toxicity information, and the abbreviations used are listed below:

A three or four-digit number is a page reference to toxicity data in "Industrial Hygiene and Toxicology," Volume II, 2nd Revised Edition, edited by Frank A. Patty, David W. Fassett and Don D. Irish, published by John Wiley and Sons, New York, 1963. This 2,377 page book represents the outstanding compilation of detailed toxicological data from chemical manufacturers, and from academic and government laboratories.

AIHA—indicates a Hygienic Guide is available from the American Industrial Hygiene Association. (Examples are reprinted, shown on pages 437–442 with permission.)

MCA—indicates there is a Chemical Safety Data Sheet available from the Manufacturing Chemists Association.

NSC—indicates a data sheet available from the National Safety Council.

HCD—indicates there is additional information in Hazardous Chemicals Data, (NFPA No. 49-1966) published by the National Fire Protection Association.

MGD—indicates reference to data in the Matheson Gas Data Book. Single data sheets are available on request from the Matheson Company.

HCP—indicates the hazard data was found in the Handbook of Chemistry and Physics, edited by Robert C. Weast, published by the Chemical Rubber Company, Cleveland, Ohio, 1966.

Merck—indicates the hazard data was found in the Merck Index, 7th Ed., edited by Paul G. Stecher, published by Merck & Co. Inc., Rahway, N.J., 1960.

STTE—indicates the material and possible treatment is listed in Symptomatology and Therapy of Toxicological Emergencies by William B. Deichmann and Horace W. Gerarde, published by Academic Press, New York and London, 1964.

Other data entered is from the open literature or from privately circulated industrial information sheets.

COLUMN 7.   N.F.P.A. HAZARD IDENTIFICATION SIGNALS

The hazard signals listed were established by the National Fire Protection Association as set forth in their publication, "Identification System for Fire Hazards of Materials," 704-M, 1966, which identifies hazards as follows:

Identification of Health Hazard

(Color Code: Blue)
4. Materials which on very short exposure could cause death or major residual injury even though prompt medical treatment were given.
3. Materials which on short exposure could cause serious temporary or residual injury even though prompt medical treatment were given.
2. Materials which on intense or continued exposure could cause temporary incapacitation or possible residual injury unless prompt medical treatment is given.
1. Materials which on exposure would cause irritation but only minor residual injury even if no treatment is given.
0. Materials which on exposure under fire conditions would offer no hazard beyond that of ordinary combustible materials.

Identification of Flammability

Susceptibility of Materials to Burning:
(Color Code: Red)
4. Materials which will rapidly or completely vaporize at atmospheric pressure and normal ambient temperature, or which are readily dispersed in air and which will burn readily.
3. Liquids and solids that can be ignited under almost all ambient temperature conditions.
2. Materials that must be moderately heated or exposed to relatively high ambient temperature before ignition can occur.
1. Materials that must be preheated before ignition can occur.
0. Materials that will not burn.

Identification of Reactivity, (Stability)

Susceptibility to Release of Energy:
(Color Code: Yellow)
4. Materials which are readily capable of detonation or of explosive decomposition or reaction at normal temperatures and pressures.
3. Materials which are capable of detonation or explosive reaction but require a strong initiating source or which must be heated under confinement before initiation or which react explosively with water.
2. Materials which are normally unstable and readily undergo violent chemical change but do not detonate. Also materials which may react violently with water or which may form potentially explosive mixtures with water.

1. Materials which are normally stable, but which can become unstable at elevated temperatures and pressures or which may react with water with some release of energy but not violently.
0. Materials which are normally stable, even under fire exposure conditions, and which are not reactive with water.

Special Symbols:
  P. Polymerizes
  W̶. Reacts violently with water
    Although these are listed in the column on flammability, they would appear in a separate place on a special label.

COLUMN 8.   EXTINGUISHING AGENTS

The information on extinguishing agents was taken from the Handbook of Industrial Loss Prevention with supplementary data supplied by the Ansul Company.

The code numbers indicate the following extinguishing agents:

1. Water (see Note A below)
2. Foam
2a. Alcohol foam
3. Carbon dioxide or dry chemical
4. See Note B below (gas fires)
5. Approved dry compound for metal fires

Note A: Water discharged from spray nozzles and standard sprinklers is finely divided and absorbs much more heat than an equal weight of water from a hose stream or an equal weight of large drops from an old-type sprinkler. The old type will extinguish fires in liquids having a flash point above 93°C (200°F) while water from spray nozzles or standard sprinkler heads will usually extinguish fire in liquids having flash points above 66°C (150°F.)

For lower-flash-point liquids, these devices reduce the intensity of burning and protect buildings and equipment from severe damage, even though they will not extinguish the fire. Automatic sprinklers are the basic protection for all indoor flammable-liquid areas, even where special fixed extinguishing systems utilizing other extinguishing agents have been installed to protect against a specific hazard.

Note B: While carbon dioxide, dry chemical, and in some instances water spray may be used to extinguish small gas fires, these agents are not generally recommended in gas fires because the discharge of gas or volatile liquid will continue unless shut off promptly and may create a more serious explosion hazard. Generally, the best procedure is to use water to keep the surroundings cool until the leak can be shut off or until the volatile liquid has completely burned.

COLUMN 9.   FLAMMABLE LIMITS IN AIR, % BY VOLUME

In the case of gases or vapors which form flammable mixtures with air or oxygen, there is a minimum concentration of vapor in air or oxygen below which propagation of flame does not occur on contact with a source of ignition. Gases and vapors may form flammable mixtures in atmospheres other than air or oxygen, as for example hydrogen in chlorine. There is also a maximum proportion of vapor or gas in air above which propagation of flame does not occur. These boundary-line mixtures of vapor or gas with air, which if ignited will just propagate flame, are known as the "lower and upper flammable or explosive limits," and are usually expressed in terms of percentage by volume of gas or vapor in air. (Flammable limits will differ greatly in oxygen and oxygen-enriched air.)

In popular terms, a mixture below the lower flammable limit is too "lean" to burn or explode and a mixture above the upper flammable limit too "rich" to burn or explode.

The flammable limit figures given in this Table are based upon normal atmospheric temperatures and pressures, unless otherwise indicated. There may be considerable variation in flammable limits at pressures or temperatures above or below normal. The general effect of increase of temperature or pressure is to lower the lower limit and raise the upper limit. Decrease of temperature or pressure has the opposite effect.

COLUMN 10. FLASH POINT

Flash point of a liquid is the temperature at which it gives off vapor sufficient to form an ignitible mixture with the air near the surface of the liquid or within the vessel used. By "ignitible mixture" is meant a mixture within the explosive range (between upper and lower limits) that is capable of the propagation of flame away from the source of ignition when ignited. By propagation of flame is here meant the spread of flame from layer to layer independently of the source of ignition. A gas or vapor mixed with air in proportions below the lower limit of flammability may burn at the source of ignition, that is, in the zone immediately surrounding the source of ignition, without propagating (spreading) away from the source of ignition. Some evaporation takes place below the flash point when vapor does not go off freely enough to meet flash point classification requirements. This term applies mostly to flammable liquids, although there are certain solids, such as camphor and naphthalene, that slowly evaporate or volatilize at ordinary room temperature and therefore have flash points while still in the solid state.

(Flash points usually measure liquid temperatures for downward propagation of flame, but upward propagation may occur at slightly lower temperatures.)

The flash point figures represent closed cup tests except where the open cup flash point is designated by the initials "OC" following the figure. Open cup flash points, determined in a different type of testing apparatus, are usually somewhat higher than the closed cup flash point figures for the same substances. Closed cup flash point figures are commonly used in determining the classification of liquids which flash in the ordinary temperature range, but for certain materials which have relatively high flash points, the open cup flash point testing is often preferred. In the case of some of the older figures quoted in this Table, there are no data to indicate whether the figures are closed or open cup tests.

Flash point values listed are reprinted with permission from the National Fire Protection Association, with many additional values supplied by the Technical Safety Laboratory of the Eastman Kodak Company.

There are several types of apparatus for determining flash point by test. The NFPA Flammable liquids Code specifies the Tag Closed Tester as authoritative in case of dispute. This tester, intended for testing liquids with a flash point below 79°C (175°F), is described by ASTM specifications D56. The Pensky-Martens Closed Tester (ASTM D93) is considered accurate for testing liquids having flash points between 66°C (150°F.) and 110°C (230°F.), and is used largely for determination of flash points of fuel oil. The Cleveland Open Tester (ASTM D92) is commonly used for high flash point liquids. The Tag Open Tester (ASTM D1310) is frequently used for low-flash liquids where it is desired to have tests more representative of conditions in open tanks of flammable liquids, or for labeling and transportation purposes. For most liquids, the numerical value in degrees Fahrenheit of the closed cup flash point is some 10 to 20 percent lower than that of the open cup flash point for the same liquid, but there are some cases where the difference is greater or smaller. Standard specifications for the open cup and other testers are published by the American Society for Testing and Materials, 1916 Race St., Philadelphia 3, Pennsylvania.

COLUMN 11. IGNITION TEMPERATURE

Ignition temperature of a substance, whether solid, liquid, or gaseous, is the minimum temperature required to initiate or cause self-sustained combustion independently of the heating or heated element.

Ignition temperatures observed under one set of condtions may be changed substantially by a change of conditions. For this reason, ignition temperatures should be looked upon only as approximations. Some of the variables known to effect ignition temperatures are percentage composition of the vapor or gas-air mixture, shape and size of the space where the ignition occurs, rate and duration of heating, kind and temperature of the ignition source, catalytic or other effect of materials that may be present, and oxygen concentration. As there are many differences in ignition temperature test methods, such as size and shape of containers, method of heating and ignition source, it is not surprising that ignition temperatures are affected by the test method.

As illustration of the effects of test methods, the ignition temperatures of hexane determined by three different methods were 225°C (437°F.), 336.1°C (637°F.), and 510°C (950°F.), respectively. The effect of percentage composition is shown by the following ignition temperatures for pentane: 547.8°C (1018.4°F) for 1.5 percent pentane in air, 501.7°C (935.6°F) for 3.75 percent pentane, and 475.6°C (888.8°F) for 7.65 percent pentane. The following ignition temperatures for carbon disulfide demonstrate the effect of size of space containing the ignitible mixture: in a 200 ml (milliliter) flask the ignition temperature was 120°C (248°F.), in a 1,000 ml flask 110°C (230°F.), and in a 10,000 ml flask 96.1°C (205°F.) That materials in which the flammable mixture is in contact may affect the ignition temperature is illustrated by ignition temperature determinations for benzene conducted in various containers: 575.6°C (1060°F.) in a quartz container, 677.8°C (1252°F.) in iron, and 721.1°C (1330°F.) in zinc.

The ignition temperature of a combustible solid is influenced by the rate of air flow, rate of heating, and size of the solid. Small sample tests have shown that as the rate of airflow or the rate of heating is increased, the ignition temperature of a solid drops to a minimum and then increases. The same appears to be the case as the size of the sample increases.

COLUMN 12. BOILING POINT

Boiling point and melting point data was generally taken from the Handbook of Chemistry and Physics, 47th Edition, edited by Robert C. Weast, published by the Chemical Rubber Company, Cleveland, Ohio, 1966.

COLUMN 13. VAPOR PRESSURE

Vapor pressure data was taken in most cases from Industrial Hygiene and Toxicology, Volume II, edited by Frank A. Patty and published by Interscience Publishers, New York, 1963.

COLUMN 14. DENSITY OR SPECIFIC GRAVITY

Density is relative to water, otherwise it has the dimensions g/ml. A superscript indicates the temperature of water to which the density is referred.

Values were taken from the Handbook of Chemistry and Physics, edited by Robert C. Weast, published by the Chemical Rubber Company, Cleveland, Ohio, 1966.

COLUMN 15. WATER SOLUBILITY

The following symbols are used to indicate the relative water solubility of the substances:

∞ miscible

v   more than 50 grams dissolve in 100 milliliters of water

s   5 to 50 grams dissolve in 100 milliliters of water

δ less than 5 grams dissolve in 100 milliliters of water

i   insoluble

d   decomposes on contact with water

Unless otherwise indicated by superscripts, solubility designations are at room temperature for organic compounds and in cold water for inorganic compounds. The superior figure h designates hot water.

EXAMPLE: American Industrial Hygiene Association
Hygienic Guides*
*Reprinted by permission

AMERICAN

Industrial

Hygiene

ASSOCIATION

HYGIENIC GUIDE SERIES

Hygienic Guide Series

Benzene
(Benzol)
(Revised 1961)

I. Hygienic Standards

A. RECOMMENDED MAXIMUM ATMOSPHERIC CONCENTRATION (8 hours): 25 parts of vapor per million parts of air, by volume (ppm).[1]

  1. *Basis for Recommendation:* Principally human experience in industry plus toxicological observations on animals.

B. SEVERITY OF HAZARDS:

  1. *Health:* Moderate for acute exposure; high for chronic. Absorption occurs chiefly by inhalation. High concentrations irritate the respiratory tract and produce narcosis. Repeated exposure to benzene may cause bone marrow damage, resulting in a decrease in the circulating white blood cells, platelets, and red blood cells. The red cells may occasionally show an increase in size in early poisoning.[7] Many serious illnesses and fatalities have occurred in association with chronic exposures to benzene. It has been stated that symptoms may occasionally occur after exposure has ceased.[5] Individual susceptibility varies widely with the remote possibility that an occasional individual may be affected by prolonged exposure to 25 ppm. A primary irritant type of dermatitis may result from repeated skin contact. Percutaneous absorption is considered insignificant. Gerarde[3, 4, 5] has shown that alkyl derivatives do not produce the bone marrow effects of benzene in rats.

  2. *Fire:* High. Explosive limits are 1.4–7.1% by volume at 212°F. Flash point is −11.1°C (12.0°F) (closed cup).

C. SHORT EXPOSURE TOLERANCE: For man on single exposure, 3000 ppm is endurable for 30 to 60 minutes; 7500 ppm is dangerous in 30 to 60 minutes.[10]

D. ATMOSPHERIC CONCENTRATION IMMEDIATELY HAZARDOUS TO LIFE: 20,000 ppm is reported fatal in 5 to 10 minutes.[10]

II. Significant Properties

Benzene is a flammable, colorless, odorous liquid

Chemical formula:	C_6H_6
Molecular weight:	78.11
Specific gravity:	0.8790 (20°/4°C)
Boiling point:	80.1°C (760 mm Hg)
Vapor pressure:	95.14 mm Hg at 25°C
Solubility:	In most organic solvents

At 25°C and 760 mm Hg:

1 ppm of vapor:	0.0032 mg/liter
1 mg/liter of vapor:	313 ppm
Saturated air concentration:	125,000 ppm
Relative density of saturated air:	1.21 (air = 1)
Relative density of vapor:	2.7 (air = 1)

Note: Benzene, an aromatic hydrocarbon, is to be distinguished carefully from benzine, a petroleum distillate containing mixed hydrocarbons (such as pentane and hexane) in uncertain proportions. However, some commercial petroleum and aromatic solvents may contain quantities of benzene of possible hygienic significance. In the past, some gasolines contained appreciable quantities of benzene, but currently the amount does not exceed 3%. An accurate estimation of the quantities of benzene present in such solvents is valuable in estimating the benzene hazard.

III. Industrial Hygiene Practice

A. RECOGNITION: Benzene may be recognized by its characteristic odor at concentrations of about 100 ppm.[10] It is an excellent solvent for a wide variety of gums, resins, fats, alkaloids, and rubber and is frequently used in combination with other

---

The Committee wishes to acknowledge the assistance of Dr, L. J. Goldwater and Dr. H. W. Gerarde in the preparation of this Hygienic Guide.

solvents. It is present, to some extent, in most gasolines and is a common ingredient of paint and varnish removers.

B. EVALUATION OF EXPOSURES:

1. *Instrumentation:* Direct determination may be made by the commercially available aromatic hydrocarbon detecting instruments, if the sensitivity is 25 ppm or lower. It may also be determined by scrubbing the air through spectroscopically pure isooctane or alcohol in an all-glass device, followed by ultraviolet spectrophotometry, or by adsorption on silica gel, with subsequent extraction and ultraviolet spectrophotometry.[8]

2. *Chemical Method:* Collection by adsorption on silica gel, followed by nitration, color development with sodium hydroxide and comparison with prepared standards,[2] or by polarography.[6] Detector tubes are available.

C. RECOMMENDED CONTROL PROCEDURES: Maintain workroom atmospheres below 25 ppm by means of process enclosure and/or ventilation. Substitute, wherever possible, a less toxic solvent. Prevent skin contact through the use of protective clothing made of nitrile type rubber (Butaprene, Chemigum, Hycar, Krynac, Paracril, Tylor) or Neoprene. Use chemical type goggles.

IV. Specific Procedures

A. FIRST AID: Remove from exposure; remove contaminated clothing, flush eyes with water, and wash contaminated skin areas with soap and water.

B. SPECIFIC MEDICAL PROCEDURES:

1. *Preplacement:* This should include a complete blood count as well as a careful history and physical examination. Particular attention should be paid to a history of any type of blood disturbance; but any chronic disturbances of general health, particularly those which might involve liver or renal diseases, should be carefully considered by the physician.

2. *Periodic Examination:* A program of routine physical examinations is essential for all persons regularly exposed to benzene and especially if exposures are at or possibly above the threshold limit. This should include a complete blood count at intervals no greater than every two months (frequency may be decreased if regular air analyses show the continued absence of hygienically significant amounts of benzene).

The measurement of urinary sulfates has been suggested as a control procedure.[13] The determination of urinary phenols has also been suggested as an index of personnel exposure, particularly useful where marked fluctuations in atmospheric concentrations occur or where the operations are intermittent or varied.[12] Values above 200 mg of total phenol per liter of urine on spot samples collected late in the day are stated to be significant.[9] While both of these methods are measures of benzene exposure, they tell nothing of the health of the individual, but may be of some use as control procedures, particularly in low-grade chronic exposures. They are of most value when obtained from groups of exposed personnel, rather than from single individuals.

V. References

1. American Conference of Governmental Industrial Hygienists: *Amer. Ind. Hyg. Assoc. J. 22:* 325 (1961).
2. Elkins, H. B.: *The Chemistry of Industrial Toxicology*, 2nd Ed., John Wiley & Sons, Inc., New York (1959).
3. Gerarde, H. W.: *AMA Arch. Ind. Health 13:* 468 (1956).
4. Gerarde, H. W.: *AMA Arch. Ind. Health 19:* 403 (1959).
5. Gerarde, H. W.: *Toxicology and Biochemistry of Aromatic Hydrocarbons.* Elsevier Publishing Co., New York (1960).
6. Gisclard, J. B.: Personal Communication (Wright-Patterson Air Force Base).
7. Goldwater, L. J.: *J. Lab. Clin. Med. 26:* 957 (1941).
8. Maffett, P. A., T. F. Doherty, and J. L. Monkman: *Amer. Ind. Hyg. Assoc. Quart. 17:* 186 (1956).
9. Pagnotto, L. D., H. B. Elkins, H. G. Brugsch, and J. E. Walkley: Industrial Benzene Exposure from Petroleum Naphtha. *Amer. Ind. Hyg. Assoc. J. 22:* 417 (1961).
10. Patty, F. A.: *Industrial Hygiene and Toxicology.* Vol. 2, page 754, Interscience Publishers, Inc., New York (1949).
11. Patty, F. A.: *Industrial Hygiene and Toxicology.* Vol. 2, page 757, Interscience Publishers, Inc., New York (1949).
12. Teisinger, J., and V. Fiserova-Bergerova: *Arch. maladies profess med. travail et securite sociale 16:* 222 (1955).
13. Yant, W. P., H. H. Schrenk, R. R. Sayers, A. A. Horvath, and W. H. Reinhart: *J. Ind. Hyg. Toxicol. 18:* 69 (1936).

EXAMPLE: American Industrial Hygiene Association
Hygienic Guides*
*Reprinted by permission

HYGIENIC GUIDE SERIES
Carbon Tetrachloride
(Revised 1961)

I. Hygienic Standards

A. RECOMMENDED ATMOSPHERIC CONCEN-TRATION (8 hours): 25 parts of vapor per million parts of air, by volume (ppm).[1] It is felt that this is a maximum which should not be exceeded for repeated daily exposures. The time weighted average should not exceed 10 ppm.[2] **These levels cannot be detected by odor.**

  1. *Basis for Recommendation:* Human experience and observation of experimental animals.[3]

B. SEVERITY OF HAZARDS:

  1. *Health:* High for both acute and chronic exposures. Narcosis with subsequent liver and kidney injury and death may occur as the result of an acute overexposure. An acute nephrosis and anuria may occur from single exposures which are insufficient to cause symptoms of narcosis or of irritation. Chronic inhalation exposure to concentrations above 25 ppm may also result in severe injury to the kidneys and liver.[3] Persons who have consumed alcohol are more susceptible to such effects than others, but individual susceptibility varies widely. Prolonged or repeated contact with the liquid may result in irritation of the skin. Liquid carbon tetrachloride can be absorbed through the intact skin of experimental animals in toxic amounts.[5] Eye contact results in pain and minimal injury to the conjunctiva. It is highly toxic if ingested.

  2. *Fire:* None. Carbon tetrachloride is used as a fire extinguishing agent. Thermal decomposition products are quite toxic.

C. SHORT EXPOSURE TOLERANCE: Animal experiments showed little or no injury from single exposures to 300 ppm for one hour, 90 ppm for four hours, or 2000 ppm for six minutes.[3]

D. ATMOSPHERIC CONCENTRATION IMMEDI-ATELY HAZARDOUS TO LIFE: Unknown, but probably about 2%. Human fatalities from acute renal damage have occurred after one-half to one hour exposure to concentrations of 1000–2000 ppm.[4]

II. Significant Properties

Carbon tetrachloride is a volatile, nonflammable, colorless liquid with a sweetish, aromatic odor. The odor threshold is about 50 ppm of vapor.[5] Odor is usually not objectionable at acutely toxic levels and may not even be apparent at concentrations which are harmful upon repeated exposure.

Chemical formula: CCl_4
Molecular weight: 153.84
Specific gravity: 1.5843 (25°/4°C)
Boiling point: 76.54°C at 760 mm Hg
Vapor pressure at 25°C: 115.25 mm Hg
Solubility: Insoluble in water, but miscible with acetone, alcohol, ether, and most organic solvents.

At 25°C and 760 mm of Hg:
  1 ppm of vapor: 0.00628 mg/liter
  1 mg/liter of vapor: 159 ppm
  Saturated air concentration: 15.2%
  Relative density of vapor: 5.3 (air = 1.0)
  Relative density of saturated air: 1.65 (air = 1.0)

III. Industrial Hygiene Practice

A. RECOGNITION: Carbon tetrachloride is used as a degreasing and cleaning agent, as a fire extinguisher, as a solvent in rubberizing fabrics, as a component in fumigant mixtures used in the grain industry, and as a solvent in the chemical processing and manufacturing industries. It is sometimes used as a component of proprietary solvent preparations to reduce their fire hazards. At elevated temperatures its vapors decompose to produce chlorine, hydrogen chloride, and (in presence of ozone, during welding etc.) possibly some phosgene.[6, 7]

B. EVALUATION OF EXPOSURES

1. *Instrumentation:* Instruments based on the Beilstein test such as leak detectors and the Davis Halide meter may be calibrated for carbon tetrachloride. Instruments employing measurement of the conductivity of water which has absorbed the combustion products of this material are available. Infrared, mass spectrographic and gas chromatographic methods can be used, especially when absolute identification is necessary. Infrared and conductivity instruments are especially suitable for continuous air analysis.

2. *Chemical Analysis:* Direct combustion followed by absorption in a basic, reducing solution and then determination of the halide ion has been used successfully,[8] as has adsorption on silica gel followed by thermal desorption and decomposition.[9] Removal from silica gel by isopropanol followed by alkaline hydrolysis can also be used.[10] Halogenated hydrocarbon "detector tubes" by various manufacturers can probably be used, but each batch should be separately calibrated. A modified Fujiwara reaction may be used.[11] None of these methods are specific for carbon tetrachloride.

C. RECOMMENDED CONTROL PROCEDURES: All employees should be instructed in the hazards and control measures. The hazard of severe renal and liver injury may be greatly reduced by substituting a less toxic solvent such as methylene chloride, trichloroethylene, perchloroethylene, or 1,1,1-trichloroethane (methyl chloroform). Petroleum solvents with flash points above 100°F may be useful in some instances as a replacement. Where carbon tetrachloride is used, local exhaust ventilation will usually be necessary, and frequently the operation should be isolated (by total enclosure) from the general work space. Concentrations to which men are repeatedly exposed should rarely or never exceed 25 ppm and should not average over 10 ppm. Air sampling is mandatory, and an automatic air monitoring system may be necessary to assure adequate control of vapor. Adequate respiratory protection should be provided for maintenance workers or in the event of spills or leaks. Gross skin contact should be prevented. Spectacles will usually offer adequate eye protection.

IV. Specific Procedures

A. FIRST AID: Remove person to uncontaminated atmosphere and apply arti-ficial respiration if indicated. Call a physician at once. Remove wet clothing and cleanse the skin with water or soap and water. Do not allow contaminated clothing to be reworn until it is thoroughly dry. Eye splashes should be treated with copious water irrigation.

Any person sustaining an accidental high level exposure should be observed by a physician for several days, even if no symptoms have been noticed at the time of exposure. Nausea and vomiting occurring twenty-four hours or so after exposure may be the first sign of renal damage.

B. SPECIFIC MEDICAL PROCEDURES: Employment should be restricted to persons free of a history of liver or kidney disorder, or of alcoholism. Treatment of acutely affected individuals should be directed toward restoring kidney and liver function. Hospitalization is essential. Oxygen therapy should be instituted if any anesthesia or cardiac failure has occurred. As with other chlorinated hydrocarbons, the use of epinephrine during any anesthetic phase may cause ventricular fibrillation. As the most serious sequelae of acute carbon tetrachloride exposure is toxic nephrosis (and death is directly related thereto), the use of an artificial kidney is indicated and may prove life-saving. Some indication of the extent of exposure may be obtained from an analysis of exhaled air.[12]

V. References

1. American Conference of Governmental Industrial Hygienists: Threshold Limit Values for 1961, *Amer. Ind. Hyg. Assoc. J. 22:* 325 (1961).
2. American Standards Association— Z37-17, 1957.
3. Adams, E. M., *et al.: AMA Arch. Ind. Hyg. & Occ. Med..6:* 50 (1952).
4. Fassett, D. W. (Eastman Kodak Co., Rochester, New York) Personal Communication.
5. The Dow Chemical Company, Unpublished Data.
6. Sjöberg, Bertil: *Svenska Kem Tid.* (Sweden) *64:* 63 (1952) (in English).
7. Crummett, W. D., and V. A. Stenger: *Ind. Eng. Chem. 48:* 434 (1956).
8. Jacobs, M. B.: *Analytical Chemistry of Industrial Poisons, Hazards and Solvents,* 2nd Ed., Interscience Publishers, Inc., New York (1949).
9. Peterson, J. E., *et al.: Amer. Ind. Hyg. Assoc. Quart. 17:* 429 (1956).
10. Fahy, J. P.: *J. Ind. Hyg. & Toxicol. 30:* 205 (1948).
11. Rogers, G. W., and K. K. Kay: *J. Ind. Hyg. and Toxicol. 29:* 229 (1947).
12. Stewart, R. D., *et al.:* To be published: *J. Am. Med. Assoc.*

EXAMPLE: American Industrial Hygiene Association
Hygienic Guides*
*Reprinted by permission

HYGIENIC GUIDE SERIES

Hydrogen Sulfide

(Revised 1962)

I. Hygienic Standards

A. RECOMMENDED MAXIMUM ATMOSPHERIC CONCENTRATION (8 hours): 20 parts of gas per million parts of air, by volume (ppm).[1,2]

   1. *Basis for Recommendation:* Human experience and animal studies.[3]

B. SEVERITY OF HAZARDS:

   1. *Health:* High for acute exposures; moderate for chronic. It may cause irritation to the eyes at concentrations above 10 ppm and to the lungs and mucous membranes at levels only moderately above 20 ppm. Because of the rapid occurrence of olfactory fatigue, its odor is an unreliable indicator of its level of exposure. Death may occur rapidly with exposures above 600 ppm. Its primary physiological effects are respiratory paralysis and pulmonary irritation. Hydrogen sulfide is noncumulative in the body, and if recovery occurs from acute exposures, there are no sequelae. The irritant effect on the eyes from repeated exposures may result in a painful conjunctivitis with some photophobia. [2,3,4,5,6,7]

   2. *Fire:* Moderate. Ignition temperature, 292°C. Explosive limits are 4.3% to 45.5% by volume.

C. SHORT EXPOSURE TOLERANCE: In 18 to 25 ppm, 25 of 78 persons exposed to hydrogen sulfide vapor complained of eye irritation.[3] 70-150 ppm produces slight symptoms after exposure of several hours. 170-300 ppm is the maximum concentration that can be inhaled for one hour without serious consequences.[5]

D. ATMOSPHERIC CONCENTRATION IMMEDIATELY HAZARDOUS TO LIFE: 400-700 ppm is dangerous to life after 30 to 60 minutes, and concentrations above 600 ppm become rapidly fatal.[5]

II. Significant Properties

Colorless gas with an extremely unpleasant odor characteristic of rotten eggs at low concentrations and a sweetish odor at higher concentrations.

Chemical formula: H_2S
Molecular weight: 34.08
Boiling point: —60.1°C
Solubility: 2.58 volumes gas per volume of water at 20°C

Vapor pressure: 19.6 atm. at 25°C
Odor threshold: 0.13 ppm[6]
Specific gravity of gas: 1.192 (dry air = 1.00)
At 25°C and 760 mm:
   1 ppm: 0.00139 mg/liter
   1 mg/liter: 717 ppm

III. Industrial Hygiene Practice

A. RECOGNITION: Odor is readily apparent at low concentrations, but it cannot be relied upon for warning against high concentrations or prolonged exposures at any level. It is readily apparent at 4.6 ppm and strong, but not intolerable, at 27 ppm.[8] Industrial exposures may occur in petroleum refining, especially of high sulfur crudes, viscose rayon manufacture, sewage disposal plants, excavating and tunneling operations, chemical laboratories, and as a side product of many chemical reactions and processes.

B. EVALUATION OF EXPOSURE:

   1. *Instrumentation:* There are commercially available a number of automatic devices suitable for the de-

tection and determination of hydrogen sulfide. For spot samples several commercial devices, operating on a color change of a solid detecting chemical, are satisfactory for approximate determinations.

2. *Chemical Methods:* Absorption in a solution of a cadmium salt and iodometric determination of the cadmium sulfide or absorption in alkaline zinc acetate solution and subsequent colorimetric determination; or, in the absence of oxidizing or reducing gases, absorption in iodine-potassium iodide solution and iodometric determination.[9]

C. RECOMMENDED CONTROL PROCEDURES: Maintain workroom atmospheres below 20 ppm by means of process enclosure and/or ventilation. Disposal must not constitute a nuisance or hazard to the public. Small amounts may be flared or burned in a furnace. Larger amounts may justify concentration and recovery as elemental sulfur. Eye and respiratory protective equipment should be used in concentrations exceeding 20 ppm.

IV. Specific Procedures

A. FIRST AID: Acute poisoning is a medical emergency. Remove victim to a noncontaminated location promptly. Summon medical assistance. Keep patient warm and at rest. Artificial respiration should be promptly instituted if breathing has ceased. Oxygen or oxygen-CO_2 inhalation is recommended, continuing after spontaneous breathing has returned, as the high oxygen concentration aids in the destruction of H_2S.[8]

B. SPECIAL MEDICAL PROCEDURES: Preplacement examinations should be thorough. Workers with any severe ailment, including eye and nervous diseases, should not be employed on operations involving H_2S exposures. Treatment for acute exposure is supportive. Eye irritation may be severe. In these cases, local anaesthetics, topical antibiotics and cycloplegic agents may be indicated.[6]

V. References

1. American Conference of Governmental Industrial Hygienists: *Amer. Ind. Hyg. Assoc. J. 23:* 419 (1962).
2. American Standards Association: "Allowable Concentration of Hydrogen Sulfide," American Standard Z37.2—1941.
3. U. S. Public Health Service: Pub. Health Reports *56:* 684 (1941) (Reprint No. 2256).
4. Fairhall, L. T.: *Industrial Toxicology,* 2nd ed., Williams and Wilkins Co., Baltimore (1957).
5. Manufacturing Chemists Association: Chemical Safety Data Sheet SD-36 (1950).
6. Milby, T. H.: *J. Occupational Med. 4:* 431 (1962).
7. Henderson, Y., and H. W. Haggard: *Noxious Gases,* 2nd rev. ed., New York, Reinhold Publishing Corp. (1943).
8. American Petroleum Institute: API Toxicological Review—Hydrogen Sulfide, New York (1948).
9. Jacobs, M. B.: *The Analytical Chemistry of Industrial Hazards, Poisons, and Solvents,* Vol. 1, 2nd ed. pp. 323-326. Interscience Publishers, Inc., New York (1949).

Chemical Safety Data Sheet SD-41

ACETIC ACID

Copyright 1951 by Manufacturing Chemists' Association, Inc.
1825 Connecticut Avenue, N.W.
Washington, D.C. 20009

(Excerpts reprinted by permission)

Chemicals in any form can be safely stored, handled or used if the physical, chemical and hazardous properties are fully understood and the necessary precautions, including the use of proper safeguards and personal protective equipment, are observed.

1. NAME

Chemical Names: Acetic Acid, Ethanoic Acid
Common Name: Acetic Acid
Formula: CH_3COOH

2. PROPERTIES

2.1. Grades and Strengths

Grades: Commercial, USP, CP
Strengths: 28%, 56%, 70%, 80%, 84%, Glacial (99.5% Minimum)

2.2. Important Physical and Chemical Properties (Glacial Grade)

Boiling Point: 118.1°C (244.6°F)
Color: Colorless
Corrosivity: Highly corrosive at dilute concentrations. Somewhat less so at glacial strength.
Explosive Limits: 4% in air (lower limit)
Flash Point: Open Cup, 43°C (110°F)
  Closed Cup, 40°C (104°F)
Ignition Temperature: 565°C (1050°F)
Freezing Point: 16.6°C (62°F)
Odor: Characteristic vinegar odor. In high concentrations pungent.
Reactivity: Reacts readily with most common metals (except aluminum), basic salts, amines, etc., to form water-soluble salts. It reacts with alcohols to form esters.
Specific Gravity: 1.049 (20°/4°C) Glacial Grade.
Vapor Density: 2.07 (air = 1).
Physical State: Liquid above 16.6°C (62°F). Solidifies at lower temperatures.

2.3. Hazardous Properties

2.3.1. *HEALTH HAZARDS*. Concentrated solutions of acetic acid (50% or more), if not promptly removed, can destroy tissue with which they come in contact, and produce severe burns, but are not as active in this way as sulfuric or nitric acid.

In the eye acetic acid or its concentrated solutions can cause severe damage. Breathing of concentrated vapor may be harmful. Swallowing may cause severe injury or death (see 8. Health Hazards and Their Control).

2.3.2. *FIRE HAZARDS*. Although acetic acid will support combustion, it is not classed as a flammable liquid. Reasonable precautions should be observed (see 6. Handling).

3. USUAL SHIPPING CONTAINERS

4. UNLOADING AND EMPTYING

5. STORAGE

6. HANDLING

7. WASTE DISPOSAL

Waste acetic acid may be flushed away with water, with due regard to local, state, and federal laws regulating health and pollution. As a safeguard against corrosion of sewer lines, the acid should first be neutralized with any readily available basic compound, such as soda ash. Acetate salts are all water-soluble and easily flushed away.

8. HEALTH HAZARDS AND THEIR CONTROL

This section includes not only recognized first aid procedures and information of interest to the layman, but also suggestions which may be of value to the attending physician.

8.1. Hazards

8.1.1. *GENERAL*. Acetic acid is dangerous when improperly handled. While concentrated solutions can be destructive to tissues with which they come in contact, producing severe burns, experience has demonstrated that it is not as hazardous in this respect as sulfuric acid or nitric acid. Contact with the eyes by concentrated solutions can cause severe damage and may result in total loss of sight. Inhalation of concentrated vapor may be harmful. Swallowing may cause severe injury or death.

8.1.2. *ACUTE TOXICITY*.

8.1.2.1. *Systemic Effects*. No general systemic effects have been noted. Inhalation of concentrated vapor from hot acid may cause serious damage to the membranes of the respiratory system. Its action is largely irritating in character. Workers exposed to low concentrations of vapor gradually become hardened to their irritant action.

8.1.2.2. *Local Effects*. Acetic acid is dangerous when improperly handled. Concentrated solutions may be destructive to any body tissues with which they come in contact.

Contact with the eyes very rapidly causes severe damage, which may be followed by total loss of sight.

Inhalation of concentrated vapor or mist from hot acid will cause damage to the upper respiratory tract and even to the lung tissue proper.

8.1.3. *CHRONIC TOXICITY*

8.1.3.1. *Systemic Effects*. No systemic effects are noted, except those secondary to tissue damage. These can be prevented by proper handling and use of the material, and minimized by prompt first aid measures.

8.1.3.2. *Local Effects.* Repeated contact with diluted solutions may cause a dermatitis. Repeated inhalation of mist may cause a chronic inflammation of the upper respiratory tract and chronic bronchitis.

8.2. Prevention and Control

Acetic acid may be handled safely, provided workers so engaged are fully and adequately instructed and are supervised in approved and safe methods of handling.

8.2.1. *EMPLOYEE EDUCATION* (see 6.4).

8.2.1.1. It is essential to provide those who handle acetic acid with instructions describing the hazards, prevention and control measures, and recommended first aid treatment.

8.2.2. *VENTILATION.* A maximum allowable concentration of 10 ppm of acetic acid vapor by volume in the air for an eight hour working day, has been suggested by some agencies. Ventilation should be adequate wherever acetic acid is handled to keep the concentration within this limit. Workers exposed to low concentrations of vapor gradually become hardened to the irritant action.

8.2.3. *SAFETY SHOWERS AND EYE BATHS.* Readily accessible, well marked, frequently inspected, rapid action safety showers must be available in the areas where acetic acid is being handled. A special eye washing fountain, a ready source of running tap water, a bubbler drinking fountain, or a hose with a soft, gentle flow of drinking water must be available for eye irritation. All of the above equipment should be inspected at frequent intervals to ensure its being in good working condition at all times (see 6.5).

8.2.4. *PREPLACEMENT PHYSICAL EXAMINATIONS.* Prior to assignment to processes involving the handling of acetic acid, all individuals should have a careful preplacement physical examination, and, in order to properly protect their health, those who have the following conditions should be excluded from such processes:

(a) Chronic skin conditions,
(b) Chronic diseases of the upper respiratory tract or chronic lung disease,
(c) Only one eye,
(d) Uncorrected severe faulty vision.

8.2.5. *PERIODIC PHYSICAL EXAMINATIONS.* All employees who work constantly in processes involving acetic acid should have a careful physical examination at least once each year.

8.3. Personal Protective Equipment (see 6.5)

8.3.1. Emergencies should be anticipated, and proper personal protective equipment in sufficient quantities should be strategically located throughout the premises to be readily available in case of need. It should be examined and tested regularly, and maintained in excellent condition at all times.

8.3.2. Personal protective equipment is not a substitute for good, safe working conditions, nor for adequate ventilation. The correct usage of personal protective equipment requires education of the worker in the proper employment of the materials available to him (see 6.5) and careful, constant supervision.

8.3.3. Employees who handle acetic acid should be provided, when indicated, with the following equipment:

(a) A brimmed felt or a treated fiber hat, rubber gloves, rubber high top safety toe shoes or boots (with tops covered by trousers), outer clothing fitted snugly at the neck and wrists, and a rubber apron. A rubber acid suit, recommended for tank car loading or unloading (see 6.5) may also be necessary. Water in ample quantity should be available immediately (see 6.5.4):

(b) Suitable gas-tight chemical safety goggles;
(c) Rescue harness and life line for those entering a tank or enclosed storage space. An outside attendant should maintain constant observation (see 6.8 Tank and Equipment Cleaning and Repairs);

Positive pressure hose masks with hose inlet originating in a vapor-free atmosphere;

Air-line masks with proper reducing valve and filter suitable for use only where conditions will permit safe escape in case of failure of the compressed air supply;

Self-contained breathing apparatus with stored oxygen or air, which allows greater mobility but usually requires more highly trained men. In tank work (see 6.8) small manholes may make this apparatus unsuitable because of its bulk, although the type known as self-generating is specially designed for entrance and egress through small openings;

Masks and breathing apparatus should be approved by the United States Bureau of Mines for use with acetic acid and should be equipped with full-face pieces;

(d) Industrial canister-type gas masks, even though approved by the United States Bureau of Mines, should never be used in emergency circumstances. They may be used only when it is certain that the concentration is less than 2% by volume (20,000 ppm), and the oxygen content is not less than 16%, and then only for exposures not exceeding one-half hour.

CAUTION: *Do not depend on creams or ointments to afford protection.*

8.4. First Aid

8.4.1. *GENERAL PRINCIPLES.* Speed in removing acetic acid is of primary importance. First aid must be started immediately in cases of contact with acetic acid, as delay in initiating treatment may result in injury.

8.4.2. *SPECIFIC ACTIONS*

8.4.2.1. *General First Aid.* Of primary importance in case of contact is the immediate and prolonged application of copious quantities of running water. All contaminated clothing must be removed immediately. This can best be accomplished while the man is under a safety shower. The application of copious quantities of running water to the affected parts must be prolonged until all traces of acetic acid have been removed. No attempt should be made to neutralize the acetic acid with mild alkaline solutions until all areas of contact have been thoroughly irrigated with copious quantities of running water.

It should be borne in mind that in cases of severe or extensive burns, shock symptoms such as rapid pulse, sweating and collapse, may appear at any time. If such symptoms should appear, have the patient lie on his back and keep him warm, not hot, until a physician arrives. No oils or ointments should be applied to the burned areas without specific direction from the attending physician. Call a physician at the earliest possible moment. Describe to him in detail the nature of the injury and the exact location of the patient.

8.4.2.2. *Contact with Eyes.* If even minute quantities of acetic acid enter the eyes, they should be immediately irrigated with copious quantities of running water for 15 minutes. The eye lids should be held apart during the irrigation to ensure contact of water with all the tissues of the surface of the eyes and lids. A physician, preferable an eye specialist, should be called in attendance at the first possible moment. If a physician is not immediately available, the eye irrigation should be continued for a second period of 15 minutes. After the first 15 minute period of irrigation is completed, it is permissible as a first aid measure to instill two or three drops of 0.5% solution of pontocaine or an equally effective aqueous topical anesthetic. No oils or oily ointments should be instilled unless ordered by the physician.

8.4.2.3. *Taken Internally.* Ingestion of concentrated acetic acid causes burns of the mucous membranes of the mouth, throat, esophagus, and stomach. Do not attempt to induce vomiting in patients who have swallowed strong solutions of acetic acid. If the patient is conscious, encourage him to wash out his mouth with copious quantities of water; then have him drink milk, if available, mixed with the whites of eggs. If they are not immediately available, have him drink as much water as possible. A physician should be notified as soon as possible. Describe the accident in detail to the physician at the time he is called and give him the exact location of the patient.

DO NOT GIVE ANYTHING BY MOUTH TO AN UNCONSCIOUS PATIENT.

The medical information in this publication has been supplied by the Medical Advisory Committee of the Manufacturing Chemists' Association.

Chemical Safety Data Sheet SD-9

CAUSTIC SODA

1. NAME

Chemical Name: Sodium Hydroxide
Common Names: Caustic Soda, Liquid Caustic Soda, Lye
Formula: NaOH

2. PROPERTIES

2.1. Grades and Strength

2.1.1. Aqueous solutions, commonly called liquid caustic, containing about 50 % and about 73 % sodium hydroxide by weight.

2.1.2. Anhydrous (dry) solid, flake, ground, and powdered forms, containing from 70 % to about 99 % sodium hydroxide.

2.1.3. CP pellet or stick.

2.2. Important Physical and Chemical Properties

2.2.1. *LIQUID CAUSTIC SODA* (*aqueous solutions*)
Color: Water white; occasionally gray.

Aqueous Solutions of Caustic Soda

	50 % Solution	73 % Solution
Boiling Point (760 mm):	142° to 148°C (288° to 298°F)	188° to 198°C (370° to 388°F)
Crystallization begins at:	+12° to +15°C (54° to 59°F)	+63°C (145°F)
Solidifies at:	+5°C (41°F)	+62°C (144°F)

Flash Point: None.
Ignition Temperature: None. Not combustible.
Hazardous to Health: Yes.
Corrosive: Non-corrosive to rubber at atmospheric temperatures. Slowly corrosive to iron, copper, and Monel metal: solutions may pick up harmful quantities of these metals. Corrodes clothing and a few metals, such as aluminum, tin, lead and zinc, and alloys containing these metals. Destroys living tissue.
Dangerously Reactive: Yes; boiling point can be attained in diluting 73 % liquid caustic with water.
Heat of dilution of liquid caustic is considerable and variable, depending upon initial and final caustic concentrations.

2.2.2. *ANHYDROUS (DRY) FORMS OF CAUSTIC SODA—Solid, Flake, Ground and Powdered.*

Color: White or light gray.
Boiling Point: White heat.
Melting Point: 310° to 320°C (590° to 608°F)
Hazardous to Health: Yes.
Corrosive: Non-corrosive to rubber at atmospheric temperatures. Slowly corrosive to iron, copper, and Monel metal: solutions may pick up harmful quantities of these metals. Corrodes clothing and a few metals, such as aluminum, tin, lead and zinc, and alloys containing these metals. Destroys living tissue.
Dangerously Reactive: Yes. Considerable heat is generated when water is added to caustic soda; boiling and spattering of hot caustic solution may result.
Hygroscopic: Yes.
Deliquescent: Yes.

2.3. Hazardous Properties

2.3.1. *HEALTH HAZARDS (see 8. Health Hazards and Their Control).* Caustic soda is a strong alkali and is dangerous when improperly handled. The solid caustic and concentrated solution are destructive to tissues with which they come in contact, producing severe burns. Contact with the eyes, either in solid form or in solution, causes severe damage to the eye. Inhalation of dust or mist of this compound is capable of causing injury to the entire respiratory tract. Swallowing usually results in severe injury.

For detailed description of health hazards and their control, see Section 8.

2.3.2. *FIRE HAZARD.* Caustic soda and its solutions will neither burn nor support combustion.

3. USUAL SHIPPING CONTAINERS

4. UNLOADING AND EMPTYING

5. STORAGE

6. HANDLING

7. WASTE DISPOSAL

7.1. Waste disposal of caustic soda depends to a great extent upon local conditions. Be sure that all federal, state, and local regulation regarding health and pollution are followed.

7.2. Waste caustic soda solution should not be discharged directly into sewers or streams. The caustic should first be converted to a neutral salt, as by neutralization with acid, and then well diluted with water to render the waste less harmful.

7.3. Strong alkali tends to diminish bacterial activity needed for proper sewage disposal by increasing the alkalinity to unfavorable levels.

8. HEALTH HAZARDS AND THEIR CONTROL

This section includes not only recognized first aid procedures and information of interest to the layman, but also suggestions that may be of value to the attending physician.

8.1. Hazards

Caustic soda is dangerous when improperly handled. Whether in solid form or in solution, marked corrosive action results from contact with all tissues of the body. Since signs and symptoms of irritation are frequently not evident immediately after contact with caustic soda, injury may result before one realizes that the chemical is in contact with the body. Therefore, adequate protection against such exposure should be provided for all parts of the body.

8.1.2. *ACUTE TOXICITY*

8.1.2.1. *Systemic Effects.* No general or systemic effects are noted, except those associated with shock or those secondary to tissue damage, as described in 8.1.2.2.

8.1.2.2. *Local Effects.* Caustic soda is a strong alkali and is dangerous when improperly handled. It exerts a marked corrosive action on those tissues with which it comes in contact, with resulting burns, frequently deep ulceration and ultimate scarring. Severe burns result not only from contact with the solid alkali, but also from solutions of this compound. Even dilute solutions, on prolonged contact, exert a destructive effect on tissues. This chemical is a strong, primary irritant. Multiple small burns may result from exposure to dust or mist of this compound. Contact with the eyes, either in solid form or in solution, very rapidly causes severe damage to the delicate eye tissues.

Ingestion, either of the solid form or of the solution, results in severe damage to the mucous membranes or deeper tissues with which contact is made. As a result, perforation of these tissues may follow, or there may occur subsequent severe and extensive scar formation. Obviously, death may result if penetration into vital areas ensues. Scarring may so constrict or destroy damaged tissues, that extensive corrective surgery may be required.

Inhalation of the dust or concentrated mist of this compound may cause damage to the upper respiratory tract and even to the lung tissue proper, depending upon the severity of the exposure. The effects of inhalation may vary accordingly from mild irritation of the nasal mucous membranes to severe pneumonitis.

8.1.3. *CHRONIC TOXICITY*

8.1.3.1. *Systemic Effects.* None, except those secondary to tissue damage. These can be minimized or prevented by prompt first aid measures.

8.1.3.2. *Local Effects.* The chronic local effects, i.e., those arising from repeated contact with dilute solutions, may be those of multiple areas of superficial destruction of the skin in the form of primary irritant dermatitis. Similarly, inhalation of the dust or mist of this compound may result in varying degrees of irritation of the respiratory tract tissues.

8.2. Prevention and Control

Caustic soda may be handled safely, provided workers so engaged are fully and adequately instructed and supervised in approved methods and in the rigid execution of proper safety precautions. If such precautions are ignored and carelessness is tolerated, caustic soda is capable of producing serious injury or death.

8.2.1. *EMPLOYEE EDUCATION.* (see 6.4 Employee Education and Training)

8.2.1.1. It is necessary to provide those who are regularly employed in processes in which caustic soda is used with instructions describing the hazards, prevention and control measures, and recommended first aid treatment.

8.2.1.2. Food should never be stored nor eaten near caustic soda nor in the work area in which it is handled.

8.2.2. *VENTILATION.* No maximum allowable concentrations for dust or mist of this compound have been established. However, ventilating equipment should be installed if there is recognizable contamination of the workroom atmosphere.

8.2.3. *SAFETY SHOWERS AND EYE BATHS.* A readily accessible, well-marked, rapid action safety shower should be available in the area where caustic soda is being handled. Special eye washing fountains or a ready source of running tap water such as a bubbler drinking fountain or a hose with a soft, gentle flow of water should be available for eye irrigation. All safety equipment should be inspected periodically at fixed intervals to insure its being in working condition at all times.

8.3. Personal Protective Equipment

8.3.1. To secure safe working conditions, personal protective equipment is an essential supplement to, but not a substitute for, other precautionary measures such as careful and intelligent conduct on the part of the employees. The correct usage of personal protective equipment requires education of the worker in the proper employment of the materials available to him. Under conditions that are sufficiently hazardous to require personal protective equipment, the use of it should be supervised.

8.3.2. Employees who may be exposed to caustic soda should be provided with proper eye, respiratory, skin, and mucous membrane protection as follows:

(a) Close fitting industrial goggles,
(b) Rubber gloves and apron,
(c) Rubber safety-toe shoes or boots (tops should be covered by the trousers),
(d) Cotton coveralls, which should fit snugly at neck and wrists,
(e) Respirator of approved type, if dust or mist is present,
(f) Do not depend on protective creams or ointments to afford protection from this compound,
(g) Rescue harness and life line for those entering tanks or closed storage spaces.

8.4. First Aid and Medical Treatment

8.4.1. *GENERAL PLAN. Speed in removing caustic soda is of primary importance.*

First aid treatment should be started at once in all cases of contact with caustic soda in any form or serious injury may result. *Refer all injured persons to a physician, even when the injury appears to be slight.* Give the physician a detailed account of the accident.

8.4.2. *SPECIFIC ACTIONS*

8.4.2.1. *General First Aid.* Of prime importance is the copious and prolonged application of water to all affected areas at the first instant after exposure. Contaminated clothing should be removed promptly. *Irrigation and prolonged application of water to the affected areas should be continued for as long as one to two hours.* If 5% ammonium chloride or 5% zinc chloride solutions are on hand, wash the affected areas promptly and thoroughly with these solutions. However, if they are not immediately available, no time should be lost awaiting them, but copious amounts of water should be used. It is generally accepted that prolonged, copious irrigations with water are less damaging to tissues than are attempts at chemical neutralization. It should be borne in mind that in severe burns and those involving a large area of body surface, shock may supervene at any time. This should be treated promptly by placing the patient in a supine position and keeping him reasonably warm until the physician arrives. No oil or ointment of any kind should be applied to burned areas within the first 24 hours after contact or subsequently without the sanction of the attending physician. A physician should be called in attendance at the earliest possible moment.

8.4.2.2. *Contact With Eyes*

8.4.2.2.1. *First Aid.* If even minute quantities of caustic soda, either in solid form or in solution, enter the eyes, they should be irrigated immediately and copiously with water for a minimum of 15 minutes. The eye lids should be held apart during the irrigation to ensure contact of water with all the tissues of the surface of the eye and lids. A physician should be called in attendance at the first possible moment, preferably an eye specialist. If a physician is not immediately available, the eye irrigation should be continued for a second period of 15 minutes. After the first 15 minute period of irrigation is completed, it is permissible as a first aid measure to instill two or three drops of a 0.5% pontocaine solution or an equally effective aqueous topical anesthetic. No oils or oily ointments should be instilled unless ordered by the physician.

8.4.2.2.2. Ophthalmologists may be interested in a method of treatment for chemical burns of the eye described by Ralph S. McLaughlin, "Chemical Burns of the Human Cornea", *American Journal of Ophthalmology, 29*: 1355, 1946.

8.4.2.3. *Taken Internally.* Ingestion of caustic soda causes severe burns of the mucous membranes of the mouth, throat, esophagus and stomach. Here, since copious irrigation is not as feasible as on surface tissue, chemical neutralization may be attempted. Dilute vinegar or a 5% solution of ammonium chloride may be administered freely, if at hand. Unless they are immediately available, their use should not be contemplated, but the patient should be encouraged to drink a large quantity of water without delay. After free caustic soda has been diluted with water or chemically neutralized, whites of eggs or mineral oil may be administered for their demulcent or soothing effect. A stomach tube should not be inserted, except by the attending physician.

8.4.2.4. *Dermatitis.* Dermatitis of the primary irritant type may occur following single or repeated exposure to dilute solutions. This merely represents a relatively mild inflammatory response of tissues to exposures of low concentration. Repeated exposures result in an eczematoid condition of the skin, which may require prolonged treatment by a dermatologist.

The medical information in this publication has been supplied by the Medical Advisory Committee of the Manufacturing Chemists' Association.

Chemical Safety Data Sheet SD-5

NITRIC ACID

Copyright 1961 by Manufacturing Chemists' Association, Inc.

(Excerpts reprinted by permission)

PREFACE

Nitric acid is a colorless to light brown liquid with an acrid odor. It is capable of causing severe skin and eye burns and its vapor, especially the oxides, can cause damage to the lungs.

It is classified by the Interstate Commerce Commission as a corrosive liquid. It is not flammable, but will react readily with most chemicals and strong nitric acid in contact with wood or other organic materials may cause fire.

The full text of this data sheet should be consulted for details of the hazards of nitric acid and suggestions as to their control.

1. NAMES

Chemical Name: Nitric Acid
Common Names: Nitric Acid, Aqua Fortis, Hydrogen Nitrate
Formula: HNO_3

2. PROPERTIES

2.1. Grades and Strengths

2.2. Properties and Characteristics

	100 %	Aqueous Solution
Physical State	Liquid	Liquid
Explosive Limits	Non-flammable	Non-flammable
Boiling Point	86°C (186.9°F)	Constant boiling mixture 68 to 68.5% HNO_3—121.6°C (251°F)
Color	Colorless	Colorless to light brown
Corrosivity	Will vigorously attack most metals	
Hygroscopicity	Yes	Yes
Light Sensitivity	Water white acid becomes amber to brown depending upon strength of acid and exposure	
Melting Point	−42°C (−43.6°F)	See 2.1.
Odor	HNO_3 vapors acrid—Oxides of Nitrogen sweet to acrid	
Reactivity	Will react readily with most chemicals. Strong nitric acid in contact with wood and some other organic materials may cause fire	
Specific Gravity	1.502	See 2.1.
Threshold Limit	10 ppm or 25mg/m³	

3. HAZARDS

3.1. Health Hazards

Nitric acid in either liquid or concentrated vapor form produces severe burns on contact with the skin or eyes. Inhalation of the vapors or of its gaseous oxides is injurious to the lungs. THE ONSET OF SYMPTOMS FOLLOWING THE INHALATION OF VAPORS MAY BE DELAYED FOR MANY HOURS.

3.1.1. *WARNING PROPERTIES*. The dangerous oxides of nitrogen are very insidious and exposure to low concentrations may not be recognized. Heavy evolution of oxides is easily recognized by the light to deep reddish-brown fumes. The vapors of nitric acid can be detected by their acrid odor.

3.2. Fire and Explosion Hazards

3.2.1. Nitric acid will vigorously attack most metals. Strong nitric acid may cause spontaneous ignition when in contact with organic materials such as sawdust, excelsior, wood scraps and shavings, paper, cotton waste and burlap bags. If fire starts, it will burn vigorously.

3.2.2. Nitric acid corrodes most metals, especially iron or steel, depending upon strength of acid.

3.2.3. Nitric acid may cause explosion when in contact with hydrogen sulfide and certain other chemicals.

3.2.4. Nitric acid or its vapor will, under certain conditions, nitrate wood, wood cellulose, cotton and similar organic materials, lowering their ignition temperature and greatly increasing their flammability. (See Section 6—Fire Fighting)

4. ENGINEERING CONTROL OF HAZARDS

4.1. Building Design

4.1.1. Operations in which large scale formation of gaseous oxides of nitrogen may occur should be housed in one-story buildings. In case this is impossible, easily accessible exits should be provided to permit rapid evacuation of the area. Special emergency ventilation may be required, arranged to be started in the event of the rapid formation of large quantities of gaseous oxides.

4.2. Equipment Design

4.2.1. The design of process piping and equipment is highly specialized. The technical problems of designing equipment, providing adequate ventilation and formulating operational procedures that insure maximum security and economy, can be handled best by engineers and safety specialists.

4.3. Ventilation

4.3.1. Workrooms where nitric acid is handled should have ventilation adequate to reduce the concentration of nitric acid vapor to less than 10 ppm, and nitrogen oxides to less than 5 ppm.

4.3.2. Since the gaseous oxides are heavier than air, hoods and downdraft exhaust systems should be used where general ventilation is inadequate. Such systems should be inspected regularly at frequent intervals.

4.4. Air Analysis

4.4.1. A portable detector for nitrogen dioxide, with a range of 1 to 500 ppm, is commercially available. It operates on the colorimetric principle with a standard color plaque for comparing results.

4.5. Electrical Equipment

4.5.1. Electrical fixtures should be of vapor-proof type to protect against corrosive action of acid vapor. All wiring and other electrical equipment should conform to the National Electrical Code and be suitable for corrosive atmospheres.

5. EMPLOYEE SAFETY

6. FIRE FIGHTING

7. HANDLING AND STORAGE

8. TANK AND EQUIPMENT CLEANING AND REPAIR

9. WASTE DISPOSAL

Dilute and neutralize before disposal.

Do NOT flush down drains where the acid will eventually pollute streams, city sewage systems, etc.

10. MEDICAL MANAGEMENT

10.1. Health Hazards

10.1.1. *GENERAL.* On contact with the skin or eyes, nitric acid produces severe burns. Inhalation of the vapor, as well as inhalation of most of the oxides, is injurious to the lung. THE ONSET OF SYMPTOMS FOLLOWING INHALATION OF THE VAPORS MAY BE DELAYED FOR SEVERAL HOURS.

10.1.2. *ACUTE TOXICITY.*

10.1.2.1. *Systemic Effects.* The toxic effects of nitric acid and of the oxides differ somewhat. It is a combination of these compounds that is commonly encountered in industry, since the oxides may be formed wherever nitric acid is used and these may be mixed with nitric acid vapor.

Gaseous oxides of nitrogen consist of nitric oxide (NO), nitrous oxide (N_2O), nitrogen trioxide (N_2O_3), and nitrogen dioxide (NO_2 and N_2O_4). The chemical relationship of these compounds is so close that they seldom occur separately in industry. Toxicologically the most important are the two forms of nitrogen dioxide, N_2O_4 (colorless) and NO_2 (dark brown), both highly toxic. The color of the gaseous oxides varies from colorless to chocolate brown, depending upon the percentage composition of the mixture, which is largely a function of temperature. A toxic concentration of the gaseous oxides may therefore be dark brown or colorless. The intensity of color is not an indicator of the degree of danger.

Gaseous oxides are formed when nitric acid comes in contact with certain metals, e.g., copper, brass, zinc, or with any organic material, e.g., wood, sawdust, cloth, paper. Gaseous oxides are also present in hazardous concentrations during various operations, e.g., burning of certain explosives, nitration processes, processes involving the use of nitric acid, electric arc welding, soldering, the operation of diesel engines, and many other operations.

Evidences of damage to the lung following exposure to the oxides of nitrogen characteristically appear after a delay of 4–30 hours. This, in the form of edema, may be severe and sometimes fatal.

The breathing of an atmosphere containing 25 ppm through an 8 hour period may cause pulmonary signs and symptoms. Pulmonary edema may follow exposure for only $\frac{1}{2}$ to 1 hour to higher concentrations (100–150 ppm). A few breaths of the gaseous oxides in a concentration of 200–700 ppm will cause severe pulmonary damage, which may prove fatal within 5 to 8 hours.

10.1.2.2. *Local Effects.* Nitric acid or its concentrated vapors will produce immediate severe and penetrating burns to the skin and membranes. Contact with the eyes will produce very severe, immediate damage and may result in permanent damage, with visual impairment.

10.1.3. *CHRONIC TOXICITY.*

10.1.3.1. *Systematic Effects.* The nature of nitric acid precludes the ingestion of sufficient quantities to produce any systemic effects except those which might be secondary to irritation to the tissues of the eyes, nose, or upper respiratory tract.

10.1.3.2. *Local Effects.* Repeated exposure of the skin or of the eyes, nose, throat, and upper respiratory tract may cause chronic irritation.

10.2. Preventive Health Measures

Nitric acid is not a serious industrial hazard if workers are adequately instructed and supervised in the proper means of handling the chemical. Contact with the skin and eyes as well as the inhalation of its vapors and oxides should be avoided. A threshold limit value of 10 ppm for nitric acid fumes and 5 ppm for nitrogen dioxide fumes has been suggested as the safe concentration in air for an eight-hour exposure, In addition, even short time exposures to higher concentrations (100–500 ppm should be avoided.

10.2.1. *PERSONAL HYGIENE.* Properly designed emergency showers and eye baths should be placed in convenient locations wherever nitric acid is used. All employees should know the location and operation of such equipment. It must be frequently inspected to make sure it is in proper working condition.

Personal protective equipment for workers who are exposed to contact with nitric acid is described in 5.2.

10.2.2. *PHYSICAL EXAMINATION.*

10.2.2.1. *Preplacement Examinations.* It may be desirable to exclude from potential exposure to nitric acid prospective employees with the following conditions:

(a) Those with only one functioning eye
(b) Those with uncorrected, severe, faulty vision
(c) Those who have chronic diseases of the upper respiratory tract or lung
(d) Those with severe preexisting skin lesions.

10.2.2.2. *Periodic Health Examination.* No specific type of periodic health examination is needed.

10.3. Suggestions to Physicians

The main feature of the clinical pattern following contact with nitric acid or its oxides is the delayed onset of pulmonary edema following sufficient vapor exposure. This may occur 4 to 30 hours after an initial contact that was virtually asymptomatic. Employees must, therefore, be placed under observation for this period after any vapor exposures that were not patently insignificant. The occurrence of tightness in the chest, rales and an elevated white blood cell and platelet counts are important premonitory signs of impending pulmonary edema.

In the treatment of vapor exposure, the use of oxygen and the corticosteroids has proved helpful. Oxygen is administered at normal or elevated pressures following exposure until the danger of pulmonary edema has past (see Section 11.5.1.1). The corticosteroids are administered whenever any signs of developing pulmonary damage appear and are continued until recovery occurs.

Chemical Safety Data Sheet SD-4*

PHENOL

Copyright 1964 by Manufacturing Chemists' Association, Inc.

(Excerpts reprinted by permission)

PREFACE

Phenol is a colorless to light pink crystalline material which melts at 40°–41°C. It is highly hazardous when not handled with care. In liquid, solid, vapor, droplet or as a solution it exerts a local corrosive effect and is readily absorbed through the skin, mucous membranes, gastrointestinal, and respiratory tracts. Massive exposure may result in collapse and death despite prompt emergency care. The Interstate Commerce Commission lists phenol as a Class B poison material.

Phenol will burn if ignited or if involved in a fire, giving off toxic vapors.

This chemical safety data sheet contains material published hitherto in MCA Manual TC-6 "Tank Cars—Unloading when filled with phenol." The full text should be consulted for details of the hazards of phenol and suggestions for their control.

1. NAME

Chemical Names:	Phenol, Monohydroxybenzene
Common Names:	Phenol, Carbolic Acid
Formula:	C_6H_5OH

2. PROPERTIES

2.1. Grades and Strengths

USP: solid and water solutions, 82 to 92%.
Technical: 82 to 92%, containing cresol.

2.2. Properties and Characteristics of USP Grade

Physical State	... Liquid or solid
Flammable limits	... Lower limit approx. 1.5%
Flash Point (Tag. open-cup) ...	... 85°C (185°F)
(closed cup) ...	... 79°C (174°F)
Boiling Point (760 mm)	... 180° to 182°C (356 to 360°F)†
Color	... Colorless to light pink solid
Deliquescent...	... Yes

*This Chemical Safety Data Sheet has been prepared as an activity of the Safety and Fire Protection Committee of the Manufacturing Chemists' Association. Other MCA committees which have co-operated in its preparation include:
Air Pollution Abatement Committee
Chemical Packaging Committee
Committee on Tank Cars, Tank Trucks, and Portable Tanks
Labels and Precautionary Information Committee
Medical Advisory Committee
Transportation and Distribution Committee
Water Resources Committee
†USP water solutions start to boil at lower temperatures; e.g., 85%: 104°C (219°F); 92%: 112°C (234°F).

Hygroscopic	...	...	...	...	Yes
Ignition Temperature		...	...	...	715°C (1319°F)
Light Sensitive	...	...	...	...	Yes. Darkens slowly on exposure to light
Melting Point	...	...	...	...	40° to 41°C (104° to 106°F)*
Odor ...	...	..:	...	...	... Characteristically sweet
Reactivity	...	...	...	...	Not dangerously reactive
Solubility	...	...	...	...	In water, 6.7 g/100 ml at 16°C (61°F), soluble in all proportions at 66°C (151°F). Also soluble in alcohol and other organic solvents.

Specific Gravity
    Solid (25°C/4°C) 1.071
    Liquid (50°C/4°C) 1.049
Threshold Limit Value (8 hr working day) 5 ppm or 19 mg/m³
Threshold (Odor)   ...   . 0.3 ppm
Vapor Density (Air = 1) 3.24

3. HAZARDS

3.1. Health Hazards (see Section 10 for details)

Phenol is highly toxic when handled improperly. It exerts locally a strong corrosive action on body tissues and produces severe systemic reactions after absorption through the skin and mucous membranes, the gastrointestinal tract, or the lungs.

3.1.1. Phenol has good warning properties due to its characteristic odor and irritation to the skin and mucous membranes.

3.2. Fire Hazards

Phenol, having a flash point of 185°F, is safe to handle at ambient temperatures, but will burn if ignited or if involved in a fire. Flammable toxic vapor will be given off at elevated temperatures, should the material become involved in fire.

4. ENGINEERING CONTROL OF HAZARDS

5. EMPLOYEE SAFETY

6. FIRE FIGHTING

7. HANDLING AND STORAGE

8. TANK AND EQUIPMENT CLEANING AND REPAIRS

9. WASTE DISPOSAL

9.1. All local and state water pollution regulations should be determined and complied with.

---

*USP water solutions start to melt at lower temperatures, e.g., 85%: 6°C (43°F); 92%: 16°C (61°F). Technical grades containing 82 to 92% phenol also melt at lower temperatures [30.5° to 36.2°C (87° to 97°F dry basis)], but have somewhat higher boiling points [at least 95% evaporates at 187°C (369°F)].

9.2. Phenol may be recovered, and possible pollution-causing discharges avoided, by charcoal absorption, solvent extraction or steam stripping. Minimum concentrations of 1% by weight are generally necessary for economical recovery.

9.3. Phenol is water soluble and is amenable to biological or chemical oxidation. Larger poundages of phenol may be more economically removed by biological oxidation. It is generally necessary to acclimate the microorganisms to phenol, so care should be taken to avoid shock loadings on biological treatment plants. Discharge to municipal sewers may offer a satisfactory solution, provided approval of the sewerage authority is obtained. Aqueous phenol solutions may be chemically oxidized by chlorine, chlorine dioxide or other oxidants. Disposal may also be accomplished by burning contaminated wastes if proper burning methods are practiced so as not to create an air pollution problem.

9.4 If phenol-bearing wastes are to be treated in oxidation ponds or disposed of in impounding basins, or otherwise discharged to the ground, necessary precautions must be taken to prevent ground water contamination. Care must also be taken to keep unprotected workers and the general public away from such disposal areas.

9.5. Phenol is classed as a taste and odor producing compound in water supplies, especially those which are chlorinated. The U. S. Public Health Service has stated in their publication, "Drinking Water Standards—1962," that the phenol content of a water supply should not be in excess of 0.001 mg/l, where other more suitable supplies are or can be made available.

10. MEDICAL MANAGEMENT

10.1. Health Hazards

10.1.1. *GENERAL.* Phenol is highly hazardous when not handled with care. In liquid, solid, vapor, droplet or as a solution it exerts a local corrosive effect. It is readily absorbed through the skin, mucous membranes, gastrointestinal, and respiratory tracts.

10.1.2. *ACUTE TOXICITY*

10.1.2.1. *Systemic Effects.* Generalized symptoms may develop rapidly after any route of exposure. These symptoms include weakness, mental confusion, rapid irregular pulse and breathing. Collapse and death may occur in a few minutes after massive exposure despite prompt emergency care.

10.1.2.2. *Local Effects.* Phenol has a marked corrosive effect on any tissue. Since it is a skin anesthetic, the first reaction is not pain, but a whitening of the exposed area. A serious burn or systemic poisoning may occur if the chemical is not removed promptly and thoroughly.

Swallowing of the liquid results in severe corrosive injury to the mouth, throat and stomach.

10.1.3. *CHRONIC TOXICITY*

10.1.3.1. *Systemic Effects.* Chronic systemic poisoning has been reported, but is probably extremely rare.

10.1.3.2. *Local Effects.* Dermatitis may result from repeated or prolonged skin contact with low concentrations of phenol in any form.

10.2. Protective Health Measures

Phenol is not an industrial hazard if workers are adequately instructed and supervised in the proper methods of handling the chemical. If these safety precautions are ignored or carelessness is tolerated, serious injury may occur.

10.2.1. *Personal Hygiene.* Emergency showers and eye baths should be placed in convenient locations wherever phenol is being used. Employees should be instructed that direct contact with the chemical requires the immediate application of large amounts of water to the contaminated area.

10.2.2. *EMPLOYEE EDUCATION* (refer to Section 5.1 for details).

10.2.3. Food should neither be stored nor eaten in a workroom where phenol is stored or used.

10.2.4. *VENTILATION.* The concentration of phenol vapor in the atmosphere under normal working conditions should never be high enough to cause irritation of the eyes and mucous membranes.

A threshold limit value of 5 ppm in air has been set by some agencies as the maximum safe concentration for daily 8 hour exposure.

10.2.5. *Personal Protective Equipment.* No personal protective equipment is an adequate substitute for safe working conditions and intelligent conduct on the part of employees who handle phenol. (Refer to Section 5.2 for details of protective equipment.)

CAUTION: *Protective creams do not afford adequate skin protection.*

10.2.6. *PHYSICAL EXAMINATIONS.*

10.2.6.1. *Preplacement Examinations.* Most employees may be assigned to processes in which the handling of phenol is carefully controlled. Since sensitivity to phenol has been reported, the physician should inquire regarding any evidence of unusual sensitivity to this chemical. Individuals with evidence of liver or kidney disease should not be assigned routinely to processes that may involve exposure to phenol.

10.3. Suggestions to Physicians

No specific antidote is known for phenol poisoning and treatment is symptomatic.

10.3.1. *OXYGEN ADMINISTRATION.* Oxygen has been found useful in the treatment of inhalation exposures of many chemicals, especially those capable of causing either immediate or delayed harmful effects in the lungs.

In most exposures, administration of 100 % oxygen at atmospheric pressures has been found to be adequate. This is best accomplished by use of a face mask having a reservoir bag of the nonrebreathing type. Inhalation of 100 % oxygen should not exceed one hour of continuous treatment. After each hour, therapy may be interrupted. It may be reinstituted as the clinical condition indicates.

Some believe that superior results are obtained when exposures to lung irritants are treated with oxygen under an exhalation pressure not exceeding 4 cm water. Masks providing for such exhalation pressures are obtainable. A single treatment may suffice for minor exposures to irritants. It is believed by some observers that oxygen under pressure is useful as an aid in the prevention of pulmonary edema after breathing irritants.

In the event of an exposure causing symptoms or in the case of a history of severe exposure, the patient may be treated with oxygen under 4 cm exhalation pressure for one-half hour periods out of every hour. Treatment may be continued in this way until symptoms subside or other clinical indications for interruption appear.

11. FIRST AID

11.1. General Principles

After severe exposure to phenol vapors or air droplets, it is important to remove the patient from the contaminated area.

In case of skin or eye exposure the chemical must be removed immediately or severe injury may result.

Call a physician.

11.2. Contact with Skin and Mucous Membranes

The most important part of the treatment is removing the chemical by large amounts of water immediately after the accident. If skin contact is extensive, the employee should get under the shower immediately. Clothing, including shoes and socks, can be removed while under the shower. Continue washing until all odor of phenol has disappeared.

Salves and ointments should not be applied to skin or mucous membrane burns.

11.3. Contact with the Eyes

If phenol in either the solid, liquid or vapor form enters the eyes, they should be irrigated immediately and copiously with water for at least 15 minutes. The eye lids should be held apart during the irrigation to ensure the removal of the chemical from all the tissues of the eye surfaces and lids. A physician, preferably an eye specialist, should be called at the first possible moment. If a physician is not immediately available, the irrigation should be continued for a further 15 minute interval. After the first 15 minute period of irrigation and if pain is still present, it is permissible as a first aid measure to instill 2 or 3 drops of a 0.5 % pontocaine solution or an equally effective aqueous topical anesthetic. No oils or oily ointments should be instilled unless ordered by the physician.

11.4. Taken Internally

If a person has swallowed phenol, the injury that occurs will be due to the corrosive action on the mouth, esophagus and stomach and to its systemic toxicity. The patient should instantly drink large quantities of water in order to reduce the concentration of the chemical. If vomiting does not occur spontaneously, induce vomiting by giving a warm salt solution (2 tablespoons to a glass of water) or tickling the back of the throat. If the patient is in shock, has severe pain or is unconscious, vomiting should not be induced.

Call a physician immediately.

Keep the patient warm, but not hot.

11.5. Inhalation

Exposed persons should be removed immediately from the contaminated atmosphere and a physician called. If breathing has ceased, effective artificial respiration should be initiated at once.

If oxygen inhalation apparatus is available, oxygen should be administered, but only by a person authorized for such duty by a physician.

If the patient is conscious, the irritation to the throat may be relieved by washing the throat with water.

Documentation of Threshold Limit Values*

American Conference of Governmental Industrial Hygienists

PREFACE

The documented evidence appearing in this publication has been taken mainly from technical papers and texts of industrial hygiene and toxicology that have appeared over the years and from the experience and knowledge of committee members. All informational sources are referenced. Occasionally, written communications to the Committee members have been used to supplement formally published material, when such information has been substantiated by valid observation. All of these materials have been used in deriving the values given in the table of threshold limits, which is published annually by the American Conference of Governmental Industrial Hygienists.

The format is designed to supply in brief, the best available technical evidence substantiating the choice of the limiting concentrations. Substances are listed in alphabetical order for ready reference, irrespective of their categoric listings in the threshold limits list. Because evidence on individual substances in this compilation is subject to additional modification as new data appear, this second edition, coming three years after the first, has been almost completely rewritten and enlarged with numerous new references on over 90 substances, bringing the text up-to-date, as well as new additions in the amount of about 130 substances.

The Committee regards these Documentations as essential to the proper implementation of the limits of the substances to which they apply. Each Documentation contains a clear statement of the response against which the limit is designed. The Committee welcomes suggestions or new data that will maintain the usefulness of this publication. Further editions will be published and offered for sale as new information justifies their appearance.

Communications should be addressed to the Chairman of the Committee, Dr. H. E. Stokinger, or to the Secretary-Treasurer of the ACGIH, 1014 Broadway, Cincinnati, Ohio 45202.

Individuals contributing to the Documentations:

Harry B. Ashe
Edward J. Baier
Allan L. Coleman
W. Clark Cooper, M.D.
Hervey B. Elkins
William G. Fredrick
Bernard Grabois
Paul Gross, M.D.

Wayland J. Hayes, Jr., M.D.
Keith H. Jacobson
Harold N. MacFarland
Ernest Mastromatteo, M.D.
William F. Reindollar
Russel G. Scovill
Ralph G. Smith
Mitchell R. Zavon, M.D.

Herbert E. Stokinger, Chairman

ACRYLAMIDE—Skin

0.3 mg/m³

Acrylamide is used as a reactive monomer and intermediate in the production of organic chemicals. It has a wide variety of uses as a polymer or copolymer in such

*The example documentations that are reprinted here by permission are excerpted from the large group of documentations published by the committee.

applications as: adhesives, fibers, paper sizes, molded parts, water coagulant aids and textiles. It is quite reactive and is known to polymerize with violence when heated.[1]

The oral LD_{50} for laboratory animals is reported to be in the range of 150–180 mg/kg, which rates it as of moderate toxicity.[2] It thus has a relatively comparable toxicity for all species of experimental animal in which it has been administered. It produces an unusual toxic effect on the central nervous system, which is manifested by muscular weakness, ataxia, incoordination, tremors, and hallucinations. The hind quarters of animals are affected more than the fore quarters. Cases of occupational poisoning in man with this type of clinical picture have been noted, according to Fassett.[3]

Toxic effects may be produced by any route of administration—ingestion, inhalation, injection, skin contact or contact with the eye. Dogs developed the neurologic syndrome within 24 hours when given single oral doses of 100 mg/kg. Cats, the most sensitive species, given 1 mg/kg daily by intravenous or intraperitoneal injection developed the same picture in about 6 months; however, dietary levels in this species between 0.3–1 mg/kg represented a "no-ill effect" dosage. The available studies indicate that the pathologic process is probably limited to the brain, but extensive histopathologic study of the central nervous system in severely poisoned cats have not revealed any significant abnormalities. Kuperman[4] reported that the toxic disturbance was probably subcortical and in the midbrain.

The toxic effects on the central nervous system appear to be quickly reversible in mild cases of poisoning, if exposure is terminated; if exposure continues, however, the recovery period may be greatly prolonged. From the results of feeding experiments in the most sensitive animal species it was recommended[2] that no more than 0.05 mg/kg/day be absorbed by workmen. Assuming a respiratory exchange of 10 m^3/day, a TLV of 0.3 mg/m^3 (0.1 ppm) may be calculated. This limit has proven to be a practical working limit for the prevention of nervous disorder.[6]

Fassett in his review[3] noted that the action of acrylamide probably arises from its conversion to a more toxic material by some metabolic process in view of the delay in onset and the need for what appeared to be a definite threshold dose before symptoms became apparent. Stokinger[5] commented on the unusual toxicity of this compound and the fact that there is an anamnestic response—following recovery from the effects of poisoning, the same syndrome is recalled with lesser amounts on reexposure.

REFERENCES

1. American Cyanamid Co., "Chemistry of Acrylamide," Technical Bulletin, Revised (March 1956).
2. McCollister, D. D., Oyen, F., Rowe, V. K., *Tox. & Appl. Pharm.*, 6, 172 (1964).
3. Fassett, D. W., in "Industrial Hygiene and Toxicology," F. A. Patty, Ed., 2nd Revised Ed., Vol. II. Toxicology, p. 1832, Interscience Publishers. J. Wiley & Sons, New York (1963).
4. Kuperman, A. S., *J. Pharm. Exptl. Therap.*, *123*, 180 (1958).
5. Stokinger, H. I., *Am. Ind. Hyg. Assn. J.*, *17*, 340 (1956).
6. Shaffer, C. B., personal communication to Committee (1966).

n-BUTYL GLYCIDYL ETHER (BGE)—Skin

50 ppm (approximately 270 mg/m^3)

The only reported data on BGE are those of Hine, et al.,[1] in which animal studies and limited human exposures are described. On animals, BGE is classified as a mild skin irritant and a mild eye irritant. The LC_{50} for mice (4 hours) was greater than 3500 ppm, and for rats (8 hours) 670 ppm. Intragastric LD_{50} values were 1.52

and 2.26 g/kg for mice and rats, respectively, while the percutaneous LD_{50} for rabbits was 4.93 g/kg. In addition, the intraperitoneal LD_{50} values were 1.14 and 0.70 g/kg, respectively, for rats and mice. Signs of delirium and depression were noted following intragastric administration. Some focal inflammatory cells were observed in livers of animals exposed to the vapors. No chronic studies were performed.

On the basis of this single publication, a TLV of 50 ppm is recommended. This level should provide an environment relatively free from irritation, but may cause some response in sensitized persons.[2]

REFERENCES

1. Hine, C. H., Kodama, J. K., Wellington, J. S., Dunlap, M. K., and Anderson, H. H.: *Arch. Ind. Health*, 14, 250 (1956).
2. Patty, F. A.: "Industrial Hygiene and Toxicology," 2nd Revised Edition, Vol. II, p. 1605, Interscience Pub., N.Y. (1962).

CARBON MONOXIDE

50 ppm (approximately 55 mg/m^3)

The inhalation of carbon monoxide, a colorless, odorless gas of specific gravity similar to that of air, causes asphyxiation (anoxia, hypoxia) by forming metastable chemical compounds primarily with hemoglobin and secondarily with other biochemical constituents, which in a complex manner reduces the availability of oxygen for the cellular systems of the body. The resulting physiologic effect is similar to, but in some respects more serious than, a simple lack of oxygen caused by a reduced partial pressure of oxygen in inspired air due to altitude or dilution with an essentially physiologically inert gas.

The health effect on exposed workers is importantly dependent upon the carbon monoxide content of the blood, the partial pressure of the oxygen in the air breathed, the duration of exposure, the ambient temperature, the work effort (oxygen demand), the metabolic efficiency of the worker, his health status, genotype and degree of and capacity for inurement to exposure.

The rate of uptake of carbon monoxide by blood, when air containing carbon monoxide is breathed, increases from three to six fold between rest and heavy work output.[1-7]

The rate of uptake is influenced by the partial pressure of oxygen in the air breathed, and increases with altitude.

The rate of uptake is dependent upon pulmonary diffusion; normal adult males, during moderate exercise show a carbon monoxide diffusion capacity ranging from 10.6 to 49.2 cc per minute.[8]

The absorption process for air containing less than 200 ppm of CO by volume will be substantially complete in from two to twelve hours, as will its elimination in normal air. At normal activity, half of the carbon monoxide in the blood may be lost in four hours, but chronically exposed people such as cigarette smokers may take 8 hours or longer.

The equilibrium concentration of carbon monoxide reacting with the hemoglobin of the blood for men at work is usually substantially complete in six to eight hours. When the air contains 100 ppm of CO, the blood at equilibrium will contain 12–14% of HbCO (carbon monoxide hemoglobin); 50 ppm of CO, 6–8% HbCO; 25–30 ppm of CO, 3–4% HbCO.[9,10,11]

The blood of cigarette smokers will contain from 2–10% HbCO and non-exposed adults will show a normal average background of 1% HbCO formed within the body.

Symptoms such as headache, fatigue and dizziness appear in healthy workers engaged in light labor near sea level when about 10% of the hemoglobin is combined with carbon monoxide. Such degree of saturation could be achieved by continuously breathing air containing 50 ppm of CO for about six to eight hours.[12-17]

Disturbance of coordination, judgment, psychomotor tasks and visual acuity appear at about 2% HbCO, but do not become importantly significant until about 5% HbCO saturation is reached. Choice discrimination errors, reaction time, etc., show a marked increase after an exposure of 4 hours to 100 ppm CO. The reduction of the threshold of light sensitivity of the eye at 5% HbCO saturation is equivalent in magnitude to that caused by an altitude of 8000–10,000 feet above sea level.[18-20]

The effect of carbon monoxide exposure on man is enhanced by many environmental factors such as heavy labor, high environmental temperatures, altitude above 2000 feet and concomitant presence of narcotic solvents in the air breathed.

Variation of individual susceptibility is great. People who show unusual sensitivity include the young and the aged, pregnant women, people with heart trouble and those with existing or potential hypoxia for whatever cause—anemia, thyroid disease, alcoholism, effects of many drugs and poor pulmonary function, including asthma and bronchitis. Variations of 25% have been noted within a given species.[1,2,6,7,11,13,14,16]

Basis genetic differences, such as glucose-6-phosphate dehydrogenase deficiency in red cells and type S hemoglobin, may be significant because of the large number of people who exhibit such abnormalities. Carbon monoxide also reacts with many pigments and enzymes, but these effects are still poorly delineated.[21,22]

Inurement to the action of carbon monoxide can occur in some people and may explain why some groups of exposed workers seem to tolerate carbon monoxide better than others and this is especially true for cigarette smokers, who frequently show HbCO values of 5–10%.[1,14,16,23]

A threshold limit value for carbon monoxide of 100 ppm has been extant for many years, but no substantial justification for its validity under ordinary circumstances of work can be found. The often quoted work of Henderson and Haggard[24] states that a 3-hour exposure at 100 ppm produces no effect, but six hours' exposure produces a perceptible effect and nine hours' exposure causes headache and nausea. Drinker,[1] pp. 87–88, states that the safety constant must be reduced to 1/3 or less under exercise or work. The Committee believes that the apparent success of the 100 ppm value was due to inurement and accumulation of individuals of low degree of susceptibility. A great deal of confusion has resulted over the years from the summation of thousands of publications on the effects of carbon monoxide on man because of serious difficulty in the determination of low levels of the compound in air and blood. It now seems reasonably certain that an equilibrium exposure of man to 50 ppm of carbon monoxide will result in a HbCO value of 8–10%, and that under usual conditions of work and rest periods an "end of work day" level of 5–6% would be expected. A crew of workers in the Holland Tunnel, who worked 2 hours in, and two hours out, for eight-hour "swing" shifts in an average tunnel CO concentration of 70 ppm, showed an average of 5% HbCO with no one above 10%. Under this average exposure of 25–50 ppm, no symptoms or health impairments were found,[25] or would be expected.

Men exposed continuously for many days in a submarine at 50 ppm CO complained of headache, but a 60-day exposure of 40 ppm CO was without effect.[26]

Schulte[27] states that exposure at 100 ppm of CO for over 4 hours is excessive and recommends a maximal exposure of 50 ppm CO for exposures over 4 hours' duration.

Studies made upon a group of healthy young men exposed for a prolonged period at 44 ppm CO produced no adverse reactions on their general health.[28]

The threshold limit value for carbon monoxide in the U.S.S.R. and in Czechoslovakia is 18 ppm.

It would appear to the Committee that for conditions of heavy labor, high temperatures or work 5000–8000 feet above sea level, the threshold limit value should be appropriately reduced to 25 ppm. No further benefit under any circumstances could be expected by reducing the level below 5–10 ppm, since at this concentration one is practically in equilibrium with the normal blood level of around 1% HbCO.

The recommended TLV for CO of 50 ppm is thus based on an air concentration that should not result in blood CO levels above 10%, a level that is just below the development of signs of borderline effects.

REFERENCES

1. Drinker, Cecil K., "Carbon Monoxide Asphyxia," pp. 81–87, Oxford Univ. Press, New York (1938).
2. "Handbook of Medical Physiology," 11th ed., pp. 594–95, C. V. Mosby Co., St. Louis, Mo. (1961).
3. Bosaeus, E., and Friberg, L., *Acta Physiol. Scand.*, *39*, 176 (1956).
4. Forbes, Sargent, and Roughton, *Am. J. Physiol.*, *143*, 594 (1945).
5. Gaensler, et al., *J. Lab. and Clin. Med.*, *49*, 945 (1957).
6. Groszkopf, Karl, "Darstellung der toxischen Wirkung des Kohlenoxyds," *Drager-Hefte*, 236, 5180–82, published by Drager Werk, Lubeck, W. Germany, (1959).
7. "Handbook of Respiratory Physiology," Air Univ. U.S.A.F. School of Aviation Medicine, Randolf Field, Texas, Chap. 5, by F. J. W. Roughton (1954).
8. Hanson, J., and Tobakin, B., *J. Appl. Physiol. 15*, 402 (1960).
9. Ringold, A., et al., *Arch. Env. Health 5*, 388 (1962).
10. Smith, Ralph G., "A Study of Carboxyhemoglobin Levels Resulting from Exposure to a Low Concentration of Carbon Monoxide." Report, Dept. of Industrial Medicine and Industrial Hygiene, Wayne State Univ., Detroit, Mich. 48207 (1960).
11. "California Standards for Ambient Air Quality and Motor Vehicle Exhaust," State of California, Dept. of Public Health, Berkeley, Calif. 94704 (1959).
12. Katz, M., *Can. Med. Assn. J.*, *78*, 182 (1958).
13. Mayers, M. R., "Carbon Monoxide Poisoning in Industry and Its Prevention," New York State Dept. of Labor, Special Bulletin No. 194 (1938).
14. von Oettingen, W. F., U. S. Public Health Service Bulletin No. 290, p. 81 (1944).
15. Lindgren, S. A., *Acta. Med. Scand. Supplement 356 to Vol. 167*, pp. 1–135 (1961).
16. Sayers, R. R., et al., U. S. Public Health Service Public Health Bulletin No. 186, p. 21 (1929).
17. Pfrender, R., *Ind. Med. Surgery*, *31*, 90 (1960).
18. Schulte, J. H., *Arch. Env. Health*, *7*, 425–30 (1963).
19. McFarland, R., *Am. Ind. Hyg. Assn. J.*, *24*, 209 (1963).
20. Halperin, M. H., et al., *J. Physiol.* (*London*), *146*, 583 (1959).
21. "The Biochemistry of Clinical Medicine," 3rd ed., p. 74, W. S. Hoffman, Year Book Medical Publishers, Chicago (1964).
22. "Biochemistry of Human Genetics," p. 76, Ciba Foundation (1959).
23. Killick, E. M., *J. Physiol.*, *108*, 27 (1948).
24. Henderson, Y. and Haggard, H. W., "Noxious Gases and the Principles of Respiration Influencing Their Action," 2nd ed., pp. 167–68, Reinhold Publishing Co., New York (1943).
25. Sievers, Rudolph, et al., "A Medical Study of Men Exposed to Measured Amounts of Carbon Monoxide in the Holland Tunnel for Thirteen Years," U. S. Public Health Service Public Health Bulletin No. 278.
26. Ebersole, J. E., *N. Eng. J. Med.*, *262*, 599 (1960).
27. Schulte, J. H., *Arch. Env. Health*, *8*, 438 (1964).
28. Schulte, J. H., *Mil. Med.*, *126*, 40 (1961).
29. Stern, A. C., "Summary of Existing Air Pollution Standards," Div. Air Pollution, Pub. Health Serv., U. S. Dept. of Health, Ed. & Welfare (1963).

DIAZOMETHANE

0.2 ppm (approximately 0.34 mg/m^3)

According to Flury and Zernik,[1] diazomethane is highly irritating to the entire respiratory tract; ten minutes' exposure at 175 ppm was fatal to cats within three days.

Fairhall states that diazomethane is undoubtedly one of the most dangerous products of the chemical laboratory, having caused irritation of eyes, dizziness, denudation of mucous membranes, and sensitization in laboratory workers. Chest pains, fever and severe asthmatic attacks have been reported.[2]

Sunderman et al.[3] report a case of diazomethane poisoning in a chemistry student. Six days after exposure, he was hospitalized in grave condition, with cyanosis and symptoms of pulmonary edema, but recovered after three weeks.

LeWinn reported a fatal case in another chemist with a clinical picture of fulminating pneumonia, who died four days after exposure.[4]

Lewis reported a case in a physician. The presenting symptoms were weakness, chest pain and severe headache.[5] A chest x-ray was normal. Subsequently, exposure to traces of diazomethane caused coughing, wheezing and malaise.

Potts et al.[6] compared the toxicity of diazomethane to that of phosgene and suggested that the mechanism of intoxication might be related to the fact that diazomethane is a strong methylating agent.

The toxicity of diazomethane seems comparable to that of phosgene, and a threshold limit of 0.2 ppm is recommended to prevent irritation of the lower respiratory passages and other toxic effects.

REFERENCES

1. Flury, F., and Zernik, F., "Schädliche Gase," p. 420, J. Springer, Berlin (1931).
2. Fairhall, L. T., "Industrial Toxicology," 2nd ed., pp. 207–208. Williams & Wilkins Co., Baltimore (1957).
3. Sunderman, F. W., Conner, R., and Fields, H., *Am. J. Med. Sci. 195*, 469 (1938).
4. LeWinn, E. B., *Am. J. Med. Sci. 218*, 556 (1949).
5. Lewis, C. E., *J. Occup. Med. 6*, 91 (1964).
6. Potts, A. M., Simon, F. P., and Gerard, P. W., *Arch. Biochem. 24*, 329 (1949).

No.	Chemical Name and Formula	Chemical Abstracts Registry Number	A.C.G.I.H. 1966-TLV (Threshold Limit Values) ppm (mg/M³)	Principal Effects of Inhalation Exposures above TLV	Eye Contact	Inhalation	Skin Penetration	Skin Irritation	Ingestion	Supplemental Effects	References to Supplementary Information on Toxicity, Flammability and Other Hazards
					(5 is high; 1 is low)						
1	Acenaphthene $C_{12}H_{10}$	83329	....	....	1x	1	1x	1x	2	....	....
—	Acetal See Acetaldehyde, diethylacetyl										
2	Acetaldehyde C_2H_4O	75070	200 (360)	Irritant	3y	5*	2	2z*	2	Sensitiz. skin, resp. tract	1966, AIHA, MCA, HCD, STTE. May deteriorate in normal storage and cause hazard.
—	Acetaldehyde cyanohydrin See Lactonitrile										
3	Acetaldehyde, diethylacetyl (Acetal) $C_6H_{11}O_2$	105577	....	....	1	3	2	1	2		1982
4	Acetamide C_2H_5NO	60355			1x	1	1x	3y	1		1828
5	4-Acetamidodiphenyl $C_8H_9NO_2$									Carcinogenic in man	....
7	4-Acetamidostilbene	841189								Carcinogenic in animals	....
—	2-Acetaminofluorene See N-Acetyl-2-aminofluorene										
8	Acetanilide C_8H_9NO	103844			Acute toxicity is very low					....	1835, STTE
9	Acetic acid $C_2H_4O_2$	64197	10 (25)	Irritant	5	3y	3	3	2		1778, AIHA, MCA, NSC, HCD, STTE
10	Acetic anhydride (Diacetyl monoxide) $C_4H_6O_3$	108247	5 (20)	Irritant	5	4	2	2y	2		1817, MCA, HCD, STTE
11	Acetoacetanilide $C_{10}H_{11}NO_2$	102012	....	....	3y	1	2x	2y	2		....
12	Acetone (2-Propanone) C_3H_6O	67641	1,000 (2,400)	Narcosis	2	3	1	1	1		1720, AIHA, MCA, NSC, STTE
—	Acetone cyanohydrin See 2-Methyl-lactonitrile										
13	Acetonitrile C_2H_3N	75058	40 (70)	Toxic	2	4z*	2	1	2	Death/hours inhalation	2013, AIHA, HCD, STTE
14	Acetophenone C_8H_8O	98862			4	1	2	1	2		1761, STTE
—	Acetylacetone See 2,4-Pentanedione										
15	N-Acetyl-2-aminofluorene (2-Acetaminofluorene; N-2-Fluorenylacetamide; 2-Acetylaminofluorene) $C_{15}H_{13}NO$	53963	....	....						Carcinogenic	2126
16	Acetyl bromide C_2H_3BrO	506967	....	....	Eye irritation; Severe tissue damage; Inhalation hazard					....	1826
17	Acetyl chloride C_2H_3ClO	75365	....	....	5x	5x	3x	4x	3x	....	HCD. May deteriorate in normal storage and cause hazard.
18	Acetylene C_2H_2	74862	....	....	....	3z	....	....	....	Simple asphyxiant	1205, MCA, NSC, HCD, MGD, STTE. Forms explosive compounds with Ag and Cu
—	Acetylene tetrabromide See 1,1,2,2-Tetrabromoethane										

No.	(7) N.F.P.A. Hazard Identification Signals (4 is high; 0 is low)			(8) Extinguishing Agents	(9) Flammable Limits in Air % by volume		(10) Flash Point °C (°F)	(11) Ignition Temperature °C (°F)	(12) Boiling Point (Melting Point, m.p.) °C	(13) Vapor Pressure mm. Hg °C	(14) Density or Specific Gravity	(15) Water Solubility	No.
	Health	Fire	Reactivity		Lower	Upper							
1	.....	.....	.....	....	....		....	....	278	....	1.024	i	1
2	2	4	2	1,3,4	4.1	55	−38 (−36)	185 (365)	20.8	740^{20}	$0.780-0.790^{20}_{20}$	∞ h	2
3	2	3	0	3	1.6	10.4	−20 (−5)	230 (446)	102	$20^{19.6}$	0.821	δ	3
4	.....	.....	.....				....	....	221.2	10^{105}	0.9986	s	4
5	.....	.....	.....				....	....	m.p. 168	....	1.293	s	5
7	.....	.....	.....		....		....	....	....	....	....		7
8	.....	.....	.....	1,3	....		169 OC (337)OC	529 (984)	304	....	1.2105	δ h	8
9	2	2	1	1,2a,3	4.0	16.0	43 (109)	524 (975)	118.5	15^{25}	$1.048-.053^{20}_{20}$	∞	9
10	2	2 W	2	3	2.9	10.3	54 (129)	380 (715)	140	10^{36}	1.082^{20}_{20}	v	10
11	2	1	0	2a	....		185 OC (365)OC	....	85	....	....	δ	11
12	1	3	0	1,2a,3	2.6	12.8	−20 (−4)	538 (1000)	56.2	226^{25}	0.7908	∞	12
13	2	3	1	2a,3	4.4	16.0	8 (45)	524 (975)	80.06	73^{20}	0.7856	s	13
14	1	1	0	1,2,3	....		105 (221)	571 (1060)	202	0.45^{25}	0.0281	i	14
15	.....	.....	.....	....	....		....	....	m.p. 192–5	....	....	i	15
16	.....	.....	.....	....	....		....	....	76.7	....	1.663	d	16
17	3	3 W	2	3	....		4 (40)	390 (734)	51.2	....	1.1039	d	17
18	1	4	3	4	2.5	81	....	299 (571)	−83.6 Sub	....	0.6181^{-82}_{4}	δ	18
	1	4	2	(When dissolved in acetone in closed cylinder)									
				Utilization of gas must be at less than 15 psi/gage (2 atmospheres).									

No.	Chemical Name and Formula	(2) Chemical Abstracts Registry Number	(3) A.C.G.I.H. 1966–TLV (Threshold Limit Values) ppm (mg/M³)	(4) Principal Effects of Inhalation Exposures above TLV	(5) Eye Contact	Inhalation	Skin Penetration	Skin Irritation	Ingestion	Supplemental Effects	(6) References to Supplementary Information on Toxicity, Flammability and Other Hazards
19	Acetyl thioglycolyl chloride	10553783	....								May deteriorate in normal storage and cause hazard
20	Acridine $C_{13}H_{11}N$	260946	(See also coal tar pitch volatiles)		4x*	1*	2x	3x	2	Irritant: eye, nose, throat	....
21	Acrolein C_3H_4O	107028	0.1 (0.25)	Irritant	5	4*	4	5*	4	Sensitiz. skin, resp. tract	AIHA, MCA, NSC, HCD, STTE
22	Acrylamide C_3H_5NO	79061	(0.3)ST	Toxic	1y	3z*	3*	1y*	3*	Sensitiz. skin, resp. tract; Chronic toxicity	1829
23	Acrylic acid $C_3H_4O_2$	79107	....	....	4	1	3	4	2	....	1719, HCD, STTE. Polymerizes at 60°C
24	Acrylonitrile C_3H_3N	107131	20 S (45) S	Toxic	2	4*	3*	1	4	Chronic toxicity	2009, AIHA, MCA, HCD, STTE. May deteriorate in normal storage and cause hazard.
25	Adipic acid $C_6H_{10}O_4$	124049	....	....	4x	1	2x	2y	2	....	1811, NSC
26	Adiponitrile (Hexane dinitrile) $C_6H_8N_2$	111693			Toxic					....	2022
27	Aldrin $C_{12}Cl_6H_8$	309002	(0.25) S	Toxic							1356
28	Alizarin dyes $C_{14}H_8O_4$	72480	....	....	3x	2x	3x	1x	3x	....	....
29	Alkyl aryl nitrosoamines									Carcinogenic in animals	....
30	Allene (Propadiene) C_3H_4	463490				3x				May have narcotic properties	MGD
31	Allyl acetate $C_5H_8O_2$	591877	....		2	4y	3	1	3	....	1876
32	Allyl alcohol C_3H_6O	107186	2 S (5) S		2*	4y*	4	1	4	Eye burn w/o pain	AIHA
33	Allylamine C_3H_7N	107119								Irritant	2038, HCD
—	Allyl bromide See 3-Bromopropene										
—	Allyl chloride See 3-Chloropropene										
—	Allyl glycidyl ether See 1-(Allyloxy)-2,3-epoxy-propane)										
—	Allyl iodide See 3-Iodopropene										
34	1-(Allyloxy)-2,3-Epoxypropane (Allyl glycidyl ether) $C_6H_{10}O_2$	106923	C 10 C (45)	Irritant	Severely irritating to the eyes					....	1598, AIHA
35	Allyl propyldisulfide	2179591	2 (12)	Irritant	....						....
36	Aluminum Al	7429905	....	....	2x	1x	....	1x	....		989, AIHA, HCD
37	Aluminum chloride, anhydrous $AlCl_3$	7446700	....	....	5*	*	....	3y	2	Irrit. eye, nose, throat	MCA, NSC

(5 is high; 1 is low)

No.	(7) N.F.P.A. Hazard Identification Signals			(8) Extinguishing Agents	(9) Flammable Limits in Air % by volume		(10) Flash Point °C (°F)	(11) Ignition Temperature °C (°F)	(12) Boiling Point (Melting Point, m.p.) °C	(13) Vapor Pressure mm. Hg °C	(14) Density or Specific Gravity	(15) Water Solubility	No.
	Health	Fire	Reactivity (4 is high; 0 is low)		Lower	Upper							
19	.....	.....	.....	....	....	....	....	....	....	....	....	......	19
20									345–6	....	1.005	δ^h	20
21	3	3P	2	2a,3	2.8	31	<18 (<0)	278 (532)	52.5–3.5	214^{20}	0.8625	v	21
22	.....						138 (280)	424 (795)	Decomp.	....	1.122	v	22
23	.....	.....		1,2a	....		49 OC (121)OC	429 (804)	141.6	3.1^{20}	1.0511	∞	23
24	4	3 P	2	2a,3	3.0	17	0 OC (32)OC	481 (898)	77.5–9	$110\text{–}115^{25}$	0.8060	$s{:}v^h$	24
25	.....	.....		1,2,3	....		196 (385)	422 (792)	337.5	....	1.360	δ	25
26	.....						93 OC (199)OC		295	2^{119}	0.965	δ	26
27	.....						....	....	m.p. 104	....	....	i	27
28												δ	28
29	.....						....	....	....	....	....	......	29
30	.....	.....		....	Extremely flammable		....	....	−34.5	....	1.787	......	30
31	.....			....	....	....	22 OC (72)OC	374 (705)	103–4	....	0.928	......	31
32	3	3	1	1,2a,3	2.5	18.0	21 (70)	378 (713)	-97	23.8^{25}	0.854^{20}_{4}	......	32
33	3	3	1	1,2a,3	2.2	22	−29 (−20)	374 (705)	58		0.7613	∞	33
—													—
—													—
34	.....	.....		....	....		57 (135)	....	154	4.7 (25)	0.9698^{20}_{4}	......	34
35	.....						....	....	....	....	....	......	35
36	0	1	1	5	....		Dust explosion hazard	....	m.p. 659.7	....	2.702	i	36
37	.....						....	....	182.7	....	2.44	$s^{h/}_{/}d$	37

No.	(1) Chemical Name and Formula	(2) Chemical Abstracts Registry Number	(3) A.C.G.I.H. 1966–TLV (Threshold Limit Values) ppm (mg/M³)	(4) Principal Effects of Inhalation Exposures above TLV	(5) Relative Hazard to Health from Concentrated Short-term Exposure (5 is high; 1 is low) Eye Contact	Inhalation	Skin Penetration	Skin Irritation	Ingestion	Supplemental Effects	(6) References to Supplementary Information on Toxicity, Flammability and Other Hazards
38	Aluminum ethoxide $C_6H_{15}AlO_3$	555759	....		....	....	....	....	....		May deteriorate in normal storage and cause hazard
—	Aluminum lithium hydride See Lithium tetrahydroalanate and Lithium hexahydro-alanate										
39	Aluminum nitrate $Al(NO_3)_3 \cdot 9H_2O$	10487599	....	....	4y	....	1x	2y	2		
—	o-Aminoazotoluene See C.I. solvent yellow										
—	4-Aminodiphenyl See 4-Biphenylamine										
40	2-Aminodiphenylene oxide $C_{12}H_9NO$	3693229			....	....	....	....	....	Carcinogenic	....
41	2-Aminoethanol (Ethanolamine) C_2H_7NO	141435	3 (6)	Toxic	....	....	....	....	....		2065
42	2-Aminoethanolamine	4747186			3x	1x	1x	3x			
43	2-Aminofluorene $C_{13}H_{11}N$	153786			....	....	....	....	....	Carcinogenic in animals	....
44	1-Amino-2-naphthol	2834926			....	....	....	....	....	Carcinogenic in animals	....
45	2-Amino-1-naphthol	606417			....	....	....	....	....	Carcinogenic in animals	....
46	o-Aminophenol C_6H_7ON	95556			Sensitization						
47	2-Aminopyridine $C_5H_6N_2$	504290	0.5 T (2) T	Toxic	....	....	....	....	....		2188
48	4-Aminostilbene	834242			....	....	....	....	....	Carcinogenic in animals	....
49	3-Aminotriazole	....			....	....	....	....	....	Carcinogenic in animals	....
50	Ammonia (anhydrous) NH_3	7664417	50 (35)	Irritant	4z	5z	....	3z	....		860, AIHA, MCA, NSC, MGD, HCD
51	Ammonium bromide NH_4Br	7789324			1x		1x	1x	1x		....
52	Ammonium dichromate $(NH_4)_2Cr_2O_7$	7789095			May cause irritation and ulceration of skin wounds. Ingestion or inhalation harmful. Recommended exposure limits 0.1 mg/m³						MCA
53	Ammonium fluoride NH_4F	7790172	....		Toxic					....	Merck
54	Ammonium hydroxide (28%) NH_4OH	1336216	....	....	4	5	3	3	3	....	STTE
—	Ammonium metavanadate See Ammonium vanadate (V)										
55	Ammonium nitrate $N_2H_4O_3$	6484522	....		2x	1x	1x	1x	2x		NSC, HCD
56	Ammonium peroxydisulfate (Ammoniumpersulfate) $(NH_4)_2S_2O_8$	7727540			1x*	2x*	2x	2x	3		....
—	Ammonium persulfate See Ammonium peroxydi-sulfate										
57	Ammonium sulfamate $N_2H_6SO_3$	10196040	(15)	Toxic	....	....	....	....	....	....	STTE
58	Ammonium thiocyanate N_2H_4CS	1762954	....		May be toxic					....	....

No.	(7) Health	Fire	Reactivity	(8) Extinguishing Agents	(9) Lower	Upper	(10) Flash Point °C (°F)	(11) Ignition Temperature °C (°F)	(12) Boiling Point (Melting Point, m.p.) °C	(13) Vapor Pressure mm. Hg °C	(14) Density or Specific Gravity	(15) Water Solubility	No.
38	.....	.....	.....									d	38
39									d 150			v	39
40	.....								m.p. 94				40
41	2	2	0	1,2a,3			85 (*185*)		170	0.4^{20}	1.0180	∞	41
42	.....	.....	.....									.....	42
43	.....	.....	.....						m.p. 129			i	43
44	.....	.....	.....										44
45	.....	.....	.....										45
46	.....	.....	.....						Sub.			δ	46
47	.....	.....	.....						204			s	47
48	.....	.....	.....										48
49	.....	.....	.....						m.p. 90			s	49
50	3	1	0	4	16	25	Gas	651 (*1204*)	−33.5		0.7716 g/l	v	50
51	.....								Sublimes		2.43		51
52	.....		.....	1	Flammable solid. May react explosively with organic materials at elevated temperatures		Decomp. begin @ 180 and becomes self-sustaining at 225					δ	52
53	.....								m.p. Sub.		1.315	v	53
54	.....								m.p. −77		2.15	s	54
55	2	1	3	1	Explosive hazard						1.725	v	55
56	.....								m.p. d/20		1.982	v	56
57	.....	.....	.....						d 160			v	57
58	.....	.....	.....										58

No.	Chemical Name and Formula	(2) Chemical Abstracts Registry Number	(3) A.C.G.I.H. 1966–TLV (Threshold Limit Values) ppm (mg/M³)	(4) Principal Effects of Inhalation Exposures above TLV	(5) Relative Hazard to Health from Concentrated Short-term Exposure — Eye Contact	Inhalation	Skin Penetration	Skin Irritation	Ingestion	Supplemental Effects	(6) References to Supplementary Information on Toxicity, Flammability and Other Hazards
					(5 is high; 1 is low)						
59	Ammonium vanadate (V) .... NH_4VO_3	7803556	....	....	1x	1x	1x	2x	3		....
—	n-Amyl acetate See Pentyl acetate										
—	sec-Amyl acetate See sec-Pentyl acetate										
—	n-Amyl alcohol See Pentyl alcohol										
—	sec-Amyl alcohol See 2-Pentanol (about 80%) and 3-Pentanol (about 20%)										
	tert-Amyl alcohol See tert-Pentyl alcohol										
—	n-Amylamine See Pentyl amine										
—	n-Amyl chloride See 1-Chloropentane										
—	n-Amyl ether See Pentyl ether										
	Amyl nitrite See Isopentyl nitrite										
60	Aniline .... C_6H_7N	62533	5 S (19) S	Toxic	3	4*	4z*	1	3	Chronic toxicity	2132, AIHA, MCA, NSC, HCD, STTE
61	Aniline HCl .... $C_6H_7N(HCl)$	142041	....	....	....	....	....	....	....		....
62	o-Anisaldehyde .... $C_8H_8O_2$	135024	....	....	....	....	....	....	....		1983
63	o-Anisidine .... (2-Methoxyaniline) C_7H_9NO	90040	(0.5) S	Toxic	4x	1x*	4z*	1x	2x	Chronic toxicity	....
64	p-Anisidine .... C_7H_9NO	104949	(0.5) S	Toxic	....	....	....	....	....		STTE
65	Anisole .... (Methoxybenzene) C_7H_8O	100663	....	....						Absence of recorded adverse effects	1681
66	Anthracene .... $C_{14}H_{10}$	120127	(See also coal tar pitch volatiles)		1x	1	1x	1x	1	....	Not carcinogenic in pure form
67	2-Anthramine .... (Anthracene amine) $C_{14}H_{11}N$	613138	....	....	....	....	....	....	....	Carcinogenic	....
68	Anthraquinone .... $C_{14}H_8O_2$	84651	....	....	....	....	....	....	....		....
69	Antimony .... Sb	7440360	(0.5)	Toxic	....	....	....	....	....		AIHA, NSC, STTE
70	Antimony compounds......... (as Sb)	....	(0.5)	Toxic	....	....	....	....	....		AIHA, STTE
71	Antimony pentachloride....... $SbCl_5$	7647189	(0.5)	Toxic	....	....	....	....	....		AIHA, NSC
72	Antimony trichloride .... $SbCl_3$	10025919	(0.5)	Toxic	Irritant					Releases hydrogen chloride in presence of moisture	AIHA, MCA, NSC
73	Antimony trioxide .... Sb_2O_3	1309644	(0.5)	Toxic	1x	1x	1x	1	1	....	AIHA, NSC
74	ANTU® .... $C_{11}H_{10}N_2S$	86884	(0.3)	Toxic	....	....	....	....	....		STTE
75	Argon .... Ar	7440371	....	....	....	....	....	....	....	Simple asphyxiant	MGD
76	Arsenic .... As	7440382	(0.5)	Toxic	....	....	....	....	....		AIHA, NSC, STTE

No.	(7) N.F.P.A. Hazard Identification Signals			(8) Extinguishing Agents	(9) Flammable Limits in Air % by volume		(10) Flash Point °C (°F)	(11) Ignition Temperature °C (°F)	(12) Boiling Point (Melting Point, m.p.) °C	(13) Vapor Pressure mm. Hg °C	(14) Density or Specific Gravity	(15) Water Solubility	No.
	Health	Fire	Reactivity		Lower	Upper							
	(4 is high; 0 is low)												
59	.....	.....	.....	....	....	....	....	....	d 200	....	2.326	s	59
—													—
—													—
—													—
—													—
—													—
—													—
—													—
—													—
60	3	2	0	1,2,2a,3	1.3	....	70 (158)	617 (1143)	184.3	15⁷⁷	1.0216	s; ∞ [h]	60
61	.....	.....	.....	1,2a,3	....	....	193 OC (380) OC		245	....	1.2215	v	61
62	2	1	0	....	....	....	118 OC (244) OC	....	249.5	1⁷³	1.1192	i	62
63	.....	.....	.....	....	....	....	....	....	224	....	1.0923	δ	63
64	.....	.....	.....	....	....	....	....	....	243	....	1.0605	s	64
65	1	2	0	....	....	....	52 OC (125) OC	....	153.8	31²⁵	0.9954	i	65
66	.....	.....	.....	1,2,3	0.6	....	121 (250)	540 (1004)	340	....	1.25	i	66
67	.....	.....	.....	....	....	....	....	....	m.p. 127	....	....	v	67
68	.....	.....	.....	1,2,3	....	....	185 (365)	....	379.8	....	1.438	i	68
69	.....	.....	.....	....	....	....	....	....	m.p. 630.5	....	6.684	i	69
70	.....	.....	.....								....		70
71	.....	.....	.....						m.p. 2.8	....	liq. 2.336	d	71
72	.....	.....	.....						223	....	3.140	v	72
73	.....	.....	.....						m.p. 656	....	5.2	δ	73
74	.....	.....	.....						m.p. 198	....	....		74
75	.....	.....	.....						−185.7	47⁻¹⁴³	1.784 g/l	δ	75
76	.....	.....	.....					....	Sub. 615	....	5.727	i	76

CHEMICAL HAZARD INFORMATION (See Explanation, page 668)

No.	(1) Chemical Name and Formula	(2) Chemical Abstracts Registry Number	(3) A.C.G.I.H. 1966–TLV (Threshold Limit Values) ppm (mg/M³)	(4) Principal Effects of Inhalation Exposures above TLV	(5) Relative Hazard to Health from Concentrated Short-term Exposure (5 is high; 1 is low) Eye Contact	Inhalation	Skin Penetration	Skin Irritation	Ingestion	Supplemental Effects	(6) References to Supplementary Information on Toxicity, Flammability and Other Hazards
77	Arsenic compounds (as As)	....	(0.5)	Toxic	....	....	..	....	....		AIHA, STTE
78	Arsenic acid $HAsO_3$	10102531	(0.5)	Toxic	....	....	..	....	....		AIHA, NSC
79	Arsenic trioxide As_2O_3	1327533	(0.5)	Toxic	....	....	..	....	....		MCA
80	Arsine (Arsenic hydride) AsH_3	7784421	0.05 (0.2)	Toxic	Highly toxic						AIHA, MGD, STTE
81	Auramine $C_{17}H_{21}N_3 \cdot HCl$	2465272								Carcinogenic in animals	
82	Azinphos-methyl (Guthion®) $C_{10}H_{12}N_3O_3PS_2$	86500	(0.2)ST	Toxic	....	....	..	....	....		
83	Aziridine (Ethylenimine) C_2H_5N	151564	0.5 ST (1)ST [skin]	Toxic	3	5y*	5	4*	4	Sensitiz. skin resp. tract. (Not a skin sensitizer, according to A. M. Thiess, *Archiv für Toxikologie 21, 67, 1965.*) *Human fatality report*	2172, AIHA, STTE, Polymerizes violently in presence of acids. Forms explosive with water.
84	2,2′-Azonaphthalene $C_{20}H_{14}N_2$	487105	....		....	....	..	....	....	Carcinogenic in animals	....
85	Azoxybenzene $C_{12}H_{10}N_2O$	495487	....		1	....	3	2	3		
86	Barium Ba	7440393	....		3x	3x	..	3x	....		AIHA, STTE
87	Barium-soluble compounds	....	(0.5)	Toxic							
88	Barium chloride $BaCl_2$	10361372	....		Extremely toxic						
89	Barium hydroxide $Ba(OH)_2 \cdot 8H_2O$	1304172	(0.5)	Toxic	4x*	....	1x	4x	3x	Irritant, eye, nose, throat	....
—	Benzal chloride See α,α-Dichlorotoluene										
90	Benzaldehyde C_7H_6O	100527	....		....	....	..	....	....		1983
91	1,2-Benzanthracene $C_{18}H_{12}$	56553								Carcinogenic in animals	....
92	Benzene C_6H_6	71432	C 25 S C(80)S	Toxic	2	4*	2	2	2	Chronic toxicity	1222, AIHA, MCA, NSC, HCD, STTE
—	Benzene hexachloride See Lindane										
93	Benzenesulfonic acid $C_6H_6SO_3$	98113	....		4	1	3x	5	3	....	1840
94	Benzidine $C_{12}H_{12}N_2$	92875	Because of high incidence of bladder tumors in man, any exposure, including skin, is extremely hazardous								2133, STTE
95	Benzoic acid $C_7H_6O_2$	65850	....		2y	1	3	2y	2	....	1838, STTE
96	Benzonitrile C_7H_5N	100470	....		2x	3	3x	3x	3	....	....
97	Benzo[a]pyrene $C_{20}H_{12}$	50328								Carcinogenic in man	....
98	p-Benzoquinone (Quinone) $C_6H_4O_2$	106514	0.1 (0.4)	(Toxic) Irritant	May produce severe irritation, erythema, swelling, dermatitis					....	1383, AIHA, STTE
99	Benzoyl chloride C_7H_5OCl	98884	....		....	....	..	....	....		STTE

728

No.	(7) N.F.P.A. Hazard Identification Signals			(8) Extin-guishing Agents	(9) Flammable Limits in Air % by volume		(10) Flash Point °C (°F)	(11) Ignition Tem-perature °C (°F)	(12) Boiling Point (Melting Point, m.p.) °C	(13) Vapor Pressure mm. Hg °C	(14) Density or Specific Gravity	(15) Water Solu-bility	No.
	Health	Fire	Reac-tivity		Lower	Upper							
	(4 is high; 0 is low)												
77	....	.....		....			....	....	....	....	....		77
78	....	.....		....					d	....	....	d	78
79	....	.....		....					m.p. 315	....	3.738	δ	79
80	....	.....		....					−55 d300	....	2.695 g/l	δ	80
81	....	.....		....					m.p. 13	....		s	81
82	....	.....		....					....	....	....		82
83	3	3	3	1,2a,3	3.6	46	−11 (*12*)	322 (*612*)	56	160^{20}	0.831_4^{20}	∞	83
84	....	.....		....			....	....	Sub.	....	....	i	84
85	....	.....		....					m.p. 87	....	1.166	i	85
86	....	.....		....					m.p. 850	....	3.51	d ev H_2	86
87	....	.....		....						....	....	......	87
88	....	.....		....					1560	....	2.856	s	88
89	....	.....		....					m.p. 78	....	2.18	δ	89
—												—	
90	2	2	0	1,2,3	....	....	64 (*148*)	192 (*377*)	178.1	1^{26}	1.0415	δ	90
91	....	.....		....			....	....	435	....		i	91
92	2	3	0	2,3	1.4	8.0	−17 (*2*)	562 (*1044*)	80.1	100^{26}	0.8787	δ	92
—												—	
93	....	.....		....					m.p. 525	....		s	93
94	....	.....		....					....	....	....	......	94
95	....	.....	.....	1,3	....	....	121 (*250*)	574 (*1065*)	249	....	1.2659_{14}^{15}	δ	95
96	....	.....		....					190.7	....	1.0102_{15}^{15}	δ^h	96
97	....	.....		....					m.p. 179	....		i	97
98	....	.....		....					560 (*1040*)	Sub.	High	1.318	δ	98
99	2	2	1	1,2,3	....	....	72 (*162*)	....	197.2	....	1.2105_4^{21}	d	99

No.	Chemical Name and Formula	(2) Chemical Abstracts Registry Number	(3) A.C.G.I.H. 1966-TLV (Threshold Limit Values) ppm (mg/M³)	(4) Principal Effects of Inhalation Exposures above TLV	(5) Relative Hazard to Health from Concentrated Short-term Exposure						(6) References to Supplementary Information on Toxicity, Flammability and Other Hazards
					Eye Contact	Inhalation	Skin Penetration	Skin Irritation	Ingestion	Supplemental Effects	
					(5 is high; 1 is low)						
100	Benzoyl peroxide .... $C_{14}H_{10}O_4$	94360	(5)T	Toxic	5x*	4x*	2x	4x	3x	Irritant, eye, nose, throat	MCA, HCD. May deteriorate in normal storage and cause hazard
101	3,4-Benzphenanthrene	195197								Carcinogenic in animals	....
103	Benzyl acetate .... $C_9H_{10}O_2$	140114	....								1860
104	Benzyl alcohol .... C_7H_8O	100516	....		4	2	2	3	2		STTE
105	Benzyl benzoate .... $C_{14}H_{12}O_2$	120514	....								1896, STTE
—	Benzyl chloride See α-Chlorotoluene										
—	Benzyl mercaptan See α-Toluenethiol										
106	Beryllium .... Be	7440417	(0.002)	Toxic	1x	5y*	....	1y*	....	Chronic toxicity. Wounds need medical care	1004, AIHA, NSC, HCD, STTE
107	Bicycloheptadiene dibromides	....	....		.4y	5z	3	4	3	....	STTE
108	Biphenyl .... (Diphenyl) $C_{12}H_{10}$	92524	0.2 (1)	Irritant	2y	1	2	2y	2	....	AIHA, STTE
109	N-4-Biphenylacetohydroxamic acid	....	....							Carcinogenic	....
110	2-Biphenylamine............ $C_{12}H_{11}N$	90415									....
111	4-Biphenylamine............ (Xenylamine; 4-Aminodiphenyl) $C_{12}H_{11}N$	92671								Carcinogenic in man & animals	....
112	3,3',4,4'-Biphenyltetramine ..... (3,3'-Diaminobenzidine) $C_{12}H_{14}N_4$	91952								(Derivative of carcinogen)	....
113	Bis(2-chloroethyl)ether .... (Dichloroethylether; Chlorex®) $C_4H_8Cl_2O$	111444	C 15 S C (90) S	Irritant; Toxic	2	4	3*	4	4	Chronic toxicity	1673, STTE
—	Bis(dimethylthiocarbamoyl) disulfide See Thiram										
114	Bis(2-ethoxyethyl)ether .... (Diethylene glycol diethyl ether; Diethyl Carbitol®) $C_8H_{18}O_3$	112367	....		....	....	....	....	....	....	1543, HCD
115	Borax $Na_2B_4O_7 \cdot 10H_2O$				3x	2x	1x	3x	2		STTE
116	Boric acid .... H_3BO_3	10043353	....		3x	2x	2x	1x	2	....	STTE
—	Boric anhydride See Boron oxide										
117	Borneol .... $C_{10}H_{18}O$										....
118	Boron .... B	7440428	....								....

No.	Health	Fire	Reactivity	(8) Extinguishing Agents	(9) Lower	Upper	(10) Flash Point °C (°F)	(11) Ignition Temperature °C (°F)	(12) Boiling Point (Melting Point, m.p.) °C	(13) Vapor Pressure mm. Hg °C	(14) Density or Specific Gravity	(15) Water Solubility	No.
	(4 is high; 0 is low)												
100	1	4	4	1,2,3	Decomposition vapors are flammable. Compound is sensitive to heat, friction shock.		....	Explodes. Self-accelerating decomp. temp. ≈68 (135)	m.p. 103.5	....	....	δ	100
101	.....	.....	.....		....		....	....	....	....	....	......	101
103	1	1	0	2a			102 (216)	461 (862)	213.5^{756}	1.9^{60}	1.057^{16}	δ	103
104	2	1	0	2a			101 (213)	436 (817)	205.35	0.15^{25}	1.0419^{20}_{4}	s	104
105	1	1	0	1,2,3			148 (298)	481 (898)	327	....	1.1121	i	105
106	4	1	1	....	Dust explosive		....	....	m.p. 1228	....	1.85^{20}	i	106
107	.....	.....	.....		....		....	....	....	....	....	......	107
108	2	1	0	1,2,3	0.6 @ 2.32	5.8 @ 311	113 (235)	540 (1004)	255.9	....	1.9896^{77}_{4}	i	108
109	.....	.....	.....		....		....	....	....				109
110	2	1	0						299			i	110
111								45 (846)	302	....		i	111
112	.....	.....									....	......	112
113	2	2	0	1,2,3	....	....	55 (131)	369 (696)	178	0.13^{20}	1.2199^{20}_{4}	i:δ[h]	113
114	1	2	0	2a	....	....	82 OC (180)OC	....	189	0.5^{25}	0.9063	v	114
115	.....	.....	.....				....	....	−10H₂O 320	....	1.73	δ	115
116	.....	.....	.....				None	None	m.p. 169	....	1.435^{15}	δ	116
117	2	2	0	1,3	....		66 (150)	....	m.p. 210 Sub.	....	1.011^{20}_{4}	i	117
118	.....	.....	.....	....	Dust ignites on contact with air	....			m.p. 230	....	2.45	i	118

No.	Chemical Name and Formula	Chemical Abstracts Registry Number	A.C.G.I.H. 1966–TLV (Threshold Limit Values) ppm (mg/M³)	Principal Effects of Inhalation Exposures above TLV	Eye Contact	Inhalation	Skin Penetration	Skin Irritation	Ingestion	Supplemental Effects	References to Supplementary Information on Toxicity, Flammability and Other Hazards
					(5 is high; 1 is low)						
119	Boron fluoride-ethyl ether complex (Ethyl ether, compound with boron trifluoride) $C_4H_{10}BF_3O$	109637	. . . .	. . . .	. . . .	. . . .		. . . .	. . . .		May deteriorate in normal storage and cause hazard
120	Boron hydrides (Decaborane; Diborane; Pentaborane)	. . . .	. . . .	. . . .							MCA
121	Boron oxide (Boric anhydride) B_2O_3	1303862	(15)	Nuisance	. . . .	. . . .		. . . .	. . . .		STTE
122	Boron trichloride BCl_3	10294545	. . . .	. . . .	Corrosive liquid, respiratory irritant					. . . .	MGD, STTE
123	Boron trifluoride BF_3	7637072	C 1 C(3)	Toxic	5x	5x		4x	. . . .	Respiratory irritant	844, MGD, STTE
124	Bromine Br_2	7726456	0.1 (0.7)	Irritant	5x	2x	1x	5x	. . . .	Extremely corrosive liquid	852, AIHA, MCA, NSC, HCD, STTE
125	Bromine pentafluoride BrF_5	7789302	. . . .	. . . .	Corrosive and irritating to eyes, skin, and mucous membranes					. . . .	MGD, STTE
126	Bromine trifluoride BrF_3	7787715	1 (3)	Toxic	. . . .	. . . .		. . . .	. . . .	Highly toxic, very corrosive	MGD, STTE
127	Bromoacetic acid $C_2H_3BrO_2$	79083	. . . .	. . . .	Toxic					. . . .	1798
128	Bromobenzene C_6H_5Br	108861	. . . .	. . . .	2x	2x	. . . .	3x	. . . .	Skin irritation. May be narcotic in high concentrations	Merck
129	1-Bromobutane (n-Butyl bromide) C_4H_9Br	109659	. . . .	. . . .	. . . .	. . . .		. . . .	. . . .		. . . .
130	Bromochloromethane CH_2BrCl	74975	200 (1050)	Narcosis	. . . .	. . . .		. . . .	. . . .		1271, AIHA
132	Bromodiethyl aluminum $C_4H_{10}AlBr$	760190	. . . .	. . . .	. . . .	. . . .		. . . .	. . . .		. . . .
133	p-Bromodiphenyl $C_{12}H_9Br$	92660	. . . .	. . . .	. . . .	. . . .		. . . .	. . . .		. . . .
134	Bromoethane (Ethyl bromide) C_2H_5Br	74964	200 (880)	Irritant toxic	. . . .	. . . .		. . . .	. . . .		1276, AIHA, STTE, MCA, HCD
135	Bromoethylene (Vinyl bromide) $CH_2:CHBr$	593602	. . . .	. . . .	. . . .	. . . .		. . . .	. . . .		MGD
—	Bromoform See Tribromomethane										
136	Bromomethane (Methyl bromide) CH_3Br	74839	C 20 S C(80)S	Toxic	1x	5x*	. . . .	4x	. . . .	Chronic toxicity	1251, MGD, STTE. May deteriorate in normal storage and cause hazard
137	Bromopentane $C_5H_{11}Br$	110532	. . . .	. . . .	. . . .	. . . .		. . . .	. . . .		. . . .
138	3-Bromopropene (Allyl bromide) C_3H_5Br	106956	. . . .	. . . .	. . . .	. . . .		. . . .	. . . .		. . . .

No.	(7) N.F.P.A. Hazard Identification Signals Health	Fire	Reactivity (4 is high; 0 is low)	(8) Extinguishing Agents	(9) Flammable Limits in Air % by volume Lower	Upper	(10) Flash Point °C (°F)	(11) Ignition Temperature °C (°F)	(12) Boiling Point (Melting Point, m.p.) °C	(13) Vapor Pressure mm. Hg °C	(14) Density or Specific Gravity	(15) Water Solubility	No.
119	3	2	0	....	....	....	64 OC (147)OC	....	101 m.p. 126.8	....	1.1	s	119
120	.....	.....	.....										120
121	.....	.....	.....						1860	....	2.46	δ	121
122	.....	.....	.....						12.5	477	1.35^{11}_{4}	......	122
123	.....	.....	.....						−99.9	27.9^{-30}	2.99 g/l	s	123
124	4	0	1	....					58.78	77.3^4	2.928^{58}	δ	124
125	.....	.....	.....						40.3	136^1	2.482	......	125
126	.....	.....	.....		Ignites paper and wood and will react violently with most organic compounds				127	18^{39}	2.843	......	126
127	.....	.....	.....		....	....	....	....	208	....	1.934	v	127
128	2	2	0	1,2,3	....	....	51.1 (124)	566 (1051)	155–6	....	1.5219^0_4	i	128
129	2	3	0	....	2–6 @ 212	6.6 @ 212	18 (65)	265 (509)	101.3	....	1.2764^{20}_4	i	129
130	.....	.....	.....		....	....	....	....	69	$155–60^{25}$	1.991	δ	130
132	.....	.....	.....										132
133	2	1	0	....	....	....	144 (291)	....	310	....	0.9327	i	133
134	2	3	0	....	6.7	11.3	....	511.11 (952)	38.4	475^{25}	1.4604	δ	134
135	.....	.....	.....	....	Flammable		....	....	15.8	....	1.4933	......	135
136	3	1	0	....	10	16	....	537 (999)	3.59	....	1.732	δ	136
137	1	3	0	....	....	....	32 (90)	....	129.6	....	1.2177^{20}_4	i	137
138	3	3	1	2a	4.4	7.3	−1 (30)	295 (563)	70	....	1.398	i	138

No.	(1) Chemical Name and Formula	(2) Chemical Abstracts Registry Number	(3) A.C.G.I.H. 1966–TLV (Threshold Limit Values) ppm (mg/M³)	(4) Principal Effects of Inhalation Exposures above TLV	(5) Relative Hazard to Health from Concentrated Short-term Exposure						(6) References to Supplementary Information on Toxicity, Flammability and Other Hazards
					Eye Contact	Inhalation	Skin Penetration	Skin Irritation	Ingestion	Supplemental Effects	
					(5 is high; 1 is low)						
139	3-Bromopropyne. . . . . (Propargyl bromide) C_3H_3Br	. . . . .	. . . .	. . . . .	. . . .						
140	o-Bromotoluene . . . . C_7H_7Br	95465	. . . .								
141	Bromotrifluoroethylene . . . . C_2BrF_3	598732	. . . .	May cause anesthesia and narcosis							MGD
142	Bromotrifluoromethane . . . . (Freon-13B1®) $CBrF_3$	75638	1000 (6100)	Narcosis	. . . .						MGD
143	Brucine $C_{23}H_{26}N_2O_4$	357573	. . . .		4x	4x	5x	3x	5x		
144	1,3-Butadiene . . . . C_4H_6	106990	1000 (2000)	Narcosis	2x	2y	1x	1x	. . . .		1203, AIHA, MCA, HCD, STTE
145	Butadiene dioxide.	1335826	. . . .		4	5*	4	4*	4	Carcinogenic in animals	. . . .
146	Butane. C_4H_{10}	106978	. . . .					3x		Simple asphyxiant	1195, MGD, STTE
147	1,4-Butanediol . . . . $C_4H_{10}O_2$	110634	. . . .								
148	2,3-Butanedione . . . . $C_4H_6O_2$	431038	. . . .								
149	1-Butanethiol . . . . (n-Butyl mercaptan) $C_4H_{10}S$	109795	10 (35)	Toxic							STTE
150	2-Butanone . . . . (Methyl ethyl ketone) C_4H_8O	78933	200 (590)	Irritant	2	4	2	1	2	. . . .	AIHA, MCA, STTE
151	1-Butene . . . . C_4H_8	106989	. . . .					3x	. . . .	Simple asphyxiant	MGD
152	2-Butene . . . . C_4H_8	107017	. . . .					3x	. . . .	Simple asphyxiant	MGD
153	3-Buten-2-one . . . . (Methyl vinyl ketone) C_4H_6O	78944	. . . .		3x	3x	2x	1x	2x	. . . .	STTE
154	2-Butoxyethanol. . . . . (Butyl Cellosolve®; Dowanol®EB; Ethylene glycol monobutyl ether) $C_6H_{14}O_2$	111762	50 S (240) S	Toxic	5	1*	3	3y*	2	Chronic toxicity	1542, AIHA, STTE
155	2-(2-Butoxyethoxy)ethanol . . . . . (Butyl Carbitol®; Dowanol®DB; Diethylene glycol monobutyl ether) $C_8H_{18}O_3$	112345	. . . .		2	1	2	1	2		1543, STTE
156	Butyl acetate. . . . . $C_6H_{12}O_2$	123864	150 T (710) T	Irritant (Narcosis)	2	3	1	1	1		1858, AIHA, STTE
157	sec-Butyl acetate . . . . $C_6H_{12}O_2$	105464	200 T (950) T	Irritant (Narcosis)	2x	3x	1x	1x	2	. . . .	1858
158	tert-Butyl acetate . . . . $C_6H_{12}O_2$	540885	200 T (950) T	Irritant (Narcosis)	. . . .						
159	Butyl alcohol. . . . . (1-Butanol) $C_4H_{10}O$	71363	100 (300)	Narcosis	3	1	2	1	2	. . . .	1441, AIHA, STTE
160	sec-Butyl alcohol . . . . $C_4H_{10}O$	78922	150 (450)	Narcosis	2	3	2x	1	2	. . . .	1445, STTE
161	tert-Butyl alcohol . . . . $C_4H_{10}O$	75650	100 (300)	Irritant (Narcosis)	. . . .					. . . .	1449, STTE
162	Butylamine . . . . $C_4H_{11}N$	109739	C 5 S C(15) S	Irritant	4	4*	3	4*	3	Sensitiz. skin, resp. tract	2038, AIHA, STTE

No.	(7) N.F.P.A. Hazard Identification Signals			(8) Extinguishing Agents	(9) Flammable Limits in Air % by volume		(10) Flash Point °C (°F)	(11) Ignition Temperature °C (°F)	(12) Boiling Point (Melting Point, m.p.) °C	(13) Vapor Pressure mm. Hg °C	(14) Density or Specific Gravity	(15) Water Solubility	No.
	Health	Fire	Reactivity		Lower	Upper							
	(4 is high; 0 is low)												
139	.....	.....	.....	....	Will detonate when heated under confinement			....	88–90	....	1.520	....	139
140	2	2	0	1	....	....	79 (174)	....	181	....	1.4222^{25}_{4}	i	140
141	.....	.....	.....	....	Non-flammable		....	....	−2 to 0	....	....		141
142	.....	.....	.....	....	Non-flammable		....	....	−59	....	....	δ	142
143	.....	.....	.....	....	....	....	....	....	m.p. 178	....	....	δ	143
144	2	4	2	4	2.0	11.5	Gas	42.9 (804)	−4.4			i	144
145	.....	.....	.....	....				....	....		....	....	145
146	1	4	0	4	1.9	8.5	Gas	405 (761)	−0.5	1823^{25}	0.6012^{0}_{4}	v	146
147	.....	.....	.....	2a	....	....	121 OC (250) OC	....	235	....	1.0171^{20}_{4}	∞	147
148	.....	.....	.....	2a	....	....	27 (80)	....	89–90	....	$0.9808^{18.5}_{4}$	v	148
149	2	3	0	2a	....	....	2 (35)	....	97.8	....	....	δ	149
150	1	3	0	2a	1.7	11,4	7 (21)	474 (885)	79.6	....	0.8054	v	150
151	1	4	0	4	1.6	9.3	−80 (−112)	384 (723)	−6.3	$760^{-6.3}$	0.5946	i	151
152	.....	.....	.....	4	1.7	9.0	....	230 (446)	3.7	760^{1}	0.6213	i	152
153	2	3	2	2,3	....	....	−2 (29)	....	$79\text{–}80^{755}$	....	0.8636^{20}_{4}	....	153
154	.....	.....	.....	1,2a,3	1.1	12.7	61 (141)	238 (460)	171	0.88^{25}	0.9027	∞	154
155	1	2	0	1,2a,3	0.85	....	78 (172)	204 (400)	231	0.023^{25}	0.9553	∞	155
156	1	3	0	2,2a,3	1.4	7.6	27 (81)	399 (750)	127	15^{25}	0.883	δ	156
157	1	3	0	2,2a,3	1.7	....	31 OC (88) OC	....	112	24 (25)	0.8758	i	157
158	.....	.....	.....						95	....	0.8620	....	158
159	1	3	0	1,2a,3	1.7	18	29 (84)	365 (689)	117.5	6.5^{25}	0.8098^{20}_{4}	s	159
160	1	3	0	1,2a,3	1.7 @ 212°	9.8 @ 212°	24 (75)	406 (763)	99.5	23.9^{30}	0.8080^{20}_{4}	s	160
161	1	3	0	1,2a,3	2.4	8.0	11 (52)	478 (892)	82.2–.3	42.0^{25}	0.7856^{26}_{4}	∞	161
162	3	3	0	1,2a,3	1.7	9.8	−12 (10)	312 (594)	77.8	72^{20}	0.764^{25}_{4}	∞	162

(1) No.	(1) Chemical Name and Formula	(2) Chemical Abstracts Registry Number	(3) A.C.G.I.H. 1966–TLV (Threshold Limit Values) ppm (mg/M³)	(4) Principal Effects of Inhalation Exposures above TLV	(5) Relative Hazard to Health from Concentrated Short-term Exposure (5 is high; 1 is low) Eye Contact	Inhalation	Skin Penetration	Skin Irritation	Ingestion	Supplemental Effects	(6) References to Supplementary Information on Toxicity, Flammability and Other Hazards
163	tert-**Butylamine** .... $C_4H_{11}N$	75649	....	....	....	....	....	....	....		....
—	n-Butyl bromide See 1-Bromobutane										
—	Butyl Carbitol® See 2-(2-Butoxyethoxy)-ethanol										
—	Butyl Cellosolve® See 2-Butoxyethanol										
—	n-Butyl chloride See 1-Chlorobutane										
164	tert-**Butyl chromate** .... $C_8H_{18}O_4Cr$	3957491	C (0.1) S	Toxic	....	....	....	....	....		STTE
165	**Butyl ether** .... (Dibutyl ether) $C_8H_{18}O$	142961	....	....	1	4	2	2	2	....	1665, STTE
166	**Butyl ethyl ether** .... (Ethyl butyl ether) $C_6H_{14}O$	628819	....	....	1	4	3	1	2	....	STTE
167	n-**Butyl formate** .... $C_5H_{10}O_2$	592847	....	....	....	....	....	....	....		....
168	n-**Butyl glycidyl ether** .... $C_7H_{14}O_2$	2426086	50 (270)	Irritant	2	2*	2	1*	2	Sensitiz. skin, resp. tract	1604, STTE
169	tert-**Butyl hydroperoxide** ...... $C_4H_{10}O_2$	75912	....	....	....	....	....	....	....		HCD, STTE
170	n-**Butyllithium** .... C_4H_9Li	109728	....	....	....	....	....	....	....	Skin contact causes caustic burns	MCA
—	n-Butyl mercaptan See 1-Butanethiol										
—	tert-Butyl mercaptan See 2-Methyl-2-propanethiol										
171	**Butyl methacrylate** .... $C_8H_{14}O_2$	97881	....	....	....	....	....	....	....		1878. Can polymerize explosively
—	n-Butyl methyl ketone See 2-Hexanone										
172	**Butyl nitrite** .... $C_4H_{9}O_2N$	544161	....	....	....	....	....	....	....		May deteriorate in normal storage and cause hazard
—	tert-Butyl perbenzoate See tert-Butyl peroxybenzoate										
173	tert-**Butyl peroxybenzoate** ...... (tert-Butyl perbenzoate) $C_{11}H_{14}O_3$	614459	....	....	....	....	....	....	....		HCD
175	**Butyl phosphate** .... (Dibutyl phosphate) $C_8H_{19}O_4P$	107664	1 T (5) T	Toxic	....	....	....	....	....		....
176	p-tert-**Butyl toluene** .... $C_{11}H_{16}$	98511	10 (60)	Toxic	....	....	....	....	....		....
177	**Butyl vinyl ether** .... (Vinyl butyl ether) $C_6H_{12}O$	111342	....	....	1	4	2	2	2	....	1671, STTE
178	**1-Butyne** .... (Ethylacetylene) C_4H_6	107006	....	....	....	....	....	....	....	Simple asphyxiant	MGD
179	**Butyraldehyde** .... C_4H_8O	123728	....	....	4	1*	2	1*	2	Sensitiz. skin, resp. tract	1966, MCA, HCD, STTE
180	**Butyric acid** .... $C_4H_8O_2$	107926	....	....	4	1	3	3	2	....	1780, HCD, STTE

No.	(7) N.F.P.A. Hazard Identification Signals			(8) Extinguishing Agents	(9) Flammable Limits in Air % by volume		(10) Flash Point °C (°F)	(11) Ignition Temperature °C (°F)	(12) Boiling Point (Melting Point, m.p.) °C	(13) Vapor Pressure mm. Hg °C	(14) Density or Specific Gravity	(15) Water Solubility	No.
	Health	Fire	Reactivity		Lower	Upper							
	(4 is high; 0 is low)												
163	3	4	0	2a	1.7 @ 212°	8.9 @ 212°	....	....	45.2	....	0.696_4^{20}	∞	163
—													—
—													—
—													—
—													—
164	.....											....	164
165	2	3	0	2a,3	1.5	7.6	25 (77)	194 (382)	141	4.8^{20}	0.769	δ	165
166	2	3	0	2a	....	....	4 (40)	....	92	....	0.7490	....	166
167	3	2	0	2,2a,3	1.7	8	18 (64)	222 (612)	107	30^{25}	0.911	δ	167
168	.....								164	3.2^{25}	0.9087	δ	168
169	1	4	4	2a,3	Flammable; heat and shock sensitive		38 (100)	Expl.	d	....	0.860	δ	169
170	.....			....	Pyrophoric if concentrated. Often dissolved in extremely flammable solvents				....	....	....		170
—													—
—													—
171	2	2	0	....	....	....	52 OC (126) OC	....	163	....	0895	i	171
—													—
172	.....			....	....	....			....	....	....	....	172
—													—
173	1	3	4	3	Flammable, limits unknown, heat sensitive		88 OC (190) OC	Self-accelerating decomp. temp. 113 (235)	113 d	....	1.035	d	173
175	.....									....		....	175
176	.....								193	0.65^{25}	0.8534	i	176
177	2	3	2	2a	....	....	−9 OC (15) OC	....	93.8	42^{20}	0.7742	i	177
178	.....			....	Flammable		....	....	8.1	....	0.669	....	178
179	2	3	1	3	1.9	....	−22 (−8)	218 (425)	75.7 m.p. −99	90^{20}	0.8170_4^{20}	s	179
180	2	2	0	1,2a,3	2.0	10.0	66 (151)	452 (846)	163.5	0.84^{20}	$0.957-.961_{20}^{20}$	∞	180

No.	Chemical Name and Formula	(2) Chemical Abstracts Registry Number	(3) A.C.G.I.H. 1966–TLV (Threshold Limit Values) ppm (mg/M³)	(4) Principal Effects of Inhalation Exposures above TLV	(5) Relative Hazard to Health from Concentrated Short-term Exposure						(6) References to Supplementary Information on Toxicity, Flammability and Other Hazards
					Eye Contact	Inhalation	Skin Penetration	Skin Irritation	Ingestion	Supplemental Effects	
					(5 is high; 1 is low)						
181	**Butyric anhydride**............ $C_8H_{14}O_2$	106310	....	....	4	3x	2	3	2	....	HCD
182	**γ-Butyrolactone**.............. $C_4H_6O_2$	96480	....	....	....	....	....	....	....	Skin irritant	....
—	Cacodylic acid See Hydroxydimethylarsine oxide										
183	**Cadmium**.................. Cd	7440439	....	....	Toxic						AIHA, NSC, STTE
184	**Cadmium sol. salts and metal dust**..................	....	(.02) T	Toxic	....	....	....	....	....	....	....
185	**Cadmium chloride**............. $CdCl_2$				Acute toxicity					....	1013
186	**Cadmium oxide fume**.......... CdO	1306190	(0.1)	Toxic	....	....	....	....	....	....	AIHA, STTE
187	**Calcium**.................. Ca	7440702	....	....	....	....	....	....	....	....	HCD
188	**Calcium arsenate**............. $Ca(AsO_4)_2$	10553692	(1)	Toxic	....	....	....	....	....	....	STTE
189	**Calcium carbide**............. CaC_2	75207	....	....	4x	2x	1x	3x	2x	....	MCA, HCD, STTE
190	**Calcium carbonate**............ $CaCO_3$	471341	....	....	2x	1x	1x	1x	1x	....	....
191	**Calcium chloride**............. $CaCl_2$	1305482	....	....	2x	....	1x	2x	2x	....	STTE
192	**Calcium hydroxide**............ $Ca(OH)_2$	1305620	(5)	Irritant	4*	3x*	1x	4y	2	Irritant, eye, nose, throat	863, STTE
193	**Calcium hypochlorite**.......... $Ca(ClO)_2$	7778543	....	....	....	....	....	....	....	Irritant	HCD, STTE
194	**Calcium oxide**............... CaO	1305788	(5)	Irritant	4x*	3x*	1x	4x	2x	Irritant, eye, nose, throat	863, HCD, STTE
195	**Camphor**.................. $C_{10}H_{16}O$	76222	(2)	Toxic	4x	2x	2x	3x	2x	....	STTE
—	Caproic acid See Hexanoic acid										
—	Caprylic acid See Octanoic acid										
—	Caprylic alcohol See Octyl alcohol										
197	**Carbazole**.................. $C_{12}H_9N$	86748	....	....	3y	....	2y	3y	1	....	....
—	Carbitol® See 2-(2-Ethoxyethoxy)ethanol										
—	Carbolic acid See Phenol										
198	**Carbon black**.............. C	7440440	(3.5) T	Toxic	....	....	....	....	....	....	....
199	**Carbon dioxide**.............. CO_2	124389	5000 (9000)	Toxic	....	3x	....	....	....	High concentration can cause resp. paralysis	937, AIHA, NSC, MGD, STTE
200	**Carbon disulfide**............. CS_2	75150	20 S (60) S	Toxic	3x	5z*	3x	2x	3x	Chronic toxicity	901, AIHA, MCA, NSC, HGD, STTE
201	**Carbon monoxide**............. CO	630080	50 T (55) T	Toxic	....	5z	....	....	....	Chemical asphyxiant	926, AIHA, MCA, NSC, MGD, STTE

No.	(7) N.F.P.A. Hazard Identification Signals			(8) Extinguishing Agents	(9) Flammable Limits in Air % by volume		(10) Flash Point °C (°F)	(11) Ignition Temperature °C (°F)	(12) Boiling Point (Melting Point, m.p.) °C	(13) Vapor Pressure mm. Hg °C	(14) Density or Specific Gravity	(15) Water Solubility	No.
	Health	Fire	Reactivity		Lower	Upper							
	(4 is high; 0 is low)												
181	3	2	1	3	....	....	77 (170)	307 (584)	195	0.3^{20}	0.969^{20}_{20}	d	181
182	.....	.....	.....	2a	....	....	79 OC (175)OC		206	....	1.049^{20}_{20}	∞	182
—													—
183	.....	.....	.....	....	Cadmium dust will burn with evolution of very hazardous fume				767 ± 2 m.p. 321	....	8.642	i	183
184	.....								....		....	.....	184
185	.....								960	....	4.047^{25}	v	185
186	.....								1559 Sub.	....	8.15	i	186
187	1	1 W	2	5	May ignite at room temp. if finely divided		....		1240	....	1.54	d to H_2 + Ca · $(OH)_2$	187
188	.....								m.p. 1.455		3.620	δ	188
189	1	4 W	2	....	Moisture liberates acetylene		....		2300	....	2.22	d	189
190	.....								d 898.6	....	2.710^{18}	i	190
191	.....								m.p. 772	....	2.15^{25}_{4}	s	191
192	.....								$-H_2O$, 580 d		2.504	v	192
193	2	1	2	....	Hazardous in contact with acids and oxidizable materials		....		m.p. d 100	....	2.35	s	193
194	1	0	1	....	Not combustible but heat of combination with water may ignite combustibles		....		m.p. 561	....	3.25–3.28	v	194
195	2	2	0	1,2,3	....	....	66 (150)	466 (871)	209		0.999^{20}_{4}	δ	195
—													—
—													—
197	.....	.....		....					355	....	....	....	197
—													—
198	.....	.....							....		....	....	198
199	.....	.....							−78.5 Sub	....	1.977 g/l	δ	199
200	2	3	0	1,3	1.3	44	−30 (−22)	100 (212)	46.3	360^{25}	1.261	i	200
201	2	4	0	4	12.5	74	....	651 (1204)	−191.5	760^{-191}	1.250 g/l	δ	201

No.	(1) Chemical Name and Formula	(2) Chemical Abstracts Registry Number	(3) A.C.G.I.H. 1966–TLV (Threshold Limit Values) ppm (mg/M³)	(4) Principal Effects of Inhalation Exposures above TLV	(5) Relative Hazard to Health from Concentrated Short-term Exposure						(6) References to Supplementary Information on Toxicity, Flammability and Other Hazards
					Eye Contact	Inhalation	Skin Penetration	Skin Irritation	Ingestion	Supplemental Effects	
					(5 is high; 1 is low)						
202	**Carbon tetrabromide** . . . . (Tetrabromomethane) CBr₄	558134	. . . .	. . . .	Highly toxic					. . . .	1270
203	**Carbon tetrachloride** . . . . CCl₄	56235	10 S (65) S	Toxic	3x	5z*	2	2x	2	Chronic toxicity (carcinogenic in mice, according to another source)	1265, AIHA, MCA, NSC, STTE
204	**Carbon tetrafluoride** . . . . (Tetrafluoromethane; Freon®-14; Genetron®-14) CF₄	75730	. . . .		Low order of toxicity; May act as simple asphyxiant					. . . .	MGD
205	**Carbonyl fluoride** . . . . COF₂	353504			Strong irritant to skin, eyes, mucous and resp.					Recommended exposure limit is 1 ppm	MGD. Special first aid preparations recommended
206	**Carbonyl sulfide** . . . . COS	463581	. . . .		Irritating to lungs, respiratory tract. Acts principally on central nervous system, with death resulting mainly from respiratory paralysis					. . . .	904, MGD, STTE
—	Caustic soda See Sodium hydroxide										
—	Cellosolve® See 2-Ethoxyethanol										
207	**Cellulose nitrate** . . . . (Pyroxylin)				Combustion produces toxic oxides of nitrogen					. . . .	HCD
208	**Cerium** . . . . Ce	7440451	. . . .		. . . .						1058, STTE
209	**Cesium** . . . . Cs	7440462	. . . .		5x			5x			HCP, STTE
210	**Chloral** . . . . C₂HCl₃O	75876	. . . .	. . . .	. . . .				. . . .	Inhalation may cause delayed fatal lung injury	. . . .
211	**Chloral hydrate** . . . . C₂H₃Cl₃O		. . . .	. . . .	3x	. . . .	3x	2x	3	. . . .	1982, STTE
212	*Chlordane* . . . . C₁₀H₆Cl₈		(0.5) S	Toxic	1	3x*	3x*	2	3	Chronic toxicity	1354, STTE
213	**Chlorinated camphene** . . . . C₁₀H₁₀Cl₈		(05) S	Toxic	1	1	2	3	1		1359, STTE
214	**Chlorinated diphenyl oxide**		(0.5)	Toxic	. . . .	. . . .	. . . .	. . . .	. . . .		STTE
215	**Chlorine** . . . . Cl₂	7782505	C 1 T C (3) T	Irritant	5z	5z	. . . .	3z	. . . .	Powerful vesicant and resp. irritant	846, MCA, NSC, HCD, MGD, STTE
216	**Chlorine dioxide** . . . . ClO₂	10049044	0.1 (0.3)	Irritant	. . . .	. . . .	. . . .	. . . .	. . . .	. . . .	AIHA, NSC, STTE, Explosive at 100 C°.
217	**Chlorine trifluoride** . . . . ClF₃	7790912	C 0.1 C (0.4)	Toxic	Highly toxic; vapor exposure may result in severe eye injury or blindness					. . . .	MGD, STTE. Special first aid preparations recommended
—	Chloroacetal See Chloroacetaldehyde, diethyl acetal										
218	**Chloroacetaldehyde** . . . . C₂H₃ClO	107200	C 1 C (3)	Irritant	. . . .	. . . .	. . . .	. . . .	. . . .		1966, STTE

No.	(7) Health	Fire	Reactivity	(8) Extinguishing Agents	(9) Lower	Upper	(10) Flash Point °C (°F)	(11) Ignition Temperature °C (°F)	(12) Boiling Point (Melting Point, m.p.) °C	(13) Vapor Pressure mm. Hg °C	(14) Density or Specific Gravity	(15) Water Solubility	No.
202	3	0	1	....	....	....	....	....	189.5	....	3.42	δ	202
203	.....	.....	.....	....	Non-flammable. Thermal decomposition products are very toxic		None	....	120.8	113^{25}	1.6311	δ	203
204	.....	.....	.....	....	Non-flammable gas		....	....	−129	760^{-128}	3.42^{0}	δ	204
205	.....			....	....	....	....	....	−83.1	....	1.139	d	205
206	3	4	1	4	12	29	....	....	−50.2	....	1.073 g/l	δ	206
—												—	—
207	Solution 1; solid 2	3; solid 3	0; 3	....	Severe fire hazard		27 (80)	....		1.66	i	....	207
208	.....	.....	.....	5	Pyrophoric		150–180	....	2417 m.p. 894	....	6.768	d	208
209	.....	.....	....	5	Reacts explosively with water; Reacts with ice above −116		....	....	690 m.p. 28.5	....	1.8785^{15}	d	209
210	.....			....	....	....	....	....	97.7 m.p. 57.5	....	1.505	....	210
211	.....								96.3^{764}	....	1.6415^{50}_{50}	v	211
212	.....						56 (132)		175^{2}	....	1.57–1.67	i	212
213	.....								65–90	....	1.66	s	213
214	.....											....	214
215	3	0	1	....	Reacts explosively with flammable materials		....	....	−34.6	3.66^{0}	....	δ	215
216	.....				....	....	....	Will detonate if heated rapidly to 100 (212)	9.9^{730}	....	3.09 g/l	δ	216
217	.....	.....	.....		Powerful oxidizing agent, ignites many organic compounds, ignites many metals at elevated temperatures, and reacts violently with water or ice.				11.3	504^{2}	1.77	d	217
—													—
218	.....	.....		....	....	....	....	....	$85–5.5^{748}$	....	1.190	v	218

No.	(1) Chemical Name and Formula	(2) Chemical Abstracts Registry Number	(3) A.C.G.I.H. 1966–TLV (Threshold Limit Values) ppm (mg/M³)	(4) Principal Effects of Inhalation Exposures above TLV	(5) Relative Hazard to Health from Concentrated Short-term Exposure — Eye Contact	Inhalation	Skin Penetration	Skin Irritation	Ingestion	Supplemental Effects	(6) References to Supplementary Information on Toxicity, Flammability and Other Hazards
					(5 is high; 1 is low)						
219	Chloroacetaldehyde, diethyl acetal (Chloroacetal) $C_6H_{13}ClO_2$	621625	....	....	Toxic					....	1982
220	Chloroacetaldehyde, dimethyl acetal (Dimethyl chloroacetal) $C_4H_9ClO_2$	97972	....	....							....
221	Chloroacetamide C_2H_4ClNO	79072			5x	4x	4x	4x	4x	....	....
222	Chloroacetic acid $C_3H_3ClO_2$				4	4z	1	4y	4x	....	1795
223	Chloroacetonitrile C_2H_2ClN	107142			2*	4*	4	1	3	Irritant, eye, nose, throat	STTE
224	α-Chloroacetophenone (Phenacylchloride) C_8H_7ClO	99912	0.05 (0.3)	Irritant	2*	4z*	4x	3	4x	Irritant, eye, nose, throat	....
225	Chloroacetyl chloride $C_2H_2Cl_2O$	79049	....		5x*	4*	4x	5	4x	Irritant, eye, nose, throat	....
226	Chlorobenzene C_6H_5Cl	108907	75 (350)	Narcosis	2x	3x*	2x*	2x	2x	Chronic toxicity	1333, AIHA, HCD, STTE
227	o-Chlorobenzylidene malonitrile	10487668	.05 T (0.4)T	Toxic	.....						....
—	2-Chloro-1,3-butadiene See Chloroprene										
228	Chlorobutane (n-Butyl chloride) C_4H_9Cl	109693			1	4*	1	2	2	Chronic toxicity	HCD, STTE
229	Chlorodiethylaluminum (Diethyl aluminum chloride) $C_4H_{10}AlCl$	96106	....		Extremely corrosive combustion products toxic					....	HCD
230	1-Chloro-1,1-difluoroethane (1,1-Difluoro-1-chloroethane; Genetron® 142b) $C_2H_3ClF_2$	75683			Relatively non-toxic						MGD
231	Chlorodifluoromethane (Freon®-22; Genetron®-22; Isotron®-22; Ucon®-22) $CHClF_2$	75456			Comparatively non-toxic; Exposure limit of 1000 ppm is generally accepted						MGD
—	Chlorodinitrobenzene See 1-Chloro-2,4-dinitro- benzene										
232	1-Chloro-2,4-dinitrobenzene (2,4-Dinitrochlorobenzene) $C_6H_3ClN_2O$	97007	....	....	5	....	4	4*	2	Sensitiz. skin, resp. tract	2109
233	Chlorodiphenyl (42% chlorine equiv.) $C_{12}H_7Cl_3$	1335928	(1)S	Toxic	1x	4z*	3x*	4x*	3x	Chronic toxicity; Acne-like eruption	1340, AIHA, STTE
234	Chlorodiphenyl (54% chlorine equiv.) $C_{12}H_5Cl_5$	1335917	(0.5)S	Toxic	1x	4z*	3x*	4x*	3x	Chronic toxicity; Acne-like eruption	1340, AIHA
235	1-Chloro-2,3-epoxypropane (Epichlorohydrin) C_3H_5ClO	106898	5 (19)	Toxic	2	4y*	3*	1	4	....	1622, AIHA, HCD, STTE
236	Chloroethane (Ethyl chloride) C_2H_5Cl	75003	1000 (2600)	Narcosis	Has narcotic properties. Over-exposure may be toxic						1273, AIHA, MCA, HCD, MGD, STTE. May deterio- rate in normal storage and cause hazard

No.	(7) N.F.P.A. Hazard Identification Signals			(8) Extin-guishing Agents	(9) Flammable Limits in Air % by volume		(10) Flash Point °C (°F)	(11) Ignition Temperature °C (°F)	(12) Boiling Point (Melting Point, m.p.) °C	(13) Vapor Pressure mm. Hg °C	(14) Density or Specific Gravity	(15) Water Solubility	No.
	Health	Fire	Reactivity		Lower	Upper							
	(4 is high; 0 is low)												
219	.....	.....	.....	....	....	....	....	....	157	20^{62}	1.026	δ	219
220	.....	2	0	3	....	....	44 (111)	232 (450)	124.5–6.5	....	....	......	220
221	.....	.....	.....	....	....	....	....	....	$224–5^{743}$	....	....	s	221
222	.....	.....	.....	....	....	....	....	....	189	1^{43}	1.4043	v	222
223	.....	.....	.....	....	....	....	....	....	123–4	....	1.193^{20}	......	223
224	.....	.....	.....	....	....	....	....	....	273	....	1.1922	i	224
225	.....	.....	.....	....	....	....	....	....	$108–10^{764}$	....	1.4177^{20}_{4}	d	225
226	2	3	0	2,3	1.3	7.1	29 (84)	638 (1180)	132	12^{25}	1.1064	i	226
227	.....	.....	.....	....	....	....	....	....	....	....	....	......	227
—													—
228	2	3	0	....	1.8	10.1	7 (15)	460 (860)	78.4	....	.08865	i	228
229	3	3 W	0	3	Pyrophoric		....	Ignites on contact with air	208	....	....	d	229
230	.....	.....	.....	4	9.0	14.8	Flam-mable	632 (1170)	−9.2	....	1.118	δ	230
231	.....	.....	.....	....	Non-flammable		....	....	−40.8	....	1.118	i	231
—													—
232	3	1	4	1,3	2.0	22	194 (382)	432 (810)	315 (m.p. 43)	....	1.4982	i:δ[h]	232
233	.....	.....	.....	....	....	....	176 to 80 OC (349 to 356) OC	....	....	30^{200}	1.378–1.388	i	233
234	.....	.....	.....	....	....	....	....	....	....	9^{200}	1.538–1.548	i	234
235	3	2	0	2a	....	....	41 OC (105) OC	....	116.5	13^{20}	1.180	δ:d[h]	235
236	2	4	1	3,4	3.8	15.4	−50 (−58)	519 (966)	131	$539^{3.5}$	0.9028	δ	236

(1) No. Chemical Name and Formula	(2) Chemical Abstracts Registry Number	(3) A.C.G.I.H. 1966–TLV (Threshold Limit Values) ppm (mg/M³)	(4) Principal Effects of Inhalation Exposures above TLV	(5) Relative Hazard to Health from Concentrated Short-term Exposure — Eye Contact	Inhalation	Skin Penetration	Skin Irritation	Ingestion	Supplemental Effects	(6) References to Supplementary Information on Toxicity, Flammability and Other Hazards
				(5 is high; 1 is low)						
237 **2-Chloroethanol** .... (Ethylene chlorohydrin) C_2H_5ClO	107073	5 S (16) S	Toxic	3	4*	4*	1	4	Chronic toxicity. Human fatality reported	1493, AIHA, MGD, STTE
238 **Chloroethylene** .... (Vinyl chloride) C_2H_3Cl	75014	C 500 C (1300)	Narcosis; Toxic	2x	3x	2x	2x	2x	....	1280, MCA, HCD, STTE
239 **Chloroform** .... $CHCl_3$	67663	C 50 C (240)	Toxic	2x	3*	2x	1	2	Chronic toxicity	1259, AIHA, MCA, STTE
240 **Chloromethane** .... (Methyl chloride) CH_3Cl	74873	C 100 C (210)	Toxic	1x	3x	....	1x	....	Chronic toxicity	1248, AIHA, MCA, HCD, MGD, STTE
241 **Chloromethyl ether** .... C_2H_5ClO	107302	....	....	....	....	....	....	....	Very irritating to eyes and nose	....
242 **3-Chloro-2-methyl propene** ..... C_4H_7Cl	563473	....	....	Irritation of eyes and resp. tract					....	STTE
243 **1-Chloronaphthalene** .... $C_{10}H_7Cl$	90131	....	....	1	1*	2*	3*	2	Chronic toxicity; Acne-like eruption	1343, AIHA
244 **Chloro-2-naphthylamine** .... $C_{10}H_8ClN$	4684122	....	....	....	....	....	....	....	Carcinogenic	....
— *o*-Chloronitrobenzene See 1-Chloro-2-nitrobenzene										
— *p*-Chloronitrobenzene See 1-Chloro-4-nitrobenzene										
245 **1-Chloro-2-nitrobenzene** .... (*o*-Nitrochlorobenzene; *o*-Chloronitrobenzene) $C_6H_4ClNO_2$	88733									HCD, STTE
246 **1-Chloro-3-nitrobenzene** .... (*m*-Chloronitrobenzene), (*m*-Nitrochlorobenzene) $C_6H_4ClNO_2$	121733									2109, HCD, STTE
247 **1-Chloro-4-nitrobenzene** .... (*p*-Chloronitrobenzene) (*p*-Nitrochlorobenzene) $C_6H_4ClNO_2$	100005	(1) S								STTE
248 **1-Chloro-1-nitropropane** .... $C_3H_6ClNO_2$	600259	20 (100)	Irritant	....	....	....	....	....		2080, STTE
249 **Chloropentafluoroethane** .... (Genetron® 115) C_2ClF_5	76153			Practically non-toxic; Exposure limit of 1000 ppm is generally accepted						MGD
250 **1-Chloropentane**.............. (*n*-Amyl chloride) $C_5H_{11}Cl$	543590									
— Chloropicrin See Trichloronitromethane										
251 **Chloroprene** .... (2-Chloro-1,3-butadiene) C_4H_5Cl	126998	25 S (90) S	Toxic	2x	4*	2x	2x*	2x	Chronic toxicity	1319, STTE
252 **1-Chloropropane** .... (*n*-Propyl chloride) C_3H_7Cl				4*	3	2	2	3	Chronic toxicity	1299
253 **1-Chloro-2-propanone** .... C_3H_5ClO	78955	....	....	Strong Irritant						....
254 **3-Chloropropene**.............. (Allyl chloride) C_3H_5Cl	107051	1 (3)	Irritant	3	3*	3	3	3	Chronic toxicity	1317, AIHA
255 **Chloropropionic acid** .... $C_3H_5ClO_2$	107948	....	....	....	....	....	....	....	....	....
256 **1-Chloropropylene**............ C_4H_5Cl				....	....	....	....	....	....	....

No.	(7) N.F.P.A. Hazard Identification Signals			(8) Extinguishing Agents	(9) Flammable Limits in Air % by volume		(10) Flash Point °C (°F)	(11) Ignition Temperature °C (°F)	(12) Boiling Point (Melting Point, m.p.) °C	(13) Vapor Pressure mm. Hg °C	(14) Density or Specific Gravity	(15) Water Solubility	No.
	Health	Fire	Reactivity										
	(4 is high; 0 is low)				Lower	Upper							
237	3	2	0	2a	4.9	15.9	60 OC (140) OC	425 (797)	128	4.9^{20}	1.204	∞	237
238	2	4	2	3	4	22	13 (55)	472 (882)	84	87^{25}	1.256	δ	238
239	.....	.....	.....	....	....	....	....	....	61.2	200^{25}	1.4916^{18}_{21}	δ	239
240	2	4	0	....	8.1	17.4	....	652 (1170)	−23.76	....	0.92	s	240
241	.....	.....	.....					....	....	....	1.07	....	241
242	.....	.....	.....				....	....	71.5–72.5	....	0.925	....	242
243	.....	.....	.....				132 OC (270) OC	>558 >(1036)	256		1.1377	i	243
244	.....	.....	.....				....	....	m.p. 60				244
—													—
245	3	1	1	1,3	....	....	127 (261)	....	242	....	1.9279^{22}_{4}	i	245
246	3	1	1	....	....	....	127 (261)	....	....	....	....	....	246
247	.....	.....	.....	....	....	....	127 (261)	....	246	....	1.368	i	247
248	.....	2	3	1,2a,3	....	....	62 OC (144) OC	....	141–3	5.8^{25}	1.209	δ	248
249	.....	.....	.....	Non-flammable at normal temperatures and pressures			....	....	−38.7		1.26	i	249
250	1	3	0	....	1.6	8.6	13 OC (55) OC	232 (450)	108.2	....	0.8828	i	250
—													—
251	.....	.....	.....	2a,4	4.0	20.0	−20 (−4)	....	59.4	215.4^{25}	0.9538^{20}_{20}	δ	251
252	2	3	0	....	2.6	11.1	<−18 (<0)	....	46.60	350^{25}	0.8923	δ	252
253	.....	.....	.....	....	....	....	....	....	179	....	1.15^{20}	s	253
254	3	3	1	2a	3.3	11.1	−32 (−25)	392 (737)	45	365^{25}	0.9397	i	254
255	.....	1	0	2a	....	....	107 (225)	....	186	....	1.28	∞	255
256	3	4	2	2a,4	4.5	16	<−6.1 (<21)	....	32.8		0.9	....	256

No.	Chemical Name and Formula	(2) Chemical Abstracts Registry Number	(3) A.C.G.I.H. 1966-TLV (Threshold Limit Values) ppm (mg/M^3)	(4) Principal Effects of Inhalation Exposures above TLV	(5) Relative Hazard to Health from Concentrated Short-term Exposure (5 is high; 1 is low) Eye Contact	Inhalation	Skin Penetration	Skin Irritation	Ingestion	Supplemental Effects	(6) References to Supplementary Information on Toxicity, Flammability and Other Hazards
257	Chlorosulfonic acid .... HClO$_3$S	7790954	....	....	5x	....	3x	5x	4x	....	HCD
258	α-Chlorotoluene............. (Benzyl chloride) C$_7$H$_7$Cl	100447	1 (5)	Toxic	....					Irritant	STTE
259	Chlorotrifluoroethylene .... (Genetron®-1113) C$_2$ClF$_3$	79389				3y	....			Can be toxic; Exposure limit of 20 ppm recommended	MGD
260	Chlorotrifluoromethane .... (Freon®-13; Genetron®-13) CClF$_3$	75729			Comparatively non-toxic					....	MGD
261	Cholanthrene .... C$_{20}$H$_{14}$	479232								Carcinogenic to animals	....
262	Chromates .... CrO$_3$ Chromic acid See Chromium trioxide		(0.1)	Irritant Toxic	....						STTE
—											
263	Chromic salts, soluble	....	(0.5)	Toxic	....						
264	Chromium............. Cr	1440473	(1)	Toxic	1x	3x	1x	1x	1x	....	1018, STTE
265	Chromium, insoluble compounds.	....	(1)	Toxic	....						
266	Chromium fluoride .... CrF$_2$	10049102	....		Strong irritant						
267	Chromium trioxide........... (Chromic acid) CrO$_3$	10553727	(0.1)	Irritant; Toxic	5x*	4y*	2x	4y	4x	Irrit. eye, nose, throat, lung	1018, AIHA, MCA, HCD, STTE
268	Cinnamaldehyde............. C$_9$H$_8$O	104552	....	....	Toxic					....	1983
269	Chromous salts, soluble		(0.5) T	Toxic	....						
270	C. I. Solvent Yellow 2......... (N,N-Dimethyl-p-phenylazo-aniline; p-Dimethylaminoazobenzene) C$_{14}$H$_{15}$N$_3$	60117								Carcinogenic in animals	2129
271	C. I. Solvent Yellow 3......... (4-(o-Tolylazo)-o-toluidine; 2-Amino-5-azotoluene) C$_{14}$H$_{15}$N$_3$	97563								Carcinogenic in animals	....
272	Citraconic anhydride .... C$_5$H$_4$O$_3$	616024				4y	1*	3	3*	2	Sensitiz. skin, resp. tract	STTE
273	Citric acid .... C$_6$H$_8$O$_7$	77929				5y	2x	2x	2x	2x	....	1814
274	Coal tar pitch volatiles .... Benzene soluble fraction Anthracene Chrysene Phenanthrene Acridine Pyrene; BaP	....	(0.2) T	Carcino-genic; Toxic	3x	2x*	1x*	2x*	1x	Chronic toxicity; Carcinogenic in animals. Not all the compounds listed are carcinogenic in pure form.	....
275	Cobalt............. Co	7440484	(0.1)	Toxic	....	....	....	....	....	....	1024, AIHA, STTE
276	Cobaltous nitrate .... Co(NO$_3$)$_2$ · 6H$_2$O	10553738	....		Releases toxic oxides of nitrogen on decomposition at 74°C (165°F)						HCD
277	Collodion	....	....	....	....	....	....	....	....		HCD
278	Copper .... Cu	7440548	....	....	2x*	....	1x	2x*	2x	....	1034, STTE
279	Copper—dust and mists	....	(1.0)	Irritant	.. ..	....	....	....	....		....

No.	(7) N.F.P.A. Hazard Identification Signals			(8) Extin-guishing Agents	(9) Flammable Limits in Air % by volume		(10) Flash Point °C (°F)	(11) Ignition Tem-perature °C (°F)	(12) Boiling Point (Melting Point, m.p.) °C	(13) Vapor Pressure mm. Hg °C	(14) Density or Specific Gravity	(15) Water Solu-bility	No.
	Health	Fire	Reac-tivity		Lower	Upper							
	(4 is high; 0 is low)												
257	3	0 W	2	....	....	....	....	....	158	....	1.766[18]	d to H_2SO_4 to HCl	257
258	2	2	0	....	1.1	....	67 (153)	585 (1085)	179.3	....	1.100	i: d[h]	258
259	3	4	2	4	8.4	38.7	....	....	....	....	1.305	......	259
260	.....	.....	.....	....	Non-flammable		....	....	−81.4	....	7.01 g/l	i	260
261	.....	.....	.....						sub. 210[0.2]	....		......	261
262	.....												262
—													—
263	.....												263
264	.....								2480	....	7.20[28]	i	264
265	.....												265
266	.....								>1300	....	4.11	δ	266
267	1	0	1	. . . .	Powerful oxidizing material		....	....	m.p. 196	....	2.70	i	267
268	.....	....	.....			246	10 (120)	....		....	1.048[20][4]	δ	268
269	.....						....		....				269
270	.....						....		m.p. 128			i	270
271	.....						....		....	....	....		271
272	.....						....		213–4	....	1.2469	d	272
273	.....						....		d	....	1.542[18]	v	273
274	.....						....		....	....	....	i	274
275	.....	.....					....		3550	....	8.9	i	275
276	1	0	1	....	Oxidizing material		....		....	....	1.87	v	276
277	1	4	0	3	....	....	<−18 (<0)	....		....	0.765–75	......	277
278	.....	....					....		2595	....	8.92	i	278
279	.....	....					....		....	....	....		279

No.	(1) Chemical Name and Formula	(2) Chemical Abstracts Registry Number	(3) A.C.G.I.H. 1966–TLV (Threshold Limit Values) ppm (mg/M³)	(4) Principal Effects of Inhalation Exposures above TLV	(5) Relative Hazard to Health from Concentrated Short-term Exposure						(6) References to Supplementary Information on Toxicity, Flammability and Other Hazards
					Eye Contact	Inhalation	Skin Penetration	Skin Irritation	Ingestion	Supplemental Effects	
					(5 is high; 1 is low)						
280	**Copper—fume**		(0.1)	Irritant	....	....	....	....	....		STTE
281	**Cupric carbonate** CuCO₃	10487704			Irritant; Corrosive						STTE
282	**Copper nitrate** Cu(NO₃)₂ · 6H₂O	10553749			3	....	1x	2y*	3	Skin stain	HCD
283	**Cotton dust (raw)**		(1)	Toxic	....						2254
284	*Crag® 1* C₈H₈Cl₂O₅S	149268	(15)	Toxic	....						STTE
285	*m*-**Cresol** C₇H₈O	108394	5 S (22) S	Toxic	4	1	3	4	3		1387, HCD, AIHA, STTE
286	*o*-**Cresol** C₇H₈O	95487	5 S (22) S	Toxic	5						1387, HCD, AIHA, STTE
287	*p*-**Cresol** C₇H₈O	106445	5 S (22) S	Toxic	....						1387, HCD, AIHA, STTE
288	**2,3-Cresotic acid** (o-Cresotinic acid) C₈H₈O₃	83409			2x	2x					....
289	**Crotonaldehyde** C₄H₆O	123739	2 T (6) T	Toxic	4	5*	3	4z*	3	Sensitiz. skin, resp. tract	1996, HCD, STTE. May deteriorate in normal storage and cause hazard.
290	**Croton oil**									Carcinogenic in animals	....
291	**Crotononitrile** C₄H₅N										
292	**Crotonyl chloride** C₄H₅O				Irritation to eyes and respiratory tract						....
292a	**Chrysene** See Coal tar pitch volatiles				1x	1x	1x	1x	1		
293	**Cumene** C₉H₁₂	98828	50 ST (245) ST	Narcosis; Toxic	1	2	2	2	2		1223, AIHA, HCD, STTE
—	Cumene hydroperoxide See α,α-Dimethylbenzyl hydroperoxide										
294	**Cupric acetate** Cu(C₂H₃O₂)₂ · H₂O				4y	....	1x	2y*	3	Skin stain	....
295	**Cuprous chloride** CuCl	1306441			Gastroenteritis, nephritis, and liver damage						Merck
296	**Cuprous cyanide** CuCN	544923			Poisoning thru ingestion; absorption thru skin						Merck
297	**Cyanamide** H₂NCN	420042			Very irritating and caustic						STTE. Merck
298	**Cyanides** CN		(5) S	Toxic	....						1991, STTE
299	**Cyanoacetamide** C₃H₄N₂O	107915			Low toxicity						2028
300	**Cyanogen** C₂N₂	460195			Highly toxic; exposure limit of 10 ppm is recommended						HCD, MGD, STTE
301	**Cyanogen bromide** CNBr	506683			Highly toxic; limit exposure to <0.5 ppm						2007. May deteriorate in normal storage and cause hazard
302	**Cyanogen chloride** CNCl	506774			Very toxic and irritating; threshold limit value of 0.5 ppm is suggested; 159 ppm is fatal to man in 10 minutes						MGD
303	**Cycloheptanone** C₇H₁₂O	502421			Moderate toxicity with central nervous system depression						Merck

No.	(7) N.F.P.A. Hazard Identification Signals (4 is high; 0 is low) Health	Fire	Reactivity	(8) Extinguishing Agents	(9) Flammable Limits in Air % by volume Lower	Upper	(10) Flash Point °C (°F)	(11) Ignition Temperature °C (°F)	(12) Boiling Point (Melting Point, m.p.) °C	(13) Vapor Pressure mm. Hg °C	(14) Density or Specific Gravity	(15) Water Solubility	No.
280	.....	.....	.....	....	....	....	....	....	....	....	....	....	280
281	.....	.....	.....	....	....	....	....	....	m.p.: d. at 200	....	4.0	i	281
282	1	0	1	....	....	....	....	....	m.p. $-3H_2O$ 26.4	....	2.074	s	282
283	.....	.....	.....	....	....	....	....	....	....	....	....	....	283
284	.....	.....	.....	....	....	....	....	....	....	....	....	s	284
285	2	1	0	1,3	1.1 @ 302°	....	94 (202)	559 (1110)	202.8	3.72^{25}	1.0336^{20}_{4}	δ:s[h]	285
286	2	2	0	1,2,3	1.4 @ 300°	....	81 (178)	599 (1038)	191.2	3.72^{25}	1.0465^{20}_{4}	s	286
287	2	1	0	1,3	1.1 @ 302°	....	94 (202)	559 (1038)	202	3.72^{25}	1.0347^{20}_{4}	δ:s[h]	287
288	.....	.....	.....	....	....	....	....	....	....	....	....	....	288
289	3	3	2	2a,3	2.1	15.5	7 (45)	207 (405)	104	19^{20}	0.869^{20}_{20}	v	289
290	.....	.....	.....	....	....	....	....	....	....	....	$0.935\text{--}0.950^{25}_{25}$	....	290
291	.....	.....	.....	....	....	....	<100 (<212)	....	122	....	0.8239	....	291
292	.....	.....	.....	....	....	....	....	....	124–5	....	1.0905	d	292
292a	.....	.....	.....	....	....	....	....	....	....	....	....	....	292a
293	0	2	0	2,3	.88	6.50	44 (111)	424 (795)	152–3	$10^{39.33}$	$.0864^{20}_{4}$	i	293
—													—
294	.....	.....	.....	....	....	....	....	....	d 240	....	1.882 Anhydr.	s	294
295	.....	.....	.....	....	....	....	....	....	1490	....	4.14	δ	295
296	.....	.....	.....	....	....	....	....	....	d	....	2.92	i	296
297	.....	.....	.....	1,3	....	....	141 (285)	....	140^{19}	....	1.0729^{18}_{4}	v	297
298	.....	.....	.....	....	....	....	....	....	....	....	....	....	298
299	.....	.....	.....	....	....	....	....	....	m.p. 118	....	....	s	299
300	.....	.....	.....	....	6	32	Gas	....	−21	2.89^{5} atm.	0.9537	....	300
301	.....	.....	.....	....	....	....	....	....	61.6	920^{20}	2.015^{20}_{4}	s	301
302	.....	.....	.....	....	....	....	....	....	13.1	2.1	1.2	....	302
303	.....	.....	.....	....	....	....	....	....	178.5-9.5	....	0.9508^{20}_{4}	δ	303

No.	Chemical Name and Formula	Chemical Abstracts Registry Number	A.C.G.I.H. 1966–TLV (Threshold Limit Values) ppm (mg/M³)	Principal Effects of Inhalation Exposures above TLV	Relative Hazard to Health from Concentrated Short-term Exposure (5 is high; 1 is low)						References to Supplementary Information on Toxicity, Flammability and Other Hazards
					Eye Contact	Inhalation	Skin Penetration	Skin Irritation	Ingestion	Supplemental Effects	
304	Cyclohexane C_6H_{12}	110827	300 T (1050) T	Narcosis	....	....	....	....	....	....	1209, AIHA, MCA, STTE
305	Cyclohexanol $C_6H_{12}O$	108930	50 (200)	Narcosis	3	1	2x	1	2	....	1477, STTE
306	Cyclohexanone $C_6H_{10}O$	108941	50 (200)	Narcosis	2	2	2y	2	2	....	1723, AIHA, STTE
307	Cyclohexene C_6H_{10}	110838	300 (1015)	Narcosis							May deteriorate in normal storage and cause hazard. STTE
308	Cyclohexylamine $C_6H_{13}N$	108918	....								2038, HCD
309	Cyclohexylbenzene (Phenylcyclohexane) $C_{12}H_{16}$	827521									
310	Cyclopentadiene C_5H_6	542927	75 (200)	Toxic	2x	3x	2x	1x	3x	....	
311	Cyclopentane C_5H_{10}	287923	....	....	May be a mild narcotic in high concentrations					....	1207
312	Cyclopentanone C_5H_8O	120923									
313	Cyclopropane C_3H_6	75194	....		Inhalation anesthetic, 400 ppm is recommended exposure limit					....	MGD, STTE
314	p-Cymene $C_{10}H_{14}$	99876	....		3x	2	2x	2x	2	....	
315	2,4-D (2,4-Dichlorophenoxyacetic acid) $C_8H_6Cl_2O_3$	94757	(10)	Toxic	4y	....	2y	3y	2	....	1837, STTE
316	DDT $C_{14}H_9Cl_5$	50293	(1) S	Toxic	4x	....	3y	3x	3	Possibly carcinogenic in animals, according to another source	1346, AIHA, NSC, STTE
317	DDVP $C_4H_7Cl_2O_4P$	62737	(1) S	Toxic	....	....	....	....	....	....	1941, STTE
318	Decaborane $B_{10}H_{14}$	1304025	0.05 S (0.3) S	Toxic	....	....	....	....	....	....	MCA, HCD, STTE. Forms shock sensitive mixtures with halogenated materials
319	Decahydronaphthalene (Decalin®) $C_{10}H_{18}$	91178	....	....	1	2	2	2	2	....	1209, STTE
—	Decalin® See Decahydronaphthalene										
320	n-Decane $C_{10}H_{22}$	124185	....	....	1x	2x	3x	1x	1x	....	....
321	n-Decyl alcohol $C_{10}H_{22}O$	112301	....	....	2	1	2	3	2	....	1467
322	Demetron® $C_8H_{14}O_3PS_2$	126750	(1) S	Toxic	....	....	....	....	....	....	1954, STTE
323	Deuterium D_2	7782390	....	....	Simple asphyxiant. Toxic when ingested in heavy water					....	MGD
—	Diacetone alcohol See 4-Hydroxy-4-methyl-2-pentanone										
—	Diacetyl monoxide See Acetic anhydride										

No.	(7) N.F.P.A. Hazard Identification Signals			(8) Extinguishing Agents	(9) Flammable Limits in Air % by volume		(10) Flash Point °C (°F)	(11) Ignition Temperature °C (°F)	(12) Boiling Point (Melting Point, m.p.) °C	(13) Vapor Pressure mm. Hg °C	(14) Density or Specific Gravity	(15) Water Solubility	No.
	Health	Fire	Reactivity		Lower	Upper							
	(4 is high; 0 is low)												
304	1	3	0	2,3	1.3	8	−20 (−4)	260 (500)	81^{755}	$103.60^{26.35}$	0.7791_4^{20}	i	304
305	1	2	0	1,2,2a,3	....	....	68 (154)	300 (572)	161.1	3.5^{34}	0.9624_4^{20}	s	305
306	1	2	0	1,2a,3	1.1 @ 212°	8.1 @ 175°	44 (111)	420 (788)	156	$4.5^{2.5}$	0.9978_4^{20}	δ	306
307	1	3	0	....	....	....	<7 (<20)	....	83	....	0.8110_4^{20}	i	307
308	2	3	0	2a,3	....	....	5 OC (40)OC	293 (560)	134.5	....	0.8668_4^{20}	s	308
309	2	1	0	....	....	....	99 OC (210)OC	....	238–9	....	0.9502	i	309
310	....	....	....						40.83^{772}	....	0.8021	i	310
311	1	3	0	....			<7 (<20)	....	49.3	....	0.7510_4^{20}	i	311
312	2	3	0	2a	....	....	26 (79)	....	130.65	....	$0.9509_4^{18.2}$	i	312
313	1	4	0	4	2.4	10.4	....	498 (928)	−33	786^{-32}	0.720_4^{-79}	δ	313
314	2	2	0	1,2,3	0.7 @ 212°	5.6	47 (117)	436 (817)	177	....	0.8569_4^{20}	i	314
315	....	....	....		....	....	....	....	$160^{0.4}$	....	....	i	315
316	....	....	....		....	....	....	....	m.p. 108.5	....	1.55	i	316
317	....	....	....					....	84	0.01^{30}	1.415	δ	317
318	3	2	1	....	....	....	~80 ~(176)	149 (300)	213 m.p. 99.7	$\sim66^{132}$	0.94^{20}	δ	318
319	2	2	0	1,2,3	0.7 @ 212°	4.9 @ 212°	58 (136)	250 (482)	185.5	$10^{47.2}$	0.8700_4^{20}	i	319
320	0	2	0	2,3	0.8	5.4	46 (115)	208 (406)	174.1	.....:....	0.7300_4^{20}	i	320
321	....	....	....	....	....	....	82 OC (180)OC	....	229	$1^{69.5}$	0.8287_4^{20}	i	321
322	....	....	....		....	....	....	....	....	0.001^{33}	1.1183	δ	322
323	0	4	0	4	5	75	Gas	....	−249.7	....	0.18 g/l	δ	323

No.	Chemical Name and Formula	(2) Chemical Abstracts Registry Number	(3) A.C.G.I.H. 1966-TLV (Threshold Limit Values) ppm (mg/M³)	(4) Principal Effects of Inhalation Exposures above TLV	(5) Eye Contact	Inhalation	Skin Penetration	Skin Irritation	Ingestion	Supplemental Effects	(6) References to Supplementary Information on Toxicity, Flammability and Other Hazards
					(5 is high; 1 is low)						
325	N,N'-Dialkyl-3,3'-dimethoxy-benzidine................	....	....							Carcinogenic	....
—	3,3-Diaminobenzidine See 3,3',4,4'-Biphenyltetramine										
327	2,2'-Diamino-1,1'-dinaphthyl ...	4488226	....							Carcinogenic in animals	....
329	Diazomethane................ CH₂N₂	463605	0.2 T (0.4) T	Toxic	Highly toxic; sensitizes; fatalities reported					Carcinogenic in animals	2213
330	1,2,5,6-Dibenzacridine........	226368								Carcinogenic in animals	....
331	1,2,7,8-Dibenzacridine........	224420								Carcinogenic in animals	....
332	1,2,5,6-Dibenzanthracene	53703								Carcinogenic in animals	....
333	1,2,5,6-Dibenzcarbazole	207841								Carcinogenic in animals	....
334	3,4,5,6-Dibenzcarbazole	194592								Carcinogenic in animals	....
335	1,2,5,6-Dibenzofluorene........	207830								Carcinogenic in animals	....
336	1,2,3,4-Dibenzophenanthrene ...	188523								Carcinogenic in animals	....
337	3,4,8,9-Dibenzpyrene	198640								Carcinogenic in animals	....
338	Diborane.................... B₂H₆	1304003	0.1 (0.1)	Toxic	Highly toxic					....	AIHA, MCA, HCD, MGD, STTE
339	Dibrom® .... (Dimethyl-1,2-dibromo-2,2-dichloroethylphosphate) C₄H₇Br₂Cl₂O₄P	300765	(3)	Toxic	....					....	....
340	2,6-Dibromo-N-chloro-p-benzoquinoneimine .... C₆H₂Br₂ClNO	537457								May deteriorate in normal storage and cause hazard	
341	Dibromodifluoromethane....... (Difluorodibromomethane) CBr₂F₂	75616	100 (860)	Toxic	Comparatively low toxicity					....	MGD
342	1,2-Dibromoethane .... (Ethylene bromide; Ethylene dibromide) C₂H₄Br₂	106934	C 25 S C (190) S	Toxic	2x	3x	....	4x		Chronic toxicity	1284, AIHA, STTE
343	1,2-Dibromotetrafluoroethane... (Freon®-114B2) C₂B₂F₄	124732	....		Comparatively non-toxic						MGD
344	Dibutylamine .... (Di-n-butylamine) C₈H₁₉N	111922			4	5	3	3	3	....	2038, STTE
345	Dibutyldichlorotin .... (Dibutyltin dichloride) (C₄H₉)₂SnCl₂	683181			5	2	3x	5	3x		STTE
346	Dibutylbis(lauroyloxy)tin ...... (Dibutyltin dilaurate) C₃₂H₆₄O₄Sn	77587			2	....	3	5	2		STTE
—	Dibutyl ether See Butyl ether										
347	Dibutyl oxalate............... C₁₀H₁₈O₄	2050604	....								
—	Dibutyl phosphate See Butyl phosphate										
348	Dibutyl phthalate .... C₁₆H₂₂O₄	84742	(5) T	Irritant	1	1y	1	1y	2	....	1900, HCD
—	Dibutyltin dichloride See Dibutyldichlorotin										
—	Dibutyltin dilaurate See Dibutylbis(lauroyloxy)tin										

No.	(7) N.F.P.A. Hazard Identification Signals			(8) Extinguishing Agents	(9) Flammable Limits in Air % by volume		(10) Flash Point °C (°F)	(11) Ignition Temperature °C (°F)	(12) Boiling Point (Melting Point, m.p.) °C	(13) Vapor Pressure mm. Hg °C	(14) Density or Specific Gravity	(15) Water Solubility	No.
	Health	Fire	Reactivity		Lower	Upper							
	(4 is high; 0 is low)												
325	.....	.....	.....	....	....		....				....	......	325
—													—
327	....			....	....		....				....	......	327
329	....	...			Explosive		Gas	Extra hazardous	−23	....	....	d	329
330													330
331	....				....		....				....		331
332	....				....		....				....		332
333	....				....		....				....		333
334	....				....		....				....		334
335	....				....		....				....		335
336	....				....		....				....	......	336
337	....				....		....				....		337
338	3	4 W	3	4	0.9	98	−90 (−130)	145 (293)	−92.5 m.p. −164.9	$760^{-9.3}$	0.445 liq.	δ, d to H_3BO_3 + H_2	338
339					....		....		....	....	....	......	339
340	....				....		....		....	....	....	......	340
341	....			....	Non-flammable		....		24.5	....	2.28	s	341
342	....								131	12^{25}	2.180_4^{20}	δ	342
343	....				Non-flammable		....		47.3	....	2.175 liq.	......	343
344	3	2	0	1,2,2a,3	....	....	52 OC (125) OC	....	159	1.9^{20}	0.501	s	344
345	....			....	m.p. 142		....				1.36	d	345
346	....							....	m.p. 27	....	1.05	i	346
—													—
347	....	1	0	1,2,3	....	....	93 (200)	....	243.4	....	1.0099_0^0	i	347
—													—
348	0	1	0	1,2,3	....	....	157 (315)	403 (757)	340	2^{150}	1.043	i	348
—													—
—													—

No.	(1) Chemical Name and Formula	(2) Chemical Abstracts Registry Number	(3) A.C.G.I.H. 1966–TLV (Threshold Limit Values) ppm (mg/M³)	(4) Principal Effects of Inhalation Exposures above TLV	(5) Eye Contact	Inhalation	Skin Penetration	Skin Irritation	Ingestion	Supplemental Effects	(6) References to Supplementary Information on Toxicity, Flammability and Other Hazards
					(5 is high; 1 is low)						
349	**Dichloroacetic acid** . . . . $C_2H_2Cl_2O$	79436	. . . .		5	1	3	5	2	. . . .	1796, STTE
350	**Dichloroacetyl chloride** . . . . C_2HCl_3O	79367	. . . .		5	4	3	5	2		HCD
351	**1,3-Dichloro-2-butene** . . . . $C_4H_6Cl_2$	926578	. . . .								May deteriorate in normal storage and cause hazard
352	**o-Dichlorobenzene** . . . . $C_6H_4Cl_2$	95501	C 50 C(300)	Toxic	3x	3x*	2x	2x	2x	Chronic toxicity	1335, AIHA, MCA, HCD, STTE
353	**p-Dichlorobenzene** . . . . $C_6H_4Cl_2$	106467	75 (450)	Toxic	. . . .	. . . .					1337, AIHA, STTE
354	**3,3′-Dichlorobenzidine**	91941	. . . .							Carcinogenic in animals	. . . .
355	**Dichlorodifluoromethane** . . . . (Freon®-12; Genetron®-12; Halon®; Isotron®-12; Ucon®-12) CCl_2F_2	75718	1000 (4950)	Narcosis	Practically non-toxic						1324, MGD, STTE
356	**1,3-Dichloro-5,5-dimethyl hydantoin** . . . . $C_5H_6Cl_2N_2O_2$	118525	(0.2)	Toxic	. . . .						. . . .
357	**1,1-Dichloroethane** . . . . (Ethylidene dichloride) $C_2H_4Cl_2$	75343	100 (400)	Toxic	1x	3y*	2	1x	1	Chronic toxicity	1279, STTE
358	**1,2-Dichloroethane** . . . . (Ethylene dichloride; Ethylene chloride) $C_2H_4Cl_2$	107062	50 (200)	Toxic	Respiratory and conjunctival irritant; Causes narcosis					. . . .	1280, HCD, AIHA, STTE
359	**1,1-Dichloroethylene** . . . . (Vinylidene chloride) $C_2H_2Cl_2$	75354	. . . .		2x	2y*	1x	2x	2x	Chronic toxicity	1305, HCD, STTE
360	*cis*-**1,2-Dichloroethylene** . . . . (Acetylene dichloride-*cis*) $C_2H_2Cl_2$	156592	200 (790)	Narcosis	1x	3y	2x	1x	1x	Chronic toxicity	1307
361	*trans*-**1,2-Dichloroethylene** . . . . . (Acetylene dichloride-*trans*) $C_2H_2Cl_2$	156605			1x	3y	2x	1x	1x	Chronic toxicity	. . . .
	Dichloroethyl ether See Bis(2-chloroethyl)ether	. . . .	. . . .								
362	**Dichlorofluoromethane** . . . . (Freon®-21; Genetron®-21) $CHCl_2F$	75434	1000 (4200)	Narcosis	. . . .	. . . .					MGD
363	**Dichloromethane** . . . . (Methylene chloride) CH_2Cl_2	75092	500 (*1740*)	Narcosis	1	1	. . . .	1	. . . .	Skin contact is painful	1257, AIHA, MCA, NSC, STTE
364	**1,2-Dichloro-1-nitroethane**	10544555	C 10 C(60)	Toxic; Irritant	. . . .						. . . .
—	2,4-Dichlorophenoxyacetic acid See 2,4-D										
365	**Dichlorophenylphosphine**		. . . .								May deteriorate in normal storage and cause hazard
366	**1,2-Dichloropropane** . . . . (Propylene dichloride) $C_3H_6Cl_2$	78875	75 (350)	Toxic	1	3*	2	1	2	Chronic toxicity	1301, STTE
367	**1,3-Dichloropropene** . . . . $C_3H_4Cl_2$	542756	. . . .		4x	3x	5x	5x	. . . .		. . . .
368	**1,2-Dichlorotetrafluoroethane** . . . (Freon®-114; Genetron®-114; Isotron®-114; Ucon®-114) $C_2F_4Cl_2$	76142	1000 (7000)	Narcosis	Low toxicity					. . . .	1329, MGD, STTE

No.	(7) N.F.P.A. Hazard Identification Signals			(8) Extin-guishing Agents	(9) Flammable Limits in Air % by volume		(10) Flash Point °C (°F)	(11) Ignition Tem-perature °C (°F)	(12) Boiling Point (Melting Point, m.p.) °C	(13) Vapor Pressure mm. Hg °C	(14) Density or Specific Gravity	(15) Water Solu-bility	No.
	Health	Fire	Reac-tivity		Lower	Upper							
	(4 is high; 0 is low)												
349	.....	.....	.....	....	....	....	....	....	192–3	1⁴⁴	1.5634	∞	349
350	3	2	1	....	....	....	66 (151)	....	107–8	....	1.5315	d	350
351	2	3	0	....	....	....	27 (80)	....	125–130	....	....	......	351
352	2	2	0	1,2,3	2.2	9.2	66 (151)	648 (1198)	179	1.56³⁵	1.3048	i	352
353	2	2	0	1,2,3	....	....	66 (150)	....	174	....	1.533	i	353
354	.....	.....	.....										354
355	.....	.....	.....		Non-flammable at normal temperatures and pressures		....	....	−29	10²³	1.292	s	355
356	.....												356
357	2	3	0	....	....	....	6 (22)	....	57 cor	234²⁵	1.1776	δ	357
358	2	3	0	....	6.2	16	13 (56)	413 (775)	84	87²⁵	1.256	δ	358
359	2	4 P	2	....	5.6	11.4	−15 OC (0) OC	458 (856)	37	....	1.3	......	359
360	2	3	2	2,3	9.7	12.8	4 (39)	....	60.3	208²⁵	12837	δ	360
361	2	3	2	2,3	9.7	12.8	2 (36)	....	47.5	324²⁵	....	......	361
—													—
362	.....	.....	.....	4	Weakly flammable		....	552 (1026)	9	....	1.426	i	362
363	2	0	0	....	(15.5 to 66 in oxygen only)		....	662 (1224)	40¹	440²⁵	1.335	δ	363
364	.....	.....	.....	....	....	....	....	....	....	....	....	......	364
—													—
365	.....	.....	.....	....	....	....	....	....	....	....	....	......	365
366	2	3	0	2,3	3.4	14.5	16 (60)	557 (1035)	96.20	50²⁵	1.1558²⁰	δ	366
367	.....	.....	.....	....	....	....	21 (70)	....	104	52° mm	1.225	i	367
368	.....	.....	.....	....	Non-flammable		....	....	3.6	....	1.440	i	368

No.	(1) Chemical Name and Formula	(2) Chemical Abstracts Registry Number	(3) A.C.G.I.H. 1966–TLV (Threshold Limit Values) ppm (mg/M^3)	(4) Principal Effects of Inhalation Exposures above TLV	(5) Relative Hazard to Health from Concentrated Short-term Exposure						(6) References to Supplementary Information on Toxicity, Flammability and Other Hazards
					Eye Contact	Inhalation	Skin Penetration	Skin Irritation	Ingestion	Supplemental Effects	
					(5 is high; 1 is low)						
369	α,α-Dichlorotoluene (Benzal-chloride) $C_7H_6Cl_2$	98873	....		1	2x*	2x	3	2	Chronic toxicity	....
370	Dicyclohexylamine $C_{12}H_{23}N$	101837	....		....	....	....	....	....	....	2038
—	Dicyclopentadiene See 3a,4,7,7a-Tetrahydro-4,7-methanoindene										
371	Dieldrin $C_{12}H_8Cl_6O$	60571	(0.25)S	Toxic	....						1358, STTE
—	Diethanolamine See 2,2-Iminodiethanol										
372	Diethyl adipate (Ethyl adipate) $C_{10}H_{18}O_4$	141286	....		2x	....	2y	2y	2	....	1882
—	Diethylaluminum bromide See Bromodiethyl aluminum										
—	Diethylaluminum chloride See Chlorodiethylaluminum										
373	Diethylamine $C_4H_{11}N$	109897	25 (75)	Irritant	5	4	3	2	3	....	2038, AIHA, STTE
374	Diethylaminoethanol $C_6H_{15}NO$	100378	10 ST (50) ST	Irritant	4	1	3	2	2	....	2062, STTE
375	Diethylaniline $C_{10}H_{15}N$	91667	....		....	....	....	....	....	....	2137, STTE
—	Diethyl carbitol® See Bis(2-ethoxyethyl)ether										
376	Diethyl carbonate $C_5H_{10}O_3$	105588	25 (77)	....	....	....	....	....	....	....	1912, STTE
377	Diethylene glycol $C_4H_{10}O_3$	111466	....		1	1	2	1	1	....	1503
—	Diethylene glycol diethyl ether See Bis(2-ethoxyethyl ether)										
—	Diethylene glycol monobutyl ether See 2-(2-butoxyethoxy)ethanol										
—	Diethylene glycol monoethyl ether See 2-(2-ethoxyethoxy)ethanol										
378	Diethylenetriamine $C_4H_{13}N_3$	111400	....		....	....	....	....	....	Irritant; corrosive; may cause sensitization	AIHA, MCA
—	Diethyl ether See Ethyl ether										
—	Diethyl ketone See 3-Pentanone										
379	Diethyl malonate $C_7H_{12}O_4$	105533	....		....	....	....	....	....	....	1881
380	Diethyl phthalate $C_{12}H_{14}O_4$	84662	....		....	....	....	....	....	....	1900
381	Diethyl sulfate $C_4H_{10}O_4S$	64675	....		3	2	3	4	3	Possibly carcinogenic in animals	1930, STTE
382	Diethylzinc $C_4H_{10}Zn$	557220	....	Combustion yields zinc oxide fumes with TLV of 5 mg/m^3							HCD
—	1,1-Difluoro-1-chloroethane See 1-Chloro-1,1-difluoroethane										
—	Difluorodibromomethane See Dibromodifluoromethane										

No.	(7) N.F.P.A. Hazard Identification Signals			(8) Extinguishing Agents	(9) Flammable Limits in Air % by volume		(10) Flash Point °C (°F)	(11) Ignition Temperature °C (°F)	(12) Boiling Point (Melting Point, m.p.) °C	(13) Vapor Pressure mm. Hg °C	(14) Density or Specific Gravity	(15) Water Solubility	No.
	Health	Fire	Reactivity		Lower	Upper							
	(4 is high; 0 is low)												
369	....	....	....	....	....	....	....	....	205.2	....	1.2557	i	369
370	....	....	....	1,2a,3	....	....	99 OC (210)OC	....	255.8	....	0.925^{18}	δ^h	370
371	....	....	....	....	....	....	....	....	m.p. 175–6	....	1.75	i	371
'372	....	....	....	....	....	....	....	....	240–45	....	1.0076	i	372
373	3	3	0	2a,3	1.8	10.1	<−26 (<−14)	312 (594)	56.3	195^{20}	0.7108	v	373
374	3	2	0	1,2a,3	....	....	60 OC (140)OC	....	163	....	0.884	∞	374
375	3	2	0	1,2a,3	....	....	85 (185)	332 (630)	216.27	....	0.93507	δ	375
376	....	3	1	2,3	....	....	....	....	126	....	0.9752	i	376
377	1	1^1	0	1,2a,3	2.0	....	124 (255)	343 (650)	245	$<0.01^{20}$	1.118^{20}_{20}	s	377
378	....	....	....	....	....	....	102 OC (215)OC	399 (750)	207.1	0.37^{20}	0.9542	∞	378
379	0	1	0	1,3	....	....	93 OC (200)OC	....	199	10^{x1}	1.0550	δ	379
380	0	1	0	1,2,3	....	....	117 (243)	....	296	0.05 (70)	1.2321^{14}_{4}	i	380
381	3	1	1	2a,3	....	....	104 (220)	436 (817)	208 δ d	....	1.1774	i: d^h	381
382	0	3 W	3	....	Ignites instantly on contact with air; Reacts violently with water				118 m.p. 28	....	1.207	d	382

No.	(1) Chemical Name and Formula	(2) Chemical Abstracts Registry Number	(3) A.C.G.I.H. 1966–TLV (Threshold Limit Values) ppm (mg/M³)	(4) Principal Effects of Inhalation Exposures above TLV	(5) Relative Hazard to Health from Concentrated Short-term Exposure						(6) References to Supplementary Information on Toxicity, Flammability and Other Hazards
					Eye Contact	Inhalation	Skin Penetration	Skin Irritation	Ingestion	Supplemental Effects	
					(5 is high; 1 is low)						
383	1,1-Difluoroethane............ (Ethylidene fluoride; Genetron® 152) C₂H₄F₂	75376	....	....	1x	2y	....	....	....	Very low toxicity but should be used with adequate ventilation to avoid suffocation from oxygen deficiency	MGD
384	1,1-Difluoroethylene.......... (Vinylidene fluoride; Genetron® 1132) C₂H₂F₂	75387	....	....	1x	1y	1x	1x	1x	....	1322, MGD
385	Diglycidyl ether............. C₆H₁₀O₃	2238075	C 0.5 C (2.8)	Toxic	3x	....	3x	4x	....	Sensitiz. skin, resp. tract	1608, STTE
—	Dihydropyran See 3,4-Dihydro-2 H-pyran										
386	3,4-Dihydro-2 H-pyran........ (Dihydropyran) C₅H₈O	110872	....	....							....
—	Diisobutyl ketone See 2,6-Dimethyl-4-heptanone										
387	Diisopropylamine............. C₆H₁₅N	108189	5 S (20) S	Irritant	4	5	2y	2	3	....	2038, HCD, STTE
388	Diisopropyl fluorophosphate .... C₈H₁₄FPO₃				Avoid inhalation and skin contact. Powerful inactivator of cholinesterase. Forms HF in presence of moisture						Merck
389	Diisopropyl peroxydicarbonate .. C₈H₁₄O₆	105646	....	....							HCD
390	3,3′-Dimethoxybenzidine....... C₁₄H₁₆N₂O₂	119904	....	....						Carcinogenic in aminals	....
391	1,2-Dimethoxyethane.......... C₄H₁₀O₂	110714	....	....							....
392	Dimethoxymethane .... (Methylal) C₃H₈O₂	109875	1000 (3100)	Toxic	....	....	....	....	....		1982, STTE
393	2,2-Dimethoxypropane .... C₅H₁₂O₂	77769	....	....	Can cause anesthesia						1667
394	N,N-Dimethylacetamide .... C₄H₉NO	127195	10 S (35) S	Toxic	2	1	2	1	2		1830, STTE
395	Dimethylamine............... C₂H₇N	124403	10 (18)	Toxic	4x	5x*	....	4x*		Sensitiz. skin, resp. tract	2038, HCD, MGD
—	p-Dimethylaminoazobenzene See C. I. Solvent Yellow 2										
396	Dimethylaminoazobenzene -1-naphthalene	1335984	....	....						Carcinogenic in animals	....
397	Dimethylaminoazobenzene-2- naphthalene...............	1335973	....	....						Carcinogenic in animals	....
—	N,N-Dimethyl-4-aminobiphenyl See N,N-Dimethyl-4-biphenyl- amine										
398	2-Dimethylaminofluorene ...... C₁₅H₁₅N	1335951	....	....						Carcinogenic in animals	....
	3,5-Dimethylaniline See 3,5-Xylidine										
400	N,N-Dimethylaniline .... C₈H₁₁N	121697	5 S (25) S	Toxic	2	1	3	2	2	Toxic fumes from decomposition	STTE
401	2,3-Dimethylazobenzene	10580548	....	....						Carcinogenic in animals	....
402	9,10′-Dimethyl-1,2- benzanthracene.............	56564	....	....						Carcinogenic in animals	....

No.	(7) N.F.P.A. Hazard Identification Signals			(8) Extin- guishing Agents	(9) Flammable Limits in Air % by volume		(10) Flash Point °C (°F)	(11) Ignition Tem- perature °C (°F)	(12) Boiling Point (Melting Point, m.p.) °C	(13) Vapor Pressure mm. Hg °C	(14) Density or Specific Gravity	(15) Water Solu- bility	No.
	Health	Fire	Reac- tivity		Lower	Upper							
	(4 is high; 0 is low)												
383	.....	.....	.....	4	3.7	18.0	....	....	−24.7	....	0.895^{25} g/cc^{25}	i	383
384	.....	.....	.....	....	5.5	21.3	....	....	−85.7	....	0.585^{25} g/cc^{25}	δ	384
385	.....	.....	.....	....	....	....	....	....	260	0.09^{25}	1.262	....	385
386	.....	3	0	2a	....	....	−18 (0)	....	86–7	....	0.922^{19}_{15}	s	386
387	3	3	0	1,2a,3	83.5	4^{743}	−1 OC (30) OC	....	....	70^{20}	0.722^{22}	δ	387
388	.....	.....	.....	....	....	....	....	....	....	....	....	....	388
389	0	4	4	....	Self-accelerating decomposition at 12°C (53°F). Shock & heat sensitive.				m.p. 8	....	....	....	389
390	.....	.....	.....	....	....	....	....	....	m.p. 137	....	....	i	390
391	.....	2	0	2a	....	....	40 (104)	....	85	....	0.8665^{20}_4	s	391
392	2	3	2	2a,3	....	....	−18 OC (0) OC	237 (459)	43.9	400^{25}	0.856	v	392
393	.....	.....	.....	....	....	....	−7 (20)	....	80	$100^{26.4}$	0.850^{20}_{20}	s	393
394	.....	.....	.....	....	1.8	13.8	62 (144)	354 (669)	165^{758}	9^{60}	0.9366^{25}_4	∞	394
395	3	4	0	3,4	2.8	14.4	<−18 (<0)	430 (806)	7.4	$2\,atm.^{10}$	0.6804^0_4	v	395
396	.....	.....	.....	....	....	....	....	....	m.p. 8–10	....	1.080	i	396
397	.....	.....	.....	....	....	....	....	....	....	....	....	....	397
398	.....	.....	.....	....	....	....	....	....	....	....	....	....	398
400	3	2	0	1,2,2a,3	....	....	63 (145)	371 (700)	194.15	....	0.9563	δ	400
401	.....	.....	.....	....	....	....	....	....	....	....	....	....	401
402	.....	.....	.....	....	....	....	....	....	....	....	....	....	402

No.	Chemical Name and Formula	(2) Chemical Abstracts Registry Number	(3) A.C.G.I.H. 1966-TLV (Threshold Limit Values) ppm (mg/M³)	(4) Principal Effects of Inhalation Exposures above TLV	(5) Eye Contact	Inhalation	Skin Penetration	Skin Irritation	Ingestion	Supplemental Effects	(6) References to Supplementary Information on Toxicity, Flammability and Other Hazards
403	α,α-Dimethylbenzyl hydroperoxide (Cumene hydroperoxide) $C_9H_{12}O_2$	80159								Skin sensitiz. toxicity	HCD, STTE
404	N,N-Dimethyl-4-biphenylamine (N,N-Dimethyl-4-aminobiphenyl)	1137797								Carcinogenic	
405	2,2-Dimethylbutane C_6H_{14}	75832									
406	Dimethyl carbonate $C_2H_6O_3$	616386									
—	Dimethyl chloroacetal See Chloroacetaldehyde, dimethyl acetal										
407	5,6-Dimethyl chrysene	3697276								Carcinogenic in animals	
408	Dimethyl-1,2-dibromo-2,2-dichloroethyl phosphate	300765	(3)	Toxic							
409	Dimethyldichlorotin (Dimethyl tin dichloride) $C_2H_6Cl_2Sn$	753731			4			4			
410	Dimethyldinitrosopropane diamine	1335848								Carcinogenic in animals	
—	Dimethyl ether See Methyl ether										
411	N,N-Dimethylformamide C_3H_7NO	68122	10 S (30) S	Toxic	2	2y	2y	2	2		1830, AIHA, STTE
412	Dimethyl fumarate $C_6H_8O_4$	624497			5	1	3	3	2		
413	2,6-Dimethyl-4-heptanone (Diisobutyl ketone) $C_9H_{18}O$	108838	50 (290)	Narcosis	1	1	1	2	2		1721, AIHA, STTE
414	1,1-Dimethylhydrazine (UDMH) $C_2H_8N_2$	57147	0.5 S (1) S	Toxic							2220, AIHA, STTE
415	Dimethylnaphthalene $C_{12}H_{12}$	1335939			2x	1	2x	1x*	2		
—	Dimethylnitrosoamine See N-Nitrosodimethylamine										
—	2,6-Dimethylphenol See Xylenol										
—	N,N-Dimethyl-p-phenylazoaniline See C.I. Solvent Yellow 2										
417	N,N-Dimethyl-p-phenylazo-o-anisidine	2438495								Carcinogenic in animals	
418	Dimethyl phthalate (DMP) $C_{10}H_{10}O_4$	131113	(5)	Nuisance	May be irritating to mucous membranes and eyes						
419	2,2-Dimethylpropane (Neopentane) C_5H_{12}	463821			Simple asphyxiant and anesthetic						MGD
420	Dimethyl sulfate $C_2H_6O_4S$	77781	1 S (5) S	Irritant	5z	5z	3x	5z	3	Possibly carcinogenic in animals	1926, MCA, STTE
—	Dimethyl sulfide See Methyl sulfide										
—	Dimethyl sulfoxide See Methyl sulfoxide										
—	Dimethyltin dichloride See Dimethyldichlorotin										
421	2,4-Dinitroaniline $C_6H_5N_3O_4$	97029			May be irritating to skin and mucous membranes. Highly toxic if absorbed.						
422	m-Dinitrobenzene $C_6H_4N_2O_4$	528290	(1) S	Toxic	Highly toxic vapors						2139, AIHA, STTE

No.	(7) N.F.P.A. Hazard Identification Signals			(8) Extinguishing Agents	(9) Flammable Limits in Air % by volume		(10) Flash Point °C (°F)	(11) Ignition Temperature °C (°F)	(12) Boiling Point (Melting Point, m.p.) °C	(13) Vapor Pressure mm. Hg °C	(14) Density or Specific Gravity	(15) Water Solubility	No.
	Health	Fire	Reactivity		Lower	Upper							
	(4 is high; 0 is low)												
403	1	2	4	....	Vapor forms explosive mixtures with air		79 (175)	221 (430)	153	....	1.048	....	403
404	.....	.....	.....	....	....	....	....	....	....	....	....	....	404
405	1	3	0	3	1.2	7.0	−48 (−54)	425 (797)	49.7	....	0.6492_4^{20}	i	405
406	.....	3	1	2a	....	....	19 OC (66) OC		90	....	0.64924_4^{20}	i	406
407	.....	.....	.....	....	....	....	....	....	....	....	....	....	407
408	.....	.....	.....	....	....	....	....	....	....	....	....	....	408
409	.....	.....	.....	....	....	....	....	....	188–90	....	....	....	409
410	.....	.....	.....	....	....	....	....	....	....	....	....	....	410
411	1	2	0	1,2,2a,3	22 @ 212°	15.2	58 (136)	445 (833)	153^{758}	3.7^{25}	0.9445_4^{25}	∞	411
412	.....	.....	.....	....	....	....	....	....	m.p. 241	....	....	∞ $\delta{:}s^h$	412
413	1	2	0	1,3	0.8 @ 212°	6.2 @ 212°	60 (140)	....	168	2.4^{25}	0.8053	i	413
414	.....	.....	.....	2a	2	95	−15 (5)	249 (480)	63^{752}	156.8^{25}	0.7914^{22}	v	414
415	.....	.....	.....	....	....	....	....	....	$262–4^{751}$	....	1.0157	i	415
417	.....	.....	.....	....	....	....	....	....	....	....	....	....	417
418	.....	.....	.....	....	....	....	157 OC (315) OC	518 (964)	282–5	....	1.1905		418
419	.....	4	0	4	1.4	7.5	gas	450 (842)	9.45	$606^{3.1}$	0.61350	i	419
420	4	2	0	1,2,3	....	....	83 (182)	....	188.5d	....	1.3322	s	420
421	3	1	3	1,3	....	....	244 (435)	....	m.p. 187.5–8	....	1.615	i	421
422	.....	.....	.....	....	Severe explosion hazard when exposed to shock or flame		150 (302)	....	291^{756} m.p. 90	....	1.575_4^{18}	i	422

No.	(1) Chemical Name and Formula	(2) Chemical Abstracts Registry Number	(3) A.C.G.I.H. 1966–TLV (Threshold Limit Values) ppm (mg/M³)	(4) Principal Effects of Inhalation Exposures above TLV	(5) Relative Hazard to Health from Concentrated Short-term Exposure (5 is high; 1 is low) Eye Contact	Inhalation	Skin Penetration	Skin Irritation	Ingestion	Supplemental Effects	(6) References to Supplementary Information on Toxicity, Flammability and Other Hazards
423	o-Dinitrobenzene $C_6H_4N_2O_4$	....	(1) S	Toxic	Highly toxic vapors						2138, HCD, STTE
424	p-Dinitrobenzene $C_6H_4N_2O_4$	100254	(1) S	Toxic	Highly toxic vapors						2139, STTE
—	2,4-Dinitrochlorobenzene See 1-Chloro-2,4-dinitrobenzene										
425	Dinitro-o-cresol $C_7H_6N_2O_5$	1335859	(0.2) S	Toxic	Highly toxic					....	2142, STTE
426	2,7-Dinitrofluorene	5405538	....		....	....	...:	....	....	Carcinogenic in animals	....
—	Dinitrogen trioxide See Nitrogen trioxide										
427	2,4-Dinitrophenol $C_6H_4N_2O_5$	51285	....	....	Readily absorbed through the skin; liver and kidney damage reported in chronic poisoning. Exposure level of 0.2 mg/m believed safe.						AIHA
428	1,4-Dinitrosopiperazine $C_4H_8N_4O_2$	140794	....		....	....	....	....	....	Carcinogenic in animals	....
429	2,4-Dinitrotoluene $C_7H_6N_2O_4$	121142	(1.5) S	Toxic	2y	1	2y	2y	2	....	MCA; HCD
430	2,6-Dinitrotoluene $C_7H_6N_2O_4$	606202	(1.5) S	Toxic	2y	1	2y	2y	2	....	MCA
431	Di-sec-octyl phthalate $C_{24}H_{39}O_4$	1335995	(5)	Toxic	1	1	1	1	1	....	1901
433	p-Dioxane (1,4-Dioxane) $C_4H_8O_2$	123911	100 S (360) S	Toxic	2	3*	2	1y	2	Chronic toxicity (Liver carcinogen in rats)	1537, AIHA, HCD, Forms peroxides. J. Nat. Cancer Inst. 35, 949, (1965).
—	Dipentene® See p-Mentha-1,8-diene										
—	Diphenyl See Biphenyl										
434	Diphenylamine (Not to be confused with Biphenylamine) $C_{12}H_{11}N$	122394	....		3x	....	3x	2x*	3x	....	2143. Toxic fumes on decomposition
435	Diphenylmethane $C_{13}H_{12}$	101815	....		....	....	....	....	....		....
—	Dipropylene glycol methyl ether See 3-(3-Methoxypropoxy)-1-propanol										
—	Disodium hydrogen phosphate See Sodium phosphate										
—	DMP See Dimethyl phthalate										
436	Dodecane $C_{12}H_{26}$	112403	....		....	....	....	....	....		....
—	Dodecanoic acid (See Lauric acid)										

No.	(7) N.F.P.A. Hazard Identification Signals			(8) Extinguishing Agents	(9) Flammable Limits in Air % by volume		(10) Flash Point °C (°F)	(11) Ignition Temperature °C (°F)	(12) Boiling Point (Melting Point, m.p.) °C	(13) Vapor Pressure mm. Hg °C	(14) Density or Specific Gravity	(15) Water Solubility	No.
	Health	Fire	Reactivity		Lower	Upper							
	(4 is high; 0 is low)												
423	3	1	4	1,3	....	....	150 (302)	....	319[773]	....	1.3119	δ	423
424	.....	.....	.....	....	....	....	150 (302)	....	299[777]	....	1.625	i	424
425	.....	.....	.....	....	....	....	....	....	85	....		i	425
426	.....	.....	.....	....	....	....	....	....	....	....	....	....	426
427	.....	.....	.....	Dangerous fire hazard when dry.					m.p. 114	....	1.681	s	427
428	.....	.....	.....	....	....	....	....	....	....	....	....	....	428
429	3	1	3	1,3	....	....	....	....	m.p. 60 Self-sustaining decomposition at ≈ 280	....	1.321[70]	i	429
430	3	1	3	....	....	....	....	....	m.p. 66 Self-sustaining decomposition at ≈ 280	....	1.2833[111]	....	430
431	0	1	0	1,3	....	....	218 OC (425) OC	410 (770)	358	0.01[20]	0.986	i	431
433	2	3	1	1,2a	2.0	22	12 (54)	366 (690)	101	37[25]	1.0336^{21}_{4}	∞	433
434	3	1	0	1,3	....	....	153 (307)	634 (1173)	302	....	1.160	i	434
435	1	1	0	1,2,3	....	....	130 (266)	486 (907)	265.6	....	1.0060	i	435
436	0	2	0	1,2,3	0.6		74 (165)	204 (399)	216.3	....	0.7487^{20}_{4}	i	436

No.	(1) Chemical Name and Formula	(2) Chemical Abstracts Registry Number	(3) A.C.G.I.H. 1966–TLV (Threshold Limit Values) ppm (mg/M³)	(4) Principal Effects of Inhalation Exposures above TLV	(5) Relative Hazard to Health from Concentrated Short-term Exposure						(6) References to Supplementary Information on Toxicity, Flammability and Other Hazards
					Eye Contact	Inhalation	Skin Penetration	Skin Irritation	Ingestion	Supplemental Effects	
					(5 is high; 1 is low)						
437	**Dodecyl sodium sulfate** . . . . (Sodium lauryl sulfate) $C_{12}H_{25}NaO_4S$	. . . .	. . . .	. . . .	4	. . . .	2x	3x	2	. . . .	1844
—	Dowanol® DB See 2-(2-Butoxyethoxy) ethanol										
—	Dowanol® DE See 2-(2-Ethoxyethoxy) ethanol										
—	Dowanol® EB See 2-Butoxyethanol										
—	Dowanol® EE See 2-Ethoxyethanol										
—	Dowanol® EM See 2-Methoxyethanol										
438	**Dysprosium**. . . . . Dy	7429916	. . . .	. . . .	. . . .	. . . .	. . . .	. . . .	. . . . . .	. . . .	1058
439	*Endrin* . . . . $C_{12}H_8Cl_6O$	72208	(0.1) S	Toxic	. . . .	. . . .	. . . .	. . . .	. . . . . .	. . . .	STTE
—	Epichlorohydrin See 1-Chloro-2,3-epoxy-propane										
440	**EPN** . . . . $C_{14}H_{14}O_4NPS$	2104645	(0.5) S	Toxic	. . . .	. . . .	. . . .	. . . .	. . . . . .	. . . .	1945
442	**Epoxy resin systems**	. . . .	. . . .	. . . .	. . . .	. . . .	. . . .	. . . .	. . . . . .	. . . .	AIHA
443	**1,2-Epoxy-3-phenoxypropane** . . . (Phenyl glycidyl ether) $C_9H_{10}O_2$	122601	10 (62)	Toxic	1	1	3	3	2	. . . .	1640, STTE
444	**Erbium** . . . . Er	7440526	. . . .	. . . .	. . . .	. . . .	. . . .	. . . .	. . . . . .	. . . .	1058
445	**Ethane**. . . . . C_2H_6	74840	. . . .	. . . .	. . . .	3x	. . . .	. . . .	. . . . . .	Simple asphyxiant; anesthetic in high concentrations.	1196, MGD, STTE
446	**Ethanethiol** . . . . (Ethyl mercaptan) C_2H_6S	75081	C 10 C (25)	Toxic	. . . .	. . . .	. . . .	. . . .	. . . . . .	. . . .	May deteriorate in normal storage and cause hazard. STTE
—	Ethanolamine See 2-Aminoethanol										
—	Ethoxy acetylene See Ethyl ethynyl ether										
447	**2-(2-Ethoxyethoxy)ethanol** . . . . . (Diethylene glycol monoethyl ether; Carbitol®; Dowanol® DE) $C_6H_{14}O_3$	.111900	. . . .	. . . .	. . . .	. . . .	. . . .	. . . .	. . . . . .	. . . .	. . . .
448	**2-Ethoxyethanol** . . . . (Ethylene glycol monoethyl ether; Cellosolve®; Dowanol® EE) $C_4H_{10}O_2$	110805	200 S (740) S	Toxic	1	2	. . . .	2	. . . . . .	. . . .	1542, AIHA, STTE
449	**2-Ethoxyethyl acetate** . . . . (Cellosolve® acetate) $C_6H_{12}O_3$	111159	100 S (540) S	Toxic	. . . .	. . . .	. . . .	. . . .	. . . . . .	. . . .	1859, STTE
450	**Ethyl acetanilide**. . . . . $C_{10}H_{13}NO$	. . . .	. . . .	. . . .	. . . .	. . . .	. . . .	. . . .	. . . . . .	. . . .	. . . .
451	**Ethyl acetate**. . . . . $C_4H_8O_2$	141786	400 (1400)	Irritant	1	3	1	1	2	. . . .	1858, AIHA, MCA, STTE
452	**Ethyl acetoacetate** . . . . $C_6H_{10}O_3$	141979	. . . .	. . . .	2	1	1	1	2	. . . .	1860, STTE
—	Ethylacetylene See 1-Butyne										

No.	(7) N.F.P.A. Hazard Identification Signals			(8) Extinguishing Agents	(9) Flammable Limits in Air % by volume		(10) Flash Point °C (°F)	(11) Ignition Temperature °C (°F)	(12) Boiling Point (Melting Point, m.p.) °C	(13) Vapor Pressure mm. Hg °C	(14) Density or Specific Gravity	(15) Water Solubility	No.
	Health	Fire	Reactivity		Lower	Upper							
	(4 is high; 0 is low)												
437	.....	...	.....	....	....	....	....	....	....	....	....	s	437
—													—
—													—
—													—
—													—
438	.....	.....	.....	....	Pyrophoric on filing		....	....	2600 m.p. 1407		8.536	i	438
439	.....	.....	.....	....	....	....			m.p. 200	....	....	i	439
—													—
440							....	....	m.p. 36	$3 \times 10^{-4\ 100}$	1.5978	δ	440
442	.....			....	....	....						......	442
443	.....			....	....	....			245	0.01^{20}	1.1092	δ	443
444	.....				Pyrophoric on filing		....		2900 m.p. 1497		9.051	i	444
445	1	4	0	4	3.0	12.5	....	515 (959)	−88.63	....	0.572	i	445
446	2	4	0	....	2.8	18.0	<27 (<80)	299 (570)	37	....	0.8391	δ	446
—													—
—													—
447	1	1	0	....	1.2	....	96 (205)	204 (400)	200	....	1.11–1.23	s	447
448	.....	.....	.....	2a	1.7	15.6	43 (110)	238 (400)	135	5.3^{25}	0.9297	∞	448
449	.....	.....	.....	....	1.2	12.7	55 (130)	382 (720)	156.4	....	0.9749	v	449
450	0	2	0	1,2,3	....	....	52 (126)	....	258^{731}	....	0.942	i	450
451	1	3	0	2a,3	2.2	11.5	4 (24)	524 (975)	77.06	100^{25}	0.9005	s	451
452	2	2	0	1,2a,3	....	....	84 OC (184)OC	....		0.8^{20}	1.03	s	452
—													—

No.	(1) Chemical Name and Formula	(2) Chemical Abstracts Registry Number	(3) A.C.G.I.H. 1966-TLV (Threshold Limit Values) ppm (mg/M³)	(4) Principal Effects of Inhalation Exposures above TLV	(5) Relative Hazard to Health from Concentrated Short-term Exposure						(6) References to Supplementary Information on Toxicity, Flammability and Other Hazards
					Eye Contact	Inhalation	Skin Penetration	Skin Irritation	Ingestion	Supplemental Effects	
					(5 is high; 1 is low)						
453	Ethyl acrylate.............. $C_5H_8O_2$	140885	25 S (100) S	Irritant	3	3z*	3	1z*	3	Sensitiz. skin, resp. tract	1876, AIHA, HCD. Poly-merizes, possibly explosive violence
—	Ethyl adipate See Diethyl adipate										
454	Ethyl alcohol............... (Ethanol) C_2H_6O	64175	1000 (1900)	Toxic; Irritant	2	1	1	1	1	....	1422, MCA, NSC, AIHA, STTE
455	Ethylamine .... (Gas) C_2H_7N	75044	10 T (18) T	Irritant	4	3y	3	4z	3	....	2038, MGD, STTE. May deteriorate in normal storage and cause hazard
456	Ethylamine .... (70% aq.) C_2H_7N	8026162	10 T (18) T	Irritant	4	3y	3	4z	3	....	2053, STTE. May deteriorate in normal storage and cause hazard
457	N-Ethylaniline .... $C_8H_{11}N$	103695	....	....	4y	....	2x	1	2	....	STTE
458	Ethyl aziridinyl formate........	1335815	....	....	....	3x*	....	4x	4x	May sensitize	....
459	Ethylbenzene .... C_8H_{10}	100414	100 T (435) T	Toxic	2	2	2	2	2	....	AIHA, HCD, STTE
460	Ethyl benzoate .... $C_9H_{10}O_2$	93890	....	....	1	1	2x	2	2	....	....
—	Ethyl bromide See Bromoethane										
461	Ethyl bromoacetate .... $C_4H_7BrO_2$	105362	....	....	....	....	....	....	....		May deteriorate in normal storage and cause hazard
—	Ethyl butyl ether See Butyl ethyl ether										
—	Ethyl chloride See Chloroethane										
462	Ethyl chloroacetate .... $C_4H_7ClO_2$	105395	....	....	4*	4x*	....	4x	....	Irritates eyes, nose, throat	....
463	Ethyl chloroformate .... $C_3H_5ClO_2$	541413	....	....	....	....	....	....	....	....	May deteriorate in normal storage and cause hazard
464	Ethyl crotonate............. $C_6H_{10}O_2$	....	....	....	2	1	2	1	2	....	STTE
465	Ethyl cyanoacetate........... $C_5H_7NO_2$	105566	....	....	....	....	....	....	....		....
466	Ethylene .... C_2H_4	74851	....	....	....	3x	....	....	....	Used as an anesthetic. Recommended exposure limit is 5500 ppm (20% of lower flammable limit)	1203, HCD, MGD
—	Ethylene bromide See 1,2-Dibromoethane										
—	Ethylene chloride See 1,2-Dichloroethane										
—	Ethylene chlorohydrin See 2-Chloroethanol										

No.	(7) N.F.P.A. Hazard Identification Signals (4 is high; 0 is low)			(8) Extinguishing Agents	(9) Flammable Limits in Air % by volume		(10) Flash Point °C (°F)	(11) Ignition Temperature °C (°F)	(12) Boiling Point (Melting Point, m.p.) °C	(13) Vapor Pressure mm. Hg °C	(14) Density or Specific Gravity	(15) Water Solubility	No.
	Health	Fire	Reactivity		Lower	Upper							
453	2	3 P	2	2a	1.8	Saturation @ 77°F	8.9 (48)	385 (524)	99.1–99.5	16.5^{10} 29.5^{20}	0.924_4^{20}	s	453
454	0	3	0	1,2a,3	3.3	19	13 (55)	423 (793)	78.5	50^{25}	0.7893_4^{20}	∞	454
455	.....	.....	.....	1,3,4	3.5	14.0	....	....	16.6^{766}	400^2	0.6892_{15}^{15}	∞	455
456	3	4	0	1,2a,3	3.5	14.0	−18 (<0)	384 (723)	17	....	0.8	∞	456
457	3	2	0	1,2,3	....	....	85 OC (185)OC	....	204.72	....	0.9625_4^{20}	i	457
458	.....	.....	.....		....	....	....		120^{23}				458
459	2	3	0	2,3	1.0	....	15 (59)	432 (810)	136	....	0.8672_4^{20}	i	459
460	1	2	0	....	....	....	>96 (>204)	....	213	....	1.0458_4^{25}	i	460
461	.....	.....	.....	....	....	....	48 (118)	....	159	....	1.5059_{20}^{20}	i	461
462	2	2	0	....	....	....	66 (151)	....	144^{740}	....	1.2570_4^1	i	462
463	.....	3	1	3	....	....	16 (61)	....	95	....	1.3577_4^{20}	d	463
464	2	3	0	....	....	....	2 (36)	....	143–7	....	0.9183_4^{20}	i	464
465	.....	1	0	....	....	....	110 (230)	....	206	....	1.063_4^{20}	i	465
466	1	4	2	4	3.1	32	....	450 (852)	−104	26^0 atm.	0.00126 at 0° 760 mm	i	466

CHEMICAL HAZARD INFORMATION (See Explanation, page 668)

No.	(1) Chemical Name and Formula	(2) Chemical Abstracts Registry Number	(3) A.C.G.I.H. 1966-TLV (Threshold Limit Values) ppm (mg/M³)	(4) Principal Effects of Inhalation Exposures above TLV	(5) Eye Contact	Inhalation	Skin Penetration	Skin Irritation	Ingestion	Supplemental Effects	(6) References to Supplementary Information on Toxicity, Flammability and Other Hazards
					(5 is high; 1 is low)						
467	**Ethylenediamine**............ $C_2H_8N_2$	107153	10 (25)	Irritant	4	4y*	3	4*	2	Sensitiz. skin, resp. tract	2039, AIHA, STTE. May deteriorate in normal storage and cause hazard
—	Ethylene dibromide See 1,2-Dibromoethane										
—	Ethylene dichloride See 1,2-Dichloroethane										
468	**Ethylene glycol**........... $C_2H_6O_2$	107211	....	....	1	1	1	1	2	....	1497, STTE
—	Ethylene glycol dinitrate See Ethylene nitrate										
—	Ethylene glycol monobutyl ether See 2-Butoxyethanol										
—	Ethylene glycol monoethyl ether See 2-Ethoxyethanol										
—	Ethylene glycol monomethyl ether See 2-Methoxyethanol										
—	Ethyleneimine See Aziridine										
469	**Ethylene nitrate** .... (Ethylene glycol dinitrate) $C_2H_4N_2O_6$	628966	C 0.2 S C(1.9)S	Toxic	....	....	....	....	....		1590
470	**Ethylene oxide** .... (Oxirane) C_2H_4O	75218	50 (90)	Irritant; Toxic	5z	3*	....	5z*	....	Sensitiz. skin, resp. tract	1626, AIHA, MCA, HCD, MGD, STTE
471	**Ethyl ether** .... (Diethyl ether) $C_4H_{10}O$ Ethyl ether, compound with Boron trifluoride See Boron fluoride ethyl ether complex	60297	400 (1200)	Narcosis	1	4	1	1	2	....	1656, MCA, NSC, STTE, HCD, AIHA
472	**Ethyl ethynyl ether**........... C_4H_6O	927800	....	....	....	....	....	....	....		STTE
—	Ethyl fluoride See Fluoroethane										
473	**Ethyl fluoroacetate**........... $C_4H_7FO_2$	459723	....	....	Extremely toxic					....	1887
474	**Ethyl formate** .... $C_3H_6O_2$	109944	100 (300)	Irritant	2	4	1	1	2	....	1854, STTE
—	Ethyl iodide See Iodoethane										
475	**Ethyl lactate** .... $C_5H_{10}O_3$	97643	....	....	....	....	....	....	....		1866
476	**Ethyl malonate** .... $C_7H_{12}O_4$	105533	....	....	....	....	....	....	....		
—	Ethyl mercaptan See Ethanethiol										
477	**N-Ethylmorpholine** .... (4-Ethylmorpholine) $C_6H_{13}NO$	100743	20 ST (94) ST	Toxic; Irritant	3	2	2x	1	2	....	2204, STTE
478	**Ethyl nitrite** .... $C_2H_5NO_2$	109955	....	....	Decomposition forms toxic oxides of nitrogen					....	2099, HCD
479	*N*-Ethyl-*N*-nitroso-*N*-butylamine				....	....	....	....	....	Carcinogenic in animals	....
480	*N*-Ethyl-*N*-nitrosovinylamine ...				....	....	....	....	....	Carcinogenic in animals	....
481	**Ethyl oxalate** .... $C_6H_{10}O_4$	95921	....	....	....	....	....	....	....		

768

No.	(7) N.F.P.A. Hazard Identification Signals			(8) Extinguishing Agents	(9) Flammable Limits in Air % by volume		(10) Flash Point °C (°F)	(11) Ignition Temperature °C (°F)	(12) Boiling Point (Melting Point, m.p.) °C	(13) Vapor Pressure mm. Hg °C	(14) Density or Specific Gravity	(15) Water Solubility	No.
	Health	Fire	Reactivity		Lower	Upper							
	(4 is high; 0 is low)												
467	3	2	0	1,2a,3	....	....	43 (110)	....	116.5	$10^{21.5}$	0.8995^{20}_{20}	v	467
468	1	1	0	1,2a,3	3.7	....	111 (232)	413 (775)	198^{760}	0.06^{20}	1.1088^{20}_4	∞	468
469	.....	......	.....	....	....	....			197 ± 3	0.049^{20}	1.4918	i	469
470	2	4	3	1,2a,3,4	3	100	<−18 (<0)	429 (804)	$13-4^{11}$	$625^{5.6}$	0.822^{10}_{10}	s	470
471	2	4	1	2a,3,4	1.9	48	−45 (−49)	180 (356)	34.6	438.9^{20}	0.714^{20}_{20}	s	471
472	.....	......	.....			....	<−7 (<20)	....	50 exp. 100	....	0.7929	Forms per-oxides	472
473	.....											....	473
474	2	3	0	2,2a,3	2.7	13.5	−20 (−4)	455 (851)	54.3	200^{21}	0.9117	s	474
475	2	2	0	1,2a,3	1.5 @ 212°	....	46 (115)	400 (752)	$69-70^{36}$	5^{30}	1.0415	s	475
476	0	1	0	....	....	....	93 OC (200) OC	....	199	....	1.0550	δ	476
477	2	3	0	1,2,2a,3	....	....	32 OC (90) OC	....	....	6.1^{20}	0.9	s	477
478	2	4	4	....	4.1	50+	−35 (−31)	d 90 d (194)	17.2	....	....	....	478
479	.....	......	.....					....	....				479
480	.....	......	.....						....				480
481	.....	2	0	1,3	....	....	76 (168)		185.7	....	1.0785	δ[h]	481

No.	Chemical Name and Formula	Chemical Abstracts Registry Number	A.C.G.I.H. 1966–TLV (Threshold Limit Values) ppm (mg/M³)	Principal Effects of Inhalation Exposures above TLV	Eye Contact	Inhalation	Skin Penetration	Skin Irritation	Ingestion	Supplemental Effects	References to Supplementary Information on Toxicity, Flammability and Other Hazards
					(5 is high; 1 is low)						
482	p-Ethylphenol . . . . $C_8H_{10}O$	123079	. . . .	. . . .	. . . .	. . . .	. . . .	. . . .	. . . .	. . . .	. . . .
—	Ethyl pyrophosphate See T.E.P.P.										
483	Ethyl vinyl ether . . . . (Vinyl ethyl ether) C_4H_8O	109922	. . . .	. . . .	. . . .	. . . .	. . . .	. . . .	. . . .	. . . .	May deteriorate in normal storage and cause hazard
485	Ferbam® . . . . $C_9H_{18}FeN_3S_6$	301053	(15)	Toxic	. . . .	. . . .	. . . .	. . . .	. . . .	. . . .	. . . .
486	Ferric chloride 60%	. . . .	. . . .		4x	. . . .	. . . .	3x	. . . .	. . . .	. . . .
487	Ferrous ammonium sulfate . . . . . $FeN_2H_8S_2O_8 \cdot 6H_2O$	10553761	. . . .		2y	2y	1x	2y	2	. . . .	. . . .
488	Ferrous chloride . . . .! . . . $FeCl_2$	7758943	. . . .		3x	. . . .	1x	3x	2x	. . . .	. . . .
489	Ferrous sulfate . . . . $FeSO_4$	. . . .	. . . .		3x	. . . .	1x	3x	2x	. . . .	1053
490	Ferro vanadium dust	. . . .	(1)	Toxic	. . . .	. . . .	. . . .	. . . .	. . . .	. . . .	STTE
491	Fluoracetic acid	. . . .	. . . .		Extremely toxic					. . . .	1773
—	N-2-Fluorenylacetamide See N-Acetyl-2-aminofluorene										
492	N,N'-Fluoren-2,7-yl bisacetamide	. . . .	. . . .		. . . .	. . . .	. . . .	. . . .	. . . .	Carcinogenic in animals	. . . .
493	Fluorides . . . . F^-	. . . .	(2.5)	Toxic	. . . .	. . . .	. . . .	. . . .	. . . .	. . . .	AIHA, STTE
494	Fluorine . . . . F_2	. . . .	0.1 (0.2)	Irritant	5z	5z	. . . .	5z	. . . .	Highly toxic	1832, AIHA, HCD, MGD. Special first preparations recommended
495	4-Fluoro-4-biphenylamine	. . . .	. . . .		. . . .	. . . .	. . . .	. . . .	. . . .	Carcinogenic in animals	. . . .
496	Fluoroethane. . . . . (Ethyl fluoride) C_2H_5F	. . . .	. . . .		. . . .	1y	. . . .	. . . .	. . . .	STTE	
497	Fluoroethylene . . . . (Vinyl fluoride) C_2H_3F	75025	. . . .		. . . .	. . . .	. . . .	. . . .	. . . .	Low toxicity	MGD
—	Fluoroform See Trifluoromethane										
498	2'-Fluoro-4-phenylacetanilide . . .	. . . .	. . . .		. . . .	. . . .	. . . .	. . . .	. . . .	Carcinogenic in animals	. . . .
498a	2'-Fluoro-4'-phenylacetanilide . . .	. . . .	. . . .		. . . .	. . . .	. . . .	. . . .	. . . .	Carcinogenic in animals	. . . .
499	4'''-Fluoro-4'-phenylacetanilide .	450555	. . . .		. . . .	. . . .	. . . .	. . . .	. . . .	Carcinogenic in animals	. . . .
500	Formaldehyde. . . . . CH_2O	50000	C 5 C (6)	Irritant	4	3z*	4	4y*	3	Sensitiz. skin, resp. tract	HCD, AIHA, STTE
501	Formalin . . . . (37% Formaldehyde; Methanol-free)		C 5 C (6)	Irritant	4	3z*	4	4y*	3	Sensitiz. skin, resp. tract	1966, AIHA, MCA, NSC, STTE
502	Formalin . . . . (37% Formaldehyde; 15% Methanol)		C 5 C (6)	Irritant	4	3z*	4	4y*	3	Sensitiz. skin, resp. tract	STTE
503	Formamide . . . . CH_3NO	75127	. . . .	. . . .	2	1	2x	1	2	. . . .	1828, STTE

No.	(7) N.F.P.A. Hazard Identification Signals			(8) Extinguishing Agents	(9) Flammable Limits in Air % by volume		(10) Flash Point °C (°F)	(11) Ignition Temperature °C (°F)	(12) Boiling Point (Melting Point, m.p.) °C	(13) Vapor Pressure mm. Hg °C	(14) Density or Specific Gravity	(15) Water Solubility	No.
	Health	Fire	Reactivity		Lower	Upper							
	(4 is high; 0 is low)												
482	.....	1	0	2a	....	....	104 (*219*)	....	219	....	1.0 @ 140°F	δ	482
483	.....	.....	.....	....	....	....	<−46 (*<−50*)	....	36	4.28	0.754	s	483
485	.....											......	485
486	.....							......	315	....		......	486
487	.....										1.864	s	487
488	.....								m.p. 670–4		3.16_4^{25}	δ	488
489	.....										2.970^{25}	δ	489
490	.....											......	490
491	.....											......	491
492	.....												492
493	.....												493
494	4	0 W	3	4	Reacts with most materials spontaneously at room temp.		Gas	....	−188	....	1.69^{15} g/l	HF+O₂	494
495	.....												495
496	.....								−37.7	....	0.00220	s	496
497	2	4	1	....	2.6±0.5	21.7±10	....	....	−72	....	....	......	497
498	.....				....							......	498
498a	.....												498a
499	.....											......	499
500	....	4	.....	1,4	7.0	73	Gas	430 (*806*)	−21	....	....	......	500
501	2	2	0	1	7.0	73	85 (*185*)	403 (*806*)	101	10^{-88}	0.815	s	501
502	....	2	.....	1	7.0	73	50 (*122*)	403 (*806*)	....	....	....	s	502
503	.....	.....	.....	....	....	....	....	....	$105\text{–}6^{11}$	10^{109}	1.134_4^{20}	∞	503

No.	(1) Chemical Name and Formula	(2) Chemical Abstracts Registry Number	(3) A.C.G.I.H. 1966-TLV (Threshold Limit Values) ppm (mg/M³)	(4) Principal Effects of Inhalation Exposures above TLV	(5) Relative Hazard to Health from Concentrated Short-term Exposure (5 is high; 1 is low)						(6) References to Supplementary Information on Toxicity, Flammability and Other Hazards
					Eye Contact	Inhalation	Skin Penetration	Skin Irritation	Ingestion	Supplemental Effects	
504	**Formic acid** CH$_2$O$_2$	64186	5 T (9)	Irritant	4	4x	2x	4	2	....	1776, STTE. May deteriorate in normal storage and cause hazard
—	Freon®-11 See Trichlorofluoromethane										
—	Freon®-12 See Dichlorodifluoromethane										
—	Freon®-13 See Chlorotrifluoromethane										
—	Freon®-13B1 See Bromotrifluoromethane										
—	Freon®-14 See Carbon tetrafluoride										
—	Freon®-22 See Chlorodifluoromethane										
—	Freon®-114 See 1,2-Dichlorotetrafluoro-ethane										
—	Freon®-C318 See Octafluorocyclobutane										
505	**Fumaric acid** C$_4$H$_4$O$_4$	110178	....	....	4y	....	2x	4x	2x	....	1811, STTE
506	**2-Furaldehyde** (Furfural) C$_5$H$_4$O$_2$	98011	5 S (20)S	Irritant	....	....	....	....	....		1983, AIHA, STTE
507	**Furan** C$_4$H$_4$O	110009	....		....	....	....	....	....		....
—	Furfural See 2-Furaldehyde										
508	**Furfuryl alcohol** C$_5$H$_6$O$_2$	98000	50 (200)	Toxic	....	....	....	....	....		1489, STTE
509	**Gadolinium** Gd	7440542	....								1058
510	**Gallic acid** C$_7$H$_6$O$_5$	149917	....		Skin irritant						....
511	**Gallium** Ga	7440553	....								1037
512	**Gasoline** C$_5$H$_{12}$ to C$_9$H$_2$O	The composition of gasoline varies greatly and thus a single TLV for all types of gasoline is no longer applicable. In general, the aromatic hydrocarbon content will determine what TLV applies. Consequently the content of benzene, other aromatics and additives should be determined to arrive at the appropriate TLV.			2x	3x*	3x*	1x*	2x	Chronic toxicity	1199, STTE. Tetraethyl lead and some other additives are toxic.
—	Genetron®-21 See Dichlorofluoromethane										
—	Genetron®-23 See Fluoroform										
—	Genetron®-142B See 1-Chloro-1,1-difluoro-ethane										
—	Genetron®-152A See 1;1-Difluoroethane										
—	Genetron®-1132A See 1,1-Difluoroethylene										
513	**Germane** (Germanium tetrahydride) GeH$_4$	7782652	....	....	Highly toxic; upper respiratory irritant; corrosive to eyes, skin, mucous membranes					....	MGD
514	**Germanium** Ge	7440564	....	....	....	....			....		1042
515	**Glass, fibrous**	....	(5)T	Toxic	....	....			....		2267

No.	(7) N.F.P.A. Hazard Identification Signals			(8) Extinguishing Agents	(9) Flammable Limits in Air % by volume		(10) Flash Point °C (°F)	(11) Ignition Temperature °C (°F)	(12) Boiling Point (Melting Point, m.p.) °C	(13) Vapor Pressure mm. Hg °C	(14) Density or Specific Gravity	(15) Water Solubility	No.
	Health	Fire	Reactivity		Lower	Upper							
	(4 is high; 0 is low)												
504	3	2	0	1,2a,3	....	....	69 (156)	601 (1114)	100.7	43^{25}	1.220_4^{20}	∞	504
—													—
—													—
—													—
—													—
—													—
—													—
505	.....	.....	.....	....				....	m.p. 286-7	....	1.635_4^{20}	δ:s^h	505
506	1	2	1	1,2,2a,3	2.1	....	60 (140)	316 (600)	161.7	1^{19}	1.1598	s:v^h	506
507	1	4	1	....	....	....	<0 (<32)	....	32^{758}	....	0.9366_4^{20}	i	507
—													—
508	1	2	1	1,2a,3	1.8 Hypergolic with fuming nitric acid	16.3	75 OC (167)OC	491 (915)	171^{750}	$1^{31.8}$	1.1296_4^{20}	∞^d	508
509	.....	.....	.....	.... Pyrophoric on filing			....	....	~3000 m.p. 1312	....	7.9	i	509
510	.....	.....	.....	....					m.p. 253 d		1.694_4^{6}	δ:v^h	510
511	.....	.....	.....						1983 m.p. 30.15	....	$5.904^{29.6}$	i	511
512	1	3	0	2,3	1.4	7.6	−43 (−45)	371 (700)		....	0.8	i	512
—													—
—													—
—													—
—													—
—													—
513	.....	.....	.....	....	Flammable		Gas	....	−90	....	3.43 g/l	....	513
514	.....	.....	.....	....				....	m.p. 947	....	5.35	i	514
515	.....	.....	.....	....				....		....	....	.....	515

No.	(1) Chemical Name and Formula	(2) Chemical Abstracts Registry Number	(3) A.C.G.I.H. 1966–TLV (Threshold Limit Values) ppm (mg/M³)	(4) Principal Effects of Inhalation Exposures above TLV	(5) Eye Contact	Inhalation	Skin Penetration	Skin Irritation	Ingestion	Supplemental Effects	(6) References to Supplementary Information on Toxicity, Flammability and Other Hazards
					(5 is high; 1 is low)						
516	Glutaraldehyde $C_5H_8O_2$	111308	....		4	1	3	3y*	3	Skin stain	1981, STTE
517	Glutaric anhydride $C_5H_6O_3$	108554			4	....	3	2y	2	....	STTE
—	Glycerine See Glycerol										
518	Glycerol (Glycerine) $C_3H_8O_3$	56815	....		1	1y	1	1y	1	....	STTE
519	Glycidol $C_3H_6O_2$	55625	50 (150)	Toxic	....	....					1634, STTE
520	Glycolic acid $C_2H_4O_3$	79141			3	2x	2y	4x	2		1803
521	Glyoxal $C_2H_2O_2$	107222			2	2y*	2y	1*	2	Skin stain; Sensitiz. skin, respiratory	1980, STTE
—	Guaiacol See o-Methoxyphenol										
522	Hafnium Hf	7440586	(0.5)	Toxic	....	....					MCA
523	Helium He	7440597	....		....	....				Can act as an asphyxiant	MGD, STTE
524	Heptachlor $C_{10}H_5Cl_7$	76448	(0.5) S	Toxic	....	....					1355, STTE
525	n-Heptane C_7H_{16}	142825	500 (2000)	Narcosis	1	2x	1x	1	1	....	1197, AIHA, STTE
526	2-Heptanone (Methyl n-amyl ketone) $C_7H_{14}O$	110430	100 T (465) T	Irritant	2x	1x	2x	1x	2x		1921, STTE
527	3-Heptanone $C_7H_{14}O$	106354	50 T (230) T	Irritant	2x	1x	2x	1x	2x	....	1721
528	n-Heptylamine $C_7H_{17}N$	111682									2038
529	Hexachlorobenzene C_6Cl_6	118741	....		1y	1	2	3y	2		STTE
—	1,2,3,4,5,6-Hexachlorocyclo- hexane Properties listed under Lindane										
530	Hexachloroethane C_2Cl_6	67721	1 S (10) S	Toxic	....	....					1297
531	Hexachloronaphthalene $C_{10}H_2Cl_6$	1335871	(0.2) S	Toxic	....	....					....
532	Hexafluoroethane (Freon® 116) C_2F_6	76164	....		Comparatively non-toxic. Avoid inhalation of decomposition products						MGD
533	Hexamethylenetetramine $C_6H_{12}N_4$	100970	....		2y	1*	2y	1y*	1		2206
534	Hexane C_6H_{14}	110543	500 (1800)	Narcosis	Impurities must be considered in evaluating health hazard						1197, AIHA, STTE
535	1,6-Hexanediamine $C_6H_{16}N_2$	124094	....		Possible skin sensitization						2061
536	Hexanoic acid (Caproic acid) $C_6H_{12}O_2$	142621	....		4	1	3	4	2	....	1783, STTE
—	1-Hexanol See Hexyl alcohol										
537	2-Hexanone (Methyl butyl ketone) $C_6H_{12}O$	591786	100 (410)	Irritant	....	....					1720, HCD
538	1-Hexene C_5H_{10}	592416	....								....

No.	(7) Health	Fire	Reactivity	(8) Extinguishing Agents	(9) Lower	Upper	(10) Flash Point °C (°F)	(11) Ignition Temperature °C (°F)	(12) Boiling Point (Melting Point, m.p.) °C	(13) Vapor Pressure mm. Hg °C	(14) Density or Specific Gravity	(15) Water Solubility	No.
516	....	....	....	....	....	....	....	....	187–9d	....	....	∞	516
517	....	....	....	....	....	....	....	....	302–4	....	....	s	517
—													—
518	1	1	0	1,2a,3	....	....	160 (320)	354 (670)	290 d	....	1.2613^{20}_{4}	∞	518
519	....	....	....	....	....	....	....	....		0.9^{25}	1.117^{20}_{4}	∞	519
520	....	....	....	....	....	....	....	....	d	....	....	s	520
521	....	....	....	....	....	....	220 (20)	....	50.4	....	1.14^{20}	v	521
—													—
522	....	....	....	5	....	....	Dust expl. hazard	20(68) for dust cloud	>3200 m.p. 2227	....	13.31	i	522
523	....	....	....	....	....	....	....	....	−269	....	0.1249 g/l liq.	i	523
524	....	....	....	....	....	....	....	....	m.p. 95–6	....	$1.57–9^{9}$	i	524
525	1	3	0	2,3	1.2	6.7	−4 (25)	223 (433)	98.42	150^{25}	0.68376^{20}_{4}	i	525
526	1	2	0	2a	....	....	49 OC (120)OC	533 (991)	151	1.6^{25}	0.8111^{20}_{4}	v	526
527	1	2	0	....	....	....	46 OC (115)OC	....	150	1.4^{25}	0.8183^{20}_{4}	i	527
528	2	2	0	2a	....	....	54 OC (130)OC	....	158.3	....	0.777^{20}_{4}	δ	528
529	....	....	....	1,3	....	....	242 (468)	....	322	....	1.569	i	529
—													—
530	....	....	....	....	....	....	....	....	186^{777}	....	2.091	i	530
531	....	....	....	....	....	....	....	....		....	....		531
532	....	....	....	....	Non-flammable		....	....	−79	....	1.590^{-78}	....	532
533	....	....	....	....	....	....	250 (482)	....	m.p. 280 sub	....	1.331^{-5}	v	533
534	1	3	0	2,3	1.1	7.5	−30 (−22)	261 (502)	68	$150^{24.8}$	0.6595^{20}_{4}	i	534
535	....	....	....	....	....	....	....	....	204	....	....	v	535
536	2	1	0	....	....	....	102 OC (215)OC	....	205	1^{72}	0.9274^{20}_{4}	δ	536
—													—
537	2	3	0	2a	1.2	8	35 OC (95)OC	533 (991)	126	3.8^{25}	0.81162	δ	537
538	1	3	0	....	....	....	< −7 (<20)	....	63.3	....	0.7^{15}_{0}	i	538

No.	(1) Chemical Name and Formula	(2) Chemical Abstracts Registry Number	(3) A.C.G.I.H. 1966–TLV (Threshold Limit Values) ppm (mg/M³)	(4) Principal Effects of Inhalation Exposures above TLV	(5) Relative Hazard to Health from Concentrated Short-term Exposure (5 is high; 1 is low)						(6) References to Supplementary Information on Toxicity, Flammability and Other Hazards
					Eye Contact	Inhalation	Skin Penetration	Skin Irritation	Ingestion	Supplemental Effects	
539	**2-Hexene** C₆H₁₂	592438	....	....	...	...	...	...	...		....
540	*sec*-**Hexyl acetate** C₈H₁₆O₂	142927	50 (295)	Irritant	2x	1x	1x	2x	2x		1858, STTE
541	**Hexyl alcohol** C₆H₁₄O	111273	....	....	4	1	2	2	2		
542	**Hexylamine** C₆H₁₅N	111262	....	....	4	3	3	4	3		2038, STTE
543	**Holmium** Ho	7440600	....	....							1058
544	**Hydrazine** N₂H₄	302012	1 S (1.3)S	Toxic	4x	4x*	3x	3x	3x		2220, AIHA, HCD, STTE
545	**Hydrazine HCl** N₂H₄ · HCl	2644704	....	....	3x		3x	3x	3		....
546	**Hydrazoic acid** HN₃	7782798	....	....	Highly toxic					Limit exposure to 1 ppm	2212
547	**Hydracrylic acid, β-lactone** (β-Propiolactone) C₃H₄O₂	57578	High acute toxicity and demonstrated skin tumor production in animals. Contact by any route should be avoided.		...				...		STTE
548	**Hydrides, Metal**		....	....							NSC
—	Hydrobromic acid See Hydrogen bromide										
549	**Hydrochloric acid** HCl (aq.) See also Hydrogen chloride	7647010	....	....	4	5z	3x	5	3x		849, HCD
550	**Hydrocyanic acid 96%** HCN (aq.) See also Hydrogen cyanide	74908	....	....	4x	5x*	4x	2x	4x	Gas or vapor penetrates skin	1996, AIHA, MCA, HCD
551	**Hydrofluoric acid** HF (aq.) See also Hydrogen fluoride	7664393	....	....	5z	5z	4x	5z	4x		MCA, NSC
552	**Hydrogen** H₂	1333740	....	....	...	3	...	...	...	May act as asphyxiant	MGD
553	**Hydrogen bromide** (Hydrobromic acid) HBr	10035106	3 (10)	Irritant	Toxic and very irritating; Corrosive to eyes, skin, mucous membrane					....	MGD, STTE
554	**Hydrogen chloride** (Gas) HCl See also Hydrochloric acid	7647010	C 5 C(7)	Irritant	4	5z	3x	5	3x	Highly toxic. Concentrations of 0.13 to 0.2% are lethal to humans in a few minutes	849, AIHA, MCA, MGD, STTE
555	**Hydrogen cyanide** HCN See also Hydrocyanic acid	74908	10 S (11)S	Toxic	4x	5z*	....	2x	....	Gas or vapor penetrates skin	1996, AIHA, MCA, HCD, STTE
556	**Hydrogen fluoride** HF See also Hydrofluoric acid	7664393	3 S (2)S	Toxic	5z	5z	....	5z	....	Toxic and violently corrosive	841, AIHA, HCD, MGD, STTE
557	**Hydrogen iodide** HI	10034852	....	....	Toxic and very irritating; corrosive to eyes, skin, mucous membrane					....	MGD
558	**Hydrogen peroxide 90%** H₂O₂	7722841	1 (1.4)	Irritant	5x	....	4x	5x	4x	....	AIHA, MCA, HCD, STTE
559	**Hydrogen peroxide 50%** H₂O₂		....	....	5x		4x	5x	4x	....	AIHA, MGA, HCD, STTE
560	**Hydrogen peroxide 35%** H₂O₂		....	....	5x		4x	5x	4x	....	AIHA, MCA, HCD, STTE

No.	(7) Health	Fire	Reactivity	(8) Extinguishing Agents	(9) Lower	Upper	(10) Flash Point °C (°F)	(11) Ignition Temperature °C (°F)	(12) Boiling Point (Melting Point, m.p.) °C	(13) Vapor Pressure mm. Hg °C	(14) Density or Specific Gravity	(15) Water Solubility	No.
	(4 is high; 0 is low)												
539	1	3	0	....	....	....	< -7 (<20)	....	68^{749}	....	0.6845^{20}_4	i	539
540	1	2	0	2a, 3	....	....	45 (113)	....	141	3.8^{20}	0.855	δ	540
541	1	2	0	1,2a,3	....	....	60 (140)	293 (559)	158	....	0.8136^{20}_4	δ	541
542	2	3	0	2a	....	....	29 OC (85)OC	....	129^{742}	6.5^{20}	0.763^{25}_4	δ	542
543	....		....	....	Pyrophoric on filing		....		2600 m.p. 1461	....	8.803	i	543
544	3	3	2	1,3	4.7	100	38 (100)	....	113.5	14.4^{25}	1.011^{15}	v	544
545	....		....	....			....	....	d 240	....	....	v	545
546	....		....	....	Highly explosive		....	....	37	....	1.09^{25}_4	∞	546
547													547
548	....		....	5	....	....	....	....					548
549	3	0	0	....	....	....	....	....	....	....		∞	549
550	4	4 P	2	1	6	41	−18 (o)	538 (1000)	26	807 (27)	....	∞	550
551	....											∞	551
552	0	4	....	4	4.0	75	....	585 (1085)	−252.8	....	0.0899 g/l	δ	552
553	....							....	−67.0	....	3.5 g/l	v	553
554	....								84.9	....	1.6397	v	554
555	4	4 P	2	3	6	41	Gas	538 (1000)	26	....	0.6884^{20}_4	∞	555
556	4	0	0	....	....	....	....	....	19.54	....	0.991	∞	556
557	....		....					....	−35.5	....	....	v	557
558	2	0	3	....	Non-flammable. Hypergolic with hydrazine		....	....	140	5^{20}	1.392^{0}_4	∞	558
559	....		....		Non-flammable		....	....	107.8	....	....	....	559
560	....		....		Non-flammable		....	....	113.9	....	....	....	560

No.	Chemical Name and Formula	(2) Chemical Abstracts Registry Number	(3) A.C.G.I.H. 1966–TLV (Threshold Limit Values) ppm (mg/M³)	(4) Principal Effects of Inhalation Exposures above TLV	(5) Relative Hazard to Health from Concentrated Short-term Exposure — Eye Contact	Inhalation	Skin Penetration	Skin Irritation	Ingestion	Supplemental Effects	(6) References to Supplementary Information on Toxicity, Flammability and Other Hazards
					(5 is high; 1 is low)						
561	**Hydrogen selenide** .... H_2Se	7783075	0.05 (0.2)	Toxic	Highly toxic					....	896, AIHA, MGD, STTE
562	**Hydrogen sulfide** .... H_2S	7783064	10 (15)	Toxic	....	5x*	....	....	....	Death/odor reliance	896, AIHA, MCA, NSC, HCD, MGD, STTE
—	Hydroquinol See Hydroquinone										
563	**Hydroquinone**.............. (Hydroquinol; Quinol) $C_6H_6O_2$	123319	(2)	Toxic	4*	3x*	2x	2y	2	Eye pigment	1380, AIHA, STTE
—	Hydroquinone monomethyl ether See *p*-Methoxyphenol										
—	*o*-Hydroxybenzaldehyde See Salicylaldehyde										
564	**Hydroxydimethylarsine oxide** ... (Cacodylic acid) $C_2H_7AsO_2$	75605	....		Acute toxicity—See Arsenic					....	STTE
565	**Hydroxylamine** .... NH_2OH	7803498	....	....	....	....		....			Hydroxylamine and its derivatives are known to break chromosomes, produce large chromosomal alterations, and induce cancer. —NIH Safety Office "Spot Hazards" Jan., 1965
566	**Hydroxylamine HCl** .... $NH_2OH \cdot HCl$	5470111	....	....	4x	3x	2x	3x	3	....	2108
567	**4-Hydroxy-4-methyl-2-pentanone** .... (Diacetone alcohol) $C_6H_{12}O_2$	123422	50 (240)	Irritant; Toxic	2	1	2	1	2	....	1470
—	8-Hydroxyquinoline See 8-Quinolinol										
568	**2,2′-Iminodiethanol** (Diethanolamine) $C_4H_{11}NO_2$	111422	....	....	3	1	2	3	2		2062
569	**5-Indanol** .... $C_9H_{10}O$	1470946	....	....	4	2x	3	4	2		STTE
570	**Indium**................... In	7440600			Toxic; Toxicity of compounds varies						1050
571	**Iodic acid** .... HIO_3	7782685			Strong irritant to skin and mucous membranes					....	
572	**Iodine** .... I_2	7553562	C 0.1 C(1)	Irritant	5x	4x	3x	4x	4x		854, NSC, AIHA, STTE
573	**Iodine monochloride**........... ClI										May deteriorate in normal storage and cause hazard
574	**Iodine pentafluoride** .... IF_5	7783666	....	....	Highly toxic and extremely corrosive					Exposure limit of 3 ppm is suggested	MGD, STTE
575	**Iodoacetic acid** .... $C_2H_3IO_2$	64697	....	....	May cause contact dermatitis					....	1799
576	**Iodoethane** .... (Ethyl iodide) C_2H_5I	75036	....	....	....	....		....			....

778

No.	(7) N.F.P.A. Hazard Identification Signals (4 is high; 0 is low) Health	Fire	Reactivity	(8) Extinguishing Agents	(9) Flammable Limits in Air % by volume Lower	Upper	(10) Flash Point °C (°F)	(11) Ignition Temperature °C (°F)	(12) Boiling Point (Melting Point, m.p.) °C	(13) Vapor Pressure mm. Hg °C	(14) Density or Specific Gravity	(15) Water Solubility	No.
561	....	....	....	....	Flammable		Gas	....	−42	....	2.004 g/ml (liquid)	δ	561
562	3	4	0	4	4.3	45	Gas	260 (500)	−60.7	5.6 atm.	1.539 g/l	v	562
563	....	....	....	1,3	....	....	165 (329)	516 (960)	285	4[150]	1.328	s:v[h]	563
564	....	....	....				....	....	m.p. 200	....	....	v	564
565	1	3	3	....	....	....		Expl. at 129 (265)	56.5	....	1.204	....	565
566	....	....	....		....	....	....	....	m.p. 153–155	....	1.67[17]	v	566
567	1	2	0	2a	1.8	6.9	64 (148)	603 (1118)	164–6[11]	0.97[20]	0.9306	∞	567
568	1	1	0	1,2a,3	....	....	152 OC (305)OC	662 (1224)	270[748]	1.09[20]	1.09664	v	568
569	....	....	....				....	....	255	....	....	δ	569
570	....	....	....				....	....	2000	....	7.3	....	570
571	....	....	....				....	....	m.p. d 110	....	4.629	v	571
572	....	....	....				....	....		309[25]	4.93	δ	572
573	....	....	....				....	....			3.18	d	573
574	....	....	....				....	....	98		3.75	....	574
575	....	....	....				....	....	d	....	....	s[h]	575
576	....	....	....	....	Flammable		....	....	72	....	1.950	δ[d]	576

No.	(1) Chemical Name and Formula	(2) Chemical Abstracts Registry Number	(3) A.C.G.I.H. 1966-TLV (Threshold Limit Values) ppm (mg/M³)	(4) Principal Effects of Inhalation Exposures above TLV	(5) Relative Hazard to Health from Concentrated Short-term Exposure — Eye Contact	Inhalation	Skin Penetration	Skin Irritation	Ingestion	Supplemental Effects	(6) References to Supplementary Information on Toxicity, Flammability and Other Hazards
577	**Iodomethane** (Methyl iodide) CH_3I	74884	5 ST (28) ST	Toxic	....	....					
578	**3-Iodopropene** (Allyl iodide) $IC_2H_3:CH_2$	556569	....		Irritation of eyes and mucous membranes						....
579	**Iron oxide fume**	....	(10) T	Toxic	....	....					STTE
580	**Isoamyl acetate** (Isopentyl alcohol, acetate) $C_7H_{14}O_2$	123922	100 T (525) T	Irritant	....						1858
—	Isobutane See 2-Methylpropane										
—	Isobutene See 2-Methylpropene										
581	**Isobutyl acetate** $C_6H_{12}O_2$	110190	150 T (700) T	Irritant	1	3	2	1	1		1858
582	**Isobutyl alcohol** (2-Methyl-1-propanol) $C_4H_{10}O$	78831	100 (300)	Irritant	3	2	2	1	2		1447, STTE
583	**Isobutylamine** $C_4H_{11}N$	78819			....	....					2038
—	Isobutylene See 2-Methylpropene										
584	**Isobutyraldehyde** C_4H_8O	78842			2	3	2	1	2		1966, STTE
585	**Isobutyric acid** $C_4H_8O_2$	79312			4	1	3	4	3		1781
586	**Isopentyl nitrite** (Amyl nitrite) $C_5H_{11}NO_2$	110463									1995. May deteriorate in normal storage and cause hazard
—	Isophorone See 3,5,5-Trimethyl 2-cyclohexene-1-one										
587	**Isoprene** (2-Methyl-1,3-butadiene) C_5H_8	78795	....		3x	3x	....	2x	....		1203
588	**Isopropenyl acetate** $C_5H_8O_2$	108225	....		1	3	1	1	2	....	STTE
589	**Isopropyl acetate** $C_5H_{10}O_2$	108214	250 T (950) T	Irritant	1	3	1	1	2		1858
590	**Isopropyl alcohol** C_3H_8O	67630	400 (980)	Narcosis; Irritant	....	....					1436, AIHA, STTE
591	**Isopropylamine** C_3H_9N	75310	5 (12)	Irritant	5	3	3	4	3		2038, MCA, STTE
592	**Isopropyl benzoate** $C_{10}H_{12}O_2$	939480	....		1	1	2	2	2		1896, STTE
593	**Isopropyl ether** $C_6H_{14}O$	108203	500 (2100)	Narcosis	1	4	2	1	2		1662, STTE. Forms peroxides
594	**Isopropyl glycidyl ether** $C_6H_{12}O_2$	4016142	50 (240)	Toxic	....	....					1637, STTE
—	Isothiocyanic acid, methyl ester See Methyl isothiocyanate										
595	**Kerosine** (Kerosene) C_9—C_{16}	....	....		1	2y	3	1	1	....	1195, STTE
596	**Ketene** C_2H_2O	463514	0.5 (0.9)	Irritant; Toxic	5x	5*	....		....	Sensitiz. skin, resp. tract	1964, STTE
597	**Krypton** Kr	7439909	....		Can act as an asphyxiant					....	MGD

No.	(7) N.F.P.A. Hazard Identification Signals			(8) Extinguishing Agents	(9) Flammable Limits in Air % by volume		(10) Flash Point °C (°F)	(11) Ignition Temperature °C (°F)	(12) Boiling Point (Melting Point, m.p.) °C	(13) Vapor Pressure mm. Hg °C	(14) Density or Specific Gravity	(15) Water Solubility	No.
	Health	Fire	Reactivity		Lower	Upper							
	(4 is high; 0 is low)												
577	.....	.....	.....					....	42.5	....	2.28	δ	577
578	.....	.....	.....					....	102–3	....	1.8454	i	578
579	.....	.....	.....								....		579
580	.....	.....	.....	2,3	1.0 @ 212°F	7.5	25 (77)	379 (715)	142	6^{25}	0.8670	δ	580
—													—
581	1	3	0	2,2a,3	1.3	7.5	18 (64)	423 (793)	117.2	20^{25}	0.8747	δ	581
582	1	3	0	1,2a,3	1.7 @ 212°F	10.9 @ 212°F	28 (82)	441 (825)	107	12.2^{25}	0.805	s	582
583	3	3	0	1,3	....	....	−9 (15)	378 (712)	65.6	100 (18.8)	0.733	∞	583
—													—
584	2	3	.....	2a,3	1.6	10.6	−16 (2)	210 (410)	62	115^{20}	$0.787–0.791^{20}_{20}$	s	584
585	.....	.....	.....	2a	....	....	56 (132)	502 (935)	154.3	$1^{14.7}$	0.9504^{20}_{4}	v	585
586	1	.....	2	....	....	....		209 (408)	....	....	0.8528^{20}_{4}	....	586
—													—
587	2	4	1	3,4	....	....	−54 (−65)	220 (428)	34	....	0.6810^{20}_{4}	i	587
588	2	3	.....	2a	....	....	16 (60)	....	$92–4^{732}$	....	0.9090	δ	588
589	1	3	0	2,2a,3	1.8	7.8	2 (35)	460 (860)	93	73^{25}	0.8732	s	589
590	1	3	0	1,2a,3	2.3	12.7	12 (53)	399 (750)	82.4	44^{25}	0.7851^{20}_{4}	∞	590
591	3	4	0	1,2a,3,4	2.0	10.4	−37 OC (−35) OC	402 (756)	33.0	460^{20}	0.691	∞	591
592	1	1	.....	....	....	....	99 (210)	....	218	....	1.0122	i	592
593	2	3	1	2,2a,3	1.4	21	−28 (−18)	443 (830)	69^{761}	119.4^{20}	0.7241	δ	593
594	.....	.....	.....	....	....	....	....	....	137	9.4^{25}	0.9186	s	594
—													—
595	0	2	0	2,3	0.7	5	38 (100)	229 (444)	170–300	....	0.81	i	595
596	.....	.....	.....	....				....	−56	....		d	596
597	.....	.....	.....	....				....	−152.9	....	....	....	597

No.	Chemical Name and Formula	Chemical Abstracts Registry Number	A.C.G.I.H. 1966–TLV (Threshold Limit Values) ppm (mg/M³)	Principal Effects of Inhalation Exposures above TLV	Relative Hazard to Health from Concentrated Short-term Exposure						References to Supplementary Information on Toxicity, Flammability and Other Hazards
					Eye Contact	Inhalation	Skin Penetration	Skin Irritation	Ingestion	Supplemental Effects	
					(5 is high; 1 is low)						
598	Lactonitrile . . . . (Acetaldehyde cyanohydrin) C_3H_5NO	78977	. . . .	. . . .	4*	5y	5	2	4	Extreme toxicity	2019
—	Lampblack See Carbon black										
599	Lanthanum . . . . La	7439910	. . . .	. . . .	. . . .	. . . .	. . . .	. . .	. . . .	. . . .	1058, STTE
600	Lauric acid . . . . (Dodecanoic acid) $C_{12}H_{24}O_2$	143077	. . . .	. . . .	2x	. . . .	1x	2x	2x	. . . .	1789
601	Lead . . . . Pb	7439921	(0.2)	Toxic	1x	5x*	1x	1x	2x	Chronic toxicity	945, AIHA, NSC, STTE
602	Lead acetate . . . . $C_4H_{12}O_7Pb$	301042	. . . .	. . . .	2x	5x*	1x	1x	2x	Chronic toxicity	945
603	Lead arsenate . . . . PbHAsO₄ (Approximate)	7784409	(0.15)	Toxic	2y	5x*	1x	2x	2	Chronic toxicity	945, STTE
604	Lead carbonate, basic . . . . $(PbCO_3)_2Pb(OH)_2$ (Approximate)	1319466	. . . .	. . . .	2x	5x*	1x	1x	2x	Chronic toxicity	945
605	Lead nitrate . . . . Pb$(NO_3)_2$	10099748									HCD
—	Limonene See p-Mentha-1,8-diene										
606	Lindane . . . . (1,2,3,4,5,6-Hexachlorocyclo-hexane) $C_6H_6Cl_6$	58899	(0.5) S	Toxic	2x	4*	5*	3y	4x*	Chronic toxicity	1351, STTE
607	Liquified petroleum gas . . . . (L.P.G.) C_3, C_4	. . . .	1000 (1800)	Narcosis	. . . .	. . . .	. . . .	. . .	. . . .	. . . .	1195, STTE
608	Lithium Li	7439932	. . . .	. . . .	5x	. . . .	. . . .	4x	. . . .	. . . .	1068, HCD, NSC, STTE
—	Lithium aluminum hydride See Lithium tetrahydroalanate and Lithium hexahydro-alanate										
609	Lithium carbonate . . . . Li_2CO_3	554132	. . . .	. . . .	4y*	3x*	2x	2y	3	Irritates eye, nose, throat	1068
610	Lithium hexahydroalanate. . . . . . $LiAlH_6$	1336249	. . . .	. . . .	. . . .	. . . .	. . . .	. . .	. . . .	. . . .	J. Am. Chem. Soc. 88, p. 858, (1966)
611	Lithium hydride . . . . LiH	7580678	(0.025)	Toxic	Irritating					. . . .	1068, AIHA, HCD, STTE, NSC
612	Lithium tetrahydroalanate. . . . . . (Lithium aluminum hydride) $LiAlH_4$	1302303	. . . .	. . . .	Highly caustic					. . . .	1068, HCD, NSC
—	L.P.G. See Liquified petroleum gas										
613	Lutetium . . . . Lu	7439943	. . . .	. . . .	. . . .	. . . .	. . . .	. . .	. . . .	. . . .	1058
614	Magnesium . . . . Mg	7439954	. . . .	. . . .	2x	1x	. . . .	3x*	. . . .	Wounds need medical care	1075, AIHA, NSC, HCD, STTE
615	Magnesium chloride. . . . . $MgCl_2$	7786303	. . . .	. . . .	2x	. . . .	1x	2x	2	. . . .	1075
616	Magnesium nitrate . . . . $Mg(NO_3)_2 \cdot 2H_2O$	10588257	. . . .	. . . .	. . . .	. . . .	. . . .	. . .	. . . .	. . . .	HCD
617	Magnesium oxide fume	. . . .	(15)	Fume fever	1x	1z*	1x	1x	2x	Metal fume fever	STTE
618	Magnesium perchlorate . . . . $MgCl_2O_8$	10034818	. . . .	. . . .	. . . .	. . . .	. . . .	. . .	. . . .	. . . .	HCD

No.	(7) Health	Fire	Reactivity	(8) Extinguishing Agents	(9) Lower	Upper	(10) Flash Point °C (°F)	(11) Ignition Temperature °C (°F)	(12) Boiling Point (Melting Point, m.p.) °C	(13) Vapor Pressure mm. Hg °C	(14) Density or Specific Gravity	(15) Water Solubility	No.
598	....	....	....	....	....	....	77 (170)	....	103^{50}	10^{74}	0.992	∞	598
599	....	....	....	Pyrophoric on filing			....	....	3470 m.p. 920	....	6.162	d	599
600	....	....	....					$1^{121.0}$	131^1		0.8679^{50}_{4}	i	600
601	....	....	....	....					1515	....	11.296	i	601
602	....	....	....						m.p. 280	....	3.25	s	602
603	....	....	....						m.p. 1042 δd 1000	....	7.80	δδ	603
604	....	....	....						m.p. d 400	....	61.4	i	604
605	1	0	1	....					m.p. d 470	....	4.53^{20}	s	605
606	....	....	....						288	0.3^{20}	1.87^{20}	i	606
607	....	....	....						....	....	....	....	607
608	1	1 W	2	5	....	....			1317	....	.534	d	608
609	....	....	....						d 1310	....	2.11	δ	609
610	....	....	....	Decomposes in air without ignition			....		d >210	....	1.26	i	610
611	1	4 W	2	5 Smother with dolomite powder	Pyrophoric		....	May ignite spontaneously in air	680	....	.82	d	611
612	3	1 W	2	....	....	....	....	Moisture causes ignition	d >125	....	0.917	d	612
613	....	....	....	Pyrophoric on filing			....	....	3327 m.p. 1652	....	9.872	....	613
614	0	1 W	2	'5	....	....	Dust explosion hazard	....	110.7		1.74	i: d	614
615	....	....	....				....		1.412	....	2.316	v	615
616	1	0	1	....	Oxidizing material		....	....	m.p. 129	....	2.0256	s	616
617	....	....	....						....	....	....	....	617
618	1	0	1	1	Forms explosive mixtures with oxidizable material		....	....				....	618

CHEMICAL HAZARD INFORMATION (See Explanation, page 668)

No.	(1) Chemical Name and Formula	(2) Chemical Abstracts Registry Number	(3) A.C.G.I.H. 1966–TLV (Threshold Limit Values) ppm (mg/M³)	(4) Principal Effects of Inhalation Exposures above TLV	(5) Eye Contact	Inhalation	Skin Penetration	Skin Irritation	Ingestion	Supplemental Effects	(6) References to Supplementary Information on Toxicity, Flammability and Other Hazards
					(5 is high; 1 is low)						
619	Malathion................. $C_{10}H_{19}O_6PS_2$	121755	(15) S	Toxic	2x	3x	2y	2x	2x	....	1947, STTE
620	Maleic acid............... $C_4H_4O_4$	110167	....	....	4	2y*	2y	4y*	2	Sensitiz. skin, resp. tract	1811, STTE
621	Maleic anhydride........... $C_4H_2O_3$	108316	0.25 (1)	Toxic	5	1*	3y	4*	2x	Sensitiz. skin resp. tract	1820, MCA, HCD, STTE
622	Malononitrile.............. $C_3H_2N_2$	109773	....	....	Toxicity similar to cyanides					....	2027
623	Manganese................ Mn	7439965	C(5)	Toxic	....	....	....	....	....		1081, AIHA, NSC, STTE
624	Manganese sulfate.......... $MnSO_4$	7785877	....		2x	4x*	1x	2x	2x	Chronic toxicity	....
—	MAPP Gas See Propyne-allene mixture										
625	p-Mentha-1,8-diene (Dipentene®) (Limonene) $C_{10}H_{16}$	138863									
626	2-Mercaptoethanol.......... C_2H_6OS	60242			3	2	3	2	3		
627	Mercuric chloride........... $HgCl_2$	7487947	....		4	4z*	2y	4y	4y	Chronic toxicity	1092
628	Mercury................. Hg	7439976	(0.1) S	Toxic	Vapor toxic. Salts are general cellular poisons					....	1090, AIHA, NSC, STTE
629	Mercury—organic compounds ..	....	(0.01) S	Toxic							1090, STTE
—	Mesityl oxide See 4-Methyl-3-penten-2-one										
—	Metal hydrides See Hydrides, metal										
630	Methacrylic acid .... (α-Methylacrylic acid) $C_4H_6O_2$	79414			4x	1x	3x	4x	2x		1793, STTE
—	Methallyl chloride See 3-Chloro-2-methyl propene										
631	Methane.................. CH_4	74828	....		....	3x	....	....	....	Simple asphyxiant	1196, MGD, STTE
632	Methanesulfonic acid.......... CH_4O_3S	75752	....		Severe skin irritant. Not absorbed.					....	....
633	Methanethiol (Methyl mercaptan) CH_4S	74931	10 (20)	Toxic	May be narcotic in high concentrations						MGD, STTE
634	Methanol .... (Methyl alcohol) CH_4O	67561	200 (260)	Narcosis Toxic	2	2	2	1	1*	Permanent injury	1409, AIHA, NSC, MCA, STTE
—	2-Methoxyaniline See o-Anisidine										
—	Methoxybenzene See Anisole										
635	Methoxychlor............... (1,1,1-Trichloro-2,2-bis(p-methoxyphenyl)ethane) $C_{16}H_{15}Cl_3O_2$	72435	(15)	Toxic	....	....	....	....	....		STTE
636	2-Methoxyethanol........... (Methyl Cellosolve®, Dowanol® EM; Ethylene glycol monomethyl ether) $C_3H_8O_2$	109864	25 S (80) S	Toxic	2	2*	3	1y	2	Chronic toxicity	1542, AIHA, STTE
637	o-Methoxyphenol............ (Guaiacol) $C_7H_8O_2$	90051	....		3x	....	2x	3x	2		1683, STTE
638	p-Methoxyphenol............. (Hydroquinone monomethyl ether) $C_7H_8O_2$	150765	....	....	4	2x	2x	1	2	....	1686, STTE

No.	(7) N.F.P.A. Hazard Identification Signals			(8) Extinguishing Agents	(9) Flammable Limits in Air % by volume		(10) Flash Point °C (°F)	(11) Ignition Temperature °C (°F)	(12) Boiling Point (Melting Point, m.p.) °C	(13) Vapor Pressure mm. Hg °C	(14) Density or Specific Gravity	(15) Water Solubility	No.
	Health	Fire	Reactivity		Lower	Upper							
	(4 is high; 0 is low)												
619	.....	.....	.....	....	....	....	....	....	$156-7^{0.7}$	4×10^{-5} (30)	1.23_4^{25}	δ	619
620									m.p. 130-5	....	....	v	620
621	3	1	1	1,2a,3	1.4	7.1	102 (215)	477 (890)	196	0.16^{20} 6.2^{70}	....	i	621
622	.....	.....	.....						218-9	....	1.0494	s	622
623	.....	.....	.....						2152		7.20	d	623
624	.....	.....	.....						d 850		3.25	v	624
—													—
625	.....	2	0	....	0.7 @ 302°	6.1 @ 302°	45 (113)	237 (458)	170.3^{695}	....	3.8402	i	625
626	2	2	.....	1,2a,3	....	δd	74 OC (165)OC	....	$157-8^{742}$	....	1.1143	s	626
627	.....	.....	.....				....		302	....	....	s	627
628	.....	.....	.....				....		356.58	0.002^{26}	13.594_4^{20}	i	628
629	.....	.....	.....						....	....	....	....	629
—													—
—													—
630	3	2	2	2a	....	....	77 OC (171)OC	....	$162-3^{757}$	$1^{25.5}$	0.9504_4^{20}	....	630
—													—
631	.....	.....	....	4	5.0	15.0	....	537 (999)	-161.49	....	0.415	δ	631
632	.....	.....	.....						167^{10}	....	1.4812	v	632
633	2	4	0	2a,4	3.9	21.8	<-18 (<0)	....	6	$760^{-6.8}$	0.8665_4^{0}	$δ^h$	633
634	1	3	0	2a,3	6.0	36	11 (52)	446 (835)	64.96	160^{30}	0.7914	∞	634
—													—
—													—
635	.....	.....	.....						m.p. 94	....	....	s	635
636	.....	.....	.....	1,3	2.3	24.5	39 (102)	288 (551)	125^{768}	9.7^{25}	0.9647	∞	636
637	.....	.....	.....	δ	....	....	82 (180)	....	205.05	10^{92}	1.1287	δ	637
638	.....	.....	.....				132 (268)	421 (790)	243		1.55	δ	638

No.	(1) Chemical Name and Formula	(2) Chemical Abstracts Registry Number	(3) A.C.G.I.H. 1966 TLV (Threshold Limit Values) ppm (mg/M³)	(4) Principal Effects of Inhalation Exposures above TLV	(5) Relative Hazard to Health from Concentrated Short-term Exposure						(6) References to Supplementary Information on Toxicity, Flammability and Other Hazards
					Eye Contact	Inhalation	Skin Penetration	Skin Irritation	Ingestion	Supplemental Effects	
					(5 is high; 1 is low)						
639	3-(3-Methoxypropoxy)-1-propanol .... (Dipropylene glycol methyl ether) $C_7H_{16}O_3$	112287	100 S (600) S	Toxic	1	1	1	1	2	....	1567, STTE
640	Methyl acetate .... $C_3H_6O_2$	79209	200 (610)	Narcosis; Toxic	2	3	2x	1		....	1858, AIHA, STTE
—	Methyl acetylene See Propyne										
—	Methyl acetylene-propadiene mixture See Propynepropadiene mixture										
641	Methyl acrylate .... $C_4H_6O_2$	96333	10 S (35) S	Irritant	2	4*	3	1*	3	Sensitiz. skin, resp. tract	1576, MCA, HCD, STTE
—	2-Methacrylic acid See Methacrylic acid										
—	Methylal See Dimethoxymethane										
642	Methylamine .... CH_5N	74895	10 T (12) T	Irritant	5x	5x*	....	4x*	....	Sensitiz. skin, resp. tract	2036, HCD, MGD
—	Methyl-n-amyl ketone See 2-Heptanone										
643	N-Methylaniline.............. C_7H_9N	100618	2 S (9) S	Toxic	....	....	....	....	....		
644	2-Methylaziridine............. (Propylenimine) C_3H_7N	75558	2 ST (5) ST	Toxic	4	4x*	4	4x*	4	Sensitiz. skin, resp. tract	2175, STTE
645	6-Methyl-1,2-benzanthracene ...	316143	....	....						Carcinogenic in animals	....
646	10-Methyl-1,2-benzanthracene ..	2381159	....	....						Carcinogenic in animals	
647	Methyl benzoate.............. $C_8H_8O_2$	93583			1	1	2x	2	2	....	1896
648	α-Methylbenzyl alcohol .... $C_8H_{10}O$	98851			2	1	3	2	3		
—	Methyl bromide See Bromomethane										
—	2-Methyl-1,3-butadiene See Isoprene										
649	2-Methyl-1-butene .... C_5H_{10}	563462									
650	2-Methyl-2-butene .... C_5H_{10}	513359									
651	3-Methyl-1-butene .... C_5H_{10}	563451			Highly toxic and irritating					....	MGD
652	N-Methylbutylamine	110689			4	4	3	4	3	....	STTE
653	Methyl butyrate .... $C_5H_{10}O_2$	623427			....	....	....	....	....		1866
—	Methyl cellosolve® See 2-Methoxyethanol										
654	Methyl cellosolve acetate .... $C_5H_{10}O_3$	110496	25 S (120) S	Toxic	1	2*	2	1	2	Chronic toxicity	1583, STTE
—	Methyl chloride See Chloromethane										
—	Methyl chloroform See 1,1,1-Trichloroethane										
655	Methylchloroformate.......... $C_2H_3ClO_2$	79221	....	....	Avoid all contact					Toxic and irritant	1885
656	3-Methyl cholanthrene	56495	....	....	....	....	....	....	....	Carcinogenic in animals	....
657	Methylcyclohexane .... C_7H_{14}	108872	500 (2000)	Narcosis	....	....	....	....	....	....	1209, STTE

No.	(7) N.F.P.A. Hazard Identification Signals			(8) Extinguishing Agents	(9) Flammable Limits in Air % by volume		(10) Flash Point °C (°F)	(11) Ignition Temperature °C (°F)	(12) Boiling Point (Melting Point, m.p.) °C	(13) Vapor Pressure mm. Hg °C	(14) Density or Specific Gravity	(15) Water Solubility	No.
	Health	Fire	Reactivity		Lower	Upper							
	(4 is high; 0 is low)												
639	.....	.....	.....	....	....	....	85 OC (185) OC	....	189	0.36 (25)	0.951	∞	639
640	.....	.....	.....	2a,3	3.1	16	−10 (15)	502 (935)	57	235^{25}	0.9723	v	640
641	2	3	2	....	2.8	25	−3 OC (27) OC	....	70^{698}	....	0.9561	s	641
642	3	4	0	3,4	4.9	20.7	Gas	430 (806)	41^{4500}	^{25}2 atm.	0.699^{-11}	v	642
643	.....	.....	.....	....	....	....	....	....	196.25	....	0.989^{12}	i	643
644	.....	.....	.....	....	....	....	....	....	63.64	....	....	s	644
645	.....	.....	.....	....	....	....	....	....	....	....	....	....	645
646	.....	.....	.....	....	....	....	....	....	....	....	....	....	646
647	0	2	0	....	....	....	83 (181)	....	199.6	....	1.0937	i	647
648	1	1	0	2a	....	....	132 OC (205) OC	....	203	....	....	δ	648
649	.....	4	0	....	....	....	<−7 (<20)	....	38.6	....	0.6623	i	649
650	.....	3	0	2a	....	....	<−7 (<20)	....	38–42	....	0.6670^{15}_{4}	δ	650
651	.....	.....	.....	....	....	....	....	....	20.06	434^{5}	0.6272	....	651
652	.....	.....	.....	....	....	....	13 OC (85) OC	....	$90.5-1.5^{764}$	....	0.7367^{15}_{4}	δ	652
653	.....	3	0	2,2a,3	....	....	14 (57)	....	102.3	40^{30}	0.8982^{20}_{4}	δ	653
654	.....	.....	.....	1,3	1.5	12.3	45 (120)	394 (740)	144.5–5	$2.0-3.7^{20}$	1.0090	s	654
655	.....	.....	.....	....	Flammable		12 (54)	504 (940)	71–2	....	1.236	d	655
656	.....	.....	.....	....	....	....	....	....	280^{80}	....	128	i	656
657	.....	3	0	2,3	1.2	....	−4 (25)	285 (545)	100.4	43^{25}	0.7695^{20}_{4}	i	657

No.	(1) Chemical Name and Formula	(2) Chemical Abstracts Registry Number	(3) A.C.G.I.H. 1966–TLV (Threshold Limit Values) ppm (mg/M³)	(4) Principal Effects of Inhalation Exposures above TLV	(5) Relative Hazard to Health from Concentrated Short-term Exposure						(6) References to Supplementary Information on Toxicity, Flammability and Other Hazards
					Eye Contact	Inhalation	Skin Penetration	Skin Irritation	Ingestion	Supplemental Effects	
					(5 is high; 1 is low)						
658	2-Methylcyclohexanol .... $C_7H_{14}O$	583595	100 (470)	Irritant	....	....	....	....	....		1480, STTE
659	2-Methcyclohexanone .... $C_7H_{12}O$	583608	100 S (460) S	Narcosis	....	....	....	....	....		STTE
660	Methylcyclohexene .... C_7H_{12}	1335860	....	....	....	....	....	....	....		....
661	3-Methyl-4-dimethylamino-azobenzene	54886	....	....						Carcinogenic in animals	....
—	Methylene bisphenyl isocyanate See Methylene di-p-phenylene isocyanate										
—	Methylene chloride See Dichloromethane										
—	Methylene dichloride See Dichloromethane										
661a	Methylene di-p-phenylene isocyanate .... Methylene-bis(4-phenyl isocyanate, MDI) $C_{15}H_{10}N_2O_2$	101688	C 0.02 C (0.2)	Toxic	....	....	....	....	....		
662	Methyl ether .... (Dimethyl ether) C_2H_6O	115106	....	....	2x	4x	2x	1x	2x		1655
663	Methyl ethyl ether .... C_3H_8O	540670			....	....	....	....	....	Anesthetic	HCD
—	Methyl ethyl ketone See 2-Butanone										
664	Methyl ethylnitrosocarbamate (Methyl nitrosoethyl carbamate)	10546233	....	....						Carcinogenic in animals	....
665	Methyl formate .... $C_2H_4O_2$	107313	100 (250)	Irritant	3x	4x	2x	2x	3x		1854, HCD, STTE
666	2-Methylfuran .... C_5H_6O	534225									
667	5-Methyl-3-heptanone .... $C_8H_{16}O$	....	25 T (130) T	Irritant	....	....	....	....	....		1721
—	Methyl-n-hexylcarbinol See 2-Octanol										
668	Methylhydrazine .... CH_6N_2	60344	Ceiling C 0.2 S C (0.35) S	Toxic	....	....	....	....	....		2220
—	Methyl iodide See Iodomethane										
—	Methyl isobutyl carbinol See 4-Methyl-2-Pentanol										
—	Methyl isobutyl ketone See 4-Methyl-2-pentanone										
669	Methyl isobutyrate........... $C_8H_{16}O_2$	547637	....	....	....	....	....	....	....		1866
670	Methyl isocyanate .... C_2H_3NO	624839	0.02 ST (0.05) ST	Irritant Toxic	5	5y	3	4	4		....
671	Methyl isothiocyanate .... (Isothiocyanic acid, methyl ester) C_2H_3NS	556616	....	....	4x	4x	3x	4x	3		
672	2-Methyllactonitrile........... (Acetone cyanohydrin) C_4H_7NO	75865	....	....	5y*	4	5	1	4	Extreme toxicity (10 ppm limit recommended by one source)	2021, HCD
—	Methyl mercaptan See Methanethiol										
673	Methyl methacrylate .... $C_5H_8O_2$	80626	100 (410)	Irritant	....	....	....	....	....		1876, HCD, STTE

No.	(7) N.F.P.A. Hazard Identification Signals			(8) Extinguishing Agents	(9) Flammable Limits in Air % by volume		(10) Flash Point °C (°F)	(11) Ignition Temperature °C (°F)	(12) Boiling Point (Melting Point, m.p.) °C	(13) Vapor Pressure mm. Hg °C	(14) Density or Specific Gravity	(15) Water Solubility	No.
	Health	Fire	Reactivity		Lower	Upper							
	(4 is high; 0 is low)												
658	.....	2	0	1,2a,2,3	....	....	296 (565)	....	167.2–7.6	1.5^{30}	0.9241_4^{26}	δ	658
659	.....	2	0	1,2,3	....	....	48 (118)	....	165^{757}	....	0.9240_4^{20}	i	659
660	.....	3	0				−1 OC (30)OC	....	102.74		0.7991_8^{14}	i	660
661	.....	.....	.....		....	....		....	....	....	....	....	661
661a	.....	.....	.....					....	....	....	....	....	661a
662	.....	4	0	3,4	3.4	18	Gas	350 (662)	−23.6	2128^{0}	0.661	δ	662
663	2	4	1	....	2	10.1	−37 (−35)	190 (374)	11				663
664	.....				....	....		....	....	....	....	....	664
665	2	4	0	2a,3,4	5.9	20	−19 (−2)	456 (853)	31.50	600^{26}	0.98674^{15}	s	665
666	.....	3	1		....	....	−30 (−22)	....	63–63.5	....	0.9159_0^{20}	δ	666
667	.....						57 OC (135)OC	....	160.5	2.0^{25}	0.850	i	667
668	.....	.....	.....	2a	....	....	<27 (<80)	....	87^{745}	49.6^{25}	0.9	δ	668
669	.....	.....	.....		....	....	13 OC (55)OC	482 (900)	147.0	10^{38}	0.8557	δ	669
670	.....	.....	.....		....	....	....	....	59.6		0.7557	δ	670
671	.....	.....	.....		....	....	....	....	119	....	1.6691	δ	671
672	4	1	2	2a	....	....	74 (165)	688 (1270)	82	0.8^{20}	0.932	s	672
673	2	3 P	2	....	2.1	12.5	10 (50)	421 (790)	100	12.9^{-10} 68.2^{20}	0.936	δ	673

No.	(1) Chemical Name and Formula	(2) Chemical Abstracts Registry Number	(3) A.C.G.I.H. 1966–TLV (Threshold Limit Values) ppm (mg/M³)	(4) Principal Effects of Inhalation Exposures above TLV	(5) Relative Hazard to Health from Concentrated Short-term Exposure — Eye Contact	Inhalation	Skin Penetration	Skin Irritation	Ingestion	Supplemental Effects	(6) References to Supplementary Information on Toxicity, Flammability and Other Hazards
					(5 is high; 1 is low)						
674	1-Methylnaphthalene.......... C₁₁H₁₀	90120	....	....	3x	4y*	2	2*	2	Photosensi- tization	....
675	3-Methyl-2-naphthalyamine .... C₁₁H₁₁N	10546244	....	....	....	....	....	....		Carcinogenic in animals	....
676	N-Methyl-N-nitrosoacetamide ..	7417676								Carcinogenic in animals	
677	N-Methyl-N-nitrosoallylamine ..	4549433								Carcinogenic in animals	
678	N-Methyl-N-nitrosoaniline	614006								Carcinogenic in animals	
679	N-Methyl-N-nitrosobenzylamine	937406								Carcinogenic in animals	
—	Methyl nitrosoethyl carbamate See Methyl ethylnitrosocar- bamate										
680	1-Methyl-1-nitrosourea	684935	....							Carcinogenic in animals	
681	N-Methyl-N-nitrosovinylamine..	454900								Carcinogenic in animals	
682	4-Methyl-2-pentanol .... (Methyl isobutyl carbinol) C₆H₁₄O	108112	25 S (100) S	Narcosis; Toxic	....						1459, STTE
683	4-Methyl-2-pentanone .... (Methyl isobutyl ketone) C₆H₁₂O	108101	....		2	3	1	1	2		1270, AIHA, HCD, STTE
684	4-Methyl-3-penten-2-one .... (Mesityl oxide) C₆H₁₀O	141797	25 (100)	Narcosis	2	3	2	1	3		1722, HCD, STTE
685	2-Methylpropane (Isobutane) C₄H₁₀	75285	....		....	3x	....				MGD, STTE
686	2-Methyl-2-propanethiol .... (tert-Butyl mercaptan) C₄H₁₀S	75661	....								
—	2-Methyl-1-propanol See Isobutyl alcohol										
687	2-Methylpropene .... (Isobutene; Isobutylene) C₄H₈	115117								Simple asphyxiant	1204, MGD
—	Methyl propyl ketone See 2-Pentanone										
—	2-Methylpyridine See 2-Picoline										
—	4-Methylpyridine See 4-Picoline										
688	1-Methylpyrrole............. C₅H₇N	96548									
—	2-Methylquinoline See Quinaldine										
689	Methyl salicylate .... C₈H₈O₃	119368	....		2	2x	3x	2x	3		1896, STTE
690	α-Methylstyrene .... (Isopropenylbenzene) C₉H₁₀	98839	C 100 C (480)	Irritant	....						STTE
691	m-,p-Methylstyrene........... (3-Vinyltoluene) C₉H₁₀	100801	100 (480)	Irritant	This compound is a mixture of meta and para isomers of methylstyrene.					....	
692	Methyl sulfate .... CH₄O₄S	75934	....		....		....	....			
693	Methyl sulfide............... (Dimethyl sulfide) C₂H₆S	75183									HCD
694	Methyl sulfoxide .... (Dimethyl sulfoxide; DMSO) C₂H₆OS	67685	....	....	1	1	2x	1	1		HCD, STTE

No.	(7) N.F.P.A. Hazard Identification Signals (4 is high; 0 is low) Health	Fire	Reactivity	(8) Extinguishing Agents	(9) Flammable Limits in Air % by volume Lower	Upper	(10) Flash Point °C (°F)	(11) Ignition Temperature °C (°F)	(12) Boiling Point (Melting Point, m.p.) °C	(13) Vapor Pressure mm. Hg °C	(14) Density or Specific Gravity	(15) Water Solubility	No.
674	2	2	0	3	....	....	....	528 (984)	240–3[759]	....	1.0287[19]	i	674
675	....	....	....	....	....	....	....	....	51–2	....	....	i	675
676	....											....	676
677	....												677
678	....												678
679	....											....	679
—													—
680	....											....	680
681	....												681
682	2	2	0	2a	1.0	5.5	41 (106)	....	130	352[20]	0.8025	δ	682
683	2	3	0	2a	1.4	7.5	23 (73)	460 (860)	116.85	7.5[25]	0.801_4^{20}	δ	683
684	3	3	0	2a,3	Flammable		31 (87)	344 (652)	130–1	9.5[25]	0.8578	s	684
685	1	4	0	4	1.8	8.4	Gas	462 (864)	−12	....	0.563	s	685
686	.....	3	0	....	Flammable		< −29 (< −20)	....	64.22	....	0.7947	i	686
—													—
687	1	4	0	4	1.8	8.8	Gas	465 (869)	−6.6	....	0.5992	i	687
—													—
—													—
—													—
688	.....	3	.....	....	....	....	16 (61)	....	115–60[756]	....	0.9145	δ	688
—													—
689	1	1	0	1,2a,3	....	....	101 (214)	454 (850)	223.3	....	1.1787_4^{25}	s	689
690	1	2	1	....	1.9	6.1	54 (129)	574 (1066)	167–70	....	$0.9139_4^{17.4}$	i	690
691	....	.....	.....	....	....	....	60 (140)	....	170–1	....	0.890	δ	691
692	....	.....	.....	....	....		....	....	....	....	....	....	692
693	4	4	0	3	2.2	19.7	< −18 (<0)	206 (403)	37.3	....	0.8458	i	693
694	1	1	0	2a	2.6	28.5	95 OC (203) OC	....	189	....	1.014	s	694

No.	Chemical Name and Formula	Chemical Abstracts Registry Number	A.C.G.I.H. 1966-TLV (Threshold Limit Values) ppm (mg/M^3)	Principal Effects of Inhalation Exposures above TLV	Eye Contact	Inhalation	Skin Penetration	Skin Irritation	Ingestion	Supplemental Effects	References to Supplementary Information on Toxicity, Flammability and Other Hazards
						(5 is high; 1 is low)					
695	N-Methyl-N,2,4,6-tetranitro-aniline . . . . $C_7H_5N_5O_8$	479458	(1.5) S	Toxic	....	....	....	....	....		2151, NSC, STTE
696	Methyl p-toluenesulfonate . . . . . $C_8H_{10}O_3S$	80488	. . . .	. . . .	Skin sensitization						1926
697	Methyl vinyl ether . . . . (Vinyl methyl ether) C_3H_6O	107255			1x	3x	2x	1x	2x		MGD
—	Methyl vinyl ketone See 3-Buten-2-one										
698	Molybdenum . . . . Mo	7439987	. . . .	. . . .	1x	2x	1x	1x	1x		1112, AIHA, STTE
699	Molybdenum compounds (soluble)	. . . .	(5)	Toxic	....	....	....	....	....		
700	Molybdenum compounds (insoluble)	. . . .	(15)	Toxic	....	....	....	....	....		
701	Morpholine . . . . C_4H_9NO	110918	20 S (70) S	Irritant Toxic	3	2	3	3	2		2202, STTE
702	Naphtha (coal tar)	. . . .	100 (400)	Narcosis	....	....	....	....	....		STTE
—	Naphtha, Petroleum See Petroleum ether										
703	Naphtha, Varnish Makers & Painters 50° Flash	. . . .	500 (2000)	Narcosis	....	....	....	....	....		1195
704	Naphtha, Varnish Makers & Painters High Flash	. . . .	500 (2000)	Narcosis	....	....	....	....	....		1195
705	Naphtha, Varnish Makers & Painters Regular . . . . [Only a few naphthas are listed since flash point, boiling range and ignition temperature will vary depending on the manufacturer.]	. . . .	500 (2000)	Narcosis	....	....	....	....	....		1195, AIHA
706	Naphthalene . . . . $C_{10}H_8$	91023	10 (50)	Toxic	2y	1	2x	2y	2		1223, MCA, NSC, HCD,
707	1-Naphthol . . . . $C_{10}H_8O$	90153	. . . .	. . . .	4	1	3	3y	2		. . . .
708	2-Naphthol . . . . $C_{10}H_8O$	135193	. . . .	. . . .	4	....	3x	2y	2		. . . .
709	1-Naphthylamine . . . . $C_{10}H_9N$	134327									
710	2-Naphthylamine . . . . (β-Naphthylamine) $C_{10}N_9N$	91598	Because of the extremely high incidence of bladder tumors in workers handling this compound and the inability to control exposures, it has been prohibited from manufacture, use and other activities that involve human contact by the State of Pennsylvania.								
711	1-Naphthyl isothiocyanate . . . . . . $C_{11}H_7NS$	551064	. . . .	. . . .	May cause dermatitis, chills, fever, and kidney damage thru skin absorption						. . . .
712	Natural gas	. . . .	. . . .	. . . .	....	3x	....	....			STTE
713	Neon . . . . Ne	7440019	. . . .		May act as an asphyxiant						MGD
714	Nickel-metal & solid compounds .	. . . .	1	Toxic; Carcinogenic	May cause dermatitis					. . . .	1121, AIHA, STTE
715	Nickel carbonyl . . . . $Ni(CO)_4$	13005317	0.001 (0.007)	Carcinogenic	Exposure may be fatal or may be carcinogenic					. . . .	1105, AIHA, MGD, STTE
716	Nickel nitrate . . . . $Ni(NO_3)_2 \cdot 6H_2O$	10580333	. . . .	. . . .	2y	....	1x	3y	2	. . . .	HCD
717	Nickel sulfate . . . . $NiSO_4$	7786814	. . . .	. . . .	2x	....	1x	3x	2x	. . . .	1120
718	Nicotine . . . . $C_{10}H_{14}N_2$	54115	(0.5) S	Toxic	....	....	....	....	....		1193, STTE

No.	(7) N.F.P.A. Hazard Identification Signals			(8) Extinguishing Agents	(9) Flammable Limits in Air % by volume		(10) Flash Point °C (°F)	(11) Ignition Temperature °C (°F)	(12) Boiling Point (Melting Point, m.p.) °C	(13) Vapor Pressure mm. Hg °C	(14) Density or Specific Gravity	(15) Water Solubility	No.
	Health	Fire	Reactivity		Lower	Upper							
	(4 is high; 0 is low)												
695	....	....	....	....	....	....	....	....	expl. 187	....	1.57	i	695
696	....	....	....	....	....	....	....	....	m.p. 27–8	....	1.230	....	696
697	2	4	2	....	....	....	Gas	....	8	....	0.7725	δ	697
—													—
698	....	....	....	....	....	....	....	....	4510	....	10.2	i	698
699	....	....	....	....	....	....	....	....	....	....	....	....	699
700	....	....	....	....	....	....	....	....	....	....	....	....	700
701	2	3	0	1,2a,3	....	....	38 OC (100)	....	128	8^{20}	0.9994	∞	701
702	2	2	0	2,3	....	....	41 (107)	277 (531)	....	....	....	....	702
—													—
703	1	3	0	2,3	0.9	6.7	10 (50)	232 (450)	116–143	....	....	....	703
704	1	3	0	2,3	1.0	6.0	29 (85)	232 (450)	139–177	....	....	....	704
705	1	3	0	2,3	0.9	6.0	−2 (28)	232 (450)	100–160	....	....	....	705
706	2	2	0	1,3	0.9	5.9	79 (174)	526 (979)	210.8^{720}	0.082^{25}	1.145^{100}_{14}	i	706
707	....	....	....	....	....	....	....	....	288	....	1.103^{101}	i:δ^h	707
708	....	....	....	1,3	....	....	153 (307)	....	295	....	1.28	i	708
709	2	1	0	1,3	....	....	157 (315)	....	300.8 Sub.	....	1.123^{25}_{25}	δ	709
710	....	....	....	....	....	....	....	....	....	....	....	....	710
711	....	....	....	....	....	....	....	....	....	....	....	....	711
712	1	4	0	4	3.8–6.5	13–17	....	482–632 (900–1170)	....	....	....	....	712
713	....	....	....	....	....	....	....	....	−246	....	Gas 0.835 g/l	i	713
714												....	714
715	....	....	....	....	May decompose violently in air at 60 (140)		....	....	43	261^{15}	1.32^{17}	δ	715
716	1	0	1	....	Oxidizing material		....	....	136.7	....	2.05	v	716
717	....	....	....	....	....	....	....	....	d 848	....	3.68	s	717
718	4	1	0	2a,3	0.7	4.0	....	244 (471)	247.3	$1^{61.8}$	1.010^{20}_4	s	718

No.	(1) Chemical Name and Formula	(2) Chemical Abstracts Registry Number	(3) A.C.G.I.H. 1966-TLV (Threshold Limit Values) ppm (mg/M³)	(4) Principal Effects of Inhalation Exposures above TLV	(5) Relative Hazard to Health from Concentrated Short-term Exposure — Eye Contact	Inhalation	Skin Penetration	Skin Irritation	Ingestion	Supplemental Effects	(6) References to Supplementary Information on Toxicity, Flammability and Other Hazards
					(5 is high; 1 is low)						
719	Niobium Nb	7440031									1058
720	Nitric acid HNO_3	7697372	2 (5)	Irritant; Toxic	4x	5x	3x	4x	4x		2238, MCA, HCD, STTE, AIHA
721	Nitric oxide NO	10102439	25 (30)	Irritant; Toxic	3x	5z*		3		Chronic toxicity	918, MGD
722	2,2',2''-Nitrilotriethanol (Triethanolamine) $C_6H_{15}NO_3$	102716			2	1	1	1	2		2065
723	3-Nitroacetophenone $C_8H_7NO_3$	121891			1y	1	2	2y	2		STTE
724	m-Nitroaniline $C_6H_6N_2O_2$	99092			Toxic					More toxic than p-Nitroaniline	2145
725	o-Nitroaniline $C_6H_6N_2O_2$	88744			Toxic						2146
726	p-Nitroaniline $C_6H_6N_2O_2$	100016	1 S (6) S	Toxic	Toxic; rapidly absorbed through skin						2145, MCA, HCD, STTE
727	Nitrobenzene $C_6H_5NO_2$	98953	1 S (5) S	Toxic							2146, AIHA, MCA, HCD, STTE
728	4-Nitrobiphenyl $C_{12}H_9NO_2$	92933								Carcinogenic	STTE
—	o-Nitrochlorobenzene See 1-Chloro-2-nitrobenzene										
—	p-Nitrochlorobenzene See 1-Chloro-4-nitrobenzene										
729	Nitrodiethanolamine	1335837								Carcinogenic in animals	
730	Nitroethane $C_2H_5NO_2$	79243	100 (310)	Irritant; Narcosis							2073, AIHA, HCD, STTE
731	2-Nitrofluorene $C_{13}H_9NO_2$	607578								Carcinogenic in animals	
732	Nitrogen N_2	7727379				3x				Can act as an asphyxiant	917, MGD, STTE
733	Nitrogen dioxide (Nitrogen peroxide) NO_2	10102440	C 5 C (9)	Irritant; Toxic	3x	5z		3x		Chronic toxicity	919, AIHA, NSC, HCD, MGD, STTE
—	Nitrogen peroxide See Nitrogen dioxide										
734	Nitrogen tetroxide (Dinitrogen tetroxide) N_2O_4	10544726			3x	5z		3x		Chronic toxicity	NSC, HCD
735	Nitrogen trifluoride NF_3	7783542	10 (29)	Toxic							STTE
736	Nitrogen trioxide N_2O_3	10544737			Highly toxic						NSC, MGD
737	Nitroglycerin $C_3H_5(NO_3)_3$	55630	C 0.2 S C (1.2) S	Toxic							AIHA, STTE
738	Nitromethane CH_3NO_2	75525	100 (250)	Irritant; Toxic							AIHA, HCD, STTE
739	α-Nitronaphthalene $C_{10}H_7NO_2$	86577									STTE
740	1-Nitropropane $C_3H_7NO_2$	108032	25 (90)	Toxic							2073, AIHA, HCD, STTE
741	2-Nitropropane $C_3H_7NO_2$	79469	25 (90)	Toxic							2073, HCD, STTE

No.	(7) N.F.P.A. Hazard Identification Signals (4 is high; 0 is low)			(8) Extinguishing Agents	(9) Flammable Limits in Air % by volume		(10) Flash Point °C (°F)	(11) Ignition Temperature °C (°F)	(12) Boiling Point (Melting Point, m.p.) °C	(13) Vapor Pressure mm. Hg °C	(14) Density or Specific Gravity	(15) Water Solubility	No.
	Health	Fire	Reactivity		Lower	Upper							
719	....	....	....	....	Pyrophoric on filing		....	....	~3300 m.p. 2415	....	8.57	i	719
720	2	0	1	....	....	....			83		1.5027_4^{25}	∞	720
721	....	....	....				....	....	−152		Gas 1.3402 g/l	s	721
722	1	1	1	1,2a,3	....	....	179 (355)		360	<0.01 (20)	1.1258	∞	722
723	....								202	....	....	v^h	723
724	3	1	1						284	....	1.442^{40}	δ	724
725	3	1	1	....			168 OC (335) OC	521 (970)	305−7	....	1.1747^{40}	δ	725
726	3	1	1	1,3	Combustible yielding toxic vapors. Dust is explosive		199 (390)	....	200^{10} m.p. 146	....	1.437^{40}	i:$δ^h$	726
727	3	2	0	1,2,3	1.8 @ 200°F	....	88 (190)	482 (900)	210.8	....	1.2037_4^{25}	δ	727
728	....	....							340	....	....	i.	728
—													—
—													—
729	....						....		....	....	....	....	729
730	1	3	3	2a,3	3.4	....	38 (100)	360 (680)	Expl. on heating 115	15.6^{20}	1.0448	s	730
731	....								m.p. 160	....		....	731
732	....								−195.8	30^{-149}	Gas 1.2506 g/l	δ	732
733	....								21.2	....	1.4494	s	733
—													—
734	3	0	1						158				734
735	....								−128.8		liq. 1.537^{-129}	δ	735
736	....								3.5				736
737	....				Extra hazardous. Explodes violently when heated or shocked		....						737
738	1	3	4	2a,3	7.3	....	35 (95)	379 (714)	108	27.8^{20}	1.1354	s	738
739	1	1	0	1,3	....	....	164 (327)		304	....	1.332	i	739
740	1	2	3	2a,3	2.6	....	49 OC (120) OC	421 (789)	130.5−1.5 m.p. 108	7.5^{20}	1.0221	δ	740
741	1	2	3	2a,3	2.6	....	39 OC (103) OC	428 (802)	May explode on heating 120	12.9^{20}	1.024	δ	741

No.	(1) Chemical Name and Formula	(2) Chemical Abstracts Registry Number	(3) A.C.G.I.H. 1966–TLV (Threshold Limit Values) ppm (mg/M³)	(4) Principal Effects of Inhalation Exposures above TLV	(5) Relative Hazard to Health from Concentrated Short-term Exposure (5 is high; 1 is low)						(6) References to Supplementary Information on Toxicity, Flammability and Other Hazards
					Eye Contact	Inhalation	Skin Penetration	Skin Irritation	Ingestion	Supplemental Effects	
742	4-Nitroquinoline-*N*-oxide	56575	....	....	....	....	....	....	....	Carcinogenic in animals	....
743	*N*-Nitrosodiethanolamine	1116547	....	....	....	....	....	....	....	Carcinogenic	....
744	*N*-Nitrosodimethylamine .... (Dimethylnitrosoamine) C₂H₆NO	62759	Because of extremely high toxicity and presumed carcinogenic potential of this compound, contact by any route should not be permitted.		....						
745	*N*-Nitroso-*N*-methylaniline ..... C₇H₈N₂O	614006	....	....	....	....	....	....	....	Carcinogenic	....
746	4-Nitrosomorpholine	59892	....	....	....	....	....	....	....	Carcinogenic in animals	....
747	1-Nitrosopiperazine	5632473	....	....	....	....	....	....	....	Carcinogenic	....
748	*N*-Nitrosopiperidine	10546346	....	....	....	....	....	....	....	Carcinogenic in animals	....
749	Nitrosyl chloride............. NOCl	2696926	....	....	5x	5x	....	5x	....		MGD
750	*m*-Nitrotoluene .... (3-Nitrotoluol) C₇H₇NO₂	99081	5 S (30) S	Toxic	....	....	....	....	....		2109, STTE
751	*o*-Nitrotoluene .... (2-Nitrotoluol) C₇H₇NO₂	88722	5 S (30) S	Toxic	....	....	....	....	....		2149, STTE
752	*p*-Nitrotoluene .... C₇H₇NO₂	99990	5 S (30) S	Toxic	....	....	....	....	....		2109, STTE
—	3-Nitrotoluol See *m*-Nitrotoluene										
—	2-Nitrotoluol See *o*-Nitrotoluene										
753	Norbornylene	498668	....	....	3	5	2	1y	1		....
754	Octachloronaphthalene .... C₁₀H₈Cl₈		(0.1) S	Toxic	....	....	....	....	....		AIHA
755	Octafluoro-2-butene........... (Perfluoro-2-butene) C₄F₈	360894	....	....	Possibly moderately toxic						MGD
756	Octafluorocyclobutane........ (Perfluorocyclobutane; Freon®-C318) C₄F₈	115253	....	....	....	....	....	....	....		MGD
757	Octafluoropropane .... (Perfluoropropane) C₃F₈	76197	....	....	....	....	....	....	....		MGD
758	*n*-Octane .... C₈H₁₈	111659	500 (2350)	Narcosis	3x	2x	1x	1x	1x		1197, STTE
759	Octanoic acid .... (Caprylic acid) C₈H₁₆O₂	124072	....	....	....	....	....	....	....	Mild irritation	1786, STTE
760	2-Octanol .... (Capryl alcohol; Methyl-hexylcarbinol) C₈H₁₈O	123966	....	....	....	....	....	....	....		2220, STTE
761	Octyl alcohol .... (Caprylic alcohol; 1-Octanol) C₈H₁₈O	111875	....	....	2	1	2	2	2		1461
762	Oil, cedar leaf.............		....	....	3x	2x	2x	2x	3x		....
763	Oil, coconut		....	....	....	....	....	....	....		....
764	Oil, lavendar		....	....	3x	2x	2x	2x	3x		....
765	Oil, mineral oil mist		(5)	Nuisance	....	....	....	....	....		....
766	Oil, olive		....	....	....	....	....	....	....		....

No.	(7) Health	Fire	Reactivity	(8) Extinguishing Agents	(9) Lower	Upper	(10) Flash Point °C (°F)	(11) Ignition Temperature °C (°F)	(12) Boiling Point (Melting Point, m.p.) °C	(13) Vapor Pressure mm. Hg °C	(14) Density or Specific Gravity	(15) Water Solubility	No.
742	....	....	....				....	....	....	....	....	....	742
743	....	....	....				....	....	....	....	....	....	743
744	....	....	....				....	....	....	....	....	....	744
745	....	....	....				....	....	225 d	....	1.1288	i	745
746	....	....	....				....	....	....	....	....		746
747	....	....	....				....	....	....	....	....	....	747
748	....	....	....				....	....	....	....	....	....	748
749	....	....	....				....	....	-5.8	....	Gas 3.0 g/l	d	749
750	....	....	....				....	....	232.6 m.p. 15	1.0^{60}	1.1571^{20}_{4}	i	750
751	....	....	....				....	....	220.4 m.p. 2.9	1.6^{60}	1.1629^{20}	i	751
752	....	1	3	1,3	....	....	106 (223)	....	238.3 m.p. 51.7	1.3^{65}	1.299	i	752
—													—
753	....	....	....				....	....	....	....	....	i	753
754	....	....	....				....	....	m.p. 185	....	2.00	......	754
755	....	....	....				....	....	1.2	....	liq. 1.5297	......	755
756	....	....	....				....	....	-6.04	....	liq. 1.513	......	756
757	....	....	....				....	....	-36.7	....	liq. 1.293	......	757
758	0	3	0	2,3	1.0	4.66	13 (56)	220 (428)	125–6	10.45^{20}	0.7025^{20}_{4}	i	758
759	....	....	....		Lower	Upper	132 OC (270)OC	....	110	1^{92}	0.8615^{80}	i	759
760	....	....	....	2a	....	....	66 (150)	....	178–9^{304}	40^{98}	0.8232^{20}_{20}	δ	760
761	1	2	0	1,2,3	....	....	81 (178)	....	194–5	....	0.8270	i	761
762	....	....	....				....	....	....	....	....	......	762
763	0	1	0	1,2,3	....	....	..216... (420)	....	....	....	0.9^{0}_{4}	i	763
764	....	....	....				....	....	....	....	....	......	764
765	0	1	0	1,3	....	....	193 OC (380)OC	....	360	....	0.809	i	765
766	0	1	0	1,2,3	....	....	225 (437)	343 (650)	....	....	0.9	i	766

797

No.	Chemical Name and Formula	(2) Chemical Abstracts Registry Number	(3) A.C.G.I.H. 1966–TLV (Threshold Limit Values) ppm (mg/M^3)	(4) Principal Effects of Inhalation Exposures above TLV	(5) Relative Hazard to Health from Concentrated Short-term Exposure — Eye Contact	Inhalation	Skin Penetration	Skin Irritation	Ingestion	Supplemental Effects	(6) References to Supplementary Information on Toxicity, Flammability and Other Hazards
					(5 is high; 1 is low)						
767	Oil, peanut										
768	Oil, soybean				1x		1x	1x	1x		
769	Oleic acid $C_{18}H_{34}O_2$	112801			1	1	1x	1	1		1791
—	Oleum See Sulfuric acid, fuming										
—	Osmic acid See Osmium tetroxide										
770	Osmium				Powdered or spongy metal slowly gives off toxic osmium tetroxide when exposed to air						1130
771	Osmium tetroxide OsO_4		(0.002)	Irritant; Toxic	Extremely irritating and toxic; Severe eye injury hazard						1130
772	Oxalic acid $C_2H_2O_4$		(1) T	Toxic	3x		2x	3x	3		1773, NSC, STTE
—	Oxirane See Ethylene Oxide										
—	Oxalic chloride See Oxalyl chloride										
773	Oxalyl chloride (Oxalic chloride) $C_2Cl_2O_2$				Severe burns; Inhalation hazard						1826
774	Oxygen O_2					2x					911, NSC, MGD, STTE
775	Oxygen difluoride OF_2		0.05 (0.1)	Toxic							831
776	Ozone O_3		0.1 (0.2)	Toxic	Very irritating; Lethal in a few minutes in concentrations over 1700 ppm						915, AIHA, MGD, STTE. Ozone should be used only in equipment cleaned for oxygen service
777	Paraffin				1	1y	1x	1	1		STTE
778	Paraldehyde $C_6H_{12}O_3$	123637			4	2	2	2	2		1966, HCD, STTE
779	Paraquat		(0.5) S	Toxic							
780	Parathion $C_{10}H_{14}NO_5PS$	56382	(0.1) S	Toxic							1949, AIHA, HCD
781	Pentaborane B_5H_9	1303975	0.005 (0.01)	Toxic							AIHA, MCA, STTE. Forms shock sensitive mixtures with halogenated materials
782	Pentachloronaphthalene $C_{10}H_3Cl_5$	1321648	(0.5) S	Toxic	1x	3z*	3x*	3x*	3	Acne-like eruption; Chronic toxicity	1343, AIHA, STTE
783	Pentachlorophenol C_6HCl_5O	87865	(0.5) S	Toxic	2x			3x*		Acne-like eruption; Chronic toxicity	1396, AIHA, STTE
—	1,3-Pentadiene See Piperylene										
784	Pentane C_5H_{12}	109660	1000 (2950)	Narcosis	May be narcotic in high concentrations						1197, AIHA, STTE
785	1,5-Pentanediol $C_5H_{12}O_2$	111295			1	1	1	1	2		STTE

No.	(7) N.F.P.A. Hazard Identification Signals			(8) Extinguishing Agents	(9) Flammable Limits in Air % by volume		(10) Flash Point °C (°F)	(11) Ignition Temperature °C (°F)	(12) Boiling Point (Melting Point, m.p.) °C	(13) Vapor Pressure mm. Hg °C	(14) Density or Specific Gravity	(15) Water Solubility	No.
	Health	Fire	Reactivity		Lower	Upper							
	(4 is high; 0 is low)												
767	0	1	0	1,2,3	....	....	282 (540)	445 (833)	....	....	0.9	i	767
768	0	1	0	1,2,3	....	....	282 (540)	445 (833)	....	....	0.9	i	768
769	0	1	0	1,2,3	....	....	189 (372)	363 (685)	286^{100}	....	0.895	i	769
—													—
—													—
770	.....	.....	.....	....			....	....	>5300 m.p. 7700	....	22.48	i	770
771	.....	.....	.....	....			None	None	130	....	4.906	s	771
772	.....	.....	.....	....					157 Sub.	....	1.90	s	772
—													—
—													—
773	.....	.....	.....	....					63–4		1.488_4^{13}	d	773
774	.....	.....	.....	....					182.96	42.2 atm. (−123)	1.429 g/l	δ	774
775	.....	.....	.....	....					−144.8	....	1.90	δ d	775
776	.....			....	Potentially explosive; Strong oxidant		....	....	−111.9	....	2.144 g/l	s	776
777	.....	.....	.....	1,3	....	....	199 (390)	245 (473)	m.p. 42–60	....	3.880 −0.915	i	777
778	2	3	1	2a,3	1.3	....	36 OC (96) OC	238 (460)	128.0	....	0.9923	v: s [h]	778
779	.....	.....	.....	....	....	....	....	....				....	779
780	4	1	0	....	....	....	....	....	375	$0.00003^{24.0}$	1.26	i	780
781	.....	.....	.....	....	0.42	....	≈ 30 ≈ (86) if pure	Ignites spontaneously in air if impure; ≈ 35(95) if pure	58.4 m.p. −46.8	60^0	0.66	d	781
782	.....	.....	.....	....	....	....	....	....	....	....	....	i	782
783	.....	.....	.....	....			....	....	$309-10^{754}$	0.00011^{20}	1.978	δ	783
—													—
784	1	4	0	2,3,4	1.4	8.0	−49 (−56)	309 (588)	36^{273}	500^{24-34}	0.6262_4^{20}	v	784
785	1	1	0	2a	....	....	129 OC (265) OC	....	260	....	0.9939_4^{25}	s	785

No.	Chemical Name and Formula	Chemical Abstracts Registry Number	A.C.G.I.H. 1966–TLV (Threshold Limit Values) ppm (mg/M³)	Principal Effects of Inhalation Exposures above TLV	Eye Contact	Inhalation	Skin Penetration	Skin Irritation	Ingestion	Supplemental Effects	References to Supplementary Information on Toxicity, Flammability and Other Hazards
					(5 is high; 1 is low)						
786	2,4-Pentanedione (Acetylacetone) $C_5H_8O_2$	123546	....	....	2	3*	2	1z*	3	Sensitiz. skin, resp. tract	....
787	2-Pentanol $C_5H_{12}O$	123524	100 (360)	Narcosis	....	....	....	....	....	....	1452
788	3-Pentanol $C_5H_{12}O$	584021	100 (360)	Narcosis	2	2	2	1	2	....	1452, STTE
789	2-Pentanone (Methyl propyl ketone) $C_5H_{10}O$	107879	200 (700)	Irritant	2	3	2	1	2	....	1452, STTE
790	3-Pentanone (Diethyl ketone) $C_5H_{10}O$	96220	....	....	2	3	1	1	2	....	STTE
791	Pentyl acetate (n-Amyl acetate) $C_7H_{14}O_2$	628637	100 (525)	Irritant	1	3	1	2	2	....	1858, NSC, STTE
792	sec-Pentyl acetate (sec-Amyl acetate) $C_{10}H_{14}O_2$	....	125 T (650) T	Irritant	....	....	....	....	....	....	1858, STTE
793	Pentyl alcohol (Amyl alcohol; 1-Pentanol) $C_5H_{12}O$	71410	100 (360)	Narcosis	4	1	2	1	2	....	1452, STTE
794	tert-Pentyl alcohol (tert-Amyl alcohol) $C_5H_{12}O$	75854	100 (360)	Narcosis	....	....	....	....	....	....	1451
795	Pentyl amine (n-Amylamine) $C_5H_{13}N$	110587	....	....							2038
796	Pentyl ether (n-Amyl ether) $C_{10}H_{22}O$	....	....	....							....
797	Peracetic acid (60% Acetic acid solution) $C_2H_4O_3$	79210	....	....					....	Irritant; Limit exposure to 10 ppm	HCD
798	Perchloric acid 70–72% $HClO_4$	7601903	....	....	5x	4x	2x	4x	4x	....	MCA, NSC, HCD
—	Perchloroethylene See Tetrachloroethylene										
—	Perchloromethyl mercaptan See Trichloromethanethiol										
799	Perchloryl fluoride ClO_3F	7616946	3 (13.5)	Irritant	Avoid inhalation, ingestion, and skin contact. Take special measures in cases of leak					....	....
—	Perfluoro-2-butene See Octafluoro-2-butene										
—	Perfluorocyclobutane See Octafluorocyclobutane										
—	Perfluoropropane See Octafluoropropane										
800	Petroleum Ether Boiling range 30–60°C	....	....	....	....	....	....	....	....	May cause narcosis. Limit exposures to 500 ppm.	1195

No.	Health	Fire	Reactivity	Extinguishing Agents	Lower	Upper	Flash Point °C (°F)	Ignition Temperature °C (°F)	Boiling Point (Melting Point, m.p.) °C	Vapor Pressure mm. Hg °C	Density or Specific Gravity	Water Solubility	No.
786	2	2	0	2a			41 OC (105) OC		139⁷⁴⁶		0.9721²⁵₄	v	786
787	1	2	0	2,2a,3	1.2	9.0	39 (103)	347 (657)	118.5–9.5		0.8103	v	787
788	1	2	0	2a	1.2	9.0	41 (105)		115.5⁷⁵⁴	2²⁰	0.8154¹⁵₁₅	δ	788
789					1.55	8.15	7.2 (45)		102	16²⁵	0.8124¹⁵₅	δ	789
790	1	3	0	2a			13 OC (55) OC	452 (846)	102.7		0.8159	v	790
791	1	3	0		1.1	7.5	25 (77)	379 (714)	148.8	5²⁵	0.8756	δ	791
792	1	3	0	2,2a,3	1.12	7.5	25 (77)		121	9²⁵	0.862	δ	792
793	1	3	0	2a	1.2	10.0 @ 212°	33 (91)	300 (572)	137.3⁷⁴⁸		0.8110	i	793
794					1.2	9.0	19 (67)	437 (819)			0.809		794
795	3	3	0	2a			7.2 OC (45) OC		103		7.614	∞	795
796	1	2	0	1,2a,3			57 OC (135) OC	171 (340)	190		0.744	i	796
797	3	2	4		Flammable liquid; shock and heat sensitive		41 (105)		105 expl. 110		1.226	v	797
798	3	0	3		Extremely, strong, active oxidizing agent and dehydrating agent at elevated temperatures (about 160°C). Vapors may form explosive perchlorates in contact with materials.				200		1.68	∞	798
799									Powerful oxidizing agent		liq. 1.434	δ	799
800	1	4	0	3	~1	–8	–57 (–70)		30–60				800

No.	(1) Chemical Name and Formula	(2) Chemical Abstracts Registry Number	(3) A.C.G.I.H. 1966–TLV (Threshold Limit Values) ppm (mg/M³)	(4) Principal Effects of Inhalation Exposures above TLV	(5) Relative Hazard to Health from Concentrated Short-term Exposure (5 is high; 1 is low)						(6) References to Supplementary Information on Toxicity, Flammability and Other Hazards
					Eye Contact	Inhalation	Skin Penetration	Skin Irritation	Ingestion	Supplemental Effects	
801	Petroleum ether (commercial hexane) .... Boiling range 60–70°C	....	500 (1800)	Narcosis	....	....	....	....	....	May cause narcosis.	1195, AIHA, STTE
802	Phenanthrene .... See also Coal tar pitch volatiles C₁₄H₁₀	85018	....	....	2y	1	2y	2y	2	....	Not carcinogenic in pure form
803	2-Phenanthreneacetamide	10546368	....	....	....	....	....	....	....	Carcinogenic in animals	....
804	3-Phenanthreneacetamide	10546379	....	....	....	....	....	....	....	Carcinogenic in animals	....
805	Phenol.................. (Carbolic acid) C₆H₆O	108952	5 S (19) S	Toxic	4	1	3	4y	2	....	1364, AIHA, MCA, NSC, HCD, STTE. Human fatality reported
806	Phenyl acetate .... C₈H₈O₂	122792	....	....	1	1	2	1	2	....	....
807	1-Phenylazo-2-naphthol........ C₁₆H₁₂N₂O Phenylcyclohexane See Cyclohexylbenzene	842079	....	....	....	....	....	....	....	Carcinogenic in animals	....
808	p-Phenylenediamine .... C₆H₈N₂	106503	(0.1) S	Irritant	....	....	....	....	....	....	2151, STTE
809	Phenyl ether .... C₁₂H₁₀O	101848	1 T (7) T	Irritant	1x	1x	....	2x	....	....	1698
810	Phenyl ether-biphenyl mixture... (Dowtherm A)	....	1 T (7) T	Irritant	....	....	....	....	....	....	1701
811	Phenyl isocyanate............ C₇H₅NO	103719	....	....	....	....	....	....	....	May deteriorate in normal storage and cause hazard	....
—	Phenylethylene See Styrene										
—	Phenyl glycidyl ether See 1,2-Epoxy-3-phenoxypropane										
812	Phenylhydrazine............. C₆H₈N₂	100630	5 S (22) S	Toxic	4x	....	....	4x	....	Eye burn w/o pain; sensitiz. skin, resp. tract	2229, STTE. May deteriorate in normal storage and cause hazard
813	Phenyl naphthylamine .... C₁₆H₁₃N	....	....	....	2x	....	2x	3*	2	Sensitiz. skin, resp. tract	....
814	o-Phenylphenol.............. C₁₂H₁₀O	90437	....	....	....	....	....	....	....	....	1407
815	Phosdrin® .... C₇H₁₃O₆P	298011	(0.1) S	Toxic	....	....	....	....	....	....	1951, STTE
816	Phosgene................... (Carbonyl chloride) CCl₂O	75445	0.1 (0.4)	Toxic	Highly toxic					....	938, AIHA, MGD, STTE
817	Phosphine .... PH₃	7803512	0.3 (0.4)	Toxic	Highly toxic					....	883, AIHA, MGD, STTE

No.	(7) N.F.P.A. Hazard Identification Signals			(8) Extinguishing Agents	(9) Flammable Limits in Air % by volume		(10) Flash Point °C (°F)	(11) Ignition Temperature °C (°F)	(12) Boiling Point (Melting Point, m.p.) °C	(13) Vapor Pressure mm. Hg °C	(14) Density or Specific Gravity	(15) Water Solubility	No.
	Health	Fire	Reactivity		Lower	Upper							
	(4 is high; 0 is low)												
801	1	4	0	3	~1	−8	−32 (−25)	....	60–70	....	....	....	801
—									Petroleum is defined in F. A. Patty's "Industrial Hygiene and Toxicology," Vol. II, p. 1195, as having a boiling range of 20–60°C and principal paraffin components C_4 to C_6; petroleum benzin as having a boiling range of 40–90°C and principal paraffin components C_5 to C_7; petroleum naphtha as having a boiling range of 65–120°C and principal paraffin components C_6 to C_8.				—
802	....	.....	.....	1,2,3	....		....	....	340	....	1.182	i	802
803	....	.....	.....	....	....		....	....	....	....	....	....	803
804	....	.....	.....	....	....		....	....	....	....	....	....	804
805	3	2	0	1,2a,3	....		79 (175)	715 (1319)	182	0.3513 (25)	1.0722^{20}_{20}	s: ∞ [h]	805
806	1	2	0	1,2a,3	....		80 (176)	....	195.7	....	1.0927^{25}_{25}	δ	806
807	....	.....	.....	....	....		....	m.p. 102–104	....	....	....	....	807
—													—
808	....	.....	.....	....	....		....	....	267	....	....	s [h]	808
809	....	.....	.....	....	....		96 OC (205) OC	646 (1195)	258–9	0.0213^{25}	1.0863	i	809
810	....	.....	.....	....	....		124 OC (255) OC	610 (1130)	257.4	0.08^{25}	1.06	i	810
811	....	.....	.....	....	....		....	....	165		1.095	d	811
—													—
—													—
812	3	2	0	1,2a,3	....		89 (192)	174 (345)	243	....	1.099^{20}_{4}	s [h]	812
813	....	.....	.....	....	....		....	....	335^{260}	....	....	i	813
814	....	1	0	1,2a,3	....		124 (255)	....	286	....	1.213	i	814
815	....	.....	.....	....	....		79 OC (175) OC	....	210–$218^{0.03}$	0.0029^{21}	1.23	∞	815
816	....	.....	.....	....	....		Gas	....	8.02	568^0	1.392	d	816
817	....	.....	.....	....	>1	....	Gas	40–60 (104–140) May ignite at room temperature	−87.4	....	liq. 0.796	δ	817

No.	Chemical Name and Formula	Chemical Abstracts Registry Number	A.C.G.I.H. 1966-TLV (Threshold Limit Values) ppm (mg/M³)	Principal Effects of Inhalation Exposures above TLV	Eye Contact	Inhalation	Skin Penetration	Skin Irritation	Ingestion	Supplemental Effects	References to Supplementary Information on Toxicity, Flammability and Other Hazards
					(5 is high; 1 is low)						
818	Phosphoric acid .... H_3PO_4	7664382	(1)	Irritant	4	....	2x	4	3x	....	2280, AIHA, MCA, STTE
—	Phosphoric anhydride See Phosphorus pentoxide										
819	Phosphorus, red .... P_4	10544464	....	....	5x	5z*	4x*	5x	4x	Chronic toxicity	880, HCD, MCA, NSC, STTE
820	Phosphorus, white or yellow .... P_1	10544464	(0.1)	Toxic	5x	5z*	4x*	5x	4x	Chronic toxicity	880, HCD, STTE
821	Phosphorus pentachloride ...... PCl_5	10026138	(1)	Irritant	5x	4x	3x	4x	3x		885, STTE
822	Phosphorus pentafluoride....... PF_5	764190	....	....	Respiratory irritant					....	MGD
823	Phosphorus pentasulfide........ P_2S_5	1314803	(1)	Irritant	Reacts with water to evolve hydrogen sulfide						MCA, HCD, STTE
824	Phosphorus pentoxide (Phosphoric anhydride) P_2O_5	1314563	....	....	Corrosive and irritant						AIHA
825	Phosphorus tribromide........ PBr_3	7789608	....	....	Toxic and corrosive					....	....
826	Phosphorus trichloride........ PCl_3	7719122	0.5 (3)	Irritant	....	....	....	....	....		885, MCA, HCD
827	Phthalic anhydride .... $C_8H_4O_3$	85449	2 (12)	Irritant	4y	....	2x	3x*	2x	Sensitized skin, resp. tract	1822, MCA, HCD, STTE
—	α-Picoline See 2-Picoline										
828	2-Picoline .... (α-Picoline; 2-Methylpyridine) C_6H_7N	109068	....		4	3	3	2	2		2188, HCD, STTE
829	4-Picoline .... (4-Methylpyridine) C_6H_7N	108894	....		4	3	3	3			STTE
830	Picric acid................. (2,4,6-Trinitrophenol) $C_6H_3N_3O_7$	88891	(0.1)S	Toxic	....	....	....	....		Toxic and irritant	2157, NSC, HCD, STTE
831	Pimelic acid................ $C_7H_{12}O_4$	111160			4		2	2y	2		1811, STTE
832	2-Pinene .... (α-Pinene) $C_{10}H_{16}$	80568			Irritates skin and mucous membranes; May cause skin eruption					....	....
833	Piperidine................. $C_5H_{11}N$	110894	....		4	3	3	4	3		2196
834	Piperylene................. (1,3-Pentadiene) C_5H_8	504609			4x	2x	2x	2x	2x		
835	*Pival* .... (2-Pivaloyl-1,3-indandion) $C_{14}H_{14}O_3$	83261	(0.1)T	Toxic	....	....	....	....	....		STTE
836	Platinum—soluble salts........	....	(0.002)	Toxic	....	....	....	....	....		1125
837	Plutonium .... Pu	7440075			....	....	....	....	....	May be ab-absorbed through inhalation; Source of ionizing radiation	STTE
838	Polonium.................. Po	7440086	....	....	Very hazardous. An alpha emitter. Possible lung carcinoma from inhalation					....	AIHA

No.	(7) N.F.P.A. Hazard Identification Signals			(8) Extin-guishing Agents	(9) Flammable Limits in Air % by volume		(10) Flash Point °C (°F)	(11) Ignition Tem-perature °C (°F)	(12) Boiling Point (Melting Point, m.p.) °C	(13) Vapor Pressure mm. Hg °C	(14) Density or Specific Gravity	(15) Water Solu-bility	No.
	Health	Fire	Reac-tivity		Lower	Upper							
	(4 is high; 0 is low)												
818	.....	.....	.....	....	....	....	....	....	$-\frac{1}{2}H_2O$; 213	0.0285^{20}	1.834^{18}	v	818
—													—
819	0	1	1	1	....	....	....	<200 (<392)	ign. 200 280	....	2.34	δ	819
820	3	3	1	1	Ignites spontaneously in air.		....	30 (86)	280.5	....	1.82^{20}	δ	820
821	.....	.....	.....	....	....	....	....	....	sub. 162	....	4.65 g/l	d	821
822	.....	.....	.....	....	....	....	....	....	−84.6	....	....	....	822
823	3	1 W	2	3	Dust is flammable		....	282 (540)	514	....	2.03	i	823
824	.....	.....	.....	....	....	....	....	....	Sub 300	....	2.39	d to H_3PO_4	824
825	.....	.....	.....	....	....	....	....	....	172.9	....	2.85^{20}	d	825
826	3	0 W	2	....	....	....	....	....	75.5^{749}	....	1.574	d	826
827	2	1	0	1,3	1.7	10.5	152 (305)	584 (1083)	Sub 284.5	....	1.527	δ	827
—													—
828	2	2	0	3	....	....	39 OC (102)	538 (1000)	128.8	....	0.9497	v	828
829	.....	.....	.....	....	....	....	....	....	143.1	....	0.9571	∞	829
830	2	4	4	1	High explosive		....	<300 (<572)	m.p. 122–3	....	....	δ: s[h]	830
831	.....	.....	.....	....	....	....	....	....	Sub 272^{100}	....	1.329	δ	831
832	1	3	0	2,3	....	....	33 (91)	....	156.2	....	0.8582	δ	832
833	2	3	3	2a	....	....	16 (61)	....	106.0	$40^{29.2}$	0.8606_4^{20}	∞	833
834	.....	.....	.....	....	....	....	....	....	418	....	0.6830	i	834
835	.....	.....	.....	....	....	....	....	....	....	....	....	....	835
836	.....	.....	.....	....	....	....	....	....	....	....	....	....	836
837	.....	.....	.....	....	....	....	....	....	....	....	....	....	837
838	.....	.....	.....	....	....	....	....	....	962 ± 2	....	9.4	δ	838

No.	Chemical Name and Formula	Chemical Abstracts Registry Number	A.C.G.I.H. 1966–TLV (Threshold Limit Values) ppm (mg/M³)	Principal Effects of Inhalation Exposures above TLV	Relative Hazard to Health from Concentrated Short-term Exposure (5 is high; 1 is low)						References to Supplementary Information on Toxicity, Flammability and Other Hazards
					Eye Contact	Inhalation	Skin Penetration	Skin Irritation	Ingestion	Supplemental Effects	
839	Polytetrafluoroethylene decomposition products ...... (Teflon decomp. products)	....	Note concerning TLV: Thermal decomposition of the fluorocarbon chain in air leads to the formation of oxidized products containing carbon, fluorine and oxygen. Because these products decompose by hydrolysis in alkaline solution, they can be quantitatively determined in air as fluoride to provide an index of exposure. No TLV is recommended pending determination of the toxicity of the products, but air concentrations should be minimal.								AIHA, STTE
840	Potassium K	7440097									864, HCD, STTE. Potassium stored under oil, but open to the atmosphere, picks up oxygen and forms KO_2 which is explosive with oil (ACS Monograph)
841	Potassium acetate $KC_2H_3O_2$	127082			1y	....	2	1y	2		
842	Potassium carbonate K_2CO_3	298146			3*	....	1x	2y	2	Irritation, eye, nose, throat	
843	Potassium chlorate $KClO_3$	3811049									HCD
844	Potassium dichromate $K_2Cr_2O_7$	7778509			Corrosive poison						MCA, STTE
845	Potassium fluoride KF	7789233			Irritating to skin, eyes, and mucous membranes						
846	Potassium hydroxide KOH	1310583			5	....	1x	5	2		864, MCA, HCD
847	Potassium nitrate KNO_3	7757791									HCD
848	Potassium permanganate $KMnO_4$	7722647			3	....	1x	3y	2		1081, HCD, STTE
849	Potassium peroxide K_2O_2	1336227			Toxic						HCD
850	Potassium persulfate $K_2S_2O_8$	7727211			3x	....	1x	2x	2x		HCD
851	Potassium sulfide K_2S	1336238			Forms H_2S in presence of water or acids						HCD
852	Praseodymium Pr	7440100									1058, STTE
853	Promethium Pm	7440122			Soft beta emitter						1058
854	Propane C_3H_8	74986	1000 (1800)	Narcosis	....	3x	....	....	....	Simple asphyxiant	1197, MGD, STTE
855	1,2-Propanediamine (Propylenediamine)	78900			4	1*	3	4*	2	Sensitiz. skin, resp. tract	2039
856	1,2-Propanediol (Propylene glycol) C_4H_8O	57556			1	1	1	1	1		1515, STTE
—	2-Propane See Acetone										
—	Propanol See Propyl alcohol										
—	Propargyl alcohol See Propyn-2-ol										
—	Propargyl bromide See 3-Bromopropyne										

No.	(7) Health	Fire	Reactivity	(8) Extinguishing Agents	(9) Lower	Upper	(10) Flash Point °C (°F)	(11) Ignition Temperature °C (°F)	(12) Boiling Point (Melting Point, m.p.) °C	(13) Vapor Pressure mm. Hg °C	(14) Density or Specific Gravity	(15) Water Solubility	No.
					(4 is high; 0 is low)								
839	.....	.....	.....	....	....		....	....	....	....	....	....	839
840	3	1 W	2	5	Contact with water releases hydrogen with sufficient heat to cause ignition or explosion. May ignite spontaneously in air.				774 m.p. 63.65	8^{432}	0.86	d to KOH	840
841	.....	.....	.....	....	....		....	....	m.p. 292	....	1.57	v	841
842	.....	.....	.....	....	....		....	....	m.p. 891	....	2.428	v	842
843	1	0	2	....	Forms explosive mixtures with oxidizable materials		....	....	400	....	2.32	δ	843
844	.....	.....	.....	....	....		....	....	d 500	....	2.676	δ	844
845	.....	.....	.....	....	....		....	....	1505	....	248	v	845
846	3	0	1	....	....		....	....	1320–44	1^{719}	2.044	v	846
847	1	0	2	....	....		....	....	d 400	....	2.109^{16}	s	847
848	0	0	1	....	....		....	....	m.p. d < 240	....	2.703	δ	848
849	3	0 W	2	....	....		Reacts vigorously with water forms explosive mixtures		d	....	....	....	849
850	1	0	1	....	....		....	....	m.p. d < 100	....	2.477	δ	850
851	2	1	0	....	Combustible in air. Dust explosive.		....	....	m.p. 470	....	....	s	851
852	.....	.....	.....	....	Pyrophoric on filing		....	....	3127 m.p. 935	....	6.64 to 6.782	d	852
853	.....	.....	.....	....	....		....	....	No stable isotopes	....	....	....	853
854	1	4	0	4	2.2	9.5	Gas	466 (871)	−44.5	8.8^{20}	0.5834^{25}_{4}	s	854
855	2	3	0	1,2a,3	....	....	33 OC (92) OC	....	117.35	8.0^{20}	0.8584	v	855
856	0	1	0	1,2a,3	2.6	12.5	99 (210)	421 (780)	189	0.19^{25}	1.0361	∞	856

No.	(1) Chemical Name and Formula	(2) Chemical Abstracts Registry Number	(3) A.C.G.I.H. 1966–TLV (Threshold Limit Values) ppm (mg/M³)	(4) Principal Effects of Inhalation Exposures above TLV	(5) Relative Hazard to Health from Concentrated Short-term Exposure (5 is high; 1 is low)						(6) References to Supplementary Information on Toxicity, Flammability and Other Hazards
					Eye Contact	Inhalation	Skin Penetration	Skin Irritation	Ingestion	Supplemental Effects	
857	Propene . . . . (Propylene) C_3H_6	115071	. . . .	. . . .	4000 ppm limit suggested					Simple asphyxiant; anesthetic	1204, HCD, MGD
—	β-Propiolactone See Hydracrylic acid, β-lactone										
858	Propionaldehyde . . . . C_3H_6O	123386			2	4*	2	1*	2	Sensitiz. skin, resp. tract	1966, HCD, STTE
859	Propionic acid . . . . $C_3H_6O_2$	79094	. . . .		4	1	3	4	2	. . . .	1779, HCD, STTE
860	Propionic anhydride . . . . $C_6H_{10}O_3$	123626			4	2	2	3	2		1818, STTE
861	Propionitrile . . . . C_3H_5N	107120			2	4	3	1	4		2016, STTE
862	Propionyl chloride . . . . C_3H_5ClO	79038			. . . .						1826
863	Propyl acetate . . . . (n-Propyl acetate) $C_5H_{10}O_2$	109604	200 (840)	Narcosis	1	3	1	1	2	. . . .	1858, STTE
864	Propyl alcohol . . . . (n-Propyl alcohol; Propanol) C_3H_8O	71238	200 (450)	Narcosis	2	2	2	1	2		1433, AIHA, STTE
865	Propylamine . . . . C_3H_9N	107108	. . . .		4	5	3	4	3		2038, HCD, STTE
866	Propylbenzene . . . . C_9H_{12}	103651			. . . .					. . . .	. . . .
—	n-Propyl chloride See 1-Chloropropane										
—	Propylene See Propene										
867	Propylene carbonate . . . . $C_4H_6O_3$	108327	. . . .		2	1	1	1	1	. . . .	1912
—	Propylene diamine See 1,2-Propanediamine										
—	Propylene dichloride See 1,2-Dichloropropane										
868	Propylene disulfate	10546448	. . . .		. . . .				. . . .	Carcinogenic in animals	. . . .
—	Propylene glycol See 1,2-Propanediol										
—	Propylenimine See 2-Methylaziridine										
869	Propylene oxide . . . . C_3H_6O	75569·	100 (240)	Toxic	2	5*	3	1*	2	Senitiz. skin, resp. tract	1643, AIHA, HCD, STTE
870	Propyl formate . . . . (n-Propyl formate) $C_4H_8O_2$	110747	. . . .		. . . .					. . . .	1854
871	n-Propyl nitrate . . . . $C_3H_7NO_3$	627134	25 (110)	Toxic	. . . .						2090, HCD, STTE
872	Propyne . . . . C_3H_4	74997	1000 (1650)	Narcosis	. . . .	3x	. . . .	. . . .	. . . .	Simple asphyxiant	MCA, HCD
873	Propyne-allene mixture . . . . (Methylacetylene-propadiene mixture; MAPP) 58% propyne and allene, with balance of mixture of paraffinic and olefinic C_3 and C_4 hydrocarbons		1000 (1650)								Forms explosive compounds with Cu and Ag
874	l-2-Propyn-1-ol . . . . (Proparyl alcohol) C_3H_4O	1071197	. . . .		Irritating to skin and mucous membranes					. . . .	. . . .

No.	(7) N.F.P.A. Hazard Identification Signals (4 is high; 0 is low)			(8) Extinguishing Agents	(9) Flammable Limits in Air % by volume		(10) Flash Point °C (°F)	(11) Ignition Temperature °C (°F)	(12) Boiling Point (Melting Point, m.p.) °C	(13) Vapor Pressure mm. Hg °C	(14) Density or Specific Gravity	(15) Water Solubility	No.
	Health	Fire	Reactivity		Lower	Upper							
857	1	4	1	....	2.4	10.3	....	927	-47.7	....	liq. 0.5139	v	857
858	2	3	1	2a	2.9	17	-9 to -7 OC (*15* to *19*) OC	207 (*405*)	48.8	300^{25}	0.807_4^{20}	s	858
859	2	2	0	2a	....	....	54 (*130*)	513 (*955*)	141	10^{40}	0.994_{20}^{20}	∞	859
860	2	2	1	....	....	....	59 (*139*)	316 (*600*)	$168.1-4^{712}$	1^{20}	$1.010-1.015_{20}^{20}$	d	860
861	.....	.....	.....				....		97.2	40^{22}	0.7720_4^{20}	d	861
862	3	3^l	1	....	....	....	12 (*54*)	....	80	....	1.0646_4^{20}	d	862
863	1	3	0	2a	2.0	8	12 (*54*)	450 (*842*)	101.6	35^{25}	0.8884_{20}^{20}	δ	863
864	1	3	0	2a	2.5	13.5	22 (*73*)	404 (*760*)	97.1	20.8^{25}	0.7796_4^{20}	v	864
865	3	3	0	2a	2.0	10.4	-37 (*-35*)	318 (*604*)	49	$400^{31.5}$	0.719_{20}^{20}	s	865
866	.....	3	0	....			30 (*86*)	....	159.2		0.8620_4^{20}	i	866
867	1	1	0	1,2a,3	....	....	135 OC (*275*) OC	....	168	....	0.9435_4^{20}	δ	867
868	.....	.....	.....									....	868
869	2	4	2	2a,3,4	2.1	21.5	-37 (*-35*)	....	33.9	445^{20}	0.8304_{20}^{20}	s	869
870	2	3	.....	2,2a,3	....	....	-3 (*27*)	455 (*851*)	81.3	85^{25}	0.9006_4^{20}	δ	870
871	2	3	3	1,2a,2,3	2	100	20 (*68*)	117 (*350*)	110.5	16^{25}	1.0580	δ	871
872	2	4	2	4	1.7	....	Gas	....	-23.2	$744^{-23.5}$	0.7062	....	872
873	.....	.....	.....					....	....	....	....	....	873
874	.....	.....	.....					....	114-5	....	0.9628	s	874

No.	Chemical Name and Formula	Chemical Abstracts Registry Number	A.C.G.I.H. 1966–TLV (Threshold Limit Values) ppm (mg/M³)	Principal Effects of Inhalation Exposures above TLV	Relative Hazard to Health from Concentrated Short-term Exposure (5 is high; 1 is low)						References to Supplementary Information on Toxicity, Flammability and Other Hazards
					Eye Contact	Inhalation	Skin Penetration	Skin Irritation	Ingestion	Supplemental Effects	
875	Protactinium. Pa				Alpha emitter and potential carcinogen						
—	Pyrene See Coal tar pitch volatiles										
876	Pyrethrum		(5)	Toxic							STTE
877	Pyridine C_5H_5N	110861	5 (15)	Toxic	3	3	3x	2	2		2189, AIHA, NSC, HCD, STTE
—	Pyrogallic acid See Pyrogallol										
878	Pyrogallol (Pyrogallic acid) $C_6H_6O_3$	87661			Toxic; sensitizer						1385, STTE
—	Pryoxylin See Cellulose nitrate										
879	Pyrrolidine (Tetrahydropyrrole) C_4H_9N	123751									2179
880	Quinaldine (2-Methylquinoline) $C_{10}H_9N$	91634			4	1	3	1	2		
881	8-Quinolinol (8-Hydroxyquinoline) C_9H_7NO	148243								May be carcinogenic	Arch. Path. 79, 245, (1965)
—	Quinol See Hydroquinone										
882	Quinoline C_9H_7N	91225			4	1	3	2	3		2207
—	Quinone See p-Benzoquinone										
883	Radium Ra	7440144								May cause sarcoma, anemia & lung carcinoma	
884	Radon Rn	10043922									STTE, AIHA
885	Resorcinol $C_6H_6O_2$	108463									1378
886	Rhodium—metal fume & dusts		(0.1) T	Toxic							1125
887	Rhodium—sol. salts		(0.001) T	Toxic							1125
888	Ronnel $C_8H_8Cl_3O_3PS$	299843	(15)	Toxic							
889	Rotenone $C_{23}H_{22}O_6$	83794	(5)	Toxic							STTE
890	Rubidium Rb	7440177									HCD
891	Ruthenium Ru	7440188									1125, STTE
892	Salicylaldehyde (o-Hydroxybenzaldehyde) $C_7H_6O_2$	90028									1983
893	Salicylic acid $C_7H_6O_3$	69727			3x		2x	3z	2		1773
894	Samarium Sm	7440199									1058
895	Selenium Se	7782492								Toxic	887, AIHA, STTE
896	Selenium compounds Se		(0.2)	Toxic							887, AIHA
897	Selenium hexafluoride SeF_6		0.05 T (0.4) T	Toxic							

No.	(7) N.F.P.A. Hazard Identification Signals (4 is high; 0 is low)			(8) Extinguishing Agents	(9) Flammable Limits in Air % by volume		(10) Flash Point °C (°F)	(11) Ignition Temperature °C (°F)	(12) Boiling Point (Melting Point, m.p.) °C	(13) Vapor Pressure mm. Hg °C	(14) Density or Specific Gravity	(15) Water Solubility	No.
	Health	Fire	Reactivity		Lower	Upper							
875	.....	.....	.....	....	....		....	....	....	....	15.37	....	875
876	.....	.....	.....	....	....		....	....	....	....		....	876
877	2	3	0	1,3	1.8	12.4	8 (47)	543 (1010)	115.5	20^{25}	$.9819^{20}$	∞	877
878	.....	.....	.....	....	....		....	....	309; d 293	....	1.453	....	878
879	2	3	1	1,2a,3	....		3 (37)	....	88.5–9	128^{139}	0.8520	∞	879
880	.....	.....	.....	....	....		....	....	246.5	....	1.0585	δ	880
881	.....	.....	.....	....	....		....	....	267 m.p. 76	....	....	.....	881
882	2	1	0	3	....		....	480 (896)	237.10	$1^{59.7}$	1.09376_4^{25}	s	882
883	.....	.....	.....	....	....		....	....	1140 m.p. 700	....	6.0	s	883
884	.....	.....	.....	....	....		....	....	61.8	....	9.72	s	884
885	.....	.....	.....	1,3	1.4 @ 392°F		127 (261)	607.8 (1126)	281	....	1.2717^{15}	s	885
886	.....	.....	.....	....	....		....	....	....	....	....	.....	886
887	.....	.....	.....	....	....		....	....	....	....	....	.....	887
888	.....	.....	.....	....	....		....	....	....	....	....	.....	888
889	.....	.....	.....	....	....		....	....	$210-20^5$	....	....	i	889
890	.....	.....	.....	5	Ignites spontaneously in air and reacts violently with water		....	700 m.p. 38.5	....	....	1.532 liq. $1.475^{38.5}$	d	890
891	.....	.....	.....	....	....		....	....	4150 m.p. 2450	....	12.30	i	891
892	.....	2	0	2a	....		78 (172)	....	197	1^{33}	1.1461_{25}^{25}	....	892
893	.....	.....	.....	....	....		157 (315)	....	211^{760}	....	1.1443	δ:v[h]	893
894	.....	.....	.....	Pyrophoric on filing			....	~150	1900 m.p. 1072	....	7.536	i	894
895	.....	.....	.....	....	....		....	....	685 m.p. 217.4	....	4.81_4^{20}	i	895
896	.....	.....	.....	....	....		....	....	....	....	....	.....	896
897	.....	.....	.....	....	....		....	....	−34.5	....	3.25 g/l	δ d	897

No.	(1) Chemical Name and Formula	(2) Chemical Abstracts Registry Number	(3) A.C.G.I.H. 1966–TLV (Threshold Limit Values) ppm (mg/M³)	(4) Principal Effects of Inhalation Exposures above TLV	(5) Relative Hazard to Health from Concentrated Short-term Exposure						(6) References to Supplementary Information on Toxicity, Flammability and Other Hazards
					Eye Contact	Inhalation	Skin Penetration	Skin Irritation	Ingestion	Supplemental Effects (5 is high; 1 is low)	
898	Sevin® . . . . (Carbaryl) C₁₂H₁₁NO₂	63252	(5)	Toxic	. . . .	. . . .	. . . .	. . . .	. . . .		
899	Silane . . . . SiH₄	7803625	. . . .		. . . .	. . . .	. . . .	. . . .	. . . .		MGD
900	Silica (free) . . . . SiO₂	1309611	. . . .		. . . .	. . . .	. . . .	. . . .	. . . .		AIHA
901	Silica, quartz	. . . . . .	TLV calculated from formula on page 420		2y	4z*	1x	1	1	Prolonged contact may injure lung	. . . .
902	Silica, cristobalite	. . . . . .	TLV calculated from formula on page 420		2y	4z*	1x	1	1	Prolonged contact may injure lung	. . . .
903	Silica, amorphous . . . . (Incl. diatomaceous earth)	. . . . . .	706 p/cc		2y	4z*	1x	1	1	Prolonged contact may injure lung	AIHA
904	Silica gel	. . . . . .	. . . .		1x	2x	1x	1x	1		. . . .
911	Silicic acid . . . . H₂SO₃	7699414	. . . .		1x	2x	1x	1x	1x		. . . .
912	Silicon tetrafluoride . . . . SiF₄	7783611	. . . .		5x	2y	. . . .	2x	. . . .	Highly toxic and irritating	844, MGD
913	Silver—metal and soluble compounds . . . . Ag	7440224	(0.01)	Toxic	. . . .	. . . .	. . . .	. . . .	. . . .		. . . .
914	Silver nitrate . . . . AgNO₃	7761888	. . . .		5x*	. . . .	2x	4x*	4x		HCD, STTE
915	Sodium . . . . Na	7440235	. . . .		5x	. . . .	. . . .	4x	. . . .	Toxic fumes from fire	865, MCA, NSC, HCD. Small quantities may be disposed of by gradual addition to ethyl or isopropyl alcohol and reacted solution poured slowly into building drains
916	Sodium acetate . . . . NaC₂H₃O₂	127093	. . . .		2x	. . . .	1x	1x	2x		. . . .
917	Sodium amide . . . . NaNH₂	. . . . . .	. . . .		. . . .	. . . .	. . . .	. . . .	. . . .	Burns skin and eyes	. . . .
918	Sodium azide . . . . NaN₃	10102473	. . . .		Highly toxic to animals. May form toxic hydrazoic acid vapors					. . . .	2210
919	Sodium benzoate . . . . NaC₇H₅O₂	532321	. . . .		2y	. . . .	3	2y	2		. . . .
920	Sodium bicarbonate . . . . NaHCO₃	144558	. . . .		2x	. . . .	1x	1x	2x		STTE
921	Sodium bisulfate . . . . NaHSO₃	7681381	. . . .		4y	. . . .	1x	3x	3x		STTE
922	Sodium carbonate . . . . Na₂CO₃	497198	. . . .		3x	. . . .	1x	2x	2x	Primary skin irritant	866
923	Sodium chlorate . . . . NaClO₃	7775099	. . . .		3x	. . . .	2x	2x	3x		AIHA, MCA, HCD, STTE
924	Sodium chloride . . . . NaCl	7647145	. . . .		2x	. . . .	1x	1x	2x		. . . .

No.	(7) N.F.P.A. Hazard Identification Signals			(8) Extinguishing Agents	(9) Flammable Limits in Air % by volume		(10) Flash Point °C (°F)	(11) Ignition Temperature °C (°F)	(12) Boiling Point (Melting Point, m.p.) °C	(13) Vapor Pressure mm. Hg °C	(14) Density or Specific Gravity	(15) Water Solubility	No.
	Health	Fire	Reactivity		Lower	Upper							
	(4 is high; 0 is low)												
898	.....	.....	.....	....	....		....	....	....	....	....		898
899	.....	.....	.....	....	Spontaneously flammable in air		....	....	−112	....	....		899
900	.....	.....	.....	....					2230 m.p. 1710	....	2.2 2.6	....	900
901	.....	.....	.....	....					....	....	....		901
902	.....	.....	.....	....									902
903	.....	.....	.....	....									903
904	.....	.....	.....	....									904
911	.....	.....	.....	....									911
912	.....	.....	.....	....					−86	$760^{94.8}$	4.69 g/l	d	912
913	.....	.....	.....	....					....	....	....		913
914	1	0	1	....	....		....	....	d 444	....	4.352^{19}	v	914
915	3	1 W	2	5	Extremely dangerous in contact with water		....	Ignites spontaneously in air	883	1^{432}	0.97^{20}	d to NaOH + H$_2$	915
916	.....	.....	.....	....	....		....	607 (1125)	m.p. 324	....	1.528	v	916
917	.....	.....	.....	Sand. Never use H$_2$O or CCl$_4$	Reacts violently with water		....	....	400	....	....	d to NaOH	917
918	.....	.....	.....	....	May form explosive hydrazoic acid vapors		....	....	....	....	1.846	....	918
919	.....	.....	.....	....	....		....	....	....	....	....	v	919
920	.....	.....	.....	....	m.p. −CO$_2$ at 270°		....	....	....	....	2.159	s	920
921	.....	.....	.....	....	....		....	....	m.p. d	....	1.48	v	921
922	.....	.....	.....	....	....		None	None	d	....	2.532	s	922
923	1	0	2	....	Forms explosive mixtures with oxidizable materials		....	....	....	Oxidizable	2.490^{15}	v	923
924	.....	.....	.....	....	....		....	....	....	....	....		924

No.	Chemical Name and Formula	Chemical Abstracts Registry Number	A.C.G.I.H. 1966–TLV (Threshold Limit Values) ppm (mg/M³)	Principal Effects of Inhalation Exposures above TLV	Eye Contact	Inhalation	Skin Penetration	Skin Irritation	Ingestion	Supplemental Effects	References to Supplementary Information on Toxicity, Flammability and Other Hazards
							Relative Hazard to Health from Concentrated Short-term Exposure (5 is high; 1 is low)				
925	Sodium chlorite $NaClO_2$	7758192	....	....	....	....	....	....	....	....	HCD
926	Sodium chromate Na_2CrO_4	7775113			3x	....	2x	1x	3x		
927	Sodium cyanide $NaCN$	3396825			4x	....	3x	3x	4	Water contact releases HCN	MCA, HCD
928	Sodium fluoride NaF	7681494			3x	....	2x	2x	3		
929	Sodium fluoroacetate $CH_2FCOONa$	62748	(0.05)S	Toxic	Extremely toxic						STTE
930	Sodium formate $NaCHO_2$	141537			3x	....	1x	2x	2x		
931	Sodium hydride NaH	7646697	....	....	....	....	....	....	....	Inhalation may cause eye, nose & throat irritation. May form NaOH	STTE, NSC
932	Sodium hydroxide (Caustic soda) $NaOH$	1310732	(2)	Irritant	5*	....	1x	5	3x	Irritation, eye, nose, throat	MCA, NSC, HCD, STTE
933	Sodium iodide NaI	7681825			3x	....	2x	2x	3x	....	
—	Sodium lauryl sulfate See Dodecyl sodium sulfate										STTE
934	Sodium nitrate $NaNO_3$	7631994			2x	....	1x	1x	2x	....	HCD, STTE
935	Sodium nitrite $NaNO_2$	7632000			2x	....	1x	1x	3		2108, STTE
936	Sodium perchlorate $NaClO_4$	7601890								Irritant	HCD
937	Sodium peroxide Na_2O_2	1313606	....	....	....	....	....	....	....	Poisonous if ingested. Avoid inhalation.	HCD, STTE
938	Sodium phosphate (Disodium hydrogen phosphate) $Na_2HPO_4 \cdot 7H_2O$	10580355			1y*	....	1x	2y	1	Irritation of eye, nose, throat may result.	STTE
939	Sodium-potassium alloys NaK									Severe skin and eye burns	HCD
940	Sodium propionate $NaC_3H_5O_2$	137406			2y	....	2y	2y	2	....	....
941	Sodium silicate $Na_2O \cdot xSiO_2 (x = 3-5)$	1344098			4x	....	1x	3x	2x		....
942	Sodium sulfide Na_2S	1313822			Hydrogen sulfide liberated by water or acids.					....	HCD
943	Sodium sulfite Na_2SO_3	7757837			4x	....	1x	3x	3x		
944	Sodium tetraborate $Na_2B_4O_7$	1330434			Acute Poison					....	
945	Sodium thiocyanate $NaSCN$	540727			2x	....	2x	1x	2x		
946	Sodium thiosulfate $Na_2S_2O_3$	7772987			2x	....	1x	1x	2x		STTE
947	Stannic chloride $SnCl_4$	7646788	(2)	Toxic	4x	....	1x	4x	4x		
948	Stearic acid $C_{18}H_{36}O_2$	57114	....		2x	....	1x	2x	1x		....
949	Stibine SbH_3	7803523	0.1 (0.5)	Toxic	Very toxic					....	994, AIHA

No.	(7) N.F.P.A. Hazard Identification Signals			(8) Extinguishing Agents	(9) Flammable Limits in Air % by volume		(10) Flash Point °C (°F)	(11) Ignition Temperature °C (°F)	(12) Boiling Point (Melting Point, m.p.) °C	(13) Vapor Pressure mm. Hg °C	(14) Density or Specific Gravity	(15) Water Solubility	No.
	Health	Fire	Reactivity		Lower	Upper							
	(4 is high; 0 is low)												
925	1	1	2	1	....	....	....	....	d 180–200	....	....	s	925
926	....	....	....								2.710 to 2.736	v 2.736	926
927	2	0	0				Water	....	1496		....	s	927
928	....		....						d		2.558^{41}	v	928
929	....											....	929
930	....								m.p. 253		1.92^{20}	v	930
931	Fire and explosion may result on contact with water						....	....	m.p. 800° d. 850	....	0.92	i	931
932	3	0	1		....		....	....	1390	....	2.130	s	932
933	....								1304		3668^{25}_{4}	v	933
—												—	
934	1	0	2	1	....		....	....	d 380		2.261	v	934
935	....								d 320		2.168^{0}	v	935
936	2	0	2	....	Powerful oxidizing material				d 482	....	2.02	s	936
937	3	0 W	2	....	Powerful oxidizing material. Reacts vigorously with water.				d 460	....	2.805	s	937
938	....	....	....	5	....		....		−5 H₂O 48.1	....	1.52	v	938
939	3	3 W	2	....	Extremely dangerous on contact with air or water			Spontaneous ignition in air.	....	....	....	....	939
940	....	....	....		....		....					s	940
941	....	....		....								s	941
942	2	1	0	....	....		....		m.p. 1180	....	1.856^{14}_{4}	s	942
943	....		....						d		2.633	s	943
944	....								d 1575		2.367	δ	944
945	....								m.p. 287	....	....	v	945
946	....										1.667	s	946
947	....								114.1	....	liq. 2.226	s	947
948	1	1	0	1,3	....	....	196 (385)	395 (743)	358–83	....	0.948	i	948
949	....	....	....	....	Flammable		Gas	....	−17.1 m.p. −88	....	530 g/l	....	949

No.	Chemical Name and Formula	(2) Chemical Abstracts Registry Number	(3) A.C.G.I.H. 1966–TLV (Threshold Limit Values) ppm (mg/M^3)	(4) Principal Effects of Inhalation Exposures above TLV	(5) Relative Hazard to Health from Concentrated Short-term Exposure (5 is high; 1 is low) Eye Contact	Inhalation	Skin Penetration	Skin Irritation	Ingestion	Supplemental Effects	(6) References to Supplementary Information on Toxicity, Flammability and Other Hazards
950	4-Stilbenamine $C_{14}H_{13}N_2$	834242	....	....	....	....	....	....	....	Carcinogenic in animals	....
951	Stoddard solvent	....	500 (2900)	Narcosis	....	....	....	....	....		1201, AIHA
952	Strontium Sr	7440246	....	....	Compounds have low to moderate toxicity.					....	1131, STTE
953	Strontium carbonate $SrCO_3$	1633052	....	....	2x	....	1x	2x	1		1133
954	Strontium nitrate $Sr(NO_3)_2$	10042769	....	....	....	....	....	....	....		HCD
955	Strontium peroxide SrO_2	1314187	....	....	....	....	....	....	....		HCD
956	Strychnine $C_{21}H_{22}N_2O_2$	57249	(0.15)	Toxic	....	....	....	....	....		STTE
957	Styrene (Phenylethylene) C_8H_8	100425	C 100 C(420)	Irritant; Narcosis	2	2	2x	2	2		1223, AIHA, MCA, HCD, STTE
958	Succinic acid $C_4H_6O_4$	110156	....	....	4	2	2	3y	2		1812, STTE
959	Succinic anhydride $C_4H_4O_3$	108305	....	....	4	....	2y	3y	2x		STTE
960	Succinonitrile $C_4H_4N_2$	110612	....	....	....	....	....	....	....		2022
961	Sulfur S_8	10544500	....	....	1x	1x	1x	1x	1x	Burning produces toxic SO_2	892, MCA, NSC, HCD, STTE
962	Sulfur chloride (Sulfur monochloride) S_2Cl_2	10025679	1 (6)	Irritant	....	....	....	....	....		MCA
963	Sulfur dichloride SCl_2	10545990	....	....	Irritant. Exposure limit of 1 ppm is suggested.					....	MCA
964	Sulfur dioxide SO_2	7446095	5 (13)	Irritant	4x	4z	....	4x			892, AIHA, HCD, MCA, MGD, STTE
965	Sulfur hexafluoride SF_6	2551624	1000 (6000)	Nuisance	Physiologically inert					Simple asphyxiant	MGD, STTE
966	Sulfuric acid H_2SO_4	76647939	(1)	Irritant	4*	*	2x	4	4y	Irritation, eye, nose, throat	895, AIHA, MCA, NSC, HCD, STTE
967	Sulfuric acid, fuming (Oleum) H_2SO_4 + Free SO_3	10580582	(1)	Irritant	4*	*	2x	4	4y	Irritation, eye, nose, throat Extremely corrosive	895, AIHA, MCA, NSC, HCD
968	Sulfur pentafluoride	10546017	0.025 (0.25)	Toxic	....	....	....	....	....		STTE
969	Sulfur tetrafluoride SF_4	7783600	....	....	Highly toxic					....	MGD
970	Sulfur trioxide SO_3	7446119	....	....	....	....	....	....	....		STTE
971	Sulfuryl fluoride SO_2F_2	2699798	5 (20)	Toxic	....	....	....	....	....		MGD
—	Tannic acid See Tannin										
972	Tannin (Tannic acid) $C_{76}H_{52}O_6$	....	....	....	....	....	....	....	....		STTE
973	Tantalum Ta	7440257	(5)	Toxic	....	....	....	....	....		1136

No.	(7) Health	Fire	Reactivity	(8) Extinguishing Agents	(9) Lower	Upper	(10) Flash Point °C (°F)	(11) Ignition Temperature °C (°F)	(12) Boiling Point (Melting Point, m.p.) °C	(13) Vapor Pressure mm. Hg °C	(14) Density or Specific Gravity	(15) Water Solubility	No.
950									m.p. 90–2				950
951				2,3	0.8	5	38–43 (100–110)	227–260 (440–500)					951
952									1500		2.6^{20}		952
953											3.70	δ	953
954	1	0	1						645		2.986		954
955	1	0	1						m.p. d 215		4.56	δ	955
956									270^5		1.36	δ[h]	956
957	2	3 P	2	2,3	1.1	6.1	32 (90)	490 (914)	145–6	4.3^{15}	0.9090_4^{20}	i	957
958									d 235		1.572	δ:v[h]	958
959									261			i	959
960				2a			132 (270)		265–7	6^{125}	0.9848_4^{45}	v	960
961	2	1	0	1,3			207 (405)	232 (450)	444.6		2.07^{20}	i	961
962	2	1	0	3			118 (245)	234 (453)	135.6 m.p. −76	6.813^{20}	1.678	d	962
963					If heated slowly may decompose at temps above 40 (104)				59 m.p. −78	7.6^{-23}			963
964	3	0	0						−10		Gas 2.3 liq. 1.5	s	964
965									63.8		Gas 6.602 g/l	δ	965
966	3	0 W	1						338	1^{146}	1.891	∞	966
967	3	0 W	1									∞	967
968													968
969									−40.4		liq. 1.9191		969
970									44.8 m.p. 16.83			d	970
971									−55.4		Gas 3.72 g/l	s	971
972				1			199 OC (390 OC)	527 (980)	m.p. 210 − δd			s	972
973									Ca 6000		16.69	i	973

No.	(1) Chemical Name and Formula	(2) Chemical Abstracts Registry Number	(3) A.C.G.I.H. 1966–TLV (Threshold Limit Values) ppm (mg/M^3)	(4) Principal Effects of Inhalation Exposures above TLV	(5) Relative Hazard to Health from Concentrated Short-term Exposure						(6) References to Supplementary Information on Toxicity, Flammability and Other Hazards
					Eye Contact	Inhalation	Skin Penetration	Skin Irritation	Ingestion	Supplemental Effects	
					(5 is high; 1 is low)						
974	*TEDP* (Tetraethyl dithionopyrophosphate) $(C_2H_5O)_4P_2OS_2$	....	(0.2) S	Toxic	....	....	....	....	....		STTE
—	Teflon decomposition products (See Polytetrafluoroethylene decomposition products)										
975	Tellurium Te	10023167	(0.1)	Toxic	....	....	....	....	....		907, AIHA, STTE
976	Tellurium hexafluoride TeF$_6$	7783804	0.02 T (0.2) T	Toxic	....	....	....	....	....		....
977	*T.E.P.P.* (Tetraethyl pyrophosphate) (Ethyl pyrophosphate) $C_8H^{20}P_2O_5$	107493	(0.05) S	Toxic	4x	5x	5	2x	5		1955, STTE
978	Terbium Tb	7440279	....	....							1058
979	*m*-Terphenyl $C_{18}H_{14}$	92068	1 (9.4)	Irritant	....	....	....	....	....		1220
980	*o*-Terphenyl $C_{18}H_{14}$	84151	1 (9.4)	Irritant	....	....	....	....	....		1220
981	*p*-Terphenyl $C_{18}H_{14}$	92944	1 T (9.4) T	Irritant	....	....	....	....	....		1220
982	*p*-Terphenyl-4-amine	....	....	....						Carcinogenic in animals	....
983	1,1,2,2-Tetrabromoethane (Acetylene tetrabromide) $C_2H_2Br_4$	79276	1 (14)	Toxic	Decomposition liberates highly toxic vapors					....	1294, HCD
—	Tetrabromomethane See Carbon tetrabromide										
984	1,2,4,5-Tetrachlorobenzene $C_6H_2Cl_4$	95943	....	....							....
985	1,1,2,2-Tetrachloroethane $C_2H_2Cl_4$	79345	5 S (35) S	Toxic	....	....	....	....	....		1292, MCA, STTE
986	1,1,1,2-Tetrachloro-2,2-difluoroethane $C_2Cl_4F_2$	76119	500 (4170)	Toxic	....	....	....	....	....		1330
987	1,1,2,2-Tetrachloro-1,2-difluoroethane $C_2Cl_4F_2$	76120	500 (4170)	Toxic	....	....	....	....	....		1330
988	Tetrachloroethylene (Perchloroethylene) C_2Cl_4	127184	100 (670)	Narcosis	2x	....	3	1	2x		1314, AIHA
989	Tetrachloronaphthalene	1335882	(2) ST	Toxic	....	....	....	....	....		....
990	Tetradecane $C_{14}H_{30}$	629594	....	....							
—	Tetraethyl dithionopyrophosphate (See TEDP)										STTE
991	Tetraethylenepentamine $C_8H_{23}N_5$	112572	....	....	3	3x	3	4*	2	Sensitiz. skin, resp. tract	2039, STTE
992	Tetraethyllead $C_8H_{20}Pb$	78002	(0.075) S	Toxic	....	....	....	....	....		945, AIHA, HCD, STTE
—	Tetraethylpyrophosphate See T.E.P.P.										
—	Tetrafluoromethane See Carbon tetrafluoride										
993	Tetrahydrofuran C_4H_8O	109999	200 (590)	Irritant Toxic	2x	4x	2x	1x	2x		AIHA, HCD, STTE. Forms peroxides
994	3a,4,7,7a-Tetrahydro-4,7-methanoindene (Dicyclopentadiene) $C_{10}H_{12}$	77736	....	....	1	2	2	3	3		1209, STTE

No.	(7) N.F.P.A. Hazard Identification Signals			(8) Extinguishing Agents	(9) Flammable Limits in Air % by volume		(10) Flash Point °C (°F)	(11) Ignition Temperature °C (°F)	(12) Boiling Point (Melting Point, m.p.) °C	(13) Vapor Pressure mm. Hg °C	(14) Density or Specific Gravity	(15) Water Solubility	No.
	Health	Fire	Reactivity		Lower	Upper							
	(4 is high; 0 is low)												
974	.....	.....	.....	....	....	....	....	....	$138–9^2$	....	1.196	i	974
975	.....	.....	.....	....	Burns slowly in air		....	....	1390	....	6.25	i	975
976	.....	.....	.....	....	....	....	....	....	35.5	....	liq. 2.56	d	976
977	.....	.....	.....	....	....	....	....	....	155^5	Volatile	1.1847	....	977
978	.....	.....	.....	....	Pyrophoric on filing		....	....	2800 m.p. 1356	....	8.272	i	978
979	.....	1	0	1,3	....	....	135 OC (275)OC	....	365	....		i	979
980	.....	1	0	1,3	....	....	163 OC (325)OC	....	332	....		i	980
981	.....												981
982	.....												982
983	3	0	1	....			....	....	d 239–42	....	2.9672	i	983
984	.....	1	0	1,3	....	....	155 (311)	....	243–6	....	1.858^{22}	i	984
985	.....	.....	.....	....	....	....	....	....	146	6^{25}	1.5984_4^{25}	δ	985
986	.....	.....	.....	....	....	....	....	....	91.5	....	....	i	986
987	.....	.....	.....	....	....	....	....	....	92.8	....	1.6447	i	987
988	.....	.....	.....	....	....	....	....	....	121	19^{25}	1.623	i	988
989	.....	.....	.....									....	989
990	.....	1	0	1,2,3	0.5	....	100 (212)	202 (396)	253.5	....	0.7627	i	990
991	2	1	0	1,3	....	....	163 OC (325)OC	....	$151–2^1$	$<0.01^{20}$	0.998	∞	991
992	3	2	3	....	....	....	93 (200)	....	d 200	....	1.659	i	992
993	2	3	1	1,2a,3	2	11.8	−14 (6)	321 (610)	64–5	....	0.888	v	993
994	1	3	1	....	....	....	32 OC (90)OC	....	170 δd	$10^{47.6}$	0.9302	i	994

(1) No. / Chemical Name and Formula	(2) Chemical Abstracts Registry Number	(3) A.C.G.I.H. 1966–TLV (Threshold Limit Values) ppm (mg/M³)	(4) Principal Effects of Inhalation Exposures above TLV	(5) Eye Contact	Inhalation	Skin Penetration	Skin Irritation	Ingestion	Supplemental Effects	(6) References to Supplementary Information on Toxicity, Flammability and Other Hazards
				(5 is high; 1 is low)						
995 Tetrahydronaphthalene .... $C_{10}H_{12}$	119642	....	....	1	1	2	2	2	....	1223, STTE
— Tetrahydropyrrole See Pyrrolidine										
996 1-Tetralone $C_{10}H_{10}O$	529340	....	....	2	1	1	2	3	....	
997 N,N,N'N'-Tetramethyl-3,3'-dimethoxybenzidine	....								Derivative of a carcinogen	....
998 N,N,N',N'-Tetramethyl-ethylenediamine .... $C_6H_{16}N_2$	110189	....	....	4	3	3	4	2		....
999 Tetramethyllead .... $C_4H_{12}Pb$	75741	(0.075) ST	Toxic	.....	....	....	....	....		HCD
1000 Tetramethyl silane $C_4H_{12}Si$	75763	....	....	5x	3y	....	4x	2		
1001 Tetramethyl succinonitrile ...... $C_8H_{12}N_2$	3333526	0.5 ST (3) ST	Toxic	.....	....	....	....	....		
1002 Tetranitromethane .... $C(NO_2)_4$	509148	1 (8)	Toxic	.....	....	....	....	....		2073, AIHA, STTE
— Tetryl See N-Methyl-N-2,4,6-tetranitroaniline										
1003 Thallium .... Tl	7440280	....		Very toxic					Limit of 0.1 mg/M³ recommended	1138, STTE
1004 Thallium—soluble compounds....	....	(0.1) S	Toxic	.....	....	....	....	....		1138
1005 Thallium sulfate .... Tl_2SO_4	7446186	....	....	Ingestion leads to convulsions, paralysis, etc.					....	
1006 Thioacetamide .... C_2H_5NS	62555	....	....	3x	....	3x	2x	3	Carcinogenic in animals	....
— Thiobenzyl alcohol See α-Toluenethiol										
1007 2,2'-Thiodiethanol .... (Thiodiglycol) $C_4H_{10}O_2S$	111488	....	....	1*	1*	2	1x	2		
— Thiodiglycol See 2,2'-Thiodiethanol										
— Thiofuran See Thiophene										
1008 Thionyl chloride .... $SOCl_2$	7719097	....	....	Decomposes above 140° forming Cl_2, SO_2, and S_2Cl_2. Hydrolyzed by H_2O forming SO_2 and HCl. Irritant. Exposure limit of 1 ppm recommended.					....	907, STTE
1009 Thionyl fluoride .... SOF_2	7783428	....	....	Highly irritating to eyes and respiratory tract.					....	
1010 Thiophene .... (Thiofuran) C_4H_4S	110021	....	....	2x	3x	2x	1x	2x	....	STTE
1011 Thiourea .... CH_4N_2S	62566	....	....	3y	....	4x	2y	4		STTE
1012 Thiram .... (Bis(dimethylthiocarbamoyl) disulfide) CH_4N_2S	137268	(5)	Toxic	....	....	....	....	....		STTE
1013 Thorium .... Th	7440291	....	....	.....	....	....	....	....	....	1145, AIHA, STTE
1014 Thorium nitrate .... $Th(NO_3)_4 \cdot 4H_2O$	10102100	....	....	Toxic oxides of nitrogen produced by fire exposure.					....	AIHA, HCD
1015 Thulium.................... Tm	7440304	....	....	.....	....	....	....	....		1058
1016 Tin—inorganic compounds	....	(2)	Toxic	.....	....	....	....	....		1148

No.	(7) N.F.P.A. Hazard Identification Signals			(8) Extinguishing Agents	(9) Flammable Limits in Air % by volume		(10) Flash Point °C (°F)	(11) Ignition Temperature °C (°F)	(12) Boiling Point (Melting Point, m.p.) °C	(13) Vapor Pressure mm. Hg °C	(14) Density or Specific Gravity	(15) Water Solubility	No.
	Health	Fire	Reactivity		Lower	Upper							
	(4 is high; 0 is low)												
995	1	2	0	1,2,3	0.8 @ 212°	5.0 @ 302°	71 (160)	384 (723)	207.3	....	0.9729	i	995
996	.....	.....	.....	....			129 (265)	....	$120\text{-}125^{10}$	0.02^{20}	1.090– 1.095^{20}_{20}	......	996
997	.....	.....	.....								....		997
998	.....	.....	.....					....	121–2	....	0.7765	s	998
999	3	3	3	....			38 (100)	....	110		1.995	i	999
1000	.....	.....	.....						26.5		0.648^{20}_{4}	i	1000
1001	.....	.....	.....						m.p. 169	....	1.070	......	1001
1002	.....	.....	.....						126	13^{25}	1.6372^{21}_{4}	i	1002
1003	.....	.....	.....						1460 (m.p. 303.5)	....	11.85	......	1003
1004	.....	.....	.....								....	.....	1004
1005	.....	.....	.....						d	....	6.77	δ	1005
1006	.....	.....	.....						m.p. 115–6		....	v	1006
1007	1	1	0	2a	....		160 OC (320) OC	....	28.2	....	1.1849	s	1007
1008	.....	.....	.....	....	....		....		75.5 m.p. 105	110^{10}	1.655^{20}_{4}	......	1008
1009	.....	.....	.....	....	....				....	....			1009
1010	.....	.....	.....				−1.1 (30)	....	84.12	....	1.0583^{20}_{4}	i	1010
1011	.....	.....	.....						m.p. 180–2	....	1.405	s	1011
1012	.....	.....	.....								1.17	i	1012
1013	.....	.....	.....	5	Pyrophoric when finely divided		Dust explosion hazard	....	4230	....	11.7	i	1013
1014	1	0	1	....	Oxidizing material		....	....	....		....	.....	1014
1015	.....	.....	.....	....	Pyrophoric on filing		....	....	1727 (1545)	....	9.332	i	1015
1016	.....	.....	.....						....	....	....		1016

(1) No. / Chemical Name and Formula	(2) Chemical Abstracts Registry Number	(3) A.C.G.I.H. 1966–TLV (Threshold Limit Values) ppm (mg/M³)	(4) Principal Effects of Inhalation Exposures above TLV	(5) Relative Hazard to Health from Concentrated Short-term Exposure (5 is high; 1 is low)						(6) References to Supplementary Information on Toxicity, Flammability and Other Hazards
				Eye Contact	Inhalation	Skin Penetration	Skin Irritation	Ingestion	Supplemental Effects	
1017 Tin—organic compounds		(0.1)	Toxic	....	....	....	....	....		1148
1018 Titanium Ti	7440326			Considered physiologically inert						1154, AIHA, NSC, STTE
1019 Titanium dioxide TiO_2	1309633	(15)	Nuisance	....	....	....	....	....		1155, AIHA, STTE
1020 Titanium tetrachloride $TiCl_4$	7550450			5x	5x	2x	4x	3x		1155
— TNT See 2,4,6-Trinitrotoluene										
1021 Toluene C_7H_8	108883	200 (750)	Irritant	4	3	2	2	2		1222, AIHA, MCA, NSC, HCD, STTE
1022 Toluene-2,4-diisocyanate (TDI) $C_9H_6N_2O_2$	86919	C 0.02 C (0.14)	Toxic	....	....	....	....	....	Irritant. Causes sensitization.	2032, AIHA, MCA, HCD, STTE
1023 α-Toluenethiol (Benzyl mercaptan) (Thiobenzyl alcohol) C_7H_8S	100538	1 (5)	Irritant	....	....	....	....	....		STTE
1024 m-Toluidine C_7H_9N	108441			4x	1x*	2x	1x	3x	Chronic toxicity	MCA
1025 o-Toluidine C_7H_9N	95534	5 S (22) S	Toxic	4	1*	2	1	3	Chronic toxicity	2155, MCA, HCD, STTE
1026 p-Toluidine C_7H_9N	106490									2154, MCA, HCD
1027 1-o-Tolylazo-2-naphthol	2646175								Carcinogenic in animals	....
— 4-(o-Tolylazo)-o-toluidine See C.I. Solvent Yellow 3										
1028 o-Tolyl phosphate (Tri-o-cresyl phosphate) $C_{21}H_{21}O_4P$	78308	(0.1)	Toxic	1x	3x	3x	1x	4z	Can cause paralysis	1919, STTE
1029 Tremolite $Ca_2Mg_5Si_8O_{24}H_2$	1305700	176.5 p/cc T	Toxic	....	....	....	....	....		STTE
1030 Tribromomethane (Bromoform) $CHBr_3$	75252	5 ST (50) ST	Toxic	....	....	....	....	....		1262, STTE
1031 Tributylamine $C_{12}H_{27}N$	102829			4x	4x	3x	3x	3x		2038, HCD
1032 Tributylchlorotin (Tributyltin chloride) $C_{12}H_{27}SnCl$				5y	....	x	5	4x		1153, STTE
1033 Tributyl phosphate $C_{12}H_{27}O_4P$	126738	(5) T	Irritant	2	3y	2x	2	2		1914
— Tributyltin chloride See Tributylchlorotin										
1034 Trichloroacetic acid $C_2HCl_3O_2$	76039			....	....	....	....	....	Irritant	1797
1035 Trichloroacetonitrile (Tritox) C_2Cl_3N	545062			5	5y	3	4	3		STTE
1036 1,2,4-Trichlorobenzene $C_6H_3Cl_3$	120821			1	2	....	2	....		....
— 1,1,1-Trichloro-2,2-bis-(p-methoxyphenyl)ethane See Methoxychlor										
1037 1,1,1-Trichloroethane (Methyl chloroform) $C_2H_3Cl_3$	71556	350 (1900)	Narcosis	1x	3x	2x	1x	2		1287, NSC, MCA, AIHA, STTE
1038 1,1,2-Trichloroethane $C_2H_3Cl_3$	79005	10 ST (45) ST	Narcosis; Toxic	1	3y*	2	1	2	Chronic toxicity	1303, STTE
1039 Trichloroethylene C_2HCl_3	79016	100 (535)	Narcosis; Toxic	2	3*	1	3	2	Chronic toxicity	1309, AIHA, NSC, MCA, STTE

No.	(7) N.F.P.A. Hazard Identification Signals			(8) Extinguishing Agents	(9) Flammable Limits in Air % by volume		(10) Flash Point °C (°F)	(11) Ignition Temperature °C (°F)	(12) Boiling Point (Melting Point, m.p.) °C	(13) Vapor Pressure mm. Hg °C	(14) Density or Specific Gravity	(15) Water Solubility	No.
	Health	Fire	Reactivity		Lower	Upper							
	(4 is high; 0 is low)												
1017	.....	.....	.....	....	....		....	....	....	....	....	....	1017
1018	.....	.....	.....	5	Can burn in nitrogen		Dust explosion hazard	....	3262	....	4.5^{20}	i	1018
1019	.....								1825	....	4.17	i	1019
1020	.....	W							136.4	....	liq. 1.726	s	1020
—													—
1021	2	3	0	2,3	1.2	7.1	4 (40)	536 (997)	110.6	$30^{26.04}$	0.8669^{20}_{4}	i	1021
1022	2	1 W	2	....			135 OC (275)OC	....	250	1^{80}	1.21	i	1022
1023	.....	.....	.....					....	94–5	....	1.058	i	1023
1024	.....	.....	.....		....	....	86 (187)	482 (900)	203.2	1^{44}	0.9916^{25}_{25}	δ	1024
1025	3	2	0	1,2,3	....		85 (185)	482 (900)	199.7	1^{44}	1.008^{20}_{20}	δ	1025
1026	3	2	0	1,2,3	....		87 (188)	482 (900)	200.4 m.p. 44	1^{42}	1.046^{20}_{4}	δ	1026
1027	.....	.....	....	....						....	....	....	1027
—													—
1028	2	1	0	1,2,3	....	....	225 (437)	385 (725)	$244^{3.5}$	10^{200}	1.247	δ	1028
1029	.....	.....	.....				....	....	....	....	....	....	1029
1030	.....								149.5	5.6^{25}	2.8899	δ	1030
1031	2	2	0	1,2,3	....	....	86 OC (187)OC	....	216–7	20^{100}	0.7782^{20}_{20}	δ	1031
1032	.....			....					....		....		1032
1033	2	1	0	1,2,3	....	....	146 OC (295) OC	....	289	....	0.9727	s	1033
—													—
1034	.....	.....						....	197.5 m.p. 57.5	....	1.6298	....	1034
1035	.....	.....						....	84.6^{741}	....	1.4403	....	1035
1036	2	1	0	1,2,3	....	....	99 OC (210)OC	....	213.5	....	1.4542	i	1036
—													—
1037	.....	.....	.....					....	74	127 (25)	1.3492^{20}_{4}	i	1037
1038	.....	.....					None	....	113.7	16.7^{20}	1.4432^{20}_{4}	δ	1038
1039	.....	.....	.....		12.5	90	32 (90)	....	87	77^{25}	1.462^{15}	δ	1039

CHEMICAL HAZARD INFORMATION (See Explanation, page 668)

No.	Chemical Name and Formula	(2) Chemical Abstracts Registry Number	(3) A.C.G.I.H. 1966–TLV (Threshold Limit Values) ppm (mg/M³)	(4) Principal Effects of Inhalation Exposures above TLV	Eye Contact	Inhalation	Skin Penetration	Skin Irritation	Ingestion	Supplemental Effects	(6) References to Supplementary Information on Toxicity, Flammability and Other Hazards
1040	Trichlorofluoromethane (Freon®) CCl₃F	75694	1000 (5600)	Narcosis	....	2y	....	....			MGD, STTE
1041	Trichloromethanethiol (Perchloromethyl mercaptan) CHCl₃S	75707	0.1 (0.8)	Irritant	....		....				STTE
1042	Trichloronaphthalene C₁₀H₅Cl₃	2437549	(5)S	Toxic							1343, STTE
1043	Trichloronitromethane (Chloropicrin) CCl₃NO₂	76062	0.1 (0.7)	Toxic; Irritant	5x*	5*	....	4x		Irritation, eye, nose, throat	2080, STTE
1044	(2,4,5-Trichlorophenoxy)acetic acid C₈H₅Cl₃O₃	93765	10 (25)	Irritant	Eye irritation, G.I. disturbance					....	STTE
1045	1,2,3-Trichloropropane C₃H₅Cl₃	96184	50 (300)	Irritant	2	3y*	3	1	3	Chronic toxicity	STTE
1046	1,1,2-Trichloro-1,2,2-trifluoroethane C₂Cl₃F₃	76131	1000 (7600)	Narcosis	....	2y	....				....
—	Tri-o-cresyl phosphate See o-Tolyl phosphate										
1047	Triethylamine C₆H₁₅N	121448	25 (100)	Irritant	4	4y	3	1	3	....	2038, STTE
1048	Triethylene glycol C₆H₁₄O₄	112276	....	....	1	1x	1	1	1		1507, STTE
1049	Trifluoroacetic acid C₂HF₃O₂	76051	....	....						Irritant	1801, STTE
—	Trifluorobromomethane See Bromotrifluoromethane										
1050	Trifluoromethane (Fluoroform; Freon®-23; Genetron®-23)	75467	....	....	May be slightly irritating to respiratory tract					....	MGD
1051	Trimethylamine C₃H₉N	75503	....	....	4x	4x	....	4x			2038, HCD, MGD. May deteriorate in normal storage and cause hazard
1052	5,9,10-Trimethyl-1,2-benzanthracene	10546391	....	....						Carcinogenic in animals	....
1053	6,9,10-Trimethyl-1,2-benzanthracene	6610408	....	....						Carcinogenic in animals	....
1054	Trimethyl borate C₃H₉BO₃	121437	....	....							
1055	3,5,5-Trimethyl-2-cyclo-hexen-1-one (Isophorone) C₉H₁₄O	78591	25 (140)	Irritant; Toxic	2	2*	3	1	2	Chronic toxicity	1723, STTE
1056	2,2,4-Trimethylpentane C₈H₁₈	540841	....	....	May cause skin irritation					....	
1057	2,4,4-Trimethyl-1-pentene C₈H₁₆	107391									
1058	2,4,4-Trimethyl-2-pentene C₈H₁₆	107404									
1059	Trinitrobenzene C₆H₃(NO₂)₃	610311			Respiratory irritant					....	2109
1060	2,4,6-Trinitrotoluene (TNT) C₇H₅N₃O₆	118967	(1.5)S	Toxic	....	....	....				2153, NSC, HCD, STTE
—	Tri-o-cresyl phosphate See o-Tolyl phosphate										
1061	S-Trioxane C₃H₆O₃	110883	....		3	....	2x	2y	2	....	....

824

No.	(7) N.F.P.A. Hazard Identification Signals Health	Fire	Reactivity	(8) Extinguishing Agents	(9) Flammable Limits in Air % by volume Lower	Upper	(10) Flash Point °C (°F)	(11) Ignition Temperature °C (°F)	(12) Boiling Point (Melting Point, m.p.) °C	(13) Vapor Pressure mm. Hg °C	(14) Density or Specific Gravity	(15) Water Solubility	No.
	(4 is high; 0 is low)												
1040	.....	.....	.....		....	....	....	....	23.77	....	liq. 1.464	δ	1040
1041	.....											....	1041
1042	.....											....	1042
1043	.....								112^{757}	16.9^{20}	1.6558	$δ^h$	1043
1044	.....								m.p. 157–8	....		δ	1044
1045	3	2	0		....	....	82 OC (180)OC	....	156		1.394^{15}	δ	1045
1046	.....								47.7	....	1.5635	i	1046
—													—
1047	2	3	0	2,2a,3	1.2	8.0	-6.7 OC (20)OC	....	89–90	$400^{31.5}$	0.7255^{20}_{4}	$s:δ^h$	1047
1048	1	1	0	1,2a,3	0.9	9.2	177 (350)	371 (700)	276	0.001^{20}	1.1274^{15}_{4}	∞	1048
1049	.....								72.4 m.p. 15.3	191^{37}	1.5351^{0}	....	1049
—													—
1050	.....								-84	....		....	1050
1051	3	4	0	4	2.0	11.6	Gas	190 (374)	3.5	$760^{2.9}$	0.6079^{0}_{4}	v	1051
1052	.....											....	1052
1053	.....											....	1053
1054	2	3	1				<27 (<80)	....	67–8		0.915^{20}	d	1054
1055	2	1	0	1,3	0.8	3.8	96 OC (205)OC	462 (864)	215.2	0.44^{25}	0.9229	δ	1055
1056	.....	3	0	3	1.1	6.0	-12 (10)	18 (784)	99.2	....	0.6918^{20}_{4}	i	1056
1057	.....	3	0				<7 (<20)		101.44	....	0.7150^{20}_{4}	i	1057
1058	.....	3	0				1.7 (35)	....	105			....	1058
1059	2	4	4	....	High explosive		....		61			δ	1059
1060	2	4	4				....	....	240 expl.	0.046^{82}	1.654	i	1060
—													—
1061	2	2	0	2,2a,3	3.6	29	45 OC (113)OC	414 (777)	114.5^{759}	....	1.17	v	1061

No.	(1) Chemical Name and Formula	(2) Chemical Abstracts Registry Number	(3) A.C.G.I.H. 1966–TLV (Threshold Limit Values) ppm (mg/M³)	(4) Principal Effects of Inhalation Exposures above TLV	(5) Relative Hazard to Health from Concentrated Short-term Exposure (5 is high; 1 is low)						(6) References to Supplementary Information on Toxicity, Flammability and Other Hazards
					Eye Contact	Inhalation	Skin Penetration	Skin Irritation	Ingestion	Supplemental Effects	
1062	Triphenyl phosphate........... $C_{18}H_{15}O_4P$	115866	(3)	Toxic	1x	1x	2x	1x	1x	....	1915, STTE
1063	Triphenylphosphine .... (Triphenylphosphorus) $(C_6H_5)_3P$	603350	....	....	2y	1	2	2y	3		1918
—	Triphenylphosphorus See Triphenylphosphine										
1064	Tripropylamine............... $C_9H_{21}N$	102692									
1065	Triton	....	....	....	2x	....	3	1x	2x		STTE
—	Tritox See Trichloroacetonitrile										
1066	Trypan blue..............	72571	....	....	....	....	....	....		Carcinogenic in animals	....
1067	Turpentine		100 (560)	Irritant; Narcosis	1	2x	2x	2	2		1209, NSC, STTE
1068	Uranium .... U	7440611	(0.25)	Toxic	....	....	....	....	....		AIHA
1069	Uranium-soluble compounds	....	(0.05)	Toxic	....	....	....	....	....		....
1070	Uranium-insoluble compounds ..	....	0.25	Toxic	....	....	....	....	....		....
1071	Urea .... CH_4N_2O	57136	....	....	2x	1x	1x	1x	2		STTE
1072	Valeric acid................ $C_5H_{10}O_2$	109524	....	....	4	1	3	4	2		1782
1073	Vanadium dust .... V_2O_5	1314621	C(0.5)	Irritant							1171, STTE
1074	Vanadium fume .... V_2O_5	1314621	(0.1)	Irritant							1171, STTE
1075	Vanadium chloride .... VCl_2	10580526	....	....	4	1	4	2y	3		....
—	Vanadium oxide dust See Vanadium dust										
1076	Vanadium pentoxide .... V_2O_5	1314621	....	....	3x	3z*	3x	2x	3	Chronic toxicity	1173, AIHA
1077	Vinyl acetate............... $C_6H_6O_2$	108054	....	....	1	3y	2y	1	2	....	1876, MCA, HCD, STTE
—	Vinyl bromide See Bromoethylene										
—	Vinyl butyl ether See Butyl vinyl ether										
—	Vinyl chloride See Chloroethylene										
1078	Vinyl ether .... C_4H_6O	109933	....	....							....
—	Vinyl ethyl ether See Ethyl vinyl ether										
—	Vinyl fluoride See Fluoroethylene										
—	Vinylidene chloride See 1,1-Dichloroethylene										
—	Vinyl methyl ether See Methyl vinyl ether										
—	Vinyl trichloride See 1,1,2-Trichloroethane										
—	Vinyl toluene See α-Methyl styrene										
1079	Warfarin.................. $C_{19}H_{16}O_4$	81812	(0.1)	Toxic	....	....	....	....	....	....	STTE
1080	Xenon .... Xe	7440633	....	....	Non-toxic					Can be an asphyxiant	MGD
—	Xenylamine See 4-Biphenylamine										
1081	m-Xylene.................. C_8H_{10}	108383	100 T (435)T	Irritant	1	2	2	3	1	....	1222, AIHA, NSC, HCD, STTE

No.	(7) N.F.P.A. Hazard Identification Signals			(8) Extinguishing Agents	(9) Flammable Limits in Air % by volume		(10) Flash Point °C (°F)	(11) Ignition Temperature °C (°F)	(12) Boiling Point (Melting Point, m.p.) °C	(13) Vapor Pressure mm. Hg °C	(14) Density or Specific Gravity	(15) Water Solubility	No.
	Health	Fire	Reactivity		Lower	Upper							
	(4 is high; 0 is low)												
1062	2	1	0	1,3	....	....	220 (428)	....	245^{11}	....	1.2055	i	1062
1063	....	....	....	....			180 OC (356) OC	....	>360	....	1.194	i	1063
—													—
1064	2	2	0	....			41 OC (105) OC		365	....	0.774^{20}_{4}	i	1064
1065	....											....	1065
—													—
1066	....												1066
1067	1	3	0	2,3	0.8	....	35 (95)	253 (488)	153–75	....	0.86–88	i	1067
1068	....	....	....	5	Pyrophoric in finely divided state		....	....	3818 m.p. 1150	....	19.05^{8} 0.02	i	1068
1069	....								....		....	....	1069
1070	....								....		....	......	1070
1071	....								d	....	1.32^{18}_{4}	v	1071
1072							96 OC (205) OC		186–7	1^{42}	0.939^{20}_{4}	s	1072
1073	....						....	....	....	....	....		1073
1074	....											......	1074
1075	....	...			....		....	....	....	....	3.23^{18}	s d	1075
—													—
1076	....	....	....	....	....	......	....	....	d 1750	....	3.357^{18}	δ	1076
1077	2	3p	2	2,2a,3	2.6	13.4	8 (18)	427 (800)	$71-2^{728}$	$115^{25.3}$	1.3941	i	1077
—													—
—													—
—													—
1078	2	3	2	....	1.8	36.5	−47 (−52)	60 (680)	39	....	0.769	s	1078
—													—
—													—
—													—
—													—
—													—
1079	....	....	....	....	....	....	....	....	m.p. 161	....	....	i	1079
1080	....	....	....				....		−108	....	Gas 5.897 g/l	......	1080
1081	2	3	0	2,3	1.1	7.0	29 (84)	528 (982)	139	$10^{28.26}$	0.8684^{0}_{4}	i	1081

No.	(1) Chemical Name and Formula	(2) Chemical Abstracts Registry Number	(3) A.C.G.I.H. 1966–TLV (Threshold Limit Values) ppm (mg/M³)	(4) Principal Effects of Inhalation Exposures above TLV	(5) Relative Hazard to Health from Concentrated Short-term Exposure (5 is high; 1 is low) Eye Contact	Inhalation	Skin Penetration	Skin Irritation	Ingestion	Supplemental Effects	(6) References to Supplementary Information on Toxicity, Flammability and Other Hazards
		416	416								430
1082	o-Xylene .... C_8H_{10}	95476	100 T (435) T	Irritant	1	2	2	3	1	....	1222, NSC, HCD, STTE
1083	p-Xylene .... C_8H_{10}	106423	100 T (435) T	Irritant	1	2	2	3	1	....	1222, NSC, HCD, STTE
—	Xenylamine See 4-Biphenylamine										
1084	Xylenol (2,6-Dimethylphenol) $C_8H_{10}O$	576261	....	....	4	1	3x	4	3	....	....
1085	3,5-Xylidine .... (3,5-Dimethylaniline) $C_8H_{11}N$	108690	5 S (25) S	Toxic	3x	3x*	4y	2x	4x	Chronic toxicity	2126, STTE
1086	Yttrium .... Y	7440655	(1)	Toxic	....	...	...	...	....	....	1061, STTE
1087	Zinc....................... Zn	7440666	....	....	Products of combustion cause fume fever					....	1184, NSC, HCD, STTE
1088	Zinc acetate .... $Zn(C_2H_3O_2)_2$	557346	....	....	2x	....	2y	2y	2	....	....
1089	Zinc chloride............... $ZnCl_2$	7046857	(1) T	Irritant	5x	....	2x	4x	3x	....	1184
1090	Zinc oxide................. ZnO	1314132	....	...:.....	2x	3z*	1x	1x	2x	....	1184, AIHA, NSC
1091	Zinc oxide fume	....	(5)	Fume fever	....	...	...	...	....	....	STTE
1092	Zinc stearate............... $Zn(C_{18}H_{35}O_2)_2$	557051	....	....	....	...	...	...	....	....	1187
1093	Zirconium .... Zr	7440677	....	....	....	...	...	...	....	....	1189, AIHA, MCA, NSC
1094	Zirconium compounds .... Zr	....	(5)	Toxic	....	...	...	...	....	....	STTE

No.	(7) N.F.P.A. Hazard Identification Signals			(8) Extin- guishing Agents	(9) Flammable Limits in Air % by volume		(10) Flash Point °C (°F)	(11) Ignition Tem- perature °C (°F)	(12) Boiling Point (Melting Point, m.p.) °C	(13) Vapor Pressure mm. Hg °C	(14) Density or Specific Gravity	(15) Water Solu- bility	No.
	Health	Fire	Reac- tivity		Lower	Upper							
	(4 is high; 0 is low)												
1082	2	3	0	2,3	1.0	6.0	32 (90)	464 (867)	144	$10^{32.11}$	0.8968_4^{20}	i	1082
1083	2	3	0	2,3	1.1	7.0	27 (81)	529 (984)	138	$10^{27.30}$	0.85667_4^{28}	i	1083
—													—
1084	.....	.....	.....	.....	....	....	....	....	218	....	....	δ	1084
1085	3	1	0	1,2,3	....	....	97 (206)	....	224^{728}	....	1.076	δ	1085
1086	.....	.....	.....	....	Pyrophoric on filing		....	....	1427 m.p. 824	....	4.34	d	1086
1087	0	1	1	5	Dust may ignite in air		Dust ex- plosion hazard	....	907 m.p. 419	....	7.14	i	1087
1088	.....	.....	.....	....	....	....	....	....	m.p. d 200	....	1.84	s	1088
1089	2	0	2	....	....	....	....	....	732	....	2.91^{25}	v	1089
1090	.....	.....	.....	....	....	....	....	....	m.p. 1975	....	5.606	δ	1090
1091	.....	.....	.....	....	....	....	....	....	....	....	....	.....	1091
1092	.....	.....	.....	1,3	....	....	276 OC (530)OC	421 (790)	m.p. 130	....	....	i	1092
1093	1	4	1	5	....	....	Dust explosion hazard	20(68) for dust cloud	>2900 m.p. 1830	....	6.49	i	1093
1094	.....	.....	.....	....	....	....	....	....	....	....	....	....	1094

INDEX

A

Abdominal wounds, 37, 39
Absorbent material for radioactive spills, 437, 438, 443, 446
Absorption, effect on flammability characteristics, 208, 209
  of the laser beam in blood, 385
  of mercury vapor, 337
  of toxic substance from GI tract into blood, 316
Accidents, 3, 97–111
  basic causes, 5
  in chemistry labs, 61, 63
  definition, 4
  electric shock, 524
  involving perchloric acid, 265, 274
  preoperational analysis, at installations using radioactive materials, 486–487, 497, 499
  prevention measures, 5, 9
  radioactive, 426E, 430, 443
Acetic acid, chemical safety data, 694
  mixtures with perchloric acid, 267
  storage, 182
Acetic anhydride, 267
Acid, eye damage, 23, 90
  protection for pouring operations, 95
  storage, 182
Acid gases, automatic monitoring, 176
Acid hematin, analytical method, 347
Acrylamide, threshold limit value, documentation, 714
Action levels, 310
Adrenals, dose from inhalation or ingestion of radionuclides, 469
Aerosols, 315
  infectious, 612, 613, 615
  produced by common microbiological techniques, 633–636
  radioactive, 438, 441, 444, 451
    inhalation, 431
    monitoring of air for contamination, 434
    production, 440
Aerosols and flammable vapors, control of, 146, 147
Air, analysis equipment, automatic, 304
  chemicals in, evaluating toxicity of, 289
  cleaning equipment, 150, 151, 153
  dewpoint of 13°C or less, 166
  gap in plumbing, 588, 592, 594
  intakes, 163–165
    location of, 173
  liquid, 570, 572, 573
  make-up requirements, 151
  monitoring requirements for radioactive materials, 493
  permissible concentrations of radionuclides in, 484–485

  pollution control, 151
  recirculation, 175
  sampling, 304–306, 308
  substances, 314, 611, 612, 670
  supply, for inhalation experiments with animals, 292
  TLV for contaminants, 302, 305, 672, 720
  toxic substances, 304
  velocity, 141, 150, 154, 583, 660
    minimum required to capture contaminants, 150 (Table 1)
    outside fume hood, 160
Air-conditioning, animal facilities, 647
Air flow, 151, 156–165, 168–173
  equation, 145
  laminar, 142
  means for measurements, 152
  minimum and maximum limits, 154, 155, 157, 160
  patterns, 142, 154
  rates, 168, 172
    at face of hood, 114
  relative, 152
  supplementary, 168
Air masks, self-contained, 18, 19
Air-moving equipment, 150, 151, 153
Air pollutants, respiratory protective equipment, 75
Air-supply systems and conditioning
  animal facilities, 647, 748
  balance in, 158, 159
  design, 166, 167
  microbiological laboratories, 623, 626
  precautions for shutdown, 662
  relation to fume hoods, 157, 158
  volume controls, 159
"Air Weave" polyethylene fabric, protective clothing from, 128
Alarm systems, 656
Alcohol-air mixtures, 113
Alcohols, removal from water, 596
Alerting personnel, 16
Alkalies, eye damage by, 23, 90
Alkyl esters of perchloric acid, 268
Alkyl hydroperoxide, 250
Alkylidene peroxides, 252
Allergic sensitizer, definition, 297
Allergies, from isocyanates, 327
Alpha emitters, 433, 448
Alpha rays, basic units of measurement, 391
  limits for surface contamination, 478–480
Alpha-selective monitor, 434
Aluminum, container for high-purity water, 601
American Conference of Governmental Industrial Hygienists, 714
American National Standards Institute, Subcommittee N6.9, 503
Americium 241, limits for surface contamination, 480

C

M

846   Handbook of Laboratory Safety

P

CRITICAL REVIEWS™

QUARTERLY JOURNALS

CRITICAL REVIEWS in ANALYTICAL CHEMISTRY
Edited by Louis Meites, Ph.D., Chairman, Department of Chemistry, Clarkson College of Technology. Associate Editors: Gunter Zweig, Ph.D., Director, Life Sciences Division, Syracuse University Research Corp., and Irving Sunshine, Ph.D., Chief Toxicologist, Cuyahoga County Coroner's Office, Ohio.

CRITICAL REVIEWS in CLINICAL LABORATORY SCIENCES
Edited by John W. King, M.D., Ph.D., Clinical Pathologist, Cleveland Clinic Foundation, and Willard R. Faulkner, Ph.D., Director, Clinical Chemistry Laboratories, Vanderbilt University Medical Center.

CRITICAL REVIEWS in ENVIRONMENTAL CONTROL
Edited by Richard G. Bond, M.S., M.P.H., Director of Environmental Health, University of Minnesota, and Conrad P. Straub, Director, Environmental Health Research and Training Center, University of Minnesota.

CRITICAL REVIEWS in FOOD TECHNOLOGY
Edited by Thomas E. Furia, Technical Development Manager, Geigy Industrial Chemicals, Division of Geigy Chemical Corporation.

CRITICAL REVIEWS in RADIOLOGICAL SCIENCES
Edited by Yen Wang, M.D., D. Sc. (Med.), University of Pittsburgh.

CRITICAL REVIEWS in SOLID STATE SCIENCES
Edited by Donald E. Schuele, Ph.D., and Richard Hoffman, Ph.D., both of the Department of Physics, Case Western Reserve University.

Condensed Explanation of
Tables of Chemical Hazard Information

COLUMN

1 CHEMICAL NAME AND FORMULA: Chemical Abstracts nomenclature is followed, except in the case of pesticides and other economic poisons. Some common and trade names are cross-referenced.

2 CHEMICAL ABSTRACTS REGISTRY NUMBER: The number listed in this column is the definitive identification of the chemical in the computer registry established by the Chemical Abstracts Service.

3 ACGIH THRESHOLD LIMIT VALUES 1966: (see page 417 for full text). Limits preceded by C are ceiling limits that should not be exceeded; other values are time-weighted average limits, below which most workers can be repeatedly exposed without adverse effect. Values for gases and vapors are given in parts of vapor or gas per million parts of air by volume at 25°C and 760 mm. Hg., followed in parentheses by approximate milligrams per cubic meter of air. Values for particulate and aerosols shown only in parentheses, in milligrams per cubic meter. **S** indicates skin absorption is possible and measures may be needed to limit cutaneous exposures. **T** indicates tentative values for which a new or changed limit is proposed.

4 PRINCIPAL EFFECTS OF INHALATION EXPOSURES ABOVE TLV (Threshold Limit Values): Designations listed in this column are judgments based on "Documentation of Threshold Limit Values," and only the principal or first major effects of inhalation exposures above TLV are listed. Many substances will produce several effects, particularly by different routes of exposure or at concentrations far above the TLV.
Fume Fever: Fever following initial or interrupted exposure, which ordinarily clears up within 24 hrs.
Toxic: Damage to some essential function, usually chronic effects produced by repeated exposures in excess of TLV.
Irritant: Irritation of tissue contacted, including skin, eyes, or entire respiratory system.
Narcosis: Depression of the central nervous system, tending to produce sleep or unconsciousness.
Nuisance: No known systemic effects or irritation are produced at concentrations several times TLV.
Carcinogenic: Substance is known to produce cancer.

5 RELATIVE HAZARD TO HEALTH FROM CONCENTRATED SHORT-TERM EXPOSURES: *Symbols Used to Describe Relative Hazard to Health*
5 Major residual injury is likely in spite of prompt treatment.
4 Major residual injury may result in spite of prompt treatment.
3 Minor residual injury may result in spite of prompt treatment.
2 Minor residual injury may result from some accidental exposures if no treatment is applied.
1 No residual injury is to be expected from accidental exposures even if no treatment is applied.
x Entry based upon analogy with a closed similar structure, or on other estimate believed sound.
y Entry based on a non-standard test (different from the tests described on page 429).
z Entry based upon human experience, superseding animal data.
***** See Supplemental Effects column.

SUPPLEMENTAL EFFECTS

Acne-like eruption: Acne-like eruption may follow prolonged contact.

Carcinogenic in animals: Cancer has resulted in animals; prolonged contact may affect humans.

Chronic toxicity: Repeated contact is more hazardous.

Death/hours inhalation: Prolonged breathing of high concentration for a few hours has killed a human.

Death/odor reliance: Odor is unreliable warning and death has resulted from relying upon odor to warn of danger.

COLUMN

Extreme toxicity: So toxic that a microdrop (as little as 0.005 ml) in the eye has killed a rabbit within a few minutes.

Eye burn w/o pain: Eye burn possible with no pain at time of contact. So-called delayed eye burns are particularly likely to result in permanent disability, because absence of pain at the time of contact reduces the likelihood that eye-washing will be undertaken early enough to limit eye injury. Contact with either vapor or fluid has injured.

Eye pigment: Long repeated contact may result in permanent eye pigmentation.

Gas or vap. penet. skin: Gas or vapor penetrates the skin freely.

Irrit. eye, nose, throat: Irritation of eye, nose, and throat may result.

Lung injury: Prolonged contact may injure lung.

Metal fume fever: Metal fume fever may follow breathing very fine fume. This effect results from a physical action in the lung, not a toxic action. It may be temporarily disabling, but has not produced permanent disability.

Nose and lung injury: Prolonged contact may injure nose and lung. Here, prolonged means a few weeks for the nose, many years for the lung.

Permanent injury: Permanent injury from non-fatal contact increases hazard. The materials bearing this entry have permanently disabled persons who appeared to recover, with the help of medical care, from the severe effect of a single contact.

Photosensitization: Photosensitization may follow exposure. Materials so marked have left a victim temporarily so sensitive to sunlight that he has received a dangerous sunburn when an unexposed companion was unaffected.

Polymer fume fever: Polymer fume fever may follow breathing very fine fume. See "metal fume fever" above.

Reac. w. HCl—hazard 5: Reaction with hydrochloric acid yields toxic irritant of hazard 5 for all contacts. Because of the hazard of the reaction product, any mixture with any amount of free hydrochloric acid at any temperature, should be considered to be in hazard grade 5.

Sensitiz. skin, resp. tract: Sensitization of skin or respiratory tract may result. Materials so marked may elicit in the skin of many people an allergic sensitization so that later contact with a very small quantity results in annoying or disabling dermatitis; or in the respiratory tract so that later contact results in disabling or fatal asthma-like interference with breathing. It is said that any material whatsoever can sensitize someone. Only those known to have sensitized a considerable proportion of frequent contactors, and close structural analogs of these, are shown.

Skin strain: Harmless stain may result (color is often specified).

Temp. visual disturbance: Temporary visual disturbance may result. The disturbance referred to has not been permanent, but it may be alarming and shows unwisely excessive contact.

Wounds need medical care: Wounds require prompt medical attention to facilitate orderly healing. Unless every fragment is removed from even a trivial wound, healing may not take place for months.

6 SUPPLEMENTARY INFORMATION ON TOXICITY, FLAMMABILITY AND OTHER HAZARDS: A three or four-digit number is a page reference to toxicity data in "Industrial Hygiene and Toxicology," Volume II, 2nd Revised Edition, 1963, published by John Wiley & Sons.
AIHA: Hygienic Guide available from the American Industrial Hygiene Ass'n.
MCA: Chemical Safety Data Sheet available from the Manufacturing Chemists Ass'n.
NSC: Data sheet available from the National Safety Council.
HCD: Additional information in Hazardous Chemicals Data (NFPA No. 49), published by the National Fire Protection Association.
MGD: Data in Matheson Gas Data Book, or single data sheets available from the Matheson Company.
HCP: Data in Handbook of Chemistry and Physics, 1966, published by the Chemical Rubber Company.
Merck: Data in the Merck Index, 7th Ed., 1960, published by Merck & Co., Inc.
STTE: Symptomatology and Therapy of Toxicological Emergencies, 1964, published by Academic Press.

this ed. still in print 4/90

DATE DUE